T0257897

Advances in Neuroregulation & Neuroprotection

Advances in Neuroregulation and Neuroprotection

Edited by
C. Collin, M. Minami, H. Parvez,
H. Saito, S. Parvez, G.A. Qureshi and C. Reiss

Leiden • Boston, 2005

VSP
an imprint of Brill Academic Publishers
P.O. Box 9000
2300 PA Leiden
The Netherlands

Tel: +31 71 535 3500
Fax: +31 71 531 7532
vsppub@brill.nl
www.vsppub.com
www.brill.nl

First published in 2005

ISBN 90-6764-412-9

Printed in The Netherlands by Ridderprint bv, Ridderkerk.

Contents

Advances in Neuroregulation and Neuroprotection (2005), pp. 1-2
Collin, C. *et al.* (Eds)
© VSP 2005

Foreword

Investigations in the field of basic and clinical neurodegeneration and neuroprotection are of vital importance to an understanding of human disease. At present, great efforts are being made to try to elucidate the physiopathology of neurogenic lesions, and also to establish the fine molecular mechanisms associated with all neurological syndromes. The present collection of papers by well known and eminent international scientists working on defects of the higher brain function was gathered during an international manifestation in Sapporo, Japan. A manuscript of nearly 900 pages overall was assembled to honour our distinguished friend, and teacher, Professor Masaru Minami, along with the participation of our master, Professor Hideya Saito. These two scientists of high distinction are leaders in neuropharmacology and neuroscience in Japan and have contributed greatly to an understanding of mechanisms of cerebral congestion, neurotoxicity of emesis, essential hypertension and brain aging. They are not only outstanding in their professional competency of neuroscience but are also very human and humble educators to open and promote widely a true international cooperation. The entire book covers most of the basic, clinical and molecular aspects of neuronal regulation. Neuroprotection, still a field in embryonic stage, has been greatly enriched by several contributions defining future active molecules to cure or stop neurodegeneration. Such a developmental neuropharmacology at present is essential to cure diseases such as Parkinsonism, Alzheimer's disease and other autoimmune neuropathologies. Experimental models of human neuropathology and cerebral congestion and their use to test the efficacy of new neurotropic factors have been widely studied in several investigations reported in this book. The possible impact of protein folding on the onset of conformational neuropathies along with the important role of amino acid propensity embedded in a context of specific usage of synonymous codons; genetic code degeneracy and folding have been correlated. Antidepressant research in the era of functional genomics calling a farewell to the monoamine hypothesis has also been introduced. Studies defining pineal gland functionality in connection to melatonin and the immune system are also presented. Beyond the classical neuropharmacological approach to different types of receptor regulation in neuroregulation, new approaches such as biofeedback training and clinical settings have also been included.

The entire collection provides an up-to-date survey of new trends in neuroscience covering basic neurophysiology, histopathology, and genetic, immune and molecular biology. We consider these manuscripts of great value and informative not only to research scientists but also to medical students and physicians. The generous help of all colleagues, especially Professor Hirafuji, to accomplish this laborious task is gratefully acknowledged to prepare this sincere and friendly homage to Professor Minami and Professor Saito.

C. COLLIN
S. H. PARVEZ
S. PARVEZ
C. REISS

Advances in Neuroregulation and Neuroprotection (2005), pp. 3-5
Collin, C. *et al.* (Eds)
© VSP 2005

Prof. Minami's contribution to neuroscientific and pharmacological research

The lifelong devotion of Professor Masaru Minami to neuroscientific research and pharmacology has had an important impact on basic and clinical research. Professor Masaru Minami (Dean of Health Sciences, University of Hokkaido, Faculty of Pharmaceutical Sciences; Emeritus member of Japanese Pharmacological Society) is one of the leading pharmacologists in Japan. After receiving his degree from Hokkaido University School of Medicine in 1964, Professor Minami continued his study of pharmacology and cardiovascular medicine under the auspices of Professor Tsuneyoshi Tanabe and Hisakazu Yasuda at the Departments of Pharmacology and Cardiovascular Medicine of Hokkaido University School of Medicine, respectively. Professor Minami later served as a postdoctoral research fellow in cardiovascular pharmacology under the auspices of Professor Benedict Lucchesi at the University of Michigan Medical School.

With this remarkable research background, Professor Minami performed outstanding research in the field of cardiovascular pharmacology in the 1980s. While an Associate Professor of Pharmacology at Hokkaido University School of Medicine, Professor Minami worked both as cardiologist and the head of the outpatient department for hypertensive patients at Hokkaido University Hospital.

Professor Minami was initially interested in catecholamine research. He published the results of his research, 'Plasma norepinephrine concentration and plasma dopamine-β-hydroxylase activity in patients with congestive heart failure' in *Circulation* **67**, 1324–1329, 1983. This manuscript summarizes his early research. Next, Professor Minami and his colleagues turned to SHR-related study. Using an Ambulo-Drinkometer, he studied the 'changes in ambulation and drinking behavior related to stroke in stroke-prone spontaneously hypertensive rats' (*Stroke* **16**, 44–48, 1985). This study pointed out that the behavioral changes in ambulation and drinking activity, including the disturbance of circadian rhythms before death in SHRSPs, may correspond to behavioral changes such as the delirium-state observed in patients with dementia. He and his colleagues point to the possibility that SHRSP may be a suitable animal model for the study of vascular dementia.

Professor Masaru Minami

During the past 18 years, Professor Minami and his colleagues discovered isatin, an endogenous MAO inhibitor, in the SHRSP brain using GC-MS. He was invited to speak about isatin at the Second International Meeting for Stress held in Melbourne, Australia in October, 1998. After he studied the antihypertensive actions of $5HT_2$ antagonist, ketanserin, Professor Minami completely changed his research direction to enter the novel field of serotonin. He studied anti-cancer drug-induced emesis from 1986 to 2004. Professor Minami was the first to import ferrets (from Marshal Co. Ltd., New York) as a model for emesis in Japan. In order to elucidate the site of action and mechanisms of anti-emetic drugs ($5\text{-}HT_3$ receptor antagonists and NK_1 receptor antagonists) against anti-cancer drug-induced emesis, Dr. Minami (and his group) undertook various studies including those involving behavioral pharmacology; electrophysiological studies with measurement of abdominal vagal afferent nerve activity *in vivo* (the measurement of isolated afferent nerve activity was also developed); biochemical techniques including the determination of 5-HT content in the gut and brain or 5-HT release from isolated ileum. Furthermore, Dr. Minami showed for the first time intracellular Ca^{2+} dynamics following the administration of various agonists in enterochromaffin (EC) cells in the isolated mouse ileal crypts by digital imaging analysis. In 1995, Professor Minami was one of 25 scientists selected from around the world to present their research at the Oxford International Symposium on 'Serotonin and the scientific basis of anti-emetic therapy'. He presented 'How do toxic emetic stimuli cause 5-HT release in the gut and brain'. Professor Minami endeavored to start a new kind of anti-emetic therapy for patients with cancer in Japan. The development of the field of 5-HT$_3$ receptor antagonists for use as anti-emetics has significantly increased the quality of life of those undergoing cancer treatment.

In 1979, Professor Hasan Parvez of the Neuropharmacology Unit, C.N.R.S., France, visited Professor Minami and Professor Hideya Saito at Hokkaido University School of Medicine, and since then Professor Minami has become Professor Parvez's life-long friend. Professor Parvez visited Professor Minami many times in Japan, and carried out several collaborative studies with him. In 1984, Professor Parvez and Professor Toshiharu Nagatsu created a new international journal, *Biogenic Amines*, published by VSP, Utrecht. Professor Minami has contributed many important manuscripts to this journal as a Deputy Chief Editor.

Professor Minami is not only outstanding in his contribution to pharmacology as a scientist, and head of an energetic and active laboratory, but is also a noble and warm-hearted person. He has educated and encouraged many graduate students in his laboratory who have gone on to become leading pharmacologists throughout Japan. Professor Minami organized the 2000 annual meeting for the Research of Biogenic Amines as well as the *14th Symposium on Defects in Higher Brain Function* in September, 2003 both in Sapporo.

We, the speakers in these symposiums, his friends, colleagues and previous students decided to pay homage to Professor Masaru Minami in the form of a collection of papers written by scientists well known in both neuroscience and pharmacology. Professor Masaru Minami's forty years of devotion to neuroscientific research and pharmacology has had an important impact on basic and clinical research. We hope that this collection of papers will increase the knowledge in neuroscience and pharmacology and that basic and clinical researchers will discover how very much we owe to Professor Minami for the advances in these neuroscientific fields.

HIDEYA SAITO
HASAN PARVEZ
MASAHIKO HIRAFUJI
CATHERINE COLLIN
KEIICHI SHIMAMURA
MITSUHIRO YOSHIOKA
SIMONE PARVEZ
CLAUDE REISS

Advances in Neuroregulation and Neuroprotection (2005), pp. 7-18
Collin, C. *et al.* (Eds)
© VSP 2005

Research on the evolution and development of autonervous system

K. NISHIHARA *

Nishihara Institute, Hara Building 301, 6-2-5 Roppongi, Minato-ku, Tokyo 106-0032, Japan

Abstract—The development of the basic construction of vertebrates, the metameric system, the sympathetic nervous system, the lymphatic vessel system, the capillary system and the bone marrow hemopoietic system have been investigated from the viewpoint of the gravity-evolutionary theory. The definitive structures and organs of the vertebrates are the bone and respiratory apparatus of the gut. All animal cells have a cellular respiratory system composed of mitochondria. The hemopoietic system in bone marrow, blood cells, and the cardiovascular system are the structures of the organ system between the outer (gut) respiration of gills and lungs and inner (cellular) respiration of mitochondria. The author has successfully developed hybrid artificial bone marrow by applying biomechanical stimuli to sintered apatite; this took on the characteristics of bone marrow hemopoiesis peculiar to higher vertebrates after migrating onto land. The author has also developed a hybrid-type artificial dental root that took on the characteristics of the gompholic tooth peculiar to mammals. By this approach, the author has suggested that evolution occurs according to the mechanical functions of the animal in response to gravitational energy. Establishment of the basic construction of mammals was verified by means of experimental evolutionary studies that showed the Heterodontus (dog shark) had developed into an archetype mammal directly during terrestrialization. These evolutionary phenomena can be seen as a revolutionary transformation of morphology due to biomechanical responses to environmental changes of energy triggering gene expression of cells. Through the establishment of the mammalian system, the tissue immune system is disclosed and a new concept for immune diseases is developed. Diseases of the immune system seem to be related to disorders of the cellular respiration of mitochondria caused by intracellular infections of non-pathogenic micro-organisms in the gut system.

Keywords: Morphology; gravity; evolution; metameric system; biomechanics; immune diseases; vertebrates; phylogenesis-ontogenesis; archetype; immune system.

INTRODUCTION

The development of the basic construction of vertebrates, this is, the metameric system as well as the sympathetic nervous system, the lymphatic system, the

---

*E-mail: nishihara-ken@a.email.ne.jp

capillary system, and the bone marrow hemopoietic system are investigated from the viewpoint of gravitational influence on evolution, that is, the so-called gravity evolutionary theory, which the author proposed and verified using ceramic artificial bone marrow chambers (Nishihara, 1987, 1999).

The definition of a vertebrate is a chordate having a bony backbone, with differing degrees of ossification. The structures and organs defined as being characteristic of vertebrates are the spine, and the gut respiratory system (the gill and lung system). The definitive structures and organs are the bone and respiratory apparatus of the gut (Torry, 1987). Moreover, all organisms have the cellular respiration system of mitochondria. What is the structure of the organ system between the outer (gut) respiration of gills and lungs and inner (cellular) respiration of mitochondria? It is the hemopoietic system in bone marrow, blood cells, and the cardiovascular system. Therefore, if the definitive substance of bones is synthesized, then, using them, the riddles of the evolution of the vertebrates can be solved by developing artificial bone marrow chambers inducing the hemopoietic system (Nishihara, 1987, 1999).

The author has successfully developed hybrid artificial bone marrow by applying biomechanical stimuli to sintered apatite (Nishihara et al., 1992, 1994). This took on the characteristics of bone marrow hemopoiesis peculiar to higher vertebrates after migrating to land.

The author has also developed a hybrid-type artificial dental root that took on the characteristics of the gompholic tooth peculiar to mammals (Nishihara, 1993, 2003a). In this way, the author has clarified that evolution occurs according to the mechanical functions of the animal in response to gravitational forces. In order to elucidate the law of evolution, the author has developed trilateral research that integrates morphology, including embryology and phylogeny, the functional study of molecular biology, and molecular genetics concerning remodeling with biomechanics (Nishihara, 1998). The author has also devised an experimental evolutionary study method that applies trilateral research to work of every phylogenetic stage representing phylogeny (Nishihara et al., 1996). From these studies the author has tried to reinterpret Wolff's Law (1870), Lamarck's Use and Disuse Theory (1809) and Haeckel's Biogenetic Law (Alberch, 1994) with the current level of science using biomechanics and molecular biology. The author used artificial bone marrow organs made of synthetic apatite and Ti-electrodes as well as using an artificial dental root inducing the cementum (fibrous bone) and hemopoietic nest by biomechanics. The author, moreover, verified Lamarck's Use and Disuse Theory. Following that, the author carried out research on the basic construction of vertebrates from the viewpoint of the Gravity Influencing Evolutionary Theory (Gravity Evolutionary Theory, Nishihara, 2003).

Establishment of the basic construction of archetypical vertebrates during the primitive evolution of vertebrates is thought to occur by means of gene duplication of the hemichordate ascidia. Prior to that, during the dawn of vertebral evolution the hemichordate ascidia, which integrated the respiration, nutrition, and excretion systems into only one tube of the gut, evolved from Bryozoa, which have a pterobranchial skin respiration system. Conventionally, primordial revolution of the establishment of the metameric system from monometameric ascidia had been completely unrecognized; also, development of the sympathetic nervous system had been only slightly investigated.

From the evolution of the vertebrate, mammals are conventionally thought to have evolved after three kinds of vertebrate revolutions. The first is the nativity of the Acantoidii; the second, terrestrialization; and the third, the birth of mammals. However, the author could find only two stages. These were verified by means of experimental evolutionary studies that showed the Heterodontus (dog shark) had developed into an archetype mammal directly during terrestrialization. These evolutionary phenomena can be seen as a revolutionary transformation of morphology due to biomechanical responses to environmental changes. Through these transitions, various kinds of concerns develop between morphology and the function of organs in the human body from the standpoint of basic construction of the vertebrates. Through this new theory of evolution, the immune system was studied and a new concept for immune disease was developed. Immune system diseases are a disorder of cellular respiration of mitochondria caused by intracellular infections of non-pathogenic micro-organisms in the gut system.

DEVELOPMENT OF BASIC CONSTRUCTION OF VERTEBRATES

Living organisms, including multi-cellular creatures and higher animals, have enormously diversified shapes. Are there any basic designs for the bodies of higher organisms? What is the initiating factor in evolution? The questions that occurred first in the sciences of life were those related to morphology and function. In modern science, the first person who noticed the presence of archetypes as the basis for higher organisms, including animals and plants, was the poet Goethe, who established the new scientific area of 'Morphologie'. He defined Morphologie as a science whose ultimate objective is to clarify the mechanisms of morphological transformation. He also studied vertebrate archetypes up to human beings. Goethe stated that even the body of a human being 'has reached today's shape through transformations every moment after the creation of life, starting from the prototype in the Archeozoic age'. Goethe's lifelong theme was the importance of 'the principles of modification of shapes', which was later called phylogeny by Haeckel.

In those days, Lamarck and Cuvier in France greatly contributed, as did Goethe, to the establishment of a new scientific field, which is called 'life sciences' today. Lamarck made 'Biologie' independent from natural history (Lamarck and Treviranus, 1802) through studies of invertebrates and proposed the evolutionary concept that the archetypes of living beings are changed by external as well as internal biomechanical factors according to the Use and Disuse Theory (Lamarck, 1809). This was a more definite scientific realization of Goethe's Morphologie of the modification of archetypes under biomechanical influences. On the other hand, Cuvier established the basics of comparative vertebrate anatomy at the age of 27, and proposed the principle of the dependence and correlation of organs (1795). However, under the patronage of Napoleon, he advocated the theory of the unchangeability of life and made efforts to abolish Lamarck's concept of evolution.

The essence of the Use and Disuse theory is biomechanics and the substance of biomechanics is dependent upon the earth's gravity (Nishihara, 2003a).

While various living beings exist on the earth, there are comparatively few basic macromolecular substances that are common to all living beings, namely, nucleic acids constituting genes and amino acids constituting proteins. There are various substances for skeletal systems related to morphology and, since the substances are not the same, the basic morphology and the modifications appropriate to each kind seem to be different.

The skeletal systems of organisms on earth can be divided approximately into five kinds: (1) the cellulose system for plants, (2) silicate system for diatoms, (3) calcium carbonate system for shellfish and coral, (4) chitin system for insects, prawns and spiders, and (5) the collagen apatite phosphate system for vertebrates. To those listed above only the last two skeletal systems have effective response mechanisms to biomechanical stimulation. Because of this, invertebrates with only chitin and vertebrates display a morphology associated with functions.

Coelenterates and Pterobranchia, having no skeleton, are able to change their morphology relatively freely in conformity with their functions. Invertebrates that developed a chitin-based skeleton evolved into organisms such as crabs and insects with social organizations, but because of their external skeleton, their remodeling system of ecdysis was limited, and therefore they failed to achieve larger sizes.

The collagen apatite-based internal skeleton of vertebrates was able to change its shape by remodeling in response to external as well as internal forces during movement under the earth's gravity. Also, they were able to grow to huge sizes and to adapt and propagate, forming considerably diversified species.

Therefore, it can be said that the bone structure is the defining characteristic of the vertebrates. For this reason, the mechanisms of bone evolution in vertebrates can be clarified by investigating bone characteristics. Bone is a connective tissue calcified by hydroxyapatite, so that collagen seems to be as important a substance as apatite for characterizing the vertebrates. The archetype of the bone structure is derived from a carapace called aspidin, which is a composite of dentine and bone forming the armor of the Pisces in the Archeozoic era. Calcium which existed in high concentrations in the Archeozoic sea was absorbed in the gut from sea water and metabolized by mitochondria in muscle and secreted from aspidin on the skin surface. The aspidin may also have had a storage function for phosphates, which were in short supply in the sea.

The starting point of the evolution of the vertebrates is the incorporation of respiratory apparatus into the gill cleft from the cutaneous respiration system of Pterobranchia, and the acquisition of an apatite bone structure. Therefore, the archetype of our mammalian human ancestors can be seen in the dog shark

(*Heterodontus japonicus*) of which constructive parts around the mouth coincide with those of the human embryo.

Since animals cannot synthesize energy sources independently from their immediate environment, the fundamental characteristic of animals is related to movement in search of food.

In morphological evolution, empirical laws exist which state that the form changes according to the method of using the body, i.e. repeated action, just as in the Use and Disuse theory. That is to say, the characteristics of constant behavior determine form and structure. Repetitive behavior can be seen as a function in many cases, and, therefore, form changes in accordance with function. Because of this, the evolution of form occurs depending on changes in the environment. Environmental change promotes change in biomechanics in a broad sense, i.e. physico-chemical stimuli including energy.

ESTABLISHMENT OF THE BASIC CONSTRUCTION OF ARCHETYPE VERTEBRATES IN EVOLUTION

The primordial revolution

In this stage, metameric organisms developed through a process of genome duplication from the ascidian (urochordate monosomite organisms). Relics of this primordial revolutionary stage are the Amphioxus and Cyclostomata. After metameric organisms (vertebrates with a homeobox), metamorphic changes occurred as the result of biochemical stimuli with no alteration to the basic structure of the homeobox genome.

The primordial vertebrate, i.e. hemichordate ascidian organism with a monosomite form, began with the acquisition of a gut for an absorbing and excreting system, that is, respiration, nutrition, hemopoiesis, generation, and urination. Ascidian has a dorsal chord, but it is a monosomite organism, while the vertebrates are multisomite organisms. Development from monosomite animals to multisomite vertebrates is thought to have occurred by geneduplication, a concept proposed by S. Ohno.

Through several cycles of gene-duplication of ascidian genomes, a cyclosalpa-type organism, which is a continuous ascidian-like chain with serial single gut of archetype independent vertebrae, evolved. This organism moved in sea water. The first front ascidia formed the head and face. Water and food were taken by the mouth while each branchial aperture opened during water aspiration to provide oxygen, which was absorbed through the gills. When the openings closed, digestion of food began in the gut. After that, food debris was discharged from the anus. Following the development of the anus, the gut of all somites in the tail part of cyclosalpa disappear completely. In this way, the tail, i.e. the serial somites without the gut developed. From this primordial vertebral evolution, archetype vertebrates like Amphioxus and Cyclostomata developed. This mechanism of development depends upon the Use and Disuse theory based on biomechanical energy. Maintenance of this system required a hard-tissue notochord, namely, the series of somites that became the vertebrae.

The first revolution

The Silurian period saw the appearance of spiny sharks (Acanthodii), with jaws, calcified teeth and a carapace of hydroxyapatite acquired through forward movement. Descendants of these are to be found today as cartilaginous fishes (Chondrichtyes), i.e. sharks and rays.

Archetype Cyclostomata continued to move forward speedily. Energy metabolism of the muscles increased and excreted large amounts of minerals from the muscles, which functioned via electron transmission in mitochondria through increased cellular respiration as a result of muscle movement. Consequently, placioids made of cartilage evolved into bone with an enamel substance and Acantoidii developed. This is the first revolution in vertebral evolution.

Use of bone made of collagen apatite compounds for hard tissue made adaptation and propagation possible for the vertebrates. The fact that the hydroxyapatite is based on calcium which is indispensable for life, and phosphate which is indispensable for cytological respiration, as well as energy and nucleic acid metabolism are considered to have far-reaching significance.

Living phenomena are systems struggling for self-continuation by utilizing energy from the incorporation and decomposition of environmental factors, so that life can be said to essentially depend on the environment. Therefore, all vital reactions can be considered biomechanical responses to environmental factors.

BASIC CONSTRUCTION OF MAMMALS

This involved development of bone marrow hemopoiesis, the parasympathetic nervous system, capillary system, lymphatic system, and the immune system.

In the second revolution of vertebrates, i.e. terrestrialization in the Devonian period, two dramatic changes occurred in the change from water to air as a habitat with a concomitant change in gravitational force (from 1/6G, due to buoyancy, to 1G), as well as in the change from branchial to pulmonary respiration (Nishihara, 2003c). In addition, the cartilaginous endoskeleton ossified. Salamanders and lungfish are relics of this evolutionary stage.

During terrestrialization, drastic changes of biomechanic stimuli in a broad sense, occurred. These include energy without mass, e.g. force of gravity, as well as substance with mass, e.g. oxygen. They are known as the following five physicochemical stimuli: (1) Gravity of 1/6G in water (buoyancy) to 1G on land or 6 times stronger energy; (2) 0.7% oxygen content in sea water to 21% in air, or 30 times greater; (3) Water as a life-medium containing minerals becomes 1/800 lighter as air without minerals; (4) Complete 'wetness' in water to complete 'dryness' in air; and (5) Highly viscous water versus extremely low viscosity of air.

As the result of gravitational action, the dog shark could adapt by increasing blood pressure as it struggled to try to escape suffocation by returning to water. Consequently, endoskeleton cartilage developed into osseous tissue with bone marrow hemopoiesis. Thirty times increased oxygen content caused the sixth gill epithelial membranes to develop air sacks of the lungs during respiration in the struggle to escape suffocation. Dryness as well as the lack of minerals in the air influenced the placoids (dermal teeth) to develop fur, which are in fact placoids made of collagen without minerals. The high viscosity of water brought about the streamlined shapes of organisms in water. The extremely low viscosity of air brought about common shapes without streamlining after terrestrialization. This is epitomized by the changes in morphology and functions in the organs of archetype vertebrates after land migration, as seen by respiratory gills, the dermal region, and inner skeletons. Metamorphosis is disclosed to be a phenomenon of metaplasia. In pathological terminology, this means a change of cells from one type into another with the same genetic code by biomechanical stimuli, i.e. via physicochemical changes, including energy.

In the second revolution of the vertebrates, the author hypothesizes that the increased gravitational action of the earth after landing affects the blood pressure of chondrichthyes through their intensive movement to escape suffocation by moving toward water (Nishihara, 2003c). With elevated blood pressure, streaming potential increases (Petrov *et al.*, 1989) and the increased currents trigger the gene expression of chondrocytes to develop osseous tissue, together with bone marrow hemopoiesis as well as erythrocyte-enucleation, the cardiovascular system, lymphatic-vessel formation, the pyramidal tract of the cerebral motor nervous system, the sympathetic nervous system along with the capillary system and major histocompatibility antigens (MHC) in conjugation with the homothermal system. The author verified these processes by means of experimental evolutionary research methods (Nishihara, 2000).

In the second revolution, mitochondria in somato-muscles required oxygen. This assures that mitochondria in muscle cells as well as in cells of all kinds of organs and viscera synthesize some kind of cytokine nerve growth factor to generate capillaries and the lymphovessal system. Consequently, capillaries and the sympathetic nervous system proliferate and develop into brain, visceral organs, as well as somato-muscles. Before the stage of archetype animals there were no nutritional capillaries even in the heart and brain. The heart was derived from branchial hematopoietic nests. Therefore, the heart itself generates hemato cells.

All vascular systems have smooth muscle cells in their walls. At this stage, the sympathetic nervous system developed from the ganglion of the parasympathetic nervous system.

But we may ask, what is the nervous system? This question is most important to solve the riddle of the brain. The nervous system was differentiated together with muscle cells. However, conventional research overlooks this. As a result, research has not been able to solve the brain system riddle. The cerebrum of archetype animals is composed of a visceral brain, which controls somato-muscles by means of the extra-pyramidal tract as well as visceral muscles.

The parasympathetic nervous system is extremely old. Even the initial developmental stage of vertebrates, such as the monosomite ascidian, had this. The parasympathetic nervous system developed together with the gut system. Therefore, the parasympathetic nervous system belongs to the visceral smooth muscle group. These systems function by cholinergic stimuli. Therefore, the preganglion nervous system of sympathetic nerves is not correct as it actually belongs to the parasympathetic nervous system because of cholinergic fibers. What then were the major factors in the development of capillaries and sympathetic nerves? The former developed from the extensive increase of movement requiring oxygen and the latter from thermal changes in the air in the environment. Thus, the surface skin of the somatic motor system has a connection with somato organs, e.g. brain, and visceral organs, such as the heart and the intestines, via the autonomic nervous system.

The sympathetic nervous system functions only by adrenalin. During the development of the sympathetic nervous system, neurons in the cerebral cortex proliferated in mammalian type reptiles and control somato motor muscles by means of the pyramidal tract. At this stage, the bone marrow hemopoietic system, together with the functioning of MHC, as well as the homothermal system are established.

The function of MHC is for remodeling and cell-metabolism as well as antibody formation of immunoglobulin A, as leukocytes depend completely upon cellular respiration and energy metabolism of mitochondria in white blood corpuscles. The body temperature of the homothermal system is generated by means of oxidative phosphorylation in mitochondria. This is the reason why the mammalian immune system of leukocytes, i.e. antibody formation as well as bacterial phagocytes and digestion strictly depend upon a body temperature of 37°C. If body temperature lowers by one degree, leukocytes lose their immune functions. The terrestrialization of Heterodontus (dog shark) led to dramatic changes in the following 12 major areas: (1) cardiovascular system, (2) capillaries, (3) lymphatic system, (4) autonomic nervous system, (5) homothermal system, (6) skeletal system, (7) bone marrow hemopoietic system, (8) tissue immune system, (9) external respiratory system, (10) pyramidal tract system, (11) erythrocyte development, and (12) placoid mineral loss (fur). These changes in evolution occur as metaplasia, which means the change of cell type with the same genetic code by means of gene expression, is triggered by physicochemical stimuli, including energy.

As mentioned above, the mammalian sympathetic nervous system developed along with capillaries derived from the ganglion of parasympathetic nerves. Therefore, it had no control over the central nervous system of the brain and spine. As a result, through capillaries and sympathetic nerves, the body surface of skin is connected to viscera as well as to the brain, spine, bone marrow, and joints.

DISCLOSURE OF THE MECHANISMS OF EVOLUTION OF THE IMMUNE SYSTEM AND THE ONSET OF HUMAN IMMUNE DISEASES

What happened to the immune system during terrestrialization? The immune system is composed of various kinds of hemocytes. During the development of bone marrow hemopoiesis, erythrocytes lose their nuclei and leukocytes differentiate with functioning HLA in their surface membrane. Erythrocytes without nuclei cannot maintain amoeboid movement, therefore, they remain in the blood vessels. Leukocytes continue to have amoeboid movement, so they can move into lymphoid vessels as well as intracellular space. Lymphoid vessels act as the major immune system of white blood corpuscles, into which the autonomous nervous system has its networks. The function of white blood corpuscles, i.e. the cytotogical digestive system, depends upon little higher temperatures than the homothermal system of each respective animal. Cytological digestion of leukocytes is carried out by the energy generated by their mitochondria.

These drastic changes in evolution occurred according to Lamarck's Use and Disuse theory, and all phenomena are based on the metaplasia of cells. Also, all drastic phenomena in evolution are sustained by the energy metabolism of mitochondria in all cells in an organism. From the standpoint of evolution, as well as from the basic construction of the vertebrates, the author questions why intractable immune diseases which comprise almost all human maladies, have risen recently in all the advanced countries.

Thinking about the establishment of the eukaryotes, in which parasitic aerobic bacteria had infected intracellularly as mitochondria, it can be easily understood that in higher animal cells, various microbes, e.g. virus, mycoplasma, bacteria, and some kinds of Protozoa can cause parasitic infections. If these microbes have pathogenicity, maladies occur in animals. Today, if we have no effective antibiotics against pathogenic bacteria, we could never overcome either epidemic or pathogenic contagious diseases. The immune system as a defence mechanism is very weak and there is almost no protective system against parasitic infections.

Self and non-self immunology is now in vogue around the world. However, this immunology is effective only for tissue immunity in organ transplants, but almost useless against infectious diseases.

Archetype vertebrate chondrichthyes (and sharks) are known to have the MHC gene. However, they have no tissue immune system; and, as their leukocytes do not phagocyte, bacteria and microbes can coexist in shark blood. They have no lymph system as their erythrocytes have a nucleus like leukocytes. Therefore, their leukocytes do not have their own vascular system of lymphoid vessels. The author successfully verified the immuno-tolerance of chondrichthyes by means of transplantation of various shark organs into mammals (Nishihara, 2004).

From this research, the author has determined that the renewal metabolism of cells in the immune system of mammals is a remodeling system by means of MHC or HLA in the cell membrane. In the human body, one trillion out of 60 trillion cells are remodeled per day during sleep. This is the function of the tissue immune system of leukocytes. The loss of this function in mammals is known as immunotolerance. This tissue immunity has developed into the most sophisticated system in mammals during evolution. The development has been in parallel with evolution of bone marrow hemopoiesis, the lung respiration system, the sympathetic nervous system, the cerebral motor of the pyamidal tract system, the cardiovascular capillary system, the completion of erythrocyte differentiation, and the homothermal system.

We must then ask what are the intractable immune diseases that are now appearing in in advanced countries, especially Japan, France and the USA? Intractable immune diseases, including self-immune maladies, develop by the intracellular contamination of parasitic non-pathogenic enteromicrobes, which are incorporated into leukocytes from lympho-adenoid follicles, i.e. GALT (gut associated lymphoid tissues). The more famous ones are Waldeyer's adeno lymphoid ring and the Peyer patch. From this, the non-pathogenic indigenous parasitic microbiota are incorporated into the follicles where microbes are digested and immunoglobulin A is generated. However, by breathing though the mouth, or by cooling the gut with cold liquids or ice cream, bacteria or viruses are incorporated into leukocytes and leukocytes disseminate these parasitic bacteria in circulating blood. This gives rise to unrecognized intracellular infections of various kinds of cells in major organs. Common pathogenic bacteria bring about diseases by contamination of the medium, i.e. blood or lymphatic fluid, or on the surface of the gut, e.g. the lungs and intestines. However, intracellular infections by non-pathogenic parasitic enterobacteria bring about no infectious diseases or toxicoses, but allergic diseases and immune system diseases instead.

These allergic diseases, e.g. atopic dermatitis as well as collagen diseases, myositis, ulcerative colitis, asthma, and mycoplasma pneumonia, are intracellular infections of non-pathogenic or only weakly pathogenic microbes.

Conventionally, intracellular infections have been known as viral infections or protozoa malaria. What happens during intracellular contamination by non-pathogenic bacteria? In intracellular infections by aerobic or anaerobic bacteria, the energy metabolism of mitochondria of the infected cells is disturbed. Aerobic bacteria consume intracellular oxygen and anaerobic bacteria consume nutrition glycolysis which allows pyruvic acid to be metabolized for the oxidative phosphorylation in mitochondria. Therefore, despite aerobic or anaerobic bacteria, facultative aerobiosis or obligate anaerobes without pathogenicity, all cells infected are caused by the disturbed cellular respiration of mitochondria. By breathing though the mouth, lower body temperature, cooling of the gut by cold drinks or ice cream, overtension of the sympathetic nerves, shortage of sheep (i.e. relief from gravity), and an overtired state, all allow parasitic non-pathogenic enterobacteria to be incorporated into leukocytes through lympho-adenoid follicles in the gut (Nishihara, 2000). These over-reactions of energy induce the intracellular infection of leukocytes, which are brought about by the mistaken use of the human body. Through the lymphatic vessels, these contaminated leukocytes are incorporated into venous vessels and the leukocytes disseminate bacteria into various cells in different organs via the bloodstream. Each specialized organ has highly differentiated cells with specialized functions. All cell functions are regulated by the energy metabolism of mitochondria. Therefore, if cells of some specialized organs, e.g. brain, kidney, spleen or pancreas, or some structure, e.g. joint, subcutaneous tissue, or bone marrow, are contaminated by parasitic enterobacteria disseminated by infected leukocytes, these organs cannot continue to function normally. And, in some cases, these contaminated structures allow the granulation of tissue to develop. These are joint rheumatitis, matrix pneumonia, histocytosis, sarcoidosis, and atopic dermatitis.

The major causal factor of evolution is energy. The modality of evolution is in accordance with Lamarck's Use and Disuse theory. The major cause of intractable immune diseases is also energy. These diseases are now occurring according to the Use and Disuse theory by the mistaken usage of the mouth for breathing, shortage of natural sleep, shortage of sunlight absorption, cooling the gut and an over-tired condition, which result in the deterioration of mitochondria in all cells of the body, as well as intracellular infections by parasitic bacteria.

SUMMARY

The basic construction of the vertebrates has been investigated from the viewpoint of gravity as a causal factor in evolutionary theory. The initial stage of the revolution of vertebral evolution, i.e. the primordial revolution in which the monosomite ascidian developed into polymetameric organisms, is thought to occur as a result of geneduplication. Lamarck's Use and Disuse theory corresponding to gravity as well as the development of the sympathetic nervous system are studied together with bone marrow hemopoiesis, the tissue immune system, capillaries and lymphovessel development, and the homothermal system. From this research, the author has determined that the causes of human-specific intractable immune diseases are intracellular contamination by parasitic nonpathogenic enteromicrobes, which are disseminated by leukocytes.

Acknowledgements

This research was supported by a Grant-in-Aid for Developmental Scientific Research (B) (1) (No. 03557107), in-part by a Grant-in-Aid for Scientific Research on Priority Area (1) (No. 05221102 and 06213102), a Grant-in-Aid for Developmental Scientific Research (B) (1) (No. 06558119), in-part by a Grant-in-Aid for Scientific Research on Priority Area (1) (No. 08233102), and a Grant-in-Aid for Co-operative Research (A) (No. 07309003) from the Ministry of Education, Science and Culture, Japan. This study also has been supported by a Grant-in-Aid for Scientific Research (A) 09309003 from the Ministry of Education, Science and Culture, Japan.

REFERENCES

Alberch, P. (1994). Heterochrony; Pattern or process? in: *Biodiversity and Evolution, 10th Intern. Symp. on Biology*, in conjunction with the awarding of the international prize of biology, pp. 26–27.

Lamarck, J. B. P. A. (1809). *Philosophie Zoologique*. France.

Nishihara, K. (1993). Studies on peri-root tissue formation around new type artificial root made of dense hydroxyapatite, *Clinical Materials* **12**, 159–167.

Nishihara, K. (1998). Development of hybrid-type artificial immune organ by means of experimental evolutionary research method using bioceramics, in: *Tissue Engineering for Therapeutic, Use 1*, Ikeda, Y. and Yamaoka, Y. (Eds), pp. 39–50.

Nishihara, K. (1999). Evidence of biomechanics-evolutionary theory by using bioceramics, *Bioceramics*, Vol. 12, Ohgushi, H., Hastings, G. W. and Yoshikawa, T. (Eds), pp. 253–256.

Nishihara, K. (2000). Evidence-based evolutionary research and development of the practical phylogenetics: verification of the gravity-corresponding evolutionary law by means of biomaterials, in: *Proc. Conf. on Ceramics, Cells and Tissues*, Faenza, Italy, pp. 167–172.

Nishihara, K. (2003a). Verification of the gravity action in the development of bone marrow hemopoiesis during terrestrialization, in: *Materials in Clinical Applications VI*, Vincenzini, P. (Ed.), pp. 277–288.

Nishihara, K. (2003b). Development of revolutionizing method for creating hybrid-type artificial organs using ceramics by means of electric energy, in: *Proc. Conf. on Ceramics, Cells and Tissues*, Faenza, Italy (in press).

Nishihara, K. (2003c). Verification of use and disuse theory of Lamarck in vertebrates using biomaterials, *Biogenic Amines* **18**, 1–17.

Nishihara, K. (2004). Establishment of a new concept of the immune system, disclosure of causes, and development of the therapeutic system of immune diseases, *Biogenic Amines* **18**, 79–93.

Nishihara, K., Tange, T., Tokumaru, H., *et al.* (1992). Study on developing artificial bone marrow made of sintered hydroxyapatite chamber, *Bioceramics* **5**, 131–138.

Nishihara, K., Tange, T. and Hirota, K. (1994). Development of hybrid type artificial bone marrow using sintered hydroxyapatite, *Bio-Medical Materials and Engineering* **4**, 61–65.

Nishihara, K., *et al.* (1996). Successful inducement of hybrid type artificial bone marrow using bioceramics in various vertebrates, in: *Bioceramics*, Vol. 9, Kokubu, T., *et al.* (Eds), pp. 69–72.

Petrov, N., Pollack, S. and Blagoeva, R. (1989). A discrete model for streaming potentials in a single osteon, *J. Biomechanics* **22**, 517–521.

Torrey, T. W. (1987). *Morphogenesis of the Vertebrates*. John Wiley, New York.

Wolff, J. (1870). Ueber die innere Arechitectur der Knochen und ihre Bedeutung für die Frage vom Knochenwachsthum, *Archiv für pathologische Anatomie und Physiologie und für Klinishe Medizin, Virchövs Archiv* **50**, 389–453.

Advances in Neuroregulation and Neuroprotection (2005), pp. 19-28
Collin, C. *et al.* (Eds)
© VSP 2005

The effect of melatonin and corticosterone on the phagocytic function of BALB/c mice macrophages

C. BARRIGA *, M. I. MARTÍN, E. ORTEGA and A. B. RODRÍGUEZ

Department of Physiology, University of Extremadura, 06071 Badajoz, Spain

Abstract—The hormones melatonin and corticosterone are able to modulate the immune response, with indications that melatonin may counteract the immunosuppressory action of the glucocorticoids. The aim of the present study is to evaluate the effects of pharmacological concentrations of melatonin and corticosterone, singly and together, on macrophage phagocytosis of antigen particles. To this end we performed an *in vitro* study using as phagocytes peritoneal macrophages from two groups of BALB/c mice, one in a basal situation and the other subjected to stress by swimming. The concentrations used were ten and one hundred times greater than the previously determined physiological levels. In particular, for incubations with macrophages from the basal group they were 0.9 pg/ml and 9 pg/ml of melatonin, and 2 μg/ml and 20 μg/ml of corticosterone, and for macrophages from the stressed group they were 0.7 pg/ml and 7 pg/ml of melatonin, and 5 μg/ml or 50 μg/ml of corticosterone. The results indicate that the immunomodulatory effect of the two hormones depends on the conditions of the study. Thus, melatonin led to increased phagocytic capacity in macrophages from both the basal and the stressed animals, while corticosterone only had an immunostimulatory effect with macrophages from the unstressed animals. With both hormones acting together, the macrophage phagocytic activity was stimulated at all the concentrations used and independently of whether or not the corresponding animals had been stressed, although the greatest activation was seen in the basal animal phagocytes incubated with the higher corticosterone concentration. In conclusion, while melatonin was definitely immunostimulatory at the concentrations tested, the immunoregulatory properties of corticosterone depended on the situation.

Keywords: Melatonin; corticosterone; phagocytosis; stress; immune.

INTRODUCTION

Stress is defined as the response of an organism to a stimulus or change, and is characterized by the activation of the autonomous nervous system and of the hypothalamic-pituitary-adrenal (HPA) axis. The results of the neuroendocrine changes affect the immune function both directly and indirectly (Felten and Felten,

---

*To whom correspondence should be addressed. E-mail: cibars@unex.es

1991; Roszman and Carlson, 1991; Ortega *et al.*, 1993). Physical exercise is one of the most relevant forms of stress for the organism (Simon, 1991). The changes that occur during physical activity include alterations in the levels of neurotransmitters such as the catecholamines, and of hormones such as cortisol, vasopressin, glucagon, prolactin, and thyroid stimulating hormone, *inter alia* (Ortega *et al.*, 1996). Some of these agents act as modulators of the immune system, and mediate the response to stress (Barriga *et al.*, 1992).

Cortisol and melatonin represent true internal chemical pacemakers of different physiological processes, and both hormones possess immunoregulatory effects (Ortega *et al.*, 1996; Skwarlo-Sonta, 1996). The secretion of the two hormones is under β-adrenergic control, with indications of the possibility that melatonin might counteract the physiological effects of the glucocorticoids released in situations of stress (Ortega *et al.*, 1996; Barriga *et al.*, 2001; Rodriguez *et al.*, 2001). Indeed, various studies indicate that the production of melatonin is altered in rats subjected to stress (Troiani *et al.*, 1988; Wu *et al.*, 1988).

Previous results from our group indicate that in situations of stress there is a change in the rate of secretion of the hormone melatonin, as well as a loss of the circadian rhythm of corticosterone. At the same time there appears an increase in the activity of the macrophages from the stressed animals in ingesting antigens (Barriga *et al.*, 2001). To look further into whether these hormones intervene directly in the activation of the non-specific immune response, we performed an *in vitro* study to evaluate the effect of melatonin and corticosterone at pharmacological concentrations, both separately and conjointly, on the phagocytic activity of peritoneal macrophages from BALB/c mice, subjected or not to situations of stress.

MATERIALS AND METHODS

Animals

Studies were performed on 24 BALB/c mice (*Mus musculus*), aged 12 ± 4 weeks, maintained at a constant temperature ($22 \pm 2\,°C$) and fed on 'Sander Mus' and water *ad libitum* in a room with an outside window, natural lighting, indirect ventilation, and temperature and humidity controlled ($23 \pm 1\,°C$ and 60%, respectively). They were kept under a light-dark cycle of approximately 14 h of light and 10 h of darkness (dark period 21:30 h) during spring (April–June) 2002. The animals were separated into two experimental groups: basal and stressed (subjected to acute physical activity).

Acute physical activity

A classical model of stress induced by physical activity was employed: swimming until exhaustion (Ferry *et al.*, 1991; Ortega *et al.*, 1993). Mice were put into individual 25×25 (height) $\times 15$ cm tanks containing 7 litres of water (at $25 \pm 2\,^\circ\text{C}$) and made to swim continuously until exhaustion, taken as the moment when the animal stopped making rapid spontaneous swimming movements. At that point, they were immediately removed from the tank. The mean duration of exercise was 30 ± 5 min. This study was approved by the Ethical Committee of the University of Extremadura (Spain). Basal animals were maintained under similar environmental conditions to the previous group but not subjected to exercise stress.

Collection of peritoneal exudate cells (PECs)

The abdomen was cleansed with 70% ethanol, the abdominal skin was carefully dissected without opening the peritoneum, and 4 ml of Hank's solution (Sigma) adjusted to pH 7.4 was injected intraperitoneally. The abdomen was massaged and the peritoneal exudate cells (PEC) removed, with recovery of 90–95% of the injected volume of fluid. The cells (macrophages and lymphocytes) were counted and adjusted to a final concentration of 5×105 macrophages/ml in Hank's solution. Cell viability was $98 \pm 1\%$ as measured by the trypan blue exclusion method. In all the determinations, the phagocytic cells were isolated at 09:00 h.

Melatonin

N-acetyl-5-methoxytryptamine (Sigma) was prepared in phosphate-buffered saline solution (PBS), starting from a base solution of 1 g/100 ml which was dissolved by heating and stirring, followed by diluting working solutions to 0.9 pg/ml and 9 pg/ml in the basal animal group, and to 0.7 pg/ml and 7 pg/ml in the stressed animal group. All determinations were accompanied by a control sample free from hormone.

Corticosterone

4-pregnene-11,21-diol-3,20-dione (Sigma) was prepared in PBS solution, starting from a base solution of 100 mg/100 ml, which was followed by diluting working solutions to 2 μg/ml and 20 μg/ml for the basal animals and to 5 μg/ml and 50 μg/ml for the stressed animal group. All determinations were accompanied by a control sample free from hormone.

Phagocytosis assay

The latex phagocytosis assay was carried out following the method described elsewhere (Ortega *et al.*, 1993). In brief, aliquots of 200 ml of the suspension of PEC were incubated (37°C and with 5% CO_2 atmosphere) on culture plates for 30 min, and the adhered monolayer was washed with PBS at 37°C. Then 20 ml of latex beads (Sigma, 1.02 mm, diluted to 1% in PBS) and 200 ml Hank's (control), or 200 ml of melatonin, corticosterone, or both hormones at different concentrations, were added, followed by another 30 min of oven incubation under the same conditions as before. Finally, the samples were fixed and stained with Diff-Quick containing methanol (5 min), eosin (five passes), and haematoxylin (five passes). The plaques were washed with tap water and dried, followed by counting under oil-immersion phase-contrast microscopy at 100×. The number of particles ingested per 100 macrophages was expressed as the latex-bead phagocytosis index (PI). The percentage of cells that had phagocytosed at least one latex bead was expressed as the phagocytosis percentage (PP). The ratio PI:PP was calculated, giving the phagocytosis efficiency (PE) which represents the mean number of latex beads phagocytosed per macrophage.

Statistical study

All data are expressed as mean (X) ± standard deviation (SD). The variables were tested for normality, and then the different groups were compared using the Scheffe ANOVA parametric F-test, with $p < 0.05$ taken as the level of significance in differences between groups.

RESULTS

Figure 1 shows the results for the latex bead Phagocytic Index (PI) of peritoneal macrophages from the two groups (basal and stressed) of BALB/c mice incubated with the different concentrations of melatonin (Fig. 1A) or corticosterone (Fig. 1B) used in this *in vitro* study. One observes that the melatonin stimulated the macrophages' capacity to ingest antigens independently of the origin of the macrophages. The maximum stimulation was for basal animal macrophages incubated with the higher melatonin concentration. Corticosterone, however, only led to an increased PI when the phagocytes were from stressed animals and had been incubated with the higher corticosterone concentration. Were the variations in the macrophages' capacity to phagocytose inert particles due to an increase in the number of cells capable of phagocytosis (measured by the Phagocytic Percentage, PP), or to the cells being more active in ingesting the antigen (measured by the Phagocytosis Efficiency, PE)? The results indicated that there were no significant differences in PP for the different concentrations of melatonin and corticosterone acting alone or together. There was, however, an increase in the effectiveness of the macrophages' antigen ingesting capacity (PE) at all the melatonin concentrations tested (Fig. 2A), and, when the macrophages were from basal animals, for each of the two corticosterone concentrations (Fig. 2B). The smallest values of PE corresponded to incubating macrophages from stressed animals with the higher corticosterone concentration.

Tables 1 and 2 list the results when the macrophages were incubated with melatonin and corticosterone together. One observes that there were variations depending on whether the phagocytes were from basal (Table 1) or stressed (Table 2) animals. Thus, for the macrophages from basal animals, both the PI and the PE were

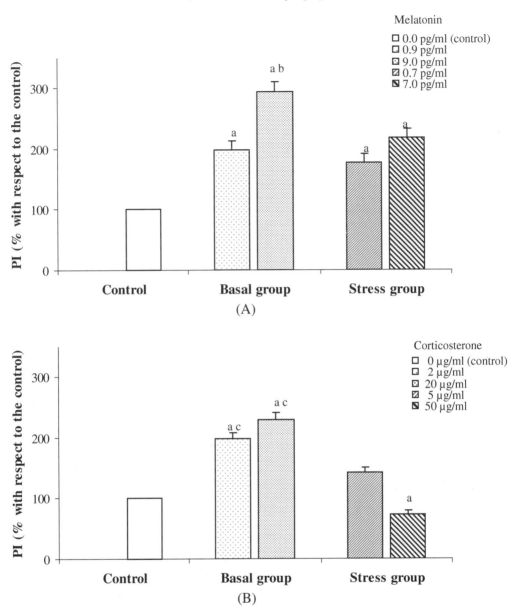

Figure 1. The *in vitro* effect of melatonin (A) and corticosterone (B) on the Phagocytic Index (PI) of peritoneal macrophages from basal and stressed BALB/c mice. Each value represents the mean ± SD of 6 determinations performed in duplicate. (a) $p < 0.05$ with respect to the control. (b) $p < 0.05$ with respect to the resting melatonin concentration. (c) $p < 0.05$ with respect to the corticosterone concentration of the stressed group.

raised by all the concentrations tested, with the greatest increase corresponding to macrophages incubated with the greatest capacity to ingest antigen corresponding to the higher concentration of corticosterone. For the macrophages from stressed

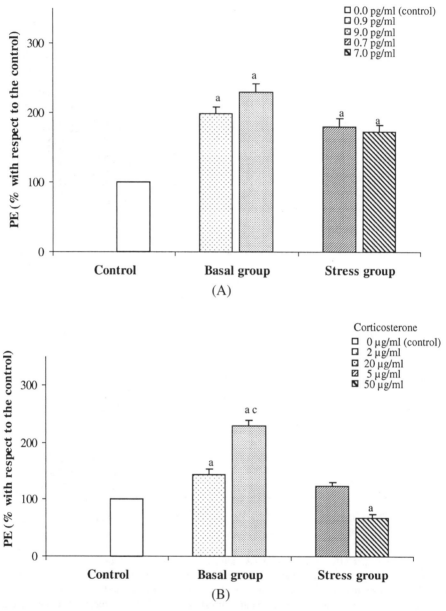

Figure 2. The *in vitro* effect of melatonin (A) and corticosterone (B) on the Phagocytic Efficiency (PE) of peritoneal macrophages from basal BALB/c mice. Each value represents the mean ± SD of 6 determinations performed in duplicate. (a) $p < 0.05$ with respect to the control. (c) $p < 0.05$ with respect to the corticosterone concentration of the stressed group.

Table 1.
Joint effect of melatonin and corticosterone on the Phagocytosis Index and Phagocytosis Efficiency in macrophages from basal BALB/c mice

Melatonin + corticosterone (concentrations)	Phagocytosis Index	Phagocytosis efficiency
0 pg/ml + 0 μg/ml	100	100
0.9 pg/ml + 2 μg/ml	276 ± 20^a	291 ± 34^a
9 pg/ml + 2 μg/ml	283 ± 15^a	317 ± 30^a
0.9 pg/ml + 20 μg/ml	356 ± 25^{ab}	363 ± 18^{ab}
9 pg/ml + 20 μg/ml	370 ± 20^{ab}	367 ± 16^{ab}

Each value represents the mean $\pm$ SD of 6 determinations performed in duplicate.
[a] $p < 0.05$ with respect to the control.
[b] $p < 0.05$ with respect to the 2 μg/ml corticosterone concentration.

Table 2.
Joint effect of melatonin and corticosterone on the Phagocytosis Index and Phagocytosis Efficiency in macrophages from stressed BALB/c mice

Melatonin + corticosterone (concentrations)	Phagocytosis Index	Phagocytosis efficiency
0 pg/ml + 0 μg/ml	100	100
0.7 pg/ml + 5 μg/ml	180 ± 20^a	175 ± 22^a
7.0 pg/ml + 5 μg/ml	188 ± 19^a	165 ± 14^a
0.7 pg/ml + 50 μg/ml	189 ± 38^a	151 ± 11^a
7.0 pg/ml + 50 μg/ml	184 ± 24^a	162 ± 12^a

Each value represents the mean $\pm$ SD of 6 determinations performed in duplicate.
[a] $p < 0.05$ with respect to the corresponding control.

animals, all of the joint concentrations of melatonin and corticosterone tested led to increases in PI and PE, with no apparent dependence on the actual concentration used.

DISCUSSION

The neuroendocrine changes that occur during a situation of stress have both a direct and indirect influence on immune function (Felten and Felten, 1991; Roszman and Carlson, 1991). In particular, corticosterone, which is considered to be the main glucocorticoid released in response to a stressing activity in the rat and the mouse (Khansari *et al.*, 1990; Ferry *et al.*, 1991; Simons, 1991), has a major effect on immune cells (Ortega *et al.*, 1996).

The mammalian pineal gland is innervated by sympathetic post-ganglionic nerve fibres that could be responsible for provoking changes in the production of melatonin at times of stress (Troiani *et al.*, 1988; Wu *et al.*, 1998). Various researchers assign melatonin a major anti-stress function in the organism (Aoyama *et al.*, 1986, 1987; Maestroni *et al.*, 1988; Maestroni and Conti, 1991). Indeed, Maestroni (1993) noted the possibility that melatonin acts as a 'buffer' in stress situations to neutralize the adverse immunological effects produced by the circadian release of glucocorticoids.

Previous results of our research team have shown that there occur changes during situations of stress in both the circulating melatonin and corticosterone levels and in the non-specific immune response (Ortega *et al.*, 1996; Barriga *et al.*, 2001), and that *in vitro* incubation with physiological concentrations of corticosterone and melatonin leads to variations in phagocyte activity (Rodríguez *et al.*, 2001; Barriga *et al.*, 2002). Given these antecedents, and that there have been relatively few published works addressing the question of whether these hormones intervene directly in the activation of the non-specific immune response, we carried out the present *in vitro* study to evaluate the effect of pharmacological concentrations of melatonin and corticosterone, separately and conjointly, on the phagocytic activity triggered by the ingestion of antigens in macrophages from mice subjected to forced swimming (the group of stressed animals), and from mice not subjected to any activity (basal animals).

We used concentrations ten and one hundred times greater than those found physiologically (Barriga *et al.*, 2001). These concentrations are in concordance with those used by Rogers *et al.* (1997). In particular, we used melatonin concentrations of 0.9 pg/ml and 9 pg/ml in the basal and of 0.7 pg/ml and 7 pg/ml in the stressed animals, and corticosterone concentrations of 2 μg/ml and 20 μg/ml in the basal and 5 μg/ml and 50 μg/ml in the stressed animals.

Melatonin at the concentrations tested led to an increase in phagocytosis by the macrophages due both to the increase in the number of activated macrophages, i.e. those with the capacity to ingest foreign material (PI), and in the number of particles ingested by each phagocyte (PE). This increase occurred in macrophages from both groups of mice — stressed and basal. These results coincide with those obtained for the phagocytic activity of ring dove (*Streptopelia risoria*) heterophils, in which there was a dose-dependent stimulatory effect (Rodríguez *et al.*, 1997).

With respect to corticosterone, our results indicate that this hormone increases the macrophages' capacity to ingest latex beads in the basal group of animals. In the stressed group, however, we observed that it had an immunosuppressory effect on phagocytosis. Similar results were found by Forner *et al.* (1995) after incubating macrophages with corticosterone concentrations ten times greater (7.2 μmol/ml) than those measured in plasma after subjecting BALB/c mice to forced swimming. At physiological concentrations, Ortega *et al.* (1996) and Rodríguez *et al.* (2001) observed an immunostimulation of phagocytic activity by corticosterone. This all seems to indicate that there is an 'optimal' range of concentrations in which glucocorticoids may induce stimulation of phagocytes.

Incubation with the two hormones conjointly at pharmacological concentrations stimulated the macrophages' antigen ingestion activity. The greatest effect was on the macrophages from basal (unstressed) animals, as had been observed in earlier work. Thus, Barriga *et al.* (2002) incubated macrophages *in vitro* with physiological concentrations of corticosterone and melatonin and observed that the greatest phagocytic stimulation was obtained in macrophages from mice in a basal situation and which had been incubated with the diurnal concentration of the two hormones (maximum corticosterone 200 ng/ml; minimum melatonin 15 pg/ml). Similar results were found by Rodríguez *et al.* (2001) in *Streptopelia risoria* peripheral blood heterophils incubated *in vitro* with physiological concentrations of melatonin and corticosterone, finding the greatest stimulation of the phagocytic capacity when the incubation was with the two hormones conjointly.

In conclusion, the findings of the present study demonstrate that pharmacological concentrations of melatonin and corticosterone possess immunoregulatory effects on the phagocytosis carried out by murine peritoneal macrophages. Whereas melatonin is immunostimulatory at the concentrations tested, the immunoregulatory properties of corticosterone depended on the situation, being immunodepressive for phagocytes from the stressed animals and immunostimulatory for those from the basal, unstressed animals. Of particular interest was the finding that when the incubation was with the two hormones together, the stimulation of the phagocytic activity in the basal group was greatest for the higher of the two concentrations of corticosterone tested, indicating, as suggested by Rodríguez *et al.* (2002) that the role of the two hormones is additive in modulating the phagocytic process *in vitro*.

Acknowledgements

This work was supported in part by a research grant from 'Junta de Extremadura-Consejeria de Sanidad y Consumo 2003–2004'.

REFERENCES

Aoyama, H., Mori, W. and Mori, N. (1986). Anti-glucocorticoid effects of melatonin in young rats, *Acta Pathol. Jpn.* **36**, 423–428.

Aoyama, H., Mori, N. and Mori, W. (1987). Anti-glucocorticoid effects of melatonin on adult rats, *Acta Pathol. Jpn.* **37**, 1143–1148.

Barriga, C., Campillo, J. E. and Ortega, E. (1992). Aspectos inmunológicos de la actividad física, in: *Fisiología de la Actividad Física*, J. Gonzalez Gallego (Ed.), pp. 161–174. Interamericana-McGraw-Hill.

Barriga, C., Martín, M. I., Tabla, R., *et al.* (2001). Circadian rhythm of melatonin, corticosterone and phagocytosis: effect of stress, *J. Pineal Res.* **30**, 180–187.

Barriga, C., Martín, M. I., Ortega, E., *et al.* (2002). Physiological concentrations of melatonin and corticosterone in stress and their relationship with phagocytic activity, *J. Neuroendocrinol.* **14**, 691–695.

Felten, S. Y. and Felten, D. L. (1991). The innervation of lymphoid tissue, in: *Psychoneuroimmunology*, Arder, R., Cohen, N. and Felten, D. L. (Eds), pp. 27–70. Academic Press, New York.

Ferry, A., Weil, B., Amiridis, I., *et al.* (1991). Splenic immunomodulation with swimming-induced stress in rats, *Immunol. Lett.* **29**, 261–264.

Forner, M. A., Barriga, C., Rodríguez, A. B., *et al.* (1995). A study of the role of corticosterone as a mediator in exercise-induced stimulations of murine macrophage phagocytosis, *J. Physiol.* **488**, 789–794.

Khansari, D. N., Murgo, A. J. and Faith, R. E. (1990). Effects of stress on the immune system, *Immunology Today* **11**, 170–175.

Maestroni, G. J. M. (1993). The immunoendocrine role of melatonin, *J. Pineal Res.* **14**, 121–126.

Maestroni, G. J. M. and Conti, A. (1991). Anti-stress role of the melatonin-immuno-opioid network: evidence for a physiological mechanism involving T-cell derived, immunoreactive-endorphin and met-enkephalin binding to thymic opioid receptors, *Intern. J. Neurosci.* **61**, 289–298.

Maestroni, G. J. M., Conti, A. and Pierpaoli, W. (1988). Role of the pineal gland in immunity. III: Melatonin antagonizes the immunosuppressive effect of acute stress via an opiatergic mechanism, *Immunology* **63**, 465–469.

Ortega, E., Forner, M. A., Barriga, C., *et al.* (1993). Effect of age and swimming-induced stress on the phagocytic capacity of peritoneal macrophages from mice, *Mech. Ageing Devel.* **70**, 53–63.

Ortega, E., Rodríguez, A. B., Barriga, C., *et al.* (1996). Corticosterone, prolactin and thyroid hormones as mediators of stimulated phagocytic capacity of peritoneal macrophages after high-intensity exercise, *Int. J. Sports Med.* **17**, 149–155.

Rodríguez, A. B., Ortega, E., Lea, R. W., *et al.* (1997). Melatonin and the phagocytic process of heterophils from ring dove (*Streptopelia risoria*), *Mol. Cell. Biochem.* **168**, 185–190.

Rodriguez, A. B., Terron, M. P., Duran, J., *et al.* (2001). Physiological concentrations of melatonin and corticosterone affect phagocytosis and oxidative metabolism in ring dove heterophils, *J. Pineal Res.* **31**, 31–38.

Rogers, N., van den Heuvel, C. and Dawaon, D. (1997). Effect of melatonin and corticosteroid on in vitro cellular immune function in humans, *J. Pineal Res.* **22**, 75–80.

Roszman, T. L. and Carlson, S. L. (1991). Neurotransmitter and molecular signalling in the immune respose, in: *Psychoneuroimmunology*, Arder, R., Cohen, N. and Felten, D. L. (Eds), pp. 311–335. Academic Press, New York.

Simon, H. B. (1991). Exercise and human immune functions, in: *Psychoneuroimmunology*, Arder, R., Felten, D. L. and Cohen, N. (Eds), pp. 869–895. Academic Press, New York.

Skwarlo-Sonta, K. (1996). Functional connections between the pineal gland and immune system, *Acta Neurobiol. Exp.* **56**, 341–357.

Troiani, M. E., Reiter, R. J., Tannenbaum, M. G., *et al.* (1988). Neither the pituitary gland nor the sympathetic nervous system is responsible for eliciting the large drop in elevated rat pineal melatonin levels due to swimming, *J. Neural Transm.* **74**, 146–160.

Wu, W. Y., Chen, Y. C. and Reiter, R. J. (1988). Day-night differences in the responses of the pineal gland to swimming stress, *Proc. Soc. Exp. Biol. Med.* **187**, 315–319.

Advances in Neuroregulation and Neuroprotection (2005), pp. 29-42
Collin, C. *et al.* (Eds)
© VSP 2005

Cobalamin deficiency and its dynamic impact on neurotoxicity and oxidative stress in Parkinson's patients in on-off status on L-dopa treatment

G. A. QURESHI [1,*], S. A. MEMON [1], C. COLLIN [2] and S. H. PARVEZ [2]

[1] *M. A. Kazi Institute of Chemistry and Biochemistry, Sind University, Jamshoro, Pakistan*
[2] *Neuroendocrine Unit, CNRS-Institute Alfred Fessard, Bat 5, Parc Chateau CNRS, 91190 Gif/Yvette, France*

Abstract—L-Dopa is considered to be one of the best drug therapies in treating Parkinson's patients with positive outcome in the beginning of its treatment. However, as long period of treatment has shown results with both on and off phenomena and high doses show various side effects. In this study, both On and Off PD patients are investigated to determine the role of biochemical markers such as cobalamin (Vitamin B12) and homocysteine in the cerebrospinal fluid (CSF). These biochemical changes in the CSF could be considered as vital information in designing new drugs in treating these patients for a longer period.

Keywords: Parkinson's patients; on-off phenomena; vitamin B12; homocysteine; L-dopa treatment; oxidative stress; neurotoxicity.

INTRODUCTION

Parkinson's disease (PD) is the second most common neurodegenerative disorder worldwide and is characterized by cardinal clinical features and specific pathological findings. It is possible to detect PD early on in the course of the disease, and certain laboratory studies may identify preclinical stages. Postulates regarding pathogenesis, such as oxidative stress and excitotoxicity, have led to the discovery of abnormal mitochondrial function in PD and a search for biochemical markers. Functional imaging studies have detected subclinical nigral dopaminergic dysfunction in individuals at risk of developing PD. Current symptomatic therapies are aimed at enhancing dopaminergic transmission. However, some commonly used PD medications may alternative reactions with both symptomatic and neuroprotec-

---

*To whom correspondence should be addressed. E-mail: ghulamali48@hotmail.com

tive consequences (Di Paolo *et al.*, 1996). There is now a wide variety of drugs and formulations available, including anticholinergics, amantidine, L-dopa, dopamine agonists including apomorphine, selegiline and soon to be available catechol-O-methyltransferase inhibitors. Disabling side-effects of treatment, fluctuations, dyskinesias and psychiatric problems require strategic use of the drugs available. There is an increasing potential for neurosurgical intervention with regard to neuroprotection. Despite these advances, until there is a better understanding of the aetiology and pathogenesis of PD, there will be no definitive long-term benefit of early diagnosis and treatment of PD. Dopamine neurons in the substantia nigra of human brain are selectively vulnerable and the number decline by aging at 5–10% per decade.

Enzymatic and non-enzymatic oxidation of dopamine generates reactive oxygen species, which induces apoptotic cell death in dopamine neurons. Parkinson's disease (PD) is also caused by selective cell death of dopamine neurons in this brain region (Bernheimer *et al.*, 1973; Hugh, 1992; Qureshi *et al.*, 2003b). The hallmark of PD is progressive loss of DA neurons in the midbrain, most pronounced in substantia nigra pars compacta even though other nerve cell populations using monoamine transmitters such as the locus coeruleus containing noradrenaline neurons also are affected (Graybiel and Ragsdale, 1979; German *et al.*, 1992; Gerfen, 1992; Qureshi *et al.*, 2003a).

The etiology of neuronal death in neurodegenerative diseases remains mysterious; however, great advances in both molecular genetics and neurochemistry have improved our knowledge of fundamental processes involved in cell death, including both excitotoxicity and oxidative damage (Wernek and Alvarenga, 1999). Several factors such as inhibition of the mitochondrial respiration, generation of hydroxyl and nitric oxide radicals and reduced free radical defense mechanisms causing oxidative stress, have been postulated to contribute to the degeneration of dopaminergic neurons (Thiessen *et al.*, 1990; Smith *et al.*, 1991; Qureshi *et al.*, 2003b). Using catecholamine staining techniques (Carlsson *et al.*, 1957), several nerve groups were identified utilizing DA as their transmitter. Oxidative stress, which results from an imbalance between oxidant production and antioxidant defense mechanisms, can promote modifications of lipids, proteins and nucleic acids. In the past decade or so, a convincing link between oxidative stress and degenerative conditions has been made and with the knowledge that oxidative changes may actually trigger deterioration in cell function, a great deal energy has focussed on identifying agents may have possible therapeutic value in combating oxidative changes with antioxidants may prevent or reduce the rate of progression of this disease (Tatton *et al.*, 2003).

The discovery of several genetic mutations associated with PD raises the possibility that these or other biomarkers of the disease may help to identify persons at risk of PD. Transcranial ultrasound has shown susceptibility factors for PD related to an increased iron load of the substantia nigra. In the early clinical phase, a number of motor and particularly non-motor signs emerge, which can be identified by the patients and physicians several years before the diagnosis is made, notably olfactory dysfunction, depression, or 'soft' motor signs such as changes in handwriting, speech or reduced ambulatory arm motion. These signs of the early, prediagnostic phase of PD can be detected by inexpensive and easy-to-administer tests (Becker *et al.*, 2002).

There is no specific cure for PD and there is limitations to current PD therapy (Rascol *et al.*, 2003). However, there are several treatments available to alleviate symptoms of the disease, such as pharmacological therapy (Korczyn *et al.*, 2002) and surgical interference (Playfer, 1997; Gullingham, 2000). The most successful candidate for drug therapy is still L-dopa that is a precursor in DA biosynthesis. The introduction of L-dopa was revolutionary in that it became an excellent treatment in an earlier essentially untreatable disorder. Mortality has been nearly normalized and L-dopa is still considered the most effective agent for treatment of PD. However, disabling motor complications occur in the late stages of PD with the most common effect that includes onset of on-off phenomena and abnormal involuntary movements called dyskinesia (Jankovic, 2002); hence the need for better, perhaps more physiologically based therapy for this complex disorder remains. Most patients treated with L-dopa develop fluctuation in motor performance. After 3–5 years of treatment one-third, after 5–7 years about half and after 10–12 years nearly all patients suffer from the motor fluctuation (Rinne, 1983). Non-motor symptoms such as pain, fatigue, anxiety and depression are more often seen in PD patients treated with L-dopa for a long time (>10 years) than are seen in patients with classical motor disturbances (Shulman *et al.*, 2002; Calon and Di Paolo, 2002).

Hence, it has been suggested that L-dopa therapy should be delayed as long as possible (Scigliano *et al.*, 1997). There is increasing evidence that free radical nitric oxide NO˙ plays an important role in pathophysiology of a variety of CNS disorders (Dawson *et al.*, 1992; Qureshi *et al.*, 2003b; Koutsilieri *et al.*, 2002). Besides, NO exposure causes cobalamin deficiency resulting from its inactivity which is known to produce impaired neurological functions (Stollhoff *et al.*, 1987; Hall, 1990). Congenital defects of cobalamine metabolism and nutritional deficiencies have been reported to produce abnormal CNS and peripheral neuronal function in humans (Hall, 1990).

The present study deals with CSF analysis of homocysteine and cobalamine in PD patients in on-off pathological conditions. These levels are reported with the assumption that CSF concentrations reflect brain or spinal cord concentrations and perhaps synaptic activity, since CSF is in constant exchange with the extracellular fluid of the CNS. Based on the facts that most of the transmitters in the brain are present in the CSF (Herkenham, 1987) and the alterations in the concentration of transmitters in the CSF can alter the same transmitter in the brain of man (Wester *et al.*, 1990), the results are compiled in two groups of PD patients showing on and off phenomena, and are compared with healthy controls.

MATERIAL AND METHODS

Patients and healthy subjects and routine analysis

Most of the patients were recruited from the Department of Neurology, Huddinge University Hospital, Sweden. 30 patients (12 women and 18 men) with PD were included in this study. The research ethics committee of Karolinska Institute, Stockholm approved the study and all patients gave written informed consent to participate. 15 patients at different stages of illness (Hoehn and Yar range 2–4) were on individual drug combinations. All 15 patients were on L-dopa (250 mg/day) and had taken it with combination of other drugs with positive response towards their motor activities (Group PD On). 15 PD patients who had been treated for 6–8 years with similar therapy showed severely 'off' response (Hoehn and Yar range of −3) (Group PD Off). These Off PD group showed motor fluctuation. Both On and Off groups were monitored one week before the samples of CSF and blood were taken and all food and fluid intake were similar in both groups and exact dose times were maintained. Blood and CSF were taken after one week of observation early in the morning between 6–8 a.m. in a fasting condition. 16 healthy individual volunteers from the institute employees were included and their blood and CSF were collected under comparable conditions. None of the healthy controls was on medication for six months. Table 1 shows the clinical data on PD patients and healthy controls.

Ten-twelve ml cerebrospinal fluid (CSF) was collected from each PD patient and healthy subject in a sitting position at the L4–L5 levels. Blood samples were collected by vanipuncture. The basic CSF analyses included cell counting by phase-contrast microscopy (Siesjö, 1967), determination of CSF/serum albumin ratio and CSF/immunoglobulin G (IgG) index (Link and Tibbling, 1977) as well as isoelectric focusing for detection of oligoclonal IgG band. Serum and CSF albumin and IgG were determined using Hitachi 737 Automatic Analyzer (Naka Works, Hitachi Ltd., Tokyo, Japan). CSF and serum samples were kept at −80°C.

Analysis of amino acids, homocysteine, vitamin B12 and nitrite

Amino acid analysis was performed based on the pre-column derivatization of OPA with amino acids as previously described (Qureshi and Baig, 1988). The reproducibility for each amino acid was <2% with this method.

Table 1.
Clinical data on the healthy subjects (HS), patients with Parkinson's disease (PD) in both On and Off groups

Patients	n	Females	Age (years)	CSF-Albumin	CSF-IgG	IgG-index
Controls	16	7	53 ± 5	218 ± 17	34 ± 4	0.44 ± 0.01
PD (On)	15	7	72 ± 12	332 ± 42*	51 ± 7**	0.43 ± 0.01
PD (Off)	15	5	74 ± 14	313 ± 37*	44 ± 7*	0.42 ± 0.01

*$p < 0.05$, **$p < 0.01$.

Because of the short half-life time of NO, its concentration was quantified indirectly by measuring the levels of its degraded product nitrite. A recent HPLC method in combination with electrochemical detection was used for this purpose (Kaku *et al.*, 1994).

Homocysteine was quantified using a previously described HPLC method (Hyland and Bottiglieri, 1992) and vitamin B12 was quantified using a competitive protein binding assay (CPBA) based on radioactivity measurement with intrinsic factor (obtain from Amershan International, Buck England) (Ikeda *et al.*, 1992).

Data are presented as mean ± SEM. Differences in concentration of vitamin B12 and homocysteine were analyzed with ANOVA and group comparisons were made with *t*-text. A *p*-value less than 0.05 was considered significant. Correlation between homocysteine vs. nitrite was submitted to a stepwise polynomial regression analysis (Zar, 1984).

RESULTS

Figure 1 shows the CSF levels of vitamin B12 in healthy controls (HS) and PD patients in On and Off groups and in comparison to Controls, the level of vitamin B12 decreases from 0.079 ± 0.006 μmol/l to 0.059 ± 0.005 ($p < 0.001$) in PD patients in the On group and 0.041 ± 0.003 ($p < 0.001$) in PD patients in Off group. Figure 2 shows the CSF levels of homocysteine (HC) which increased from 1.09 ± 0.096 μmol/l in Controls to 1.62 ± 0.13 ($p < 0.001$) and 2.26 ± 0.23

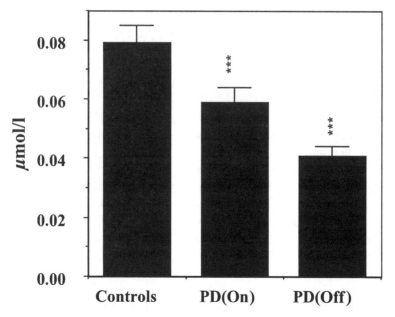

Figure 1. CSF levels of vitamin B12 in healthy controls and in On and Off groups of PD patients on L-dopa therapy.

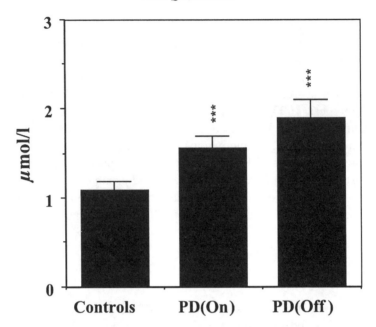

Figure 2. CSF levels of homocysteine in PD patients (On and Off groups) and healthy controls. All the values are expressed as mean ± SEM. *** $p < 0.001$.

($p < 0.001$) in PD patients. In On and Off groups, respectively. Figure 3 shows the relationship between CSF homocysteine and nitrite levels in Controls and patients in both On and Off groups.

DISCUSSION

Tremor, akinesia, rigidity and postual instability are key signs of Parkinson's disease. The most important one is akinesia, which includes decreased spontaneous locomotor activity, slowness of movement, awkwardness and freezing. The main pathophysiology of Parkinson's disease is neurodegeneration of nigrostriatal dopaminergic neurons.

A great deal of interest has been focused on the possibility that oxidative demage plays an important role in the pathogenesis of various degenerative disorders (Smith *et al.*, 1991; Dexter *et al.*, 1994; Koroshetz *et al.*, 1994; Qureshi *et al.*, 1997). Hence, the role of free radical in cell death induced by activation of EAA receptors is an area of expanding interest (Kontos, 1989; Halliwell, 1992; Coyle *et al.*, 1993; Qureshi *et al.*, 1995; Qureshi *et al.*, 2003b). Biological systems have developed a comprehensive array of defense mechanisms to protect against free radicals. These include enzymes to decompose peroxides, proteins to sequester transition metal ions and a range of compounds to scavenge free radicals. Oxidative stress is caused by accumulation of free radicals, and is defined as a disturbance in prooxidant-antioxidant balance in favour of the former, leading to potentially

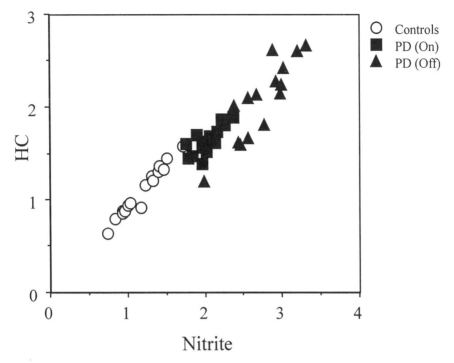

Figure 3. Relationship between homocysteine and nitrite in both On and Off PD patients and healthy controls.

damaging reactions with biological molecules (Sies, 1985). The results indicate a possible participation of oxidative stress in the neuropathology of PD patients, especially during a crisis, when the metabolites are highly increased, and point to the use of antioxidant drugs as a possible adjuvant therapy in such situations to improve the neurological status of the patients and to prevent sequelae.

All of the vitamin B12 or cobalamin in nature is of microbial origin. Cobalamin is a large polar molecule that must be bound to specialized transport proteins to gain entry into cells. Entry from the lumen of the intestine under physiological conditions occurs only in the ileum and only when bound to intrinsic factor. It is transported into all other cells only when bound to another transport protein, transcobalamin II. Congenital absence or defective synthesis of intrinsic factor or transcobalamin II result in megaloblastic anemia. In most bacteria and in all mammals, cobalamin regulates DNA synthesis indirectly through its effect on a step in folate metabolism, the conversion of N5-methyltetrahydrofolate to tetrahydrofolate, which in turn is linked to the conversion of homocysteine to methionine (Lokk, 2003). This reaction occurs in the cytoplasm, and it is catalyzed by methionine synthase, which requires methyl cobalamin (MeCbl), one of the two coenzymes forms of the vitamin, as a cofactor. Defects in the generation of MeCbl result in homocystinuria; affected infants present with megaloblastic anemia, retardation, and neurological and ocular defects. 5′-Deoxyadenosyl cobalamin (AdoCbl), the other coenzyme form of cobalamin is present within mitochondria, and it is an essential cofactor for the enzyme methylmalonyl-CoA mutase, which converts L-methylmalonyl CoA to succinyl CoA. This reaction is used in the pathway for the metabolism of odd chain fatty acids via propionic acid, as well as that of amino acids iosleucine, methionine, threonine, and valine (Lokk, 2003). Impaired synthesis of AdoCbl (cobalamin A or B disease) results in infants with methylmalonic aciduria who ar mentally retarded, hypotonic, and who present with metabolic acidosis, hypoglycemia, ketonemia, hyperglycemia and hyperammonemia. Megaloblastic anemia does not develop in these children because adequate amounts of MeCbl are present, but the effect of methylmalonic acid on marrow stem cells may give rise to pancytopenia. Congenital absence of reductases in the cytoplasm, which normally reduce the cobalt atom in cobalamin from its oxidized to its reduced state (cobalamin C and D diseases), results in impaired synthesis of both MeCbl and AdoCbl. Both methylmalonic aciduria and homocystinuria therefore develop in these children, and they present with megaloblastosis, mental retardation, a host of neurological and ocular disorders, and failure to thrive; however, they do not have hyperglycinemia or hyperammonemia. A similar biochemical profile and clinical presentation is also seen in cobalamin F disease, which results from a defect in the release of cobalamin from lysosomes, following receptor-mediated endocytosis of the transcobalamin II-cobalamin complex into cells (Lokk, 2003). It is important to recognize these inborn errors of cobalamin absorption, transport, or function as soon after birth as possible, because most respond (in some patients more fully than others) to parenteral administration of cobalamin. Delays in diagnosis can lead to grave clinical consequences (Kapadia, 1995; Qureshi *et al.*, 1998).

In our study, significant decrease in CSF levels of vitamin B12 and increased in homocysteine were observed in both groups of PD patients (Figs 1 and 2). Folate, cobalamin and pyridoxine deficiency are associated with psychiatric and neurological symptomatology. Disturbances in sulfur amino acid metabolism leading to accumulation of homocysteine occurs in all three conditions, as the metabolism of homocysteine depends on enzymes requiring these vitamins as cofactors (Santosh-Kumar *et al.*, 1994). Oxidation products of homocysteine (homocysteine sulfinic acid and homocysteic acid) and cysteine (cysteine sulfinic acid and cysteic acid) are excitatory sulfur amino acids and may act as excitatory neurotransmitters, whereas taurine and hypotaurine (decarboxylation products of cysteic acid and cysteine sulfinic acid) may act as inhibitory transmitters. Homocysteic acid and cysteine sulfinic acid have been considered as endogenous ligands for N-methyl-D-aspartate (NMDA) type of glutamate receptors. The profile of these sulfur amino acid neurotransmitters could be altered in a similar fashion in states of decreased availability of folate, cobalamin or pyridoxine. It is proposed that the mechanism of neuropsychiatric manifestations in all three conditions result from a combination of two insults to homocysteine catabolism in the brain. Serum vitamin B12 levels are often low in human immunodeficiency virus (HIV)-infected patients (Remacha *et al.*, 1999).

The biochemical and clinical aspects of methylation in neuropsychiatric disorders and the clinical potential of their treatment with ademethionine (S-adenosylmethionine; SAMe). SAMe is required in numerous transmethylation reactions involving nucleic acids, proteins, phospholipids, amines and other neurotransmitters. The synthesis of SAMe is intimately linked with folate and vitamin B12 (cyanocobalamin) metabolism, and deficiencies of both these vitamins have been found to reduce CNS SAMe concentrations. Both folate and vitamin B12 deficiency may cause similar neurological and psychiatric disturbances, including depression, dementia, myelopathy and peripheral neuropathy. SAMe has a variety of pharmacological effects in the CNS, especially on monoamine neurotransmitter metabolism and receptor systems. SAMe has antidepressant properties, and preliminary studies indicate that it may improve cognitive function in patients with dementia. Treatment with methyl donors (betaine, methionine and SAMe) is associated with remyelination in patients with inborn errors of folate and C-1 (one-carbon) metabolism. These studies support a current theory that impaired methylation may occur by different mechanisms in several neurological and psychiatric disorders (Bottiglieri *et al.*, 1994). An abnormal uptake and/or utilization of the vitamin B12 in the brain leads to deficiency and may contribute to a number of neurologicaland psychiatric abnormalities (Ikeda *et al.*, 1992). Lack of Ca^{+2} in the food can also reduce the uptake and so can heavy metals (Shevell and Rosenblatt, 1992). Vitamin B12 deficiencies are followed by neurological and psychiatric disorders such as disturbed sense of coordination, loss of memory, abnormal reflexes, weakness, loss of muscle strength, exhaustion, confusion, low self-confidence, incontinenece, impaired vision, abnormal gait, frequent need to pass water and physiological deviance (Metz, 1992).

It is also known that free radical nitrous oxide $NO^{\bullet}$ interacts with vitamin B12 resulting in selective inhibition of methionine synthase, a key enzyme in methionine and folate metabolism. Thus, nitrous oxide may alter one-carbon and methyl-group transfer, which is most important for DNA, purine and thymidylate synthesis. There is correlation in the levels of nitrite and homocysteine suggesting that both these parameters are interrelated (Fig. 3). HC accumulates in methylcobalamin deficiency and is thought to be neurotoxic for CNS (Qureshi *et al.*, 1998).

Recent studies seem to suggest a correlation between nitrous oxide anaesthesia and hyperhomocysteinaemia which is accepted to be an independent risk factor for coronary artery disease (Weimann, 2003). Although the mechanisms responsible for the neurological lesion of vitamin B12 deficiencies are less well understood, vitamin B12 analogues including methylcobalamin (MCbl) have been widely used in therapy of neurological diseases (Chanarin *et al.*, 1988). Vitamin B12 was shown to improve memory, emotional function and communication ability in Alzheimer's patients (Ikeda *et al.*, 1992). MCbl is an active coenzyme of vitamin B12 analogue that is essential for cell growth and replication (Hall, 1990). In fact, detection of vitamin B12 deficiency in humans is a problem to be solved in the clinical area, because data on its levels in blood do not always correlate with CSF levels (Ikeda *et al.*, 1992). However, there is now good evidence that the majority (>90%) of the patients who are deficient in vitamin B12 accumulate high levels of methylmalonic acid and homocysteine in the blood. Hence, both these substances can be considered a better marker for the clinical diagnosis of vitamin B12 deficiency (Chandefaux et al., 1994).

NO toxicity due to its interaction with vitamin B12-dependent enzyme (methionine synthase) produces haematological and less frequently neurological symptoms (Yagiela, 1991).

NO is known to oxidize active reduced Cob(II)alamin to inactive Cob(III)alamin (Banks et al., 1968). As Cbl in the form of reduced MCbl is required as cofactor for methionine synthetase, exposure to NO causes rapid inactivation of this enzyme (Deacon et al., 1978). The inactive CBl is excreted, so that repeated exposure to NO results in depletion of body Cbl stores, with reduced Ado Cbl, so that the activity of the Ado Cbl dependent enzyme MMCoA is also affected (Deacon et al., 1978). Long-term Cbl deficiency resulting from its inactivation following NO exposure can produce impaired neurological function. Congenital defects of Cbl metabolism and nutritional deficiencies have also been reported to produce abnormal CNS and peripheral neuronal function in humans. NO interaction with cobalamin results in the liberation of hydroxy free radicals responsible for inactivation of methionine synthase (Drummond and Methews, 1994). Since our endogenous antioxidant defenses are not always completely effective, and since exposure to damaging environmental factors is increasing, it seems reasonable to propose that exogenous antioxidants could be very effective in diminishing the cumulative effects of oxidative damage. Antioxidants of widely varying chemical structures have been investigated as potential therapeutic agents. However, the therapeutic use of the most of these compounds is limited since they do not cross the blood brain barrier (BBB). Although a few of them have shown limited efficiency in animal models or in small clinical studies, none of the currently available antioxidants proven efficacious in a large-scale controlled study. Therefore, any novel antioxidant molecules designed as potential neuroprotective treatments in acute or chronic neurological disorders should have the mandatory prerequisite that they can cross the BBB after systemic administration (Hunot et al., 2003).

Future therapy in PD is likely to include a 'cocktail' of neuroprotective compounds to interfere with several molecular pathways that lead to neuronal injury. In using therapeutic strategies aimed towards retarding or arresting neuronal death, as observed in PD, close attention will need to be paid to quality of life issues. One of the main problems during the many years of research on PD is to detect the disease early. The disease is diagnosed on a clinical basis and there are no diagnostis tests. However, a positive response to levodopa administration is obvious and separates PD from other neurological disorders, which present with parkinsonism. New methods for early detection and monitoring of the disease progression to help clinicians in their clinical diagnosis are important, particularly in future pharmacological treatments that can halt further neurodegeneration and nerve cell death. One of the tools that has been used during last few decades is to examine the patient's brain with positron emission tomography (PET). Combining this technique with analysis of biochemical, pathological and behavioral changes, more information on the PD patients could be monitored.

There is probably no single intracellular pathway that is responsible for the cell death of neurological diseases. However, hyperactivity of the glutaminergic innervation, the associated activation of NMDA-receptor ion channels selective for Ca^{+2} and ensuing increase in the intracellular concentration of Ca^{+2} may contribute to the neurodegenerative process (Rothman and Olney, 1995). Besides, oxygen-free radicals, nitric oxide (NO) in particular is thought to mediate NMDA-toxicity (Dawson and Dawson, 1996). The mechanism by which NO kills brain cells in unknown, however, free radical formation has been generally implicated in various forms of neurotoxicity (Whetsell, 1996). Our previous results (Qureshi *et al.*, 2003b) do indicate increased production of nitrite in both PD patients i.e. On and Off groups and decreased CSF level of vitamin B12 in both groups, as we have shown in this study. There is clear evidence that of a linear correlation exists between the level of homocysteine and nitrite in PD patients. On the basis of these results on degenerative disorders, the role of various neurotransmitters, and the role of free radical NO is clearly defined in order to develop effective drug therapy. This may include glutamate release inhibitors, EAA antagonists, agents to improve mitochondrial function, free radical scavengers and neurotropic factors, intake of antioxidants along with vitamin B12. Furthermore, the purpose of selective exogenous antioxidants could be that they should be very effective in diminishing the cumulative effects of oxidative damage, as the currently available antioxidants have proven efficacious in a large-scale controlled study.

CONCLUSIONS

Advances in symptomatic therapy for PD patients include the development of specialist clinics, with input from multidisciplinary teams, as well as hospice care in the late stages of the disease. A number of recent therapeutic trials of potential neuroprotective drugs have been conducted. These results undoubtedly provided us with a possibility in understanding the interrelationship between different neurotransmitters in the brain as well as their distribution and functions. Therefore, any novel antioxidant molecules designed as potential neuroprotective treatment in acute or chronic neurological disorders such as PD should have the mandatory prerequisite that they can cross the BBB after systemic administration.

REFERENCES

Bank, R. G. S., Henderson, J. R. and Pratt, J. M. (1968). Reaction of gases in solution. III. Some reactions of nitric oxide and transition metal complexes, *J. Chem. Soc. A*, 2886–2890.

Becker, G., Muller, A., Braune, S., *et al.* (2002). Early diagnosis of Parkinson's disease, *J. Neurol.* **249**, 40–48.

Berheimer, H., Birkmayar, W., Hornkeiwics, O., *et al.* (1973). Brain dopamine and syndromes of Parkinson's and Huntington. Clinical, morphological and neurochemical correlations, *J. Neurosci.* **17**, 6761–6768.

Bottiglieri, T., Hyland, K. and Reynold, E. H. (1994). The clinical potential of ademethionine in neurological disorders, *Drugs* **48**, 137–152.

Calon, F. and Di Paolo, T. (2002). Levodopa response motor complications — GABA receptors and preproenkephalin expression in human brain, *Parkinsonism Relat. Disord.* **6**, 449–454.

Carlsson, A., Lindquist, M. and Magnusson, T. (1957). 3,4-dihydroxyphenylalanine and 5-hydroxy-tryptophan as reserpine antagonists, *Nature* **180**, 1200–1201.

Chanarin, I., Deacon, R., Lumb, M., *et al.* (1988). Cobalamin folate interrations: A critical review, *Blood* **66**, 479–482.

Chandefaux, B., Cooper, B. A., Gilfix, B. M., *et al.* (1994). Homocysteine relationship to serum cobalamin, serum folate, erythrocyte folate, and lobation of neutrophils, *Clin. Invest. Med.* **17**, 540–550.

Coyle, J. T. and Puttfarcken, P. (1993). Oxidative stress, glutamate and neurodegenerative disorders, *Science* **262**, 689–694.

Dawson, T. M., Dawson, V. L. and Snyder, S. H. (1992). A novel neuronal messenger molecule in brain: The free radical, nitric oxide, *Ann. Neurol.* **32**, 297–311.

Dawson, V. L. and Dawson, T. M. (1996). Free radicals and neuronal cell death, *Cell Death and Differentiation* **3**, 71–78.

Deacon, R., Lamb, M. and Perry, J. (1978). Selective inactivation of vitamin B12 in rats by nitrous oxide, *Lancet* **2**, 1023–1024.

Dexter, D. T., Holley, A. E. and Flitter, W. D. (1994). Increased levels of lipid hydroperoxides in Parkinsonian substantia nigra: An HPLC and ESR study, *Mov. Disord.* **9**, 92–97.

Di Paolo, R. and Uitti, R. J. (1996). Early detection of Parkinson's disease. Implication for treatment, *Drug Aging* **9**, 159–168.

Drummond, J. T. and Methews, R. G. (1994). Nitrous oxide degradation by Cobalamin dependent synthase: Characterization of reactants and products in the inactivation reaction, *Biochemistry* **33**, 3732–3741.

Grefen, C. R. (1992). The neostriatal mosaic: multiple levels of compartmental organization, *Trend Neurosci.* **15**, 133–139.

German, D. C., Manaye, K. F., White, C. L., *et al.* (1992). Disease-specific patterns of locus coeruleus cell loss, *Ann. Neurol.* **32**, 667–676.

Graybiel, A. M. and Rogsdale, C. W. (1979). Fiber connections of the basal ganglia, *Prog. Brain Res.* **51**, 279–283.

Hall, C. A. (1990). Function of vitamin B12 in central nervous system as revealed by congenital defects, *Amer. J. Hemat.* **34**, 121–127.

Halliwell, B. (1992). Reactive oxygen species and the central nervous system, *J. Neurochem.* **59**, 1609–1614.

Halliwell, B. and Gutteridge, J. M. C. (1989). *Free Radicals in Biology and Medicine*. Clarendon Press, Oxford.

Herkenham, M. (1987). Mismatches between neurotransmitters and receptor localizations in brain: observations and implications, *Neuroscience* **23**, 1–38.

Hornykeiwicz, O. (1966). Dopamine and brain function, *Pharmacol. Rev.* **18**, 925–964.

Hughes, A. J. D. S., Kilford, L. and Lees, A. J. (1992). Accuracy of clinical diagnosis of idiopathic Parkinson's disease: a clinico-pathological study of 100 cases, *J. Neurol. Neurosurg. Psychiatry* **55**, 181–185.

Hunot, S. and Hirsch, E. C. (2003). Neuroinflammatory processes in Parkinson's disease, *Ann. Neurol.* **53** (Suppl. 3), S49–S60.

Hyland, K. and Bottiglieri, T. (1992). Measurement of total plasma and CSF homocysteine by fluorescence detection followed HPLC and pre-column derivatization with OPA, *J. Chromatogr.* **579**, 55–62.

Ikeda, T., Yamamoto, K., Takahashi, K., *et al.* (1992). Treatment of Alzheimer-type dementia with intravenous cobalamin, *Clin. Therapeutics* **14**, 426–437.

Jancovic, J. (2002). Levadopa strengths and weaknesses, *Neurology* **58**, S19–S32.

Kaku, S., Tanaka, M., Muramatsu, M., *et al.* (1994). Determination of nitrite by HPLC with EC: Measurement of nitric oxide synthase activity in rat cerebellum cytosol, *J. Biomed. Chromatogr.* **8**, 24–28.

Kapadia, C. R. (1995). Vitamin B12 in health and disease. Part 1. Inherited disorders of function, absorption and transport, *Gastroenterologist* **3**, 329–344.

Kontos, H. A. (1989). Oxygen radicals in CNS damage, *Chem. Biol. Interact.* **72**, 229–236.

Koutsilleri, E., Scheller, C., Grunblatt, E., *et al.* (2002). Free radicals in Parkinson's Disease, *J. Neurol.* **249**, 1–5.

Koroshetz, W. J., Jenkins, B. and Rosen, B. (1994). Evidence from metabolic disorder in Huntington's disease, *Neurology* **44** (Suppl. 2), A338.

Link, H. and Tibbling, G. (1977). Principles of albumin and IgG analyses in neurological disorders, III. Evaluation of IgG synthesis within the central nervous system in multiple sclerosis, *Scand. J. Clin. Lab. Invest.* **37**, 297–301.

Lokk, J. (2003). Treatment with levodopa can effect latent vitamin B12 and folic acid deficiency. Patients with Parkinson disease run the risk of elevated homocysteine levels, *Lakatidningen* **100**, 2674–2677.

Playfer, J. R. (1997). Parkinson's disease, *Postgrad. Med. J.* **73**, 257–264.

Metz, J. (1992). Cobalamin deficiency and the pathogenesis of nervous system disease, *Ann. Ren. Nub.* **12**, 59–79.

Qureshi, G. A. and Baig, S. M. (1988). Quantitation of amino acids in biological samples by HPLC, *J. Chromatogr.* **459**, 237–244.

Qureshi, G. A., Baig, S., Bednar, I., *et al.* (1995). Increased cerebrospinal fluid concentration of nitrite in Parkinson's disease, *NeuroReport* **6**, 1642–1644.

Qureshi, G. A., Halawa, A., Baig, S. M., *et al.* (1997). The neurochemical markers in cerebrospinal fluid to differentiate between viral and bacterial meningitis, *J. Neurochem.* **32**, 197–203.

Qureshi, G. A. and Baig, S. M. (1998). The relationship between the deficiency of vitamin B12 and neurotoxicity of homocysteine with nitrite level in cerebrospinal fluid of neurological patients, *Biogenic Amines* **14**.

Qureshi, G. A., Memon, S. A., Memon, D. and Parvez, H. (2003a). Levels of catecholamines and 5-hydroxy-tryptamine and eating behavior in lateral hypothalamic aphagic rats, *Biogenic Amines* **18**, 19–28.

Qureshi, G. A., Memon, S. A., Qureshi, A. A., Collin, C. and Parvez, S. H. (2003b). Neurotoxicity and dynamic impact of oxidative stress in neural regulation of Parkinson's patients in on-off phenomena, *Biogenic Amines* **18**, 55–78.

Rascol, O., Payous, P., Ory, F., *et al.* (2003). Limitation of current Parkinson's disease therapy, *Annals Neurology* **53**, S3–S-15.

Remacha, A. F. and Cadafalch, J. (1999). Cobalamin deficiency in patients infected with human immunodeficiency virus, *Semin. Hematol.* **36**, 75–87.

Rinne, U. K. (1983). Problems associated with long term levodopa treatment of Parkinson's disease, *Acta Neurolog. Scand. Suppl.* **95**, 19–26.

Rothman, S. M. and Olney, J. W. (1995). Excitatory and NMDA receptors — still lethal after eight years, *Trends in Neurosci.* **18**, 57–58.

Santhosh-Kumar, C. R., Hassell, K. L., *et al.* (1994). Are neurophychiatric manifestations of folate, cobalamin and pyridoxine deficiency mediated through imbalances in excitatory sulfur amino acids? *Med. Hypotheses* **43**, 239–244.

Scigliano, G., Girotti, F., Soliveri, P., *et al.* (1997). *Ital. J. Neurol. Sci.* **18**, 69–72.

Shevell, M. I. and Rosenblatt, D. S. (1992). The neurology of cobalamin, *Can. J. Neurol. Sci.* **19**, 472–486.

Shulman, L. M., Taback, R. L., Rabinstein, A. A., *et al.* (2002). Non-recognition of depression and other non-motor symptoms in Parkin's disease, *Parkinsonism Relat. Disord.* **8**, 193–197.

Sies, H. (1985). *Oxidative Stress.* Academic Press, London.

Siesjö, R. (1967). A new method for the cytological examination of the cerebrospinal fluid, *J. Neurol. Neurosurg. Psych.* **30**, 568–577.

Smith, C. D., Carney, J. M. and Strake-Reed, P. E. (1991). Excess brain protein oxidation and enzyme dysfunction in normal aging and Alzehamier's disease, *Proc. Natl. Acad. Sci. (USA)* **88**, 10540–10543.

Stollhoff, K. and Schulte, F. J. (1987). Vitamin B12 and brain development, *Eur. J. Paediatr.* **146**, 201–205.

Tatton, W. G., Chalmer-Redman, R., Brown, D., *et al.* (2003). Apoptosis in Parkinson's disease: Signal for neuronal degradation, *Ann. Neurol.* **53** (Suppl. 3), S61–S72.

Thiessen, B., Rajput, A. H., Loverty, W., *et al.* (1990). Age, environments and the number of substancia nigra neurons, *Adv. Neurol.* **53**, 201–206.

Weimann, J. (2003). Toxicity of nitrous oxide, *Best Pract. Res. Clin. Anaesthesiol.* **17**, 46–61.

Wernek, A. L. and Alvarenga, H. (1999). Genetics, drugs and environmental factors in Parkinson's disease: A control study, *Arq. Neurosiquiatr.* **57**, 347–355.

Wester, P., Bergström, U. and Eriksson, A. (1990). Venticular cerebrospinal fluid monoamine transmitter and their metabolite concentrations refect human brain neurochemistry in autopsy cases, *J. Neurochem.* **54**, 1148–1156.

Zar, J. H. (1984). *Biostatistical Analysis.* Prentice Hall, Englewood Cliffs, London.

Advances in Neuroregulation and Neuroprotection (2005), pp. 43-61
Collin, C. *et al.* (Eds)
© VSP 2005

Alzheimer's disease: from bench to bedside

MOHAMMAD SAEED [1,*], PHILIPPE M. FROSSARD [1]
and HASAN S. PARVEZ [2]

[1] *Department of Biological and Biomedical Sciences, The Aga Khan University, Stadium Road, Karachi 74800, Pakistan*
[2] *Developmental Neuroendocrinology and Neuropharmacology Unit, Institute Alfred Fessard for Neurosciences, CNRS, Gif sun Yrelte, France*

Abstract—Alzheimer's disease (AD) is the most common cause of dementia worldwide. It causes major debility for the elderly with medical, social and economic implications. The pathology of AD was described in early twentieth century. Significant scientific progress has been made to date in understanding the pathophysiology and genetics as well as developing therapeutic strategies for slowing disease progression and providing symptomatic relief to AD patients. This chapter reviews these advances and explores the links, molecular genetics and cellular biology has provided us today to tackle this devastating disorder.

Keywords: Alzheimer's; amyloid; tau; apolipoprotein E; genetics.

'I can't think. I can't plan. I feel as though my feet were in sand . . . and I have no solid ground to stand on' — Impressions of an Alzheimer's patient.

INTRODUCTION

Alzheimer's disease (AD) is an enigma. We have learned much in the last decade about biological markers, genes, and other possible factors relating to the development of this disease, yet there is still much more to understand. The seminal work of Blessed and colleagues (Blessed *et al.*, 1968) recognized the disease, not as a rare neurological disorder, but as the most common cause of dementia. AD is a progressive neurologic disease that results in the irreversible loss of neurons in the cerebral cortex and affects about 12 million people worldwide at present (Hy and Keller, 2000). The prevalence of AD is the highest in people 85 years

---

*To whom correspondence should be addressed. E-mail: saeed.mahmood@aku.edu

of age and older (approximately 26%), which is the fastest growing segment of western populations (CSHA Working Group, 1994). The clinical hallmarks are progressive impairment in memory, judgment, decision making, orientation to physical surroundings, and language. Diagnosis is based on neurologic examination and the exclusion of other causes of dementia; a definitive diagnosis can only be made at autopsy.

CLINICAL FEATURES

Alzheimer's disease is the most common cause of dementia, accounting for 55–65%. In AD, there is an overall shrinkage of brain tissue. The grooves or furrows in the brain, called sulci, are noticeably widened and there is shrinkage of the gyri, the well-developed folds of the brain's outer layer. In addition, the ventricles, or chambers within the brain that contain cerebrospinal fluid, are noticeably enlarged.

AD has several stages as described by Braak and Braak (1991). In the early stages short-term memory begins to decline. Hippocampus, which is part of the limbic system that is involved in processing experiences, prior to storage as permanent memories, appears to be preferentially affected on histopathological examination. This correlates with the clinical deficits observed in the early stages of AD in learning and in the creation of new memories, as well as with the relative preservation of established memories. The neurons at the basal forebrain that provide most of the cholinergic innervation to the cortex are also prominently affected; the resulting cholinergic neurotransmitter deficits are often treated with cholinesterase inhibitors such as donepezil.

As the disease spreads through the cerebral cortex, judgment as well as the ability to perform routine tasks declines, emotional outbursts may occur and language is impaired. Progression of the disease leads to the death of more nerve cells and subsequent behavioral changes, such as wandering and agitation. The ability to recognize faces and to communicate is completely lost in the final stages. Patients lose bowel and bladder control, and eventually need constant care. This stage of complete dependency may last for years before the patient dies. The average length of time from diagnosis to death is 4 to 8 years, although it can take 20 years or more for the disease to run its course.

PLAQUES AND TANGLES

The pathological hallmarks of the disease are neuronal loss, extracellular senile plaques and intracellular neurofibrillary tangles, first observed by the German neurologist, Dr. Alois Alzheimer in 1907 (Alzheimer, 1907). Nearly 100 years later there is convincing evidence that these plaques and tangles do indeed lead to the development of damaged brain tissue, producing the characteristic dementia of the disease. In 1984, a protein was isolated from the plaques in the parenchyma of

AD brains and its sequence determined (Glenner and Wong, 1984). Subsequently it was discovered that this protein formed the core of the neuritic plaques and was also their major constituent (Masters *et al.*, 1985). This protein, β-amyloid (Aβ), is so called because it aggregates into a β-pleated sheet structure. Aβ is a 40 to 42 amino acid long peptide cleaved from a larger molecule, the amyloid precursor protein (APP). Aβ forms fibrils that interact with stains such as Congo red, giving amyloid its characteristic appearance on fluorescent microscopy. Plaques start as innocuous deposits of nonaggregated, putatively non-neurotoxic β-amyloid (diffuse plaques). However, they later undergo an orderly sequential transformation into the mature senile neuritic plaques that are associated with the development of AD (Mackenzie, 1994).

Neurofibrillary tangles are intracellular aggregates containing tau protein, which is a 55 kDa protein abundantly expressed in the brain. In contrast to senile plaques, neurofibrillary tangle formation is a late event. In AD, the areas forming neurofibrillary tangles precisely match those exhibiting neuronal loss. Moreover, the abundance of tau inclusions correlates well with the extent of neuronal loss as well as with the degree of dementia (Gomez-Isla *et al.*, 1997). Tau is normally involved in cross-bridging microtubules, which are important cytoskeletal components of cellular architecture for maintenance of synaptic integrity and of intracellular transport system for molecules and organelles along an axon or dendrite. Phosphorylation of tau is an essential and dynamic process that regulates the ability of the protein to facilitate the assembly and stabilization of microtubules (Hasegawa *et al.*, 1998). In AD brains tau aggregates in a hyperphosphorylated form as homodimers, which later form helical filaments that in turn form neurofibrillary tangles (Kondo *et al.*, 1988; Lee *et al.*, 1991).

Tau and Aβ are not unique to AD brains but are found in normal brains as well, being secreted constitutively from various types of cells (Grundke-Iqbal *et al.*, 1986; Haass *et al.*, 1992). Moreover, neuritic plaques and neurofibrillary tangles are similar in chemical composition and regional distribution to those in normal aging brains (Arriagada *et al.*, 1992; Fukumoto *et al.*, 1996; Wang and Munoz, 1995). However, they are found in much greater abundance in AD. Thus, plaques and tangles bear a relation to dementia similar to that of atherosclerosis and infarcts (in atherosclerosis, the primary lesions are common in aging, but clinical manifestations will appear after a certain density of these lesions is reached).

Plaques and tangles form as a result of protein aggregation. When proteins are transcribed from mRNA, they are in the form of primary structures — a string of amino acids. When they are released into solution (intracellular or extracellular fluid), proteins undergo conformational changes depending on the pH, osmolality, chemical and electrical potentials, temperature and other physical conditions of the solution. Protein folding follows a funnel-like pathway in which the conformational intermediates merge into a final species depending upon the Gibbs free energy levels of interactions in the protein structure (Merlini and Bellotti, 2003). Folding occurs under the influence of van der Waals forces, hydrogen bonds, hydrophobic

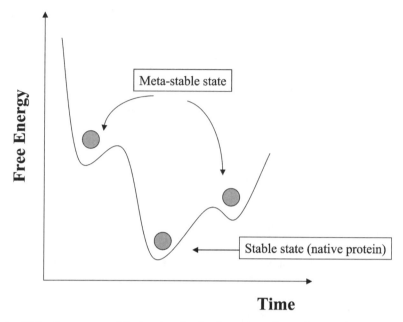

Figure 1. Gibbs free energy of interaction for the formation of a stable protein structure. Note: Protein folding results in the storage of energy in conformational bonds and lowering of free energy level of the protein molecule. The native protein state is the most stable with minimum free energy. In the course of folding, there are energy pockets in which the conformational intermediates are in meta-stable states having mis-folded structures, prone to aggregation.

interactions and other weak electrostatic forces, resulting in the storage of energy in these bonds and lowering the free energy level of the protein. The native protein state is the most stable with minimum free energy. Just before the protein can fold into its most stable native state, there is an alternative energy path available which allows the protein to achieve a stable, yet 'misfolded' low free-energy state, prone to aggregation (Fig. 1).

There is an acute energy barrier between the 'misfolded' (or metastable) state and the native biologically functional state, which is not easily overcome. If this barrier could be surpassed the protein could theoretically assume its native state. This could only happen if the energy barrier is either lowered by a catalytic process, or the 'misfolded' state is energized to overcome it (Fig. 1).

Misfolding can result due to several reasons. The protein may have an intrinsic propensity to misfold, which becomes evident with aging, such as in the case of normal transthyretin in senile systemic amyloidosis (Saraiva, 2001). High serum concentrations of a protein may alter the physical conditions needed for its proper folding and result in a pathological conformation, as with β_2-microglobulin in patients undergoing long-term hemodialysis for chronic renal failure (Verdone *et al.*, 2002). Mutations and proteolytic remodeling of the synthesized protein by enzymes can also give rise to incorrect conformations. Some mutations in APP increase the production of β-amyloid, while others favor the formation of long (42 amino acid) forms of β-amyloid, by altering the cleavage sites of processing enzymes (Hardy and Selkoe, 2002; Nussbaum and Ellis, 2003). Aβ42 (long form) aggregates more readily than the 40 amino acid short form, Aβ40 (Esler and Wolfe, 2001).

GENETICS

Early-onset AD

In 1987, the gene encoding APP was mapped on chromosome 21q and cloned (Kang *et al.*, 1987). This caused great excitement and was the beginning of the subsequent intensive molecular and cell biological studies of AD. The finding was supported by the fact that patients with Down's syndrome (trisomy 21) invariably developed AD neuropathology (senile plaques and neurofibrillary tangles) much earlier, in their second to fourth decades of life. The pathogenic significance of chromosome 21 was confirmed by the identification of a missense mutation at codon 717 of APP in two families affected by autosomal dominant familial AD (Goate *et al.*, 1991). Thus, the APP gene was the first identified causative gene for familial AD. In 1995, two additional causative genes for familial AD, *presenilin (PS)1* on chromosome 14 (Sherrington *et al.*, 1995) and PS2 on chromosome 1 (Levy-Lahad *et al.*, 1995; Rogaev *et al.*, 1995), were successively identified. Phenotypic analysis of these genes revealed that all are associated with increased Aβ42 production or its enhanced deposition in the brain (Borchelt *et al.*, 1996; Scheuner *et al.*, 1996; Suzuki *et al.*, 1994).

The familial forms of AD account for only 4–8% of cases. Most individuals with sporadic AD become affected after the age of 65; the disease is thus considered late-onset. Several point mutations in the gene coding for APP on chromosome 21 can cause early-onset autosomal dominant familial AD with complete penetrance. To date, 12 AD-related mutations have been discovered in the APP gene, while more than 70 mutations in the *PS1* gene and six mutations in the PS2 gene have been reported (Borchelt *et al.*, 1996; Scheuner *et al.*, 1996). Families carrying mutations in the PS1 gene have the earliest ages of onset, largely between 35 and 55 years (Holmes, 2002). One such family member has been reported to develop AD at 24 years (Wisniewski *et al.*, 1998). The mean duration of illness in families with PS1 mutations is also significantly shorter (5.8–6.8 years) than in families with APP (9–16 years) or PS2 (4.4–10.8 years) mutations, reflecting the severity of *PS1*-associated AD (Holmes, 2002). Ages of onset for patients having APP mutations is 40–65 years with the maximum reported age of onset for APP carriers being 67 years. PS2 gene mutation carriers have onset of AD largely between 40 and 70 years, thus showing some overlap with late-onset AD. Thus, there is evidence that the mutations that cause AD also affect onset of disease and its progression and their mechanism of pathogenesis is largely driven through Aβ42 modulation.

Similarly, mutations in the tau gene have also been characterized and shown to be sufficient to cause neuronal loss on their own, leading to dementia (Hutton *et al.*, 1998; Poorkaj *et al.*, 1998). In the human brain, six isoforms of tau are produced from a single gene on chromosome 17q by alternative mRNA splicing. In its C-terminal portion, three or four repeats composing the microtubule-binding domain are located in tandem (Morishima-Kawashirma and Ihara, 2002). Tau binds to tubulin through the repeats and promotes microtubule assembly. The splicing out of the second 31-residue repeat, encoded by exon 10, gives rise to three-repeat isoforms which are less stable and dissociate from microtubules leading to tangle formation (Lee *et al.*, 1991).

Late-onset AD

The inheritance of late-onset AD is more complex than that of the early-onset familial autosomal dominant form. Both twin and family studies suggest a polygenic and multifactorial mode of inheritance. Thus late-onset AD is what is known as a complex disorder, where disease results due to mutations in multiple genetic loci interacting with each other and creating only subtle changes in the function of the gene-products. These changes lead to the variability and heightened complexity of such disorders. The genes involved in complex disorders are often referred to as 'susceptibility genes', since they increase the risk and predispose individuals towards a particular disease. This distinguishes them from causative genes in which a mutation directly leads to a particular phenotype, as in most Mendelian disorders.

The genes causing early-onset familial autosomal dominant AD were discovered using the statistical genetic method of linkage analysis. It does not require information regarding underlying biochemical defects, which remain unknown for most diseases and has thus become a key strategy in identifying disease-genes. Linkage allows the determination of regions of chromosomes that are likely to contain a risk gene, and rule out areas where there is a low chance of finding a risk gene. A family having members with the disease of interest as well as healthy individuals (known as a pedigree), is the first and foremost requirement for linkage analysis. The method attempts to search for a genetic marker (from a known set of well characterized markers) that is consistently found in the affected individuals of the family more commonly than in their healthy counterparts. In such a case, the marker and the disease-causing gene are said to be linked, and are thought to be in close physical proximity to each other. This is based on the assumption that alleles on the same chromosome should segregate together at a rate that is proportional to the distance between them. Thus, this approach identifies disease genes by their chromosomal location, gradually zeroing in on them among their neighbors in the genome.

Linkage analysis is indeed the most powerful method for identifying disease-genes in monogenic disorders but there are several downsides to using linkage for complex (multifactorial and polygenic) disorders, the most important being the need to use DNA from several pedigrees harboring the disease and a huge sample size (Risch and Merikangas, 1996) to tease out the subtle effects of a particular gene on the disease phenotype. In 1991, Margaret Pericak-Vance and her colleagues, however, were successfully able to identify a region on the proximal long arm of chromosome 19, harboring the disease susceptibility gene for late-onset AD, using a modified approach to linkage analysis (Pericak-Vance *et al.*, 1991), following which association studies dominated the scene on the Alzheimer's front. The gene involved was thus shown to be Apolipoprotein E (APOE) (Corder *et al.*, 1993; Saunders *et al.*, 1993). Subsequently, APOE was localized to senile plaques, vascular amyloid and neurofibrillary tangles and its high avidity binding in vitro to Aβ was also demonstrated (Strittmatter *et al.*, 1993).

APOE is a plasma protein synthesized by the liver that is involved in cholesterol transport, but it is also produced and secreted in the brain by astrocytes and microglia. There are seven isoforms of APOE (ε1 through ε7), which differ from each other in one or two amino acids. Most common (>99%) isoforms, however, are ε2 (7%), ε3 (79%) and ε4 (14%). It has been shown that APOE ε4 is strongly associated with late-onset familial and sporadic AD (Corder *et al.*, 1993; Saunders *et al.*, 1993). The risk of developing AD increases from 20% to 90% and the mean age at onset of AD decreases by almost two decades with increasing number of APOE ε4 alleles (Corder *et al.*, 1993). APOE ε4/ε4 genotypes are rare in human populations (the frequency of APOE ε4 is in the range of 10–20%), but are found in high frequency (>90%) in AD patients. Each additional APOE ε4 allele nearly triples the lifetime risk of AD; patients having APOE ε4/ε4 genotypes were shown to be eight times more likely to be affected than subjects having ε3/ε3 genotypes (Corder *et al.*, 1993).

The elucidation of the genetics of APOE in AD is a landmark contribution, recognizing association studies as powerful genetic tools for exploring complex disorders. Association is based on the concept of linkage disequilibrium (LD). LD occurs when a marker allele lies in close proximity of the disease susceptibility allele, such that these two alleles are inherited together over several generations, in the population studied. Thus the same allele will be detected in affected individuals from multiple unrelated families, but belonging to the same population and thus to the same genetic pool. The concept is similar to that which underlies linkage analysis, except that linkage is studied within families, and that the chromosomal region mapped is much larger, owing to greater genetic homogeneity of families compared to populations. Thus the markers in close proximity of the mutation will appear to co-segregate with the mutation. In such a case where the marker allele and the mutation are in *linkage disequilibrium*, there is an *association* between the marker allele and the disease phenotype. This is observed as a significant change in the frequency of the marker allele for the disease phenotype compared to the control population, resulting in deviations from the expected random occurrence of the alleles in the case and control subsets of the population. In the case of AD, APOE ε4 has been shown to be directly responsible for a biological function — binding with high avidity to Aβ. Thus markers in close physical proximity to APOE ε4 were found to be associated with AD and the strongest association was obtained for APOE ε4 itself (Martin *et al.*, 2000).

PATHOPHYSIOLOGY

APOE, as a plasma protein, transports cholesterol from lipoproteins into the liver. In the brain it also appears to have a transport function for complex molecules such as Aβ, which would otherwise be insoluble making inter-compartmental transport difficult. APOE is produced in the brain mainly by astrocytes, which also synthesize APP. APP undergoes enzymatic processing to form Aβ in the cytoplasm and is later secreted into the extracellular fluid in association with APOE, which carries it to the neuron. On the neuronal cell membrane are receptors for APOE, called LDL-related protein (LRP) receptors. APOE-Aβ complex binds LRP receptors and the complex is endocytosed in vesicles, to whose membrane APOE remains attached while liberating Aβ for a trophic function in the neuron. APOE is released extracellularly for performing further transport functions (Strittmatter and Roses *et al.*, 1996).

Synthesis of Aβ in the brain is increased following injury and during developmental phases of the nervous system (Smith *et al.*, 2003; Xu *et al.*, 1996). This finding implicates Aβ in the growth and repair of the nervous system during development or after injury. APOE is also increased in chronic neurodegenerative disorders. It strongly binds to and co-precipitates with Aβ (Strittmatter *et al.*, 1993). APOE ε4 has a very high affinity for Aβ and therefore binds Aβ strongly (Strittmatter *et al.*, 1993). This prevents endocytosis of the complex and release of Aβ into the cytosol of the neuron. ApoE-Aβ complexes aggregate outside the cell membranes of neurons and form plaques (Strittmatter *et al.*, 1996) (Fig. 2).

In humans, APOE is also synthesized by the neurons (Xu *et al.*, 1996). *In vitro* studies have shown that APOE binds tau protein as well (Strittmatter *et al.*, 1994). The binding varies according to the isoform; it is strongest for the ε3-isoform and lowest for the ε4-isoform (Strittmatter *et al.*, 1994). APOE binds tau and carries it to a microtubule assembly where APOE is displaced, freeing tau for its normal function. Tau has a β-tubulin (a constituent of microtubules) binding domain, with which it binds to the microtubule assembly and forms stabilizing cross-bridges (Kondo *et al.*, 1988).

At the β-tubulin domain, two tau proteins can bind to each other as well, forming intertwined homodimers called paired helical filaments (PHF), which aggregate and eventually form neurofibrillary tangles (Kondo *et al.*, 1988; Lee *et al.*, 1991). After homodimerization, tau is unable to bind APOE. APOE ε3 is hypothesized to protect the β-tubulin binding site until tau is delivered to a microtubule assembly. APOE ε4, due to its reduced affinity for tau, prematurely releases tau in the cytosol and prevents it from forming cross-bridges at the microtubule assembly. Instead, it makes it available for homodimerization with other tau molecules in the cytosol (Roses, 1997). The effects are enhanced in APOE ε4 homozygote individuals,

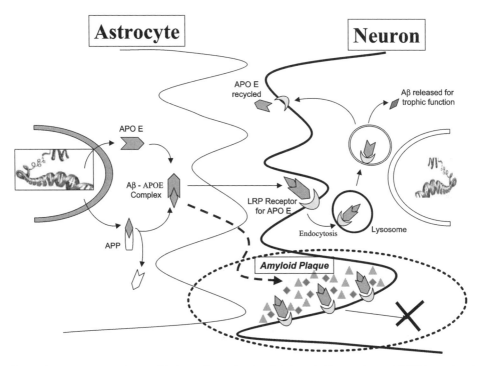

Figure 2. Pathogenesis of amyloid plaque formation in Alzheimer's disease. Note: APOE is produced in the brain mainly by astrocytes, which also synthesize APP which is processed to form Aβ and transported to neurons in association with APOE. On the neuronal cell membrane are receptors for APOE, called LDL-related protein (LRP) receptors. APOE-Aβ complex binds LRP receptors and the complex is endocytosed in lysosomal vesicles. Aβ is liberated to perform a trophic function, as yet unknown, in the neuron. APOE is subsequently released extracellularly for performing further transport functions. Encircled area on the diagram shows plaque formation. APOE ε4 isoform has a very high affinity for Aβ and therefore binds Aβ strongly, preventing endocytosis of the complex and release of Aβ into the cytosol of the neuron. APOE-Aβ complexes aggregate outside the cell membranes of neurons and form plaques.

in whom tau is being predominantly released into the cytosol of neurons forming PHF, rather than stabilizing microtubule assemblies. On a timescale, microtubule maintenance, repair and remodeling become progressively less efficient, thereby hampering neuronal transport and synaptic integrity (Fig. 3).

Phosphorylation plays an essential role in the regulation of the interactions of tau and microtubule assemblies (Hasegawa *et al.*, 1998). In AD brains, tau is hyperphosphorylated; this may also cause it to dissociate from microtubules, besides modifying its interactions with different isoforms of APOE. More than 25 phosphorylation sites have been identified in PHF-tau purified from AD brains (Hanger *et al.*, 1998; Morishima-Kawashima *et al.*, 1995). Both the number of phosphorylation sites on tau and the extent of their phosphorylation, far exceed those in normal brain. Most of the phosphorylation sites are clustered in the flanking region of the microtubule binding domain, which forms the core of PHFs

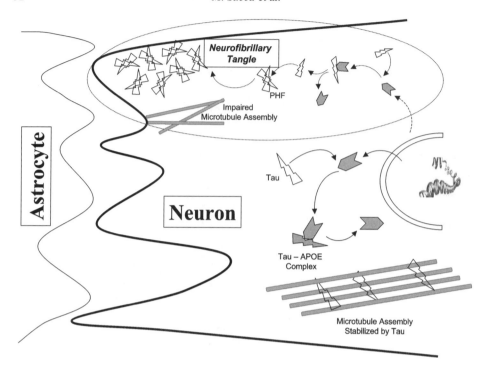

Figure 3. Neurofibrillary tangle formation. Note: In humans, APOE is also synthesized in the neurons and binds tau protein as well. The binding varies according to the isoform; it is strongest for the $\varepsilon3$-isoform and lowest for the $\varepsilon4$-isoform. APOE binds tau and carries it to a microtubule assembly where APOE is displaced, freeing tau to form stabilizing cross-bridges. APOE $\varepsilon3$ is hypothesized to protect the binding site of tau, whereas APOE $\varepsilon4$, due to its reduced affinity for tau, prematurely releases it in the cytosol. Two tau proteins can then bind, forming intertwined homodimers called paired helical filaments (PHF), which aggregate and eventually form neurofibrillary tangles.

(Kondo *et al.*, 1988). Additionally, multiple protein kinases appear to be involved in hyperphosphorylation of tau (Morishima-Kawashima *et al.*, 1995) which indicates that multiple phosphorylation cascades may be activated simultaneously by tangle formation.

Proteolytic processing is another important determinant of AD pathology. Its major role has been established in the development of neuritic plaques. $A\beta$ is a 39–43-residue protein with a molecular weight of 4 kDa (Glenner and Wong, 1984). The major $A\beta$ species found *in vivo* are $A\beta40$, which ends at Val40, and $A\beta42$, which has two additional hydrophobic residues, Ile and Ala. As a result, $A\beta42$ is more hydrophobic and has higher aggregation potential (Morishima-Kawashirma and Ihara, 2002).

After synthesis, APP undergoes N-linked and O-linked glycosylation in the endoplasmic reticulum and Golgi apparatus, respectively, and is transported via secretory vesicles to the plasma membrane (Morishima-Kawashirma and Ihara, 2002). APP is a type-I membrane protein, and $A\beta$ consists of its ectodomain just outside the membrane (28 residues) and the luminal half of the transmembrane

APP

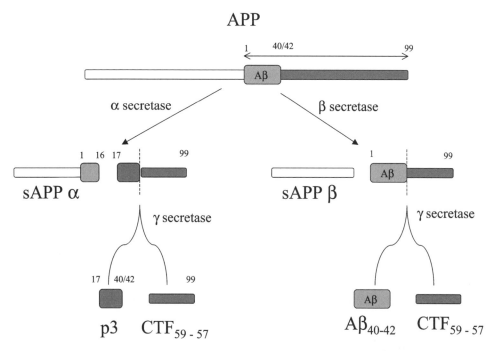

Figure 4. Amyloid precursor protein (APP) processing and beta-amyloid (Aβ) formation. Note: There are two proteolytic pathways operating in the processing of APP. The α-secretase pathway results in the formation of a soluble sAPPα component and a membrane bound 99-amino acid long C-terminal fragment (CTF) which is further cleaved by γ-secretase to yield a 59 to 57 amino acid long CTF and a p3 fragment. APP processing through the β-secretase pathway results in the formation of Aβ. Cleavage of APP by β-secretase results in a membrane bound 99 amino acid long CTF, which is further cleaved by γ-secretase at position 40 or 42 to yield Aβ40 or Aβ42, and a 59 to 57 amino acid long fragment respectively. Thus the processing of APP results in the formation of toxic Aβ42 fragment depending on the type of secretase involved and the location of the cleavage.

domain (12 or 14 residues) (Morishima-Kawashirma and Ihara, 2002). During the trafficking from intracellular organelles to the cell membrane, APP undergoes various sequential proteolytic processing steps, resulting in the production of Aβ as well.

There are two proteolytic pathways operating in the processing of APP (Bayer *et al.*, 2001; Jarrett *et al.*, 1993) (Fig. 4). APP consists of a much larger molecule with a 99 amino acid long segment towards its C-terminal end. The α-secretase pathway results in the formation of a soluble APP-α component (sAPPα) which is shed into the lumen and a membrane bound 99-amino acid long C-terminal fragment (CTF). Gamma-secretase further cleaves the CTF in the middle of the luminal region between residues Lys-16 and Leu-17, creating a 59 to 57 amino acid long CTF and a 23 to 25 amino acid long p3 fragment. This pathway precludes the production of amyloidogenic Aβ. APP can also be processed through the β-secretase pathway. Cleavage of APP by β-secretase results in a membrane bound 99 amino acid long CTF, which is further cleaved by γ-secretase at position 40 or

42 to yield Aβ40 or Aβ42, and a 59 to 57 amino acid long fragment respectively. Several mutations of APP, known to segregate with familial AD, are located in the flanking regions of the Aβ molecule, and one can speculate that these locations affect either β-secretase or γ-secretase cleavage sites, thereby increasing total Aβ or Aβ42 production. Thus the processing of APP results in the formation of toxic Aβ42 fragment depending on the type of secretase involved and the location of the cleavage.

Aβ deposited in the extracellular space of brain parenchyma undergoes a maturation process to form neuritic plaques. Initially amyloid is in the form of a granular amorphous material and is known as diffuse plaques. Later, an orderly sequential transformation of the innocuous deposits into Aβ fibrils takes place which accompany dystrophic neuritis and glial reaction that is characteristic of neuritic plaques. It is thought that the enzyme butyrylcholinesterase (BChE) may play an essential role in this maturation process (Guillozet *et al.*, 1997). It has been suggested that BChE becomes associated with amyloid plaques at approximately the same time that the Aβ deposit assumes a compact β-pleated conformation. Like amyloid elsewhere in the body, complex sugar polymer components (glycosaminoglycans) are also thought to be critical in the assembly and maturation of these deposits. Aβ binds to GM1 ganglioside that helps Aβ molecules to be converted to β-sheets leading to fibril formation (Choo-Smith *et al.*, 1997; Kakio *et al.*, 2001). BChE and glycosaminoglycans may therefore participate in the transformation of Aβ from an initially benign form to an eventually malignant form associated with neuritic tissue degeneration and clinical dementia.

DIAGNOSTICS AND THERAPEUTICS

With advances in molecular and cellular biology of AD, the field has entered into the futuristic arena of 'Predictive Medicine', where we will be able to prevent disease by predicting its onset in susceptible individuals. The patient's medical history and findings on physical examination, however, will continue to direct diagnostic evaluation for AD (Clark and Jason, 2003). Additionally, most patients in general would undergo a complete blood count, thyroid function screening, and routine blood studies for liver, kidney, and endocrine function. Serum B12 level determinations are also recommended by some AD specialists, as they may be low despite the absence of anemia or macrocytosis (Carmel *et al.*, 1988). However, hardly any clinical benefit of B12 replacement occurs in patients with dementia (Cunha, 1990; Eastley *et al.*, 2000). Imaging studies (computed tomography (CT) or magnetic resonance imaging (MRI)) may identify conditions other than neurodegeneration that could explain or contribute to the dementia symptoms (Chui and Zhang, 1997). Behavior and mood problems often occur as AD progresses. Depression, agitation, irritability, restlessness, aggression, delusions, and paranoia develop and worsen with disease progression. Usually, these problems are best handled by educating the caregiver about behavioral management strategies that do not include medication.

Over the past several years, there has been an intensive search for structural and biochemical markers that can serve as diagnostic tests during the earliest stage of the disease. Cerebral atrophy is an invariant consequence of the neuronal death that marks AD. Therefore the greatest promise of MRI may lie in its ability to document and quantitate the presence of regional and whole-brain atrophy that exceeds age-adjusted norms. This may allow it to serve as an important 'anatomical' biomarker of neurodegenerative pathology (DeCarli, 2001; Fox *et al.*, 1996, 1999; Jack *et al.*, 2000).

Biochemical changes that reflect the presence of disease-related pathology also have the potential to serve as diagnostic biomarkers of AD (Growdon, 1999). To date, the most extensively studied biochemical markers are the cerebrospinal fluid (CSF) proteins tau and Aβ (Andreasen *et al.*, 2001). Both are particularly relevant to the pathology of AD and thus may provide diagnostically useful information. CSF tau levels are elevated in most patients with AD, found during the earliest clinically detectable stage of the disease (Arai *et al.*, 1997; Galasko *et al.*, 1997). CSF Aβ levels are lowered in patients with AD. Some studies suggest that patients with both an elevated tau level and a reduced Aβ level in their CSF are most likely to have AD (Galasko *et al.*, 1998; Kanai *et al.*, 1998; Motter *et al.*, 1995).

APOE genotyping been shown to have tremendous clinical value in patients who have 'possible AD'. Genotyping suspected AD patients for APOE raises the accuracy of diagnosis to greater than 97% if an ε4 allele is identified (van Gool and Hijdra, 1994; Mayeux *et al.*, 1998). This can save on the cost of multiple brain imaging studies, which are commonly performed. Additionally, it can help predict the duration of the disease (Corder *et al.*, 1993) allowing the patient to participate in planning his or her care, and in deciding how to spend the time remaining.

APOE genotyping in healthy populations, however, has not proved to be useful in predicting the likelihood of developing AD, nor it is able to rule out individuals who will not develop AD, i.e. the test has low positive and negative predictive values (NIAA Working Group, 1996). Although determining APOE genotype in a blood sample is technically simple and routinely done in a number of laboratories in the Europe and North America, the consensus has been that it should not be offered as a routine presymptomatic test (Brodaty *et al.*, 1995; NIAA Working Group, 1996). The rationale is that there is no preventive intervention and that some ε4 carriers will not get AD. Genetic counseling, based on APOE genotyping, is therefore not only very complex, but also so far unsatisfactory.

Rational therapies derived from increased understanding of AD pathogenesis are now being developed. Many laboratories are investigating the potential methods for blocking plaque and tangle formation. In this context, seeking the secretases has been one of the keenest topics in AD research. BACE1 (β-site APP cleaving enzyme; also termed Asp2 or memapsin 2) was identified as the long-sought β-secretase in 1999, based on the cleavage site in APP and enhanced cleavage in a Swedish mutation of APP (Hussain *et al.*, 1999; Sinha *et al.*, 1999; Vassar *et al.*, 1999; Yan *et al.*, 1999). BACE1 is a type-I membrane bound aspartyl protease belonging to the pepsin protease family. In the cell, BACE1 is preferentially located in the Golgi apparatus and endosomes and might cleave APP at the β-site in these organelles.

On the other hand, the identity of γ-secretase, which cleaves the C-terminus of Aβ either at Val-40 or at Ala-42 within the transmembrane domain, is still ambiguous. PS1 and PS2 (PSs) are highly homologous multispanning membrane proteins: an eight-span transmembrane (TM) model is most preferred (Morishima-Kawashirma, and Ihara, 2002). PSs play a critical role in γ-cleavage of APP. The cells from PS1-deficient mice (De Strooper *et al.*, 1998) and the cells from PS1 and PS2 doubly deficient mice (Herreman *et al.*, 2000; Zhang *et al.*, 2000) show markedly reduced and almost undetectable Aβ production, respectively, which accompanies a remarkable accumulation of APP CTFs, C83 and C99, the immediate substrates for γ-secretase. If specific blockers of BACE1 or γ-secretase are discovered, AD would have very targeted therapeutics whereby the cause of the disease, i.e. Aβ production, would be eliminated or at least kept in check.

Cellular architecture and synaptic integrity as well as intracellular transport of molecules, is affected by the formation of plaques and tangles (Gomez-Isla *et al.*, 1997). Cholinergic input to the cortex is prominently affected as the neurons at the basal forebrain become dysfunctional. When vesicular traffic is interrupted the delivery of synaptic vesicles to presynaptic terminals is impaired and neurotransmitter deficit may exceed neuronal loss. Cholinesterase inhibitors increase low levels of acetylcholine in the cortex and are considered efficacious as antidementia drugs on the basis of improvements seen on a standard (Rosen *et al.*, 1984). An additional (and initially unexpected) benefit of cholinesterase inhibitors is their ability to reduce some of the neuropsychiatric symptoms associated with AD, especially apathy and visual hallucinations (Cummings, 2000).

Four cholinesterase inhibitors have so far been approved for the symptomatic treatment of patients with AD. These are tacrine (Cognex, Warner-Lambert, Morris Plains, New Jersey); donepezil (Aricept, Eisai, Inc., Teaneck, New Jersey, and Pfizer, Inc., New York, New benefit York) (Doody *et al.*, 2001b); rivastigmine (Exelon, Novartis, Basel, Switzerland); and galantamine (Reminyl, Janssen, Titusville, New Jersey) (Davis *et al.*, 1992 ; Raskind *et al.*, 2000; Rogers *et al.*, 1998; Rösler *et al.*, 1999; Tariot *et al.*, 2000). All produce essentially the same degree of modest improvement in approximately 30% to 40% of patients with mild to moderate (Mini-Mental State Examination score between 10 and 26) AD. The drugs have different organ toxicities, side effects, and dosing frequencies.

The major effect of cholinesterase inhibitors, which remains to be confirmed in prospective longitudinal clinical trials, lies in their potential to slow the rate of progression of AD. One retrospective analysis suggested that long-term use of donepezil may slow the rate of disease progression (Doody *et al.*, 2001a). If so, the drug might be indicated, even in the absence of observable improvement in cognition or function. To date, only high-dose vitamin E has been shown to have this effect (Sano *et al.*, 1997). The American Academy of Neurology now recommends 1000 IU of vitamin E twice daily as a standard care for the treatment of patients with AD (Doody *et al.*, 2001c).

The emergence of diagnostically specific molecular and genetic markers offer the potential for achieving an accurate diagnosis at the earliest stage of the disease. This in turn offers the best opportunity to initiate disease-modifying treatment. Therapies available at present provide at most a modest benefit in improving symptoms and possibly slowing progression. The promise for truly effective therapy, however, remains a realistic expectation and the identification of a medication and therapeutic modality that could prevent or delay the progression of the underlying neurodegenerative pathologic cascade would be of considerable benefit.

CONCLUSION

Much progress has been made in the field of AD, especially in the last decade. Elucidation of the pathogenesis of AD, the discovery of several genes and the identification of the secretases has brought Medicine closer to finding the cure for AD. What is it, however, that is processing at the core of the mind, the heart and the soul of one affected by Alzheimer's disease? The question is yet to be answered.

REFERENCES

Alzheimer, A. (1907). Uber eine eignartige Erkrankung der Birnrinde, *All. Z. Psychiat.* **64**, 146–148.

Andreasen, N., Minthon, L., Davidsson, P., *et al.* (2001). Evaluation of CSF-tau and CSF-Abeta42 as diagnostic markers for Alzheimer disease in clinical practice, *Arch. Neurol.* **58**, 373–379.

Arai, H., Nakagawa, T., Kosaka, Y., *et al.* (1997). Elevated cerebrospinal fluid tau protein level as a predictor of dementia in memory-impaired individuals, *Alzheimer's Research* **3**, 211–213.

Arriagada, P. V., Marzloff, K. and Hyman, B. T. (1992). Distribution of Alzheimer-type pathologic changes in nondemented elderly individuals matches the pattern in Alzheimer's disease, *Neurology* **42**, 1681–1688.

Bayer, T. A., Wirths, O., Majtényi, K., *et al.* (2001). Key factors in Alzheimer's disease: beta-amyloid precursor protein processing, metabolism and intraneuronal transport, *Brain Pathol.* **11**, 1–11.

Blessed, G., Tomlinson, B. E. and Roth, M. (1968). The association between quantitative measures of dementia and of senile change in the cerebral gray matter of elderly subjects, *Br. J. Psychiatry* **114**, 797–811.

Borchelt, D. R., Thinakaran, G., Eckman, C. B., *et al.* (1996). Familial Alzheimer's disease-linked presenilin1 variants elevate Aβ1–42/1–40 ratio in vitro and in vivo, *Neuron* **17**, 1005–1013.

Braak, H. and Braak, E. (1991). Neuropathological staging of Alzheimer-related changes, *Acta Neuropathol. (Berl.)* **82**, 239–259.

Brodaty, H., Conneally, M., Gauthier, S., *et al.* (1995). Consensus statement on predictive testing for Alzheimer disease, *Alzheimer Dis. Assoc. Disord.* **9**, 182–187.

Canadian Study of Health and Aging Working Group (1994). The Canadian Study of Health and Aging: study methods and prevalence of dementia, *CMAJ* **150**, 899–913.

Carmel, R., Sinow, R. M., Siegel, M. E., *et al.* (1988). Food cobalamin malabsorption occurs frequently in patients with unexplained low serum cobalamin levels, *Arch. Intern. Med.* **148**, 1715–1719.

Choo-Smith, L. P., Garzon-Rodriguez, W., Glabe, C. G., *et al.* (1997). Acceleration of amyloid fibril formation by specific binding of A_-(1–40) peptide to ganglioside-containing membrane vesicles, *J. Biol. Chem.* **272**, 22987–22990.

Chui, H. and Zhang, Q. (1997). Evaluation of dementia: a systematic study of the usefulness of the American Academy of Neurology's practice parameters, *Neurology* **49**, 925–935.

Clark, C. M. and Jason, H. T. K. (2003). Alzheimer Disease: Current concepts and emerging diagnostic and therapeutic strategies, *Ann. Intern. Med.* **138**, 400–410.

Corder, E. H., Saunders, A. M., Strittmatter, W. J., *et al.* (1993). Gene dose of apolipoprotein E type 4 allele and the risk of Alzheimer's disease in late onset families, *Science* **261**, 921–923.

Cummings, J. L. (2000). Cholinesterase inhibitors: A new class of psychotropic compounds, *Am. J. Psychiatry* **157**, 4–15.

Cunha, U. G. (1990). An investigation of dementia among elderly outpatients, *Acta Psychiatr. Scand.* **82**, 261–263.

Davis, K. L., Thal, L. J., Gamzu, E. R., *et al.* (1992). A double-blind, placebo-controlled multicenter study of tacrine for Alzheimer's disease. The Tacrine Collaborative Study Group, *New Engl. J. Med.* **327**, 1253–1259.

DeCarli, C. (2001). The role of neuroimaging in dementia, *Clin. Geriatr. Med.* **17**, 255–279.

De Strooper, B., Annaert, W., Cupers, P., *et al.* (1999). A presenilin-1-dependent γ-secretase-like protease mediates release of Notch intracellular domain, *Nature* **398**, 518–522.

Doody, R. S., Dunn, J. K., Clark, C. M., *et al.* (2001a). Chronic donepezil treatment is associated with slowed cognitive decline in Alzheimer's disease, *Dement. Geriatr. Cogn. Disord.* **12**, 295–300.

Doody, R. S., Geldmacher, D. S., Gordon, B., *et al.* (2001b). Openlabel, multicenter, phase 3 extension study of the safety and efficacy of donepezil in patients with Alzheimer disease, *Arch. Neurol.* **58**, 427–433.

Doody, R. S., Stevens, J. C., Beck, C., *et al.* (2001c). Practice parameter: management of dementia (an evidence-based review). Report of the Quality Standards Subcommittee of the American Academy of Neurology, *Neurology* **56**, 1154–1166.

Eastley, R., Wilcock, G. K. and Bucks, R. S. (2000). Vitamin B12 deficiency in dementia and cognitive impairment: the effects of treatment on neuropsychological function, *Int. J. Geriatr. Psychiatry* **15**, 226–233.

Esler, W. P. and Wolfe, M. S. (2001). A portrait of Alzheimer secretases — new features and familiar faces, *Science* **293**, 1449–1454.

Fox, N. C., Warrington, E. K., Freeborough, P. A., *et al.* (1996). Presymptomatic hippocampal atrophy in Alzheimer's disease. A longitudinal MRI study, *Brain* **119**, 2001–2007.

Fox, N. C., Scahill, R. I., Crum, W. R., *et al.* (1999). Correlation between rates of brain atrophy and cognitive decline in AD, *Neurology* **52**, 1687–1689.

Fukumoto, H., Asami-Odaka, A., Suzuki, N., *et al.* (1996). Amyloid beta protein deposition in normal aging has the same characteristics as that in Alzheimer's disease. Predominance of A beta 42(43) and association of A beta 40 with cored plaques, *Am. J. Pathol.* **148**, 259–265.

Galasko, D., Clark, C., Chang, L., *et al.* (1997). Assessment of CSF levels of tau protein in mildly demented patients with Alzheimer's disease, *Neurology* **48**, 632–635.

Galasko, D., Chang, L., Motter, R., *et al.* (1998). High cerebrospinal fluid tau and low amyloid beta42 levels in the clinical diagnosis of Alzheimer disease and relation to apolipoprotein E genotype, *Arch. Neurol.* **55**, 937–945.

Glenner, G. G. and Wong, C. W. (1984). Alzheimer's disease: initial report of the purification and characterization of a novel cerebrovascular amyloid protein, *Biochem. Biophys. Res. Commun.* **120**, 885–890.

Goate, A., Chartier-Harlin, M. C., Mullan, M., *et al.* (1991). Segregation of a missense mutation in the amyloid precursor protein gene with familial Alzheimer's disease, *Nature* **349**, 704–706.

Gomez-Isla, T., Hollister, R., West, H., *et al.* (1997). Neuronal loss correlates with but exceeds neurofibrillary tangles in Alzheimer's disease, *Ann. Neurol.* **41**, 17–24.

Growdon, J. H. (1999). Biomarkers of Alzheimer disease, *Arch Neurol.* **56** 281–283.

Grundke-Iqbal, I., Iqbal, K., Quinlan, M., *et al.* (1986). Microtubule-associated protein tau. A component of Alzheimer paired helical filaments, *J. Biol. Chem.* **261**, 6084–6089.

Guillozet, A. L., Smiley, J. F., Mash, D. C., *et al.* (1997). Butyrylcholinesterase in the life cycle of amyloid plaques, *Ann. Neurol.* **42**, 909–918.

Haass, C., Schlossmacher, M. G., Hung, A. Y., *et al.* (1992). Amyloid-peptide is produced by cultured cells during normal metabolism, *Nature* **359**, 322–325.

Hanger, D. P., Betts, J. C., Loviny, T. L., *et al.* (1998). New phosphorylation sites identified in hyperphosphorylated tau (paired helicalfilament-tau) from Alzheimer's disease brain using nanoelectrospray massspectrometry, *J. Neurochem.* **71**, 2465–2476.

Hardy, J. and Selkoe, D. J. (2002). The amyloid hypothesis of Alzheimer's disease: progress and problems on the road to therapeutics, *Science* **297**, 353–356.

Hasegawa, M., Smith, M. J. and Goedert, M. (1998). Tau proteins with FTDP-17 mutations have a reduced ability to promote microtubule assembly, *FEBS Lett.* **437**, 207–210.

Herreman, A., Serneels, L., Annaert, W., *et al.* (2000). Total inactivation of γ-secretase activity in presenilin-deficient embryonic stem cells, *Nature Cell Biol.* **2**, 461–462.

Holmes, C. (2002). Genotype and phenotype in Alzheimer's disease, *Br. J. Psych.* **180**, 131–134.

Hussain, I., Powell, D., Howlett, D. R., *et al.* (1999). Identification of a novel aspartic protease (Asp 2) as β-secretase, *Mol. Cell Neurosci.* **14**, 419–427.

Hutton, M., Lendon, C. L., Rizzu, P., *et al.* (1998). Association of missense and 5′-splice-site mutations in tau with the inherited dementia FTDP-17, *Nature* **393**, 702–705.

Hy, L. X., Keller, D. M. (2000). Prevalence of AD among whites: a summary by levels of severity, *Neurology* **55**, 198–204.

Jack, C. R., Jr., Petersen, R. C., Xu, Y., *et al.* (2000). Rates of hippocampal atrophy correlate with change in clinical status in aging and AD, *Neurology* **55**, 484–489.

Jarrett, J. T., Berger, E. P. and Lansbury, P. T., Jr. (1993). The carboxy terminus of the beta amyloid protein is critical for the seeding of amyloid formation: implications for the pathogenesis of Alzheimer's disease, *Biochemistry* **32**, 4693–4697.

Kakio, A., Nishimoto, S. I., Yanagisawa, K., *et al.* (2001). Cholesterol-dependent formation of GM1 ganglioside-bound amyloid β-protein, anendogenous seed for Alzheimer amyloid, *J. Biol. Chem.* **276**, 24985–24990.

Kanai, M., Matsubara, E., Isoe, K., *et al.* (1998). Longitudinal study of cerebrospinal fluid levels of tau, A beta1-40, and A beta1-42(43) in Alzheimer's disease: a study in Japan, *Ann. Neurol.* **44**, 17–26.

Kang, J., Lemaire, H. G., Unterbeck, A., *et al.* (1987). The precursor of Alzheimer's disease amyloid A4 protein resembles a cell-surface receptor, *Nature* **325**, 733–736.

Kondo, J., Honda, T., Mori, H., *et al.* (1988). The carboxyl third of tau is tightly bound to paired helical filaments, *Neuron* **1**, 827–834.

Lee, V. M., Balin, B. J., Otvos, L., Jr., *et al.* (1991). A68: a major subunitof paired helical filaments and derivatized forms of normal tau, *Science* **251**, 675–678.

Levy-Lahad, E., Wasco, W., Poorkaj, P., *et al.* (1995). Candidate gene for the chromosome 1 familial Alzheimer's disease locus, *Science* **269**, 973–977.

Mackenzie, I. R. (1994). Senile plaques do not progressively accumulate with normal aging, *Acta Neuropathol. (Berlin)* **87**, 520–525.

Martin, E. R., Lai, E. H., Gilbert, J. R., *et al.* (2000). SNPing away at complex diseases: analysis of single-nucleotide polymorphisms around APOE in Alzheimer disease, *Am. J. Human Genet.* **67**, 383–394.

Masters, C. L., Simms, G., Weinman, N. A., *et al.* (1985). Amyloid plaque core protein in Alzheimer disease and Down syndrome, *Proc. Natl. Acad. Sci. USA* **82**, 4245–4249.

Mayeux, R., Saunders, A. M., Shea, S., *et al.* (1998). Utility of the apolipoprotein E genotype in the diagnosis of Alzheimer's disease. Alzheimer's Disease Centers Consortium on Apolipoprotein E and Alzheimer's Disease, *New Engl. J. Med.* **338**, 506–511.

Merlini, G. and Bellotti, V. (2003). Molecular mechanisms of amyloidosis, *New Engl. J. Med.* **349**, 583–596.

Morishima-Kawashima, M., Hasegawa, M., Takio, K., *et al.* (1995). Proline-directed and non-proline-directed phosphorylation of PHF-tau, *J. Biol. Chem.* **270**, 823–829.

Morishima-Kawashirma, M. and Ihara, Y. (2002). Alzheimer's disease: ß-amyloid protein and tau, *J. Neurosci. Res.* **70**, 392–401.

Motter, R., Vigo-Pelfrey, C., Kholodenko, D., *et al.* (1995). Reduction of beta-amyloid peptide42 in the cerebrospinal fluid of patients with Alzheimer's disease, *Ann. Neurol.* **38**, 643–648.

National Institute on Aging/Alzheimer's Association Working Group (1996). Apolipoprotein E genotyping in Alzheimer's disease, *Lancet* **347**, 1091–1095.

Nussbaum, R. L. and Ellis, C. E. (2003). Alzheimer's disease and Parkinson's disease, *New Engl. J. Med.* **348**, 1356–1364.

Pericak-Vance, M. A. *et al.* (1991). Linkage studies in familial Alzheimer disease: Evidence for chromosome 19 linkage, *Am. J. Human Genet.* **48**, 1034–1049.

Poorkaj, P., Bird, T. D., Wijsman, E., *et al.* (1998). Tau is a candidate gene for chromosome 17 frontotemporal dementia, *Ann Neurol.* **43**, 815–825.

Raskind, M. A., Peskind, E. R., Wessel, T., *et al.* (2000). Galantamine in AD: A 6-month randomized, placebo-controlled trial with a 6-month extension. The Galantamine USA-1 Study Group, *Neurology* **54**, 2261–2268.

Risch, N. and Merikangas, K. (1996). The future of genetic studies of complex human disease, *Science* **273**, 1516–1517.

Rogaev, E. I., Sherrington, R., Rogaeva, E. A., *et al.* (1995). Familial Alzheimer's disease in kindreds with missense mutations in a gene on chromosome 1 related to the Alzheimer's disease type 3 gene, *Nature* **376**, 775–778.

Rogers, S. L., Farlow, M. R., Doody, R. S., *et al.* (1998). A 24-week, double-blind, placebo-controlled trial of donepezil in patients with Alzheimer's disease. Donepezil Study Group, *Neurology* **50**, 136–145.

Rosen, W. G., Mohs, R. C. and Davis, K. L. (1984). A new rating scale for Alzheimer's disease, *Am. J. Psychiatry* **141**, 1356–1364.

Roses, A. D. (1997). Alzheimer's Disease: The Genetics of Risk. Hospital Practice (http://www.hosppract.com/genetics/9707gen.htm).

Rösler, M., Anand, R., Cicin-Sain, A., *et al.* (1999). Efficacy and safety of rivastigmine in patients with Alzheimer's disease: international randomised controlled trial, *Brit. Med. J.* **318**, 633–638.

Sano, M., Ernesto, C., Thomas, R. G., *et al.* (1997). A controlled trial of selegiline, alpha-tocopherol, or both as treatment for Alzheimer's disease. The Alzheimer's Disease Cooperative Study, *New Engl. J. Med.* **336**, 1216–1222.

Saraiva, M. J. (2001). Transthyretin amyloidosis: a tale of weak interactions, *FEBS Lett.* **498**, 201–203.

Saunders, A. M., Strittmatter, W. J., Schmechel, D., *et al.* (1993). Association of apolipoprotein E allele epsilon 4 with late-onset familial and sporadic Alzheimer's disease, *Neurology* **43**, 1467–1472.

Scheuner, D., Eckman, C., Jensen, M., *et al.* (1996). Secreted amyloid β-protein similar to that in the senile plaques of Alzheimer's disease is increased in vivo by the presenilin 1 and 2 and APP mutations linked to familial Alzheimer's disease, *Nature Med.* **2**, 864–870.

Sherrington, R., Rogaev, E. I., Liang, Y., *et al.* (1995). Cloning of a gene bearing missense mutations in early-onset familial Alzheimer's disease, *Nature* **375**, 754–760.

Sinha, S., Anderson, J. P., Barbour, R., *et al.* (1999). Purification and cloning of amyloid precursor protein β-secretase from human brain, *Nature* **402**, 537–540.

Smith, D. H., Uryu, K., Saatman, K. E., *et al.* (2003). Protein accumulation in traumatic brain injury, *Neuromolecular Med.* **4**, 59–72.

Strittmatter, W. J., Saunders, A. M., Schmechel, D., *et al.* (1993). Apolipoprotein E: high-avidity binding to beta-amyloid and increased frequency of type-4 allele in late-onset familial Alzheimer disease, *Proc. Natl. Acad. Sci. USA* **90**, 1977–1981.

Strittmatter, W. J., Saunders, A. M., Goedert, M., *et al.* (1994). Isoform-specific interactions of apolipoprotein E with microtubule-associated protein tau: implications for Alzheimer disease, *Proc. Natl. Acad. Sci. USA* **91**, 11183–11186.

Strittmatter, W. J. and Roses, A. D. (1996). Apolipoprotein E and Alzheimer's disease, *Annu. Rev. Neurosci.* **19**, 53–77.

Suzuki, N., Cheung, T. T., Cai, X. D., *et al.* (1994). An increased percentage of long amyloid βprotein secreted by familial amyloid β protein precursor (βAPP717) mutants, *Science* **264**, 1336–1340.

Tariot, P. N., Solomon, P. R., Morris, J. C., *et al.* (2000). A 5-month, randomized, placebo-controlled trial of galantamine in AD. The Galantamine USA-10 Study Group, *Neurology* **54**, 2269–2276.

van Gool, W. A. and Hijdra, A. (1994). Diagnosis of Alzheimer's disease by apolipoprotein Egenotyping [Letter], *Lancet* **344**, 275.

Wang, D. and Munoz, D. G. (1995). Qualitative and quantitative differences in senile plaque dystrophic neurites of Alzheimer's disease and normal aged brain, *J. Neuropathol. Exp. Neurol.* **54**, 548–556.

Vassar, R., Bennett, B. D., Babu-Khan, S., *et al.* (1999). β-Secretase cleavage of Alzheimer's amyloid precursor protein by the transmembrane aspartic protease BACE, *Science* **286**, 735–741.

Verdone, G., Corazza, A., Viglino, P., *et al.* (2002). The solution structure of human beta2-microglobulin reveals the prodromes of its amyloid transition, *Protein Sci.* **11**, 487–499.

Wisniewski, T., Dowjat, W. K., Buxbaum, J. D., *et al.* (1998). A novel Polish presenlin-2 mutation (PII7L) is associated with familla Alzhemer's disease and leads to death as early as the age of 28 years, *Neuroreport* **9**, 217–221.

Xu, P.-T., *et al.* (1996). Human apolipoprotein E2, E3, and E4 isoform specific transgenic mice: Human-like pattern of glial and neuronal immunoreactivity in central nervous system not observed in wild-type mice, *Neurobiol. Dis.* **3**, 229.

Yan, R., Bienkowski, M. J., Shuck, M. E., *et al.* (1999). Membrane-anchored aspartyl protease with Alzheimer's disease β-secretase activity, *Nature* **402**, 533–537.

Zhang, Z., Nadeau, P., Song, W., *et al.* (2000). Presenilins are required for γ-secretase cleavage of β-APP and transmembrane cleavage of Notch-1, *Nat. Cell Biol.* **2**, 463–465.

Advances in Neuroregulation and Neuroprotection (2005), pp. 63-81
Collin, C. *et al.* (Eds)
© VSP 2005

Neuronal responses, sensitivity and fates as a consequence of hypoxia in the immature brain

J. L. DAVAL *

Inserm EMI 0014, Faculté de Médecine, 9 avenue de la Forêt de Haye, B. P. 184, 54505 Vandoeuvre-les-Nancy Cedex, France

Abstract—Hypoxia is the primary event of neonatal asphyxia, which remains a major cause of brain injury and subsequent neurological disabilities. The effects of a hypoxic episode of varying duration were investigated *in vitro* by using primary neuronal cell cultures and *in vivo* in newborn rat pups. Severe hypoxia was shown to induce an age-dependent delayed neuronal death that mainly reflects apoptosis. Before brain cells displayed morphological hallmarks of apoptotic death, hypoxia was shown to promote re-entry into the cell cycle followed by the activation of pro-apoptotic proteins, such as Bax and caspases. By contrast, mild hypoxia was able to trigger neuronal proliferation by stimulating the expression of neurogenic and survival-associated proteins, including proliferating cell nuclear antigen (PCNA) and Bcl-2. Moreover, *in vivo* studies revealed that hypoxia-induced neuronal apoptosis observed in the vulnerable CA1 subfield of the hippocampus was then followed by an apparent anatomical recovery, which suggests that neurogenesis might occur as a self-repair mechanism in the developing brain.

Keywords: Cultured neurons; newborn rat; apoptosis; necrosis; cell cycle; gene regulation; neurogenesis; excitotoxicity; oxidative stress.

PATHOPHYSIOLOGICAL BACKGROUND AND OBJECTIVES

The normal development of the fetus is known to be connected with the sufficient oxygen requirement. However, hypoxic episodes are particularly frequent in the antenatal period as well as in premature and full-term human newborns (Volpe, 1995), and it remains unclear to what extent they are deleterious for the developing brain.

The primary cause of pre- or postnatal hypoxia in human pathology is the reduced placental or pulmonary gas exchange (Carter, 1989; Volpe, 1995). This leads to a decrease in fetal or neonatal blood and tissue pO_2 and an increase in pCO_2

---

*E-mail: Jean-Luc.Daval@nancy.inserm.fr

(hypercapnia) which may compensate for the oxygen deficit up to a certain degree by increasing cerebral blood flow and pulmonary ventilation. Severe and/or long-lasting hypoxia impairs cerebral vascular autoregulation and suppresses cardiac function manifested in a bradycardia and a decrease in blood pressure. Ultimately, hypoxia will lead to ischemia, with a reduction of brain tissue concentration of glucose (Brierley and Graham, 1984; Vannucci and Plum, 1975; and for reviews, see Nyakas et al., 1996 and Volpe, 1995).

The immature brain has been repeatedly reported to be more resistant to oxygen deprivation than the adult one (Fazekas et al., 1941; Haddad and Donelly, 1990). The underlying mechanisms for this apparent tolerance are not clear. Higher resistance of the newborn to hypoxia has been attributed to specific properties of the cardiovascular system (Stattford and Wheatherhall, 1960) and, more recently, to differences in metabolic pathways (Xia et al., 1992), such as the ability of the developing brain to decrease its overall metabolic rate in order to preserve ATP levels (Duffy et al., 1975; Kass and Lipton, 1989) and utilization of lactate and ketone bodies as alternative fuels for energy metabolism (Dombrowski et al., 1989). In this respect, the perinatal age has particular significance if we consider the process of delivery and the associated sudden adaptation to postnatal life, which are both demanding events for metabolic homeostasis (Kohle and Vannucci, 1977; Vannucci and Duffy, 1974). Moreover, vulnerable periods may be delineated during brain development regarding the various steps of maturation of distinct cellular and tissue processes (Alling, 1985; Morgane et al., 1993), and developing neurons probably require more energy for enhanced synthesis of structural proteins and other macromolecules, differentiation processes, such as axon and dendrite elongation, synapse formation or signal transmission. At least during critical periods of the brain formation and maturation, the relative resistance to oxygen deprivation is certainly limited, and it has been confirmed that the rapid rate of differentiation and synaptogenesis may contribute to an increased neuronal susceptibility to hypoxic injury (Slotkin et al., 1986). Finally, it must be emphasized that certain neuronal groups in the perinatal animal, particularly at a developmental period in the rat analogous to human brain at term, appear to be more vulnerable to injury than similar neurons in the adult, a phenomenon which has been associated with an increased sensitivity of excitatory amino acid receptors (Ikonomidou et al., 1989; McDonald et al., 1988).

In spite of marked improvements in obstetric care and neonatal practice, hypoxic brain damage still accounts for a major part of perinatal neuropathology. Clinically, neonatal asphyxia may result in complex cerebral dysfunctions like cerebral palsy or seizure disorders. It has been associated with long term handicaps, including motor impairments, visual and hearing deficits as well as cognitive and learning disabilities (Berger and Garnier, 1999; Maneru et al., 2001; Mutch et al., 1992; Patel and Edwards, 1997; Simon, 1999). Because of its clinical implications, hypoxic brain injury in neonates has attracted considerable attention and accelerated research investigations.

Since the lack of oxygen is the primary event in case of brain asphyxia, there was a controversy as to whether hypoxia *per se* was able to induce neuronal damage and could subsequently alter brain functions (Banasiak and Haddad, 1998; Pearigen *et al.*, 1996; Vannucci, 1990). Whereas alterations in neuronal metabolic activities after hypoxic stress may share common properties with those underlying ischemia-induced neuronal damage, only few studies have focused on the specific consequences of transient hypoxia by itself, and it has been proposed that different mechanisms for neuronal injury in the forebrain may occur between hypoxia and ischemia (Taniguchi *et al.*, 1994). In our own work, we aimed to address the following questions: does oxygen deprivation induce neuronal death? If the hypoxic stress accounts for brain injury, what are the underlying cellular and molecular mechanisms? What is the influence of hypoxia duration/severity? What is the role of age when hypoxia occurs? What new neuroprotective strategies could be developed?

EXPERIMENTAL MODELS

Much of our knowledge concerning the pathophysiology of perinatal brain damage caused by hypoxia and/or ischemia originates from studies using various animal models. The rat is most frequently used in these studies and also our experiments were performed in this species. Compared to the human, the rat is born prematurely, and the maturity level of the newborn rat brain is considered to reflect the human fetal brain during the third trimester of pregnancy — and thus possibly premature babies — whereas the full-term human infant's brain would correspond to a 7- to 10-day-old rat pup (Alling, 1985; Morgane *et al.*, 1993).

To study the effects of hypoxia on developing brain neurons, we used two distinct models consisting of (i) *in vitro* neuronal cell cultures isolated from the fetal rat brain and (ii) *in vivo* newborn rat pups.

Culture model

Although neuronal cells in primary culture serve as a simplified model, they appear to be very useful in elucidating cellular and molecular mechanisms involved in various diseases, including cerebral hypoxia.

Primary cultured neurons were obtained from 14-day-old rat embryo forebrains (Daval *et al.*, 1991). Whole embryos excised by caesarian section were placed in culture medium equilibrated at 37°C and consisting of a mixture of Dulbecco's modified Eagle's medium (DMEM) and Ham's F12 medium (50 : 50) supplemented with 5% inactivated fetal calf serum. Forebrains were carefully collected, dissected free of meninges and gently dispersed in culture medium. After centrifugation at 700*g* for 10 min, the pellet was redispersed in the same medium and passed through a 46 μm-pore size nylon mesh. Aliquots of the cell suspension were transferred into 35 mm Petri dishes precoated with poly-L-lysine in order to obtain a final density

Table 1.
Evolution of physiological parameters in the culture medium

	Duration of hypoxia		
	0 (control)	3 hours	6 hours
pH	7.27 ± 0.02	7.23 ± 0.03	$7.17 \pm 0.03^{**}$
pO_2	138.6 ± 3.9	$30.8 \pm 2.9^{**}$	$29.6 \pm 3.1^{**}$
pCO_2	32.6 ± 1.8	37.1 ± 2.4	35.25 ± 3.1

For each measurement, data were obtained from 10 samples. Significantly different from controls: ** $p < 0.01$ (Student's t-test).

of 10^6 cells/dish. Cultures were then placed at 37°C in a humidified atmosphere of 95% air/5% CO_2. The following day, the culture medium was replaced with a fresh hormonally defined serum-free medium consisting of the DMEM/Ham's F12 mixture enriched with human transferrin (1 mM), bovine insulin (1 mM), putrescine (0.1 mM), progesterone (10 nM), estradiol (1 pM), Na selenite (30 nM), and also containing fibroblast growth factor (bFGF, 2 ng/ml) and epidermal growth factor (EGF, 10 ng/ml). After two additional days, the culture medium was renewed with serum-free medium in the absence of growth factors.

Hypoxic conditions were produced after 6 days *in vitro* by transferring neuronal cell cultures to a humidified incubation chamber thermoregulated at 37°C and flushed by a gas mixture corresponding to 95% N_2/5% CO_2. Hypoxia was achieved for various durations, and, for subsequent analyses, culture dishes were then returned for the ensuing 96 h to normal atmosphere, whereas matched controls were constantly maintained under standard normoxic conditions. Reduction in oxygen delivery to the neurons was scored by measuring O_2 content in samples of extracellular medium sheltered from ambient air, by means of a gas analyzer (Table 1).

Cell morphology was routinely assessed by phase-contrast microscopic observations. Purity of neuronal cultures was regularly evaluated by using monoclonal antibodies against neuron-specific enolase (NSE, a neuronal marker) and glial fibrillary acidic protein (GFAP, a marker for glial cells). Cell viability was monitored by Trypan blue exclusion and by the spectrophotometric method using 3-[4,5-dimethylthiazol-2-yl]-2,5-diphenyltetrazolium bromide (MTT), according to Hansen *et al.* (1989).

Morphological hallmarks of apoptosis, necrosis, and mitosis were analyzed by staining fixed cultured cells with the fluorescent dye 4,6-diamidino-2-phenylindole (DAPI, 0.5 μg/ml) (Wolvetang *et al.*, 1994). Using this procedure, healthy neurons exhibit intact round-shaped nuclei with diffuse fluorescence, indicative of homogeneous chromatin. Necrotic neurons are characterized by highly refringent smaller nuclei with uniformly dispersed chromatin, while condensation and fragmentation of chromatin lead to shrunken nuclei in apoptotic neurons (Park *et al.*, 1997). Typical morphological features of mitosis are easily recognizable.

Table 2.
Evolution of blood gases in rat pups exposed to birth hypoxia

	Duration of hypoxia		Time post-hypoxia
	0 (control)	20 min	20 min
pH	7.40 ± 0.03	6.62 ± 0.03[**]	7.32 ± 0.04
pO_2	60.7 ± 7.4	11.8 ± 4.8[**]	63.0 ± 6.8
pCO_2	35.6 ± 7.3	130.5 ± 6.9[**]	30.6 ± 7.0

For each measurement, data were obtained from 10 rats. Significantly different from controls: [**] $p < 0.01$ (Student's t-test).

In vivo *model*

In vivo models classically use combined hypoxia and ischemia in rat pups at $\sim$1 week of postnatal life (Renolleau *et al.*, 1998; Rice *et al.*, 1981; Schwartz *et al.*, 1992). Such an insult produces a delineated regional infarction, and thus these models do not allow investigation of the consequences of global brain hypoxic conditions around the birth period. We have therefore used a model approximating birth asphyxia (Nyakas *et al.*, 1996) by the transient exposure of newborn rats to nitrogen (Grojean *et al.*, 2003b). Between 8 to 24 h after delivery, half of the neonates were placed for 20 min in a thermostated plexiglas chamber flushed with 100% N_2, whereas the remaining pups were taken as controls and exposed to 21% O_2/79% N_2 (a mixture corresponding to air) for the same time. The temperature inside the chamber was adjusted to 36°C to maintain body temperature in the physiological range. All pups were allowed to recover for 20 min in normoxic conditions, and they were then returned to their dams. In such experimental conditions, the overall mortality was 4% in hypoxic rats, and the litter size was finally reduced to 10 pups, corresponding to 5 controls and 5 hypoxic rats, for homogeneity in subsequent experiments. In some experiments, blood samples were rapidly withdrawn by decapitation and sheltered from air for measurement of pH, pO_2 and pCO_2 (Andiné *et al.*, 1990) (Table 2).

To evaluate the extent of cell loss resulting from hypoxia, brain sections were generated at the level of anterior hippocampus according to the developing rat brain atlas of Sherwood and Timiras (1970); they were fixed and stained with thionin, and morphometric analyses were conducted by means of a microscope coupled to a computerized image-processing system. Adjacent brain sections were stained with DAPI for the measurement of cell density by counting cell nuclei, and for monitoring of apoptosis and necrosis.

EFFECTS OF SEVERE HYPOXIA

Cellular alterations

When neuronal cell cultures were subjected to hypoxia for 6 h, cell viability progressively declined, and started to be significantly different from normoxic

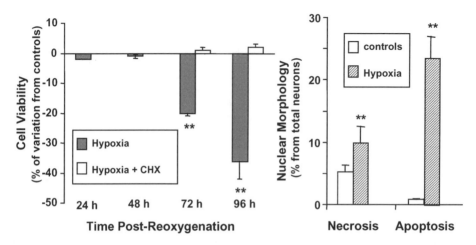

Figure 1. *Left panel*: Evolution of cell viability in cultured neurons exposed to hypoxia for 6 h, and effects of cycloheximide (CHX). *Right panel*: Proportions of necrosis and apoptosis in cultured neurons at 96 h post-hypoxia. Signicantly different from controls: ** $p < 0.01$ (Dunnett's test).

control cultures by 72 h post-reoxygenation, to reach -36% at 96 h (Fig. 1, left panel). The presence of typical apoptotic nuclei could be detected by 48 h post-hypoxia. The amount of these nuclei increased thereafter, reaching 23% at 96 h, whereas necrosis was only slightly enhanced (Fig. 1, right panel).

As illustrated, addition to the culture medium of 1 μM cycloheximide, a protein synthesis inhibitor, prevented hypoxia-associated cell death, and therefore significantly reduced apoptosis in cultured cells. These observations indicate that oxygen deprivation, if severe enough, induces delayed neuronal death reflected by an apoptotic mechanism that requires protein synthesis, as previously documented in other experimental models (Martin *et al.*, 1988; Papas *et al.*, 1992).

Subsequent analysis of the temporal profile of the incorporation of radioactive leucine into cultured cells revealed that cell death was accompanied by a biphasic increase in the rate of global protein synthesis (Bossenmeyer *et al.*, 1998). A strong increase (77% above controls) was first recorded 1 h after the onset of hypoxia, and a second one (72% above controls) was then measured at 48 h after reoxygenation. Parallel changes in RNA synthesis, as reflected by the incorporation of radiolabeled uridine, were also shown, supporting that hypoxia triggers the activation of specific sets of genes implicated in the apoptotic pathway.

In vivo, exposure of newborn rats to N_2 for 20 min appears to mimic birth asphyxia in human infants (Amiel-Tison and Ellison, 1986), since such a treatment rapidly elicited pronounced hypoxemia associated with hypercapnia and subsequent acidosis which persisted after reoxygenation. Regarding general development, neonatal hypoxia in our model induced a long lasting reduction of body weight, and only a transient alteration of brain weight.

Cell damage was repeatedly observed in various brain regions of hypoxic rats, including those which are known to be particularly sensitive to oxygen supply

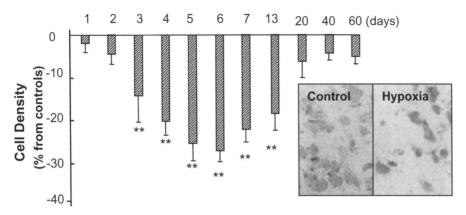

Figure 2. Changes in cell density in the infragranular part of the parietal cortex (layers III-IV) after birth hypoxia. Number of cells was measured with an occular grid after nuclear staining by DAPI ($n = 5$), and compared to controls ($^{**}p < 0.01$). Microphotograph: thionin coloration ($\times 40$).

such as the cerebral cortex and the CA1 subfield of the hippocampus. Indeed, it has been well established that neuronal vulnerability to oxygen deprivation varies according to brain regions, both in developing and adult brains (Pulsinelli *et al.*, 1982; Rivkin and Volpe, 1996). For illustration, Fig. 2 shows that cell density was progressively reduced after hypoxia in the parietal cortex. The decline was maximal around 6–7 days post-reoxygenation, reaching 25% as compared to controls. Immunohistological analyses and cell counts revealed that NSE-positive cells were predominantly affected, suggesting that the reduction of cell density mostly reflects the loss of neurons. Accordingly, neuronal cells are known to be more sensitive than other brain cells to oxygen reduction (Sochocka *et al.*, 1994), but it cannot be totally excluded that other cell types may be altered to a lesser extent.

As previously reported *in vitro*, DAPI staining and electronic miscroscopy observations showed the accumulating presence of neurons with nuclear condensation, clumping and fragmentation into spherical entities corresponding to apoptotic bodies; in parallel, specific DNA fragmentation patterns on agarose gels suggested that birth hypoxia may transiently reactivate the developmental apoptotic programme of neuronal elimination (Grojean *et al.*, 2003b).

Interestingly, time course studies showed that cell deficits were followed by an apparent gradual recovery, starting from day 20 after reoxygenation (Fig. 2). Such observations contrast with those previously reported in studies associating middle cerebral artery occlusion to oxygen deprivation in older rats (i.e. by using the Rice-Vannucci model), where permanent lesions were found (Mitsufugi *et al.*, 1996; Renolleau *et al.*, 1998; Towfighi *et al.*, 1995). Both severity of the insult and age of the subject certainly account for this difference in the brain outcome (Haddad and Jiang, 1993), and these findings ask the fundamental question of whether the immature brain may dispose of efficient compensatory repair processes.

Nevertheless, additional neurodevelopmental studies revealed long-term functional impairments (Grojean et al., 2003b). Indeed, although they successfully achieved various behavioral tests such as the righting reflex, negative geotaxis, locomotor coordination, and the eight-arm maze tasks (reflecting spatial orientation, learning and memory capacities), both developing and adult hypoxic rats were repeatedly slower than controls, suggesting that transient hypoxia around birth may be associated with moderate but persistent sequelae. In this respect, neurotransmitter systems may be a target of hypoxia, and their disturbance might participate in such long-term effects.

UNDERLYING MECHANISMS

Expression of specific proteins

As already mentioned above, neuronal death triggered by severe hypoxia was associated with a specific temporal pattern of total protein synthesis. Further investigations by immunohistochemistry and Western blotting indicated that the expression of various apoptosis-related proteins was temporally regulated in our models (Bossenmeyer-Pourié et al., 1999b, 2000b, 2002; Chihab et al., 1998b; Grojean et al., 2003b).

We first examined in cultured neurons the expression profile of AP-1 (activator protein-1) which regulates the transcription of various genes by DNA binding to the TRE sequence (Rauscher et al., 1988). Electrophoretic mobility shift assays showed that AP-1 DNA-binding activities were present in both control and hypoxic neurons. Moreover, all AP-1 components were markedly overexpressed by one hour after the onset of hypoxia. They include c-Jun, Jun B, Jun D, c-Fos and Fos-related antigens (FRA). Whereas only c-Jun remained elevated for up to 96 h post-reoxygenation, time at which neurons were severely injured, other gene products showed patterned induction/repression as hypoxia progressed and then during the post-reoxygenation period, with Fos-related antigens being finally induced at 96 h. Among the kinases known to phosphorylate AP-1 components, only JNK1 (c-Jun N-terminal kinase-1) was constitutively detected in cultured neurons, and its expression was inhibited during hypoxia. Nonetheless, both JNK1 and JNK3 were then markedly, but transiently, induced at 48 h post-reoxygenation, when apoptosis-related morphological features became evident (Fig. 3). Such data support the hypothesis that transient hypoxia, independently of ischemia, may trigger apoptosis in developing brain neurons through JNK signaling pathway, and emphasize the potential role of JNK3 in neuronal outcome.

In parallel, proteins which are known to promote apoptosis were stimulated in cultured neurons following hypoxia, whereas anti-apoptotic components were gradually repressed. Bax, the death-inducing prototype member of the Bcl-2 family, was highly expressed within 24 h after the insult (Fig. 4), its induction preceding apparent cell damage. This protein appears as an early component of the signaling

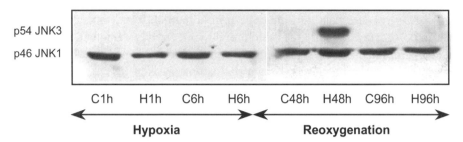

Figure 3. Temporal expression of JNK1 and JNK3 studied by Western blotting in cultured neurons during and after hypoxia (C = control, H = hypoxia).

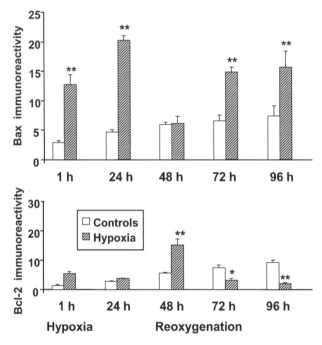

Figure 4. Temporal expression of Bax and Bcl-2 in cultured neurons exposed to hypoxia for 6 h and their matched controls ($n = 4$). Immunoreactivity is expressed as arbitrary units; Statistically significant difference from controls: $^*p < 0.05$ and $^{**}p < 0.01$.

pathways leading to brain apoptosis. Also, Bax has been demonstrated to activate caspase-3, the activity of which is strongly stimulated as a consequence of hypoxia in our culture model (Bossenmeyer-Pourié *et al.*, 1999b). This enzyme has been reported as the main effector in the execution of apoptosis in central neurons (Chen *et al.*, 1998). Caspase-1 was also stimulated in hypoxic neurons (Bossenmeyer-Pourié *et al.*, 2000b). Interestingly, defense proteins like Bcl-2 and HSP 70 were significantly increased at 48 h post-reoxygenation, at which time Bax expression transiently remained in the range of basal values. Another study has shown that cultured neurons can be rescued from hypoxia-induced apoptosis when they are treated by caspase inhibitors during hypoxia or even 24 h later (Bossenmeyer-Pourié *et al.*, 1999b). These data indicate that an important shift may occur at 48 h post-insult in our *in vitro* model, time which might correspond to a potential therapeutic window before neurons are irreversibly committed to die.

In other respects, it has been shown that apoptosis is closely related to the cell cycle, including in postmitotic neurons, and would be associated with abnormal or conflicting signals for cell cycle activation and cell cycle arrest (Meikrantz and Schlegel, 1995; Timsit *et al.*, 1999). Accordingly, a treatment by olomoucine (50 μM), a cell cycle inhibitor, protected neurons from hypoxia-induced injury (Bossenmeyer-Pourié *et al.*, 2002). In our culture model, a robust stimulation of PCNA (proliferating cell nuclear antigen) was observed by 1 h after the onset of hypoxia. This protein is required for DNA replication and repair, and is known to be synthesized during the late G1-early S phase of the cell cycle (Kurki *et al.*, 1988). These data are in good agreement with other reports that have shown that apoptotic death in various paradigms is preceded by increased expression of cyclins, such as cyclin D1, which are key markers of cell cycle activation (Timsit *et al.*, 1999). A strong and persistent overexpression of the tumor suppressor p53 was also observed in hypoxic neurons, and may reflect a subsequent blockade of the progression of the cell cycle. Induction of p53 has previously been associated with neuronal apoptotic death following ischemia (Renolleau *et al.*, 1997). After its translocation to the nucleus, p53 binds to DNA to transcriptionally activate its various response genes (Agarwal *et al.*, 1998), among which are p21, Gadd45 (growth arrest and DNA damage inducible gene 45), mdm2 (murine double minute-2) and the gene bax, whereas bcl-2 is known to be downregulated by p53 (Miyashita and Reed, 1995).

Although not illustrated herein, similar profiles of specific protein expression were also recorded in the rat brain after *in vivo* neonatal hypoxia (Grojean *et al.*, 2003b).

Oxidative stress

Numerous reports have suggested that reactive oxygen species (ROS) may play an important role in the processes leading to neuronal cell damage consecutively to hypoxia/reoxygenation (Berger and Garnier, 1999; Halliwell, 1992; Mishra and Delivoria-Papadopoulos, 1999). Additionally, it is known that the immature brain is rather deficient in intracellular defense systems against ROS (Mishra and Delivoria-Papadopoulos, 1988; Mover and Ar, 1997), possibly rendering the premature infant's brain more susceptible to oxidative stress. Whereas oxygen radicals may serve as effectors of cell death, resulting in oxidative damage of major cellular components (e.g. DNA, lipids and proteins) as well as in alterations of membrane functions, they were also implicated as signaling molecules capable of participating in the apoptotic cascade (Bredesen, 1995; Sastry and Rao, 2000; Suzuki *et al.*, 1997).

In an attempt to address the participation of oxidative pathways in the process of *in vitro* cell death induced by transient hypoxia, we monitored the production of superoxide radicals in the extracellular cell medium (Daval *et al.*, 1995; Oillet *et al.*, 1996) as well as the intracellular production of ROS in individual cells by

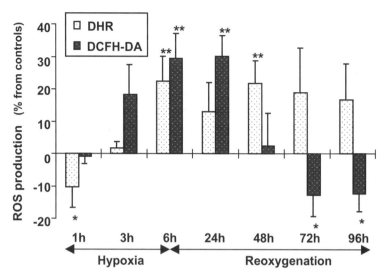

Figure 5. Intracellular production of mitochondrial (DHR) and cytosolic (DCFH-DA) ROS in cultured neurons exposed to hypoxia for 6 h ($n = 6$). Significantly different from normoxic controls: $^*p < 0.05$ and $^{**}p < 0.01$.

flow cytometry using the fluorogenic dyes dihydrorhodamine 123 (DHR, mostly reflecting intra-mitochondrial ROS) and 2′,7′-dichlorofluorescein-diacetate (DCFH-DA, detecting oxidative stress in the cytoplasm). In parallel, temporal changes in the main endogenous defense systems (i.e. superoxide dismutase, glutathione peroxidase, catalase, intracellular reduced glutathione and glutathione reductase) were analyzed in cultured neurons, and the influence of various antioxidants was evaluated (Lièvre *et al.*, 2000, 2001).

Transient cell exposure to a superoxide generating system (1 mU/ml xanthine oxidase + 50 μM xanthine) induced marked cytotoxicity that was mostly reflected by apoptosis, suggesting that oxidative stress triggers apoptotic pathways. Moverover, our data showed that neuronal production of free radicals in response to hypoxia/reoxygenation did not appear as a short and transient event, but was detectable over several days after the onset of reoxygenation (Fig. 5). The parallel sequential alterations of the expression and activities of endogenous defense enzymes, coupled with the neuroprotective effects of the glutathione precursor N-acetyl-cysteine (NAC), suggested that, even if ROS overproduction seems to be of moderate magnitude, it may be sufficient to overwhelm the antioxidant capacities of neurons, mainly glutathione-dependent elimination of hydroperoxide, to finally participate to apoptotic cell death in neuronal cultures.

Participation of excitotoxicity and influence of age

Beyond their functions as neurotransmitters, excitatory amino acids (EAAs) like glutamate are considered as mediators of neuronal injury in various pathological conditions, including hypoxia-ischemia (Lipton and Rosenberg, 1994). Neurotoxicity is classically achieved through the 'excitotoxic' process involving excessive glutamate release followed by overactivation of specific EAA receptors, especially the N-methyl-D-aspartate (NMDA) receptor subtype (Mishra *et al.*, 2001). Specifically in the immature brain, however, the role of excitotoxicity mediated by NMDA receptors was controversial. First, and as documented above, hypoxia mainly results in apoptosis, whereas excitotoxicity was mostly related to necrosis (Dessi *et al.*, 1993). Second, it has been shown that brain sensitivity to EAAs only develops around one week of postnatal age in the rat (Ikonomidou *et al.*, 1989), in line with the ontogenetic profile of cerebral NMDA receptors (Insel *et al.*, 1990).

To delineate the actual participation of excitotoxicity in our experimental models, we monitored the effects of exogenously applied glutamate in cultured neurons, and evaluated the influence of selective EAA receptor antagonists on hypoxia effects, both *in vitro* and *in vivo*. Also, we compared the consequences of hypoxia at different brain maturational stages (Chihab *et al.*, 1998a, 1998c; Grojean *et al.*, 2003a).

Exposure to glutamate or other EAAs (at concentrations up to 100 μM) remained without any effect on 6-day-old cultured neurons, and addition of EAA receptor antagonists (MK-801 for NMDA receptors or NBQX for AMPA and kainate receptors) did not protect cultured cells against hypoxic injury. However, *in vitro* neuronal cells became vulnerable to exogenous glutamate when they were co-exposed to staurosporine (30 nM), a protein kinase C inhibitor (Chihab *et al.*, 1998a, 1998c). In accordance with the *in vitro* studies, a pre-hypoxic administration of MK-801 (5 mg/kg) to the newborn rat pup did not prevent hypoxia-induced neuronal damage *in vivo* (Grojean *et al.*, 2003a).

Collectively, these findings suggest that (i) hypoxia effects in the immature brain are not mediated by excitotoxicity, (ii) immature neurons are rather resistant to excitotoxicity, and (iii) this resistance to glutamate toxicity may be explained, at least partly, by the inability of neuronal cells at this developmental stage to trigger inhibition of membrane protein kinase C activity, an early and necessary step for glutamate to mediate cell injury (Durkin *et al.*, 1997).

By contrast, more mature brain neurons, as reflected by neuronal cells maintained in culture for 13 days or those present in the brain of 7-day-old rats *in vivo*, were sensitive to excitotoxicity. Accordingly, when hypoxia was induced in rats at 7 days of age (100% N_2, 8 min), morphological analyses, DNA fragmentation studies on agarose gels as well as monitoring of the expression of specific proteins, all were indicative of high levels of necrosis which were associated with permanent brain lesions and resemble excitotoxic manifestations. Moreover, a pre-treatment by MK-801 almost completely protected the brain against hypoxic damage.

As summarized in Fig. 6, our results have demonstrated that the characteristics of cell death associated with transient hypoxia in the developing central nervous system are strongly influenced by the degree of brain maturation, with apoptosis being largely predominant in the rat brain around birth, and massive necrosis occurring at one week of age. Such a remarkable shift in the rat brain response to

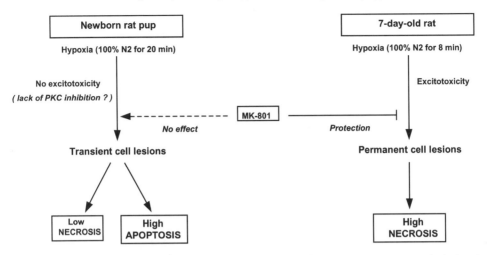

Figure 6. Summary of main conclusions on the participation of excitotoxicity in hypoxia-induced cell injury in the developing brain.

oxygen deprivation appears to correlate with a parallel recruitment of the excitotoxic processes.

EFFECTS OF MILD HYPOXIA

It is well admitted that severity and/or duration of the hypoxic episode may strongly influence brain cell final outcome (Banasiak and Haddad, 1998; Haddad and Jiang, 1993). In this respect, various reports have documented that a potentially damaging stimulus, when applied below the threshold of injury, can activate protective mechanisms, to finally reduce the deleterious impact of subsequent, more severe stimuli (Dirnagl *et al.*, 2003). Such a tolerance phenomenon may be induced by mild hypoxia, both *in vitro* (Bossenmeyer-Pourié and Daval, 1998; Bossenmeyer-Pourié *et al.*, 2000a) and *in vivo* (Cantagrel *et al.*, 2003; Gidday *et al.*, 1994). In an attempt to depict cell mechanisms which underlie hypoxic tolerance, we have studied the effects of a shorter hypoxia (3 h) in our culure model (Bossenmeyer-Pourié *et al.*, 1999a, 2002).

By contrast to hypoxia imposed for 6 h, a 3-h hypoxic stress did not damage cultured neurons, and several arguments are in favor of a stimulation of neurogenesis by sublethal hypoxia. Indeed, the final viability score in hypoxic neurons was repeatedly found to be above control values ($\sim$15%), along with an increased number of mitotic nuclei depicted by DAPI staining. Moreover, intracellular incorporation of thymidine, which is considered to reflect DNA synthesis, was significantly increased in response to short hypoxia, and a persistent overexpression of PCNA, a co-factor for DNA polymerase, was also recorded. Finally, the addition to the culture medium of the cell cycle blocker olomoucine blocked the beneficial effects of

J. L. Daval

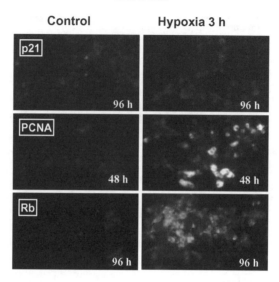

Figure 7. Effects of sublethal hypoxia on the expression of various proteins involved in the regulation of the cell cycle.

hypoxia, indicating that the neuronal response to sublethal hypoxia actually also implies cell cycle activation.

Though neuronal cultures still contain proliferating cells, evidence has been provided that differentiated neurons still possess mitotic capacity and that progenitor cells which are normally present in the adult brain can differentiate into neurons (Hastings *et al.*, 2001). Further analysis of cell cycle-related proteins associated with sublethal (i.e. 3 h) hypoxia in cultured neurons revealed that p53 and its target gene product p21 were only transiently stimulated in response to mild hypoxia. Between 24 and 48 h post-reoxygenation, their expression levels were no longer different from controls. PCNA was highly overexpressed in hypoxic neurons, and then gradually declined when the amounts of Rb protein (the retinoblastoma susceptibility gene product) progressively increased (Fig. 7). It is conceivable that, due to the lack of induction of p21, cyclin-dependent kinases can phosphorylate Rb, and thus allow the cell cycle to progress.

CONCLUSIONS AND FUTURE DIRECTION

Our *in vitro* and *in vivo* investigations have shown that neurons of the immature brain are very sensitive to hypoxia as a single source for initiating the pathophysiological response. Depending on the severity of the insult, specific patterns of gene induction/repression appear to play a pivotal role for neuronal destiny in response to hypoxia.

Severe hypoxia around the birth period in the rat triggers delayed apoptotic neuronal death in the most vulnerable brain regions. Apoptosis corresponds to a ubiquitous form of naturally occurring cell death that plays a fundamental role in brain development. These observations confirm that the developing brain is more prone to apoptosis, probably by retaining a part of the developmental cell death programme, and suggest the existence of a potential therapeutic window before neurons are irreversibly committed to die. Age certainly represents a key factor in the brain response to hypoxic episodes, since the more mature neurons, by contrast to the very immature ones, can die through an excitotoxic process normally typified by cell homeostasis failure and rapid lysis of neuronal membranes.

A striking observation is the possible induction of neurogenesis by sublethal hypoxia. As already mentioned, it has been reported that cells destined to die attempt to re-enter the cell cycle before proceeding to death. Accordingly, numerous genes that have been implicated in apoptosis, including c-myc, c-fos, c-jun or p53, are known to play a role in modulating cell proliferation. Our findings support the hypothesis that mild hypoxia promotes cell cycle activation and would finally allows neuronal proliferation.

Another salient feature is the apparent repair capacities of the immature brain after a hypoxic insult *in vivo*. Based on the above considerations and in accordance with recent reports (for review, see Kokaia and Lindvall, 2003), it is tempting to speculate that neurogenesis may account for the gradual anatomical recovery observed in rat pups exposed to transient hypoxia, and this warrants future exciting investigations.

Acknowledgements

Many thanks are due to all the talented PhD students who contributed to the different parts of the reported work: F. Nicolas, J. Oillet, C. Bossenmeyer-Pourié, R. Chihab, V. Lièvre, and S. Grojean as well as to P. Vert and V. Koziel for their help.

REFERENCES

Agarwal, M. L., Taylor, W. R., Chernov, M. V., *et al.* (1998). The p53 network, *J. Biol. Chem.* **273**, 1–4.

Alling, C. (1985). Biochemical maturation of the brain and the concept of vulnerable period, in: *Alcohol and the Developing Brain*, Rydberg, U. (Ed.), pp. 5–10. Raven Press, New York.

Amiel-Tison, C. and Ellison, P. (1986). Birth asphyxia in the fullterm newborn: early assessment and outcome, *Dev. Med. Child Neurol.* **28**, 671–682.

Andiné, P., Thordstein, M., Kjellmer, I., *et al.* (1990). Evaluation of brain damage in a rat model of neonatal hypoxic-ischemia, *J. Neurosci. Meth.* **35**, 253–260.

Banasiak, K. J. and Haddad, G. G. (1998). Hypoxia-induced apoptosis: effect of hypoxic severity and roles of p53 in neuronal cell death, *Brain Res.* **797**, 295–304.

Berger, R. and Garnier, Y. (1999). Pathophysiology of perinatal brain damage, *Brain Res. Rev.* **30**, 107–134.

Bossenmeyer-Pourié, C. and Daval, J. L. (1998). Prevention from hypoxia-induced apoptosis by preconditioning: A mechanistic approach in cultured neurons from fetal rat forebrain, *Mol. Brain Res.* **58**, 237–239.

Bossenmeyer, C., Chihab, R., Muller, S., *et al.* (1998). Hypoxia/reoxygenation induces apoptosis through biphasic induction of protein synthesis in central neurons, *Brain Res.* **787**, 107–116.

Bossenmeyer-Pourié, C., Chihab, R., Schroeder, H., *et al.* (1999a). Transient hypoxia may lead to neuronal proliferation in the developing mammalian brain: from apoptosis to cell cycle completion, *Neuroscience* **91**, 221–231.

Bossenmeyer-Pourié, C., Koziel, V. and Daval, J. L. (1999b). CPP32/caspase-3-like proteases in hypoxia-induced apoptosis in developing brain neurons, *Mol. Brain Res.* **71**, 225–237.

Bossenmeyer-Pourié, C., Koziel, V. and Daval, J. L. (2000a). Effects of hypothermia on hypoxia-induced apoptosis in cultured neurons from the developing rat forebrain — Comparison with preconditioning, *Pediatr. Res.* **47**, 385–391.

Bossenmeyer-Pourié, C., Koziel, V. and Daval, J. L. (2000b). Involvement of caspase-1 proteases in hypoxic brain injury. Effects of their inhibitors in developing neurons, *Neuroscience* **95**, 1157–1165.

Bossenmeyer-Pourié, C., Lièvre, V., Grojean, S., *et al.* (2002). Sequential expression patterns of apoptosis- and cell cycle-related proteins in neuronal response to severe or mild transient hypoxia, *Neuroscience* **114**, 869–882.

Bredesen, D. E. (1995). Neuronal apoptosis, *Ann. Neurol.* **38**, 839–851.

Brierley, J. B. and Graham, D. I. (1984). Hypoxia and vascular disorders of the central nervous system, in: *Greenfield's Neuropathology*, 4th edn, Adams, J. H., Corsellis, J. A. N. and Duchen, L. W. (Eds), pp. 125–207. Edward Arnold, London.

Cantagrel, S., Krier, C., Ducrocq, S., *et al.* (2003). Hypoxic preconditioning reduces apoptosis in a rat model of immature brain hypoxia-ischemia, *Neurosci. Lett.* **347**, 106–110.

Carter, A. (1989). Factors affecting gas transfer across the placenta and the oxygen supply to the fetus, *J. Dev. Physiol.* **12**, 305–322.

Chen, J., Nagayama, T., Jin, K., *et al.* (1998). Induction of caspase-3-like protease may mediate delayed neuronal death in the hippocampus after transient cerebral ischemia, *J. Neurosci.* **18**, 4914–4928.

Chihab, R., Bossenmeyer, C., Oillet, J., *et al.* (1998a). Lack of correlation between the effects of transient exposure to glutamate and those of hypoxia/reoxygenation in immature neurons *in vitro*, *J. Neurochem.* **71**, 1177–1186.

Chihab, R., Ferry, C., Koziel, V., *et al.* (1998b). Sequential activation of activator protein-1-related transcription factors and JNK protein kinases may contribute to apoptotic death induced by transient hypoxia in developing brain neurons, *Mol. Brain Res.* **63**, 105–120.

Chihab, R., Oillet, J., Bossenmeyer, C., *et al.* (1998c). Glutamate triggers cell death specifically in mature central neurons through a necrotic process, *Mol. Gen. Metab.* **63**, 142–147.

Daval, J. L., Koziel, V. and Nicolas, F. (1991). Functional changes in cultured neurons following transient asphyxia, *NeuroReport* **2**, 97–100.

Daval, J. L., Ghersi-Egea, J. F., Oillet, J., *et al.* (1995). A simple method for evaluation of superoxide radical production in neural cells under various culture conditions: Application to hypoxia, *J. Cereb. Blood Flow Metab.* **15**, 71–77.

Dessi, F., Charriaut-Marlangue, C., Khrestchatisky, M., *et al.* (1993). Glutamate-induced neuronal death is not a programmed cell death in cerebellar cultures, *J. Neurochem.* **60**, 1953–1955.

Dirnagl, U., Simon, R. P. and Hallenbeck, J. M. (2003). Ischemic tolerance and endogenous neuroprotection, *Trends Neurosci.* **26**, 248–254.

Dombrowski, G. J., Jr., Swiatek, K. R. and Chao, K. L. (1989). Lactate, 3-hydroxybutyrate, and glucose as substrates for the early postnatal rat brain, *Neurochem. Res.* **14**, 667–675.

Duffy, T. E., Kohle, S. J. and Vannucci, R. C. (1975). Carbohydrate and energy metabolism in perinatal rat brain, *J. Neurochem.* **24**, 271–276.

Durkin, J. P., Tremblay, R., Chakravarthy, B., *et al.* (1997). Evidence that the early loss of membrane protein kinase C is a necessary step in the excitatory amino acid-induced death of primary cortical neurons, *J. Neurochem.* **68**, 1400–1412.

Fazekas, J. F., Alexander, F. A. D. and Himwich, H. E. (1941). Tolerance of the newborn to hypoxia, *Am. J. Physiol.* **134**, 281–287.

Gidday, J. M., Fitzgibbons, J. C., Shah, A. R., *et al.* (1994). Neuroprotection from ischemic brain injury by hypoxic preconditioning in the neonatal rat, *Neurosci. Lett.* **168**, 221–224.

Grojean, S., Pourié, G., Vert, P., *et al.* (2003a). Differential neuronal fates in the CA1 hippocampus following hypoxia in newborn and 7-day-old rats — Effects of a pre-treatment by MK-801, *Hippocampus* **13**, 970–977.

Grojean, S., Schroeder, H., Pourié, G., *et al.* (2003b). Histopathological alterations and functional brain deficits after transient hypoxia in the newborn rat pup: a long term follow-up, *Neurobiol. Dis.* **14**, 265–278.

Haddad, G. G. and Donelly, D. F. (1990). O_2 deprivation induces a major depolarization in brainstem neurons in the adult but not in the neonate, *J. Physiol., London* **429**, 411–428.

Haddad, G. G. and Jiang, C. (1993). O_2 deprivation in the central nervous system: on mechanisms of neuronal response, differential sensitivity and injury, *Prog. Neurobiol.* **40**, 277–318.

Halliwell, B. (1992). Reactive oxygen species and the central nervous system, *J. Neurochem.* **59**, 1609–1623.

Hansen, M. B., Nielsen, S. E. and Berg, K. (1989). Re-examination and further development of a precise and rapid dye method for measuring cell growth/cell kill, *J. Immunol. Meth.* **119**, 203–210.

Hastings, N. B., Tanapat, P. and Gould, E. (2001). Neurogenesis in the adult mammalian brain, *Clin. Neurosci. Res.* **1**, 175–182.

Ikonomidou, C., Mosinger, J. L., Salles, K. S., *et al.* (1989). Sensitivity of the developing rat brain to hypobaric/ischemic damage parallels sensitivity to N-methyl-aspartate neurotoxicity, *J. Neurosci.* **9**, 2809–2818.

Insel, T. R., Miller, L. P. and Gelhard, R. E. (1990). The ontogeny of excitatory amino acid receptors in rat forebrain. I. N-methyl-D-aspartate and quisqualate receptors, *Neuroscience* **35**, 31–43.

Kass, I. S. and Lipton, P. (1989). Protection of hippocampal slices from young rats against anoxic transmission damage is due to better maintenance of ATP, *J. Physiol., London* **413**, 1–11.

Kohle, S. J. and Vannucci, R. C. (1977). Glycogen metabolism in fetal and postnatal rat brain: influence of birth, *J. Neurochem.* **28**, 441–443.

Kokaia, Z. and Lindvall, O. (2003). Neurogenesis after ischaemic brain insults, *Curr. Opin. Neurobiol.* **13**, 127–132.

Kurki, P., Ogata, K. and Tan, E. M. (1988). Monoclonal antibodies to proliferating cell nuclear antigen (PCNA)/cyclin as probes for proliferating cells by immunofluorescence microscopy and flow cytometry, *J. Immunol. Meth.* **109**, 49–59.

Lièvre, V., Bécuwe, P., Bianchi, A., *et al.* (2000). Free radical production and changes in superoxide dismutases associated with hypoxia/reoxygenation-induced apoptosis of embryonic rat forebrain neurons in culture, *Free Radicals Biol. Med.* **29**, 1291–1301.

Lièvre, V., Bécuwe, P., Bianchi, A., *et al.* (2001). Intracellular generation of free radicals and modifications of detoxifying enzymes in cultured neurons from the developing rat forebrain in response to transient hypoxia, *Neuroscience* **105**, 287–297.

Lipton, S. A. and Rosenberg, P. A. (1994). Excitatory amino acids as a final common pathway for neurologic disorders, *New Engl. J. Med.* **330**, 613–622.

Maneru, C., Junque, C., Botet, F., *et al.* (2001). Neuropsychological long-term sequelae of perinatal asphyxia, *Brain Injury* **15**, 1029–1039.

Martin, D. P., Schmidt, R. E., DiStefano, P., *et al.* (1988). Inhibitors of protein synthesis and RNA synthesis prevent neuronal death caused by nerve growth factor deprivation, *J. Cell. Biol.* **106**, 829–844.

McDonald, J. W., Silverstein, F. S. and Johnston, M. V. (1988). Neurotoxicity of N-methyl-D-aspartate is markedly enhanced in developing rat central nervous system, *Brain Res.* **459**, 200–203.

Meikrantz, W. and Schlegel, R. (1995). Apoptosis and the cell cycle, *J. Cell. Biol.* **58**, 160–174.

Mishra, O. P. and Delivoria-Papadopoulos, M. (1988). Anti-oxidant enzymes in fetal guinea pig brain during development and the effect of maternal hypoxia, *Dev. Brain Res.* **42**, 173–179.

Mishra, O. P. and Delivoria-Papadopoulos, M. (1999). Cellular mechanisms of hypoxic injury in the developing brain, *Brain Res. Bull.* **48**, 233–238.

Mishra, O. P., Fritz, K. I. and Delivoria-Papadopoulos, M. (2001). NMDA receptor and neonatal hypoxic injury, *Mental Retard. Dev. Disabil. Res. Rev.* **7**, 249–253.

Mitsufigi, N., Yoshioka, H., Okano, S., *et al.* (1996). A new model of transient cerebral ischemia in neonatal rats, *J. Cereb. Blood Flow Metab.* **16**, 237–243.

Miyashita, T. and Reed, J. (1995). Tumor suppressor p53 is a direct transcriptional activator of the human bax gene, *Cell* **80**, 293–299.

Morgane, P. J., Austin-LaFrance, R. J., Bronzino, J. D., *et al.* (1993). Prenatal malnutrition and development of the brain, *Neurosci. Biobehav. Rev.* **17**, 91–128.

Mover, H. and Ar, A. (1997). Antioxidant enzymatic activity in embryos and placenta of rats chronically exposed to hypoxia and hyperoxia, *Comp. Biochem. Physiol.* **117C**, 151–157.

Mutch, L., Alberman, E., Hagberg, B., *et al.* (1992). Cerebral palsy epidemiology: where are we and where are we going? *Dev. Med. Child Neurol.* **34**, 547–551.

Nyakas, C., Buwalda, B. and Luiten, P. G. M. (1996). Hypoxia and brain development, *Prog. Neurobiol.* **49**, 1–51.

Oillet, J., Koziel, V., Vert, P., *et al.* (1996). Influence of post-hypoxia reoxygenation conditions on energy metabolism and superoxide production in cultured neurons from the rat forebrain, *Pediatr. Res.* **39**, 598–603.

Papas, S., Crépel, V., Hasboun, D., *et al.* (1992). Cycloheximide reduces the effects of anoxic insult *in vivo* and *in vitro*, *Eur. J. Neurosci.* **4**, 758–765.

Park, D. S., Morris, E. J., Greene, L. A., *et al.* (1997). G1/S cell cycle blockers and inhibitors of cyclin-dependent kinases suppress camptothecin-induced neuronal apoptosis, *J. Neurosci.* **17**, 1256–1270.

Patel, J. and Edwards, A. D. (1997). Prediction of outcome after perinatal asphyxia, *Curr. Opin. Pediatr.* **9**, 128–132.

Pearigen, P., Gwinn, R. and Simon, R. P. (1996). The effects *in vivo* of hypoxia on brain injury, *Brain Res.* **725**, 184–191.

Pulsinelli, W. A., Brierley, J. B. and Plum, F. (1982). Temporal profile of neuronal damage in a model of transient forebrain ischemia, *Ann. Neurol.* **11**, 491–498.

Rauscher, F. J., III, Sambucetti, L. C., Curran, T., *et al.* (1988). A common DNA binding site for Fos protein complexes and transcription factor AP-1, *Cell* **52**, 471–480.

Renolleau, S., Benjelloun, N., Ben-Ari, Y., *et al.* (1997). Regulation of apoptosis-associated proteins in cell death following transient focal ischemia in rat pups, *Apoptosis* **2**, 368–376.

Renolleau, S., Aggoun-Zouaoui, D., Ben-Ari, Y., *et al.* (1998). A model of transient unilateral focal ischemia with reperfusion in the P7 neonatal rat: morphological changes indicative of apoptosis, *Stroke* **29**, 1454–1461.

Rice, J. E., Vannucci, R. C. and Brierley, J. B. (1981). The influence of immaturity on hypoxic-ischemic brain damage in the rat, *Ann. Neurol.* **9**, 131–141.

Rivkin, M. J. and Volpe, J. J. (1994). Asphyxia and brain injury, in: *Intensive Care of the Fetus and Neonate*, Spitzer, A. R. (Ed.), pp. 685–695. Mosby, St. Louis.

Sastry, P. S. and Rao, K. S. (2000). Apoptosis and the nervous system, *J. Neurochem.* **74**, 1–20.

Schwartz, P. H., Massarweh, W. F., Vinters, H. V., *et al.* (1992). A rat model of severe neonatal hypoxic-ischemic brain injury, *Stroke* **23**, 539–546.

Sherwood, N. M. and Timiras, P. S. (1970). *A Stereotaxic Atlas of the Developing Rat Brain.* University of California Press, Berkeley.

Simon, N. P. (1999). Long-term neurodevelopmental outcome of asphyxiated newborns, *Clin. Perinatol.* **26**, 767–778.

Slotkin, T. A., Cowdery, T. S., Orband, L., *et al.* (1986). Effects of neonatal hypoxia on brain development in the rat: immediate and long-term biochemical alterations in discrete regions, *Brain Res.* **374**, 63–74.

Sochocka, E., Juurlink, B., Code, W., *et al.* (1994). Cell death in primary cultures of mouse neurons and astrocytes during exposure to and 'recovery' from hypoxia, substrate deprivation and simulated ischemia, *Brain Res.* **638**, 21–28.

Stattford, A. and Wheatherhall, J. A. C. (1960). The survival of young rats in nitrogen, *J. Physiol.* **153**, 457–472.

Suzuki, Y. J., Forman, H. J. and Sevanian, A. (1997). Oxidants as stimulators of signal transduction, *Free Radic. Biol. Med.* **22**, 269–285.

Taniguchi, T., Fukunaga, R., Matsuoka, Y., *et al.* (1994). Delayed expression of c-fos protein in rat hippocampus and cerebral cortex following transient *in vivo* exposure to hypoxia, *Brain Res.* **640**, 119–125.

Timsit, S., Rivera, S., Ouaghi, P., *et al.* (1999). Increased cyclin D1 in vulnerable neurons in the hippocampus after ischaemia and epilepsy: a modulator of *in vivo* programmed cell death? *Eur. J. Neurosci.* **11**, 263–278.

Towfighi, J., Zec, N., Yager, J., *et al.* (1995). Temporal evolution of neuropathologic changes in immature rat model of cerebral hypoxia: a light microscopic analysis, *Acta Neuropathol.* **90**, 375–386.

Vannucci, R. (1990). Experimental biology of cerebral hypoxia-ischemia: relation to perinatal brain damage, *Pediatr. Res.* **27**, 317–326.

Vannucci, R. C. and Duffy, T. E. (1974). Influence of birth on carbohydrate and energy metabolism in rat brain, *Am. J. Physiol.* **226**, 933–940.

Vannuci, R. C. and Plum, F. (1975). Pathophysiology of perinatal hypoxic-ischemic brain damage, in: *Biology of Brain Dysfunction*, Gaull, G. E. (Ed.), pp. 163–178. Plenum Press, New-York.

Volpe, J. J. (1995). *Neurology of the Newborn*. W. B. Saunders, Philadelphia, PA.

Wolvetang, E. J., Johnson, K. L., Krauer, K., *et al.* (1994). Mitochondrial respiratory chain inhibitors induce apoptosis, *FEBS Letters* **339**, 40–44.

Xia, Y., Jiang, C. and Haddad, G. G. (1992). Oxidative and glycolytic pathways in rat (newborn, adult) and turtle brain: role during anoxia, *Am. J. Physiol.* **262**, R595–R603.

Advances in Neuroregulation and Neuroprotection (2005), pp. 83-102
Collin, C. *et al.* (Eds)
© VSP 2005

Neurodegeneration and neuroprotection: epilepsy and Parkinsonism

DEBJANI GUHA *

S. N. Pradhan Centre for Neurosciences, Calcutta University, Kolkata, India

Abstract—Mounting evidence suggests that neurological dysfunctions have accelerated among the world patient population. The present chapter focuses on pathogenesis and protection of two such prevailing disorders of the central nervous system — epilepsy and Parkinsonism. While epilepsy is more prevalent among the young, the older generation of the population is affected by Parkinsonism. Although a wide variety of drugs are available for treatment of these nervous system disorders, neither any prophylaxis nor cure is available; therapy is symptomatic. There is a need for new drugs that can halt the course of progression of the underlying disorder. Mechanisms that underlie the disease process constitute potential target areas for development of newer therapeutic drugs. Inflammation in the brain has traditionally been recognised as playing an important role in the pathogenesis of several neurological disorders. However, newer sources like neurochemical, environmental toxicants and oxidative stress have been implicated as risk factors in such neurological diseases and a combination of these factors is likely to be responsible for disease onset and progression. Effort is being devoted to novel approaches elucidating neurochemical, genetic, cellular and molecular mechanisms of such neurologicaldisorders. Researchers are now aiming to design therapeutic agents free from toxic side effects and, in the search for an ideal therapy, focus has shifted to development of therapeutic agents from herbal sources. Neurotherapeutics from natural reserves seems to be the emerging strategy.

Keywords: Neurodegeneration; Parkinsonism; epilepsy; neurotransmitters; aging; oxidative stress; nitric oxide; neuroprotection.

EPILEPSY

Introduction

The brain occupies the key position in the central nervous system and is affected by various diseases in youth and old age. Epilepsy is encountered more among the younger generation and Parkinson's and Alzheimer's disease are observed increasingly as diseases of old age. Epilepsy, one of the commonest disorders of

---

*E-mail: debjaniguha@rediffmail.com

the CNS has been with us for centuries, yet today it is still a mystery. The common description of epilepsy is 'seizure discharge'.

Epilepsy is a major neurological problem all over the world. Incidence and prevalence studies of epilepsy have been reported from many countries but comparisons are often difficult because investigators have adopted different definitions of epilepsy, classification schemes, and selection bias (Shorvon, 1990). Nevertheless, most studies have found incidence rates of 20–70 cases per 100 000 per year (range 11–134 cases per 100 000 per year) and point prevalence rates of 0.4 to 0.8% (Chadwick, 1990).

Incidence varies with age. The incidence rate is highest in children under 2 years of age and in persons over 65 years of age. About 30% of patients with seizures have an identifiable neurological disorder and the remainder have idiopathic epilepsy (Elwes et al., 1985). Males are more likely to have a new diagnosis of epilepsy than females. The seizure type and the cause of the seizure changes with age.

Epilepsy is hyperactivity of a group of nerve cells and is characterized by physical manifestation of behaviour.

The brain cells (neurons) like many other cells in the body have the built-in capacity to generate potentials or electric signals. It is through these signals that the message is transmitted to a far-off part of the body. A brain cell may be receiving several messages while sending out different messages at the same time. It is normally done in a very meticulous, orderly fashion. Therefore, different functional signals or messages are not jumbled up. The brain acts like a telephone switch-board. If the incoming and outgoing calls are plugged in wrongly or there is something else wrong with the switch-board, then there is chaos and impairment of communications. All functional activities are represented in the brain not at the same place but in different areas and at different depths. Which brain cells are impaired will be reflected in the disturbed or altered functions of that area of the brain. Malfunction of the switch-board can happen only once or frequently at different intervals and for different durations. Epilepsy is an electric disturbance resulting from a disorderly discharge of brain cells. The discharge is sudden and recurrent in nature. The discharge may or may not be accompanied by alteration in consciousness, or in motor, sensory, psychological and behavioural functions: the clinical manifestations depend on the nature of the discharge. In epilepsy, as a disease, the discharge has to be recurrent. A single convulsion should not be taken as equivalent to epilepsy. Motor manifestations of a seizure are called convulsions. A single episode is called a seizure. The other common names for seizures are blackouts, attacks, and fits. Thus epilepsy is the disease of the brain (Benardo and Pedley, 1984).

For many years epilepsy was not considered a brain disease. Hippocrates suggested that the seat of the disease is in the brain which overflows with phlegm that clogs the veins and causes convulsions. Many people consider that epileptic seizures are caused by supernatural power while many others blame evil spirits, devils, demons or black magic and so on. Some people also regard epilepsy a mental illness and a person suffering from epileptic fits is considered insane (Maheswari, 1993).

Since epilepsy is not a homogeneous entity, various factors influence prognosis. Patients with a good prognosis are those with seizures precipitated by alcohol, drugs, or metabolic disturbance (Shorvon, 1984). Patients with the poorest prognosis are those with evidence of diffuse cerebral disorder (often with intellectual or behavior disturbance); early onset seizures; partial or mixed seizure types; progressive neurological disorders; or severe epileptic syndromes (e.g. Lennox Gastaut syndrome, West syndrome). The length of active epilepsy is also important — the longer the seizures continue after the onset of treatment the worse the ultimate prognosis (Elwes *et al.*, 1984).

Seizure activity is characterized by paroxysmal discharges occurring synchronously in a large population of cortical neurons. This is characterized in the EEG as sharp wave or spike. The physiology of a seizure is traceable to an unstable cell membrane or its surrounding, supportive cells (Meldrum, 1990). An abnormality of potassium conductance, a defect in the voltage-sensitive calcium channels, or a deficiency in the membrane ATPases linked to ion transport may result in neuronal membrane instability and a seizure (Benardo and Pedley, 1984). Neurotransmitters (e.g. acetylcholine (Ach), norepinephrine (NE), histamine (HA) and corticotrophin-releasing factor) enhance the excitability and propagation of neuronal activity, whereas GABA and dopamine inhibit neuronal activity and propagation. Normal neuronal activity also depends on an adequate supply of glucose, oxygen, sodium, potassium, calcium, and amino acids. Systemic pH is also a factor in precipitating seizures. There may be primary defects in the GABAergic inhibitory system or in the sensitivity or arrangement of the receptors involved in excitatory neurotransmission that result in a seizure (Benardo and Pedley, 1984).

Dendritic degeneration is a non-specific finding, but may be associated with membrane changes, including receptor hypersensitivity, that could contribute to epileptogenesis. Another common finding associated with both generalized seizures and complex partial seizures is an abnormality of cortical maturation often called microdysgenesis. This may manifest as clusters of abnormally large neurons in the cortex, or as dystrophic groups of neurons in the sub-cortical white matter. It has been proposed that such abnormalities predispose to diverse types of epilepsy including primary generalized epilepsy, West's syndrome, and temporal lobe epilepsy (Ellenberg, 1986). However, their relevance remains uncertain.

Anything that disrupts the normal homeostasis of neuronal cell and disturbs its stability may trigger seizures. The most clearly established of these factors are severe head trauma, infections of the central nervous system (CNS), and stroke, although many other factors are also important antecedents. A hereditary predisposition to seizures has been suggested. Epilepsy frequently occurs in families. The parents, siblings, and offspring of a person with epilepsy are more likely than the general population to have epilepsy. This familial aggregation does not necessarily imply a genetic mechanism. In addition to genes, families

Table 1.
Common causes of seizures

Mechanical	Metabolic disturbances
• Head trauma	• Electrolytes
• Birth injury	• Water
• Neoplasm	• Glucose
• Vascular	• Amino acids
• Vascular abnormalities	• Lipids
	• pH
Sudden withdrawal of CNS drugs	Toxins
• Alcohol	• Fever
• Street drugs	• Infection
• Antipsychotics	
• Antidepressants	
• Antiepileptic drugs	
Heredity	Idiopathic

share environmental exposures that may also increase the risk of epilepsy. Primary generalized epilepsies have a strong genetic contribution (Anderson *et al.*, 1982; Meierkord, 1989). Patients with mental retardation and cerebral palsy are at increased risk for seizures. The more profound the degree of mental retardation as measured by IQ, the greater the incidence of epilepsy (Sundaram, 1989). The causes of seizures in the elderly are cerebrovascular disease, tumor, head trauma, metabolic disorder, and CNS infections (Scheuer and Pedley, 1990). Hyperventilation may precipitate absence seizures. Sleep, sleep deprivation, sensory stimuli and emotional stress may initiate seizures. Hormonal changes occurring around the time of menses, puberty, or pregnancy have been associated with the onset of, or an increase in, seizure activity. Other precipitating factors include fever, lack of food, and drugs. Also, antiepileptic drugs (AEDs) in excessive concentrations may cause seizures. Table 1 summarizes the common causes of seizures.

Classification

The International League Against Epilepsy has developed a classification system (Table 2) that combines clinical description with EEG findings. Over 90% of seizure patients may be classified using this system. Using the international classification scheme, seizures may be divided into partial, generalized, or unclassified.

Neurotransmitter involvement in basic mechanisms of epilepsy

Evidence is accumulating to support a relationship between central nervous system transmitters and convulsions (Purkayastha and Guha, 2003; Schlichter *et al.*, 1986).

Table 2.
Classification of epileptic seizures (Chadwick *et al.*, 1989)

Traditional classification	New nomenclature

Focal motor; Jacksonian seizure	I. Partial seizures (seizures begin locally)
	 (1) Simple (without impairment of consciousness)
	      • With motor symptoms
	      • With special sensory or somatosensory symptoms
	      • With autonomic symptoms
	      • With psychic symptoms

Temporal lobe or psychomotor seizures	(2) Complex (with impairment of consciousness)
	      • Simple partial onset followed by impairment of consciousness — with or without automatisms
	      • Impaired consciousness at onset-with or without automatisms
	 (3) Secondarily generalized (partial onset evolving to generalized tonic-clonic seizures)

Petit mal	II. Generalized seizures (bilaterally symmetrical and without local onset) (1) Absence

Minor motor
 Limited grand mal
	(2) Myoclonic
	      (3) Clonic
	      (4) Tonic

Grand mal
 Drop attacks
	(5) Tonic-clonic
	      (6) Atonic
	      (7) Infantile spasms

—	III. Unclassified seizures IV. Status Epilepticus (prolonged partial or generalized seizures without recovery between attacks)

Considerable emphasis has been given in studying the cerebral metabolism of the biogenic amines in view of the possibility that affective neuronal and psychiatric disorders may be associated with abnormalities in their metabolism. Currently a great deal of interest is centred on the possible involvement of certain amino acids in the mechanism of epilepsy. This is partly the result of an awakening in neurophysiology to the paradox that a range of simple and ubiquitous amino acids which are involved in a wide range of metabolic pathways in the CNS are likely also to function as major synaptic transmitters in the brain and spinal cord.

It has been hypothesized that there is an inverse relation between seizure predisposition and levels of noradrenergic activity in brain (Jobe *et al.*, 1994). Brain stem seizures (tonic and clonus extensor convulsions) are characterized by innate noradrenergic deficits and from selective lesioning of noradrenergic neurons and/or pathways (Mishra *et al.*, 1994).

The increase or decrease of norepinephrine level with epilepsy is area specific. An increase in seizure severity is always associated with marked depletion of NE in the midbrain excluding the inferior colliculus (Wang *et al.*, 1994).

Interestingly norepinephrine (NE) has been proposed to have both pro- and anti-convulsant properties (Rutecki, 1995). On the other hand, dopamine (DA) has an antiepileptic action. It inhibits most hippocampal neurons. The traditional anticonvulsant action of DA was attributed to D-2 receptor stimulation in the forebrain, while the advent of selective D-1 agonist with proconvulsant properties revealed that DA could also lower the seizure threshold from the mid-brain (Benardo and Pedley, 1985; Starr, 1996). The inhibitory effects of DA are derived from induction or enhancement of a calcium activated K^+ conductance (Benardo and Pedley, 1985).

Cavalheiro *et al.* (1994) reported that after pilocarpine administration in rats, hippocampal NE level was decreased whereas dopamine (DA) content increased. Utilization rate measurement of monoamines showed increased NE consumption and decreased DA consumption. A considerable body of evidence indicates that the noradrenergic system provides the forebrain with substantial protection against the development of seizure activity.

Low levels of dopamine in lumbar CSF in epileptics (Hiramatsu *et al.*, 1987) and in epileptic foci in human brain have been observed (Mori *et al.*, 1987). Again, in focal epileptic cases it has been shown that in the temporal neocortex, the focal area has increased levels of NE, DA, dihydroxyphenylalanine (DOPA), 5-hydroxyindole acetic acid (5-HIAA) and homovalinic acid (HVA) (Goldstein *et al.*, 1988; Louw *et al.*, 1989).

Histamine also is believed to affect neurobehavioural disorders such as Alzheimer's disease, Down's syndrome, attention deficit hyperactive disease, epilepsy, Parkinson's disease, etc. (Onodera and Miyazaki, 1999). CNS HA has been suggested to participate in seizure control. Intracerebroventricular (ICV) ad-ministration of HA decreased seizure susceptibility on electrically and pentylene-tetrazol-induced convulsions significantly and dose dependently, while centrally acting HA H1 antagonists such as pyrilamine (or mepyramine) and ketotifen an-tagonized the inhibitory effect of HA (Yokoyama *et al.*, 1994a, b).

The histaminergic neuron system is also believed to be involved in inhibition of seizures associated with febrile illness in childhood. The increased susceptibility to seizures during fever is hypothesized to be connected to the lack of increase in CSF HA in the febrile convulsive group (Kiviranta *et al.*, 1995).

Upon stimulation of the GABAergic system, decreases in cellular excitability are observed, which leads to control of seizures. GABA receptor agonists have a wide spectrum as they antagonize not only seizures that are dependent on decreased

GABA synaptic activity but also convulsion states that are apparently independent of alterations in GABA mediated events (Bartholini, 1985).

Seizure disorders have always been associated with a complex mixture of psychopathology. However, insufficient attention has been paid to the extent of structural damage that is associated with the disorder.

The observation, made by Bouchet and Cazauvieihlh in 1825 of a palpable hardening and atrophy of the uncus and the mesial temporal lobe in patients with epilepsy did not attract much attention until Sommer (1880), some 50 years later, described neuron loss in a particular area of the pyramidal cells of the hippocampus in relation to epilepsy. However, Sommer concluded that sclerosis was the cause of epilepsy. The concept of the nature of ammonshorn sclerosis (Sommer, 1880) reached a new stage with the delineation of psychomotor epilepsy (Bailey and Gibbs, 1951; Earle *et al.*, 1953; Falconer *et al.*, 1955; Gibbs *et al.*, 1948; Penfield and Jasper, 1954).

Subsequent studies (Seitelberger, 1969) have tried to reverse this hypothesis but still the older idea of epilepsy as a functional disorder without any consequent structural damage prevailed (Townsend, 1976).

It has been reported that recurrent limbic seizures caused a massive, delayed, and reversible reduction in levels of the kainite receptor mRNA in dentate gyrus; lesser decreases were found in pyramidal cell fields of hippocampus and superficial cortex (Gall *et al.*, 1990).

These specific neurotransmitters take an active part in the control of brain activity. There may be a release of excitatory and inhibitory transmitters into the brain regions in response to nerve signals. The excitation passes down the cortex to widespread areas of brain and excites all the neurotransmitter systems differentially causing liberation of the neurotransmitters at different terminals that maintain the normal integrity of the brain functioning. Thus seizure activity is associated with a wide range of local biochemical changes affecting various monoamines such as NE, DA, 5-HT and HA.

Neuroprotection

The diagnosis of epilepsy is made primarily on history obtained from patient, relatives and witnesses. In the present era of diagnostic medicine with major reliance and dependence on modern equipment, there is a tendency to bypass or shortcut the clinical history and examination. The most important aspect of the management of epilepsy is to remember that (1) whatever form of therapy is considered, epilepsy is a disease where therapy once started is continued for a long time, at times for the whole life either continuously or with breaks. Most of the epilepsies with some exceptions are controllable but not curable. (2) The second point of importance is that epilepsy in the majority of cases affects an individual at a young age in the early part of life when one starts acquiring education, vocational training, new career and social responsibility. In 75% of patients epilepsy begins before the age of seventeen years. (3) The third point of great importance is that prescribing medicines is only one of the several facets of total management of patients with epilepsy. The different forms of treatment, which are available for the treatment of epilepsies, are (a) surgical (b) general (c) medical (Maheswari, 1993).

All forms of seizures should be treated with antiepileptic drugs, but an attempt should be made simultaneously to identify the difficult-to-treat other variations of the treatable epilepsies. Surgically treatable epilepsies should be identified. It should be appreciated, however, that only a very small number of patients are fit for surgical treatment.

Among dietary measures the Ketogenic Diet (high fat) has been found useful in severe childhood epilepsies. How it helps is not well understood. Also the Ketogenic Diet is not very palatable.

Electromagnetic therapy has been detected around each individual person. However, there are few reports suggesting its influence on epilepsy. This needs further exploration.

Sleep disorders and epileptogenesis are very interlinked. Sleep deprivation has been used in the laboratory as provoking factor. Normalisation of sleep disorders may help the EEG abnormality and control the seizures in some patients.

If there are stress-induced seizures then it is sensible to reduce or avoid those stresses. Psychotherapy is effective in this direction in eliminating stress. Some change in life-style and occupation may also help in controlling seizures. Abstinence from alcohol and smoking and avoidance of toxic substances may at times help some patients.

Control of seizures by medication, though important, is only a part of total care of the patient. If social and environmental problems are not taken care of, then good control of seizures cannot be achieved. Better education, self-confidence and careful life-style restriction are important goals (Maheswari, 1993).

The aim of treatment with antiepileptic drug is to stop all seizures and to achieve complete suppression of all epileptic activity in EEG without producing side-effects for a long enough duration that the tendency to epilepsy ceases and treatment can be stopped. It is unlikely that this aim can be achieved. At present in a large number of cases, suppression of seizures is largely achieved by the development of newer and more effective drugs in the 20th century. Bromides were used for many years in the 19th century until phenobarbitone was made available for clinical use in the year 1912. Formerly, the aim of treating epilepsy was often regarded to be merely keeping the sufferer free of convulsions, but the ever-growing contemporary hope for cure often entails a more vigorous and meticulous approach to drug therapy combined with social and rehabilitative measures. However, the aim remains restrictive and unrealistic in epilepsies due to progressive neuronal diseases and other developmental disorders. Impaired cognitive functions, i.e. learning and behaviour are effects of the condition causing epilepsy. Epilepsy itself may cause changes through its metabolic and neurotransmitter interactions of excitation and inhibition. Although numerous drugs are available in the market for treatment of epilepsies, the existence and use of other forms of treatment in the management of epilepsies is itself an indication that all types of epilepsy are not easily controllable.

Herbal therapies are on the rise all over the world. Promising herbs like *Acorus calamus* (Hazra and Guha, 2003) and *Moringa oleifera* (Ray *et al.*, 2003) can be developed into therapeutic agents. These drugs, once developed will be free from unpleasant side effects, and they will be comparatively cheaper than those drugs currently available.

The ideal antiseizure drug would suppress all seizures without causing any unwanted side effects. Unfortunately, the drugs used currently not only fail to control seizure activity in some patients, but they frequently cause side effects that range in severity from minimal impairment of the CNS to death from aplastic anaemic or hepatic failure. Physicians who treat a patient with epilepsy are thus faced with the task of selecting the appropriate drug or combination of drugs that best controls seizures in an individual patient at an acceptable level of untoward effects. It is generally held that complete control of seizures can be achieved in up to 50% of patient and that another 25% can be improved significantly. The degree of success is greater in newly diagnosed patients and is dependent on such factors as the type of seizure, the family history, and the extent of associated neurological abnormalities. All anti-convulsant drugs have side-effects, trivial or major.

No system of the human body is spared by these drugs and common adverse reactions of commonly used antiepileptic drugs are given in Table 3.

Searches in different directions have been conducted with the sole aim of helping the patient in controlling seizures. These various methods are not the only ones undertaken but are regarded as complementary to drug therapy.

Table 3.
Some antiepileptic drugs and their side effects

Drug	Side effects
Phenobarbitone (PHEN)	Drowsiness, aggression, irritability, poor concentration, hyperkinetic, impaired learning, skin rash
Diphenylhydantoin (DPH)	Gum hyperplasia, nystagmus, ataxia, acne, hirsutism, coarsening of facial features, dysarthria, skin rash, involuntary movement
Carbamazepine (CAB)	Skin rash, giddiness, double vision or blurred vision
Sodium valproate (SV)	Nausea, vomiting, weight gains, loss or thinning of hair, tremors
Ethosuximide (ETHO)	Vomiting, anorexia, euphoria, dizziness, headache, hiccup, anxiety, aggressiveness, inability to concentrate
Primidone (PRIM)	Sedation, vertigo, dizziness, nausea, vomiting, ataxia, diplopia
Benzodiazepine	Lethargy, hypnotic, dysarthria, behavioural disturbance in children
Diazepam	Sedation, respiratory depression
Gabapentin	Somnolence, dizziness, ataxia
Nitrazepam	Sedation, drooling of saliva, dizziness

PARKINSON'S DISEASE

Parkinson's disease (PD) is a disease of progressive neurodegeneration. It was first described in 1817 by the physician James Parkinson. Previous to this, accounts of the symptoms were remarkably scarce, which led many researchers to theorize whether this disease may have been a product of the beginning of the early 19th century and the Industrial Revolution in England. Some speculated that, because certain environmental neurotoxins cause Parkinson's-like ('Parkinsonian') symptoms, some contaminant in the new industrial environment may have increased its prevalence. The present concept is in favour of oxidative causation linked to environmental toxins.

Parkinson's disease is widespread in Westernized countries. The disease is highly age-dependent: it can manifest as early as the mid-30s, but becomes more common past the age of 50, with 57 being the average age of diagnosis. It is visible as tremor in a limb, and as it progresses three other symptoms arise — bradykinesia (slowness of movement), rigidity (both 'cogwheel' jerkiness and 'leadpipe' stiffness) and posture instability with impaired gait, associated with the stooped stance. Bradykinesia causes the patient to feel glued to the ground or to the chair in which they sit, and progressively erases body language and facial expressiveness. The disease is not restricted to motor degeneration; as many as 35% of PD cases also develop dementia.

Parkinson's disease is now recognized to be a widespread degenerative illness that affects not just the central nervous system, but also the peripheral and enteric systems. Formerly the disease was typecast as motor system degeneration, yet sensory fields, association areas, and premotor fields become damaged throughout the brain. Various lines of evidence has suggested that PD is primarily an oxidative disease, fueled by endogenous susceptibility and driven by the cumulative contributions of endogenous and exogenous (environment) oxidant stressors. The limbic, autonomic, and neurosecretory control fields (hypothalamus) all show micro-anatomic damage. At the cellular level, neuron death in PD is more systemic than previously assumed: non-invasive imaging recently demonstrated that the nerve supply to the heart degenerates in PD subjects. Biochemically, abnormalities of liver detoxification and mitochondrial oxidative phosphorylation also occur (Kidd, 2000).

The pathological process that underlies PD typically is slow-paced but relentlessly progressive, the clinical symptoms tending to manifest relatively late in the pathological progression. Classically, the hallmark of PD has been degeneration of dopamine-producing neurons in the relatively small substantia nigra (SN), most intensely localized in the zonal compacta.

Normally, dopamine produced in the SN is moved to the caudate nucleus and the putamen, where it is involved in stimulating and coordinating the body's motor movements. In PD, neurons producing dopamine in the SN die, reducing the overall supply of dopamine and compromising the brain's capacity to effect movement. Curiously, dopamine-producing neurons outside the SN tend not to be affected, though many other neuronal types — glutamatergic, cholinergic, tryptaminergic, GABAergic, noradrenergic, adrenergic — show 'grievous cytoskeletal damage'.

The characteristic pattern of nerve cell destruction in SN neurons appears to be linked to abnormalities that develop in the cytoskeleton. The pathognostic Lewy bodies and Lewy neurites are composed mainly of abnormal cytoskeletal neurofilament proteins. Neurons afflicted with Lewy formations remain viable for a relatively long period, but are functionally compromised and die prematurely. As a rule, projection neurons with long axons are more vulnerable than local circuit and projection neurons with short axons, which tend to be spared.

The broad spectrum of potential etiologic factors

In spite of the extensive studies performed on postmortem substantia nigra from Parkinson's disease patients, the etiology of the disease has not yet been established. Nevertheless, studies have demonstrated that, at the time of death, a cascade of events had been initiated that may contribute to the demise of the melanin-containing nigro-striatal dopamine neurons. These events include increased levels of iron and monoamine oxidase (MAO)-B activity, oxidative stress, inflammatory processes, glutamatergic excitotoxicity, nitric oxide synthesis, abnormal protein folding and aggregation, reduced expression of trophic factors, depletion of endogenous antioxidants such as reduced glutathione, and altered calcium homeostasis (Mandal *et al.*, 2003).

As with almost all disease states, a broad spectrum of both genetic and environmental factors have been suggested as contributing to the initiation and progression of PD. Aging is also implicated, with advanced age being the single most important risk factor for the disease.

Role of aging

Parkinson's disease is clearly age-dependent. Several of the neurodegenerative syndromes documented in the elderly — gait slowing, for example — resemble those seen in PD and may be prodromal for the disease while the inexorable downhill slide of Parkinson's disease is unmistakably a disease process. Aging undoubtedly contributes to PD progression, perhaps because of its accumulative oxidative damage and steady decrease of antioxidant capacity.

Heritability and genetic susceptibilities

There appears to be an inherited component to PD, and a number of family pedigrees with multiple cases of PD have been extensively studied. To date, seven loci on four chromosomes are reliably linked to PD and/or to neurodegeneration of the Parkinsonian type, not always with the presence of the Lewy structures. Non-familial PD subjects carrying a specific alpha-synuclein allele and ApoE4 have a 13-fold increased risk of developing PD. Parkin, localized to chromosome 6, causes early-onset of parkinsonism without the Lewy bodies that define PD. Most PD cases, however, do not have other affected family members and have no apparent familial contribution.

With many key studies now done, the preponderance of the evidence indicates the general PD population has no more than a mild genetic contribution. Still, overall absence of defined heritabilty does not necessarily rule out subpopulations with higher heritability, or subtle genetically conditioned vulnerabilities. Various authors have described how various specific gene mutations and deletions might potentially contribute to PD. The genes involved could act with varying degrees of penetrance, or polygenically contribute to disease vulnerability with no single gene being wholly responsible. For early onset PD, the cumulative evidence is consistent with a strong heritability component.

Some individuals with PD have impaired liver detoxification. The P450 IID6 enzyme, which was characterised based on its capacity to metabolize debrisoquine, was found to be dysfunctional more frequently in PD subjects than in non-PD controls. Of the PD subjects, those with very early onset (less than age 40) are most likely to have this problem. The gene for P450 IID6 was localised to chromosome 22, and efforts are underway to develop it into a biomarker for early-onset PD, but this may not prove practical. Conducting such assays *in vivo* is expensive and laborious and, in the studies, some of the medications being taken for PD may have complicated the outcomes.

Interestingly, P450 IID6 is present in the nigrostriatal system and the notorious Parkinsonian toxin, 1-methyl-4-phenyl-1,2,3,6-tetrahydropyridine (MPTP) is a substrate for this complex. MPTP came to the forefront of PD research in 1982, when drug addicts in northern California began to develop severe Parkinsonism after intravenous injection of a synthetic heroin that was contaminated with MPTP as a by-product of its synthesis. The identification of three genes and several additional loci associated with inherited forms of levodopa-responsive PD has confirmed that this is not a single disorder. Yet, analysis of the structure and function of these gene products point to the critical role of protein aggregation in dopaminergic neurons of the substantia nigra as the common mechanism leading to neurodegeneration in all known forms of this disease. The three specific genes identified to date — alpha-synuclein, Parkin, and ubiqutin C terminal hydrolase L1 — are either closely involved in the proper functioning of the ubiquitin–proteasome pathway or are degraded by the protein-clearing machinery of cells. Knowledge gained from genetically transmitted PD also has clear implications for nonfamilial forms of the disease. Lewy bodies, even in sporadic PD, contain these three gene products, and particularly abundant amounts of fibrillar alpha-synuclein. Increased aggregation of alpha-synuclein by oxidative stress, as well as oxidant-induced proteasomal dysfunction, link genetic and potential environmental factors in the onset and progression of the disease. The biochemical and molecular cascades elucidated from genetic studies in PD can provide novel targets for curative therapies.

The MPTP poisonings led to a serendipitous finding: that the symptoms resulting from exposure to MPTP very closely matched the many features of PD. This shifted the focus toward environmental factors as potential PD initiators or contributors. As the search progresses, a single toxic cause remains elusive but a role for environmental factors seems almost certain.

Although the contribution from environmental toxins cannot be denied, currently the cumulative evidence suggests PD is a multifactorial oxidative disease. The main causal, oxidative contributors indicated to date are: (1) measurable amplification of the endogenous oxidative load by constitutive impairments of mitochondrial energy transformations, (2) innate vulnerability of the brain's substantia nigra region to oxidative challenge, and (3) initiation or promotion by toxic exposure(s) that further deplete antioxidants. These factors combine to initiate a downhill course for the neurons of the SN and elsewhere in the brain, the end result of which appears to be a slow-acting yet long-term progressing, inflammatory process. This eventually results in the micro-anatomic degeneration and clinical symptomatologies of Parkinson's disease.

The oxidative stress (OS) theory has implicated the involvement of reactive oxygen species (ROS) in both aging and age-dependent neurodegenerative diseases. The dopaminergic system is particularly vulnerable to ROS and dopamine (DA) itself can be an endogenous source of ROS. These results suggest that a neurochemical deficit and not cell loss *per se* within the nigrostriatal system underlies the motor behavioural deficits (Cantutu-Castelvetre *et al.*, 2003; Choi *et al.*, 2003).

It is becoming widely accepted that the inflammatory response is involved in neurodegenerative disease (Castano *et al.*, 2002). Elevated synaptic levels of dopamine may induce striatal neurodegeneration in I-DOPA-unresponsive Parkinsonism subtype of multiple system atrophy (MSA-P subtype), multiple system atrophy, and metham-phetamine addiction (Chen *et al.*, 2003).

Parkinson's disease is most commonly a sporadic illness, and is characterized by degeneration of substantia nigra dopamine neurons and abnormal cytoplasmic aggregates of alpha-synuclein. Rarely, PD may be caused by missense mutations in alpha-synuclein (Dauer *et al.*, 2002). It is characterized by focal microglial activation and progressive dopaminergic neurodegeneration in substantia nigra compacts (SNc) (DeGiorgio *et al.*, 2002) with no effective protective treatment characterized by a massive degeneration of dopaminergic neurons in the substantia nigra and the subsequent loss of their projecting nerve fibers in the striatum (Delgado and Ganea, 2003).

The motor disturbances occurring in Parkinson's disease have been partially attributed to a hyperactivity of gamma-aminobutyric acid (GABA)-ergic nigral cells, largely in the substantia nigra pars reticulata (SNr) secondary to the degeneration of dopaminergic nigrostriatal neurons. However, some aspects of this response remain unclear. Findings indicate a complex regulation of nigral GABAergic activity after nigrostriatal dopaminergic degeneration that probably involves local mechanisms, the nigro-striato-nigral loop as well as interhemispheric mechanisms whose anatomical basis remains unstudied (Diaz et al., 2003).

Nitric oxide and Parkinsonism

Nitric Oxide (NO), in excess, behaves as a cytotoxic substance mediating the pathological processes that cause neurodegeneration. The NO-induced dopaminergic cell loss causing Parkinson's disease has been postulated to include the following: an inhibition of cytochrome oxidase, ribonucleotide reductase, mitocondrial complex I, II and IV in the respiratory chain, superoxide dismutase, glyceraldehyde-3-phosphate dehydrogenase; activation or initiation of DNA strand breakage, poly (ADP-ribose) synthase, lipid peroxidation, and protein oxidation; release of iron; and increased generation of toxic radicals such as hydroxyl radicals and peroxynitrite. NO is formed by the conversion of L-arginine to L-citrulline by NO synthase (NOS). At least three NOS isoforms have been identified by molecular cloning and biochemical studies: a neuronal NOS or type 1 NOS (nNOS), an immunologic NOS or type 2 NOS (iNOS), and an endothelial NOS or type 3 NOS (eNOS). The enzymatic activities of eNOS or nNOS are induced by phosphorylation triggered by $Ca^{(2+)}$ entering cells and binding to calmodulin. In contrast, the regulation of iNOS seems to depend on *de novo* synthesis of the enzyme in response to a variety of cytokines, such as interferon-γ and lipopolysaccharide. Selegiline, an irreversible inhibitor of monoamine oxidase B, is used in PD as a dopaminergic function-enhancing substance. Selegiline and its metabolite, desmethylselegiline, reduce apoptosis by altering the expression of a number of genes, for instance, superoxide dismutase, Bcl-2, Bcl-xl, NOS, c-Jun, and nicotinamide adenine nucleotide dehydrogenase. The selegiline-induced antiapoptotic activity is associated with prevention of a progressive reduction of mitochondrial membrane potential in preapoptotic neurons. As apoptosis is critical to the progression of neurodegenerative disease, including PD, selegiline or selegiline-like compounds to be discovered in the future may be efficacious in treating PD (Ebadi and Sharma, 2003; Pang et al., 2000; Pardini et al., 2003).

Inflammation in the brain has increasingly been recognized to play an important role in the pathogenesis of several neurodegenerative disorders, including Parkinson's disease and Alzheimer's disease. Inflammation-mediated neurodegeneration involves activation of the brain's resident immune cells, the microglia, which produce proinflammatory and neurotoxic factors, including cytokines, reactive oxygen intermediates nitric oxide, and eicosanoids that impact on neurons to induce neurodegeneration (Gao et al., 2002; Liu et al., 2003).

Spiny neurons in the neostriatum are highly vulnerable to cerebral ischemia. Recent studies have shown that the postischemic cell death in the right striatum was reduced after ipsilateral dopamine denervation whereas no protection was observed in the left striatum after dopamine denervation in the left side. It is observed that after ipsilateral dopamine denervation, the depression of excitatory synaptic transmission and neuronal excitability therefore, might play an important role in neural protection after ischemic insult.

Current management of Parkinson's disease

Currently, PD is managed mainly through dopamine replacement therapy — pharmaceutical agents are aimed at replacing dopamine in the brain or mimicking its actions at dopamine receptors. Most commonly used is the dopamine precursor levodopa in combination with carbodopa (Sinemest (R) and Sinemet (CR)). The vast majority of patients experience benefits initially, but rarely do the benefits persist. Typically, after 2–5 years on levodopa drugs, the patient's responses become erratic. Nausea is a constant threat, and dyskinesias develop that feature excessive and uncontrollable movements. Other adverse effects are mental confusion, 'freezing' and inability to move, dystonia, low blood pressure episodes, sleep disturbances, and hallucinations. Some classes of prescription drugs that can precipitate Parkinsonian symptoms are given in Table 4.

Adverse side effects usually pose a major ongoing challenge to the PD patient. For example, the combined effects of the disease and the drugs used to treat it produce sleep problems in an estimated 70% of patients and daytime hallucinations in about 30%. Levodopa is usually effective for motor symptoms at the beginning, but over time tends to cause motor fluctuations, dyskinesias, and other adverse side effects. These can become so disabling that surgical treatment becomes the only apparent option for restoring any quality of life.

Other drugs used for PD symptom management include amantadine (Symmetrel (R), selegiline (Eldepryl (R), deprenyl), dopamine agonists (bromocriptine, pergolide, pramipexole, ropinirole), and several anticholinergic drugs. All these have major adverse effects and generally are less effective than Sinemet (R) in suppressing symptoms. Tolcapone, an inhibitor of COMT (catechol-O-methyltransferase, the enzyme which normally inactivates dopamine), became available in 1998. It caused several deaths from liver failure. Another COMT inhibitor — entacapone

Table 4.
Classes of prescription drugs that can precipitate Parkinsonian symptoms: from worst pills to best pills

---

Antihypertensive, diuretic: Diupres, Enduronyl, Hydropres, Regroton, Demi-Regroton, Salutensin, Ser-AP-Es.
Antihypertensive, non-diuretic: Aldomet.
Antidepressant: Asendin, Aventyl/Pamelor, Desyrel, Elavil, Limbitrol, Ludiomil, Luvox, Norpramin, Paxil, Prozac, Sinequan, Tofranil, Triavil, Wellbutrin, Zoloft.
Antipsychotic: Compazine, Haldol, Mellaril, Navane, Prolixin, Risperdal, Stelazine, Thorazine, Triavil, Zyprexa.
Other: Reglan, Zyban

— was released in 1999 which was not liver-toxic but still caused dyskinesias, nausea, diarrhoea, abdominal pain, and urine discoloration.

As PD progresses, in addition to the adverse effects accruing from levodopa therapy, the ever-worsening loss of dopamine neurons causes progressively crippling damage to motor control circuits throughout the brain. The control may shift, so the pathways that normally inhibit movement come to dominate those that activate movement. The increasing desperation of the patient can become the rationale for risky surgical intervention: for example, whether to remove inhibitory zones or to implant electrodes aimed at restoring a healthy balance of circuits.

Surgical destruction of brain tissue was tried prior to the advent of levodopa therapy, but produced inconsistent results. More recently, microelectrodes are being used to detect signals from individual brain cells, using these signals as 'signposts' to arrive at more precise locations in the brain. Pallidal and subthalamic nuclear surgery can improve motor symptoms and levodopa-induced dyskinesias, but only unilateral pallidotomy is acceptable since the bilateral procedure carries unacceptably high risk. The unilateral procedure, however, probably the most common surgery for advanced PD, unfortunately does not allow for postoperative reduction in levodopa doses. Postsurgical mortality is 1–1.8%, risk of permanent neurological deficit is about 5%, and benefits tend to dissipate within 1–4 years.

Deep-brain electrical stimulation (DBS) by way of surgically implanted electrodes has the advantages over ablation of being regulatable and reversible. DBS also has been bilaterally performed in many patients with marked benefit and little permanent morbidity. DBS post-operative morbidity and mortality is less than for ablation, and the stimulation side effects are relatively mild. On the negative side, infection may ensue, mechanical failure occurs in 3–4% of cases, batteries must be replaced at regular intervals, and the device is expensive. On the positive side, successful bilateral stimulation can allow medication dosing to be reduced, providing the patients a better quality of life. Although results from randomized studies are not yet available, surgeons who have done both ablation and DBS agree that DBS is better for the patient.

Epidemiological and clinical studies provide growing evidence for marked sex difference in the incidence of certain neurological disorders that are largely attributed to the neuroprotective effects of estrogen. Thus there is a keen interest in the clinical potential of estrogen-related compounds to act as novel therapeutic agents in conditions of neuronal injury and neurodegeneration such as Parkinson's disease.

Obviously, the need to broaden the options for therapy in Parkinson's disease is urgent. The urgency compels renewed focus on the etiology and pathogenesis of the disease. Deeper scientific understanding of PD would lead to better preclinical detection and prophylaxis, validation of biomarkers, confirmation of genetic and environmental risk factors, and more prolonged symptom control with fewer adverse effects.

SUMMARY

Parkinson's disease (PD) is characterized pathologically by preferential degeneration of the dopaminergic neurons in the substantia nigra pars compacta (SNc). Nigral cell death is accompanied by the accumulation of a wide range of poorly degraded proteins and the formation of proteinaceous inclusions (Lewy bodies) in dopaminergic neurons. Mutations in the genes encoding alpha-synuclein and two enzymes of the ubiquitin-proteasome system, parkin and ubiquitin C-terminal hydrolase L1, are associated with neurodegeneration in some familial forms of PD (McNaught *et al.*, 2003). This disorder is characterized by tremor, muscular rigidity, difficulty in initiating motor activity, and loss of postural reflexes.

Acknowledgements

This work is dedicated to my beloved teacher, Professor S. N. Pradhan, who kindled in me the quest for science: I wish his early recovery and good health.

I wish to express my gratitude to my senior PhD scholar Miss Sudarshana Purkayastha who helped me in preparation of the chapter. She read the manuscript and offered numerous comments on content and style. Without her help, writing of this manuscript would have been difficult.

REFERENCES

Anderson, V. E., Hauser, W. A. and Penry, J. K. (1982). In: *The Genetic Basis of the Epilepsies*, p. 120. Raven Press, New York.

Bailey, P. and Gibbs, F. A. (1951). The surgical treatment of psychomotor epilepsy, *J. Am. Med. Assoc.* **145**, 365–370.

Bartholini, G. (1985). GABA receptor agonists: pharmacological spectrum and therapeutic actions, *Med. Res. Rev.* **5**, 55–75.

Benardo, L. S. and Pedley, T. A. (1984). Basic mechanisms of epileptic seizures, *Cleve Clin.* **Q 51**, 195.

Benardo, L. S. and Pedley, T. A. (1985). Cellular mechanisms of focal epileptogenesis, in: *Recent Advances in Epilepsy*, No. 2, Pedley, T. A. and Meldrum, B. (Eds), pp. 1–17. Churchill Livingstone, UK.

Bouchet, C. and Cazauvieihlh, Y. (1825). Epilépsie et l'aliénation mentale. *Arch. Gen. Med.* **9**, 510–542. Cited by Dam, M. A. (1980). Epilepsy and neuron loss in the hippocampus, *Epilepsia* **21**, 617–629.

Cantutu-Castelvetre, I., Shukitt-Hale, B., *et al.* (2003). Dopamine neurotoxicity : age-dependent behavioral and histological effects, *Neurobiol. Aging.* **24**, 697–706.

Castano, A., Herrera, A. J., *et al.* (2002). The degenerative effect of a single intranigral infection of LPS on the dopaminergic system prevented by dexamethasone and not mimicked by rhTNF-alpha, IL-beta and IFN-gamma, *J. Neurochem.* **81**, 150–157.

Cavalheiro, E. A., Fernandez, M. J. and Turski, L. (1994). Spontaneous recurrent seizures in rats: amino acid and monoamine determination in the hippocampus, *Epilepsia* **35**, 1–11.

Chadwick, D. W. (1990). Diagnosis of epilepsy, *Lancet* **336**, 291–295.

Chadwick, D., Cartlidge, N. and Bates, D. (1989). In: *Medical Neurology*, pp. 152–185. Churchill Livingstone, UK.

Chen, J., Wersinger, C., *et al.* (2003). Chronic stimulation of D1dopamine receptors in human SK-N-MC neuroblastoma cells induces nitric-oxide synthase activation and cytotoxicity, *J. Biol. Chem.* **278**, 28089–28100.

Choi, H. J., Kim, S. W., *et al.* (2003). Dopamine-dependent cytotoxicity of tetrahydrobiopterin: a possible mechanism for selective neurodegeneration in Parkinson's disease, *J. Neurochem.* **86**, 143–152.

Dauer, W., Kholodilov, N., *et al.* (2002). Resistance of alpha-synuclein null mice to the parkinsonian neurotoxic MRTP, *Proc. Natl. Acad. Sci. USA* **99**, 14524–14529.

DeGiorgio, L. A., Shimizu, Y., *et al.* (2002). APP knockout attenuates microglial activation and enhances neuron survival in substantia nigra compacta after axotomy, *Gila* **38**, 174–178.

Delgado, M. and Ganea, D. (2003). Neuroprotective effect of vasoactive intestinal peptide (VIP) in a mouse model of Parkinson's disease by blocking microglial activation, *Faseb. J.* **17**, 944–946.

Diaz, M. R., Barroso-Chinea, P., *et al.* (2003). Effects of dopaminergic cell degeneration on electrophysiological characteristics and GAD65/GAD67 expression in the substantia nigra: different action on GABA cell subpopulations, *Mov. Disord.* **18**, 254–266.

Earle, K. M., Baldwin, M. and Penfield, W. (1953). Incisural sclerosis and temporal lobe seizures produced by hippocampal herniation at birth, *Arch. Neurol. Psych.* **69**, 27–42.

Ebadi, M. and Sharma, S. K. (2003). Peroxynitrite and mitochondrial dysfunction in the pathogenesis of Parkinson's disease, *Antioxid. Redox Signal.* **5**, 319–335.

Ellenberg, J. H., Hirtz, P. G. and Nelson, A. (1986). Do seizures in children cause intellectual deterioration? *New Engl. J. Med.* **314**, 1085–1088.

Elwes, R. D. C., Chesterman, P. and Reynolds, E. H. (1985). Prognosis after a first untreated tonic clonic seizure, *Lancet* **ii**, 752–826.

Elwes, R. D. C., Johnson, A. L., Shorvon, S. D., *et al.* (1984). The prognosis for seizure control in newly diagnosed epilepsy, *New Engl. J. Med.* **311**, 944–947.

Falconer, M. A., Hill, D., Meyer, A., *et al.* (1955). Treatment of temporal lobe epilepsy by temporal lobectomy, *Lancet* **1**, 827–835.

Gall, C., Sumikawa, K. and Lynch, G. (1990). Levels of mRNA for a putative kainate receptor are affected by seizures, *Proc. Natl. Acad. Sci.* **87**, 7643–7647.

Gao, H. M., Jiang, J., *et al.* (2002). Microglial activation-mediated delayed and progressive degeneration of rat nigral dopaminergic neurons: relevance to Parkinson's disease, *J. Neurochem.* **81**, 1285–1297.

Gibbs, E. L., Gibbs, F. A. and Fuster, B. (1948). Psychomotor epilepsy, *Arch. Neurol. Psych.* **60**, 331–339.

Goldstein, D., Nadi, N. S. and Stull, R. (1988). Levels of catechols in normal and epileptic regions of the human brain, *J. Neurochem.* **50**, 225–229.

Hazra, R. and Guha, D. (2003). Effect of chronic administration of *Acorus calamus* on electrical activity and regional monoamine levels in rat brain, *Biogenic Amines* **17**, 161–169.

Hiramatsu, M., Fujimoto, N. and Mori, A. (1987). Catecholamine level in cerebrospinal fluid of epileptics, *Neurochem. Res.* **7**, 157–171.

Jobe, P. C., Mishra, P. K., Browning, R. A., *et al.* (1994). Noradrenergic abnormalities in the genetically epilepsy-prone rat, *Brain Res. Bull.* **35**, 493–496.

Kidd, P. M. (2000). Parkinson's disease as multifactorial oxidative neurodegeneration: Implications for integrative management, *Altern. Med. Rev.* **5**, 502–529.

Kiviranta, T., Tuomisto, L. and Airaksinen, E. M. (1995). Histamine in cerebrospinal fluid of children with febrile convulsions, *Epilepsia* **36**, 276–280.

Klepser, T. B. and Klepser, M. E. (1999). Unsafe and potentially safe herbal therapies, *Am. J. Health Syst. Pharm.* **56**, 125–138; Quiz, 139–141.

Liu, Y., Quin, L., *et al.* (2003). Dextramethorphan protects dopaminergic neurons against inflammation mediated degeneration through inhibition of microglial activation, *J. Pharmacol. Exp. Ther.* **305**, 212–218.

Louw, D., Sutherland, G. R., Blavin, G. B., *et al.* (1989). A study of monoamine metabolism in human epilepsy, *Can. J. Neurol. Sci.* **16**, 394–397.

Maheswari, M. C. (1993). All you want to know about epilepsy, ed. 2, NBT India.

Mandal, S., Grunblalt, E., *et al.* (2003). Neuroprotective strategies in Parkinson's disease: an update on progress, *CNS Drugs.* **17**, 729–762.

McNaught, K. S., Belizaire, R., *et al.* (2003). Altered proteasomal function in sporadic Parkinson's disease, *Exp. Neurol.* **179**, 38–46.

Meierkord, H. (1989). Advances in genetics and their application to epilepsy, in: *Chronic Epilepsy: its Prognosis and Management*, Trimble, M. R. (Ed.), p. 243. Wiley, Chichester.

Meldrum, B. S. (1990). Anatomy, physiology and pathology of epilepsy, *Lancet* **336**, 231–233.

Mishra, P. K., Burger, R. L., Bettendorf, A. F., *et al.* (1994). Role of norpeinephrine in fore brain and brain-stem seizures: Chemical lesioning of locus ceruleus with DSP4, *Exp. Neurol.* **125**, 58–61.

Mori, A., Hiramatsu, M., Nishimoto, A., *et al.* (1987). Decreased dopamine levels in the epileptic focus, *Res. Commun. Chem. Pathol. Pharmacol.* **56**, 157–164.

Onodera, K. and Miyazaki, S. (1999). The roles of histamine H3 receptors in the behavioural disorders and neuropsychopharmacological aspects of its ligands in the brain, *Nippon Yakurigaku Zasshi.* **114**, 89–106.

Pang, Z. P., Ling, G. Y., *et al.* (2000). Asymmetrical changes of excitatory synaptic transmission in dopamine-denervated striatum after transient forebrain ischemia, *Neuroscience* **114**, 317–326.

Pardini, C., Vaglini, F., *et al.* (2003). Dose-dependent induction of apoptosis by R-apomorphine in CHO-K1 cell line in culture, *Neuropharmacology* **45**, 182–189.

Penfield, W. and Jasper, H. (1954). *Epilepsy and the Functional Anatomy of the Brain*. Little Brown, Boston.

Purkayastha, S. and Guha, D. (2003). Monoaminergic correlates of penicillin induced epilepsy, *Biogenic Amines* **18**, 29–40.

Ray, K., Hazra, R. and Guha, D. (2003). Central inhibitory effect of *Moringa oleifera* root extract: possible role of neurotransmitters, *Indian J. Exp. Biol.* **41**, 1279–1284.

Rutecki, P. A. (1995). Noradrenergic modulation of epileptiform activity in the hippocampus, *Epilepsy Res.* **20**, 125–129.

Scheuer, M. L. and Pedley, T. A. (1990). The evaluation and treatment of seizures, *New Engl. J. Med.* **323**, 1048–1074.

Schlichter, W., Bristow, M. F. and Schultz, R. S. (1986). Seizures occurring during intensive chlorpromazine therapy, *Can. Med. Assoc. J.* **74**, 364–366.

Seitelberger, F. (1969). General neuropathology of degenerative processes of the nervous system, *Neurosci. Res.* **2**, 253–299.

Sharma, S. K. and Ebadi, M. (2003). Metallothionein attenuates 3-morpholinosydnonimine (SIN-1)-induced oxidative stress in dopaminergic neurons, *Antioxid. Redox Signal.* **5**, 251–264.

Shorvon, S. D. (1984). The temporal aspects of prognosis in epilepsy, *J. Neurol. Neurosurg. Psychiatry* **47**, 1157–1165.

Shorvon, S. D. (1990). Epidemiology, classification, natural history, and genetics of epilepsy, *Lancet* **336**, 93–96.

Sommer, W. (1880). Erkrankung des Ammonshorns als aetiologisches Moment der Epilepsie, *Arch. Psychiatr. Nervenkr.* **10**, 631–675. Cited by Dam, M. A. (1980). Epilepsy and neuron loss in the hippocampus, *Epilepsia* **21**, 617–629.

Starr, M. S. (1996). The role of dopamine in epilepsy, *Synapse* **22**, 159–194.

Sundaram, M. B. M. (1989). Etiology and patterns of seizures in the elderly, *Neuroepidemiology* **8**, 234–238.

Townsend, H. R. A. (1976). Epilepsy — a clinician's view, in: *Biochemistry and Neurology*, Bradford, H. F. and Marsden, C. D. (Eds), p. 175. Academic Press, London.

Wang, C., Mishra, P. K., Dailey, J. W., *et al.* (1994). Noradrenergic terminal fields as determinants of seizure predisposition in GEPR-3S: a neuroanatomic assessment with intracerebral microinjections of 6-hydroxydopamine, *Epilepsy Res.* **18**, 1–6.

Yokoyama, H., Onodera, K., Maeyama, K., *et al.* (1994a). Clobenpropit (VUF -9153), a new histamine H_3 receptor antagonist, inhibits electrically induced convulsions in mice, *Eur. J. Pharmacol.* **260**, 23–28.

Yokoyama, H., Sato, M., Onodera, K., *et al.* (1994b). 2-Thiazolylethylamine, a selective histamine H1 agonist, decreased seizure susceptibility in mice, *Pharmacol Biochem. Behav.* **47**, 503–507.

Advances in Neuroregulation and Neuroprotection (2005), pp. 103-126
Collin, C. *et al.* (Eds)
© VSP 2005

Oxidative stress and neurotoxicity in Parkinson's patients in on-off phenomena

G. A. QURESHI [1,*], S. A. MEMON [1], C. A. COLLIN [3], A. A. QURESHI [2] and S. H. PARVEZ [3]

[1] *M. A. Kazi Institute of Chemistry and Biochemistry, Sind University, Jamshoro, Pakistan*
[2] *Department of Neurosurgery, Liquat Medical University of Health Science, Jamshoro, Pakistan*
[3] *Alfred Essard de Neurosciences, Bat 5, Parc Chateau CNRS, 91190 Gif/Yvette, France*

Abstract—L-Dopa is considered to be one of the best therapies in treating Parkinson's patients, with a positive outcome at the start of treatment; however, long-term treatment has shown both on and off phenomena and even high doses have shown various clinical symptoms. In this chapter, both on and off PD patients on L-dopa therapy are investigated to determine the role of various neurotransmitters, such as aspartic acid, glutamic acid, homocysteine, catecholamines, vitamin B12 and free radical nitric oxide resulting in neurotoxicity and oxidative stress. All these biochemical changes may constitute vital information in designing multiple antioxidant therapy to avoid neurotoxicity and oxidative stress in PD patients.

Keywords: Parkinson's disease; L-dopa therapy; oxidative stress; neurotoxicity.

INTRODUCTION

Parkinson's disease (PD) is a brain disease involving neurodegeneration in one or a few selected regions of the central nervous system (CNS) in which one or few cell types are affected. PD is one of the most common degeneration diseases, and affects approximately 1% of the population over 65 years of age. Many different insults appear to be involved in etiology of the disease, among these, environmental toxins and mitochondrial dysfunction. The disease is characterized by slowing of emotional and voluntary involvement, muscular rigidity, postural abnormality and tremor (Bernheimer *et al.*, 1973; Hugh, 1992). Motor symptoms are the predominant features of PD, however, other symptoms may occur i.e. neuropsychiatric problems, autonomic dysfunction and sleep disorders. Some of these complications might be induced by dopaminomimetic therapy whereas some

---

*To whom correspondence should be addressed. E-mail: ghulamali48@hotmail.com

features are only linked to disease progression. The diagnostic might canceal a wide pattern of similar disorders with different causes and courses; however, a few criteria and a clear response to dopaminomimeric therapy constitute idiopathic PD. The symptoms of PD have been linked with deficiency of dopamine (DA), both in the midbrain substantia nigra and the forebrain caudate-putman or neotriatum (Ehringer and Hornykiewicz, 1960; Hornykiewicz, 1966). Using catecholamine staining techniques (Carlsson *et al.*, 1957), several nerve cell groups utilizing DA as their transmitter were identified.

The DA projection from the nerve cell bodies in the substantia nigra to terminals in the forebrain is part of nerve cell loops called the basal ganglia, where DA is known to have several physiological and biochemical effects (Gerfen, 1992; Graybiel and Ragsdale, 1979). The hall mark of PD is progressive loss of DA neurons in the midbrain, most pronounced in substantia nigra pars compacta even though other nerve cell populations using monoamine transmitters such as the locus coeruleus containg noradrenaline neurons also are affected (German *et al.*, 1992).

The cause of PD is still unknown, although both genetic (Werneck and Alvarenga, 1999) and environmental risk factors have been examined (Thiessen *et al.*, 1990). Increasing age is generally accepted as a risk factor for PD, with an increasing prevalence in ages over 60 years. In the last few years, mutation in five different genes have been linked with PD and these mutated proteins are shown implicated in synaptic plasticity and vesicle functions (Smith *et al.*, 2003) and dysfunction of these proteins are held responsible cytoplasmic DA levels that are readily prone to autooxidation and enzymatic for increasing degradation, which can be a common pathway in the pathogenesis of the disease.

There is no specific cure for PD and there are limitation to current PD therapy (Rascol *et al.*, 2003). However, there are several treatments available to alleviate symptoms of the disease, such as pharmacological therapy (Korczyn and Nuss-baum, 2002) and surgical interference (Gullingham, 2000). The most successful candidate among drug therapy is still L-DOPA, which as a precursor in DA biosyn-thesis was considered to be the most appropriate drug therapy. The introduction of L-DOPA was revolutionary in that it became an excellent treatment in an ear-lier essentially untreatable disorder. Mortality has been nearly normalized and L-DOPA is still considered the most effective agent for treatment of PD. How-ever, disabling motor complications occur in the late stages of PD with the most common effect that includes onset of on-off phenomena and abnormal involuntary movements called dyskinesia (Jankovic, 2002). Hence, the need for better, per-haps more physiologically based therapy for this complex disorder remains. The Increasing knowledge of monoamines led trails with monoamine oxidase inhibitors in the late 1960, and in 1978 the facilitation of dopamine by selegiline was pre-sented (Knoll, 1978). During the 1990s, inhibitors of another enzyme, catechol-O-methyltransferase (COMT), was further developed to further increase the utilization of L-DOPA (Mannisto and Kaakkola, 1999). Apart from these advancements in PD therapy, bromocriptine, the oldest dopamine receptor agonist was accompanied by several new agonists in the late 1990s (Diamond *et al.*, 1987). However, comparing the rate of mortality among PD patients, L-DOPA showed markedly reduced mortality and it has been concluded that L-DOPA treatment is far superior regarding life expectancy to other treatments (Diamond *et al.*, 1987). Most patients treated with L-DOPA develop fluctuation in motor performance. After 3–5 years of treatment, one-third, after 5–7 years about half and after 10–12 years nearly all patients suffer from the motor fluctuation (Rinne, 1983; Markham and Diamond, 1986). Non-motor symptoms such as pain, fatigue, anxiety and depression are often seen in PD patients treated with L-DOPA for a long time (>10 years) than are seen in the patients with classical motor distrubances (Sulman *et al.*, 2002).

Transmitter and its metabolite concentrations in brain tissue are reflected by density of cells using specific transmitter in a brain region and the activity of the neurons within the area. The heterogeneity in the chemical features of various degenerative diseases highlights the importance of biological markers as diagnostic tools. Even more important is the possibility of following each disease during its different stages, and to understand the metabolic aspects of the disease identified by the biological marker. Establishing a primary role of a biochemical marker in neurodegenerative disease is an important strategy in the design of new pharmacological treatments. Quantitative analyses of biochemical constituents of cerebrospinal fluid (CSF) represent an initial strategy in attempts to link neurological and psychiatric disorders with pathology within specific neurotransmitters and/or neuromodulator systems in the central nervous system (CNS).

In addition, oxygen free radicals may play a role in triggering neuronal destruction, and nitric oxide (NO) has been implicated in a number of diverse physiological processes (Snyder, 1992). There is increasing evidence that NO plays an important role in pathophysiology of a variety of CNS disorders, including cerebral ischemia and seizures and in acute and chronic inflammation (Dawson *et al.*, 1992). Endogenous NO is synthesized in neurons during the conversion of arginine (Arg) to citrulline (Cit) by cleavage of the terminal guanidino nitrogen by nitric oxide synthase (NOS). Besides, NO exposure causes cobalamin deficiency resulting from its inactivity, which is known to produce impaired neurological functions. Congenital defects of cobalamin metabolism and nutritional deficiencies have been reported to produce abnormal CNS and peripheral neuronal function in humans (Stollhoff, 1987; Hall, 1990). This review deals with CSF analysis of amino acids, nitrite (a metabolite of free radicals, nitric oxide), its precursor arginine (Arg), and catecholamines in PD patients in on-off pathological conditions.

These levels are reported with the assumption that CSF concentrations reflect brain or spinal cord concentrations and perhaps synaptic activity, since CSF is in constant exchange with the extracellular fluid of the CNS. Based on the facts that most of the transmitters in the brain are present in the CSF (Herkenham, 1987) and that alterations in the concentration of transmitters in the CSF can alter the same transmitters in the brain of man (Wester *et al.*, 1990), the results are compiled in two groups of PD patients and compared with healthy controls.

BLOOD-BRAIN BARRIER

The presence of the blood-brain barrier (BBB) has major implications for the passage of relatively large and hydrophilic compounds in brain. However, essential nutrients are transported into the brain by means of (selective) carrier mechanisms. Several transport systems have been characterized varying from passive transport (diffusion) to active and energy-requiring processes. Under healthy conditions, the BBB not only regulates the entry of drug or endogenous compounds into the brain, but also diminishes cellular infiltration compounds to peripheral organ. However, the barrier function of BBB can change dramatically during various CNS diseases; inflammatory reactions underlie changes in the integrity of BBB during cerebral diseases such as cerebrovascular disorder (CVD) and bacterial meningitis (Yu, 1994). In some neurologic disorders, such as multiple sclerosis (MS) and tuberculus meningitis (TBM), the common consequence is an alteration of blood brain barrier permeability (Saez-Llorens *et al.*, 1990) and leakage of serum proteins and macromolecules into the CSF which results in the development of vasogenic edema. Additionally, larger numbers of leukocytes enter the subarachnoid space and release toxic substances that contribute to the production of cytotoxic edema (Talan *et al.*, 1988). Antidiuretic hormone secretion also contributes to the development of cytotoxic edema by decreasing the toxicity of brain parenchymal fluid and increasing the permeability of the brain to water (Kaplan, 1989). All these events if not modulated promptly cause alterations of CSF dynamics of brain metabolisms and of cerebrovascular autoregulation which results in irreversible brain injury (Lipton *et al.*, 1993; Saez-Llorens *et al.*, 1990).

L-DOPA is transported across the intestinal mucosa and BBB by a large neutral amino acids (LNAA) transport sytem, which means that some amino acids competitively inhibit L-DOPA membrane transport (Wade *et al.*, 1973). Doses of L-DOPA taken with a meal, particularly with high protein content, will be less effective than doses taken on an empty stomach (Nutt *et al.*, 1984). This has led to the use of a protein distribution diet for patients with motor fluctuations (Pinus and Barry, 1988). By restricting most of the daily protein intake to evening meals, daytime plasma L-DOPA levels are more predictable and motor performance is improved, including so called L-DOPA resistant 'Off' periods (Karstead and Pincus, 1992). It has also been concluded that plasma L-DOPA levels predict motor response to a great extent (Frankel *et al.*, 1990) under controlled condition regarding diet and medication and L-DOPA response was found to be predictable.

In order to study these changes, CSF obtained by lumbar puncture technique is the most important and widely used diagnostic tool in evaluating the levels of neurotransmitters and their metabolites. Lumbar CSF neurochemical measurements aim to serve as indirect *in vivo* markers of human central neurotransmission and these markers can be used in the diagnosis of neurodegeneration process.

EXCITOTOXICITY AND OXIDATIVE STRESS IN CNS DISORDERS

The etiology of neuronal death in neurodegenerative diseases remains mysterious; however, great advances in both molecular genetics and neurochemistry have improved our knowledge of fundamental processes involved in cell death, including both excitotoxicity and oxidative damage. Excitotoxicity refers to neuronal cell death caused by activation of excitatory amino acid (EAA) receptors N-methyl-D-aspartate (NMDA), and other excitatory amino acid receptors have been implicated in neuronal death sometimes. These receptors not only mediate EAA neurotransmission within the hippocampus, basal ganglia and cerebral cortex but when activated for a sufficient period of time, may also result in neurodegeneration and seizures (Meldrum *et al.*, 1990). An increase body of evidence has implicated excitotoxicity as a mechanism of cell death in both acute and chronic neurological diseases (Beal, 1992a; Choi, 1988; Coyle, 1993). The tvs concept of slow or weak excitotoxicity that evolves slowly over many years has been proposed in neurodegenerative diseases (Albin *et al.*, 1992; Beal, 1992a). Calcium (Ca^{+2}) influx via NMDA receptors is much more effective in mediating cell death than that occurring through non-NMDA receptors or voltage dependent Ca^{+2} channels (Tymiaski *et al.*, 1993). This suggests compartmentalization of Ca^{+2}-dependent neurotoxic processes within neurons with a preferential localization in the submembrane space adjacent to NMDA receptors. Increased intracellular Ca^{+2} levels can initiate a number of deleterious processes, including activation of nitric oxide synthase and free radical generation (Beal, 1995). NMDA receptors appear to have a critical role in excitotoxicity, although non-NMDA receptors could also mediate prolonged stimulation (Choi *et al.*, 1987). NMDA receptor activation is able to trigger neuronal injury more readily than can non-NMDA receptors such as kainic acid (KA) and 3-hydroxy-5-methyl-4-isoxazolepropionic acid (AMPA) activation due to a greater ability to induce Ca^{+2} overload within a cell. Glutamate (Glu) and aspartate (Asp) are considered as neurotoxic when present in higher amounts in brain (Rothman *et al.*, 1986). The neurotoxic effects of Asp and Glu have been known for some time and these have been found responsible for neuronal death in a wide range of neurological disorders (Farber *et al.*, 1981; Rothman, 1984; Rothman *et al.*, 1986).

An association between excitotoxic mechanisms and the slow, often insidious development of human neurodegenerative disorders has undergone widespread scrutiny. The relatively recent recognition of interaction between excitotoxicity and neuronal metabolic stress underscores the likelihood that these two biological mechanisms may act in nature to produce neuropathological conditions (Albin *et al.*, 1992). Cell injury or cell death may be induced by several processes. It has been suggested (Choi, 1992) that neuronal injury following brief intense exposure to Glu involves two components, distinguishable by differences in time-course and ionic dependence. The first component marked by neuronal swelling probably reflects Na^+ ion influx with subsequent passive Cl^- and water entry leading to an expansion of cell volume. It occurs within minutes and depends on the presence of extracellular Na^+ and Cl^- ions and may or may not be lethal. The second and possibily more significant component may involve Ca^{+2} influx mediated by NMDA subtype of Glu receptors (Collingridge *et al.*, 1989). The process leads to a delayed neuronal disintegration (Choi, 1992) occurring over a period of hours after exposure to Glu in cultured cells and is dependent on the presence of extracellular Ca^{+2} ions. Glu is known to be released in excess during cerebral ischemia and Glu receptors are enriched in brain regions susceptible to ischemic injury (Nakanishi, 1992). Neurotoxic mechanisms mediated by both free radicals and neuronal death follow hypoxic or ischemic injury to brain (Choi, 1990; Floyd, 1990). Generation of toxic free radicals is also thought to play an important role in the pathogenesis of ischemic induced neuronal death. Recent evidence has suggested that excitotoxicity and oxidative stress (the cytotoxic effect of oxygen radicals) may be sequential and interactive mechanisms leading to neuronal degeneration (Coyle, 1993). The most direct evidence linking excitotoxicity to oxidative stress is the finding that NMDA exposure leads to superoxide generation in cultures of cerebellar neurons (Lofton-Cazal *et al.*, 1993). The most accepted theory dependent on various biochemical markers leading to neurodegeneration is as follows;

The cycle of neurodegeneration = Free radicals (oxidative stress) + Cytosolic Ca^{+2} + Mitochondrial damage + Excitotoxicity of EAA and perhaps the role of homocysteine (HC) and vitamin B12, vitamin B6 and folate deficiency. All these factors one way or another result in cell death processes in the brain.

MATERIAL AND METHODS

Patients and healthy subjects and routine analysis

Most of the patients were recruited from the department of Neurology, Huddinge University Hospital, Sweden. 30 patients (12 women and 18 men) with PD were included in this study. The reseach ethics commitee of Karolinska Institute, Stockholm approved the study and all patients gave written informed consent to participate. 15 patients each at different stages of the illness (Hoehn and Yar range 2–4) were medicated with individual drug combinations. All 15 patients were on L-DOPA dose (250 mg/day) and had been taking it with combination of other drugs with positive response towards their motor activities (Group PD On). 15 PD patients who had been treated for 6–8 years with the similar drug therapy showed severely 'off' response (Hoehn and Yar range of −3) (Group PD Off) were included. These Off PD patients showed motor fluctuation. Both On and Off groups were monitored one week before the samples of CSF and blood were taken and all food and fluid intake were similar in both groups and exact dose times were maintained. Blood and CSF were taken after one week of observation early in the morning between 6–8 a.m. in a fasting condition. 16 healthy individual volunteers from the institute employees were included and their blood and CSF were collected under comparable conditions. None of the healthy controls was on medication for six months. Table 1 shows the clinical data on PD patients and healthy controls.

Table 1.
Clinical data on the healthy subjects (HS), patients with Parkinson's disease (PD) in both On and Off group

Patients	n	Females	Age (years)	CSF-Abumin	CSF-IgG	IgG-index
Controls	16	7	53 ± 5	218 ± 17	34 ± 4	0.44 ± 0.01
PD (On)	15	7	72 ± 12	$332 \pm 42^*$	$51 \pm 7^{**}$	0.43 ± 0.01
PD (Off)	15	5	74 ± 14	$313 \pm 37*$	$44 \pm 7^*$	0.42 ± 0.01

$^*p < 0.05, ^{**}p < 0.01, ^{***}p < 0.001.$

10–12 ml cerebrospinal fluid (CSF) was collected from each patient and healthy subject in a sitting position at the L4–L5 levels. Blood samples were collected by venipuncture. The basic CSF analyses included: cell counting by phase-contrast microscopy (Siesjö, 1967), determination of CSF/serum albumin ratio and CSF/immunoglobulin G (IgG) index (Link *et al.*, 1977) as well as isoelectric focusing for detection of oligoclonal IgG band. Serum and CSF albumin and IgG were determined using Hitachi 737 Automatic Analyzer (Naka Works, Hitachi Ltd., Tokyo, Japan). CSF and serum samples were kept at $-80°C$.

Analysis of amino acids and nitrite

Amino acid analysis was performed based on the pre-column derivatization of OPA with amino acids as previously described (Qureshi *et al.*, 1988). The reproducibility for each amino acid was <2% with this method.

Since the half-life time for NO is short, its concentration is quantitated indirectly by measuring the levels of its degraded product nitrite. Recent HPLC method in combination with electrochemical detection is used for this purpose (Kaku *et al.*, 1994).

Analysis of catecholamines and metabolites

The CSF was prepared as previously described (Qureshi *et al.*, 1988) and the concentration of DA, homovanillic acid (HVA), 3,4-dihydroxyphenylacetic acid (DOPAC), noradrenaline (NA), 3-methoxy-4-hydroxyphenylglycol (MHPG), 5-hydroxytryptamine (5-HT) and 5-hydroxyindoleacetic acid (5-HIAA) was measured using previously described methods (Baig *et al.*, 1991). Correlation between Glu versus nitrite methods were submitted to a stepwise polynomial regression analysis (Zar, 1984).

NEUROTOXICITY

Excitatory amino acids

Glutamate (Glu) is the main excitatory neurotransmitter apart from aspartate (Asp) in mammalian brain. In normal physiological conditions Glu acts as a neurotrans-

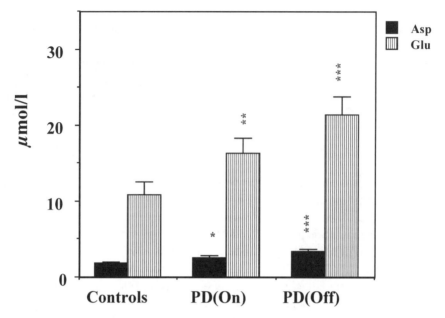

Figure 1. CSF levels of glutamate and aspartate in PD patients (On and Off groups) and healthy controls. All the values are expressed as mean ± SEM. $^*p < 0.05$, $^{**}p < 0.01$, $^{***}p < 0.001$.

mitter in several brain regions, regulates neuronal outgrowth and synaptogenesis during normal development and adult plasticity, and has a special function in learning and memory processes (Gasic et al., 1992; Lipton et al., 1989; Monaghan et al., 1989). Figure 1 shows elevated CSF levels of Asp and Glu, respectively, in both PD patient groups (On and Off). The concept of excitotoxicity and the observations collectively suggested that an excessive amount of the intrinsic EAA, specifically Glu, is capable of exciting CNS neurons to such an extent as to become toxic, ultimately killing Glu sensitive neurons through over-excitation. Apart from PD, an excitotoxic mechanism is involved in the pathogenesis of several other neurodegenerative disorders, including eplepsia, hypoglycemia, concussive brain injury and certain dementia and metabolic disorders (Benveniste, 1991; Choi, 1992; Lipton et al., 1994; Qureshi et al., 1996b). If the Glu concentration is elevated over physiological level, the consequence is overestimulation of glutamate receptors, resulting in neuronal injury — the so-called excitotoxic damage. The excitotoxic mechanism is probably involved in several neuropathological conditions. These include

(a) energy deficiency disorders such as hypoxia and ischemia (Benveniste, 1984; Rothman, 1984), hypoglycemia and prolonged seizures (Schwarcz et al., 1984);

(b) metabolic disorders such as olivopontocerebellar degeneration (Plaitakis et al., 1984), amyotropic lateral sclerosis (ALS) (Plaitakis et al., 1987) and sulfide oxide deficiency (Olney et al., 1975);

(c) disorders induced by food excitotoxicitys, such as endemically occurring neurolathyrism (Ross et al., 1987). In addition, an excitotoxic mechanism has been hypothesized to be involved in several other neuropathological conditions, including Huntington's disease (Schwarcz *et al.*, 1978), Alzheimer's disease (Geddes *et al.*, 1986), parkinsonism (Sonsall *et al.*, 1989) and AIDS encephalopathy (Lipton, 1992).

An elevated Glu (or related excitatory compound) concentration is common to all of these conditions. Subsequently it was observed that another intrinsic EAA, the Glu analogue aspartate (Asp), is also capable of inducing an excitotoxic effect (Olney, 1978). More recently, however, it has been recognized that excitotoxic neurodegeneration is considerably enhanced when intracellular energy levels are reduced (McMaster *et al.*, 1991; Zeevalk *et al.*, 1991), and evidence now indicates that even in the face of relatively low levels of excitotoxic substances, when neuronal energy metabolism is impaired, characterstic excitotoxic neurodegeneration can be demonstrated. Recognition of the interaction between excitatory effects of EAA and reduced intracellular levels has led to studies on more slowly-evolving or chronic forms of excitotoxic neurodegeneration. These investigations have brought into sharper focus the posssibility that certain slowly-evolving neurodegenerative diseases of the CNS may in fact have a chronic excitotoxic neurodegenerative basis or component. Evidence which came to light more than a decade ago showed that various types of iGlu-receptors responded differentially when stimulated by their respective ligands so that different subtypes of iGlu-receptors are selective in allowing influx of Na^+, Cl^-, and Ca^{+2}. These differential responses bear significantly on iGlu-receptor function in both normal and pathological conditions in the CNS (Monaghan *et al.*, 1989). Cell death could result from disturbances of the metabolism or physiology that maintains EAA concentrations at sub-toxic levels. These may include defective catabolism or uptake or excessive production of EAA, particularly Glu (Dodd *et al.*, 1994). A number of excitotoxic hypotheses have been formed to account for regional variations in the pathology of different disorders. It has also been proposed (Albin *et al.*, 1992) that neurons in circumscribed brain areas could possess altered receptor subtypes, which could make them more susceptible to the excitotoxic effects of endogenous EAA.

Activation of EAA receptor leads to increased intracellular Ca^{+2} followed by activation of protein kinases, phospholipases, nitric oxide synthase, impaired mitochondrial function and the generation of free radicals (Beal, 1992b). Our results show the levels of Asp and Glu were increased in the CSF of PD patients who have positive response from L-DOPA therapy; however, it is interesting to note that these levels are much more elevated when the L-DOPA treatment does not show any response at all (Off situation).

It is known that enhanced Glu release increases NO production, causes DNA damage, energy depletion in neurons and neuronal death as occurs in PD (Zhang *et al.*, 1994). This hypothesis is also supported by our recent study (Qureshi *et al.*, 2002b) that showed high levels of arginine and nitrite (metabolite of NO) in CSF of PD patients. Similar to CVD patients (Qureshi *et al.*, 2002a), PD patients also show elevated levels of Glu and Asp. Local application of Glu or Asp on experimental animal neurons causes spreading depression (Murphy *et al.*, 1989). This process is blocked by Glu antagonist (Siesjö *et al.*, 1989). In diseases with acute neuronal damage caused by ischemia, hypoxia or seizures, it has been suggested that EAAs are released in excess and after specific receptor stimulation, EAAs can promote or intensify neuronal destruction (Buchan, 1990). Both of these EAAs are considered to be responsible for fast excitatory postsynaptic potentials (EPSP) in brain and spinal cord. A major impetus in understanding how the excitotoxic potential of endogenous Glu may be unleashed to cause neuronal degeneration came with the discovery (Benveniste *et al.*, 1984) that conditions such as hypoxia/ischemia and head trauma cause a marked outpouring of Glu and Asp from the intracellular to extracellular compartment of brain where they interact with EAA receptors to trigger an excitotoxic cascade. It was also demonstrated that cerebral ischemia causes marked elevation in the extracellular concentration of the endogenous excitotoxins Glu and Asp in rat hippocampus, and *in vivo* hippocampal cell culture (Kochhar *et al.*, 1988), EAA antagonists can prevent anoxic neuronal degeneration. The latter finding was confirmed suggesting that NMDA antagonist can protect from hypoxic/ischemic brain damage (Buchan, 1990; Rothman *et al.*, 1995). The excitatory action of Glu is mediated by membrane receptors resulting in increased sodium permeability of neurons. Tissue culture studies had also shown that postsynaptic amino acid antagonist 6-glutamylglycine, which blocks Glu and Asp receptors, reduced EPSP and Asp and Glu responses. These studies further indicate that Glu- and Asp-antagonists blocked neural death when culture was exposed to anoxic conditions. Although the detailed molecular basis of Glu neurotoxicity is not known, it appears that Ca^{+2} influx plays a critical role (Goldberg *et al.*, 1988; Siesjö *et al.*, 1991).

Glu, Asp and related EAA account for most of the excitatory synaptic activity in the CNS; they are released via a Ca^{+2}-dependent mechanism by approximately 40% of all synapses (Fonnum, 1984). A transient intense influx of Ca^{+2} may lead to uncontrolled activation of one or more of these potentially lethal processes. It has been demonstrated that acute excitotoxicity experimentally induced by excessive amount of Glu or Glu-analogues can produce neuropathological changes of acute CNS infarction like those seen with ischemia, hypoxia or head trauma (Choi *et al.*, 1988; Meldrum *et al.*, 1990). It is also established that excessive release of excitatory neurotransmitter Glu and sustained activation of Glu receptors may also be responsible for neuronal degeneration associated with cerebral ischemia and other neurodegenerative diseases (Choi, 1988; Meldrum *et al.*, 1990; Rothman *et al.*, 1987). It is now emerging that free radical formation and Glu receptor activation may act in concert, cooperating in the genesis and propagation of neuronal damage (Bondy *et al.*, 1993; Chio, 1990; Coyle *et al.*, 1993).

Free radicals and oxidative stress

Most free radicals are unstable species due to one or more unpaired electrons that can extract an electron from neighbouring molecules leading to oxidative damage.

Although there are number of intracellular sources of free radicals, the mitochondria are thought to be the most important. The most reliable and robust risk factor for neurodegenerative disease is normal aging, since there is substantial evidence that mitochondrial function declines with age (Beal, 1995). A great deal of interest has been focused on the possibility that oxidative damage plays an important role in the pathogenesis of various degenerative disorders (Dexter *et al.*, 1994; Koroshetz *et al.*, 1994; Smith *et al.*, 1991). Hence, the role of free radicals in cell death induced by activation of EAA receptors is an area of expanding interest (Coyle *et al.*, 1993; Halliwell, 1992; Kontos, 1989). The initial report linking free radicals to excitotoxicity (Dykens *et al.*, 1987) showed that kianate-induced damage to cerebellar neurons could be attenuated by superoxide dismutase, allopurinol and hydroxyl radical scavengers such as mannitol. The brain is known to have high oxygen consumption needs and is rich in oxidizable substrates, especially unsaturated lipids and catecholamines. It is a highly oxygenated organ, consuming almost one-fifth of the body's total oxygen, and it derives most of its energy from oxidative metabolism of mitochondrial respiratory chain. Oxygen-centred free radicals are the main types of radicals formed in cells as normal attributes of aerobic life, either as accidental by-products of metabolism or as selectively generated species commonly named as reactive oxygen species (ROS).

ROS are also involved in intracellular signalling processes, modifying the activity of proteins at the post-translational level and regulating gene expression (Halliwell *et al.*, 1989; Sies, 1991). As a result, it is not surprising that several CNS disorders are related to oxygen radicals either as primary process or a secondary one. In this latter case, secondary free radical formation may result in increased damage over and above that caused by the primary disease process, since the brain is sensitive to hyperbaric oxygen exposure. Although the fundamental mechanism is not fully understood, excessive oxygen radical production is suggested by the protective action of liposome entrapped superoxide dismutase and catalase (Yusa *et al.*, 1984). One of the prerequisites of aerobic life is a competent defence system against ROS (Yu, 1994). As with atherosclerosis and acute myocardial infarction, oxygen radicals are probably equally important in neurological degenerative disorders. In addition to these conditions, various studies have shown increased radical formation and lipid peroxidation in most neurologic and neuropsychiatric disorders including schizophrenia, MS, CVD and AD (Halawa *et al.*, 1996; Hunter *et al.*, 1985; Qureshi, 1997a).

Oxygen-derived free radicals are very important mediators of cell injury and death. Not only are these highly reactive chemical species important in the aging process but directly or indirectly these species are involved in a wide variety of clinical disorders (Floyd, 1990; Knight, 1995). In addition, these species play an important role in chronic granulomatous diseases and act as secondary sources of cellular injury in chronic inflammatory processes. Biological systems have developed a comprehensive array of defence mechanisms to protect against free radicals. These includes enzymes to decompose peroxides, proteins to sequester transition metal ions and a range of compounds to scavenge free radicals. The term 'oxidative stress' is caused by accumulation of free radicals, and is defined as a distrubance in prooxidant-antioxidant balance in favour of former, leading to potentially damaging reactions with biological molecules (Sies, 1985). It has been suggested that oxidative stress is involved in the etiology of various biological processes such as inflammation, aging, carcinogenesis and ischaemia/reperfusion damage (Halliwell et al., 1989; Sies, 1991). However, establishing the involvement of ROS in the pathogenesis of a disease is extremely difficult due to the short lifetime of these species. The generation of ROS in vitro or in vivo, in sufficient amounts to overwhelm normal defence mechanisms, can result in serious cell and tissue damage. All of the major classes of biological macromolecules may be attacked, but membrane lipids are probably the most susceptible. The overall effect of lipid peroxidation is to decrease membrane fluidity, thus destabilising membrane proteins such as receptors. Lipid peroxidation has been implicated in a wide range of tissue injuries and diseases (Cross et al., 1987; Hallliwell et al., 1989). On the other hand, the protective effects of iron chelation in several disorders, superoxide dismutase and catalase in ischemia reperfusion and the protection by various vitamins and other oxidants under a variety of experimental conditions, along with other examples clearly indicate a key role for free radicals in many, if not most of these disorders (Halliwell et al., 1985; McCord, 1985). Although free radicals can damage cells directly through oxidation of biological molecules, they can also inhibit mitochondrial activity, induce the release of EAA and promote a rise in the level of cytosolic free Ca^{+2}, thus promoting a cascade of events leading to neuronal degeneration (Dawson et al., 1996). The two most actively investigated oxygen radicals at present are the superoxide anion (O_2^-.) and nitric oxide (NO$^{\bullet}$). The mammalian brain may be exceptionally vulnerable to oxidative stress through attack of these radicals (Cross et al., 1987; Dawson et al., 1996).

The free radical nitric oxide (NO) has been implicated in a range of neurodegenerative diseases (Dodd et al., 1992; Halawa et al., 1996; Qureshi et al., 1995, 1996a, b). Arg the precursor of NO and nitrite (a metabolite of NO) are significantly increased in both groups of PD patients, as in other neurological diseases (Halawa et al., 1996; Qureshi et al., 1995, 1997b) (Figs 2 and 3).

It has been shown previously (Qureshi et al., 1996b) that there is a linear relationship between Glu and nitrite in TBM, CVD and MS patients and, similar to these findings, both groups of PD patients show linear relations between these parameters (Fig. 4). There is probably no single intracellular pathway that is responsible for neurotoxicity; however, this shows that these neurotoxic generating processes based on Glu and NO$^{\bullet}$ are interrelated in these patients (Qureshi et al., 2002a, 2002b). Accumulation of reactive oxygen species in cerebral tissues may be prevented by agents that have antioxidant activity (Rice-Evans and Diplock, 1993).

NO functions as neuronal messanger in the nervous system and in situations of excessive production, it may act as a neurotoxic. It has been clearly demonstrated that Glu and related amino acids such as NMDA markedly stimulate NO synthase

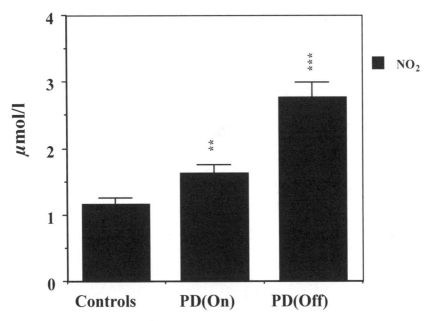

Figure 2. CSF levels of nitrite (metabolite of NO) in PD patients (On and Off groups) and healthy controls. All the values are expressed as mean ± SEM. ** $p < 0.01$, *** $p < 0.001$.

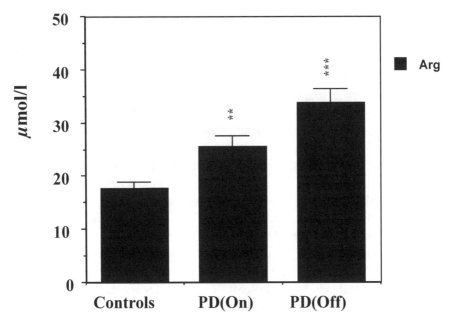

Figure 3. CSF levels of arginine (Arg) in PD patients (On and Off groups) and healthy controls. All the values are expressed as mean ± SEM. ** $p < 0.01$, *** $p < 0.001$.

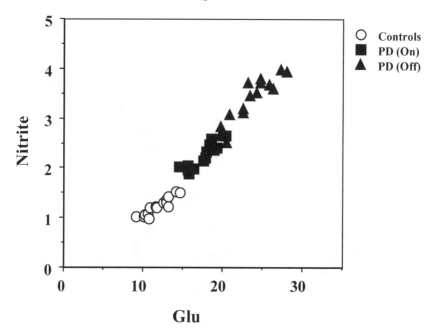

Figure 4. Relationship between nitrite and glutamic acid in PD patients (On and Off groups) and healthy controls.

(NOS) activity in cerebellar slices monitored by the conversion of Arg to citrulline (Cit). NMDA receptor activation by EAA and Glu is known as a major trigger for NO formation in several areas of the brain (Dawson *et al.*, 1992; Garthwaite, 1991).

Our results clearly show the linear interrelationship between CSF nitrite and Glu in healthy subjects as well as in PD patients (Fig. 5). An abundance of experimental evidence supports the concept that an acute event causing sudden focal, regional or global loss of oxygen and energy in the CNS can induce a state of neuronal membrane depolarization which allows ion and water shift characterstic of excitoxicity. While these changes appear to be quite consistent with the immediate effects of ion and water influx, the compounding effects of mitochondrial damage due to direct Ca^{+2} toxicity and impairment of mitochondrial function must also be taken into account. Neuroleptics are the major class of drugs used to treat PD; however, these drugs are associated with a wide variety of extrapyramidal side effects that include tardive dyskinesia as a hyperkinetic syndrome consisting of choriform, athetoid or rhythmically abnormal involuntary movements (Kulkami and Naidu, 2003). Numerous reports have indicated that induction of free radicals by neuroleptic drug treatments that increase the free radical production and cause structural damage eventually leads to oxidative stress and resultant structural abnormalities in the pathophysiology of tardive dyskinesia. Oxidative stress contributes to the cascade, to develop cell degeneration in PD (Jenner, 2003). This is linked to other degenerative processes such as mitochondrial

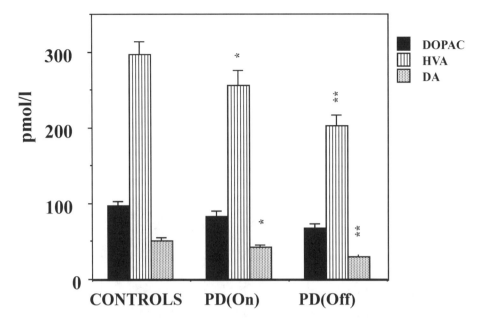

Figure 5. CSF levels of dopamine (DA) and its metabolites HVA and DOPAC in PD patients (On and Off groups) and healthy controls. All the values are expressed as mean ± SEM. $*p < 0.05$, $**p < 0.01$.

dysfunction, excitotoxicity, nitric oxide toxicity and inflammation; however, the mechanism that makes the major contribution has still not been elucidated.

Monoamines and their metabolites

Abnormalities of biogenic amine metabolism in the central nervous system (CNS) have been implicated in various psychiatric and neurological disorders (Davis *et al.*, 1991; Gottfries, 1990). The estimation of monoamine and indoleamine are shown to be helpful for the diagnosis and interpretation of these disorders (Burns *et al.*, 1985). Both DA and NA are synthesized from the amino acid tyrosine derived from food intake or protein breakdown. The main metabolites of DA are homovalinic acid (HVA) and 3,4-dihydroxyphenylacetic acid (DOPAC) and of NA, 3-methoxy-4-hydroxyphenylglycol (MHPG). The major inactivation of monoamines is due to an active reuptake of the parent substances from the synaptic cleft into the presynaptic bouton. Apart from this reuptake, neurotransmitters and their metabolites in the synaptic cleft are cleared to the cerebrovenous blood. However, a small fraction diffuses into the CSF (Björklund *et al.*, 1984).

In PD patients, the CSF levels of NA and its metabolite MHPG were increased whereas DA and its metabolites, HVA, and DOPAC were decreased significantly in both On and Off groups of PD patients (Figs 5 and 6).

The therapeutic effect of L-DOPA is very promising in the course of the disease and normal mobility can be acheived. A single dose of L-DOPA gives a fast motor response, which can be maintained for several hours due to preserved dopamine

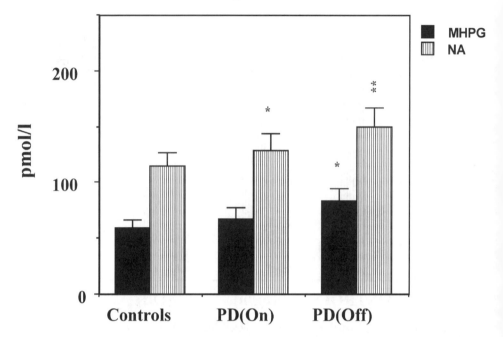

Figure 6. CSF levels of noradrenaline (NA) and its metabolite MHPG in PD patients (On and Off groups) and healthy controls. All the values are expressed as mean ± SEM. *p < 0.05, ** p < 0.01.

buffering capacity in the neurons; the fluctuating pharmokinetics is no problem in PD (On) patients. As the disease progresses within a few years, the response duration gets shorter (wearing-off phenomenon) and the therapeutic window is narrowed (Mouradian *et al.*, 1988). This is clearly reflected by CSF levels of NA and DA and their metabolites (Figs 5 and 6). In the advancing stages of the PD disease, patients take small and frequent doses of L-DOPA as continuously as possible to reduce fluctuations in plasma L-DOPA levels and motor performance (Obeso *et al.*, 1994).

Apart from catecholamines, serotonin (5-hydroxy tryptamine, 5-HT) plays an important role in the CNS. 5-HT is synthesized from the amino acid tryptophan (Trp) and is metabolized mainly to 5-hydroxyindoleacetic acid (5-HIAA). 5-HT neurons originate from the raphe nuclei in the brain stem and adjacent nuclear groups and, as the noradrenergic system, project widely to most areas in the brain (Copper *et al.*, 1991). The highest concentration of 5-HT is found in subcortical nuclei (Wester *et al.*, 1990). In both groups of PD patients, there is a decrement of 5-HT; however, it is more significant in the off group (Fig. 7). 5-HIIA and 5-HT are known to pass through the BBB but the rate of Trp entry in the brain is dependent on its availility in the blood and its relationship with other large neutral amino acids through the transport system (Fernstrom *et al.*, 1972).

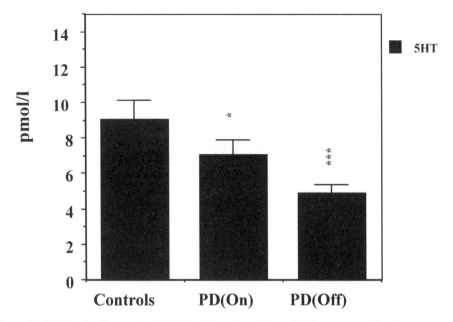

Figure 7. CSF levels of serotonin (5-HT) in PD patients (On and Off groups) and healthy controls.

CONCLUSIONS

One of the main problems during the many years of research on PD is to detect the disease early, since clinical diagnosis of PD will not appear until the DA depletion reaches a critical threshold in neostratum (Bernheimer *et al.*, 1973). The disease is diagnosed on a clinical basis and there are no diagnostic tests. However, a positive response to levodopa administration is obvious and separates PD from other neurological disorders, which present with parkinsonism. New methods for early detection and monitoring of the disease progression to help the clinician in clinical diagnosis are important, particularly if future pharmacological treatments can halt further neurodegeneration and nerve cell death. One of the tools that have been used during the last few decades is to examine the patient's brain with positron emission tomography (PET). Combining this technique with analysis of biochemical, pathological and behavioral changes, more information on PD patients could be monitored. There is probably no single intracellular pathway that is responsible for the cell death of neurological diseases (Rothman *et al.*, 1995). However, hyperactivity of the glutaminergic innervation, the associated activation of NMDA-receptor ion channels selective for Ca^{+2} and ensuing increase in the intracellular concentration of Ca^{+2} may contribute to the neurodegenerative process (Rothman *et al.*, 1995). Besides, oxygen-free radicals, nitric oxide (NO) in particular, is thought to mediate NMDA-toxicity (Dawson *et al.*, 1996). The evidence is accumulating to support the view that besides its beneficial actions, NO production from Arg could contribute and even initiate the cytotoxic effect and brain cell death. The mechanism by which NO kills brain cells is unknown;

however, free radical formation has been generally implicated in various forms of neurotoxicity (Whetsell, 1996). Our results do indicate increased accumulation of Arg and increased production of nitrite in both patient groups, and there is clear evidence that there is a linear correlation between the level of Glu and nitrite in PD patients. These results undoubtedly provide us with a possibility of understanding the interrelationship between different neurotransmitters in the brain as well as their distribution and functions. On the basis of these results on degenerative disorders, the role of various neurotransmitters, and the role of free radical NO must be clearly defined in order to develop effective drug therapy. If such a scenario indeed occurs, then a variety of therapeutic approach are possible. This includes glutamate release inhibitors, EAA antagonists, agents to improve mitochondrial function, free radical scavengers and neurotropic factors.

REFERENCES

Albin, R. L. and Greenamyre, J. T. (1992). Alternative excitatoxic hypotheses, *Neurology* **42**, 733–738.

Baig, S. M. and Qureshi, G. A. (1995). Homocysteine and vitamin B12 in multiple sclerosis, *Biogenic. Amine* **11**, 479–485.

Baig, S. and Qureshi, G. A. (2003). Catecholamines, tryptophan and their metabolites in CSF of patients with neurologic disorders, *J. Biochemistry* (in press).

Baig, S., Halawa, I. and Qureshi, G. A. (1991). HPLC as a tool in definition of abnormalities in monoamines and tryptophan metabolites in CSF from patients with neurological disorders, *Biomed. Chromatogr.* **5**, 108–112.

Beal, M. F. (1992a). Does impairment of energy metabolism result in excitotoxic neuron al death in neurodegenerative illness? *Ann. Neurol.* **31**, 119–130.

Beal M. F. (1992b). Mechanism of excitotoxicity in neurological disease, *FASEB J.* **6**, 3338–3344.

Beal, M. F. (1995). Aging, energy and oxidative stress in neurodegenerative disease, *Ann. Neurol.* **38**, 357–366.

Benveniste, H. (1991). The excitotoxin hypothesis in relation to cerebral ischemia, *Cerebrovasc. Brain Metab. Rev.* **3**, 213–245.

Benveniste, H., Drejer, J., Schousboe, A. and Diemer, N. M. (1984). Elevation of the extracellular concentrations of glutamate and aspartate in hippocampus during transient cerebral ischemia monitored by intracerebral microdialysis, *J. Neurochem.* **43**, 1369–1374.

Berheimer, H., Birkmayar, W., Hornkeiwics, O., Jellinger, K. and Seitelberger, F. (1973). Brain dopamine and syndromes of Parkinson's and Huntington. Clinical, morphological and neurochemical correlations, *J. Neurosci.* **17**, 6761–6768.

Björklund, A. and Lindvall, O. (1984). Dopamine-containing systems in the CNS, in: *Handbook of Chemical Neuroanatomy*, Björklund, A. and Hĸfelt, T. (Eds). Elsevier, Amsterdam, The Netherland.

Bondy, S. C. and LeBele, C. P. (1993). The relationship between excitotoxicity and oxidative stress in the central nervous system, *Free Rad. Biol. Med.* **14**, 633–642.

Buchan, A. M. (1990). Do antagonist protect against cerebral ischemia. Are clinical trails warranted? *Cerebrovasc. Brain Metab. Rev.* **2**, 1–26.

Burns, R. S., LeWitt, P. A., Ebert, M. H., Pakkenberg, H. and Kopin, J. (1985). The clinical syndrome of striatal dopamine deficincy. parkinsonism induced by 1-methyl 4-phenyl 1,2,3,6-tetra hydropyridine (MPTP), *N. Engl. J. Med.* **32**, 1418–1421.

Carlsson, A., Lindquist, M. and Magnusson, T. (1957). 3,4-dihydroxyphenylanaline and 5-hydroxytryptophan as reserpine antagonists, *Nature* **180**, 1200–1201.

Chanarin, I., Deacon, R., Lumb, M., Muir, M. and Perry, J. (1988). Cobalamin folate interrations: A critical review, *Blood* **66**, 479–482.

Chandefaux, B., Cooper, B. A., Gilfix, B. M., Lue-Shing, H., Carson, W., Gavsie, and Rosenbladt, D. S. (1994). Homocysteine: relationship to serum cobalamin, serum folate, erythrocyte folate, and lobation of neutrophils, *Clin. Invest. Med.* **17**, 540–550.

Chan-Palay, V. (1988). Galanin hyperinnervates surviving neurons of human basal nucleus of meynert in dementias of Alzheimer's and Parkinson's disease: A hypothesis for the role of galanin in accentuating cholinergic dysfunction in dementia, *J. Comp. Neurol.* **273**, 543–557.

Choi, D. W. (1988). Glutamate neurotoxicity and diseases of nervous system, *Neuron* **1**, 623–634.

Choi, D. W. (1990). Cerebral hypoxia: Some new approaches and unanswered questions, *J. Neurosci.* **10**, 2493–2496.

Choi, D. W. (1992). Excitoxic cell death, *J. Neurobiol.* **23**, 1261–1276.

Choi, D. W. and Rothman, S. M. (1987). The role of glutamate neurotoxicity in hypoxic-ischemic neuronal death, *Ann. Rev. Neurosci.* **13**, 171–182.

Collingridge, G. L. and Lester, R. A. J. (1989). Excitatory amino cid receptors inthe vertebrate central nervous sytem, *Pharmac. Rev.* **40**, 143–210.

Cooper, J. R., Bloom, F. E. and Roth, R. H. (1991). *The Biochemical Basis of Neuropharmacology*. Oxford University Press, New York.

Coyle, J. T. and Puttfarcken, P. (1993). Oxidative stress, glutamate and neurodegenerative disorders, *Science* **262**, 689–694.

Crellin, R. F., Bottiglieri, T. and Reynolds, E. H. (1990). Multiple sclerosis and macrocytosis, *Acta Neurol. Scand.* **81**, 388–391.

Cross, C. E., Halliwell, B., Borish, E. T., Pryor, W. A., Ames, B. N., Saul, R. L., McCord, J. M. and Harman, D. (1987). Oxygen radicals and human disease, *Ann. Int. Med.* **107**, 526–545.

Davis, K., Kahn, R., Ko, G. and Davidson, M. (1991). Dopamine in schizophrenia: Coexistence between neuropeptides and catecholamines — a review and reconceptaulization, *Am. J. Psychiatry* **148**, 1474–1486.

Dawson, T. M., Dawson, V. L. and Snyder, S. H. (1992). A novel neuronal messager molecule in brain: The free radical, nitric oxide, *Ann. Neurol.* **32**, 297–311.

Dawson, V. L. and Dawson, T. M. (1996). Free radicals and neuronal cell death, *Cell Death and Differentiation* **3**, 71–78.

Deacon, R., Lamb, M. and Perry, J. (1978). Selective inactivation of vitamin B12 in rats by nitrous oxide, *Lancet* **2**, 1023–1024.

Dexter, D. T., Holley, A. E. and Flitter, W. D. (1994). Increased levels of lipid hydroperoxides in Parkinsonian substantia nigra: An HPLC and ESR study, *Mov. Disord.* **9**, 92–97.

Diamond, S. G., Markham, C. H., Hoehn, M. M., McDowell, F. H. and Muenter, M. D. (1987). Multicenter study of Parkinson mortality with early versus later dopa treatment, *Ann. Neurol.* **22**, 8–12.

Dodd, P. R., Williams, S. H., Gundlach, A. L., Harper, P. A. W., Healy, P. J., Dennis, J. A. and Johnston, G. A. R. (1992). Glutamate and GABA neurotransmitter systems in the acute phase of maple syrup urine disease and citrullinemia encephalopathies in new born calves, *J. Neurochem.* **59**, 582–590.

Dodd, P. R., Scott, H. L. and Westphalen, R. I. (1994). Excitotoxic mechanism in the pathogenesis of dementia, *Neurochem. Int.* **25**, 203–219.

Dykens, J. A., Stern, A. and Trenkner, E. (1987). Mechanisms of kainate toxicity to cerebellar neurons *in vitro* is analogous to reperfusion tissue injury, *J. Neurochem.* **49**, 1222–1228.

Ehringer, H. and Hornykiewics, O. (1960). Noradrenalin und Dopamin im gehrin des Menschen und ihr Verhalten bei Erkrankungen des extrapyramidalen Systems, *Klin. Wochensch.* **38**, 1236–1239.

Farber, J. I., Chein, K. R. and Mittnachi, S. (1981). The pathogenesis of irreversible cell injury in ischemia, *Am. J. Pathol.* **102**, 271–281.

Ferstrom, J. D. and Wurtman, R. J. (1972). Brain serotonin content: Physiological regulation of plasma neutral amino acids, *Science* **178**, 414–416.

Floyd, R. A. (1990). Role of oxygen free radicals in brain ischemia, *FASEB J.* **4**, 2587–2592.

Fonnum, F. (1984). Glutamate: neurotransmitters in mammalian brain, *J. Neurochem.* **42**, 1–6.

Frankel, J. P., Pitrosek, Z., Kempster, P. A., Bovingdon, M., Webster, R. and Lees, A. J. (1990). Diurnal differences in response to oral levodopa, *J. Neurol. Neurosurg. Psychiatry* **53**, 948–950.

Garthwaite, J. (1991). Glutamate, nitric oxide and cell-cell signalling in the nervous system, *Trends Neurosci.* **14**, 60–67.

Gasic, G. P. and Hollmann, M. (1992). Molecular neurobiology of glutamate receptors, *Ann. Rev. Physiol.* **54**, 507–536.

Geddes, J. M., Chang-Chiu, J., Cooper, S. M., Lott, I. T. and Cotman, C. W. (1986). Density and distribution of NMDA receptors in the human hippacampus in Alzheimer's disease, *Brain Res.* **3-9**, 156–161.

Gerfen, C. R. (1992). The neostriatal mosiac: multiple levels of compartmental organization, *Trend. Neurosci.* **15**, 133–139.

German, D. C., Manaye, K. F., White, C. L., Woodward, D. J., McIntire, D. D., Smith, W. K., Kalaria, R. N. and Mann, D. M. (1992). Disease-specific patterns of locus coeruleus cell loss, *Ann. Neurol.* **32**, 667–676.

Goldberg, M. P., Monyer, H. and Chio, D. W. (1988). Hypoxic neuronal injury *in vitro* depends on extracellular glutamine, *Neurosci. Lett.* **94**, 52–57.

Gottfries, C. (1990). Brain monoamines and their metabolites in dementia, *Acta Neurol. Scand.* (Suppl.) **129**, 8–11.

Graybiel, A. M. and Rogsdale C. W. (1979). Fiber connections of the basal ganglia, *Prog. Brain Res.* **51**, 239–283.

Halawa, I., Baig, S. and Qureshi, G. A. (1991). Application of HPLC in defining the abnormalities in patients with aseptic meningitis, *J. Biomed. Chromatogr.* **5**, 216–222.

Halawa, A., Baig, S. and Qureshi, G. A. (1996). Amino acids and nitrite in ischemic brain stroke, *Biogenic Amines* **12**, 27–36.

Hall, C. A. (1990). Function of Vitamin B12 in the central nervous system as revealed by congenital defects, *Amr. J. Hemalt.* **34**, 121–127.

Halliwell, B. (1992). Reactive oxygen species and the central nervous system, *J. Neurochem.* **59**, 1609–1614.

Halliwell, B. and Gutteridge, J. M. C. (1989). *Free Radicals in Biology and Medicine.* Clarendon Press, Oxford.

Herkenham, M. (1987). Mismatches between neurotransmitter and receptor localizations in brain: observations and implications, *Neuroscience* **23**, 1–38.

Hornykeiwicz, O. (1966). Dopamine and brain function, *Pharmacol. Rev.* **18**, 925–964.

Hughes, A. J. D. S., Kilford, L. and Lees, A. J. (1992). Accuracy of clinical diagnosis of idiopathic Parkinson's disease: a clinico-pathological study of 100 cases, *J. Neurol. Neurosurg. Psychiatry* **55**, 181–185.

Hunter, M. I. S., Niemadim, B. C. and Davidson, D. L. W. (1985). Lipid peroxidation products and antioxidant proteins in plasma and CSF from multiple sclerosis patients, *Neurochem. Res.* **10**, 1645–1652.

Jancovic, J. (2002). Levodopa strengths and weaknesses, *Neurology* **58**, S19–S32.

Jenner, P. (2003). Oxidative stress in Parkinson's Disease, *Ann. Neurol.* **53** (3), S26–S36.

Kaakkola, S. and Wurtman, R. J. (1993). Effects of catechol-O-methyltransferase inhibitors and L-3,4-dihydroxyphenylanaline with or without carbidopa on extracellular dopamine in rat striatum, *J. Neurochem.* **60**, 137–144.

Kaku, S., Tanaka, M., Muramatsu, M. and Otomo, S. (1994). Determination of nitrite by HPLC with EC: Measurement of nitric oxide synthase activity inrat cerebellum cytosol, *J. Biomed. Chromatogr.* **8**, 24–28.

Kaplan, S. L. (1989). Recent advances in bacterial meningitis, *Adv. Pediatr. Infect. Dis.* **4**, 83–110.

Knight, J. A. (1995). Diseases related to oxygen-derived free radicals, *Annals Clin. Lab. Sci.* **25** (2), 111–121.

Kochhar, A., Zivin, J. A., Lyden, P. D. and Mazzarella, V. (1988). Glutamate antagonist theraphy reduces neurologic deficit produced by focal nervous sytem ischemia, *Arch. Neurol.* **45**, 148–153.

Korczyn, A. and Nussbaum, M. (2002). Emerging therapies in pharmacological treatment of Parkinson's disease, *Drugs* **62**, 775–786.

Kontos, H. A. (1989). Oxygen radicals in CNS damage, *Chem. Biol. Interact.* **72**, 229–236.

Koroshetz, W. J., Jenkins, B. and Rosen, B. (1994). Evidence from metabolic disorder in Huntington's disease, *Neurology* **44** (Suppl. 2), A338.

Kulkami, S. K. and Naidu, P. S. (2003). Pathophysiology and drug therapy of tardive dyskinesia; Current concept and future perspective, *Drugs Today (Barcelona)* **39** (1), 19–49.

Link, H. and Tibbling, G. (1977). Principles of albumin and IgG analyses in neurological disorders, III. Evaluation of IgG synthesis within the central nervous system in multiple sclerosis, *Scand. J. Clin. Lab. Invest.* **37**, 297–301.

Lipton, S. A. (1992). Models of neuronal injury in AIDS: another role for the NMDA receptor? *Trends Neurosci.* **15**, 75–79.

Lipton, S. A. and Kater, S. B. (1989). Neurotransmitter regulation of neuronal outgrowth, plasticity and survival, *Trends Neurosci.* **12**, 265–270.

Lipton, J. D. and Schafermeyer, R. W. (1993). Evolving concepts in pediatric bacterial meningitis: Part1-pathophysiology and diagnosis, *Ann. Emerg. Med.* **22**, 1602–1615.

Lipton, S. A. and Rosenberg, P. A. (1994). Excitatory amino acids as a final common pathway for neurologic disorders, *New Engl. J. Med.* **330**, 613–622.

Lofton-Cazal, M., Pietri, S., Culcasi, M. and Bockaert, J. (1993). NMDA-dependent superoxide production and neurotoxicity, *Nature* 364, 535–537.

Lundberg, J. M. and Hökfelt, T. (1983). Coexistence of peptides and classical transmitters, *TIN* **6**, 325–333.

Mannisto, P. T. and Kaakkola, S. (1999). Catechol-O-methyltransferase (COMT): Biochemistry, molecular biology, pharmacology and clinical efficacy of new selective COMT inhibitors, *Pharmacol. Rev.* **51**, 503–628.

McCord, J. M. (1985). Oxygen-derived free radicals in post-ischemic tissue injury, *New Engl. J. Med.* **321**, 159–163.

McMaster, O. G., Du, F., French, E. D. and Schwarcz, R. (1991). Focal injection of aminooxyacetic acid produces seizures and lesions in rat hippocampus: Evidence for mediation by NMDA receptors, *Exp. Neurol.* **113**, 378–385.

Meldrum, B. and Garthwaite, J. (1990). Excitatory amino acid neurotoxicity and neurodegeneration disease, *Trends in Pharmacological Science* **11**, 79–86.

Monaghan, D. R., Bridges, R. J. and Cotman, C. W. (1989). The excitatory amino acid receptors: Their classes, pharmacology and distinct properties in the function of central nervous system, *Ann. Rev. Pharmacol. Toxicol.* **29**, 365–402.

Montague, P. R., Cancayco, C., Winn, M. J., Marchase, R. B. and Friedlander, M. J. (1994). Role of NO production in NMDA receptor-mediated neurotransmitter release in cerebral cortex, *Science* **263**, 973–977.

Mouradian, M. M., Juncos, J. L., Fabbrini, G., Schlegel, J. and Chase, T. N. (1988). Motor fluctuation in Parkinson's disease: central pathophysiological mechanisms, Part II, *Ann. Neurol.* **24**, 372–378.

Murphy, T. H., Miyamoto, M., Saste, A., Schnaar, R. L. and Coyle, J. T. (1989). Glutamate toxicity in neuronal cell line involves inhibition of cystine transport leading to oxidative stress, *Neuron* **2**, 1547–1548.

Nakanishi, S. (1992). Molecular diversity of glutamate receptors and implications for brain function, *Science* **258**, 597–664.

Nutt, J. G., Woodward, W. R., Hammerstad, J. P., Carter, J. H. and Anderson, J. L. (1984). The on-off phenomenon in Parkinson's disease: relation to levodopa absorption and transport, *New Engl. J. Med.* **310**, 483–488.

Obeso, J. A., Grandas, F., Herrero, M. T. and Horowski, R. (1994). The role of pulsatile versus continuous dopamine receptor stimulation for functional recovery in Parkinson's disease, *Eur. J. Neurosci.* **6**, 889–897.

Olney, J. W. (1978). Neurotoxicity of excitatory amino acids, in: *Kainic Acid as a Tool in Neurobiology*, McGeer, E. G., Olney, J. W. and McGeer, P. L. (Eds), pp. 37–69. New York, Raven Press.

Olney, J. W. (1990). Excitatory amino acids and neuropsychiatric disorders, *Annu. Rev. Pharmacol. Toxicol.* **30**, 47–71.

Olney, J. W., Misra, C. H. and deGubareff, T. (1975). Cysteine-S-sulfate: Brain damaging metabolite in sulfite oxidase defiency, *J. Neuropathol. Exp. Neurol.* **34**, 167–176.

Pincus, J. H. and Barry, K. (1988). Protein distribution diet restores motor function in patients with dopa-resistant off periods, *Neurology* **38**, 481–483.

Plaitakis, A. and Caroscio, J. T. (1987). Abnormal glutamate metabolism in amyotropic lateral sclerosis, *Ann. Neurol.* **22**, 575–579.

Plaitakis, A., Berl, S. and Yahr, M. D. (1984). Neurobiological disorders associated with deficiency of glutamate dehydrogenase, *Ann. Neurol.* **7**, 297–303.

Qureshi, G. A. and Baig, S. M. (1988). Quantitation of amino acids in biological samples by HPLC, *J. Chromatogr.* **459**, 237–244.

Qureshi, G. A. and Baig, S. (1993). Role of neurotransmitter amino acids in multiple sclerosis in exacerbation, remission, and progressive course, *Biogenic. Amines* **10**, 39–48.

Qureshi, G. A. and Baig, S. M. (1998). The relationship between the deficiency of vitamin B12 and neurotoxicity of homocysteine with nitrite level in cerebrospinal fluid of neurological patients, *Biogenic Amines* **14** (1).

Qureshi, G. A., Bednar, I., Min, Q., Södersten, P., Silberring, J., Nyberg, F. and Thörnwall, M. (1993). Quantitation and identification of two cholecystokinin peptides, CCK-4 and CCK-8s in rat brain by HPLC and fast atom bombardment mass spectroscopy, *Biomed. Chromatogr.* **7**, 251–255.

Qureshi, G. A., Baig, S., Bednar, I., Söderersten, P., Forsberg, G. and Siden (1995). Increased cerebrospinal fluid concentration of nitrite in Parkinson's disease, *Neuro Report* **6**, 1642–1644.

Qureshi, G. A., Baig, S. M. and Parvez, S. H. (1996a). Biochemical markers in tuberculosis meningitis, *Biogenic Amines* **12** (6), 499–519.

Qureshi, G. A., Halawa, A. and Baig, S. (1996b). Multiple sclerosis and neurotransmission, *Biogenic Amines* **12** (5), 353–376.

Qureshi, G. A., Halawa, A., Baig, S. M., Bednar, I. and Parvez S. H. (1997). The neurochemical markers in cerebrospinal fluid to differentiate between viral and bacterial meningitis, *J. Neurochem.* **32**, 197–203.

Qureshi, G. A., Baig, S. M. and Parvez, S. H. (1998a). Neuroxicity and the role of aspartic acid, glutamic acid and GABA in neurologic disorders, *Biogenic Amines* **13** (6), 537–546.

Qureshi, G. A., Ansari, A. F., Halawa, A., Minami, M. and Parvez, S. H. (1998b). Cholecystokinins and their coexistence with catecholamines in CSF of patients with some neurological disorders, *Biogenic Amines* **14** (2), 101–116.

Qureshi, G. A., Baig, S. and Parvez, S. H. (1998c). Neuropeptide Y, noradrenaline and their existence in CSF of patients with ischemic brain stroke, *Biogenic Amines* **14** (6), 615–624.

Qureshi, G. A., Baig, S. and Parvez, S. H. (2002a). Cerebrospinal fluid levels of neuropeptides in neurologic patients, *Biogenic Amines* **17** (2), 81–90.

Qureshi, G. A., Baig, S. and Parvez, S. H. (2002b). Coexistence of SP with catecholamines in CSF of healthy and patients with various neurologic disorders, *Biogenic Amine* **17** (2), 91–100.

Rascol, O., Payous, P., Ory, F., Ferreira, J. J., Berefel-Courbon, C. and Montastrue, J. L (2003). Limitation of current Parkinson's disease therapy, *Annals of Neurology* **53** (3), S3–S-15.

Rinne, U. K. (1983). Problems associated with long-term levodopa treatment of Parkinson's disease, *Acta Neurolog. Scand. Suppl.* **95**, 19–26.

Rice-Evans, C. A. and Diplock, A. T. (1993). Current status of antioxidant therapy, *Free Rad. Biol. Med.* **15**, 77–96.

Ross, S. M., Seelig, M. and Spencer, P. S. (1987). Specific antagonism of excitotoxic action of uncommon amino acids assayed in organotypic mouse cortical cultures, *Brain Res.* **425**, 120–127.

Rothman, S. M. (1984). Synaptic release of excitatory amino acids mediates anoxic neuronal death, *J. Neurosci.* **4**, 1884–1891.

Rothman, S. M. and Olney, J. W. (1986). Glutamate and pathophysiology of hypoxic-ischemic brain damage, *Ann. Neurol.* **19**, 105–111.

Rothman, S. M. and Olney, J. W. (1987). Excitotoxicity and the NMDA receptor, *Trends Neurosci.* **10**, 299–303.

Rothman, S. M and Olney, J. W. (1995). Excitatory and NMDA receptors-still lethal after eight years, *Trends Neurosci.* **18**, 57–58.

Saez-Llorens, X., Ramilo, O. and Mustafa, M. M. (1990). Molecular Phathophysiology of bacterial meningitis: Current concepts and therapeutic implications, *J. Pediatr.* **116**, 671–684.

Schwarcz, R., Scholz, D. and Coyle, J. T. (1978). Structure-activity relations for the neurotoxicity of kainic acid derivatives and glutamate analogues, *Neuropharmacology* **17**, 145–151.

Schwarcz, R., Foster, A. C., French, E. D., Whetsell, W. O., Jr. and Köhler, C. (1984). Excitotoxic models for neurodegenerative disorders, *Life Sci.* **35**, 19–32.

Shulman, L. M., Taback, R. L., Rabinstein, A. A. and Weiner, W. J. (2002). Non-recognition of depression and other non-motor symptoms in Parkinson's disease, *Parkinsonism Relat. Disord.* **8**, 193–197.

Sies, H. (1985). *Oxidative Stress*. Academic Press, London.

Sies, H. (1991). *Oxidative Stress: Oxidants and Antioxidants*. Academic Press, London.

Siesjö R. (1967). A new method for the cytological examination of the cerebrospinal fluid, *J. Neurol. Neurosurg. Psych.* **30**, 568–577.

Siesjö, B. K. and Bengtsson, F. (1989). Calcium fluxes, calcium antagonists and calcium related pathology in brain ischemia, hypoglycemia and spreading depression, a unifying hypothesis, *Cereb. Blood Flow Metab.* **9**, 127–140.

Smith, C. D., Carney, J. M. and Strake-Reed, P. E. (1991). Excess brain protein oxidation and enzyme dysfunction in normal aging and Alzeheimer's disease, *Proc. Natl. Acad. Sci. USA* **88**, 10540–10543.

Snyder, S. H. and Bredt, D. S. (1992). Biological roles of nitric oxide, *Science* **299**, 68–71.

Sonsall, P. K., Niklas, W. J. and Heikkila, R. E. (1989). Role of excitatory amino acids in metamphetamine-induced nigrostriatal dopaminergic toxicity, *Science* **243**, 398–400.

Stollhoff, K. and Schulte, F. J. (1987). Vitamin B12 and brain development, *Eur. J. Paediatr.* **146**, 201–205.

Talan, D. A., Guterman, J. J. and Overturf, G. D. (1988). Influence of granulocytes of brain edema, intracranial pressure and CSF concentrations of lactate and protein in experimental meningitis, *J. Infect. Dis.* **157**, 456–459.

Thiessen, B., Rajput, A. H., Loverty, W. and Desai, H. (1990). Age, environments and the number of substantia nigra neurons, *Adv. Neurol.* **53**, 201–206.

Tymiaski, M., Charlton, M. P. and Carlen, P. L. (1993). Source specificity of early calcium neurotoxicity in cultured embryonic spinal neurons, *J. Neurosci.* **13**, 2085–2104.

Wade, D. N., Mearrick, P. T. and Morris, J. L. (1973). Active transport of L-dopa in intestine, *Nature* **242**, 463–465.

Walker, J. E. (1983). Glutamate, GABA and CNS diseases, *Neurochem. Res.* **8**, 521–550.

Wernek, A. L. and Alvarenga, H. (1999). Genetics, drugs and environmental factors in Parkinson's Disease: A conrol study, *Arq. Neurosiquiatr.* **57**, 347–355.

Wester, P., Bergström, U. and Eriksson, A. (1990). Venticular cerebrospinal fluid monoamine transmitter and metabolite concentrations reflect human brain neurochemistry in autopsy cases, *J. Neurochem.* **54**, 1148–1156.

Whetsell, W. O. (1996). Current concept of excitotoxicity, *J. Neuropath. Expt. Neurology* **55**, 1–13.

Yu, B. P. (1994). Cellular defences against damage from reactive oxygen species, *Physiol. Rev.* **74**, 139–162.

Yusa, T., Crapo, J. D. and Freeman, B. A. (1984). Lipase mediated augmentation of brain SOD and catalase inhibits CNS oxygen toxicity, *J. Physiol.* **57**, 1674–1681.

Zar, J. H. (1984). *Biostatistical Analysis*. Prentice Hall, Englewood Cliffs.

Zeevalk, G. D. and Nicklas, W. J. (1991). Mechanism underlying initiation of excitatoxicity associated with metabolic inhibition, *J. Pharmacol. Exp. Ther.* **257**, 870–878.

Zhang, J., Dawson, V. L., Dawson, T. M. and Snyder, S. H. (1994). Nitric oxide activation of poly (ADP-ribose) synthetase in neurotoxicity, *Science*, 687–689.

Advances in Neuroregulation and Neuroprotection (2005), pp. 127-163
Collin, C. *et al.* (Eds)
© VSP 2005

Modulation of dopamine release in the striatum: physiology, pharmacology and pathology

V. LEVIEL *

*CNRS, Physiologie Integrative Cellulaire et Moléculaire, UMR 5123, Bat Rene Dubois,
8 rue Dubois, Campus La Doua, 43 Boulevard du 11 novembre, 69622 Villeurbanne, France*

Abstract—In response to the depolarization of the neuronal membrane, dopamine is released into synapses and binds to postsynaptic specific receptors. It is likely that some molecules can also diffuse out of the synapse, and reach distant targets. Alongside this process, active transport also occurs, outside the synapses, and involving a carrier protein able to act bi-directionally. These two releasing processes are concerned with the intraterminal pools of dopamine located or not in vesicles. From this complexity of storing and releasing mechanisms, it emerges that regulation of dopamine release involves the intraterminal amine metabolism. Indeed, various pharmacological treatments and physiological situations seem to modulate basal or evoked processes of release by alterations of dopamine synthesis, storage or catabolism before the action of the natural triggers, the ion gradients. Pathological situations, including the Parkinson's syndrome for instance, could result from such tonic modifications of dopamine metabolism in the terminals. In the present review, the variety of dopamine releasing agents or experimental situations is considered for their involvement in the amine metabolism.

Keywords: Dopamine synthesis; dopamine metabolism; dopamine transport; dopamine release; physiological regulation; Parkinson's disease; experimental models.

INTRODUCTION

From the beginning of the 1960s, dopamine (DA) has been identified as a neurotransmitter in the brain. This implied that DA could be released from cellular structures to be recognized at the level of specific receptors on homologous or heterologous cells. This relation between release and linkage substantiates the neurotransmission in specialized structures called the synapses. During the last decades, technological developments have shown that neurotransmission is not limited to

---

*E-mail: Vincent.Leviel@univ-lyon1.fr

synapses and that DA, like the majority of neurotransmitters, is engaged in the whole extracellular space.

According to this new concept, DA produced inside the synapses could migrate outside and produce both local and distant effects. In addition, a unique mechanism of release has not been characterized, and it is not strictly localized in synapses. The monoamine can be produced on axons but outside of the synapses, and even in the somatic or dendritic cellular region.

The presence of multiple mechanisms of release, diffusion and binding make it difficult to understand the regulatory processes occurring in the extracellular space, because extracellular space is not a homogeneous medium but a medium of several diffusing substances. It is a crucial challenge in physiological and pathological research to clearly identify these mechanisms that can produce various, simultaneous and even antagonistic effects.

The present article will focus on the mechanisms of DA release that are responsible for the alterations in the extracellular DA concentration. They will be presented and characterized first. Natural regulatory processes and the main disposable pharmacological tools will be reviewed. Finally, the consequences of experimental situations constituting biochemical models of Parkinson's disease will be presented.

TWO MAIN PROCESSES RELEASING DA

This section will be devoted to a rapid description of the two main mechanisms releasing DA in the extracellular space. Particular attention will be focused on reverse transport, the more recently characterized process, still under debate.

Exocytosis

There is general agreement in the literature about one of the mechanisms used by neurons to transfer information. It involves the association of specialized structures in both the emitting and the receiving neurons. This complex is constituted of dozens of proteins with specific functions and is called the synapse. Neurotransmittory molecules are synthesized in the terminal buttons or are picked up in the environmental medium of the presynaptic neuron. They are stored in cytoplasmic vesicles whose membrane can merge with the presynaptic membrane excreting their content in a narrow synaptic space (some hundreds of Ångstroms). Then, neurotransmitter molecules bind to specific receptors located on the postsynaptic membrane.

The successive steps of the exocytose have been the subject of many studies over the two last decades. The identification and the function of the proteins constituting the synaptic complex in both the pre- and post-neuron are still under study. In the presynaptic part of the synapse, vesicles are imbedded in a network of actin and microtubules that prevent contact with the cytosolic membrane. Cohesion is maintained by a dephosphorylated form of synapsin Ib. When this enzyme

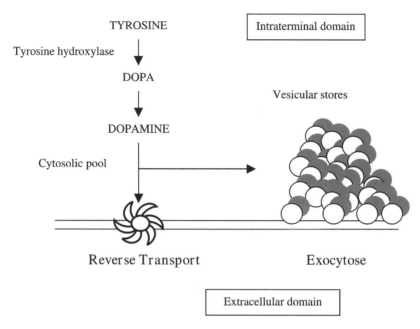

Figure 1. Two main mechanisms are responsible for the release of DA in the extracellular space. Exocytose involves the vesicularly stored amine and the reverse transport involves the cytosolic amine.

is phosphorylated following membrane depolarization and calcium entry, this cytosqueleton breaks up and the vesicles can reach the presynaptic membrane (Petrucci and Morrow, 1987). Due to the presence of various proteins that are also sensitive to the presence of calcium, vesicles are anchored in the presynaptic membrane (Greengard *et al.*, 1993). This linkage could be due to the binding of the synaptotagmine (on the vesicular membrane) with the α-latrotoxin receptor, or a family of proteins called neurexines (on the cytosolic membrane). Various other proteins could be involved in this presynaptic process, such as the synaptobrevin on the vesicular membrane or synaptophysine, a transmembrane enzyme (Bennet *et al.*, 1993). Also, proteins of the Rab type are involved in the mechanisms addressing the vesicles to the membrane.

Two protein complexes are successively aggregated: the first between synaptotagmine, synaptobrevin, syntaxin and SNAP25 and the last associated with cytosolic membrane. The dissociation of synaptotagmin in the presence of calcium ions allows the binding of a soluble protein, the α-SNAP, on which the fusion protein NSF can be loaded, relapsing the first complex and allowing the fusion of the membranes. This step still requires the presence of a large amount of calcium ions. For some authors, these complexes are directly associated with the calcium channels located in the presynaptic membrane. For instance, syntaxin, a tetrameric enzyme, could be associated with the complex synaptobrevin/NSF/SNAP and thus constitute a second complex with a calcium channel, synaptotagmine and neurexine (O'Connor *et al.*, 1993).

The Rab3 protein is located into the cytoplasm, bound to GDP and GDI (an inhibitor of the GDP hydrolysis). The dissociation of this complex allows Rab3 to bind to vesicular membrane through rabphilin-3α and GTP (Zerial and Stenmark, 1993). GTP is thus hydrolysed by GAP (Pfeffer, 1992). This could constitute the addressing mechanism. As mentioned, most of the processes constituting this mechanism are activated by large amounts of calcium ions. Thus exocytosis is obviously triggered by the massive calcium entry induced by the spike induced membrane depolarization.

Reverse transport (DA-RT)

Neurochemical observations. Paton (1973a, b) proposed a carrier mediated process of the monoamine efflux. The ability of NE, metaraminol and tyramine to increase pre-loaded [^{3}H]NE efflux was consistent with an exchange-diffusion model (Stein, 1967) and in line with previously described observations. Secondly, ouabain addition or K^+ omission in the superfusing medium increased [^{3}H]efflux likely by inhibiting the coupled Na^+-K^+ pump. This phenomenon, inhibited by cocaine and desipramine (inhibitors of NE uptake carrier) was sensitive to temperature. Thus the mechanism of the NE efflux appears to be compatible with the Na^+ gradient model, a hypothesis proposed by Bogdanski and Brodie (1969).

More recently, the mechanisms responsible for NE and γ-aminobutyric acid (GABA) release were analyzed by Levi *et al.* (1976) and Raiteri *et al.* (1977) with similar conclusions concerning the two neurotransmitters. Raiteri *et al.* (1979) also observed, as others, that nomifensine (a DA carrier inhibitor) had no effect on the spontaneous release of [^{3}H]DA from rat striatal synaptosomes (Hunt *et al.*, 1974; Schacht *et al.*, 1977). However, it prevented the increased release of [^{3}H]DA produced by the superfusion with a Na^+ free medium. As in the earlier work of Paton (1973a), inhibition of Na^+-K^+ ATPase with ouabain and superfusion with free K^+ also produced and increased [^{3}H]DA release that was blocked by nomifensine. A major point of these works was the inability of carrier inhibition to reduce the spontaneous release of [^{3}H]DA from synaptosomes. This suggests that this mode of release is involved in a transient process and not in a simple continuous leakage. It was initially proposed that depolarization-induced DA release could be independent of the DA transport (Raiteri *et al.*, 1979); however, the recent model described by Wheeler *et al.* (1993) is consistent with the participation of the transporter in depolarization induced release.

Direct observation of the reverse transport in vivo. Using two different methods simultaneously to evaluate DA release, the two mechanisms of release can be differentiated (Olivier *et al.*, 1995). A push-pull canula was implanted in the striatum of anesthetized rats. Basal extracellular DOPAC and DA were assayed in the superfusate using HPLC and electrochemical detection. In order to detect simultaneously the firing-evoked DA release, a carbon fiber electrode was implanted in close proximity to the canula. The DA and DOPAC were detected by differential pulse amperometry (DPA) allowing a measurement every second. The firing-evoked DA released was obtained by electrical stimulation of the DA fibers passing through the medial forebrain bundle in the region of the lateral hypothalamus. Limited periods of stimulation (20 seconds) were imposed, consisting of a train of 20 pulses (0.5 ms, 40 Hz). The push-pull superfusion was chosen in preference to microdialysis since it allows both inflow of exogenous drugs and the measurement of released substances, simultaneously. This experimental set-up (Fig. 2) allowed continuous measurement of the extracellular amount of DA (spontaneous release) and the response to an increase of the firing (evoked release).

Each 20 s stimulation train of the medial forebrain bundle resulted in a large increase in the amperometric signal in the ipsilateral caudate nucleus. The signal returned to baseline immediately after cessation of the stimulation. The amplitude was constant over time but correlated with the duration and the frequency of the stimulation as described previously (Gonon and Buda, 1985). The DPA is unable to differentiate between DA and DOPAC since the oxidative potentials are too close for these two substances. However, it can be ascertained that, in these experimental conditions, the signal observed is due to DA oxidation rather than to DOPAC oxidation. This is evidenced by an increase of the signal in the presence of pargyline, a substance inhibiting DOPAC formation. This could be due to the higher sensitivity of this method for DA oxidation than for DOPAC.

The conditions of stimulation we used produced an increase in the DPA signal but remained ineffective in increasing the amount of DA detected by push–pull superfusion. This is not surprising since the total time of the stimulation constituted only 0.06% of the total superfusion time. An evoked increase of the amine remained undetectable, as the experimental time was too short.

To block the voltage dependent calcium channels, 100 μmol cadmium were added in the superfusing fluid (Fig. 3). The stimulation-evoked release of DA disappeared rapidly after the beginning of the application. The same effect resulted from cobalt application. The uptake-blocker GBR12909 was without effect on these inhibitions. In contrast, the amount of DA collected with the push-pull canula was largely increased for at least 40 minutes. In the presence of GBR12909 no further increase could be detected in the extracellular space. These observations confirmed the dependence of the evoked DA release on calcium ions. In contrast, they also showed that blocking the calcium channel induced a dramatic increase of the DA release involving the carrier protein responsible for the uptake.

The glutamate-agonists NMDA and kainate produced the same type of effect (Fig. 4). A large reduction in the evoked DA release was associated with a large increase in the amount of DA in the extracellular space. This increase was also sensitive to the presence of GBR12909 showing that these effects of the glutamate also involve the transport protein. The blockade of the firing-evoked DA released was suspected to result from an autoregulation by local DA on the D2 receptors, but sulpiride (D2 antagonist) remained without effect on these alterations.

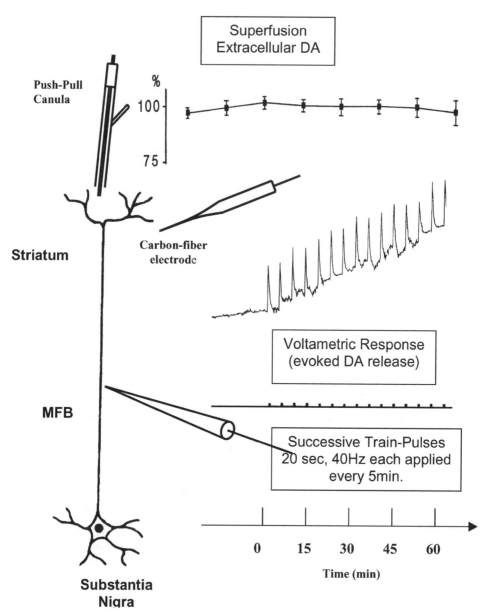

Figure 2. Experimental set-up used to visualize simultaneously the two releasing processes *in vivo*. Each train of pulses (20 s, 40 Hz) applied to DA axons at the level of the medial forebrain bundle (lateral hypothalamic) results in a DA release in the ipsilateral striatum strictly limited at the stimulation period (medial panel). The amount of DA released during the train pulses (every 5 min) is too low to modify extracellular DA evaluated every 20 min (top panel).

Amphetamine was added during 20 min to the superfusing fluid and during this treatment (1 μmol) the amount of DA collected was enhanced (250%) and the evoked DA release was also increased (200%). It is not surprising that amphetamine

Cadmium

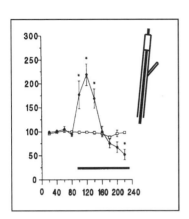
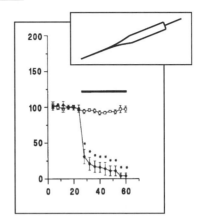

Figure 3. Local application of cadmium ions in the striatum on the basal release of DA (left panel) measured by push-pull superfusion. The evoked DA release is measured by DPA and presented in the right panel.

NMDA

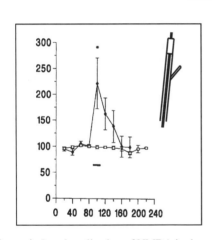
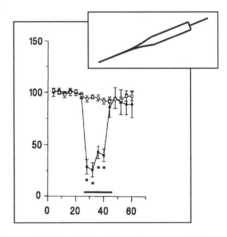

Figure 4. Local application of NMDA in the striatum. The basal DA release is measured after push-pull superfusion and is expressed in percent of spontaneous release (left panel). The evoked DA release is reduced by the treatment as visualized by DPA.

appears as enhancing the two modes of DA release. However, the only mechanism increased *per se* is likely to be the DA-RT. Indeed, exocytosis is probably insensitive to amphetamine but this substance exhibits uptake-blocker properties by a direct competition with DA for transport. Thus the amount of DA uptake between the pulses constituting the stimulation is reduced in the presence of amphetamine. The consequence is a virtual increase of the efficacy of the firing evoked release. These observations were the first direct *in vivo* demonstration of the duality of the DA releasing mechanism. In addition they showed the probable antagonistic nature of these two processes. This is not surprising when we consider their biochemical mechanisms.

The mechanism of the reverse transport.   Often evoked by pharmacological agents, DA-RT comprises many steps. First, the releasing molecules enter the terminal buttons. Second, many factors combine to produce a rise in the concentration of neurotransmitters in the terminal cytosol. Third, transmitter molecules are transported outside the cell through an exchange-diffusion process, which has been extensively studied and various mathematical models have been proposed over the course of the last three decades (Chubb *et al.*, 1972; Thoenen *et al.*, 1969; Ziance *et al.*, 1972; see Stein, 1990 for a review).

The first step.   DA-RT is often evoked in response to a releaser agent but not exclusively.   Many psychotropic substances, known to induce DA release, act through this process. Amphetamine, for example, has been shown to accumulate in the DA-terminals.   This accumulation is inhibited by DA uptake inhibitors (Arnold *et al.*, 1977; Azzaro *et al.*, 1974).   Various other sympathomimetic amines (tyramine, parahydroxyamphetamine, etc.) are substrates for uptake into neuron terminals (Horn, 1973; Ross and Renyi, 1964). A large set of substances competing with DA for uptake remain without effect on the release. Benztropine, nomifensine, methylphenidate only exhibit uptake blocker function (Bonnet *et al.*, 1984; Hunt *et al.*, 1974; Miller and Shore, 1982).   Some of these uptake-blocker substances are also blocking agents of the amphetamine effect (Arbuthnott *et al.*, 1990). Synaptosomal sequestration of [^{3}H]methylenedioxyamphetamine and [^{3}H]amphetamine was inhibited by permeant cations (Na$^+$, K$^+$), is saturable and temperature sensitive (Zaczek *et al.*, 1990). The accumulation of these releasing agents is thus likely submitted to an active transport but that remains a question of debate.

The second step.   The intraterminal mechanism leading to the amine release is now better understood.   The DA transport protein is located on the cytoplasmic membrane; thus DA efflux should derive from a presumably cytoplasmic compartment. The cytoplasmic compartment is maintained by the DA synthesis, replenished by the displacement from the vesicular stores and limited by the catabolic enzyme, the mono-amino-oxidase (MAO) that produces the dihydroxyphenyl acetic acid (DOPAC). DOPAC is generally considered as passively diffusing through the cytoplasmic membrane. The natural uptake of extracellular DA also supplies the cytosolic DA compartment.

Various physiological or pharmacological situations lead to changes in the level of the intraterminal DA pools.

1. Reducing DA catabolism.   Amphetamine enhances the cytoplasmic amine compartment by reducing the action of the MAO (Miller and Shore, 1982). The IMAO effect of amphetamine is only one of its pharmacological actions.

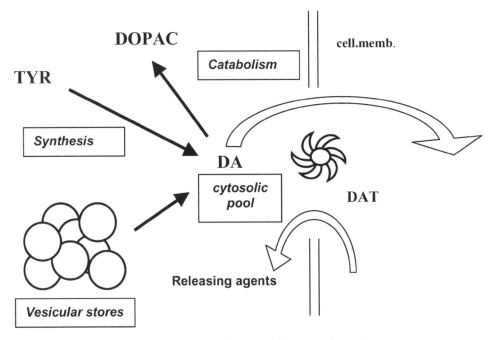

Figure 5. The various steps of DA metabolism on which most of the releasing agents are acting. Some of them enter the terminal using the transporter. Displacing vesicular DA *or* increasing DA synthesis *or* inhibiting DA catabolism leads to the presence of a significative cytosolic DA pool. This pool becomes a substrate for the reverse transport.

2. Displacing DA from intraterminal stores. Amphetamine displaces DA from intraterminal stores and its action is not synergistic with a K^+ application or an electrical stimulation (Kamal *et al.*,1981; Langer and Arbilla, 1984). Kamal *et al.* (1983) proposed an intraterminal transfer of DA by amphetamine from an electrically releasable pool (probably the vesicular one) to a compartment only releasable by amphetamine and related substances (probably the cytoplasmic one). Displacement from vesicular stores may involve the vesicular transport mechanism: Fairbrother *et al.* (1990) reported a reduction of the tyramine effect by reserpine, a substance inhibiting the vesicular DA transport. A *weak base model* could explain the vesicular displacement of DA. Indeed, a vesicular pH gradient is required for monoamine accumulation (Henry *et al.*, 1998; Johnson, 1988). Weak bases could abolish the intracellular pH gradient causing a huge increase of cytosolic DA and giving rise to the DA overflow (Schuldiner *et al.*, 1993). This mechanism was evoked to explain the releasing action of amphetamine, 3,4-methylendioxy-methamphetamine ('ectasy'), and fenfluramine on serotonin (5HT) (Rudnick and Wall, 1992).

3. Increasing DA synthesis. Increased DA synthesis could increase intraterminal pool of DA leading to increase amine molecules disposable for DA-RT (see below).

The exchange-diffusion transfer. Fischer and Cho (1979) proposed a model taking place at the binding site of the uptake carrier and involving cytoplasmic DA. They observed that this exchange is temperature dependent, saturable, Na^+-dependent, stereoselective, and cocaine-sensitive. Raiteri *et al.* (1979), tested amphetamine, octopamine and β-phenylethylamine as DA-releasing substances and observed the same blockade with another uptake inhibitor, nomifensine. Using *in vivo* or *in vitro* experiments as well, many studies conducted during the last 15 years have confirmed the role played by the protein responsible for DA uptake (DAT) in the releasing mechanism (Butcher *et al.*, 1988). Thus, Nash and Brodkin (1991) reported that methylene-dioxymethamphetamine released DA through a carrier mediated process blocked by mazindol and GBR12909 (two uptake blockers) and also activated by 5HT but insensitive to ketanserin (5HT2 antagonist). The finding by Giros *et al.* (1996) that mice lacking the DA carrier protein by disruption of the transporter gene are insensitive to amphetamine, thus unable to increase extracellular DA, is the most recent and direct demonstration of the involvement of DAT in the amphetamine-releasing effect.

Thus, it may be said that general agreement exists about the fact that amphetamine-like substances release DA, at least partly, by inducing DA-RT from a cytosolic DA compartment, even if the molecular mechanisms involved are not completely understood.

Why two mechanisms of release? The physiological significance of the presence at the terminal level of several mechanisms of release remains an opened question that has been considered in the literature (Bjelke *et al.*, 1996; Grace, 1991; Langer and Arbilla, 1984; Parker and Cubeddu, 1986a, b; Vizi and Labos, 1991). The extracellular DA does not simply constitute a feed-back mechanism by which DA terminals regulate the DA release, even if this mechanism also occurs. It is likely that these two mechanisms constitute two separated processes with their own physiological significance. The fast synaptic release of the amine is involved in connectionist transmission of the signals. The slow DA release constitutes a modulation of homologous and heterologous neurons.

These two releasing functions could be brought about by different types of dopaminergic neurons. The degeneration process occurring during Parkinson's disease will be considered in evaluating such a hypothesis. It is admitted that during the presymptomatic phases of the disease, a metabolic compensation is realized by the spared neurons. Such compensation should be more efficacious if the degenerating neurons are only weakly involved in the connectionist neural transmission. In this hypothesis, the syndrome should only appear when the remaining cells become unable to compensate for the lesion or when a second class of neurons directly involved in the motor integration begin to degenerate. The existence of various types of neurons with different morphology was often reported (Gerfen *et al.*, 1987). It should be of interest to define if each type of DA-releasing mechanism described here could be relevant to just one particular type of dopaminergic neuron.

PHYSIOLOGICAL CONTROLS OF THE RELEASE

As described in the previous section, extracellular DA is regulated by at least two main process of release. Until now, diffusion has been considered as a passive phenomenon. The re-uptake of the extracellular molecule by the carrier protein is closely associated with the mechanism leading to the release. Some arguments suggest that the regulation of the DA-RT could be different from the regulation of the uptake, but this question will not be debated here. In this section, the physiological controlling processes will be considered.

The intraterminal origin of the amine released

As stated previously, DA cells can release their neurotransmitter through two different mechanisms, one depending directly on calcium entry and triggered by the membrane depolarization and the other using the cytosolic pool. The presence of two pools of DA in the DA terminal of the striatum has been known for at least three decades (see Glowinski, 1973, for a review). It has emerged from the original studies (Besson *et al.*, 1969, 1973) that the intraterminal DA is located mainly in the vesicles (about 70–80%) with a half-life of two hours. The cytoplasmic compartment constituted by the neosynthesized amine (the tyrosine hydroxylase, TH, being a soluble enzyme) presents a short half-life not exceeding 15 or 20 minutes.

From these metabolic differences it has been found that a local superfusion of the DA terminal with the radioactive precursor of the DA ([^{3}H]tyrosine) leads to a highly labeled cytoplasmic DA and to poorly labeled vesicles stores. This principle

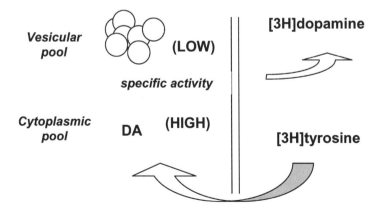

Figure 6. A highly specific activity labels the DA originating from the cytoplasmic compartment with fast turnover rate. A low specific activity characterizes the previously stored amines.

was used in some of the next studies we describe to identify the intraterminal origin of the released DA. The method consists in evaluating the specific activity of the DA released. An example is given in Fig. 9. In that case, the specific activity of DA was decreased when DA release was evoked by electrical stimulations of the DA axons. On the contrary, in the presence of glutamate the specific activity of DA increased. Thus, electrical depolarization evoked preferentially the previously stored amine and GLU enhanced DA release by an increase of the cytosolic pool (Leviel *et al.*, 1991).

Synthesis

The role of synthesis in the processes of release was pinpointed in several studies. Indeed, the amount of DA released by the spiking activity is obviously smaller than the amount of DA involved in the DA-RT. The synapse can be considered as a two-dimensional space and the extracellular medium a three-dimensional space. Thus non-synaptic diffusion probably requires a greater production of molecules. This implies that synthesis is more probably involved in the reverse transport than in the synaptic neurotransmission.

Regulation of synthesis by calcium ions. The push-pull superfusion technique is based on a liquid/liquid exchange at the tip of the canula and allows a stable and reproducible labeling of the tissues with the tritiated precursors ([³H]tyrosine). That cannot be obtained with a dialysis probe. The present results were obtained

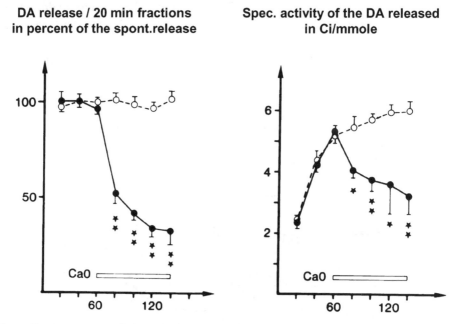

Figure 7. The local superfusion with 0 calcium induces a sharp decrease of the DA released and a decrease of its specific activity. This is the index of a reduced synthesis.

mainly by using the push-pull superfusion technique. Alterations in synthesis were revealed by the dynamic variations of [³H]DA and [³H]DOPAC efflux. The specific activity of DA can be considered as an index of the DA synthesis (Herdon *et al.*, 1985) since the newly synthesized DA has a higher specific activity than the stored amine (Leviel *et al.*, 1989). The total amount of DOPAC in extracellular space being in the range of the micromolar, it cannot be directly related to DA release (in the nanomolar range). This could be due to the various origins of this substance. However, [³H]DOPAC efflux can be considered as a good index of the amine synthesis, since, according to many authors, extracellular DOPAC comes from the metabolism of an unreleased and recently synthesized pool of dopamine (Herdon *et al.*, 1985; Leviel *et al.*, 1989; Soares-Da-Silva, 1987; Soares-Da-Silva and Garrett, 1990; Zetterström *et al.*, 1988).

Removing the calcium ions from the superfusing fluid induced a partial decrease in the extracellular DA that reached 30% of the basal values. It can be argued from this observation that under these conditions the DA synthesis was largely impaired. Indeed, it has been demonstrated some time ago (Glowinski, 1973) that the DA, spontaneously released, is originating from an intraterminal compartment constituted by newly synthesized molecules (in our case with a high specific activity). The intraterminal compartment originating the DA released was probably different, since [³H]DA release and [³H]DOPAC efflux were reduced indicating

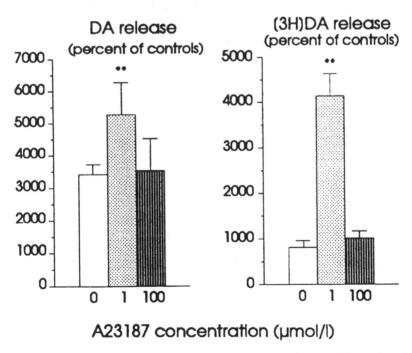

Figure 8. The calcium ionophore A23187 applied at low concentrations amplifies preferentially the action of amphetamine on the [³H]DA release (right panel) rather than on the total DA (left panel). This is the index of an increased synthesis by the forced entry of calcium ions.

the involvement of the intraterminally stored amine with a low specific activity. Thus, even to maintain a low rate of DA release, the stored amine had to be involved. Secondly, the addition of a TH inhibitor, α-mpt, did not further decrease the extracellular DA or DA specific activity also suggesting that the activity of TH was already maximally inhibited by Ca^{2+} removal.

Adding cadmium ions (calcium channel blocker) in the superfusing fluid produced rapid increase of the DA release (threefold that of the spontaneous level) followed (120 min later) by a dramatic decrease in the DA release and a drastic reduction in the [^{3}H]DA. The [^{3}H]DOPAC and the total DOPAC were also largely reduced. The delayed effect of the Cd^{2+} treatment (120 min) is in agreement with the well known half-life of the storage compartment of DA in the striatum (Javoy and Glowinski, 1971). Thus, reducing the entry of extracellular calcium favoured in a first step the DA-RT, but simultaneously impaired the DA synthesis.

As a corollary, a forced Ca^{2+} entry could have the opposite effect. Indeed a low dose of Ca^{2+} ionophore A23187 (1 μmol) slightly increased basal [^{3}H]DOPAC but dramatically potentiated the amphetamine-induced [^{3}H]DA release. The effect particularly concerned the newly synthesized DA showing that the synthesis was mainly affected. On control rats, 1 mmol amphetamine produced a 34-fold increase in extracellular DA and 5.6-fold increase in [^{3}H]DA. One hour after superfusion with A23187, the amphetamine effect was still more pronounced (53-fold increase of DA and 37-fold increase for [^{3}H]DA). This increased synthesis was not affecting the spontaneous DA-release but was able to potentiate DA-RT when it was evoked.

The mechanism of the synthesis activation is complex since cytoplasmic DA potently inhibits TH activity; thus an accumulation of the cytoplasmic DA in terminals, could only be rendered possible by a simultaneous reduction in the feedback inhibition of TH by DA. Such a reduction could have occurred as a consequence of the calcium entry. Indeed, changes in the inhibitory constant (K_i) for DA was reported to be well correlated with the phosphorylation states of the TH protein that are clearly dependent on the calcium ions (Pasqualini et al., 1994, 1995). Table 1 illustrates the changes in TH activity that appear to be directly correlated with the increasing phosphorylation state of the enzyme. The DA-inhibitory constant (K_i DA) is the better index of these phosphorylation states. Indeed the end product inhibition is known to be inversely proportional to the level of phosphorylation of the protein.

How the synthesis regulates the release.

1. Amphetamine. In 1975, Chieuh and Moore considered that the sustained release of DA by amphetamine appears to be dependent on an activated synthesis. In the same year, Kuczenski (1975), observed an increased rate of DA synthesis after amphetamine treatment. For these authors this increase was mainly a consequence of the release induced by the drug. However, for Uretsky and Snodgrass (1977), the synthesis induced by amphetamine was not only considered to be compensatory of the release.

Table 1.
Activation of TH enzyme following increasing levels of phosphorylation

On tissue slices	Activity	K_i-DA
Controls	100	30
Oestradiol (10^{-11} M, 30 min)	60	—
Oestradiol + Ac. Okadaique	100	30
Prolactine (10 μg/ml, 2 h)	180	60
Prolactine + PKC inhibitor (GF109203x)	100	30
TPA (100 nM)	180	60
db cAMP(2 mM)	450	120
On tissue homogenate + triton X100		
Controls	100	30
cAMP (2 mM)	500	120

2. Glutamate. The case of glutamate is of particular interest since the releasing action of glutamate is likely to involve DA-RT (see below). Moghaddam and Bolinao (1994), proposed that the stimulatory effect of GLU on DA release may be due to an activation of DA synthesis rather than release. It was also proposed that the glutamatergic cortico-striatal pathway could exert a tonic inhibition of the DA synthesis in the striatum (Leviel *et al.*, 1990). This was deduced from the observation of an increased efflux of [^{3}H]DA during cortical xylocaine application. It should be noted however that this treatment did not significantly affect the basal DA release, which led to the suggestion of a dual potency of glutamate, and a tonic inhibition on the DA synthesis (direct or indirect) associated with a phasic ability to induce release (direct). This was later confirmed by others (Castro *et al.*, 1996; Desce *et al.*, 1992; Fillenz, 1993).

Can increased synthesis induce increased extracellular DA through the activation of the DA-RT? Alterations in DA synthesis were generally expected to occur as a consequence of the DA release, not to originate a release. By the use of simultaneous monitoring of both neosynthesized and endogenous DA release it is possible to differentiate alterations of the synthesis from alterations of the release. Neosynthesized amine was evaluated by measuring the neoformed [^{3}H]DA during a continuous labeling of the tissues with [^{3}H]TYR. The total DA released was also measured using electrochemical detection after HPLC analysis. The ratio between these two parameters (the specific activity of the released amine) could give a continuous index of the activity of the DA synthesis *in vivo*. It was confirmed directly with this method that electrically evoked neuronal firing evoked the release of previously stored amine with a low specific activity (Leviel *et al.*, 1991). In contrast, local application of glutamate or amphetamine, also increasing the DA release, simultaneously increased the specific activity of DA. This showed a preferential release of neosynthesized amine. Other treatments previously considered to be able to increase DA release were clearly more efficient on the

DA Specific activity following different releasing situations

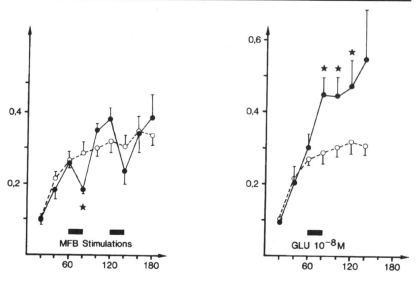

Figure 9. Direct electrical stimulations of the DA axons (left panel) produce a decrease of the DA specific activity. In contrast, local GLU application increases DA specific activity.

DA synthesis (Gobert *et al.*, 1992; Leviel *et al.*, 1989; Pasqualini *et al.*, 1995). It can thus be proposed that intraterminal DA concentration may be the target of presynaptic regulations in the caudate nucleus leading *in fine* to a modulation of the amount of DA releasable by DA-RT (Olivier *et al.*, 1999).

Humoral controls

Autoregulation.

Effects on the electrical activity of DA cells. Presynaptic control of DA release was shown some time ago (Farnebo and Hamberger, 1971). It was observed that DA agonists reduce release while an antagonist favours it. Various mechanisms have been proposed to explain this observation. Electrophysiological studies have shown that DA-cell activity is well regulated through activation of DA receptors located on dendrites and/or cell bodies (Aghajanian and Bunney, 1973; Groves *et al.*, 1981). In turn, activation of these autoreceptors led to a reduction of the DA release in the striatum by a reduction of the spontaneous activity. The presence of DA receptors on the DA terminals in the striatum was clearly established. Their direct role on the DA release was also generally accepted (Cubeddu and Hoffman, 1982; Gonon and Buda, 1985). The receptors involved are of the D2 type. The presence of D1 autoreceptors in the striatum is still a subject of debate. However, local application of D1 agonist was shown to modulate DA release (Diana *et al.*, 1991; Imperato and di Chiara, 1988). The mechanisms involved in this regulation have not been completely clarified. It seems that cAMP is not implicated and this regulation interferes with only the calcium-dependent DA release, while the amphetamine effect is not affected by DA receptor agonists and antagonists.

Effects on the reverse transport. Substrate concentration affects the potency of a carrier. Recently, Zahniser *et al.* (1999) reported that the transport velocity of DA carrier and thus the DA clearance is regulated by the extracellular DA concentration. This phenomenon, called facilitated transfer, was described a while ago by Stein (1990). This may explain the non-linearity of the transport when the amount of substrate increases. The DA carrier is not regulated in the same way as a receptor. It is not subjected to up-regulation and supersensitization by treatments with antagonists (Kula and Baldessarini, 1991). The DA receptor of D2 type (DA-D2) mediates a decrease in the rate of DA uptake (Cass and Gerhardt, 1994; Meiergerd *et al.*, 1993; Parsons *et al.*, 1993). For others, the uptake kinetic parameters remained unchanged by the presence of metoclopramide and sulpiride (May and Wightman, 1989a) or were increased after chronic deprenyl (Weiner *et al.*, 1989). Using genetically engineered mice devoid of DA-D2, Dickinson *et al.* (1999) observed a reduction in DA clearance not related to a reduction in DAT expression. This is consistent with the idea that DA-D2 could modulate the activity of the DA carrier.

The case of DA-RT is more complex: some authors reported that the releasing action of amphetamine is modified by the presence of DA-D2 agonists or antagonists (Wieczorek and Kruk, 1994), while others have reported the opposite (Kamal *et al.*, 1981; Kuczenski *et al.*, 1990). Vizi reported that idazoxan a pure $\alpha2$ blocking agent, failed to enhance DA release evoked by 1-phenylephrine which was proved to be Ca^{++} independent (Vizi *et al.*, 1986). The question of a DA-D2 regulated DA-RT and uptake has not been completely resolved yet.

Serotonin. A close relationship between 5HT and DA neurotransmission was often reported. An increased extracellular level of 5HT in the striatum led to an increased DA. The 5HT reuptake blockers, including alaproclate (Yadid *et al.*, 1994), fluoxetine (Sills *et al.*, 1999) and fenfluramine (De Deuwaerdere *et al.*, 1995), all enhance extracellular DA. There is no clear consensus about the mechanism linking 5HT and DA releases. The 5HT receptor was evoked in mediating an action of 5HT on the DA release (Benloucif *et al.*, 1993; Bonhomme *et al.*, 1995). The effect of 5HT on basal DA overflow was not altered by the serotonin antagonist, methysergide, or the serotonin re-uptake blocker, chlorimipramine, but was reversed by the DA re-uptake carrier inhibitors nomifensine and benztropine. In addition, 5HT does not modulate K^{+}-induced release of DA in the nucleus accumbens or striatum (Nurse *et al.*, 1988). Using phenylbiguanide (PBG), a 5HT type-3 receptor agonist, it was observed that the PBG-evoked DA outflow from striatal slices can be inhibited by nomifensine, is insensitive to reserpine and does not require the presence of Ca^{++} ions in the superfusing medium (Nurse *et al.*, 1988; Schmidt and Black, 1989). These authors thus concluded that PBG could promote the efflux of DA by a carrier dependent mechanism. The possibility that 5HT enhances DA efflux by a process of exchange diffusion was also proposed by Yi *et al.* (1991) but they considered that 5HT could act as amphetamine or tyramine by entering the DA terminals. Jacocks and Cox (1992) and Crespi *et al.* (1997) confirmed this latter.

Glutamate. DA uptake velocity was found to be increased by the glutamate agonist NMDA (Welch and Justice, 1996) and this effect was antagonized by AP5 (NMDA antagonist). The same activation was obtained with kainic acid. These effects were proposed to be receptor-mediated, direct and independent from action potentials. NMDA however also causes arachidonic and nitric oxide release which inhibit DA uptake in striatal synaptosomes (Lonart and Johnson, 1994; Pogun *et al.*, 1994; L'Hirondel *et al.*, 1995). This could explain the reduction of DA uptake by NMDA also reported by Lin and Chai (1998). Finally, NMDA facilitates an inward calcium movement that in turn could activate the PKc enzyme and it was observed that PKc activators decrease the DA transport in the COS cells (Kitayama *et al.*, 1994).

Considering that the corticostriatal pathway is likely to be glutamatergic in nature (at least in part), the relationship between an afferent glutamate path to the striatum and the local DA release is of primary interest and has been extensively studied. DA release can be elicited in the striatum by glutamate or glutamate agonist and this effect is clearly independent of DA neuron firing. Grace reviewed the role of glutamate in DA release in 1991. It is likely that glutamate could interfere with DA mainly through NMDA receptors located on the DA terminals even if kainate and/or quisqualate receptors were also evoked. The way in which glutamate increases extracellular DA could be a reduction of the DA uptake.

Recently some observations showed that simultaneously glutamate may promote DA-RT. Extensive studies were devoted to the regulation of DA release by excitatory amino-acids and particularly glutamate (Robert and Sharif, 1978; Clow and Jhamandas, 1989). Early observations revealed that this effect of Glutamate was insensitive to TTX and thus unrelated to the depolarization-induced firing. This was recently confirmed by Keefe *et al.* (1993) who observed that NMDA applied directly in the striatum of unanesthetized rats produced a significant elevation in extracellular DA during the blockade of impulse activity in DA neurons with TTX. For these authors, the glutamate does not tonically regulate the extracellular concentration of DA in resting conditions (Keefe *et al.*, 1992).

Lonart and Zigmond (1991), suggested that one component of the action of glutamate on DA release could involve its uptake into DA terminals, as proposed for GABA by Bonanno and Raiteri (1987). It seems likely that Na^+ co-transported with glutamate acts to depolarize the terminal (McMahon *et al.*, 1989) promoting the DA-RT. The DA release under these conditions is both Ca^{++} independent and blocked by nomifensine (Lonart and Zigmond, 1990). Olivier *et al.* (1995) studied the dual effect of glutamate through NMDA and Kainate receptors. It appeared from these experiments that NMDA and kainate induce a blockade of the ability of

DA terminal to release DA by a firing dependent process. In contrast, extracellular DA was enhanced by the presence of these two substances, an effect blocked by systemic injection of GBR12909. These results were substantially confirmed one year later with a different method by Iravani and Kruk (1996).

Ascorbic acid. This endogenous substance was often considered as a modulator of the DA uptake (Berman *et al.*, 1996; Debler *et al.*, 1988; Sershen *et al.*, 1987). In a study of the DA-RT on mouse striatal synaptosomes it was revealed that external ascorbic acid enhanced the efflux of preloaded [^{3}H]DA (Debler *et al.*, 1991). Since ascorbic acid does not appear to be a substrate for the carrier it may be more likely a modulator of the carrier translocation, promoting the conformation of the binding site oriented toward the side with the lowest ascorbic acid concentration (Debler *et al.*, 1991).

Oestradiol. It as been known for a long time, that steroid hormones affect the DA system (Barber *et al.*, 1976; Ramirez *et al.*, 1985). Oestradiol was reported to affect receptors coupled to adenylate cyclase (Maus *et al.*, 1990) and the DA synthesis (Pasqualini *et al.*, 1995). Recently an inhibition of the DA transport was also reported (Disshon *et al.*, 1998). The action of oestradiol on DA-RT remains thus questionable but this substance is a good candidate as DA-RT modulator.

PHARMACOLOGICAL CONTROLS OF THE RELEASE

This section reviews the exogenous substances altering extracellular DA. Some of them are used as therapeutic agents. The understanding of their mode of action is a key step in discovering the natural mechanism regulating extracellular DA.

Ethanol

From the various pharmacological situations interfering with DA transporter, the case of ethanol is interesting. The effect of ethanol on the DA transmission remains controversial (Lynch *et al.*, 1983; Snape and Engel, 1988). Using *in vivo* voltammetry, Lin and Chai (1995) observed that local pressure ejection of ethanol in the rat striatum did not elicit significant changes in spontaneous DA release. However, this treatment augmented the time course of NMDA-evoked DA release, as did nomifensine. These results have been confirmed by others (Brown *et al.*, 1992; Woodward and Gonzales, 1990) as the fact that the rate of DA clearance was reduced after ethanol treatment (Daoust *et al.*, 1986; Smolen *et al.*, 1984). These data demonstrate that ethanol simultaneously inhibits NMDA-evoked DA uptake, suggesting a role for ethanol on the translocation of the carrier. Recently, using *in vivo* voltammetry on urethane anesthetized rats, it was observed that local ethanol application in the striatum reduced the potassium-evoked DA overflow, when, in contrast, the same treatment remained ineffective on the tyramine-induced

DA overflow (Wang et al., 1997). The reduction in the effect of potassium for these authors is relevant of an activated DA clearance through a facilitated DA uptake by ethanol. These controversial observations show however that the uptake function of the DA transport may be regulated differently from the releasing function.

Amphetamine

It was proposed from a long time that amphetamine and related substances increase extracellular DA through multiple mechanisms, including DA synthesis, release and re-uptake. To better characterize these mechanisms, a push-pull canula was implanted in the striatum of anesthetized rats. The push-pull superfusion was chosen in preference to the microdialysis since it allows both the inflow of exogenous drugs and the measurement of released substances. The superfusion was realized with an artificial cerebrospinal fluid containing the tritiated precursor of DA the [^{3}H]tyrosine. Total DA and dihydroxyphenyl acetic acid (DOPAC) were measured in successive 20-min fractions using HPLC and electrochemical detection. Radioisotopic counting of the peaks permitted the calculation of the specific activity of both DA and DOPAC released into the extracellular space. This method is based on the fact that DA is synthesized in the cytoplasm and constitutes a pool of newly synthesized molecules with a short (15 min) half life. Due to this rapid turnover rate, the specific activity of this compartment reaches half of the specific activity of the precursor. In contrast, the vesicular compartment is replenished from cytosolic amine with a long turnover rate (2 hours). In consequence, during the first hours of superfusion, the newly synthesized DA pool presents a high specific activity and the vesicular compartment a low specific activity. This property was used to characterize the origin of the DA released, an increase of the specific activity being the index of the involvement of the neosynthesized molecules, and a decrease of the specific activity being the index of the previously stored molecules.

When applied locally during a 20-min period, 1 μmol amphetamine produced an increased DA release. As a matter of fact, the increase in total DA was only transitory in contrast with the neosynthesized amine whose the level remained elevated for hours (Leviel and Guibert, 1987). The amplitudes of the two effects were similar. The specific activity of the amine released was also elevated as the tritiated (neosynthesized) DOPAC. The same discrepancy was noticeable between the kinetics of both forms of DOPAC. The total DOPAC increase was only transitory when the tritiated DOPAC was enhanced for hours.

When high concentrations were used (1 mM), amphetamine increases extracellular DA by 3000%! The neosynthesized amine was unable to follow such an increase, and the specific activity was largely reduced. The DOPAC production is interrupted by amphetamine (IMAO effect), and the neosynthesized DOPAC (tritiated) falls down rapidly after the drug application.

These observations show the different roles of the two pools of terminal DA in the release. The vesicular amine, constituting 80% of the total amine content, is only displaced by large amounts of amphetamine (200% vs 3000% for 1 μM and 1 mM,

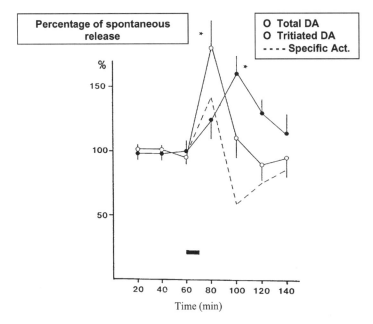

Figure 10. The two-steps of amphetamine action. An increase of the release of the cytosolic amine (high specific activity) followed by the displacement from the vesicles of the previously stored molecules (low specific activity).

respectively). The mechanism activated by amphetamine is also an intraterminal displacement of the stored amine toward cytosolic DA. In the normal situation, the cytosolic concentration of DA is reduced and low doses of amphetamine mainly affect this compartment.

Methylphenydate

After 1 μM methylphenidate (ritalin), the two forms of DA are increased (total and neosynthesized). Even in the presence of a 230% increase of the DA release, the specific activity remains constant. No alterations in the DOPAC production were pin-pointed. High dose methylphenidate (1 mM) dramatically enhances the amplitude of the effect. This produces a large reduction of the specific activity of DA. Surprisingly an increase of the total DOPAC results from the ritalin application simultaneously with a reduction in the neosynthesized DOPAC. This observation was interpreted as an argument in favour of an intraterminal release of DA, inducing both an increased DOPAC production (ritalin being not an IMAO) and an inhibition of the synthesis (due to the high DA concentration).

Potassium

Various concentrations of potassium have been tested on the DA release. A relative dose response effect was observed: the higher the dose of potassium, the higher is the increase of DA. The specific activity is reduced during the time of application of the potassium suggesting that the amine released is mainly coming from a direct exocytose of the vesicular DA (with a low specific activity). Simultaneously, the potassium-evoked depolarization of the DA terminals enhances the DOPAC production, ascertaining an increased synthesis (Leviel *et al.*, 1989).

GAP

Gonadotropin associated peptide (GAP) applied at a concentration of 1 μM produces a stable increase of the extracellular neosynthesized DA (Gobert *et al.*, 1992). The absence of effect of GAP on the total DA was also observed and this discrepancy induces an increase of the specific activity of DA. An enhancement of the tritiated (neosynthesized) DOPAC was also noticeable. The opposite effect is observed when GAP was replaced by an immunoserum directed against GAP: a reduced neosynthesized DA and a reduced DOPAC production. It has been concluded that GAP could activate DA synthesis and indirectly favor DA release. The same observation was done more recently with 17-B-oestradiol (Pasqualini *et al.*, 1995).

Reserpine

Reserpine is a blocking agent of the vesicular uptake of DA. When reserpine is injected into rats (10 mg/kg), a rapid reduction of DA release (total and neosynthesized) and a reduction of DOPAC production was observed. The simultaneous decrease of the DA synthesis and DA release could be due to the inhibitory action of the cytosolic DA. Indeed, DA potently inhibits the tyrosine hydroxylase enzyme by an end-product retro-control. The blockade of DA uptake in vesicles makes every DA molecule become a TH inhibitor.

Nomifensine

Nomifensine represents a typical uptake blocker agent. The extracellular DA was increased under its two states (total and neosynthesized). No modifications of the specific activity of DA or DOPAC can be observed. This is because of the relatively pure extracellular action of this substance.

From these experimental observations, a functional model of the DA terminal can be drawn (Fig. 11, Leviel *et al.*, 1989). The cytoplasmic DA synthesized from the tyrosine can be used as a substrate for DA-RT. This constitutes a significant increase in the extracellular amine, when the intraterminal concentration is elevated, following activated synthesis, reduced catabolism or displacement of the vesicular amine. A part of this cytosolic amine is vesicularized in a presynaptic vesicle immediately disposable for exocytoses but also in storage vesicles. This last compartment releases the DA inside the terminal under the influence of the releasing agents entering the vesicles or disrupting the Ph gradient.

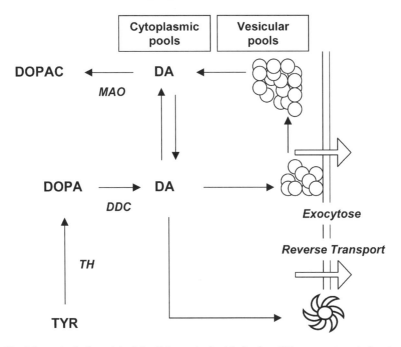

Figure 11. A hypothetical model of the DA terminal with the four DA compartments (see text).

PATHOLOGICAL CONTROLS: PARKINSON' DISEASE

Parkinson's disease constitutes a natural pathological situation that illustrates the plasticity of the neural system. In spite of a progressive degeneration of a population of DA cells (constituting the striatal DA denervation), no symptom can be detected until the damage reaches 60 to 80% of the initial population. The metabolic compensation of the DA-cells death has thus been the target of many studies during the last four decades. Indeed it is a crucial challenge to understand how DA cells re-organize their metabolism with a view to maintaining an apparently normal DA function. This section will review two studies carried out on the rat and showing that the reduction in the number of DA cells in the substantia nigra led to a specific increase of the DA-RT, involving the DA synthesis and the glutamate afference to the DA terminals.

Partial lesions increase extracellular DA

A toxin specific to DA cells, 6-hydroxydopamine (6OHDA), was injected in the restricted lateral region of the rat substantia nigra. Three weeks later, tissue DA, DOPAC and homovanillic acid (HVA) were measured using LCED. The extracellular DA and DOPAC were measured by electrochemical detection. The amount of TH enzyme was quantified by immuno-autoradiography, and the DA synthesis evaluated by measuring the DOPA/DA ratio after a metabolic blockade of TH. Finally, a binding study of the GBR12935 (ligand of the DA transport protein) was used as an index of the DA denervation in the terminal region. The toxin injection was restricted to the lateral part of the substantia nigra and in the terminal region, comparison was done between the lateral and the medial part of the striatum.

As expected, a lateral lesion in the substantia nigra corresponded to a lateral denervation in the striatum. The tissue DA, DOPAC and HVA were reduced by 80%, as the TH amount and the GBR12935 binding. In contrast, TH activity was unchanged and extracellular DA was slightly increased (130%). The TH activity and extracellular DA amount was apparently unchanged in spite of the important DA-cells denervation. Furthermore, the important reduction of the DA storage pin-pointed a large increase of the DA turnover in this region. This is in agreement with many reports about severe DA denervation of the CP (Agid et al., 1973; Altar et al., 1987; Hefti et al., 1980, 1985).

In the neighbouring region of the striatum (medial part), the DA terminals were clearly spared. In this region DA, DOPAC and HVA, TH immuno-autoradiography and GBR12935 binding were only reduced by about 20% indicating a very weak DA denervation. However, using in vivo voltammetry, extracellular basal DA levels were found to be greatly elevated when compared to unoperated animals (up to 235%). In the median regions, TH activity was also significantly increased (161%). These results show an over-compensation in the spared DA neurons. This could be due to an increased efficacy of the firing/release on DA terminal. However, the direct electrical stimulation of the DA fibers produced the same DA overflow when comparing control and lesioned animals. These results indicate that elevated basal DA levels in this region cannot be attributed to the reduced DA uptake and/or an increased ability of DA neurons to release DA in response to impulse flow.

The nature of the mechanism of release, tonically enhanced in the medial region of the striatum, needs to be questioned, particularly if we consider that the denervation in this region did not exceed 20%.

The enhanced extracellular DA is unlikely to be due to a tonic increase in the electrical activity of the DA neurons. Indeed, more than 80% denervation is needed to modify the firing activity of DA neurons after nigro-striatal partial lesion (Hollerman and Grace, 1990).

The high level of extracellular DA could also be the result of an enhanced amount of amine released per pulse and produced by the spontaneous activity of the DA neurons. In contrast with this idea, GBR12909 reduced the evoked DA overflow when compared to the same stimulation on controls. This effect of GBR12909 was already reported by others (Horne et al., 1992; Wang et al., 1994) and suggests a reduction of the DA release per pulse on the lesioned animals, as was already proposed (Garris et al., 1997). Furthermore, in the absence of GBR12909, the amount of the amine released by electrical stimulation remained unchanged on the lesioned animals. Thus, on these animals, a reduced uptake of the extracellular amine should occur as often proposed (Earl et al., 1998; Gerhardt et al., 1996; Horne et al., 1992) but only compensating the reduced release. Extracellular DA could be responsible for the reduced release per pulse through an action on the

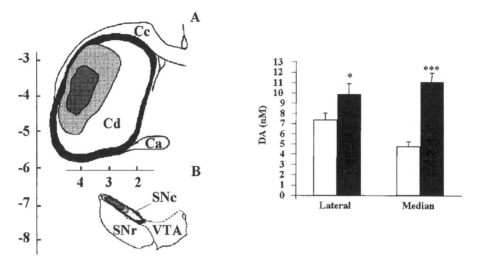

Figure 12. A lesion in the lateral part of the substantia nigra results in a lateral lesion in the striatum. In the extracellular space, the DA concentration is slightly increased. In the median spared region the extracellular DA is greatly enhanced.

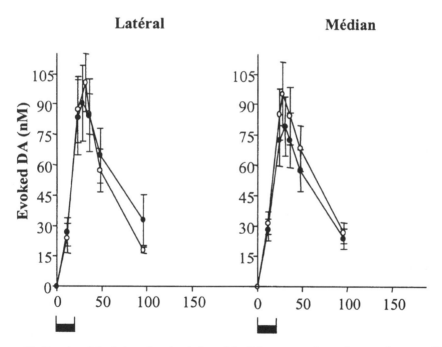

Figure 13. In spite of the lesion, the stimulation of the DA axons produces the same increase of DA release in the lateral and the medial region of the striatum.

preterminal D2 receptors (May and Wightman, 1989b). Thus the firing evoked DA release is unlikely to be responsible for the sustained increase of the extracellular DA we observed after the weak DA denervation.

The enhanced extracellular DA release is more likely to be related to a modified DA metabolism. Several authors have reported that in 6-OHDA lesioned rat the DA release is more sensitive to synthesis inhibition than in controls (Heffner *et al.*, 1977; Marshall, 1979; Snyder *et al.*, 1990). This is in line with a reduced amine storage in the DA terminals and the observed increased TH activity, largely confirmed in the literature (Hefti *et al.*, 1985; Stackowiack *et al.*, 1987; Snyder *et al.*, 1990).

The increased extracellular basal DA level could thus be related to another mechanism of the classical firing-induced DA release. RT of DA by the uptake DA carrier could be proposed as mediating this effect. Indeed its high sensitivity to extracellular glutamate has been reported (Kitayama *et al.*, 1994; Welch and Justice, 1996) and an activated glutamatergic neurotransmission seems to take place in striatum after partial lesions. It was recently reported that daily treatment with glutamate receptor antagonists is able to block the neurochemical compensation of a partial 6-OHDA lesion (Emmi *et al.*, 1996). Also, a local application of NMDA or kainate in the CP increased levels of extracellular DA simultaneously with reduced ability of DA terminals to release DA in response to MFB stimulation (Olivier *et al.*, 1995). Finally, glutamate was observed to enhance TH enzyme activity in various models (Castro *et al.*, 1996; Desce *et al.*, 1993; Leviel *et al.*, 1991). The result that will be described in the following paragraph will sustain the hypothesis that the partial lesion of DA pathway enhances through glutamate neurotransmission and the cortico-striatal path the RT of DA in the striatum.

Partial lesion activates reverse transport

An acute lesion and the observation of further reorganizations is often used to investigate the consequences of nigral neurodegeneration. However, Parkinsonian degeneration is not actually of this type. Particularly, the progressivity should be considered. For this reason, a progressive denervation of the rat striatum was achieved by repeated (weekly) intraventricular injections of the toxin 6-OHDA. The extent of the lesion was monitored after each of the toxin injections by measuring tissue DA, DOPAC and TH protein. After reaching approximately 40% of denervation, the mechanism of transport was investigated on preparations of synaptosomes. It was expected that following this protocol a greater similarity with the natural process of reorganization could be obtained.

Reduction of the tissue DA and DOPAC was detectable immediately after beginning the injections. The DOPAC/DA ratio, reflecting TH protein activity (Lavielle *et al.*, 1978), and the tissue amount of TH protein remained the same during the first six weeks. This reduction of intraterminal DA stores not associated with changes in DA synthesis suggests that TH activity is under the control of cytoplasmic amine levels through an end-product regulatory process (Ames *et al.*, 1978; Fillenz, 1993; Mann and Gordon, 1979; Zigmond *et al.*, 1989). Therefore increased TH activity, often reported after partial lesion of the DA pathway (Zigmond *et al.*, 1990) could be, at least in part, a consequence of reduced intraterminal DA pools.

In these experiments, the quantification of TH immunoreactivity and GBR12935 binding suggested that about 40% of the DA terminals have degenerated after 6 injections of the toxin. However, on these animals, the global DA uptake was only reduced by 11% (insignificant decrease). In addition, the V_{max} an index of transport capacity, was only 12% of the value in control rats (reduced by 88%). This observation shows that a large number of sites, still linking [^{3}H]GBR12935, have become unable to carry DA. Thus the number of sites carrying DA was drastically reduced but the amount of molecules carried only reduced by 11%. Furthermore, the K_m value was also reduced showing a large increased affinity of the DA carrier protein for the amine.

These results show that a moderate (very partial) DA denervation of the CP complex results in a drastic reduction in the number functional DAT able to carry DA, but with an increase in their affinity, the results producing in turn an increased transport rate.

Which local phenomenon could be responsible for these paradoxical effects on the DA transport? Chronic treatments with MK801 (antagonist of the NMDA receptors) have already been reported as able to counteract consequences of 6-OHDA treatment.

An activated Glu neurotransmission in the striatum has been observed on animal models and Parkinsonian patients as well (Iwasaki *et al.*, 1992; Lindefors and

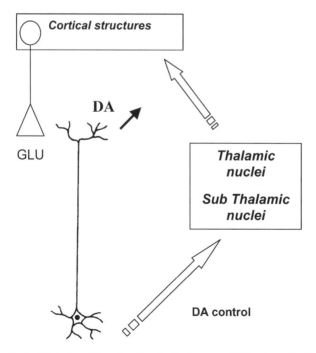

Figure 14. Hypothetical functional loop leading to the increased DA release in the striatum after a partial lesion of the substantia nigra. The reduced DA control of the thalamic nuclei could activate the indirect thalamocortico-striatal loop producing the increase of the DA reverse transport in the striatum.

Ungerstedt, 1990; Samuel *et al.*, 1990; Wullner *et al.*, 1994). It has been proposed that Glu could maintain a high level of DA release in CP by promoting burst firing of DA cells (Overton and Clark, 1992). The activation of DA neurotransmission by Glu is more likely to be a consequence of the altered DA transport affinity that we report here. Glu can both reduce the DA uptake (Lin and Chai, 1998) and/or favor DA-RT of amine (Leviel, 2001; Lonart and Zigmond, 1991). DA uptake velocity was reported to be increased by Glu (Welch and Justice, 1996) also known to activate DA release independent of the electrical activity of the nigral DA neurons (Keefe *et al.*, 1992; Olivier *et al.*, 1995). Finally Glu also activates DA synthesis in DA terminals of CP (Castro *et al.*, 1996; Desce *et al.*, 1992; Fillenz, 1993; Leviel *et al.*, 1990, 1991). The active transport of DA is well known to be bi-directional and responsible for the so-called DA-RT (Leviel, 2001). GLU, activating intraterminal DA synthesis, could also favor DA-RT involving the cytoplasmic DA compartment. An increased basal release could explain why extracellular DA concentration is maintained in CP complex after partial DA denervation.

CONCLUSIONS

To conclude this overview of the biochemical process regulating extracellular DA through the terminal release of the amine, the question can be raised as to the physiological nature of DA. Indeed, DA like other monoamines, is often considered as a neuromodulator more than a neurotransmitter. Its most often evoked function in striatum is a filtering of the afferent cortico-striatal (glutamatergic) path on the GABAergic neurons constituting the efferent system of striatum. The large extension of the DA terminal field when compared to the restricted number of originating cells in midbrain also supports this concept of a rather diffuse and general regulatory process. Finally, the late survey of clinical symptoms in Parkinson's disease when at least 40–50% of DA neurons have degenerated in the SN, could lead to minimize the role of the neuronal connections in the motor control realized by dopaminergic innervation of the striatum. The control of the extracellular DA concentration by DA cells metabolic activity are well in line with this concept of DA as a neuromodulator responsible for so called volume transmission.

However, the anatomical and electrophysiological approaches of the midbrain dopaminergic pathways innervating the striatum argue against this simplistic view. Indeed, the nigro striatal pathway appears to be rather strictly organized and a precise somatotopy of the neurons making this pathway is now established (Maurin *et al.*, 1999). In addition this anatomical organization seems to well correspond to the various functionalities recognized for the extrapyramidal system (Alexander and Crutcher, 1990). Electrophysiological analysis of the activity of DA neurons during behavioral test have also demonstrated that these neurons are activated by very specific situations (Mirenowicz and Schultz, 1996; Schultz, 1986). Finally, the importance of neuronal connections in the occurrence of motor symptoms in

Parkinson's could have been underestimated in humans since motor dysfunctions were reported to occur on MPTP chronically treated monkeys as early as in the first weeks of the treatment (Hantraye *et al.*, 1990). It could be that in humans non extrapyramidal compensation could occur.

From these considerations, DA modulation could operate in the following two ways in the striatum: a diffuse and long lasting action on the terminal environment, probably realized through the DA-RT, regulated by preterminal afferences, and a more targeted influence involving synaptic contacts.

REFERENCES

Aghajanian, G. K. and Bunney, B. S. (1973). Central dopaminergic neurons: neurophysiological identification and responses to drugs, in: *Frontiers in Catecholamine Research*, Usdine, E. and Snyder, S. H., pp. 643–663. Pergamon Press, New York.

Agid, Y., Javoy, F. and Glowinski, J. (1973). Hyperactivity of remaining dopaminergic neurons after partial destruction of the nigrostriatal dopaminergic system in the rat, *Nature* **245**, 150–151.

Alexander, G. E. and Crutcher, M. D. (1990). Functional architecture of basal ganglia circuits: neural substrates of parallel processing, *Trends in Neurosci.* **13**, 266–271.

Altar, C. A., Marien, M. R. and Marshall, J. F. (1987). Time course of adaptations in dopamine biosynthesis, metabolism, and release following nigrostriatal lesions: Implications for behavioral recovery from brain injury, *J. Neurochem.* **48**, 390–399.

Ames, M. M., Lerner, P. and Lovenberg, W. (1978). Tyrosine hydroxylase: activation by protein phosphorylation and end product inhibition, *J. Biol. Chem.* **253**, 27–31.

Arbuthnott, G. W., Fairbrother, I. S. and Butcher, S. P. (1990). Dopamine release and metabolism in the rat striatum: an analysis by 'in vivo' brain microdyalisis, *Pharmac. Ther.* **48**, 281–293.

Arnold, E. B., Molinoff, P. B. and Rutledge, C. O. (1977). The release of endogenous norepinephrine and dopamine from cerebral cortex by amphetamine, *J. Pharm. Exp. Ther.* **202**, 544–557.

Azzaro, A. J., Ziance, R. J. and Rutledge, C. O. (1974). The importance of neuronal uptake of amines for amphetamine-induced release of 3H-norepinephrine from isolated brain tissue, *J. Pharm. Exp. Ther.* **189**, 110–118.

Barber, P. V., Arnold, A. G. and Evans, G. (1976). Recurrent hormone-dependent chorea: effects of oestrogens and progesterone, *Clin. Endocrinol.* **5**, 291–293.

Benloucif, S., Keegan, M. J. and Galloway, M. P. (1993). Serotonin-facilitated dopamine release *in vivo*: Pharmacological characterization, *J. Pharm. Exp. Therap.* **265**, 373–377.

Bennet, M. K., Garcia-Arraras, J. E., Elferink, L. A., *et al.* (1993). The syntaxin family of vesicular transport receptors, *Cell* **74**, 863–873.

Berman, S. B., Zigmond, M. J. and Hastings, T. (1996). Modification of dopamine transporter function: Effect of reactive oxygen species and dopamine, *J. Neurochem.* **67**, 593–600.

Besson, M. J., Cheramy, A., Feltz, P. and Glowinski, J. (1969). Release of newly synthesized dopamine from dopamine-containing terminals in the striatum of the rat, *Proc. Natl. Acad. Sci. USA* **62**, 741–748.

Besson, M. J., Chéramy, A., Gauchy, C., *et al.* (1973). In vivo continuous estimation of 3H-dopamine synthesis and release in the cat caudate nucleus, *Naunyn Schmiedeberg's Arch. Pharmacol.* **278**, 101–105.

Bjelke, B., Goldstein, M., Tinner, B., *et al.* (1996). Dopaminergic transmission in the rat retina: evidence for volume transmission, *J. Chem. Neuroanat.* **12**, 37–50.

Bogdanski, D. F. and Brodie, B. B. (1969). The effects of inorganic ions on the storage and uptake of [3H]norepinephrine by rat heart slices, *J. Pharm. Exp. Ther.* **165**, 181–189.

Bonanno, G. and Raiteri, M. (1987). Coexistence of carriers for dopamine and GABA uptake on a same nerve terminal in the rat brain, *Br. J. Pharmacol.* **91**, 237–243.

Bonhomme, N., De Deurwaerdere, P., Le Moal, M., *et al.* (1995). Evidence for 5HT$_4$ receptor subtype involvement in the enhancement of striatal dopamine release induced by serotonin: a microdialysis study in the halothane-anesthetized rat, *Neuropharmacology* **34**, 269–279.

Bonnet, J. J., Lemasson, M. H. and Costentin, J. (1984). Simultaneous evaluation by a double labelling method of drug-induced uptake inhibition and release of dopamine in synaptosomal preparation of rat striatum, *Biochem. Pharmacol.* **33**, 2129–2135.

Brown, L. M., Trent, R. D., Jones, T. W., *et al.* (1992). Alcohol inhibition of NMDA-stimulated catecholamine efflux in aging brain, *Alcohol* **9**, 555–558.

Butcher, S. P., Fairbrother, I. S., Kelly, J. S., *et al.* (1988). Amphetamine induced dopamine release in the rat striatum: An in vivo microdialysis study, *J. Neurochem.* **50**, 346–355.

Cass, W. A. and Gerhardt, G. A. (1994). Direct in vivo evidence that D2 dopamine receptors can modulate dopamine uptake, *Neurosci. Lett.* **176**, 259–263.

Castro S. L., Sved, A. F. and Zigmond, M. J. (1996). Increased neostriatal tyrosine hydroxylation during stress: role of extracellular dopamine and excitatory amino acids, *J. Neurochem.* **66**, 824–833.

Chiueh, C. C. and Moore, K. E. (1975). d-amphetamine induced release of newly synthesized and stored dopamine from the caudate nucleus in vivo, *J. Pharm. Exp. Ther.* **192**, 642.

Chubb, I. W., De Porter, W. P. and De Schaepdryver, A. F. (1972). Tyramine does not release noradrenaline from splenic nerves by exocytosis, *Naunyn Schmiedeberg's Arch. Pharmacol.* **274**, 281–286.

Clow, D. W. and Jhamandas, K. (1989). Characterization of L-glutamate action on the release of endogenous dopamine from the rat caudate putamen, *J. Pharm. Exp. Ther.* **248**, 722–728.

Crespi, D., Mennini, T. and Gobbi, M. (1997). Carrier-dependent and Ca(2+)-dependent 5-HT and dopamine release induced by (+)amphetamine, 3,4-methylendioxymethamphetamine, p-chloroamphetamine and (+)-fenfluramine, *Br. J. Pharmacol.* **121**, 1735–1743.

Cubeddu, L. X. and Hoffman, I. S. (1982). Operational characteristics of the inhibitory feedback mechanism for regulation of dopamine release, via presynaptic receptors, *J. Pharmacol. Exp. Ther.* **223**, 497–501.

Daoust, M., Moore, N., Saligaut, C., *et al.* (1986). Striatal dopamine does not appear involved in the voluntary intake of ethanol by rats, *Alcohol* **3**, 15–17.

De Deurwaerdere, P., Bonhomme, N., Le Moal, M., *et al.* (1995). d-Fenfluramine increases striatal extracellular dopamine in vivo independently of serotoninergic terminals or dopamine or dopamine uptake sites, *J. Neurochem.* **65**, 1100–1108.

Debler, E. A., Hashim, A., Lajtha, A., *et al.* (1988). Ascorbic acid and striatal transport of [3H]1-methyl-4-phenylpyridine (MPP+) and [3H]dopamine, *Life Sci.* **42**, 2553–2559.

Debler, E. A., Sershen, H., Hashim, A., *et al.* (1991). Carrier mediated efflux of [3H]dopamine and [3H]1-methyl-4-phenylpyridine: effect of ascorbic acid, *Synapse* **7**, 99–105.

Desce, J. M., Godeheu, G., Galli, T., *et al.* (1992). L-Glutamate evoked release of dopamine from synaptosomes of the rat striatum: involvement of AMPA and N-methyl-D-aspartate receptors, *Neuroscience* **47**, 333–339.

Desce, J. M., Godeheu, G., Galli, T., *et al.* (1993). Opposite presynaptic regulations by glutamate through NMDA receptors of dopamine synthesis and release in rat striatal synaptosomes, *Brain Res.* **640**, 205–214.

Diana, M., Young, S. J. and Groves, P. M. (1991). Modulation of dopaminergic terminal excitability by D1 selective agents: further characterization, *Neuroscience* **42**, 441–449.

Dickinson, S. D., Sabeti, J., Larson, G. A., *et al.* (1999). Dopamine D2 receptor deficient mice exhibit decreased dopamine transporter function but no changes in dopamine release in dorsal striatum, *J. Neurochem.* **72**, 148–156.

Disshon, K. A., Boja, J. W. and Dluzen, D. E. (1998). Inhibition of striatal dopamine transporter activity by 17 β-estradiol, *Eur. J. Pharm.* **345**, 207–211.

Earl, C. D., Sautter, J., Xie, J., *et al.* (1998). Pharmacological characterization of dopamine overflow in the striatum of the normal and MPTP-treated common marmoset, studied in vivo using fast cyclic voltammetry, nomifensine and sulpiride, *J. Neurosci. Methods* **85**, 201–209.

Emmi, A., Rajabi, H. and Stewart, J. (1996). Behavioral and neurochemical recovery from partial 6-hydroxydopamine lesions of the substantia nigra is blocked by daily treatment with glutamate receptor antagonists MK-801 and CPP, *J. Neurosci.* **16**, 5216–5224.

Fairbrother, I. S., Arbuthnott, G. W., Kelly, J. S., *et al.* (1990). In vivo mechanisms underlying dopamine release from rat nigrostriatal terminals: II Studies using potassium and tyramine, *J. Neurochem.* **54**, 1844–1851.

Farnebo, L. O. and Hamberger, B. (1971). Drug-induced changes in the release of 3H-monoamines from field stimulated rat brainslices, *Acta Physiol. Scand. Suppl.* **371**, 35–44.

Fillenz, M. (1993). Short-term control of transmitter synthesis in central catecholaminergic neurones, *Prog. Biophys. Molec. Biol.* **60**, 29–46.

Fischer, J. F. and Cho, A. (1979). Chemical release of dopamine from striatal homogenate: evidence for an exchange diffusion model, *J. Pharm. Exp. Ther.* **208**, 203–209.

Garris, P. A., Walker, Q. D. and Wightman, R. M. (1997). Dopamine release and uptake rates both decrease in the partially denervated striatum in proportion to the loss of dopamine terminals, *Brain Res.* **753**, 225–234.

Gerfen, C., Herkenham, M. and Thibault, J. (1987). The neostriatal mosaïc: II Patch-and matrix-directed mesostriatal dopaminergic and non-dopaminergic systems, *J. Neurosci.* **7**, 3915–3934.

Gerhardt, G. A., Cass, W. A., Hudson, J., *et al.* (1996). In vivo electrochemical studies of dopamine overflow and clearance in the striatum of normal and MPTP-treated rhesus monkeys, *J. Neurochem.* **66**, 579–587.

Giros, B., Jaber, M., Jones, S. R., *et al.* (1996). Hyperlocomotion and indifference to cocaine and amphetamine in mice lacking the dopamine transporter, *Nature* **379**, 606–612.

Glowinski, J. (1973). Some characteristics of the functionnal and main storage compartments in central catecholaminergic neurons, *Brain Res.* **62**, 489–493.

Gobert, A., Guibert, B., Lenoir, V., *et al.* (1992). GnRH-associated peptide (GAP) is present in the rat striatum and affects the synthesis and release of dopamine, *J. Neurosci. Res.* **31**, 354–359.

Gonon, F. and Buda, M. J. (1985). Regulation of dopamine release by impulse flow and by autoreceptors as studied by in vivo voltammetry in the rat striatum, *Neuroscience* **14**, 765–774.

Grace, A. A. (1991). Phasic versus tonic dopamine release and the modulation of dopamine system responsivity: a hypothesis for the etiology of schizophrenia, *Neuroscience* **41**, 1–24.

Greengard, P., Valtorta, F., Czernik, A. J., *et al.* (1993). Synaptic vesicle phosphoproteins and regulation of synaptic function, *Science* **259**, 780–785.

Groves, P. M., Fenster, G. A., Tepper, J. M., *et al.* (1981). Changes in dopaminergic terminal excitability induced by amphetamine and haloperidol, *Brain Res.* **221**, 425–431.

Heffner, T. G., Zigmond, M. J. and Stricker, E. M. (1977). Effects of dopamine agonists and antagonists on feeding in intact and 6-hydroxydopamine treated rats, *J. Pharm. Exp. Therap.* **201**, 386–399.

Hefti, F., Melamed, E. and Wurtman, R. J. (1980). Partial lesion of the dopamine nigrostriatal system in rat brain: biochemical characterization, *Brain Res.* **195**, 123–137.

Hefti, F., Enz, A. and Melamed, E. (1985). Partial lesions of the nigrostriatal pathway in the rat. Acceleration of transmitter synthesis and release of surviving dopaminergic neurones by drugs, *Neuropharmacology* **24**, 19–23.

Henry, J. P., Sagné, C., Bedet, C., *et al.* (1998). The vesicular monoamine transporter: from chromaffin granule to brain, *Neurochem. Int.* **32**, 227–246.

Hentraye, P., Khalili-Varasteh, M., Peschanski, M., *et al.* (1990). MPTP: mécanismes biologiques et limites d'un modèle expérimental de la maladie de Parkinson, *Circul. et métab. du Cerveau* **7**, 15–19.

Herdon, H., Strupish, J. and Nahorski, S. R. (1985). Differences between the release of radiolabelled and endogenous dopamine from superfused rat brain slices: effects of depolarizing stimuli, amphetamine and synthesis inhibition, *Brain Res.* **348**, 309–320.

Hollerman, J. R. and Grace, A. A. (1990). The effects of dopamine-depleting brain lesions on the electrophysiological activity of rat substantia nigra dopamine neurons, *Brain Res.* **533**, 203–212.

Horn, A. S. (1973). Structure-activity relations for the inhibition of catecholamine uptake into synaptosomes from noradrenaline and dopaminergic neurones in rat brain homogenates, *Br. J. Pharmacol.* **47**, 332–338.

Horne, C., Hoffer, B. J., Strömberg, I., *et al.* (1992). Clearance and diffusion of locally applied dopamine in normal and 6-hydroxydopamine-lesioned rat striatum, *J. Pharmacol. Exp. Therap.* **263**, 1285–1291.

Hunt, P., Kennengiesser, M. H. and Raynaud, J. P. (1974). Nomifensine: A new potent inhibitor of dopamine uptake into synaptosomes from rat brain corpus striatum, *J. Pharm. Pharmacol.* **26**, 370–371.

Imperato, A. and Di Chiara, G. (1988). Effects of locally applied D-1 and D-2 receptor agonists and antagonists studied with brain dialysis, *Eur. J. Pharmacol.* **156**, 385–393.

Iravani, M. M. and Kruk, Z. L. (1996). Real-time effects of N-methyl-D-aspartic acid on dopamine release in slices of rat caudate putamen: a study using fast cyclic voltammetry, *J. Neurochem.* **66**, 1076–1085.

Iwasaki, Y., Ikeda, K., Shiojima, T., *et al.* (1992). Increased plasma concentrations of aspartate, glutamate and glycine in Parkinson's disease, *Neurosci. Letters* **145**, 175–177.

Jacocks, H. M. and Cox, B. M. (1992). Serotonin-stimulated release of [3H]dopamine via reversal of the dopamine transporter in rat striatum and nucleus accumbens: A comparison with release elicited by potassium, N-methyl-D-aspartic acid, glutamic acid and D-amphetamine, *J. Pharm. Exp. Therap.* **262**, 356–364.

Javoy, F. and Glowinski, J. (1971). Dynamic characteristic of functional compartment of dopamine in dopaminergic terminals of the rat striatum, *J. Neurochem.* **18**, 1305–1311.

Johnson, R. G. (1988). Accumulation of biological amines into chromaffin granules; a model for hormone and neurotransmitter transport, *Physiol. Rev.* **68**, 232–307.

Kamal, L. A., Arbilla, S. and Langer, S. Z. (1981). Presynaptic modulation of the release of dopamine from the rabbit caudate nucleus: differences between electrical stimulation, amphetamine and tyramine, *J. Pharm. Exp. Ther.* **216**, 592–598.

Kamal, L. A., Arbilla, S., Galzin, A. M., *et al.* (1983). Amphetamine inhibits the electrically evoked release of [3H]dopamine from slices of the rabbit caudate, *J. Pharm. Exp. Ther.* **227**, 446–458.

Keefe, K. A., Zigmond, M. J. and Abercrombie, E. D. (1992). Extracellular dopamine in striatum: influence of nerve impulse activity in medial forebrain bundle and local glutamatergic input, *Neuroscience* **47**, 325–332.

Keefe, K. A., Zigmond, M. J. and Abercrombie, E. D. (1993). In vivo regulation of extracellular dopamine in the neostriatum: influence of impulse activity and local excitatory amino acids, *J. Neural. Transm.* **91**, 223–240.

Kitayama, S., Dohi, T. and Uhl, G. R. (1994). Phorbolesters alters functions of the expressed dopamine transporter, *Eur. J. Pharmacol.* **268**, 115–119.

Kuczenski, R. (1975). Effects of catecholamine releasing agents on synaptosomal dopamine biosynthesis: multiple pools of dopamine or multiple forms of tyrosine hydroxylase, *Neuropharmacology* **14**, 1–10.

Kuczenski, R., Segal, D. S. and Manley, L. D. (1990). Apomorphine does not alter amphetamine induced dopamine release measured in striatal dialysates, *J. Neurochem.* **54**, 1492–1499.

Kula, N. S. and Baldessarini, R. J. (1991). Lack of increase in dopamine transporter binding or function in rat b rain tissue after treatment with blockers of neuronal uptake of dopamine, *Neuropharmacology* **30**, 89–92.

L'Hirondel, M., Cheramy, A., Godeheu, G., *et al.* (1995). Effects of arachidonic acid on dopamine synthesis, spontaneous release, and uptake in striatal synaptosome from the rat, *J. Neurochem.* **64**, 1406–1409.

Langer, S. Z. and Arbilla, S. (1984). The amphetamine paradox in dopaminergic neurotransmission, *Trends in Pharm. Sci.* **84** (Sept.), 387–390.

Lavielle, S., Tassin, J. P., Thierry, A. M. *et al.* (1978). Blockade by benzodiazepines of the selective high increase in dopamine turn over induced by stress in mesocortical dopaminergic neurons of the rat, *Brain Res.* **168**, 585–594.

Levi, G., Rusca, G. and Raiteri, M. (1976). Diaminobutyric acid: A tool for dicriminating between carrier-mediated and non carrier-mediated release of GABA from synaptosomes, *Neurochem. Res.* **1**, 581–590.

Leviel, V. (1991). The glutamate-mediated release of dopamine in the rat striatum: further characterization of the dual excitatory-inhibitory function, *Neuroscience* **39**, 305–312.

Leviel, V. (2001). The reverse transport of DA, what physiological significance, *Neurochem. Int.* **38**, 83–106.

Leviel, V. and Guibert, B. (1987). Involvement of intraterminal dopamine compartments in the amine release in the cat striatum, *Neurosc. Lett.* **76**, 197–202.

Leviel, V., Gobert, A. and Guibert, B. (1989). Direct observation of dopamine compartmentation in striatal nerve terminal by 'in vivo' measurement of the specific activity of released dopamine, *Brain Res.* **499**, 205–213.

Leviel, V., Gobert, A. and Guibert, B. (1990). The glutamate-mediated release of dopamine in the rat striatum: further characterization of the dual excitatory-inhibitory function, *Neuroscence* **39**, 305–312.

Leviel, V., Gobert, A. and Guibert, B. (1991). Specifically evoked release of newly synthesized or stored dopamine by different treatments, in: *The Basal Ganglia III*, Bernardi, G., *et al.* (Eds), pp. 407–416. Plenum, New York.

Lin, A. M. and Chai, C. Y. (1995). Dynamic analysis of ethanol effects on NMDA-evoked dopamine overflow in rat striatum, *Brain Res.* **696**, 15–20.

Lin, A. M. and Chai, C. Y. (1998). Role of dopamine uptake in NMDA-modulated K(+)-evoked dopamine overflow in rat striatum: an in vivo electrchemical study, *Neurosci. Res.* **31**, 171–177.

Lindefors, N. and Ungerstedt, U. (1990). Bilateral regulation of glutamate tissue ans extracellular levels in caudta putamen by midbrain dopamine neurons, *Neurosci. Letters* **115**, 248–252.

Lonart, G. and Johnson, K. M. (1994). Inhibitory effects of nitric oxide on the uptake of [3H]dopamine and [3H]glutamate by striatal synaptosomes, *J. Neurochem.* **63**, 2108–2117.

Lonart, G. and Zigmond, M. J. (1990). N-methyl-D-aspartate evokes dopamine release from rat striatum via dopamine uptake system, *Soc. Neurosci. Abs.* **16**, 552.

Lonart, G. and Zigmond, M. J. (1991). High glutamate concentrations evoke Ca^{++} independent dopamine release from striatal slices, a possible role of reverse dopamine transport, *J. Pharm. Exp. Therap.* **256**, 1132–1138.

Lynch, M. A. and Littleton, J. M. (1983). Possible association of alcohol tolerance with increased synaptic Ca^{++} sensitivity, *Nature* **303**, 175–176.

Mann, S. P. and Gordon, J. I. (1979). Inhibition of guinea pig tyrosine hydroxylase by catechols and biopterin, *J. Neurochem.* **33**, 133–138.

Marshall, J. F. (1979). Somatosensory inattention after dopamine depleting intracerebral 6-OHDA injections: spontaneous recovery and pharmacological control, *Brain Res.* **177**, 311–324.

Maurin, Y., Banzeres, B., Menetrey, A., *et al.* (1999). Three-dimensional distribution of nigrostriatal neurons in the rat: relation to the topography of striatonigral projections, *Neuroscience* **91**, 891–909.

Maus, M., Premont, J. and Glowinski, J. (1990). In vitro effects of 17 β-oestradiol on the sensitivity of receptors coupled to adenylate cyclase on striatal neurons in primary culture, in: *Steroids and neuronal activity, CIBA Foundation symposium 153*, pp. 145–153. Wiley, New York.

May, L. J. and Wightman, R. M. (1989a). Effects of D2 antagonists on frequency dependent stimulated dopamine overflow in nucleus accumbens and caudate putamen, *J. Neurochem.* **53**, 898–906.

May, L. J. and Whitman, R. M. (1989b). Heterogeneity of stimulated dopamine overflow within rat striatum as observed with in vivo voltammetry, *Brain Res.* **487**, 311–320.

McMahon, H. T., Barrie, A. P., Lowe, M., *et al.* (1989). Glutamate release from guinea-pig synaptosomes: stimulation by reuptake-induced depolarization, *J. Neurochem.* **53**, 71–79.

Meiergerd, S. M., Patterson, T. A. and Schenk, J. O. (1993). D2 receptors may modulate the function of the striatal transporter for dopamine: kinetic evidence from studies in vitro and in vivo, *J. Neurochem.* **61**, 764–767.

Miller, H. H. and Shore, P. A. (1982). Effects of amphetamine and amfonelic acid on the disposition of striatal newly synthesyzed dopamine, *Eur. J. Pharmacol.* **78**, 33–44.

Mirenowicz, J. and Schultz, W. (1996). Preferential activation of midbrain dopamine neurons by appetitive rather than aversive stimuli, *Nature* **379**, 449–451.

Moghaddam, B. and Bolinao, M. L. (1994). Glutamatergic antagonists attenuate ability of dopamine uptake blockers to increase extracellular levels of dopamine: implications for tonic influence of glutamate on dopamine release, *Synapse* **18**, 337–342.

Nash, J. F. and Brodkin, J. (1991). Microdialysis studies on 3,4-methylenedioxymethamphetamine induced dopamine release: effect of dopamine uptake inhibitors, *J. Pharm. Exp. Ther.* **259**, 820–825.

Nurse, B., Russel, V. A. and Taljaard, J. J. (1988). Characterization of the effects of serotonin on the release of [3H]dopamine from rat nucleus accumbens and striatal slices, *Neurochem. Res.* **13**, 403–407.

O'Connor, V. M., Shamotienko, O., Grishin, E., *et al.* (1993). On the structure of the synapto-secretosome evidence for a neurexin/syntaxin/Ca^{2+} channel complex, *FEBS Letters* **326**, 255–260.

Olivier, V., Guibert, B. and Leviel, V. (1995). Direct in vivo comparison of two mechanisms releasing dopamine in the rat striatum, *Brain Res.* **695**, 1–9.

Olivier, V., Gobert, A., Guibert, B., *et al.* (1999). The in vivo modulation of dopamine synthesis by calcium ions: Influences on the calcium independent release, *Neurochem. Int.* **35**, 431–438.

Overton, P. and Clark, D. (1992). Iontophoretically administered drugs acting at the N-methyl-D-aspartate receptor modulate burst firing in A9 dopamine neurons in the rat, *Synapse* **10**, 131–140.

Parker, E. M. and Cubeddu, L. X. (1986a). Effects of d-amphetamine and dopamine synthesis inhibitors on dopamine and acetylcholine neurotransmission in the striatum. I Release in the absence of vesicular transmitter stores, *J. Pharm. Exp. Ther.* **237**, 179–192.

Parker, E. M. and Cubeddu, L. X. (1986b). Effects of d-amphetamine and dopamine synthesis inhibitors on dopamine and acetylcholine neurotransmission in the striatum. II Release in the presence of vesicular transmitter stores, *J. Pharm. Exp. Ther.* **237**, 193–203.

Parsons, L. H., Schad, C. A. and Justice, J. B., Jr. (1993). Co-administration of the D2 antagonist pimozide inhibits up-regulation of dopamine release and uptake induced by repeated cocaine, *J. Neurochem.* **60**, 376–379.

Pasqualini, C., Guibert, B., Frain, O., *et al.* (1994). Evidence for protein kinase C involvement in the short term activation by prolactin of tyrosine hydroxylase in tuberoinfundibular dopaminergic neurons, *J. Neurochem.* **62**, 967–977.

Pasqualini, C., Olivier, V., Guibert, B., *et al.* (1995). Acute stimulatory effect of estradiol on striatal dopamine synthesis, *J. Neurochem.* **65**, 1651–1657.

Paton, D. M. (1973a). Evidence for a carrier-mediated efflux of noradrenaline from the axoplasm of adrenergic nerves in rabbit atria, *J. Pharm. Pharmacol.* **25**, 265–267.

Paton, D. M. (1973b). Mechanism of efflux of noradrenaline from adrenergic nerves in rabbit atria, *Br. J. Pharm.* **49**, 614–627.

Petrucci, T. C. and Morrow, J. S. (1987). Synapsin I: an acting bundling protein under phosphorylation control, *J. Cell. Biol.* **105**, 1355–1363.

Pfeffer, S. R. (1992). GTP-binding proteins in intracellular transport, *Trends Cell Biol.* **2**, 41–45.

Pogun, S., Baumann, M. H. and Kuhar, M. J. (1994). Nitric oxide inhibits [3H]dopamine uptake, *Brain Res.* **641**, 83–91.

Raiteri, M., Del Carmine, R., Bertollini, A., *et al.* (1977). Effects of desmethyl imipramine on the release of [3H]norepinephrine induced by various agents in hypothalamic synaptosomes, *Mol. Pharmacol.* **13**, 746–758.

Raiteri, M., Cerrito, F., Cervoni, A. M., *et al.* (1979). Dopamine can be released by two mechanisms differentially affected by the dopamine transport inhibitor nomifensine, *J. Pharm. Exp. Ther.* **208**, 195–202.

Ramirez, V. D., Kim, K. and Dluzen, D. E. (1985). Progesterone action on the LHRH and the nigrostriatal dopamine neuronal systems: in vitro and in vivo studies, *Recent Prog. Hormone Res.* **41**, 421–472.

Roberts, P. J. and Shariff, N. A. (1978). Effects of L-glutamate and related amino acids upon the release of [3H]dopamine from rat striatal slices, *Brain Res.* **157**, 391–395.

Ross, S. B. and Renyi, A. L. (1964). Blocking action of sympathomimetic amines on the uptake of NE by mouse cerebral cortex, *Acta Pharmacol. Toxicol.* **31**, 226–239.

Rudnick, G. and Wall, S. C. (1992). The molecular mechanism of "ecstasy" [3,4-methylenedioxy-methamphetamine (MDMA)]: serotonin transporters are targets for MDMA induced serotonin release, *Proc. Natl. Acad. Sci. USA* **89**, 1817–1821.

Samuel, D., Errami, M. and Nieoullon, A. (1990). Localization of N-methyl-D-aspartate receptors in the rat striatum: effects of specific lesions on the [3H]3-(2-carboxypiperazin-4)propyl-1-phosphonic acid binding, *J. Neurochem.* **54**, 1926–1933.

Schacht, U., Leven, M. and Backer, G. (1977). Studies on brain metabolism of biogenic amines, *Brit. J. Clin. Pharmacol.* **4**, 77S–87S.

Schmidt, C. J. and Black, C. K. (1989). The putative 5HT$_3$ agonist phenylbiguanide induces carrier-mediated release of [^{3}H]dopamine, *Eur. J. Pharmac.* **167**, 309–310.

Schuldiner, S., Steiner-Mordoch, S., Yelin, R., *et al.* (1993). Amphetamine derivatives interact with both plasma membrane and secretory vesicle biogenic amine transporters, *Mol. Pharmac.* **44**, 1227–1231.

Schultz, W. (1986). Response of midbrain dopamine neurons to behavioral trigger stimuli in the monkey, *J. Neurophysiol.* **56**, 1439–1461.

Sershen, H., Debler, E. A. and Lajhta, A. (1987). Effect of ascorbic acid on the synaptosomal uptake of [3H]MPP+, [3H]dopamine, and [3H]GABA, *J. Neurosci. Res.* **17**, 298–301.

Sill, T. L., Greenshaw, A. J., Baker, G. B., *et al.* (1999). Acute fluoxetine treatment potentiates amphetamine hyperactivity and amphetamine-induced nucleus accumbens dopamine release: possible pharmacokinetic interaction, *Psychopharmacology* **141**, 421–427.

Smolen, T. N., Howerton, T. C., *et al.* (1984). Effects of ethanol and salsolinol on catecholamine function in LS and SS mice, *Pharmacol. Biochem. Behav.* **20**, 125–131.

Snape, B. M. and Engel, J. A. (1988). Ethanol enhances the calcium-dependent stimulus induced release of endogenous dopamine from slices of rat striatum and nucleus accumbens in vitro, *Neuropharmacology* **27**, 1097–1101.

Snyder, G., Keller, R. W. and Zigmond, M. J. (1990). Dopamine efflux from striatal slices after intracerebral 6-hydroxydopamine: evidence for compensatory hyperactivity of residual terminals, *J. Pharmacol. Exp. Therap.* **253**, 867–876.

Soares-Da-Silva, P. (1987). Does brain 3,4-dihydroxyphenylacetic reflect dopamine release, *J. Pharm. Pharmacol.* **39**, 127–129.

Soares-Da-Silva, P. and Garrett, M. C. (1990). A kinetic study of the rate of formation of dopamine, 3,4-dihydroxyphenylacetic acid (DOPAC) and homovanillic acid (HVA) in the rat brain: implications for the origin of DOPAC, *Neuropharmacol.* **29**, 869–874.

Stackowiack, M. K., Keller, R. W., Stricker, E. M., *et al.* (1987). Increased dopamine efflux from striatal slices during development and after nigrostriatal bundle damage, *J. Neurosci.* **7**, 1648–1654.

Stein, W. D. (1967). *The Movement of Molecules Across Cell Membranes.* Academic Press, New York, pp. 177–206.

Stein, W. D. (1990). *Channels, Carriers and Pumps: An Introduction to Membrane Transport.* Hartcourt, Brace, Jovanovich, New York.

Thoenen, H., Hurlimann, A. and Haefely, W. (1969). Cation dependence of the noradrenaline releasing action of tyramine, *Eur. J. Pharmacol.* **6**, 29–37.

Uretsky, N. J. and Snodgrass, S. R. (1977). Studies on the mechanism of stimulation of dopamine synthesis by amphetamine in striatal slices, *J. Pharm. Exp. Ther.* **202**, 565–580.

Vizi, E. S. and Labos, E. (1991). Nonsynaptic interactions at presynaptic level, *Prog. Neurobiol.* **37**, 145–163.

Vizi, E. S., Bernath, S., Kapocsi, J., *et al.* (1986). Transmitter release from the cytoplasm is of physiological importance but no subject to presynaptic modulation, *J. Physiol., Paris* **81**, 283–288.

Wang Y., Wang, S. D., Lin, S. Z., *et al.* (1994). Restoration of dopamine overflow and clearance from the 6-hydroxydopamine lesioned rat striatum reinnervated by fetal mesencephalic grafts, *J. Pharmacol. Exp. Therap.* **270**, 814–821.

Wang, Y., Palmer, M. R., Cline, E. J., *et al.* (1997). Effects of ethanol on striatal dopamine overflow and clearance: an in vivo electrochemical study, *Alcohol* **14**, 593–601.

Welch, S. M. and Justice, J. B. (1996). Regulation of dopamine uptake in rat striatal tissue by NMDA receptors as measured using rotating disk electrode voltammetry, *Neurosci. Letters* **217**, 184–188.

Wheeler, D. D., Edwards, A. M., Chapman, B. M., *et al.* (1993). A model of the sodium dependence of dopamine uptake in rat striatal synaptosomes, *Neurochem. Res.* **18**, 927–936.

Wieczorek, W. J. and Kruk, Z. L. (1994). A quantitative comparison on the effects of benztropine, cocaine and nomifensine on electrically evoked dopamine overflow and rate of re-uptake in the caudate putamen and nucleus accumbens in the rat brain slice, *Brain Res.* **657**, 42–50.

Wiener, H. L., Hashim, A., Lajtha, A., *et al.* (1989). Chronic L-deprenyl-induced up-regulation of the dopamine uptake carrier, *Eur. J. Pharmacol.* **163**, 191–194.

Woodward, J. J. and Gonzales, R. A. (1990). Ethanol inhibition of NMDA-stimulated endogenous dopamine release from rat striatal slices: reversal by glycine, *J. Neurochem.* **54**, 712–715.

Wullner, U., Testa, C. M., Catania, M. V., *et al.* (1994). Glutamate receptors in striatum and substantia nigra: effects of medial forebrain bundle lesions, *Brain Res.* **645**, 98–102.

Yadid, G., Pacak, K., Kopin, I. J., *et al.* (1994). Endogenous serotonin stimulates striatal dopamine release in conscious rats, *J. Pharm. Exp. Therap.* **270**, 1158–1165.

Yi, S., Gifford, A. N. and Johnson, K. M. (1991). Effect of cocaine and $5HT_3$ receptor antagonists on 5HT-induced [^{3}H]dopamine release from rat striatal synaptosomes, *Eur. J. Pharmac.* **199**, 185–189.

Zaczek, R., Culp, S. and De Souza, E. B. (1990). Intrasynaptosomal sequestration of [3H]amphetamine and [3H]methylenedioxy-amphetamine: Characterization suggests the presence of a factor responsible for maintaining sequestration, *J. Neurochem.* **54**, 195–204.

Zahniser, N. A., Larson, G. A. and Gerhardt, G. A. (1999). In vivo dopamine clearance rate in rat striatum: regulation by extracellular dopamine concentration and dopamine transporter inhibitors, *J. Pharm. Exp. Ther.* **289**, 266–277.

Zerial, M. and Stenmark, H. (1993). RabGTPases in vesicular transport, *Curr. Opin. Cell. Biol.* **5**, 613–620.

Zetterström, T., Sharp, T., Collin, A. K., *et al.* (1988). In vivo measurement of extracellular dopamine and DOPAC in rat striatum after various dopamine-releasing drugs; implications for the origin of extracellular DOPAC, *Eur. J. Pharmacol.* **148**, 327–334.

Ziance, R. J., Azzaro, A. J. and Rutledge, C. O. (1972). Characteristic of amphetamine induced release of norepinephrine from rat cerbral cortex in vitro, *J. Pharm. Exp. Ther.* **182**, 284–294.

Zigmond, M. J., Berger, T. W., Grace, A. A., *et al.* (1989). Compensatory response to nigrostriatal bundle injury. Studies with 6-hydroxydopamine in an animal model of parkinsonism, *Mol. Chem. Neuropathol.* **10**, 185–200.

Zigmond, M. J., Abercrombie, E. D., Berger, T. W., *et al.* (1990). Compensatory changes after lesions of central dopaminergic neurons: some clinical and basis implications, *Trends Neurol. Sci.* **13**, 290–295.

Advances in Neuroregulation and Neuroprotection (2005), pp. 165-187
Collin, C. *et al.* (Eds)
© VSP 2005

Conditions for culture of different types of neural cells: growth regulation

F. ROBERT [1], S. CONSTANTIN [2], A. H. DUITTOZ [2] and T. K. HEVOR [1,*]

[1] *Laboratoire de Métabolisme Cérébral et Neuropathologies — U.P.R.E.S. E.A. 2633, Université d'Orléans, B.P. 6759, F-45067 Orléans Cedex 2, France*
[2] *INRA, Station de Physiologie de la Reproduction et du Comportement, Equipe de Neuroendocrinologie Sexuelle, F-37380 Nouzilly, France*

Abstract—The central nervous system is composed of a variety of cell types. Among them neuronal and astroglial cells have been the subjects of numerous studies. Normal cerebral functions require a great degree of cooperation between neuronal and glial cells. Due to the fact that neural cells are highly mixed in the nervous system, it is not easy to know the role played by each cell type *in vivo*. The cell culture approach is an interesting alternative for research on a given cell type. In the present paper, we review different methods of obtaining the different kinds of neural cells that exist in the normal nervous system. Neuronal cells were easily cultured by seeding single cells in a defined Sato-type medium when the dish was coated by polylysine. A monolayer of protoplasmic astrocytes was obtained when culturing brain cells in a Dulbecco's modified Eagle's medium without a previous coating of the dishes. Fibrous astrocytes were obtained using dibutyryl cyclic AMP or a defined medium. Neural precursors were obtained and maintained in an undifferentiated state using epidermal growth factor and basic fibroblast growth factor. Starting from the sheep embryo brain, it was possible to culture brain stem cells for six months. These cells differentiated into radial glial cells. All the kinds of cells of the *in vivo* nervous system can be cultured. This offers the possibility of getting models for studying all the types of neural cells. Now, the factors inducing the development of the different phenotypes have to be studied.

Keywords: Neurons; neurospheres; protoplasmic astrocytes; fibrous astrocytes; radial astrocytes; stem cells.

INTRODUCTION

Neuronal cells, glial cells, and endothelial cells are highly mixed in the brain tissue to ensure the brain functions. A fundamental feature of these cells is the high degree of cooperation to such a level that it is not easy to always identify the role played

---

*To whom correspondence should be addressed. E-mail: tobias.hevor@univ-orleans.fr

by each cell type. Another feature is the existence of many different processes, which make up a complex network in the brain tissue. Knowledge of each cell role needs to delimit the exact functional frontiers of these cells. This approach is very tedious and its efficiency is limited *in vivo*. During the last decades, the development of cell culture techniques has helped to overcome some of the difficulties of the nervous system studies. Indeed, it was possible to culture separately neuronal cells, astrocytes, oligodendrocytes, microglial cells, brain endothelial cells, etc. As these cultures are made in a medium which is not the extracellular fluid of the brain, cooperation between different neural cells no longer exists in the culture, and the blood brain barrier is disrupted, it is obvious that the cultured neural cells cannot completely mimic the *in vivo* neural cell. So, it is necessary to characterize the behaviour of the cultured neural cells. Among the characteristics, their growth must be considered apart. Indeed, according to the medium utilized to culture a given cell, the latter could develop a given phenotype and a given behaviour. In the present work, we propose to review some culture conditions and their influence on the neural cell growth.

NEURONAL CELLS

All the experiments were done according to the EEC ethical regulations for animal research.

Monolayer culture of neuronal cells

Pregnant rats of Sprague Dawley strain were anaesthetized at the stage of 16 days of gestation. The foetuses were taken and the cerebral cortices were dissected out. The tissues were dissociated in single cells by trituration through a needle of 1.4 mm (i.d.) in a Hanks solution. The cell suspension was poured on the top of foetal calf serum contained in a tube and the latter was centrifuged at $3000g$ for 5 min. The cells of the pellet were suspended in Neurobasal™ medium (GIBCO, Invitrogen, France) and the seeding was done at the rate of 8×10^5 cells per one 60-mm-diameter Petri dish, previously coated with poly-L-lysine. The incubation took place in a CO_2 incubator (Forma Scientific, Marietta, OH) at 37°C, 95% air, 5% CO_2 in a water saturated atmosphere. The medium was renewed by half once per week (Fig. 1).

The cells obtained were characterized by immunolabelling using anti-neuro-filaments antibodies. Astrocytic proliferation was negligible, even when antimitotic agents were not used. Within 24 h after the start of culture, many cells adhered to coated coverslips and had begun to develop neurites. From one day to 7 days after seeding, the processes were steadily growing and showed a progressive complexity with the beginning of a ramification. Individual processes displayed varicosities and made frequent contact with proximal or distant neuronal soma or neurites. From 14 days after the cell seeding, neurons displayed a second level of ramification

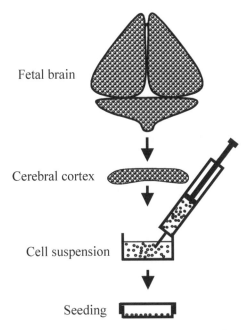

Figure 1. Simplified scheme of the method for the primary culture of neural cells from the rodent brain.

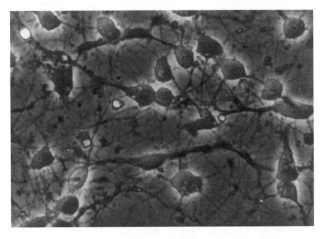

Figure 2. Primary culture of neurons from rat cerebral cortex. The dissociated cells were plated onto a polylysine-coated Petri dish and then incubated. The Neurobasal™ medium was renewed by half every week. The cells were photographed on the 20th day after plating. Note the individual cell bodies, the processes and their network. Magnification ×280.

consisting of a fine and dense network. Cultured cells showed mature neuronal features with little perinuclear cytoplasm. These soma bore multiple, long, thin, varicose and branched processes extending from the perinuclear cytoplasm. Three weeks after the seeding, the frequency of close contacts between neighbouring cells and processes increased and the network density was enhanced (Fig. 2).

Neural stem cells

It is well accepted now that several neural cell type originate from common precursors. *In vitro* studies (Murphy *et al.*, 1990; Reynolds *et al.*, 1992; Reynolds and Weiss, 1992; Kilpatrick and Bartlett, 1993) and *in vivo* studies (Wetts and Fraser, 1988; Turner *et al.*, 1990) have now established that neurons, astrocytes and oligodendrocytes can differentiate from a common precursor: neural stem cells. These stem cells were first described in embryonic rodents and have none of the characteristics of neuronal or glial cell types (Reynolds and Weiss, 1992). They have the ability to divide in response to epidermal growth factor (EGF) and they form spheres of clonal cells *in vitro*. Stem cells have also been found in adult central nervous system in the ventricular/periventricular zone and in the hippocampus (Weiss *et al.*, 1996). As embryonic stem cells, these EGF-responsive cells lack the antigenic properties of neuronal and glial cells and expressed the intermediate filament, nestin. Numerous studies rely only on immuno-cytochemical characterization of neural cells obtained from neural stem cells. However, one must ensure that neurons obtained from differentiated stem cells are functional, i.e. they must be excitable and they must be able to release their neurotransmitter after stimulation.

In the present study, we described different culture conditions allowing the propagation of ovine neural stem cells and their differentiation into neuronal or astrocytic cell types. Characterization of neuronal cells was made using an immunocytochemical approach, electrophysiological recordings and intracellular calcium measurement.

Cell culture conditions. Neural stem cells were obtained from 26-30-day-old sheep embryos removed by cesarean section. The brain was dissected out in chilled sterile Hank's balanced salt solution (HBSS). The meningeal membranes were carefully removed as far as possible. After mechanical dissociation, the cell suspension was filtered through 100 and 40 μm sterile nylon mesh, and gently centrifuged ($100g$ for 5 min). HBSS was replaced by a Sato-type defined medium (Bottenstein and Sato, 1979): Neurobasal^{TM} and N2 supplement (GIBCO, Invitrogen, France) supplemented with 2 μg epidermal growth factor (EGF), 400 ng basic fibroblast growth factor (bFGF) (GIBCO, In Vitrogen, France) for 100 ml. The cell suspension was distributed in Petri dishes at the final concentration of 2×10^5 cells/ml. Two-thirds of the medium was replaced every two days. Most cells did not adhere onto plastic and started to divide to form floating colonies called neurospheres. When changing the medium, neurospheres were mechanically dissociated. Dissociated cells from these neurospheres formed secondary neurospheres. Neurospheres were composed of neural precursors: neural stem cells, neuronal progenitors and glial progenitors. Differentiation into neuronal or glial types was achieved by plating the cells and the use of different culture conditions. Once plated, neural precursors could be maintained in Neurobasal and

N2 supplement with EGF bFGF only for a limited period not exceeding two weeks, since cell adhesion initiated differentiation.

For neuronal differentiation, dissociated neurospheres were plated onto poly-L-lysine coated glass coverslips and maintained into Neurobasal supplemented with B27 complement or Sato medium supplemented with 2 ng/ml brain derived neurotrophic factor (BDNF, GIBCO, In Vitrogen France).

For glial differentiation, dissociated neurospheres were plated onto poly-L-lysine coated glass coverslips into DMEM/Ham's F12 (1/1) supplemented with 10% fetal calf serum or in Neurobasal and G5 supplement. Cultures were then examined at different times: 7, 14, 21, 28 and 53 days after differentiation using immunocyto-chemistry and electrophysiological techniques to characterize each cell type.

One hundred ml of all the media were supplemented as follows: 5000 IU penicillin, 50 000 μg streptomycine, 50 μg fungizone, 2 mM glutamine.

Immunocytochemical characterization. For immunocytochemical characterization, cell cultures were rinsed with phosphate buffer saline (PBS) and fixed with paraformaldehyde 4% for 30 min. The cultures were then incubated with PBS, 0.3% triton X100, 1 : 15 (V/V) normal horse serum or normal goat serum according to the species in which the secondary antibody had been made. Neural precursors were characterized using the monoclonal anti-nestin antibody diluted to 1 : 1000. Neuronal cells were characterized using the monoclonal anti-β-III-tubulin antibody (diluted to 1 : 1000) or using the monoclonal anti-microfilament associated protein 2 (MAP2) antibody (diluted to 1/1000). Glial cells were characterized using the anti-GFAP (glial fibrillary acidic protein) (diluted to 1/3000). All primary antibodies were from Chemicon, Temecula, USA, except GFAP from Dako Carpintaria, USA. Single and double staining were performed using fluorescent secondary antibodies: CY-3 anti-rabbit and FITC anti-mouse (Jackon immunoresearch, West Grove, USA) or Alexa-546 anti rabbit and Alexa-488 anti-mouse from Molecular Probes (USA). Cells were counterstained using DAPI (Sigma, St. Louis USA).

Electrophysiological characterization and calcium measurement. For electrophysiological characterization, cells were recorded using the patch-clamp technique in the whole-cell configuration. Borosilicate electrodes (GC150-F, Harvard Apparatus, Edenbridge, UK), pulled with a two-stage vertical puller (PC-10, Narishige, Japan), had a resistance of about 5 MΩ. For sodium current recordings, pipettes were back-filled with a solution containing (in mmol): CsCl 145, MgCl$_2$ 2, glucose 10 and HEPES 10 (pH 7.3), and the cells were maintained in Hank's balanced salt solution (HBSS). For calcium currents, intrapipette solution were filled with: CsCl 120, TEA-Cl 20, MgCl$_2$ 4, EGTA 10, HEPES 10, glucose 9, TrisGTP 0.3, MgATP 4, creatine-PO4 14 and creatine phosphokinase 50 U/ml (pH 7.2) and extracellular media were composed with TMA-Cl 100, TEA 20, CaCl$_2$ 10, MgCl$_2$ 1 and HEPES 10 (pH 7.4). Cells, maintained in a mini-Petri dish, were visualized through a 20$\times$ objective on an inverted microscope (Leica Microsystems, Wetzlar,

Germany). Currents, low-pass filtered at 10 kHz, were recorded with an amplifier (RK-400, Bio-Logic, Science Instrument SA Claix, France) and digitised at 20 kHz by an analogic-digital converter (Digidata 1200A, Axon Instruments, Foster City, CA, USA). Experiments were piloted by the pClamp 6 software (Axon Instruments). For intracellular calcium measurement, cells were loaded with Fura 2-AM (Molecular Probes, Eugene, USA) for 30 min at 37°C and then thoroughly rinsed. Calcium transients were recorded at 510 nm following excitation at 340 and 380 nm. Time lapse images were recorded through a Coolsnap fx™ camera and Metafluor™ software (Roper Scientific, Evry, France).

Results

Maintenance of undifferentiated stem cells. The definition of neural stem cells relies on their property to differentiate into neuronal or astrocytic progeny according to the cell culture conditions. To maintain neural stem cells, they must not adhere onto plastic and the floating clonal neurospheres must not be allowed to grow too big, otherwise differentiation and cell death occur in the center of the sphere.

In our hands, mechanical dissociation using blue and yellow tips was sufficient if one does not want to maintain neurospheres over 1 to 3 months. This mechanical dissociation leaves some neurospheres intact and it is sometimes preferable to filter the cell suspension using 100 μm mesh. Non-enzymatic methods using low calcium and low magnesium solutions (cell dissociation solution) did not give better results than mechanical dissociation and were time consuming. Once plated, neural precursors will keep their properties for only two weeks, since cell adhesion will induce differentiation.

Neuronal differentiation. Based on the immunocytochemical identification oriented by the morphology of the living cells, neuronal differentiation needed a time delay of two weeks in the appropriate medium. However, we worked on very young sheep embryos (26-day-old embryo (E26) in sheep is equivalent to E11 in mouse and E13 in rats). Using rat embryos at E14, we obtained neuronal differentiation in less than one week (data not shown). However, electrophysiological differentiation took a longer time, since the first Na currents were recorded after 3 weeks of differentiation and Ca-currents and Ca transients were recorded after 4 weeks of differentiation. These results suggested that (i) our culture conditions allowed good neuronal differentiation and (ii) that immunocytochemical markers are not sufficient to estimate neuronal differentiation.

Astrocytic differentiation. Astrocytes were easy to obtain from neural stem cells. They appeared during the first week of differentiation in the appropriate medium. One must note that GFAP positive cells could be detected in cultures under neuronal differentiation, when the stem cells were not properly handled and the differentiation started (bad dissociation of neurospheres, cellular adhesion, EGF and bFGF at low concentration) (Fig. 3).

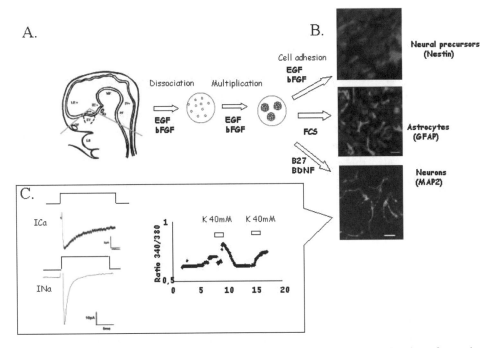

Figure 3. (A) Preparation and propagation of neural stem cells. Schematic drawing of an ovine embryo's head of 26-day of gestation. Grey lines indicate the limits of the dissection. Neural stem cells culture and propagation under epidermal growth factor (EGF) and basic fibroblast growth factor (bFGF) are represented by schematised neurospheres in Petri dish. (B) Plating and differentiation of neural stem cells according to the culture conditions. Platted neural precursors can be maintained undifferentiated with EGF and bFGF factors for two weeks. Neural precursors are immunoreactive to nestin (FITC, DAPI counterstain). Addition of fetal calf serum to the medium induced astrocytic differentiation as assessed by GFAP immunoreactivity (CY-3, DAPI counterstain). Neuronal differentiation was obtained after two weeks in Sato type medium supplemented with B27 or BDNF (MAP2 FITC). (C) Electrophysiological recordings and intracellular calcium measurements. Sodium currents and calcium currents were recorded in neurons differentiated from neural precursors after three weeks of culture in the differentiation medium. Calcium transients were also recorded after a non specific 56 mM K^+-induced depolarization. FCS, foetal calf serum; B27, trade mark culture medium supplement from GIBCO Invitrogen (France); BDNF, brain-derived neurotrophic factor, GFAP, glial fibrillary acidic protein; MAP2, microtubule-associated protein 2.

Conclusion

The use of stem cells from embryonic or adult animals can represent a good model to study the intimate mechanisms underlying cellular differentiation. It should be noted however that the cultures often are not pure (stem cells with neuronal or astrocytic progenitors). Care must be taken when culturing undifferentiated cells since once differentiation has started, it cannot be stopped. Note also that neuronal differentiation can be impaired by inappropriate substrate or medium and that immunocytochemical markers are not always sufficient to ensure the functionality of a neuron.

ASTROGLIAL CELLS

On the basis of their morphology, the astrocytes are commonly separated into protoplasmic astrocytes and fibrous astrocytes. Regarding other characteristics, such as the presence of some antigens and the pattern of development, it was proposed to distinguish type I astrocytes and type II astrocytes (for review, see Raff, 1989). Type I astrocytes are often associated with protoplasmic astrocytes and type II astrocytes with fibrous ones. However, Raff (1989) mentioned that some protoplasmic astrocyte antigens were not present in the type I astrocytes and *vice versa*. This is also the case for fibrous astrocytes and type II astrocytes. In the present study, we investigated the conditions of culture of protoplasmic astrocytes and fibrous astrocytes using the cerebral cortex of the brain.

Astrocytes from rodent brain

Conditions of primary culture of protoplasmic astrocytes.

Methodology. Newborn rats from Sprague Dawley strain were used for these cultures. The cerebral cortex was dissected out in a laminar flow hood. The tissues were placed in DMEM supplemented with 10% inactivated fetal calf serum, penicillin (100 units/ml) and streptomycin (100 μg/ml). The meninges were carefully taken off, and then the culture was performed using the method of Booher and Sensenbrenner (1972) with slight modifications. The main steps were the dissociation of the tissue into single cells by trituration through a needle, the sieving of the cell suspension through a nylon mesh of 48-μm-diameter pores, the plating in 60-mm-diameter Petri dishes at a density of 2×10^5 cells/dish. The dishes were kept in a CO_2 incubator (Forma Scientific, Marietta, OH) regulated at 37°C, 95% air, 5% CO_2, and water saturated atmosphere. The medium was changed after 3–4 days and thereafter twice a week.

Primary cultures resulted from the preceding procedure (Fig. 4). Different fields were found in the cultures from the cortex of the rat brain. These cells have already been described (Hansson *et al.*, 1982; Saneto and De Vellis, 1987). Briefly, when the culture medium used was DMEM supplemented with 10% fetal calf serum, most cells were flat and roughly polygonal astrocytes without important processes. These cells were referred to as protoplasmic astrocytes. Rarely, in some fields of some Petri dishes, cells bore long processes, which were parallel or not. They were referred to as fibrous astrocytes (data not shown). Another kind of cell was easily identifiable because of the high darkness of their bodies and their position above the astrocyte layer. They were oligodendrocytes (Saneto and De Vellis, 1987). Microgliocytes and neuron-like cells were very rare. In our hands, no ependymocyte was observed in the culture, presumably because of the careful elimination of the subcortical tissues.

Characterization. The astrocytes were characterised by the presence of Glial Fibrillary Acidic Protein (GFAP) inside their cytoplasm. For this characterisation,

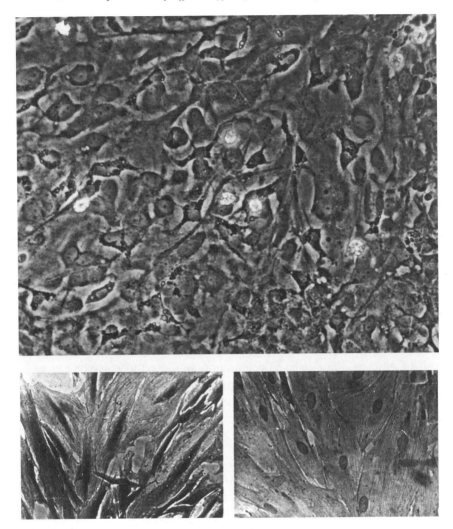

Figure 4. Astrocytes can display different morphologies. Top figure: a micrograph of protoplasmic astrocytes from one-month primary culture. The rat cerebral cortex was used for plating. Subconfluent monolayer of astrocytes were photographed in order to distinguish the single cell bodies. Lower figures: characterization of astrocytes using the immunolabelling of glial fibrillary acidic protein in another culture. Left: the labelling with the anti-serum; right: the control with a normal serum. Magnification ×120 everywhere.

cover slips were put into some dishes before plating. After the cell growth, the cover-slips were washed with PBS at pH 7.4. The cells were then fixed with 10% paraformaldehyde buffered solution (pH 7.4) containing 0.25% Triton X100 for 1 h. For the immunochemical study of GFAP, the cells were successively incubated with anti-GFAP antibodies, intensively washed with PBS, and incubated again with anti-IgG antibodies that were raised in goat. The second antibodies were coupled

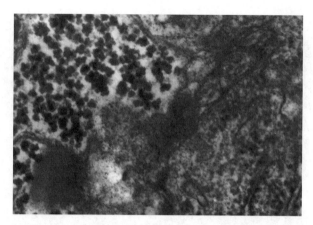

Figure 5. Glycogen particles in astrocytes cultured in a medium containing 5 g/l of glucose. Note the presence of alpha and beta glycogen particles. Magnification ×35 000.

to peroxidase and the processing was made using diaminobenzidine and hydrogen peroxide. The labelling was marked inside the astrocytes.

Astrocytes are the only neural cells that are able to store glycogen (Phelps, 1972, 1975). So, the presence of glycogen particles is also used to characterize astrocytes. For this, glycogen was searched for in cells cultured in a medium containing 1 g/l or 5 g/l of glucose. The latter concentration was used to enhance the glycogen concentration in the cells. When the astrocytes were harvested 6 h after renewing the medium containing 5 g/l of glucose, glycogen particles were clearly visible in the cytoplasm (Fig. 5).

Purifying astrocyte culture. The preceding primary cultures can be purified by a physical method to eliminate non-astrocytic cells. For this, the cells were allowed to grow for about two to three weeks after their seeding in the incubator. Then the dishes were shaken inside the incubator using a shaker during one night. The detached cells were eliminated, the medium was renewed, and the dishes were incubated as before (Verge *et al.*, 1996). The examination of these cultures showed a homogeneous field of protoplasmic astrocytes (Fig. 6). Another way of purifying primary cultures of astrocytes is a general method of cell culture. When a cell culture is replated many times, the number of slow-growing cells decreases and these disappear after a number of passages. Only the rapid-growing cells remain in the culture. Astrocytes are slow-growing cells. However, their velocity of growing is significantly larger than those of the other cells that are present in the primary culture. So, after a number of passages, the cultures contained only astrocytes. For this way of purification, primary cultures of astrocytes were obtained as described above. When the cells reached the subconfluence, they were detached by trypsinization as follows: a solution of 0.05% trypsin and 0.02% EDTA in phospho-buffered saline was added. A few minutes after (5–10 min), when the cells begin to detach, the complete DMEM containing serum was added to inhibit trypsin. The cell suspension was centrifuged at the lower speed of 3000*g* and

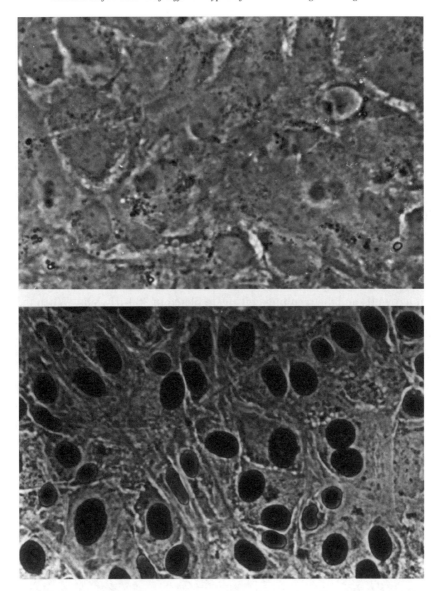

Figure 6. Purified cultures of astrocytes after shaking. The single cells was not easily observable because the extracellular space was reduced. The coloration of the nucleus by Giemsa allowed each cell to be recognised. Magification ×380.

the plating was performed as before. After the growing of the obtained cells, the subculture protocol could be repeated many times. After about three passages, the cultures contained only protoplasmic astrocytes, as in the case of purification using the shaking method. However, the morphology of the cells was different in the same dish, whatever the purification method utilized. Therefore, we decided to separate each category of cell by a cloning method.

Cloning cells. The primary culture of astrocytes was performed in a flask. The culture was purified by shaking, using the method of Saneto and De Vellis (1987). The cells were harvested by trypsinization and plated in 60-mm-diameter Petri dishes. They were kept in the incubator until they reached subconfluence. They were harvested again by trypsinization, and then the final cell suspension was greatly diluted. The latter dilution was made using a cloning medium that was prepared as follows. The media from primary cultures were collected 24 h after their renewal and were added to an equal volume of fresh medium. The resulting mixture was supplemented with fetal calf serum up to 25%. Trypan blue was added when counting the cells in order to detect unviable cells. A cell suspension of 100 viable cells/ml was made and was poured at the rate of 1 ml per well into 24-well cloning dishes. Each well was subdivided into sixteen 20-μl-volume microwells that all overflowed with the volume of 1 ml. The cloning dish was incubated for 24 h, and then the 384 microwells were checked and those having only one cell were registered. Half of the medium of each well was changed once per week. When colonies developed in the 1-cell microwells, a tripsinization was made in these microwells and the resulting cells were plated in 300-μl wells. When the cells reached a sizeable growth, they were again harvested after tripsinization and they were plated in 16-mm-diameter well. Then, the passages were repeated successively in a 35-mm-diameter dish and in a 60-mm-diameter dish. Subcultures were made from the latter dish (Mbarek *et al.*, 1998). Throughout the study, the cells were observed using an inverted microscope (Nikon, model TMS, Tokyo) equipped with an automatic camera (Nikon, model F-601M) (Fig. 7).

The cloning procedure required successive replatings. In order to know if the passages disturbed the characteristics of the cells, the primary cultured astrocytes

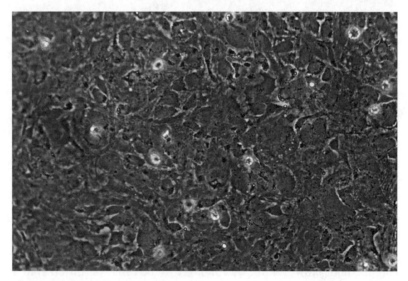

Figure 7. Cloned protoplasmic astrocytes from the rat cerebral cortex in culture. Note the high homogeneity of the cells. Magnification ×140.

were subcultured and their positivity toward GFAP was investigated. From the early step of 4 days after the second replating, some cells from the primary culture, including the isolated clusters, were positively immunostained with anti-GFAP antibodies. When the number of cells replated was low (10^4 cells per 60-mm Petri dish), the astrocytes underwent a singular development since they became large, flat and very hyaline. These hyaline astrocytes were however GFAP-positive. The labelling was more intense around the nuclei. When the number of passages increased, the astrocyte labelling was always well marked. So, the passages did not alter the positivity of astrocytes to GFAP.

The cloning was done in 32-well-cloning dishes, each well containing 1 ml volume. In order to obtain one cell per microwell, we calculated that 1 ml of astrocyte suspension might contain 16 viable cells/ml, since a well was made of 16 microwells. This calculated cell dilution was not appropriate because the microwells bearing one cell were very rare. The experiments showed that the optimal dilution was 100 viable cells/ml. In these conditions, 1 to 4 single cells adhered alone to the bottom of some microwells in one well. The adhesion was followed by one or two weeks without cell division. Many cells died. During this 'lag' period, the living cells underwent a very important development. They became giant and very hyaline. These cells looked like the big cells, which were observed when a low number of primary cultured astrocytes were subcultured. One to two weeks after the seeding, two nuclei were visible in the giant isolated cells and a straight line of division separated them into two cells. Each of these cells underwent the same kind of division. This pattern of division was the same for the subcultured cells from primary cultured astrocytes. After the first divisions, all the astrocytes underwent an exponential growth. The result of such a proliferation was the formation of colonies of very large and hyaline cells. An important feature of the colonies was the presence of numerous cells bearing two visible nuclei. This was an indication of active cell division, since these mitotic figures disappeared after some hours. The simultaneous presence of two nuclei in many cells may also be considered as a sign of synchrony in astrocyte proliferation. When the colonies were growing, the cells in the center became smaller and smaller and looked like subcultured cells from primary cultured astrocytes. However, the bordering cells of the colony remained large and many of them bore mitotic figures. These figures showed that the growing of the colony took place at the borders. When such a colony was large enough (100–500 cells), a passage was undertaken. The cells quickly adhered to the dishes (some hours) and cell division rapidly started, from the second day. Each cell generated one colony, the growth of which followed the same pattern as that of the preceding colony. Finally, all the colonies joined. After the other passages, the cell growth followed the same pattern. The cells of the replated primary cultures also grew by the increasing of many colonies, each resulting from some of the plated cells. These primary cultured cells were much smaller than cloned cells of the same step.

All the steps of the cloning procedure were checked for the presence of GFAP in the cells. Single cells which were alone in the microwells were markedly immunolabelled with the anti-GFAP antibodies. In the large cells of the colonies, the labelling was important and was localized around the nuclei, as already shown in the case of subcultured astrocytes from primary cultures. All the cloned cells were GFAP-positive.

The morphology of the cloned cells was the same as that of the common cultured astrocytes from the rat cerebral cortex, as already described above. The morphological homogeneity of the cultures was tremendous. In most cases, the cells remained flat and polygonal indefinitely. However, unexpectedly, the morphology of some clones underwent a transformation during the growth. They became long and spindle-shaped and they peculiarly looked like fibroblasts. This transformation was absolutely spontaneous. Although fibroblasts were not routinely cultured in the laboratory, an accidental contamination by the latter cells when changing the medium might be possible. To verify this possibility, the dishes containing the fibroblast-like cells were immunostained with anti-GFAP antibodies. All the cells were GFAP-positive and the labelling was around the nuclei, as in the case of typical clones. A further verification was the immunocytochemical study of true fibroblasts cultured from the meninges and from the head of the newborn rat by anti-GFAP antibodies. No labelling was observed.

Cultured astrocytes provide a convenient material for the study of many functions in the central nervous system and many progresses were made using these cultures (Pentreath *et al.*, 1986; Swanson *et al.*, 1989; Wiesinger *et al.*, 1991). The availability of homogenous populations of astrocytes would improve these investigations. The present results show that a single astroglial cell is able to generate many millions of cells that are typical clones. The morphology of these cells changed during the cloning procedure. The final shape of most clones was the common flat and polygonal form, although some other clones were spindle-shaped. All along the cloning process, the general aspect of the cells was roughly the same as that of subcultured cells from primary cultured astrocytes. Our results show that the eukaryotic cells astrocytes are suitable for cloning without a preliminary immortalization. The yield of the cloning is convenient, thus showing that astrocytes have a high potentiality of proliferation. This proliferation is in agreement with the biology of the astrocytes. Indeed, *in vivo*, astrocytes are able to greatly proliferate in many instances (Cavanagh, 1970; Eng, 1988; Nathaniel and Nathaniel, 1977). Our results are also in agreement with the observations of Lee *et al.* (1992) who have made 17 passages from primary cultured astrocytes without apparent degeneration.

Conditions of primary culture of fibrous astrocytes. The fibrous astrocytes can also be cultured up to a high level of purity. This depends on the technique of culture. One possibility is to first culture protoplasmic astrocytes, as described above. When the astrocytes reach confluence or subconfluence, they are submitted to the cyclic nucleotide AMP. The problem is that this nucleotide does not easily cross the plasma membrane. There is a soluble form, dibutyryl cyclic AMP, that is lipophilic enough to cross the plasma membrane. When dibutyryl cyclic AMP

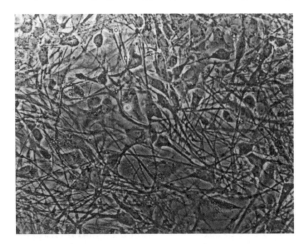

Figure 8. A typical culture of fibrous astrocytes from the rat cerebral cortex. Note the long processes making a complex network. Magnification ×100.

was added to the medium up to a concentration of 1 μmol/litre, the protoplasmic astrocytes underwent a deep change in their morphology. From about three days after adding the cyclic nucleotide, the astrocyte cell body volume steadily decreased. The processes that were not clearly visible became longer and longer. The spaces between the astrocytes increased more and more. Less than ten days after adding the cyclic nucleotide, the astrocytes became typical fibrous astrocytes, i.e. they bore long processes on a small cell body with large extracellular spaces (Fig. 8).

These astrocytes were checked for the presence of the common astrocyte markers, GFAP and glycogen. Like the situation *in vivo*, the astrocytes that underwent the cyclic AMP transformation were intensively immunostained with the anti-GFAP antibodies. That means the presence of a high number of GFAP filaments in their cytoplasm. Moreover, glycogen was measurable in these astrocytes. Note however that the concentration of glycogen was less than in protoplasmic astrocytes (Hevor and Delorme, 1991). So, fibrous astrocytes obtained by transformation are probably of the same kind as *in vivo*.

There is another way to obtain fibrous astrocytes in culture. The cerebral cortex was prepared as above to obtain single cells. The seeding is performed in the same DMEM as above. When the cells adhered to the bottom of the Petri dishes, the medium was changed using the defined medium of Bottenstein and Sato (1979), which was not supplemented by foetal calf serum. The feedings of the cultures continued using the last medium. The obtained astrocytes were fibrous astrocytes that singularly looked like the preceding ones.

Astrocytes from sheep brain

Avian materials, principally chick eggs, are often used for cultures because of the ease of accession to the embryo neural cells and the possibility of a thorough control of developmental stages (Kentroti and Vernadakis, 1996). It is highly probable that there are differences between the neural cells of birds and that of mammals. So, many other investigations were performed using mouse or rat brains. Beyond the group of birds and the group of mammals, there are differences between species in each group. Unfortunately, the results that are obtained using rodent neural cell culture are often extrapolated to other mammals. Here, we investigate the conditions that are necessary to obtain suitable neural cell culture from a non-conventional laboratory animal. Our aim is to utilize sheep neural cells as an experimental model. A disease appeared in many breeding animals during recent years. Termed prion disease, this illness is characterized by a severe encephalopathy, which leads to behavior impairments and death. Neural cell necrosis is abundant in the brain of sick animals. The appearance of prion disease in man, presumably following ingestion of tissue from a sick animal, resulted in a general panic because it was stated that a species barrier existed between animal and man regarding this disease. Many investigations started on the disease during the few last years. A convenient biological material for such investigations could be neural cell from ovine brain, because sheep was the first species known to develop a pathology closely related to what is termed prion disease. The frontiers of the neural cell development are more and more disclosed since many characteristics of the stem cells have already been described (Cattaneo and McKay, 1990; Gage et al., 1995). However, one difficulty in studying the early stages of neural cell development is the size of the biological material, since the smallness of the embryo from common laboratory rodents constitutes one disadvantage for easy manipulation. Contrarily, even at the stage of the neural tube, the central nervous system of sheep can be easily manipulated. Finally, the interest in agronomy has to be mentioned. Tillet and Thibault (1987) have studied the ontogeny of the catecholaminergic system in vivo in sheep regarding the central control of reproduction pathways. The utilization of cultured neural cell would improve this kind of study as previously mentioned (Duittoz et al., 1997).

For the preceding reasons, we thought to make available data for an easy way of getting neural cells from sheep. The step of the neural tube was particularly investigated to search for stem cells or progenitors that differentiate into final cells.

Conditions of culture of astrocytes from sheep brain.   The animals came from a local flock (INRA, Tours-Nouzilly, France). The ewes from cross-bred Ile de France ovine were used throughout the experiments. The embryos were obtained by caesarean section. For this, on the chosen embryogenesis or fetal stage, the pregnant ewes were anaesthetized using pentobarbitone and halothane. The embryos were aseptically removed from the uterus in an animal surgery and they were immersed in DMEM at 4°C. When the size of the fetus was large, the encephalon was immediately dissected out after the caesarean section and immersed in the DMEM medium. Some experiments were performed on newborn lamb encephalon (Richard et al., 1998).

For glial cell cultures, the tissue was dissociated by passing it successively through a needle previously filled with the medium. The cell suspension was directly used for seeding at the rate of 5.10^5 cells/ml in a non coated 60-mm-diameter Petri dish. The culture medium was DMEM supplemented with 2 mM glutamine, 100 units/ml of penicillin, 100 μg/ml of streptomycin and 10% fetal calf serum. The medium was renewed 3 to 4 days after plating, and then once per week. Sometimes, subcultures were made. For this, when the cells almost reached confluence, they were harvested by trypsinization and replated. The dishes were incubated at 37°C in a water-saturated atmosphere containing 5% CO_2.

The cultures obtained from the cerebral cortex of 40-day-old embryo (E40) to newborn lamb were very similar to the astrocytes cultured from the cerebral cortex of newborn mouse brain. Flat and polygonal cells developed quickly and gave confluent culture within two weeks. The main difference from the rodent cultured astrocytes was that the growth velocity of sheep astrocytes seemed to be higher when the number of cells seeded was the same. As in the case of rodent astrocyte culture, some fibrous astrocytes were present in the dishes. The presence of GFAP was searched for in sheep cultured astrocytes using anti-GFAP antibodies. Most cells cultured from sheep cerebral cortices were obviously stained. Glycogen was sought in these cells as a second way of characterization. We observed many glycogen particles in the astrocyte slices. This was particularly important when the culture medium was enriched in glucose.

Radial astrocytes and stem cells.   A major property of adult neural cells is that they are highly differentiated *in vivo*. In recent years, it has been shown that cells from special brain regions, such as the circumventricular regions, are able to undergo differentiation up to neural cells. They behave as neural stem cells. Because of the potential interest of such cells in many studies on the nervous system, it is of interest to obtain models of stem cells in culture. We had an opportunity of culturing such cells using sheep embryos.

Because of its large size as compared to that of rodents, sheep brain offers the possibility of getting enough material at a low stage of embryogenesis. So, we dissected out sheep embryos at 26 days of gestation, a stage when the neural tube is not completely closed. The culture was done as described above for sheep brain tissue. The neural tube was dissociated into single cells and plated either in polylysine-coated or in non-coated Petri dishes. The culture medium was fetal calf serum-supplemented or defined medium. During the first ten days, when the dishes were coated, neuronal cells developed and then died. Many morphological types of cells developed when the incubation of the Petri dishes was prolonged. Between these cells, some were astrocytes. There was a family of cells that looked like epithelial cells since they were flat and so close to each other that extracellular space was not visible. These epitheloid cells were together, thus forming islets inside the other cells. They were actively dividing, since the islets were very small at the

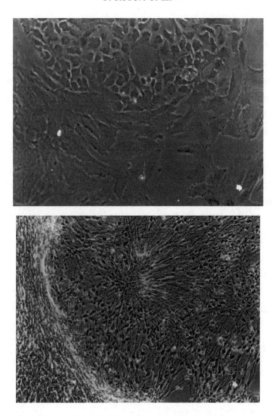

Figure 9. Neural stem cells (top) and radial astrocytes (bottom) from sheep cerebral cortex in culture. The conditions of the culture are explained in the text. Magnification: left photograph ×180; right photograph ×100.

beginning and they grew actively. They were not stained by anti-GFAP antibodies or by antineurofilament antibodies or by antineuron specific enolase antibodies.

Depending on an unknown factor, the epitheloid cells began to differentiate. They were less close together. Some of them developed a singular figure: from a focus, the cells aligned along straight radii on a relatively important area. These cells were labeled by anti-GFAP antibodies. Some other islet cells changed in common fibrous astrocytes, especially when the medium was a defined one. The most important feature of the islet cells was their long life span, since they were cultured during 7 months without change in their undifferentiated state. All along this life span, some of them differentiated into 'radial' or common fibrous astrocytes. Another feature was that after a subculture, these islets were always present in the Petri dishes, even after a second passage (Fig. 9).

In our hands, the epitheloid islet cells give principally astrocytes. This is not surprising, on the basis of the data published by many authors. It was shown that the phenotype of a given neural cell is largely influenced by the presence of some factors. The presence of platelet-derived growth factor stimulates the formation of oligodendrocytes (Raff *et al.*, 1988). Nerve growth factor and basic fibroblast growth factor promote the proliferation of neuronal precursors and their differentiation (Cattaneo and McKay, 1990). The influence of many factors has been reported (Gage *et al.*, 1995). In our experiments, no factor was added to the culture medium of epitheloid islet cells. In these conditions, only astrocytes develop. The utilization of different factors will help to understand the nature of our islet cells. Whatever, for the first time, the culture of sheep neural cells provides the possibility of making available stem neural cells for a long period, more than seven months, without a 'spontaneous' differentiation. This constitutes a suitable biological material easy to study for researchers who investigate cellular development in central nervous system as Kentroti and Vernadakis (1997) did in birds. Indeed, a large number of such cells can be easily obtained and the life span of these stem cells is no longer a limitation for the different studies.

NEURAL CELL ULTRASTRUCTURE

The ultrastructure of neural cells was studied *in vivo* when the electron microscope became available for biological research about five decades ago. When neural cells in culture became common models for the study of the brain, no thorough ultrastructural observation was made in these cultured cells, so we undertook the ultrastructural study of neurons and astrocytes.

Neurons

The cultures were obtained as described above. They were rinsed using PBS at pH 7.4 and fixed in 2.5% buffered glutaraldehyde. They were washed in PBS and were post-fixed for one hour in a solution containing 1% osmium tetroxide and 1.5% potassium ferricyanide. The cell layer was then dehydrated in graded alcohols and was pre-embedded in three mixtures of epoxypropane and Epon 812 resin in the ratio 2 : 1, 1 : 1, and 1 : 2, respectively. The embedding was done in 100% Epon and the polymerization was done at 60°C and lasted for 24 h. The samples were then sliced into ultrathin sections of 90- to 100-nm-thickness. These sections were contrasted with 4% uranyl acetate for 20 min followed by 0.4% lead acetate for 10 min. The sections were covered with carbon powder, examined at an accelerating power of 80 Kev, and photographed using a transmission electron microscope (Philips CM 10).

The ultrastructure of cultured neurons was similar to that *in vivo* (Fig. 10). Neurons showed abundant organelles in the perinuclear cytoplasm such as free ribosomes, polysomes, endoplasmic reticular, Golgi complex and mitochondria. Near these conventional organelles, two kinds of abnormal organelles were observed: (1) vesicles similar to myeloid bodies and (2) disfigured mitochondria. The neuronal nuclei contained homogenous and clear chromatin and a prominent nucleolus. The processes displayed microtubules, intermediate filaments and mitochondria. Myelin was not observed anywhere. From the 14th day after the seeding, neurons showed

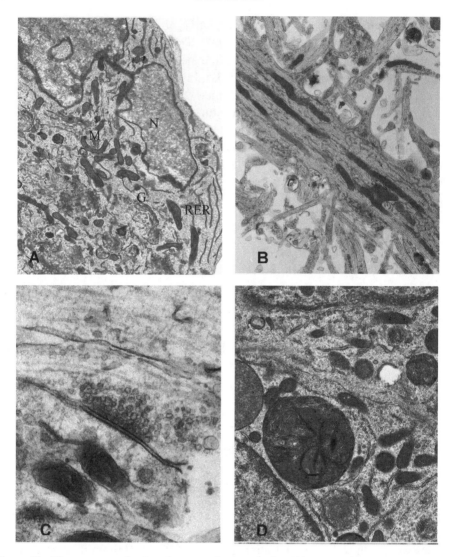

Figure 10. Ultrastructure of cultured neurons from rat brain. (A): soma with common organelles; (B): neurite displaying microtubules and mitochondria; (C): a typical asymetric synapse containing clear vesicles; (D): a multilamellar vesicle is observed in neuronal cytoplasm. N, nucleus; RER, rough endoplasmic reticulum; M, mitochondrion; G, Golgi apparatus Magnification: A and B ×6700; C ×34 500; D ×10 000.

synapses that had the morphological features of *in vivo* synapses. Indeed, synapses displayed asymmetric membranes densities, synaptic clefts and clear synaptic vesicles of 30–50 nm average diameter in bulbous presynaptic endings. Between 2 and 3 weeks after seeding, neuron processes developed numerous synaptic connections and displayed a variety of complex synapses. For example, a single spine might receive contacts from two presynaptic terminals.

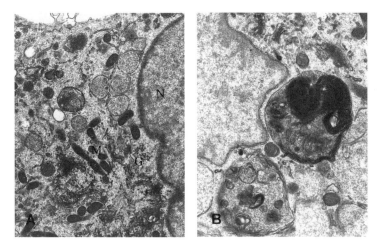

Figure 11. Ultrastructure of cultured astrocytes from rat brain. The cytoplasm of primary cultured astrocytes contain current organelles (A) and abnormal organelles constituted by multilamellar vesicles (B). N, nucleus; M, mitochondrion; G, Golgi apparatus. Magnification: (A) ×8400 and (B) ×15 000.

The description given above shows that neuronal culture is a simple method of producing convenient neurons *in vitro*. This method allows cells to be maintained in culture for many weeks without any passage. So, neuron cultures should be considered as a valuable tool which could be particularly useful for many studies such as neurodegeneration, neuroprotection, pharmacology, investigations on neuron growth and neuron injury, etc.

Astrocytes

The observation of slices of rat astrocytes from primary culture or from subcultures revealed the presence of normal organelles. Cytomembrane bilayers were clearly observed when a high magnification was set for the electron microscope observations. Endoplasmic reticular, bearing ribosomes or not and Golgi cisternae were abundant inside the cytoplasms. Long bundles of filaments, which were GFAP, lay among the organelles. The nuclei were apparently normal, since compact chromatin was near the nuclear envelope and the largest part of these nuclei contained euchromatin. In the neighbourhood of these normal organelles, there were two kinds of abnormal organelle: disfigured mitochondria and vesicles that looked like myeloid bodies (Fig. 11). These abnormal organelles are under investigation currently.

CONCLUSION

The present work shows that it is possible to produce a great variety of neural cells in culture. Depending on the conditions of culture, all kinds of neurons and astrocytes observed *in vivo* could be cultured for the purpose of many kinds of studies on the nervous system. For the future, the factors that induce a given phenotype must be investigated. In this area some factors such as growth factors are already published. However, as defined media are also able to give different kinds of phenotypes, there may exist other factors than growth factors that influence the phenotype. These studies may be extended to oligodendrocytes and microgliocytes.

Acknowledgements

This work has benefited from financial support from the French Region Centre, Conseil Général du Loiret, and l'ARC. F. Robert is a scholarship recipient of the Region Centre. We thank M. C. Nicolle, Mrs. B. Delaleu from INRA of Tours-Nouzilly, and M. P. Moreau for their technical help.

REFERENCES

Booher, J. and Sensenbrenner, M. (1972). Growth and cultivation of dissociated neurons and glial cells from embryonic chick, rat and human brain in flask cultures, *Neurobiology* **2**, 97–105.

Bottenstein, J. E. and Sato, G. H. (1979). Growth of a rat neuroblastoma cell line in serum-free supplemented medium, *Proc. Natl. Acad. Sci. USA* **76**, 514–517.

Cattaneo, E. and McKay, R. (1990). Proliferation and differentiation of neuronal stem cells regulated by nerve growth factor, *Nature* **347**, 762–765.

Cavanagh, J. B. (1970). The proliferation of astrocytes around a needle wound in the rat brain, *J. Anat.* **106**, 471–487.

Duittoz, A. H., Batailler, M. and Caldani, M. (1997). Primary cell culture of LHRH neurones from embryonic olfactory placode in the sheep (Ovis aries), *J. Neuroendocrinol.* **9**, 669–675.

Eng, L. F. (1988). Astrocytic response in injury, in: *Current Issues in Neural Regeneration Research*, Reir, P., Bunge, R. and Seil, F. (Eds), pp. 247–255. A. R. Liss, New York.

Gage, F. H., Ray, J. and Fisher, L. J. (1995). Isolation, characterization, and use of stem cells from the CNS, *Annu. Rev. Neurosci.* **18**, 159–192.

Hansson, E., Ronnback, L., Lowenthal, A., *et al.* (1982). Brain primary culture — a characterization (part II), *Brain Res.* **231**, 173–183.

Hevor, T. K. and Delorme, P. (1991). Biochemical and ultrastructural study of glycogen in cultured astrocytes submitted to the convulsant methionine sulfoximine, *Glia* **4**, 64–69.

Kentroti, S. and Vernadakis, A. (1996). Immunocytochemical and biochemical characterization of glial phenotypes in normal and immortalized cultures derived from 3-day-old chick embryo encephalon, *Glia* **18**, 79–91.

Kentroti, S. and Vernadakis, A. (1997). Differential expression in glial cells derived from chick embryo cerebral hemispheres at an advanced stage of development, *J. Neurosci. Res.* **47**, 322–331.

Kilpatrick, T. J. and Bartlett, P. F. (1993). Cloning and growth of multipotential neural precursors: requirements for proliferation and differentiation, *Neuron* **10**, 255–265.

Lee, K., Kentroti, S., Billie, H., *et al.* (1992). Comparative biochemical, morphological, and immunocytochemical studies between C-6 glial cells of early and late passages and advanced passages of glial cells derived from aged mouse cerebral hemispheres, *Glia* **6**, 245–257.

Mbarek, O., Verge, V. and Hevor, T. (1998). Direct cloning of astrocytes from primary culture without previous immortalization, *In Vitro Cell Dev. Biol. Anim.* **34**, 401–411.

Murphy, T. H., Schnaar, R. L. and Coyle, J. T. (1990). Immature cortical neurons are uniquely sensitive to glutamate toxicity by inhibition of cystine uptake, *Faseb J.* **4**, 1624–1633.

Nathaniel, E. J. and Nathaniel, D. R. (1977). Astroglial response to degeneration of dorsal root fibers in adult rat spinal cord, *Exp. Neurol.* **54**, 60–76.

Pentreath, V. W., Seal, L. H., Morrison, H. J., *et al.* (1986). Transmitter mediated regulation of energy metabolism in nervous tissue at the cellular level, *Neurochem. Int.* **9**, 1–10.

Phelps, C. H. (1972). Barbiturate-induced glycogen accumulation in brain. An electron microscopic study, *Brain Res.* **39**, 225–234.

Phelps, C. H. (1975). An ultrastructural study of methionine sulphoximine-induced glycogen accumulation in astrocytes of the mouse cerebral cortex, *J. Neurocytol.* **4**, 479–490.

Raff, M. C. (1989). Glial cell diversification in the rat optic nerve, *Science* **243**, 1450–1455.

Raff, M. C., Lillien, L. E., Richardson, W. D., *et al.* (1988). Platelet-derived growth factor from astrocytes drives the clock that times oligodendrocyte development in culture, *Nature* **333**, 562–565.

Reynolds, B. A. and Weiss, S. (1992). Generation of neurons and astrocytes from isolated cells of the adult mammalian central nervous system, *Science* **255**, 1707–1710.

Reynolds, B. A., Tetzlaff, W. and Weiss, S. (1992). A multipotent EGF-responsive striatal embryonic progenitor cell produces neurons and astrocytes, *J. Neurosci.* **12**, 4565–4574.

Richard, O., Duittoz, A. H. and Hevor, T. K. (1998). Early, middle, and late stages of neural cells from ovine embryo in primary cultures, *Neurosci. Res.* **31**, 61–68.

Saneto, R. P. and De Vellis, J. (1987). Neuronal and glial cells: cell culture of the central nervous system, in: *Neurochemistry: A Practical Approach*, Turner, A. J. and Bachelard, H. S. (Eds), pp. 27–63. IRL Press, Oxford.

Swanson, R. A., Yu, A. C., Sharp, F. R., *et al.* (1989). Regulation of glycogen content in primary astrocyte culture: effects of glucose analogues, phenobarbital, and methionine sulfoximine, *J. Neurochem.* **52**, 1359–1365.

Tillet, Y. and Thibault, J. (1987). Early ontogeny of catecholaminergic structures in the sheep brain. Immunohistochemical study, *Anat. Embryol. (Berlin)* **177**, 173–181.

Turner, D. L., Snyder, E. Y. and Cepko, C. L. (1990). Lineage-independent determination of cell type in the embryonic mouse retina, *Neuron* **4**, 833–845.

Verge, V., Legrand, A. and Hevor, T. (1996). Isolation of carbohydrate metabolic clones from cultured astrocytes, *Glia* **18**, 244–254.

Weiss, S., Dunne, C., Hewson, J., *et al.* (1996). Multipotent CNS stem cells are present in the adult mammalian spinal cord and ventricular neuroaxis, *J. Neurosci.* **16**, 7599–7609.

Wetts, R. and Fraser, S. E. (1988). Multipotent precursors can give rise to all major cell types of the frog retina, *Science* **239**, 1142–1145.

Wiesinger, H., Schuricht, B. and Hamprecht, B. (1991). Replacement of glucose by sorbitol in growth medium causes selection of astroglial cells from heterogeneous primary cultures derived from newborn mouse brain, *Brain Res.* **550**, 69–76.

Advances in Neuroregulation and Neuroprotection (2005), pp. 189-205
Collin, C. *et al.* (Eds)
© VSP 2005

Behavioural and kinetic analysis of locomotion in anxious and non-anxious strains of mice tested on a rotating beam

J. NEGRONI*, P. VENAULT, C. BENSOUSSAN, C. COHEN-SALMON,
E. LEPICARD, F. PEREZ-DIAZ, R. JOUVENT and G. CHAPOUTHIER

*'Vulnérabilité, Adaptation et Psychopathologie', CNRS UMR 7593, Hôpital Pitié-Salpêtrière,
Université Paris VI, 91 Boulevard de l'Hôpital, 75634 Paris Cedex 13, France*

Abstract—The model of the rotating beam was used to analyse detailed patterns of locomotor behaviour in anxious and non-anxious strains of mice. Comparisons involved the non-anxious B6D2F1 strain, the anxious ABPLe strain and the DBA/2J strain vulnerable to stressors. Detailed analysis was made of several parameters, both postural (height of the trunk, tail angle, number of imbalances and falls and head movements) and kinetic (mean speed on the beam, distance covered for large and small movements, plus time spent in no-motion episodes). The findings provided evidence of numerous differences between the strains: DBA/2J mice were moderately sensitive to the task and ABPLe subjects were extremely reactive. Signs of increased anxiety included a reduction in the tail angle, an increase in the number of imbalances, falls and head movements, a decrease in the mean speed and distance covered in large movements, and an increase in the number of no-motion episodes, as well as time spent and distance covered with small movements. All these data validate the use of the model to study the relationship between anxiety and balance control.

Keywords: Anxiety; balance; posture; locomotion; mice.

INTRODUCTION

Previous studies have reported a strong association between dizziness and anxiety in patients with agoraphobia and panic attack (Yardley *et al.*, 1994). Because vestibular system dysfunction leads to dizziness and disorientation, it has been suggested that a mismatch in the neural integration of sensory input may generate disturbances in space perception and balance control (i.e. vertigo and motion sickness) (Jacob *et al.*, 1985; Reason, 1978; Treisman, 1977). However, it is difficult to demonstrate a cause-and-effect relationship between pathological anxiety and vestibular system dysfunction.

---

*To whom correspondence should be addressed. E-mail: negroni@ext.jussieu.fr

Very few experimental animal procedures have been designed to study the functional link between anxiety and balance control; consequently, the mechanisms underlying the relationship between anxiety and balance control are still unclear. In a previous article (Lepicard *et al.*, 2000) attention was drawn to the possible relationship in mice between anxiety as measured in several classical anxiety tasks and balance control tested on a rotating beam. In this study, animals motivated by partial food deprivation had to move along a rotating beam, trying to cross without falling. The procedure gave an accurate assessment of balance control and posture, using sensitive measurements such as the number of imbalances and postural positions, the angle of the tail and elevation of the trunk. Striking differences were observed between the two inbred strains of mice known to have radically different anxiety-related behaviour. In the open-field test (Crusio and Schwegler, 1987; Goodrick, 1976; Hallam and Hinchcliffe, 1991) and the elevated plus-maze test (Cole *et al.*, 1995; Conti *et al.*, 1994), BALB/cByJ is considered an emotive strain and C57BL/6J is considered a non-emotive strain. In the rotating beam test, the highly anxious strain, BALB/cByJ, performed poorly compared to the non-anxious strain, C57BL/6J (Treisman, 1977).

Behavioural data from the rotating beam test have been confirmed by pharmacological studies. Balance control and postural control of anxious BALB/cByJ mice were improved by acute anxiolytic treatment with benzodiazepine diazepam. Poorer behavioural performance for both balance control and posture was observed in non-anxious C57BL/6J mice given acute anxiogenic treatment with the benzodiazepine receptor inverse agonist, methyl β-carboline-3-carboxylate (β-CCM).

Another study using the rotating beam test assessed the effect of two chronically administered selective serotonin reuptake inhibitors (SSRIs), fluoxetine and paroxetine. For the three behavioural parameters (number of imbalances, angle of tail and elevation of trunk) observed in anxious BALB/cByJ mice, both compounds had the same diazepam-like effects, i.e. a reduction in the number of imbalances, higher elevation of the trunk and an increase in the tail angle (Venault *et al.*, 2001). These data suggested that SSRIs could be useful in the treatment of anxiety-induced balance impairments and confirmed the findings of Lepicard *et al.* (2000) and their hypothesis that balance and posture response to a sensory challenge are related to anxiety processes in mice.

Measurements on the rotating beam used in all previous articles by the present authors were limited to easily observable behavioural parameters, such as the height of the trunk, tail angle and number of imbalances and head movements. The present study observed three different strains and was designed to validate highly sensitive kinetic recordings of behavioural data so as to compare the different behavioural strategies on the beam. The purpose therefore was not to draw a direct correlation between anxiety and balance control, but to assess new techniques in order to improve the behavioural recording of locomotion on the beam. To evaluate both balance control and locomotor performance in the three strains, a detailed analysis was made of several parameters including both the postural measurements used in previous reports and the new, more specific and accurate kinetic measurements, i.e. the mean speed along the beam, the distance covered for large and small movements, plus time spent in no-motion episodes. The three strains used were the B6D2F1 strain, which has been shown to be resistant to stressors (Pardon *et al.*, 2000), DBA/2J, which has shown a high level of vulnerability to stressors (Gilad and Gilad, 1995; Shanks and Anisman, 1988), and ABPLe, an anxious strain displaying certain anxiety traits controlled by genes which have been located (Clément *et al.*, 1994, 1995). The three strains appeared to be sufficiently different for a convincing validation of the kinetic parameters used.

MATERIALS AND METHODS

All procedures complied with the directive from the European Community council on the ethical treatment of animals (86/609/EEC).

Animals

Twenty four-month-old female B6D2F1/JICO mice [Iffa-Credo, Lyon, France], twenty four-month-old ABP/Le mice (10 females and 10 males) [Jackson Institute, Bar. Harbor, ME, USA] and twelve four-month-old DBA/2 mice (6 females and 6 males) [Iffa-Credo, Lyon, France] were used. The mice were brought into the laboratory two months before the beginning of the experiment. On arrival, the animals were housed in groups of five per cage ($20 \times 32 \times 12$ cm) in the animal research facility and maintained under standard laboratory conditions: $12:12$ h light:dark cycle (with lights on at 07.30 h), $22 \pm 2°C$, 60% relative humidity, *ad libitum* access to food and water.

The animals were tested on the rotating beam task after 1 hour of familiarisation with the test room.

Characteristics of mice strains used

B6D2F1. Because genetic factors are thought to play a role in inter-individual variability to stress responses (Chrousos and Gold, 1992), the study was performed on B6D2F1 mice, the first generation produced by crossing C57BL/6J with DBA/2, both inbred strains; observations of these two strains have shown contrasts in a number of behavioural and biological parameters measured as a response to stress. The impact of genetic influences in B6D2F1 mice was reduced by crossing the alleles that may be involved in vulnerability or resistance to stressors. All B6D2F1 subjects were genetically identical and displayed resistance to stressors with a heterosis effect (Shanks and Anisman, 1988).

DBA/2J. DBA/2J is an inbred mouse strain. The light/dark procedure used by different author has shown DBA/2J to have a lower level of anxiety than BALB/cByJ, but it is still an emotive strain (Crawley and Davis, 1982; Trullas and

Skolnick, 1993). DBA/2J also shows a higher level of vulnerability to stressors than the inbred strain C57BL/6J which has shown resistance to stressors (Gilad and Gilad, 1995; Shanks and Anisman, 1988). In spatial-specific learning tasks (Morris and place learning-set tasks), DBA/2J mice perform poorly compared to C57BL/6J mice (Wehner *et al.*, 1990).

ABP/Le. ABP/Le is also an inbred mouse strain with six easily identifiable traits, e.g. colour of fur and size of ears. These traits are controlled by genes located on small chromosomal fragments previously transferred into the ABP/Le strain and their location has been identified (Clément *et al.*, 1994, 1995). Some of these genes are associated with 'anxious' behaviour. By using Mendelian crossings (intercrosses and backcrosses) between ABP/Le mice and the inbred mouse strain, C57BL/6By, it was possible to dissociate the different loci to see if any were associated with behaviour patterns classically related to anxiety. In previous studies, using the open-field test, Clément *et al.* found that three loci were associated with anxious behaviour: the regions containing the brown (b) and pink-eyed dilution (p) loci, located respectively on chromosomes 4 and 7, were associated with an increase in peripheral activity in the open-field (Clément *et al.*, 1995); the fragment containing the short-ear locus (*se*) was associated with an increase in grooming (Clément *et al.*, 1994), a behaviour pattern often observed after anxious episodes (D'Aquila *et al.*, 2000; Rodgers and Cole, 1993). The tendency to avoid moving in the centre of an open area and an increase in grooming activity are classically linked to an 'anxious' process in rodents. It has also been shown that one locus on chromosome 9 is associated with seizures induced by methyl β–carboline-3-carboxylate (β-CCM, a benzodiazepine receptor inverse agonist) (Clément *et al.*, 1996). Analysis of (3H) flumazenil binding to brain GABA$_A$ receptors has suggested a possible involvement of a decrease in B$_{max}$ in both β-CCM-induced seizures and anxiogenic processes (Clément *et al.*, 1997).

Rotating beam test

Apparatus. The apparatus consisted of a beam, set 40 cm above the floor, operated by a motor at one end, rotating along its axis at a constant speed of 2.25 rpm, (Fig. 1a). At the other end, a food reward was placed on a platform. The beam could be removed and replaced by another. A layer of sawdust was provided to break any possible fall.

Behavioural procedure.

Learning phase (LP). Two days before the beginning of the experiment, the mice were partially deprived of food, but had free access to water. Their weight was recorded every day during the experiment to ensure that body weight did not fall below 20% of the initial value. During a 4-day training session (LP1 to LP4), adult mice were trained to move along a non-rotating horizontal cylindrical wooden beam

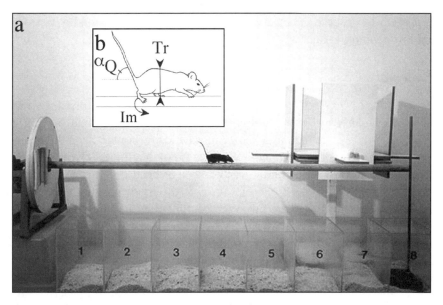

Figure 1. (a) Rotating beam apparatus, (b) measurement of balance control and posture: imbalances (Im) in numbers, elevation of trunk (Tr) as a ratio (height/length of paw). Angle of tail (α_Q) in degrees. (Modified from Lepicard *et al.* (2000).)

(w) ($D = 1.5$ cm; $L = 105$ cm), to a platform (Fig. 1a). The distance covered by the mice was increased each day. After each trial, the mice could easily climb onto the platform where they were given a food reward for 5 min.

Test phase.   On the first day of the test (T1), the wooden beam (w) was replaced by a metal beam (m) which did not rotate ($D = 1.5$ cm; $L = 105$ cm). The variation in the texture of the beam was intended to disturb the tactile system of the mice trained with the wooden beam over the 4 days of learning (Lepicard *et al.*, 2000). On the second day of the test (T2) the same metal beam was used, but was rotated at a constant speed. The variation with the movement of the beam was intended to disturb the balance control of the mice so as to stimulate the vestibular system. On T3, the rotating beam was longer than on the previous day ($D = 1.5$ cm; $L = 190$ cm). The variation in length was designed as a greater challenge motivating the mice and changing the visual reference marks for locating the platform previously used by the mice. On T4, the beam was a metal rod ($D = 1.2$ cm) extended with a second metal rod ($D = 1.0$ cm) fitted into the end of the first. The total length was 190 cm with smaller diameters than the previous beams. The two rods rotated together in the same direction. T4 was the most difficult test day; the variation in diameter challenged the motor functions and stability of the mice. On T5, the same beam with two different diameters was used; the same task was repeated on this last day of the test so as to study the ability of the mice to cope with the task. All the experiments were recorded in full using a VHS video camera to observe balance control and posture analysis and a Video-Track system for locomotion analysis.

Behavioural analysis.

Balance control and posture analysis. Balance control performance was assessed by the number of mice falling and the number of imbalances (Fig. 1b). Imbalances (Im) were defined as at least one paw losing contact with the beam outside the normal locomotion context. The elevation of the trunk (Tr), the tail position (α_Q) and the head position of the mice (number of head dippings) were observed and recorded for each mouse: the elevation of the trunk was defined as the distance between the trunk and the beam (in centimetres), i.e. the distance between the middle of the top of the animal's back and the surface of the beam. To avoid bias due to inter-individual differences, the measurement was expressed as a ratio (height/length of foot). The position of the tail was defined as the angle between the horizontal line of the beam and the tail (in degrees).

Locomotion analysis. A computer-operated digital video acquisition system was used to record different rates of locomotion (inactivity, slow and fast motion). The Video-track® 512 system, developed by View Point, a company based in Champagne au Mont d'Or (France), monitored the movement of the laboratory animals in real time. The position of the animal was recorded with a sample frequency of 25 Hz. An operator entered two cut-off values for different rates: inactivity/slow movements and slow/fast movements. The two cut-off values were recorded by 20 C57BL/6J non-anxious mice tested when moving along a non-rotating horizontal cylindrical wooden beam.

The system recorded the total time animals spent on the beam, the number of movements, the distance covered, the duration of inactivity, plus the number of movements, distance and duration at the slow rate as well as the number of movements, distance and duration at the fast rate. The mean velocity was calculated as the distance covered (length of beam used) / total time.

In the findings, the description of locomotion variables focused on mean velocity, the number of pauses (inactivity), the number of movements at a slow rate, the distance covered at a slow rate, the number of movements at a fast rate and the distance covered at a fast rate. Other variables, such as the duration of pauses and the duration of movements at both slow and fast rates were not significant.

Statistical analysis

To assess possible inter-strain behavioural differences, all measurements were assessed using an ANOVA procedure. An analysis of repeated values (ANOVA) using the SAS software mixed procedure was carried out for each variable: mean values of elevation of the trunk, angle of the tail, number of falls, number of imbalances, number of head dippings, velocity, inactivity, slow rate and fast rate.

For each subject, variables were measured for the 'day' factor (six conditions LP4, T1, T2, T3, T4 and T5), the 'group' factor (three conditions: B6D2F1, DBA/2J and ABPLe). The 'strain X day' interaction was also included in the analysis. Two-by-two comparisons of strains (B6D2F1 and DBA/2J, B6D2F1 and ABPLe, DBA/2J and ABPLe) were made for each day, as well as day-to-day comparisons for each strain, all being analysed using a contrast procedure.

RESULTS

After the learning phase, the rotating beam test procedure was comprised of sequential tasks over five test days. From test day 1 (T1) to test day 4 (T4), the difficulty of the task was increased. Observations on T1 and T2 showed the three strains had intra-strain performance levels similar to the levels reached on the last day of the learning phase, LP4. In the present study, LP4, T1 and T2 are therefore considered as simple tasks; T3 was considered an intermediate task and T4 the most difficult. T5 is the same task as T4, repeated on the next day. The findings therefore present only LP4, T3, T4 and T5, so as to compare the performance of the mice dealing with the increasing difficulty between LP4 and T5.

Balance control performance and posture analysis

Balance control performance. This was assessed by the number of falls and the number of imbalances. Statistical analysis of the number of falls and the number of imbalances showed significant inter-strain differences. For the entire procedure, the number of ABPLe mice falling was significantly higher than for DBA/2J (t_{28} = 2.75, p = 0.0103) and B6D2F1 (t_{18} = 5.50, p < 0.0001) (Fig. 2A). The mean number of imbalances (Fig. 2B) showed significant differences between ABPLe and B6D2F1 (t_{18} = 9.79, p < 0.0001) and between ABPLe and DBA/2J (t_{28} = 5.63, p < 0.0001). In relation to the day factor (Fig. 2C), the difference in imbalances between ABPLe and B6D2F1 (t_{107} = 2.81, p = 0.0059: LP4; t_{107} = 3.24, p = 0.0016: T3, t_{107} = 5.76, p < 0.0001: T4, t_{107} = 4.64, p < 0.0001: T5) and between ABPLe and DBA/2J (t_{166} = 2.31, p = 0.0223: LP4; t_{166} = 5.22, p < 0.0001: T4) was significant for two experimental conditions: LP4 (last day of learning phase) and T4 (change in diameter of the rotating beam).

Balance control performance can be used to rank the strains in order of vestibular system dysfunction: ABPLe > DBA/2J > B6D2F1. The anxious strain, ABPLe, fell most often and recorded the highest number of imbalances.

Posture. Measurements of the elevation of the trunk relative to the beam were recorded for each strain (Fig. 3A). Results showed that trunk elevation in DBA/2J and ABPLe was significantly lower than in B6D2F1 (t_{14} = 8.94, p < 0.0001, t_{18} = 5.74, p < 0.0001, respectively). Most DBA/2J and ABPLe mice kept their trunk close to the beam, whereas most B6D2F1 mice had high arched trunks.

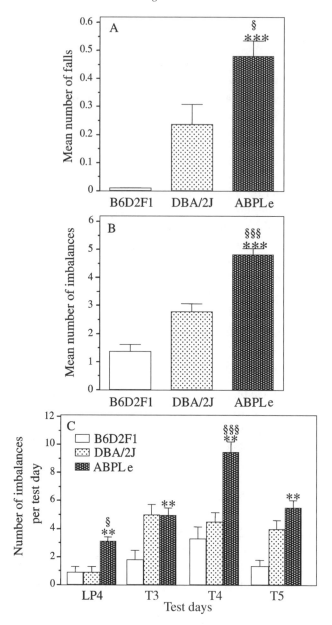

Figure 2. Mean number (±SEM) of falls (A) and imbalances (B) for the total procedure for each strain. Number of imbalances per test day (C). $^{§}p < 0.05$ between ABPLe and DBA/2J, $^{**}p < 0.01$ between ABPLe and B6D2F1, $^{§§§}p < 0.0001$ between ABPLe and DBA/2J, $^{***}p < 0.0001$ between ABPLe and B6D2F1.

Differences in tail elevation relative to the beam were observed (Fig. 3B) between ABPLe and DBA/2J ($t_{28} = 19.61$, $p < 0.0001$), ABPLe and B6D2F1 ($t_{18} = 10.36$, $p < 0.0001$) and B6D2F1 and DBA/2J ($t_{14} = 8.70$, $p < 0.0001$). Most ABPLe mice kept their tails low and close to the beam (mean 5°), whereas most

B6D2F1 mice maintained their tails raised for proper stability. However, the tail angle of DBA/2J mice (mean 60°) was twice the angle of B6D2F1 (mean 30°). During locomotion on the beam, many DBA/2J mice displayed jerky, swinging tail movements, which was not the case for B6D2F1 mice.

The mean number of head dippings was recorded for each strain (Fig. 3C). The ABPLe strain recorded significantly more head dippings during locomotion on the beam compared to DBA/2J (t_{28} = 6.40, p < 0.0001) and B6D2F1 (t_{18} = 11.58, p < 0.0001).

Kinetic analysis of locomotion

Velocity. The mean speed of mice on the beam was analysed for each strain (Fig. 4A) and differences were observed between strains: B6D2F1 moved fast on the beam, differing significantly from DBA/2J (t_{14} = 3.49, p = 0.0036) and ABPLe (t_{18} = 9.38, p < 0.0001). The mean velocity of DBA/2J was intermediate and the mean velocity of ABPLe was slowest (t_{28} = 5.53, p < 0.0001). The fast motion of B6D2F1 mice was steady over the total distance covered, but DBA/2J and ABPLe moved at variable speeds with many stops and pauses. In relation to the day factor, the most striking significant differences between ABPLe and B6D2F1 and between DBA/2J and B6D2F1 (Fig. 4B) were observed for the last day of the learning phase LP4 (t_{108} = 5.95, p < 0.0001; t_{82} = 4.13, p < 0.0001, respectively) and for the last test day T5 between B6D2F1 and ABPLe, (t_{108} = 5.28, p < 0.0001) and between B6D2F1 and DBA/2J (t_{82} = 2.49, p = 0.0147).

Inactivity. Values measured for inactivity for each strain (Fig. 5A) tallied with the observations on velocity and showed the mean number of pauses by ABPLe to be significantly higher than B6D2F1 (t_{18} = 5.78, p < 0.0001) and DBA/2J (t_{28} = 2.54, p = 0.0171). The mean number of pauses by DBA/2J was at an intermediate level and significantly different from values recorded for B6D2F1 (t_{14} = 2.72, p = 0.0167). In relation to the day factor, inactivity values (Fig. 5B) were significantly different for LP4 between B6D2F1 and DBA/2J (t_{82} = 3.01, p = 0.0034); for T4 between B6D2F1 and ABPLe (t_{108} = 2.91, p = 0.0044) and between DBA/2J and ABPLe (t_{166} = 3.00, p = 0.0031); and for T5 between B6D2F1 and DBA/2J and between B6D2F1 and ABPLe (t_{82} = 4.04, p = 0.0001; t_{108} = 4.34, p < 0.0001, respectively).

Slow motion. The mean number of movements at a slow rate measured (Fig. 6A) for DBA/2J and ABPLe was significantly higher than for B6D2F1 (t_{14} = 4.81, p = 0.0003; t_{18} = 2.62, p < 0.01, respectively). In relation to the day factor, the number of movements at a slow rate (Fig. 6B) by ABPLe was significantly higher than B6D2F1 for LP4 (t_{108} = 2.68, p = 0.0082), T3 (t_{108} = 2.30, p = 0.0228), T4 (t_{108} = 3.92, p < 0.0001) and T5 (t_{108} = 2.84, p < 0.0001). No significant difference was observed between B6D2F1 and DBA/2J for T3 and T4, but the

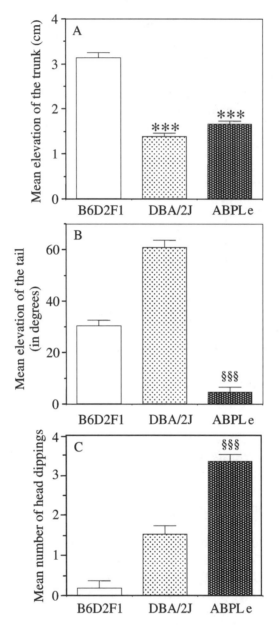

Figure 3. Mean elevation of the trunk (A), of the tail (B), values (±SEM) and mean number of head dippings (±SEM) for each strain. $^{§§§}p < 0.0001$ between ABPLe and DBA/2J, $^{***}p < 0.0001$ between ABPLe and B6D2F1, and between DBA/2J and B6D2F1.

number of slow movements by DBA/2J on LP4 and T5 was significantly higher than for B6D2F1 ($t_{82} = 4.97$, $p < 0.0001$; $t_{82} = 4.12$, $p < 0.0001$, respectively).

Measurements of the mean distance travelled at a slow rate (Fig. 6C) by DBA/2J and ABPLe were significantly higher than for B6D2F1 ($t_{14} = 7.78$, $p < 0.0001$;

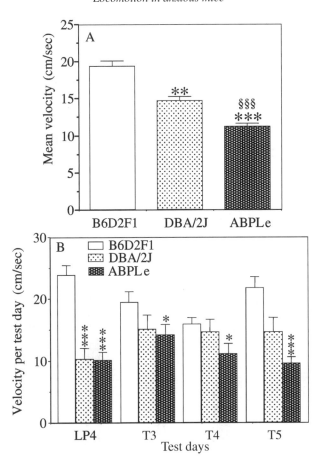

Figure 4. Mean velocity (A) and velocity per test day (B) values (±SEM) for each strain. $^{*}p < 0.05$ between ABPLe and B6D2F1, between DBA/2J and B6D2F1, $^{**}p < 0.01$ between DBA/2J and B6D2F1, $^{\S\S\S}p < 0.0001$ between ABPLe and DBA/2J, $^{***}p < 0.0001$ between ABPLe and B6D2F1, and between DBA/2J and B6D2F1.

$t_{18} = 8.02$, $p < 0.0001$, respectively). A significant difference was also observed between ABPLe and DBA/2J ($t_{28} = 4.21$, $p = 0.0002$).

Fast motion. In DBA/2J, the mean number of movements at a fast rate (Fig. 7A) was higher than B6D2F1 and ABPLe ($t_{14} = 3.23$, $p = 0.0061$; $t_{28} = 4.77$, $p < 0.0001$, respectively). The mean number of movements at a fast rate by B6D2F1 was at an intermediate level with no significant difference when compared to ABPLe. ABPLe made few movements at a fast rate.

Measurements for the mean distance travelled at a fast rate (Fig. 7B) by ABPle were significantly lower than for B6D2F1 ($t_{18} = 10.89$, $p < 0.0001$) and DBA/2J ($t_{28} = 7.46$, $p < 0.0001$). B6D2F1 had the highest values for the distance travelled at a fast rate; DBA/2J was at an intermediate level which was significantly different from B6D2F1 ($t_{14} = 3.85$, $p = 0.0018$).

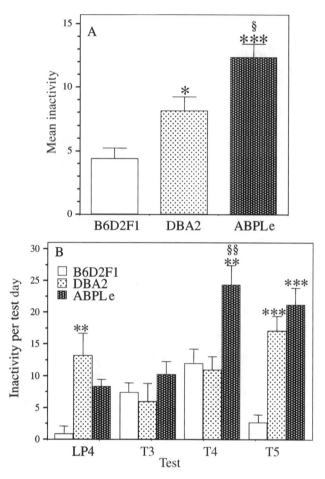

Figure 5. Mean inactivity (A) and inactivity per test day (B) values (±SEM) for each strain. $^{§}p < 0.05$ between ABPLe and DBA/2J, $^{*}p < 0.05$ between DBA/2J and B6D2F1, $^{§§}p < 0.01$ between ABPLe and DBA/2J, $^{**}p < 0.01$ between DBA/2J and B6D2F1, between ABPLe and B6D2F1, $^{§§§}p < 0.0001$ between ABPLe and DBA/2J, $^{***}p < 0.0001$ between ABPLe and B6D2F1, and between DBA/2J and B6D2F1.

DISCUSSION

The balance control values, similarly to those used in previous articles (Lepicard *et al.*, 2000), showed clear differences in balance between the three strains: ABPLe mice recorded the highest number of falls and imbalances, particularly on test day 4 (T4), when the diameter of the beam was changed. The variation in beam diameter was designed to challenge motor abilities and stability in mice. On the last day of the learning phase (LP4), the balance control of ABPLe was disturbed compared to B6D2F1 and DBA/2J. In line with the conclusions of the earlier studies (Lepicard *et al.*, 2000), the anxiety profile of ABPLe mice seems to delay the process of learning how to move along the non-rotating wooden beam. The ranked order of balance control placed B6D2F1 ahead of DBA/2J. Under the stressful

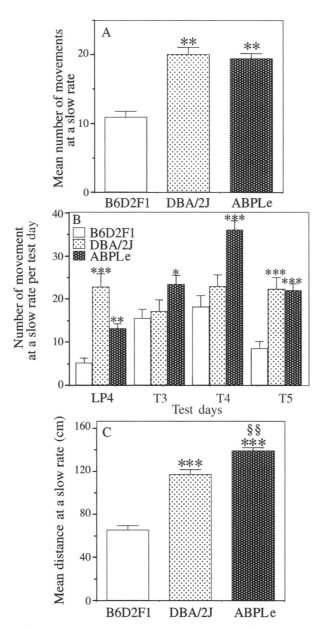

Figure 6. Mean number of movements at a slow rate (A), number of movements at a slow rate per test day (B) and mean distance covered at a slow rate (C) values (±SEM) for each strain. *$p < 0.05$ between DBA/2J and B6D2F1, §§$p < 0.01$ between ABPLe and DBA/2J, **$p < 0.01$ between DBA/2J and B6D2F1, between ABPLe and B6D2F1, ***$p < 0.0001$ between ABPLe and B6D2F1, and between DBA/2J and B6D2F1.

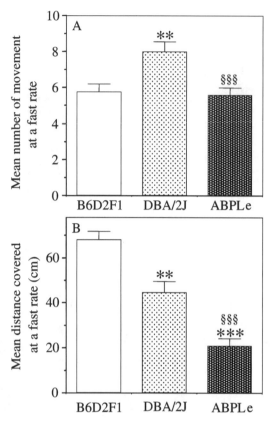

Figure 7. Mean number of movements at a fast rate (A) and mean distance covered at a fast rate (B) values (±SEM) for each strain. ** $p < 0.01$ between DBA/2J and B6D2F1, §§§ $p < 0.0001$ between ABPLe and DBA/2J, *** $p < 0.0001$ between ABPLe and B6D2F1.

environmental conditions of the experiment, balance control in DBA/2J was mildly disturbed, suggesting that the level of anxiety of the mice might be increased; this was not the case for the B6D2F1 strain which is resistant to stressors.

Postural values recorded for each strain showed a similar postural profile for ABPLe and DBA/2J, differing from the profile of the non-anxious B6D2F1. ABPLe and DBA/2J had a lower centre of gravity with the trunk closer to the beam. The angle of elevation of the tail was low for ABPLe (5°), high for DBA/2J (60°) and intermediate for B6D2F1 (30°). The tail of ABPLe mice was positioned closer to the beam, in a position comparable to observations of BALB/cByJ mice, a strain reported as being anxious (Cole *et al.*, 1995; Conti *et al.*, 1994; Crusio and Schwegler, 1987; Lepicard *et al.*, 2000). The dynamic use of the tail differed between DBA/2J and B6D2F1: VHS video recordings show that DBA/2J mice maintained equilibrium by normal use of their tail as a balancer when moving, whereas in B6D2F1, the tail remained unusually steady throughout the movement of the animal, with an average measurement of 30 degrees to the beam. This was the best position for stability as a 30° angle produces an optimal centre of gravity and forms a level line extending from the body. While the steady position of the tail is unusual, it did not seem to make the B6D2F1 mice lose their balance and as their postural profile was satisfactory, they obviously used their tail efficiently.

The posture of ABPLe and DBA/2J, however, was not efficient and failed to counter imbalances induced by anxiety and reaction to stress. Balance control and postural profile data confirmed that ABPLe and DBA/2J were more sensitive to a high-risk environment and more anxious than B6D2F1.

The postural phenomena observed were similar to observations made for the initial articles (Lepicard *et al.*, 2000; Venault *et al.*, 2001) and the detailed kinetic measurements here clearly validate them. Directed exploration activity was measured by the number of head dippings; the ABPLe strain recorded the highest number of head movements during locomotion along the beam. The number of head dippings in DBA/2J was at an intermediate level. Dipping, being exploration activity directed to the floor means higher risk assessment behaviour (Rodgers and Cole, 1993) in ABPLe and DBA/2J, a behavioural factor which is no doubt linked to the level of anxiety and vulnerability to stressors. The lowest number of head dippings was recorded by B6D2F1.

The mean velocity of mice along the beam tallied with the number of imbalances and the number of pauses which obviously reduced velocity. The comparisons produced a ranked order for mean velocity (ABPLe < DBA/2J < B6D2F1) which was the opposite of inactivity (ABPLe > DBA/2J > B6D2F1). The ABPLe strain recorded the lowest velocity on each day of the procedure (LP4, T1 to T5) and the highest number of interruptions in movement. Comparisons of DBA/2J and B6D2F1 showed differences in velocity and inactivity but these were significant on two days only: LP4 and T5. The anxiety profile and vulnerability to stressors, in ABPLe and DBA/2J, respectively, affected locomotor performance, particularly on the last day of the learning phase and on the last day of the procedure.

The velocity of ABPLe was reduced by the high number of imbalances on LP4, while the velocity of DBA/2J was reduced by the large number of pauses on both LP4 and T5. The ABPLe strain recorded the highest values for inactivity when the beam changed diameter (T4, challenging the vestibular system). No significant differences in the number of imbalances were observed on T4 for DBA/2J or B6D2F1.

ABPLe and DBA/2J had higher values for the number of movements and the distance covered at a slow rate. B6D2F1 adopted a fast rate of locomotion. The results confirm that ABPLe and DBA/2J were more anxious and were more sensitive to the rotating beam task than B6D2F1. Locomotion using small movements, as observed with ABPLe, is no doubt related to efforts to maintain balance. The vulnerability to stressors of DBA/2J may be related to the behavioural hesitation (slowed performance) observed in the experiment with locomotion by small movements.

These results combine to validate with kinetic data the conclusions of our previous articles (Lepicard *et al.*, 2000) suggesting that the rotating beam model could be a good model for measuring the correlation between anxiety and balance control in rodents. Indeed, anxious ABP/le mice were highly sensitive to the beam rotation, whereas the B6D2F1 population, resistant to stressors, were quite resistant to the beam rotation. Observations of the DBA/2J strain vulnerable to stressful situations showed locomotor and postural performances that were often at an intermediate level compared to ABPLe and B6D2F1. DBA/2J mice were thus moderately sensitive to the rotating beam test. Further experiments should extend the validity of these results to other strains and other populations of mice with different levels of anxiety and reaction to stress.

Acknowledgements

We wish to thank Shan Benson for her help in improving this manuscript.

REFERENCES

Chrousos, G. P. and Gold, P. W. (1992). The concepts of stress and stress system disorders. Overview of physical and behavioral homeostasis, *JAMA* **267**, 1244–1252.

Clément, Y., Adelbrecht, C., Martin, B., *et al.* (1994). Association of autosomal loci with the grooming activity in mice observed in open-field, *Life Sci.* **55**, 1725–1734.

Clément, Y., Martin, B., Venault, P., *et al.* (1995). Involvement of regions of the 4th and 7th chromosomes in the open-field activity of mice, *Behav. Brain Res.* **70**, 51–57.

Clément, Y., Launay, J. M., Bondoux, D., *et al.* (1996). A mouse mutant strain highly resistant to methyl beta-carboline-3-carboxylate-induced seizures, *Exp. Brain Res.* **110**, 28–35.

Clément, Y., Bondoux, D., Launay, J. M., *et al.* (1997). Convulsive effects of a benzodiazepine receptor inverse agonist: are they related to anxiogenic processes? *J. Physiol., Paris* **91**, 21–29.

Cole, J. C., Burroughs, G. J., Laverty, C. R., *et al.* (1995). Anxiolytic-like effects of yohimbine in the murine plus-maze: strain independence and evidence against alpha 2-adrenoceptor mediation, *Psychopharmacology (Berlin)* **118**, 425–436.

Conti, L. H., Costello, D. G., Martin, L. A., *et al.* (1994). Mouse strain differences in the behavioral effects of corticotropin-releasing factor (CRF) and the CRF antagonist alpha-helical CRF9-41, *Pharmacol. Biochem. Behav.* **48**, 497–503.

Crawley, J. N. and Davis, L. G. (1982). Baseline exploratory activity predicts anxiolytic responsiveness to diazepam in five mouse strains, *Brain Res. Bull.* **8**, 609–612.

Crusio, W. E. and Schwegler, H. (1987). Hippocampal mossy fiber distribution covaries with open-field habituation in the mouse, *Behav. Brain Res.* **26**, 153–158.

D'Aquila, P. S., Peana, A. T., Carboni, V., *et al.* (2000). Exploratory behaviour and grooming after repeated restraint and chronic mild stress: effect of desipramine, *Eur. J. Pharmacol.* **399**, 43–47.

Gilad, G. M. and Gilad, V. H. (1995). Strain, stress, neurodegeneration and longevity, *Mech. Ageing Dev.* **78**, 75–83.

Goodrick, C. L. (1976). Mode of inheritance of emotionality in the mouse (Mus Musculus): sex differences and the effects of trials and illumination, *Psychol. Rep.* **39**, 247–256.

Hallam, R. S. and Hinchcliffe, R. (1991). Emotional stability: its relationship to confidence in maintaining balance, *J. Psychosom. Res.* **35**, 421–430.

Jacob, R. G., Moller, M. B., Turner, S. M., *et al.* (1985). Otoneurological examination in panic disorder and agoraphobia with panic attacks: a pilot study, *Am. J. Psychiatry* **142**, 715–720.

Lepicard, E. M., Venault, P., Perez-Diaz, F., *et al.* (2000). Balance control and posture differences in the anxious BALB/cByJ mice compared to the non anxious C57BL/6J mice, *Behav. Brain Res.* **117**, 185–195.

Pardon, M. C., Perez-Diaz, F., Joubert, C., *et al.* (2000). Influence of a chronic ultramild stress procedure on decision-making in mice, *J. Psychiatry Neurosci.* **25**, 167–177.

Reason, J. T. (1978). Motion sickness adaptation: a neural mismatch model, *J. Royal Soc. Med.* **71**, 819–829.

Rodgers, R. J. and Cole, J. C. (1993). Influence of social isolation, gender, strain, and prior novelty on plus-maze behaviour in mice, *Physiol. Behav.* **54**, 729–736.

Shanks, N. and Anisman, H. (1988). Stressor-provoked behavioral changes in six strains of mice, *Behav. Neurosci.* **102**, 894–905.

Treisman, M. (1977). Motion sickness: an evolutionary hypothesis, *Science* **197**, 493–495.

Trullas, R. and Skolnick, P. (1993). Differences in fear motivated behaviors among inbred mouse strains, *Psychopharmacology (Berlin)* **111**, 323–331.

Venault, P., Rudrauf, D., Lepicard, E. M., *et al.* (2001). Balance control and posture in anxious mice improved by SSRI treatment, *Neuroreport* **12**, 3091–3094.

Wehner, J. M., Sleight, S. and Upchurch, M. (1990). Hippocampal protein kinase C activity is reduced in poor spatial learners, *Brain Res.* **523**, 181–187.

Yardley, L., Luxon, L., Bird, J., *et al.* (1994). Vestibular and posturographic test result in people with symptoms of panic and agoraphobia, *J. Audio Med.* **3**, 48–65.

Advances in Neuroregulation and Neuroprotection (2005), pp. 207-215
Collin, C. *et al.* (Eds)
© VSP 2005

Effect of adrenergic and cholinergic agonists on vasoactive intestinal polypeptide release from pancreatic islets of normal and diabetic rats

ERNEST ADEGHATE [1,*], ABDULSAMAD PONERY [1] and HASAN PARVEZ [2]

[1] *Department of Anatomy, Faculty of Medicine and Health Sciences, United Arab Emirates University, PO Box 17666, Al Ain, United Arab Emirates*
[2] *Neuroendocrinologie et Neuropharmacologie du Développement, Institut Alfred Fessard, Parc Chateau CNRS, France*

Abstract—*Background and aims*: Vasoactive intestinal polypeptide (VIP) is present in pancreatic islet cells and stimulates insulin and glucagon release from normal and diabetic rats. The aim of this study was to examine the effect of noradrenaline (NA) and acetylcholine (ACh) on VIP release from pancreatic islets of normal and diabetic rats.

Methods: NA- and ACh-induced VIP release from isolated pancreatic tissue fragments was measured using radioimmunoassay technique.

Results: The basal VIP release from the islets of diabetic rats was significantly ($p = 0.02$) higher compared to normal. NA at and 10^{-6} M also evoked large and significant ($p = 0.04$) increases in VIP release from the islets of normal rat. Moreover, NA at 10^{-8} M, significantly ($p < 0.05$) stimulated VIP release from the islets of diabetic rats. ACh (10^{-4} M), induced up to 46.6% increase in VIP secretion from the islet cells of normal ($n = 6$) rats. However, a more dilute concentration of ACh (10^{-8} M) was sufficient for the induction of maximal VIP secretion from the isolated islets of diabetic ($n = 6$) rat. ACh (10^{-6} M) evoked 32.3% increase in VIP secretion in diabetic rat islets.

Conclusion: NA and ACh stimulated VIP secretion from pancreatic islets of normal rat. NA and ACh can also induce increases in VIP secretion from the islets of diabetic rat. The increase in VIP in the islets of diabetic rats may also play a role in the pathogenesis of diabetic complications.

Keywords: Noradrenaline; acetylcholine; vasoactive intestinal polypeptide secretion; islets; pancreas; diabetes mellitus; rat.

INTRODUCTION

Vasoactive intestinal polypeptide (VIP) belongs to a peptide family that includes glucagon, PACAP, secretin and glucagon-like-peptide-1 (Arimura, 1992). VIP

---

*To whom correspondence should be addressed. E-mail: eadeghate@uaeu.ac.ae

was first isolated from the pig duodenum by Said and Mutt in 1970. A few years later it was also shown to be present in the brain (Palkovits, 1980) and in many other parts of the gastrointestinal system (Balaskas et al., 1995). VIP is present in pancreatic islet cells (Adeghate and Donath, 1990; Adeghate et al., 2001a) where it co-localizes with glucagon (Sheu et al., 1989). The number of VIP positive cells is increased in diabetes (Adeghate et al., 2001a). The close morphological relation of VIP with other pancreatic hormones indicates that VIP may play a role in the regulation of pancreatic hormone secretion. In fact, it has been shown that VIP stimulates insulin release from the pancreas of normal and diabetic rats (Adeghate et al., 2001a). In addition to the effect of VIP on hormone release from pancreatic islets, VIP has also been shown to stimulate acetylcholine synthesis in rat hippocampal tissue (Lapchak and Collier, 1988). VIP is also capable of inducing nitric oxide in the myenteric neurons of the opossum internal anal sphincter (Chakder and Rattan, 1996).

If VIP is co-localized with neurotransmitters and peptides such as NA and ACh, it is pertinent to suggest a role for noradrenaline (NA) and acetylcholine (ACh) in the physiology of VIP. The effect of NA and ACh on VIP release from the isolated rat pancreatic islets has never been reported. The aim of this study was to determine whether NA and ACh would have any effect on VIP release from rat pancreatic islets.

MATERIALS AND METHODS

Experimental animals

Twelve week-old male Wistar rats weighing approximately 250 grams were used in this study. Rats were obtained from the United Arab Emirates University breeding colony and the Animal Research Group's guidelines for the care and use of laboratory animals were followed. The rats were divided into two groups, streptozotocin (STZ)-induced diabetics and age-matched controls. Diabetes was induced by a single intraperitoneal injection of STZ (Sigma, Poole, UK) at 60 mg kg^{-1} prepared in 5 mM citrate buffer pH 4.50 (Adeghate, 1999). The animals were kept in plastic cages and maintained on standard laboratory animal diet with food and water ad libitum. One-Touch II® Glucometer (LifeScan, Johnson and Johnson, Milpitas, USA) was used to measure the blood glucose for each individual animal. The animals were considered diabetic if the random blood glucose levels were equal to or more than 300 mg dl^{-1}. After six weeks from the date of induction of diabetes, all of the animals from both groups were sacrificed under chloral hydrate general anesthesia (7% chloral hydrate 6-ml kg^{-1} of body weight, injected intraperitoneally). A mid-line abdominal incision was made, and the pancreas was rapidly removed and placed in ice-cold Krebs solution. Representative fragments were taken from the tail end of the pancreas.

Isolation of pancreatic islets

The pancreas of six rats were removed and placed in ice-cold Krebs buffer (KB). The pancreas was trimmed free of adherent fat and connective tissue and minced into small fragments (0.5–1 mm^3). The minced pancreatic tissue fragments were placed in vials incubated in collagenase solution (1 mg ml^{-1}) in a warm waterbath ($37°$C) for 30 min. After the incubation period, the islets were handpicked and centrifuged to get rid of excess acinar tissue. The islet mass retrieved was aliquoted for experiment on VIP release.

Estimation of in vitro effect of adrenaline and acetylcholine on VIP release from pancreatic islets

The islet mass was placed in 2 ml glass vials containing 1 ml of KB and pre-incubated for 30 min in a waterbath at $37°$C, in order to wash away any hormones due to destruction of the cells by collagenase. After the pre-incubation period, the KB solution was drained and the islets were subsequently incubated for 1 h with different concentrations of either NA (10^{-8}–10^{-4} M) or ACh (10^{-8}–10^{-4} M). In control experiments, the islets were incubated in KB solution alone for the same duration of time. During the incubation period, each vial was gassed with a mixture of 95% oxygen and 5% carbon dioxide every 2 min. At the end of the experiment the tissues were removed, blotted, weighed and the effluent stored at $-20°$C for radioimmunoassay.

Radioimmunoassay of VIP

VIP was determined using a modified method of Fahrenkrug (1979). Briefly, 200 μl of the control and unknown samples of the supernatant were added to 100 μl of VIP antiserum with the exception of T and NSB tubes. The mixture was vortexed and incubated for 24 h at $4°$C. After this, 100 μl of ^{125}I-VIP was added into all tubes and vortexed before incubation for another 24 h. The samples were incubated at $4°$C for 2 h after the addition of 0.85% saline and the precipitating complex. The samples were centrifuged for 20 min and decanted, before counting for 1 min in a Beckman gamma counter (Beckman, Fullerton, CA, USA). VIP RIA kit was purchased from Diasorin® (Stillwater, Minnesota, USA). Results were analyzed using a Beckman Immunofit EIA/RIA analysis software, version 2.00.

Statistical analysis

All values were expressed as mean $\pm$ standard deviation (STD). Statistical significance was assessed using a two-tailed Student's t-test. Only values with $p < 0.05$ were accepted as significant.

RESULTS

The basal VIP release from isolated pancreatic islets of diabetic rats [157.87 $\pm$ 27.54 pg ml^{-1} (200 μl of islet mass)$^{-1}$] was significantly ($p = 0.02$) higher compared to that of normal rat [9.68 $\pm$ 1.4 pg ml^{-1} (200 μl of islet mass)$^{-1}$].

Noradrenaline and VIP secretion

Noradrenaline (NA) stimulated VIP secretion from isolated pancreatic islets of normal rat. NA at 10^{-6} M evoked large and significant ($p < 0.04$) increases in VIP release from the islets of normal rat (Fig. 1). Moreover, NA at 10^{-8} M significantly ($p < 0.05$) stimulates VIP release from the islets of Langerhans of diabetic rats

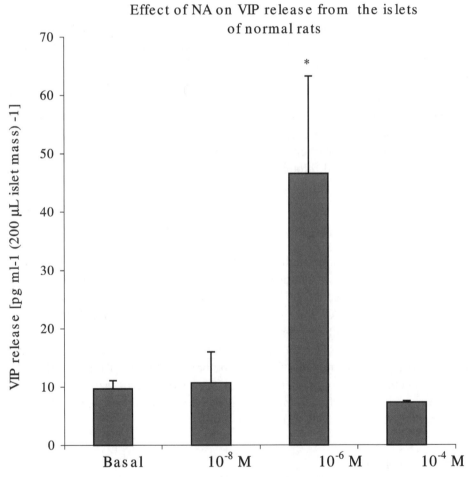

Figure 1. Illustrating the effect of noradrenaline (NA) (10^{-8}–10^{-4} M on VIP release from isolated pancreatic islet of normal rats ($n = 6$). NA (10^{-8} M and 10^{-6} M) evoked significant ($p < 0.05$) increases in VIP release from the pancreatic islets cells of normal rats. *($p < 0.05$ compared to basal).

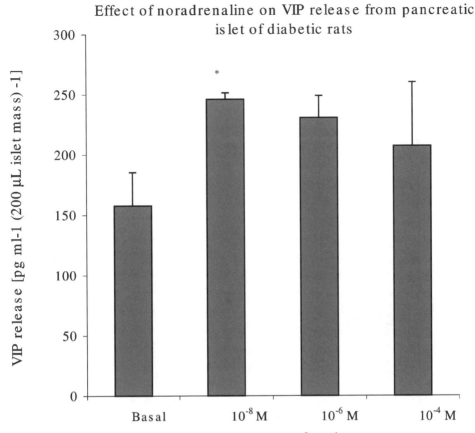

Figure 2. Illustrating the effect of noradrenaline (NA) (10^{-8}–10^{-4} M) on VIP release from the isolated pancreatic islet of diabetic ($n = 6$) rats. NA (10^{-8} M) evoked significant (*$p < 0.05$) increases in VIP release from the pancreatic islets of diabetic rats compared to basal.

(Fig. 2). NA exerts its maximal effect on VIP secretion from the islet of normal rats at 10^{-6} M while in diabetic rat pancreas 10^{-8} M was enough to elicit maximal VIP release.

Acetylcholine and VIP secretion

ACh at 10^{-4} M, induced up to 46.6% increase in VIP secretion from the islet cells of normal ($n = 6$) rat. All other concentrations (10^{-8} M and 10^{-6} M) failed to stimulate or inhibit VIP secretion from the islets of normal rat (Fig. 3). However, a more dilute concentration of ACh (10^{-8} M) was sufficient for the induction of maximal VIP secretion from the isolated islets of diabetic ($n = 6$) rat. ACh (10^{-6} M) evoked 32.3% increase in VIP secretion in diabetic rat islets (Fig. 4). The probability of this slight increase was $p = 0.07$. The maximal effect of ACh on VIP secretion from the islets of normal rats was achieved at 10^{-6} M. In contrast, ACh induced the largest amount of VIP release from the isolated islets of diabetic rat at 10^{-8} M.

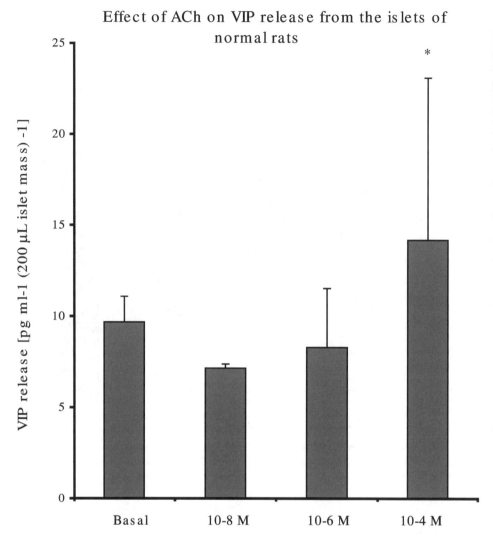

Figure 3. A family of histograms showing the effect of acetylcholine (ACh) (10^{-8}–10^{-4} M) on VIP secretion from the isolated pancreatic islet of normal ($n = 6$). ACh (10^{-4} M) caused 46.6% increase in VIP release from the pancreatic islets cells of normal rat.

DISCUSSION

The results of this study showed that adrenergic (NA) and cholinergic (ACh) neurotransmitters could stimulate VIP release from the isolated pancreatic islets of normal rat. The observations of this study agree with those of Cheung (1988) and Fahrenkrug *et al.* (1978) who reported that ACh at 50 μM stimulated VIP release from fetal adrenal gland. It was also shown that VIP is released into the cerebrospinal fluid by a mechanism involving nicotine-sensitive cholinergic pathway (Kaji *et al.*, 1983). Our observation on the effect of NA on VIP release

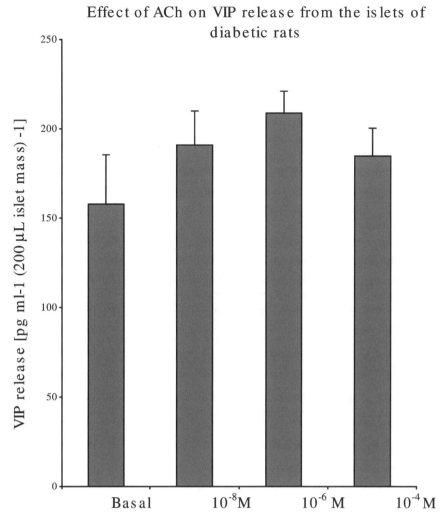

Figure 4. A family of histograms showing the effect of acetylcholine (ACh) (10^{-8}–10^{-4} M) on VIP secretion from the isolated pancreatic islet of diabetic ($n = 6$) rats. ACh (10^{-6} M) induced 32.3% increase in VIP release from the islets of diabetic rat.

did not corroborate that of Cheung (1988), who reported that NA did not have any effect on VIP release from fetal adrenal gland.

NA and ACh exerted their maximal effects on VIP secretion from the islets of Langerhans of diabetic rats were achieved at a much lower concentration. Moreover, there are no literature data regarding the effect of adrenergic and cholinergic and transmitters of VIP release from the pancreatic islets of diabetic rat.

It was also of interest to note that pancreatic islets of diabetic rats contained a significantly larger quantity of VIP compared to basal. The large quantity of VIP measured in the basal solution confirms previous observation (Adeghate *et al.*, 2001a) that the pancreatic islets of diabetic rats contained larger number of VIP positive cells compared to normal. Moreover, VIP level in the plasma of diabetic rats is higher than that of control (Adeghate et al., 2001b). The relatively high quantity of immunoreactive VIP in diabetic rats might have rendered the islets cells of diabetic rats more susceptible to the secretory effect of NA and ACh. The other possible reason why the islets of diabetic rats were more susceptible to the actions of NA and ACh might be that the structure of the receptors of these chemical transmitters might be altered in diabetes and hence the changes in signal transduction within these islet cells.

While ACh enhances VIP release from the adrenal gland (Cheung, 1988), brain (Kaji et al., 1983) and from enteric nerves (Adeghate et al., 2001b), VIP itself has been shown to increase ACh synthesis in rat hippocampal slices (Lapchak and Collier, 1988). VIP in this manner may also play a role in the metabolic activities of other pancreatic islet cells.

In conclusion, NA or ACh stimulated VIP secretion from the isolated pancreatic islets of normal and diabetic rat. However, NA induced larger increases in VIP release from islets of diabetic rat compared to ACh. The large quantity of VIP in the islets of diabetic rats may play a factor in the pathogenesis of diabetic complications. The actions of these neurotransmitters may indirectly regulate hormone secretion from the endocrine pancreas.

Acknowledgement

This work was supported by a Research Grant from the Faculty of Medicine and Health Sciences, United Arab Emirates University, Al Ain, United Arab Emirates.

REFERENCES

Adeghate, E. (1999). Effect of subcutaneous pancreatic tissue transplants on streptozotocin-induced diabetes in rats. II. Endocrine and metabolic functions, *Tissue Cell* **31**, 73–83.

Adeghate, E. and Donáth, T. (1990). Distribution of neuropeptide-Y and vasoactive intestinal polypeptide immunoreactive nerves in normal and transplanted pancreatic tissue, *Peptides* **11**, 1087–1092.

Adeghate, E., Ponery, A. S., Pallot, D. J., et al. (2001a). Distribution of vasoactive intestinal polypeptide, neuropeptide-Y and substance P and their effects on insulin secretion from the in vitro pancreas of normal and diabetic rats, *Peptides* **22**, 99–107.

Adeghate, E., Ponery, A. S., Sharma, A. K., et al. (2001b). Diabetes mellitus is associated with a decrease in vasoactive intestinal polypeptide content of gastrointestinal tract of rat, *Archiv Physiol. Biochem.* **109**, 246–251.

Arimura, A. (1992). Pituitary adenylate cyclase activating polypeptide (PACAP): discovery and current status of research, *Regul. Pept.* **37**, 287–303.

Balaskas, C., Saffrey, M. J. and Burnstock, G. (1995). Distribution and co-localization of NADPH-diaphorase activity, nitric oxide synthase immunoreactivity, and VIP immunoreactivity in the newly hatched chicken gut, *Anat. Rec.* **243**, 10–18.

Belai, A., Ralevic, V. and Burnstock, G. (1987). VIP release from enteric nerves is independent of extracellular calcium, *Regul. Pept.* **19**, 79–89.

Chakder, S. and Rattan, S. (1996). Evidence for VIP-induced increase in NO production in myenteric neurons of opossum internal anal sphincter, *Am. J. Physiol.* **270**, G492–497.

Cheung, Y. (1988). Ontogeny of adrenal VIP content and release from adrenocortical cells in the ovine fetus, *Peptides* **9**, 107–111.

Fahrenkrug, J. (1979). Vasoactive intestinal polypeptide: Measurement, distribution and putative neurotransmitter function, *Digestion* **19**, 149–156.

Fahrenkrug, J., Galbo, H., Holst, J. J., *et al.* (1978). Influence of the autonomic nervous system on the release of vasoactive intestinal polypeptide from the porcine gastrointestinal tract, *J. Physiol.* **280**, 405–422.

Kaji, H., Chihara, K., Minamitani, N., *et al.* (1983). Release of vasoactive intestinal polypeptide into the cerebrospinal fluid of the fourth ventricle of the rat: involvement of cholinergic mechanism, *Brain Res.* **269**, 303–310.

Lapchak, P. A. and Collier, B. (1988). Vasoactive intestinal polypeptide increases acetylcholine synthesis in rat hippocampal slides, *J. Neurochem.* **50**, 58–64.

Palkovits, M. (1980). Topography of chemically identified neurons in the central nervous system. Progress in 1977–1979, *Med. Biol.* **58**, 188–227.

Said, S. I. and Mutt, V. (1970). Polypeptide with broad biological activity: Isolation from small intestine, *Science* **169**, 1217–1218.

Sheu, H. W., Chou, S. Y., Yang, K. C., *et al.* (1989). Pancreatic vasoactive intestinal polypeptide-secreting tumor: report of a case, *Taiwan I Hsueh Hui Tsa Chih* **88**, 931–935.

Advances in Neuroregulation and Neuroprotection (2005), pp. 217-237
Collin, C. *et al.* (Eds)
© VSP 2005

Conformational diseases and the protein folding problem: role of amino-acid propensity to be embedded in context of specific usage of synonymous codons. Genetic code degeneracy and folding

JEAN SOLOMOVICI [1], THIERRY LESNIK [1], RICARDO EHRLICH [2], HASAN PARVEZ [3], SIMONE PARVEZ [3] and CLAUDE REISS [1,*]

[1] *Alzheim' R&D, 2 rue de la Noue, F91190 Gif, France*
[2] *Seccion Bioquimica, Facultad de Ciencias, Universidad de la Republica, 4225 Igua, U11400 Montevideo, Uruguay*
[3] *Laboratoire de Neuroendocrinologie, IAF-CNRS, F91198 Gif, France*

Abstract—Translation was long thought to be a smooth process, producing invariably native proteins. However, it appeared recently that this may not always be the case. First, a family of apparently unconnected diseases, including important neurodegenerative conditions, like Alzheimer, Parkinson, Huntington, and prion disorders, was shown to be linked to misfolded proteins ('conformational diseases'). Secondly, up to 70% of nascent proteins are recognized as misfolded and tagged for proteolysis whilst still attached to the ribosome. Misfolding seems therefore to be a quite common event, which is however most often corrected by efficient proteolysis of the misconformers produced. Misfolding therefore usually remains unnoticed, and safe when proteolysis is unable to remove the misconformers, which is the case in conformational diseases. Since the latter usually onset in the second half of human life, the mechanism of misfolding is likely to be linked to defects in the cellular translation and proteolysis machinery. This speculation is further substantiated by the observation that the amount of nascent proteins, recognized as misconformers whilst still attached to the ribosome, depends on the cellular state. On the other hand, the large amount of work devoted to prion proteins has highlighted the fact that certain proteins are prone to become easily misfolded. This implies that particular motifs in the primary structure of proteins enter metastable or 'soft' conformations upon synthesis, which can easily flip from one state to another.

The present study investigates a possible link between particular primary structures and soft conformations. The study is based on experimental evidence, showing that silent codon exchange can affect protein conformation. Synonymous codons correspond to cognate tRNAs that may be present in the cell in rather different concentrations. Since tRNA availability determines the cognate codon translation rate, it follows that synonymous codons may have rather different translation rates. Taken together, part at least of the factors determining the native protein conformation may be

---

*To whom correspondence should be addressed. E-mail: cjreiss@yahoo.com

encoded in the protein translation rate, i.e. in the selection of synonymous codons. Finally, the codon translation rate can be approximately estimated from the relative codon usage. Indeed, since the cellular concentration of charged tRNA on one hand controls the translation rate of the cognate codon, and on the other correlates closely with the usage of that codon in the cell, a reasonable approximation is to take as relative rate of codon translation the inverse of the usage of this codon. These various assumptions allow us to compute the local translation rate at any codon of a given messenger RNA. We were struck by the observation that 'fast' (frequent) or 'slow' (rare) codons tend to cluster in most mRNAs, and that the clusters tend to harbour preferentially certain amino acids. We therefore set out to analyse all the genes of *E. coli*, identify the 'fastest' and 'slowest' codon window in each and check if certain amino-acids are favored, or disfavored, within these windows, with respect to the general amino-acid usage in the genome of the species under investigation. We find that indeed, most aminoacids are constrained within these windows, and more so as the size of the window is reduced from 21 to 5 codons. Significant correlations are found between the type (excess or depletion) of constraint and the physical characteristics of the amino acids (hydrophobic, aromatic, charged or polar) or their propensity to enter particular secondary structure. It is therefore likely that synonymous codon selection in the mRNA locally sets kinetic factors contributing to the native conformation of the protein. Modifications of these kinetic factors, whether by silent codon exchange or by changes of the translation rate due to modifications of the cellular concentrations of components of the translation machinery (charged tRNA, elongation factor, foldases, etc.) would then possibly result in misconformation of the protein. We surmise that conformational diseases may result from spontaneous protein misfolding resulting from the incompatibility of the translation machinery components with the folding-specific translation rate. The misfolded proteins would either adopt a conformation that cannot be proteolyzed, or aggregate so as to be unable to enter the proteasome.

Keywords: Conformational disease; intracellular mechanism; translation kinetics.

INTRODUCTION

Quite recently, it was realized that several important, apparently disparate pathologies, actually share a common cause: they result from misconformation of some type of cellular proteins, despite the fact that they carry the wild type primary sequence (Bross *et al.*, 1999). In general, the vast majority of misconformed proteins are tagged for ubiquitination during their synthesis or early during quality control and directed to the proteasome (Schubert *et al.*, 2000), but some manage to escape proteolysis and remain in the cell or its environment (Gregersen *et al.*, 2000). The result can be loss of protein function, as is the case for instance for misconformed p53[1], a much-celebrated tumor supressor, or accumulation and aggregation of misconformers, which can possibly form bodies, fibers or plaques (Johnson, 2000). These in turn can induce apoptosis or preclude cell to cell communication. If such events take place in neural cells or in other susceptible tissues, they can elicit 'conformational diseases', such as Alzheimer's and prion disease, or type II diabetes (DeMager *et al.*, 2002).

In order to understand the general mechanism leading to misconformation-based pathologies, two questions must be addressed: how can misconformation occur, that is, how can folding of a particular protein escape the Anfinsen principle? And how can a misfolded protein escape immediate proteolysis despite being ubiquitinated? Here we address the first question.

A long-standing, yet unsolved problem in biology addresses protein folding. This problem is of basic importance, for fundamental reasons, biotechnological applications (heterologous gene expression), and the mechanisms leading to the devastating diseases just mentioned. Though the primary sequence of a protein was shown to be the main determinant of folding, many proteins nevertheless require the help of a large set of factors (foldases) (Baneyx and Palumbo, 2003; Sakahira *et al.*, 2002; Schiene-Fischer *et al.*, 2002). Even so, a large proportion of nascent proteins have been detected as misconformers and ubiquinated while still attached to the ribosome or in the process of maturation in the rough ER (Schubert *et al.*, 2000). Obviously, other factors participate also in protein folding (Sato *et al.*, 1998).

Here we focus on co-translational events that could affect protein folding (Mayor *et al.*, 2003). Given the architecture of secondary structure elements, it is clear that folding of an alpha helix for instance can be fast, since only neighbouring amino acids contribute, but formation of beta structures in general would require much more time, because of the lag between synthesis of the distant aminoacid sequences involved (Bodenreider *et al.*, 2002; Doig *et al.*, 2001; Maness *et al.*, 2003; Williams *et al.*, 1996). Similarily, disulfate bond formation, or *cis-trans* isomerization of proline residues and, more generally, 'reactions' between protein parts and foldases, require some time to be completed. It is therefore plausible that the rate of translation may come into play for protein folding.

Careful biometric studies carried out in *E. coli* cells have shown that the rate of translation of a codon present at the P site of the ribosome is limited by the availability of the cognate tRNA ternary complex. On the other hand, it was shown that the relative cellular concentration in *E. coli* of a given tRNA correlates closely whith the frequency with which the cognate codon is used in the *E. coli* genome. Taken together, one might expect that the local translation rate is controlled by the frequencies with which the local codons are used in *E. coli* (Berg and Kurland, 1997).

We may therefore take the relative (%) frequency of usage of a given codon as a measure of its relative translation rate. Inspection of the actual codon usage (ACU below) in any gene reveals that frequent or rare codons have a propensity to come in clusters of variable size (on average, less than 10 codons) (Ma *et al.*, 2002). Closer inspection of the clusters shows that by far the major contribution to their average ACU is made by appropriate selections of synonymous codons. ACU clustering is not a consequence of clustering of aminoacids encoded anyway by frequent (or rare) codons, but of a deliberate and extended selection of synonymous codons in the cluster. It follows that the modulation of the translation rate along the messenger RNA, which is phased by the codon clusters, is to a large extent set by appropriate selections of synonymous codons in the clusters.

Local codon selection could then presumably contribute to folding of the nascent protein, at certain places (ACU clusters) at least (Oresic and Shalloway, 1998).

We had previously obtained experimental results that support this hypothesis. It was for instance shown *in vitro* that the substitution of a set of 16 minor codons of the chloramphenicol acetyl transferase (CAT) mRNA by a set of synonymous major codons resulted, when expressed in a cell-free S30 extract of *E. coli*, in a 20% loss of the enzyme's specific activity (Komar *et al.*, 1999). Similarily, *in vivo*, the replacement of five consecutive minor codons by their major synonyms in the EgFABP1 gene expressed in *E. coli*, markedly decreased the cytoplasmic solubility of the protein and strongly activated an intracellular reporter gene designed to respond to misfolded proteins (Cortazzo *et al.*, 2002).

In both examples, the silent mutations surround turns in the wild type conformation of the proteins. Could it be then that the local translation rate is slowed down in order to favour these turns? More generally, are sequences characterized by their abundance in rare codons constrained in amino acids, which have propensities to participate in particular secondary structures, or to interact with some foldase? Conversely, would document frequently found in alpha helices have a tendency to be located in sequences rich in frequent codons, since the assembly of that secondary structure involves consecutive amino acids and appears to depend less on foldases.

We report below our preliminary results, showing that within the coding sequences of *E. coli,* the codon islets of various extent, having on average the highest constraint in frequent, or rare, codons, display charcteristic amino acid constraints. This observation lends strong support to the hypothesis that the translation rate may participate in steps of co-translational folding (Kolb *et al.*, 2000; Purvis *et al.*, 1987).Work in progress shows that other species share similar propensities.

MATERIAL AND METHODS

The gene sequences used in this work were taken from the Colibri database provided by the Pasteur Institute (http://genolist.pasteur.fr/Colibri). The computer programs used were written in Microsoft Quickbasic and can be provided on request to JS.

RESULTS

Codon translation rate

For a cell in a steady growth state, the global mass of protein synthesis varies linearly with time, at a constant average rate. Therefore, per unit time, the average number of codons translated in the cell is constant also and corresponds to the average codon demand. It follows that, at the level of individual codons (i), the time t_i it takes to translate that codon is on average proportional to the inverse of its demand, the actual codon usage (not the frequency with which it appears in the genome, see below) of codon (i) in all the genes of the species under consideration). Hence $t_i \propto K/v_i$. Within a given cell in a given, steady state, to a good approximation the rate of translation of a given codon is proportional to the frequency with which that codon is used in the global protein synthesis.

Codon usage

The number of times a given codon appears in the genes of a given species, or codon *frequency*, which can be derived from the gene sequences, does not reflect the actual codon *usage* in the cell. Results obtained by quantitative proteomic studies of *E. coli* for instance show that for certain proteins, ten- to hundred-thousand copies are present in the cell (large excess of actual codon usage over codon frequency), whereas for others, only a few copies are found (comparable actual codon usage and frequency), and some are not expressed at all (no codon usage). Biometric data (Cordwell *et al.*, 1999; VanBogelen *et al.*, 1999) are available that allow an estimation of the proportion of proteins present in the cell at a given time, derived from high- or low-expressed genes, respectively. Some one hundred genes are heavily expressed and contribute of the order of 1.2×10^6 proteins in an exponential growing cell. These genes code mainly for proteins involved in the essential biosynthesis pathways (namely, amino acid, nucleotide, fatty acid) and energy metabolism, for ribosomal proteins, nucleic acid polymerases and factors assisting transcription and translation, for enzymes in charge of controlling and maintaining protein conformation (foldases, chaperons, hsp), proteases and nucleases. The remaining genes of *E. coli* (over 4000) contribute together some 2×10^4 proteins only. On average then, roughly 85% of the proteins present in the cell come from highly expressed genes, 15% from the remainder.

It follows that the codon usage in the set of highly expressed genes is predominant in the actual codon usage in the cell. In order to quantify this predominance, we assume that the actual codon *usage* is the linear combination of 85% of the codon *frequency* in highly expressed genes, and 15% of the codon *frequency* found in the remaining genes (we take the latter as the codon frequency in all genes). These frequencies are readily derived from the coding sequences.

Although in *E. coli,* highly expressed genes other than ribosomal genes have been listed and their biometry established, this is not the case for most other species, for which only approximate ribosomal biometry is usually available. In order to allow simple comparison of results obtained from different species by the study reported, we therefore take for the codon frequency in highly expressed genes simply the codon frequency in ribosomal protein genes. Therefore, the actual codon usage (ACU below) is taken as: ACU = 0.85(codon frequency in 'ribo') + 0.15(codon frequency 'total'), where codon frequency 'total' is that in all genes.

The ACU for the 64 codons in *E. coli* is given in Table 1.

Searching genes for segments with highest, or lowest, actual codon frequency

In order to see whether, within coding sequences, some amino acids tend to be associated with particular local translation rates, we set out looking for such

Table 1.
Actual codon usage (ACU) in *E. coli* (%)

AAA	6.267	TAA	0.566	GAA	4.892	CAA	0.85
AAT	0.691	TAT	0.557	GAT	1.867	CAT	0.61
AAG	2.458	TAG	0.003	GAG	1.687	CAG	2.531
AAC	2.986	TAC	1.288	GAC	2.726	CAC	1.429
ATA	0.066	TTA	0.387	GTA	2.382	CTA	0.058
ATT	1.679	TTT	0.929	GTT	4.635	CTT	0.463
ATG	2.518	TTG	0.41	GTG	1.382	CTG	5.83
ATC	3.675	TTC	2.23	GTC	0.952	CTC	0.371
AGA	0.057	TGA	0.073	GGA	0.145	CGA	0.08
AGT	0.31	TGT	0.18	GGT	4.434	CGT	4.938
AGG	0.018	TGG	0.73	GGG	0.252	CGG	0.107
AGC	1.202	TGC	0.336	GGC	3.326	CGC	2.387
ACA	0.243	TCA	0.184	GCA	2.767	CCA	0.466
ACT	2.158	TCT	1.632	GCT	4.692	CCT	0.488
ACG	0.412	TCG	0.176	GCG	2.026	CCG	2.21
ACC	2.255	TCC	1.073	GCC	1.165	CCC	0.1

correlations in the regions within the coding sequence having the highest, or the lowest, average ACU values. To this end, a window of defined size z (ranging from 5 to 21 codons, as stated) is passed through the coding sequence by steps of one codon. The arithmetic average of the ACU of the z codons in the window is computed for each window position, and the positions in the sequence of the windows with the highest, zMAX, or the lowest, zMIN, scores, are memorized. In order to see whether constraints are present in the surrounding of the zMIN or zMAX windows, the sequences of the 50 document including these windows and such that the window starts at position 26, are stored, and the whole operation is repeated for all the coding sequences of the species under investigation. The 50-amino acid sequences translated slowest or fastest in each gene were respectively listed in table TMAX (largest translation time) or TMIN (shortest translation time). The number of times each of the 20 document is present at positions $1, 2, \ldots, 50$ in the TMIN or TMAX lists, is counted, normalized with respect to the average occurrence of the particular amino acid in all coding sequences analyzed, and displayed for the 20 document (Fig. 1, computed for window size $z = 13$). Bars 1–25 show the relative occurrence of the aminoacid on the N-terminal side of window z, bars 26 to 38 the propensity within the window, and bars 39 to 50 the propensity on the C-terminal side of z, for window z including the gene sequence translated slowest (TMAX, lowest average ACU), left side of Fig. 1 or fastest (TMIN, highest average ACU), right side of Fig. 1). The dotted horizontal line (ordinate 1) in each figure indicates the no-constraint or neutral value (relative average occurrence of the amino-acid in all the analyzed genes of the species.

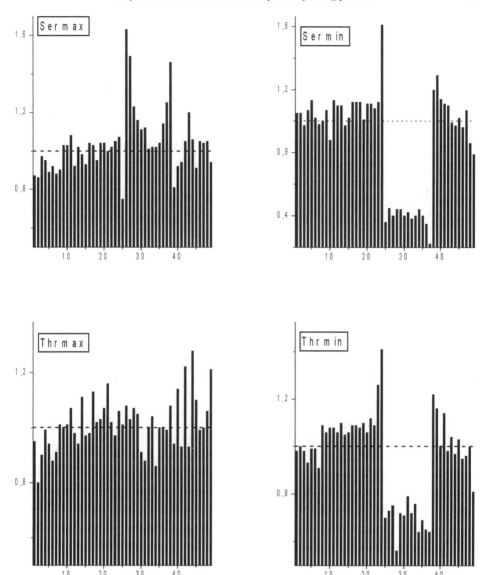

Figure 1. Constraints, averaged over all coding sequences of *E. coli*, of polar amino-acids in the 50-codon sequences ($5' \rightarrow 3'$) surrounding the 13-codon windows (position 27 to 39) translated slowest (left) and fastest (right). Dotted line: average occurrence of amino-acid in the genome.

Strong amino-acid constraints in both TMAX and TMIN windows

For each species, we computed the ratio B of the average value of occurrence of the aminoacid inside windows z corresponding to TMAX and TMIN, to its average occurrence in all analyzed genes of the species. B represents the constraints (if any) experienced by the amino-acid (avoidance < 1 or preference > 1) within the TMAX (or TMIN) window. The ratios b corresponding to amino-acid constraints

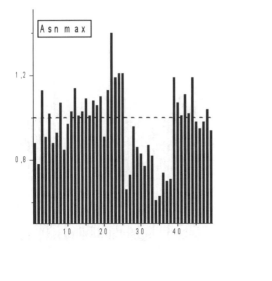

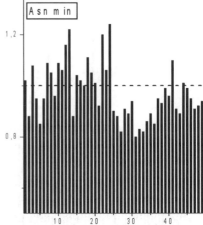

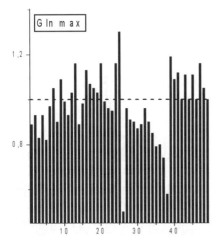

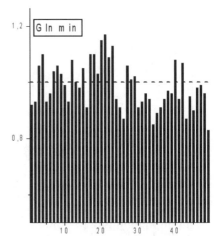

Figure 1. (Continued).

in the 26 residues on the N-terminal side of z have been computed in a similar way. Figure 2 shows the classification of document for decreasing values of B and b, for TMAX and TMIN.

Taking as significant constraints those departing from neutrality (B = 1) by 15% (dotted line in Fig. 2) at least, the TMAX window (translated slowest) favours most the basic aminoacid R, the special amino acids C and P, the polar aminoacid S and the hydrophobic L. It disfavours strongly both acidic amino acids D and E, the basic amino acid K, the polar amino acids N and Q, the hydrophobic aminoacids V, A and M and the aromatic aminoacid Y. Most of these amino acids, with the significant exception of R, experience the opposite constraint in TMIN, the window translated fastest, in which the acidic amino acids E, D and the basic amino acid K are most

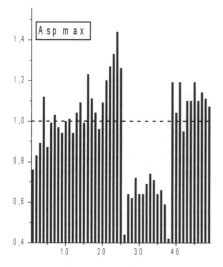
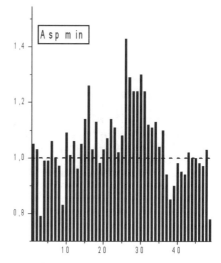
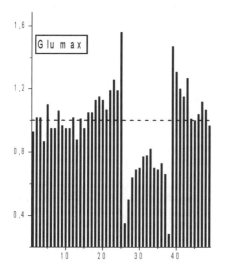
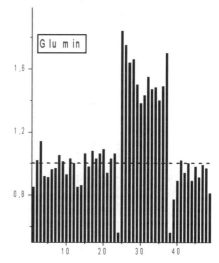

Figure 1. (Continued).

prominent, together with A and I, whereas the special amino acid C and P, the polar amino acids S and T, the aromatic amino acids F, Y and W and the basic amino acid H are disfavoured. Most hydrophobic amino acids are subjected to moderate or barely significant constraints.

Figure 2 shows also the constraint of the amino acids within the 25 codons in front of TMIN or TMAX windows. With few exceptions, the constraint deviates from the

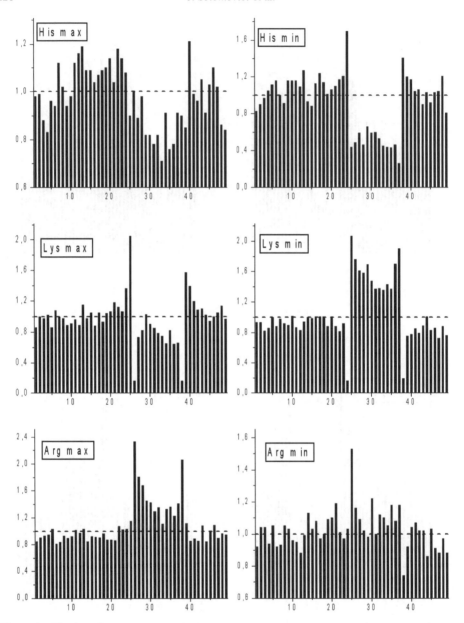

Figure 1. (Continued).

neutral value B = 1 by less than 10%. As noticed previously (Lesnik *et al.*, 2000), over three out of four genes, the TMAX windows are located close to the gene start. This may explain why, for M, B exceeds the average value upstream of the TMAX windows, since the initiation codon was frequently included in the gene sequences analyzed.

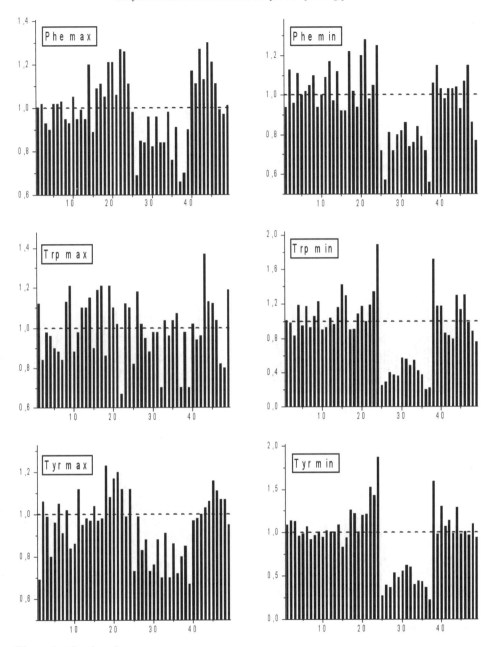

Figure 1. (Continued).

Statistical significance

No significant constraint on amino acid occurrence was observed in windows of 13 codons located at a fixed distance (codon 100) from the initiation codon of the 4258 genes of *E. coli*, nor in front of these windows (Fig. 3, lower left and right diagrams,

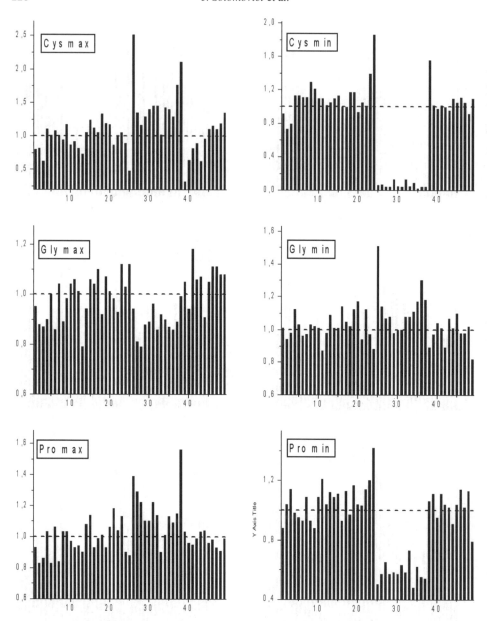

Figure 1. (Continued).

respectively), meaning that on average, amino acid occurrence is not correlated with codon position in the genes.

The same computation was carried out using instead of the actual codon usage defined as above, the codon *frequency* as derived from the coding sequences of *E. coli*. The results are displayed for the TMIN window in Fig. 3 (upper left). Comparison with Fig. 2 shows that in general rather similar trends are seen to those

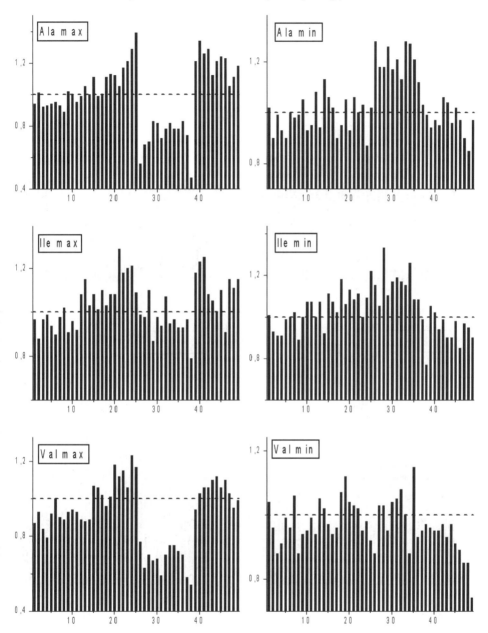

Figure 1. (Continued).

shown in Fig. 2, with sometimes more extreme values of B. Notable differences are L and R, respectively slightly favored and discriminated, whilst no significant constraint was observed in TMIN computed with ACU. The constraint classification in TMAX windows using codon frequency conforms to that observed in Fig. 2, but again with a broader distribution of B values.

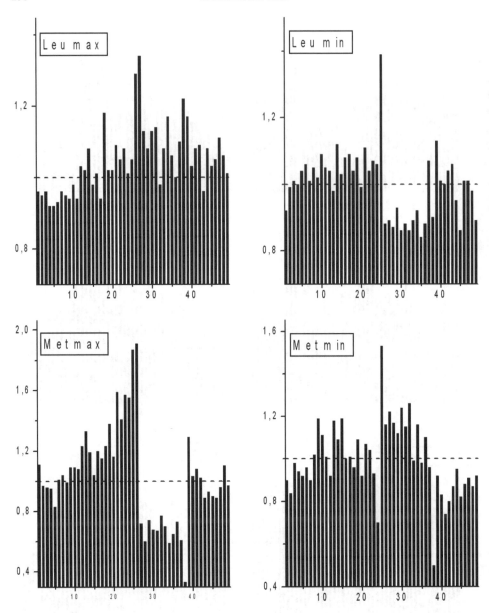

Figure 1. (Continued).

Effect of window size on the amino acid constraints

To study the effect of z, the window size, on amino acid constraint, the latter were studied in TMIN and TMAX windows of 5 to 21 consecutive codons. The results are displayed on Fig. 3. The amino acids were classified according to their general characteristics, i.e. hydrophobic, polar, acidic, basic, special and aromatic. For TMAX and TMIN windows of z = 9 codons and above, all document constraints

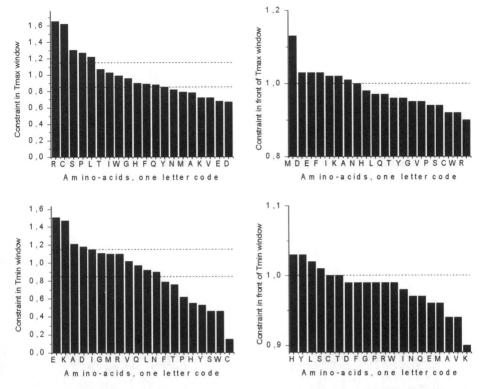

Figure 2. Average amino acid constraints (relative to amino acid occurrence in all genes) in 13-codon TMAX (upper left) and TMIN (lower left) windows, classified by decreasing constraints. Dotted lines: B = 0.85 and B = 1.15, the interval within which constraints are not considered significant. Upper right and lower right: as left, but for the 25-codon sequences in front of TMAX and TMIN windows (no significant constraint).

vary monotonously with z. With few exceptions, the constraints (whether favoring or discriminating the amino acid) increase as z is reduced. For window size 5 codons, in TMAX windows the constraints become strongly enhanced (except for T, G and W, for which the constraint is low anyway). In TMIN windows, for most of the document experiencing low constraints in larger windows, the constraint becomes reversed in the window of 5 codons, compared to the constraint in larger windows.

It must be stressed that, as a rule, TMAX (or TMIN) windows in a given gene, computed for different window sizes, have different locations, depending on the local density of unfrequent (or frequent) codons.

The data show that the amino acid constraints associated with TMAX increase steadily as the size of the window is reduced, meaning that the local selection of unfrequent codons becomes more stringent as the codon specifying the amino acid is approached. The same holds in general for TMIN for z = 9 or more. A few exceptions to this rule are observed at z = 5, where in TMIN A, D, H, I and M become discriminated (they were favored or neutral for z = 9) and L is favored

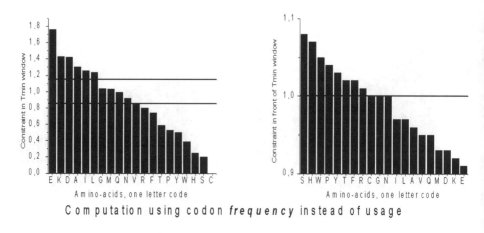

C o m p u t a t i o n   u s i n g   c o d o n   *f r e q u e n c y*   i n s t e a d   o f   u s a g e

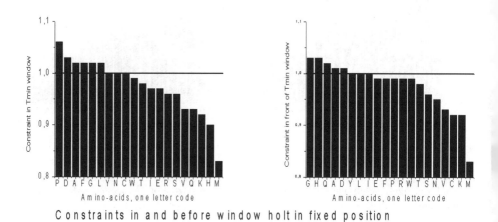

C o n s t r a i n t s   i n   a n d   b e f o r e   w i n d o w   h o l t   i n   f i x e d   p o s i t i o n

Figure 3. Controls. Amino acid constraints in TMIN windows, computed using the codon *frequency* instead of the *actual codon usage* (upper left), and amino-acid constraints in the 25 codon sequences in front of the TMIN window (upper right, no significant constraint). Lower left: amino-acid constraint in windows locked at fixed position (codon 100) in the 4258 genes examined, and in the 25 codons in front of these windows (lower right). No significant constraint is observed in the latter two cases.

(neutral for z = 9). In close vicinity of the codons of this document, the codon frequency strategy departs from neutrality, and can even reverse abruptly.

DISCUSSION AND CONCLUSION

Each coding sequence of *E. coli* has been explored for codon islets of 5 to 21 codons having, averaged over the islet, the highest, or lowest, actual codon usage frequency in the sequence. Within these islets, the content of most document is subjected to significant bias. With very few exceptions, the strength of the bias increases as the

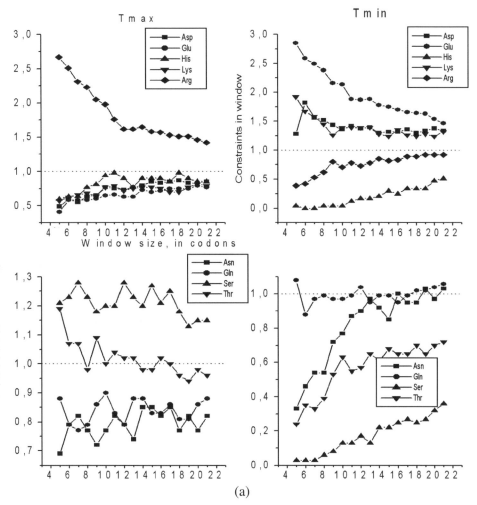

Figure 4. Dependence of acidic and basic amino acids (upper), and polar amino acids (lower), constraints, on window size (left TMAX, right TMIN windows).

size of the islet is reduced, and opposite bias is observed in islets with other extreme codon usage frequency.

In *E. coli*, the actual codon usage correlates with the relative cellular concentration of cognate tRNA (Ikemura, 1981a; 1981b). In addition, the codon translation rate is limited by the availability of the ternary complex with the cognate tRNA (Berg and Kurland, 1997; Dong *et al.*, 1996). It is therefore very likely that the actual codon usage frequency is proportional to the inverse of the codon translation rate. The islet of the highest, or lowest, actual codon usage frequency in the sequence would then be the mRNA sequences translated fastest or slowest, respectively. Experimental proof fully supports this conclusion (Komar *et al.*, 1999).

It is quite remarkable that the translation rate can be closely controlled by means of the degree of freedom granted by the degeneracy of the genetic code. The many

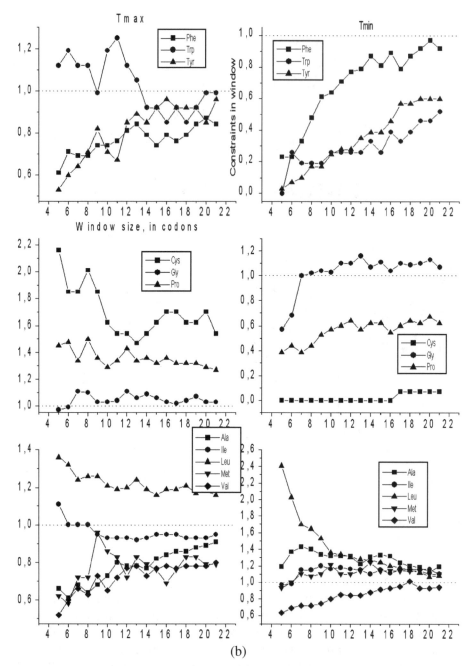

Figure 4. (Continued). Dependence of aromatic amino acids (upper), special amino acids (middle), and hydrophobic amino-acids (lower), constraints on window size (left TMAX, right TMIN windows).

genome sequences presently available show that each species has its own selection of synonymous codon usage, and probably selects the population of its tRNA pool accordingly. It follows that each species might precisely place its synonymous codons in its genes and mRNAs so as to be able to control specifically their local translation rate.

Specific constraints in the selection of most amino acids are found within both islet types. Interestingly, document involved in turns, beta-structures and especially disulphide bond formation are strongly avoided (favoured) in islets translated fastest (slowest), whereas document found in islets translated fastest have a propensity to enter alpha helices. We therefore surmise that more generally, the local modulations of the translation rate assist the folding of secondary structure, and even the completion of tertiary and quaternary structures. Again, experimental support for this hypothesis exists (Komar *et al.*, 1999). A given gene in a given species will be translated according to a rate schedule set by the species-specific codon usage frequency. It follows that the translation rate in a given host of a heterologous gene, which carries synonymous codons according to the selection in the species from which it has been derived, will be translated in the host-specific rate, which may not allow for the rate modulation required by native folding. This could one of the reasons why heterologous expression often lead to misconformation of the gene product, despite the fact that it carries the authentic primary sequence.

The degeneracy of the genetic code allows each species to adopt a 'dialect' of its own, by encoding its proteins with a particular selection of synonymous codons (corresponding to the population of its tRNA pool), thereby modulating the local translation rate as required by the protein folding. This could in addition provide the species with a strategy to avoid invasion of foreign genes, since the latter would be expressed mostly as misfolded proteins and tagged for degradation even before completion of their synthesis.

Studies similar to that presented here on *E. coli* are in progress in our laboratory. Preliminary results show that most constraints in the islets described in this paper are shared by *B. subtilis*, *S. cerevisiae* and *H. sapiens*. This observation strongly suggest that indeed the translation rate–folding dependence may be universal. It would mean that protein folding is mainly determined by the codon sequence of its mRNA, rather than by its primary sequence, which does not take account of the modulation of the translation rate. Each species is free to make specific use of synonymous codons, but must introduce translation rate modulations at places as required by the folding of the protein, by introducing locally constraints in the codon usage frequency and by setting its tRNA population accordingly. Protein folding must obey physical and chemical laws that are necessarily species-independent. The precise chronology of protein synthesis contributes to these laws.

With few exceptions, the constraint for, or against, the presence of an amino acid in the codon islets considered is enhanced as the size of the islet shrinks. The smaller the islet surrounding the codon of the aminoacid under consideration, the stronger the average density of frequent (or rare) codons in the islet, i.e. the faster (or the slower) the average translation rate of the islet. However, within a given mRNA, the position of the islets may change, depending on window size. It can therefore not be concluded that the translation rate increases (or decreases) as a particular amino acid in the islet under consideration is approached, although in most instances explored, this has been the case. In the latter event, the shorter the islet surrounding a particular amino acid, the stronger the codon frequency constraint in that islet and the stronger the constraint for that document, in other words there is a rising constraint gradient as the amino acid is approached, which corresponds to a positive (or negative) acceleration of the translation rate as the ribosome approaches its codon.

Here we have only examined the extreme islets in each gene. It would be of interest to look at amino acid constraints in less extreme islets, and determine the codon usage constraint limits for which the amino acid constraints vanish. This would allow assessment of the codon frequency constraint, or local translation rate modulation, significant for the control of folding. Work is in progress along this line.

What is the significance of these findings for conformational diseases? We understand that changes in the translation rate can induce misconformation of the nascent protein. Such changes will of course not come from a change in the actual codon usage within the coding sequences, which is stably encoded and will only be affected by the evolutionary, mutational drift. Rather, the changes could be introduced in real time by modification of the population in the tRNA pool, i.e. its size, the relative amounts of the different tRNA species and their base modification status, known to affect the kinetics of the codon-anticodon 'reaction', hence the translation rate. In particular, change in the growth status is known to deeply modify the cell's tRNA pool. The large amount of misfolded proteins, found ubiquitinated (tagged for proteolysis) while still attached to the ribosome could be the consequence of the imbalance of the tRNA pool with the codon usage. In this view, conformational diseases would ensue as the imbalance exceeds a critical value, at which the cellular content of misfolded proteins become large enough for aggregation to take place, simply because of the mass action law. This hypothetical mechanism, which depends on intracellular imbalances, could as well be the consequence of extracellular signals, endocrine, exocrine or environmental in origin, which force the cell to change its growth status (hormones) or to overload its translation machinery (excess demand on document, tRNAs, foldases) which cannot be dealt with by the cell and thereby shifts the translation rate–folding relationship towards misfolding. This hypothesis can be easily subjected to experimental proof.

REFERENCES

Baneyx, F. and Palumbo, J. L. (2003). Improving heterologous protein folding via molecular chaperone and foldase co-expression, *Methods Mol. Biol.* **205**, 171–197.
Berg, O. G. and Kurland, C. G. (1997). Growth rate-optimised tRNA abundance and codon usage, *J. Mol. Biol.* **270**, 544–550.
Bodenreider, C., Kellershohn, N., Goldberg, M. E., *et al.* (2002). Kinetic analysis of R67 dihydrofolate reductase folding: from the unfolded monomer to the native tetramer, *Biochemistry* **41**, 14988–14999.

Bross, P., Corydon, T. J., Andresen, B. S., *et al.* (1999). Protein misfolding and degradation in genetic diseases, *Human Mutation* **14**, 186–198.

Cordwell, S. J., Nouwens, A. S., Verrills, N. M., *et al.* (1999). The microbial proteome database — an automated laboratory catalogue for monitoring protein expression in bacteria, *Electrophoresis* **20**, 3580–3588.

Cortazzo, P., Cervenansky, C., Marin, M., *et al.* (2002). Silent mutations affect in vivo protein folding in *Escherichia coli*, *Biochem. Biophys. Res. Commun.* **293**, 537–541.

DeMager, P. P., Penke, B., Walter, R., *et al.* (2002). Pathological peptide folding in Alzheimer's disease and other conformational disorders, *Curr. Med. Chem.* **9**, 1763–1780.

Doig, A. J., Andrew, C. D., Cochran, D. A., *et al.* (2001). Structure, stability and folding of the alpha-helix, in: *Biochem. Soc. Symp.*, pp. 95–110.

Dong, H., Nilsson, L. and Kurland, C. G. (1996). Co-variation of tRNA abundance and codon usage in Escherichia coli at different growth rates, *J. Mol. Biol.* **260**, 649–663.

Gregersen, N., Bross, P., Jorgensen, M. M., *et al.* (2000). Defective folding and rapid degradation of mutant proteins is a common disease mechanism in genetic disorders, *J. Inherit. Metab. Dis.* **23**, 441–447.

Ikemura, T. (1981a). Correlation between the abundance of *Escherichia coli* transfer RNAs and the occurrence of the respective codons in its protein genes, *J. Mol. Biol.* **146**, 1–21.

Ikemura, T. (1981b). Correlation between the abundance of *Escherichia coli* transfer RNAs and the occurrence of the respective codons in its protein genes: a proposal for a synonymous codon choice that is optimal for the *E. coli* translational system, *J. Mol. Biol.* **151**, 389–409.

Johnson, W. G. (2000). Late-onset neurodegenerative diseases — the role of protein insolubility, *J. Anat.* **196**, 609–616.

Kolb, V. A., Makeyev, E. V. and Spirin, A. S. (2000). Co-translational folding of an eukaryotic multidomain protein in a prokaryotic translation system, *J. Biol. Chem.* **275**, 16597–16601.

Komar, A. A., Lesnik, T. and Reiss, C. (1999). Synonymous codon substitutions affect ribosome traffic and protein folding during in vitro translation, *FEBS Lett.* **462**, 387–391.

Lesnik, T., Solomovici, J., Deana, A., *et al.* (2000). Ribosome traffic in E. coli and regulation of gene expression, *J. Theor. Biol.* **202**, 175–185.

Ma, J., Zhou, T., Gu, W., *et al.* (2002). Cluster analysis of the codon use frequency of MHC genes from different species, *Biosystems* **65**, 199–207.

Maness, S. J., Franzen, S., Gibbs, A. C., *et al.* (2003). Nanosecond temperature jump relaxation dynamics of cyclic beta-hairpin peptides, *Biophys. J.* **84**, 3874–3882.

Mayor, U., Guydosh, N. R., Johnson, C. M., *et al.* (2003). The complete folding pathway of a protein from nanoseconds to microseconds, *Nature* **421**, 863–867.

Oresic, M. and Shalloway, D. (1998). Specific correlations between relative synonymous codon usage and protein secondary structure, *J. Mol. Biol.* **281**, 31–48.

Purvis, I. J., Bettany, A. J., Santiago, T. C., *et al.* (1987). The efficiency of folding of some proteins is increased by controlled rates of translation in vivo. A hypothesis, *J. Mol. Biol.* **193**, 413–417.

Sakahira, H., Breuer, P., Hayer-Hartl, M. K., *et al.* (2002). Molecular chaperones as modulators of polyglutamine protein aggregation and toxicity, *Proc. Natl. Acad. Sci. USA* **99**, 16412–16418.

Sato, S., Ward, C. L. and Kopito, R. R. (1998). Cotranslational ubiquitination of cystic fibrosis transmembrane conductance regulator in vitro, *J. Biol. Chem.* **273**, 7189–7192.

Schiene-Fischer, C., Habazettl, J., Schmid, F. X., *et al.* (2002). The hsp70 chaperone DnaK is a secondary amide peptide bond cis-trans isomerase, *Nat. Struct. Biol.* **9**, 419–424.

Schubert, U., Anton, L. C., Gibbs, J., *et al.* (2000). Rapid degradation of a large fraction of newly synthesized proteins by proteasomes, *Nature* **404**, 770–774.

Van Bogelen, R. A., Schiller, E. E., Thomas, J. D., *et al.* (1999). Diagnosis of cellular states of microbial organisms using proteomics, *Electrophoresis* **20**, 2149–2159.

Williams, S., Causgrove, T. P., Gilmanshin, R., *et al.* (1996). Fast events in protein folding: helix melting and formation in a small peptide, *Biochemistry* **35**, 691–697.

Advances in Neuroregulation and Neuroprotection (2005), pp. 239-268
Collin, C. *et al.* (Eds)
© VSP 2005

The pineal gland: Functional connection between melatonin and immune system in birds

C. BARRIGA [1,*], J. A. MADRID [2], M. P. TERRÓN [1], R. V. RIAL [3], J. CUBERO [1],
S. D. PAREDES [1], S. SÁNCHEZ [1] and A. B. RODRÍGUEZ [1]

[1] *Department of Physiology, University of Extremadura, 06071 Badajoz, Spain*
[2] *Department of Physiology, Faculty of Biology, University of Murcia, 30100 Murcia, Spain*
[3] *Department of Biology, F. i C.S., University Illes Balears, Spain*

Abstract—Research into the activity of the epiphysial complex and its hormone, melatonin, has been part of the growing general interest in chronobiology and in the potential therapeutic applications of melatonin. However, the studies have been almost exclusively oriented to humans and, by extension, to mammals, although humans are diurnal animals, while most other mammals are either nocturnal or crepuscular. The aim of this review is to direct attention to epiphysial complex in birds, diurnal species, that have an immense commercial interest, and the connection between melatonin and the immune system, system responsible for the survival of species. We also go more deeply into biosynthesis, metabolism, regulation and receptors of melatonin, a truly exceptional substance which is perhaps the only biomolecule which has undergone no structural change in all living beings and has probably always been dedicated to biological time measurement and control.

Keywords: Melatonin; pineal gland; immune system; birds.

INTRODUCTION

Both the epiphysial complex and melatonin have an extraordinary phylogenetic antiquity. Unequivocal signs have been found of the existence of the parietal foramen in the skulls of fossils of certain vertebrates of the Devonian and Silurian, ancestors of modern-day fish, amphibians and lacertilians, suggesting that the pineal organ already existed then (Roth and Roth, 1980). However, in spite of the early appearance of the pineal in vertebrates, melatonin shows an even greater antiquity and a degree of conservation and ubiquity that is highly conspicuous in a

---

*To whom correspondence should be addressed. Professor Carmen Barriga Ibars, Department of Physiology, Faculty of Science, University of Extremadura, Avda Elvas s/n, 06071 Badajoz, Spain. E-mail: cibars@unex.es

biological molecule. Such conservation in a biomolecule that apparently has neither a metabolic nor a structural function is difficult to match.

The presence of melatonin, identical in structure to that found in all vertebrates, has been described in: (a) unicellular organisms, including Monera (*Rodospirillum rubrum*) (Pooggeler, 1995), Protista (*Gonyaulax polyedra*) (Poeggeler *et al.*, 1991; Hardeland, 1999) and Fungi (*Saccharomyces cerevisiae*) (Hardeland, 1999); (b) plants (Hattori *et al.*, 1995; Dubbels *et al.*, 1995); and (c) invertebrates (for a review, see Vivien-Roels and Arent, 1983; Hardeland and Poeggeler, 2003) such as the migratory locust, *Locusta migratoria*, the crab, *Carcinus maenas*, the cuttlefish, *Sepia officinalis*, the face fly, *Musca autumnalis* and the flatworm, *Dugesia dorotocephala*. In plants, melatonin's biological role has yet to be clarified. In other organisms, however, its role has been related to the transduction of seasonal and circadian photoperiod information (Hardeland, 1993).

The aforementioned ubiquity and conservation of a molecule such as melatonin attracted the attention of some researchers to attempt to shed some light on how its functions might have developed on an evolutionary scale. From a chemical point of view, the same reaction seems to be the cause of melatonin's extreme instability and sensitivity to light: an oxidation reaction, catalysed by iron, of the pyrrole ring contained in reactive oxygen species, the superoxide anions. These products can be induced either by light or by other types of energy transfer reactions (Hardeland, 1993).

The fact that melatonin acts as a powerful antioxidant reflects a fundamental property of potential importance for understanding its role as a chemical mediator of darkness. In the beginning and for millions of years melatonin would originally have been used as part of the antioxidant protection system. In general, all the indoleamines could be used for this purpose. If it is compared with other antioxidants at a physiological pH, melatonin could be regarded as the most efficient. With these properties, the melatonin present in an organism would be destroyed by light but would stay relatively stable in darkness. Its effective concentration would hence undergo oscillations in coincidence with the light-dark cycle. This alternation could be seen as the primary reason that melatonin, somewhat later, was adopted as a mediator of information on darkness. There also must have arisen a cycle in the concentration of reactive oxygen species, and any cell exposed to light would have found itself subjected to this strong rhythm. This could have been made use of by the ancestors of modern-day organisms whose cells, and most especially their retinal and pineal photoreceptors, would have had to be protected from oxygen free radicals. It is possible, therefore, that the role of melatonin evolved polyphyletically on the basis of the molecule's exceptional properties and availability. In a last evolutionary step, melatonin synthesis could have been coupled to a circadian oscillator, a connection that would have had the advantage of the melatonin rhythm being a predictor of oscillations in levels of both light and free oxygen radicals (Hardeland, 1993; Hardeland *et al.*, 1995). For the animal, the oscillating levels of melatonin would serve to determine with great precision both the time of day (the melatonin clock) and the season of the year (the melatonin calendar), allowing it to make the physiological and behavioural changes needed to anticipate with advantage the dramatic fluctuations in the environment (Reiter, 1993).

In the following sections of this paper, particular attention will be paid to the synthesis and receptors of this indoleamine, and the effects that this hormone has on different functions of vital importance for the survival of birds, especially the connection between melatonin and the immune system.

HISTORICAL OVERVIEW

> "Let us then conceive here that the soul has its principal seat in the little gland which exists in the middle of the brain, from whence it radiates forth through all the remainder of the body by means of the animal spirits, nerves and even the blood, which, participating in the impressions of the spirits, can carry them by the arteries into all the members ..." (René Descartes, 1640).

The pineal gland or epiphysis cerebri has been known for more than 2000 years. The most ancient existing written description is that of Galen (130–200 AD) who refers to Herophilus of Alexandria (325–280 BC). Herophilus apparently thought that the pineal was a valve regulating the flow of *pneuma* (or *spiritus* in Latin) from the 3rd to the 4th ventricle. *Pneuma* was considered to be derived from air and transformed into *pneuma psychikon* (*spiritus animalis*) in the brain's ventricular system. *Pneuma psychikon* was the direct cause of the development of knowledge. Galen, however, considered that the pineal was a gland, at that time regarded as an organ which filled gaps between blood vessels and supported them. Classical Indian literature refers to the pineal as an organ of clairvoyance and meditation that also permits men to remember their past lives. There is a profusion of representations of the third eye in oriental imagery (Fig. 1). The studies of the classical authors of Greco-Roman medicine considered it to be a structure capable of materializing and transporting the flow of thought from the third to the fourth ventricle (Zrenner, 1985).

The first good description of the pineal derives originally from the 15th century, but comes to us from Vesalius in the 16th century with a description of its anatomical situation in his celebrated work *De Humanis Corporis Fabrica*. Probably the most famous of ancient texts on the pineal is that of the influential French philosopher René Descartes (1596–1656) who embellished the existing physiological and anatomical conceptions by proposing in his book *De Homine* that the pineal is a centre where the soul receives information from the body (Fig. 2). Descartes considered that, by moving, the pineal controlled the flow of 'animal spirits' into motor nerves and thus influenced movements of the body. He thought that the stimulus for the pineal function came from visual input to the retina, a most remarkable insight, as effectively this is seen as true today (Descartes, 1662). It is usually Descartes to whom is attributed the concept of the pineal as the seat of the soul, although this idea probably derived from Herophilus. These ancient

Figure 1. Representation of the Buddha. One observes the *third eye* corresponding to the sixth chakra. This third eye symbolizes total sight.

Figure 2. In 'De Homine', Descartes describes light as transmitting images to the retina and stimulating the animal spirit which travels through the nerves and activates the pineal gland.

ideas persisted for many centuries and to some extent have coloured 20th century perceptions of the pineal. They may have been responsible for the slow progress of early pineal research in rendering the gland somehow not quite scientifically respectable.

Voltaire described the epiphysis as the director of the brain, guiding the operation of the hemispheres by means of two bands of nerves. Ahlborn in 1884 was the first to notice the remarkable resemblance between the pineal/parietal organ of some poikilotherms and the structure of the lateral eyes. This was certainly a crucial observation. It led to extensive interest in the evolutionary history of the pineal in that century and to the statement by Studnicka in the next (in 1905) that the photosensory organ of poikilotherms became the secretory mammalian pineal (Arendt, 1995). This is firmly supported by modern comparative anatomy and underpins our current knowledge of the influence of light on this secretory organ, whether it be direct or indirect.

Early observations of pineal tumours by Heubner and Marburg, together with the work of others, led to the important conclusion that the gland influenced reproductive function (Heubner, 1898). In the mid 20th century, Holmgren observed that the pineal cells with a secretory activity, the pinealocytes, of a certain type of elasmobranch were sensory in nature, similar to the photoreceptor cells of the retina (Holmgren, 1959). In amphibians, lacertids and some fish, the pineal gland is of an extracranial parietal nature with photoreceptive structures resembling the lens or the retina that can act as a 'third eye'. This anatomical and physiological characteristic led the pineal gland, including that of humans, to be regarded as a vestige of some such primitive visual organ (Collin, 1972).

Our primary understanding of pineal control mechanisms derives from Ariëns Kappers' classical studies on the innervation pathway of the pineal from the superior cervical ganglion of the sympathetic nervous system (Ariëns Kappers, 1965). In all of the history of pineal research there is, however, nothing more extraordinary or more important than the discovery of the structure of its principle hormone melatonin by Aaron Lerner and his co-workers in 1958. They processed chromatographically thousands of bovine pineal glands and identified the bioactive compound N-acetyl-5-methoxytryptamine (Lerner *et al.*, 1958). They gave it the name melatonin because of its effects on the skin pigmentation of the frog and because it was chemically related to serotonin (5-hydroxytryptamine) (Lerner *et al.*, 1959, 1960). It was observed that melatonin had very powerful effects on the skin pigmentation of many animals, although the most important discoveries were those related to its antigonadal effects on mammalian reproductive systems (Lerner and Nordlund, 1975). Despite its name, however, it should be emphasized that melatonin has no effect on melanophores in mammals.

ANATOMY OF THE EPIPHYSIAL COMPLEX IN BIRDS

The pineal organ of birds, unlike that of fish, amphibians and reptiles, sometimes (in 20% of the birds that have been studied) has a bipartite structure: a primary pineal organ and secondary or accessory pineal tissue. For this reason, some authors speak of the pineal system for birds.

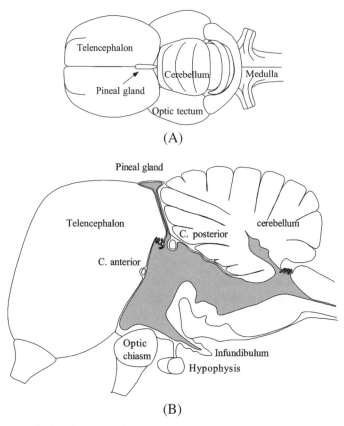

Figure 3. The avian brain. A: Dorsal view. B: Midsagittal section showing the position of the pineal gland between the anterior and posterior commisures.

The primary pineal organ can appear as a structure that extends from the intercommissural region to the roof of the third ventricle in the brain (Fig. 3). There are references to species with atrophied pineals in two avian orders, procellariiforms and strigiforms. These species apparently could have evolved from nocturnal predecessors (Kappers, 1979).

The parenchyma of the avian pineal organ presents several types of cell, but the most clearly defined are the receptor pinealocytes which constitute a group of cells similar to the pineal photoreceptor cells of poikilotherms (Fig. 4). This type of pinealocyte seems to be restricted to pineal organs of a follicular or saccular type and they are distributed on the edge of the lumen.

The principal characteristics of this type of photoreceptor cell that distinguish it from other pineal cell types is the presence of short external segments with concentric laminations. These structures were first described by González and Valladolid (1966) in *Gallus gallus*. Transmission electron microscopy shows a set of rudimentary structures in the form of concentric zones corresponding to the outer segments of photoreceptor cells.

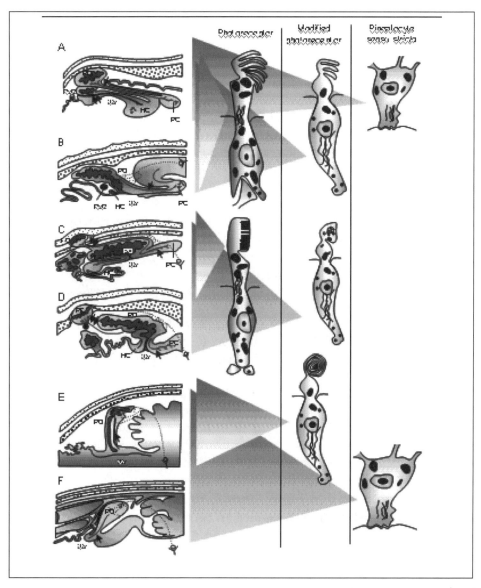

Figure 4. Evolution of the epiphyseal complex in various vertebrate classes — (A) lampreys; (B) teleosts; (C) anurans; (D) lacertilians; (E) birds; and (F) mammals — correlated with the evolution of ultrastructural features of the receptor line. Except in mammals, the pineal of the different groups of vertebrates receives light information through the skin and the bones of the skull. The pineal organ (PO) is situated on the roof of the diencephalon, dorsally to the third ventricle (IIIv) and between the habenular commissure (HC) and the posterior commissure (PC). In poikilotherms, it is associated with another photosensitive structure — the fish parapineal organ (PaO), the amphibian frontal organ (FO), and the reptile parietal eye (PE).

Scanning electron microscopy studies on *Urolancha domestica* showed morphological differences in these outer segments, with forms that were bulbous, cup-shaped, or elongated. The internal structure of these outer segments and their relationship with the inner segments has yet to be clearly defined. Their concentric zones may be chaotic in appearance and have perhaps undergone some degenerative process. The basal prolongation of these cells and their perikarya possess the same characteristics as vertebrate pineal photoreceptors, including the presence of synaptic layers. There is still some doubt about the existence of afferent nerve connections, although in *Gallus gallus* Collin *et al.* (1976) and González and Valladolid (1996) describe at an ultrastructural level the presence of an aminergic innervation with dark vesicles. The existence and nature of these photoreceptor cells in chicken pineal glands was confirmed in the work of Binkley *et al.* (1977) and Goto *et al.* (1990).

The presence of secretion granules in the receptor-type pinealocytes of birds would seem to be a sign that there indeed exists a secretory process in these cells. They have been described in the pinealocytes of *Gallus gallus* and *Columba livia*. These granules are formed in the Golgi apparatus and accumulate at various sites: in the basal and distal prolongations as in the case of *Gallus gallus*, *Columba livia* and *Passer domesticus*; or in the perikarya, as in the case of *Melopsittacus undulatus* and *Anas platyrhynchos*. The quantity of these secretion granules is highly variable, although they are observed to be abundant in *Pica pica* and *Melopsittacus undulatus*. They possibly have a serotoninergic function (González and Valladolid, 1996).

With respect to the presence of synaptic layers in this cell type, there seems no reason to think that they play a relevant part in the activity of nerve transmission since no afferent innervation has as yet been observed in these cells. These layers are frequent in *Vanellus vanellus*, *Coturnix coturnix japonica* and *Gallus gallus*, but are absent in *Pica pica* (González and Valladolid, 1966).

BIOSYNTHESIS AND METABOLISM OF MELATONIN

Biosynthesis of melatonin

The precursor of melatonin is the essential amino acid L-tryptophan (L-TRP), which is taken up from the blood by pinealocytes through an active transport mechanism under adrenergic control, as has been observed in particular in mammals (Urbanqui, 2000), but also more generally in lampreys and teleosts, lizards and birds (Collin *et al.*, 1986, 1989; Falcón *et al.*, 1992). Within the cell, L-TRP is converted into 5-hydroxytryptophan by the enzyme tryptophan hydroxylase (TPOH). The enzyme aromatic L-amino acid decarboxylase (AAAD) acts on the 5-hydroxytryptophan to form 5-hydroxytryptamine or serotonin. The concentration of serotonin in the pineal is very high, exceeding that in any other organ. From that point, the most important pathway in the pineal metabolism of serotonin involves its transformation into

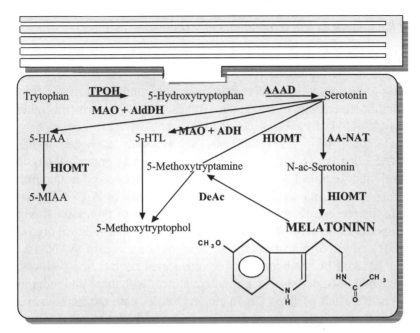

Figure 5. Pathways of the indole metabolism in the photosensitive pineal cells. Enzymes: AAAD, aromatic L-amino acid decarboxylase; ADH, alcoholdeydrogenase; Ald DH, aldehyde dehydrogenase; AA-NAT, arylalkylamine N-acetyltransferase; DeAc, deacetylase; HIOMT, hydroxyindole-*O*-methyltransferase; MAO, monoamine oxidase; TPOH, tryptophan hydroxylase. Indoles: N-acserotonin, N-acetylserotonin; 5-HIAA, 5-hydroxyindole acetic acid; 5-HTL, 5-hydroxytryptophol; 5-MIAA, 5-methoxyindole acetic acid. The chemical structure of melatonin is given in the lower part of the figure. The upper part, in which there appear thick dark lines, represents the outer segment of the photosensitive pinealocytes. Redrawn from Falcón, 1999.

N-acetylserotonin by the action of arylalkylamine N-acetyl transferase (AA-NAT), the enzyme that constitutes the rate-limiting step in the synthesis of melatonin. Finally, N-acetylserotonin is converted by HIOMT into melatonin (N-acetyl-5-methoxytryptamine) which is released into the circulation by the pinealocytes (Falcón, 1999). In addition, melatonin itself may be deacetylated *in situ* to produce 5-methoxytryptamine and 5-methoxytryptophol, as has been observed in the pineal organs of two fish and two lizard species (Falcón *et al.*, 1985; Grace and Besharse, 1994; Yanez and Meissl, 1995). The pathways of the indole metabolism in the photosensitive pineal cells.is shown in Fig. 5.

Other indole compounds, 5-hydroxyindole acetic acid (5-HIAA) and 5-hydroxytryptophol (5-HTL), are also synthesized after the oxidative deamination of serotonin, catalysed by monoamine oxidase (MAO). Methylation of these compounds by HIOMT leads to the formation of 5-methoxyindole acetic acid (5-MIAA) and 5-methoxytryptophol, respectively. Serotonin can also be directly methylated to 5-methoxytryptamine by HIOMT (Falcón, 1999).

Effect of light on the pineal biosynthesis of melatonin

The synthesis and release of melatonin from the pineal into the circulation is under the influence of the light/dark cycle (Liebmann *et al.*, 1997). Large changes in melatonin production are typically associated with similar changes in the activity of the penultimate enzyme in melatonin synthesis, AA-NAT, which is inhibited by light. Two AA-NATs have been identified in fish: AANAT1, more closely related to AANATs found in higher vertebrates, is specifically expressed in the retina; and AANAT2 that is specifically expressed in the pineal organ. A physiological day/night rhythm in pineal AANAT2 protein has been observed in the pike, where the light exposure at midnight decreases the abundance of AANAT2 protein and activity. In the present, se apunta la posibilidad de que of proteasomal proteolysis play an important role in the regulation of AANAT2 in the fish pineal organ (Falcón *et al.*, 2001), since in pineal organ cultures, the decrease of AANAT2 activity by light is blocked by inhibitors of the proteasomal degradation pathway, and if glands are maintained under light at night, treatment with these inhibitors increases AANAT2 activity and protein. Organ culture studies with the trout and seabream also indicate that the light-induced decrease of AANAT2 activity is prevented when proteosomal proteolysis is blocked, indicating that proteosomal proteolysis is a conserved element in the regulation of AANAT in vertebrates (Falcón *et al.*, 2001).

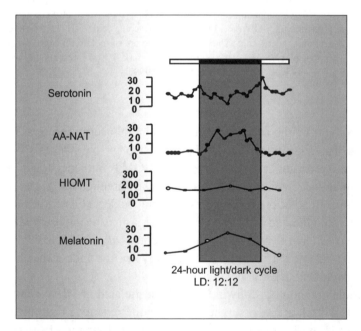

Figure 6. Daily measurements of arylalkylamine (AA-NAT) and hydroxyindole-O-methyltransferase (HIOMT) activity, and serotonin and melatonin content, in the chicken pineal gland. The measurements are expressed as: serotonin and melatonin (ng/gland); AA-NAT and HIOMT (activity/ng/gland/h). Modified from Hadley (1997).

This inhibition of AANAT by light puede llevarse a cabo too by the participation of enzyme phosphodiesterase 6 (PDE6, the primary effector of retinal phototransduction) actuando sobre the AA-NAT (the rate-limiting enzyme of melatonin synthesis). In fact, in culture, chick pineal glands were protected from inhibition by light when selective phosphodiesterase 6 (PDE6) inhibitors (zaprinast and 1,3-dimethyl-6-(2-propoxy-5-methanesulfonylamidophenyl)pyrazolo[3,4d]-pyrimidin-4-(5H)-one (DMPPO)), were added to the culture medium. PDE6 inhibitors did not affect AA-NAT activity in the dark. Moreover, a general PDE inhibitor (IBMX, 3-isobutyl-1-methylxanthine) increased AA-NAT in a light-independent manner (Morin *et al.*, 2001). Figure 6 shows the changes day/night in AANAT and HIOMT activity, as well as serotonin and melatonin content, in the chicken pineal gland.

If the conditions of external lighting are inverted, the enzymatic activities and the pineal biosynthesis of indoleamines are also inverted. There is therefore a daily rhythm of pineal biosynthetic activity that is controlled by the daily changes in the duration of light. This rhythm is lost if animals are subjected to conditions of continuous illumination, but is maintained, although at a lower intensity, if they are kept in darkness.

Intracellular regulation of melatonin production

In all vertebrates that have been studied, cyclic AMP (cAMP) is actively involved in the regulation of melatonin biosynthesis (Fig. 7). *In vitro*, an increase in the intracellular cAMP content always results in an increase in AA-NAT activity and melatonin production. Neither cyclic GMP analogues, nor agents that specifically increase intracellular cGMP levels are able to modulate melatonin secretion (Falcón *et al.*, 1990). A fall in cAMP might mediate the acute inhibitory effects of light on melatonin biosynthesis. Indeed, the light-induced inhibition of AA-NAT activity and melatonin secretion is partially prevented in the presence of cAMP analogues or of agents which increase intracellular cAMP levels (Falcón *et al.*, 1992). Similarly, lighting at midnight also induces a 40% reduction in cAMP content in isolated trout photoreceptors in culture (Falcón *et al.*, 1992). This suggests that light might inhibit adenylate cyclase and stimulate cAMP-phosphodiesterase. Light also reduces cGMP levels by 40%, but cGMP analogues do not prevent the light-induced decrease in melatonin production (Falcón *et al.*, 1992). The observation that stable analogues of cAMP never completely prevent the inhibitory effects of light suggests that cAMP is not the only intracellular messenger involved in the control of melatonin production.

Typical (fish) and modified (sauropsid) photoreceptor cells are not only light sensitive but also their activity is modulated by temperature and/or neurotransmitters. Many of these effects are mediated through cAMP. In fish, cAMP accumulation and AA-NAT activity respond similarly to different temperatures, displaying a peak at 12–15°C in the trout (Thibault *et al.*, 1993a) and at 18–25°C in the pike (Thibault *et al.*, 1993b). In the chicken pineal, norepinephrine and vasoactive intestinal peptide (VIP) act on cell surface receptors to inhibit or stimulate melatonin biosynthesis, re-

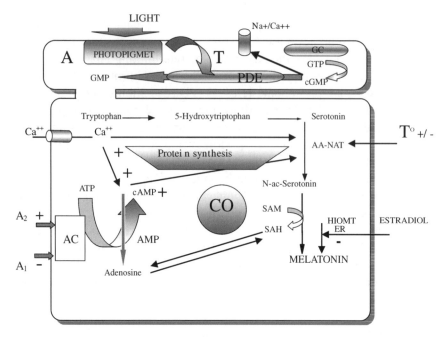

Figure 7. Intracellular regulation of melatonin secretion in the fish pineal photoreceptor. A, arrestin; A1, adenosine A1 receptor subtype; A2, adenosine A2 receptor subtype; AC, adenylate cyclase; Ca^{++}, calcium associated to a calcium binding protein; CO, circadian oscillator; ER, oestradiol receptor; GC, guanylate cyclase; HIOMT, hydroxyindole-O-methyltransferase; 5'-N, 5'-nucleotidase; na/ca, cationic channel; N-ac-serotonin, N-acetylserotonin; NE, norepinephrine; AA-NAT, arylalkylamine N-acetyltransferase; SAM, S-adenosylmethionine; SAH, S-adenosylhomocysteine; T, transducin; T°, temperature; PDE, phosphodiesterase. Redrawn from Falcón, 1999.

spectively (Zatz *et al.*, 1990). The most complete picture of the mechanisms linking cAMP to the modulation of AA-NAT has been obtained from studies on rodents, indicating that in mammals the norepinephrine released by the postganglionic neurons of the superior cervical ganglion during the dark phase (night) interact with the adrenergic receptors of pinealocytes, which gives rise to an increase in cAMP production (Pevet, 1998). The rise in cAMP levels activates protein kinase A (PKA) that phosphorylates the transcription factor CREB into P-CREB, leading to a massive expression of the gene coding for AA-NAT (Roseboom and Klein, 1995; Roseboom *et al.*, 1996). Once synthesized and active (with a very rapid turnover), the AA-NAT converts serotonin into N-acetyl-serotonin, which is immediately methylated to form melatonin. In rodents, a down regulation carried out by ICER (inducible cAMP early repressor) in camp rhythm inside the pineal gland has been also observed. Thus, it has been seen that while the P-CREB accumulates endogenously at the beginning of the dark period and declines during the second half of the night; concomitant with this decline, the amount of ICER rises, and these changes in the ratio between P-CREB and ICER shapes the *in vivo* dynamics in mRNA and, thus, protein levels of AA-NAT. A silenced ICER expression in pinealocytes leads to a desinhibited AA-NAT transcription and a primarily enhanced melatonin synthesis (Maronde *et al.*, 1999). The role of ICER on the AA-NAT es llevado a cabo mediante la regulation of gene expression in rat pineal gland. Asi, inhibition of specific kinase in primary pinealocyte cultures show that ICER induction depends pivotally on the activation of cAMP-dependent protein kinase II. Eliminating ICER´s impact by transfecting antisense constructs into pinealocytes revealed a predominant beta-adrenergic mechanism in regulating a cotransfected CRE-inducible reporter AA-NAT gene (Pfeffer *et al.*, 2000).

Adenosine, a neuromodulator produced locally by the pineal organ of vertebrates, also modulates AA-NAT activity and/or melatonin production in the chicken, the trout and the pike pineal organ (Falcón, 1999). Inhibitory effects are observed at low concentrations of this nucleoside and stimulatory effects at higher concentrations. The former is a result of the activation of A1 adenosine receptors which are negatively coupled to adenylate cyclase and the latter to activation of A2 adenosine receptors which are positively coupled to adenylate cyclase.

In isolated trout photoreceptor cells and in chicken pineal cells, low extracellular Ca^{2+} concentrations decrease melatonin secretion and high concentrations increase melatonin production (Begay *et al.*, 1994b; Meissl *et al.*, 1996). Agents that increase Ca^{2+} influx through voltage sensitive L-type Ca^{2+} channels are also able to increase melatonin production in trout or chicken pineal cells, whereas antagonists of these channels counteract these effects (Bégay *et al.*, 1994b). The presence of voltage-gated L-type Ca^{2+} channels was demonstrated using patch-clamp recordings and measurements of Ca^{2+} in trout photoreceptors (Bégay *et al.*, 1994; Kroeber *et al.*, 1997). The channels are activated at potentials ranging from -30 to $+40$ mV with a maximum amplitude observed at 0 mV (Bégay *et al.*, 1994). These data are consistent with the idea that the depolarized state, as observed during darkness, favours entry of Ca^{2+} through activation of the voltage-dependent L-type channels. A reduction in Ca^{2+} might participate in the light-induced reduction of melatonin secretion as a consequence of hyperpolarization of the cell (-80 mV) and closure of the channels. Studies in chicken and trout pineal cells suggest that Ca^{2+} acts through modulation of cAMP (Zatz, 1992), although the mechanism of this modulation is still unclear.

A catecholaminergic control of melatonin production occurs in some, but not all, fish species (Cahill, 1996; Falcón *et al.*, 1991; Martín and Meissl, 1992). Whereas the effects of noradrenaline are bimodal in the pike, they are inhibitory in the chicken and stimulatory in the rat (Falcón, 1999). Since the intracellular pathways of activation via norepinephrine have as yet not been well established in non-mammalian vertebrates, Fig. 8 presents the intracellular mechanisms activating cAMP for the subsequent activation of AA-NAT by norepinephrine as have been defined in mammals.

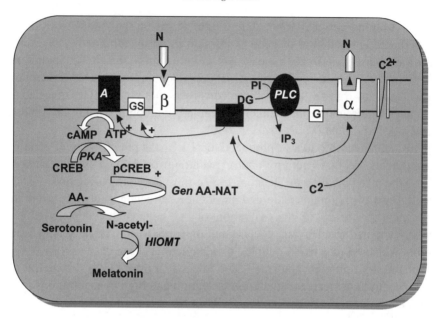

Figure 8. Regulation of the synthesis of melatonin by the pineal gland. The binding of noradrenaline (NA) to the β-adrenergic receptors provokes the activation of adenylate cyclase (AC) via a stimulated G protein (Gs). The active AC hydrolyses on adenosine tryptophan (ATP) converting it into cyclic adenosine monophosphate (cAMP). The cAMP, with the intervention of protein kinase A (PKA) fosforilan el factor de transcripción CREB en pCREB, el cual provoca una expression masiva del gen que codifica la arylalkylamine N-acetyltransferase (gen AA-NAT) to transform serotonin into N-acetylserotonin, which is methylated by hydroxyindole-*O*-methyltransferase (HIOMT) to melatonin. The binding of noradrenaline to alpha-1 receptors stimulates the β-adrenergic receptors. The alpha-1 receptors are paired with a G protein that activates phospholipase C (PLC). This enzyme hydrolyses phosphatidylinositol (Pi) into inositolphosphate (IP) and diacylglycerol (DG). These compounds, via protein kinase C (PKC) and an increase in intracellular calcium, enhance the activation of NAT induced by the stimulation of the β-adrenergic receptors. The stimulation of the alpha-1 receptors likewise increases the entry of extracellular calcium through voltage-dependent channels. Redrawn from Pevet, 1998.

MELATONIN RECEPTORS AND THEIR REGULATION

Melatonin is involved in a great variety of processes, including synchronization of the circadian system, regulation of sleep, seasonal synchronization and reproduction, regulation of blood pressure, oncogenesis, osmoregulation, thermogenesis, retinal physiology, etc. This broad spectrum of functions may be explained on the basis of various mechanisms (for review, see Witt-Enderby *et al.*, 2003): (i) The melatonin levels undergo circadian and seasonal fluctuations which determine changes in the activation of melatonin's own receptors *in vivo* (Reiter, 1991). (ii) Melatonin may act on transduction cascades directly, without acting on the receptors. This property could be attributed to its lipophilic nature and/or to the existence of active capture mechanisms (Benitez-King and Antón-Tay, 1993; Finocchiaro and Glikin, 1998; Menéndez-Peláez and Reiter, 1993). (iii) The existence

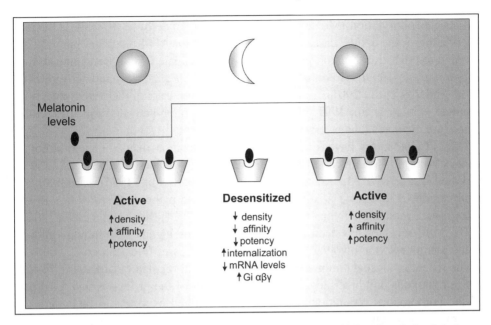

Figure 9. Homologous regulation of melatonin receptors. Continuous high melatonin levels induce a desensitised state in melatonin receptors. Redrawn from Enderby *et al.*, 2003.

of diverse high (Reppert *et al.*, 1994; Reppert *et al.*, 1995) and low (Nosjean *et al.*, 2001) affinity melatonin receptors. (iv) The regulation of the expression of melatonin receptors and possibly modifications in their affinity, as an effect of a multiplicity of factors (light-dark cycle, endogenous pacemaker, melatonin, or other hormones) (Witt-Enderby *et al.*, 2003) (Fig. 9).

On the basis of the pharmacological and kinetic differences in the binding of $2[^{125}I]$iodomelatonin, melatonin receptors were initially classified into two types: high affinity, initially denominated ML1 and low affinity or ML2. To date, three subtypes of ML1 receptors have been cloned and characterized. One is the MT1 receptor, which corresponds to the subtype previously known as ML1A or Mel1a (Chan *et al.*, 2002) and is encoded by the *Mel1a* gene. Another is the MT2 receptor, corresponding to the subtype previously known as ML1B or Mel1b (Clemens *et al.*, 2001) encoded by *Mel1b* gene. And finally, the Mel1c receptor has been cloned from *Xenopus laevis* (Conway *et al.*, 1997), but this gene is not expressed in mammals. As well as these high affinity receptors, there is evidence for another melatonin binding site of lower affinity for melatonin (type ML2), which was purified from the brain and kidney of the hamster (Dubocovich, 1995; Molinari *et al.*, 1996; Nosjean *et al.*, 2000) and is known as MT3.

The MT1 receptors are members of the superfamily of seven transmembrane domain G-protein coupled receptors. On stimulation, they produce responses that inhibit the cAMP transduction cascade (Brydon *et al.*, 1999a,b; Reppert *et al.*, 1994; Witt-Enderby and Dubocovich, 1996) and hence lead to the inhibition of PKA (Morgan *et al.*, 1994, Witt-Enderby *et al.*, 1998) and to a decrease in

CREB phosphorylation (McNulty *et al.*, 1994; Witt-Enderby *et al.*, 1998). Besides intervening in the cAMP cascade, the activation of the MT1 receptors may also be coupled to other transduction processes. It may stimulate PLC-dependent cascades (Brydon *et al.*, 1999a; Ho *et al.*, 2001; MacKenzie *et al.*, 2002) and activate PKC (Witt-Enderby *et al.*, 2001). It may also connect with calcium-activated potassium channels (Geary *et al.*, 1997, 1998) and with G-protein-activated inward rectifier potassium channels (Jiang *et al.*, 1995).

The MT2 receptors have a more restricted localization than MT1 and although their activation also induces a decline in the cAMP cascade (Brydon *et al.*, 1999a, b; Jones *et al.*, 2000; MacKenzie *et al.*, 2002; Reppert *et al.*, 1995), it additionally inhibits cGMP levels via a soluble guanylyl cyclase pathway (Petit *et al.*, 1999). They also stimulate the hydrolysis of PI (Ho *et al.*, 2001; MacKenzie *et al.*, 2002).

The MT3 receptor has been found to be 95% homologous with human quinone reductase, an enzyme involved in detoxification processes (Nosjean *et al.*, 2000).

The receptors of melatonin are subjected to several regulatory processes, including homologous and heterologous regulation. Four of the processes that form part of the homologous regulation produced by prolonged exposure to melatonin are G-protein uncoupling, phosphorylation, internalization and down-regulation.

(i) The persistence of high levels of melatonin for long periods leads to increases in the proportion of inactive forms of G proteins, with a concomitant loss of potency of melatonin on its receptor (Gauer *et al.*, 1993, 1994; Tenn and Niles, 1993; Witt-Enderby *et al.*, 1998).

(ii) Another form of regulation of the melatonin receptors is their desensitization through their phosphorylation produced by PKA, PKC, or G-protein receptor kinases (GRKs) (Hosey *et al.*, 1995; Krupnick and Benovic, 1998; Witt-Enderby *et al.*, 2003). This phosphorylation process impedes the interaction between G-protein and the receptor (Witt-Enderby *et al.*, 2003).

(iii) The prolonged presence of melatonin could induce another form of desensitization, the internalization of the receptors, although this process has not yet been described for melatonin (Witt-Enderby *et al.*, 2003). According to the description for other GPC receptors, this process would take place via the elimination of receptors through the formation of clathrin-coated vesicles.

(iv) Finally, the fourth mechanism of homologous regulation is down-regulation. The $2[^{125}I]$iodomelatonin affinity diminishes after exposure of the receptors to nocturnal melatonin levels (Schuster *et al.*, 2001). This effect might be explained by the decline in mRNA levels (Schuster *et al.*, 2001) and by the action of the cAMP-dependent signaling pathway (Barrett *et al.*, 1996).

With respect to heterologous regulation, various stimuli can change the density of melatonin receptors. In the SCN, this density varies according to the phase of the light/dark cycle, being lower at night, even in pinealectomized animals in which the daily cycle of melatonin is absent (Masson-Pevet *et al.*, 2000), so that the change could be produced by the light conditions. It is possible that the action of the light-dark cycle on the melatonin receptors is mediated by the clock genes of the circadian clock (SCN), although it is as yet still not known whether the genes of the melatonin receptors are clock-controlled genes (Witt-Enderby *et al.*, 2003).

Other factors that affect receptor density are the circulating levels of oestradiol and scheduled arousal. Oestradiol levels seem to be inversely correlated with the density of MT1Rs receptors in the cerebral and caudal arteries of female rats (Seltzer *et al.*, 1992). Scheduled arousal in hamsters, elicited either by handling or by subcutaneous injections of saline solution, is also capable of synchronizing the expression of melatonin receptors in the SCN (Hastings *et al.*, 1997).

FUNCTIONAL CONNECTION BETWEEN THE PINEAL ORGAN AND THE IMMUNE SYSTEM

Interpretation of the melatonin message within the body is essential for the physiological functions of an animal to adapt to environmental conditions and needs, an adaptation that would increase the probability of survival. Immune system activity is one of the physiological capabilities most responsible for the survival of an individual animal and reproductive system function guarantees the survival of the species. There is no longer any doubt about the interaction between the immune, nervous and endocrine systems (Homo-Delarche and Dardenne, 1993; Fabris, 1994; Skwarlo-Sonta, 1996) (Fig. 10). The efficacy of the immune system function in defending against harmful microorganisms, foreign molecules, or tumour cells requires it to be coordinated into periods of about 24 h. As with other body functions, the activity of the immune system undergoes circadian changes and reciprocal synchrony is of great importance to homeostasis. Illnesses, however, can alter these rhythms and modify their temporal coordination (Levi *et al.*, 1992).

Corticosteroids were the first humoural factors recognized as regulators of the daily rhythm of the immune system (Skwarlo-Sonta, 1996). There exists clear evidence, however, that certain parameters and some immune cells fluctuate differentially over a 24-h period and exhibit different phase relationships with circulating corticosteroid levels (Levi *et al.*, 1992). The implication is hence that there are one or more other factors involved in regulating the circadian rhythm of the immune function and one of the main candidates would seem to be the pineal gland and its secretion of melatonin. There are three reasons for this statement: (i) the circadian and seasonal periodicity of the pineal gland function; (ii) the strong dependence of the circadian rhythm of melatonin synthesis on light conditions; and (iii) the participation of melatonin in the control of different biological rhythms, including those associated with ageing and with affective and psychosomatic diseases, which in turn, are related to an increased incidence of infections, autoimmune disorders and cancer (Skwarlo-Sonta, 1996).

There have been few papers published on the functional connection between melatonin and the immune system of poikilotherms and birds, in spite of the

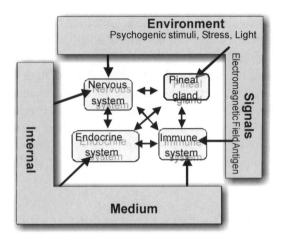

Figure 10. The pineal gland is involved in the transduction signal between environmental information and the immune, nervous, and endocrine systems. Modified from Skwarlo-Sonta, 1996.

interesting immuno-pineal relationships recognized in these animals. A chrono-haematological study was performed on the blood of pinealectomized and sham-operated lizards over a 48-h time period. The removal of the pineal significantly inhibited leukopoiesis and erythropoiesis (as reflected in the reduced number of cells in the circulation) and led to hypoglycaemia. In the sham-operated lizards, however, there was an evident circadian rhythm in the white and red blood cell and glucose levels (Haldar-Misra and Thaphliyal, 1981a). In mammals and birds, developmental and age-related changes in pineal function appear to be at least partially related to immune system efficiency. The mechanisms by which melatonin influences immune system function are complex, but are known to involve the participation of such mediators as endogenous opioids, cytokines, hormones, etc. As melatonin is a highly lipophilic compound, it may easily penetrate immune cells without the mediation of any specific receptors and act within the cells as a potent free-radical scavenger and as an anti-ageing and oncostatic factor. The immune system may, in turn, via the synthesis and secretion of soluble factors, such as cytokines, influence pineal gland function, thereby closing the information loop to maintain homeostasis in order to face the harmful environment (Skwarlo-Sonta, 1996).

Pinealectomy, bursectomy and the immune system

The avian embryo offers an excellent model for the study of the reciprocal interdependence between pineal gland and the immune system, especially because the thymus and pineal start to develop at the same time (Skwarlo-Sonta, 1996, 1999, 2002). In chicken embryos pinealectomized at 96 h of incubation, Jankovic *et al.* (1994) observed a delay in the development of the primary lymphoid organs (thymus and bursa of Fabricius) and a decrease in the humoural immune response and in the activities of different types of cell-mediated immunity. These results clearly indicate the necessity of an intact pineal gland for normal immune system development and suggest that the pineal gland may influence the lymphoid organs directly and/or indirectly via other neuroendocrine secretions (Skwarlo-Sonta, 2002). Nevertheless, one can not reject the possibility that the extrapineal synthesis of melatonin exerts a compensatory effect on the late stages of development.

A strategic relationship between the immune system and the pineal gland was also demonstrated in experiments involving the bursa of Fabricius, a primary lymphoid gland found exclusively in avian species and responsible for the maturation of B-cells. It was demonstrated that early embryonic bursectomy not only diminished the chicken immune response but also influenced the circadian rhythm of the pineal gland function in reducing the nocturnal peaks in pineal NAT activity and serum melatonin levels (Youbicier-Simo *et al.*, 1996). Also, pinealectomy performed in early post-hatched chickens abolished the circadian rhythm of non-specific immune parameters, which were restored by prolonged treatment with very low, physiological doses of melatonin (Rosolowska-Huszcz *et al.*, 1991).

With respect to the specific immune response, pinealectomy in the turtle dove increased the number of white blood cells and the total protein level and altered heterophil phagocytosis (Rodríguez and Lea, 1994) with the heterophils presenting an augmented oxidative metabolism. This effect was related to a reduction in circulating melatonin (Rodríguez *et al.*, 1999a).

Immunomodulatory action of melatonin

General observations suggest that the effects exerted by melatonin on aspects of the function of the immune system not only depend on the species, age and sex, but also on the experimental protocol (including the season), melatonin dose and the route of administration. One of the most interesting relationships between melatonin and the immune system is represented by the season-dependent changes in immunity observed in wild-living animals, both in nature and under laboratory conditions, where the animals can be kept under different lighting regimes (Markowska *et al.*, 2000).

It is surprising that there have been no studies of the influence of melatonin on the immune system of fish, given the great applied interest in knowledge of the possible immunostimulatory role of melatonin for fish-farming. Abundant indirect evidence, however, supports such a role. Environmental factors, particularly temperature and photoperiod, are known to be immunomodulatory in the lower vertebrates such as fish (Zapata *et al.*, 1992). With respect to non-specific immune mechanisms (e.g. phagocytosis), in the tench *Tinca tinca*, a warm-water cyprinid, the different stages of the phagocytic process of blood granulocytes show the highest level of activity during the winter. It remains high in the spring and declines in the summer, when the lowest level of activity is found (Collazos *et al.*, 1995).

The complement system is one of the main routes by which the inflammatory process is effected. There are two different but convergent pathways of complement activation: the immunoglobulin-dependent, or classical pathway and the immunoglobulin-independent, or alternative pathway. In *Tinca tinca*, it was found that the alternative complement pathway activity was greater in winter in both males and females than in the other three seasons (Collazos *et al.*, 1994). When the seasonal variations in specific immunity in this fish were analysed through the mitogen-induced proliferative response of lymphocytes, it was found that the lowest levels of this response to the different mitogens (phytohaemagglutinin, concavalin A, *Escherichia coli* lipopolysaccharide and pokeweed mitogen) occurred during winter and the highest during summer (Collazos *et al.*, 1996). Hence one can presume that the high winter melatonin concentrations activate the non-specific immune response, doubtless to counteract the stress-mediated winter suppression of the specific immune response, i.e. as a compensatory mechanism for the activity of the specific immune system which appears principally to operate in summer. In sum, while there has been no work directly approaching the influence of melatonin on the immune response in fish, it is probable that the seasonal changes in melatonin levels affect that response.

In birds, melatonin has been shown to modulate several immune functions, namely, antibody production, lymphocyte proliferation, ADCC activity, NK cell cytotoxicity, cytokine synthesis and release, etc. (Skwarlo-Sonta, 1996). It is known that melatonin enhances mitogen-induced T-cell blood lymphocyte and T-cell and B-cell splenocyte proliferation in male broiler chickens (Kliger *et al.*, 2000). It has also been suggested that melatonin inhibits PHA-stimulated chicken lymphocyte proliferation *in vitro* (Markowska *et al.*, 2001)

In the chicken, the circadian rhythm of different immune parameters was found to be strongly dependent on the presence of an intact pineal gland (Rosolowska-Huszcz *et al.*, 1991). In seven-week-old chickens immunized three times at 9-day intervals with T-dependent porcine antigen (Youbicier-Simo *et al.*, 1996), the diurnal serum melatonin concentration increased after the second antigen challenge. Also, Moore *et al.* (2002) have shown that exogenous melatonin can reconstitute a deficient cellular and humoural immune response in pinealectomized Japanese quail.

Melatonin added to avian lymphocyte cultures over a wide range of concentrations did not influence cell proliferation, as measured by 3H-thymidine incorporation. However, when the culture was stimulated with mitogens, the addition of melatonin generally diminished cell proliferation. Also, when melatonin was added to splenocytes pre-treated with T-cell mitogens, blast formation was almost completely blocked. This effect was best seen in cells isolated from the youngest (5 days old) group of studied chickens (Skwarlo-Sonta, 1999).

In recent years, many reports have shown that melatonin is a broad-spectrum antioxidant due to its ability to scavenge free radicals and to stimulate antioxidant enzymes (Tan *et al.*, 2000). It has also been shown that it is a significant scavenger *in vitro* of both free radicals and other reactive oxygen species and, at both physiological and pharmacological concentrations, reduces oxidative stress *in vivo* (Reiter *et al.*, 2000).

Nevertheless, although melatonin is secreted by the avian pineal and eyes, there is no literature establishing a source for the melatonin involved in the immune responses. Research has concentrated on the pineal gland and there is no evidence linking the eye to immune functions in birds. Most of the literature concerning photoperiod effects in the avian immune system is indirect, describing growth, performance and reproductive changes (Lewis *et al.*, 1996). However, Kirby and Froman (1991) reported a suppressed cellular immunity and secondary antibody response in immature cockerels reared under LL (light-light) as compared to LD (light-dark) 12:12 daily cycles. A clear dose-response immunoenhancement was observed after melatonin administration to birds (Japanese quail) with immunosuppression induced by constant light (Moore and Siopes, 2000). Moore *et al.* (2002) have demonstrated that pineal melatonin, but not ocular melatonin, is sufficient to maintain the normal cellular and humoural immune response in adult Japanese quail. Furthermore, melatonin administered within physiological levels had important immunoreconstituting effects on surgically immunocompromised birds. These data support the role of melatonin as an immunoregulatory hormone in birds and establish the pineal gland as the primary source of this melatonin-mediated immune regulation.

Melatonin as an antioxidant in phagocytic cells

Phagocytosis is an important element of the non-specific immune response and represents a fundamental mechanism of defence against infection. Phagocytic cells engulf their target (antigen) and then destroy it by the action of enzymes which form oxygen-derived free radicals by means of a series of oxidation-reduction reactions which lead to what is known as the 'respiratory burst'. In this process, various chemically aggressive species are formed, such as superoxide anions, hydrogen peroxide, hydroxyl radicals, or hypochlorite. Their function is to destroy the invading microorganism. The presence of free radicals in phagocytes is beneficial for the organism, since it is thanks to their formation within those cells that pathogenic microorganisms are destroyed. It is clearly an effective adaptation and solid defence adopted by the organism in its natural habitat. What would really be an advantage would be if once the radicals had fulfilled their goal they were then sequestered and/or eliminated from the phagocytes, as this would have the effect of guaranteeing the integrity of those cells.

Rodríguez *et al.* (1997) observed a decrease in superoxide anion levels (O_2^-) in heterophils of *Streptopelia risoria* after the phagocytosis of inert particles when the phagocytes had been incubated with pharmacological doses of melatonin. Also, in the same species, Rodríguez *et al.* (1998) found that incubation with pharmacological doses of melatonin led to the disappearance of the antigen-produced rise in the activity of superoxide dismutase (SOD), a metalloenzyme which catalyses the dismutation of superoxide anion into oxygen and hydrogen peroxide. These workers also observed that the same melatonin dose induced an increase in the concentration of myeloperoxidase stored in heterophils, this being the major component of the bactericidal armoury of phagocytes (Barriga *et al.*,

1998) and a decrease in the production of malonaldehyde (MDA), an indicator of induced oxidative damage to lipid membranes (Rodríguez et al., 1999b). All these data confirm the existence of a negative correlation between serum melatonin levels over a 24-h period and the superoxide anion levels in heterophils, with minimum and maximum levels coinciding with the diurnal oscillations of melatonin (Rodríguez et al., 1999a). In addition, Terrón et al. (2001) observed that melatonin acts as an antioxidant in phagocytic cells even at physiological doses, favouring phagocytic activity at the same time as neutralizing free radical levels after the digestion of the antigen.

The effect of melatonin on phagocytosis has also been studied in the ring dove (*Streptopelia risoria*) using isolated heterophils (Rodríguez et al., 2001). Melatonin, at pharmacological concentrations, enhanced both the chemoattractant capacity of these cells and their capacity to phagocytose antigen particles and reduced the intensity of the respiratory burst (Rodríguez et al., 1997). It also modulated the superoxide dismutase activity in the same species (Rodríguez et al., 1998) and increased the concentration of myeloperoxidase, an enzyme used as an indicator of the bactericidal capacity of heterophils (Barriga et al., 1998). These results were confirmed when the circadian changes in plasma melatonin were found to be positively correlated with the phagocytic capacity of the heterophils and negatively with their oxidative metabolism (Rodríguez et al., 1999a).

On this same theme, Terrón et al. (2002) studied in vitro the effect of the physiological melatonin concentrations found in young and mature ring doves (300 pg/ml as the maximum nocturnal concentration and 50 pg/ml as the minimum diurnal concentration) on the heterophils obtained from old animals, evaluating the capacity for ingestion and destruction of *Candida albicans* and the oxidative metabolism associated with phagocytosis by determining the superoxide anion levels. Melatonin induced a dose-dependent increase in both phagocytosis and candidicide index. Also, a decline in superoxide anion levels was found after incubation with both concentrations. These results thus confirm the physiological effects of melatonin on phagocytic function.

In sum, melatonin is a significant endogenous antioxidant for bird heterophils as, even at physiological concentrations, it is an effective free-radical scavenger, yielding protection from the oxidative stress that accompanies phagocytosis.

Direct action of melatonin on immune function

Consistent with the effects of melatonin and pinealectomy on immune function and disease, receptors for melatonin have been isolated on specific immune cells (Nelson and Demas, 1996). Yu et al. (1991) demonstrated the presence of melatonin receptors at a greater density in birds (duck and chicken) than in mice and suggested that they were high-affinity melatonin receptors (ML1) (Dubocovich, 1988). Specific high-affinity binding sites for melatonin have been described in membrane preparations from the guinea pig and chicken spleen (Poon and Pang, 1996) and the bursa of Fabricius and thymus of duck (Liu and Pang, 1992).

Other workers (Skwarlo-Sonta *et al.*, 1994) have shown the presence of abundant $2[^{125}I]$iodomelatonin binding sites on brain, gonad and lymphoid tissues (although at a lower densities in the last two) in four-week-old cockerels. Of the lymphoid tissues examined, there was weak binding of iodomelatonin in preparations of membrane isolated from the bursa of Fabricius, less from the spleen and only a small trace in the thymus (Liu and Pang, 1992).

In birds, melatonin receptors are present not only in different central nervous system structures mainly connected with sensory functions (Cassone *et al.*, 1995), but also in many peripheral organs such as the lung (Pang *et al.*, 1993), spleen (Yu *et al.*, 1991) and gastrointestinal tract (Lee and Pang, 1992). The extensive distribution of melatonin receptors may reflect a broader sensitivity of different physiological processes to circadian variations of melatonin in birds when compared with mammals. Melatonin binding sites have also been localized in the spleen of duck, chickens and pigeons, as well as the thymus and bursa of Fabricius (Calvo *et al.*, 1995; Wang *et al.*, 1993). In almost all cases reported, the Kd values for these receptors are between 10 and 100 pM. This range is similar to the values found in rat and hamster (Calvo *et al.*, 1995). The low Kd values reported for melatonin binding on immune targets suggest that these receptors serve a physiological role rather than merely being an artifact of non-specific binding. Although very little is known about the adaptive significance of melatonin receptors on peripheral lymphoid tissue, it is likely that its presence provides an adaptive advantage by allowing for a potentially rapid increase in immune function during times when fitness may be compromised (e.g. winter, migration, malnutrition).

Influence of the immune system on the function of the pineal gland

To close the regulatory loop between the immune system and the function of the pineal gland, the messages sent by the activated immune system have to be understood by the pineal, i.e. they have to influence pineal gland activity and hence melatonin synthesis. Some evidence for that has come from studies performed on mammals and birds (Skwarlo-Sonta, 2002). The cytokines seem to be the best candidates for this function, with effects mediated by endogenous opiates (Di Stefano and Paules, 1994).

The pineal gland of 2- to 8-week-old chickens contains infiltrated lymphocytes forming a well-organized peripheral accumulation similar to that of the lymphoid organs (Olah, 1995). Using monoclonal antibodies, it was observed that the intrapineal lymphoid cells have characteristics of T cells and that there are also immunoglobulin-producing cells, forming a true lymphopineal tissue. The function of these lymphocytes is still unknown, but nonetheless they provide evidence in favour of immuno-neuro-endocrine interactions.

While embryonic chicken bursectomy has been found to diminish the humoural immune response, the melatonin level increases after multiple immunization and both effects are reversed by two injections of low doses of bursin into the embryo (Youbicier-Simo *et al.*, 1996). Single immunization with sheep red blood cells caused season, age and gender dependent changes in nocturnal AA-NAT activity in the chicken (Markowska *et al.*, 2000). Therefore, there is no doubt that, besides information on external lighting conditions, the pineal gland is able to perceive messages coming from the immune system, most probably via soluble factors (e.g. cytokines) and therefore, to close the regulation loop (Skwarlo-Sonta, 2002).

Acknowledgements

The authors extend thanks to Elena Circujano Vadillo for her secretarial assistance. This work has been supported by grants Junta de Extremadura 03/10, Consejería de Sanidad y Consumo and Ministerio de Ciencia y Tecnología BFI 2002-04583-C02-01, Spain.

REFERENCES

Arendt, J. (1995). Melatonin and the mammalian pineal gland, in: *Melatonin and the Mammalian Pineal Gland*, Arendt, J. (Ed.), pp. 1–5. Chapman and Hall, London.

Ariëns Kappers, J. A. (1965). Survey of the innervation of the epiphysis cerebri and the accessory pineal organs of vertebrates, *Prog. Brain. Res.* **10**, 87–153.

Barriga, C., Nogales, G., Marchena, J. M., *et al.* (1998). Myeloperoxidase activity in ring dove heterophils after latex bead ingestion. Effect of melatonin, *J. Physiol.* **509**, 95.

Barrett, P., MacLean, A., Davidson, G., *et al.* (1996). Regulation of the Mel 1a melatonin receptor mRNA and protein levels in the ovine pars tuberalis: evidence for a cyclic adenosine $3',5'$-monophosphate-independent Mel 1a receptor coupling and an autoregulatory mechanism of expression, *Mol. Endocrinol.* **10**, 892–902.

Bégay, V., Collin, J. P. and Falcón, J. (1994). Calciproteins regulate cyclic AMP content and melatonin secretion in trout pineal photoreceptors, *Neuroreport.* **5**, 2019–2022.

Benítez-King, G. and Antón-Tay, F. (1993). Calmodulin mediates melatonin cytoskeletal effects, *Experientia.* **49**, 635–641.

Binkley, S., Stephens, J. L., Riebman, J. B., *et al.* (1977). Regulation of pineal rhythms in chickens: photoperiod and dark-time sensitivity, *Gen. Comp. Endocrinol.* **32**, 411–416.

Binkley, S. A., Riebman, J. B. and Reilly, K. B. (1978). The pineal gland: a biological clock in vitro, *Science* **202**, 1198–1200.

Brydon, L., Roka, F., Petit, L., *et al.* (1999a). Dual signaling of human Mel1a melatonin receptors via G(i2), G(i3), and G(q/11) proteins, *Mol. Endocrinol.* **13**, 2025–2038.

Brydon, L., Petit, L., de Coppet, P., *et al.* (1999b). Polymorphism and signalling of melatonin receptors, *Reprod. Nutr. Dev.* **39**, 315–324.

Cahill, G. M. (1996). Circadian regulation of melatonin production in cultured zebrafish pineal and retina, *Brain Res.* **708**, 177–181.

Calvo, J. R., Rafii-el-Idrissi, M., Pozo, D., *et al.* (1995). Immunomodulatory role of melatonin: specific binding sites in human and rodent lymphoid cells, *J. Pineal Res.* **18**, 119–126.

Cassone, V. M., Brooks, D. S. and Kelm, T. A. (1995). Comparative distribution of 2[125]iodomelatonin binding in the brains of diurnal birds: outgroup analysis with turtles, *Brain Behav. Evol.* **45**, 241–256.

Chan, A. S., Lai, F. P., Lo, R. K., *et al.* (2002). Melatonin mt1 and mt2 receptors stimulate c-Jun N-terminal kinase via pertussis toxin-sensitive and -insensitive G proteins, *Cell. Signal.* **14**, 249–257.

Clemens, J. W., Jarzynka, M. J. and Witt-Enderby, P. A. (2001). Down-regulation of mt1 melatonin receptors in rat ovary following estrogen exposure, *Life Sci.* **69**, 27–35.

Collazos, M. E., Barriga, C. and Ortega, E. (1994). Optimum conditions for the activation of the alternative complement pathway of a cyprinid fish (*Tinca tinca* L.). Seasonal variations in the titres, *Fish Shellfish Immun.* **4**, 499–506.

Collazos, M. E., Barriga, C. and Ortega, E. (1995). Seasonal variations in the immune system of the cyprinid *Tinca tinca* phagocytic function, *Comp. Immunol. Microbiol. Infect. Dis.* **18**, 105–113.

Collazos, M. E., Barriga, C. and Ortega, E. (1996). Seasonal variations in the immune system of the tench, *Tinca tinca* (Cyprinidae). II Proliferative response of lymphocytes induced by mitogens, *J. Comp. Phys.* **B165**, 592–595.

Collin, J. P. (1972). Differentiation and regression of the cells of the sensory line in the epiphysis cerebri, in: *The Pineal Gland*, Wolstenhome, G. E. W. and Knight, J. (Eds), pp. 79–125. Churchill Livingstone, Edinburgh.

Collin, J. P., Calas, A. and Juillard, M. T. (1976). The avian pineal organ. Distribution of exogenous indoleamines: a qualitative study of the rudimentary photoreceptor cells by electron microscopic radioautography, *Exp. Brain Res.* **25**, 15–33.

Collin, J. P., Falcón J., Voisin, P., *et al.* (1986). The pineal organ: ontogenetic differentiation of photoreceptor cells and pinealocytes, in: *The Pineal Gland from Fetus to Adult*, Gupta, A. and Reiter, R. J. (Eds), pp. 14–30. Croom Helm, London.

Collin, J. P., Voisin, P., Falcón, J., *et al.* (1989). Pineal transducers in the course of evolution: molecular organization, rhythmic metabolic activity and role, *Arch. Histol. Cytol.* **52**, 441–449.

Conway, S., Canning, S. J., Barrett, P., *et al.* (1997). The roles of valine 208 and histidine 211 in ligand binding and receptor function of the ovine Mel1a beta melatonin receptor, *Biochem. Biophys. Res. Commun.* **239**, 418–423.

Descartes, R. (1662). *De Homine. Figuris et Latinitate Donatus a Schuyl. F. Lugduni Batavorum.*

Di Stefano, A. and Paulesu, L. (1994). Inhibitory effect of melatonin on production of INF gamma or TNF alpha in peripheral blood mononuclear cells of some blood donors, *J. Pineal Res.* **17**, 164–169.

Dubbels, R., Reiter, R. J., Klenke, E., *et al.* (1995). Melatonin in edible plants identified by radioimmunoassay and by high performance liquid chromatography-mass spectrometry, *J. Pineal Res.* **18**, 28–31.

Dubocovich, M. L. (1988). Pharmacology and function of melatonin receptors, *FASEB J.* **2**, 2765–2773.

Dubocovich, M. L. (1995). Melatonin receptors: are there multiple subtypes? *Trends Pharmacol. Sci.* **16**, 50–56.

Fabris, N. (1994). Neuroendocrine regulation of immunity, *Adv. Pineal Res.* **7**, 41–56.

Falcón, J. (1999). Cellular circadian clocks in the pineal, *Prog. Neurobiol.* **58**, 121–162.

Falcón, J., Balemans, M. G., van Benthem, J., *et al.* (1985). *In vitro* uptake and metabolism of (3H)indole compounds in the pineal organ of the pike. I. A radiochomatographic study, *J. Pineal Res.* **2**, 341–356.

Falcón, J., Thibault, C., Blázquez, J. L., *et al.* (1990). Atrial natriuretic factor increases cyclic GMP and cyclic AMP levels in a directly photosensitive pineal organ, *Pflügers Arch.* (*Eur. J. Physiol.*) **417**, 243–245.

Falcón, J., Thibault, C., Martin, C., *et al.* (1991). Regulation of melatonin production by cate-cholamines and adenosine in a photoreceptive pineal organ. An *in vitro* study in the pike and the trout, *J. Pineal Res.* **11**, 123–134.

Falcón, J., Thibault, C. Bégay, V., *et al.* (1992). Regulation of the rhythmic melatonin secretion by fish pineal photoreceptor cells, in: *Rhythms in Fishes*, Ali, M. A. (Ed.), pp. 167–198. Plenum Press, New York.

Falcón, J., Galarneau, K. M., Weller, J. L., et al. (2001). Regulation of arylalkylamine N-acetyl-transferase-2 (AANAT2, EC 2.3.1.87) in the fish pineal organ: evidence for a role of proteasomal proteolysis, *Endocrinology* **142**, 1804–1813.

Finocchiaro, L. M. and Glikin, G. C. (1998). Intracellular melatonin distribution in cultured cell lines, *J. Pineal Res.* **24**, 22–34.

Gauer, F., Masson-Pévet, M., Skene, D. J., et al. (1993). Daily rhythms of melatonin binding sites in the rat pars tuberalis and suprachiasmatic nuclei; evidence for a regulation of melatonin receptors by melatonin itself, *Neuroendocrinology* **57**, 120–126.

Gauer, F., Masson-Pevet, M., Stehle, J., et al. (1994). Daily variations in melatonin receptor density of rat par tuberalis and suprachiasmatic nuclei are distinctly regulated, *Brain Res.* **641**, 92–98.

Geary, G. G., Krause, D. N. and Duckles, S. P. (1997). Melatonin directly constricts rat cerebral arteries through modulation of potassium channels, *Am. J. Physiol.* **273**, H1530–1536.

Geary, G. G., Duckles, S. P. and Krause, D. N. (1998). Effect of melatonin in the rat tail artery: role of K+ channels and endothelial factors, *Br. J. Pharmacol.* **123**, 1533–1540.

González, G. and Valladolid, M. (1996). Ultraestructura de la glándula pineal de aves, *Trab. Inst. Cajal. Inv. Biol.* **58**, 55–67.

Goto, K., Yamagata, K., Miki, N., et al. (1990). Direct photosensitivity of chick pinealocytes as demonstrated by visinin immunoreactivity, *Cell Tissue Res.* **262**, 501–505.

Grace, M. S. and Besharse, J. C. (1994). Melatonin deacetylase activity in the pineal gland and brain of the lizards Anolis carolinensis and Sceloporus jarrovi, *Neuroscience* **62**, 615–623.

Haldar-Misra, C. and Thapliyal, J. P. (1981). Effect of melatonin on the testes and on the renal sex segment in the garden-lizard *Calotes versicolor*, *Can J. Zool.* **59**, 70–74.

Hardeland, R. (1993). The presence and function of melatonin and structurally related indoleamines in a dinoflagellate, and a hypothesis on the evolutionary significance of these tryptophan metabolites in unicellulars, *Experientia* **49**, 614–622.

Hardeland, R. (1999). Melatonin and 5-methoxytryptamine in non-metazoans, *Reprod. Nutr. Dev.* **39**, 399–408.

Hardeland, R. and Poeggeler, B. (2003). Non-vertebrate melatonin, *J. Pineal Res.* **34**, 233–241.

Hardeland, R., Balzer, I., Poeggeler, B., et al. (1995). On the primary functions of melatonin in evolution: mediation of photoperiodic signals in a unicell, photooxidation, and scavenging of free radicals, *J. Pineal Res.* **18**, 104–111.

Hastings, M. H., Duffield, G. E., Ebling, F. J., et al. (1997). Non-photic signalling in the suprachiasmatic nucleus, *Biol. Cell.* **89**, 495–503.

Hattori, A., Migitaka, H., Iigo, M., et al. (1995). Identification of melatonin in plants and its effects on plasma melatonin levels and binding to melatonin receptors in vertebrates, *Biochem. Mol. Biol. Int.* **35**, 627–634.

Heubner, O. (1898). Tumor der glandula pinealis, tsch, *M. Tsch. Med. Wschr.* **24**, 214–220.

Ho, M. K., Yung, L. Y., Chan, J. S., et al. (2001). Galpha(14) links a variety of G(i)- and G(s)-coupled receptors to the stimulation of phospholipase C, *Br. J. Pharmacol.* **132**, 1431–1440.

Homo-Delarche, F. and Dardenne, M. (1993). The neuroendocrine-immune axis, *Springer Semin. Immunopathol.* **14**, 221–238.

Hosey, M. M., Benovic, J. L., DebBurman, S. K., et al. (1995). Multiple mechanisms involving protein phosphorylation are linked to desensitization of muscarinic receptors, *Life Sci.* **56**, 951–955.

Jankovic, B. D., Knezevic, Z., Kojic, L., et al. (1994). Pineal gland and immune system. Immune functions in the chick embryo pinealetomized at 96 hours of incubation, *Ann. N.Y. Acad. Sci.* **719**, 398–409.

Jiang, Z. G., Nelson, C. S. and Allen, C. N. (1995). Melatonin activates an outward current and inhibits Ih in rat suprachiasmatic nucleus neurons, *Brain Res.* **687**, 125–132.

Jones, M. P., Melan, M. A. and Witt-Enderby, P. A. (2000). Melatonin decreases cell proliferation and transformation in a melatonin receptor-dependent manner, *Cancer Lett.* **151**, 133–143.

Kappers, J. A. (1979). Short history of pineal discovery and research, in: *Progress in Brain Research, The Pineal Gland of Vertebrates including Man*, Ariens-Kappers, J. and Pevet, P. (Eds), Vol. 52, pp. 3–22. Elsevier/North Holland Biomedical Press.

Kirby, J. D. and Froman, D. P. (1991). Research note: evaluation of humoral and delayed hypersensitivity responses in cockerels reared under constant light or a twelve hour light twelve hour dark photoperiod, *Poultry Sci.* **70**, 2375–2378.

Kliger, C. A., Gehad, A. E., Hulet, R. M., *et al.* (2000). Effects of photoperiod and melatonin on lymphocyte activities in male broiler chickens, *Poultry Sci.* **79**, 18–25.

Kroeber, S., Schomerus, C. and Korf, H. W. (1997). Calcium oscillations in a subpopulation of S-antigen-immunoreactive pinealocytes of the rainbow trout (*Oncoryhynchus mykiss*), *Brain Res.* **744**, 68–76.

Krupnick, J. G. and Benovic, J. L. (1998). The role of receptor kinases and arrestins in G protein-coupled receptor regulation, *Annu. Rev. Pharmacol. Toxicol.* **38**, 289–319.

Lee, P. P. and Pang, S. F. (1992). Identification and characterization of melatonin bindings sites in the gastrointestinal tract of ducks, *Life Sci.* **50**: 117–125.

Lerner, A. B. and Nordlund, J. J. (1975). Comment. Administration of melatonin to human subjects, in: *Frontiers of Pineal Physiology*, Altschule, M. D. (Ed.), pp. 42–43. Cambridge Press.

Lerner, A. B., Case, J. D., Takahashi, Y., *et al.* (1958). Isolation of melatonin, the pineal gland factor that lightens melanocytes, *J. Amer. Chem. Soc.* **80**, 2587–2594.

Lerner, A. B., Case, J. D. and Heinzelmann, R. V. (1959). Structure of melatonin, *J. Amer. Chem. Soc.* **81**, 6084–6085.

Lerner, A. B., Case, J. D. and Takahashi, Y. (1960). Isolation of melatonin and 5-methoxyindole-3-acetic acid from bovine pineal gland, *J. Biol. Chem.* **235**, 1992–1997.

Levi, F., Canon, C., Depres-Brummer, P., *et al.* (1992). The rhythmic organization of the immune network: implication for the chronopharmacologic delivery of inteferons, interleukins and cyclosporin, *Adv. Drug. Deliv. Rev.* **9**, 85–112.

Lewis, P. D., Morris, T. R. and Perry, G. C. (1996). Lighting and mortality rates in domestic fowl, *Brit. Poultry Sci.* **37**, 295–300.

Liebmann, P. M., Wolfler, A., Felsner, P., *et al.* (1997). Melatonin and the immune system, *Int. Arch. Allergy Immunol.* **112**, 203–211.

Liu, Z. M. and Pang, S. F. (1992). [125I]iodomelatonin-binding sites in the bursa of Fabricius of birds: biding characteristics, subcellular distribution, diurnal variations and age studies, *J. Endocrinol.* **138**, 51–57.

MacKenzie, R. S., Melan, M. A., Passey, D. K., *et al.* (2002). Dual coupling of MT(1) and MT(2) melatonin receptors to cyclic AMP and phosphoinositide signal transduction cascades and their regulation following melatonin exposure, *Biochem. Pharmacol.* **63**, 587–595.

Markowska, M., Bialecka, B., Ciechanowska, M., *et al.* (2000). Effect of immunization on nocturnal NAT activity in chicken pineal gland, *Neuroendocrinol. Lett.* **21**, 367–373.

Markowska, M., Waloch, M. and Skwarlo-Sonta, K. (2001). Melatonin inhibits PHA-stimulated chicken lymphocyte proliferation in vitro, *J. Pineal Res.* **30**, 220–226.

Maronde, E., Pfeffer, M., Olcese, J., *et al.* (1999). Transcription factors in neuroendocrine regulation: rhythmic changes in pCREB and ICER levels frame melatonin synthesis, *J. Neurosci.* **19**, 3326–3336.

Martín, C. and Meisal, H. (1992). Effects of dopaminergic and noradrenergic mechanisms on the neuronal activity of the isolated pineal organ of the trout, *Oncorhynchus mykiss, J. Neural. Transmt. Gen. Sect.* **79**, 81–91.

McNulty, S., Ross, A. W., Barret, P., *et al.* (1994). Melatonin regulates the phosphorylation of CREB in ovine pars tuberalis, *J. Neuroendocrinol.* **6**, 523–532.

Meissl, H., Kroeber, S., Yanez, J., *et al.* (1996). Regulation of melatonin production and intracellular calcium concentrations in the trout pineal organ, *Cell. Tissue Res.* **286**, 315–323.

Menéndez-Peláez, A. and Reiter, R. J. (1993). Distribution of melatonin in mammalian tissues: the relative importance of nuclear versus cytosolic localization, J. Pineal Res. 15, 59–69.

Molinari, E. J., North, P. C. and Dubocovich, M. L. (1996). 2-[125]iodo-5-methoxycarbonylamino-N-acetyltriptamine: a selective radioligand for the characterization of melatonin ML2 binding sites, Eur. J. Pharmacol. 301, 159–168.

Moore, C. B. and Siopes, T. D. (2000). Effects of lighting conditions and melatonin supplementation on the cellular and humoral immune responses in Japanese quail (Coturnix coturnix japonica), Gen. Comp. Endocrinol. 119, 95–104.

Moore, C. B., Siopes, T. D., Steele, C. T., et al. (2002). Pineal melatonin secretion, but not ocular melatonin secretion, is sufficient to maintain normal immune responses in Japanese quail (Coturnix coturnix japonica), Gen. Comp. Endocrinol. 126, 352–358.

Morgan, P. J., Barret, P., Howell, H. E., et al. (1994). Melatonin receptors: localization, molecular, pharmacology and physiological significance, Neurochem. Int. 24, 101–146.

Morin, F., Lugnier, C., Kameni, J., et al. (2001). Expression and role of phosphodiesterase-6 in the chicken pineal gland, J. Neurochem. 78, 88–99.

Nelson, R. J. and Demas, G. E. (1996). Seasonal changes in immune function, Quart. Rev. Biol. 71, 511–548.

Nosjean, O., Ferro, M., Coge, F., et al. (2000). Identification of the melatonin-binding site MT3 as the quinine reductase-2, J. Biol. Chem. 275, 31311–31317.

Nosjean, O., Nicolas, J. P., Klupsch, F., et al. (2001). Comparative pharmacological studies of melatonin receptors: MT1, MT2 and MT3/QR2, Tissue distribution of MT3/QR2, Biochem. Pharmacol. 61, 1369–1379.

Olah, I. (1995). The pineal gland is a transitory lymphoid organ in the chicken. Abstract of Conference on Neuroendocrine-Immune Interactions, September 8–10, Ladek Zdrój near Wrocław, Poland.

Pang, C. S., Brown, G. M., Tang, P. L., et al. (1993). 2-[125I]iodomelatonin binding sites in the lung and heart: a link between the photoperiodic signal, melatonin and the cardiopulmonary system, Biol. Signals 2, 228–236.

Petit, L., Lacroix, I., de Coppet, P., et al. (1999). Differential signaling of human Mel1a and Mel1b melatonin receptors through the cyclic guanosine 3′-5′-monophosphate pathway, Biochem. Pharmacol. 58, 633–639.

Pevet, P. (1998). Melatonin and biological rhythms. Therapie 53, 411–420.

Pfeffer, M., Maronde, E., Korf, H. W., et al. (2000). Antisense experiments reveal molecular details on mechanisms of ICER suppressing cAMP-inducible genes in rat pinealocytes, J. Pineal Res. 29, 24–33.

Poeggeler, B. (1993). Introduction. Melatonin and the light-dark zeitgeber in vertebrates, invertebrates and unicellular organisms, Experientia 49, 611–613.

Poeggeler, B., Balzer, I., Hardeland, R., et al. (1991). Pineal hormone melatonin oscillates also in the dinoflagellate Gonyaulax polyedra, Naturwisseschaften. 78, 268–269.

Poon, A. M. S. and Pang, S. F. (1996). Pineal melatonin-immune system interaction, in: Melatonin: A Universal Photoperiodic Signal with Diverse Actions, Pang, P. L. and Reiter, R. J. (Eds), pp. 73–74. Karger, Basel.

Reiter, R. J. (1991). Pineal melatonin: cell biology of its synthesis and of its physiological interactions, Endocr. Rev. 12, 151–180.

Reiter, R. J. (1993). The melatonin rhythm: both a clock and a calendar, Experientia 49, 654–664.

Reiter, R. J., Tan, D. X., Qi, W., et al. (2000). Pharmacology and physiology of melatonin in the reduction of oxidate stress in vivo, Biol. Signals Recept. 9, 160–171.

Reppert, S. M., Weaver, D. R. and Ebisawa, T. (1994). Cloning and characterization of a mammalian melatonin receptor that mediates reproductive and circadian responses, Neuro. 13, 1177–1185.

Reppert, S. M., Godson, C., Mahle, C. D., et al. (1995). Molecular characterization of a second melatonin receptor expressed in human retina and brain: the Mel1b melatonin receptor, Proc. Natl. Acad. Sci. USA 92, 8734–8738.

Rodríguez, A. B. and Lea, R. W. (1994). Effect of pinealectomy upon nonspecific immune response of the ring dove (*Streptopelia risoria*), *J. Pineal Res.* **16**, 159–166.

Rodríguez, A. B., Ortega, E., Lea, R. W., *et al.* (1997). Melatonin and the phagocytic process of heterophils from ring dove (*Streptopelia risoria*), *Mol. Cell. Biochem.* **168**, 185–190.

Rodríguez, A. B., Nogales, G., Ortega, E., *et al.* (1998). Melatonin controls of superoxide anion level: modulation of superoxide dismutase activity in ring dove heterophils, *J. Pineal Res.* **24**, 9–14.

Rodríguez, A. B., Marchena, J. M., Nogales, G., *et al.* (1999a). Correlation between the circadian rhythm of melatonin, phagocytosis, and superoxide anion levels in ring dove heterophils, *J. Pineal Res.* **26**, 35–42.

Rodríguez, A. B., Nogales, G., Marchena, J. M., *et al.* (1999b). Suppression of both basal and antigen-induced lipid peroxidation in ring dove heterophils by melatonin, *Biochem. Pharmacol.* **58**, 1301–1306.

Rodríguez, A. B., Terron, M. P., Duran, J., *et al.* (2001). Physiological concentrations of melatonin and corticosterone affect phagocytosis and oxidative metabolism of ring dove heterophils, *J. Pineal Res.* **31**, 31–38.

Roseboom, P. H. and Klein, D. C. (1995). Norepinephrine stimulation of pineal cyclic AMP response element-binding protein phosphorylation: primary role of a beta-adrenergic receptor/cyclic AMP mechanism, *Mol. Pharmacol.* **47**, 439–449.

Roseboom, P. H., Coon, S. L., Baler, R., *et al.* (1996). Melatonin synthesis: analysis of the more than 150-fold nocturnal increase in serotonin N-acetyltransferase messenger ribonucleic acid in the rat pineal gland, *Endocrinology* **137**, 3033–3044.

Rosolowska-Huszcz, D., Thaela, M. J., Jagura, M., *et al.* (1991). Pineal influence on the diurnal rhythm of nospecific immunity indices in chickens, *J. Pineal Res.* **10**, 190–195.

Roth, J. J. and Roth, E. C. (1980). The parietal-pineal complex among paleovertebrates, in: *A Cool Look at the Warm-Blooded Dinosaurs*. AAAS Selected Symposium, Thomas, R. D. K. and Olson, E. C. (Eds), pp. 189–231. Weshiew Press, Boulder, CO.

Schuster, C., Gauer, F., Malan, A., *et al.* (2001). The circadian clock, light/dark cycle and melatonin are differentially involved in the expression of daily and photoperiodic variations in mt(1) melatonin receptors in the Siberian and Syrian hamsters, *Neuroendocrinology* **74**, 55–68.

Seltzer, A., Viswanathan, M. and Saavedra, J. M. (1992). Melatonin-binding sites in brain and caudal arteries of the female rat during the estrous cycle and after estrogen administration, *Endocrinology* **130**, 1896–1902.

Skwarlo-Sonta, K. (1996). Functional connections between the pineal gland and immune system, *Acta Neurobiol. Exp.* **56**, 341–357.

Skwarlo-Sonta K. (1999). Reciprocal interdependence between pineal gland and avian immune system, *Neuroendocrinol. Lett.* **20**, 151–156.

Skwarlo-Sonta K. (2002). Melatonin in immunity: comparative aspects, *Neuroendocrinol. Lett.* **23**, 61–66.

Skwarlo-Sonta, K., Raczynska, J., Zujko, J., *et al.* (1994). Lack of the anti-glucocorticoid activity of melatonin in chicken immune system, *Adv. Pineal Res.* **7**, 137–142.

Tan, D. X., Chen, L. D., Poeggeler, B., *et al.* (2000). Melatonin: a potent endogenous hydroxyl radical scavenger, *Endocrinology J.* **1**, 57–60.

Tenn, C. and Niles, L. P. (1993). Physiological regulation of melatonin receptors in rat suprachiasmatic nuclei: diurnal rhythmicity and effect of stress, *Mol. Cell. Endocrinol.* **98**, 43–48.

Terrón, M. P., Marchena, J. M., Shai, F., *et al.* (2001). Melatonin: An antioxidant at physiological concentrations, *J. Pineal Res.* **31**, 95–96.

Terrón, M. P., Cubero, J., Marchena, J. M., *et al.* (2002). Melatonin and aging: in vitro effect of young and mature ring dove physiological concentrations of melatonin on the phagocytic function of heterophils from old ring dove, *Exp. Gerontol.* **37**, 421–426.

Thibault, C., Falcón, J., Greenhouse, S. S., *et al.* (1993a). Regulation of melatonin production by pineal photoreceptor cells: role of cyclic nucleotides in the trout (*Oncorhynchus mykiss*), *J. Neurochem.* **61**, 332–339.

Thibault, C., Collin, J. P. and Falcón, J. (1993b). Intrapineal circadian oscillator(s), cyclic nucleotides and melatonin production in pike pineal photoreceptor cells, in: *Melatonin and Pineal Gland: From Basic*, Touitou, Y. (Ed.), pp. 11–18. Elsevier, Amsterdam.

Urbanski, H. F. (2000). Influence of light and the pineal gland on biological rhythms, in: *Neuroendocrinology in Physiology and Medicine*, Conn, P. M. and Freeman, M. E. (Eds), pp. 405–420. Humana Press.

Vivien-Roels, B. and Arendt, J. (1983). How does the indoleamine production of the pineal gland respond to variations in the environment in a non-mammalian vertebrate, *Testudo gemelin*, *Psychoneuroendocrinology* **8**, 327–332.

Wang, X. L., Yuan, H. and Pang, S. F. (1993). Specific biding of [125I]iodomelatonin in pigeon and quail spleen membrane preparations and effect with hydrocortisone-treatment, *Zhongguo Yao Li Xue Bao.* **14**, 292–295.

Witt-Enderby, P. A. and Dubocovich, M. L. (1996). Characterization and regulation of human ML1A melatonin receptor stably expressed in Chinese hamster ovary cells, *Mol. Pharmacol.* **50**, 166–174.

Witt-Enderby, P. A., Masana, M. I. and Dubocovich, M. L. (1998). Physiological exposure to melatonin supersensitizes the cyclic adenosine $3',5'$-monophosphate-dependent signal transduction cascade in Chinese hamster ovary cells expressing the human mt 1 melatonin receptor, *Endocrinology* **139**, 3064–3071.

Witt-Enderby, P. A., Jarzynka, M. J. and Melan, M. A. (2001). Microtubules modulate melatonin receptors function, *Neuroscience Abstracts* **27**, 142.

Witt-Enderby, P. A., Bennet, J., Jarzynka, M. J., *et al.* (2003). Melatonin receptors and their regulation: biochemical and structural Mechanism, *Life Sci.* **72**, 2183–2198.

Yanez, J. and Meissl, H. (1995). Secretion of methoxyindoles from trout pineal organs in vitro: indication for a paracrine melatonin feedback, *Neurochem. Int.* **27**, 195–200.

Youbicier-Simo, B. J., Boudard, F., Mekaouche, M., *et al.* (1996). A role of bursa fabricii and bursin in the ontogeny of the pineal biosynthetic activity in the chicken, *J. Pineal Res.* **21**, 35–43.

Yu, Z. H., Lu, Y. and Pang, S. F. (1991). [125I]iodomelatonin binding sites in spleens of birds and mammals, *Neurosci. Lett.* **125**, 175–178.

Zapata, A. G., Varas, A. and Torroba, M. (1992). Seasonal variations in the immune system of lower vertebrates, *Immunol. Today* **13**, 142–147.

Zatz, M. (1992). Agents that affect calcium influx can change cyclic nucleotide levels in cultured chick pineal cells, *Brain Res.* **583**, 304–307.

Zatz, M., Kasper, G. and Marquez, C. R. (1990). Vasoactive intestinal peptide stimulates chick pineal melatonin production and interacts with other stimulatory and inhibitory agents but does not show alpha 1-adrenergic potentiation, *J. Neurochem.* **55**, 1149–1153.

Zrenner, C. (1985). Theories of pineal function from classical antiquity to 1900: A history, *Pineal Res. Rev.* **3**, 1–40.

Advances in Neuroregulation and Neuroprotection (2005), pp. 269-286
Collin, C. *et al.* (Eds)
© VSP 2005

Parkinson's disease and neurotransmitters interplay in processes of neurotoxicity and oxidative stress in On and Off phenomena

G. A. QURESHI [1,*], C. COLLIN [2], S. A. MEMON [1] and S. H. PARVEZ [2]

[1] *M. A. Kazi Institute of Chemistry, University of Sindh, Jamshoro, Pakistan*
[2] *Neuroendocrine Unit, Institut Alfred Fessard de Neurosciences, Bat 5, Parc Chateau CNRS, 91190 Gif Yvette, France*

Abstract—L-Dopa is considered to be one of the best therapies in treating Parkinson's disease (PD), with positive outcome in the beginning of its treatment; however, long treatment periods with this drug have shown both On and Off phenomena and, when it is used in high doses, produces various clinical symptoms. In this chapter, both On and Off PD patients on L-dopa therapy are investigated to determine the role of various neurotransmitters, such as aspartic and glutamic acid, homocysteine, catecholamines, vitamin B12 and free radical nitric oxide, in neurotoxicity and oxidative stress. All these biochemical changes might provide vital information in designing multiple therapy consisting of antioxidants to prolong the avoidance of neurotoxicity and oxidative stress in PD patients.

Keywords: Parkinson's disease; L-dopa therapy; On and Off phenomena; oxidative stress; neurotoxicity.

INTRODUCTION

Parkinson's disease (PD) is the most frequently encountered disease among neurological disorders in old age that neurologists meet in their clinical practice. PD affects about 1.5% of the population over 65 years of age. In recent years, several genes that cause certain forms of inherited PD have been identified, and great progress has been made in elucidating their molecular mechanism. From a clinical point of view, PD is characterized by motor disturbances, such as tremor, rigidity and akinesia, and from a pathological point of view by loss of dopaminergic neurons in the substantia nigra pars compacta (Marsden, 1994). The hallark of PD is progressive loss of dopamine neurons in the midbrain, most pronounced in the sub-

---

*To whom correspondence should be addressed. E-mail: ghulamali48@hotmail.com

stantia nigra pars compacta, though other nerve cell populations using monoamine transmitters, such as the locus coeruleus containing noradrenaline neurons are also affected (Gerfen, 1992; German *et al.*, 1992; Graybiel and Rogsdale, 1979; Qureshi *et al.*, 2003a). Pathologically, PD is associated with preferential degeneration of nigrostriatal dopamine neurons, but the degeneration also affect other parts of the central and peripheral nervous system, such as serotoninergic, noradrenergic, cholinergic and peptidergic pathways (Agid *et al.*, 1986; Dubois *et al.*, 1985; Jenkovic *et al.*, 2001; Pillon *et al.*, 1989).

The etiology of neuronal death in neurodegenerative disease remains unclear; however, great advances in both molecular genetics and neurochemistry have improved our knowledge of fundamental processes involved in cell death, including both excitotoxicity and oxidative damage (Werneck and Alvarenga, 1999). Several factors such as inhibition of the mitochondrial respiration, generation of hydroxyl and nitric oxide radicals and reduced free radical defence mechanisms causing oxidative stress have been postulated to contribute to the degeneration of dopaminergic neurons (Qureshi *et al.*, 2003b; Smith *et al.*, 2003; Thiessen *et al.*, 1990). In the past decade or so, a convincing link between oxidative stress and degenerative conditions has been made and with the knowledge that oxidative changes may actually trigger deterioration in cell function, a great deal of energy has been focussed on identifying agents that may have therapeutic value in combating oxidative changes with antioxidants which may prevent or reduce the rate of progression of this disease (Tatton *et al.*, 2003). On the other hand, the discovery of several genetic mutations associated with PD raises the possibility that these or other biomarkers may help to identify persons at risk of PD (Mandir *et al.*, 2001).

The treatment of choice is still mainly symptomatic, either by means of substances acting on dopaminergic system such as L-dopa and dopamine (DA) agonists or by means of drugs that modify the metabolism of L-dopa or DA with MAO B and COMT inhibitors (Jankovic *et al.*, 1998). Because of its diagnosis at the later stage, the actual treatment strategies have pitfalls and caveats.

The discovery of L-dopa constitutes a major milestone in modern neuropharmacology. Among many pharmacological therapies adopted in the past two or three decades, the most effective therapy during the first years of PD is with L-dopa. However, in L-dopa treatment, more than 80% of PD patients after 5–10 years show On and Off phenomena in the PD patients with dyskinesia. Its limitation is most noteworthy in the advanced stages, where the drug cannot improve motor and non-motor parkinsonian features; it produces side effects and it does not halt PD progression. Despite the disadvantages, L-dopa remains the so-called 'gold standard' for the treatment of PD because of its impressive efficacy in treating the motor symptoms and its cheap price as compared to other treatments. However, several motor features typically do not respond to L-dopa; on the contrary, it tends to deteriorate speech, gait, posture and balance over time. Furthermore, L-dopa therapy tends to aggravate nonmotor functions such as hallucinations, cognitive impairment and orthostatic hypotension. In a recent study (Koller *et al.*, 1999), it was reported that motor fluctuations and dyskinesias could be minimized by using the lowest possible L-dopa dosage throughout the several years of treatment.

This is one of the main reasons for the development of dopamine agonists as new treatment strategies (Rinne, 1983, 1986). Various drugs such as Bromcriptine, Lisurid and Pergolid have been introduced, but these drugs have limited efficiency in the long-term treatment of *denovo* patients, as patients have more side effects and the clinical outcome was poorer compared to patients treated with L-dopa (Rinne, 1983, 1986). Other studies using early combination of L-dopa with dopamine agonists reduced dyskinesia and fluctuations in the long-term (Olsson, 1985; Rinne, 1985). Most treatment strategies suggest that the decision as to whether to treat with dopamine agonist alone or combine L-dopa with dopamine agonists depends upon the age of the patient. However, one recent study (Schoenfeld *et al.*, 2003) could not find any relationship between the age and the treatment type with regard to the clinical outcome.

This paper deals with analysis of amino acids, nitrite (a metabolite of free radical, nitric oxide), vitamin B_{12}, homocysteine, catechoamines and their metabolites in the cerebrospinal fluid (CSF) in On and Off PD patients. These levels are reported with the assumption that CSF concentrations reflect brain or spinal cord concentrations and perhaps synaptic activity, since CSF is in constant exchange with the extracellular fluid of the CNS. Based on the fact that most of the transmitters in the brain are present in the CSF (Herkenham, 1987) and that alterations in the concentration of transmitters in the CSF can alter the same transmitters in the brain of man (Wester *et al.*, 1990), the results are compiled in PD patients (On and Off groups) and compared with healthy subjects.

MATERIAL AND METHODS

Patients, healthy subjects and routine analysis

Most of the patients were recruited from the department of Neurology, Huddinge University Hospital, Stockholm. The research Ethics Committee of Karolinska Institute, Stockholm, approved the study and all patients gave written informed consent to participate. 30 patients (12 women, 18 men) with PD were included in this study. 15 patients were at different stages of illness (Hoehn and Yar range 2–4) and on individual drug combinations of L-dopa (250 mg/day) combined with other drugs so as to produce a positive response for motor activities (PD On group). The other 15 PD patients had been treated for 6–8 years with similar drug therapy but showed a severely Off response (Hoehn and Yar range −3) (PD Off group) with motor fluctuation. Both On and Off groups were monitored one week before the samples of blood and CSF were taken and all food and fluid intake were similar in both groups and exact dose times were maintained. Blood and CSF were taken early in the morning between 6–8 a.m. in a fasting condition. 16 healthy volunteers from Karolinska Institute employees were included and their blood and

Table 1.
Clinical data on the healthy controls and PD patients in both On and Off groups

Patients	n	Females	Age (years)	CSF-Albumin	CSF-IgG	IgG-index
Controls	16	7	53 ± 7	218 ± 17	34 ± 4	0.44 ± 0.01
PD (On)	15	7	72 ± 12	$332 \pm 42^*$	$51 \pm 7^{**}$	0.43 ± 0.01
PD (Off)	15	5	74 ± 14	$313 \pm 37^*$	$44 \pm 6^*$	0.42 ± 0.01

* $p < 0.05$.
** $p < 0.01$.

CSF were collected under comparable conditions. None of the healthy controls was on medication for six months. Table 1 shows the clinical data on PD patients and healthy controls.

10–12 ml CSF was collected from each PD patient and healthy control in sitting position at the L4–L5 levels. Blood samples were collected by venipuncture. The basic CSF analyses included cell counting by phase-contrast microscopy (Siesjo, 1967), determination of CSF/serum/albumin ratio and CSF/immunoglobulin G (igG) index (Link and Tibbling, 1977) as well as isoelectric focusing for the detection of oligoclonal IgG band. Serum and CSF albumin and IgG were determined using Hitachi 737 Automatic Analyzer (Naka Works, Hitachi Ltd., Tokyo, Japan). CSF and serum samples were kept at $-80°C$ if not analyzed immediately.

Analysis of amino acids, catecholamines, tryptophan, vitamin B_{12} and nitrate

Amino acid analysis was performed based on the pre-column derivatization of OPA with amino acids as previously described (Qureshi and Baig, 1988). The degree of error for each amino acid was $<2\%$ with this method.

Because of the short half-life time for NO, its concentration is quantified indirectly by measuring the levels of its degraded product nitrite. A recent HPLC method in combination with electrochemical detection is used for this purpose (Kaku et al., 1994). Homocysteine was quantified using a previously described HPLC method (Hyland and Bottiglieri, 1992) and vitamin B_{12} was quantified using a competitive protein binding assay (CPBA) based on radioactivity measurement with intrinsic factor (obtained from Amersham International, Buckinghamshire, UK) (Ikeda et al., 1990).

Data are presented as mean $\pm$ SEM. Differences in concentration of vitamin B_{12} and homocysteine were analyzed with ANOVA, and group comparisons were made with t-text. A p-value less than 0.05 was considered significant.

The concentrations of dopamine (DA), homovanillic acid (HVA), 3,4-dihydroxy-phenylacetic acid (DOPAC), noradrenaline (NA), 3-methoxy-4-hydroxyphenyl-glycol (MHPG), 5-hydroxytryptamine (5-HT) and 5-hydroxyindoleacetic acid (5-HIAA) were measured using previously described methods (Baig and Qureshi,

1991). Measurements were also made in serum but since no differences were found these will not be considered further.

RESULTS AND DISCUSSION

On and Off phenomena in PD

Parkinson's disease is an inexorably progressive disorder that worsens over time. The rate of nigral cell death is not exactly known, but neuro-imaging techniques estimate cell death occurs at a rate of approximately 10% per year (Olanow, 1997). PD symptoms are considered to be a consequence of an imbalance between stimulatory and inhibitory impulses in the extrapyramidal system, and this is mainly due to DA transmission in the nigrostriatal pathway. Disease progression mainly affects presynaptic terminals by reducing not only their buffer capacity but also their feedback control on striatal neurons (Carlsson, 1983). Various studies suggest that PD may develop in subjects exposed to pesticides, herbicides, contaminated well water or other hazards of rural living and the disease is not dependent on cigarette smoking or caffeine consumption (Olanow, 1997; Palmer, 2001). The diagnosis of PD generally relies on the clinical observation of the combination of four cardinal motor signs, namely, tremor, rigidity, bradykinesia and balance impairment or postual instability (Agid *et al.*, 1989). These symptoms, and especially the first three, are typically improved by dopamine replacement therapies, and a positive response to L-dopa is mandatory for the diagnosis of PD. However, not all motor features in PD are adequately controlled with dopaminergic medication. Furthermore, PD is not just a motor disorder and dysfunction of autonomic, cognitive and psychiatric system frequently accompany the classic motor features of PD (Rascol *et al.*, 2003). These nonmotor features frequently represent an important source of disability for PD patients and severely impact on their quality of life. There is no specific cure for PD and there is limitation of current PD therapy (Rascol *et al.*, 2003). However, there are several treatments available to alleviate symptoms of the disease, such as pharmacological therapy (Korczyn and Nassbaum, 2002) and surgical interference (Playfer, 1997). The treatment of choice is still mainly symptomatic, either by means of substances acting on dopaminergic system such as L-dopa and dopamine (DA) agonists or by means of drugs that modify the metabolism of L-dopa or DA with MAO B and COMT inhibitors (Jankovic and Kapadia, 2001). The most successful drug in treatment of PD is still L-dopa, which is a precursor in dopamine biosynthesis, and mortality has been nearly normalized. However, disabling motor complications occur in the late stages of PD: the most common effect includes onset of On and Off phenomena and abnormal involuntary movements called dyskinesia (Jankovic, 2002). Most PD patients treated with L-dopa develop fluctuation in motor performance. After 3–5 years of treatment, one-third; after 5–7 years, about one-half; and after 10–12 years, nearly all patients suffer from the motor fluctuation (Rinne, 1983). Furthermore, non-motor symptoms such as pain, fatigue, anxiety and depression are seen more often in PD patients treated with L-dopa for a long time (>10 years) than are seen in patients with classical motor disturbances (Calon and Di Paulo, 2002; Shulman et al., 2002), so it has been suggested that L-dopa therapy should be delayed as long as possible (Scigliano et al., 1997).

Most cases of PD, however, appear to be sporadic, and these are likely to represent an interplay between both genetic and environmental factors. To date, the main risk factors for developing PD in an individual apart from biochemical brain defects, are increasing age and presence of another affected family member (Werner and Schapira, 2003). It is also proposed that an earlier age of onset of PD implies a greater likelihood that genetic factors play a dominant role.

Pharmacokinetics of L-dopa and the blood-brain barrier

L-Dopa is transported across the intestinal mucosa and blood-brain barrier (BBB) by the large neutral amino acid (LNAA) transport system, which means that some amino acids competitively inhibit L-dopa membrane transport (Wade et al., 1973). Doses of L-dopa taken with a meal rich in proteins will be less effective than doses taken on an empty stomach (Nutt et al., 1984). By restricting most daily protein intake to evening meals, daytime plasma L-dopa levels are more predictable and motor performance is improved. Hence, it is concluded that plasma L-dopa levels predict motor response to a greater extent under controlled conditions regarding diet and medication (Frankel et al., 1990). In healthy individuals, the BBB regulates the entry of any drug or endogenous compound into peripheral organs. However, the barrier functions of BBB can change dramatically during various CNS diseases. The most common consequence under inflammation and bacterial disorders is an alteration of BBB permeability (Saez-Llorens et al., 1990). Antidiuretic hormone secretion and accumulation of toxic substances also change BBB transport properties (Lipton and Kater, 1989).

L-Dopa is susceptible to a large number of pharmacokinetic interactions (Furlanut et al., 2001), and it has a short half-life. It produces remarkable blood fluctuations of the drug with unimportant consequences in the early stages of the disease but of much greater significance in the later more advanced stages. This is due to the complicated kinetics in the brain as compared to the kinetics in the blood. It makes it clear that PD is an evolving disease, not only from a neuropathological and pharmacodynamic point of view but also from a pharmacokinetic aspect (Furlanut et al., 2001). This may be the reason why there are different responses to L-dopa at the early stage (On phenomena) and at the advanced stages (Off phenomena), where the number of dopaminergic neurons are destroyed with time (Fearnley and Lees, 1991). In order to study these changes, CSF obtained by lumbar puncture technique is the most important and widely used diagnostic tool in evaluating the levels of neurotransmitters and their metabolites. Lumbar CSF neurochemical measurements aim to serve as indirect in vivo markers of human central neurotransmission and these markers can be used in the diagnosis of neurodegeneration processes.

Neurotoxicity in PD

Excitotoxicity refers to neuronal cell death caused by activation of excitatory amino acid (EAA) receptors, mainly N-methyl-D-aspartate (NMDA), and other EAA receptors have been implicated in neuronal death. An increasing body of evidence has implicated such a pathway as a mechanism of cell death in both acute and chronic neurological diseases (Beal, 1992; Choi, 1988). In the past decade or so, a convincing link between oxidative stress and degenerative conditions has been made and with the knowledge that oxidative changes may actually trigger deterioration in cell function, a great deal of energy has been focussed on identifying agents that may have possible therapeutic value in combating oxidative changes with antioxidants and which thereby may prevent or reduce the rate of progression of this disease (Tatton *et al.*, 2003). On the other hand, the discovery of several genetic mutations associated with PD raises the possibility that these or other biomarkers may help to identify persons at risk of PD (Mandir *et al.*, 2001). Cell injury may be introduced by several processes. It was proposed (Beal, 1992; Choi, 1988) that neuronal injury following brief intense exposure to glutamic acid (Glu) involves two components, distinguishable by differences in time course and ionic dependence. The first component marked by neuronal swelling probably reflects Na^+ ion influx with subsequent passive Cl^- and water entry leading to an expansion of cell volume. It occurs within minutes and depends on the presence of extracellular Na^+ and Cl^- and may or may not be lethal. The second and possibly more significant component may involve calcium (Ca^{2+}) influx mediated by NMDA subtype of Glu receptors (Collingridge and Lester, 1989). This second component leads to a delayed neuronal disintegration occurring over a period of hours after exposure to Glu. Table 2 shows the CSF levels of glutamate and aspartate in both On and Off groups of PD patients, and these are significantly higher than those in healthy controls.

Glu acts as a neurotransmitter in several brain regions, regulates neuronal outgrowth and synaptogenesis during normal development and adult plasticity, and has a special function in learning and memory processes (Lipton and Kater, 1989). An elevated Glu concentration is common in both On and Off groups in PD. It is also observed that another intrinsic EAA, the Glu analogue aspartate (Asp), is also in-

Table 2.
CSF levels of glutamic and aspartic acid in healthy controls and patients with PD On and Off groups. All values are expressed as mean $\pm$ SEM in μmol/l

Patients	Glutamic acid (Glu)	Aspartic acid (Asp)
Healthy controls	10.86 ± 1.63	1.82 ± 0.13
PD (On)	$16.32 \pm 1.89^{**}$	$2.60 \pm 0.19^{*}$
PD (Off)	$21.37 \pm 2.38^{***}$	$3.32 \pm 0.32^{***}$

$^{*}p < 0.05.$
$^{**}p < 0.01.$
$^{***}p < 0.001.$

creased in both groups of PD. It is also known that Asp is also capable of inducing an excitotoxic effect (Olney, 1990). Activation of EAA receptors leads to increase intracellular Ca^{2+} followed by activation of protein kinases, phospholipases, nitric acid synthase, impaired mitochondrial function and generation of free radicals (Beal, 1992).

Free radicals and oxidation stress in PD

Most free radicals are unstable species due to one or more unpaired electrons that can extract an electron from neighboring molecules leading to oxidative damage. The role of free radicals in cell death induced by activation of EAA receptors is an area of expanding interest (Coyle and Puttfarcken, 1993). The brain is known as a highly oxygenated organ. Oxygen-centred free radicals are the main types of radical that are formed in neurons as accidental by-products of metabolism or as selectively generated species commonly known as reactive oxygen species (ROS). There is increasing evidence that free radical nitric oxide ($NO^{\bullet}$) plays an important role in pathophysiology of a variety of CNS disorders (Dawson et al., 1992; Koutsilieri et al., 2002; Qureshi et al., 2003b).

ROS are very important mediators of cell injury and death. Not only are these highly reactive chemical species important in aging processes but they are also directly or indirectly involved in a wide variety of pathological conditions (Floyd, 1990; Knight, 1995; Qureshi et al., 1995, 1997, 1998, 2003b, 2004). The generation of ROS in excessive amounts is enough to overwhelm normal defence mechanisms, resulting in serious cell and tissue damage. Nearly all the major classes of biological materials can undergo destruction of their structure and biological activity, especially membrane lipids which are most susceptible. Among, the ROS, superoxide anion ($O_2^{\bullet-}$) and nitric oxide ($NO^{\bullet}$) are the most studied free radicals, and the mammalian brain may be exceptionally vulnerable to oxidative stress through attack of these radicals (Dawson and Dawson, 1996). Nitric oxide has been shown to be involved in various neurodegenerative disorders (Dodd et al., 1994; Halawa et al., 1996; Qureshi et al., 1996). It is also known that enhanced Glu release increases free radical nitric oxide ($NO^{\bullet}$) production, causes DNA damage, energy depletion in neurons and neuronal death in PD (Zhang et al., 1994). This hypotheses is also supported by our recent study (Qureshi et al., 1995), which showed high levels of arginine and NO in patients with PD. Table 3 shows CSF levels of arginine (Arg), the precursor of NO and nitrite (a metabolite of NO) in healthy controls and On and Off groups of PD patients: both Arg and NO are significantly increased in both PD groups.

In our recent study (Qureshi et al., 2003b), a linear correlation between NO and Glu for both these groups was shown. Neuroleptics are the major class of drugs used to treat PD; however, these drugs are associated with wide variety of extrapyramidal side effects which includes tardive dyskinesia (Kulkami and Naidu, 2003). Numerous reports have indicated that induction of free radicals in neuroleptic drug treatment increases free radical production and causes structural

Table 3.

CSF levels of arginine and nitrite in healthy controls and patients with PD On and Off groups (All values are expressed as mean $\pm$ SEM in μmol/l)

Patients	Arginine	Nitrite
Healthy Controls	17.62 ± 1.32	1.16 ± 0.06
PD (On)	25.41 ± 1.96[**]	1.62 ± 0.13[*]
PD (Off)	33.82 ± 2.62[***]	2.26 ± 0.23[***]

[*] $p < 0.05$.
[**] $p < 0.01$.
[***] $p < 0.001$.

damage that eventually leads to oxidative stress (Jenner, 2003). Besides, NO exposure causes cobalamin deficiency resulting from its inactivity which is known to produce impaired neurological functions (Hall, 1990; Qureshi *et al.*, 1998; Stollhof and Schulte, 1987).

Naturally occurring molecules such as bioflavinoids, polyphenols, vitamins, terpenes, alkaloids and enzymes constitute the barrier that plants, animals and humans use to counter the damaging effects of free radicals and provide effective natural remedies for several radical-related diseases. The mechanism of the damage that free radicals produce on biomolecules, such as on polyunsaturated fatty acids or nucleic acids is known. It is essential to maintain the balance between pro-oxidants and anti-oxidants *in vitro* and *in vivo*, so that free radicals from both reactive oxygen species and reactive nitrogen species that cause oxidative stress damage are not removed or converted into harmless species.

Biological systems in humans have developed a comprehensive array of defence mechanisms to protect against free radicals. These include enzymes to decompose peroxides, proteins to sequester transition metal ions and a range of compounds such as antioxidants to scavenge free radicals. The condition of oxidative stress is caused by accumulation of free radicals and is defined as a disturbance in pro-oxidant and anti-oxidant balance in favour of the former leading to potentially damaging reactions with biological molecules (Qureshi *et al.*, 1995; Sies, 1985). The results indicate a possible participation of oxidative stress in the neuropathology of PD patients, especially during the crisis, i.e. in the Off situation, when the metabolites are highly increased, and to the point where the use of antioxidant drugs is a possible adjuvant therapy in such situations to improve the neurological status of PD patients and to prevent sequelae. An unbalanced overproduction of reactive oxygen species (ROS) may give rise to oxidative stress which can induce neuronal damage, ultimately leading to neuronal death by apoptosis or necrosis. Oxidative stress is a ubiquitously observed hallmark of neurodegenerative disorders. Neuronal cell dysfunction and cell death due to oxidative stress may contribute to the pathogenesis of progressive neurodegenerative disorders, such as Alzheimer's disease (AD) and Parkinson's disease (PD), as well as acute syndromes of neurodegeneration, such as ischaemic and haemorrhagic stroke (Dawson and

Dawson, 1996).   A large body of evidence indicates that oxidative stress is involved in the pathogenesis of AD and PD. An increasing number of studies show that nutritional antioxidants (especially vitamin E and polyphenols) can block neuronal death *in vitro*, and may have therapeutic properties in animal models of neurodegenerative diseases including AD, PD, and ALS. Moreover, clinical data suggest that nutritional antioxidants might exert some protective effect against AD, PD, and ALS. Neuroprotective antioxidants are considered a promising approach to slowing the progression and limiting the extent of neuronal cell loss in these disorders. The clinical evidence demonstrates that antioxidant compounds can act as protective drugs in neurodegenerative disease, but the data are still relatively scarce. The major challenges for drug development are the slow kinetics of disease progression, the unsolved mechanistic questions concerning the final causes of cell death, the necessity of attaining an effective permeation of the blood-brain barrier and the need to reduce the high concentrations currently required to evoke protective effects in cellular and animal model systems.   Finally, an outlook as to which direction antioxidant drug development and clinical practice may be leading to in the near future will be provided. Antioxidants of widely varying chemical structures have been investigated as potential therapeutic agents.   However, the therapeutic use of most of these compounds is limited since they do not cross the blood brain barrier (BBB). Although a few of them have shown limited efficiency in animal models or in small clinical studies, none of the currently available antioxidants has proven effective in a large-scale controlled study. Therefore, any novel antioxidant molecules designed as potential neuroprotective treatment in acute or chronic neurological disorders should have the mandatory prerequisite that they can cross the BBB after systemic administration.

Monoamines and their metabolites

The estimation of monoamines and indoleamines have been shown to be helpful for the diagnosis and interpretation of various neurological and psychiatric disorders (Burns *et al.*, 1985; Davis *et al.*, 1991; Gottfries, 1990).   Both noradrenaline (NA) and dopamine (DA) are synthesized from the amino acid tyrosine derived from food intake or protein breakdown products.   The main metabolites of DA are homovanilic acid (HVA) and 3,4-dihydroxyphenylacetic acid (DOPAC), and of NA is 3-methoxy-4-hydroxyphenylglycol (MHPG). In PD patients, the CSF NA and its metabolite, MHPG, are increased in both On and Off groups whereas DA and its metabolites, DOPAC and HVA are significantly decreased in both groups. Table 4 shows CSF levels of DA, NA, and their metabolites in On and Off PD patients and are compared with healthy controls. A single dose of L-dopa gives a consistent response to motor activity in the On situation whereas once the disease has progressed for a few years, the response duration gets shorter, which is when DA and its metabolites are further decreased in the Off situation as the therapeutic window is narrowed.

Table 4.
CSF levels of noradrenaline and dopamine and their metabolites in healthy controls and patients with PD On and Off groups (All values are expressed as mean $\pm$ SEM in pmol/l)

Patients	NA	MHPG	DA	HVA	DOPAC
Healthy controls	114 ± 13	59 ± 7	51 ± 4	296 ± 18	97 ± 6
PD (On)	$129 \pm 15^{*}$	$67 \pm 10^{*}$	$42 \pm 3^{*}$	$244 \pm 19^{*}$	$83 \pm 7^{*}$
PD (Off)	$150 \pm 17^{**}$	$83 \pm 11^{**}$	$30 \pm 3^{***}$	$202 \pm 15^{**}$	$67 \pm 6^{**}$

$^{*} p < 0.05.$
$^{**} p < 0.01.$
$^{***} p < 0.001.$

Table 5.
CSF levels of tryptophan and its metabolites in healthy controls and patients with PD On and Off groups (All values are expressed as mean $\pm$ SEM in pmol/l)

Patients	TRP	5-HT	5-HIAA
Healthy controls	2115 ± 105	9.7 ± 1.2	296 ± 51
PD (On)	$1635 \pm 99^{**}$	$7.3 \pm 0.85^{*}$	$241 \pm 39^{*}$
PD (Off)	$1378 \pm 78^{***}$	$4.8 \pm 0.52^{***}$	$202 \pm 23^{**}$

$^{*} p < 0.05.$
$^{**} p < 0.01.$
$^{***} p < 0.001.$

Apart from catecholamines, serotonin (5-HT), which is synthesized from the amino acid tryptophan, plays a very important role in the CNS. 5-HT metabolizes to 5-hydroxy indole acetic acid (5-HIIA). The estimation of 5-HT is related to depression, and Table 5 shows the levels of tryptophan, 5-HT and 5-HIIA, which are significantly decreased in both groups as both these groups are known to be depressed and inactive.

Cobalamin and its role in PD

It is also known that free radical nitric oxide NO$^{\bullet}$ interacts with vitamin B_{12} resulting in selective inhibition of methionine synthase — a key enzyme in metabolism of methionine and folate. Thus, NO may altar one-carbon and methyl group transfer, which is most important for DNA purine and thymidylate biosynthesis. Our results show the correlation between nitrite and homocysteine, suggesting that both these parameters are interrelated (Qureshi et al., 2004). Although the mechanisms responsible for the neurological lesions caused by vitamin B_{12} deficiency are less well understood, vitamin B_{12} analogues including methylcobalamin (MCbl) have been widely used in therapy of neurological diseases (Canarin et al., 1988). Vitamin B_{12} has been shown to improve memory, emotional function and communication ability in Alzheimer's patients. MCbl is an active coenzyme of vitamin B_{12} analogues that is essential for cell growth and replication (Hall, 1990). In fact,

Table 6.
CSF levels of vitamin B_{12} and homocysteine in healthy controls and patients with PD On and Off groups (All values are expressed as mean $\pm$ SEM in μmol/l)

Patients	Vitamin B_{12}	Homocysteine
Healthy controls	0.079 ± 0.006	1.09 ± 0.096
PD (On)	0.059 ± 0.005[**]	1.56 ± 0.130[**]
PD (Off)	0.041 ± 0.003[***]	1.89 ± 0.21[***]

[*] $p < 0.05$.
[**] $p < 0.01$.
[***] $p < 0.001$.

detection of vitamin deficiency in humans is a problem that needs to be solved in clinical medicine, because data on levels in blood do not always correlate with CSF levels. However, there is now good evidence that the majority ($>90\%$) of patients who are vitamin B_{12} deficient accumulate high levels of methylmalonic acid and homocysteine in the blood. Hence, both these substances can be considered better markers for clinical diagnosis of vitamin B_{12} deficiency (Chandefaux et al., 1994). NO toxicity due to its interaction with vitamin B_{12} dependent enzyme, i.e. methionine synthase, produces hematological and less frequently neurological symptoms (Yagiela, 1991). The NO· radical is also known to oxidize active Cob(II)alamin to inactive Cob(III)alamin (Banks et al., 1968). As Cbl in the form of reduced MCbl is required as cofactor for methionine synthase, exposure to NO· radical causes rapid inactivation of this enzyme (Deacon et al., 1978). Hence, the inactive Cbl is excreted, so that repeated exposure to NO· radical results in depletion of body Cbl stores, with reduced AdoCbl so that the activity of the AdoCbl dependent enzyme methylmalonic acid CoA (MMCoA) is also affected, which results in impaired neurological functions (Deacon et al., 1978). NO· radical interaction with cobalamin results in the liberation of hydroxyl free radicals responsible for inactivation of methionine synthase (Drummond and Methews, 1994).

Table 6 shows the CSF levels of cobalamin and homocysteine in PD patients (On and Off groups) statistically compared with those in healthy controls. Cobalamin is significantly decreased and increases in homocysteine are observed in both On and Off groups of PD.

In patients with cardiac disorders, the high levels of blood homocysteine translate into a significant increase in hardening of arteries known as arteriosclerosis. For a high-risk person who has moderate or severe arteriosclerosis, this increase in homocysteine could be enough to trigger a heart attack (Nappo, 1999). These patients are advised to take supplementary vitamins B_6, B_{12} and folic acid along with antioxidants such as vitamins C and E so that the level of homocysteine can be reduced and in addition they can also protect the blood from clotting (Nappo, 1999).

Pharmacological strategy in PD

There should be two major strategies in trying to improve on PD patients on L-dopa therapy. One involves developing specific therapies for each of the problems that are unresponsive to or aggravated by L-dopa. These may include symptomatic treatments for motor fluctuations or dyskinesias, antidementia or antipsychotic agents and drugs to control orthostatic hypotension, impotence, constipation and abnormal daytime somnolence. A second strategy should by to device diagnostic methods in identifying PD in its early stages, blocking disease progression with effective and safe neuroprotective agents in preventing the disease from reaching the advanced stage in which there are new features that do not respond to current treatment, i.e. Off phenomena. More recent therapeutic interventions, including DA agonists, MAO-B inhibitors, COMT inhibitors and modern functional surgery such as deep brain stimulation, have been developed to help control L-dopa therapy shortcomings. Although helpful, such complementary interventions are not fully safe or effective (Green *et al.*, 2002). Symptomatic orthostatic hypotension is present in about 20% of PD patients and can be worsened by dopaminergic drugs (Senard *et al.*, 2001). Constipation, neurogenic bladder with urinary frequency, urgency and incontinence, sexual dysfunction and abnormal sweating and salivation are also frequent (Mathias, 1996).

Rational, integrative management of PD requires: (1) dietary revision, especially to lower calories; (2) rebalancing of essential fatty acid intake away from pro-inflammatory and toward anti-inflammatory prostaglandins; (3) aggressive repletion of glutathione and other nutrient antioxidants and cofactors; (4) energy nutrients like acetyl L-carnitine, coenzyme Q10, NADH, and the membrane phospholipid phosphatidylserine (PS), (5) chelation as necessary for heavy metals; and (6) liver P450 detoxification support.

CONCLUSIONS

Neurodegeneration is the main consequence of PD, which is multi-factorial, and there seems to be a cycle of steps involved. From these results one can conclude that the neurodegeneration in PD patients involves Free radicals (oxidative stress) + Cytostatic Ca^{2+} + Mitochondrial damage + Excitotoxicity of EAA and homocysteine + Deficiency of vitamin B_{12}, B_6 and folate + Role of transition metals, especially iron and copper. All these factors in some way result in cell death processes.

Since our endogenous antioxidant defences are not always completely effective, and exposure to damaging environmental factors is increasing, it seems reasonable to propose that exogenous antioxidants could be very effective in diminishing the cumulative effects of oxidative damage. Antioxidants of widely varying chemical structures have been investigated as potential therapeutic agents. However, the therapeutic use of most of these compounds is limited because they do not cross the blood brain barrier (BBB). Therefore, any novel antioxidant molecules designed as potential neuroprotective treatments in acute or chronic neurological disorders should have the mandatory prerequisite that they can cross the BBB after systemic administration (Hunot *et al.*, 2003). Neuroprotection is a key issue in modern management of PD. However, none of the currently available antiparkinsonian treatments has succeeded in retarding disease progression or providing a neuroprotective effect.

It is concluded that the future therapy in PD is likely to include a combination of neuroprotective compounds to interfere with several molecular pathways that lead to neuronal injury. In using therapeutic strategies aimed towards retarding or arresting neuronal death, close attention will need to be paid to quality of life issues. One of the main problems despite many years of research on PD is to detect the disease early. New methods for early detection and monitoring of the disease progression to help clinicians in their clinical diagnosis are important, particularly in future pharmacological treatments which can halt further neurodegeneration and nerve cell death (Becker *et al.*, 2002). One of the tools that have been used during the last few decades is to examine the patient's brain with positron emission tomography (PET). Combining this technique with analysis of biochemical, pathological and behavioral changes, more information on PD patients can be monitored. On the basis of our study, the results of degenerative disorder, the role of various neurotransmitters, and the role of free radical NO•, cobalamin and homocysteine, is clearly defined in order to develop effective drug therapy. This may include glutamate- releasing inhibitors, excitatory amino acid antagonist agents to improve mitochondrial function, free radical scavengers and neuroprotective agents as antioxidants along with vitamin B_{12}. Mortality remains abnormally high in PD, and improving life expectancy is the major objective for future antiparkinsonian treatment.

REFERENCES

Agid, Y., Taquet, H. and Cesselin, F. (1986). Neuropeptides and Parkinson's Disease, *Prog. Brain Res.* **66**, 107–116.

Agid, Y., Cerveta, P. and Hitch, E. (1989). Biochemistry of Parkinson's Disease 28 years later: A critical review, *Mov. Disor* **4**, S126–S144.

Baig, S. M. and Qureshi, G. A. (1991). HPLC as a tool in defining abnormalities in monoamines and tryptophan metabolites in CSF from patients with neurological disorders, *Biomed. Chromatogr.* **8**, 108–112.

Beal, M. F. (1992). Mechanism of excitotoxicity in neurological disease, *FASEB J.* **6**, 3338–3344.

Burns, R. S., Lewitt, P. A., Ebert, M. H., *et al.* (1985). The clinical syndrome of striatal dopamine deficiency parkinsonism induced by MPTP, *New Engl. J. Med.* **32**, 1418–1421.

Calon, F. and Di Poalo, T. (2002). Levadopa response to motor complications — GABA receptors and preproenkephalin expression in human brain, *Parkinson Relat. Disord.* **6**, 449–454.

Carlsson, A. (1983). Are On-Off effects during chronic L-dopa treatment due to faulty feedback control of nigrostriatal dopamine pathway? *J. Neural Transmission* **19**, 153–161.

Chanarin, I., Deacon, R., Lumb, M., *et al.* (1988). Cobalamin folate interrelations: A critical review, *Blood* **66**, 479–482.

Chandefaux, B., Copper, B. A., Gilfix, B. M., *et al.* (1994). Homocysteine: Relationship to serum cobalamin, serum folate, erthrocyte folate and lobation of neutrophils, *Clin. Invest. Med.* **17**, 540–550.

Choi, D. W. (1988). Glutamate neurotoxicity and diseases of nervous system, *Neuron* **1**, 623–634.

Choi, D. W. (1992). Excitotoxic cell death, *J. Neurobiol.* **23**, 1261–1276.

Collingridge, G. L. and Lester, R. A. J. (1989). Excitatory amino acid receptors in the vertebrate central nervous system, *Pharmac. Rev.* **40**, 143–210.

Coyle, J. T. and Puttfarcken, P. (1993). Oxidative stress, glutamate and neurodegenerative disorders, *Science* **262**, 689–694.

Davis, K., Kahn, R., Ko, G., *et al.* (1991). Dopamine in schizophrenia: Coexistence between neuropeptides and catecholamines. A review and reconceptaulization, *Am. J. Psychiatry* **148**, 1474–1486.

Dawson, T. M., Dawson, V. L. and Snyder, S. H. (1992). A novel neuronal messager molecule in brain: The free radical, nitric oxide, *Ann. Neurol.* **32**, 297–311.

Dawson, V. L. and Dawson, T. M. (1996). Free radicals and neuronal cell death, *Cell Death and Differentiation* **3**, 71–78.

Deacon, R., Lamb, M. and Perry, J. (1978). Selective inactivation of vitamin B_{12} in rats by nitrous oxide, *Lancet* **2**, 1023–1024.

Dodd, P. R., Scott, H. L. and Wesphalen, R. I. (1994). Excitotoxic mechanism in the pathogenesis of dementia, *Neurochem. Int.* **25**, 203–219.

Drummond, J. T. and Methews, R. G. (1994). Nitrous oxide degradation by cobalamin dependent synthase: Characterization of reactants and products in the inactivation reaction, *Biochemistry* **33**, 3732–3741.

Dubois, B., Hauw, J. J. and Ruberg, M. (1985). Dementia and Parkinson's disease: Biochemical and anatomo-clinical correlation, *Rev. Neurol.* **141**, 184–193.

Fearnley, J. U. and Lees, A. J. (1991). Ageing and Parkinson's disease: Substantia nigra regional selectivity, *Brain* **114**, 2283–2301.

Floyd, R. A. (1990). Role of oxygen free radicals in brain ischemia, *FASEB J.* **4**, 2587–2592.

Frankel, J. P., Pitrosek, Z., Kempster, P. A., *et al.* (1990). Diurnal differences inresponse to oral levodopa, *J. Neurol. Neurosurg. Psychiatry* **53**, 948–950.

Furlanut, M., Furlanut, M., Jr. and Benetello, P. (2001). Monitoring of L-dopa concentrations in Parkinson's disease, *Pharmaco. Res.* **43**, 423–427.

German, D. C., Manaye, K. F., White, C. L., *et al.* (1992). Disease-specific patterns of locus coeruleus cell loss, *Ann. Neurol.* **32**, 667–676.

Gottfries, C. (1990). Brain monoamines and their metabolites in dementia, *Acta Neurol. Scand. (suppl.)* **129**, 8–11.

Graybiel, A. M. and Rogsdale, C. W. (1979). Fiber connections of the basal ganglia, *Prog. Brain Res.* **51**, 279–283.

Green, C., Koller, W. and Poewe, W. (2002). Management of Parkinson disease: an evidence based review, *Mov. Disord.* **17**, 1–166.

Grefen, C. R. (1992). The neostriatal mosiac: Multiple levels of compartmental organization, *Trend Neurosci.* **15**, 133–139.

Halawa, A., Baig, S. and Qureshi, G. A. (1996). Amino acids and nitrite in ischemic brain stroke, *Biogenic Amines* **12**, 27–36.

Hall, C. A. (1990). Function of vitamin B_{12} in CNS as revealed by congenital defects, *Am. J. Hemat.* **34**, 121–127.

Herkenham, M. (1987). Mismatches between neurotransmitter and receptor localization in brain: Observation and implications, *Neuroscience* **23**, 1–38.

Hunot, S. and Hirsch, E. C. (2003). Neuroinflammatory processes in Parkinson's disease, *Ann. Neurol.* **53**, 181–185.

Hyland, K. and Bottiglieri, T. (1992). Measurement of total plasma and CSF homocysteine by fluorescence detection followed HPLC and pre-column derivatization with OPA, *J. Chromatogr.* **579**, 55–62.

Ikeda, T., Yamamoto, K., Takahashi, K., *et al.* (1992). Treatment of Alzheimer-type dementia with intravenous cobalamin, *Clin. Therapeutics* **14**, 426–437.

Jankovic, J. and Kapadia, A. S. (2001). Functional decline in Parkinson's disease, *Arch. Neurol.* **58**, 1611–1615.

Jankovic, J. (2002). Levadopa strengths and weaknesses, *Neurology* **58**, S19–S32.

Jenner, P. (2003). Oxidative stress in Parkinson's disease, *Ann. Neurol.* **53**, S26–S36.

Kaku, S., Tanaka, M., Muramatsu, M., *et al.* (1994). Determination of nitrite by HPLC with EC: Measurement of Nitric oxide sythase activity in rat cerebellum cytosol, *J. Biomed. Chromatogr.* **8**, 24–28.

Knight, J. A. (1995). Diseases related to oxygen derived free radicals, *Annal Clin. Lab. Sci.* **25** (2), 111–121.

Koller, W. C., Hutton, J. T., Tolosa, E., *et al.* (1999). Immediate release and controlled release carbidopa-levodopa in PD: A 5 year randomized multicenter study, *Neurology* **53**, 1012–1019.

Korczyn, A. and Nassbaum, M. (2002). Emerging therapies in pharmacological treatment of Parkinson's disease, *Drugs* **62**, 775–786.

Kulkami, S. K. and Naidu, P. S. (2003). Pathophysiology and drug therapy of tardive dyskinesia: Current concest and future prospectiev, *Drug Today (Barcelona)* **39**, 19–49.

Link, H. and Tibbling, G. (1977). Principles of Albumin and IgG analyses in Neurological disorders, *Scand. J. Clin. Lab. Invest.* **37**, 297–301.

Lipton, S. A. and Kater, S. B. (1989). Neurotransmitter regulation of neuronal outgrowth, plasticity andsurvival. *Trend Neurosci.* **12**, 265–270.

Mathias, C. J. (1996). Disorders affecting autonomic function in Parkinsonian patients, *Adv. Neurol.* **69**, 383–391.

Marsden, C. D. (1994). Parkinson's Disease, *J. Neurol. Neurosurg. Psychiatry* **57**, 672–681.

Nappo, F. (1999). Impairment of endothelial functions by acute hyperhomocysteinemia and reversal by antioxidant vitamins, *JAMA* **281**, 2113–2118.

Nutt, J. G., Woodward, W. R., Hammarstad, J. P., *et al.* (1984). The on and off phenomena in Parkinson's disease: Relation to levodopa absorption and transport, *N. Eng. J. Med.* **310**, 483–488.

Olanow, C. W. (1997). Attempts to obtain neuroprotection in Parkinson's disease, *Neurology* **49**, S26–S33.

Olney, J. W. (1990). Excitatory amino acids and neuropsychiatric disorders, *Annu. Rev. Pharmacol. Toxicol.* **30**, 47–71.

Olsson, J. E. (1985). Bromcriptine and levodopa in early combination in Parkinson's disease; first results of the collobrative European Multicenter Trail, in: *Parkinson's Disease and Role of Dopamine Agonists*, Leiberman, A. and Lataste, X. (Eds), pp. 77–82. Partheon publishing group, UK.

Palmer, G. C. (2001). Neuroprotection by NMDA receptor antagonists in a variety of neuropathologies, *Curr. Drug Targets* **2**, 241–271.

Pillon, B., Dubois, B. and Cusimano, G. (1989). Does cognitive impairment in Parkinson's disease result from non-dopaminergic lesions? *J. Neurol. Neurosurg. Psychiatry* **52**, 201–206.

Playfer, J. R. (1997). Parkinson's disease, *Postgrad. Med. J.* **73**, 257–264.

Qureshi, G. A. and Baig, S. (1988). Quantitation of amino acids in biological samples by HPLC, *J. Chromatogr.* **459**, 237–244.

Qureshi, G. A., Baig, S., Bednar, I., *et al.* (1995). Increased cerebrospinal fluid concentration of nitrite in Parkinson's disease, *Neuro Report* **6**, 1642–1644.

Qureshi, G. A., Baig, S. and Parvez, S. H. (1996). Biochemical markers in tuberculosis meningitis, *Biogenic Amines* **12**, 499–519.

Qureshi, G. A., Halawa, A., Baig, S., *et al.* (1997). The neurochemical markers in cerebrospinal fluid to differentiate between viral and bacterial meningitis, *J. Neurochem.* **32**, 197–203.

Qureshi, G. A., Baig, S. and Minami, M. (1998). The relationship between deficiency of vitamin B_{12} and neurotoxicity of homocysteine with nitrite level in cerebrospinal fluid of neurological patients, *Biogenic Amines* **14**, 1–14.

Qureshi, G. A., Memon, S. A., Memon, D., *et al.* (2003a). Levels of catecholamines and 5-hydroxy-tryptamine and eating behavior in lateral hypothalamic aphagic rats, *Biogenic Amines* **18**, 19–28.

Qureshi, G. A., Memon, S. A., Collin, C., *et al.* (2003b). Neurotoxicity and dynamic impact of oxidative stress in neural regulation of Parkinson's patients in On- and Off phenomena, *Biogenic Amines* **18**, 55–77.

Qureshi, G. A., Sarwar, M., Baig, S., *et al.* (2004). Neurotoxicity, oxidative stress and Cerebrovascular disorders, *Neurotoxicity* **25**, 121–138.

Rascol, O., Payous, P., Ory, F., *et al.* (2003). Limitation of current Parkinson's disease therapy, *Annals of Neurology* **53**, S3–S15.

Rinne, U. K. (1983). Problems associated with long-term levodopa treatment of Parkinson's disease, *Acta Neurol. Scand.* **68** (Suppl. 95), 19–26.

Rinne, U. K. (1985). Combined bromcriptine-levodopa therapy early in Parkinson's disease, *Neurology* **35**, 1196–1198.

Rinne, U. K. (1986). Dopamine agonists as primary treatment, in: *Parkinson's Disease. Advances in Neurology 45*, Yahr, M. D. and Bergmann, K. J. (Eds), pp. 519–523. Raven Press, New York.

Saez-Llorens, X., Ramilo, O. and Mustafa, M. M. (1990). Molecular pathophysiology and bacterial meningitis: Current concepts and therapeutic implications, *J. Pediatr.* **116**, 671–684.

Santosh-Kumar, C. R., Hassell, K. L., Deutsch, J. C., *et al.* (1994). Are neurophychiatric manifestation of cobalamin, folate and pyridoxine deficiency mediated through imbalances in excitatory sulfur amino acids? *Med. Hypotheses* **43**, 239–244.

Schoenfeld, M. A., Pantelie, C. M. and Schwartz, B. (2003). Clinical criteria for switch of treatment strategies in Parkinson's disease, *Clin. Neurol. Neurosurg.* (in press).

Scigliano, G., Girotti, F., Soliveri, P., *et al.* (1997). Parkinson's disease and Oxidative Stress, *Ital. J. Neurol. Sci.* **18**, 69–72.

Senard, J. M., Brefel-Courbon, C., Rascol, O., *et al.* (2001). Orthostatic hypotension in patients with Parkinson's disease: pathophysiology and management, *Drugs Aging* **18**, 495–505.

Shevell, M. I. and Rosenblatt, D. S. (1992). The neurology of cobalamin, *Can. J. Neurol. Sci.* **19**, 472–486.

Shulman, L. M., Taback, R. L., Robinstein, A. A., *et al.* (2002). Non-recognition of depression and other non-motor symptoms in Parkinson's disease, *Parkinsonism Relat. Disord.* **8**, 193–197.

Sies, H. (1985). *Oxidative Stress.* Academic Press, London.

Siesjo, R. (1967). New Method for the cytological examination of cerebrospinal fluid, *J. Neurol. Neurosurg. Psych.* **30**, 568–577.

Smith, C. D., Carney, J. M. and Strake-Reed, P. E. (1991). Excess brain protein oxidation and enzyme dysfunction in normal aging and Alzheimer's disease, *Proc. Natl. Acad. Sci. USA* **88**, 10540–10543.

Stollhoff, K. and Schulte, F. J. (1987). Vitamin B_{12} and brain development, *Eur. J. Paediatr.* **146**, 201–205.

Tatton, W. G., Chalmer-Redman, R., Brown D., *et al.* (2003). Apoptosis in Parkinson's disease: Signal for neuronal degradation, *Ann. Neurol.* **53**, S61–S72.

Thiessen, B., Rajput, A. H., Loverty, W., *et al.* (1990). Age, environments, and the number of substantia nigra neurons, *Adv. Neurol.* **53**, 201–206.

Wade, D. N., Mearick, P. T. and Morris, J. L. (1973). Active transport of L-dopa in intestine, *Nature* **242**, 463–465.

Weimann, J. (2003). Toxicity of nitrous oxide, *Best Pract. Res. Clin. Anaesthesiol.* **17**, 46–61.

Werner, T. T. and Schapira, A. H. V. (2003). Genetic and environmental factors in the cause of Parkinson disease, *Annals of Neurology* **53**, S16–S25.

Wernek, A. L. and Alvarenga, H. (1999). Genetics, drugs and environmental factors in Parkinson's disease, *Arq. Neurosiquiatr.* **57**, 347–355.

Wester, P., Bergstrom, U. and Eriksson, A. (1990). Venticular cerebrospinal fluid monoamin transmitters and their metabolites concentrations reflect human brain neurochemistry in autopsy cases, *J. Neurochem.* **54**, 1148–1156.

Yagiela, J. A. (1991). Health hazards and nitric oxide: a time for reappriasal, *Anesth. Prog.* **38**, 1–11.

Zar, J. H. (1984). *Biostatistical Analysis.* Prentice Hall, Englewood Cliffs, NJ.

Advances in Neuroregulation and Neuroprotection (2005), pp. 287-301
Collin, C. *et al.* (Eds)
© VSP 2005

Anxiety-related behavior in juvenile stroke-prone spontaneously hypertensive rats — an animal model of attention-deficit/hyperactivity disorder

HIROKO TOGASHI [1,*], KEN-ICHI UENO [1], TAKU YAMAGUCHI [1], MACHIKO MATSUMOTO [1], KASANE HIGUCHI [2], HIDEYA SAITO [3] and MITSUHIRO YOSHIOKA [1]

[1] *Department of Neuropharmacology, Hokkaido University Graduate School of Medicine Kita-15, Nishi-7, Kita-ku, Sapporo 060-8638, Japan*
[2] *Department of Oral Science Function, Hokkaido University Graduate School of Dental Medicine, Kita-15, Nishi-7, Kita-ku, Sapporo 060-8638, Japan*
[3] *Department of Basic Sciences, Japanese Red Cross Hokkaido College of Nursing, Kitami 090-0011, Japan*

Abstract—Attention-deficit/hyperactivity disorder (AD/HD) is defined as a developmental disorder, manifested by deficit sustained attention (inattention), and/or hyperactivity-motor impulsiveness. Neuropsychological evidence indicates the comorbidity of anxiety disorder with AD/HD. The aim of the present study was to characterize anxiety-related behavior of the juvenile stroke-prone spontaneously hypertensive rat (SHRSP), an animal model of AD/HD, and compare with genetic and/or normotensive controls. Our hypothesis, that low susceptibility to fear/anxiety stimuli might underlie impulsive behavior in juvenile SHRSP, was assessed by a contextually conditioned fear paradigm. In order to examine whether contextual fear stimuli would affect the synaptic efficacy in the hippocampus, changes in field potentials of perforant path-dentate gyrus (DG) synapses were recorded in the freely behaving rat. Aversive footshock (FS) stimuli elicited intense freezing behavior in genetic and/or normotensive controls. Re-exposing to the FS chamber in the 30-min retention period also produced freezing behavior, as contextually conditioned fear response, in genetic and/or normotensive controls. In contrast, SHRSP exhibited a significant attenuation in freezing behavior both in the 5-min post FS period (immediately after FS) and in the 30-min retention period (24 hours after FS) as compared to genetic and/or normotensive controls. Pain perception as measures of behavioral responses to electric FS, jumping and/or vocalization, indicated that less anxiety-related response in SHRSP did not simply result from low susceptibility to FS stimuli. During the re-exposure to contextual fear stimuli, SHRSP and a normotensive control rat exhibited a decrease in the amplitude of the evoked population spikes in perforant path-DG synapses, accompanying freezing behavior. The synaptic response in this hippocampal subfield was mimicked by low frequency stimulation (1 Hz). These synaptic responses induced by behavioral and electrophysiological manipulations, freezing and

---

*To whom correspondence should be addressed. E-mail: thiro@med.hokudai.ac.jp

LFS, were less pronounced in SHRSP than those in the control. Our findings indicate that the impaired responsiveness to contextually conditioned fear stimuli, less anxiety-related freezing behavior, in juvenile SHRSP, might explain the impulsive behavior in this AD/HD animal model.

Keywords: Stroke-prone spontaneously hypertensive rats (SHRSP); attention-deficit/hyperactivity disorder (AD/HD); anxiety-related behavior; contextual fear conditioning; freezing behavior; hippocampal synaptic transmission.

INTRODUCTION

Attention-deficit/hyperactivity disorder (AD/HD) is a developmental disorder, manifested by inattention and/or hyperactivity-impulsivity (American Psychiatric Association, 1994), with an incidence of 3–5% in school-aged children (Barkley, 1990, 1998; Szatmari *et al.*, 1989). AD/HD is a heterogenous disorder, and some AD/HD children show comorbid psychological disorders, such as conduct, cognitive and/or anxiety disorders (Anderson *et al.*, 1987; Pliszka, 1989; Tannock *et al.*, 1995). However, the causes of AD/HD, including the reason why it is predominant among males (The ratio of male to female children with AD/HD is currently 9 : 1 (Lahey *et al.*, 1994; Taylor, 1998)), are poorly understood, although inherent or genetic mechanisms are also postulated (Barkley, 1998).

To explore the neurobiological basis and to evaluate therapeutic drugs for treating AD/HD, the spontaneously hypertensive rat (SHR) (Okamoto and Aoki, 1963) has been widely used as an animal model. They exhibit behavioral characteristics resemble to AD/HD patients, such as hyperactivity (Cierpial *et al.*, 1989; Knardahl and Sagvolden, 1979; Myers *et al.*, 1982; Wultz *et al.*, 1990), and impulsivity/inattention (Sagvolden, 2000; Sagvolden *et al.*, 1992, 1993a, b). In addition, central dopaminergic hypofunctions such as a decrease in dopamine (DA) release (Russell *et al.*, 1995; Tsuda *et al.*, 1991) and an increase in DAT density (Watanabe *et al.*, 1997) have been found in SHR as in AD/HD. However, the behaviors are not male preponderant in SHR; female, not male, SHR exhibit impulsivity/inattention, whereas both are hyperactive (Berger and Sagvolden, 1998; Sagvolden and Berger, 1996). Furthermore, the effectiveness of psychostimulants in SHR is not consistent with the behavioral paradigms examined. Namely, the hyperactive behavior in SHR is ameliorated by high doses of *d*-amphetamine and methylphenidate (Myers *et al.*, 1982; Wultz *et al.*, 1990), while impulsivity and inattention are not (Evenden and Meyerson, 1999; Sagvolden *et al.*, 1992). SHR do not exhibit impaired acquisition and performance in the differential low rate test (Bull *et al.*, 2000). Thus, SHR does not fulfill the characteristics as an AD/HD animal model in terms of etiology, symptomatology, pathophysiology and therapeutics.

The stroke-prone spontaneously hypertensive rat (SHRSP) is a strain that has been bred from SHR (Okamoto *et al.*, 1974). Based on the symptomatic relevance, we have recently proposed juvenile SHRSP as an animal model of AD/HD (Ueno *et al.*, 2002a, b; 2003). Namely, SHRSP exerts higher motor activity than SHR (Minami *et al.*, 1985; Togashi *et al.*, 1982). Dopaminergic hypofunction such as decreased neural transmission was observed in the prefrontal cortex, nucleus accumbens shell, striatum, and basolateral amygdala of SHRSP (Nakamura *et al.*, 2001), as reported in SHR (Russell *et al.*, 1995; Tsuda *et al.*, 1991; Watanabe *et al.*, 1997). The 5-HT levels in cerebrospinal fluid are significantly decreased in SHRSP, but not in SHR, as compared with Wistar-Kyoto rats (WKY) (Togashi *et al.*, 1994). Moreover, regional cerebral blood flow in the frontal cortex significantly reduced in SHRSP (Shibayama *et al.*, 2004; Yamori and Horie, 1977). Cognitive impairment accompanied with central cholinergic dysfunction has also been reported in SHRSP (Minami *et al.*, 1997; Togashi *et al.*, 1996). Particularly of note is that male, but not female, SHRSP showed cognitive dysfunction assessed by spontaneous alternation behavior. These behavioral symptoms were ameliorated by several pharmacological interventions, including the psychostimulant methylphenidate, a first choice drug for the treatment of AD/HD (Togashi *et al.*, 2002; Ueno *et al.*, 2002a, b; 2003).

Impulsive behavior is an important factor in the symptomatology of AD/HD, which is increasing as the symptom of greatest significance among the AD/HD symptoms (Sagvolden and Sergeant, 1998; Taylor, 1998). AD/HD is thought to be a heterogenous disorder. Some children with AD/HD exhibit the comorbidity of anxiety or phobic disorder (Anderson *et al.*, 1987; Pliszka, 1989; Tannock *et al.*, 1995). It has also been reported that AD/HD children with comorbid anxiety showed less impulsive behavior and exerted poorer psychostimulant effects on cognition than those without anxiety (Pliszka, 1989; Tannock *et al.*, 1995). Thus, it is hypothesized that the altered susceptibility to fear/anxiety might underlie the impulsive behavior of an AD/HD animal model, SHRSP.

The present study has investigated characteristic impulsive behavior in juvenile SHRSP from the viewpoint of susceptibility to fear or anxiety, using a contextually conditioned fear paradigm. In order to examine whether anxiety-related behavior would be accompanied by altered synaptic efficacy in the hippocampus, we also recorded the field potentials of the freely moving animal in perforant path-dentate gyrus (DG) synapses during behavioral and electrophysiological manipulations.

MATERIALS AND METHODS

Animals

Juvenile (6 weeks of age) male-/female-SHRSP/Ezo rats were used. The original SHRSP were donated in 1979 by the late professor Kozo Okamoto, Department of Pathology, Kinki University School of Medicine, Osaka, Japan, and inbred in our laboratory (SHRSP/Ezo, F55-57). They were weaned at postnatal weeks 3 to 4, and housed in a room maintained at a temperature of $22 \pm 2°C$ and relative humidity of $55 \pm 10\%$ with a 12-h light and dark cycle (lights on 06:00–18:00). Age- and sex-matched SHR and Wistar-Kyoto rats (WKY) or Wistar rats (Slc:Wistar/ST, Shizuoka Laboratory Animal Center, Hamamatsu, Japan) were used as genetic and/or normotensive controls. The animals had free access to food and water. All experimental procedures conformed to the Guidelines for the Care and Use of Laboratory Animals published by the Animal Research Committee at Hokkaido University School of Medicine.

Contextual fear conditioning and behavioral analysis

The rats were placed in a conditioning chamber and allowed to acclimate to the novel environment for 5 min. After exploration of their environment, the rats received 5 footshocks (FS: 2 s, 0.5 mA, every 30 s) as an aversive and unconditioned stimulus, which was delivered through a grid floor. Immediately after FS, freezing as behavioral response to unconditioned stimuli was measured for 5 min (post-FS period). Thereafter, rats were returned to their home cage. Twenty-four hours after FS, rats were re-exposed to the conditioning chamber (FS chamber) for 30 min as retention test. Freezing was defined as the absence of all movement except that related to respiration, and determined each 5 s for the corresponding period of 5 or 30 min. Freezing behavior was also monitored and scored with automated detection of pixel differences (FreezeFrame, Actimetrics, IL).

In another series of experiments, pain perception was evaluated as a measure of behavioral responses to the electric FS. Rats were placed into the FS chamber, and stimulated by scrambled electric FS of rising intensity (starting from 0.01 mA). The corresponding FS intensity to elicit behavioral responses, jumping and/or vocalization, was defined as pain threshold.

Electrophysiological analysis

To examine the synaptic response in the hippocampus to contextual fear conditioning, evoked field potentials in peforant path-DG subfield were recorded in freely behaving rats with implanted electrodes. Under ketamine anesthesia (100 mg/kg, i.p.), a stainless steel bipolar stimulation electrode was stereotaxically implanted into the perforant path and a monopolar recording electrode was implanted into the granule cell layer of the DG of the right hemisphere; coordinates were based on the atlas of Paxinos and Watson (1998) and modified by the length between bregma and lambda. Each electrode consisted of an insulated stainless-steel wire 125 μm in diameter. During preparation, test pulses were delivered to optimize the population-spike amplitude (PSA). These electrodes were fixed in place with dental cement on the skull. Rats were allowed 7–10 days in their home cage to recover from surgery. The electrodes were connected to swivel via a flexible cable. This allowed the rat freedom of behaviour. The responses were pre-amplified by a differential head amplifier (JB-220J, NIHON KOHDEN, Tokyo, Japan), transformed by an analog-to-digital interface (Power Lab System; AD Instruments Japan, Nagoya, Japan), and stored on a personal computer. Rectangular constant current pulses (250 μs) were applied to the perforant path to evoke DG field potentials of ~50% of the maximum PSA. After registering a stable baseline for 30 min in their home cage, rats were re-exposed to the FS chamber for 30 min during the retention period. Test stimuli were delivered every 20 s, and the 5-min recording was averaged.

To characterize the functional synaptic changes in the hippocampus accompanied with contextually conditioned fear, electrophysiological manipulations, low frequency stimuli (LFS: 1 Hz for 15 min) or high frequency stimuli (tetanus: 400 Hz, 8 train $\times$ 10) were used. After tetanization, the time course changes in the evoked PSA were measured for 1 hour.

Statistics

All results are expressed as means $\pm$ SEM. Two-tailed Student's t-test analyzed differences between two groups. When more than two groups were compared, the significance of the difference among groups was evaluated by one-way analysis of variance (ANOVA), and for further statistical *post-hoc* comparisons two-tailed Dunnett's multiple comparison was used. Differences in pain threshold were evaluated by the Mann–Whitney U-test after an overall comparison with the Kruskal–Wallis H-test. All tests were two-tailed, and the level of significance was set at $p < 0.05$. The statistical software was SPSS for Windows (Ver. 7.5.1J, SPSS Inc., Japan).

RESULTS

Behavioral response in the post-footshock period

As presented in Fig. 1, electric FS (2 s/30 s, 5 times) resulted in severe freezing behavior in SHR and normotensive controls WKY and Wistar/ST rats. Freezing behavior during a 5-min post-FS period was comparable, and the average response in SHR was almost identical to that in normotensive controls. In contrast, freezing behavior was less distinct in SHRSP, and the average response was markedly and significantly attenuated, when compared with SHR, normotensive controls WKY and Wistar/ST. The average response in the post FS period in SHRSP ($25.4 \pm 7.9\%$, $n = 4$) was significantly smaller than that in SHR ($79.6 \pm 4.0\%$, $n = 8$), WKY ($90.0 \pm 3.3\%$, $n = 7$) and Wistar/ST animals ($87.2 \pm 2.8\%$, $n = 6$).

Behavioral response to contextually conditioned fear stimuli

The behavioral response to contextually conditioned fear stimuli determined 24 hour after FS, was summarized in Fig. 2. A 30-min re-exposure to the FS chamber elicited freezing behavior both in WKY and Wistar/ST. The freezing behavior in SHR and SHRSP was attenuated, although the impairment was more pronounced in SHRSP. The average response during the retention period in SHRSP ($4.2 \pm 1.4\%$, $n = 5$) and SHR ($26.70 \pm 3.9\%$, $n = 8$) was significantly smaller than that in WKY ($47.1 \pm 7.7\%$, $n = 7$) and Wistar/ST rats ($61.5 \pm 6.5\%$, $n = 8$).

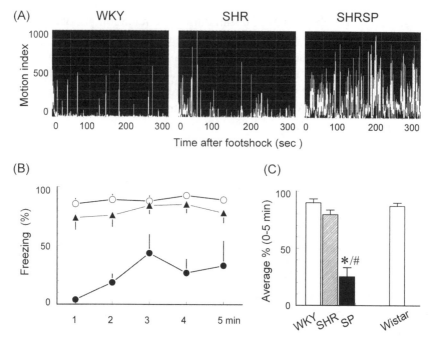

Figure 1. Behavioral analysis in the post-footshock period. Freezing behavior was evaluated immediately after FS for 5 min. (A) Specimen recordings obtained from WKY, SHR and SHRSP for 300 s (5 min) in the post-FS period. (B) Time course changes in freezing (%) of WKY (open circles), SHR (closed triangles), and SHRSP (closed circles), in the post-FS period. For clarity, time course data for Wistar–ST were not presented. (C) Averaged freezing response (%) of WKY, SHR, SHRSP and Wistar–ST during the 5-min post FS period. Means ± SEM are given. *$p < 0.01$ *vs.* WKY, #$p < 0.01$ *vs.* SHR.

Pain perceptivity to electrical footshock

To eliminate the possibility that the behavioral responses to contextual fear conditioning simply reflect the altered pain susceptibility in SHRSP, pain threshold was determined as measures of behavioral responses, jumping or vocalization, to a rising electric FS. As shown in Fig. 3, the corresponding FS intensity, defined as pain threshold, was 0.16 ± 0.01 mA ($n = 8$) in SHRSP, 0.16 ± 0.01 mA ($n = 8$) in SHR, 0.28 ± 0.02 mA ($n = 8$) in WKY and 0.19 ± 0.02 mA ($n = 8$) in Wistar/ST. Thus, pain perceptivity in SHRSP was almost comparable to both SHR and Wistar/ST, while was significantly lower than that in WKY (Fig. 3).

Effects of contextual fear stimuli on the evoked field potentials in perforant path-dentate gyrus synapses

In order to examine whether the altered behavioral response to contextually conditioned fear stimuli would accompany the changes in synaptic transmission in the hippocampus, field potentials of DG region were recorded in the freely moving SHRSP and compared with a normotensive control, Wistar/ST. Re-exposure in the

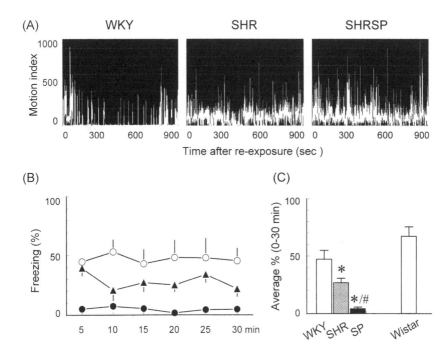

Figure 2. Behavioral analysis in the retention period. Freezing behavior elicited by contextually conditioned fear was evaluated 24 hours after FS for 30 min. (A) Specimen recordings obtained from WKY, SHR and SHRSP in the retention period (0–900 s). (B) Time course changes in freezing (%) of WKY (open circles), SHR (closed triangles), and SHRSP (closed circles), in the retention period for 30 min. For clarity, time course data for Wistar–ST were not presented. (C) Averaged freezing response (%) of WKY, SHR, SHRSP and Wistar–ST in the 30-min retention period. Means ± SEM are given. $*p < 0.05$ *vs.* WKY, $^\#p < 0.05$ *vs.* SHR.

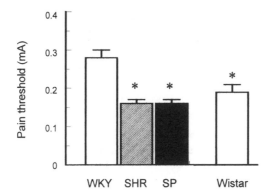

Figure 3. Pain perceptivity to electric footshock. Pain thresolds were evaluated as measures of behavioral responses to the electric FS. Rats were stimulated by electric FS (starting from 0.01 mA). The corresponding FS intensity to elicit animals jumping and/or vocalization, was defined as pain threshold (mA). Means ± SEM are given. $*p < 0.01$ *vs.* WKY.

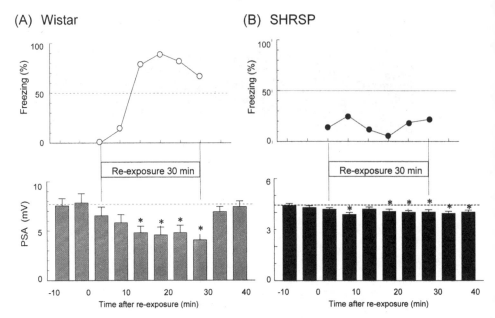

Figure 4. Effects of contextually conditioned fear stimuli on hippocampal neurotransmission in perforant path-dentate gyrus synapses. The evoked population spike amplitude (PSA) before, during and after re-exposing into the FS-chamber was determined in the freely moving rat. (A) Wistar–ST and (B) SHRSP. Upper parts of the figure show simultaneously determined freezing behavior (%) in the retention period (30 min). Data are expressed means ± SEM of 5-min recordings obtained every 20 s. Asterisks indicate significant time point differences ($p < 0.05$) *vs.* the baseline obtained before re-exposure to the FS chamber, which was indicated by the dashed line.

FS chamber, where animals had received the aversive FS 24 hours before, led to a suppression of the PSA below baseline; rats produced a rapid decrease in the evoked PSA in this synapse, accompanying freezing behavior as contextually conditioned fear response. Returning to the home cage, the inhibition of the evoked PSA gradually returned to the baseline obtained before re-exposing (Fig. 4).

Effects of low frequency stimulation on the evoked potentials in perforant path-dentate gyrus synapses

As shown in Fig. 5, LFS produced a transient but significant decrease in the PSA of perforant path-DG synapses, both in SHRSP and Wistar/ST. The response was less potent in SHRSP, when compared with Wistar/ST.

Effects of high frequency stimulation on the evoked potentials in perforant path-dentate gyrus synapses

As shown in Fig. 6, HFS (tetanus) produced a long-lasting potentiation in the PSA of DG region, both in SHRSP and Wistar/ST rats. No difference could be found between their synaptic responses to HFS.

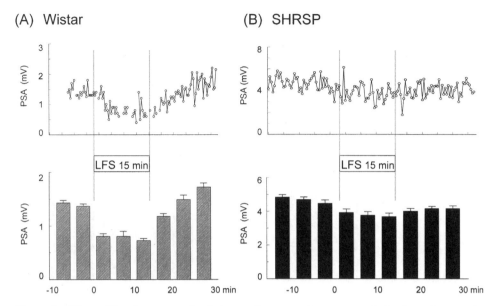

Figure 5. Effects of low frequency stimulation on hippocampal neurotransmission in perforant path-dentate gyrus synapses. Low frequency stimulation (LFS: 1 Hz) was applied for 15 min, as indicated by dashed lines. The evoked population spike amplitude (PSA) before, during and after LFS was determined in the freely moving rat. (A) Wistar–ST and (B) SHRSP. Upper parts of the figure show the value determined every 20 s (mV). In lower parts of the figure, data are expressed means ± SEM of 5-min recordings obtained every 20 s. Asterisks indicating significant time point differences were not presented for clarity.

DISCUSSION

We demonstrated here that juvenile SHRSP rats, an animal model of AD/HD, were less sensitive to contextual fear stimuli and to aversive physical stimuli FS, in a contextually conditioned fear paradigm than control animals. The behavioral responsiveness of SHRSP in this paradigm was not explained by the altered susceptibility to an input of aversive stimuli, since pain threshold was lower rather than higher as compared to genetic and/or normotensive control, WKY. Synaptic response in DG regions to contextually conditioned fear stimuli was also attenuated in SHRSP, which was mimicked by an electrophysiological manipulation, LHS. Our findings suggest that abnormal anxiety-related behavior, less anxiousness, might underlie the impulsive behavior in this AD/HD animal model.

Fear and anxiety as emotional responses are generally classified into two forms: one is innate response, and the other is based on cognition. The elevated plus-maze is a behavioral paradigm that has widely been used to evaluate the anxiolytic properties of drugs by determining the behavioral response when exposing animals to unconditioned fear, as innate emotional response (Pellow and File, 1986). We have recently reported, in this paradigm, juvenile male and female SHRSP showed less anxiety; SHRSP entered significantly more often into the danger-zone open arms than WKY (Ueno *et al.*, 2002b), implying that SHRSP was less sensitive

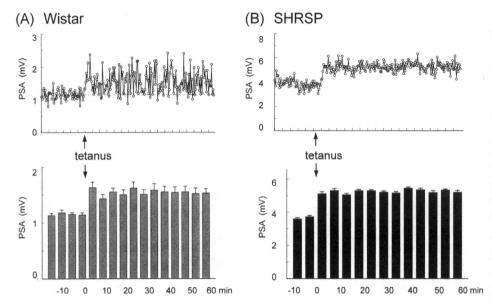

Figure 6. Effects of contextually conditioned fear stimuli on hippocampal neurotransmission in the perforant path-dentate gyrus synapses. High frequency stimulation (tetanus: 400 Hz, 8 train × 5) was applied as indicated by the arrows. The evoked population spike amplitude (PSA) before and after tetanus was determined in the freely moving rat. (A) Wistar–ST and (B) SHRSP. Upper parts of the figure show the value determined every 20 s (mV). In lower parts of the figure, data are expressed means ± SEM of 5-min recordings obtained every 20 s. Asterisks indicating significant time point differences were not presented for clarity.

to unconditioned fear. In addition, we revealed here that SHRSP exerted the attenuated freezing during the post-FS period in a contextual fear-conditioning paradigm. These findings indicate that SHRSP exhibited low susceptibility to two types of unconditioned fear — innate/situational stress-induced fear in the elevated plus maze and aversive/physical stress-induced fear in the contextual fear conditioning. In addition, the present study demonstrated that SHRSP showed less behavioral response to contextually conditioned fear stimuli as well; re-exposure to the FS chamber without FS produced much less anxiety-related behavior in SHRSP, compared with control animals. Of note is that juvenile female SHRSP also exerted less freezing to contextual fear stimuli (data not shown) as previously reported in the elevated plus maze paradigm (Ueno *et al.*, 2002b). Thus, SHRSP exhibited low sensitivity to conditioned as well and unconditioned fear stimuli.

In many animal species, including rodents, the behavioral response to fear and anxiety is characterized by increased reactivity to environmental stimuli and cessation of ongoing behavior. In a contextually conditioned fear paradigm, animals exhibit a profound immobility called freezing both in the post-FS as the unconditional response and in the retention test periods as the conditional response, indicative short-term and long-term memory, respectively (Wallace and Rosen, 2001). Therefore, the present observations in SHRSP, that freezing was severely attenuated in both the post FS and the retention test periods, possibly indicate that cognitive dysfunction involving both short- and long-term memory processes is responsible for the impaired expression of anxiety-related behavior in SHRSP. Our speculations appear to be supported by our previous observations that the spontaneous alternation performance in the Y-maze test, a paradigm evaluating shot-term and working memory (Sarter *et al.*, 1988) based on attention (Katz and Schmaltz, 1980), was attenuated in juvenile male SHRSP (Ueno *et al.*, 2002a, b). Instead, it may not be surprising that SHR exhibited the dissociated freezing responses in this conditioned fear paradigm, no change in the post-FS period and the reduced freezing response in the retention test period, since they did not exhibit the Y-maze performance impairment in our preliminary study (unpublished observation). Hence, the different severity in the impaired expression of anxiety-related freezing behavior between SHRSP and SHR in the retention test period is likely to be due, in part, to the impaired short-term memory processes based on inattention of SHRSP. In other words, the reduced freezing during the retention period in SHR might simply reflect the deficit in long-term memory, including consolidation and retrieval processes.

It is well known that a behavioral parameter of short-term memory, spontaneous alternation behavior in the Y-maze, positively correlates with the hippocampal LTP in perforant path-DG synapses (Mori *et al.*, 2001; Nakao *et al.*, 2001). Moreover, some types of contextual fear conditioning are thought to be hippocampal-dependent (Fanselow, 2000; Phillips and LeDoux, 1992). To assess whether the impaired anxiety-related response to contextually conditioned fear in SHRSP is implicated with the altered synaptic function, we recorded electrophysiological responses in the evoked field potentials to behavioral and electrophysiological manipulations. Freezing behavior as a contextually conditioned fear response was accompanied with a suppression of the PSA below baseline in perforant path-DG synapses. The synaptic response was less potent in SHRSP than in Wistar rats. Of particular interest is the fact that the synaptic response was mimicked by electrophysiological manipulation, LFS. Thus, the reduced synaptic response in this hippocampal sub-region might be attributable to the impaired behavioral response to contextual fear stimuli in SHRSP, although it remains unclear what is the biological basis of these synaptic changes. Concerning the LTP, critically involved in a cellular mechanism of learning and memory (Bliss and Lømo, 1973), the difference in hippocampal LTP formation in perforant path-DG synapses was not observed between SHRSP and a normotensive control Wistar/ST, as previously reported in anesthetized animals (Ueno *et al.*, 2002a, b), suggesting that the apparent memory impairment in SHRSP is not due to disability of hippocampal LTP formation.

A key concern for contextual fear conditioning is pain perception, i.e. hyperalgesia or hypoalgesia, since hypertensive humans and animals are known to have higher pain thresholds and smaller pain responses, indicating changes in nociceptive processing in hypertensive patients (Bruehl *et al.*, 1992; Zamir and Shuber, 1980) and inherited hypertensive rats (Sitsen and de Jong 1984; Taylor *et al.*, 2001).

To eliminate the possibility that difference in pain perception is responsible for the attenuated anxiety-related response observed in SHRSP, we compared behavioral nociceptive responses to FS among SHRSP, SHR and two normotensive control strains. Both SHRSP and SHR exhibited comparable nociceptive responses, jumping or vocalization, by lower intensity FS stimuli than that in a genetic normotensive control, WKY. The difference was also noted between two normotensive controls, WKY and Wistar/ST rats. If the altered nociceptive response in SHRSP was contributable to the difference in behavioral responses in post- FS and/or during re-exposing to the contextual stimuli, both SHRSP and SHR with lower pain thresholds should exhibit more marked behavioral responses than their genetic/normotensive control WKY. However, they did not, but inversely exhibited much less freezing. Meanwhile, freezing in the post-FS period in SHRSP was significantly less than that in SHR, although their pain thresholds were almost the same. Moreover, the behavioral responses were unaltered between two normotensive controls, though their pain thresholds and nociceptive responses to FS were significantly different from each other. Hence, it is unlikely that the altered pain perception, consistently affected the anxiety-related behavior in SHRSP as well as SHR.

Impulsiveness is a symptomatologically important factor for AD/HD, and is increasingly seen as the symptom of greatest significance (Sagvolden and Sergeant, 1998; Taylor, 1998). In a heterogenous disorder, AD/HD, some patients exert the comorbidity of anxiety or phobic disorder (Anderson *et al.*, 1987; Pliszka, 1989; Tannock *et al.*, 1995). AD/HD children with comorbid anxiety are known to show less impulsive behavior than those without anxiety (Pliszka, 1989), suggesting the close relation between impulsivity and anxiety. It has also been reported that AD/HD children with comorbid anxiety showed poorer effects of psychostimulants on cognition than those without anxiety (Pliszka, 1989; Tannock *et al.*, 1995). Our previous studies demonstrated that SHRSP showed an increased number of entries into open arms that are a danger zone in the elevated plus-maze (Ueno *et al.*, 2002b). These findings indicate less anxiety in SHRSP, since the elevated plus maze is a paradigm widely used for evaluating anxiolytic properties of drugs. Methylphenidate was not effective in ameliorating impulsive-like behavior in SHRSP, and this coincides well with the observations on premature lever pressing response in SHR (Sagvolden *et al.*, 1992; Evenden and Meyerson, 1999). Our current findings that juvenile SHRSP exerted low susceptibility to conditioned and unconditioned fear stimuli in a contextual fear conditioning paradigm strongly support the contention that their impulsive-like behavior is based on their lower anxiety, as reported in children with AD/HD (Pliszka, 1989). However, it is still open to question whether methylphenidate would ameliorate the impaired anxiety-related behavior in juveniles.

Considered together, the present findings that behavioral responses in a contextual fear conditioning paradigm were impaired in SHRSP possibly indicate that less anxiety might underlie impulsive behavior, which is in part explainable by the short-term memory impairment in this animal model for AD/HD.

Acknowledgements

The authors thank for Ms. Yukie Makio, Ms. Asako Kaku and Mr. Tetsuya Ono for their technical assistance.

REFERENCES

American Psychiatric Association (1994). Attention-deficit and disruptive behavior disorders, in: *Diagnostic and Statistical Manual of Mental Disorders*, DSM-IV. American Psychiatric Association, Washington DC, pp. 78–85.

Anderson, J. C., Williams, S., McGee, R., *et al.* (1987). DSM-III disorders preadolescent children: prevalence in a large sample from general population, *Arch. Gen. Psychiatry* **44**, 69–76.

Barkley, R. A. (1990). *Attention Deficit Hyperactivity Disorder: A Handbook for Diagnosis and Treatment*. Guilford Press, New York.

Barkley, R. A. (1998). Attention-deficit hyperactivity disorder, *Sci. Am.* **279**, 43–50.

Berger, D. F. and Sagvolden, T. (1998). Sex differences in operant discrimination behaviour in an animal model of attention-deficit hyperactivity disorder, *Behav. Brain Res.* **94**, 73–82.

Bliss, T. V. and Lømo, T. (1973). Long-lasting potentiation of synaptic transmission in the dentate area of the anaesthetized rabbit following stimulation of the perforant path, *J. Physiol.* **232**, 331–356.

Bruehl, S., Carlson, C. R. and McCubbin, J. A. (1992). The relationship between pain sensitivity and blood pressure in normotensives, *Pain* **48**, 463–467.

Bull, E., Reavill, C., Hagan, J. J., *et al.* (2000). Evaluation of the spontaneously hypertensive rat as a model of attention deficit hyperactivity disorder: acquisition and performance of the DRL-60s test, *Behav. Brain Res.* **109**, 27–35.

Cierpial, M. A., Shasby, D. E., Murphy, C. A., *et al.* (1989). Open-field behavior of spontaneously hypertensive and Wistar-Kyoto normotensive rats: effects of reciprocal cross-fostering, *Behav. Neural Biol.* **51**, 203–210.

Evenden, J. and Meyerson, B. (1999). The behavior of spontaneously hypertensive and Wistar Kyoto rats under a paced fixed consecutive number schedule of reinforcement, *Pharmacol. Biochem. Behav.* **63**, 71–82.

Fanselow, M. S. (2000). Contextual fear, gestalt memories, and the hippocampus, *Behav. Brain Res.* **110**, 73–81.

Katz, R. J. and Schmaltz, K. (1980). Dopaminergic involvement in attention: a novel animal model, *Prog. Neuropsychopharmacol.* **4**, 585–590.

Knardahl, S. and Sagvolden, T. (1979). Open-field behavior of spontaneously hypertensive rats, *Behav. Neural Biol.* **27**, 187–200.

Lahey, B. B., Applegate, B., McBurnett, K., *et al.* (1994). DSM-IV field trials for attention deficit hyperactivity disorder in children and Adolescents, *Am. J. Psychiatry* **151**, 1673–1685.

Minami, M., Togashi, H., Koike, Y., *et al.* (1985). Changes in ambulation and drinking behavior related to stroke in stroke-prone spontaneously hypertensive rats, *Stroke* **16**, 44–48.

Minami, M., Kimura, S., Endo, T., *et al.* (1997). Dietary docosahexaenoic acid increases cerebral acetylcholine levels and improves passive avoidance performance in stroke-prone spontaneously hypertensive rats, *Pharmacol. Biochem. Behav.* **58**, 1123–1129.

Mori, K., Togashi, H., Ueno, K.-I., *et al.* (2001). Aminoguanidine prevented the impairment of learning behavior and hippocampal long-term potentiation following transient cerebral ischemia, *Behav. Brain Res.* **120**, 159–168.

Myers, M. M., Musty, R. E. and Hendley, E. D. (1982). Attenuation of hyperactivity in the spontaneously hypertensive rat by amphetamine, *Behav. Neural Biol.* **34**, 42–54.

Nakamura, K., Shirane, M. and Koshikawa, N. (2001). Site-specific activation of dopamine and serotonin transmission by aniracetam in the mesocorticolimbic pathway of rats, *Brain Res.* **897**, 82–92.

Nakao, K., Ikegaya, Y., Yamada, M. K., *et al.* (2001). Spatial performance correlates with long-term potentiation of the dentate gyrus but not of the CA1 region in rats with fimbria-fornix lesions, *Neurosci. Lett.* **307**, 159–162.

Okamoto, K. and Aoki, K. (1963). Development of a strain of spontaneously hypertensive rat, *Jap. Circ. J.* **27**, 282–293.

Okamoto, K., Yamori, Y. and Nagaoka, A. (1974). Establishment of the stroke-prone spontaneously hypertensive rat (SHR), *Circ. Res.* **34/35** (Suppl. I), 143–153.

Paxinos, G. and Watson, C. (1998). *The Rat Brain in Stereotaxic Coordinates*, 3nd edn. Academic Press, New York.

Pellow, S. and File, S. E. (1986). Anxiolytic and anxiogenic drug effects on exploratory activity in an elevated plus-maze: a novel test of anxiety in the rat, *Pharmacol. Biochem. Behav.* **24**, 525–529.

Phillips, R. G. and LeDoux, J. E. (1992). Differential contribution of amygdala and hippocampus to cued and contextual fear conditioning, *Behav. Neurisci.* **106**, 274–285.

Pliszka, S. R. (1989). Effect of anxiety on cognition, behavior, and stimulant response in ADHD, *J. Am. Acad. Child Adolesc. Psychiatry* **28**, 882–887.

Russell, V., de Villiers, A., Sagvolden, T., *et al.* (1995). Altered dopaminergic function in the prefrontal cortex, nucleus accumbens and caudate-putamen of an animal model of attention-deficit hyperactivity disorder — the spontaneously hypertensive rat, *Brain Res.* **676**, 343–351.

Sagvolden, T. (2000). Behavioral validation of the spontaneously hypertensive rat (SHR) as an animal model of attention-deficit/hyperactivity disorder (AD/HD), *Neurosci. Biobehav. Rev.* **24**, 31–39.

Sagvolden, T. and Berger, D. F. (1996). An animal model of attention deficit disorders: The female shows more behavioral problems and is more impulsive than the male, *Eur. Psychologist* **1**, 113–122.

Sagvolden, T. and Sergeant, J. A. (1998). Attention deficit/hyperactivity disorder- from brain dysfunctions to behaviour, *Behav. Brain Res.* **94**, 1–10.

Sagvolden, T., Metzger, M. A., Schiorbeck, H. K., *et al.* (1992). The spontaneously hypertensive rat (SHR) as an animal model of childhood hyperactivity (ADHD): changed reactivity to reinforcers and to psychomotor stimulants, *Behav. Neural Biol.* **58**, 103–112.

Sagvolden, T., Metzger, M. A. and Sagvolden, G. (1993a). Frequent reward eliminates differences in activity between hyperkinetic rats and controls, *Behav. Neural Biol.* **59**, 225–229.

Sagvolden, T., Pettersen, M. B. and Larsen, M. C. (1993b). Spontaneously hypertensive rats (SHR) as a putative animal model of childhood hyperkinesis: SHR behavior compared to four other rats strains, *Physiol. Behav.* **54**, 1047–1055.

Sarter, M., Bodewitz, G. and Stephens, D. N. (1988). Attenuation of scopolamine-induced impairment of spontaneous alternation behavior by antagonist but not inverse agonist and agonist β-carbolines, *Psychopharmacology* **94**, 491–495.

Shibayama, A., Yamaguchi, T., Togashi, H., *et al.* (2004). Age-related changes in regional cerebral blood flow of the prefrontal cortex in stroke-prone spontaneously hypertensive rats, *Japan. Pharmacol. Sci.* **89** (suppl. I) (in press).

Sitsen, J. M. and de Jong, W. (1984). Observations on pain perception and hypertension in spontaneously hypertensive rats, *Clin. Exp. Hypertens.* **6**, 1345–1356.

Szatmari, P., Boyle, M. N. and Offord, D. R. (1989). ADHD and conduct disorder: degree of diagnostic overlap and differences among correlates, *J. Am. Acad. Child Adolesc. Psychiatry* **28**, 865–872.

Tannock, R., Ickowicz, A. and Schachar, R. (1995). Differential effects of methylphenidate on working memory in ADHD children with and without comorbid anxiety, *J. Am. Acad. Child Adolesc. Psychiatry* **34**, 886–896.

Taylor, B. K., Robyn, R. E., Lezin, E. S. T., *et al.* (2001). Hypoalgesia and hyperalgesia with inherited hypertension in the rat, *Am. J. Physiol. Regulatory Integrative Comp. Physiol.* **280**, R345–R354.

Taylor, E. (1998). Clinical foundations of hyperactivity research, *Behav. Brain Res.* **94**, 11–24.

Togashi, H., Minami, M., Bando, Y., *et al.* (1982). Effects of clonidine and guanfacine on drinking and ambulation in spontaneously hypertensive rats, *Pharmacol. Biochem. Behav.* **17**, 519–522.

Togashi, H., Matsumoto, M., Yoshioka, M., *et al.* (1994). Neurochemical profiles in cerebrospinal fluid of stroke-prone spontaneously hypertensive rats, *Neurosci. Lett.* **166**, 117–120.

Togashi, H., Kimura, S., Matsumoto, M., *et al.* (1996). Cholinergic changes in the hippocampus of stroke-prone spontaneously hypertensive rats, *Stroke* **27**, 520–526.

Togashi, H., Ueno, K.-I., Ohashi, S., *et al.* (2002). Cholinergic intervention improves shot-term memory impairment in an animal model of attention-deficit/hyperactivity disorder, Program No. 684.7, *Abstract in Society for Neuroscience 32nd Annual Meeting.*

Tsuda, K., Tsuda, S., Masuyama, Y., *et al.* (1991). Alteration in catecholamine release in the central nervous system of spontaneously hypertensive rats, *Jpn. Heart J.* **32**, 701–709.

Ueno, K.-I., Togashi, H., Matsumoto, M., *et al.* (2002a). $\alpha 4 \beta 2$-Nicotine acetylcholine receptor activation ameliorates impairment of spontaneous alternation behavior in stroke-prone spontaneously hypertensive rats, an animal model of attention-deficit/hyperactivity disorder, *J. Pharmacol. Exp. Ther.* **302**, 95–100.

Ueno, K.-I., Togashi, H., Mori, K., *et al.* (2002b). Behavioral and pharmacological relevance of stroke-prone spontaneously hypertensive rats as an animal model of a developmental disorder, *Behav. Pharmacol.* **13**, 1–13.

Ueno, K.-I., Togashi, H., Matsumoto, M., *et al.* (2003). Juvenile stroke-prone spontaneously hypertensive rats as an animal model of attention-deficit/hyperactivity disorder, *Biogenic Amines* **17**, 293–312.

Wallace, K. J. and Rosen, J. B. (2001). Neurotoxic lesions of the lateral nucleus of the amygdala decrease conditioned fear but not unconditioned fear of a predator odor: comparison with electrolytic lesions, *J. Neurosci.* **21**, 3619–3627.

Watanabe, Y., Fujita, M., Ito, Y., *et al.* (1997). Brain dopamine transporter in spontaneously hypertensive rats, *J. Nucl. Med.* **38**, 470–474.

Wultz, B., Sagvolden, T., Moser, E. I., *et al.* (1990). The spontaneously hypertensive rat as an animal model of attention-deficit hyperactivity disorder: effects of methylphenidate on exploratory behavior, *Behav. Neural Biol.* **53**, 88–102.

Yamori, Y. and Horie, R. (1977). Developmental course of hypertension and regional cerebral blood flow in stroke-prone spontaneously hypertensive rats, *Stroke* **8**, 456–461.

Zamir, N. and Shuber, E. (1980). Altered pain perception in hypertensive humans, *Brain Res. 201*, 471–474.

Advances in Neuroregulation and Neuroprotection (2005), pp. 303-313
Collin, C. *et al.* (Eds)
© VSP 2005

Characterization of behavioral responses to a 5-HT releasing drug, *p*-chloroamphetamine, in mice and the involvement of 5-HT receptor subtypes in *p*-chloroamphetamine-induced behavior

J. YAMADA *, Y. SUGIMOTO, M. OHKURA and K. INOUE

Department of Pharmacology, Kobe Pharmaceutical University, Motoyamakita-machi, Higashinada-ku, Kobe 658-8558, Japan

Abstract—Behavioral effects of a serotonin (5-hydroxytryptamine, 5-HT)-releasing drug, *p*-chloro-amphetamine (PCA) were investigated in mice. We found that PCA elicited 5-HT syndrome including head weaving, hindlimb abduction and tremor and increased locomotor activity in mice. PCA-induced 5-HT syndrome was strongly reduced by both the 5-HT depleter *p*-chlorophenylalanine (pCPA) and the 5-HT transporter inhibitor fluoxetine. pCPA and fluoxetine significantly reduced PCA-induced hyperactivity. PCA-induced 5-HT syndrome and hyperactivity were antagonized by the 5-HT_{1A} receptor antagonist WAY100635. Furthermore, 5-HT syndrome was also reduced by the $5\text{-HT}_{2A/2B/2C}$ receptor antagonist LY 53857 and the $5\text{-HT}_{2B/2C}$ receptor antagonist, SB 206553. However, the 5-HT_{2A} receptor antagonist ketanserin was without effect. The 5-HT_2 receptor antagonists did not affect hyperactivity elicited by PCA. Since PCA can elicit dopamine release, the involvement of dopamine in PCA-induced behavior was also studied. The dopamine depleter α-methyl-*p*-tyrosine did not affect PCA-induced 5-HT syndrome, although it strongly reduced PCA-induced hyperactivity. These results suggest that PCA is incorporated into nerve terminals via 5-HT transporter and that PCA facilitates 5-HT release, which induces 5-HT syndrome and hyperactivity. The 5-HT syndrome elicited by PCA is caused by stimulation of the $5\text{-HT}_{2B/2C}$ receptor in addition to the 5-HT_{1A} receptor. It also suggests that PCA-induced dopamine releasing effects are not related to the 5-HT syndrome in mice, while dopamine is related to hyperactivity induced by PCA.

Keywords: *p*-chloroamphetamine; 5-HT syndrome; 5-HT_{1A} receptor; 5-HT_{2A} receptor; $5\text{-HT}_{2B/2C}$ receptor; dopamine.

---

*To whom correspondence should be addressed. E-mail: j-yamada@kobepharma-u.ac.jp

INTRODUCTION

An amphetamine analog, *p*-chloroamphetamine (PCA) has been shown to release serotonin (5-hydroxytryptamine, 5-HT) from nerve terminals, inhibit 5-HT reuptake and tryptophan hydroxylase activity, which are similar to those of the other 5-HT-releasing amphetamine analogue 3,4-methylenedioxymethamphetamine (MDMA) (Fuller, 1992; Green *et al.*, 1995). PCA is known to deplete brain 5-HT levels for a long time by inhibiting 5-HT reuptake, and the activity of tryptophan hydroxylase, which regulates 5-HT synthesis (Fuller, 1992; Sanders-Bush *et al.*, 1972; Steranka and Sanders-Bush, 1978). PCA induces several pharmacological effects, by facilitation of 5-HT release to the synaptic cleft. PCA stimulates secretions of several hormones such as corticosterone or prolaction (Fuller, 1992; Van de Kar, 1991). Amphetamine derivatives, including methamphetamine and MDMA, induce hyperthermic responses in rodents (Albers and Sonsalla, 1995; Green *et al.*, 1995). We previously reported that PCA induces an apparent hyperthermia in mice similar to other amphetamine analogs and such hyperthermia is elicited by the 5-HT$_{2A}$ receptor (Sugimoto *et al.*, 2000). Moreover, our previous results demonstrated that blockade of the 5-HT$_{2B/2C}$ receptor enhanced hyperthermic responses to PCA (Sugimoto *et al.*, 2000).

Characteristic 5-HT syndrome including head weaving, hindlimb abduction, tremor or forepaw treading is caused in rats and mice by L-tryptophan given with MAO inhibitors by elevation of brain 5-HT concentration (Green and Backus, 1990; Jacobs, 1976). It is known that the 5-HT syndrome is elicited by stimulation of 5-HT receptors following the administration of 5-HT receptor agonists, including 8-hydroxy-2-(di-*n*-propylamino) tetralin (8-OH-DPAT) or 5-methoxy-N,N-dimethyltryptamine (5-MeODMT) (Green and Backus, 1990; Tricklebank *et al.*, 1984). It has been reported that the 5-HT syndrome is induced in humans treated with serotonergic drugs such as antidepressants, selective serotonin reuptake inhibitors (SSRIs) or an anxiolytic, buspirone (Goldberg and Huk, 1992; Lane and Baldwin, 1997; Sternbach, 1991). In many cases, 5-HT syndrome in humans is caused by administration of these serotonergic drugs concomitant with MAO inhibitors (Sternbach, 1991). Frequent clinical features of 5-HT syndrome in humans include changes in mental status, restlessness, myoclonus, shivering and tremor (Lane and Baldwin, 1997; Sternbach, 1991). Recognition of 5-HT receptor subtypes has suggested that the 5-HT$_{1A}$ receptor may be involved in the 5-HT syndrome (Green and Backus, 1990; Tricklebank *et al.*, 1984). We also reported that, in mice, 5-HT syndrome induced in mice by the 5-HT$_{1A}$ receptor agonist 8-OH-DPAT and the nonselective 5-HT receptor agonist tryptamine is mediated by the 5-HT$_{1A}$ receptor (Yamada *et al.*, 1988, 1989). However, it has been reported that the 5-HT$_2$ receptor antagonists can reduce 5-HT syndrome in rats (Smith and Peroutka, 1986).

Following acute systemic administration, PCA elicits typical 5-HT syndrome including head weaving, hindlimb abduction, tremor or forepaw treading in rats (Colado *et al.*, 1993; Trulson and Jacobs, 1976). However, the behavioral effects of PCA in mice remain unclear and the involvement of 5-HT receptor subtypes in PCA-induced behavior has not yet been clarified. In this paper, we studied the effects of PCA on behavior of mice and the involvement of 5-HT receptor subtypes in PCA-induced behavior.

MATERIALS AND METHODS

Animals

Male ddY mice weighing 24–30 g were obtained from SLC Japan, Inc. (Japan). Mice were given free access to food and water and were housed under a controlled 12-h/12-h light-dark cycle (light from 7:00 a.m. to 7:00 p.m.), with room temperature at $23\pm1°C$ and humidity at $55\pm5\%$. All experiments were performed under the same ambient conditions. The experimental procedure was approved by the Kobe Pharmaceutical University Animal Care and Use Committee.

Drug treatment

p-Chloroamphetamine HCl (PCA), fluoxetine HCl, p-chlorophenylalanine methyl-ester HCl (pCPA), WAY 100635 maleate, LY 53857 maleate, SB 206553 HCl and ketanserin tartrate were obtained from Sigma (USA). α-Methyl-p-tyrosine methylester HCl (α-MT) was purchased from Nacalai Tesque(Japan). Drugs were dissolved in saline and injected i.p. pCPA at 400 mg/kg was injected i.p. 5, 3 and 1 days before PCA. α-MT at 250 mg/kg was administered 4 h before PCA. Other 5-HT receptor antagonists were given 30 min before PCA.

Behavioral observation

Mice were observed individually, in clear polycarbonate cages. Immediately after administration of PCA, mice were placed in the test cages and observed for 60 min. Each mouse was evaluated for the presentation of the head weaving, hindlimb abduction and tremor. Head weaving, hindlimb abduction and tremor were rated for 1 min per 5 min according to the ranked intensity scale (0 = absent, 1 = occasional, 2 = frequency, 3 = continuous). Results are shown as means of total scores.

Measurement of locomotor activity

Locomotor activity was measured by a digital counter with an infrared sensor (Neuroscience Inc., Japan) for 120 min after the injection of PCA. The apparatus detects, and gives a digital count of, horizonal movements of animals.

Statistics

The statistical significance was evaluated with Mann–Whitney's U-test.

RESULTS

Effects of PCA on behavior of mice

PCA i.p. dose-dependently elicited the apparent head weaving, hindlimb abduction and tremor, which are components of the 5-HT syndrome (Fig. 1). Other 5-HT syndrome components, forepaw treading and head twitch response, which are elicited by the 5-HT_{2A} receptor (Green and Backus, 1990) were observed in mice given PCA, but these behaviors were not intense and were of short duration. Therefore, these behaviors were excluded from rating. Effects of PCA on locomotor activity are shown in Fig. 2. PCA increased locomotor activity dose-dependently.

Effects of pCPA and fluoxetine on PCA-induced behavior

Effects of the 5-HT depleter pCPA and the 5-HT transporter inhibitor fluoxetine on PCA-induced 5-HT syndrome are shown in Figs 3 and 4. pCPA and fluoxetine strongly reduced 5-HT syndrome. pCPA and fluoxetine also inhibited PCA-induced hyperactivity.

Effects of WAY 100635, LY 53857, ketanserin and SB206553 on PCA-induced behavior

Effects of the 5-HT_{1A} receptor antagonist WAY 100635 at 3 mg/kg on PCA-induced behavior are shown in Fig. 5. WAY 100635 significantly reduced PCA-induced head weaving and hindlimb abduction and tremor. WAY 100635 also inhibited PCA-induced hyperactivity (Fig. 5). Effects of the $5\text{-HT}_{2A/2B/2C}$ receptor antagonist

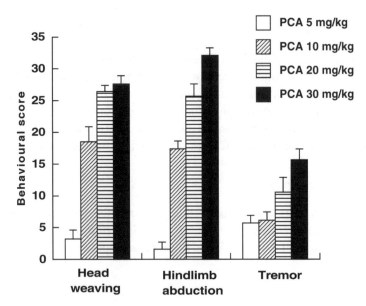

Figure 1. Effects of PCA on behavior of mice. Results are shown as mean ± S.E. ($n = 7$–9).

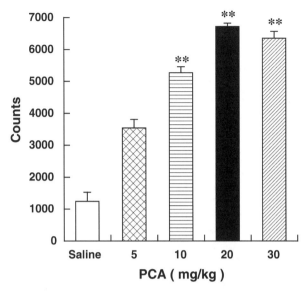

Figure 2. Effects of PCA on locomotor activity of mice. Results are shown as mean ± S.E. ($n = 7$–9).

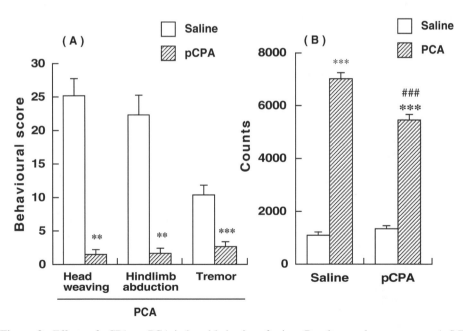

Figure 3. Effects of pCPA on PCA-induced behavior of mice. Results are shown as mean ± S.E. ($n = 6$–8). (A) 5-HT syndrome (B) Hyperactivity. PCA at 20 mg/kg was injected i.p. pCPA at 400 mg/kg was given i.p. 5, 3 and 1 days before PCA. *** $p < 0.001$ *vs.* saline of respective group. ### $p < 0.001$ *vs.* saline + PCA-treated group.

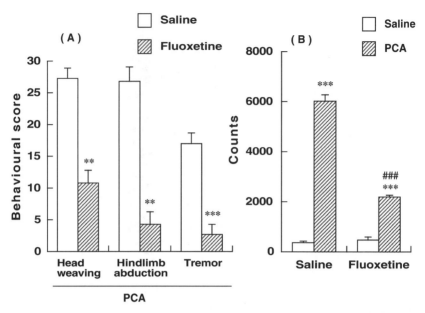

Figure 4. Effects of fluoxetine on PCA-induced behavior of mice. Results are shown as mean ± S.E. (n = 6–9). (A) 5-HT syndrome (B) Hyperactivity. PCA at 20 mg/kg was injected i.p. Fluoxetine at 10 mg/kg was given i.p. 30 min before PCA. ** p < 0.01, *** p < 0.001 *vs.* saline of respective group. ### p < 0.001 *vs.* saline + PCA-treated group.

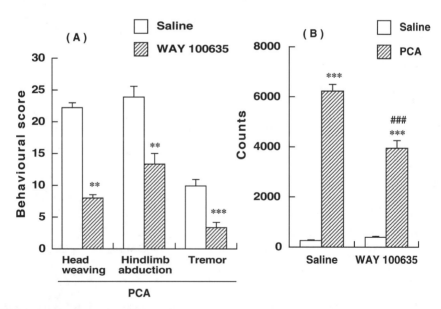

Figure 5. Effects of WAY 100635 on PCA-induced behavior of mice. Results are shown as mean ± S.E. (n = 5–9). (A) 5-HT syndrome (B) Hyperactivity. PCA at 20 mg/kg was injected i.p. WAY 100635 at 3 mg/kg was given i.p. 30 min before PCA. ** p < 0.01, *** p < 0.001 *vs.* saline of respective group. ### p < 0.001 *vs.* saline + PCA-treated group.

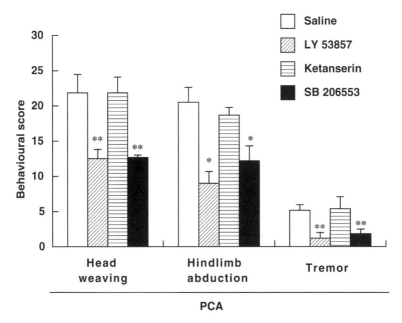

Figure 6. Effects of 5-HT$_2$ receptor antagonists on PCA-induced 5-HT syndrome in mice. Results are shown as mean ± S.E. (n = 5–9). PCA at 20 mg/kg was injected i.p. LY 53857 at 5 mg/kg, ketanserin at 1 mg/kg and SB 206553 at 5 mg/kg were given i.p. 30 min before PCA. $^*p$ < 0.05, $^{**}p$ < 0.01 *vs.* saline of respective group.

LY 53857 at 5 mg/kg, the 5-HT$_{2A}$ receptor antagonist ketanserin at 1 mg/kg and the 5-HT$_{2B/2C}$ receptor antagonist SB 206553 at 5 mg/kg on PCA-induced 5-HT syndrome are demonstrated in Fig. 6. As shown in the results, LY 53857 and SB 206553 apparently inhibited PCA-induced behavioral syndrome. Effects of 5-HT$_2$ receptor antagonists on hyperactivity elicited by PCA are shown in Fig. 7. LY 53857, ketanserin and SB 206553 did not affect it.

Effects of α-MT on PCA-induced behavior

Effects of the dopamine depleter α-MT on 5-HT syndrome and hyperactivity induced by PCA are shown in Fig. 8. α-MT did not affect 5-HT syndrome, head weaving, hindlimb abduction or tremor. However, α-MT significantly reduced PCA-induced hyperactivity.

DISCUSSION

It has been reported that in rats, PCA caused 5-HT syndrome including head weaving, hindlimb abduction, and tremor (Fuller, 1992; Trulson and Jacobs, 1976). However, it is not well known whether PCA induces 5-HT syndrome in mice. The present results demonstrate that PCA induces several components of 5-HT syndrome, head weaving, hindlimb abduction, and tremor in mice. PCA also caused

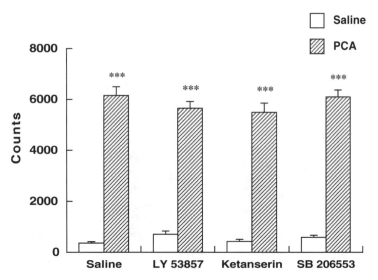

Figure 7. Effects of 5-HT$_2$ receptor antagonists on PCA-induced hyperactivity in mice. Results are shown as mean ± S.E. (n = 5–6). PCA at 20 mg/kg was injected i.p. LY 53857 at 5 mg/kg, ketanserin at 1 mg/kg and SB 206553 at 5 mg/kg were given i.p. 30 min before PCA. *** $p < 0.001$ *vs.* saline of respective group.

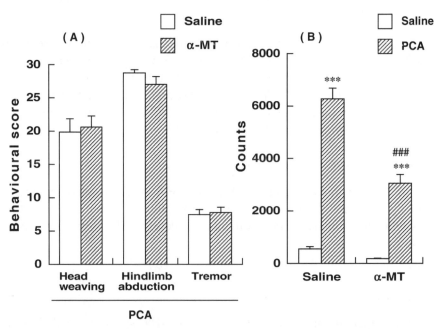

Figure 8. Effects of α-MT on PCA-induced behavior of mice. Results are shown as mean ± S.E. (n = 5–8). (A) 5-HT syndrome (B) Hyperactivity. PCA at 20 mg/kg was injected i.p. α-MT at 250 mg/kg was given i.p. 4 h before PCA. *** $p < 0.001$ *vs.* saline of respective group. ### $p < 0.001$ *vs.* saline + PCA-treated group.

hyperactivity in mice. Therefore, in mice, PCA elicits 5-HT syndrome comparable to that in rats. It was reported that in rats, forepaw treading clearly appeared after treatment with PCA (Trulson and Jacobs, 1976). However, in mice treated with PCA, the intensity of forepaw treading is not strong, differing from findings in rats.

Previous findings reported that the pharmacological effects of PCA are closely associated with the release of 5-HT from nerve terminals (Fuller, 1992). The depletion of 5-HT can inhibit several PCA-induced pharmacological effects. The 5-HT neurotoxin 5,7-dihydroxytryptamine, or the 5-HT depleter pCPA, decreases ejaculation or corticosterone release elicited by PCA (Fuller, 1992; Renyi, 1985; Trulson and Jacobs, 1976). It is well known that PCA is incorporated into nerve terminals via the 5-HT transporter, resulting in facilitation of 5-HT release (Fuller, 1992). Previous findings indicated that inhibitors of the 5-HT transporter can inhibit PCA-induced neurotoxicity (Fuller, 1980). To examine serotonergic mechanisms in PCA-induced behavior, we investigated the effects of the 5-HT depleter pCPA and the 5-HT transporter inhibitor fluoxetine on PCA-induced behavior.

Pretreatment with the 5-HT depleter PCPA strongly inhibited PCA-induced 5-HT syndrome. pCPA significantly reduced the hyperactivity induced by PCA, although its inhibitory effects were weaker than those on 5-HT syndrome. Furthermore, the 5-HT transporter inhibitor fluoxetine powerfully reduced both 5-HT syndrome and hyperactivity induced by PCA. It is reported that PCA-induced 5-HT syndrome is reduced by pCPA in rats, which is consistent with the present results (Fuller, 1992; Trulson and Jacobs, 1976). Our results indicate that 5-HT releasing effects of PCA after uptake via 5-HT transporter participate in PCA-induced 5-HT syndrome in mice as well as in rats.

It is recognized that 5-HT syndrome in rats is mediated by the postsynaptic 5-HT_{1A} receptor, since the 5-HT_{1A} receptor agonists induce 5-HT syndrome and the 5-HT_{1A} receptor antagonists inhibit it (Green and Backus, 1990; Tricklebank *et al.*, 1984). In mice, we demonstrated that the 5-HT_{1A} receptor is strongly related to 5-HT syndrome elicited by the 5-HT receptor agonist 8-OH-DPAT and tryptamine (Yamada *et al.*, 1988, 1989). In the present study, the selective 5-HT_{1A} receptor antagonist WAY 100635 inhibited PCA-induced 5-HT syndrome in mice. Therefore, it is confirmed that the 5-HT_{1A} receptor plays an important role in the occurrence of the 5-HT syndrome in rats and mice. Hyperactivity elicited by PCA was significantly reduced by pretreatment with WAY 100635. Thus, hyperactivity is also related to the 5-HT_{1A} receptor.

5-HT receptor is divided into many subtypes and the 5-HT_2 receptor is subdivided into 5-HT_{2A}, 5-HT_{2B}, 5-HT_{2C} receptors (Baxter *et al.*, 1995). Although 5-HT_{1A} receptor activation is important in 5-HT syndrome, the involvement of the 5-HT_2 receptor was also reported. Smith and Peroutka (1986) reported antagonism of the 5-HT_2 receptor antagonist on the 5-HT syndrome. To clarify the involvement of the 5-HT_2 receptor, we examined the 5-HT_2 receptor antagonists on PCA-induced behavior in mice. The 5-HT_2 receptor antagonist LY 53857 inhibited PCA-induced 5-HT syndrome. Since LY 53857 blocks the all 5-HT_2 receptor subtypes, that is, the 5-HT$_{2A}$, 5-HT$_{2B}$ and 5-HT$_{2C}$ receptor (Baxter *et al.*, 1995), we further studied the 5-HT$_{2A}$ receptor antagonist ketanserin and the 5-HT$_{2B/2C}$ receptor antagonist SB 206553 on PCA-induced behavior. As shown in the results, SB 206553 significantly reduced PCA-induced 5-HT syndrome, while ketanserin did not affect it. SB 206553 inhibited all components of the 5-HT syndrome studied in this study. These results suggest that the 5-HT$_{2B/2C}$ receptor participates in PCA-induced 5-HT syndrome in addition to the 5-HT$_{1A}$ receptor. Hyperactivity induced by PCA was not affected by the 5-HT$_2$ receptor antagonists. Therefore, hyperactivity is not connected to the 5-HT$_{2B/2C}$ receptor. This is the first report demonstrating the involvement of the 5-HT$_{2B/2C}$ receptor in 5-HT syndrome. It is recognized that the 5-HT$_{1A}$ receptor is important in inducing the 5-HT syndrome (Green and Backus, 1990; Tricklebank *et al.*, 1984; Yamada *et al.*, 1988, 1989). Therefore, our findings suggest that the 5-HT$_{2B/2C}$ receptor may enhance 5-HT syndrome.

Although PCA releases 5-HT from nerve terminals, it can also release dopamine. By microdialysis, it was demonstrated that PCA increases dopamine release in the rat brain (Sharp *et al.*, 1986). We recently reported that dopamine is involved in hyperthermia elicited by PCA, since the dopamine depleter α-MT inhibits hyperthermia in mice (Sugimoto *et al.*, 2001). Since dopamine is closely associated with regulation of behavior and motor activity, dopamine may be related to PCA-induced behavior. Therefore, we studied the effects of α-MT on 5-HT syndrome and hyperactivity induced by PCA. α-MT did not affect 5-HT syndrome induced by PCA. This indicates that dopamine does not play a role in 5-HT syndrome elicited by PCA. In contrast, α-MT significantly inhibited PCA-induced hyperactivity. Since PCA can induce dopamine release, dopamine may be related to PCA-induced hyperactivity.

In conclusion, the present findings indicate that PCA-induced 5-HT syndrome in mice and that this syndrome is elicited by 5-HT release mediated by uptake via 5-HT transporter. Furthermore, PCA-induced 5-HT syndrome in mice is mediated by the 5-HT$_{1A}$ receptor and the 5-HT$_{2B/2C}$ receptor is also related to PCA-induced 5-HT syndrome. It is suggested that dopamine is not involved in PCA-induced 5-HT syndrome. In addition, it is indicated that hyperactivity elicited by 5-HT is related to the 5-HT$_{1A}$ receptor but not the 5-HT$_2$ receptor and that dopamine may play a role in hyperactivity induced by PCA. In humans, 5-HT syndrome shows serotonergic hyperstimulation following the administration of serotonergic antidepressants or anxiolytics and has recognized side effects (Goldberg and Huk, 1992; Lane and Baldwin, 1997; Sternbach, 1991). Our findings obtained with PCA in mice, may be helpful in the study and treatment of 5-HT syndrome in humans.

REFERENCES

Albers, D. S. and Sonsalla, P. K. (1995). Methamphetamine-induced hyperthermia and dopaminergic neurotoxicity in mice: Pharmacological profile of protective and nonprotective agents, *J. Pharmacol. Exp. Ther.* **275**, 1104–1114.

Baxter, G., Kennett, G. A., Blaney, F., *et al.* (1995). 5-HT$_2$ receptor subtypes: a family re-united? *Trends Pharmacol. Sci.* **16**, 105–110.

Colado, M. I., Murray, T. K. and Green, A. R. (1993). 5-HT loss in rat brain following 3,4-methylenedioxymethamphetamine(MDMA), *p*-chloroamphetamine and fenfluramine administration and effects of chlormethiazole and dizocilpine, *Brit. J. Pharmacol.* **108**, 583–589.

Fuller, R. W. (1980). Mechanism by which uptake inhibitors antagonize *p*-chloroamphetamine-induced depletion of brain serotonin, *Neurochem. Res.* **5**, 241–245.

Fuller, R. W. (1992). Effects of *p*-chloroamphetamine on brain serotonin neurons, *Neurochem. Res.* **17**, 449–456.

Goldberg, R. J. and Huk, M. (1992). Serotonin syndrome from trazodone and buspirone, *Psychosomatics* **33**, 235–236.

Green, A. R. and Backus, L. I. (1990). Animal models of serotonin behavior, *Ann. N. Y. Acad. Sci.* **600**, 237–249.

Green, A. R., Cross, A. J. and Goodwin, G. M. (1995). Review of the pharmacology and clinical pharmacology of 3,4-methylenedioxymethamphetamine (MDMA or Ecstasy), *Psychopharmacology* **119**, 247–260.

Jacobs, B. L. (1976). An animal behavior model for studying central serotonergic synapses, *Life Sci.* **19**, 777–786.

Lane, R. and Baldwin, D. (1997). Selective serotonin reuptake inhibitor-induced serotonin syndrome: review, *J. Clin. Psychopharmacol.* **17**, 208–221.

Renyi, L. (1985). Ejaculations induced by *p*-chloroamphetamine in the rat, *Neuropharmacology* **24**, 697–704.

Sanders-Bush, E., Bushing, J. A. and Sulser, F. (1972). Long-term effects of *p*-chloroamphetamine on tryptophan hydroxylase activity and on the levels of 5-hydroxytryptamine and 5-hydroxyindoleacetic acid in brain, *Eur. J. Pharmacol.* **20**, 385–388.

Sharp, T., Zetterström, T., Christmanson, L., *et al.* (1986). *p*-Chloroamphetamine releases both serotonin and dopamine into rat brain dialysates in vivo, *Neurosci. Lett.* **72**, 320–324.

Smith, L. M. and Peroutka, S. J. (1986). Differential effects of 5-hydroxytryptamine 1a selective drugs on the 5-HT behavioral syndrome, *Pharmacol. Biochem. Behav.* **24**, 1513–1519.

Steranka, L. R. and Sanders-Bush, E. (1978). Long-term reduction of brain serotonin by *p*-chloroamphetamine: Effects of inducers and inhibitors of drug metabolism, *J. Pharmacol. Exp. Ther.* **206**, 460–467.

Sternbach, H. (1991). The serotonin syndrome, *Am. J. Psychiatry* **148**, 705–713.

Sugimoto, Y., Ohkura, M., Inoue, K., *et al.* (2000). Involvement of the 5-HT$_2$ receptor in hyperthermia induced by *p*-chloromaphetamine, a serotonin-releasing drug in mice, *Eur. J. Pharmacol.* **403**, 225–228.

Sugimoto, Y., Ohkura, M., Inoue, K., *et al.* (2001). Involvement of serotonergic and dopaminergic mechanisms in hyperthermia induced by a serotonin-releasing drug, *p*-chloroamphetamine in mice, *Eur. J. Pharmacol.* **430**, 265–268.

Tricklebank, M. D., Forler, C. and Fozard, J. R. (1984). The involvement of subtypes of the 5-HT$_1$ receptor and of catecholaminergic systems in the behavioural response to 8-hydroxy-2-(di-*n*-propylamino)tetralin in the rat, *Eur. J. Pharmacol.* **106**, 271–282.

Trulson, M. E. and Jacobs, B. L. (1976). Behavioral evidence for the rapid release of CNS serotonin by PCA and fenfluramine, *Eur. J. Pharmacol.* **36**, 149–154.

Van de Kar, L. D. (1991). Neuroendocrine pharmacology of serotonergic(5-HT) neurons, *Ann. Rev. Pharmacol. Toxicol.* **31**, 289–320.

Yamada, J., Sugimoto, Y. and Horisaka, K. (1988). The behavioral effects of 8-hydroxy-2-(di-*n*-propylamino)tetralin (8-OH-DPAT) in mice, *Eur. J. Pharmacol.* **154**, 229–304.

Yamada, J., Sugimoto, Y. and Horisaka, K. (1989). The evidence for the involvement of the 5-HT$_{1A}$ receptor in 5-HT syndrome induced in mice by tryptamine, *Jpn. J. Pharmacol.* **51**, 421–424.

Advances in Neuroregulation and Neuroprotection (2005), pp. 315-330
Collin, C. *et al.* (Eds)
© VSP 2005

Antidepressant research in the era of functional genomics: Farewell to the monoamine hypothesis

MISA YAMADA [1], KOU TAKAHASHI [1], MIKA TSUNODA [1], TOMOKO IWABUCHI [1], SHINYA KOBAYASHI [1], NATSUKO TSUKAHARA [1], TOMOYUKI NAKAGAWA [1], MARI AWATSU [1], SATORU YAMAZAKI [1], MIHO HIRANO [1], HISAYUKI OHATA [1], GENTARO NISHIOKA [2], KENTARO KUDO [2], SATOSHI TANAKA [2], KUNITOSHI KAMIJIMA [2], TERUHIKO HIGUCHI [3], KAZUTAKA MOMOSE [1] and MITSUHIKO YAMADA [4,*]

[1] *Department of Pharmacology, School of Pharmaceutical Sciences, Showa University, Tokyo 142-8555, Japan*
[2] *Department of Psychiatry, School of Medicine, Showa University, Tokyo 142-8555, Japan*
[3] *Musashi Hospital, National Center of Neurology and Psychiatry, Tokyo 187-8551, Japan*
[4] *Division of Psychogeriatrics, National Institute of Mental Health, National Center of Neurology and Psychiatry, 1-7-3 Kohuodai, Ichikawa, Chiba 272-0827, Japan*

Abstract—Although blockade by antidepressants of monoamine uptake into nerve endings is one of the cornerstones of the monoamine hypothesis of depression, there is a clear discrepancy between the rapid effects of antidepressants in increasing synaptic concentrations of monoamine and the lack of immediate clinical efficiency of antidepressant treatment. Novel biological approaches beyond the 'monoamine hypothesis' are definitely expected to cause paradigm shifts in depression research. Functional genomics are powerful tools that can be used to identify genes affected by antidepressants or by other effective therapeutic manipulations. Using RNA fingerprinting technique, we have previously identified several cDNA fragments as antidepressant related genes and from these, original cDNA microarrays were developed. Some of these candidate genes may encode common functional molecules induced by chronic antidepressant treatment. Defining the roles of these genes in drug-induced neural plasticity is likely to transform the course of research on the biological basis of depression. Such detailed knowledge will have profound effects on the diagnosis, prevention, and treatment of depression.

Keywords: Depression; antidepressant; neural plasticity; differential cloning; microarray.

---

*To whom correspondence should be addressed. E-mail: mitsu@ncnp-k.go.jp

INTRODUCTION

Depression is one of the major psychiatric diseases, represent abnormality of emotional, cognitive, autonomic and endocrine functions. Therapeutic manipulations used to treat depressed patients are listed in Table 1. Among them, antidepressants are very effective agents for the prevention and treatment of depression, and have been used clinically for more than 50 years. In addition, a large choice of therapeutic medications are now available, including various newer inhibitors of monoamine reuptake, the selective serotonin reuptake inhibitors (SSRIs), and the selective serotonin noradrenaline reuptake inhibitors (SNRIs). Although the therapeutic action of these antidepressants most likely involves the regulation of serotonergic and noradrenergic signal transduction pathways, to date, no consensus has been reached concerning the precise molecular and cellular mechanism of action of these drugs. Many antidepressants acutely regulate monoaminergic signal transduction, resulting in a significant increase in synaptic concentrations of the monoamines noradrenaline or serotonin within a few hours of initial treatment. But at the same time, the onset of the clinical effect of these drugs lags by several weeks. A satisfying explanation for the discrepancy in the acute increase of synaptic monoamines and delayed clinical effect remains elusive. Theories that postulate long-term changes in receptor sensitivity have unsuccessfully tried to bridge this gap (Siever and Davis, 1985). Consequently, the monoamine hypothesis does not fully explain this clear discrepancy. Novel biological approaches beyond the 'monoamine hypothesis' are definitely expected to cause paradigm shifts in the future of depression research.

There are several preclinical investigations showing that the delayed action of antidepressants on mood, motivation and cognition is not linked to their primary mechanism of action but rather to the development of various modifications (Duman and Vaidya, 1998; Hyman and Nestler, 1996). Hyman and Nestler proposed an 'initiation and adaptation' model to describe the drug-induced neural plasticity associated with the long-term actions of antidepressants in the brain (Hyman and Nestler, 1996). However, the detailed mechanisms underlying such drug-induced adaptive neuronal changes are unknown. The delay of clinical effect

Table 1.
Therapeutic manipulations used to treat depression

---

(1) Antidepressants
   Tricyclic antidepressants (TCA)
   Serotonin selective reuptake inhibitor (SSRI)
   Serotonin and noradrenaline reuptake inhibitor (SNRI)
   Noradrenaline selective reuptake inhibitor (NRI)
   Reversible inhibitor of monoamine oxidase type A (RIMA)
(2) Other manipulations
   Electroconvulsive therapy (ECT)
   Transcranial magnetic stimulation (TMS)
   Vagus nerve stimulation (VNS)

---

from antidepressants could be the result of indirect regulation of neural signal transduction systems or changes at the molecular level by an action on gene transcription following chronic treatment. Indeed, there are selective effects of antidepressants on specific immediate early genes and transcription factors including c-fos (Dahmen *et al.*, 1997; Torres *et al.*, 1998), zif268 (Dahmen *et al.*, 1997), NGFI-A (Bjartmar *et al.*, 2000; Johansson *et al.*, 1998) and the phosphorylation of CRE binding protein (Pei *et al.*, 2000). These molecules activate or repress genes encoding specific proteins by binding to a regulating element of DNA. These functional proteins may be involved in critical steps in mediating treatment-induced neural plasticity.

In the present review, we demonstrated that certain novel candidate genes may underlie the mechanism of action of antidepressants.

CHANGES IN GENE EXPRESSION ELICITED BY ANTIDEPRESSANTS

Recent developments in molecular neurobiology provide new conceptual and experimental tools to investigate and facilitate understanding of the mechanisms by which antidepressants produce long-lasting alterations in brain function. Pharmacogenomic tools, such as differential display PCR, serial analysis of gene expression (SAGE), total gene expression analysis (TOGA), representational difference analysis (RDA), cDNA microarrays, and GeneChip® (Affymetrics, Santa Clara, CA), are now being used to study antidepressant-elicited changes in gene expression. The emerging techniques and powerful tools derived from the relatively new subfields of genomics and proteomics hold great promise for the identification, independent of any preconceived hypotheses, of genes and gene products that are altered by chronic antidepressant treatment or other effective therapeutic manipulations, such as electroconvulsive treatment (ECT).

Using a differential cloning strategy, we and other groups have isolated genes that are differentially expressed in the brain after chronic antidepressant treatment (Huang *et al.*, 1997; Nishioka *et al.*, 2003; Wong *et al.*, 1996; Yamada *et al.*, 1999, 2000, 2001, 2003). Independent of any preconceived hypothesis, these genes and proteins have been implicated in a physiological or pathophysiological process. For example, in the amygdala of rats that received daily treatment with the TCA imipramine for 3 weeks, the gene encoding a mutation suppressor for the Sec4-8 yeast (Mss4) transcript was overexpressed (Andriamampandry *et al.*, 2002). This overexpression was also found in the hippocampus of rats treated chronically with two antidepressants having opposite molecular mechanisms of action — the serotonin reuptake enhancer tianeptine and the SSRI fluoxetine.

We employed the RNA fingerprinting technique, a modified differential display PCR, to identify biochemical changes induced by chronic antidepressant treatments. To date, we have cloned several cDNA candidates as ESTs from rat frontal cortex and hippocampus. Some of these candidate cDNAs should be affected by antidepressants and are thus named antidepressant related genes (ADRGs).

Control group (Cy5-labeled)

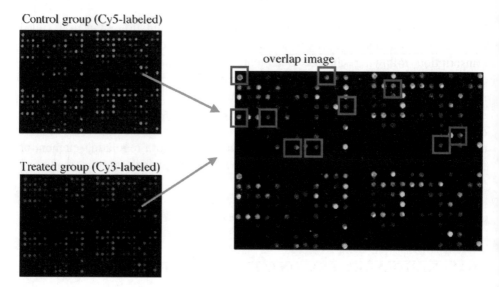

Treated group (Cy3-labeled)

Figure 1. Pseudo-color image of ADRG microarray after hybridization with fluorescent probes. One hundred ninety-six spots representing ADRGs, house keeping genes and other control genes are shown here. Each of the cDNA inserts were amplified by vector primers and spotted in duplicated on the glass slide using GMS417 Arrayer. The pseudo-color images of control group data (green) and antidepressant-treated group (red) were overlapped. The spots within the pink rectangles represent genes up-regulated by chronic antidepressant treatment. The spots within the blue rectangles represent genes down-regulated by chronic antidepressant treatment.

Each candidate molecule must then be evaluated with traditional approaches (e.g. Northern analysis, quantitative RT-PCR, *in situ* hybridization, immunoblotting analysis, immunohistochemistry, etc.). However, large labors and time would be needed to evaluate all candidate ADRGs one by one. Therefore, for high throughput secondary screening of these candidate genes, we developed an original cDNA microarray (ADRG microarray) (Yamada and Higuchi, 2002).

Each of the ADRGs was first amplified using PCR and then spotted in duplicate onto glass slides using the modified method of Salunga (Salunga *et al.*, 1999). Messenger RNA was purified from pooled total RNA using oligo-dT columns, and then converted to cDNA in the presence of either Cy-5 or Cy-3-dUTP to make fluorolabeled probes. Hybridization of the probes to the microarray was done competitively. The probes were mixed and placed on an array, overlaid with coverglass, and hybridized. After the hybridization and washing procedures, each slide was analyzed and gene expression levels were quantified using a PC and ImaGene software (Bio-Discovery Ltd., Swansea, UK). As expected, we obtained low background and consistent results in duplicate experiments. A pseudo-color image of an ADRG microarray is shown in Fig. 1. After normalizing the fluorescent signals using both negative and positive controls, several spots of interest on our ADRG microarray showed increased or decreased fluorescence intensities following

Table 2.

Composition of gene expression profile after chronic antidepressant treatmnent and repeated ECT

ADRG#	Antidepressant	ECT	ADRG#	Antidepressant	ECT
1	↑	↑	13	—	—
2	—	↑	14	—	—
3	↑	↑	15	—	↓
4	↑	↑	16	↓	↓
5	↑	—	17	↑	—
6	↑	—	18	—	—
7	—	↑	19	↑	—
8	↓	—	20	—	—
9	↓	—	21	—	—
10	—	↑	22	↑	—
11	↑	↑	23	↑	—
12	—	—	24	↑	—

chronic antidepressant treatment (Nishioka *et al.*, 2003; Yamada *et al.*, 2000, 2001, 2002). The functional implications of these gene expression changes are currently under investigation.

More recently, we used ADRG microarray to search for some genes commonly induced after chronic antidepressant treatment and repeated ECT. ECT is another therapy for depressive patients that is widely used, particularly in the treatment of patients who are resistant to antidepressant medications. It is an efficient treatment modality, although the basis for its therapeutic mechanism is still unknown. Interestingly, we found some genes were commonly induced or decreased after chronic antidepressant treatment and repeated ECT (Table 2). These genes might be important in therapeutic treatment of depression. In the era of functional genomics, the ADRG microarray we developed seemed to be a powerful tool for the discovery of novel therapeutic targets for future drug development with a new class of action in the brain.

NEW CANDIDATE MOLECULAR SYSTEMS IN ANTIDEPRESSANT RESEARCH

Many of the previous reports describing mechanisms of antidepressant action have focused on acute changes in synaptic pharmacology, especially on neurotransmitter turnover and neurotransmitter receptor changes. To advance our understanding of the therapeutic actions of antidepressants, we must now extend our efforts beyond theories based on the simple pharmacology of the synapse. This new effort must seek a deeper understanding of cellular and molecular neurobiology as well as examine the architecture and function of relevant neural systems. Many now believe that changes in brain gene expression, which are elicited after chronic antidepressant treatment, might underlie the drug-induced neural plasticity associated with the

Table 3.
New candidate molecular systems in antidepressant research

---

(1) Heat shock proteins and molecular chaperons
(2) Survival of neurons and neurogenesis in the hippocampus
(3) Neurotransmitter release and vesicular exocytotic machinery
(4) Neurite outgrowth and sprouting

---

long-term actions of antidepressants in the brain and their clinical effects. Here, we introduce some of the new candidate molecules and molecular systems in antidepressant research (Table 3).

Molecular chaperons and protein-protein interactions

Heat shock proteins (HSPs) are ubiquitous cellular proteins with a highly conserved structure, mode of regulation, and function, indicating their important role in cellular functions. The family of ∼70 kDa HSPs includes heat-shock-inducible HSP70 and the constitutively expressed heat shock cognate protein 70, HSC70 (O'Malley et al., 1985). HSPs are induced by physical and chemical insults, and confer cellular resistance to subsequent lethal stressors. For example, heat and ischemia are well known stimuli that induce the HSP70 family in the central nervous system (Nowak et al., 1990). In mammals, the HSP70 family is also stimulated by stress mediators such as adrenocorticotropic hormone and catecholamines (Blake et al., 1993). The expression of the HSP70 family may therefore be associated with stress responses involving the endocrine, nervous, and immune systems. Interestingly, HSC70 immunoreactivity is distributed throughout the normal rat brain in neuronal somata, dendrites and axons; moreover, postsynaptic chaperon activity may be mediated by HSC70 (Suzuki et al., 1999). However, the physiological role of constitutively expressed HSC70 in the central nervous system is still unclear.

The HSP70 family, including heat-shock-inducible HSP70 and HSC70, may be closely linked to the pathophysiology of depression. Deletions of 29 bp nucleotides in the 5′ non-coding region and subsequent 133 bp in the coding sequences of HSP70 mRNA occur in some patients with depression, but not in normal subjects, or patients with other psychiatric diseases (Shimizu et al., 1996). Given the association between ECT and depression mentioned earlier, it is notable that electroconvulsive seizures induce HSC70 mRNA, but not that of HSP70 in the mouse brain (Kaneko et al., 1993). Glucocorticoid levels are also increased in depressed patients (Carroll and Duman, 1982), and glucocorticoid receptor function is regulated by HSPs (Brady et al., 1991).

We previously identified a novel splice variant of HSC70 while screening differentially expressed molecules in rat brain after chronic antidepressant treatment (Yamada et al., 1999). This clone, named HSC49, lacked the 470 bp of nucleotides found in rat HSC70. HSC49 encoded 442 amino acid residues with a calculated molecular mass of 48.6 kDa. Previously, a genomic clone encoding the rat HSC70 gene was isolated and characterized (Sorger *et al.*, 1987). The HSC70 gene was approximately 4.3 kb long and had eight introns. DNA sequence analysis revealed that HSC49 lacked the entire Exon 7 and Exon 8 of the HSC70 gene. Interestingly, the deletion resulted in a novel termination codon not found in HSC70 itself. Therefore, the amino acid sequence deduced from the nucleotide sequence was terminated at this deletion site, resulting in a 204 amino acid truncation. Recently, Demand and colleagues reported that the carboxy-terminal domain of rat HSC70 mediates interactions with the chaperon cofactors, HSP40 and HOP (HSC70-HSP90-organizing protein) (Demand *et al.*, 1998). Therefore, the specific induction of HSC49, which lacks the carboxy-terminal domain of HSC70 responsible for HSP40 and HOP binding, may play an important role in the regulation of the binding and folding of these peptides. On the other hand, HSC49 still contains the ATPase domain, a putative binding domain for BAG-1 and HIP. These data suggest that HSC49 may be one of the common molecules induced after chronic antidepressant treatment.

Using the ADRG microarray, we identified a gene, ADRG34, that was significantly increased in rat hippocamps treated with antidepressants (Yamada *et al.*, 2000). More recently, we reported that this gene was also induced after repeated ECT (Nishioka *et al.*, 2003). This cDNA encoded 685 amino acid residues yielding a mass of 79 kDa and contained a RING-H2 finger motif at the carboxy-terminus. Using the GeneBank_EMBL database for comparison, homology analysis revealed that ADRG34 was significantly homologous to mouse and human kf-1 gene. Kf-1 was originally identified as a gene that exhibited augmented expression in the cerebral cortex of a sporadic Alzheimer's disease patient (Yasojima *et al.*, 1997). The carboxy-terminal domain of ADRG34 has the same structure as defined by the RING-H2 finger motif. Various proteins containing the RING-H2 finger motif have diverse functions. For example, *Drosophila* goliath plays a developmental role in mesoderm formation or differentiation (Bouchard and Cote, 1993), and RAPsyn is involved in the clustering and aggregation of acetylcholine receptors (Frail *et al.*, 1988). In addition, Neurodap1 mediates synaptic communication and plasticity by controlling the formation of postsynaptic densities and thus maintains vital functions of nerve cells (Nakayama *et al.*, 1995). The precise physiological function of ADRG34 protein is as yet unclear, though current evidence suggests that it may be involved in protein-protein interactions and may play a role in the assembly of large multiprotein complexes.

As described above, we previously reported that the expression of HSC49 is increased by chronic antidepressant treatment (Yamada *et al.*, 1999). Although the relationship between the coordinated up-regulation of HSC49 and ADRG34 by chronic antidepressant treatment is still unclear, the possibility exists that protein-protein interactions may be related to the therapeutic action of antidepressants.

Survival of neurons and neurogenesis in the hippocampus

Although depression involves many psychological and social factors, it also represents a biological process: the effects of repeated exposure to stress on a vulnerable brain. Preclinical and clinical research has focused on the interactions between stress and depression and their effects on the hippocampus (Duman *et al.*, 1999; McEwen, 2000). The hippocampus is one of several brain regions that, when exposed to stressful stimuli, can contribute to the emotional, cognitive, and vegetative abnormalities found in depressed patients. This region of the brain is also involved in the feedback regulation of the hypothalamus-pituitary-adrenal axis, the dysfunction of which is associated with depression (Young *et al.*, 1991). Recent studies suggest that stress-induced atrophy and loss of hippocampal neurons may contribute to the pathophysiology of depression. Interestingly, hippocampal volume is decreased in patients with stress-related psychiatric illnesses, including depression and post-traumatic stress disorder (Sapolsky and Duman, 2000; Sheline *et al.*, 1996).

In vitro and *in vivo* data provide direct evidence that brain-derived neurotrophic factor (BDNF) is one of the key mediators of the therapeutic response to antidepressants (D'Sa and Duman, 2002). BDNF promotes the differentiation and survival of neurons during development and in the adult brain, as well as in cultured cells (Memberg and Hall, 1995; Palmer *et al.*, 1997; Takahashi *et al.*, 1999). Stress decreases the expression of BDNF, and reduced levels could contribute to the atrophy and compromised function of stress-vulnerable hippocampal neurons. In contrast, antidepressant treatment increases the expression of BDNF in the hippocampus, and could thereby reverse the stress-induced atrophy of neurons or protect these neurons from further damage (Duman, 1998; Duman *et al.*, 1997). These findings have resulted in the development of a novel model of the mechanism of antidepressant action and have suggested new targets for the development of therapeutic agents.

While hippocampal volume can decrease in disease, the hippocampus is also one of only a few brain regions where the production of neurons normally occurs throughout the lifetime of several species of animals, including humans (Eriksson *et al.*, 1998). Hippocampal neurogenesis is influenced by several environmental factors and stimuli (Gould and Tanapat, 1999; Nilsson *et al.*, 1999; van Praag *et al.*, 1999). For example, both acute and chronic stress cause decreases in cell proliferation (Fuchs and Flugge, 1998). On the other hand, administration of several different classes of antidepressant, as opposed to non-antidepressant, agents increases the number of BrdU-labeled cells, indicating that this is a common and selective action of antidepressants (Malberg *et al.*, 2000). In addition, recent evidence indicates that electroconvulsive seizures (an animal model of ECT in humans) can also enhance neurogenesis in rat hippocampus (Hellsten *et al.*, 2002; Madsen *et al.*, 2000; Scott *et al.*, 2000). These findings raise the possibility that increased cell proliferation and increased neuronal number may be a mechanism by which antidepressant treatment mitigates stress-induced atrophy and loss of hippocampal neurons, and thus may contribute to the therapeutic actions of antidepressant treatment. Furthermore, increased formation of new neurons in the hippocampus related to antidepressant treatment may lead to altered expression of genes specifically expressed in immature neurons. Therefore, observed changes in gene expression may reflect alterations in cell composition of the tissue rather than changes in individual neurons.

Neurotransmitter release and vesicular exocytotic machinery

Considerable evidence indicates that VAMP-2 is a key component of the synaptic vesicle docking/fusion machinery that forms the SNARE (soluble N-ethylmaleimide-sensitive fusion protein attachment protein receptor) complex (Bennett *et al.*, 1992; Oyler *et al.*, 1989; Trimble *et al.*, 1988; Weis and Scheller, 1998). The SNARE complex consists of proteins on the target membrane, called t-SNARE, and proteins on the vesicular membrane, called v-SNARE. SNAP-25 and syntaxin are t-SNARE proteins, whereas VAMP-2 is a v-SNARE protein. Fusion of vesicles with the plasma membrane leads to exocytosis, which mediates the release of neurotransmitter into the synapse. Previously, we demonstrated a significant increase of both VAMP2 mRNA and protein levels in rat frontal cortex after chronic treatment with two different classes of antidepressants, imipramine and sertraline, and also with repeated ECT (Yamada *et al.*, 2002). In this context, pharmacological modulation of VAMP2 gene expression would also be predicted to alter neurotransmitter release. Our data suggest that VAMP2 may be one of the common functional molecules induced after chronic antidepressant treatment. Interestingly, the work of others shows that acute and chronic administration of antidepressants diminishes the release of glutamate and aspartate, and inhibits veratridine-evoked 5-HT release (Golembiowska and Dziubina, 2000). An important feature of the action of antidepressants and ECT is that they do not globally alter the expression of other membrane-trafficking proteins. In contrast to the enhanced expression of VAMP2, we detected no significant change in the expression of other synaptic vesicle proteins (syntaxin-1 and SNAP-25). Although there are more than a dozen synaptic vesicle proteins (Sudhof, 1995), we chose to investigate the expression of syntaxin-1 and SNAP-25 because they make a SNARE-complex with VAMP2 and mediate the synaptic vesicle docking/fusion machinery. We reasoned that a coordinated change of VAMP2 and the expression of syntaxin-1 and SNAP-25 might signal a change in the overall number of SNARE complexes. An antidepressant-induced change in the expression of syntaxin-1 and SNAP-25, associated predominantly with the presynaptic plasma membrane, would have been indicative of more complex changes in the transmitter secretory pathway, such as an increase in the number of active zones. Instead, the absence of such a coordinated change in syntaxin-1 and SNAP-25 expression indicates that antidepressants or ECT produces a more selective modification of the regulated secretory machinery. Additional work will be necessary to understand the role of selective VAMP2 induction in rat frontal cortex.

On the other hand, post-mortem depressive suicide brain samples were investigated to test the hypothesis that the regulation of SNARE proteins could be abnormal in depression (Honer *et al.*, 2002). Interestingly, the immunoreactivity of

VAMP2 was increased in the depressive group. Further, the correlation between VAMP2 and other SNARE protein or synaptophysin were remarkably weak, and in some cases clearly nonsignificant. Of course, there were limitations on the availability of tissue for investigation and drug treatment history; nevertheless, the authors concluded that the abnormalities of SNARE complex could represent a molecular substrate for abnormalities of neural connectivity in depression.

In addition, the cyclic adenosine monophosphate (cAMP) second messenger system is another pathway that could be involved in antidepressant action. Chronic administration of antidepressants up-regulates the cAMP pathway at several levels, including at points that increase expression and phosphorylation of the cAMP response element binding protein (CREB) (Thome *et al.*, 2000). Among the multiple target genes that could be regulated by CREB is BDNF (Duman *et al.*, 1997; Duman and Vaidya, 1998). BDNF promotes long-term potentiation at hippocampal CA1 synapses via a presynaptic enhancement of synaptic transmission during high-frequency stimulation (HFS). Pozzo-Miller *et al.* showed that heterozygous mice with BDNF knockout display more pronounced synaptic fatigue at CA1 hippocampal synapses upon HFS, suggesting an impairment in transmitter release (Pozzo-Miller *et al.*, 1999). This was associated with a decrease in the number of synaptic vesicles docked to the active zones (with no changes in the reserve pool of vesicles), and a decrease in the synaptic (not total) expression of VAMP2 and synaptophysin, respectively. Therefore, BDNF appears to be directly involved in the regulation of protein machinery at presynaptic terminals, an action consistent with its short- and long-term modulation of synaptic transmission. Recently, the same authors have found that long-term treatment of hippocampal slices with BDNF increases the number of docked vesicles at CA1 synapses, without altering the reserve pool, and greatly increases the expression of synaptotagmin, synaptophysin, and VAMP2 (Tartaglia *et al.*, 2001).

Previously, we demonstrated that a unique cysteine-rich protein, called cysteine string protein (CSP), is clearly elevated in rat brain after chronic antidepressant treatment (Yamada *et al.*, 2001). CSP was originally identified as a family of nervous system-specific antigens in *Drosophila*. CSP is localized to synaptic vesicle membranes (Gundersen and Umbach, 1992). Several reports indicate that CSP functions in the central nervous system to modulate the activity of presynaptic calcium channels, resulting in neurotransmitter release at the nerve terminal (Chamberlain and Burgoyne, 1998; Umbach *et al.*, 1994). Consistent with these findings is a recent report showing that antibodies against CSP inhibit evoked neurotransmitter release at the *Xenopus* neuromuscular junction (Poage *et al.*, 1999). In rat brain, CSP interacts with VAMP2 in synaptic vesicle membranes (Chamberlain and Burgoyne, 2000; Leveque *et al.*, 1998). Taken together, this coordinated induction of two presynaptic molecules suggests that the number of secretory organelles, which includes both small clear vesicles as well as large dense-core granules, might be increased after chronic antidepressant treatment.

Popoli and coworkers have demonstrated that the long-term treatment with antidepressants induced presynaptic CaM Kinase II activity, one of the kinases present involved in the modulation of transmitter release. Further, phosphorylation of synapsin I and synaptotagmin, the presynaptic substrates of CaMK II were also increased after these treatments (Celano *et al.*, 2003; Consogno *et al.*, 2001; Popoli *et al.*, 1997). In addition, as described above, the expression of Mss4 is increased in the rat amygdala and hippocampus after chronic antidepressant treatment (Andriamampandry *et al.*, 2002). Mss4 protein has the properties of a guanine nucleotide exchange factor, and interacts with several members of the Rab family implicated in Ca^{2+}−dependent exocytosis of neurotransmitters. Interestingly, Mss4 transcripts were specifically down-regulated in the hippocampus and amygdala of rats after exposure to chronic, mild stress. These findings suggest that gene expression-dependent alterations of neuronal transmitter release may be an important component of the pharmacological action of antidepressants.

From these data, it is reasonable to assume that alterations of mood, neurovegetative signs, or even social behavior of depressed patients reflect some changes in patterns of synaptic activity in the brain. Thus, it will be of interest to determine whether these changes in the brain contribute to clinical effects in patients treated with antidepressants. Additional work will be necessary to test this hypothesis.

Neurite outgrowth and sprouting

Interestingly, vesicular docking/fusion at the plasma membrane is responsible not only for the release of neurotransmitters, but also for surface expression of plasma membrane proteins and lipids. Therefore, exocytosis plays a fundamental role in axonal and dendritic outgrowth because both processes involve major increases in the surface area of the plasma membrane (Martinez-Arca *et al.*, 2001). Several reports demonstrate that the SNARE complex has an important role in neurite outgrowth. For instance, inhibition of SNAP-25 expression by antisense oligonucleotides prevents neurite elongation in rat cortical neurons and neural crest-derived rat phenochromocytoma (PC12) cells *in vitro* (Morihara *et al.*, 1999; Osen-Sand *et al.*, 1996). Cleavage of SNAP-25 with botulinum neurotoxin A inhibits axonal growth (Morihara *et al.*, 1999; Osen-Sand *et al.*, 1996). Overexpression of SNAP-25 increases the number of neurites in nerve growth factor (NGF)-differentiated PC12 cells (Shirasu *et al.*, 2000), and it has been reported that overexpression of syntaxin 1A inhibits NGF-induced neurite extension (Zhou *et al.*, 2000). On the other hand, it has also been reported that overexpression of syntaxin 1A neither promotes nor inhibits neurite outgrowth in NGF-differentiated PC12 cells (Shirasu *et al.*, 2000). This latter point still needs to be resolved. Inhibition of syntaxin 1A with antisense oligonucleotides or antibodies increases neurite sprouting and neurite length in rat dorsal root ganglion neurons, as well as in retinal ganglion neurons (Yamaguchi *et al.*, 1996). It has also been reported that cleavage of syntaxin by botulinum neurotoxin C1 inhibits axonal growth (Igarashi *et al.*, 1996). Several reports demonstrate that VAMP-2 also has an important role in neurite outgrowth; however, there are several discrepancies in these studies, and the detailed mechanism of VAMP-2 in neurite outgrowth is still unclear.

Treatment with chronic antidepressant increases the expression of GAP-43 in the rat dentate gyrus (Chen et al., 2003). Laifenfeld et al. have also demonstrated that noradrenaline treatment results in an increase in GAP-43 in human neuroblastoma SH-SY5Y cells (Laifenfeld et al., 2002). Because GAP-43 regulates growth of axons and modulates the formation of new connections, these findings suggest that chronic antidepressant treatment may have an effect on structural neuronal plasticity in the central nervous system. Laifenfeld et al. have also reported an increase in the expression of two neurite-outgrowth promoting genes: neural cell adhesion molecule L1 and laminin (Laifenfeld et al., 2002). Along with these effects, SH-SY5Y cells treated with noradrenaline had elongated granule-rich somas and increased numbers of neurites, when compared with non-treated cells. Moreover, cell survival was enhanced in the presence of noradrenaline, while proliferation was inhibited. Taken together, the results support a role for noradrenaline in processes of synaptic connectivity, and may point to a role of this neurotransmitter in mediating the hypothesized neuronal plasticity in antidepressant treatment. More recently, Dihne et al. reported that L1 influences proliferation and differentiation of neural precursor cells (Dihne et al., 2003).

As mentioned above, ECT is a safe and the most effective treatment for severely depressed patients who are resistant to antidepressant medications. Interestingly, the common effects of antidepressants and ECT on connectivity and synaptic plasticity in the dentate gyrus are likely to relate to affective functions of depression (Stewart and Reid, 2000). Consistent with these findings are data demonstrating that chronic electroconvulsive seizure administration in animals induces sprouting of the granule cell mossy fiber pathway in the hippocampus (Vaidya et al., 1999).

CONCLUSION

In the present review, we demonstrated that certain novel candidate genes may underlie the mechanism of action of antidepressants. Defining the roles of these genes in drug-induced neural plasticity is likely to transform the course of research on the biological basis of depression. Identification of such targets will advance future efforts in the quest to develop effective therapeutics that have a new mode of action in the brain. Such detailed knowledge will have profound effects on the diagnosis, prevention, and treatment of depression. In conclusion, in the era of functional genomics, novel biological approaches beyond the 'monoamine hypothesis' are expected to evoke paradigm shifts in the future of depression research.

Acknowledgements

This work was in part supported by Health Science Research Grants from the Ministry of Health, Labour and Welfare, Ministry of Education, Culture, Sport,

Science, and Technology, the Japan Society for the Promotion of Science, Showa University School of Medicine Alumni Association, and the Mitsubishi Pharma Research Foundation.

REFERENCES

Andriamampandry, C., Muller, C., Schmidt-Mutter, C., *et al.* (2002). Mss4 gene is up-regulated in rat brain after chronic treatment with antidepressant and down-regulated when rats are anhedonic, *Mol. Pharmacol.* **62**, 1332–1338.

Bennett, M. K., Calakos, N. and Scheller, R. H. (1992). Syntaxin: a synaptic protein implicated in docking of synaptic vesicles at presynaptic active zones, *Science* **257**, 255–259.

Bjartmar, L., Johansson, I. M., Marcusson, J., *et al.* (2000). Selective effects on NGFI-A, MR, GR and NGFI-B hippocampal mRNA expression after chronic treatment with different subclasses of antidepressants in the rat, *Psychopharmacology* **151**, 7–12.

Blake, M. J., Buckley, D. J., Buckley, A. R., *et al.* (1993). Dopaminergic regulation of heat shock protein-70 expression in adrenal gland and aorta, *Endocrinology* **132**, 1063–1070.

Bouchard, M. and Cote, S. (1993). The Drosophila melanogaster developmental gene g1 encodes a variant zinc-finger-motif protein., *Gene* **125**, 205–209.

Brady, L. S., Whitfield, H. J., Fox, R. J., *et al.* (1991). Long-term antidepressant administration alters corticotropin-releasing hormone, tyrosine hydroxylase, and mineralocorticoid receptor gene expression in rat brain. Therapeutic implications, *J. Clin. Invest.* **87**, 831–837.

Carroll, B. J. and Duman, R. S. (1982). The dexamethasone suppression test for melancholia, *Br. J. Psychiatry* **140**, 292–304.

Celano, E., Tiraboschi, E., Consogno, E., *et al.* (2003). Selective regulation of presynaptic calcium/calmodulin-dependent kinase II by psychotropic drugs, *Biol. Psychiatry* **53**, 442–449.

Chamberlain, L. H. and Burgoyne, R. D. (1998). The cysteine-string domain of the secretory vesicle cysteine-string protein is required for membrane targeting, *Biochem. J.* **335**, 205–209.

Chamberlain, L. H. and Burgoyne, R. D. (2000). Cysteine-string protein: the chaperone at the synapse, *J. Neurochem.* **74**, 1781–1789.

Chen, B., Wang, J. F., Sun, X., *et al.* (2003). Regulation of GAP-43 expression by chronic desipramine treatment in rat cultured hippocampal cells, *Biol. Psychiatry* **53**, 530–537.

Consogno, E., Tiraboschi, E., Iuliano, E., *et al.* (2001). Long-term treatment with S-adenosylmethionine induces changes in presynaptic CaM kinase II and synapsin I, *Biol. Psychiatry* **50**, 337–344.

D'Sa, C. and Duman, R. S. (2002). Antidepressants and neuroplasticity, *Bipolar Disord.* **4**, 183–194.

Dahmen, N., Fehr, C., Reuss, S., *et al.* (1997). Stimulation of immediate early gene expression by desipramine in rat brain, *Biol. Psychiatry* **42**, 317–323.

Demand, J., Luders, J. and Hohfeld, J. (1998). The carboxy-terminal domain of Hsc70 provides binding sites for a distinct set of chaperone cofactors, *Mol. Cell Biol.* **18**, 2023–2028.

Dihne, M., Bernreuther, C., Sibbe, M., *et al.* (2003). A new role for the cell adhesion molecule L1 in neural precursor cell proliferation, differentiation, and transmitter-specific subtype generation, *J. Neurosci.* **23**, 6638–6650.

Duman, R. S. (1998). Novel therapeutic approaches beyond the serotonin receptor, *Biol. Psychiatry* **44**, 324–335.

Duman, R. S. and Vaidya, V. A. (1998). Molecular and cellular actions of chronic electroconvulsive seizures, *J. ECT.* **14**, 181–193.

Duman, R. S., Heninger, G. R. and Nestler, E. J. (1997). A molecular and cellular theory of depression, *Arch. Gen. Psychiatry* **54**, 597–606.

Duman, R. S., Malberg, J. and Thome, J. (1999). Neural plasticity to stress and antidepressant treatment, *Biol. Psychiatry* **46**, 1181–1191.

Eriksson, P. S., Perfilieva, E., Bjork-Eriksson, T., *et al.* (1998). Neurogenesis in the adult human hippocampus, *Nat. Med.* **4**, 1313–1317.

Frail, D., McLaughlin, L., Mudd, J., *et al.* (1988). Identification of the mouse muscle 43,000-dalton acetylcholine receptor-associated protein (RAPsyn) by cDNA cloning., *J. Biol. Chem.* **263**, 15602–15607.

Fuchs, E. and Flugge, G. (1998). Stress, glucocorticoids and structural plasticity of the hippocampus, *Neurosci. Biobehav. Rev.* **23**, 295–300.

Golembiowska, K. and Dziubina, A. (2000). Effect of acute and chronic administration of citalopram on glutamate and aspartate release in the rat prefrontal cortex, *Polish J. Pharmacol.* **52**, 441–448.

Gould, E. and Tanapat, P. (1999). Stress and hippocampal neurogenesis, *Biol. Psychiatry* **46**, 1472–1479.

Gundersen, C. B. and Umbach, J. A. (1992). Suppression cloning of the cDNA for a candidate subunit of a presynaptic calcium channel, *Neuron* **9**, 527–537.

Hellsten, J., Wennstrom, M., Mohapel, P., *et al.* (2002). Electroconvulsive seizures increase hippocampal neurogenesis after chronic corticosterone treatment, *Eur. J. Neurosci.* **16**, 283–290.

Honer, W. G., Fralkai, P., Bayer, T. A., *et al.* (2002). Abnormalities of SNARE mechanism proteins in anterior frontal cortex in severe mental illness, *Cerebral Cortex* **12**, 349–356.

Huang, N. Y., Strakhova, M., Layer, R. T., *et al.* (1997). Chronic antidepressant treatments increase cytochrome b mRNA levels in mouse cerebral cortex, *J. Mol. Neurosci.* **9**, 167–176.

Hyman, S. E. and Nestler, E. J. (1996). Initiation and adaptation: a paradigm for understanding psychotropic drug action, *Am. J. Psychiatry* **153**, 151–162.

Igarashi, M., Kozaki, S., Terakawa, S., *et al.* (1996). Growth cone collapse and inhibition of neurite growth by Botulinum neurotoxin C1: a t-SNARE is involved in axonal growth, *J. Cell Biol.* **134**, 205–215.

Johansson, I. M., Bjartmar, L., Marcusson, J., *et al.* (1998). Chronic amitriptyline treatment induces hippocampal NGFI-A, glucocorticoid receptor and mineralocorticoid receptor mRNA expression in rats, *Brain Res. Mol. Brain Res.* **62**, 92–95.

Kaneko, M., Abe, K., Kogure, K., *et al.* (1993). Correlation between electroconvulsive seizure and HSC70 mRNA induction in mice brain, *Neurosci. Lett.* **157**, 195–198.

Laifenfeld, D., Klein, E. and Ben-Shachar, D. (2002). Norepinephrine alters the expression of genes involved in neuronal sprouting and differentiation: relevance for major depression and antidepressant mechanisms, *J. Neurochem.* **83**, 1054–1064.

Leveque, C., Pupier, S., Marqueze, B., *et al.* (1998). Interaction of cysteine string proteins with the alpha1A subunit of the P/Q-type calcium channel, *J. Biol. Chem.* **273**, 13488–13492.

Madsen, T. M., Treschow, A., Bengzon, J., *et al.* (2000). Increased neurogenesis in a model of electroconvulsive therapy, *Biol. Psychiatry* **47**, 1043–1049.

Malberg, J. E., Eisch, A. J., Nestler, E. J., *et al.* (2000). Chronic antidepressant treatment increases neurogenesis in adult rat hippocampus, *J. Neurosci.* **20**, 9104–9110.

Martinez-Arca, S., Coco, S., Mainguy, G., *et al.* (2001). A common exocytotic mechanism mediates axonal and dendritic outgrowth, *J. Neurosci.* **21**, 3830–3838.

McEwen, B. S. (2000). The neurobiology of stress: from serendipity to clinical relevance, *Brain Res.* **886**, 172–189.

Memberg, S. P. and Hall, A. K. (1995). Proliferation, differentiation, and survival of rat sensory neuron precursors in vitro require specific trophic factors, *Mol. Cell Neurosci.* **6**, 323–335.

Morihara, T., Mizoguchi, A., Takahashi, M., *et al.* (1999). Distribution of synaptosomal-associated protein 25 in nerve growth cones and reduction of neurite outgrowth by botulinum neurotoxin A without altering growth cone morphology in dorsal root ganglion neurons and PC-12 cells, *Neuroscience* **91**, 695–706.

Nakayama, M., Miyake, T., Gahara, Y., *et al.* (1995). A novel RING-H2 motif protein downregulated by axotomy: its characteristic localization at the postsynaptic density of axosomatic synapse, *J. Neurosci.* **15**, 5238–5248.

Nilsson, M., Perfilieva, E., Johansson, U., *et al.* (1999). Enriched environment increases neurogenesis in the adult rat dentate gyrus and improves spatial memory, *J. Neurobiol.* **39**, 569–578.

Nishioka, G., Yamada, M., Kudo, K., *et al.* (2003). Induction of kf-1 after repeated electroconvulsive treatment and chronic antidepressant treatment in rat frontal cortex and hippocampus, *J. Neural. Transm.* **110**, 277–285.

Nowak, T. S., Bond, U., Schlesinger, M. J., *et al.* (1990). Heat shock RNA levels in brain and other tissues after hyperthermia and transient ischemia, *J. Neurochem.* **54**, 451–458.

O'Malley, K., Mauron, A., Barchas, J. D., *et al.* (1985). Constitutively expressed rat mRNA encoding a 70-kilodalton heat-shock-like protein, *Mol. Cell Biol.* **5**, 3476–3483.

Osen-Sand, A., Staple, J. K., Naldi, E., *et al.* (1996). Common and distinct fusion proteins in axonal growth and transmitter release, *J. Comp. Neurol.* **367**, 222–234.

Oyler, G. A., Higgins, G. A., Hart, R. A., *et al.* (1989). The identification of a novel synaptosomal-associated protein, SNAP-25, differentially expressed by neuronal subpopulations, *J. Cell Biol.* **109**, 3039–3052.

Palmer, T. D., Takahashi, J., Gage, F. H., *et al.* (1997). The adult rat hippocampus contains primordial neural stem cells, *Mol. Cell Neurosci.* **8**, 389–404.

Pei, Q., Lewis, L., Sprakes, M. E., *et al.* (2000). Serotonergic regulation of mRNA expression of Arc, an immediate early gene selectively localized at neuronal dendrites, *Neuropharmacology* **39**, 463–470.

Poage, R. E., Meriney, S. D., Gundersen, C. B., *et al.* (1999). Antibodies against cysteine string proteins inhibit evoked neurotransmitter release at Xenopus neuromuscular junctions, *J. Neurophysiol.* **82**, 50–59.

Popoli, M., Venegoni, A., Vocaturo, C., *et al.* (1997). Long term blockade of serotonin reuptake affects synaptotagmin phosphorylation in the hippocampus, *Mol. Pharmacol.* **51**, 19–26.

Pozzo-Miller, L. D., Gottschalk, W., Zhang, L., *et al.* (1999). Impairments in high-frequency transmission, synaptic vesicle docking, and synaptic protein distribution in the hippocampus of BDNF knockout mice, *J. Neurosci.* **19**, 4972–4983.

Salunga, R., Guo, H., Lu, L., *et al.* (1999). Gene expression analysis via cDNA microarrays, in: *DNA Microarrays. A Practical approach*, Schena, M. (Ed.), pp. 121–137. Oxford University Press, New York, USA.

Sapolsky, R. M. and Duman, R. S. (2000). Glucocorticoids and hippocampal atrophy in neuropsychiatric disorders, *Arch. Gen. Psychiatry* **57**, 925–935.

Scott, B. W., Wojtowicz, J. M. and Burnham, W. M. (2000). Neurogenesis in the dentate gyrus of the rat following electroconvulsive shock seizures, *Exp. Neurol.* **165**, 231–236.

Sheline, Y., Wany, P., Gado, M., *et al.* (1996). Hippocampal atrophy in recurrent major depression, *Proc. Natl. Acad. Sci. USA* **93**, 3908–3913.

Shimizu, S., Nomura, K., Ujihara, M., *et al.* (1996). An allele-specific abnormal transcript of the heat shock protein 70 gene in patients with major depression, *Biochem. Biophys. Res. Commun.* **219**, 745–752.

Shirasu, M., Kimura, K., Kataoka, M., *et al.* (2000). VAMP-2 promotes neurite elongation and SNAP-25A increases neurite sprouting in PC12 cells, *Neurosci. Res.* **37**, 265–275.

Siever, L. J. and Davis, K. L. (1985). Overview: toward a dysregulation hypothesis of depression, *Am. J. Psychiatry* **142**, 1017–1031.

Sorger, P. K., Pelham, H. R., Brady, L. S., *et al.* (1987). Cloning and expression of a gene encoding hsc73, the major hsp70-like protein in unstressed rat cells, *Embo J.* **6**, 993–998.

Stewart, C. A. and Reid, I. C. (2000). Repeated ECS and fluoxetine administration have equivalent effects on hippocampal synaptic plasticity, *Psychopharmacology* **148**, 217–223.

Sudhof, T. C. (1995). The synaptic vesicle cycle: a cascade of protein-protein interactions, *Nature* **375**, 645–653.

Suzuki, T., Usuda, N., Murata, S., *et al.* (1999). Presence of molecular chaperones, heat shock cognate (Hsc) 70 and heat shock proteins (Hsp) 40, in the postsynaptic structures of rat brain, *Brain Res.* **816**, 99–110.

Takahashi, J., Palmer, T. D., Gage, F. H., *et al.* (1999). Retinoic acid and neurotrophins collaborate to regulate neurogenesis in adult-derived neural stem cell cultures, *J. Neurobiol.* **38**, 65–81.

Tartaglia, N., Du, J., Tyler, W. J., *et al.* (2001). Protein synthesis-dependent and -independent regulation of hippocampal synapses by brain-derived neurotrophic factor, *J. Biol. Chem.* **276**, 37585–37593.

Thome, J., Sakai, N., Shin, K., *et al.* (2000). CAMP response element-mediated gene transcription is upregulated by chronik antidepressant treatment, *J. Neurosci.* **20**, 4030–4036.

Torres, G., Horowitz, J. M., Laflamme, N., *et al.* (1998). Fluoxetine induces the transcription of genes encoding c-fos, corticotropin-releasing factor and its type 1 receptor in rat brain, *Neuroscience* **87**, 463–477.

Trimble, W. S., Cowan, D. M. and Scheller, R. H. (1988). VAMP-1: a synaptic vesicle-associated integral membrane protein, *Proc. Natl. Acad. Sci. USA* **85**, 4538–4542.

Umbach, J. A., Zinsmaier, K. E., Eberle, K. K., *et al.* (1994). Presynaptic dysfunction in Drosophila csp mutants, *Neuron* **13**, 899–907.

Vaidya, V. A., Siuciak, J. A., Du, F., *et al.* (1999). Hippocampal mossy fiber sprouting induced by chronic electroconvulsive seizures, *Neuroscience* **89**, 157–166.

van Praag, H., Christie, B. R., Sejnowski, T. J., *et al.* (1999). Running enhances neurogenesis, learning, and long-term potentiation in mice, *Proc. Natl. Acad. Sci. USA* **96**, 13427–13431.

Weis, W. I. and Scheller, R. H. (1998). Membrane fusion. SNARE the rod, coil the complex, *Nature* **395**, 328–329.

Wong, M. L., Khatri, P., Licinio, J., *et al.* (1996). Identification of hypothalamic transcripts upregulated by antidepressants, *Biochem. Biophys. Res. Commun.* **229**, 275–279.

Yamada, M. and Higuchi, T. (2002). Functional genomics and depression research. Beyond the monoamine hypothesis, *Eur. Neuropsychopharmacol.* **12**, 235–244.

Yamada, M., Kiuchi, Y., Nara, K., *et al.* (1999). Identification of a novel splice variant of heat shock cognate protein 70 after chronic antidepressant treatment in rat frontal cortex, *Biochem. Biophys. Res. Commun.* **261**, 541–545.

Yamada, M., Yamada, M., Yamazaki, S., *et al.* (2000). Identification of a novel gene with RING-H2 finger motif induced after chronic antidepressant treatment in rat brain, *Biochem. Biophys. Res. Commun.* **278**, 150–157.

Yamada, M., Yamada, M., Yamazaki, S., *et al.* (2001). Induction of cysteine string protein after chronic antidepressant treatment in rat frontal cortex., *Neurosci. Lett.* **301**, 183–186.

Yamada, M., Takahashi, K., Tsunoda, M., *et al.* (2002). Differential expression of VAMP2/synaptobrevin-2 after antidepressant and electroconvulsive treatment in rat frontal cortex, *Pharmacogenomics J.* **2**, 377–382.

Yamaguchi, K., Nakayama, T., Fujiwara, T., *et al.* (1996). Enhancement of neurite-sprouting by suppression of HPC-1/syntaxin 1A activity in cultured vertebrate nerve cells, *Brain Res.* **740**, 185–192.

Yasojima, K., Tsujimura, A., Mizuno, T., *et al.* (1997). Cloning of human and mouse cDNAs encoding novel zinc finger proteins expressed in cerebellum and hippocampus, *Biochem. Biophys. Res. Commun.* **231**, 481–487.

Young, E. A., Haskett, R. F., Murphy-Weinberg, V., *et al.* (1991). Loss of glucocorticoid fast feedback in depression, *Arch. Gen. Psychiatry* **48**, 693–699.

Zhou, Q., Xiao, J. and Liu, Y. (2000). Participation of syntaxin 1A in membrane trafficking involving neurite elongation and membrane expansion, *J. Neurosci. Res.* **61**, 321–328.

Advances in Neuroregulation and Neuroprotection (2005), pp. 331-346
Collin, C. *et al.* (Eds)

Acute and chronic effects of tryptophan and alcohol on serotonin in the locus coeruleus in rats

K. HOSHI *, M. HAYASHI and T. BANDOH

Department of Clinical Pharmacology, Hokkaido College of Pharmacy, 7-1 Katsuraoka-cho, Otaru, Hokkaido 047-0264, Japan

Abstract—The present review will concentrate on a discussion of recent investigations that demonstrated whether the levels of 5-hydroxytryptamine (serotonin; 5-HT) and its metabolite, 5-hydroxyindoleacetic acid (5-HIAA), in the locus coeruleus are influenced by tryptophan alone or simultaneous administration of tryptophan and ethanol. The impetus for this line of investigation derives from a report that mice treated with tryptophol plus alcohol increased brain tryptophol level and became highly susceptible to convulsions. A deficiency in 5-HT synthesis and metabolism (turnover) has been implicated in several neuropsychiatric disorders, including depression, obsessive-compulsive disorder, bulimia nervosa, late luteal phase dysphoric syndrome, alcoholism, impulsive violence and aggression. Emphasis will be placed on demonstration of a rodent model in which the tryptophan administration induces the increased 5-HIAA level through the activation of serotonergic neuronal system. The use of *in vivo* microdialysis techniques clearly identifies, in this model, that acutely- or consecutively-treated tryptophan plus ethanol results in focal increases in extracellular levels of 5-HIAA within the locus coeruleus. This study provides evidence to indicate that the 5-HIAA level in the locus coeruleus after tryptophan alone or tryptophan plus ethanol is indicative of changes in the activity of serotonergic neurotransmission in locus coeruleus neurons and the participation of 5-HIAA as a general phenomenon in behavioral activation.

Keywords: Locus coeruleus; dorsal raphe nucleus; tryptophan; 5-HIAA (5-hydroxyindoleacetic acid); alcohol; microdialysis.

INTRODUCTION

The locus coeruleus and the raphe nuclei are part of a venerable concept called the ascending reticular activating system, which implicates the reticular 'gcore' of the brain stem in processes that arouse and awaken the forebrain (Bear *et al.*, 1996). The serotonergic afferents to the locus coeruleus originate from the dorsal and caudal aspects of the dorsal raphe nucleus (Cedarbaum and Aghajanian, 1978; Morgane

---

*To whom correspondence should be addressed. E-mail: hoshi@hokuyakudai.ac.jp

and Jacobs, 1979; Vertes and Kocsis, 1994), from the median raphe nucleus (Luppi *et al.*, 1995; Maeda *et al.*, 1991; Morgane and Jacobs, 1979) and the supralemniscal area (Luppi *et al.*, 1995; Maeda *et al.*, 1991). Serotonergic afferents from the periventricular gray of the pons and from serotonergic perikarya of the pericoerulear region (Aston-Jones *et al.*, 1991) also extend to the locus coeruleus. Serotonergic perikarya are located in the locus coeruleus (Iijima, 1993; Kaehler *et al.*, 1999a). The noradrenergic neurons located in the locus coeruleus modulate the activity of serotonergic neurons in the dorsal raphe nucleus via excitatory α_1-adrenoceptors (Baraban and Aghajanian, 1980; Szabo *et al.*, 1999). In turn, the activity of serotonergic neurons in the locus coeruleus is modulated by 5-HT_{1A} receptors located within the dorsal raphe nucleus (Singewald *et al.*, 1995, 1997). Moreover, the stimulation of serotonergic nerve could play a functional role in the modulation of the noradrenergic system in the locus coeruleus (Cedarbaum and Aghajanian, 1978; Mateo *et al.*, 2000; Vertes and Kocsis, 1994). Stimulation of raphe 5-HT receptors reduces the firing rate of serotonergic neurons, 5-HT synthesis and 5-HT release in projection areas, such as the hippocampus (Hjorth and Magnusson, 1988; Sharp *et al.*, 1989; Sinton and Fallon, 1988; Sprouse and Aghajanian, 1986, 1987). Especially, in an area with serotonergic and noradrenergic nerve terminals such as the hippocampus, 5-HT release is modulated by noradrenaline (NA) through presynaptic α_2-adrenergic heteroreceptors. In addition, Szabo *et al.* (1999) also demonstrated that a decreased firing activity of locus coeruleus neurons induced by long-term selective serotonin reuptake inhibitor (SSRI) gives support to the contribution of the noradrenergic system in the therapeutic activity of these drugs. Recently, Singewald *et al.* (1997) demonstrated that the serotonergic system of the locus coeruleus is implicated in behavioral activation, pain, noxious and stress stimuli, and modulation of cardiovascular activity (Bear *et al.*, 1996). In this connection, 5-HT and NA also are involved in both the pathogenesis and the recovery from depression (Blier *et al.*, 1990; Caldecott *et al.*, 1991; Delgado *et al.*, 1993).

Abnormal behaviors associated with or perhaps due to a serotonergic dysfunction are often part of the clinical picture of alcohol abuse and dependence. There are the findings that some alcoholic individuals may have lowered central serotonergic neurotransmission (LeMarquand *et al.*, 1994) and that brain 5-HT has been implicated in obsessions and compulsions (Insel *et al.*, 1990). In addition, lower levels of cerebrospinal fluid (CSF) 5-hydroxyindoleacetic acid (5-HIAA), the primary metabolite of 5-HT, have been reported in clinical studies of aggression (Insel *et al.*, 1990), suicide (Mann *et al.*, 1990; Roy *et al.*, 1990), and impulsivity (Linnoila *et al.*, 1983). On the contrary, compounds that increase serotonergic neurotransmission (uptake inhibitors, releasers, direct agonists) have been shown to decrease food intake (Curzon, 1990). Thus, it is possible that the serotonergic system that controls tryptophan intake may regulate alcohol consumption (LeMarquand *et al.*, 1994). However, the neuroanatomical and functional interaction between the locus coeruleus and the dorsal raphe nucleus that release NA and 5-HT is still not clarified.

TRYPTOPHAN AND SEROTONERGIC FUNCTIONING

Tryptophan is the dietary precursor of 5-HT and a deficiency in either tryptophan intake or its metabolism could result in behavioral disorders (Tiihonen *et al.*, 2001). Dietary tryptophan augmentation is known to increase brain 5-HT synthesis and content (Fadda *et al.*, 2000; Stancampiano *et al.*, 1997). The syndrome of 5-HT excess in laboratory animals and humans has been termed 5-HT syndrome. Patients with 5-HT syndrome had abnormalities in cognitive and behavioral functions, autonomic nervous system function, and neuromuscular function (Mills, 1997). Moreover, compounds that increase serotonergic neurotransmission (uptake inhibitors, releasers, direct agonists) have been shown to decrease food intake (Curzon, 1990). Conversely, a deficiency in 5-HT synthesis and metabolism has been implicated in several neuropsychiatric disorders, including alcoholism, aggression (LeMarquand *et al.*, 1994) and depression (Meltzer, 1990). In laboratory animals, ingestion of a tryptophan-free diet alters behavioral indices of serotonergic function, increasing pain sensitivity, acoustic startle, motor activity, and aggression and reducing rapid eye movement sleep (Schaechter and Wurtman, 1989; Stancampiano *et al.*, 1997). These behavioral changes are reversed once brain tryptophan is restored (Schaechter and Wurtman, 1989). The result of acute plasma tryptophan depletion, in rats, is a decrease in brain tryptophan and 5-HT concentration (Moja *et al.*, 1989) and a decrease in cerebrospinal fluid (CSF) 5-HT (Stancampiano *et al.*, 1997) and 5-HIAA concentration (Bel and Artigas, 1996; Stancampiano *et al.*, 1997). CSF 5-HIAA, the principal metabolite of 5-HT, reflects central 5-HT turnover and correlates with brain 5-HIAA in humans (Stanley *et al.*, 1985; Williams *et al.*, 1999). In addition, CSF 5-HIAA has been proposed as a valid indicator of general chnages in 5-HT metabolism in the central nervous system (CNS) (Aizenstein and Korf, 1979; Banki and Molnar, 1981; LeMarquand *et al.*, 1994). Furthermore, CSF 5-HIAA levels also correlate with 5-HIAA levels in the cerebral cortex of humans at autopsy (Stanley *et al.*, 1985).

The usual metabolic pathway to the formation of serotonin involves the conversion of tryptophan, an essential amino acid, to 5-hydroxytryptophan (5-HTP) via tryptophan hydroxylase in a rate-limiting fashion. Once formed, 5-HTP is constitutively converted to serotonin by aromatic L-amino acid decarboxylase (Gwaltney-Brant *et al.*, 2000). 5-HT is oxidatively deaminated by monoamine oxidase to 5-hydroxyindoleacetaldehyde (5-HIAL). The 5-HIAL formed can be metabolized in two ways: dehydrogenation by aldehyde dehydrogenase to 5-HIAA, or reduced by aldehyde reductase to 5-hydroxytryptophol (5-HTPL) (Gwaltney-Brant *et al.*, 2000; Youdim and Ashkenazi, 1982) (Fig. 1).

ALCOHOL AND SEROTONERGIC FUNCTIONING

Alcohol intake has been suggested to be directly affected by the status of serotonergic neurotransmission (LeMarquand *et al.*, 1994). An acute alcohol dose in

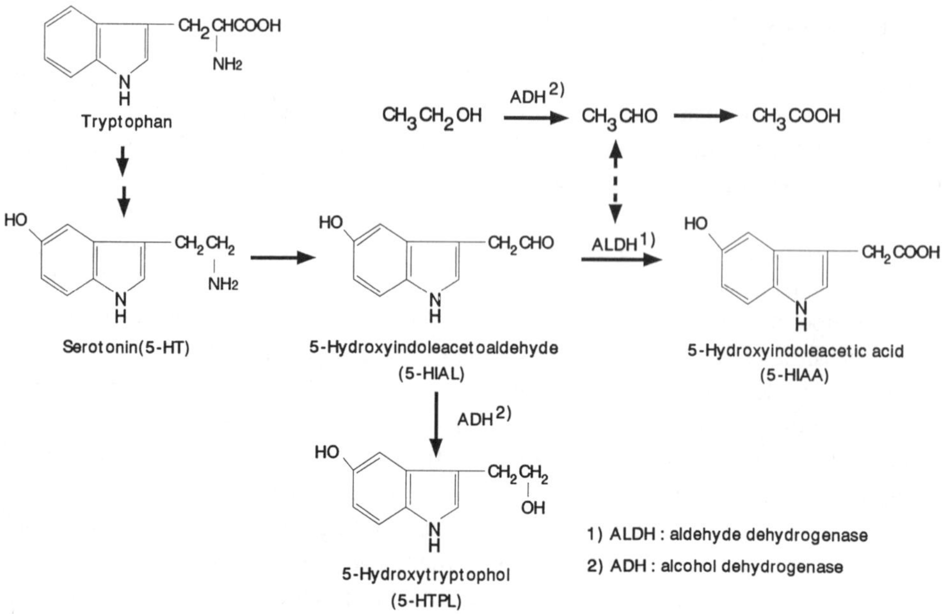

Figure 1. Inhibition of the metabolism of hydroxyindoleacetaldehyde derived from tryptophan by acetaldehyde.

healthy individuals lowers blood tryptophan and CSF tryptophan and decreased central 5-HT levels (LeMarquand *et al.*, 1994; Morgan and Badawy, 2001). Acute alcohol inhibits 5-HIAA transport from the brain by the choroid plexus, a mechanism that could account for the increase in brain 5-HIAA in rodents. This process may be responsible for the lower levels of CSF 5-HIAA found in alcoholics, particularly if it persists into abstinence (LeMarquand *et al.*, 1994; Tabakoff *et al.*, 1975). In this connection, acute ethanol may facilitate 5-HT reuptake in the hippocampus and decrease 5-HT$_{1A}$ receptor functioning in the cortex (Dillon *et al.*, 1991; LeMarquand *et al.*, 1994). On the other hand, chronic alcohol treatment increases 5-HT synthesis and turnover (Yamane *et al.*, 2003). Autoradiographic experiments after chronic alcohol treatment revealed the up-regulation of the somatodendritic 5-HT$_{1A}$ autoreceptors in the dorsal raphe, and the down-regulation of the postsynaptic 5-HT$_{1A}$ receptors in some projection areas (frontal cortex, entorhinal cortex and hippocampus) (Nevo *et al.*, 1995). The 5-HT$_{1A}$ agonist buspirone reduces self-report alcohol craving and drinking behavior in alcohol abusers (Tollefson *et al.*, 1991). In mammals, alcohol is metabolized into acetaldehyde by alcohol dehydrogenase, cytochromes P-450 2E1 (CYP2E1) and catalase, and then acetaldehyde is further metabolized into acetic acid by aldehyde dehydrogenase and CYP2E1 (Lieber, 1999). The most important enzymes catalyzing the conversion of alcohol to acetaldehyde and acetate are alcohol dehydrogenase and aldehyde dehydrogenase, respectively. Alcohol dehydrogenase was the only significant pathway for alcohol metabolism. The multiple forms of this cytosolic enzyme catalyze the conversion of alcohol to acetaldehyde, coupled with the reduction of NAD^+ to NADH. Moreover, alcohol had no inducing effect on hepatic alcohol dehydrogenase in either male or female rats (Aasmoe and Aarbakke, 1999). On the one hand, aldehyde dehydrogenase activity is detectable in practically all tissues (Karamanakos *et al.*, 2001; Pappas *et al.*, 1997), and acute alcohol intake inhibits the metabolism of other drugs by competition for biogenetic enzymes in liver or brain (Lieber, 1999). Contrasting with the acute effects of alcohol consumption, chronic alcohol intake increases the activity of the microsomal ethanol-oxidizing system (MEOS) with an associated rise in cytochrome P450, especially CYP2E1. The MEOS is distinct from alcohol dehydrogenase and catalase and dependent on cytochromes P450 (Lieber, 1999), and the alcohol-inducible form is designated as CYP2E1 (Nelson *et al.*, 1993).

LOCUS COERULEUS ADRENERGIC AND SEROTONERGIC NEURONS

The noradrenergic neurons originating in the locus coeruleus exert a tonic facilitatory influence on serotonergic neurons in the dorsal raphe nucleus via activation of α_1-adrenoceptors (Baraban and Aghajanian, 1980; Szabo *et al.*, 1999). The activity of serotonergic neurons in the locus coeruleus is implicated in behavioral activation, pain, noxious and stress stimuli, and modulation of cardiovascular activity (Bear *et al.*, 1996; Singewald *et al.*, 1997) and is modulated by 5-HT1A receptors located within the dorsal raphe nucleus (Kaehler *et al.*, 1999a; Singewald *et al.*, 1995, 1997). In this connection, the serotonergic system seems to provide inhibitory inputs to the principal noradrenergic nucleus, the locus coeruleus (Aston-Jones *et al.*, 1991; Goddard *et al.*, 1994). This is consistent with the finding that electrical stimulation of the dorsal raphe reduces locus coeruleus neuronal activity (Mateo *et al.*, 2000; Segal, 1979). On the other hand, lesioning of serotonergic neurons with a selective 5-HT neurotoxin has been shown to produce an elevation of firing rate of noradrenergic neurons (Haddjeri *et al.*, 1997; Szabo *et al.*, 1999). Moreover, there is the finding that stimulation of raphe 5-HT1A receptors reduces the firing rate of serotonergic neurons, 5-HT synthesis and 5-HT release in projection areas, such as the hippocampus (Hjorth and Magnusson, 1988; Sharp *et al.*, 1989; Sinton and Fallon, 1988; Sprouse and Aghajanian, 1986, 1987). Especially, in an area with serotonergic and noradrenergic nerve terminals such as the hippocampus, 5-HT release has been shown to be modulated by NA through presynaptic α_2-adrenergic heteroreceptors (Mongeau *et al.*, 1998). Thus, several lines of evidence indicate that the dorsal raphe is the target of noradrenergic projections from the locus coeruleus (Anderson *et al.*, 1977; Baraban and Aghajanian, 1981; Sakai *et al.*, 1977).

Actually, Mongeau *et al.* (1998) demonstrated that the inhibitory effect of milnacipran, called a 5-HT and NA reuptake inhibitor (SNRI), on serotonergic neuronal firing might be mediated via the noradrenergic system. This is confirmed by the finding that the suppressant effect of intravenous milnacipran was abolished by a 6-hydroxydopamine (6-OH-DA) lesion of noradrenergic neurons. Szabo *et al.* (1999) also demonstrated that a decreased firing activity of locus coeruleus neurons induced by long-term selective 5-HT reuptake inhibitor (SSRI) gives support to the contribution of the noradrenergic system in the therapeutic activity of these drugs. Long-term administration of paroxetine (SSRI), but not short-term treatment, significantly decreased firing activity (Szabo *et al.*, 1999). It appears that enhancing serotonergic neurotransmission by sustained SSRI administration leads to a reduction of the firing rate of noradrenergic neurons. The increased concentration of 5-HT at the somatodendritic level of noradrenergic neurons would regulate NA release in the area by acting on 5-HT receptors located at this level (Mateo *et al.*, 2000). This fits well with the inhibition of locus coeruleus firing activity induced by the 5-HT1A receptor antagonist WAY100635 through an increased serotonergic neural input to the locus coeruleus (Haddjeri *et al.*, 1997). Therefore, the stimulation of tryptophan treatment to locus coeruleus serotonergic neurons may lead to a transient decrease of noradrenergic function in rat locus coeruleus.

EXPERIMENTAL PARADIGM FOR MICRODIALYSIS STUDIES OF DRUG TREATMENT

Our recent studies have utilized a model produced by coadministration of tryptophan plus alcohol for a single and three days that has been well described (Hoshi *et al.*, 2000). In those investigations in which microdialysis has been used to measure 5-HIAA concentrations within the locus coeruleus, rats were first prepared by implantation of two guide canula to permit access into the CNS. A CMA/11 guide cannula (Bioanalytical Systems Inc., West Lafayette, IN) was stereotaxically implanted in the locus coeruleus -9.8 mm posterior to the bregma, 1.1 mm lateral, and 6.8 mm below the skull surface. A stainless steel guide cannula (outer diameter 0.35 mm, inner diameter 0.15 mm) with its stylet was stereotaxically inserted until the tip of the cannula was 2 mm above the dorsal raphe nucleus. Coordinates were (in mm) AP 7.8 posterior to bregma; L 0.0; V 5.0 from the skull surface, according to the atlas of Paxinos and Watson (1986). One week following stereotaxic surgery and the day before the beginning of the experiment, a freshly calibrated microdialysis probe was placed into the locus coeruleus guide cannula. The probe was then perfused with filtered Ringer's solution at a low rate (0.2 μl/min) overnight. On the morning of the following day, the flow rate was increased to 2 μl/min. After 2 h of equilibration, 3 or 4 samples were collected for determination of basal values. After administration of each drug, serial 15-min locus coeruleus dialysate samples were collected for analysis for 2–3 h. Baseline values for calculation of drug-induced changes in 5-HIAA levels were determined as the average of the 3–4 serial samples prior to drug challenge. Probes were dialyzed with Ringer's solution, and dialysate samples were analyzed by high-performance liquid chromatography (HPLC; BAS LC-4C Detector) with electrochemical detection using a Waters Spherisorb Column (150 × 4.6 mm; 5 μm ODS2) with electrochemical detection.

ACUTE AND SHORT-TERM EFFECTS OF COADMINISTRATION OF TRYPTOPHAN AND ETHANOL ON SEROTONIN METABOLITES IN THE LOCUS COERULEUS IN RATS

Fadda *et al.* (2000) reported that although a tryptophan-supplemented diet did not significantly change extraneuronal 5-HT content, the significant increase in extracellular 5-HIAA in the frontal cortex nonetheless suggested a stimulation of 5-HT metabolism in serotonergic neurons. In this connection, 5-HIAA, the principal metabolite of 5-HT, reflects central 5-HT turnover (Fernstrom and Wurtman, 1971; Stark and Scheich, 1997; Williams *et al.*, 1999). Moreover, when tryptophan is given to normal rats, there is a much greater accumulation of 5-HIAA than of 5-HT (Curzon *et al.*, 1978).

To investigate whether the tryptophan-induced release of 5-HT in the locus coeruleus is altered by ethanol, we examined alterations in 5-HT and its metabolites (5-HIAA and 5-hydroxytryptophol (5-HTPL)) in the locus coeruleus after administration of tryptophan in combination with ethanol to rats. Moreover, behavioral signs during treatment with each drug were also investigated (Hayashi *et al.*, 2003). These studies provided evidence to indicate that the 5-HIAA and 5-HTPL levels in the locus coeruleus after tryptophan alone or tryptophan plus ethanol is indicative of changes in the activity of serotonergic neurotransmission in locus coeruleus neurons.

We noted increases in the extracellular fluid levels of 5-HIAA in the locus coeruleus of rats which had been i.p. injected with tryptophan (50 mg/kg) alone. In our experiment, increases in 5-HIAA were observed only when the microdialysis probe was properly located in the locus coeruleus; no increase was seen in adjacent regions. This may indicate that the increased 5-HIAA levels in the locus coeruleus after tryptophan administration are due to an increase in the innervation of locus coeruleus serotonergic neurons (Kaehler *et al.*, 1999a, 1999c). The serotonergic afferents to the locus coeruleus originate for more than 50% from cell bodies located in the dorsal raphe nucleus (Morgane and Jacobs, 1979; Mateo *et al.*, 2000; Reneric *et al.*, 2002). In the present study, however, in spite of the damage to serotonergic neurons caused by dorsal raphe nucleus injection of 5,7-dihydroxytryptamine (5,7-DHT), which destroys presynaptic serotonergic nerve fibers (Kaehler *et al.*, 1999a), 5-HIAA levels in the locus coeruleus decreased by no more than about 35%. In addition, the levels of 5-HIAA in the locus coeruleus during 5,7-DHT infusion into the dorsal raphe nucleus after pretreatment with tryptophan were approximately 21% lower than those after tryptophan pretreatment alone. This indicates that about 70–80% of the increased 5-HIAA levels in the locus coeruleus seems to be from other distant serotonergic cell groups or be due to activation of locus coeruleus serotonergic neurons (Aston-Jones *et al.*, 1986; Luppi *et al.*, 1995; Maeda *et al.*, 1991).

Coadministration of tryptophan (50 mg/kg, i.p.) and alcohol (1.25 g/kg, i.p.) produced a delayed increase in 5-HIAA, which reached a peak level at 90 min in the locus coeruleus, and the dose-response curve for 5-HIAA after combined

administration was significantly shifted to the right. That is, there was a time
lag in the levels of 5-HIAA after tryptophan alone and after tryptophan plus
alcohol. This may imply that acute alcohol inhibits the metabolism of tryptophan by
competing for shared enzymes. The oxidative metabolite of alcohol, acetaldehyde,
may block the metabolism of 5-hydroxyindoleacetaldehyde (5-HIAL) to 5-HIAA
by competing with 5-HIAL for brain aldehyde dehydrogenase, which may lead
to an elevation of 5-HIAL in the brain (Fig. 1). Moreover, there is the finding
that concurrent administration of tryptophan and alcohol may increase not only
level of 5-HIAL itself but also that of 5-HTPL in the locus coeruleus (Fukumori
et al., 1981). Beck *et al.* (1980) also reported that the 5-HTPL level in the
cerebrospinal fluid of alcoholic inpatients is increased. In fact, in our study, 5-HTPL
levels in the locus coeruleus were significantly elevated by coadministration of
tryptophan and alcohol (Fig. 2). In addition, a significant increase in 5-HIAA level
induced by concurrent administration was also observed. This may be expained by
the fact that an increased 5-HIAA level affects not only aldehyde dehydrogenase
but also CYP2E1, which is capable of metabolizing 5-HIAL to 5-HIAA (Lieber,
1999). Thus the stimulation of 5-HT metabolism in locus coeruleus serotonergic
neurons by tryptophan was strengthened by the simultaneous administration of
alcohol. Moreover, the stimulation of tryptophan treatment to locus coeruleus
serotonergic neurons may lead to a transient decrease of noradrenergic function
in rat locus coeruleus. This is supported by the finding that the serotonergic system
seems to provide inhibitory inputs to the principal noradrenergic nucleus, the locus
coeruleus (Aston-Jones *et al.*, 1991; Goddard *et al.*, 1994). Actually, our studies

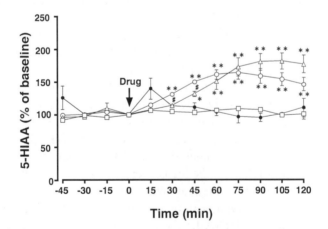

Figure 2. Extracellular fluid levels of 5-HIAA in the locus coeruleus induced by coadministration
of tryptophan and alcohol. Rats were given i.p. a single dose of tryptophan ($\bigcirc$, 50 mg/kg; $n = 5$),
alcohol ($\square$, 1.25 g/kg; $n = 5$) alone or tryptophan plus alcohol ($\triangle$, 50 mg/kg + 1.25 g/kg; $n = 5$),
respectively. The control animals received an equivalent volume of saline ($\bullet$, 50 mg/kg; $n = 4$).
Values are expressed as percentage change (mean $\pm$ S.E.M.) from basal values. $^*p < 0.05$ and
$^{**}p < 0.01$, control *vs.* tryptophan or alcohol plus tryptophan; $^\#p < 0.05$, tryptophan *vs.* alcohol
plus tryptophan.

have identified that the levels of NA in the locus coeruleus remain unchanged after treatment with tryptophan alone or tryptophan plus alcohol (Fig. 3). This result suggests that the doses used in this study may have been insufficient to cause a decrease in NA levels.

Our most recent studies examined even more specifically the participation of 5-HIAA levels within the locus coeruleus after a 3-day treatment with tryptophan (50 mg/kg) plus alcohol (1.25 g/kg). We noted increases in the extracellular fluid levels of 5-HIAA in the locus coeruleus of rats which had been i.p. injected twice daily for 3 consecutive days with tryptophan alone. This may imply that the increased 5-HIAA levels within the locus coeruleus are due to more increase in the innervation of locus coeruleus serotonergic neurons (Kaehler *et al.*, 1999a, c). Combined administration caused much more increase in 5-HIAA level in the locus coeruleus than those in tryptophan alone, but had no effect on 5-HTPL level. In addition, a time lag in the increased 5-HIAA levels between tryptophan alone and tryptophan plus alcohol for 3 consecutive days was not observed (Fig. 4). This may imply that aldehyde dehydrogenase induced by repeated administration of alcohol for 3 days accelerated the metabolism of 5-HIAL to 5-HIAA. Moreover, unchanged 5-HTPL level by combined treatment may be explained by the finding that brain CYP2E1 and/or alcohol dehydrogenase, which is capable of metabolizing 5-HIAL to 5-HTPL, were not induced sufficiently by alcohol for 3 consecutive days. In this study, a single administration of alcohol had no effect on the levels of 5-HIAA. The presence of alcohol in the blood at autopsy was related to decreased binding of the 5-HT1A agonist to 5-HT1A receptors in several cortical regions at the frontal-parietal level (Dillon *et al.*, 1991). Moreover, chronic alcohol treatment induced

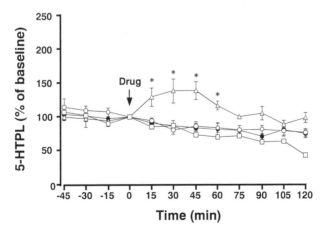

Figure 3. Extracellular fluid levels of 5-HTPL in the locus coeruleus induced by coadministration of tryptophan and alcohol. Rats were given i.p. a single dose of tryptophan (○, 50 mg/kg; $n = 4$), alcohol (□, 1.25 g/kg; $n = 4$) alone or tryptophan plus alcohol (△, 50 mg/kg + 1.25 g/kg; $n = 4$), respectively. The control animals received an equivalent volume of saline (●, 50 mg/kg; $n = 4$). Values are expressed as percentage change (mean ± S.E.M.) from basal values. $^*p < 0.05$ and $^{**}p < 0.01$, control *vs.* alcohol plus tryptophan; $^#p < 0.05$, tryptophan *vs.* alcohol plus tryptophan.

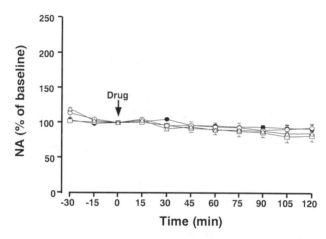

Figure 4. Extracellular fluid levels of NA in the locus coeruleus induced by coadministration of tryptophan and alcohol. Rats were given i.p. a single dose of tryptophan (○, 50 mg/kg; $n = 5$), alcohol (□, 1.25 g/kg; $n = 5$) alone or tryptophan plus alcohol (△, 50 mg/kg + 1.25 g/kg; $n = 6$), respectively. The control animals received an equivalent volume of saline (●, 50 mg/kg; $n = 6$). Values are expressed as percentage change (mean ± S.E.M.) from basal values.

a significant increase of 5-HT synthesis in descending serotonergic cell bodies, nigrostriatal structures, the hippocampus and cortices, but no significant changes were observed in the dorsal and median raphe nuclei (Yamane *et al.*, 2003). In our study, the doses of alcohol (1.25 g/kg) applied for 3 consecutive days had no effect on the levels of 5-HT and its metabolites, 5-HIAA as well as a single administration of it. It may reflect the lower concentration of alcohol used in this study that central 5-HIAA level was not altered by alcohol, despite the fact that alcohol changed 5-HIAA level in the periphery (Beck *et al.*, 1980; Fukumori *et al.*, 1981). These results could be explained by both the lower doses of alcohol and its short-term administration. In addition, this may imply that the unaltered 5-HIAA levels within the locus coeruleus after alcohol treatment are due to the decrease in the function of locus coeruleus serotonergic neurons via dorsal raphe 5-HT$_{1A}$ receptors (Singewald *et al.*, 1995, 1997).

Disulfiram is used in the treatment of chronic alcoholism. However, the disulfiram–alcohol reaction is characterized by nausea, vertigo and cardiovascular effects such as flushing of the face and neck, hypotension and tachycardia (Karamanakos *et al.*, 2001). Since disulfiram inhibits hepatic aldehyde dehydrogenase, the accumulation of acetaldehyde is suggested as the basic mechanism for the produced unpleasant symptoms (Richie, 1985). Aldehyde dehydrogenase activities are inhibited by disulfiram in a dose-dependent way and the lowest dose of disulfiram (25 mg/kg) produces no effect, either on dopamine or noradrenaline (Karamanakos *et al.*, 2001). In our study, the enhancement of 5-HIAA levels by combined administration of tryptophan plus alcohol was markedly suppressed by pretreatment with disulfiram (25 mg/kg, i.p.). This suggests that the inhibition of aldehyde dehydrogenase by disulfiram fails to increase the levels of 5-HIAA in the locus coeruleus. Therefore, it is conceivable that the stimulation of 5-HT metabolism in locus coeruleus serotonergic neurons by tryptophan is strengthened by the simultaneous administration of alcohol.

BEHAVIORAL SIGNS DURING TRYPTOPHAN ALONE AND IN COMBINATION WITH ETHANOL

Neurotransmission in locus coeruleus neurons is closely related to behavioral signs (Aghajanian *et al.*, 1994; Hoshi *et al.*, 1997; Rasmussen *et al.*, 1990). A deficiency in 5-HT synthesis and metabolism has been implicated in several neuropsychiatric disorders, including alcoholism, aggression (LeMarquand *et al.*, 1994) and depression (Meltzer, 1990). In laboratory animals, ingestion of a tryptophan-free diet alters behavioral indices of serotonergic function, increasing pain sensitivity, acoustic startle, motor activity, and aggression and reducing rapid eye movement sleep (Schaechter and Wurtman, 1989; Stancampiano *et al.*, 1997). A deficiency in either tryptophan intake or its metabolism could result in behavioral disorders (Tiihonen *et al.*, 2001). These behavioral changes are reversed once brain tryptophan is restored (Schaechter and Wurtman, 1989). Therefore, change in 5-HT metabolites in the locus coeruleus may play an important role in mediating the changes in behavioral activity (Beck *et al.*, 1980; Fukumori *et al.*, 1981). We noted increases in the extracellular fluid levels of 5-HIAA in the locus coeruleus of rats which had been i.p. injected with tryptophan alone. Thus, tryptophan stimulation may activate serotonergic fibers which influence 5-HT release in the locus coeruleus (Kaehler *et al.*, 1999b), and the spontaneous release of 5-HT in the locus coeruleus may depend in part on the activity of serotonergic neurons of the dorsal raphe nucleus. Morgan and Badawy (2001) showed that acute alcohol consumption by human volunteers decreases circulating tryptophan concentration and availability to the brain. In our experiment, the elevation of 5-HIAA and 5-HTPL in the locus coeruleus after acute tryptophan plus alcohol treatment resulted in an increase in teeth-chattering. The levels of both 5-HTPL and 5-HIAA increased markedly 15–30 min after substance administration and were sustained for about 1.0–1.5 h. Conversely, sniffing and rearing scores after tryptophan plus alcohol were lower than those after tryptophan alone (Table 1). On the other hand, the elevation of 5-HIAA in the locus coeruleus after repeated coadministration of tryptophan and alcohol for 3 days resulted in an increase in teeth-chattering and locomotion. However, the enhancement of behavioral signs in animals was perfectly suppressed by pretreatment with disulfiram. Conversely, salivation score in combination with combined treatment and disulfiram was lower than that after combined treatment. The alterations in neurotransmitter levels and behavioral signs after tryptophan alone or in combination with alcohol showed virtually identical time courses. Thus the stimulation of 5-HT metabolism in locus coeruleus serotonergic neurons resulted from the simultaneous administration of tryptophan and alcohol.

Table 1.
Acute effect of simultaneous administration of tryptophan and ethanol on behavior in rats

Withdrawal signs	Control	Tryptophan	25% EtOH	25% EtOH + tryptophan
Teeth-chattering	19.5 ± 6.2	20.8 ± 6.3	18.8 ± 5.6	$42.8 \pm 3.8^{a,c,d}$
Wet-dog shakes	1.8 ± 1.0	2.5 ± 2.2	1.8 ± 1.4	1.3 ± 0.6
Penis-licking	3/4	4/5	2/5	3/5
Stretching	0.8 ± 0.5	0.5 ± 0.3	0.4 ± 0.2	0.8 ± 0.8
Locomotion	8.8 ± 3.7	14.8 ± 4.0	10.6 ± 2.6	7.6 ± 4.6
Scratching	8.8 ± 7.4	0.0 ± 0.0	3.3 ± 3.3	6.5 ± 6.2
Sniffing	1.3 ± 1.0	$12.6 \pm 3.6^{b,e}$	3.8 ± 1.1	1.6 ± 0.5^{d}
Rearing	1.5 ± 0.9	9.8 ± 3.7^{e}	1.2 ± 0.5	1.2 ± 0.6^{d}
Salivation	2.0 ± 0.4	2.3 ± 0.6	2.8 ± 0.5	1.3 ± 0.8
Abnormal pasturing	0.0 ± 0.0	5.6 ± 2.8	2.8 ± 1.9	0.8 ± 0.4

[a] $p < 0.05$. Control *vs.* 25% EtOH + tryptophan.
[b] $p < 0.05$. Control *vs.* tryptophan.
[c] $p < 0.01$. 25% EtOH *vs.* EtOH + tryptophan.
[d] $p < 0.05$. Tryptophan *vs.* 25% EtOH + tryptophan.
[e] $p < 0.05$. Tryptophan *vs.* 25% EtOH.

Our experimental results may be explained by the finding that alterations in 5-HT turnover rate or serotonergic activity (firing rates) are responsible for behavioral signs. This is supported by clinical evidence showing that several neuro-psychiatric disorders such as stress and aggression seen after simultaneous dietary tryptophan augmentation and acute alcohol intake in healthy individuals may be caused by active metabolites of 5-HT (Fadda *et al.*, 2000; LeMarquand *et al.*, 1994; Meltzer, 1990; Williams *et al.*, 1999).

CONCLUSIONS

Examination of the investigation that has been reviewed leads to the conclusion that the increased levels of 5-HIAA in the locus coeruleus in response to tryptophan alone are potentiated by administration of alcohol. There was a time lag in the levels of 5-HIAA after acute tryptophan alone and after acute tryptophan plus alcohol. This may suggest that aldehyde dehydrogenase is inhibited by acute alcohol treatment. Acute coadministration of tryptophan and alcohol may increase the levels of 5-HIAL and 5-HTPL in the locus coeruleus. This suggests that acetaldehyde may block the metabolism of 5-HIAL to 5-HIAA by competing with 5-HIAL for aldehyde dehydrogenase, which may lead to an elevation of 5-HIAL in the locus coeruleus. On the other hand, combined administration for three consecutive days may increase only 5-HIAA level because of alcohol-induced aldehyde dehydrogenase, which did not cause a time lag in the increased 5-HIAA levels between tryptophan alone and tryptophan plus alcohol. In addition, the levels of these alterations may be partly responsible for behavioral activation. This strongly suggests the participation of serotonergic neurons as a general phenomenon in the biochemistry of an essential amino acid, tryptophan. Although tryptophan has been suggested to directly and/or indirectly affect the status of serotonergic neurotransmission (LeMarquand *et al.*, 1994), alcohol differs markedly from tryptophan in that it acts directly on the serotonergic neurons to alter endogenous 5-HT level. Brain tryptophan availability controls the rate at which neurons synthesize and release 5-HT. The coadministration of tryptophan diet with alcohol elicits higher 5-HT levels than tryptophan or alcohol alone. This should indicate that novel pharmacological strategies may be exploited to prevent an/or manage the psychiatric disorder at alcohol intake with tryptophan diet. The specific nature, location and synaptic function of serotonergic neurons within the locus coeruleus remain an area for fruitful research and will undoubtedly provide new facets to the study of tryptophan with respect to the role of the noradrenergic neurons in regulation of serotonergic neurons.

REFERENCES

Aasmoe, L. and Aarbakke, J. (1999). Sex-dependent induction of alcohol dehydrogenase activity in rats, *Biochem. Pharmacol.* **57**, 1067–1072.

Aghajanian, G. K., Kogan, J. H. and Moghaddam, B. (1994). Opiate withdrawal increases glutamate and aspartate efflux in the locus coeruleus: an in vivo microdialysis study, *Brain Res.* **636**, 126–130.

Aizenstein, M. L. and Korf, J. (1979). On the elimination of centrally formed 5-hydroxyindoleacetic acid by cerebrospinal fluid and urine, *J. Neurochem.* **32**, 1227–1233.

Anderson, C., Pasquier, D., Forbes, W., *et al.* (1977). Locus coeruleus-to-dorsal raphe input examined by electrophysiological and morphological methods, *Brain Res. Bull.* **2**, 209–221.

Aston-Jones, G., Ennis, M., Pieribone, V. A., *et al.* (1986). The brain nucleus locus coeruleus: restricted afferent control of a broad efferent network, *Science* **234**, 734–737.

Aston-Jones, G., Shipley, M. T., Chouvet, G., *et al.* (1991). Afferent regulation of locus coeruleus neurons: anatomy, physiology and pharmacology, *Prog. Brain Res.* **88**, 47–75.

Banki, C. M. and Molnar, G. (1981). Cerebrospinal fluid 5-hydroxyindoleacetic acid as an index of central serotonergic processes, *Psychiatry Res.* **5**, 23–32.

Baraban, J. M. and Aghajanian, G. K. (1980). Suppression of firing activity of 5-HT neurons in the dorsal raphe by alpha-adrenoceptor antagonists, *Neuropharmacology* **19**, 355–363.

Baraban, J. M. and Aghajanian, G. K. (1981). Suppression of serotonergic neuronal firing by α-adrenoceptor antagonists: evidence against GABA mediation, *Eur. J. Pharmacol.* **66**, 287–294.

Bear, M. F., Connors, B. W. and Paradiso, M. A. (1996). In: *Neuroscience: Exploring the Brain*, Bear M. F., *et al.* (Eds), pp. 404–430. Williams and Wilkins, Baltimore, MD, USA.

Beck, O., Borg, S., Holmsted, B., *et al.* (1980). Levels of 5-hydroxytryptophol in cerebrospinal fluid from alcoholics determined by gas chromatography-mass spectrometry, *Biochem. Pharmacol.* **29**, 693–696.

Bel, N. and Artigas, F. (1996). In vivo effects of the simultaneous blockade of serotonin and norepinephrine transporters on serotonergic function. Microdialysis studies, *J. Pharmacol. Exp. Ther.* **278**, 1064–1072.

Blier, P., de Montigny, C. and Chaput, Y. (1990). A role for the serotonin system in the mechanism of action of antidepressant treatments: Preclinical evidence, *J. Clin. Psychiatry* **51** (suppl.), 14–20, discussion 21.

Caldecott, H. S., Morgan, D. G., Deleon, J. F., *et al.* (1991). Clinical and biochemical aspects of depressive disorders: II. Transmitter/receptor theories, *Synapse* **9**, 251–301.

Cedarbaum, J. M. and Aghajanian, G. K. (1978). Afferent projections to the rat locus coeruleus as determined by retrograde tracing technique, *J. Comp. Neurol.* **178**, 1–16.

Curzon, G. (1990). Serotonin and appetite, *Ann. NY Acad. Sci.* **600**, 521–531.

Curzon, G., Fernando, J. and Marsden, C. (1978). 5-Hydroxytryptamine: the effects of impaired synthesis on its metabolism and release in rat, *Br. J. Pharmacol.* **63**, 627–634.

Delgado, P. L., Miller, H. L., Salomon, R. M., *et al.* (1993). Monoamines and the mechanism of antidepressant action: effects of catecholamine depletion on mood of patients treated with antidepressants, *Psychopharmacol. Bull.* **29**, 389–396.

Dillon, K. A., Gross-Isseroff, R., Israeli, M., *et al.* (1991). Autoradiographic analysis of serotonin 5-HT1A receptor binding in the human brain postmortem: Effects of age and alcohol, *Brain Res.* **554**, 56–64.

Fadda, F., Cocco, S., Rossetti, Z. L., *et al.* (2000). A tryptophan-free diet markedly reduces frontocortical 5-HT release, but fails to modify ethanol preference in alcohol-preferring (sP) and non-preferring (sNP) rats, *Behavioural Brain Res.* **108**, 127–132.

Fernstrom, J. and Wurtman, R. (1971). Brain serotonin content: physiological dependence on plasma tryptophan levels, *Science* **173**, 149–151.

Fukumori, R., Minegishi, A., Satoh, T., *et al.* (1981). Changes in seizure susceptibility after successive treatments of mice with tryptophol and ethanol, *J. Pharm. Pharmacol.* **33**, 586–589.

Goddard, A. W., Sholomskas, D. E., Walton, K. E., *et al.* (1994). Effects of tryptophan depletion in panic disorder, *Biol. Psychiatry* **36**, 775–777.

Gwaltney-Brant, S. M., Albretsen, J. C. and Khan, S. A. (2000). 5-Hydroxytryptophan toxicosis in dogs: 21 cases (1989–1999), *J. Am. Vet. Med. Assoc.* **216**, 1937–1940.

Haddjeri, N., De Montigny, C. and Blier, P. (1997). Modulation of the firing activity of noradrenergic neurones in the rat locus coeruleus by the 5-hydroxytryptamine system, *Br. J. Pharmacol.* **120**, 865–875.

Hayashi, M., Nakai, T., Bandoh, T., *et al.* (2003). Acute effect of simultaneous administration of tryptophan and ethanol on serotonin metabolites in the locus coeruleus in rats, *Eur. J. Pharmacol.* **462**, 61–66.

Hjorth, S. and Magnusson, T. (1988). The 5-HT1A receptor agonist, 8-OH-DPAT, preferentially activates cell body 5-HT autoreceptors in rat brain in vivo, *Naunyn-Schmiedeberg's Arch. Pharmacol.* **338**, 463–471.

Hoshi, K., Ma, T., Oh, S., *et al.* (1997). Increased release of excitatory amino acids in rat locus coeruleus in κ-opioid agonist dependent rats precipitated by nor-binaltorphimine, *Brain Res.* **753**, 63–68.

Hoshi, K., Yamamoto, A., Ishizuki, S., *et al.* (2000). Excitatory amino acid release in the locus coeruleus during naloxone-precipitated morphine withdrawal in adjuvant arthritic rats, *Inflamm. Res.* **49**, 36–41.

Iijima, K. (1993). Chemocytoarchitecture of the rat locus coeruleus, *Histol. Histopathol.* **8**, 581–591.

Insel, T. R., Zohar, J., Benkelfat, C., *et al.* (1990). Serotonin in obsessions, compulsions, and the control of aggressive impulses, *Ann. NY Acad. Sci.* **600**, 574–586.

Kaehler, S. T., Singewald, N. and Philippu, A. (1999a). Dependence of serotonin release in the locus coeruleus on dorsal raphe neuronal activity, *Naunyn-Schmiedeberg's Arch. Pharmacol.* **359**, 386–393.

Kaehler, S. T., Singewald, N. and Philippu, A. (1999b). Release of serotonin in the locus coeruleus of normotensive and spontaneously hypertensive rats (SHR), *Naunyn-Schmiedeberg's Arch. Pharmacol.* **359**, 460–465.

Kaehler, S. T., Sinner, C., Chatterjee, S. S., *et al.* (1999c). Hyperforin enhances the extracellular concentrations of catecholamines, serotonin and glutamate in the rat locus coeruleus, *Neurosci. Lett.* **262**, 199–202.

Karamanakos, P. N., Pappas, P., Stephanou, P., *et al.* (2001). Differentiation of disulfiram effects on central catecholamines and hepatic ethanol metabolism, *Pharmacol. Toxicol.* **88**, 106–110.

LeMarquand, D., Pihl, R. and Benkelfat, C. (1994). Serotonin and alcohol intake, abuse, and dependence: clinical evidence, *Biol. Psychiatry* **36**, 326–337.

Lieber, C. S. (1999). Microsomal ethanol-oxidizing system (MEOS): The first 30 years (1968–1998) — A review, *Alcohol Clin. Exp. Res.* **23**, 991–1007.

Linnoila, M., Virkkunen, M., Scheinin, M., *et al.* (1983). Low cerebrospinal fluid 5-hydroxy-indoleacetic acid concentration differentiates impulsive from nonimpulsive violent behavior, *Life Sci.* **33**, 2609–2614.

Luppi, P. H., Aston-Jones, G., Akaoka, H., *et al.* (1995). Afferent projections to the rat locus coeruleus demonstrated by retrograde and anterograde tracing with choleratoxin B subunit and phaseolus vulgaris leukoagglutinin, *Neuroscience* **65**, 119–160.

Maeda, T., Kojima, Y., Arai, R., *et al.* (1991). Monoaminergic interaction in the central nervous system: a morphological analysis in the locus coeruleus of the rat, *Comp. Biochem. Physiol.* **C98**, 193–202.

Mann, J. J., Arango, V. and Underwood, M. D. (1990). Serotonin and suicidal behavior, *Ann. NY Acad. Sci.* **600**, 476–485.

Mateo, Y., Ruiz-Ortega, J. A., Pineda, J., *et al.* (2000). Inhibition of 5-hydroxytryptamine reuptake by the antidepressant citalopram in the locus coeruleus modulates the rat brain noradrenergic transmission in vivo, *Neuropharmacol.* **39**, 2036–2043.

Meltzer, H. Y. (1990). Role of serotonin in depression, *Ann. NY Acad. Sci.* **600**, 486–500.

Mills, K. C. (1997). Serotonin syndrome. A clinical update, *Crit. Care Clin.* **13**, 763–783.

Moja, E., Cipolla, P., Castoldi, D., *et al.* (1989). Dose-response decrease in plasma tryptophan and in brain tryptophan and serotonin after tryptophan-free amino acid mixtures in rats, *Life Sci.* **44**, 971–976.

Mongeau, R., Michel, W., Montigny, C., *et al.* (1998). Effect of acute, short- and long-term milnacipran administration on rat locus coeruleus noradrenergic and dorsal raphe serotonergic neurons, *Neuropharmacology* **37**, 905–918.

Morgane, P. J. and Jacobs, M. S. (1979). Raphe projections to the locus coeruleus in the rat, *Brain Res. Bull.* **4**, 519–534.

Morgan, C. J. and Badawy, A. A.-B. (2001). Alcohol-induced euphoria: exclusion of serotonin, *Alcohol. Alcoholism.* **36**, 22–25.

Nelson, D. R., Kamataki, T., Waxman, D. J., *et al.* (1993). The P-450 superfamily: Update on new sequences, gene mapping, accession numbers, early trivial names of enzymes, and nomenclature, *DNA Cell Biol.* **12**, 1–51.

Nevo, I., Langlois, X., Laporte, A.-M., *et al.* (1995). Chronic alcoholization alters the expression of 5-HT1A and 5-HT1B receptor subtypes in rat brain, *Eur. J. Pharmacol.* **281**, 229–239.

Pappas, P., Stephanou, P., Vasiliou, V., *et al.* (1997). Ontogenesis and expression of ALDH activity in the skin and the eye of the rat, *Adv. Exp. Med. Biol.* **414**, 73–80.

Paxinos, G. and Watson, C. (1986). *The Rat Brain in Stereotaxic Coordinates*, 2nd edn. Academic Press, Orlando, FL, USA.

Rasmussen, K., Beitner-Johnson, D. B., Krystal, J. H., *et al.* (1990). Opiate withdrawal and the rat locus coeruleus: Behavioral, electrophysiological, and biochemical correlates, *J. Neurosci.* **10**, 2308–2317.

Reneric, J.-P., Bouvard, M. and Stinus, L. (2002). In the rat forced swimming test, NA-system mediated interactions may prevent the 5-HT properties of some subacute antidepressant treatments being expressed, *Eur. Neuropsychopharmacol.* **12**, 159–171.

Richie, J. M. (1985). The aliphatic alcohols, in: *The Pharmacological Basis of Therapeutics*, Gilman, A. G., Goodman, L. S., Rall, T. W. and Murad, F. (Eds), pp. 372–386. McMillan, New York, USA.

Roy, A., Virkkunen, M. and Linnoila, M. (1990). Serotonin in suicide, violence, and alcoholism, in: *Serotonin in Major Psychiatric Disorders*, Coccaro, E. F. and Murphy, D. L. (Eds), pp. 187–208. American Psychiatric Press, Washington, DC, USA.

Sakai, K., Salvert, D., Touret, M., *et al.* (1977). Afferent connections of the raphe dorsalis in the cat as visualized by the horseradish peroxidase technique, *Brain Res.* **137**, 11–35.

Schaechter, J. D. and Wurtman, R. J. (1989). Tryptophan availability modulates serotonin release from rat hypothalamic slices, *J. Neurochem.* **53**, 1925–1933.

Segal, M. (1979). Serotonergic innervation of the locus coeruleus from the dorsal raphe and its action on responses to noxious stimuli, *J. Physiology* **286**, 401–415.

Sharp, T., Bramwell, S. R. and Grahame-Smith, D. G. (1989). 5-HT$_1$ agonists reduce 5-hydroxy-tryptamine release in rat hippocampus in vivo as determined by brain microdialysis, *Br. J. Pharmacol.* **96**, 283–290.

Singewald, N., Guo, L., Schneider, C., *et al.* (1995). Serotonin outflow in the hypothalamus of conscious rats: origin and possible involvement in cardiovascular control, *Eur. J. Pharmacol.* **294**, 787–793.

Singewald, N., Kaehler, S., Hemeida, R., *et al.* (1997). Release of serotonin in the rat locus coeruleus: effects of cardiovascular, stressful and noxious stimuli, *Eur. J. Neurosci.* **9**, 556–562.

Sinton, C. M. and Fallon, S. L. (1988). Electrophysiological evidence for a functional differentiation between subtypes of the 5-HT1 receptor, *Eur. J. Pharmacol.* **157**, 173–181.

Sprouse, J. S. and Aghajanian, G. K. (1986). Propranolol blocks the inhibition of serotonergic dorsal raphe cell firing by 5-HT1A selective agonists, *Eur. J. Pharmacol.* **128**, 295–298.

Sprouse, J. S. and Aghajanian, G. K. (1987). Electrophysiological response of serotonergic dorsal raphe neurons to 5-HT$_{1A}$ and 5-HT$_{1B}$ agonist, *Synapse* **1**, 3–9.

Stancampiano, R., Melis, F., Sarais, L., *et al.* (1997). Acute administration of a tryptophan-free amino acid mixture decreases 5-HT release in rat hippocampus in vivo, *Am. J. Physiol.* **272**, R991–R994.

Stanley, M., Traskman-Bendz, L. and Dorovini-Zis, K. (1985). Correlations between aminergic metabolites simultaneously obtained from human CSF and brain, *Life Sci.* **37**, 1279–1286.

Stark, H. and Scheich, H. (1997). Dopaminergic and serotonergic neurotransmission systems are differentially involved in auditory cortex learning: A long-term microdialysis study of metabolites, *J. Neurochem.* **68**, 691–697.

Szabo, S. T., Demontigny, C. and Blier, P. (1999). Modulation of noradrenergic neuronal firing by selective serotonin reuptake blockers, *Br. J. Pharmacol.* **126**, 568–571.

Tabakoff, B., Ritzmann, R. F. and Boggan, W. O. (1975). Inhibition of the transport of 5-hydroxy-indoleacetic acid from brain by ethanol, *J. Neurochem.* **24**, 1043–1051.

Tiihonen, J., Virkkunen, M., Rasanen, P., *et al.* (2001). Free L-tryptophan plasma levels in antisocial violent offenders, *Psychopharmacology* **157**, 395–400.

Tollefson, G. D., Lancaster, S. P. and Montague-Clouse, J. (1991). The association of buspirone and its metabolite 1-pyrimidinylpiperazine in the remission of comorbid anxiety with depressive features and alcohol dependency, *Psychopharmacol. Bull.* **27**, 163–170.

Vertes, R. P. and Kocsis, B. (1994). Projections of the dorsal raphe nucleus to the brainstem: PHA-L analysis in the rat, *J. Comp. Neurol.* **340**, 11–26.

Williams, W. A., Shoaf, S. E., Hommer, D., *et al.* (1999). Effects of acute tryptophan depletion on plasma and cerebrospinal fluid tryptophan and 5-hydroxyindoleacetic acid in normal volunteers, *J. Neurochem.* **72**, 1641–1647.

Yamane, F., Tohyama, Y. and Diksic, M. (2003). Continuous ethanol administration influences rat brain 5-hydroxytryptamine synthesis non-uniformly: α-[^{14}C]methyl-L-tryptophan autoradiographic measurements, *Alcohol. Alcoholism.* **38**, 115–120.

Youdim, M. B. H. and Ashkenazi, R. (1982). Regulation of 5-HT catabolism, in: *Serotonin in Biological, Psychiatry*, Ho, B. T., *et al.* (Eds). Raven Press, NY, USA.

Advances in Neuroregulation and Neuroprotection (2005), pp. 347-357
Collin, C. *et al.* (Eds)
© VSP 2005

Perospirone hydrochloride: the novel atypical antipsychotic agent with high affinities for 5-HT$_2$, D$_2$ and 5-HT$_{1A}$ receptors

TADASHI ISHIBASHI * and YUKIHIRO OHNO [†]

*Discovery Research Laboratories I, Research Division, Sumitomo Pharmaceuticals Co. Ltd.,
3-1-98, Kasugade-naka, Konohana-ku, Osaka 554-0022, Japan*

Abstract—We reviewed the pharmacological properties of perospirone hydrochloride (perospirone), a novel atypical antipsychotic agent, and discussed the potential role of 5-HT$_2$ and 5-HT$_{1A}$ receptors in its action for the treatment of schizophrenia. Perospirone shows potent 5-HT$_2$ and D$_2$ receptor blocking activities in various animal models *in vivo*. Unlike other serotonin and dopamine antagonists (SDA), such as risperidone and olanzapine, perospirone exhibits a high affinity for 5-HT$_{1A}$ receptors, and acts as a partial agonist. Perospirone has efficacies not only for classical animal models for schizophrenia, but also in the models of negative symptoms and mood disorders, where the typical antipsychotics were inactive. In addition, perospirone has a weaker propensity to induce extrapyramidal side effects, such as catalepsy and bradykinesia in rodents, compared with haloperidol and risperidone. Depressant actions of the central nervous system (e.g. potentiation of anesthesia and inhibition of motor coordination) and a lower propensity to induce orthostatic hypotension seemed to be low. Our studies revealed that serotonergic mechanisms of perospirone (i.e. 5-HT$_2$ receptor and 5-HT$_{1A}$ receptor action) are involved in its atypical properties such as weaker propensities to induce extrapyramidal side-effects or improvement of mood disturbance. These findings suggest that perospirone is a new type of antipsychotic agent with a broad therapeutic spectrum and fewer side effects.

Keywords: Antipsychotic agents; perospirone; 5-HT$_2$ receptors; D$_2$ receptors; 5-HT$_{1A}$ receptors; extrapyramidal side effects.

---

*To whom correspondence should be addressed. E-mail: isibasi@sumitomopharm.co.jp

[†]Present address; Product Development Division, Sumitomo Pharmaceuticals Co. Ltd., 2-8, Doshomachi 2-chome, Chuo-ku, Osaka 541-8510, Japan.

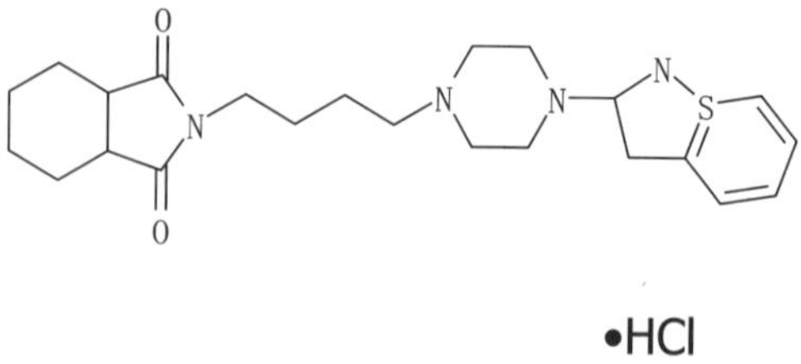

•HCl

Figure 1. Chemical structure of perospirone hydrochloride.

INTRODUCTION

Schizophrenia is a heterogeneous disorder with diverse symptoms, such as positive symptoms (e.g. hallucination and delusion), negative symptoms (e.g. apathy, social and emotional withdrawal) and mood disturbances (e.g. anxiety and depression). Since the discovery of the efficacy of chlorpromazine for schizophrenia in the early 1950s, a number of dopamine D_2 receptor antagonists have been developed as antipsychotic agents, based on the 'dopamine hypothesis' that the overexcitation of dopaminergic neurons is involved in the etiology of schizophrenia, especially positive symptoms. However, dopamine antagonists are ineffective against negative symptoms and mood disturbances, and frequently induce extrapyramidal side effects (EPS) like parkinsonisms, akathesia and tardive dyskinesia (TD). On the other hand, there is accumulating evidence that suggests that the blockade of central $5\text{-}HT_2$ receptors can provide beneficial effects in the treatment of schizophrenia by ameliorating negative symptoms and antipsychotic-induced EPS (Meltzer and Nash, 1991). Based on this idea, several $5\text{-}HT_2$ and D_2 antagonists are being developed as SDA-type antipsychotic agents.

Perospirone (Fig. 1) is a novel succimide derivative and has been launched as a new SDA-type antipsychotic agent in Japan. Perospirone has a high affinity for D_2 and $5\text{-}HT_2$ receptors but, unlike other SDA-type antipsychotics, it also has a high affinity for $5\text{-}HT_{1A}$ receptors. In this article, we review the pharmacological profiles of perospirone in comparison with other SDA and typical antipsychotics, and discuss the potential role of $5\text{-}HT_2$ and $5\text{-}HT_{1A}$ receptors its action for the treatment of schizophrenia.

PHARMACOLOGICAL PROFILES OF PEROSPIRONE: COMPARISON WITH OTHER ATYPICAL AND TYPICAL ANTIPSYCHOTICS

Receptor binding studies

Perospirone binds with high affinities to dopamine D_2 and $5\text{-}HT_2$ receptors. The Ki values for D_2 and $5\text{-}HT_2$ receptors were 1.4 nM and 0.61 nM, respectively, in radioligand binding assay in rat brain membrane preparation (Hirose *et al.*, 1990). The relative potency ratio of the $5\text{-}HT_2$ binding affinity to the D_2 binding affinity ($5\text{-}HT_2$ selectivity) of perospirone was 2.3. Other atypical antipsychotics,

Table 1.
Receptor binding affinities of perospirone and other antipsychotics in rat synaptic membrane

	Perospirone	Risperidone	Olanzapine	Haloperidol
D_2 receptors[a]	1.4	3.7	14	1.8
5-HT$_2$ receptors[a]	0.61	0.66	5.8	120
	(2.3)	(5.6)	(2.4)	(0.015)
5-HT$_{1A}$ receptors[a]	2.9	270[b]	>1000[c]	>1000[b]
	(0.48)	(0.014)	(<0.014)	(<0.0018)

Values in parentheses show the relative potency ratio of Ki value for D_2 receptors to that for 5-HT$_2$ receptors (5-HT$_2$ selectivity) or that for 5-HT$_{1A}$ receptors (5-HT$_{1A}$ selectivity).
[a] Ki values (nM).
[b] Leysen *et al.*, 1992.
[c] Bymaster *et al.*, 1996.

risperidone and olanzapine, also have high affinities for 5-HT$_2$ receptors, and the 5-HT$_2$ selectivity of these agents is similar to that of perospirone. In contrast, typical antipsychotic haloperidol shows a much lower affinity for 5-HT$_2$ receptors (Ki = 120 nM), and the order of 5-HT$_2$ selectivity of these antipsychotics is as follows, risperidone > perospirone = olanzapine ≫ haloperidol (Table 1).

Unlike risperidone or olanzapine, perospirone also has a high affinity for 5-HT$_{1A}$ receptors (Kato *et al.*, 1990; Yabuuchi *et al.*, 1998). The Ki value of perospirone for 5-HT$_{1A}$ receptors was 2.9 nM in rat hippocampal membrane preparation. In CHO cells expressing human 5-HT$_{1A}$ receptors, perospirone showed a high affinity (Ki = 0.72 nM), and exhibited partial agonistic efficacy (Emax 40% of 10 μM 5-HT) (Yabuuchi *et al.*, 1998). Risperidone, olanzapine and haloperidol have only weak affinities for rat 5-HT$_{1A}$ receptors (270, > 1000 and > 1000 nM, respectively). The order of the relative potency ratios of these agents to 5-HT$_{1A}$ receptors are as follows: perospirone ≫ risperidone > olanzapine = haloperidol. In addition, a recent electrophysiological study demonstrated that perospirone activates 5-HT$_{1A}$ receptors to induce hyperpolarization in the raphe nucleus neurons *in vitro* (Shiwa *et al.*, 2003). Therefore, the 5-HT$_{1A}$ agonistic action of perospirone is a unique property distinguishable from that of other atypical antipsychotics, and this property could have a significant role in the treatment of schizophrenia (Millan, 2000; discussed below).

Animal studies

Antipsychotic actions. Table 2 summarizes the *in vivo* pharmacological actions of perospirone in comparison with other antipsychotics. Perospirone inhibited various dopaminergic behaviors (e.g. methamphetamine-induced hyperactivity and apomorphine-induced stereotypy or climbing behavior) in rodents (Hirose *et al.*, 1990; Tokuda *et al.*, 1997). These D_2 blocking actions are slightly weaker than those of risperidone and haloperidol and are similar to those of olanzapine (Table 2). Perospirone also inhibited the rat conditioned avoidance response with about one-

Table 2.
Comparison of the pharmacological actions of perospirone with other antipsychotic agents

Pharmacological actions	Atypical antipsychotics			Typical antipsychotics
	Perospirone	Risperidone	Olanzapine	Haloperidol
D_2 blocking actions				
MAP-induced hyperactivity (rats)[b]	2.2	1.1	3.3	0.56
APO-induced stereotypy (rats)[b]	5.8	11	5.1	2.0
APO-induced climbing behavior (mice)[b]	3.5	0.17	1.1	0.67
$5\text{-}HT_2$ blocking actions				
Tryptamine-induced clonic seizures (rats)[b]	1.4	0.20	1.4	14
p-CAMP-induced hyperthermia (rats)[b]	1.8	0.098	0.62	>30
EPS liability				
Catalepsy induction (mice)[c]	57 (16)	0.85 (5.0)	>10 (>9)[d]	3.1 (4.6)
Bradykinesia (mice)[e]	100 (29)	3 (18)	10 (9.1)	1 (1.5)
General behavior				
Inhibition of motor coordination (mice)	34	1.1	5.2	2.7
Potentiation of HEX-induced anesthesia (mice)	37	0.55	8.3	11

MAP; methamphetamine, APO; apomorphine, p-CAMP; p-chloroamphetamine, EPS; extrapyramidal side effects, HEX; hexobarbital, numbers in parentheses show the potency ratio of D_2-blocking action to EPS induction (ED_{50} for catalepsy induction/ID_{50} for APO-climbing behavior).
[a] Ki values (nM).
[b] ID_{50} values (mg/kg, p.o.).
[c] ED_{50} values (mg/kg, p.o.).
[d] Unable to be tested due to the prominent muscle relaxation at higher doses.
[e] Minimal effective doses (mg/kg, p.o.).

fourth the potency of haloperidol (Tokuda *et al.*, 1997). These findings suggest that perospirone has a clinical efficacy for the positive symptoms of schizophrenia via dopamine D_2 receptor antagonism.

Perospirone also has a high affinity for $5\text{-}HT_2$ receptors like the atypical antipsychotics risperidone and olanzapine. In behavioral tests, perospirone, risperidone and olanzapine markedly inhibited serotonergic behavior (e.g. tryptamine-induced clonic seizures, and p-chloroamphetamine-induced hyperthermia) in rats, whereas the actions of haloperidol were weak (Table 2). The relative potencies of these drugs are as follows: risperidone > perospirone = olanzapine ≫ haloperidol. Since the blockade of $5\text{-}HT_2$ receptors has been suggested to improve the negative or deficit symptoms of schizophrenia (Meltzer and Nash, 1991), these agents are expected to be effective in treating negative symptoms in patients with schizophrenia. In fact, a double-blind comparative study with haloperidol revealed that perospirone is as potent as haloperidol in improving positive symptoms, but was superior to haloperidol against negative symptoms (Murasaki *et al.*, 1997).

Extrapyramidal side effect liability. Table 2 shows propensities of the antipsychotics to induce EPS in rodents. All the SDA-type antipsychotics including perospirone were much weaker than haloperidol in inducing catalepsy or bradykinesia, and showed higher therapeutic indices as revealed by the potency ratios of D_2 blocking actions to EPS induction. Interestingly, the therapeutic index of perospirone was considerably higher than that of risperidone even though its 5-HT_2 selectivity is lower than that of risperidone. The rank order of the therapeutic index was as follows: perospirone > olanzapine = risperidone > haloperidol.

Repeated administration of dopamine D_2 antagonists, e.g. haloperidol, easily causes supersensitivity of dopamine receptors in rats (Jeste and Caligiuri, 1993). It is thought that this phenomenon is related to the etiology of TD, which occurs after long-term antipsychotic treatment in humans. Unlike haloperidol, subchronic treatment with the D_2 blocking dosage of perospirone did not alter dopaminergic behavioral sensitivity (i.e. apomorphine-induced stereotypy and SKF-38393-induced vacuous chewing movement) or the density of striatal D_2 receptors in rats (Ohno *et al.*, 1995, 1997). These findings suggest that perospirone may have a reduced propensity to induce TD in humans. Similar actions of other SDA (e.g. clozapine) have also been reported (see and Ellison, 1990).

Expression of *c-fos* mRNA or Fos-like immunoreactivity is widely used as a neural marker for exploring the site of antipsychotic action in the brain and for differentiating atypical from typical antipsychotics. In fact, whereas typical antipsychotics increase *c-fos* expression in the nucleus accumbens and the striatum, atypical antipsychotics show the action only in the nucleus accumbens, not in the dorsolateral striatum (Robertson *et al.*, 1994). Perospirone induced only a marginal increase in striatal *c-fos* mRNA expression, which was in contrast to the action of haloperidol (Ishibashi *et al.*, 1996). Furthermore, Robertson *et al.* (1994) proposed the atypical index, as revealed by the difference in numbers of Fos-positive neurons between the nucleus accumbens and the dorsolateral striatum, which would serve as a useful marker for discrimination of atypical from typical antipsychotics. Perospirone showed a dose-dependent increase in the atypical index, which remained positive over the dose range tested (1–10 mg/kg, p.o.) (Ishibashi *et al.*, 1999) (Fig. 2). Risperidone also produced a positive atypical index, but it reduced with an increasing dose. In contrast, haloperidol showed a negative or near zero atypical index in Fos expression except for the lowest dose. These findings suggest that perospirone has a preferential action on the mesocortical (*vs.* nigrostriatal) dopaminergic system. This was also supported by its contrasting effects on dopamine metabolism in rats (Maruoka *et al.*, 1993). All these results suggest that perospirone has an atypical antipsychotic property characterized by

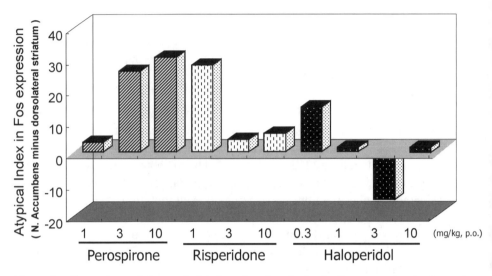

Figure 2. Comparison of the atypical index of antipsychotics in Fos expression in the nucleus accumbens and the dorsolateral striatum in rats. The atypical indexes of perospirone, risperidone and haloperidol were obtained by subtracting the number of Fos-positive neurons in the dorsolateral striatum from that of the nucleus accumbens in the same animals (Ishibashi et al., 1999).

Table 3.
Mood stabilizing actions of perospirone in rats

Model	Perospirone[a]	Haloperidol
CFS-induced freezing behavior	3.0	no effect
Conditioned defensive burying behavior	0.3	no effect
Social interaction	0.1	no effect

[a] Values show minimal effective doses in rats (mg/kg, p.o.).
CFS; conditioned fear stress.

low EPS liability in humans. Clinical study also demonstrated that the EPS score of the perospirone-treated group was significantly lower than that of the haloperidol-treated group.

Mood stabilizing effects. Perospirone has anxiolytic-like effects and mood stabilizing effects in various animal models (Table 3). Perospirone significantly improved rat freezing behavior observed in an environment previously paired with inescapable foot shock (Ishida-Tokuda et al., 1996). The inhibitory effect of perospirone on conditioned fear stress (CFS)-induced freezing behavior resembled those of anxiolytic diazepam and antidepressants, imipramine and desipramine, suggesting that perospirone has anxiolytic and/or antidepressant activities. Furthermore, perospirone but not haloperidol significantly inhibited conditioned defensive burying behavior and increased the time spent in active social interaction between naive rats, which also mimicked the effects of diazepam (Table 3) (Sakamoto et al.,

Table 4.

Comparison of $\alpha 1$-blockade-related behavior of perospirone with risperidone

		Perospirone	Risperidone
Norepinephrine-induced lethality[a]	rats	5.1	1.7
Ptosis[a]	mice	8.7	0.68
Orthostatic hypotension[b]	rabbits	>1.0	0.05

[a] ED_{50} value (mg/kg, p.o.).

[b] Minimal effective dose (mg/kg, i.p.).

1998). These findings support the clinical findings that perospirone significantly alleviated anxiety in patients with schizophrenia (Murasaki *et al.*, 1997).

Perospirone shows an efficacy for a negative symptoms model of schizophrenia, which was induced by PCP (10 mg/kg/day). Repeated treatment of PCP increases the immobility time in a forced swim test in rats, and this effect is inhibited by pretreatment with atypical antipsychotics, but not the typical antipsychotics (Noda *et al.*, 2000). Perospirone also antagonized the prolongation of immobility time induced by repeated treatment of PCP, suggesting that perospirone could ameliorate negative symptoms of schizophrenia in the clinic (Ohno, 2002). When these findings are taken together, perospirone is suggested to have a broader clinical profile than typical antipsychotics with regard to the efficacy for the emotional or motivational axis of schizophrenic symptomatology.

Other pharmacological characteristics. Perospirone inhibits motor coordination in a rota-rod test, and potentiates the duration of hexobarbital-induced anesthesia with ED_{50} values of 34 and 37 mg/kg (p.o.), respectively (Table 2). The other antipsychotics, risperidone, olanzapine and haloperidol, also show impairment of motor coordination and potentiation of hexobarbital-induced anesthesia with higher potencies compared to perospirone (Table 2). These results suggest that the central nervous system (CNS) depressant actions of perospirone are relatively mild.

It is thought that the noradrenergic $\alpha 1$ blocking action of antipsychotics is involved in orthostatic hypotension in schizophrenic patients with drug treatment. Perospirone shows a relatively weak *in vivo* $\alpha 1$ blocking action (i.e. norepinephrine-induced lethality and ptosis) compared with risperidone (Table 4). Furthermore, propensity to induce the orthostatic hypotension of perospirone is much weaker than that of risperidone (Table 4; Ohno *et al.*, 2002). These results suggest that perospirone possesses a weaker cardiovascular side effect than risperidone.

A POTENTIAL ROLE OF 5-HT$_2$ AND 5-HT$_{1A}$ RECEPTORS IN THE ATYPICAL ANTIPSYCHOTIC PROPERTY OF PEROSPIRONE

Efficacy for negative symptoms and mood disorders

Clinical studies have suggested that selective blockade of 5-HT$_2$ receptors can ameliorate the negative symptoms of schizophrenia (Meltzer and Nash, 1991). This was supported by the findings that only the SDA-type antipsychotics including perospirone, which share the potent 5-HT$_2$ blocking actions, as well as selective 5-HT$_2$ antagonists, were effective in an animal model of negative symptoms (Noda *et al.*, 2000; Ohno, 2002). Furthermore, since selective 5-HT$_2$ receptor antagonists improved CFS-induced freezing behavior in rats (Ishida-Tokuda *et al.*, 1996), 5-HT$_2$ blocking activity is suggested to be involved in the mood stabilizing action of perospirone. In addition, perospirone shows efficacy in various animal models of anxiety and depression, such as social interactions and conditioned defensive burying models (Sakamoto *et al.*, 1998). Since previous studies demonstrated that 5-HT$_{1A}$ receptor partial agonists can ameliorate anxiety and depressive disorder

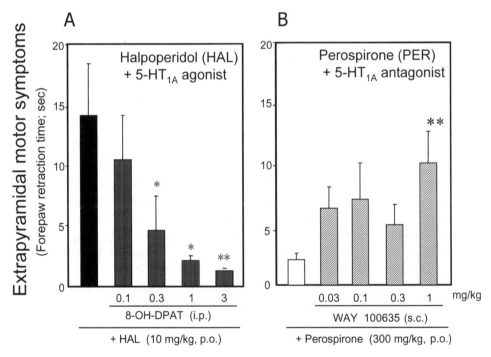

Figure 3. Effects of 5-HT$_{1A}$ agonist (8-OH-DPAT, A) and antagonist (WAY-100635, B) on the action of haloperidol and perospirone in the rat paw-test, respectively. Haloperidol (10 mg/kg, p.o.) or perospirone (300 mg/kg, p.o.) were administrated 1 hour before the test. WAY-100635 or 8-OH-DPAT were simultaneously injected with perospirone or haloperidol, respectively. Forepaw retraction time was evaluated according to Ellenbroek *et al.* (1987). $*p < 0.05$, $**p < 0.01$ *vs.* haloperidol or perospirone alone (Dunnett's test).

in these animal models, it is suggested that the action of perospirone for 5-HT$_{1A}$ receptors plays a significant role in these effects.

Reduced extrapyramidal side effects

As mentioned above, the propensity of SDA-type antipsychotics (e.g. perospirone) for EPS induction is weaker than conventional antipsychotics such as haloperidol (Table 2). Several experiments revealed that selective 5-HT$_2$ receptor antagonists, such as ritanserin and ketanserin, attenuated D$_2$ antagonist-induced bradykinesia (Ohno *et al.*, 1994), striatal *c-fos* expression (Ishibashi *et al.*, 1996), and supersensitivity of dopaminergic neurons after repeated administration (Ohno *et al.*, 1997). These results suggest that the 5-HT$_2$ blocking action of perospirone is involved in the reduced actions of EPS.

Furthermore, several studies revealed that the 5-HT$_{1A}$ agonistic action of perospirone is also involved in the lower propensity to induce EPS (Ohno *et al.*, 2001). In the paw-test, forepaw rigidity induced by haloperidol was attenuated by the coadministration of a selective 5-HT$_{1A}$ agonist, 8-OH-DPAT (Fig. 3A). In contrast, the forepaw rigidity induced by perospirone at relatively higher doses (300 mg/kg, p.o.) was significantly aggravated by the coadministration of a 5-HT$_{1A}$ antagonist, WAY-100635 (Fig. 3B). These findings suggest that the 5-HT$_{1A}$ agonistic action of perospirone contributes to the lower propensity of perospirone to induce extrapyramidal side effects.

CONCLUSION

Perospirone shows high affinities to dopamine D$_2$ and 5-HT$_2$ receptors, and acts as an antagonist for these receptors. Furthermore, unlike the other atypical antipsychotics, perospirone has a partial agonistic efficacy for 5-HT$_{1A}$ receptors. These characteristics may be involved in the anxiolytic/mood stabilizing action or reduced EPS profiles of perospirone. Because recent findings revealed that the 5-HT$_2$ blocking action and 5-HT$_{1A}$ agonistic action work cooperatively for the anxiolytic/mood stabilizing action and reduction of EPS, perospirone is thought to be more useful in the treatment of schizophrenic patients compared with the other atypical antipsychotics (Fig. 4). Furthermore, it is reported that 5-HT$_{1A}$ agonistic action could improve cognitive function in schizophrenic patients (Sumiyoshi *et al.*, 2000). These findings indicate that perospirone has a broad therapeutic spectrum and good tolerability.

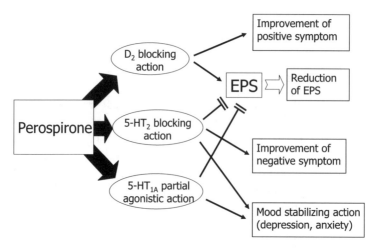

Figure 4. Mechanism of the action of perospirone.

REFERENCES

Bymaster, F. P., Calligara, D. O., Falcone, J. F., *et al.* (1996). Radioreceptor binding profile of the atypical antipsychotic olanzapine, *Neuropsychopharmacology* **14**, 87–96.

Ellenbroek, B. A., Peeters, B. W., Honig, W. M., *et al.* (1987). The paw test: a behavioural paradigm for differentiating between classical and atypical neuroleptic drugs, *Psychopharmacology (Berlin)* **93**, 343–348.

Hirose, A., Kato, T., Ohno, Y., *et al.* (1990). Pharmacological actions of SM-9018, a new neuroleptic drug with both potent 5-hydroxytryptamine$_2$ and dopamine$_2$ antagonistic actions, *Jpn. J. Pharmacol.* **53**, 321–329.

Ishibashi, T., Ikeda, K., Ishida, K., *et al.* (1996). Contrasting effects of SM-9018, a potential atypical antipsychotic, and haloperidol on *c-fos* mRNA expression in the rat striatum, *Eur. J. Pharmacol.* **303**, 247–251.

Ishibashi, T., Tagashira, R., Nakamura, M., *et al.* (1999). Effects of perospirone, a novel 5-HT$_2$ and D$_2$ receptor antagonist, on Fos protein expression in the rat forebrain, *Pharmacol. Biochem. Behav.* **63**, 535–541.

Ishida-Tokuda, K., Ohno, Y., Sakamoto, H., *et al.* (1996). Effects of perospirone (SM-9018), a novel serotonin-2 and dopamine-2 receptor antagonist, and other antipsychotics in the conditioned fear stress-induced freezing behavior model in rats, *Jpn. J. Pharmacol.* **72**, 119–126.

Jeste, D. V. and Caligiuri, M. P. (1993). Tardive dyskinesia, *Schizophrenia Bull.* **19**, 303–315.

Kato, T., Hirose, A., Ohno, Y., *et al.* (1990). Binding profile of SM-9018, a novel antipsychotic candidate, *Jpn. J. Pharmacol.* **54**, 478–481.

Leysen, J. E., Janssen, P. M. F., Gommeren, W., *et al.* (1992). *In vitro* and *in vivo* receptor binding and effects on monoamine turnover in rat brain regions of the novel antipsychotics risperidone and ocaperidone, *Mol. Pharmacol.* **41**, 494–508.

Maruoka, Y., Ohno, Y., Kato, T., *et al.* (1993). Effects of SM-9018, a potential atypical neuroleptic, on the central monoaminergic system in rats, *Jpn. J. Pharmacol.* **62**, 419–422.

Meltzer, H. Y. and Nash, J. F. (1991). Effects of antipsychotic drugs on serotonin receptors, *Pharmacol. Rev.* **43**, 587–604.

Millan, M. J. (2000). Improving the treatment of schizophrenia: Focus on serotonin (5-HT)$_{1A}$ receptors, *J. Pharmacol. Exp. Ther.* **295**, 853–861.

Murasaki, M., Koyama, T., Machiyama, Y., *et al*. (1997). Clinical evaluation of a new antipsychotic, perospirone HCl, on schizophrenia — A comparative double-blind study with haloperidol, *Clin. Eval.* **24**, 159–205.

Noda, Y., Kamei, H., Mamiya, T., *et al*. (2000). Repeated phencyclidine treatment induces negative symptom-like behavior in forced swimming test in mice: imbalance of prefrontal serotonergic and dopaminergic functions, *Neuropsychopharmacology* **23**, 375–387.

Ohno, Y. (2002). Pharmacological characteristics of the new atypical antipsychotic agents, serotonin-dopamine antagonists (SDA), *Jpn. J. Clin. Psychopharmacol.* **5**, 147–153.

Ohno, Y., Ishida, K., Ikeda, K., *et al*. (1994). Evaluation of bradykinesia induction by SM-9018, a novel 5-HT$_2$ and D$_2$ receptor antagonist, using the mouse pole test, *Pharmacol. Biochem. Behav.* **49**, 19–23.

Ohno, Y., Ishibashi, T., Okada, K., *et al*. (1995). Effects of subchronic treatments with SM-9018, a novel 5-HT$_2$ and D$_2$ antagonist, on dopamine and 5-HT receptors in rats, *Prog. Neuropsychopharmacol. Biol. Psychiatry* **19**, 1091–1101.

Ohno, Y., Ishida-Tokuda, K., Ishibashi, T., *et al*. (1997). Effects of perospirone (SM-9018), a potential atypical neuroleptic, on dopamine D$_1$ receptor-mediated vacuous chewing movement in rats: A role of 5-HT$_2$ receptor blocking activity, *Pharmacol. Biochem. Behav.* **57**, 889–895.

Ohno, Y., Matsumoto, K., Horisawa, T., *et al*. (2001). *Jpn. J. Neuropsychopharmacol.* **21**, 345 (in Japanese).

Ohno, Y., Matsumoto, K. and Nakamura, M. (2002). *Jpn. J. Neuropsychopharmacol.* **22**, 253 (in Japanese).

Robertson, G. S., Matsumura, H. and Fibiger, H. C. (1994). Induction patterns of Fos-like immunoreactivity in the forebrain as predictors of atypical antipsychotic activity, *J. Pharmacol. Exp. Ther.* **271**, 1058–1066.

Sakamoto, H., Matsumoto, K., Ohno, Y., *et al*. (1998). Anxiolytic-like effects of perospirone, a novel serotonin-2 and dopamine-2 antagonist (SDA)-type antipsychotic agent, *Pharmacol. Biochem. Behav.* **60**, 873–878.

See, R. E. and Ellison, G. (1990). Comparison of chronic administration of haloperidol and the atypical neuroleptics, clozapine and raclopride, in an animal model of tardive dyskinesia, *Eur. J. Pharmacol.* **181**, 175–186.

Shiwa, T., Amano, T., Matsubayashi, H., *et al*. (2003). Perospirone, a novel antipsychotic agent, hyperpolarizes rat dorsal raphe neurons via 5-HT$_{1A}$ receptor, *J. Pharmacol. Sci.* **93**, 114–117.

Sumiyoshi, T., Matsui, M., Tamashita, I., *et al*. (2000). Effect of adjunctive treatment with serotonin$_{1A}$ agonist tandospirone on memory functions in schizophrenia, *J. Clin. Pharmacol.* **20**, 386–388.

Tokuda, K., Ohno, Y., Ishibashi, T., *et al*. (1997). Studies on central actions of perospirone hydrochloride, a novel antipsychotic agent, *Clin. Rep.* **31**, 853–878.

Yabuuchi, K., Tagashira, R., Horisawa, T., *et al*. (1998). *Jpn. J. Psychopharmacol.* **18**, 374 (in Japanese).

Advances in Neuroregulation and Neuroprotection (2005), pp. 359-368
Collin, C. *et al.* (Eds)
© VSP 2005

Effects of tandospirone, a novel anxiolytic agent, on human 5-HT$_{1A}$ receptors expressed in Chinese hamster ovary cells (CHO cells)

KAZUKI YABUUCHI *, RIE TAGASHIRA and YUKIHIRO OHNO

Discovery Research Laboratories I, Research Center, Sumitomo Pharm. Co., Ltd., Konohana-ku, Osaka 554-0022, Japan

Abstract—Tandospirone is a novel anxiolytic agent which selectively binds to 5-HT$_{1A}$ receptors. In this study, we established a stable CHO cell line expressing human 5-HT$_{1A}$ receptors and examined the action of tandospirone in this cell line. The human 5-HT$_{1A}$ receptor gene was transfected into CHO cells. Several clones showed specific binding of [^{3}H]8-OH-DPAT and one (5HT1A/CHO-No.5) with high density of 5-HT$_{1A}$ receptors was isolated. Scatchard analysis revealed the dissociation constant (Kd) for [^{3}H]8-OH-DPAT of 2.8 nM and the expression level (Bmax) of 956 fmol/mg protein. Competition experiments of [^{3}H]8-OH-DPAT binding showed that affinities (Ki) of several compounds to human 5-HT$_{1A}$ receptors were as follows; 5-HT (8.8 nM), 8-OH-DPAT (3.9 nM), tandospirone (72 nM) buspirone (32 nM), WAY-100635 (0.22 nM) and ritanserin (IC50 > 1 μM). 5-HT, 8-OH-DPAT and tandospirone (0.1 nM–10 μM) concentration-dependently stimulated [^{35}S]GTPγS binding in 5-HT$_{1A}$/CHO-No.5 cells, but WAY-100635 (10 nM–10 μM) did not. Furthermore, treatment with 5-HT, 8-OH-DPAT, tandospirone and buspirone (0.1 nM–10 μM), but not WAY-100635 (10 nM–10 μM), inhibited forskolin-induced cAMP production in a concentration-dependent manner. These findings suggest that 5-HT$_{1A}$/CHO-No.5 cells retain 5-HT$_{1A}$ receptor and Gi/o-protein coupling and that tandospirone acts as an agonist at human 5-HT$_{1A}$ receptors.

Keywords: Tandospirone; 5-HT$_{1A}$ receptor; GTPγS binding; adenylate cyclase; CHO cells.

INTRODUCTION

5-Hydroxytryptamine (5-HT) is an important neurotransmitter in the central and peripheral nervous system and its involvement in a variety of functions such as sleep, pain, appetite, sexual behavior and control of mood have been evaluated (Jacobs and Azmitia, 1992). In accordance with multiple functions of 5-HT, many types of 5-HT receptor have been identified and their subtypes can be roughly

---

*To whom correspondence should be addressed. E-mail: yabuuchi@sumitomopharm.co.jp

divided into seven subfamilies (Hoyer *et al.*, 1994). Except for the 5-HT$_3$ receptor, all other 5-HT receptors are shown to belong to G-protein coupled receptor. Of these, 5-HT$_1$ receptor is of particular interest because newly developed compounds targeting to this type of receptors have multiple effects in the nervous system (Hoyer *et al.*, 1994). 5-HT$_1$ receptors are subdivided into 5 subclasses (Hoyer *et al.*, 1994) (5-HT$_{1A}$, 5-HT$_{1B}$, 5-HT$_{1D}$, 5-HT$_{1E}$ and 5-HT$_{1F}$) and agonists for 5-HT$_{1A}$ receptors have been developed as anxiolytics and agonists for 5-HT$_{1B/1D}$ receptors as drugs for migraine (Hoyer *et al.*, 1994).

Tandospirone (3aα, 4β, 7β, 7aα-hexahydro-2-(4-(4-(2-pyrimidinyl)-1-piper-azinyl)-butyl-4, 7-methano-1H-iso-indole-1,3 (2H)-dione dihydrogen citrate) is a novel anxiolytic that has been marketed in Japan. Pharmacological profiles of tandospirone reveal that it exhibits few of the side effects, especially sedative effects on the central nervous system that are often seen by use of benzodiazepine derivatives (Shimizu *et al.*, 1987). Previously, we reported that tandospirone induced flat-body posture in rat (Shimizu *et al.*, 1992) and suppressed 5-HT metabolism (Tatsuno *et al.*, 1989) in the rat brain. Furthermore, tandospirone displaced the specific binding of [^{3}H]8-OH-DPAT to rat hippocampal membrane (Shimizu *et al.*, 1988) and suppressed forskolin-stimulated cAMP in the rat hippocampus (Tanaka *et al.*, 1995). These data strongly suggest that tandospirone is an agonist for rat 5-HT$_{1A}$ receptor. Although the intrinsic activity of tandospirone to rat 5-HT$_{1A}$ receptor is shown to be 0.86 by cAMP assay in the rat hippocampal membrane (Tanaka *et al.*, 1995), it is well-known that several types of 5-HT receptor (5HT$_{1A}$, 5HT$_4$, 5HT$_7$ etc.) are expressed in the hippocampus (Hoyer *et al.*, 1994). In this context, it is important to elucidate the pure effects of tandospirone on the 5-HT$_{1A}$ receptor. Recently, several groups reported the sequence of 5-HT$_{1A}$ receptor (Albert *et al.*, 1990; Fargin *et al.*, 1988). In this study, we established a stable CHO cell line expressing human 5-HT$_{1A}$ receptors. In addition, we examined the effects of tandospirone on GTPγS binding and forskolin-induced cAMP accumulation in this cell line.

MATERIALS AND METHODS

Materials

Tandospirone and WAY-100635 were synthesized in our company. 5-HT HCl, 8-OH-DPAT, ritanserin, buspirone HCl and IBMX were purchased from Research Biochemicals International (Natick, USA). Forskolin, GDP and GTPγS were from Sigma Chemical (St. Louis, USA). [^{3}H]8-OH-DPAT, [^{35}S] GTPγS and cAMP assay kit were from Amersham (Buckinghamshire, UK). All materials for cell culture were from GIBCO (Grand Island, USA).

Transfection of 5-HT1A receptor gene into CHO-K1 cells

A gene of human 5-HT$_{1A}$ receptor (G-21) was supplied by Duke University (Fargin *et al.*, 1988). The *Bam* HI-*Hind* III fragment was excised from the plasmid pSP64-G21 and inserted into the same digested arm of pBluescript II KS+ (Stratagene). The constructed plasmid was cut by *Xba* I and its fragment was inserted into the pcDNA3.1+ (Invitrogen) digested by *Xba* I. The resultant plasmid (5HT1A/pcDNA3.1+) was introduced into *E. coli* (XL-1Blue) and purified by QIAGEN plasmid purification MEGA kit.

CHO-K1 cells were growing in Ham's F-12 medium (GIBCO) supplemented with 10% heat inactivated fetal calf serum (GIBCO), penicillin-streptomycin (GIBCO). After seeding the cells into a 60 mm dish at a density of 2×10^6 cells/ml, cells were growing for 24 h and changed the medium to serum-free F-12 medium. The plasmid containing human 5-HT$_{1A}$ receptor gene (5HT1A/pcDNA3.1+) was transfected using lipofectin (GIBCO). After the plasmid was introduced, the medium was changed to one containing serum that maintained the cells for 3 days. Then, the medium was removed and cells were cultured in the medium containing G-418 (50 μg/ml, GIBCO) for 1 week. Transfected cells were harvested by trypsinize and seeded into the plate at the density of about 20 cells/10 ml medium. The cells carrying the introduced plasmid survived in the medium containing G-418 and formed small colonies after 5 days from seeding. These colonies were removed by trypsin-EDTA and transferred to a 24-well plate. In this study, we picked 48 colonies. After cells were growing enough to conduct a binding assay, the cell membrane of each colony was prepared by osmotic shock. The membrane was tested for expression of 5-HT$_{1A}$ receptor by binding of 0.5 nM concentration of 5-HT$_{1A}$ specific agonist [^{3}H]8-OH-DPAT. A monoclonal cell line (No. 5) that highly expressed 5-HT$_{1A}$ receptor was isolated and used for the following experiments.

Radioligand binding assay

Cell membrane was prepared by the following method. Briefly, 80-90% confluently grown cells were scraped and homogenized in ice-cold assay buffer (Tris-HCl, 50 mM, pH 7.4). Crude cell membrane was centrifuged at 48 000g (4°C, 20 min), then the pellet was resuspended in assay buffer and stored at −80°C. Protein concentration was measured using protein assay kit (BIO-RAD).

For saturation studies, 0.4 ml of membrane suspension was incubated with increased concentration of [^{3}H]8-OH-DPAT (Amersham 223 Ci/mmol; final concentration, 0.1–4 nM) in a volume of 0.5 ml at room temperature. For the inhibition experiment, the membrane was incubated for 30 min at room temperature with 0.5 nM [^{3}H]8-OH-DPAT in the presence of various concentrations of compounds. In both experiments, incubation was carried out with 0.4 mM CaCl$_2$, and 8-OH-DPAT (1 μM) was used to define the non-specific binding. After the incubation, the reaction was terminated by adding ice-cold assay buffer and immediately passed through glass filters (GF/B, Whatman) treated with 0.3% polyethyleneimine. Radioactivity was counted by liquid scintillation counter. Kd value and Bmax value of [^{3}H]8-OH-DPAT to 5-HT1A/CHO membrane was analyzed by Scatchard analysis. IC$_{50}$ values were calculated as concentrations that each drug required to inhibit [^{3}H] 8-OH-DPAT binding to 50% of their maximum inhibition by Hill's plot. Ki values of each drug for competition binding were calculated by the following equation:

$$Ki = IC_{50}/(1 + S/Kd),$$

where S is the radioligand concentration.

[^{35}S] GTPγS binding assay

[^{35}S] GTPγS binding assay was carried out following the method of Newman-Tancredi with minor modifications (Newman-Tancredi et al., 1998). Briefly, the cells were scraped and suspended in buffer A (20 mM HEPES pH 7.4, 5 mM MgSO$_4$). This membrane was centrifuged at 50 000g for 30 min and pellet was resuspended in buffer A. 5HT1A/CHO cell membranes (50 μg) were incubated in a buffer (1 ml) containing 20 mM HEPES (pH 7.4), 3 μM GDP, 3 mM MgSO$_4$, and 0.05 nM [^{35}S] GTPγS for 20 min at 22°C. Cold GTPγS (10 μM) were used to define the non-specific binding. Incubation was terminated by rapid filtration through the glass filter (GF/B, Whatman) and radioactivity was determined by liquid scintillation counter. Results were represented as percentage of 5-HT binding at 10 μM. EC$_{50}$ values were calculated as concentrations that each drug required to stimulate the [^{35}S] GTPγS binding to 50% of their maximum response by the Litchfield-Wilcoxon method.

cAMP assay

5HT1A/CHO cells were seeded in 24-well plates (1×10^5 cells/well) one day before the assay. The cells were washed by PBS and briefly rinsed with cAMP assay buffer (118 mM NaCl, 4.6 mM KCl, 1 mM CaCl$_2$, 1 mM MgCl$_2$, 10 mM D-glucose and 20 mM HEPES, pH 7.2). After 10 min treatment with cAMP assay buffer containing 200 μM IBMX at 37°C, cells were incubated with 10 μM forskolin and 200 μM IBMX in the presence or absence of various concentrations of compounds at 37°C for 15 min. The incubation was terminated by the replacement of buffer to cAMP assay buffer containing 0.05% Triton X-100. The Triton X-100 containing buffer was recovered and centrifuged for 5 min to remove the cellular debris. Sample solution was stored at −80°C until cAMP assay was conducted.

Concentrations of cAMP were measured using enzyme immune assay kit (Amersham). Sample solution was appropriately diluted and acetylated. All assays were carried out following the manufacturer's instruction manual. IC$_{50}$ values were calculated as concentrations that each drug required to inhibit forskolin-induced cAMP to 50% of their maximum response by Litchfield-Wilcoxon method.

RESULTS

Binding characteristics of 5-HT and 5-HT$_{1A}$ ligands in the 5HT$_{1A}$/CHO cells

In this study, we obtained 8 clones that showed specific binding of [^{3}H]8-OH-DPAT. Of these cell lines, we chose No.5 cell line because this cell line showed

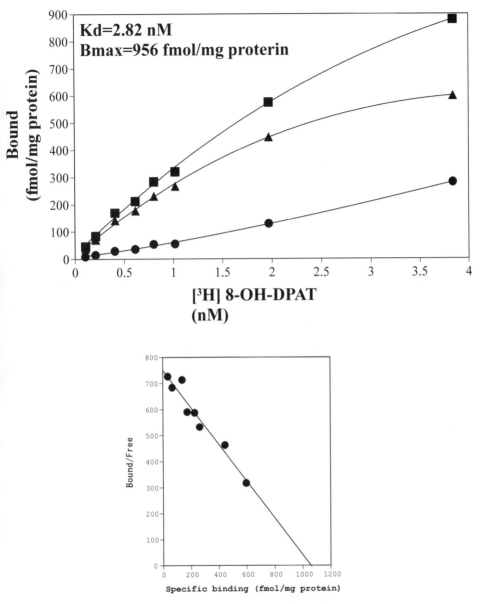

Figure 1. Specific binding of [^{3}H]8-OH-DPAT to CHO cells expressing the human 5-HT$_{1A}$ receptor. Radioligand concentrations are plotted as a function of total (■), nonspecific (●) or specific (▲) binding of radioligand (Upper panel). Nonspecific binding was defined by 1 μM 8-OH-DPAT. Lower panel showed the Scatchard plot of upper panel.

high binding sites to [^{3}H]8-OH-DPAT. As shown in Fig. 1, cell membrane from No.5 cells showed high affinity, saturable [^{3}H]8-OH-DPAT binding (Kd = 2.8 nM, Bmax = 956 fmol/mg protein). No specific binding of [^{3}H]8-OH-DPAT was observed in the cell that was transfected with pcDNA3.1+ alone (data not shown).

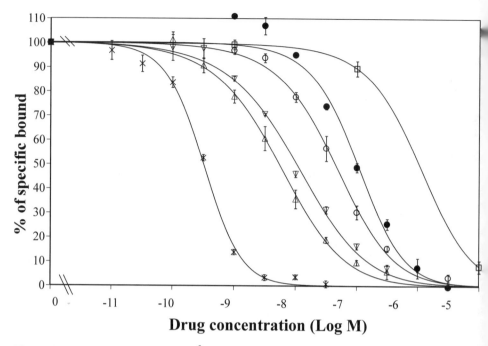

Figure 2. Inhibition of binding of [³H]8-OH-DPAT to 5HT1A/CHO cells. Membranes from 5HT1A/CHO cells were incubated with [³H]8-OH-DPAT (1 nM). Each point represents the mean ± S.E.M. of triplicate determinations. 5-HT (▽), 8-OH-DPAT (△), tandospirone (●), buspirone (○), WAY-100635 (×) and ritanserin (□).

Table 1.
Pharmacological properties of 5-HT, 8-OH-DPAT, tandospirone, buspirone and WAY-100635 to the CHO cells expressing the human 5-HT$_{1A}$ receptors

Drug	Binding assay	cAMP assay		GTPγS binding	
	Ki (nM)	IC$_{50}$ (nM)	I.A.	EC$_{50}$ (nM)	I.A.
5-HT	8.8	2.4	1.00	3.2	1.00
8-OH-DPAT	3.9	4.2	0.90	1.9	1.03
Tandospirone	72	42	0.87	24	0.84
Buspirone	32	22	0.84	ND	ND
WAY-100635	0.22	—	—	—	—

The values of intrinsic activity (I.A.) represent the ratio of Emax value of each drug to that of 5-HT. Each value was obtained from the average of 3–5 separate experiments. ND Not Determined, — No effect.

As shown in Fig. 2, [³H]8-OH-DPAT binding to 5-HT1A/CHO cell membrane was displaced by 5-HT and other 5-HT$_{1A}$ ligands in a concentration-dependent manner. The Ki values of these 5-HT$_{1A}$ ligands are shown in Table 1. On the other hand, ritanserin (5-HT$_{2/7}$ ligand) did not displace the specific [³H]8-OH-DPAT binding up to 100 nM.

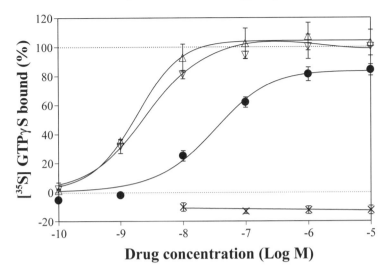

Figure 3. Activation of [^{35}S] GTPγS binding to 5HT1A/CHO membranes. Each point represents the mean ± S.E.M. of triplicate determinations and is expressed as a percentage of the maximal [^{35}S] GTPγS binding given by 5-HT. 5-HT (∇), 8-OH-DPAT ($\triangle$), tandospirone ($\bullet$) and WAY-100635 ($\times$).

Effects of 5-HT and 5-HT$_{1A}$ ligands on [^{35}S] GTP-γS binding in 5HT1A/CHO cells

5-HT stimulated the specific [^{35}S] GTPγS binding in a concentration-dependent manner (Fig. 3). The specific binding of [^{35}S] GTPγS showed 180.7 ± 2.4% ($n = 3$) increase over basal by 10 μM 5-HT with an EC$_{50}$ of 3.2 nM. 8-OH-DPAT and tandospirone also showed concentration-dependent increase in [^{35}S] GTPγS binding. The EC$_{50}$ values of 8-OH-DPAT and tandospirone were 1.9 nM ($n = 3$) and 24 nM ($n = 3$), respectively. Maximal activation (Emax) by 8-OH-DPAT and tandospirone amounted to 103% and 84% relative to 5-HT (10 μM), respectively. However, WAY-100635 did not activate the specific [^{35}S] GTPγS binding but rather inhibited it even at a concentration of 10 μM ($n = 3$).

Effects of 5-HT and other 5-HT$_{1A}$ ligands on forskolin-stimulated cAMP accumulation in 5HT1A/CHO cell

Tandospirone, as well as 5-HT and 8-OH-DPAT, suppressed forskolin (10 μM)-induced cAMP accumulation in a concentration-dependent manner. Buspirone also inhibited this accumulation (Fig. 4). On the other hand, WAY-100635 did not modify the accumulation of cAMP. These compounds maximally inhibited the cAMP productions to 83.8% (tandospirone), 96.0% (5-HT), 86.0% (8-OH-DPAT) and 81.1% (buspirone) of the control. The IC$_{50}$ values of tandospirone, 5-HT, 8-OH-DPAT and buspirone were 42 nM, 2.4 nM, 4.2 nM and 22 nM, respectively.

The inhibitory effect of tandospirone on the forskolin-induced cAMP accumulation was concentration-dependently inhibited by WAY-100635 (Fig. 5). The IC$_{50}$ values of tandospirone were shifted to 4.9 nM, 16.4 nM, 348 nM and 11.1 μM by WAY-100635 at concentrations of 0 nM, 1 nM, 10 nM and 100 nM, respectively.

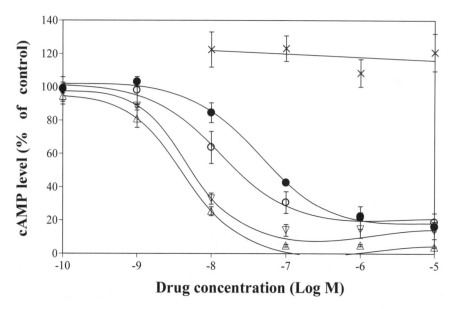

Figure 4. Inhibition of forskolin-stimulated adenylate cyclase activity by 5-HT ($\triangledown$), 8-OH-DPAT ($\triangle$), tandospirone ($\bullet$), buspirone ($\circ$) and WAY-100635 ($\times$). Values are means $\pm$ S.E.M. of 4–5 separate measurements.

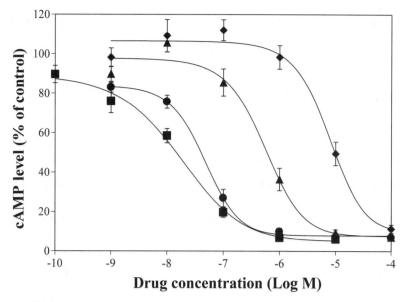

Figure 5. Concentration-response curve of tandospirone in the absence ($\blacksquare$) and presence of 1 nM ($\bullet$), 10 nM ($\blacktriangle$) and 100 nM WAY-100635 ($\blacklozenge$) for inhibition of cAMP accumulation in 5-HT1A/CHO cells. Values are means $\pm$ S.E.M. of 5 separate measurements.

DISCUSSION

In this study, we tried to establish a stable CHO cell line expressing human 5-HT$_{1A}$ receptors. The cell line that we constructed showed specific binding of [^{3}H]8-OH-DPAT which is considered to be a specific ligand to 5-HT$_{1A}$ receptor (Hoyer *et al.*, 1994). The saturation binding experiment clearly exhibited concentration dependent saturable binding to the membrane of CHO cells and this binding was displaced by WAY-100635 with the Ki value of 0.22 nM. WAY-100635 is reported to be a very specific antagonist to the 5-HT$_{1A}$ receptor *in vivo* and *in vitro* (Forster *et al.*, 1995). Furthermore, reitanserin, which is known to have no affinity to 5-HT$_{1A}$ receptor (Bourson *et al.*, 1997) but to be an antagonist to 5-HT$_2$ or 5-HT$_7$ receptor, did not displace the specific [^{3}H]8-OH-DPAT binding up to 100 nM. These binding profiles clearly showed the expression of 5-HT$_{1A}$ receptor in this CHO cell line.

It is reported that 5-HT$_{1A}$ receptor is coupled with Gi or Go protein in cultured mammalian cells and artificial cell lines (Hoyer *et al.*, 1994). The receptors coupled with Gi proteins have common properties that are characterized by the inhibition of adenylate cyclase. In this study we clearly showed that both 8-OH-DPAT and 5-HT concentration-dependently stimulated the binding of [^{35}S] GTPγS, which is shown to be a good parameter of agonist character of a ligand to Gi or Go coupled receptor (Newman-Tancredi *et al.*, 1998). In addition to this observation, 8-OH-DPAT inhibited the forskolin-stimulated cAMP accumulation in a concentration-dependent manner in this cell line. These data clearly demonstrate that our CHO cell line is expressing 5-HT$_{1A}$ receptor that is functionally coupled with Gi/o protein.

Using this CHO cell line, we evaluated the pharmacological properties of tandospirone to the human 5-HT$_{1A}$ receptor. In our previous reports, tandospirone had specific preference to rat 5-HT$_{1A}$ receptor and showed agonistic properties (Tanaka *et al.*, 1995). In this study, we showed that tandospirone displaced the specific [^{3}H]8-OH-DPAT binding in a concentration-dependent manner to the membrane expressing human 5-HT$_{1A}$ receptor in CHO cells. The Ki value of tandospirone to the human 5-HT$_{1A}$ receptors in CHO cells was calculated as 72 nM. Previously, we have reported that tandospirone showed no affinity to dopamine D$_2$, 5-HT$_{1B/1D,2}$, adrenaline, benzodiazepine and GABA receptors (Shimizu *et al.*, 1988). These reports suggest that tandospirone has specific preference to 5-HT$_{1A}$ receptors.

As mentioned above, agonists for 5-HT$_{1A}$ receptor stimulate the [^{35}S] GTPγS binding in the CHO cells expressing 5-HT$_{1A}$ receptors. We conducted the [^{35}S] GTPγS binding experiment to evaluate if tandospirone had agonistic properties to this receptor. Tandospirone stimulated the specific [^{35}S] GTPγS binding in a dose dependent manner and its intrinsic activity was 0.84. In addition to the [^{35}S] GTPγS experiment, we also examine the effects of tandospirone on the forskolin-induced cAMP accumulation. It was observed that tandospirone dose-dependently inhibited the accumulation of cAMP induced by forskolin. In this assay system, tandospirone showed high intrinsic activity of 0.87. Furthermore, these agonistic properties of tandospirone to 5-HT$_{1A}$ receptors were canceled by WAY-100635, a specific antagonist to 5-HT$_{1A}$ receptor and its concentration-effect curve was shifted to the right with increasing concentration of WAY-100635. These data suggest that tandospirone is an agonist with high intrinsic activity to the human 5-HT$_{1A}$ receptor.

As shown by accumulated reports, 5-HT$_{1A}$ agonists such as buspirone, ipsapirone and geprione have anxiolytic effects in both animal models and humans (De Vry, 1995). Furthermore, it is reported that mice lacking 5-HT$_{1A}$ receptor showed anxiogenic behavior (Heisler *et al.*, 1998; Parks *et al.*, 1998; Ramboz *et al.*, 1998). The present study clearly demonstrated that tandospirone behaves as a nearly full agonist at human 5-HT$_{1A}$ receptors as it does in rat membranes. These facts support the clinical efficacy of tandospirone for the treatment of anxiety and depressive disorders in human.

REFERENCES

Albert, P. R., Zhou, Q. Y., Van Tol, H. H. M., *et al.* (1990). Cloning, functional expression, and mRNA tissue distribution of the rat 5-hudroxytryptamine$_{1A}$ receptor gene, *J. Biol. Chem.* **265**, 5825–5832.

Bourson, A., Kapps, V., Zwingelstein, C., *et al.* (1997). Correlation between 5-HT$_7$ receptor affinity and protection against sound-induced seizures in DBA/2J mice, *Naunyn-Schmiedebergs Arch. Pharmacol.* **56**, 820–826.

De Vry, J. (1995). 5-HT$_{1A}$ receptor agonists: recent developments and controversial issues, *Psychopharmacol.* **121**, 1–26.

Fargin, A., Raymond, J. R., Lohse, M. J., *et al.* (1988). The genomic clone G-21 which resembles a β-adrenergic receptor sequence encodes the 5-HT$_{1A}$ receptor, *Nature* **335**, 358–360.

Forster, E. A., Cliffe, I. A., Bill, D. J., *et al.* (1995). A pharmacological profile of the selective silent 5-HT$_{1A}$ receptor antagonist, WAY-100635, *Eur. J. Pharmacol.* **281**, 81–88.

Heisler, L. K., Chu, H. M., Brennan, T. J., *et al.* (1998). Elevated anxiety and antidepressant-like responses in serotonin 5-HT$_{1A}$ receptor mutant mice, *Proc. Natl. Acad. Sci. USA* **95**, 15049–15054.

Hoyer, D., Clarke, D. E., Fozard, J. R., *et al.* (1994). International Union of Pharmacology classification of receptors for 5-hydroxytryptamine (Serotonin), *Pharmacol. Rev.* **46**, 157–203.

Jacobs, B. L. and Azmitia, E. C. (1992). Structure and function of the brain serotonin system, *Physiol. Rev.* **72**, 165–229.

Newman-Tancredi, A., Gavaudan, S., Conte, C., *et al.* (1998). Agonist and antagonist actions of antipsychotic agents at 5-HT$_{1A}$ receptors: a [^{35}S] GTPγS binding study, *Eur. J. Pharmacol.* **355**, 245–256.

Parks, C. L., Robinson, P. S., Sibille, E., *et al.* (1998). Increased anxiety of mice lacking the serotonin$_{1A}$ receptor, *Proc. Natl. Acad. Sci. USA* **95**, 10734–10739.

Ramboz, S., Oosting, R., Amara, D. A., *et al.* (1998). Serotonin receptor 1A knockout: An animal model of anxiety-related disorder, *Proc. Natl. Acad. Sci. USA* **95**, 14476–14481.

Shimizu, H., Hirose, A., Kato, T., *et al.* (1987). Pharmacological properties of SM-3997: a new anxioselective anxiolytic candidate, *Jpn. J. Pharmacol.* **54**, 493–500.

Shimizu, H., Karai, N., Hirose, A., *et al.* (1988). Interaction of SM-3997 with serotonin receptors in rat brain, *Jpn. J. Pharmacol.* **46**, 311–314.

Shimizu, H., Tatsuno, T., Tanaka, H., *et al.* (1992). Serotonergic mechanisms in anxiolytic effect of tandospirone in the Vogel conflict test, *Jpn. J. Pharmacol.* **59**, 105–112.

Tanaka, H., Tatsuno, T., Shimizu, H., *et al.* (1995). Effects of tandospirone on second messenger systems and neurotransmitter release in the rat brain, *Gen. Pharmacol.* **26**, 1765–1772.

Tatsuno, T., Shimizu, H., Hirose, A., *et al.* (1989). Effects of the putative anxiolytic SM-3997 on central monoaminerguc systems, *Pharmacol. Biochem. Behav.* **32**, 1049–1055.

Advances in Neuroregulation and Neuroprotection (2005), pp. 369-378
Collin, C. *et al.* (Eds)
© VSP 2005

Antiparkinsonian actions of a selective 5-HT$_{1A}$ agonist, tandospirone, in rats

TADASHI ISHIBASHI * and YUKIHIRO OHNO †

Discovery Research Laboratories I, Research Division, Sumitomo Pharmaceuticals Co., Ltd., 3-1-98 Kasugade-naka, Konohana-ku, Osaka 554-0022, Japan

Abstract—Antiparkinsonian effects of tandospirone, a selective 5-HT$_{1A}$ receptor agonist, were evaluated using rat models of Parkinson's disease. Tandospirone reversed catalepsy induced by the D$_2$ antagonist haloperidol, in a dose-dependent manner. The anti-cataleptic action of tandospirone was comparable to that of bromocriptine and greater than that of L-DOPA. In rats with unilateral dopaminergic lesion by 6-hydroxydopamine, tandospirone markedly induced contralateral rotation. Furthermore, tandospirone dose-dependently restored spontaneous locomotor activity in reserpine-treated rats. These antiparkinsonian effects of tandospirone were abolished by coadministration of WAY-100635, a selective 5-HT$_{1A}$ antagonist, but not by haloperidol. The present results suggest that tandospirone has a therapeutic potential in treating parkinsonian symptoms, which is brought about through activation of 5-HT$_{1A}$ receptor.

Keywords: Tandospirone; 5-HT$_{1A}$ agonist; catalepsy; rotation behavior; L-DOPA; Parkinson's disease.

INTRODUCTION

Tandospirone (3aa,4b,7b,7aa-hexahydro-2-(4-(4-(2-pyrimidinyl)-1-piperazinyl)-butyl)-4,7-methano-1H-isoindole-1,3(2H)-dione dihydrogen citrate) is a novel serotonergic anxiolytic agent that lacks the benzodiazepine-like side effects. Tandospirone selectively bound to and acted as an agonist at 5-HT$_{1A}$ receptor (Ki = 25 nM) without significantly interacting with many other receptors such as GABA$_A$, benzodiazepine, noradrenergic α_1, β and dopamine D$_2$ receptors (Shimizu *et al.*, 1988). Tandospirone exerted significant anxiolytic actions in various animal models of anxiety (e.g. Vogel's and Geller-Seifter's conflict tests in rats) and ameliorated

---

*To whom correspondence should be addressed. E-mail: isibasi@sumitomopharm.co.jp

†Present address; Product Development Division, Sumitomo Pharmaceuticals Co., Ltd., 2-8, Doshomachi 2-chome, Chuo-ku, Osaka 541-8510, Japan.

the stress-induced behavioral responses (e.g. food intake reduction and peptic ulcer formation associated with physical and/or psychological stress). Furthermore, unlike benzodiazepines (e.g. diazepam), it showed antidepressant-like actions in animal models of depression (e.g. forced swim and muricide tests in rats) and did not induce muscle-relaxant, hypnotic and anticonvulsant actions even at high doses (Shimizu et al., 1987, 1992). As compared to buspirone, the prototype of serotonergic anxiolytic agent, tandospirone was much weaker in interacting D_2 receptors (Shimizu et al., 1987). Thus, tandospirone seems to have a broad clinical efficacy and a favorable safety profile in treating anxiety and depression.

Recent advances in research on 5-HT receptor function have suggested that 5-HT$_{1A}$ receptor plays a role not only in reducing anxiety and depression, but also in ameliorating extrapyramidal motor symptoms due to a deficit of nigro-striatal dopaminergic system (Meltzer and Nash, 1991; Millan, 2000). Several studies have shown that selective 5-HT$_{1A}$ agonists (e.g. 8-hydroxy-2-(di-n-propylamino)-tetraline, 8-OH-DPAT), as well as 5-HT$_2$ antagonists (e.g. ritanserin), can relieve the antipsychotic-induced extrapyramidal side effects (e.g. catalepsy and bradykinesia) associated with striatal D_2 receptor blockade (Andersen and Kilpatrick, 1996; Lucas et al., 1997; Neal-Beliveau et al., 1993). 5-HT$_{1A}$ receptor is now recognized to be a potential target site of the certain atypical antipsychotic agents (e.g. perospirone and ziprasidone) with reduced extrapyramidal side effects (Hirose et al., 1990; Meltzer and Nash, 1991; Millan, 2000). Furthermore, Nomoto et al. reported that tandospirone ameliorated walking stability and activity of daily life in patients with Parkinson's disease (Nomoto et al., 1997). These findings suggest that the 5-HT$_{1A}$ agonist tandospirone may have a benefit not only for psychotic symptoms (e.g. anxiety and depression), but also for extrapyramidal motor symptoms in Parkinson's disease. To evaluate the antiparkinsonian effect of tandospirone and explore its mechanism, we studied actions of tandospirone in rat models of Parkinson's disease, i.e. the haloperidol-induced catalepsy model, rotation model in animals with unilateral dopaminergic lesion and the reserpine-induced hypolocomotion model.

MATERIALS AND METHODS

Animals and drugs

Male Sprague-Dawley rats (SLC, Japan) weighing 170–280 g were used. The animals were kept in air-conditioned rooms at $23 \pm 2°C$ and $55 \pm 10\%$ relative humidity under a 12 h-light/12 h-dark cycle (dark period: 20:00–08:00) and given standard rat chow and tipped water ad libitum.

Tandospirone citrate and haloperidol were synthesized in our laboratory. WAY-100635 (N-[2-[4-(2-methoxyphenyl)-1-piperazinyl]ethyl]-N-2-pyridinyl-cyclo-hexanecarboxamide maleate), reserpine (APOPLON®) and 6-hydroxydopamine (6-OHDA) hydrobromide were purchased from RBI (Natick, MA), Daiichi Pharmaceuticals (Tokyo, Japan) and Aldrich Chemical (Milwaukee, WI), respectively.

Haloperidol-induced catalepsy model

Catalepsy was evaluated by the method of Hirose *et al.* (1990) with a slight modification. Haloperidol (30 mg/kg, p.o.) was administered to rats in conjunction with an intraperitoneal injection of tandospirone (2.5–20 mg/kg), bromocriptine (3–30 mg/kg), L-DOPA (25–100 mg/kg) or vehicle. One hour after the drug treatment, the forepaws of each rat were placed on a horizontal bar positioned 9 cm above the bench, and the duration of catalepsy, which defined as the time in seconds while the rat kept this unusual posture, was measured with a maximum cut-off time of 180 s. In the experiments with L-DOPA, benserazide (50 mg/kg, i.p.), a peripheral inhibitor of aromatic amino acid decarboxylase, was injected 15 min before the administration of L-DOPA.

Rotation behavior model in unilateral 6-OHDA-lesioned rats

Unilateral dopaminergic lesion was carried out according to the method of Robertson *et al.* (1991) with a slight modification. Rats were injected with desipramine hydrochloride (25 mg/kg, i.p.) and 30 min later anesthetized with sodium pentobarbital (40 mg/kg, i.p.). After the animals were fixed in a stereotaxic apparatus, unilateral lesions of the right medial forebrain bundle (MFB) were made by injecting 8 μg of 6-OHDA hydrobromide in 4 μl of saline containing 0.05% ascorbic acid. The solution was injected into the MFB (P 4.5, L: 1.3, H: 8.0) over a period of 10 min. Behavioral screening was carried out 4 weeks after the surgery, and the animals that rotated 60 turns/15 min or more in response to an apomorphine (0.5 mg/kg, i.p.) injection were selected. To test the effects of compounds on rotation behavior, rats were placed in a stainless bowl (400 mm in diameter, 200 mm height) and the number of rotations was measured for 60 min immediately after tandospirone injection. In the experiments with bromocriptine, rotation measurement was started 40 min after the injection because bromocriptine usually exhibited a lag of about 40 min to induce the rotation behavior in our preliminary experiments. Each rat was used more than once with a drug-free interval of at least two days.

Reserpine-induced hypolocomotion model

Rats were pretreated with reserpine (1 mg/kg, s.c.) 24 h before the evaluation. The rats were placed in observation cages for a period of 30 min. After habituation, the rats were intraperitoneally injected with each dose of tandospirone (2.5–10 mg/kg) and locomotor activity was measured for 60 min using locomotor activity measurement apparatus (SCANET, Toyo Sangyo, Japan). When the effects of WAY-100635 (3 mg/kg, s.c.) or haloperidol (3 mg/kg, p.o.) were studied, these drugs were administered 30 min before the tandospirone injection.

Statistical evaluation

The results of the catalepsy experiments were analyzed by Kruskal-Wallis analysis of variance, followed by Steel's test. In the reserpine model and the rotation model,

T. Ishibashi and Y. Ohno

the results were analyzed by one-way ANOVA, followed by Dunnett's test. Data are represented as means ± S.E.M. In the catalepsy model, ED_{50} values were calculated using log-probit analysis from values expressed as a percentage of control values.

RESULTS

Haloperidol-induced catalepsy model

Normal animals usually removed one of their forepaws from the horizontal bar within 1 s, and no cataleptic behavior was observed (data not shown). Administration of haloperidol (30 mg/kg, p.o.), a dopamine D_2 antagonist, induced catalepsy in all animals tested and the catalepsy time increased to 175.8 ± 3.0 s. Tandospirone (2.5–20 mg/kg, i.p.) dose-dependently attenuated the haloperidol-induced catalepsy (Fig. 1). The effects of tandospirone at doses of 10 and 20 mg/kg were statistically significant as compared to the control group treated with vehicle alone ($p < 0.01$; Steel's test). The antiparkinsonian agents, bromocriptine (3–30 mg/kg, i.p.) and L-DOPA (25–100 mg/kg, i.p.), also reversed the haloperidol-induced catalepsy in a dose-dependent manner. The effects of bromocriptine were statistically significant at 10 and 30 mg/kg and those of L-DOPA at 100 mg/kg (Fig. 2). The ED_{50} values, which reversed the haloperidol-increased catalepsy time by 50%, were 12.2 mg/kg for tandospirone, 6.3 mg/kg for bromocriptine and 62.4 mg/kg for L-DOPA. In addition, pretreatment of animals with a selective 5-HT_{1A} antagonist, WAY-100635 (3 mg/kg, s.c.) completely abolished the ameliorating effect of tandospirone (10 mg/kg, i.p.) on the haloperidol-induced catalepsy (Fig. 3).

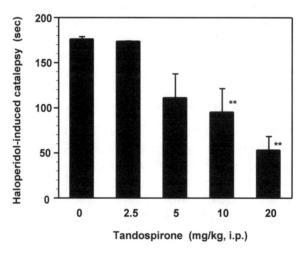

Figure 1. Effect of tandospirone on haloperidol-induced catalepsy in rats. Tandospirone (2.5–20 mg/kg, i.p.) was injected simultaneously with haloperidol (30 mg/kg, p.o.) 60 min before the catalepsy evaluation. Results represent the means ± S.E.M. (*bars*) of the time that animals displayed cataleptic posture (max. 180 s). ** $p < 0.01$ *vs.* vehicle treatment (Steel's test, $n = 8$).

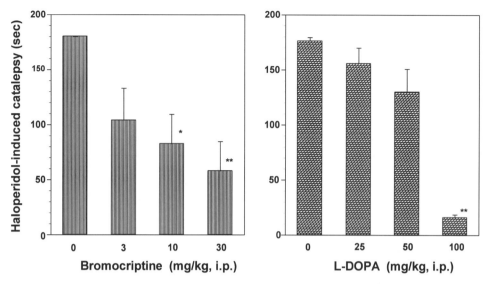

Figure 2. Effect of bromocriptine or L-DOPA on haloperidol-induced catalepsy in rats. Bromocriptine or L-DOPA was injected simultaneously with haloperidol (30 mg/kg, p.o.) 60 min before the evaluation. Benserazide (50 mg/kg, i.p.) was pretreated 15 min prior to L-DOPA injection. Results represent the means ± S.E.M. (*bars*) of the catalepsy time (max. 180 s). *$p < 0.05$, **$p < 0.01$ *vs.* vehicle treatment (Steel's test, $n = 8$).

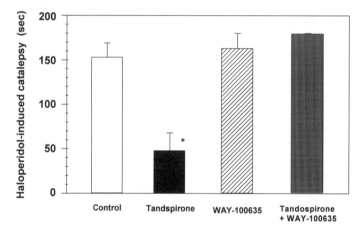

Figure 3. Involvement of 5-HT$_{1A}$ receptor on attenuation of haloperidol-induced catalepsy by tandospirone. WAY-100635 (3 mg/kg, s.c.), a selective 5-HT$_{1A}$ receptor antagonist, or saline were injected simultaneously with tandospirone (10 mg/kg, i.p.) and haloperidol (30 mg/kg, p.o.). *$p < 0.05$ *vs.* control (Steel's test, $n = 8$).

Rotation behavior in 6-OHDA-lesioned rats

In rats with unilateral dopaminergic lesion by 6-OHDA (8 μg/head), vehicle treatment caused only a few rotations (3.3 ± 1.1 turns/60 min). However, treatment with tandospirone (10 and 20 mg/kg, i.p.) markedly enhanced the contralateral

Table 1.
Tandospirone-induced rotation behavior in 6-OHDA-lesioned rats

Drugs	Dose (mg/kg, i.p.)	Contralateral rotation (turns /60 min)
Vehicle		3.3 ± 1.1
Tandospirone	10	170.3 ± 38.5
	20	354.2 ± 53.0
Bromocriptine	10	361.2 ± 85.9

Tandospirone or vehicle (0.5% methylcellulose) were administered just before the evaluation of rotation behavior. Bromocriptine was administered 40 min before the evaluation. Results represented the means $\pm$ S.E.M. of round of contralateral rotation for 60 min ($n = 6$).

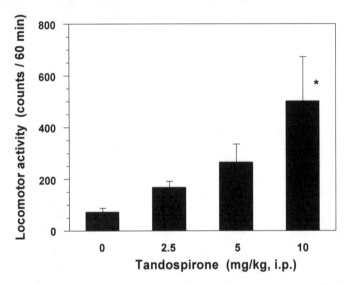

Figure 4. Effect of tandospirone on hypolocomotion induced by reserpine in rats. Tandospirone (2.5–10 mg/kg, i.p.) was injected to rats pretreated with reserpine (1 mg/kg, s.c.) 24 hours before. Results represent the means $\pm$ S.E.M. (*bars*) of total locomotor activity for 60 min. *$p < 0.05$ *vs.* vehicle treatment (Dunnett test, $n = 8$).

rotations in the 6-OHDA-lesioned rats. The number of rotations increased to 170.3 ± 38.5 and 354.2 ± 53.0 turns/60 min in response to the treatment with tandospirone at 10 and 20 mg/kg (i.p.), respectively (Table 1). Bromocriptine (10 mg/kg, i.p.) also significantly increased the contralateral rotations to 361 ± 85.9 turns/60 min. No ipsilateral rotation was observed either with tandospirone or bromocriptine under our experimental conditions.

Reserpine-induced hypolocomotion

Treatment of animals with reserpine (1 mg/kg, s.c.; injected 24 h before locomotor measurement) produced a marked reduction in their spontaneous locomotor activity (saline-control; 1318.5 ± 434.9 counts/60 min, reserpine treament; 95.5 ± 35.7

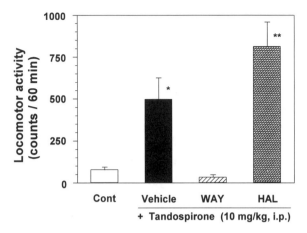

Figure 5. Involvement of dopamine D_2 or 5-HT_{1A} receptor on tandospirone-induced locomotor stimulation in reserpinized rats. Haloperidol (3 mg/kg, p.o.), WAY-100635 (3 mg/kg, s.c.) or vehicle were injected 30 min before tandospirone injection (10 mg/kg, i.p.). Results represent as the means ± S.E.M. (bars) of total locomotor activity for 60 min. $^*p < 0.05$, $^{**}p < 0.01$ *vs.* control (Dunnett's test, $n = 8$).

counts/60 min). As shown in Fig. 4, tandospirone (2.5–10 mg/kg, i.p.) reversed the reserpine-induced hypolocomotion in a dose-dependent manner, and the effect at 10 mg/kg (i.p.) was statistically significant as compared with control ($p < 0.05$, Dunnett's test). Pretreatment with WAY-100635 (3 mg/kg, s.c.) markedly antagonized the stimulant effect of tandospirone (10 mg/kg, i.p.) in the reserpine-treated rats, whereas haloperidol (3 mg/kg, p.o.) did not influence on the effect of tandospirone (Fig. 5).

DISCUSSION

Parkinson's disease is a prevalent and serious neurological disease that is accompanied with specific loss of the nigro-striatal dopamine neurons and extrapyramidal motor symptoms (e.g. akinesia/bradykinesia, tremor, rigidity and postural defect) (Lang and Lozano, 1998). The parkinsonian symptoms can be induced in animals by agents that block striatal dopamine D_2 receptors (e.g. antipsychotic agents such as haloperidol) or cause dopamine deficiency (e.g. reserpine and 6-OHDA) (Jolicoeur and Rivest, 1992). The present study examined the effects of tandospirone, a new serotonergic anxiolytic agent, in animal models of Parkinson's disease and demonstrated that the agent has a significant antiparkinsonian action.

The antipsychotic-induced catalepsy is a sign of extrapyramidal side effects which are known to be brought about through inhibition of the nigro-striatal dopaminergic neurotransmission (Elliott *et al.*, 1990; Jolicoeur and Rivest, 1992; Kobayashi *et al.*, 1997). In this study, tandospirone significantly reversed the haloperidol-induced catalepsy. The potency of tandospirone in ameliorating catalepsy was similar to that of bromocriptine but greater than that of L-DOPA. In addition, the effect of tandospirone was eliminated by simultaneous administration with a selective 5-HT$_{1A}$ antagonist WAY-100635. Consistent with previous findings (Andersen and Kilpatrick, 1996; Neal-Beliveau et al., 1993), our results suggest that activation of 5-HT$_{1A}$ receptor by tandospirone can reduce catalepsy.

Reserpine interferes with the storage of monoamines in the nerve terminals, results in depletion of monoamines and induces parkinsonism-like behaviors (e.g. hypolocomotion and muscular rigidity) (Colpaert, 1987; Jolicoeur and Rivest, 1992). The rotation model in the dopaminergic hemilesioned rat by 6-OHDA is also widely used in order to evaluate antiparkinsonian activity (Jolicoeur and Rivest, 1992; Ungerstedt and Arbuthnott, 1970). In both animal models, tandospirone significantly improved motor deficits, in that the agent dose-dependently reversed the reserpine-induced hypolocomotion and induced the contralateral rotation in the 6-OHDA-lesioned rats. Furthermore, the ameliorative effects of tandospirone on the reserpine-induced hypolocomotion was fully antagonized by the co-administration with WAY-100635, but not with haloperidol. These results strongly suggest that tandospirone improved parkinsonian symptoms through direct activation of 5-HT$_{1A}$ receptor, but not through secondary activation of D$_2$ receptor.

It is reported that serotonergic neurons negatively regulate the firing of dopaminergic neurons in the midbrain probably through 5-HT$_2$ receptors and inhibit dopamine release in the striatum (Saller et al., 1990; Ugedo et al., 1989). Since the microinjection of 8-OH-DPAT, a selective 5-HT$_{1A}$ receptor agonist, into the medial raphe nucleus can reduce the raclopride-induced catalepsy in rats (Wadenberg and Hillegaart, 1995), it is suggested that inhibition of 5-HT neurons by activation of somatodendric 5-HT$_{1A}$ autoreceptor might disinhibit the activity of dopamine neurons and thereby reduce catalepsy. On the other hand, several studies have suggested that postsynaptic 5-HT$_{1A}$ receptor play a crucial role in the antiparkinsonian action of 5-HT$_{1A}$ agonists. In early studies, Costall et al. (1976) demonstrated that intrastriatal injection of 5-HT at doses compatible to those of dopamine (5-HT, 25–100 μg; dopamine, 50–100 μg) caused a contralateral circling behavior in rats, suggesting that activation of striatal 5-HT receptors is involved in the antiparkinsonian effects. Furthermore, the anti-cataleptic action of 8-OH-DPAT has been shown to persist even in rats of which 5-HT contents had been depleted by treatment with p-chlorophenylalanine (Neal-Beliveau et al., 1993). Our finding that the ameliorative effect of tandospirone is mediated by 5-HT$_{1A}$ receptor activation, but not by D$_2$ receptor, supports the latter possibility and implies a potential role of postsynaptic 5-HT$_{1A}$ receptor in its antiparkinsonian effects. Further studies are required to elucidate the precise mechanism of antiparkinsonian action of tandospirone.

Previous studies demonstrated that tandospirone shows significant anxiolytic and antidepressant effects without producing benzodiazepine-like side effects, such as sedation, muscle relaxation, potentiation of anesthesia, impairment of motor coordination, cognitive disturbances and drug dependence. In addition, tandospirone is much weaker than buspirone, a prototype of 5-HT$_{1A}$ receptor-related anxiolytic agent, in inducing dopamine D$_2$ blocking actions. Since anxiety ind depression are the symptoms frequently observed in patients with Parkinson's lisease (Menza *et al.*, 1993; Menza and Mark, 1994), tandospirone would be useful for the treatment of anxiety and depression in Parkinson's disease. Furthermore, :he present study demonstrated that tandospirone directly improved motor deficits n an animal model of Parkinson's disease. The dosage of tandospirone for the antiparkinsonian effects was comparable to that for anxiolytic activities reported an previous studies (Shimizu *et al.*, 1987, 1992) and was similar to that of promocriptine (D_2 agonist). Our results support the recent clinical findings (Nomoto *et al.*, 1997) that tandospirone ameliorated walking stability in about 40% of patients with Parkinson's disease, and strongly suggest that tandospirone has a therapeutic potential in treating parkinsonian symptoms.

Acknowledgement

The authors are very grateful to Mrs. H. Ohki for her skillful technical assistance.

REFERENCES

Andersen, H. L. and Kilpatrick, I. C. (1996). Prevention by ($\pm$)-8-hydroxy-2-(di-n-propyl-amino)tetralin of both catalepsy and the rises in rat striatal dopamine metabolism caused by haloperidol, *Br. J. Pharmacol.* **118**, 421–427.

Colpaert, F. C. (1987). Pharmacological characteristics of tremor, rigidity and hypokinesia induced by reserpine in rat, *Neuropharmacology* **26**, 1431–1440.

Costall, B., Naylor, R. J. and Pycock, C. (1976). Non-specific supersensitivity of striatal dopamine receptors after 6-hydroxydopamine lesion of the nigrostriatal pathway, *Eur. J. Pharmacol.* **35**, 275–283.

Elliott, P. J., Close, S. P., Walsh, D. M., *et al.* (1990). Neuroleptic-induced catalepsy as a model of Parkinson's disease. I. Effect of dopaminergic agents, *Neural Transm. Park. Dis. Dement. Sect.* **2**, 79–89.

Hirose, A., Kato, T., Ohno, Y., *et al.* (1990). Pharmacological actions of SM-9018, a new neuroleptic drug with both potent 5-hydroxytryptamine$_2$ and dopamine$_2$ antagonistic actions, *Jpn. J. Pharmacol.* **53**, 321–329.

Jolicoeur, F. B. and Rivest, R. (1992). Rodent models of Parkinson's disease, in: *Neuromethods, Vol. 21: Animal Models of Neurological Disease I*, Boulton, A., Baker, G. and Butterworth, R. (Eds), pp. 135–158. The Humana Press, New Jersey, USA.

Kobayashi, T., Araki, T., Itoyama, Y., *et al.* (1997). Effects of L-DOPA and bromocriptine on haloperidol-induced motor deficits in mice, *Life Sci.* **61**, 2529–2538.

Lang, A. E. and Lozano, A. M. (1998). Parkinson's disease. First of two parts, *New Engl. J. Med.* **339**, 1044–1053.

Lucas, G., Bonhomme, N., De Deurwaerdère, P., *et al.* (1997). 8-OH-DPAT, a 5-HT$_{1A}$ agonist and ritanserin, a 5-HT$_{2A/C}$ antagonist, reverse haloperidol-induced catalepsy in rats independently of striatal dopamine release, *Psychopharmacology* **131**, 57–63.

Meltzer, H. Y. and Nash, J. F. (1991). Effects of antipsychotic drugs on serotonin receptors, *J. Pharmacol. Exp. Ther.* **295**, 853–861.

Menza, A. M. and Mark, M. H. (1994). Parkinson's disease and depression: the relationship to disability and personality, *J. Neuropsychiatry Clin. Neurosci.* **6**, 165–169.

Menza, A. M., Robertson-Hoffman, D. E. and Bonapace, A. S. (1993). Parkinson's disease and anxiety: comorbidity with depression, *Biol. Psychiatry* **34**, 465–470.

Millan, M. J. (2000). Improving the treatment of schizophrenia: Focus on serotonin (5-HT)$_{1A}$ receptors, *J. Pharmacol. Exp. Ther.* **295**, 853–861.

Neal-Beliveau, B. S., Joyce, J. N. and Lucki, I. (1993). Serotonergic involvement in haloperidol-induced catalepsy, *J. Pharmacol. Exp. Ther.* **265**, 207–217.

Nomoto, M., Iwata, S., Kaseda, S., *et al.* (1997). A 5-HT$_{1A}$ receptor agonist, tandospirone improves gait disturbance of patients with Parkinson's disease, *J. Mov. Disord. Disability* **7**, 65–70.

Robertson, G. S., Fine, A. and Robertson, H. A. (1991). Dopaminergic grafts in the striatum reduce D1 but not D2 receptor-mediated rotation in 6-OHDA-lesioned rats, *Brain Res.* **539**, 304–311.

Saller, C. F., Czupryna, M. J. and Salama, A. I. (1990). 5-HT$_2$ receptor blockade of ICI 169,369 and other 5-HT$_2$ antagonist modulates the effects of D-2 dopamine receptor blockade, *J. Pharmacol. Exp. Ther.* **253**, 1162–1170.

Shimizu, H., Hirose, A., Tatsuno, T., *et al.* (1987). Pharmacological properties of SM-3997: a new anxioselective anxiolytic candidate, *Jpn. J. Pharmacol.* **45**, 493–500.

Shimizu, H., Karai, N., Hirose, A., *et al.* (1988). Interaction of SM-3997 with serotonin receptors in rat brain, *Jpn. J. Pharmacol.* **46**, 311–314.

Shimizu, H., Kumasaka, Y., Tanaka, H., *et al.* (1992). Anticonflict action of tandospirone in a modified Geller-Seiffer conflict test in rats, *Jpn. J. Pharmacol.* **58**, 283–289.

Ugedo, L., Grenhoff, J. and Svensson, T. H. (1989). Ritanserin, a 5-HT$_2$ receptor antagonist, activates midbrain dopamine neurons by blocking serotonergic inhibition, *Psychopharmacology* **98**, 45–50.

Ungerstedt, U. and Arbuthnott, G. W. (1970). Quantitative recording of rotational behavior in rats after 6-hydroxy-dopamine lesions of the nigrostriatal dopamine system, *Brain Res.* **24**, 485–493.

Wadenberg, M.-L. and Hillegaart, V. (1995). Stimulation of median, but not dorsal, raphe 5-HT$_{1A}$ autoreceptors by the local application of 8-OH-DPAT reverses raclopride-induced catalepsy in the rat, *Neuropharmacology* **34**, 495–499.

Advances in Neuroregulation and Neuroprotection (2005), pp. 379-387
Collin, C. et al. (Eds)
© VSP 2005

Post-encephalitic parkinsonism: clinical features and experimental model

AKIHIKO OGATA [1,*], NAOYA HAMAUE [2], MASARU MINAMI [2],
ICHIRO YABE [1], SEIJI KIKUCHI [1], HIDENAO SASAKI [1]
and KUNIO TASHIRO [1]

[1] Department of Neurology, Hokkaido University Graduate School of Medicine, Kita-15, Nishi-7,
Kita-ku, Sapporo 060-8638, Japan
[2] Department of Pharmacology, Faculty of Pharmaceutical Sciences, Health Scienses University of
Hokkaido, Ishikari-Tobetsu, Hokkaido 061-0293, Japan

Abstract—Post-encephalitic parkinsonism has been well documented by now. Recently, an Indian group reported some post-encephalitic parkinsonism patients that resembled Japanese encephalitis cases with bilateral substantia nigra lesions that were detected by MRI. Post-encephalitis parkinsonism has been described following Coxsackie B virus, influenza A, poliovirus and measles virus infections. The possible involvement of virus infection has also been supported by experimental animal models. We have demonstrated pathological and to a certain extent clinical features consistent with Parkinson's disease following infection of rats with Japanese encephalitis virus (JEV). We evaluated new treatments with Parkinson's disease using this model. It was reported that tremor primarily involving the fingers, tongue and eyelids, muscle rigidity and masked face as the clinical features by JEV. Recently brainstem lesions produced by West Nile virus, which is a flavivirus like JEV, were documented by MRI. In the future, post-encephalitic parkinsonism may be found elsewhere in the world.

In adult Fischer rats sacrificed 12 weeks after infection with JEV at the age of 13 days, neuronal loss with gliosis was confined mainly to the zona compacta of the bilateral substantia nigra, without lesions in the cerebral cortex and cerebellum. Furthermore, the most severe lesions were in the central part of zona compacta; the lateral cell groups were less affected. Thus, the distribution of the pathologic lesions in infected rats resembled those found in Parkinson's disease. JEV-infected rats showed marked bradykinesia. Significant behavioral improvement was observed upon administration of L-DOPA and monoamine oxidase (MAO) inhibitor. The findings suggest that JEV infection of rats under the conditions described may serve as a model of virus induced Parkinson's disease.

Keywords: Post-encephalitic parkinsonism; Parkinson's disease; dopamine; Japanese encephalitis virus; Parkinson's disease model.

*To whom correspondence should be addressed. E-mail: a-ogata@med.hokudai.ac.jp

INTRODUCTION

One of the models of Parkinson's disease is that induced by the metabolites of 1-methyl-4-phenyl 1,2,3,6-tetrahydropyridine (MPTP) (Ballard *et al.*, 1985; Burns *et al.*, 1983; Langston *et al.*, 1983; Vingerhoets *et al.*, 1994). The neuropathologic findings in MPTP-treated monkeys are similar to those of Parkinson's disease, in that neurons in the substantia nigra are primarily involved, while other basal ganglia structures are spared. Despite these remarkable similarities, MPTP-induced parkinsonism is different from Parkinson's disease in several respects (Ballard *et al.*, 1985; Langston, 1989; Langston *et al.*, 1983). Furthermore, MPTP acts only minimally or not at all in several animal species, including rats (Heikkila *et al.*, 1984). There is another model of parkinsonism, induced by stereotactic injection of 6-hydroxydopamine into the substantia nigra (Kelly *et al.*, 1975). We have reported the distribution of Japanese encephalitis virus (JEV) antigens in developing rat brain after intracerebral virus inoculation (Ogata *et al.*, 1991). We concluded that the susceptibility to JEV infection in the rat brain was closely associated with neuronal immaturity; the neurons of basal ganglia and substantia nigra remained susceptible to infection longer than did those of the cerebral cortex. In 10-day-old rats infected with JEV, JEV antigen was almost undetectable in the cerebral cortex, and in 14-day-old rats JEV antigen was confined to the basal ganglia and substantia nigra (Ogata *et al.*, 1991). The analysis of the neuropathologic findings in rats infected with JEV on days 12, 13 and 14 after birth suggested that it would be possible to induce pathologic lesions similar to those of Parkinson's disease. These results strengthen the hypothesis that JEV might be a causal agent for some cases of Parkinsonian syndrome. Pradhan *et al.* (1999) reported clinical states like this model. Recently, brainstem lesions produced by West Nile virus, which is a flavivirus like JEV, were documented by MRI (Agid *et al.*, 2003).

Monoamine oxidase (MAO) is an enzyme that metabolizes a wide range of monoamines and plays an important role in control of the concentration of neurotransmitters. However, very little has been elucidated about detailed CNS functions of isatin. Some experiments suggested that isatin, an endogenous MAO inhibitor, can serve as marker for stress and anxiety. Exogenously administered isatin increased DA levels in the rat striatum. We recently discovered that a significant increase in urinary isatin excretion was observed in patients with Parkinson's disease (Yahr III, IV, V) (Hamaue *et al.*, 2000). These results suggested that isatin may be associated with processes in the pathomechanism of Parkinson's disease. Many therapies for Parkinson's disease are related to overcoming the declining dopamine levels that occur in the striatum as a result of the disease.

MATERIALS AND METHODS

Animals and virus

Albino rats of the Fischer strain were obtained from SLC Japan. The virus strain used was the JaGAr-01 strain of JEV. The supernates from 10% homogenates of infected mouse brains (10^9 PFU/ml) were diluted with 20% Hemaccell (Hoechst) in Eagle's minimum essential medium and stored at $-70°C$ until use. Virus (0.03 ml containing 3×10^6 PFU) was inoculated intracerebrally with a specially designed two-step thin 27 gauge needle (Hoshimori Iryoki KK, Tokyo, Japan) with a stopper 3 mm from the tip. The site of inoculation was located at the midpoint of the line connecting the left eye to the midpoint between the right and left ears. Hemaccel (20%) in Eagle's minimum essential medium was injected into control rats.

Neuropathologic study

The topographical distribution of JEV antigen in the developing rat brain was determined 3 days after JEV inoculation. The neuropathologic changes in rats infected with JEV on days 12, 13, and 14 after birth were examined. Animals were sacrificed 3 days, 10 days, 12 weeks, or one year after inoculation under ether anesthesia by perfusion fixation via the aorta with 4% freshly prepared paraformaldehyde in 0.1 M phosphate buffer. Coronal brain sections were taken from the frontal tip to the medulla, embedded in paraffin, and stained with hematoxylin-eosin, Luxol fast blue-cresyl violet (Klüver-Barrera method), anti-JEV antibody, and anti-tyrosine hydroxylase (TH) monoclonal antibody (Chemicon). The avidin-biotin-peroxidase comlex (ABC) method was used in this immunohistochemical study (Hsu *et al.*, 1981). After deparaffinization, the specimens were treated with 0.3% H_2O_2-methanol to suppress endogenous peroxidase activity, incubated with 10% normal goat serum, and allowed to react with anti-JEV rabbit serum or anti-TH monoclonal antibody, diluted in 1% bovine serum albumin (BSA) at 4°C overnight. Incubation with 1% BSA only was used as a negative control. The sections reacted with anti-JEV antibody and anti-TH monoclonal antibody were reacted with biotinylated goat anti-rabbit IgG and biotinylated goat anti-mouse IgG (Vecstain), respectively. ABC reaction products were visualized with 3,3'-diaminobenzidine tetrahydrochloride (Sigma), and counterstained with hematoxylin.

Assessment of motor function

A pole test (Ogawa *et al.*, 1985) was performed as a way of evaluating bradykinesia in the rats. The time for the rats to descend from the top of a rough-surfaced pole (2.5 cm in diameter and 100 cm in height) to the floor was recorded in JEV-infected adult rats and control adult rats. To assess the efficiency of L-DOPA, selegiline, and isatin, the same procedure was repeated immediately after the successive intraperitoneal injection of L-DOPA (10 mg/kg/day), selegiline and isatin (100 mg/kg per day) for 7 days. Comparisons among mean values for groups were done by Student's *t*-test.

RESULTS

The mortality of the rats infected with JEV

The mortality of rats infected with JEV is shown in Fig. 1. All rats infected when they were from 2 to 12 days old died. However, the mortality rate of animals infected when they were older than 12 days decreased with age, with 13-day-old rats having 50% mortality and 14-day-old rats having 8.3% mortality. Animals inoculated when they were older than 17 days showed 0% mortality. This showed that the older the rat was at the time of infection, the longer it survived.

Neuropathologic study

Pathological examination at higher magnifications clearly showed that, in rats infected with JEV at the age of 7 days, JEV antigen was present in the apical cytoplasmic portion of the neurons in the superficial areas of layer II. Since the neuronal maturation of the cerebral cortex is known to proceed from the deeper layers (layers V and VI) to the upper layers (layers III and II), the results indicate that neurons in the cerebral cortex become resistant to JEV infection with maturation. In 15-day-old animals sacrificed 3 days after JEV infection, JEV antigen was found mainly in the substantia nigra and the caudate putamen in a distribution similar to that previously described (Ogata *et al.*, 1991). The expression of JEV antigen was strikingly localized to neurons in the substantia nigra. However,

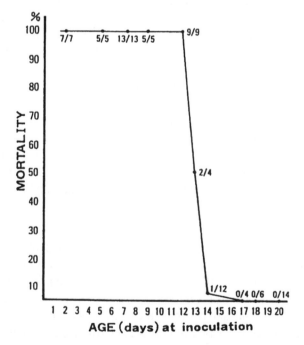

Figure 1. Relationship of age at inoculation and mortality in rats infected intracerebrally with Japanese encephalitis virus. A striking age-dependent decrease in susceptibility is evident.

JEV antigen had disappeared from the brains of the rats sacrificed 10 days, 12 weeks, and one year after infection. In the adult rats previously infected with JEV at the age of 13 days, neuronal loss was confined mainly to the zona compacta of the bilateral substantia nigra without lesions in the cerebral cortex or cerebellum. Furthermore, the most severe lesions were found in the central part of the zona compacta, whereas the lateral cell groups were less affected. In the JEV-treated rats, marked neuronal loss with gliosis was found in the zona compacta of the substantia nigra, in contrast to the controls. In addition, immunohistochemical analysis with anti-tyrosine hydroxylase (TH) antibody showed that TH-positive neurons were decreased in the substantia nigra. However, many TH-positive neurons were found in the caudate putamen in JEV-infected rats as well as an age-matched control rat. These pathologic findings imply rather selective vulnerability of neurons of the substantia nigra and to a lesser extent caudate putamen of rats to the destructive effects of JEV following intracerebral inoculation. In the JEV-infected rat brains, the same pathologic features were observed in more than 80% of the rats. In the rats sacrificed one year after JEV inoculation, almost no TH-positive neurons could be found in the substantia nigra. This finding suggests that the function of the dopaminergic system may deteriorate further with age. Although no Lewy bodies could be observed and immunohistochemical staining with an anti-tau antibody was not seen, the pathologic distribution found in JEV-treated rats was similar to that found in Parkinson's disease.

Motor function

We measured the motor activity of the adult rats (12 weeks after infection) infected with JEV at the age of 13 days and age-matched control rats. The difference between the JEV-infected rats and the control rats was significant ($p < 0.001$). The pole test results are consistent with marked bradykinesia of the JEV-infected rats. Masked faces or tremor could not be assessed in the rats. Significant behavioral improvement was observed after each treatment with L-DOPA, selegiline (Hamaue *et al.*, 2001) and isatin (Ogata *et al.*, 2003).

DISCUSSION

The pathogenesis of Parkinson's disease currently is thought to depend upon hereditary, aging, and environmental factors (Adams and Odunze, 1991; Ballard *et al.*, 1985; Burns *et al.*, 1983; Calne and Langston, 1983; Cohen, 1986; Dexter *et al.*, 1989; Nagatsu and Yoshida, 1988; Riederer *et al.*, 1989; Yoshida *et al.*, 1990). Among the environmental factors, attention has largely focused on the role of toxic and infectious agents. Among the toxic factors, exogenous toxins such as MPTP and endogenous toxins such as free radicals or tetrahydroisoquinoline (Yoshida *et al.*, 1990) have been implicated. Viruses can also selectively attack the substantia nigra and induce parkinsonism (Duvoisin and Yahr, 1965; Kristensson, 1992). Post-encephalitic parkinsonism has been well documented (von Economo, 1917; Yahr,

1978). The world pandemic of encephalitis lethargica, von Economo encephalitis, from 1916 to 1927 resulted in the death or disability of approximately half a million people. Walters (1960) described a 54-year-old woman who manifested symptoms of parkinsonism while convalescing from meningoencephalitis due to Coxsackie B virus. Influenza A virus (Hudson and Rice, 1990), poliovirus (Bojinov, 1971), and measles virus (Alves *et al.*, 1992) have also been suspected from case reports as possible viral causes. Fishman *et al.* (1985) reported an experimental model with a selective attack on the substantia nigra and subthalamic nucleus by a strain of mouse hepatitis virus which is a coronavirus that causes persistent CNS infection. Recently, similar features have been described in rats infected with influenza virus (Takahashi *et al.*, 1995).

JEV is a positive-stranded enveloped RNA virus that belongs to the family of the flaviviruses and is the most common cause of arthropod-borne human encephalitis worldwide. JEV is also occasionally the cause of post-encephalitis parkinsonism. Dickerson *et al.* (1952), in a study of 200 acute cases of Japanese encephalitis, noted tremors chiefly involving the fingers, tongue, and eyelids in 90%, muscular rigidity in more than 40%, and a mask-like face in 75% of the cases. Goto (1962) detected parkinsonian sequelae in 11.6% of 143 unselected patients five years after they had Japanese encephalitis.

The parkinsonian syndrome after Japanese encephalitis differs from that which follows encephalitis lethargica in several respects. In general, parkinsonism after Japanese encephalitis is mild, develops in the acute phase, and occasionally improves slightly over a long period. Recently, it was reported that MRI abnormalities were seen mainly in the substantia nigra and putamen in a case of typical parkinsonism after Japanese encephalitis (Shoji *et al.*, 1993). Patients without a clear history of encephalitis who follow the clinical course and have pathologic findings consistent with postencephalitic parkinsonism have been reported (Geddes *et al.*, 1993; Gibb and Lees, 1987). Although there were no Lewy bodies found in the substantia nigra in the JEV-treated rats, the pathologic findings otherwise resembled those of idiopathic Parkinson's disease. Furthermore, the immunohistochemical data using anti-TH antibody suggested that the function of the dopaminergic system might have deteriorated with age in the absence of ongoing or persistent JEV infection. McGeer *et al.* (1988) showed that the rate of neuronal cell degeneration was considerably higher in Parkinsonian patients than could be accounted for on the basis of normal age-related neuronal degeneration alone. It seems likely that neuronal cell degeneration progressed more rapidly in the JEV-treated rats than in the controls. This observation raises the possibility that post-encephalitic parkinsonism as well as Parkinson's disease is a continuing degenerative process rather than an acute illness on which the effects of aging or decompensation are superimposed. Such late deterioration might be due to a resurgence of viral-mediated damage (Appel *et al.*, 1992), though our model does not support this view.

The complete nucleotide sequence of JEV genome RNA has been determined (Sumiyoshi *et al.*, 1987). Our RT-PCR study for NS3 region amplification (Morita *et al.*, 1991) of the JEV genome showed that JEV genome was undetectable in rats sacrificed 12 weeks after JEV infection at the age of 13 days. Moreover, JEV antigen as well as the JEV genome disappeared from the brain. These findings indicate that there is no persistent infection in the brain and suggest that, following the acute phase, JEV-infected rats are a safe model for researchers.

Thus far, no virus has been isolated from patients with Parkinson's disease, and there are no data that directly link known viruses to idiopathic Parkinson's disease. However, our findings support the possibility that as yet unidentified specific pathogens could cause similar pathologic lesions in man resulting in Parkinson's disease. Why neurons of the subtantia nigra remain susceptible to JEV infection longer than those in other parts of the brain is unclear. One possibility is that virus receptors on the substantia nigra neurons persist longer. Certainly, the capacity of viruses to attack specific tissues selectively depends on an interaction between viral genes or proteins and host factors. The immune mechanism following an infection or other factors could be associated with the destruction of the substantia nigra. A more detailed understanding of JEV tropism for the substantia nigra in this experimental model might reveal mechanisms that aid in unraveling the degeneration of nigral dopaminergic neurons that is central to Parkinson's disease.

The JEV-induced parkinsonism in rats is characterized by selective destruction of neurons in the bilateral substantia nigra, especially in the zona compacta of the substantia nigra, similar to the lesions found in Parkinson's disease. Furthermore, significant behavioral improvement was observed upon administration of L-DOPA (Ogata *et al.*, 1997), selegiline and isatin. Accordingly, this JEV-induced parkinsonian model is expected to be useful in the assessment of new anti-parkinsonian drugs (Hamaue *et al.*, 2001; Ogata *et al.*, 1998, 2003) the efficiency of neuronal transplantation therapy, and in the study of the pathophysiology of parkinsonism.

Isatin was found to be a potent MAO inhibitor that is more active against MAO-B than MAO-A. Moreover, exogenously administered isatin increased both ACh and DA levels in the rat striatum. Kumar *et al.* (1994) reported that isatin inhibits ACh esterase activity in rat brain. Thus, isatin may ameliorate dementia such as that associated with Alzheimer's disease. Although the biosynthetic and metabolic pathways of isatin have not been established, we reported that a significant increase in urinary isatin excretion was observed in patients with Parkinson's disease (Yahr Stages III, IV, and V) (Hamaue *et al.*, 2000). The increase in urinary isatin excretion may be a compensatory response to a lower level of cerebral DA. Thus, the results support that isatin will ameliorates Parkinson's disease as well as parkinsonism induced by JEV.

REFERENCES

Adams, J. D. and Odunze, I. N. (1991). Oxygen free radicals and Parkinson's disease, *Free Radical Biol. Med.* **10**, 161–169.

Agid, R., Ducreuz, D., Halliday, W. C., *et al.* (2003). MR diffusion-weighted imaging in a case of West Nile virus encephalitis, *Neurology* **61**, 1821.

Alves, R. S. C., Barbosa, E. R. and Scaff, M. (1992). Postvaccinial parkinsonism, *Movement Disorders* **7**, 178–180.

Appel, S. H., Le, W.-D., Tajti, J., *et al.* (1992). Nigral damage and dopaminergic hypofunction in mesencephalon-immunized guinea pigs, *Ann. Neurol.* **32**, 494–501.

Ballard, P. A., Tetrud, J. W. and Langston, J. W. (1985). Permanent human parkinsonism due to 1-methyl-4-phenyl-1,2,3,6-tetrahydropyridine (MPTP): seven cases, *Neurology* **35**, 949–956.

Bojinov, S. (1971). Encephalitis with acute parkinsonian syndrome and bilateral inflammatory necrosis of the substantia nigra, *J. Neurol. Sci.* **12**, 383–415.

Burns, R. S., Chiueh, C. C., Markey, S. P., *et al.* (1983). A primate model of parkinsonism: Selective destruction of dopaminergic neurons in the pars compacta of the substantia nigra by N-methyl-4-phenyl-1,2,3,6-tetrahydropyridine, *Proc. Natl. Acad. Sci. USA* **80**, 4546–4550.

Calne, D. B. and Langston, J. W. (1983). Aetiology of Parkinson's disease, *Lancet* **2**, 1457–1459.

Cohen, G. (1986). Monoamine oxidase, hydrogen peroxidase and Parkinson's disease, *Adv. Neurol.* **45**, 119–125.

Dexter, D. T., Wells, F. R., Lees, A. J., *et al.* (1989). Increased nigral iron content and alterations in other metal ions occurring in brain in Parkinson's disease, *J. Neurochem.* **52**, 1830–1836.

Dickerson, R. B., Newion, J. R. and Hansen, J. E. (1952). Diagnosis and immediate prognosis of Japanese B encephalitis, *Am. J. Med.* **12**, 277–288.

Duvoisin, R. C. and Yahr, M. D. (1965). Encephalitis and parkinsonism, *Arch. Neurol.* **12**, 227–239.

Fishman, P. S., Gass, J. S., Swoveland, P. T., *et al.* (1985). Infection of the basal ganglia by a murine coronavirus, *Science* **229**, 877–879.

Geddes, J. F., Hughes, A. J., Lees, A. J., *et al.* (1993). Pathological overlap in cases of parkinsonism associated with neurofibrillary tangles. A study of recent cases of postencephalitic parkinsonism and comparison with progressive supranuclear palsy and Guamanian parkinsonism-dementia complex, *Brain* **116**, 281–302.

Gibb, W. R. G. and Lees, A. J. (1987). The progression of idiopathic Parkinson's disease is not explained by age-related changes. Clinical and pathological comparisons with post-encephalitic parkinsonian syndrome, *Acta. Neuropathol. (Berl.)* **73**, 195–201.

Goto, A. (1962). Follow-up study of Japanese B encephalitis, *Psychiat. Neurol. Jpn. (Tokyo)* **64**, 236–266.

Hamaue, N., Yamazaki, N., Terado, M., *et al.* (2000). Urinary isatin concentrations in patients with Parkinson's disease determined by a newly developed HPLC-UV method, *Res. Commun. Mol. Pathol. Pharmacol.* **108**, 63–73.

Hamaue, N., Ogata, A., Terado, M., *et al.* (2001). Selegiline effects on bradykinesia and dopaminelevels in a rat model of Parkinson's disease induced by the Japanese encephalitis virus, *Biogenic Amines* **16**, 523–530.

Heikkila, R. E., Hess, A. and Duvoisin, R. C. (1984). Dopaminergic neurotoxicity of 1-methyl-4-phenyl-1,2,5,6-tetrahydropyridine in mice, *Science* **224**, 1451–1453.

Hsu, S. M., Raine, L. and Fanger, H. (1981). Use of avidin-biotin-peroxidase complex (ABC) in immunoperoxidase techniques: a comparison between ABC and unlabeled antibody (PAP) procedures, *J. Histochem. Cytochem.* **29**, 577–580.

Hudson, A. J. and Rice, G. P. A. (1990). Similarities of Guamanian ALS/PD to post-encephalitic parkinsonism/ALS: possible viral cause, *Can. J. Neurol. Sci.* **17**, 427–433.

Kelly, P. H., Seviour, P. W. and Iversen, S. D. (1975). Amphetamine and apomorphine responses in the rat following 6-OHDA lesions of the nucleus accumbens septi and corpus striatum, *Brain Res.* **94**, 507–522.

Kristensson, K. (1992). Potential role of viruses in neuro-degeneration, *Mol. Chem. Neuropathol.* **16**, 45–58.

Kumar, R., Bansal, R. C. and Mahmood, A. (1994). In vivo effects of isatin on certain enzymes, lipids and serotonergic system of rat brain, *Indian J. Med. Res.* **100**, 246–250.

Langston, J. W. (1989). Current theories on the cause of Parkinson's disease, *J. Neurol. Neurosurg. Psychiatry* **52**, 13–17.

Langston, J. W., Ballard, P., Tetrud, J. W., *et al.* (1983). Chronic parkinsonism in humans due to a product of meperidine-analog synthesis, *Science* **219**, 979–980.

McGeer, P. L., Itagaki, S., Akiyama, H., *et al.* (1988). Rate of cell death in parkinsonism indicates active neuropathological process, *Ann. Neurol.* **24**, 574–576.

Morita, K., Tanaka, M. and Igarashi, A. (1991). Rapid identification of dengue virus serotypes by using polymerase chain reaction, *J. Clin. Microbiol.* **29**, 2107–2110.

Nagatsu, T. and Yoshida, M. (1988). An endogenous substance of the brain, tetrahydroiso-quinoline, produces parkinsonism in primates with decreased dopamine, tyrosine hydroxylase and biopterin in the nigrostriatal regions, *Neurosci Lett.* **87**, 178–182.

Ogata, A., Nagashima, K., Hall, W. W., *et al.* (1991). Japanese encephalitis virus neurotropism is dependent on the degree of neuronal maturity, *J. Virol.* **65**, 880–886.

Ogata, A., Tashiro, K., Nukuzuma, S., *et al.* (1997). A rat model of Parkinson's disease induced by Japanese encephalitis virus, *J. Neurol. Virol.* **3**, 141–147.

Ogata, A., Nagashima, K., Yasui, K., *et al.* (1998). Sustained release dosage of thyrotropin-releasing hormone improves experimental Japanese encephalitis virus-induced parkinsonism in rats, *J. Neurol. Sci.* **159**, 135–139.

Ogata, A., Hamaue, N., Terado, M., *et al.* (2003). Isatin, an endogenous MAO inhibitor, improves bradykinesia and dopamine levels in a rat model of Parkinson's disease induced by Japanese encephalitis virus, *J. Neurol. Sci.* **206**, 79–83.

Ogawa, N., Hirose, Y., Ohara, S., *et al.* (1985). A simple quantitative bradykinesia test in MPTP-treated mice, *Res. Commun. Chem. Pathol. Pharmacol.* **50**, 435–441.

Pradhan, S., Pandey, D. M., Shanshank, S., *et al.* (1999). Parkinsonism due to predominant involvement of substantia nigra in Japanese encephalitis virus, *Neurology* **53**, 1781–1786.

Richter, R. W. and Shimojyo, S. (1961). Neurologic sequelae of Japanese B encephalitis, *Neurology* **11**, 553–559.

Riederer, P., Sotic, E., Rausch, W. D., *et al.* (1989). Transition metals ferritin, glutathione and ascorbic acid in Parkinsonian brain, *J. Neurochem.* **52**, 515–520.

Shoji, H., Watanabe, M., Itoh, S., *et al.* (1993). Japanese encephalitis and parkinsonism, *J. Neurol.* **240**, 59–60.

Sumiyoshi, H., Mori, C., Fuke, I., *et al.* (1987). Complete nucleotide sequence of the Japanese encephalitis virus genome RNA, *Virology* **161**, 497–510.

Takahashi, M., Yamada, T., Nakajima, S., *et al.* (1995). The substania nigra is a major target f′ or neurovirulent influenza A virus, *J. Exp. Med.* **181**, 2161–2169.

Vingerhoets, F. J. G., Snow, B. J., Tetrud, J. W., *et al.* (1994). Positron emission tomographic evidence for progression of human MPTP-induced dopaminergic lesions, *Ann. Neurol.* **36**, 765–770.

von Economo, C. (1917). Encephalitis lethargica, *Wjen kljn Wschr* **30**, 581–585.

Walters, J. H. (1960). Post-encephalitic parkinson syndrome after meningoencephalitis due to coxsackie virus group B, type 2, *New Engl. J. Med.* **263**, 744–747.

Yahr, M. D. (1978). Encephalitis lethargica (von Economo's disease, epidemic encephalitis), in: *Handbook of Clinical Neurology*, Vinken, P. J. and Bruyn, G. W. (Eds), Vol. 34, pp. 451–457. Elsevier, Amsterdam, North-Holland.

Yoshida, M., Miwa, T. and Nagatsu, T. (1990). Parkinsonism in monkeys produced by chronic administration of an endogenous substance of the brain, tetrahy droisoquinoline: the behavioral and biochemical changes, *Neurosci. Lett.* **119**, 109–113.

Advances in Neuroregulation and Neuroprotection (2005), pp. 389-399
Collin, C. *et al.* (Eds)
© VSP 2005

Visual searching impairment in patients with major depressive disorder: performance in the Raven coloured progressive matrices test

N. NAKANO [1,*], N. OKUMURA [2], S. HAYASHI [3], Y. HAYASHI [3], S. SAITO [3], N. SASAKI [3] and S. MURAKAMI [2]

[1] *Health Sciences University of Hokkaido, School of Psychological Science, Department of Clinical Psychology, Ainosato 2-5, Kita-ku, Sapporo, 002-8072, Japan*
[2] *Sapporo Medical University, School of Health Science, Department of Occupational Therapy, Minami-1, Nishi-17, Chuo-ku, Sapporo, 060-8556, Japan*
[3] *Sapporo Medical University, School of Medicine, Department of Neuropsychiatry, Minami-1, Nishi-16, Chuo-ku, Sapporo, 060-8543, Japan*

Abstract—It is well documented that major depressive disorders (MDD) frequently cause mild cognitive deficits, but it is often difficult to distinguish them from dementia. We examined visual searching impairments in nine MDD patients and nine healthy controls (HC), using the well-known non-verbal cognitive test of Raven's coloured progressive matrices (RCPM). All subjects were shown the slides of RCPM, which contained both symmetrical pattern matching and analogical reasoning tasks, while eye movements were recording by a Free View-DTS (TKK2920). The results showed: (1) The MDD patients showed a significantly longer response times than HC in both tasks, but no major impairment on the saccades between the incomplete figures and six response alternatives. (2) In the analogical reasoning task, MDD patients had a tendency to fixate clearly false alternatives during the problem solving, which evidenced impairment of narrowing down the alternatives to the correct piece.

These findings may show that patients with MDD exhibit visual searching impairments that are similar to visual cognitive deficits suggestive of early dementia.

Keywords: Eye movements; major depressive disorder; Raven's coloured progressive matrices; pseudodementia; analogical reasoning task.

INTRODUCTION

Depression is a common neuropsychiatric problem in old age. The importance of the problem will become greater as the geriatric segment of the population

---

*To whom correspondence should be addressed. E-mail: nakanon@hoku-iryo-u.ac.jp

grows. It is well documented that major depressive disorders (MDD) frequently cause mild cognitive deficits (Elderkin-Thompson *et al.*, 2003; Fossati *et al.*, 2001). There is some evidence that depression may be a risk factor of Alzheimer's disease (AD) in later life (Green *et al.*, 2003). Reversible cognitive deficits related to MDD are summarized under the reversible neuropsychiatric condition termed 'pseudodementia' (Wells, 1979), yet the extent to which mild MDD influences cognition has not been studied in detail. Cognitive studies suggest that the patients with AD, who show an attentional disengagement deficit, should exhibit impairment when asked to perform a visual search task in which repeated shifts of spatial attention are required (Daffner *et al.*, 1992; Parasuraman *et al.*, 2000). The syndromes associated with dementia and depression in old age show considerable overlap and even coincidence.

We sought to identify the influence of MDD on visual cognitive performance by using Raven's coloured progressive matrices (RCPM) for studying the visual searching performance in MDD. RCPM is a non-verbal cognitive test which consists of visual pattern matching and simple analogical reasoning problems (Lezak, 1995). In this paper, we studied the eye movements of patients with MDD and healthy controls (HC) using a Free View-DTS (TKK2920b, Takei & Co), and found characteristic findings for eye movements of visual cognitive impairments in the patients with MDD.

MATERIALS AND METHODS

The subjects consisted of 9 patients (5 males, 4 females) with MDD based on criteria established by Diagnostic and Statistical Manual of Mental Disorders IV Criteria (APA, 1994), and 9 age-matched cognitively intact HC (7 males and 2 females). The patients were recruited from the outpatients at recovery in Sapporo Medical University Hospital. HC were recruited from the local residents. All subjects had almost normal findings on bedside neurological examination and no other disease, particularly no additional neurologic, psychiatric, or systemic medical disorder. In addition, they were given a bedside clinical neuro-ophthalmologic examination evaluating visual fields, pursuit movements, saccadic movements, and partial-field optokinetic nystagmus to test for elementary abnormalities of eye movement control. No subjects exhibited significant impairment in ocular motility. No subject was taking hypnotics, sedatives, or major tranquilizers. The patients with MDD were taking low dose antidepressants (fluvoxamine $<$ 50 mg/day or trazodone $<$ 50 mg/day). The following psychological assessment tool had been administered to the patients with MDD and HC: Center for Epidemiologic Studies Depression Scale (CES-D, Radloff, 1977). Informed consent was obtained from all subjects after the nature of the procedures had been explained.

Eye movement equipment

Eye movement data in this study were collected by means of a Free View-DTS employing an infrared-LED cornea reflex system. The pupil and corneal bright (Purkinje-Sanson's image) spot were imaged with a CCD camera sensitive in the near-infrared range (850 nm). The subject's eye-line gaze is determined by the changes in the reflected light quantity in relation to the movements of the iris. Calibration was accomplished employing a five-point array and using the software supplied with the system. The system does not require subject restraints or attachment while viewing the stimulus presentation. Subjects were excluded if they could not cooperate in the calibration procedure. Eye calibration was recurrently checked between experiments to ensure the accuracy and stability of calibration. The system has an accuracy of $\pm 0.5°$. The criterion for eye fixation is that the eye be moving at a velocity of less than 5 deg/s (Collewin and Tamminga, 1984; Yamada and Fukuda, 1986). The system samples at a rate of 30 Hz, with a temporal resolution of 33 ms. The subjects were seated in an experimental room 1.2 m from a 21-inch monitor. The total field of the monitor screen subtended a visual angle of 20° horizontally and 15° vertically. Eye position was recorded and saved to an IBM compatible personal computer for further analysis.

Procedure

At the start of the experiment, the subject was seated in the experimental room and a calibration procedure was run to adjust the measurement system to each subject's eye. At the beginning of the calibration procedure, the subjects were instructed to fixate on the center circle on the calibration slide on the screen while eye position was recorded.

Practice task

The first easy task was considered practice. The slide was a subset of the A5 slide in RCPM (Raven, 1962, 1965). The subjects were instructed to 'look at the slide, and choose the correct one from the lower six pieces to complete the upper pattern'. If the subjects did not choose the correct one, the experiments were cancelled at this point.

Symmetrical pattern matching task

The subjects were shown two slides of RCPM (sets B3 and A_B4) that involved completing a symmetrical pattern matching task (Fig. 1A). The subjects were instructed to 'look at the slide, and choose the correct one from the lower six pieces to complete the upper pattern'.

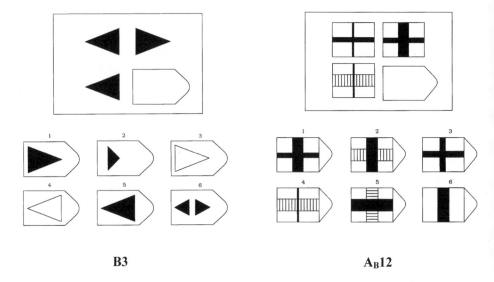

B3   **A$_B$12**

A. Symmetrical pattern matching task   B. Analogical reasoning task

Figure 1. Two examples from the symmetrical pattern matching and analogical reasoning task. (From *Coloured Progressive Matrices*, Oxford Psychologists Press Ltd. ©JC Raven Ltd.)

Analogical reasoning task

The subjects were shown two slides of RCPM (sets B9 and A$_B$12) that involved completing an analogical reasoning task (Fig. 1B). The subjects were instructed to 'look at the slide, and choose the correct one from the lower six pieces to complete the upper pattern.'

Except for the practice task, eye movements and voice were recorded while the subjects solved the task. The scan path of fixation and voice of the subjects were recorded on videotape. The subjects were instructed to fixate on the center circle of the screen before beginning each task.

Data analysis

For each RCPM slide, the visual field was divided into several areas of interest (AOIs). In the upper part of the slide (incomplete figure), the AOIs were $10° \times 5°$ boxes. Six AOIs ($2.5° \times 2°$) were located in the lower part of the slide (response alternatives). All raw eye-position data were recorded for processing and analysis. Processing of raw eye-position data yielded eye movement fixation files that contained information on numbers of fixations, mean fixation duration, and velocity of eye movements. Dependent variables for the experiments consisted of mean duration of fixations, response time of problem solving, number of saccades between the incomplete figure and six response alternatives, and percent dwell time with respect to the AOIs. For each experiment, the main dependent variables were percent dwell time (duration of fixations/response time), expressed as a percentage of total viewing subjects spent fixated on the AOIs, the numbers of AOIs viewed (0–6), and the total number of AOIs viewed (number/response time). The MDD and HC groups were compared for these measures using the Mann–Whitney U-test.

RESULTS

The characteristics of the 9 patients with MDD and the 9 HC are summarized in Table 1. These groups were not significantly different in age and sex ratio, but patients with MDD were significantly more impaired on the CES-D (25.8 *vs.* 5.6, $p < 0.01$). Scores on the CES-D were normal for HC (range 0–10) and in the mood disorder range for subjects with MDD (5–48). All subjects performed all activities of daily living normally. The clinical neuro-ophthalmologic examination did not reveal any major abnormalities in patients with MDD and HC. Data for individual subjects with MDD and HC are summarized in Table 2 and Table 3. Group mean scores for each dependent variable are given in Table 4.

The MDD patients had significantly longer response times for both symmetrical pattern matching and the analogical reasoning task than the HC ($p < 0.01$) (Table 4). In contrast, no significant differences were found in the percentage of correct trials between those with MDD and the HC (Table 4). In the symmetrical pattern matching task, there were no significant differences between two groups for the mean percent dwell time on the incomplete figure, number of saccades between the incomplete figure and lower 6 response alternatives, number and total number of AOIs viewed in the lower region and overall mean percent dwell time. For the analogical reasoning task, the MDD patients looked at the incomplete figure a significantly shorter time than at the lower region involving the 6 AOIs ($51.6\% \pm 16.2$ *vs.* 62.6 ± 9.6, $p < 0.05$) (Table 4). The MDD patients looked at significantly more AOIs than the HC (4.72 ± 1.23 *vs.* 3.33 ± 1.19, $p < 0.01$) (Table 4). The MDD patients had a tendency to look at the same AOIs repeatedly, and exhibited wide visual exploration compared to the HC. (Figure 2 presents an example of the scan path of the subjects with MDD and HC.) In contrast, the number of saccades between the incomplete figure and 6 response alternatives, and the mean percent dwell time overall were not significantly different between the two groups (Table 4).

Table 1.
Characteristics of patients with major depressive disorder and healthy control subjects (mean ± SD)

Variables	Patients with major depressive disorder	Control subjects	p
Age (years)	53.4 ± 7.1	51.2 ± 7.4	NS
Sex (male/female)	5/4	7/2	NS
CES-D score	25.8 ± 15.3	5.6 ± 3.3	<0.01

Table 2.
Summary of the data (symmetrical pattern matching tasks)

Subjects	Age	CES-D	Response time	DT-U	No. SUL	Total no. AOIs	No. AOIs	DT-O
MDD1	60	35	9.9	38.9	0.27	0.79	4.5	35
MDD2	45	5	4.2	48.8	0.35	1.00	4.0	47
MDD3	65	48	7.5	29.6	0.21	0.88	4.0	68
MDD4	49	41	4.5	49.4	0.56	0.67	3.0	36
MDD5	59	15	4.7	50.1	0.32	0.77	3.0	47
MDD6	47	35	5.5	41.1	0.35	1.27	5.0	54
MDD7	53	29	9.1	33.2	0.17	0.89	4.5	48
MDD8	57	17	8.2	37.4	0.43	1.15	4.5	44
MDD9	46	7	8.7	24.5	0.29	0.90	3.0	31
HC1	55	10	4.1	46.5	0.36	0.73	3.0	33
HC2	52	5	5.1	40.0	0.41	0.77	3.0	58
HD3	45	0	4.1	30.7	0.51	1.08	2.5	56
HD4	45	5	3.7	45.0	0.20	0.40	1.5	8
HD5	46	8	3.4	37.6	0.30	1.62	4.0	51
HD6	45	4	4.3	47.6	0.36	1.30	4.0	57
HD7	48	2	4.9	57.0	0.40	1.15	4.5	59
HD8	60	7	4.3	36.0	0.24	0.79	3.0	39
HD9	65	9	10.5	36.9	0.47	1.15	5.0	56

CES-D: Center for Epidemiologic Studies Depression Scale (0–60), DT-U: Mean percent dwell time on the upper part (incomplete figure), No. SUL: Number of saccades between the upper part and lower region of the slide (number/s), Total No. AOIs: Total number of areas of interest fixated on the lower region of the slide (number/s), No. AOIs: Number of areas of interest fixated on the lower region of the slide (0–6), DT-O: Mean percent dwell time overall on the slide, MDD: Major depressive disorder patients, HC: Healthy controls.

DISCUSSION

Our study demonstrated that persons with MDD exhibited significant impairments of spatial attention during visual searching that differed from the HC as measured by eye movements. In this study, the visual cognitive impairments of those with MDD revealed the following: (1) the MDD patients showed a significant increase in response time compared with the HC. (2) In the analogical reasoning tasks, which were more difficult than the symmetrical pattern matching, the subjects with MDD spent more time on the response alternatives to choose the correct piece, and went back and forth between the 6 AOIs in the lower region. (3) In the analogical reasoning task, those with MDD viewed more AOIs than the HC.

As described above, depressive symptoms before the onset of AD are associated with the development of AD, and are a risk factor for late development of AD (Green *et al.*, 2003). Slowing of the speed of information processing in MDD has been reported as described below. MDD cause a significant increase in reaction time when decision processes and the visuospatial pattern matching process were involved (Hoffman *et al.*, 2000; Thomas *et al.*, 1999). In the Emotional Stroop task,

Table 3.
Summary of the data (analogical reasoning tasks)

Subjects	Age	CES-D	Response time	DT-U	No. SUL	Total no. AOIs	No. AOIs	DT-O
MDD1	60	35	30.2	37.8	0.37	0.61	4.5	36
MDD2	45	5	25.8	73.5	0.28	0.50	4.0	54
MDD3	65	48	19.5	45.7	0.38	0.93	6.0	70
MDD4	49	41	9.1	61.3	0.58	0.67	3.5	36
MDD5	59	15	15.7	50.7	0.45	0.73	4.5	39
MDD6	47	35	16.6	59.0	0.46	0.85	6.0	56
MDD7	53	29	14.8	31.0	0.42	0.41	4.5	59
MDD8	57	17	22.2	64.7	0.37	0.83	5.0	46
MDD9	46	7	10.3	40.8	0.47	1.27	4.5	34
HC1	55	10	6.0	62.8	0.15	0.34	2.0	32
HC2	52	5	12.6	68.4	0.43	0.73	4.0	51
HD3	45	0	11.8	61.3	0.76	1.13	4.0	53
HD4	45	5	12.8	57.1	0.18	0.20	2.0	9
HD5	46	8	6.4	66.7	0.48	0.65	3.0	52
HD6	45	4	11.0	68.1	0.64	0.91	4.0	53
HD7	48	2	6.1	59.5	0.62	1.06	3.5	53
HD8	60	7	6.4	66.2	0.39	0.56	2.5	36
HD9	65	9	13.2	52.9	0.57	0.80	5.0	60

CES-D: Center for Epidemiologic Studies Depression Scale (0–60), DT-U: Mean percent dwell time on the upper part (incomplete figure), No. SUL: Number of saccades between the upper part and lower region of the slide (number/s), Total No. AOIs: Total number of areas of interest fixated on the lower region of the slide (number/s), No. AOIs: Number of areas of interest fixated on the lower region of the slide (0–6), DT-O: Mean percent dwell time overall on the slide, MDD: Major depressive disorder patients, HC: Healthy controls.

MDD patients were slower than controls and AD patients (Dudley *et al.*, 2002). Our data were similar to these reports, and showed a slowing of central information processing. We predicted that MDD might increase the saccades between the incomplete figure and 6 response alternatives because these patients exhibit deficits in mental speed and visual perceptual abilities required for successful completion of RCPM. However, the number of saccades between the upper and lower region showed no significant difference between the two groups. The MDD patients showed no major impairment of the saccades between the incomplete figures and six response alternatives. In analogical reasoning tasks, it was necessary for the MDD patients to increase the dwell time on the lower region because of the increase of trial and error over the response alternative. The MDD patients had a strategy to increase the saccades between the 6 pieces in the lower region, and consequently increase the number of AOIs viewed.

The patients with MDD did not exhibit difficulty in shifting their line of gaze to the points on the screen during the calibration procedure and exhibited normal findings on a bedside neuro-ophthalmologic examination. Thus, these abnormalities

Table 4.

Data on the symmetrical pattern matching and analogical reasoning task (mean ± SD)

	MDD patients	Controls	p
(1) Symmetrical pattern matching task			
% correct trials	94.4 ± 23.6	100 ± 0	NS
Response time (s)	6.8 ± 2.6	4.9 ± 2.5	<0.01
% dwell time on the upper part	39.2 ± 12.2	41.9 ± 14.6	NS
No. saccades between upper and lower	0.32 ± 0.14	0.36 ± 0.15	NS
Total No. AOIs viewed	0.92 ± 0.26	1.00 ± 0.42	NS
No. AOIs viewed	3.94 ± 1.06	3.39 ± 1.58	NS
% dwell time overall	45.2 ± 11.3	45.9 ± 16.8	NS
(2) Analogical reasoning task			
% correct trials	55.6 ± 51.1	83.3 ± 38.3	NS
Response time (s)	18.2 ± 12.2	9.6 ± 5.2	<0.01
% dwell time on the upper part	51.6 ± 16.2	62.6 ± 9.6	<0.05
No. saccades between upper and lower	0.42 ± 0.11	0.47 ± 0.22	NS
Total No. AOIs viewed	0.71 ± 0.33	0.71 ± 0.23	NS
No. AOIs viewed	4.72 ± 1.23	3.33 ± 1.19	<0.01
% dwell time overall	46.8 ± 12.3	44.2 ± 15.9	NS

of spatial attention cannot be attributed to primary oculomotor dysfunction. The effect of antidepressant treatment on visual attention tasks in this study seemed not to be significant because the patients with MDD took low doses of antidepressants. In other cognitive and attentional tasks, continuous performance tests, Koetsier *et al.* (2002) reported that imipramine and fluvoxamine treatments did not increase the reaction time and the number of errors.

Processing of visual attention has been hypothesized to be controlled by an integrated network of three cerebral regions, the posterior parietal region, frontal cortex, and cingulate gyrus (Mesulam, 1981). The frontal cortex coordinates the motor programs for exploration, scanning, reaching, and fixating (Mesulam, 1981). It is well documented that frontal lobe dysfunction causes the disinhibition of habitual responses (Milner, 1963; Perret, 1974). On the other hand, a decrease in regional cerebral blood flow in the frontal cortex has been described in major depression (Dolan *et al.*, 1992; Mayberg *et al.*, 1994). In our data, the MDD patients had a tendency to fixate more AOIs than the HC, and the number of AOIs viewed increased. These findings may reflect the hypofunction of the frontal cortex, since this behavior is analogous to the disinhibition of habitual responses.

Immediately after the slide was presented, the HC subjects made a quick determination of the gist of the pattern by peripheral vision and used information located in the periphery to guide fixation choices (Antes, 1974). After acquiring the gist, subsequent fixations may have the purpose of verifying the presence of various objects that belong in the slide (Parker, 1978). Then, the HC subjects fixated on a piece with relatively high informativeness. In the HC, a brief scan of the clearly false pieces (AOIs) was not necessary to determine the correct one. On the other hand, the sub-

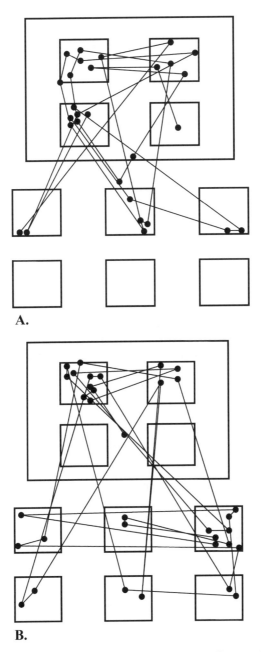

A.

B.

Figure 2. Eye fixation patterns superimposed on the experimental diagram. (A) An example of eye fixation locations (circles) and scan paths (lines) from HC (HC5) on the $A_B 12$ slide. The subject fixated the 3 AOIs. (B) Similar data from a MDD patient (MDD6). The subject fixated all 6 AOIs. In the lower regions, the scan path of the HC was located in the 3 AOIs including the correct piece. However, the scan path of the MDD patient was widely distributed and located other false pieces.

jects with MDD have an impairment of the peripheral editing process, and clearly false regions are also investigated by central vision. Thus, they look at the AOIs that are easily ruled out as false pieces by the HC. These data suggest that the subjects with MDD have an impairment in narrowing down the candidates for the correct piece, especially in the more difficult task (analogical reasoning). As mentioned above, our data may show that MDD patients exhibit visual searching impairments that are similar to visual cognitive deficits suggestive of early dementia.

However, more research is needed, particularly in patients with mild cognitive impairment, to determine whether eye movement analysis can be used as a psychophysiological marker of visual cognitive impairment suggestive of early dementia.

REFERENCES

American Psychiatric Association (1994). *Diagnostic and Statistical Manual of Mental Disorders*, 4th edn. Washington, DC: American Psychiatric Association.

Antes, J. R. (1974). The time course of picture viewing, *J. Exp. Psychol.* **103**, 62–70.

Collewin, H. and Tamminga, E. P. (1984). Human smooth and saccadic eye movements during voluntary pursuit of different backgrounds, *J. Physiol.* **351**, 217–250.

Daffner, K. R., Scinto, L. F. M., Weintraub, S., *et al.* (1992). Diminished curiosity in patients with probable Alzheimer's disease as measured by exploratory eye movements, *Neurology* **42**, 320–328.

Dolan, R. J., Bench, C. J., Brown, R. G., *et al.* (1992). Regional cerebral blood flow abnormalities in depressed patients with cognitive impairment, *J. Neurol. Neurosurg. Psychiatry* **55**, 768–773.

Dudley, R., O'Brien, J., Barnett, N., *et al.* (2002). Distinguishing depression from dementia in later life: a pilot study employing the Emotional Stroop task, *Int. J. Geriatr. Psychiatry* **17**, 48–53.

Elderkin-Thompson, V., Kumar, A., Bilker, W. A., *et al.* (2003). Neuropsychological deficits among patients with late-onset minor and major depression, *Arch. Clin. Neuropsychol.* **18**, 529–549.

Farrer, L. (2003). Depression as a risk factor for Alzheimer disease: the MIRAGE study, *Arch. Neurol.* **60**, 753–759.

Fossati, P., Ergis, A. M. and Allilaire, J. F. (2001). Problem-solving abilities in unipolar depressed patients: comparison of performance on the modified version of the Wisconsin and the California sorting tests, *Psychiatry. Res.* **104**, 145–156.

Green, R. C., Cupples, L. A., Kurz, A., *et al.* (2002). CPT performance in major depressive disorder before and after treatment with imipramine or fluvoxamine, *J. Psychiatr. Res.* **36**, 391–397.

Hoffman, M., Seifritz, E., Krauchi, K., *et al.* (2000). Alzheimer's disease, depression and normal ageing: merit of simple psychomotor and visuospatial tasks, *Int. J. Geriatr. Psychiatry* **15**, 31–39.

Lezak, M. D. (1995). Raven's coloured progressive matrices, in: *Neuropsychological Assessment*, 3rd edn, pp. 615–616. Oxford University Press, New York.

Mayberg, H. S., Lewis, P. J., Regenold, W., *et al.* (1994). Paralimbic hypoperfusion in unipolar depression, *J. Nucl. Med.* **35**, 929–934.

Mesulam, M. M. (1981). A cortical network for directed attention and unilateral neglect, *Ann. Neurol.* **10**, 309–325.

Milner, B. (1963). Effect of different brain lesions on card sorting, *Arch. Neurol.* **9**, 90–100.

Parasuraman, R., Greenwood, P. M. and Alexander, G. E. (2000). Alzheimer disease constricts the dynamic range of spatial attention in visual search, *Neuropsychologia* **38**, 1126–1135.

Parker, R. E. (1978). Picture processing during recognition, *J. Exp. Psychol. Hum. Percept. Perform.* **4**, 284–293.

Perret, E. (1974). The left frontal lobe of man and the suppression of habitual responses in verbal categorical behaviour, *Neuropsychologia* **12**, 323–330.

Radloff, L. S. (1977). A self-report depression scale for research in the general population, *Appl. Psychol. Measur.* **1**, 385–401.

Raven, J. C. (1962). *Colored Progressive Matrices: Sets A, A_B, B*. Lewis, London.

Raven, J. C. (1965). *Guide to Using to the Colored Progressive Matrices Sets A, A_B, B*. Lewis, London.

Thomas, P., Goudemand, M. and Rousseaux, M. (1999). Attentional resources in major depression, *Eur. Arch. Psychiatry. Clin. Neurosci.* **249**, 79–85.

Wells, C. E. (1979). Pseudodementia, *Am. J. Psychiatry.* **136**, 895–900.

Yamada, M. and Fukuda, T. (1986). Quantitative evaluation of eye movements as judged by sight-line displacements, *SMPTE Journal* **95**, 1230–1241.

Advances in Neuroregulation and Neuroprotection (2005), pp. 401–408
Collin, C. *et al.* (Eds)
© VSP 2005

Salivary concentration of 5-hydroxytryptamine in patients with bulimia nervosa

N. TAKAHASHI [1,*], N. HAMAUE [2], N. KURONUMA [1], T. YOSHIHARA [1],
S. ANDO [1], M. HIRAFUJI [2], T. ENDO [2], M. SENJO [3], S. H. PARVEZ [4]
and M. MINAMI [2]

[1] *School of Psychological Science, Health Sciences University of Hokkaido,*
   Sapporo 002-8072, Japan
[2] *Department of Pharmacology, Health Sciences University of Hokkaido, Hokkaido 061-0293, Japan*
[3] *Goryokai Medical Corporation Hospital, Sapporo 002-8029, Japan*
[4] *Alfred Fressard Institute of Neuroscience, CNRS UPR 2212-Bâts, Chateau CNRS 91190,*
   Gif Sur Yvette, France

Abstract—Patients with bulimia nervosa (BN) have disturbances of mood and behavior related to monoamine activities. There have been no reports concerning salivary concentrations of 5-hydroxytryptamine (5-HT) in patients with eating disorders such as anorexia nervosa (AN) and BN. In order to elucidate the involvement of 5-HT in eating disorders, 5-HT levels in the saliva of patients with BN were measured. Salivary 5-HT levels in patients with avoidant personality disorder (APD) were also measured as an active control. 5-HT levels in the saliva of patients with BN and APD were compared with those of 32 healthy volunteers. Simultaneously, salivary 3,4-dihydroxyphenylacetic acid (DOPAC) was measured using HPLC-ECD. 5-HT levels were 41.04 ± 14.41 in 11 patients with BN (10 women and 1 man, average age 19.6 years), 78.55 ± 45.55 in 5 patients with APD (3 women and 2 men, average age 19.2 years) and 8.91 ± 1.10 ng/ml in 32 healthy controls (25 women and 7 men, average age 21.7 years), respectively. This is the first report that salivary concentrations of 5-HT were significantly higher in BN ($p < 0.05$) and in APD than those in healthy controls. We found no significant differences among the salivary DOPAC levels in BN, APD and those of healthy controls. There were no significant differences in body mass index among the three groups. 5-HT concentrations in the saliva may be a useful marker of stress or eating disorders.

Keywords: Eating disorder; bulimia nervosa; avoidant personality disorder; salivary 5-HT; salivary DOPAC; human study.

*To whom correspondence should be addressed. E-mail: noritaka@hoku-iryo-u.ac.jp

INTRODUCTION

Various evidences suggest that a disturbance of 5-HT neuronal pathways may contribute to the pathogenesis of anorexia nervosa (AN) (Frank *et al.*, 2002). Patients with AN show disturbances in appetite and behavior, such as dysphoria, inhibition and obsessions, that could be related to altered 5-HT activity (Frank *et al.*, 2001). Along with mood and behavior disturbances, women with bulimia nervosa (BN) display changes in 5-HT activity (Kaye *et al.*, 1998). It is not known whether these changes are secondary to pathological eating behavior or traits that could contribute to the pathogenesis of BN. Although considerable progress has been made in the understanding and treatment of anorexia and bulimia nervosa, a substantial proportion of people with these disorders display a limited response to treatment (Frank *et al.*, 2002).

Tryptophan, an essential amino acid in the diet, is the precursor of 5-HT. Diet-induced changes in the plasma tryptophan ratio can cause increases or decreases in brain tryptophan levels and in the synthesis of 5-HT (Wurtman and Wurtman, 1986). Diets poor in tryptophan affect the tissue concentration of 5-HT (Culley *et al.*, 1963). Administration of tryptophan will increase the tissue concentration of 5-HT (Moir, 1971). Thus, 5-HT disturbances may be secondary to dietary abnormalities. However, disturbances of 5-HT activity appear to persist after long-term weight recovery from AN (Kaye and Weltzin, 1991a). Brain 5-HT activity contributes to satiety. Binge-eating (bingeing) behavior is consistent with reduced 5-HT function, whereas AN is consistent with increased 5-HT activity (Kaye and Weltzin, 1991a).

5-HT release in the saliva could be a promising, non-invasive parameter for the study of stress or eating disorders. Salivation is controlled by serotonergic and dopaminergic innervation. 5-HT stimulates the secretion of a protein-rich saliva; dopamine (DA) causes the production of a saliva without proteins (Baumann *et al.*, 2002). Baumann *et al.* (2002) suggest that DA is released on the acinar surface, close to the peripheral cells, and along the entire duct system. 5-HT is probably released close to the peripheral and central cells, and at the initial segments of the duct system. The determination of salivary 5-HT and DOPAC, a metabolite of DA, therefore, were made simultaneously.

In order to elucidate the relationship between eating disorders and 5-HT activity, 5-HT levels in the saliva of BN were determined. 5-HT levels in the saliva of patients with avoidant personality disorders (APD) were also determined using HPLC-ECD within 2 hours after sampling to serve as an active control.

MATERIALS AND METHODS

Subjects

Eleven subjects with eating disorders and 5 patients with avoidant APD were selected in this study. The psychiatric diagnosis and symptoms to determine eating

Table 1.
Characteristics of subjects

Subjects	*n* (Gender)	Age (year)	BMI	SBP (mmHg)
APD	5 (3F2M)	19.20 ± 1.18	20.06 ± 1.75	113.60 ± 6.68
BN	11 (10F1M)	19.64 ± 0.77	19.52 ± 1.85	112.91 ± 2.77
Healthy	32 (25F7M)	21.72 ± 0.34	19.66 ± 0.23	97.20 ± 3.06

Mean ± SE, BMI: body mass index, SBP: systolic blood pressure, APD: patients with avoidant personality disorders, BN: patients with bulimia nervosa (7 of 10: purging type, 3 of ten: non-purging type).

disorders and APD were assessed according to the guideline of DSM-IV (1994) by a psychiatrist. The eleven BN patients were 10 women and 1 man with an average age of 19.6 years (Table 1), and the five patients with APD consisted of 3 women and 2 men with an average age of 19.2 years. Thirty-two healthy volunteers acting as a control group consisted of 25 women and 7 men with an average age of 21.7 years. To exclude the possibility of drug interference with the serotonergic pathway, no drugs such as selective serotonin re-uptake inhibitor (SSRI) were administered. All subjects including patients and healthy volunteers gave their informed consent to participate in this study.

Sampling of saliva

Saliva samples were taken while subjects were seated, using self-drainage into sterilized dishes from which cotton containing the saliva was collected into plastic tubes (Salivette: Aktiengesellschaft and Co., Germany). Saliva specimens were immediately cooled on crushed-ice, EDTA-2Na was added to samples before centrifuging at $3000g$ at 3°C for 10 min to obtain the supernatant. A quantitative and qualitative desynchrony was found in the circadian rhythms of 5-HT in patients with BN or APD and controls as well. Saliva samples were collected hourly from 13:00 to 15:00. Following analysis of variance with a repeated measures design, it was determined that a 3-min collection was the best fit. An additional goal of this study was to develop a standardized protocol for sampling and processing saliva for the laboratory assessment of 5-HT.

The determination of 5-HT concentration in saliva

Salivary 5-HT levels were measured using high performance liquid chromatography (HPLC) (EP-30, Eicom, Kyoto, Japan) with an electrochemical detector (ECD) (ECD-300, Eicom, Kyoto, Japan) (Matsumoto *et al.*, 1990; Shintani *et al.*, 1993). HPLC was performed with a column Eicompak (SC-5 ODS: 3.0 mm × 150 mm, Eicom, Kyoto, Japan) containing EDTA-2Na (5 mg/l) (Sigma Chemical Co. Ltd., St. Louis, MO, USA), sodium 1-octanesulfonate (190 mg/l) (Nakarai Tesque Co.,

Kyoto, Japan), and methanol (17%) (Wako Pure Chemical Industries Ltd., Osaka, Japan) as the mobile phase. The oxidation potential of the ECD was set at +750 mV.

Statistical analysis

All values are expressed as mean $\pm$ SE. The statistical significant differences between two groups was assessed using Student's t-test. Analysis of variance, followed by the Bonferroni-modified t-test was used to compare more than two groups (Wallenstein *et al.*, 1980). Values of $p < 0.05$ were considered statistically significant.

RESULTS

5-HT levels in saliva

The concentration of 5-HT in the saliva of patients with BN was 41.04 ± 14.41 ng/ml (Table 2). The salivary 5-HT levels of BN were significantly higher than those of healthy volunteers ($p < 0.05$). The patients with APD had higher 5-HT levels as well.

DOPAC levels in saliva

The concentration of DOPAC in the saliva was 15.95 ± 5.03 ng/ml in BN subjects (Table 2). There were no significant differences in salivary DOPAC levels among those of BN, APD and healthy volunteers.

Factors affecting 5-HT concentration in saliva

As the 5-HT concentrations in saliva might be affected by the circadian rhythms of salivation and the autonomic nervous system, we collected salivary samples in the afternoon (13:00–17:00). Factors affecting 5-HT concentrations in the saliva were examined including the stability of 5-HT in saliva, blood pressure, age and sex differences. Mean values ($n = 6$) of 5-HT levels in the saliva gradually go down at 2 hours ($94.94 \pm 0.61\%$) and 4 hours ($83.13 \pm 1.88\%$) from the control values at 0 hour. We determined salivary 5-HT within 2 hours after collection.

Table 2.
5-HT and DOPAC levels in saliva

Subjects	5-HT	DOPAC
Healthy	8.91 ± 1.10 (32)	8.97 ± 1.77 (28)
BN	41.04 ± 14.41 (11)*	15.95 ± 5.03 (11)
APD	78.55 ± 45.55 (5)	6.81 ± 1.19 (5)

Mean $\pm$ SE, (n), *$p < 0.05$ *vs* healthy control, BN: patients with bulimia nervosa, APD: patients with avoidant personality disorders.

There were no significant differences in BMI between BN and healthy controls (Table 1). The BMI of BN may have been kept within normal limits by purging using fingers after bingeing (seven of ten BN: purging type). There were no significant differences in systolic blood pressure and age between BN and the healthy controls (Table 1). A significant correlation between salivary 5-HT and systolic blood pressure in normal healthy volunteers was observed ($r = 0.4253$, $p < 0.05$). A significant relationship was also seen between salivary 5-HT and DOPAC ($r = 0.344$, $p < 0.05$) in these volunteers.

DISCUSSION

There have been many reports concerning the cerebrospinal fluid (CSF) levels of 5-HT in patients with eating disorders (Kaye and Weltzin, 1991b). There have been no reports, however, concerning salivary concentrations of 5-HT in patients with eating disorders such as AN and BN. In this study, for the first time, we demonstrated that salivary concentrations of 5-HT were higher in patients with BN than those of the healthy controls. It also has been reported that patients with BN have impaired satiety and secretion of cholecystokinin, a peptide known to induce satiety and reduce food intake in animals and humans (Kaye and Weltzin, 1991b). Most data show that females with BN show alterations in 5-HT and norepinephrine (NE) activity. In animals, increased 5-HT activity appears to inhibit eating behavior due to induced nausea, anorexia and vomiting (Endo *et al.*, 1990; Minami *et al.*, 1995). Endogenous NE, on the other hand, activates eating behavior via the alpha-2 receptors in the hypothalamus. Binge-eating (bingeing) behavior is consistent with overactivity of the hypothalamic alpha-noradrenergic system, an underactivity of hypothalamic serotonergic systems, or a combination of both defects (Kaye and Weltzin, 1991b).

Important depots for 5-HT in mammals are the enterochromaffin (EC) cells in the gastrointestinal mucosa, the brain, the pineal gland and the platelets (Tyce *et al.*, 1983). Significant amounts of 5-HT have been detected in other mammalian tissues, notably in the heart, the kidney, the spleen, and the thyroid, although the significance of the presence of 5-HT in these tissues is not yet understood. 5-HT is synthesized in most tissues in which it is stored and has a rapid turnover in most of these tissues. The half-life in the gastrointestinal tract has been reported to be 10 to 17 hours (Erspamer and Testini, 1959).

Little information is available on the uptake of tryptophan from the plasma by peripheral tissues that synthesize 5-HT, such as the salivary glands. The storage granules of platelets are viewed as a major potential source of 5-HT, that affects the precapillary resistance vessels, as platelets can reach all parts of the body including the salivary glands. 5-HT stored in the platelets is not synthesized in them but originates in other tissues, notably the EC cells (Erspamer and Testini, 1959). The turnover of 5-HT in the platelets is relatively slow, the reported half-life being 33 and 48 hours (Udenfriend and Weissbach, 1958). Daily in their urine, healthy individuals excrete large amounts of 5-hydroxyindoleacetic acid (5-HIAA), the major metabolite of 5-HT (Tyce *et al.*, 1983), and small amounts of 5-HT (10 to 120 μg) (Korf, 1969). In similar fashion, 5-HT may be excreted daily from the salivary glands.

As the 5-HT concentrations in saliva might be affected by the circadian rhythms of salivation and the autonomic nervous system, we collected salivary samples at fixed times (13:00–17:00). Factors affecting 5-HT concentrations in the saliva were examined, including the stability of the 5-HT in saliva, blood pressure, age and sex differences.

The concentration of 5-HT in venous blood decreased gradually after insertion of a probe but fluctuated markedly. An electron microscopic study of the probe membrane after dialysis showed platelet adhesion and aggregation on part of the membrane (Yoshioka *et al.*, 1993). The measurement of free 5-HT concentrations in the blood might be inaccurate due to contamination by 5-HT derived from the adhesion and aggregation of platelets. Furthermore, the invasion of the syringe might increase sympathetic tone in patients and healthy volunteers, thereby affecting 5-HT levels. Since the collection of saliva is non-invasive and convenient, it may be a suitable substance for the study of neurotransmitter release under emotional stress.

5-HT has been shown to be released into the CSF and cerebral venous blood from an infarcted brain (Welch *et al.*, 1972). The relatively increased concentrations of 5-HT found in the CSF, which had been released by platelets shortly after bleeding, suggests that 5-HT may be active in the early stages of vasospasms. Ruwe *et al.* (1985) reported that the 5-HT detected in brain perfusates may not be neuronal in origin. Freed *et al.* (1985) showed that brain 5-HIAA increased during phenylephrine-induced hypertension and during reflex hypertension, following a period of hypertension. These changes were blocked by sino-aortic denervation, indicating that these central serotonergic neurons were responding to increased pressors sensed by the baroreceptors. Freed *et al.* (1985) concluded that brain 5-HT has a role as a neurotransmitter in the homeostatic control of blood pressure. In our study, the salivary 5-HT significantly correlated with systolic blood pressure in healthy volunteers ($p < 0.05$).

Malmgren and Hasselmark (1988) proposed that migraine is associated with a lowered threshold for stimulus response in both the platelets and the serotonergic neurons, and that alterations in platelet function reflect central serotonergic disturbances. Shimomura and Takahashi (1990) measured platelet 5-HT levels in patients with tension-type headache during cold pressor tests and showed a significant reduction in platelet 5-HT compared to levels in the controls. They suggested that, under stress, the absorbance of 5-HT into the platelets in patients with tension-type headache is reduced due to abnormalities in both 5-HT in the platelets and in factors which cause their release. They proposed that low levels of 5-HT in the platelets reflect a decrease in 5-HT in the central nervous system that results from the release of 5-HT from activated platelets in patients with tension-type headache. Marukawa *et al.* (1996) reported that the high levels of salivary substance P and 5-HT in tension-type headache patients during headache periods might reflect the release of substance P from the pain sensory system. They suggested that during periods of active headache in tension-type headache, substance P is released from the afferent fibers of the trigeminal nerves and is secreted from the salivary glands. Our results were in agreement with the data of Marukawa *et al.* (1996). We also speculate that, since patients with eating disorders show abnormalities in 5-HT uptake by the platelets and other factors which cause release of 5-HT from the platelets, their 5-HT release into the saliva may be increased.

Concerning serotonergic and behavioral abnormalities in the patients with BN, Kaye *et al.* (1998) speculated that these psychological alterations might be trait-related and contribute to the pathogenesis of BN. Frank *et al.* (2002) reported that females with AN had significantly reduced 5-HT_{2A} receptor binding on PET imaging in the mesial temporal (amygdala and hippocampus), as well as cingulate cortical regions. It is well known that humans and animals vomit after 5-HT levels increase during anti-cancer drug-induced emesis (Minami *et al.*, 2003). The cause of BN purging behavior might be a reaction to severe nausea induced by the increased 5-HT levels precipitated by over-eating and the resultant 5-HT release from the EC cells on the gut lumen. The possibility that BN is purging to avoid nausea cannot be excluded.

In conclusion, we first demonstrated that salivary concentrations of 5-HT were higher in patients with BN than those of healthy controls. Further studies are necessary concerning the relationship between eating disturbances and elevated levels of 5-HT in the saliva.

REFERENCES

Baumann, O., Dames, P., Kuhnel, D., *et al.* (2002). Distribution of serotonergic and dopaminergic nerve fibers in the salivary gland complex of the cockroach *Periplaneta Americana*, *BMC Physiol.* **2**, 9.

Culley, W. J., Saunders, R. N., Mertz, E. T., *et al.* (1963). Effect of a tryptophan deficient diet on brain serotonin and plasma tryptophan level, *Proc. Soc. Exp. Biol. Med.* **113**, 645–648.

DSM-IV (1994). *Diagnostic and Statistical Manual of Mental Disorders*, 4th edn. American Psychiatric Association, Washington, DC, USA.

Endo, T., Minami, M., Monma, Y., *et al.* (1990). Effect of GR38032F on cisplatin- and cyclophosphamide-induced emesis in the ferret, *Biogenic Amines* **8**, 79–86.

Erspamer, V. and Testini, A. (1959). Observations of the release and turnover rate of 5-hydroxy-tryptamine in the gastrointestinal tract, *J. Pharm. Pharmacol.* **11**, 618–623.

Essman, W. B. (1978). Serotonin distribution in tissues and fluids, in: *Serotonin in Health and Disease: Vol. 1, Availability, Localization and Distribution*, Essman, W. B. (Ed.), pp. 15–180. S. P. Medical and Scientific Books, New York.

Frank, G. K., Kaye, W. H., Weltzin, T. E., *et al.* (2001). Altered response to meta-chlorophenyl-piperazine in anorexia nervosa support for a persistent alteration of serotonin activity after short-term weight restoration, *Int. J. Eat. Disord.* **30**, 57–68.

Frank, G. K., Kaye, W. H., Meltzer, C. C., *et al.* (2002). Reduced 5-HT2A receptor binding after recovery from anorexia nervosa, *Biol. Psychiatry* **52**, 896–906.

Freed, C. R., Echizen, H. and Bhaskaran, D. (1985). Brain serotonin and blood pressure regulation: studies using in vivo electrochemistry and direct tissue assay, *Life Sci.* **37**, 1783–1793.

Kaye, W. H. and Weltzin, T. E. (1991a). Serotonin activity in anorexia and bulimia nervosa: relationship to the modulation of feeding and mood, *J. Clin. Psychiatry* **52** (Suppl.), 41–48.

Kaye, W. H. and Weltzin, T. E. (1991b). Neurochemistry of bulimia nervosa, *J. Clin. Psychiatry* **52** (Suppl.), 21–28.

Kaye, W. H., Greeno, C. G., Moss, H., *et al.* (1998). Alterations in serotonin activity and psychiatric symptoms after recovery from bulimia nervosa, *Arch. Gen. Psychiatry* **55**, 927–935.

Korf, J. (1969). The determination of 5-hydroxytryptamine in human urine, *Clin. Chim. Acta* **23**, 483–487.

Malmgren, R. and Hasselmark, L. (1988). The platelet and the neuron: two cells in focus in migraine, *Cephalalgia* **8**, 7–24.

Marukawa, H., Shimomura, T. and Takahashi, K. (1996). Salivary substance P, 5-hydroxytryptamine, and γ-aminobutyric acid levels in migraine and tension-type headache, *Headache* **36**, 100–104.

Matsumoto, M., Togashi, H., Yoshioka, M., *et al.* (1990). Simultaneous high-performance liquid chromatographic determination of norepinephrine, serotonin, acetylcholine and their metabolites in the cerebrospinal fluid of anesthetized normotensive rats, *J. Chromatogr.* **526**, 1–10.

Minami, M., Endo, T., Nemoto, M., *et al.* (1995). How do toxic emetic stimuli cause 5-HT release in the gut and brain? in: *Serotonin and the Scientific Basis of Anti-emetic Therapy*, Reynolds, D. J. M., Andrews, P. L. R. and Davis, C. J. (Eds), pp. 68–76. Oxford Clinical Communications, Oxford.

Minami, M., Endo, T., Hirafuji, M., *et al.* (2003). Pharmacological aspects of anticancer drug-induced emesis with emphasis on serotonin release and vagal nerve activity, *Pharmacol. Ther.* **99**, 149–165.

Moir, A. T. (1971). Interaction in the cerebral metabolism of the biogenic amines: effect of intravenous infusion of L-tryptophan on the metabolism of dopamine and 5-hydroxyindoles in brain and cerebrospinal fluid, *Br. J. Pharmacol.* **43**, 715–723.

Murakawa, H., Shimomura, T. and Takahashi, K. (1996). Salivary substance P, 5-hydroxytryptamine, and g-aminobutyric acid levels in migraine and tension-type headache, *Headache* **36**, 100–104.

Ruwe, W. D., Naylor, A. M., Bauce, L., *et al.* (1985). Determination of the endogenous and evoked release of catecholamines from the hypothalmus and caudate nucleus of the conscious and unrestrained rat, *Life Sci.* **37**, 1749–1756.

Shimomura, T. and Takahashi, K. (1990). Alteration of platelet serotonin in patients with chronic tension-type headache during cold pressor test, *Headache* **30**, 581–583.

Shintani, F., Kanda, S., Nakaki, T., *et al.* (1993). Interleukin-1 beta augments release of norepinephrine, dopamine, and serotonin in the rat anterior hypothalamus, *J. Neurosci.* **13**, 3574–3581.

Tyce, G. M., Stockard, J., Sharpless, N. S., *et al.* (1983). Excretion of amines and their metabolites by two patients in hepatic coma treated with L-dopa, *Clin. Pharmacol. Ther.* **34**, 390–398.

Udenfriend, S. and Weissbach, H. (1958). Turnover of 5-hydroxytryptamine (serotonin) in tissues, *Proc. Soc. Exp. Biol. (NY)* **97**, 748–751.

Wallenstein, S., Zucker, C. L. and Fleiss, J. L. (1980). Some statistical methods useful in circulation research, *Circ. Res.* **47**, 1–9.

Welch, K. M., Meyer, J. S., Teraura, T., *et al.* (1972). Ischemic anoxia and cerebral serotonin levels, *J. Neurol. Sci.* **16**, 85–92.

Wurtman, R. J. and Wurtman, J. J. (1986). Carbohydrate craving, obesity and brain serotonin, *Appetite* **7** (Suppl.), 99–103.

Yoshioka, M., Kikuchi, A., Matsumoto, M., *et al.* (1993). Evaluation of 5-hydroxytryptamine concentration in portal vein measured by microdialysis, *Res. Commun. Chem. Pathol. Pharmacol.* **79**, 370–376.

Advances in Neuroregulation and Neuroprotection (2005), pp. 409-417
Collin, C. *et al.* (Eds)
© VSP 2005

Central and peripheral sympathetic controls on catecholamine release in male 2-month-old SHRSP and WKY investigated by means of 6-hydrodydopamine injections

HIDEAKI HIGASHINO *, MASAZUMI KAWAMOTO, HIROSHI ENDO
and KANA OOSHIMA

*Department of Pharmacology, Kinki University School of Medicine,
Osaka-Sayama, 589-8511, Japan*

Abstract—In order to investigate the role of central and peripheral sympathetic activities in stroke-prone spontaneously hypertensive rats (SHRSP), maximum dose of 6-OHDA was injected into a lateral cerebral ventricle (i.c.v.) or a peritoneal cavity (i.p.), and changes of catecholamine metabolites in the brain, adrenal glands, plasma, and urine were measured and analyzed. 6-OHDA i.c.v. showed no changes in blood pressure, but showed a tendency to lower heart rates, and a significant decrease in NE in the brain and plasma in SHRSP. On the other hand, 6-OHDA i.p. showed a significant decline of blood pressure, an increase in heart rate, and a decrease of Ep in plasma without effects on NE level in SHRSP. These findings indicate that (1) 6-OHDA i.p. treatment destroys the accessory or extra-adrenal glands besides peripheral sympathetic nerves; (2) Ep is released from the extra-secretary organs, and contributes to maintaining the hypertension in SHRSP; (3) 6-OHDA i.c.v. destroys the central sympathetic nerves and inhibits temporally the peripheral sympathetic activity, adrenal glands and the accessory or extra-adrenal glands; and (4) peripheral sympathetic nerve function works to maintain hyper-sympathetic states in SHRSP.

Keywords: SHRSP; catecholamine; 6-hydroxydopamine; hypertension; chemical sympathectomy.

INTRODUCTION

Our previous studies (Maeda *et al.*, 1999, 2000) concerning catecholamine (CA) metabolites in plasma and urine between stroke-prone spontaneously hypertensive rats (SHRSP) (Okamoto *et al.*, 1974) and normotensive Wistar rats (WKY) concluded that excessive doses of norepinephrine (NE) and epinephrine (Ep) were more

---

*To whom correspondence should be addressed. E-mail: pharm@med.kindai.ac.jp

rapidly released in the blood stream from the sympathetic nerve endings and adrenal medulla in SHRSP than those in WKY at night and under cold stress. The following experiment (Higashino and Ooshima, 2003) using bilateral adrenalectomized rats showed that almost all of the methoxy-hydroxyphenyl glycol (MHPG) and vanillic acid (VA) in blood were released from adrenal glands in SHRSP, and that the final catecholamine products of Ep and metanephrine (MN) must have been secreted from the extra-adrenal glands much more than original release from the adrenal glands in SHRSP. Consequently, exploring the differences of central and peripheral role on the sympathetic regulatory systems between SHRSP and WKY has become an object of our interest. In this connection, 6-hydroxydopamine (6-OHDA) was used as a tool to destroy the sympathetic nerves, chemical sympathectomy, existing in the central and peripheral tissues separately by two different routs of injection, and the control systems between the central and peripheral sympathetic activities in SHRSP were investigated.

MATERIALS AND METHODS

Animals and blood pressure and heart rate measurements

Male 6-week-old SHRSP and WKY were taken from the Center for Animal Experiments in our school. Those were raised in our laboratory with commercial chow (SP, Funahashi, Chiba, Japan) and tap water *ad libitum* under conditions of $22 \pm 2°C$, 60% humidity, and the light cycle consisting of 7:00–19:00 in the dark and 19:00–7:00 in the light for 2 weeks, and 6-OHDA loading experiments were carried out for a period of 3 days. Rats were at the 124–125th generation in SHRSP and 87–88th generation in WKY, respectively, from the establishment of SHRSP strain by Okamoto *et al.* (1974). These animals were divided into three groups of 7–8. One was a sham intracerebral ventricle (i.c.v.) puncture group (control), a second was a 6-OHDA (6-hydroxydopamine hydrobromide, Sigma Co., St. Louis, MO, US) i.c.v. injection group (200 μg/rat, 6-OHDA i.c.v. group), and the other was a 6-OHDA intraperitoneal injection group (50 mg/kg, 6-OHDA i.p. group). These doses given to the rats was thought to be maximal, because 2 times these doses caused an epileptic seizure or sudden death, respectively. Systolic blood pressure (SBP) and pulse rate (HR) were measured with a volume-oscillometric manometer (TK-370A, UNICOM, Japan) as follows. Each rat, while conscious, was pre-warmed at 35°C for 5 min in a warming box, and then put into a folder made of stainless steel net, and the tail was inserted into a cuff with a laser blood pulse sensor. SBP and HR were measured in 1 min by an automatic up and down control on the cuff pressure.

Procedure for i.c.v. injection, methods of sampling, catecholamine determination, and data analysis

I.c.v. injection was performed as follows according to the method of Noble *et al.* (1967). After sagittal cut of the head skin in rats under sodium pentobarbital anesthesia (35 mg/kg i.p.) at 9:00, the cross point (Bregma) of sagittal line and coronal suture was exposed. A hole in the bone was rotatedly punched by a needle sized 24G at the point of 0.8 mm lateral and 1.4 mm posterior from Bregma, and 200 μg per 20 μlitres of 6-OHDA solution was injected at a depth of 4–5 mm in the lateral cerebral ventricle at 10 s after confirmation of CSF sampling through a microsyringe. The blood samples (0.5 ml) were taken from the tail vein of the rats bound in a rat holder at 10:00 in dark in the conscious condition, and rapidly centrifuged to collect plasma, and then acidified by adding the same volume of 1.0 N perchloric acid solution (PCA) containing 75 ng of 3,4-dihydroxybenzylamine hydrobromide (DHBA) for recovery calculation. The urine samples were collected from a urine reservoir containing 0.1 N HCl solution in metabolic cages where the rats were put for 24 hours. The filtrated samples were diluted with 0.1 N PCA solution by 400 times before measurement. The tissue samples were taken after decapitation on the 2nd day after treatments, and homogenized for 10 s at a middle round speed, using a Polytron®-homogenizer with 10 times volume of 0.5 N PCA. Every sample was filtered through Ultrafree-MC®. Twenty-one catecholamine metabolites in the filtrates were assayed with a Neurochem®-HPLC system. CA metabolite concentrations were expressed as ng/mg protein in the tissues, ng/ml in plasma, and mg/day/rat in urine, respectively. All the experiments were conducted in accordance with the guidelines for the Care and Use of animals approved by the Kinki University School of Medicine.

All the data were presented as mean ± S.E.M., and compared by the method of multiple statistical analysis of ANOVA-Scheffe between the groups. Statistical difference between the groups was determined when *p* value was below 0.05.

RESULTS

Effects of 6-OHDA treatments on systolic blood pressures (BP) and heart rates (HR) in SHRSP and WKY

The values of the control group of SHRSP without treatments were significantly higher than those of WKY. Through 6-OHDA i.c.v. administration, BP changed little in SHRSP and WKY, but 6-OHDA i.p. administration decreased BPs in both of SHRSP and WKY at days 1 and 2, especially in SHRSP (Fig. 1). HR did not change in WKY, though showed a decreasing tendency by 16% in SHRSP at day 2 through 6-OHDA i.c.v. injection. On the other hand, 6-OHDA i.p. administration into SHRSP significantly increased HR at days 1 by 28% and 2 by 30%, but did not in WKY (Fig. 2).

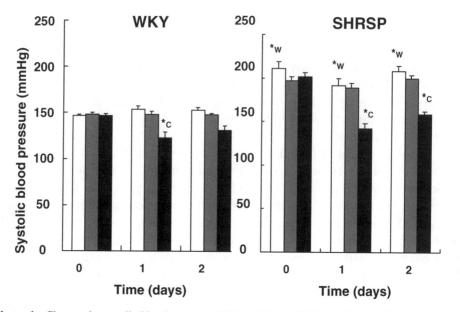

Figure 1. Changes in systolic blood pressures (BP, mmHg) in WKY and SHRSP before and after the chemical sympathectomy using 6-hydroxydopamine (6-OHDA). White, gray, and black columns represent each BP value of control, 6-OHDA i.c.v., 6-OHDA i.p. treatments, respectively, at days 0, 1, and 2. *C and *W represent the significant differences between the values of control and treated groups, and between the values of WKY and SHRSP at $p < 0.01$, respectively.

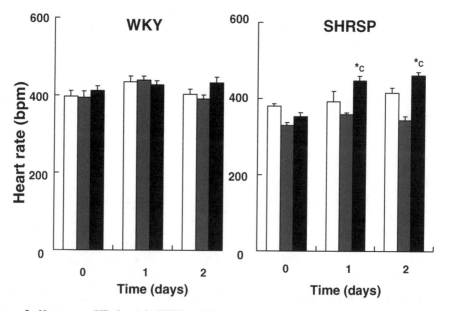

Figure 2. Heart rates (HR, bpm) in WKY and SHRSP before and after the chemical sympathectomy using 6-hydroxydopamine (6-OHDA). White, gray, and black columns represent each HR value of control, 6-OHDA i.c.v., 6-OHDA i.p. treatments, respectively, at days 0, 1, and 2. *C represents the significant difference between the values of control and treated groups at $p < 0.01$.

Effects of 6-OHDA treatments on the levels of catecholamine metabolites in the tissue, plasma and urine

Catecholamine levels in the brain stem and medulla oblongata before and after 6-OHDA administration. L-DOPA concentrations converted from L-tyrosine by tyrosine hydroxylase, a rate limiting enzyme on the course of CA production, in the brain stem, were 22.3 ± 15.4 and 31.8 ± 13.2 ng/mg protein in SHRSP and WKY, respectively. In both of the rat groups, this metabolite tended to decrease to the values of 6.25 ± 0.92 and 7.47 ± 1.28 ng/mg protein in SHRSP and WKY, respectively, by 6-OHDA i.c.v. treatment, but did not by 6-OHDA i.p. treatment. However, NE contents in the brain stem also changed the same as those of L-DOPA. That is, they changed from 701 ± 82 to 345 ± 20 ng/mg protein in SHRSP and from 742 ± 81 to 213 ± 69 ng/mg protein in WKY by 6-OHDA i.c.v. treatment, but did not by i.p. treatment in either of the rats. In the medulla oblongata, almost all the same changes were observed as those of the former findings. NE levels changed from 846 ± 83 to 379 ± 63 in SHRSP and from 973 ± 83 to 416 ± 31 ng/mg protein in WKY, respectively, by 6-OHDA i.c.v. treatment, but did not by 6-OHDA i.p. injection (figure not shown).

Catecholamine levels in the adrenal glands before and after 6-OHDA administration. Ep and NE contents in the adrenal glands were not different between SHRSP and WKY as 214 ± 51 and 151 ± 22 ng/mg protein in Ep, and 55.7 ± 11.8 and 37.9 ± 4.0 ng/mg protein in NE, respectively. Neither 6-OHDA i.c.v. nor i.p. injections caused any changes in Ep and NE concentrations in the adrenal glands of SHRSP and WKY (figure not shown).

Catecholamine basal levels in plasma between SHRSP and WKY. L-DOPA levels in plasma were not different between SHRSP and WKY at 3.03 ± 1.22 and 2.69 ± 0.91 ng/ml, respectively. NE levels were not also different between the two rat groups, at 40.7 ± 14.9 in SHRSP and 46.1 ± 16.8 ng/ml in WKY. Ep levels of SHRSP showed an increasing tendency than those of WKY, at 9.57 ± 3.19 in SHRSP and 6.29 ± 1.49 ng/ml in WKY, as reported by Jablonskis and Howe (1994). MHPG, a metabolite from NE and Ep, was also at the same levels between the two rat groups, at 27.0 ± 12.3 in SHRSP and 24.8 ± 8.8 ng/ml in WKY (figure not shown).

Catecholamine basal levels in urine between SHRSP and WKY. Normethanephrine (NMN), a metabolite from NE, excretions in urine were not different between SHRSP and WKY, at 2.13 ± 0.42 and 2.41 ± 0.74 mg/day, respectively. MN contents, a metabolite from Ep, in urine were also similar between the two rat groups, at 4.19 ± 0.77 mg/day in SHRSP and 4.22 ± 1.55 mg/day in WKY. VA excretions in urine, the final metabolite from NE and Ep, were also at similar levels between SHRSP and WKY, at 22.6 ± 1.5 and 25.7 ± 1.6 mg/day, respectively (figure not shown).

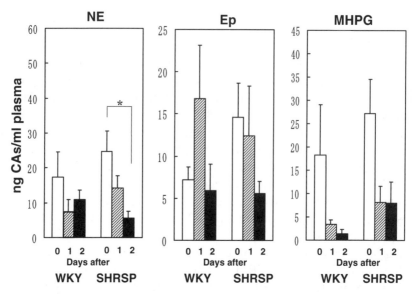

Figure 3. Norepinephrine (NE), epinephrine (Ep), and methoxy-hydroxyphenyl glycol (MHPG) levels in plasma before and after the chemical sympathectomy using 200 μg/rat 6-hydroxydopamine (6-OHDA) i.c.v. injection. Open, shaded, and closed columns represent each catecholamine value at days 0, 1, and 2. *represents the significant difference between the values before and after 6-OHDA i.c.v. treatment.

Catecholamine level changes in plasma before and after 6-OHDA administration. By the treatment with 6-OHDA i.c.v., L-DOPA levels in plasma changed little in either SHRSP or WKY: from 2.29 ± 0.56 to 2.70 ± 0.92 ng/ml at day 2 in SHRSP and from 3.73 ± 0.92 to 2.73 ± 0.67 ng/ml at day 1 in WKY. By the treatment with 6-OHDA i.p., L-DOPA levels in plasma showed a slightly decrease in both rat groups, from 4.08 ± 1.35 to 1.71 ± 0.92 ng/ml at day 1 in SHRSP and from 2.44 ± 0.89 to 1.67 ± 0.71 ng/ml at day 2 in WKY (figure not shown).

By the treatment with 6-OHDA i.c.v., NE levels decreased more significantly in SHRSP than those of pretreatment, but not in WKY. Ep and MHPG levels changed little in either SHRSP or WKY as shown in Fig. 3. On the other hand, Ep levels significantly decreased in SHRSP by the treatment with 6-OHDA i.p., and NE and MHPG levels changed little in either of the rat groups as shown in Fig. 4.

Catecholamine level changes in urine before and after 6-OHDA administration. Though there was no significant change in NMN excretions between before and after 6-OHDA i.c.v. injection in either of the rat groups, VA excretions significantly decreased at day 1 compared with those of pretreatment in both of the rat groups, and recovered at day 2. Regarding MN, the same tendencies as those of VA were found in both rat groups. Though there was no significant change in NMN excretions between before and after 6-OHDA i.p. treatment in either of the rat groups, VA excretions significantly decreased at day 1 compared with those of

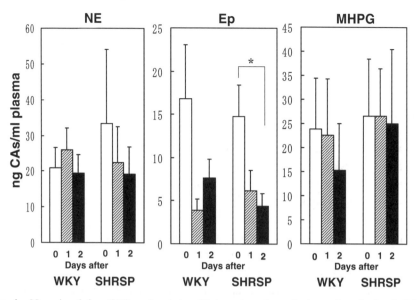

Figure 4. Norepinephrine (NE), epinephrine (Ep), and methoxy-hydroxyphenyl glycol (MHPG) levels in plasma before and after the chemical sympathectomy using 50 mg/kg 6-hydroxydopamine (6-OHDA) i.p. injection. Open, shaded, and closed columns represent each catecholamine value at days 0, 1, and 2. *represents the significant difference between the values before and after 6-OHDA i.p. treatment.

pretreatment in both of the rat groups, and recovered at day 2. MN excretions changed as well as those of VA, especially in WKY (figure not shown).

DISCUSSION

As we and others (Ely *et al.*, 1997; Kumai *et al.*, 1995) have clarified so far, SHRSP, one of pathophysiological animal model for human essential hypertension, is on a hyper-sympathetic state as compared with the normotensive WKY. That is, three major enzymes contributing to higher catecholamine production — phenylalanine-4-monooxygenase, tyrosine hydroxylase and phenylethanolamine-N-methyltransferase — might increase more in SHRSP than in WKY (Maeda *et al.*, 1999, 2000). Therefore, NE and Ep levels in plasma or urine go up more in young SHR or SHRSP than in WKY. Our former report (Higashino and Ooshima, 2003) clarified that every CA metabolite content in the adrenal glands of SHRSP was not different from those in WKY in the dark, but every CA increased more in the light in SHRSP than in WKY, since the adrenal glands of SHRSP released much more CAs into the fluid than those of WKY in the dark, for an active period. On the other hand, MHPG and VA, metabolites from Ep, in the urine significantly decreased after adrenalectomy (Adrex) in SHRSP, because the total amount of Ep secretion in the fluid decreased more by the Adrex in SHRSP than in WKY. Therefore, those findings clearly showed that sympathetic activity with CA productions from the adrenal glands in SHRSP was higher than that of WKY as known so far (Kumai *et al.*, 1994). Moreover, it showed that Ep synthesis and release in SHRSP must have occurred in the accessory or extra-adrenal glands such as Zuckerkandle's body besides the adrenal glands, in order to compensate at the time of lack in it (Higashino and Ooshima, 2003). Then, our interest in investigation was focused on the differences between the central and peripheral sympathetic activities in SHRSP. Therefore, a maximum dose of 6-OHDA was injected into a lateral cerebral ventricle (i.c.v.) in order to destroy the central sympathetic nerves, and the central sympathetic activity in SHRSP was investigated. Likewise, a maximum dose of 6-OHDA was injected into a peritoneal cavity (i.p.) in order to destroy the peripheral nerves, and the peripheral sympathetic activity (Nyborg *et al.*, 1986) was investigated. That is what we call a chemical sympathectomy.

Treatment of 6-OHDA i.c.v. showed no changes in BP, but showed a decreasing tendency of HR, L-DOPA, and a significant decrease in NE in the brain stem and medulla oblongata in SHRSP. On the other hand, treatment of 6-OHDA i.p. showed a significant decline of BP and an increase of HR, but showed no changes in L-DOPA and NE levels in the brain stem and medulla oblongata in SHRSP. These findings indicate that the peripheral sympathetic control system is dominant compared with the central sympathetic control system in SHRSP, though CAs in the brain affect largely the peripheral sympathetic nerve activity as Unger *et al.* (1984) pointed out in the past. Levels of catecholamines, such as NE and Ep, in the adrenal glands did not change with either 6-OHDA i.c.v. or i.p. treatments even by submaximal doses. That indicates that the adrenal glands work outside of the central and peripheral sympathetic control systems, although they are slightly dependent on both of the sympathetic control systems.

NE level in plasma fell greatly through 6-OHDA i.c.v. treatment, and Ep level in plasma fell significantly through 6-OHDA i.p. treatment in SHRSP, but not in WKY. MHPG levels showed a decreasing tendency by the treatment of 6-OHDA i.c.v., but did not through the treatment of 6-OHDA i.p. in either SHRSP or WKY. MN and VA levels in urine declined and then reversed by both treatments with 6-OHDA. 6-OHDA i.c.v. showed no changes in BP, but 6-OHDA i.p. showed a significant decline of BP and an increase of HR.

These findings indicate that (1) 6-OHDA i.p. treatment destroys the accessory or extra-adrenal glands (Murakami *et al.*, 1989) besides peripheral sympathetic nerves, (2) Ep released from these extra-secretary organs affects largely the contraction of peripheral resistant arteries, and contributes to elevating and maintaining the hypertension in SHRSP, (3) Though 6-OHDA i.c.v. treatment destroys the central sympathetic nerves and inhibits temporally the peripheral sympathetic activity, adrenal glands and the accessory or extra-adrenal glands compensate for the lack of sympathetic activity, and (4) peripheral sympathetic nerve function works mainly to maintain hyper-sympathetic states in SHRSP as Ikeda *et al.* (1979) pointed out in the past. Therefore, Adrex or 6-OHDA i.c.v. treatments cannot overcome the cardiovascular control system through the peripheral sympathetic nerve function in SHRSP.

Essential hypertension in humans might also have these characteristics. If so, it might be useful and important to depress the peripheral sympathetic activity by means of sympathetic blockers in order to inhibit the progression of arteriosclerotic disorders.

REFERENCES

Ely, D., Caplea, A., Dunphy, G., *et al.* (1997). Spontaneously hypertensive rat Y chromosome increases indexes of sympathetic nervous system activity, *Hypertension* **29**, 613–618.

Higashino, H. and Ooshima K. (2003). Central and peripheral sympathetic characteristics in hypertension considered from differences of catecholamine metabolite concentrations in plasma before and after adrenalectomy in male 2-month-old SHRSP and WKY, *Biogenic Amines* **17**, 389–399.

Ikeda, H., Shino, A. and Nagaoka, A. (1979). Effects of chemical sympathectomy on hypertension and stroke in stroke-prone spontaneously hypertensive rats, *Eur. J. Pharmacol.* **53**, 173–179.

Jablonskis, L. T. and Howe, P. R. (1994). Elevated plasma adrenaline in spontaneously hypertensive rats, *Blood Pressure* **3**, 106–111.

Kumai, T., Tanaka, M., Watanabe, M., *et al.* (1995). Influence of androgen on tyrosine hydroxylase mRNA in adrenal medulla of spontaneously hypertensive rats, *Hypertension* **26**, 208–212.

Kumai, T., Tanaka, M., Watanabe, M., *et al.* (1994). Elevated tyrosine hydroxylase mRNA levels in the adrenal medulla of spontaneously hypertensive rats, *Jpn. J. Pharmacol.* **65**, 367–369.

Maeda, K., Azuma, M., Nishimura, Y., *et al.* (1999). Comparison of catecholamine metabolite levels in plasma drawn at daytime, nighttime, and acute cold stress between 2-month-old SHRSP and WKY, *Clin. Exp. Hypertension* **21**, 464–465.

Maeda, K., Azuma, M., Nishimura, Y., *et al.* (2000). Catecholamine metabolites levels in plasma, and its synthetic and secretary activities in adrenal glands between 2-month-old SHRSP and WKY, *Clin. Exp. Hypertension* **22**, 359–360.

Murakami, T., Oukouchi, H., Uno, Y., *et al.* (1989). Blood vascular beds of rat adrenal and accessory adrenal glands, with special reference to the corticomedullary portal system: a further scanning electron microscopic study of corrosion casts and tissue specimens, *Arch. Histol. Cytol.* **52**, 461–476.

Noble, E. P., Wurtman, R. J. and Axelrod, J. (1967). A simple method for injecting ^{3}H-norepinephrine into the lateral ventricle of the rat brain, *Life Sci.* **6**, 281–291.

Nyborg, N. C., Korsgaard, N. and Mulvany, M. J. (1986). Neonatal sympathectomy of normotensive Wistar-Kyoto and spontaneously hypertensive rats with 6-hydroxydopamine: effects on resistance vessel structure and sensitivity to calcium, *J. Hypertension* **4**, 455–461.

Okamoto, K., Yamori, Y. and Nagaoka A. (1974). Establishment of the stroke-prone spontaneously hypertensive rats (SHR), *Circ. Res.* **34/35** (Suppl. 1), 143–153.

Unger, T., Ganten, D. and Lang, R. E. (1984). The central nervous system, regulation of blood pressure and hypertension. Recent advances in the role of the central nervous system in the regulation of blood pressure, *Fortschr. Med.* **14**, 606–608.

Advances in Neuroregulation and Neuroprotection (2005), pp. 419-430
Collin, C. *et al.* (Eds)
© VSP 2005

Effects of serotonin (5-HT) on catecholamine secretion from cultured bovine chromaffin cells

Y. SUGIMOTO *, F. NISHIKAWA, T. YOSHIKAWA, T. NOMA and J. YAMADA

Department of Pharmacology, Kobe Pharmaceutical University, Motoyamakita-machi, Higashinada-ku, Kobe 658-8558, Japan

Abstract—Effects of serotonin (5-hydroxytryptamine, 5-HT) on catecholamine (CA) secretion were examined in cultured bovine adrenal chromaffin cells. Although 5-HT itself did not increase CA secretion, 5-HT significantly inhibited acethylcholine-induced CA secretion and 5-HT blocked nicotine-evoked CA release in a dose-dependent manner. Co-incubation with muscarine and 5-HT did not increase CA secretion. At a high dose only, 5-HT inhibited high K^+ (56 mM) and veratridine-induced CA release, but 5-HT did not affect secretion induced by Ca^{2+}-ionophore A23187. Furthermore, 5-HT dose-dependently inhibited the nicotine-induced rise of intracellular free Ca^{2+} ($[Ca^{2+}]_i$), as well as the CA secretion elicited by nicotine. Lineweaver–Burk Plot analysis demonstrated that 5-HT non-competitively inhibited nicotine-induced CA release. These findings suggest that 5-HT inhibits nicotine-induced CA secretion and increases intracellular free Ca^{2+} in adrenal chromaffin cells by non-competitive inhibition of nicotinic acetylcholine-receptor.

Keywords: 5-HT; catecholamine secretion; adrenal chromaffin cells; nicotinic receptor; intracellular free Ca^{2+}.

INTRODUCTION

Serotonin (5-hydroxytryptamine, 5-HT) is a neurotransmitter and regulates several physiological functions (Murphy *et al.*, 1991). It is well established that 5-HT is an important factor controlling hormone secretion and that 5-HT increases secretions of several hormones in animals and humans. 5-HT stimulates corticosterone, rennin, and prolactin (Van de Kar, 1991; Van de Kar *et al.*, 1996). It was reported that 5-HT stimulates corticosterone secretion from the adrenal cortex mediated by the 5-HT₄ receptor (Delarue *et al.*, 1998; Lefebvre *et al.*, 1998).

Adrenal chromaffin cells synthesize and store catecholamines (CA) (Slotkin and Kirshner, 1971). Immunohistochemical study demonstrated that an indoleamine,

---

*To whom correspondence should be addressed. E-mail: yumisugi@kobepharma-u.ac.jp

5-HT is present in adrenal chromaffin cells (Holzwarth *et al.*, 1984; Verhofstad and Jonsson, 1983). It was reported that adrenaline and 5-HT coexist in adrenal chromaffin cells of rats (Delarue *et al.*, 1992; Holzwarth *et al.*, 1984, 1985). It has been suggested that 5-HT may be synthesized within chromaffin cells or taken up from the blood circulation (Kent and Coupland, 1984; Slotkin and Kirshner, 1971). Therefore, the evidence of 5-HT in chromaffin cells raises the possibility that 5-HT may modulate CA secretion from chromaffin cells. In the present study, therefore, we investigated the effects of 5-HT on CA secretion from cultured bovine adrenal chromaffin cells. Since CA secretion is stimulated by activation of nicotinic acetylcholine receptor or voltage-dependent Ca^{2+} or Na^+ channels (Livett, 1984), the effects of 5-HT on CA secretion elicited by activation of these receptors and channels were examined.

MATERIALS AND METHODS

Chromaffin cell preparation and culture

Adrenal medullary chromaffin cells were prepared from bovine adrenal medulla according to the method of Kilpatrick *et al.* (1980). Briefly, chromaffin cells were collagenase digestion and purified by Percoll gradient. Isolated cells were plated on 24-well plastic cluster plates at a density of 5×10^5 cells/well and maintained for 3–4 days as monolayer cultures at 37°C in humidified atmosphere containing 5% CO_2 in 1 ml of Dulbecco's modified Eagle medium (DMEM, Gibco, USA) containing 10% heat-inactivated fetal calf serum (Gibco, USA), penicillin G (170 U/ml), streptomycin (100 μg/ml), tetracycline (5 μg/ml), gentamicin (10 μg/ml), nystatin (25 U/ml), 5-fluoro-2-deoxyuridine (10^{-5} M), cytosinearabinoside (10^{-5} M) and uridine (10^{-5} M).

Determination of CA secretion

Cells were washed with 1 ml of Krebs-Ringer bicarbonate (KRB) buffer (pH 7.4) and then incubated by several drugs at 37°C for 10 min in 250 μl of KRB buffer. After incubation, the medium was withdrawn and the cells were lysed by adding 10% acetic acid followed by a freeze-thawing. Determination of CA in the medium and cells were determined according to the method based on trihydroxyindole method (Kelner *et al.*, 1985). CA secretion was calculated as the percentage of total cellular CA concentration. In experiments using various CA stimulants, results were calculated by subtracting the basal secretion form the secretion stimulated by various stimulants.

Determination of intracellular free $Ca^{2+}([Ca^{2+}]_i)$

Determination of $[Ca^{2+}]_i$ was performed by the method of Grynkiewicz *et al.* (1985). Isolated chromaffin cells in bacteriological dishes were incubated in 10 ml of Krebs-HEPES buffer (pH 7.4) containing fura2/AM 2×10^{-6} M (Dojin, Japan) at 37°C for 30 min. Cells loaded with fura2/AM were washed with Krebs-HEPES buffer was prepared at a concentration of 2×10^6 cells/ml. 480 μl cell suspension was transferred into a cuvette and challenged with 20 μl stimulants at 37°C under continuous stirring. The fluorescence ratio obtained by dividing the fluorescence excited at 340 nm by that at 380 nm was determined Ca^{2+} analyzer (CAF-100, Jasco, Japan). The emission wavelength was 500 nm. Changes in the fluorescence ratio were calibrated to changes in $[Ca^{2+}]_i$ using the methods of Grynkiewicz *et al.* (1985).

Chemicals

Drugs used were purchased from the following sources: serotonin creatinine sulfate (5-hydroxytytpamine, 5-HT, Merck, Germany); acetylcholine chloride, nicotine tartrate, muscarine chloride, veratridine (Sigma, USA); A23817 (Calbiochem, USA).

Statistical analysis

The statistical significance was evaluated with Dunnett test followed by analysis of variance (ANOVA).

RESULTS

Effects of 5-HT on secretion of CA

Effects of 5-HT on CA secretion in cultured chromaffin cells were studied. However, 5-HT (10^{-5}–10^{-3} M) did not show significant effects (Fig. 1).

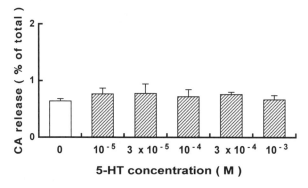

Figure 1. Effects of 5-HT on CA release from cultured adrenal chromaffin cells. Results are shown as mean ± S.E. ($n = 5$–8). Cells were incubated at 37°C for 10 min.

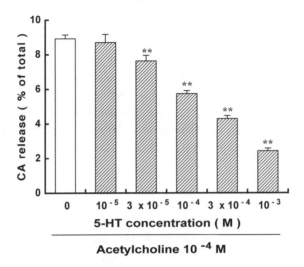

Figure 2. Effects of 5-HT on CA release induced by acetylcholine from cultured adrenal chromaffin cells. Results are shown as mean ± S.E. ($n = 6$–7). Cells were incubated at 37°C for 10 min. Spontaneously released CA was subtracted in each group **$p < 0.01$.

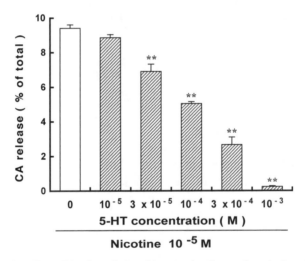

Figure 3. Effects of 5-HT on CA release induced by nicotine from cultured adrenal chromaffin cells. Results are shown as mean ± S.E. ($n = 5$–7). Cells were incubated at 37°C for 10 min. Spontaneously released CA was subtracted in each group **$p < 0.01$.

Effects of 5-HT on acetylcholine-induced CA secretion

Figure 2 demonstrates effects of 5-HT on acetylcholine-stimulated CA secretion. Acetylcholine (10^{-4} M) caused an apparent CA secretion. 5-HT dose-dependently inhibited CA secretion evoked by acetylcholine.

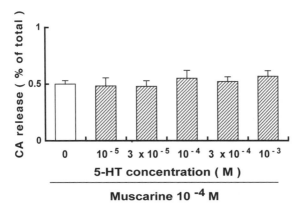

Figure 4. Effects of 5-HT on CA release induced by muscarine from cultured adrenal chromaffin cells. Results are shown as mean ± S.E. (n = 5–9). Cells were incubated at 37°C for 10 min. Spontaneously released CA was subtracted in each group.

Effects of 5-HT on nicotine- and muscarine-induced CA secretion

Effects of 5-HT on CA secretion induced by nicotine and muscarine are shown in Figs. 3 and 4, respectively. Nicotine (10^{-5} M) caused the secretion of CA corresponding to 9.4% of total CA. 5-HT (3×10^{-5}–10^{-3} M) apparently reduced nicotine-induced CA secretion (Fig. 3). Muscarine (10^{-4} M) induced a small CA secretion from cultured chromaffin cells. 5-HT did not affect CA secretion induced by muscarine (Fig. 4).

Effects of 5-HT on 56 mM K$^+$- and veratridine-induced CA secretion

As shown in Figs. 5 and 6, 56 mM K$^+$ and veratridine (10^{-5} M) induced the CA secretion of 12.0% and 7.0% of total cellular CA, respectively. 5-HT at the only concentration of 10^{-3} M inhibited CA secretion evoked by 56 mM K$^+$ and veratridine.

Effects of 5-HT on A23187-induced CA secretion

Effects of 5-HT on A23187 (10^{-5} M)-induced CA secretion are shown in Fig. 7. 5-HT did not affect A23187-induced CA secretion.

Effects of 5-HT on nicotine-, muscarine- and 56 mM K$^+$-induced intracellular Ca^{2+} elevation

Nicotine-induced $[Ca^{2+}]_i$ rises were inhibited by 5-HT dose-dependently (Fig. 8). However, 5-HT did not affect muscarine-induced $[Ca^{2+}]_i$ rise (Fig. 9). 5-HT at a high concentration (10^{-3} M) reduced 56 mM K$^+$-induced $[Ca^{2+}]_i$ elevation (Fig. 10). 5-HT itself did not change $[Ca^{2+}]_i$ (data is not shown).

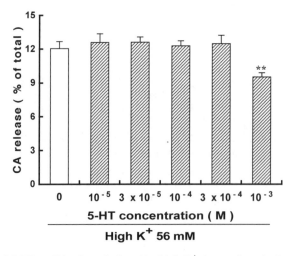

Figure 5. Effects of 5-HT on CA release induced by high K^+ from cultured adrenal chromaffin cells. Results are shown as mean $\pm$ S.E. ($n = 6$–7). Cells were incubated at 37°C for 10 min. Spontaneously released CA was subtracted in each group $**p < 0.01$.

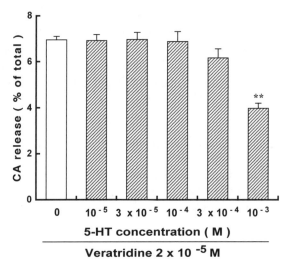

Figure 6. Effects of 5-HT on CA release induced by veratridine from cultured adrenal chromaffin cells. Results are shown as mean $\pm$ S.E. ($n = 5$–8). Cells were incubated at 37°C for 10 min. Spontaneously released CA was subtracted in each group $**p < 0.01$.

Non-competitive blockade of nicotine-induced CA release by 5-HT

Dose-response curves of nicotine or nicotine plus 5-HT on CA secretion are shown in Fig. 11. Over the range of 3–100 μM nicotine, 5-HT 300 μM inhibited maximal CA secretion elicited by nicotine. Lineweaver–Burk plot analysis is shown in Fig. 11. As shown in results, the inhibitory effects of 5-HT on nicotine-induced CA secretion are noncompetitive.

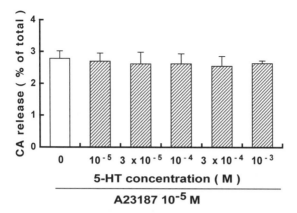

Figure 7. Effects of 5-HT on CA release induced by A23187 from cultured adrenal chromaffin cells. Results are shown as mean ± S.E. ($n = 5$–6). Cells were incubated at 37°C for 10 min. Spontaneously released CA was subtracted in each group.

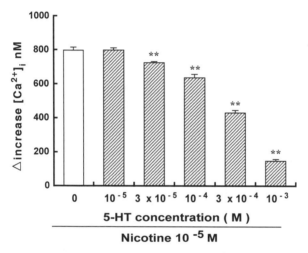

Figure 8. Effects of 5-HT on nicotine-induced increase in $[Ca^{2+}]_i$ in cultured adrenal chromaffin cells. Results are shown as mean ± S.E. ($n = 5$–8). Adrenal chromaffin cells were loaded with fura-2/AM at 37°C for 30 min $**p < 0.01$.

DISCUSSION

Although it has been reported that 5-HT is present in adrenal chromaffin cells (Holzwarth *et al.*, 1984, 1985; Kent and Coupland, 1984; Slotkin and Kirshner, 1971; Verhofstad and Jonsson, 1983), it remains unclear whether 5-HT affects CA secretion from adrenal chromaffin cells. As shown in results, 5-HT inhibited acetylcholine-induced CA secretion, although 5-HT alone did not affect it. CA secretion is physiologically regulated by nicotinic acetylcholine receptors, the activation of which leads to Na^+ and Ca^{2+} influx in adrenal chromaffin cells (Kilpatrick, 1984; Livett, 1984). It has been reported that muscarinic acetylcholine receptors are also present in chromaffin cells and can increase CA secretion (Livett, 1984;

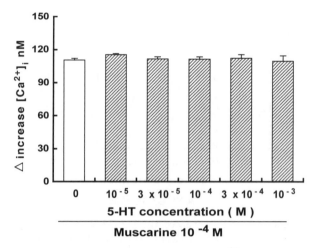

Figure 9. Effects of 5-HT on muscarine-induced increase in $[Ca^{2+}]_i$ in cultured adrenal chromaffin cells. Results are shown as mean ± S.E (n = 5–7). Adrenal chromaffin cells were loaded with fura-2/AM at 37°C for 30 min.

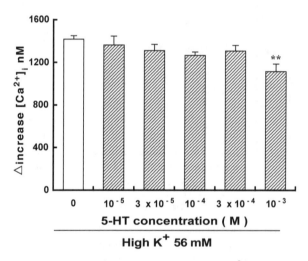

Figure 10. Effects of 5-HT on high K^+-induced increase in in $[Ca^{2+}]_i$ cultured adrenal chromaffin cells. Results are shown as mean ± S.E (n = 6–9). Adrenal chromaffin cells were loaded with fura-2/AM at 37°C for 30 min **$p < 0.01$.

Misbahuddin *et al.*, 1985). Therefore, we studied effects of 5-HT on nicotine- and muscarine-induced CA secretion. Although 5-HT apparently decreased nicotine-evoked CA secretion in a dose-dependent manner, it did not affect muscarine-induced CA secretion. These findings suggest that 5-HT inhibits nicotinic acetyl-choline receptor-mediated CA secretion without affecting muscarinic acetylcholine receptors.

Nicotine-induced CA secretion is elicited by stimulation of nicotinic acetylcholine receptor-coupled Na^+ channels, resulting in activation of voltage-dependent Na^+

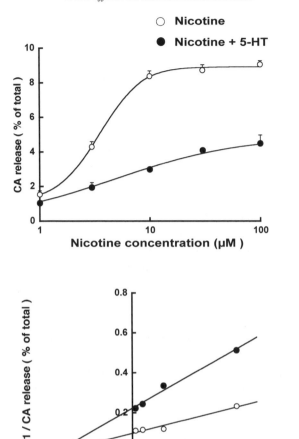

Figure 11. Inhibitory effects of 5-HT on CA release induced by nicotine and Lineweaver–Burk plot analysis on inhibitory effects of 5-HT on nicotine-induced CA release. Results are shown as mean ± S.E. ($n = 6$–10). Cells were incubated at 37°C for 10 min. Spontaneously released CA was subtracted in each group.

and Ca^{2+} channels. Consequently, Ca^{2+} entry into the cells is markedly raised, thus inducing CA secretion (Livett, 1984). It has been reported that 5-HT blocks voltage-dependent Na^+ channels in the vertebrate nervous system (Anwyl, 1990). It indicates that these channels may mediate the inhibitory effects of 5-HT on nicotine-induced CA secretion. Thus, we studied the effects of 5-HT on CA secretion induced by stimulation of voltage-dependent Ca^{2+} and Na^+ channels. The concentration of 5-HT 10^{-3} M reduced CA secretion induced by high K^+, which is mediated by voltage-dependent Ca^{2+} channels (Kilpatrick, 1984; Livett, 1984). Veratridine, which activates the voltage-dependent Na^+ channels, induced CA secretion (Kilpatrick, 1984). 5-HT also inhibited veratridine-induced CA secretion at a 5-HT concentration of 10^{-3} M. These results suggest that at a high concentration only, 5-HT inhibited CA secretion associated with both voltage-dependent Ca^{2+} and Na^+ channels. Therefore, it is likely that the inhibitory effects of 5-HT on CA secretion caused by both voltage-dependent Ca^{2+} and Na^+ channels are not strong. It suggests that the effects of 5-HT on these channels are not attributed to the blocking ability of 5-HT on nicotine-induced CA secretion. In addition, 5-HT did not affect CA secretion elicited by Ca^{2+} ionophore A23187, which is an agent that increases the intracellular secretory system by introducing Ca^{2+} into the cell (Morita *et al.*, 1985). This suggests that 5-HT did not affect Ca^{2+}-activated secretory mechanisms within the cells. Thus, the inhibitory effects of 5-HT on nicotine-induced CA secretion may be derived by blockade of nicotinic acetylcholine receptor-induced depolarization at the cell membrane.

Co-administration of 5-HT dose-dependently decreased nicotine-evoked marked elevation of intracellular free Ca^{2+} concentration ($[Ca^{2+}]_i$), while 5-HT alone was without effect. The slight elevation of $[Ca^{2+}]_i$ induced by muscarine was not affected by 5-HT. In addition, 5-HT at a high concentration (10^{-3} M) inhibited high K^+-induced rise in $[Ca^{2+}]_i$, which is mediated by the voltage-dependent Ca^{2+} channel. These findings regarding 5-HT effects on $[Ca^{2+}]_i$ are very similar to those on CA secretion. Therefore, it is suggested that 5-HT inhibits nicotine-induced CA secretion by inhibiting nicotinic acetylcholine receptor-associated Ca^{2+} entry into the cells.

As indicated, 5-HT inhibits nicotinic acetylcholine receptor-mediated CA secretion. A previous report demonstrated that 5-HT blocks nicotinic acetylcholine receptors in mouse myotubes, human cloned TE671/RD cells and *Xenopus laevis* oocytes and that this blockade is noncompetitive (Blanton *et al.*, 2000; Garcia-Colunga and Miledi, 1999; Grassi *et al.*, 1993). Therefore, we further examined the interaction between 5-HT and nicotinic acetylcholine receptors in CA secretion. Dose-response curves of nicotine or nicotine plus 5-HT on CA secretion were drawn from the results. Over the range of 3–100 μM nicotine, 5-HT 300 μM inhibited maximal CA secretion elicited by nicotine. Lineweaver–Burk plot analysis demonstrated that the inhibitory effects of 5-HT on nicotine-induced CA secretion are noncompetitive, which is in agreement with previous findings using other cells (Blanton *et al.*, 2000; Garcia-Colunga and Miledi, 1999; Grassi *et al.*, 1993).

CONCLUSION

Our results demonstrate that 5-HT inhibits nicotinic acetylcholine receptor-mediated CA secretion and rise in $[Ca^{2+}]_i$ in cultured adrenal chromaffin cells. Furthermore, it is indicated that blockade of 5-HT on nicotinic acetylcholine receptor is noncompetitive. In other cultured cells, it is suggested that 5-HT blocks nicotinic acetylcholine receptor noncompetitively (Blanton *et al.*, 2000; Garcia-Colunga and Miledi, 1999; Grassi *et al.*, 1993). Therefore, in adrenal chromaffin cells, 5-HT may interact with nicotinic acetylcholine receptor in this manner. Since it is known that 5-HT is present in adrenal chromaffin cells (Delarue *et al.*, 1992; Holzwarth *et al.*, 1984, 1985; Kent and Coupland, 1984; Slotkin and Kirshner, 1971), 5-HT may modulate CA secretion mediated via nicotinic acetylcholine receptor.

REFERENCES

Anwyl, R. (1990). Neurophysiological actions of 5-hydroxytryptamine in the vertebrate nervous system, *Prog. Neurobiol.* **35**, 451–468.

Blanton, M. P., McCardy, E. A., Fryer, J. D., *et al.* (2000). 5-Hydroxytryptamine interaction with the nicotinic acetylcholine receptor, *Eur. J. Pharmacol.* **389**, 155–163.

Delarue, C., Becquet, D., Idres, S., *et al.* (1992). Serotonin synthesis in adrenochromaffin cells, *Neuroscience* **46**, 495–500.

Delarue, C., Contesse, V., Lefebvre, H., *et al.* (1998). Pharmacological profile of serotonergic receptors in the adrenal gland, *Endocr. Res.* **24**, 687–694.

Garcia-Colunga, J. and Miledi, R. (1999). Blockage of mouse muscle nicotinic receptors by serotonergic compounds, *Exp. Physiol.* **84**, 847–864.

Grassi, F., Polenzani, L., Mileo, A. M., *et al.* (1993). Blockage of nicotinic acetylcholine receptors by 5-hydroxytryptamine, *J. Neurosci. Res.* **34**, 562–570.

Grynkiewicz, G., Poenie, M. and Tsien, R. Y. (1985). A new generation of Ca^{2+} indicators with greatly improved fluorescence properties, *J. Biol. Chem.* **260**, 3440–3450.

Holzwarth, M. A., Brownfield, M. S. and Poff, B. C. (1985). Serotonin coexists with epinephrine in rat adrenal medullary cells. Ultrastructural immunocytochemical co-localization of serotonin and PNMT in adrenal medullary vesicles, *Neuroendocrinology* **41**, 230–236.

Holzwarth, M. A., Sawetawan, C. and Brownfield, M. S. (1984). Serotonin-immunoreactivity in the adrenal medulla: distribution and response to pharmacological manipulation, *Brain Res. Bull.* **13**, 299–308.

Kelner, K. L., Levine, R. A., Morita, K., *et al.* (1985). A comparison of trihydroxyindole and HPLC/electrochemical methods for catecholamine measurement in adrenal chromaffin cells, *Neurochem. Int.* **7**, 373–383.

Kent, C. and Coupland, R. E. (1984). On the uptake and storage of 5-hydroxytryptamine, 5-hydroxytryptophan and catecholamines by adrenal chromaffin cells and nerve endings, *Cell Tissue Res.* **236**, 189–195.

Kilpatrick, D. L. (1984). Ion channels and membrane potential in stimulus-secretion coupling in adrenal paraneurons, *Can. J. Physiol. Pharmacol.* **62**, 477–483.

Kilpatrick, D. L., Ledbetter, F. H., Carson, K. A., *et al.* (1980). Stability of bovine adrenal medulla cells in culture, *J. Neurochem.* **35**, 679–692.

Lefebvre, H., Contesse, V., Delarue, C., *et al.* (1998). Serotonergic regulation of adrenocortical function, *Horm. Metab. Res.* **30**, 398–403.

Livett, B. G. (1984). Adrenal medullary chromaffin cells in vitro, *Physiol. Rev.* **64**, 1103–1161.

Misbahuddin, M., Isosaki, M., Houchi, H., *et al.* (1985). Muscarinic receptor-mediated increase in cytoplasmic free Ca^{2+} in isolated bovine adrenal medullary cells. Effects of TMB-8 and phorbol ester TPA, *FEBS Lett.* **190**, 25–28.

Morita, K., Brocklehurst, K. W., Tomares, S. M., *et al.* (1985). The phorbol ester TPA enhances A23187 — but not carbachol — and high K^+-induced catecholamine secretion from cultured bovine adrenal chromaffin cells, *Biochem. Biophys. Res. Commun.* **129**, 511–516.

Murphy, D. L., Lesch, K. P., Aulakh, C. S., *et al.* (1991). Serotonin-selective arylpiperazines with neuroendocrine, behavioral, temperature, and cardiovascular effects in humans, *Pharmacol. Rev.* **43**, 527–552.

Slotkin, T. A. and Kirshner, N. (1971). Uptake, storage, and distribution of amines in bovine adrenal medullary vesicles, *Mol. Pharmacol.* **7**, 581–592.

Van de Kar, L. D. (1991). Neuroendocrine pharmacology of serotonergic (5-HT) neurons, *Ann. Rev. Pharmacol. Toxicol.* **31**, 289–320.

Van de Kar, L. D., Rittenhouse, P. A., Li, Q., *et al.* (1996). Serotonergic regulation of renin and prolactin secretion, *Behav. Brain Res.* **73**, 203–208.

Verhofstad, A. A. J. and Jonsson, G. (1983). Immunohistochemical and neurochemical evidence for the presence of serotonin in the adrenal medulla of the rat, *Neuroscience* **10**, 1443–1453.

Advances in Neuroregulation and Neuroprotection (2005), pp. 431-442
Collin, C. *et al.* (Eds)
© VSP 2005

Enhanced inhibitory effect of 5-hydroxytryptamine on nitric oxide production by vascular smooth muscle cells derived from stroke-prone spontaneously hypertensive rats

MASAHIKO HIRAFUJI [1,*], YUKITATSU KANAI [1], AKI KAWAHARA [1], TAKUJI MACHIDA [1], NAOYA HAMAUE [1], HIDEYA SAITO [2] and MASARU MINAMI [1]

[1] *Department of Pharmacology, Faculty of Pharmaceutical Sciences, Health Sciences University of Hokkaido, Ishikari-Tobetsu, Hokkaido 061-0293, Japan*
[2] *Department of Basic Sciences, Japanese Red Cross Hokkaido College of Nursing, Kitami, Hokkaido 090-0011, Japan*

Abstract—We investigated the effect of 5-hydroxytryptamine (5-HT) on nitric oxide (NO) production by vascular smooth muscle cells derived from 7–8 weeks old stroke-prone spontaneously hypertensive rats (SHRSP) and age-matched normotensive Wistar Kyoto rats (WKY). 5-HT significantly inhibited NO production and inducible NO synthase (iNOS) expression induced by interleukin-1β (IL-1β), which effect was greater in SHRSP cells than in WKY cells. The inhibitory effect of 5-HT was mimicked by α-methyl-5-HT, a 5-HT$_2$ receptor agonist, but not by 5-HT$_1$, 5-HT$_3$ or 5-HT$_4$ receptor agonists. 5-HT inhibition of NO production was dose-dependently reversed by sarpogrelate, a 5-HT$_2$ receptor antagonist. Staurosporin, a protein kinase C (PKC) inhibitor, dose-dependently reversed the inhibitory effect of 5-HT, and phorbol 12-myristate 13-acetate, a PKC activator, dose-dependently inhibited the NO production in both WKY and SHRSP cells. Thus, 5-HT had an inhibitory effect on NO production and iNOS expression by vascular smooth muscle cells via 5-HT$_2$ receptor subtype involving PKC activation, which effect was greater in SHRSP cells than in WKY cells. The enhanced inhibitory effect of 5-HT may have a pathophysiological relevance to vascular diseases in SHRSP.

Keywords: 5-hydroxytryptamine; nitric oxide; vascular smooth muscle cells, spontaneously hypertensive rats.

INTRODUCTION

Nitric oxide (NO) produced by NO synthase (NOS) in the vascular wall is now well recognized to have important pathophysiological roles in cardiovascular disorders. As well as in the endothelium, NO can be produced by inducible NOS

---

*To whom correspondence should be addressed. E-mail: hirafuji@hoku-iryo-u.ac.jp

(iNOS) in vascular smooth muscle cells following stimulation with cytokines such as interleukin-1β (IL-1β) or lipopolysaccharide (Busse and Mülsch, 1990). Besides the vasodilating action, NO regulates platelet aggregation and adhesion, leukocyte adhesion, and vascular smooth muscle cell proliferation and migration (Dusting, 1995). Therefore, alteration in NO production by iNOS in the vascular wall may contribute to the development of cardiovascular disorders such as atherosclerosis, vasospasm, thrombosis, or hypertension. Several studies have reported the difference in NO production and iNOS expression by cultured vascular smooth muscle cells between spontaneously hypertensive rats (SHR)/stroke-prone SHR (SHRSP) and normotensive Wistar Kyoto rats (WKY). The IL-1β-induced NO production or iNOS expression are significantly lower in SHR/SHRSP cells than in age-matched WKY cells (Dubois, 1996; Hirafuji et al., 2002; Malinski et al., 1993; Singh et al., 1996). However, the iNOS protein content in the homogenates of lipopolysaccharide- or IL-1β-stimulated aorta isolated from SHR in the established stage of hypertension has been demonstrated to be higher than that from WKY (Junquero et al., 1993; Wu et al., 1996; Chou et al., 1998; Vaziri et al., 1998). Therefore, the in situ up-regulation in NO production may be an adaptive response to the elevated blood pressure and a consequence of changes in levels of humoral factors. Indeed, a higher plasma level of IL-1β in patients with essential hypertension (Dalekos et al., 1997) and of TNF-α in SHR (Wu et al., 1996) than controls has been reported.

5-hydroxytryptamine (5-HT) is a major secretory product released from activated platelets. In the cardiovascular system, 5-HT is known to enhance vasoconstriction, smooth muscle cell proliferation, and platelet aggregation. Therefore, 5-HT is considered to play important roles in the pathogenesis of various cardiovascular diseases such as vasospasm, thrombosis, atherosclerosis or hypertension (Frishman et al., 1995). 5-HT has been shown to inhibit IL-1β-induced NO production by rat vascular smooth muscle cells (Shimpo et al., 1997). Therefore, 5-HT may also be critically involved in the progression of local vascular injury through the modulatory effect on the NO production. A selective increase in sensitivity and affinity for 5-HT has been shown at the 5-HT$_{2A}$ receptor subtype on the aorta in SHR as compared with that in WKY (Doggrell, 1995). In a previous paper, we reported that the resting levels of intracellular Ca^{2+} concentration ([Ca^{2+}]$_i$) without stimulation and the peak levels induced by 5-HT were higher in vascular smooth muscle cells derived from SHRSP than in cells from WKY (Hirafuji et al., 1998). These results suggest that there may be genetic differences in the affinity or number of 5-HT receptors in vascular smooth muscle cells between WKY and SHRSP. SHRSP is an experimental model of not only essential hypertension but also cerebrovascular disorders (Okamoto et al., 1974). In the present study, therefore, we investigated the effect of 5-HT on IL-1β-induced NO production and iNOS expression by vascular smooth muscle cells derived from WKY and SHRSP.

MATERIALS AND METHODS

Materials

Dulbecco's modified Eagle medium (DMEM) and fetal bovine serum (FBS) were purchased from GIBCO BRL; bovinve serum albumin (BSA) from Boehringer Mannheim; interleukin-1β (IL-1β) from Collaborative Biomedical Products; 5-HT creatinine sulfate and 5-methoxytryptamine (5-MT) from Sigma Chemical Co.; ($\pm$)-8-hydroxy-dipropylaminotetralin hydrobromide (8-OH-DPAT), α-methyl-5-HT maleate (α-Me-5-HT), 2-methyl-5-HT (2-Me-5-HT), staurosporin and phorbol 12-myristate 13-acetate (PMA) from Research Biochemicals International; anti-iNOS mouse monoclonal antibody from Transduction Laboratory; rabbit anti-mouse IgG conjugated to horseradish peroxidase from Zymed Laboratories; enhanced chemiluminescence detection system from NEN Life Science Products. Sarpogrelate hydrochloride was a kind gift from Mitsubishi Pharma Corporation. All other chemicals used were of the highest grade commercially available.

Animals

SHRSP and normotensive WKY rats as controls were originally donated by the late Dr. Okamoto (Okamoto *et al.*, 1974), and maintained in our laboratory. They were fed water and food freely, and kept at room temperature of $22 \pm 2°C$ and humidity of $50 \pm 10\%$. The systolic blood pressure of WKY and SHRSP that we used is around 120 mmHg in both strains at 6 weeks old, and, thereafter, hypertension progressively develops in SHRSP with around 200 mmHg at 20 weeks old (Minami *et al.*, 1997).

Cell culture

Vascular smooth muscle cells were enzymatically isolated from aortic media of 7–8 weeks old WKY and SHRSP, and cultured in DMEM containing 10% FBS, as described previously (Hirafuji *et al.*, 1998). Primary culture cells were used throughout the experiments. Cells grown to sub-confluence were rinsed 3 times with phenol red-free DMEM containing 0.1% BSA and then treated with IL-1β and/or test agents for the indicated time period.

NO determination

NO was measured as its stable oxidative metabolite, nitrite, by colorimetric method. At the end of the incubation, the culture medium was mixed with an equal volume of Griess reagent (0.1% naphthylethylenediamine dihydrochloride:1% sulfanilamide in 5% phosphoric acid $= 1:1$). The absorbance at 540 nm was measured, and the nitrite concentration was determined using sodium nitrite as the standard. Results were expressed as NO nmol per mg protein normalized to the protein content of cells in each dish.

Western blotting analysis

Protein expression of iNOS was determined by Western blotting analysis as described previously (Hirafuji *et al.*, 2002). After stimulation with IL-1β (3 ng/ml) for 24 hours, cells were washed with phosphate-buffered saline (PBS) and lysed with sonication in a buffer containing 50 mM Tris-HCl (pH 7.8), 150 mM NaCl, 10 μg/ml aprotinin, 10 μg/ml pepstatin A, 10 μg/ml leupeptin, 1 μM phenylmethylsulphonyl fluoride, 10 mM EDTA, 0.5% Nonidet P-40, 0.01% sodium dodecyl sulfate (SDS), and 0.5% sodium deoxycholic acid. The lysate was centrifuged at 10 000g for 30 min and the resultant clear supernatant was stored at $-80°$C until further analysis.

Aliquots of cell lysate containing 30–50 μg of protein was reduced and separated on 7.5% SDS-polyacrylamide gel electrophoresis. The separated proteins were electrophoretically transferred to a polyvinylidene difluoride transfer membrane. The blot was incubated with anti-iNOS mouse monoclonal antibody (1 : 1000) for 2 hours. It was finally incubated for 45 min with rabbit anti-mouse IgG conjugated to horseradish peroxidase (1 : 5000). The immunocomplexes were developed using an enhanced chemiluminescence detection system on an X-ray film. For densitometric analysis of protein expression, the iNOS (130 kDa) single bands corresponnding to appropriate positive controls were analyzed using NIH Image (v.1.61).

Statistical analysis

Results were expressed as means $\pm$ SE or representative of replicate experiments. Statistical analysis of the results was made by Student's t-test or ANOVA, and values of $p < 0.05$ were considered as significant.

RESULTS

Effect of 5-HT on NO production

Vascular smooth muscle cells derived from WKY and SHRSP were stimulated with 3 ng/ml IL-1β for 48 hours in the absence or the presence of 5-HT. Figure 1 demonstrates the effect of 5-HT on the NO production as a function of 5-HT concentration. The NO production by cells derived from SHRSP was lower than that by cells from age-matched normotensive WKY, in agreement with a previous paper (Hirafuji *et al.*, 2002). 5-HT dose-dependently inhibited the IL-1β-induced NO production in both strains. However, the magnitude of inhibition was greater in SHRSP cells than in WKY cells. 5-HT at a concentration of 10 μM inhibited the NO production by 23.9 $\pm$ 2.2 and 37.8 $\pm$ 5.2%, mean $\pm$ SE, $n = 5$, in WKY and SHRSP cells, respectively, between which there was a significant difference ($p < 0.05$).

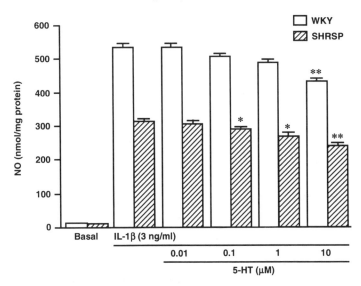

Figure 1. Effect of 5-HT on IL-1β-induced NO production by vascular smooth muscle cells derived from WKY and SHRSP. Cells were stimulated with 3 ng/ml IL-1β in the absence or the presence of various concentrations of 5-HT for 48 hours. Each column represents mean ± SE ($n = 4$). *$p < 0.05$, **$p < 0.01$ *versus* IL-1β alone in each strain.

Effects of 5-HT receptor agonists and antagonist on NO production

Figure 2 illustrates the effect of several 5-HT receptor agonists on IL-1β-induced NO production by vascular smooth muscle cells derived from SHRSP. Among these agonists, 5-HT and α-Me-5-HT, a 5-HT$_2$ receptor agonist, at a concentration of 10 μM significantly inhibited the NO production. A 5-HT$_{1A}$ receptor agonist 8-OH-DPAT, a 5-HT$_3$ receptor agonist 2-Me-5-HT, and a 5-HT$_4$ receptor agonist 5-MT had no significant inhibitory effect.

Figure 3 shows the effect of sarpogrelate, a 5-HT$_2$ receptor antagonist, on the inhibitory effect of 5-HT. The 5-HT inhibition of IL-1β-stimulated NO production was dose-dependently and completely reversed by sarpogrelate.

Effects of PKC inhibitor and activator on NO production

The results with 5-HT receptor agonists and 5-HT$_2$ receptor antagonist strongly suggest that the inhibitory effect of 5-HT on IL-1β-induced NO production is mediated via 5-HT$_2$ receptor subtype coupled to Gq protein involving the activation of protein kinase C (PKC). Therefore, we then investigated the effect of staurosporin, a potent PKC inhibitor, on the inhibitory effect of 5-HT in WKY and SHRSP cells. As demonstrated in Fig. 4, staurosporin dose-dependently reversed the effect of 5-HT, which effect seemed to be more potent in WKY cells. Furthermore, as demonstrated in Fig. 5, PMA, an activator of PKC, dose-dependently inhibited the NO production in both WKY and SHRSP cells. SHRSP cells have a tendency to be more

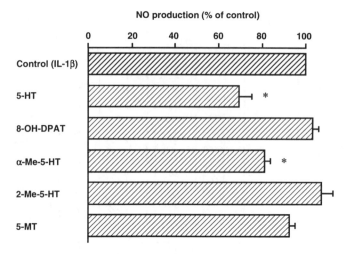

Figure 2. Effects of several 5-HT receptor agonists on IL-1β-induced NO production by vascular smooth muscle cells derived from SHRSP. Cells were stimulated with 3 ng/ml IL-1β in the absence or the presence of each 5-HT receptor agonist at a concentration of 10 μM for 48 hours. Results are expressed as percentage of NO production induced by IL-1β (Control). Each column represents mean $\pm$ SE ($n = 3$). *$p < 0.05$ *versus* Control.

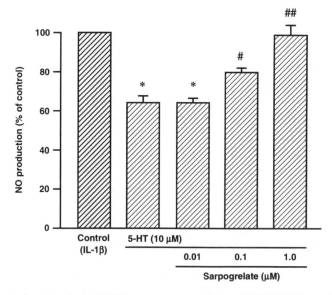

Figure 3. Effect of sarpogrelate, a 5-HT$_2$ receptor antagonist, on 5-HT inhibition of IL-1β-induced NO production by vascular smooth muscle cells derived from SHRSP. Cells were stimulated with 3 ng/ml IL-1β in the absence or the presence of 10 μM 5-HT and sarpogrelate for 48 hours. Results were expressed as percentage of NO production induced by IL-1β (Control). Each column represents mean $\pm$ SE ($n = 4$). *$p < 0.01$ *versus* Control; #$p < 0.05$, ##$p < 0.01$ *versus* 5-HT alone.

sensitive to PMA than WKY cells, although the extent of inhibition by the highest concentration (10 nM) was almost the same in both strains.

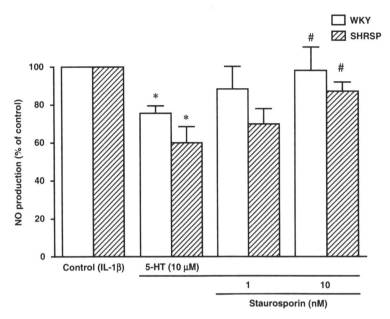

Figure 4. Effect of staurosporin, a PKC inhibitor, on 5-HT inhibition of IL-1β-induced NO production by vascular smooth muscle cells derived from WKY and SHRSP. Cells were stimulated with 3 ng/ml IL-1β in the absence or the presence of 10 μM 5-HT and staurosporin for 48 hours. Results were expressed as % of NO production induced by IL-1β (Control) in each strain. Each column represents mean $\pm$ SE ($n = 3$). *$p < 0.01$ *versus* Control; #$p < 0.05$ versus 5-HT alone in each strain.

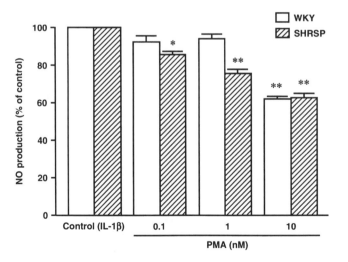

Figure 5. Effect of PMA, a PKC activator, on IL-1β-induced NO production by vascular smooth muscle cells derived from WKY and SHRSP. Cells were stimulated with 3 ng/ml IL-1β in the absence or the presence of PMA for 48 hours. Results were expressed as percentage of NO production induced by IL-1β (Control) in each strain. Each column represents mean $\pm$ SE ($n = 3$). *$p < 0.05$, **$p < 0.01$ *versus* each Control.

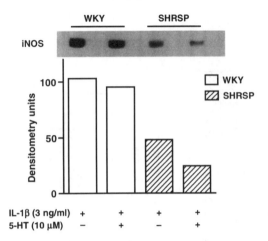

Figure 6. Effects of 5-HT on IL-1β-induced iNOS protein expression in vascular smooth muscle cells derived from WKY and SHRSP. Cells were stimulated with 3 ng/ml IL-1β in the absence or the presence of 10 μM 5-HT for 24 hours. Result shows a representative Western blotting image and densitometric analysis.

Effect of 5-HT on iNOS expression

Figure 6 demonstrates the effect of 10 μM 5-HT on IL-1β-induced iNOS expression in cells derived from WKY and SHRSP. The expression of iNOS protein in SHRSP cells was remarkably lower than that in WKY cells. In both WKY and SHRSP cells, 5-HT caused an inhibition in the iNOS protein expression, which effect was greater in SHRSP cells than in WKY cells. iNOS protein was not detectable in unstimulated cells derived from both WKY and SHRSP.

DISCUSSION

In a previous paper, we demonstrated that NO production following stimulation with IL-1β was significantly lower in vascular smooth muscle cells derived from young SHRSP in the early stage of evolution of hypertension than in cells from age-matched WKY (Hirafuji *et al.*, 2002). The expression of iNOS protein induced by IL-1β stimulation was also significantly lower in SHRSP cells than in WKY cells. These differences are a genetic property, since isolated vascular smooth muscle cells were used after 7–9 days culture in this study.

In the present study, we first investigated the effects of 5-HT on NO production and iNOS expression in IL-1β-stimulated vascular smooth muscle cells derived from SHRSP and WKY. 5-HT was found to inhibit the NO production in a dose-dependent manner, which effect was greater in SHRSP cells than in WKY cells. 5-HT was also found to inhibit IL-1β-induced iNOS protein expression more strongly in SHRSP cells than in WKY cells. The present study also indicated that among 5-HT receptor agonists, only α-Me-5-HT, a 5-HT$_2$ receptor agonist, mimicked the inhibitory effect of 5-HT on IL-1β-induced NO production. Further, the inhibitory effect of 5-HT was completely reversed by sarpogrelate, a selective 5-HT$_2$ receptor antagonist. These results strongly suggest that 5-HT inhibits IL-1β-induced NO production and iNOS expression via the activation of 5-HT$_2$ receptors in vascular smooth muscle cells derived from both WKY and SHRSP.

Stimulation of 5-HT$_2$ receptor, a G protein-coupled receptor, activates phospholipase C resulting in the formations of 1,2-diacylglycerol which activates conventional PKC in concert with Ca^{2+}, and inositol 1,4,5-triphosphate which mobilizes Ca^{2+} from intracellular stores (Doyle *et al.*, 1986). The present study demonstrated that staurosporin, a potent PKC inhibitor, completely reversed the inhibitory effect of 5-HT on NO production. In agreement with a previous study (Nakayama *et al.*, 1994), PMA, a direct activator of PKC, dose-dependently inhibited the NO production in both WKY and SHRSP cells. These results suggest that the inhibitory effect of 5-HT on the NO production is mediated, at least in part, via 5-HT$_2$ receptor involving PKC activation. Although the precise mechanism has been not yet clarified, PKC activation also seems to mediate the inhibitory effect of 5-HT on the iNOS expression (Shimpo *et al.*, 1997). SHRSP cells had a tendency to be more sensitive to PMA than WKY cells, while the extent of inhibition by the highest concentration was almost the same in both strains. We have shown that [Ca^{2+}]$_i$ in resting and 5-HT-stimulated vascular smooth muscle cells derived from young SHRSP is significantly higher than in age-matched WKY (Hirafuji *et al.*, 1998). Furthermore, it is suggested that the greater vascular contractile sensitivity to 5-HT, at least in part, is attributable to an increased turnover of phosphoinositides in SHR (Huzoor-Akbar *et al.*, 1989) and SHRSP (Turla and Webb, 1990). The content of 1,2-diacylglycerol after the exposure to noradrenaline as a stimulant has been shown to be higher in the aorta of young SHR than in that of age-matched WKY (Okumura *et al.*, 1990). A selective increase in sensitivity and affinity for 5-HT has been shown at the 5-HT$_{2A}$ receptors on the aorta in SHR (Doggrell, 1995). Therefore, the enhanced inhibitory effects of 5-HT on NO production and iNOS expression in SHRSP cells may be, at least in part, due to the increased receptor sensitivity and PKC activation mechanism in the 5-HT$_2$ receptor signaling pathway. Further studies are required to clarify the precise mechanism.

5-HT is implicated in numerous cardiovascular diseases, and vascular reactivity to 5-HT is increased in arteries from different animal models of hypertension including SHRSP (Nishimura, 1996; Satoh *et al.*, 1994). In SHRSP, serum 5-HT level does not differ from that in control WKY, while serum level of catecholamines is higher than in WKY (Minami *et al.*, 1988). However, 5-HT is known to amplify the responses to diverse vasoconstrictor agents such as noradrenaline (Meehan *et al.*, 1988; Xiao and Rand, 1989), angiotensin II (Van Neuten *et al.*, 1982) or endothelin (Yang *et al.*, 1992). Therefore, the 5-HT inhibition of vasodilating NO production that is up-regulated in vascular smooth muscle cells in the established stage of hypertension may play a crucial role in the further development of hypertension.

On the other hand, at the site of local vascular lesions, such as those that are found in atherosclerosis, IL-1β and TNF-α are actively produced and released from activated macrophages, which may induce iNOS expression in vascular smooth muscle cells (Ross, 1999). A large amount of 5-HT locally released from aggregated platelets may act on the sub-endothelial smooth muscle cells. Besides the vasodilating action, NO has inhibitory effects on platelet aggregation and adhesion, leukocyte adhesion, and vascular smooth muscle cell proliferation/migration (Dusting, 1995). These physiological properties of NO suggest that iNOS expressed in vascular smooth muscle cells may primarily function as a defensive and compensatory mechanism at the site of vascular injury. Therefore, the inhibition of iNOS induction by 5-HT at the site of vascular injury may produce a critical consequence in the development of a vascular lesion. However, excess amounts of NO may cause cytotoxicity or apoptosis as potent biogenic radicals by reacting with oxygen radical species such as superoxide anions to form peroxynitrite (Jeremy *et al.*, 2002). In this context, the inhibitory effect of 5-HT on cytokine-induced NO production may be favorable to prevent tissue injury due to the cytotoxic radicals. The pathophysiological significance of enhanced effect of 5-HT on NO production by vascular smooth muscle cells derived from SHRSP remains to be clarified.

CONCLUSION

The present study has indicated that 5-HT had an enhanced inhibitory effect on IL-1β-induced NO production and iNOS expression in vascular smooth muscle cells derived from SHRSP as compared with cells from age-matched WKY. Our results further suggest that the inhibitory effect is mediated via 5-HT$_2$ receptor subtype involving PKC activation, and these signaling pathways are up-regulated in SHRSP cells. The enhanced inhibitory effect of 5-HT may have a pathophysiological relevance to the development of hypertension and vascular disease in SHRSP.

Acknowledgements

This work was supported in part by Grant-in-Aid for Science Research from Japan Society for the Promotion of Science, and the Academic Science Frontier Project of the Ministry of Education, Culture, Sports, Science and Technology, Japan.

REFERENCES

Busse, R. and Mülsch, A. (1990). Induction of nitric oxide synthetase by cytokines in vascular smooth muscle cells, *FEBS Lett.* **275**, 87–90.

Chou, T. C., Yen, M. H., Li, C. Y., *et al.* (1998). Alterations of nitric oxide synthase expression with aging and hypertension in rats, *Hypertension* **31**, 643–648.

Dalekos, G. N., Elisaf, M., Bairaktari, E., *et al.* (1997). Increased serum levels of interleukin-1β in the systemic circulation of patients with essential hypertension: Additional risk factor for atherogenesis in hypertensive patients? *J. Lab. Clin. Med.* **129**, 300–308.

Doggrell, S. A. (1995). Increase in affinity and loss of 5-hydroxytryptamine2A-receptor reserve for 5-hydroxytryptamine on the aorta of spontaneously hypertensive rats, *J. Auton. Pharmacol.* **15**, 371–377.

Doyle, V. M., Creba, J. A., Ruegg, U. T., *et al.* (1986). Serotonin increases the production of inositol phosphates and mobilises calcium via the 5-HT$_2$ receptor in A7r5 smooth muscle cells, *Naunyn-Schmiedeberg's Arch. Pharmacol.* **333**, 98–103.

Dubois, G. (1996). Decreased L-arginine-nitric oxide pathway in cultured myoblasts from spontaneously hypertensive versus normotensive Wistar-Kyoto rats, *FEBS Lett.* **392**, 242–244.

Dusting, G. J. (1995). Nitric oxide in cardiovascular disorders, *J. Vasc. Res.* **32**, 143–161.

Frishman, W. H., Huberfeld, S., Okin, S., *et al.* (1995). Serotonin and serotonin antagonism in cardiovascular and non-cardiovascular disease, *J. Clin. Pharmacol.* **35**, 541–572.

Hirafuji, M., Ebihara, T., Kawahara, F., *et al.* (1998). Effect of docosahexaenoic acid on intracellular calcium dynamics in vascular smooth muscle cells from normotensive and genetically hypertensive rats, *Res. Commun. Mol. Pathol. Pharmacol.* **102**, 29–42.

Hirafuji, M., Tsunoda, M., Machida, T., *et al.* (2002). Reduced expressions of inducible nitric oxide synthase and cyclooxygenase-2 in vascular smooth muscle cells of stroke-prone spontaneously hypertensive rats, *Life Sci.* **70**, 917–926.

Huzoor-Akbar, Chen, N. Y., Fossen, D. V., *et al.* (1989). Increased vascular conractile sensitivity to serotonin in spontaneously hypertensive rats is linked with increased turnover of phosphoinositide, *Life Sci.* **45**, 577–583.

Jeremy, J. Y., Yim, A. P., Wan, S., *et al.* (2002). Oxidative stress, nitric oxide, and vascular disease, *J. Card. Surg.* **17**, 324–327.

Junquero, D. C., Schini, V. B., Scott-Burden, T., *et al.* (1993). Enhanced production of nitric oxide in aortae from spontaneously hypertensive rats by interleukin-1β, *Am. J. Hypertens.* **6**, 602–610.

Malinski, T., Kapturczak, M., Dayharsh, J., *et al.* (1993). Nitric oxide synthase activity in genetic hypertension, *Biochem. Biophys. Res. Commun.* **194**, 654–658.

Meehan, A. G., Medgett, I. C. and Story, D. F. (1988). Involvement of Ca^{2+} mobilization in the amplifying effect of serotonin on responses of rabbit isolated ear artery to exogenous noradrenaline, *Naunyn-Schmiedeberg's Arch. Pharmacol.* **337**, 500–503.

Minami, M., Sano, M., Togashi, H., *et al.* (1988). Stroke-related plasma noradrenaline, angiotensin II, arginine-vasopressin and serotonin concentrations in stroke-prone spontaneously hypertensive rats, in: *Progress in Hypertension, Vol. 1.* Saito, H., Parvez, H., Parves, S. and Nagatsu, T. (Eds), pp. 89–114. VSP BV, Zeist, Utrecht, The Netherlands.

Minami, M., Kimura, S., Endo, T., *et al.* (1997). Dietary docosahexaenoic acid increases cerebral acetylcholine levels and improves passive avoidance performance in stroke-prone spontaneously hypertensive rats, *Pharmacol. Biochem. Behav.* **58**, 1123–1129.

Nakayama, I., Kawahara, Y., Tsuda, T., *et al.* (1994). Angiotensin II inhibits cytokine-stimulated inducible nitric oxide synthase expression in vascular smooth muscle cells, *J. Biol. Chem.* **269**, 11628–11633.

Nishimura, Y. (1996). Characterization of 5-hydroxytryptamine receptors mediating contractions in basilar arteries from stroke-prone spontaneously hypertensive rats, *Br. J. Pharmacol.* **117**, 1325–1333.

Okamoto, K., Yamori, Y. and Nagaoka, A. (1974). Establishment of the stroke-prone SHR, *Circ. Res.* **34/35** (Suppl. I), 143–153.

Okumura, K., Kondo, J., Shirai, Y., *et al.* (1990). 1,2-Diacylglycerol content in thoracic aorta of spontaneously hypertensive rats, *Hypertension* **16**, 43–48.

Ross, R. (1999). Atherosclerosis — an inflammatory disease, *New Engl. J. Med.* **14**, 115–126.

Satoh, S., Kreutz, R., Wilm, C., *et al.* (1994). Augmented agonist-induced Ca^{2+}-sensitization of coronary artery contraction in genetically hypertensive rats. Evidence for altered signal transduction in the coronary smooth muscle cells, *J. Clin. Invest.* **94**, 1397–1403.

Shimpo, M., Ikeda, U., Maeda, Y., *et al.* (1997). Serotonin inhibits nitric oxide synthesis in rat vascular smooth muscle cells stimulated with interleukin-1, *Eur. J. Pharmacol.* **338**, 97–104.

Singh, A., Sventek, P., Larivière, R., *et al.* (1996). Inducible nitric oxide synthase in vascular smooth muscle cells from prehypertensive spontaneously hypertensive rats, *Am. J. Hypertens.* **9**, 867–877.

Turla, M. B. and Webb, R. C. (1990). Augmented phosphoinositide metabolism in aortas from genetically hypertensive rats, *Am. J. Physiol.* **258**, H173–H178.

Van Neuten, J. M., Janssen, P. A., De Ridder, A., *et al.* (1982). Interaction between 5-hydroxy-tryptamine and other vasoconstrictor substances in the isolated femoral artery of the rabbit; effect of ketanserin (R41 468), *Eur. J. Pharmacol.* **77**, 281–287.

Vaziri, N. D., Ni, Z. and Oveosi, F. (1998). Up-regulation of renal and vascular nitric oxide synthase in young spontaneously hypertensive rats, *Hypertension* **31**, 1248–1254.

Wu, C.-C., Hong, H.-J., Chou, T.-C., *et al.* (1996). Evidence for inducible nitric oxide synthase in spontaneously hypertensive rats, *Biochem. Biophys. Res. Commun.* **228**, 459–466.

Xiao, X. H. and Rand, M. J. (1989). Amplification by serotonin of responses to other vasoconstrictor agents in the rat tail artery, *Clin. Exp. Pharmacol. Physiol.* **16**, 725–736.

Yang, B. C., Nichols, W. W., Lawson, D. L., *et al.* (1992). 5-Hydroxytryptamine potentiates vasoconstrictor effect of endothelin-1, *Am. J. Physiol.* **262**, H931–H936.

Advances in Neuroregulation and Neuroprotection (2005), pp. 443-457
Collin, C. *et al.* (Eds)
© VSP 2005

Influence of occlusal support on learning memory and cholinergic neurons in the rat: a mini review

Y. IKEDA [1,*], T. HIRAI [1], T. MAKIURA [1], T. TERASAWA [1] and N. HAMAUE [2]

[1] *Department of Prosthodontics, School of Dentistry, Health Sciences University of Hokkaido, 1757 Kanazawa, Ishikari-Tobetsu, Hokkaido 061-0293, Japan*
[2] *Department of Pharmacology, Faculty of Pharmaceutical Sciences, Health Sciences University of Hokkaido, 1757 Kanazawa, Ishikari-Tobetsu, Hokkaido 061-0293, Japan*

Abstract—In order to verify the influence of tooth loss and brain function in the rat, Wistar male aged rats were divided into the following groups: a control group (fed with solid diet), a soft diet group (fed with powder diet containing the same components as solid one), and a molar-crownless group (all molars were removed at 25 weeks and then fed with a powder diet).

Experiment 1: To evaluate both learning ability and memory, rats were tested with a one-way step through type of passive avoidance apparatus divided into light and dark chambers at 40-weeks. There was no significant difference between the molar-crownless group and the control group in the response latency before the acquisition trails (non-stimulated period). At day 4 and 7 after the acquisition trials, the response latency of the molar-crownless group was significantly shorter than that in the control group ($p < 0.05$). After the passive avoidance test, determination of acetylcholine (ACh) concentration of the cerebral cortex and hippocampus was performed. The ACh levels of the molar-crownless group in the cerebral cortex and hippocampus were significantly lower than those of the control group ($p < 0.05$).

Experiment 2: At 15 and 35 weeks, the number of choline acetyltransferase (ChAT)-positive neurons in the nucleus of the diagonal band/medial septal nucleus (NDB/MS) was significantly smaller in the molar-crownless group than in the control group ($p < 0.01$).

The results of the study suggested that the soft diet and the decease of occlusal-masticatory function caused by occlusal support loss could accelerate learning ability and memory; moreover, this decrease of oral sensory information may have caused a reduction in the number of ChAT-positive neurons selectively in NDB/MS, which in turn caused a decline of ACh concentrations in the hippocampus.

Keywords: Tooth loss; learning ability and memory; cholinergic neurons; nucleus of the diagonal band; pedunculopontine tegmental nucleus.

---

*To whom correspondence should be addressed. E-mail: ikeda@hoku-iryo-u.ac.jp

INTRODUCTION

Loss of teeth is reported to be significantly associated with Alzheimer's disease (Isse *et al.*, 1991). Moreover, it is reported that the activity of choline acetyltransferase (ChAT) was remarkably reduced in the cerebral cortex and hippocampus of patients with Alzheimer-type dementia (Iizuka, 1986).

Occlusal-masticatory function or malfunction may affect the higher brain function. In fact, an impairment of spatial memory due to a decrease in acetylcholine (ACh) level in the cerebral cortex was caused in 135 weeks old rats by tooth extraction (Kato *et al.*, 1997). A similar impairment of spatial memory has been reported to occur in aged mice by extracting or cutting the molar teeth at young ages (Onozuka *et al.*, 1999), and also in adult rats by feeding soft-diet after the weaning period (Yamamoto and Hirayama, 2001). In these studies, both the neuronal density in the CA1 hippocampus (Onozuka *et al.*, 1999) and the synaptic formation in the hippocampus and the parietal cortex (Yamamoto and Hirayama, 2001) have been reported to decrease.

The nucleus of the diagonal band and the medial septal nucleus (NDB/MS), together with the nucleus basalis of Meynert, are known to be involved in learning and/or memory. Cholinergic neurons in NDB/MS project to both the hippocampus and the cerebral cortex (Davies and Feisullin, 1982; Houser *et al.*, 1983; Johnston *et al.*, 1981; McKinney *et al.*, 1983; Mesulam *et al.*, 1983a, b; Pearson *et al.*, 1983; Sofroniew *et al.*, 1987), whereas those in the nucleus basalis of Meynert send fewer axons to the hippocampus in comparison with the cerebral cortex (Lehmann *et al.*, 1980; Rye *et al.*, 1984).

We investigated the influence of dietary- and occlusal loss-related changes on the central nerves system by the following two experiments. In the first experiment, learning ability and memory were evaluated by the passive avoidance response, and changes in ACh concentration in the cerebral cortex and hippocampus were assessed using biochemical methods (Makiura *et al.*, 2000).

In the second experiment, the immunohistochemical results were compared with those obtained from cholinergic neurons in the pedunculopontine tegmental nucleus (PPT) and the trigeminal motor nucleus (Vmo) (Terasawa *et al.*, 2002).

MATERIALS AND METHODS

Animals

In the first experiment, 36 male Wistar rats (25-weeks-old) were divided into three groups with approximately equal mean body weights: (1) a control group (fed a solid diet), (2) a soft diet group (fed a powder diet containing the same components as the solid one) and (3) a molar-crownless group (fed the powder diet) as shown in Fig. 1 subjected to a passive avoidance test and biochemical studies In the second experiment, 42 male Wistar rats (25 weeks old) were also divided in a similar way

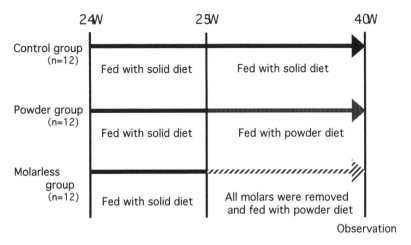

Figure 1. Time schedule of the experiment.

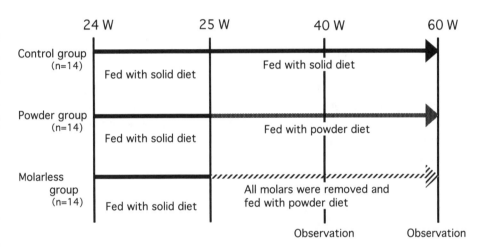

Figure 2. Time schedule of the experiment.

to that in the first experiment as shown in Fig. 2 and immunohistochemical and biochemical studies were performed.

In the molar-crownless group, at the 25th postnatal week, rats were anesthetized with sodium pentobarbital (35 mg/kg), and all maxillary and mandibular molar crowns were cut off at the gingival margin with a forceps. In the control and the soft diet group, animals were given anesthesia alone, without undergoing removal of the molar crown. The rats were kept on a light-dark cycle (light up through 08:00–20:00) in an air-conditioned room (22 ± 2°C, 55–65% humidity). This study was conducted in accordance with the Guidelines for the Care and Use of Laboratory Animals by the Animal Research Committee in Health Sciences at the University of Hokkaido.

Passive avoidance test

In the first experiment, to evaluate learning ability and memory, rats were tested with use of a one-way step through type of passive avoidance apparatus divided into light and dark chambers at 40-weeks as described previously (Togashi *et al.*, 1996). Acquisition trials were performed 24 hours before retention trials. Each rat received a footshock (75 mV, 0.2 m seconds) for 3 seconds upon entering the dark chamber. This trial was repeated until the rats eventually remained in the light chamber for more than 600 s. In the retention trials, rats were placed in the light chamber, and the time taken to enter the dark chamber, denoted as the response latency, was measured up to 600 s. The retention trials were performed 1, 2, 3, 4 and 7 days after the acquisition trials. The passive avoidance test was performed in the dark phase. The passive avoidance performance was evaluated by the response latency in each retention trial and by the area under the retention curve for seven days after the acquisition trials.

Determination of ACh concentration in cerebral cortex and hippocampus

After the passive avoidance test, rats were sacrificed by microwave irradiation (5 kW for 1.5 s). Their brains were removed, and the right hemispheres were dissected into seven regions according to the method of Glowinski and Iversen (1966) to allow for the measurement of ACh contents. The tissues were stored at $-80°C$ until they were assayed. Extraction of the ACh tissue was carried out with aliquots of 0.1 M perchloric acid including ethylhomocholine (EHC) as an internal standard. The homogenates were centrifuged for 15 min (10,000 rpm, 2–4°C). The supernatant was neutralized with 0.1 M potassium hydrogen carbonate and was filtered (0.45 μm, Millipore filter) and injected into a high performance liquid chromatography system (EP-10, Eicom, Japan) connected with an immobilized enzyme reactor and an electrochemical detector (ECD-100, Eicom, Japan) (Matumoto *et al.*, 1990). ACh concentrations were corrected with tissue weight.

Immunohistochemistry

In the second experiment, at 15 weeks post-treatment (40 weeks old) and at 35 weeks post-treatment (60 weeks old), the rats were deeply anesthetized by intraperitoneal injection of sodium pentobarbital (35 mg/kg) and perfused transcardially with 0.1 M phosphate buffered saline (pH 7.2–7.4) followed by 4% paraformaldehyde in 0.1 M phosphate buffer (pH 7.2–7.4). The brain was removed and serial transverse sections were made with Microslicer (Dosaka EM, Kyoto, Japan) at 50 μm thickness. The sections were processed with 3% H_2O_2 in 0.1 M phosphate buffered saline (PBS) for 30 min to block endogenous peroxidase activity. The sections were rinsed 3 times (10 min each) with 0.1 M PBS and then incubated in 5% normal goat serum (NGS) for 1 hour to reduce background staining. Following this incubation, the sections were incubated with rabbit anti-sera against choline acetyltransferase

(CHEMICON, Temecula, CA, USA; diluted at $1:1000$) in 0.1 M PBS containing 1% NGS and 0.5% Triton x-100 at 4°C for 2 days. The sections were washed with 0.1 M PBS 3 times (10 min each), and then incubated overnight at 4°C in biotinylated goat anti-rabbit IgG ($1:200$, Nichirei, Tokyo, Japan) in 1% NGS containing 0.5% Triton X-100. The sections were washed twice with 0.1 M PBS, twice (10 min each) with 0.05 M Tris buffered saline (TBS, pH 7.6), and then incubated for 2 h in streptavidin conjugated with horseradish peroxidase for 2 h ($1:400$, Nichirei, Tokyo, Japan) in 0.05 M TBS containing 1% Triton X-100. After several rinses with TBS, sections were reacted with 0.05% 3,3'-diaminobenzidine tetrahydrochloride containing 0.003% H_2O_2. After rinsing with TBS and PBS, these sections were mounted on a glass slide, counterstained with 1% Neutral Red, dehydrated with graded alcohol, and coverslipped.

The locus and the number of labeled neurons were examined by light microscopy. In every third section, a count was made of cells whose nucleus was seen under 10 objective (Fig. 3). No corrections were made for double count because the maximal diameter of labeled somata was less than 50 mm.

Statistical analysis

In the first experiment, the one-way analysis of variance (ANOVA) and multiple range tests (Dunkan test) were used to analyze differences between the groups.

In the second experiment, the one-way analysis of variance (ANOVA) and Scheffe multiple range test were used to analyze differences between the groups.

RESULTS

Effects of diet and occlusal support on the passive avoidance test

There were no significant differences in the response latency and the frequency of foot shocks needed to acquire avoidance task (acquisition performance) between the soft diet group and the control group. The response latency of the control group, the soft diet group and the molar-crownless group at the 1st, 2nd and 3rd day were almost the same. However, the response latency of the molar-crownless group measured before the acquisition trails (non-stimulated period) was 22.3 ± 7.9 s and that of control group was 36.7 ± 19.2 s, although statistically not significant. On the other hand, at days 4 and 7 after the acquisition trials, the response latency of the molar-crownless group was 274.7 ± 190.0 s and 227.7 ± 177.1 s and those of the control group was 587.9 ± 36.3 s and 550.4 ± 83.1 s, respectively. The response latency of the molar-crownless group was significantly shorter than that in the control group at the 4th and 7th day ($p < 0.05$) (Fig. 4).

The concentration of ACh in cerebral cortex and hippocampus

ACh levels of the molar-crownless group in the cerebral cortex (40.3 ± 9.8 pmol/mg wet tissue) were significantly lower than that of the control group ($53.8 \pm$

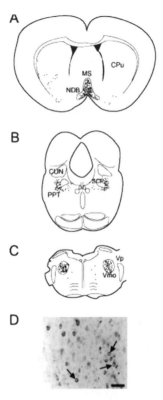

Figure 3. Illustrations summarizing the distribution of ChAT immunoreactive neurons in forebrain (A), midbrain (B) and pons (C). Abbreviations are as follows: CPu, caudate/putamen; CUN, cuneiform nucleus; NDB, nucleus of the diagonal band; MS, medial septum; PPT, pedunculopontine tegmental nucleus; SCP, superior cerebellar peduncle; Vmo, trigeminal motor nucleus; Vp, trigeminal main sensory nucleus. D, ChAT immunoreactive neurons in NDB/MS. The neurons whose nucleus could be seen (arrows) were counted. Scale bar = 50 mm.

7.5 pmol/mg wet tissue) (p < 0.05) (Fig. 5). In the molar-crownless group, hippocampal ACh levels (18.3 ± 1.6 pmol/mg wet tissue) were also significantly lower than those of the control group (25.1 ± 0.8 pmol/mg wet tissue) (p < 0.05) (Fig. 6).

The number of ChAT-positive neurons in NDB/MS

ChAT immunohistochemistry showed that the numbers of ChAT-positive neurons in NDB/MS of the soft diet group and the molar-crownless group were smaller than those of the control group, at 15 and 35 weeks post-treatment as shown in Fig. 7 and in Table 1. There was no significant age-related change in the number of ChAT-positive neurons in NDB/MS of the control group when compared between 15 and 35 weeks post-treatment. However, the numbers of ChAT-positive neurons in the molar-crownless groups at 15 and 35 weeks post-treatment were significantly smaller than those of the control group (p < 0.01). At 15 and 35 weeks post-

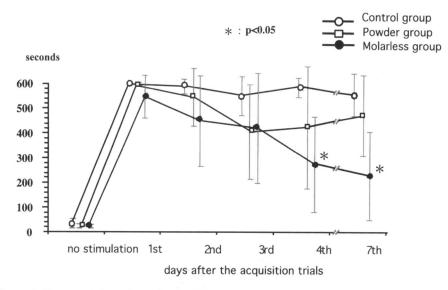

Figure 4. Response latency in passive avoidance test.

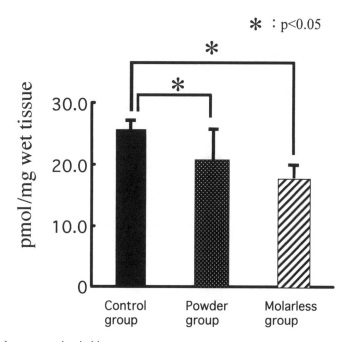

Figure 5. Ach concentration in hippocampus.

treatment, there was no significant difference in the number of ChAT-positive neurons either between the control and soft diet groups or between the soft diet and molar-crownless groups. The remaining ChAT-positive neurons in NDB/MS of the molar-crownless group, showed prominent shrinkage of dendrites. In the

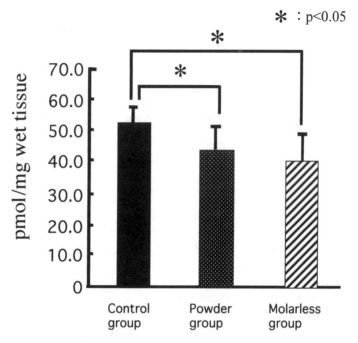

Figure 6. Ach concentration in brain cortex.

Table 1.
The number of ChAT immunoreactive neurons in NDB/MS

Post-operative week (Postnatal week)	0 (25th)	15th (40th)	35th (60th)
Control	1496 ± 176.1	1427 ± 212.9	1594 ± 305.1
Soft diet		1207 ± 97.1	1319 ± 139.5
Molar-crownless		$1040 \pm 91.1^{**}$	$964 \pm 226.3^{**}$

Values are mean ± S.D.
NDB/MS: nucleus of the diagonal band and the medial septal nucleus.
** $p < 0.01$ compared to the control group.

following sets of experiments, whether or not these changes in cholinergic neurons were selective for NDB/MS was investigated.

The pedunculopontine tegmental nucleus (PPT) is another source of major cholinergic projection neurons in the central nervous system. The influence of dietary changes and occlusal loss was examined on cholinergic neurons in the PPT as well as those in the trigeminal motor nucleus (Vmo). At 15 and 35 weeks post-treatment, there were no significant differences in the number of ChAT-positive neurons in the Vmo and in the PPT any two groups (e.g. Figs 8 and 9, Table 2), although there appeared to be a slight shrinkage in dendrites of the molar-crownless group of the Vmo, but not in PPT, in comparison with the control group.

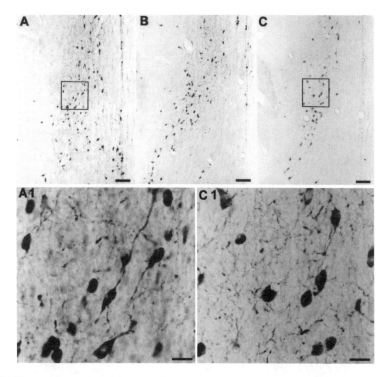

Figure 7. Photomicrographs of ChAT-positive neurons in the nucleus of the diagonal band/medial septal nucleus (NDB/MS) at 15 weeks post-treatment. A, control group. B, soft diet group. C, molar-crownless group. Scale bar = 100 mm. A1, a higher magnification view of the boxed area in A. C1, a higher magnification view of the boxed area in C. Scale bar = 30 mm. Note that the number of ChAT-positive neurons in B and C is smaller than that in A.

Table 2.
The number of ChAT immunoreactive neurons in PPT and Vmo

Post-operative week (Postnatal week)	15th (40th)
PPT	
Control	564 ± 33.8
Soft diet	475 ± 18.9
Molar-crownless	503 ± 77.5
Vmo	
Control	1215 ± 263.4
Soft diet	1095 ± 83.4
Molar-crownless	1209 ± 68.1

Values are mean $\pm$ S.D.
PPT: pedunculopontine tegmental nucleus. Vmo: trigeminal motor nucleus.
[**] $p < 0.01$ compared to the control group.

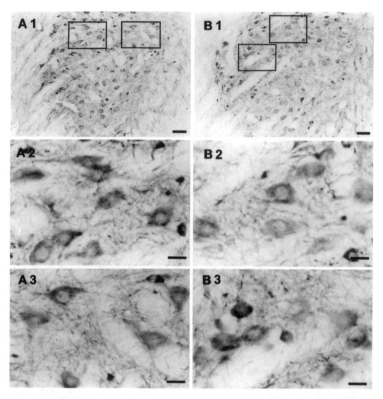

Figure 8. Photomicrographs of ChAT-positive neurons in the trigeminal motor nucleus (Vmo) at 15 weeks post-treatment. A1, the control group. B1, the molar-crownless group. Scale bar = 100 mm. A2 and 3, higher magnification views of the boxed areas in A1. B2 and 3, higher magnification views of the boxed areas in B1. Scale bar = 30 mm. Note that no apparent difference is observed in the number of ChAT-positive neurons between A and B.

DISCUSSION

The response latency of the molar-crownless group was significantly shorter than that in the control group. Simultaneously, the ACh levels of the molar-crownless group in the cerebral cortex and hippocampus were significantly lower than those of the control group. As for the study on learning memory disorder by the passive avoidance response, it is reported that the response latency of SHRSP (Stroke-Prone-Spontaneously Hypertensive Rats) was shorter than that of the normotensive control Wistar Kyoto rats (WKY). Furthermore, cerebral ACh levels correlated positively with the total response latency in the passive avoidance task (Kimura *et al.*, 1998). SHRSP is one of the animal models of vascular dementia and exhibits histopathological, behavioral and biochemical abnormality similar to those found in the patients with vascular dementia (Saito *et al.*, 1995). It is well known that cerebral blood flow carries the oxygen and glucose to brain neurons and excretes waste matter and carbon dioxide. The hippocampal neurons are reported to be weak when a deficiency of oxygen exists.

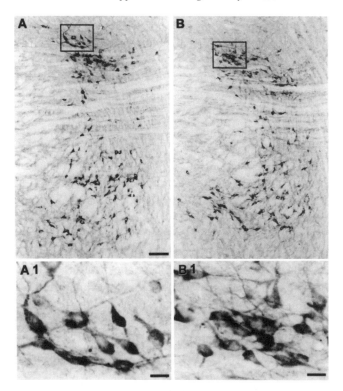

Figure 9. A and B show low power photomicrographs of ChAT-positive neurons in the pedunclopontine tegmental nucleus (PPT) at 15 weeks post-treatment of the control and molar-crownless groups, respectively. Scale bar = 100 mm. A1, a higher magnification view of the boxed area in A. B1, a higher magnification view of the boxed area in B. Scale bar = 30 mm. Note that no apparent difference is observed in the number of ChAT-positive neurons between A and B.

On the other hand, interference with cholinergic function in the hippocampus and cerebral cortex produce impairments of memory and cognitive performance (Bartus *et al.*, 1983; Drachman, 1977; Olton *et al.*, 1982), as seen in Alzheimer's disease (Smith and Swash, 1978). A neurochemical investigation of patients with Alzheimer's disease and senile dementia has demonstrated a reduction in the density of presynaptic nerve terminals of cholinergic neurons in the hippocampus and cerebral cortex (Whitehouse *et al.*, 1982). The behavioral impairment in the passive avoidance task was observed in the molar-crownless group with a positive correlation being found in the hippocampal ACh level of this study. The ChAT-positive neurons in the nucleus of basalis of Meynert (NDB/MS in rats) are reported to have a vasodilative effect in the cerebral cortex and hippocampus (Biesold *et al.*, 1989). The brain cholinergic system is generally thought to be critical for memory function and a disturbance in the central cholinergic system is reported in the patients experiencing vascular dementia (Wallin *et al.*, 1989). Accordingly, the decrease in the number of ChAT-positive neurons in the basal forebrain is supposed to cause a decrease of blood flow volume and the ACh synthesis in the hippocampus and the cerebral cortex. The aging process of the central cholinergic neuron is known to involve major changes in ACh synthesis, storage and release mechanism Perry, 1980.

A distinct decrease in the number of ChAT-positive neurons in NDB/MS was observed following a decrease in sensory information from intact proprioceptors in the periodontal ligament in the present study. Sensory information arising from the periodontal ligament, muscle spindle and temporomandibular joint play important roles in modulating masticatory and/or occlusal functions (Li *et al.*, 1995; Mizuno *et al.*, 1983; Shigenaga *et al.*, 1988a, b). Loss of posterior occlusal stop may affect the musculature, temporomandibular joints, teeth, and periodontium (Osborn and Lammie, 1974). Therefore, a wide variety of sensory information would have been altered by the removal of molar crowns, which might have consequently resulted in an unusual neuronal cell death in NDB/MS, although the causal neuronal connections or pathways are not clear. Tooth extraction can induce cell death in the trigeminal mesencephalic neurons (Kimoto, 1993), and tooth pulp extraction induces degenerative changes in primary trigeminal axons and in their target neurons of the nucleus caudalis (Gobel and Binck, 1977). The causal relationship is clear in these studies. By contrast, the causal neuronal relationship to the cell death of cholinergic neurons in NDB/MS is not clear. Cell death can be caused either by excitotoxicity through activation of glutamate receptors and/or free radical nitric oxide (NO). NO is produced through the activity of nitric oxide synthetase (NOS) (Vincent, 1994). Especially, cholinergic neurons in NDB/MS are reported to be 300 times more susceptible to NO than those in PPT (Fass *et al.*, 2000). This is presumably due to the sparse presence of superoxide dismutase, preventing the formation of peroxynitrite, in neurons of NDB/MS in comparison with those of PPT (Kent *et al.*, 1999). Cholinergic neurons in NDB/MS receive cholinergic inputs from PPT (Sarter and Bruno, 2000; Semba *et al.*, 1988; Vertes, 1988), and cholinergic neurons in PPT are considered to exert presynaptic inhibition onto glutamatergic inputs to NDB/MS (Rassmussen *et al.*, 1994). On the other hand, trigeminal sensory neurons are reciprocally connected with parabrachial regions, including parabrachial nucleus, cuneiform nucleus and PPT (Hayashi and Tabata, 1989, 1990, 1991; Yoshida *et al.*, 1997). Taken together with their vulnerability to NO, it may not be unreasonable to assume that decreases in the excitability of PPT cholinergic neurons possibly induced by decreases in trigeminal sensory inputs would allow excessive glutamatergic activation to induce a production of NO in cholinergic neurons in NDB/MS, consequently causing cell death of these cholinergic neurons.

The present studies showed that relationship of tooth loss and memory function can be explained by the following possibilities. First, ChAT activity remained at a low level due to a permanent loss of cholinergic neurons in NDB/MS while the activity of acetylcholine esterase (AChE) in the postsynaptic membrane of the hippocampus neurons might also be decreased in compensation for decreases in ACh synthetic ability. Second, the remaining or surviving cholinergic neurons in

NDB/MS might have increased their ability to synthesize ACh in their presynaptic terminals.

Based on the present studies, it is speculated that the deprivation of the oral sensory input caused by the decreased mechanical stress of teeth causes a decrease in the number of ChAT-positive neurons in NDB/MS, and a decline in ACh concentrations in the hippocampus and acetylcholine synthesis resulting in a learning memory disorder similar to the symptom displayed by dementia. It was suggested that there is a close relationship between the masticatory function and the learning and/or memory ability. Moreover, it is emphasized that maintenance of oral health is essential for the elderly to retain their own quality of life.

REFERENCES

Bartus, R. T., Dean, R. L. and Beer, B. (1983). An evaluation of drugs for improving memory in aged monkeys: Implications for clinical trials in humans, *Psychopharmacol. Bull.* **19**, 168–184.

Biesold, D., Inanami, O., Sato, A., *et al.* (1989). Stimulation of the nucleus basalis of Meynert increases cerebral cortical blood flow in rats, *Neurosci. Lett.* **98**, 39–44.

Davies, P. and Feisullin, S. (1982). A search for discrete cholinergic nuclei in the human ventral forebrain, *J. Nurochem.* **39**, 1743–1747.

Drachman, D. A. (1977). Memory and cognitive function in man: Does the cholinergic system have a specific role? *Neurology* **27**, 783–790.

Fass, U., Panickar, K., Personett, D., *et al.* (2000). Differential vulnerability of primary cultured cholinergic neurons to nitric oxide excess, *Neuroreport* **11**, 931–936.

Glowinski, J. and Iversen, L. L. (1966). Regional studies of catecholamines in the rat brain, I. The disposition of [^{3}H]norepinephrine, [^{3}H]dopamine and [^{3}H]dopa in various regions of the brain, *J. Neurochem.* **13**, 655–669.

Gobel, S. and Binck, J. M. (1977). Degenerative changes in primary trigeminal axons and in neurons in nucleus caudalis following tooth pulp extractions in the cat, *Brain Res.* **132**, 347–354.

Hayashi, H. and Tabata, T. (1989). Physiological properties of sensory trigeminal neurons projecting to mesencephalic parabrachial area in the cat, *J. Neurophysiol.* **61**, 1153–1160.

Hayashi, H. and Tabata, T. (1990). Pulpal and cutaneous inputs to somatosensory neurons in the parabrachial area of the cat, *Brain Res.* **511**, 177–179.

Hayashi, H. and Tabata, T. (1991). Distribution of trigeminal sensory nucleus neurons projecting to the mesencephalic parabrachial area of the cat, *Neurosci. Lett.* **122**, 75–78.

Houser, C. R., Crawford, G. D., Barber, R. P., *et al.* (1983). Organization and morphological characteristics of cholinergic neurons: An immuno-cytochemical study with a monoclonal antibody to choline acetyltransferase, *Brain Res.* **266**, 97–119.

Iizuka, R. (1986). Neurotransmitters in Alzheimer's disease, *Sinkei-Shinnpo* **30**, 697–710 (in Japanese).

Isse, K., Kanamori, M., Uchiyama, M., *et al.* (1991). A case-control study of risk factors associated with Alzheimer type dementia in Japan, in: *Studies in Alzheimer's disease: Epidemiology and Risk Factors. Proceedings of the Third International Symposium on Dementia.* E. Satoyoshi (Ed.), pp. 63–67. National Center of Neurology and Psychiatry Publishers, Tokyo.

Johnston, M. V., McKinney, M. and Coyle, J. T. (1981). Neocortical cholinergic innervation: A description of extrinsic and intrinsic components in the rat, *Exp. Brain Res.* **43**, 159–172.

Kato, T., Usami, T., Noda, Y., *et al.* (1997). The effect of the loss of molar teeth spatial memory and acetylcoline release from the partial cortex in aged rat, *Behav. Brain Res.* **83**, 239–242.

Kent, C., Sugaya, K., Bryan, D., *et al.* (1999). Expression of superoxide dismutase messenger RNA in adult rat brain cholinergic neurons, *J. Mol. Neurosci.* **12**, 1–10.

Kimoto, A. (1993). Change in trigeminal mesencephalic neurons after teeth extraction in guinea pig, in Japanese, English abstract, *J. Stomatol. Soc. Jpn.* **60**, 199–212.

Kimura, S., Minami, M., Endo, *et al.* (1998). Methylcobalamine (V-B$_{12}$) increase cerebral acetylcholine levels and improves passive avoidance response in stroke-prone spontaneously hypertesive rats, *Biogenic Amines* **14**, 15–24.

Lehmann, J., Nagy, J. I., Atmadia, S., *et al.* (1980). The nucleus basalis magnocellularis: The origin of a cholinergic projection to the neocortex of the rat, *Neuroscience* **5**, 1161–1174.

Li, Y. Q., Takada, M., Kaneko, T., *et al.* (1995). Premotor neurons for trigeminal motor nucleus neurons innervating the jaw-closing and jaw-opening muscles: differential distribution in the lower brainstem of the rat, *J. Comp. Neurol.* **356**, 563–579.

Makiura, T., Ikeda, Y., Hirai, T., *et al.* (2000). Influence of diet and occlusal support on learning memory in rats behavioral and biochemical studies. *Res. Commun. Mol. Pathol. Pharmacol.* **107**, 269–277.

Matumoto, M., Togashi, H., Yoshioka, M., *et al.* (1990). Simultaneous high-performance liquid chromatographic determination of norepinephrine, serotonin, acetylcholine and their metabolites in the cerebrospinal fluid of anaesthetized normotensive rats, *J. Chromatography* **526**, 1–10.

McKinney, M., Coyle, J. T. and Hedreen, J. C. (1983). Topographic analysis of the innervation of the rat neocortex and hippocampus by the basal forebrain cholinergic system, *J. Comp. Neurol.* **217**, 103–121.

Mesulam, M. M., Mufson, E. J., Levey, A. I., *et al.* (1983a). Cholinergic innervation of cortex by the basal forebrain: Cytochemistry and cortical connections of the septal area, diagonal band nuclei, nucleus basalis, substantia innominata, and hypothalamus in the Rhesus monkey, *J. Comp. Neurol.* **214**, 170–197.

Mesulam, M. M., Mufson, E. J., Wainer, B. H., *et al.* (1983b). Central cholinergic pathways in the rat: An overview based on an alternative nomenclature, (Ch1-Ch6), *Neuroscience* **10**, 1185–1201.

Mizuno, N., Yasui, Y., Nomura, S., *et al.* (1983). A light and electron microscopic study of premotor neurons for the trigeminal motor nucleus, *J. Comp. Neurol.* **215**, 290–298.

Olton, D. S., Walker, J. A. and Wolf, W. A. (1982). A disconnection analysis of hippocampal function, *Brain Res.* **233**, 241–253.

Onozuka, M., Watanabe, K., Mirbod, S.M., *et al.* (1999). Reduced mastication stimulates impairment of spatial memory and degeneration of hippocampal neurons in aged SAMP8 mice, *Brain Res.* **826**, 148–153.

Osborn, J. and Lammie, G. (1974). in: *Partial Dentures*, Osborn, J. and Lammie, G. (Eds), pp. 53–56. Blackwell Science, Oxford.

Pearson, R. C. A., Gatter, K. C. and Powell, T. P. S. (1983). The cortical relationships of certain basal ganglia and the cholinergic basal forebrain nuclei, *Brain Res.* **261**, 327–330.

Perry, E. K. (1980). The cholinergic system in old age and Alzheimer's disease. *Age and Aging* **9**, 1–8.

Rasmusson, D., Clow, K. and Szerb, J. C. (1994). Modification of neocortical acetylcholine release and electroencephalogram desynchronization due to brainstem stimulation by drugs applied to the basal forebrain, *Neuroscience* **60**, 665–677.

Rye, D. B., Wainer, B. H., Mesulam, M. M., *et al.* (1984). Cortical projections arising from the basal forebrain: a study of cholinergic and noncholinergic components employing combined retrograde tracing and immunohistochemical localization of choline acetyltransferase, *Neuroscience* **13**, 627–643.

Saito, H., Togashi, H., Yohioka, M., *et al.* (1995). Animal models of vascular dementia with emphasis on stroke-prone spontaneously hypertensive rats, *Progress in Hypertension* **3**, 159–171.

Sarter, M. and Bruno, J. P. (2000). Cortical cholinergic inputs mediating arousal, attentional processing and dreaming: differential afferent regulation of the basal forebrain by telencephalic and brainstem afferents, *Neuroscience* **95**, 933–952.

Semba, K., Reiner, P. B., McGeer, E. G., *et al.* (1988). Brainstem afferents to the magnocellular basal forebrain studied by axonal transport, immunohistochemistry, and electrophysiology in the rat, *J. Comp. Neurol.* **267**, 433–453.

Shigenaga, Y., Yoshida, A., Mitsuhiro, Y., *et al.* (1988a). Morphology of single mesencephalic trigeminal neurons innervating periodontal ligament of the cat, *Brain Res.* **448**, 331–338.

Shigenaga, Y., Mitsuhiro, Y., Yoshida, A., *et al.* (1988b). Morphology of single mesencephalic trigeminal neurons innervating masseter muscle of the cat, *Brain Res.* **445**, 392–399.

Sofroniew, M. V., Pearson, R. C. A. and Powell, T. P. S. (1987). The cholinergic nuclei of the basal forebrain of the rat: normal structure, development and experimentally induced degeneration, *Brain Res.* **411**, 310–331.

Smith, C. M. and Swash, M. (1978). Possible biochemical basis of memory disorder in Alzheimer disease, *Ann. Neurol.* **3**, 471–473.

Terasawa, H., Hirai, T., Ninomiya, T., *et al.* (2002). Influence of tooth-loss and concomitant masticatory alterations on cholinergic neurons in rats: immunohistochemical and biochemical studies, *Neurosci. Res.* **43**, 373–379.

Togashi, H., Kimura, S., Matsumoto, M., *et al.* (1996). Cholinergic changes in the hippocampus of stroke-prone spontaneously hypertensive rats, *Stroke.* **27**, 520–525.

Vertes, R. P. (1988). Brainstem afferents to the basal forebrain in the rat, *Neuroscience* **24**, 907–935.

Vincent, S. R. (1994). Nitric oxide: a radical neurotransmitter in the central nervous system, *Prog. Neurobiol.* **42**, 129–160.

Wallin, A., Alafuzoff, I., Carlsson, A., *et al.* (1989). Neurotransmitter deficits in a non-multi-infarct category of vascular dementia, *Acta. Neurol. Scand.* **79**, 397–406.

Whitehouse, P. J., Price, D. L., Struble, R. G., *et al.* (1982). Alzheimer's disease and senile dementia: loss of neurons in the basal forebrain. *Science* **215**, 1237–1239.

Yamamoto, T. and Hirayama, A. (2001). Effects of soft-diet feeding on synaptic density in the hippocampus and parietal cortex of senescence-accelerated mice, *Brain Res.* **902**, 255–263.

Yoshida, A., Chen, K., Moritani, M., *et al.* (1997). Organization of the descending projections from the parabrachial nucleus to the trigeminal sensory nuclear complex and spinal dorsal horn in the rat, *J. Comp. Neurol.* **383**, 94–111.

Advances in Neuroregulation and Neuroprotection (2005), pp. 459-474
Collin, C. *et al.* (Eds)
© VSP 2005

The ferret: a cytotoxic drug-induced emesis model

T. ENDO [1,*], M. MINAMI [1], M. HIRAFUJI [1], N. HAMAUE [1] and S. H. PARVEZ [2]

[1] *Department of Pharmacology, Health Sciences University of Hokkaido, Ishikari-Tobetsu,*
Hokkaido 061-0293, Japan
[2] *Alfred Fressard Institute of Neuroscience, CNRS UPR 2212-Bât 5, Chateau CNRS 91190,*
Gif Sur Yvette, France

Abstract—The ferret (*Mustela putorius furo*), a carnivore weighting 1–1.5 kg, is highly valued as a small animal model for determining the emetic activity of cytotoxic drugs. Because of its sensitivity to emetic stimuli, the ferret is now being used routinely for testing novel cytotoxic drugs and antiemetic agents. In our laboratory, ferrets were evaluated on their behavioral changes, pathological changes and pharmacological responses to cytotoxic drug-induced emesis. Cisplatin-induced emesis in ferrets was inhibited by 5-HT$_3$ receptor antagonists in a dose dependent manner. 5-HT levels in the area postrema and in the ileum were increased by cytotoxic agents. Both abdominal vagotomy and ondansetron, a 5-HT$_3$ receptor antagonist, remarkably inhibited cisplatin-induced emesis and cisplatin-induced increase in 5-HT levels in the area postrema. The cisplatin-induced increase of the ileal 5-HT levels, however, was not inhibited by vagotomy or ondansetron. Furthermore, compared with intraperitoneal administration, oral administration of ondansetron blocked cyclophosphamide-induced emesis more effectively and significantly. These results suggest that cytotoxic drugs induce emesis mainly through actions on the gastrointestinal tract. Cytotoxic drugs may cause 5-HT release from the enterochromaffin (EC) cells of the intestinal mucosa to stimulate 5-HT$_3$ receptors on the afferent vagal fibers. Stimulation from the afferent vagal nerves appear to produce an increase in 5-HT levels in the area postrema. The increased 5-HT levels in the area postrema might trigger the emetic response induced by cytotoxic agents. In this review, we also discuss abdominal afferent vagus nerve activity and 5-HT release from EC cells using data from our laboratory.

Keywords: Ferret; 5-HT; vagus afferent; EC cells; cisplatin.

INTRODUCTION

Nausea and vomiting are side effects of cancer chemotherapy involving anti-neoplastic agents. Cisplatin and cyclophosphamide, among others, induce severe spells of nausea and vomiting. Animal models for testing the emetic activity of anti-cancer drugs have been limited to the dog, cat and monkey. Rats and guinea-

---

*To whom correspondence should be addressed. E-mail: toruendo@hoku-iryo-u.ac.jp

pigs do not vomit. *Suncus murinus* can vomit but the size is too small to examine the mechanisms by sampling the blood or tissues. The ferret, a carnivore weighting 1–1.5 kg, was evaluated as a small animal for determining the emetic activity of cisplatin (Florczyk *et al.*, 1982). The study of the physiology of emesis is an area in which the use of the ferret has become firmly established as an alternative animal model to the cat and dog (Andrews, 1988; Fox, 1992). Since the 1980s when the ferret became widely used for the study of emesis, it has been shown to respond to various stimuli, including the cytotoxic agents used in chemotherapy: cisplatin (Endo *et al.*, 1990a; Higgins *et al.*, 1989); cyclophosphamide (Hawthorn *et al.*, 1988); radiation X-ray (Andrews, 1988); vagal afferent stimulation under urethane anesthesia (Andrews and Davidson, 1990; Minami *et al.*, 1995a) and intragastric irritants ($CuSO_4$) (Endo *et al.*, 1991). The ferret exhibits all three phases of human emesis. Before retching, they show characteristic behavior, including licking, chin rubbing, walking backward and slit eyes, which are thought to parallel nausea in humans (Andrews, 1988). The search for an alternative for testing antiemetic drugs on cancer patients receiving cisplatin led to the use of this animal model (Florczyk *et al.*, 1982). As in the cat and dog, retching is characterized by large negative oscillations in intrathoracic pressure but no ejection of stomach contents (Andrews, 1988). Vomiting is associated with the expulsion of the upper gastrointestinal contents. Studies of the mechanisms by which various stimuli evoke vomiting have concentrated on the role of the area postrema, the 'chemoreceptor trigger zone (CTZ)' for vomiting located in the forth ventricle (Andrews, 1988). However, abdominal vagotomy produced a marked reduction in vomiting induced by cytotoxic drugs such as cyclophosphamide, cisplatin (Andrews, 1988; Endo *et al.*, 1992) and copper sulfate (Endo *et al.*, 1992), suggesting an important role for abdominal innervation in triggering vomiting by these stimuli (Andrews, 1988; Endo *et al.*, 1992). Because of its sensitivity to a number of clinically important emetic stimuli, the ferret is now being used routinely for testing novel antiemetic agents.

BEHAVIORAL CHANGES

Behavioral changes in anti-cancer drug-induced acute emesis

A large proportion of the action of cisplatin is thought to be peripheral, and related to the activation of 5-HT_3 receptors positioned on distal terminals of the gastric vagal afferent nerve fibers (Andrews *et al.*, 1988). Adult, castrated, fifteen to twenty-week-old male fitch ferrets (Marshall Research Animals Iinc., NY, USA) weighing 0.8–1.5 kg were best suited for investigation. The emetic episodes induced by cisplatin were usually characterized by retching and expulsion, as described by McCarthy and Borison (1974). Our behavioral study was performed using the method of Stables *et al.* (1987) in order to obtain more digital behavioral data. All animals moved freely about in cages. Following administration of the cytotoxic

Table 1.
The mode of behavioral changes induced by cytotoxic drugs in male ferrets (Endo *et al.*, 1990b, 1991)

Dose (mg/kg, i.p.)	No. of animals	No. of retches	No. of vomits	Latency (min)	Duration (min)
Cisplatin (7)	2/3	76.0 ± 61.6	7.0 ± 6.6	105.5 ± 3.5	133.0 ± 85.0
Cisplatin (10)	6/6	107.2 ± 33.9	20.3 ± 8.7	100.3 ± 5.8	132.3 ± 41.9
Cyclophosphamide (200)	6/6	119.2 ± 20.7	17.5 ± 2.9	21.2 ± 3.2	167.2 ± 34.5

Mean ± SE, $n = 6$.

agent, ferrets were placed in individual cages and observed continuously for 6 hours (acute emesis). We administered cisplatin (10 mg/kg, i.p.) to six ferrets, and all animals displayed emetic behavior (Table 1). Antiemetic drugs were administered 30 min before the cytotoxic drugs and the number of vomits and retches, as well as the latency of the first vomit and duration of emesis were carefully recorded (Endo *et al.*, 1990a). Cisplatin produced a dose-dependent emetic response in the ferret at doses of 10 (no. of animals vomit: 6/6), 7 (2/3) but not 5 mg/kg, i.p. Injection of cisplatin (10 mg/kg, i.p.) was followed after a latency period of approximately 100 min by 100 retches and 20 vomits (Table 1). After the 6-hour observation period, emetic episodes were infrequent. The body weight of all ferrets receiving cisplatin (10 mg/kg) decreased by 10%. At a dose of 10 mg/kg, cisplatin produced diarrhea in ferrets by the second day. All doses (5–10 mg/kg) of cisplatin were lethal within 3–5 days, except for one ferret, which received cisplatin (7 mg/kg) that died 14 days after testing. Cyclophosphamide-induced emesis occurred with shorter latency and longer duration than that of the acute emesis induced by cisplatin (Table 1) (Endo *et al.*, 1991).

As shown in Table 2, ondansetron, a selective 5-HT$_3$ receptor antagonist, significantly and dose-dependently delayed the onset of cisplatin-induced emesis. At 1 mg/kg, ondansetron significantly reduced the incidence of both cisplatin-induced emesis and cyclophosphamide-induced emesis (Endo *et al.*, 1990a).

Copper sulfate-induced emesis and the effect of ondansetron

Copper sulfate is generally considered to be a peripherally acting emetic agent that causes gastrointestinal irritation which, in turn, stimulates the emetic reflex through unspecified transmitter mechanisms (Wang and Borrison, 1951). Copper sulfate, administered intra-gastrically in a volume of 5.0 ml/kg at doses of 20 mg/kg (3/6), 40 mg/kg (6/6), produced a dose-related increase in the number of retches and vomits (Endo *et al.*, 1991). Ondansetron reduced the number of retches and vomits induced by copper sulfate (40 mg/kg) by half (Table 3). The results presented in this study demonstrated the effectiveness of 5-HT$_3$ receptor antagonist against emesis evoked by copper sulfate. Our findings indicate that copper sulfate may mediate

Table 2.
The inhibitory effects of ondansetron, a selective 5-HT$_3$ receptor antagonist, on anticancer drug-induced acute emesis in the ferret (Endo et al., 1990a, 1991)

Dose (mg/kg, i.p.)	No. of animals	No. of retches	No. of vomits	Latency (min)	Duration (min)
Ondansetron (0.1)	6/6	95.2	11.0	154.7[**]	118.7
+ cisplatin (10)		± 13.4	± 1.4	± 13.8	± 33.2
Ondansetron (1.0)	4/7	11.0[*]	1.0[*]	303.8[***]	24.5
+ cisplatin (10)		± 5.7	± 0.4	± 23.2	± 18.9
Ondansetron (0.1)	6/6	72.3	16.0	145.5	169.3
+ cyclophosphamide (200)		± 17.8	± 4.2	± 26.7	± 41.5
Ondansetron (1.0)	4/6	13.3[###]	2.0[###]	274.3[###]	79.5
+ cyclophosphamide (200)		± 7.8	± 0.8	± 43.7	± 46.1

Mean ± SE.
[*] $p < 0.05$.
[**] $p < 0.01$.
[***] $p < 0.001$ vs. cisplatin (10) in Table 1.
[###] $p < 0.001$ vs. cyclophosphamide (200) in Table 1.

Table 3.
The inhibitory effects of ondansetron on copper sulfate-induced emesis in the ferret (Endo et al., 1991)

Dose (mg/kg)	Animals retching/tested	No. of retches	No. of vomits	Latency (min)	Duration (min)
CuSO$_4$ (20)	3/6	19.5	1.7	9.3	136.3
		± 12.1	± 0.9	± 5.4	± 43.9
CuSO$_4$ (40)	6/6	81.3[**]	10.2[**]	3.0	190.7
		± 13.9	± 2.2	± 1.2	± 27.9
Ondansetron (1)					
+ CuSO$_4$ (40)	6/6	44.0	5.5	7.2[#]	72.5
		± 10.4	± 1.9	± 0.9	± 47.3
Ondansetron (5)					
+ CuSO$_4$ (40)	7/7	40.7	4.3[#]	7.3[#]	102.9
		± 15.2	± 1.3	± 1.1	± 51.8

Mean ± SE.
[**] $p < 0.01$ vs. CuSO$_4$ (20).
[#] $p < 0.05$ vs. CuSO$_4$ (40).

this effect through a 5-HT system. However, Rudd et al. (1990) reported that the 5-HT$_3$ receptor antagonist ICS205-930 failed to prevent or significantly reduce the emesis evoked by copper sulfate.

In copper sulfate-treated ferrets, there were significant increases in the ileal mucosal levels of 5-HT and norepinephrine, but not of dopamine (Endo et al., 1991). Copper sulfate-treated ferrets had significant increases in aromatic L-amino acid decarboxylase (AADC) activity, and reductions in monoamine oxidase (MAO) activity in the ileum (Endo *et al.*, 1991; Minami *et al.*, 1995a). These increased levels of 5-HT might play a triggering role in the vomiting induced by copper sulfate. It is proposed that copper sulfate may stimulate the release and synthesis of 5-HT from the gut. Abdominal vagal afferents can be activated by 5-HT release.

Cytotoxic drug-induced delayed emesis

Acute emesis induced by anticancer agents in humans is classified as that occurring during the first 24 hours after anticancer chemotherapy; delayed emesis is classified as that occurring 1 to 4 days after administration of chemotherapy (Fukunaka *et al.*, 1998; Rudd and Naylor, 1994). Delayed emesis was recently recognized as a separate emetic entity with a different pathological mechanism (Grunberg and Hesketh, 1993). 5-HT$_3$ receptor antagonists such as ondansetron and granisetron have been observed to be almost completely effective in inhibiting emesis during a 4 to 6 hour period in animal models (Bermudez *et al.*, 1988; Higgins *et al.*, 1989). Delayed emesis in animals showed different sensitivities to antiemetic treatment (Marr *et al.*, 1992; Rudd and Naylor, 1994). Cancer chemotherapeutic agent-induced emesis may last for several days in humans and there remains a population of patients resistant to therapy with 5-HT$_3$ receptor antagonists (Gandara *et al.*, 1993; Kris *et al.*, 1992). Use of dexamethasone in combination with 5-HT$_3$ receptor antagonist has decreased both acute and delayed emesis in humans (Rath *et al.*, 1993; Roila, 1993). However, animal experiments using cisplatin with 4 to 6 hour observation periods have failed to show that antiemetic control by 5-HT$_3$ receptor antagonists can be reliably enhanced with dexamethasone (Marr *et al.*, 1992). Using Rudd *et al.*'s 72-hour observation model for low dose cisplatin (5 mg/kg, i.p.) (Rudd and Naylor, 1994), we demonstrated that while both granisetron and its combination with dexamethasone are effective in cisplatin-induced emesis, a combination treatment was more effective than granisetron alone for the duration of emesis in the delayed phase (Fukunaka *et al.*, 1998).

Food intake in the granisetron plus dexamethasone combined group showed a three-fold significant increase as compared with the cisplatin group (Fukunaka *et al.*, 1998). As shown in Table 4, the cisplatin-induced increase in blood urea nitrogen (BUN) and plasma creatinine levels had a tendency to decrease in the combined group. No significant differences have been observed in the 5-HT levels between the granisetron group and the combined group. Since Rosenberg *et al.* (1969) reported cisplatin to be a potent anticancer drug, the major dose-limiting factor in its use has been nephrotoxicity. Dexamethasone prevented the increase of BUN and plasma creatinine concentration (Table 4), which are reliable parameters for early renal lesions induced by cisplatin. BUN is an well known emetogenic factor (Andrews, 1993). Delayed emesis of ferrets induced by cisplatin might not be due to the increase in the tissue 5-HT levels, but rather due to another mechanism.

Table 4.
Biochemical data 72 hour after cisplatin (5 mg/kg, i.p.) administration in ferrets (Fukunaka *et al.*, 1998)

	BUN (mg/dl)	s-Creatinine (mg/dl)	s-Na+ (mEq/l)	s-K+ (mEq/l)
Control (*n* = 6)	25.0 ± 3.9	0.7 ± 0.04	151.3 ± 0.9	4.9 ± 0.2
Group A (*n* = 6)	145.9 ± 13.6*	3.1 ± 0.53*	150.7 ± 1.1	4.8 ± 0.2
Group B (*n* = 6)	150.9 ± 29.3*	4.4 ± 1.04*	145.8 ± 1.2*#	4.9 ± 0.3
Group C (*n* = 6)	111.2 ± 41.2	2.5 ± 1.13	145.0 ± 1.5*#	4.6 ± 0.3

Mean ± SE. BUN: blood urea nitrogen, Control: nondrug control ferrets, Group A: Cisplatin plus vehicle group, Group B: Cisplatin plus granisetron treatment group, Group C: Cisplatin plus granisetron plus dexamethasone treatment group.
* $p < 0.05$ *vs.* Control.
# $p < 0.05$ *vs.* cisplatin group.

Animal models of nausea

Nausea generally is not life threatening, but it can have a significant negative impact on clinical procedures (Wetchler, 1991) and chronic nausea leads to a marked reduction in the quality of life (Stewart, 1991). A valid animal model of nausea would contribute importantly to the study of the neural and physiology systems involved in this state. Development and verification of an animal model of nausea is difficult, however, for both practical and theoretical reasons. Practical difficulties arise because there is no accepted physiological method for identifying the subjective state of nausea in either animals or humans. The inability to directly measure nausea creates a significant problem for the validation of animal models. Theoretical interpretations of the neurophysiological mechanisms of vomiting indicate another problem for the development of models. The traditional concept that effector activation of vomiting center is coordinated by a localized group of neurons in a vomiting 'center' (Borison and Wang, 1953) is now questioned (Miller and Wilson, 1983a).

Current interpretations of possible mechanisms for the nausea-emetic syndrome propose that this state may be mediated via multiple pathways (Miller and Wilson, 1983b) rather than a single emetic center. Such interpretations suggest the involvement of predominant pathways for a given emetic stimulus or species (Harding, 1990) or a cascade of effector systems that may vary for different animals (Lawes, 1991). A wide variety of individual responses have been used to assess sickness in animals. Some of these are listed in typical rating scales, while others are not. A partial summary of the responses used with various species is outlined in Table 5 (Fox, 1992). Several of these responses such as pica are observed in humans during activation of the emetic syndrome and were adopted as criteria to be used with species that do not possess a complete emetic reflex. Although the use of species that do not show a complete emetic reflex to assess emetic mechanisms is question-

Table 5.
Several putative measures of sickness (nausea) and the species that have been tested with each measure (Fox, 1992)

Measure	Species
Arginine vasopressin	human, monkey, cat, rat
Burrowing and backing	ferret
Cardiac rhythm	human, squirrel monkey
Conditioned taste aversion	human, squirrel monkey, cat, rat, guinea pig, mouse
Defecation	cat, ferret, rat
Gastric rhythm	human, dog
Pica	human, rat
Reduced activity	ferret, rat
Reduced intake of food or water	human, rat
Skin color changes	human, squirrel monkey

able, they continue to receive consideration as multiple, or supplemental indices of sickness (Ossenkopp and Ossenkopp, 1985).

AREA POSTREMA AND VOMITING CENTER

Area postrema and vomiting center in the ferret

The area postrema, which lies on the wall of the fourth ventricle at the level of the obex in the medulla oblongata, is a major neural component of the emetic reflex (Andrews *et al.*, 1988b). Area postrema is a circumventricular organ which lacks a typical blood-brain-barrier. In ferrets, 5-HT$_3$ receptor within the area postrema participate in the central regulation of emesis (Barnes *et al.*, 1988). The CTZ lies within the area postrema (Borison and Wang, 1953). Its histological, vascular and neurochemical organization is adapted to the role of a detector (sensory) organ capable of sensing and transmitting emetic stimuli either from the blood, the cerebrospinal fluid or from other brain regions to the vomiting center. CTZ responds to endogenous substances of pathophysiologic origin as well as to exogenous chemicals. On the other hand, within the vomiting center a drug action affects synaptic transmission and must, therefore, involve one or more neurotransmitters (Costello and Borison, 1977).

Integration of emetic stimuli and synchronization of the appropriate responses occur in the lateral reticular formation of the brainstem, termed the 'vomiting center'. This region is not a rigidly defined anatomical or histological structure.

Effects of vagotomy and 5-HT$_3$ receptor antagonist on cisplatin-induced emesis and 5-HT increase in the area postrema and ileum in the ferret

The 5-HT levels in the area postrema of the cyclophosphamide- or cisplatin-treated ferrets were significantly higher than those of the saline-treated control group

Table 6.

The inhibitory effects of vagotomy on cisplatin-induced emesis and 5-HT levels in the ferret (Endo *et al.*, 1992; Minami *et al.*, 1995a)

Treatment	No. of retches	No. of vomits	5-HT in AP (ng/mg protein)	5-HT in ileum (ng/mg protein)
Cisplatin (10 mg/kg, i.p.)	107.2	20.3	9.7	50.6
(n = 6)	± 33.9	± 8.7	± 1.3	± 3.3
Vagotomy	37.4[*]	2.7[*]	6.5[*]	54.6
+ cisplatin (10 mg/kg, i.p.)	± 10.8	± 1.4	± 0.4	± 3.3
(n = 7)				
Ondansetron (1 mg/kg)	11.0[*]	1.0[*]	6.2[*]	44.0
+ cisplatin (10 mg/kg, i.p.)	± 5.7	± 0.4	± 0.7	± 6.9
(n = 6)				
Saline-treated control			4.9[*]	25.5[*]
(n = 6)			± 0.4	± 2.9

Mean ± SE.

[*] $p < 0.05$ *vs.* Cisplatin (10 mg/kg, i.p.).

AP: area postrema.

(Table 6) (Endo *et al.*, 1992; Minami *et al.*, 1995a). Vagotomy or ondansetron significantly reduced the number of retches and vomits induced by cisplatin. Pretreatment with ondansetron or abdominal vagotomy also significantly inhibited the increase of 5-HT concentration in the area postrema. Intestinal 5-HT levels, on the other hand, were not inhibited by vagotomy or ondansetron. We have demonstrated that electrical stimulation of the abdominal vagal afferents induces an increase in the concentration of 5-HT in the area postrema (Minami *et al.*, 1995a). Taken together, these results suggest that cytotoxic agents induce emesis mainly through actions on the gastrointestinal tract. In ferrets, the area postrema contains a high concentration of 5-HT$_3$ receptors (Higgins *et al.*, 1989). It receives visceral afferent innervation from the gastrointestinal tract (Andrews *et al.*, 1990) and its 5-HT levels are increased and decreased by cytotoxic agents and 5-HT$_3$ receptor antagonists, respectively (Endo *et al.*, 1992). The area postrema, therefore, acts as a relay between the visceral afferents of the vagus and the vomiting center (Higgins *et al.*, 1989). Cytotoxic agents may cause 5-HT release from the EC cells of the intestinal mucosa to stimulate 5-HT receptors on the afferent vagal fibers.

HISTOPATHOLOGICAL CHANGES AFTER CISPLATIN ADMINISTRATION

Following cisplatin administration, slight hyaline droplets in the tubular epithelium and cystic changes in segments of the kidney were observed. Significant changes were noted in the ileum and spleen taken from ferrets treated with cisplatin. Karyorrhexis of epithelial cells was seen in the ileum (Endo *et al.*, 1990b) as well as increased extra-medullary hematopoiesis hemorrhage and karyorrhexis in the lymph follicles of the spleen. No histopathological abnormalities were observed in the heart, liver or adrenal glands taken from ferrets six hours after cisplatin administration. Numerous investigators have reported that the major toxicities induced by cisplatin are nephrotoxicity, nausea with vomiting, myelosuppression and ototoxicity. Other toxicities reported to occur less frequently include anaphylactic-like reaction, neurotoxicity, cardiac abnormalities, protracted anorexia and elevated liver enzymes (Prestayko *et al.*, 1979). Doses larger than 2.5 mg/kg cisplatin produced prompt and severe emesis, anorexia, abdominal tenderness and diarrhea in both dogs and monkeys (Prestayko *et al.*, 1979). The toxic effects of cisplatin in the ferret, characterized by weight loss, anorexia, renal impairment and low white blood cell count, were qualitatively similar to those observed in other species (Miner and Sanger, 1986). Stables *et al.* (1987) reported marked congestion of the lamina propria, hemorrhage and glandular cell necrosis of the ileum, but not the stomach in ferrets following cisplatin administration. We evaluated and observed the manifest lesions in the spleen and intestine that occurred within 6 h (Endo *et al.*, 1990b). The renewal rate of the epithelium of the intestinal mucosa villi is rapid and includes differentiation, mobilization and exfoliation. It is postulated that cisplatin is distributed to organs with active mitosis where it causes lesions. The gastrointestinal system is an important source of sensory stimuli, sending impulses through the vagii and sympathetic nerves. This suggests that cisplatin-induced emesis is associated with intestinal lesions and may act through this pathway. Emetic responses are accomplished through the vomiting center, located in the dorsolateral reticular formation of the medulla. Afferent input to this region comes from a number of sources.

5-HT IN THE GUT

Significance of 5-HT

Large quantities of 5-HT exist in the epithelial tissue of the gastrointestinal tract. Over 80% of the 5-HT in the body is localized in the gut, and 90% of this 5-HT is derived from EC cells (Erspamer, 1966). The vomiting or diarrhea induced by 5-HT seems to be a defence or protective reaction of the body to the outer environment; both responses purge ingested toxic substances from the body (Andrews *et al.*, 1990). Therefore, both the EC cells and the vagal afferent nerve appear to be the anatomical mucosal chemoreceptor of the gastrointestinal tract (Newson *et al.*, 1982).

5-HT levels and 5-HT-related enzyme activities in the ileum of the ferret

5-HT is synthesized by tryptophan hydroxylase (TPH), which is a rate limiting enzyme for the synthesis of 5-HT (Dey and Hoffpsiur, 1984; Nilsson *et al.*, 1985). Administration of cisplatin to ferrets results in histological damage to the gut, which is maximal in the ileal mucosa (Endo *et al.*, 1990a). Simultaneously, cisplatin and cyclophosphamide induced significant increases in TPH activity in the ileum,

Table 7.

The levels of 5-HT, NE and DA in the ileum of ferrets (Endo *et al.*, 1990a; Minami *et al.*, 1991)

Treatment	5-HT (ng/mg protein)	NE (ng/mg protein)	DA (ng/mg protein)
Ileum mucosa			
Control	25.5 ± 2.9	3.23 ± 0.37	0.12 ± 0.02
Cisplatin (10 mg/kg, i.p.)	$50.6 \pm 3.3^*$	$5.59 \pm 0.25^*$	0.14 ± 0.04
Ondansetron (1.0 mg/kg)			
+ cisplatin (10 mg/kg, i.p.)	44.0 ± 6.9	$5.67 \pm 0.59^*$	0.13 ± 0.02
Gastric mucosa			
Control	41.2 ± 10.9	5.93 ± 0.74	0.21 ± 0.07
Cisplatin (10 mg/kg, i.p.)	31.4 ± 4.0	7.13 ± 0.53	0.23 ± 0.04
Ondansetron (1.0 mg/kg)			
+ cisplatin (10 mg/kg, i.p.)	48.2 ± 7.1	6.87 ± 0.61	0.21 ± 0.03

Mean $\pm$ SE.

$^*p < 0.05$ *vs.* Control. Ferrets received vehicle or cisplatin (10 mg/kg, i.p.) on two consecutive days and were killed 3 hours after the second dose.

as compared with control animals (Endo *et al.*, 1993). EC cells synthesize and secrete 5-HT in the gut. Cisplatin-treated ferrets showed significant increase in ileal mucosal levels of 5-HT and norepinephrine (NE) but not dopamine (DA) (Table 7) (Endo *et al.*, 1990a; Minami *et al.*, 1991). There were no significant changes in the levels of 5-HT and DA in the gastric mucosa (Table 7). Cyclophosphamide and copper sulfate also produce a significant increase in ileal 5-HT levels in the ferret (Endo *et al.*, 1990a, 1991, 1992). It is proposed that cytotoxic agents stimulate the release and synthesis of neuroactive agents including 5-HT from the gut that directly sensitize gastrointestinal afferent axons. Aromatic L-amino acid decarboxylase (AADC) activity also increased after administration of these drugs. On the other hand, these cytotoxic drugs induce a significant decrease in MAO activity in the ileum (Endo *et al.*, 1993). Thus, cytotoxic agents may activate ileal 5-HT biosynthesis and further enhance the concentration of 5-HT by reducing its degeneration. Through this mechanism, the life-span of 5-HT might be prolonged. However, cisplatin administered to ondansetron-pretreated ferrets induced no significant changes in TPH or MAO activity (Endo *et al.*, 1993). Although 5-HT$_3$ receptors are widely distributed on peripheral neurons including those of the enteric nervous system (Bradley *et al.*, 1986), 5-HT$_3$ receptor antagonists may have no significant effects on the activities of 5-HT-related enzyme.

In addition to the study of emesis, the ferret has been widely used as a laboratory animal in physiological, immunological, pathological and pharmacological research. Yao *et al.* (1992) reported that there were no significant differences in plasma renin activity between the manually restrained (6.5 ± 1.0 ng/ml/hour, $n = 6$) and device-restrained conscious ferrets (6.5 ± 1.0 ng/ml/hour, $n = 6$). The average mean arterial pressures were stable and within normal limits (106 ± 8 mmHg,

$n = 6$) in the conscious ferret using the restraint device. The blood chemistry and physiology of the ferret was reviewed by Andrews (1988).

5-HT release from EC cells

Examination of the ileum of cisplatin-treated ferrets using electron microscopy demonstrates that 5-HT is stored in large electron-dense granules that move toward the base of the EC cells and are released from the basal surface (Minami *et al.*, 1995a). In the forming-phase of the EC cells, secretory granules (5-HT) are being created in the Golgi apparatus. The basal granules are decreased in number. We examined the EC cells 6 h after cisplatin administration and found them empty of 5-HT. Mast cells were still intact 6 h after cisplatin administration. This suggests that the source of 5-HT that is released during emesis originates from the EC cells.

In species that vomit, elevated intestinal 5-HT may stimulate abdominal vagal afferent nerve fibers which, in turn, evoke the vomiting reflex. The release of 5-HT from intestinal EC cells is regulated by polymodal mechanisms. 5-HT release from the EC cells is triggered or regulated by multiple autoreceptors and heteroreceptors present on the EC cells (Hirafuji *et al.*, 2000; Racké *et al.*, 1988). The effects of many substances on 5-HT release have been investigated extensively in a variety of animals (Minami *et al.*, 2003). 2-Methyl-5-HT, a selective 5-HT_3 receptor agonist, induced a concentration-dependent increase of 5-HT in the ferret and the rat ileum (Minami *et al.*, 1995b). This increase in 5-HT release was significantly reduced by granisetron, a selective 5-HT_3 receptor antagonist.

Previous studies point to the possibility that 5-HT_3 receptors (ferrets: Minami *et al.*, 1995b; rats: Endo *et al.*, 1998a; guinea-pigs: Gebauer *et al.*, 1993) stimulate 5-HT release by acting as ligand-gated cation channels that evoke action potential (Albuqueque *et al.*, 1995).

VAGAL AFFERENT ACTIVITY

In vivo *afferent vagal nerve activity*

Afferent abdominal vagal nerve activity was recorded from the peripheral cut end of the dorsal abdominal vagus nerve, using bipolar platinum-iridium wire electrodes. Analysis of nerve activity was performed after the conversion of raw data to standard pulses by a window discriminator that distinguished the discharge of afferent fibers from background noise (Yoshioka *et al.*, 1992). Cisplatin produced a significant increase in afferent abdominal vagal activity approximately 90 min after its administration (Endo *et al.*, 1995). The time-course of cisplatin-induced emesis in another group of ferrets paralleled the changes in the afferent vagal nerve activity. Intravenous bolus injection of 5-HT and 2-methyl-5-HT, a selective 5-HT_3 receptor agonist, produced a dose-dependent increase in abdominal afferent vagus nerve activity in ferrets. N-3389 (Hagihara *et al.*, 1994), a new 5-HT_3 receptor antagonist, significantly inhibited the 2-methyl-5-HT-induced increase in vagus nerve activity (Minami et al., 1997). After copper sulfate administration, afferent vagus nerve activity increased immediately (Endo et al., 1995). Emetic episodes were also evoked immediately after copper sulfate administration in another group of ferrets. We also found that granisetron, a 5-HT$_3$ receptor antagonist, blocked the increase in afferent vagal nerve activity induced by ouabain (Endo et al., 1998b). These findings demonstrate that the activation of vagal afferent activity may be relevant to cytotoxic agent-induced emesis in the ferret. This excitatory response may be mediated by the action of 5-HT on 5-HT$_3$ receptors located on the vagal afferent fibers.

It is well known that 5-HT$_3$ receptor antagonists block anti-cancer drug-induced emesis in several species (Costall et al., 1986; Endo et al., 1990a; Hawthorn et al., 1988). The issue of the involvement of 5-HT$_4$ receptors in emesis, however, is controversial. 5-Methoxytryptamine (5-MT), a 5-HT$_4$ receptor agonist, caused vomiting in dogs (Fukui et al., 1994) and the 5-MT-induced emesis was inhibited by vagotomy or ICS205-930 (a 5-HT$_3$ and 5-HT$_4$ receptor antagonist). However, 5-MT failed to increase the discharge of vagal afferent fibers in rats in vitro (Tonini, 1995). We previously reported that a 5-MT-induced increase in abdominal afferent vagal nerve activity in anesthetized rats was inhibited by GR113808, a 5-HT$_4$ receptor antagonist, but not by granisetron (Tamakai et al., 1996).

In vitro *isolated vagus nerve measurement*

A 5-HT-induced depolarization reaction of isolated cervical vagus nerves in rats, rabbits, and ferrets has been reported previously (Bentley and Barnes, 1998; Elliott and Wallis, 1990; Ireland, 1987; Newberry et al., 1992). These reports suggest that the 5-HT-induced depolarization occurs mainly due to the reaction mediated by 5-HT$_3$ receptors (Peters et al., 1995) and, at least in part, to the reaction mediated by 5-HT$_4$ receptors. We found that the 5-HT-induced depolarization response of the isolated rat abdominal vagal nerves was approximately 170% higher than that observed in the cervical vagus nerves (Nemoto et al., 2001). 5-HT and 2-methyl-5-HT produced a similar tendency.

CONCLUSIONS

The ferret was evaluated as a small animal for determining the emetic activity of cytotoxic drugs. The cisplatin-induced emesis of ferrets was inhibited by 5-HT$_3$ receptor antagonists in a dose dependent manner. 5-HT levels in the area postrema and in the ileum were increased by cytotoxic agents. Both abdominal vagotomy and ondansetron, a 5-HT$_3$ receptor antagonist, remarkably inhibited cisplatin-induced emesis and the cisplatin-induced increase in 5-HT levels in the area postrema. On the other hand, cisplatin-induced increase in the ileal 5-HT levels was not inhibited by vagotomy or ondansetron. Compared with intraperitoneal administration, oral administration of ondansetron blocked cyclophosphamide-induced emesis more effectively and significantly. These results suggest that cytotoxic drugs induce emesis mainly thorough actions on the gastrointestinal tract.

Initially, cytotoxic drugs activate a peripheral mechanism. The hypothesis that $5-HT_3$ receptors located on the peripheral vagus in the gut are important sites for the $5-HT_3$ receptor antagonists to exert their antiemetic effects has also been based on the ability of cisplatin synthesis/turnover in the mucosa following administration in the ferret. Cisplatin- and cyclophosphamide-treated ferrets produced significant increases in TPH activity, a rate limiting enzyme of 5-HT in the ferret ileum, as compared to control value. Copper sulfate, on the other hand, did not modify TPH activity significantly. Cisplatin-, cyclophosphamide and copper sulfate-treated ferrets showed significant decreases in MAO activities in the ileum as compared with that of the control group. The cytotoxic drugs cause degenerative changes and affect the levels of 5-HT in the gastrointestinal tract. Cytotoxic drugs may cause 5-HT release from the EC cells of the intestinal mucosa to stimulate $5-HT_3$ receptors on the afferent vagal fibers. Finally, stimuli from afferent vagal nerves appear to produce an increase of 5-HT levels in the area postrema. The increased 5-HT levels of the area postrema might trigger the emetic response induced by cytotoxic agents. These results suggest that cytotoxic drugs induce emesis mainly through actions on the gastrointestinal tract. A different mechanism from that of acute emesis involving a higher center may be involved in cytotoxic drug-induced delayed emesis.

REFERENCES

Albuquerque, E. C., Pereira, E. F., Castro, N. G., *et al.* (1995). Nicotinic receptor function in the mammalian central nervous system, *Ann. N.Y. Acad. Sci.* **757**, 48–72.

Andrews, P. L. R. (1988). The physiology of the ferret, in: *Biology and Diseases of the Ferret*, Fox, J. G. (Ed.), pp. 100–134. Lea and Febiger, Philadelphia, USA.

Andrews, P. L. R. (1993). The mechanism of emesis induced by anti-cancer therapies, in: *Emesis in Anticancer Therapy. Mechanisms and Treatment*, Andrews, P. L. R. and Sanger, G. J. (Eds), pp. 111–161. Chapman and Hall, London.

Andrews, P. L. R. and Davidson, H. I. M. (1990). A method for the induction of emesis in the conscious ferret by abdominal vagal stimulation, *J. Physiol.* **422**, 5.

Andrews, P. L. R., Rapeport, W. G. and Sanger, G. J. (1988). Neuropharmacology of emesis induced by anti-cancer therapy, *Trend Pharm. Sci.* **9**, 334–341.

Andrews, P. L. R., Davis, C. J., Bingham, S., *et al.* (1990). The abdominal visceral innervation and the emetic reflex: pathways, pharmacology and plasticity, *Can. J. Physiol. Pharmacol.* **68**, 325–345.

Barnes, N. M., Costall, R. J., Naylor, F. D., *et al.* (1988). Identification of $5-HT_3$ recognition sites in the ferret area postrema, *J. Pharm. Pharmacol.* **40**, 586–588.

Bentley, K. R. and Barnes, N. M. (1998). 5-Hydroxytryptamine-3 ($5-HT_3$) receptor-mediated depolarization of the rat isolated vagus nerve: modulation by trichloroethanol and related alcohols. *Eur. J. Pharmacol.* **354**, 25–31.

Bermudez, J., Boyle, E. A., Miner, W. D., *et al.* (1988). The antiemetic potential of the 5-hydroxytryptamine receptor antagonist BRL43694, *Br. J. Pharmac.* **58**, 644–650.

Borison, H. L. and Wang, S. C. (1953). Physiology and pharmacology of vomiting, *Pharmacol. Rev.* **5**, 193–230.

Bradley, P. B., Engel, G., Feniuk, W., *et al.* (1986). Proposals for the classification and nomenclature of functional receptors for 5-hydroxytryptamine, *Neuropharmacol.* **25**, 563–576.

Costall, B., Domeney, A. M., Naylor, R. J., *et al.* (1986). 5-Hydroxytryptamine M-receptor antagonism to prevent cisplatin-induced emesis, *Neuropharmacol.* **25**, 959–961.

Costello, D. J. and Borison, H. L. (1977). Naloxone antagonizes narcotic self-blockade of emesis in the cat, *J. Pharmacol. Exp. Ther.* **203**, 222–230.

Dey, R. D. and Hoffpsuir, J. (1984). Ultrastructural immunocytochemical localization of 5-hydroxytryptamine in gastric enterochromaffin cells, *J. Histochem. Cytochem.* **32**, 661–666.

Elliott, P. and Wallis, D. I. (1990). Analysis of the actions of 5-hydroxytryptamine on the rabbit isolated vagus nerve, *Naunyn-Schmiedeberg's Arch. Pharmacol.* **341**, 494–502.

Endo, T., Minami, M., Monma, Y., *et al.* (1990a). Effect of GR38032F on cisplatin- and cyclophosphamide-induced emesis in the ferret, *Biogenic Amines* **7**, 525–533.

Endo, T., Minami, M., Monma, Y., *et al.* (1990b). Emesis-related biochemical and histopathological changes induced by cisplatin in the ferret, *J. Toxicol. Sci.* **15**, 235–244.

Endo, T., Minami, M., Monma, Y., *et al.* (1991). Effect of ondansetron, a 5-HT$_3$ antagonist, on copper sulfate-induced emesis in the ferret, *Biogenic Amines* **8**, 79–86.

Endo, T., Minami, M., Monma, Y., *et al.* (1992). Vagotomy and ondansetron (5-HT$_3$ antagonist) inhibited the increase of serotonin concentration induced by cytotoxic drugs in the area postrema of ferrets, *Biogenic Amines* **9**, 163–175.

Endo, T., Takahashi, M., Minami, M., *et al.* (1993). Effects of anticancer drugs on enzyme activities and serotonin release from ileal tissue in ferret, *Biogenic Amines* **9**, 479–489.

Endo T., Nemoto M., Minami M., *et al.* (1995). Changes in the afferent abdominal vagal nerve activity induced by cisplatin and copper sulfate in the ferret, *Biogenic Amines* **11**, 399–407.

Endo, T., Ogawa, T., Hamaue, N., *et al.* (1998a). Granisetron, a 5-HT$_3$ receptor antagonist, inhibited cisplatin-induced 5-hydroxytryptamine release in the isolated ileum of ferrets, *Res. Commun. Mol. Pathol. Pharmacol.* **100**, 243–253.

Endo, T., Sugawara, J., Nemoto, M., *et al.* (1998b). Effects of granisetron, a 5-HT$_3$ receptor antagonist, on ouabain-induced emesis in ferrets, *Res. Commun. Mol. Pathol. Pharmacol.* **102**, 227–239.

Erspamer, V. (1966). Occurrence of indolalkyl amines in nature, in: *Handbook of Experimental Pharmacology*, Feichler, O. and Farah, A. (Eds), Vol. 19, pp. 132–181. Springer, Berlin.

Florczyk, A. P., Schurig, J. E. and Bradner, W. T. (1982). Cisplatin-induced emesis in the ferret: a new animal model, *Cancer Treat. Rep.* **66**, 187–189.

Fox, R. A. (1992). Current status: animal models of nausea, in: *Mechanisms and Control of Emesis*, Colloque INSERM, vol. 223, Bianchi, A. L., Grelot, L., Miller, A. D. and King, G. I. (Eds), pp. 341–350. John Libbey Eurotext Ltd.

Fukui, H., Yamamoto, M., Sasaki, S., *et al.* (1994). Possible involvement of peripheral 5-HT$_4$ receptors in copper sulfate-induced vomiting in dogs, *Eur. J. Pharmacol.* **257**, 47–52.

Fukunaka, N., Sagae, S., Kudo, R., *et al.* (1998). Effects of granisetron and its combination with dexamethasone on cisplatin-induced delayed emesis in the ferret, *Gen. Pharmacol.* **31**, 775–781.

Gandara, D. R., Harvey, W. H., Monaghan, G. G., *et al.* (1993). Delayed emesis following highdose cisplatin-a double blind randomised comparative trial of ondansetron (GR-38032F) versus placebo, *Eur. J. Cancer* **29A**(Suppl.), s35–s38.

Gebauer, A., Merger, M. and Kilbinger, H. (1993). Modulation by 5-HT$_3$ and 5-HT$_4$ receptors of the release of 5-hydroxytryptamine from the guinea-pig small intestine, *Naunyn-Schmiede Arch. Pharmacol.* **347**, 137–140.

Grunberg, S. M. and Hesketh, P. J. (1993). Control of chemotherapy induced emesis, *New Engl. J. Med.* **329**, 1790–1796.

Hagihara, K., Hayakawa, T., Arai, H., *et al.* (1994). Antagonistic activities of N-3389, a newly synthesized diabicyclo derivative, at 5-HT$_3$ and 5-HT$_4$ receptors, *Eur. J. Pharmacol.* **271**, 159–166.

Harding, R. K. (1990), Concepts and conflicts in the mechanism of emesis, *Can. J. Physiol. Pharmacol.* **68**, 218–220.

Hawthorn, J., Ostler, K. J. and Andrews, P. L. R. (1988). The role of the abdominal visceral innervation and 5-hydroxytryptamine M-receptors in vomiting induced by the cytotoxic drugs cyclophosphamide and cisplatin in the ferret, *Quart. J. Exp. Physiol.* **173**, 7–21.

Higgins, G. A., Kilpatrick, G. J., Bunce, K. T., *et al.* (1989). 5-HT$_3$ receptor antagonists injected into the area postrema inhibit cisplatin-induced emesis in the ferret, *Br. J. Pharmacol.* **97**, 247–255.

Hirafuji, M., Minami, M., Endo, T., *et al.* (2000). Intracellular regulatory mechanisms of 5-HT release from enterochromaffin cells in intestinal mucosa, *Biogenic Amines* **16**, 29–52.

Ireland, S. J. (1987). Origin of 5-hydroxytryptamine-induced hyperpolarization of the rat superior cervical ganglion and vagus nerve, *Br. J. Pharmacol.* **92**, 407–416.

Kris, M. G., Tyson, L. B., Clark, R. A., *et al.* (1992). Oral ondansetron for the control of delayed emesis after cisplatin. Report of a phase-II study and a review of combined trials to manage delayed emesis, *Cancer* **70**, 1012–1016.

Lawes, I. N. C. (1991). The central connections of area postrema define the paraventricular system involved in antinoxious behaviors, in: *Nausea and Vomiting: Recent Research and Clinical Advances*, Kucharczyk, J., Stewart, D. J. and Miller, A. D. (Eds), pp. 77–101. CRC Press, Boca Raton, FL.

Marr, H. E., Davey, P. D. and Blower, P. (1992). The effect of dexamethasone alone or in combination with granisetron, on cisplatin-induced emesis in the ferret, *Br. J. Pharmacol.* **104**, 371.

McCarthy, L. E. and Borison, H. L. (1974). Respiratory mechanics of vomiting in decerebrate cats, *Am. J. Physiol.* **226**, 738–743.

Miller, A. D. and Wilson, V. J. (1983a). 'Vomiting center' reanalyzed: An electrical stimulation study, *Brain Res.* **270**, 154–158.

Miller, A. D. and Wilson, V. J. (1983b). Vestibular-induced vomiting after vestibulocerebellar lesions, *Brain Behav. Evol.* **23**, 26–31.

Minami, M., Endo, T., Monma, Y., *et al.* (1991). Pharmacology of emesis induced by anti-cancer drugs, *J. Toxicol. Sci.* **16** (Suppl. II), 35–39.

Minami, M., Endo, T., Nemoto, M., *et al.* (1995a). How do toxic emetic stimuli cause 5-HT release in the gut and brain? in: *Serotonin and the Scientific Basis of Anti-emetic Therapy*, Raynold, D. J. M., Andrews, P. L. R. and Davis, C. J. (Eds), pp. 68–76. Oxford Clinical Communications, Oxford.

Minami, M., Saito, H. and Yoshioka, M. (1995b). Toxicological aspects of cisplatin-induced emesis with emphasis on serotonin release, *J. Toxicol. Sci.* **20**, 77–85.

Minami, M., Endo, T., Tamakai, H., *et al.* (1997). Antiemetic effects of N-3389, a new synthesized 5-HT$_3$ and 5-HT$_4$ receptor antagonist, in ferrets, *Eur. J. Pharmacol.* **321**, 333–342.

Minami, M., Endo, T., Hirafuji, M., *et al.* (2003). Pharmacological aspects of anticancer drug-induced emesis with emphasis on serotonin release and vagal nerve activity, *Pharmacol. Ther.* **99**, 149–165.

Miner, W. D. and Sanger, G. J. (1986). Inhibition of cisplatin-induced vomiting by selective 5-hydroxytryptamine M-receptor antagonism, *Br. J. Pharmacol.* **88**, 497–499.

Nemoto, M., Endo, T., Minami, M., *et al.* (2001). 5-Hydroxytryptamine (5-HT)-induced depolarization in isolated abdominal vagus nerves in the rat: involvement of 5-HT$_3$ and 5-HT$_4$ receptors, *Res. Commun. Mol. Pathol. Pharmacol.* **109**, 217–230.

Newberry, N. R., Watkins, C. J., Reynolds, D. J., *et al.* (1992). Pharmacology of the 5-hydroxy-tryptamin-induced depolarization of the ferret vagus nerve in vitro, *Eur. J. Pharmacol.* **221**, 157–160.

Newson, B., Ahlman, H., Dahlström, A., *et al.* (1982). Ultrastructural observations in the rat ileal mucosa of possible epithelial 'taste cells' and submucosal sensory neurons, *Acta Physiol. Scand.* **114**, 161–164.

Nilsson, O., Ericson, L. E., Dahlström, A., *et al.* (1985). Subcellular localization of serotonin immunoreactivity in rat enterochromaffin cells, *Histochemistry* **82**, 351–355.

Ossenkopp, K.-P. and Ossenkopp, M. D. (1985). Animal models of motion sickness: Are nonemetric species an appropriate choice? *The Physiologist* **28**, s61–s62.

Peters, J. A., Lawbert, J. J., Hope, A. G., *et al.* (1995). The electrophysiology of vagal and recombinant 5-HT$_3$ receptors, in: *Serotonin and the Scientific Basis of Anti-emetic Therapy*, Raynold, D. J. M., Andrews, P. L. R. and Davis, C. J. (Eds), pp. 95–105. Oxford Clinical Communications, Oxford.

Prestayko, A. W., D'Aoust, J. C., Issell, B. F., *et al.* (1979). Cisplatin (cis-diamminedichloroplatinum II), *Cancer Treat. Rev.* **6**, 17–39.

Racké, K., Schwörer, H. and Kilringer, H. (1988). Adrenergic modulation of the release of 5-hydroxytryptamine from the vascularly perfused ileum of the guinea-pig, *Br. J. Pharmacol.* **95**, 923–931.

Rath, U., Upadhyaya, B. K., Arechavala, E., *et al.* (1993). Role of ondansetron plus dexamethasone in fractionated chemotherapy, *Oncology* **50**, 168–172.

Roila, F. (1993). Ondansetron plus dexamethasone compared to the standard metocloplamide combination, *Oncology* **50**, 163–167.

Rosenberg, B., VanCamp, L., Trosko, J. E., *et al.* (1969). Platinum compounds: a new class of potent antitumour agents, *Nature* **222**, 385–386.

Rudd, J. A., Costall, B., Naylor, R. J., *et al.* (1990). The emetic action of copper sulfate in the ferret, *Eur. J. Pharmacol.* **183**, 1213.

Rudd, J. A. and Naylor, R. J. (1994). Effects of 5-HT$_3$ receptor antagonists on models of acute and delayed emesis induced by cisplatin in the ferret, *Neuropharmacology* **33**, 1607–1608.

Stables, R., Andrews, P. L. R., Bailey, H. E., *et al.* (1987). Antiemetic properties of the 5-HT$_3$ receptor antagonist, GR38032F, *Cancer Treat Rev.* **14**, 333–336.

Stewart, D. J. (1991). Nausea and vomiting in cancer patients, in: *Nausea and vomiting: Recent research and clinical advances*, Kucharczyk, J., Stewart, D. J. and Miller, A. D. (Eds), pp. 177–203. CRC Press, Boca Raton FL.

Tamakai, H., Sugahara, J., Ogawa, T., *et al.* (1996). Effects of 5-HT$_4$ receptor agonist on abdominal afferent nerve activity in rats (abstract), *Jpn. J. Pharmacol.* **71** (Suppl. 1), 245.

Tonini, M. (1995). 5-HT$_4$ receptor involvement in emesis, in: *Serotonin and the Scientific Basis of Anti-emetic Therapy*, Raynold, D. J. M., Andrews, P. L. R. and Davis, C. J. (Eds), pp. 192–199. Oxford Clinical Communications, Oxford.

Yao, Z., Adler, A., Kovar, P., *et al.* (1992). A restraint device for blood sampling and direct blood pressure measurement in conscious ferrets, *Contemporary Topics of the American Association for Laboratory Animal Sciences* **31**, 19–21.

Yoshioka, M., Ikeda, M., Abe, H., *et al.* (1992). Pharmacological characterization of 5-hydroxytryptamine-induced excitation of afferent cervical vagus nerve in anesthetized rats, *Br. J. Pharmacol.* **106**, 544–549.

Wang, S. C. and Borrison, H. L. (1951). Copper sulfate emesis: a study of afferent pathways from the gastrointestinal tract, *Am. J. Physiol.* **164**, 520–526.

Wetchler, B. V. (1991). Outpatient anesthesia: What are the problems in the recovery room? *Can. J. Anesth.* **38**, 890–894.

Advances in Neuroregulation and Neuroprotection (2005), pp. 475-490
Collin, C. *et al.* (Eds)

Na$^+$-K$^+$ pump and β-adrenergic responses in smooth muscle

K. SHIMAMURA [1,2,*], S. KIMURA [1], A. OHASHI [1], M. TOBA [1], S. OKUNO [1], Y. KANAMARU [3] and H. SAITO [4]

[1] *Department of Clinical Pharmacology, Faculty of Pharmaceutical Sciences, Health Sciences University of Hokkaido, 1757 Kanazawa, Ishikari-Tobetsu, Hokkaido 061-0293, Japan*
[2] *The Research Institute of Personalized Health Sciences, Health Sciences University of Hokkaido, 1757 Kanazawa, Ishikari-Tobetsu, Hokkaido 061-0293, Japan*
[3] *Aeromedical Laboratory, JASDF, 1-2-10 Tachikawa, Tokyo 190-8585, Japan*
[4] *Department of Basic Sciences, Japanese Red Cross Hokkaido College of Nursing, Kitami, Hokkaido 090-0011, Japan*

Abstract—Changes in Na$^+$-K$^+$ pump activity influence intracellular Na$^+$ and K$^+$ concentrations. Activation of the electrogenic Na$^+$-K$^+$ pump hyperpolarizes plasma membrane and inhibits voltage-dependent Ca^{2+} current. Na$^+$-K$^+$ pump activity has been shown to be inhibited in the absence of extracellular K$^+$ or by cardiac glycosides. Involvement of Na$^+$-K$^+$ pump in regulation of cellular functions has been identified by inhibition of responses under inhibition of the Na$^+$-K$^+$ pump. In some smooth muscle tissues, responses to β-agonists were inhibited by K$^+$-free medium or ouabain, indicating that responses to β-adrenoceptor agonists are mediated by the Na$^+$-K$^+$ pump. However, in other smooth muscle tissues, responses to β-agonists were not inhibited by Na$^+$-K$^+$ pump inhibition. The Na$^+$-K$^+$ pump plays an important role in the regulation of circulatory function, and the levels of endogenous Na$^+$-K$^+$ pump inhibitors have been shown to be elevated in some types of hypertension. Endogenous Na$^+$-K$^+$ pump inhibitors possibly affect some of β-adrenoceptor-mediated responses. Regulation of Na$^+$-K$^+$ pump may be important in the regulation of smooth muscle activity in normal conditions, as well as under pathophysiological conditions, such as hypertension.

Keywords: Smooth muscle; Na-K-ATPase; adrenergic; hyperpolarization.

INTRODUCTION

It has been reported that low temperature (2–3°C), K$^+$-free medium, or ouabain causes intracellular Na$^+$ increase and K$^+$ loss in cardiac muscle (Page *et al.*, 1964). As cellular electrical activity depends mainly on the electrochemical gradient of

---

*To whom correspondence should be addressed. E-mail: shimamu@hoku-iryo-u.ac.jp

Na$^+$ and K$^+$ across the cell membrane, maintenance of intracellular concentrations of Na$^+$ and K$^+$ is important for excitable cells. Na$^+$-K$^+$-ATPase (Na$^+$-K$^+$ pump) plays a major role in the maintenance of intracellular Na$^+$ and K$^+$ concentrations by using ATP. It has been reported that in resting skeletal muscle, the Na$^+$-K$^+$ pump utilizes around 5–10% of total energy turnover (Clausen, 2003). Intracellular Na$^+$ is also important in the maintenance of intracellular Ca^{2+} concentration through a Na$^+$/Ca^{2+} exchange mechanism (Miura and Kimura, 1989).

Na$^+$-K$^+$ pump current

Activation of the Na$^+$-K$^+$ pump is electrogenic because ATP-driven pumping of 3Na$^+$ out of and 2K$^+$ into the cell creates a net efflux of positive charge. Na$^+$-K$^+$ pump current is thought to be voltage-dependent, to be small at hyperpolarized membrane and to be large at depolarized membrane (De Weer *et al.*, 1988). When ionic channel currents and Na$^+$/Ca^{2+} exchange currents were minimized, Na$^+$-K$^+$ pump current was isolated and shown to be sensitive to strophantidine, a cardiac glycoside, in guinea-pig ventricular muscle cells. The current was voltage-dependent and its amplitude increased almost linearly between -100 and 0 mV and stayed stable between 0 and $+50$ mV (Gadsby and Nakao, 1989). It has also been shown that an intracellular regulatory molecule connected to G protein is important for the current because GTPγS restored the current recorded by patch clamping in human endothelial cells (Oike *et al.*, 1993).

The electrogenic component of the Na$^+$-K$^+$ pump has been shown to contribute to the resting membrane potential of smooth muscle with a magnitude of around -10 mV (Bolton, 1973; Fleming, 1980). In isolated smooth muscle cells from guinea-pig mesenteric artery, outward current induced by the Na$^+$-K$^+$ pump was inhibited by ouabain, K$^+$-free medium and low temperature. It was shown that when temperature increased, the current amplitude increased with Q$_{10}$ of 1.87. The current amplitude increased almost linearly as the membrane potential was made more positive up to about 0 mV (Nakamura *et al.*, 1999).

Na$^+$-K$^+$ pump-induced hyperpolarization and relaxation in smooth muscle

Na$^+$-loading by storage in cold K$^+$-free medium has been used to induce electrogenic Na$^+$-K$^+$ pump activation and the activation resulted in membrane hyperpolarization. Resting membrane potential of pregnant rat uterine smooth muscle cells, which were stored in K$^+$-free medium at 4°C for 18 h was around -15 mV and was increased to -70 mV by K$^+$ reintroduction at 37°C for 2 min (Taylor *et al.*, 1969). Similar effects of Na$^+$-K$^+$ pump activation on membrane potential were observed in rabbit uterine smooth muscle (Kao and Nishimura, 1969). In arterial preparations incubated in K$^+$-free medium, K$^+$-induced relaxation was associated with ouabain-sensitive membrane hyperpolarization (Bonaccorsi *et al.*, 1977a). The effect of ouabain on smooth muscle Na$^+$-K$^+$ pump has been well studied (Matthews and Sutter, 1966). Na$^+$-K$^+$ pump activity was shown to be also important for the development of membrane potential slow wave in cat intestine (Connor, 1984). The electrogenic activity of the Na$^+$-K$^+$ pump was also observed in guinea pig taenia coli (Casteels *et al.*, 1971; Tomita and Yamamoto, 1971) and ileum (Bolton, 1973).

We examined the effect of Na$^+$-K$^+$ pump activation on membrane potential in smooth muscle of circular muscle layer in rat stomach. Male Wistar rats weighing 200–300 g were anesthetized with CO_2 and circular muscle preparations were made. Animals were treated according to the Guiding Principles for the Care and Use of Animals in the Field of Physiological Sciences approved by The Physiological Society of Japan (March 29, 2002). Isometric tension was recorded by a force-displacement transducer in modified Tyrode's solution. Modified Tyrode's solution was composed of (in mmol): NaCl 137, KCl 5.4, $CaCl_2$ 2.0, $MgCl_2$ 1.0, NaH_2PO_4 0.4, glucose 5.6, and was equilibrated with a gas mixture of 95% O_2 and 5% CO_2 (pH 7.3) at 37°C. Nominally K$^+$-free or Ca^{2+}-free medium was made by omitting KCl or $CaCl_2$, respectively, from the modified Tyrode's solution. After preincubation in normal medium for 2 h, the preparations were applied with passive tension, which stretched the length of preparations by around 50% in Ca^{2+}-free medium. Application of 2 mM $CaCl_2$ induced phasic and tonic contraction. When Tyrode's solution was changed to a K$^+$-free medium, increase in the tonic contraction was observed. Application of 5.4 mM KCl during the tonic contraction caused marked relaxation (data not shown). Membrane potential was measured intracellularly using a conventional microelectrode technique. Briefly, the circular muscle layer strips were mounted on a silicon rubber bed in a chamber perfused continuously with warmed Tyrode's solution at 5 ml/min and were impaled with glass capillary electrodes filled with 3 M KCl (tip resistance 40–50 MΩ) to record transmembrane potential. As shown in Fig. 1, after incubation in K$^+$-free medium, application of 5.4 mM K$^+$ caused transient but high-magnitude hyperpolarization of the plasma membrane. This hyperpolarization appeared to be responsible for the relaxation because the membrane hyperpolarized enough to close voltage-dependent Ca^{2+} channels. The K$^+$-free incubation time-dependent increase in hyperpolarization amplitude indicated that electrogenic Na$^+$-K$^+$ pump transported Na$^+$ and K$^+$ continuously and that incubation time-dependent intracellular accumulation of Na$^+$ stimulated the Na$^+$-K$^+$ pump in circular muscle cells of rat stomach fundus.

Vascular smooth muscle and Na$^+$-K$^+$ pump

Inhibition of electrogenic Na$^+$-K$^+$ pump activity by cooling, K$^+$-free solution, or ouabain caused membrane depolarization in guinea pig portal vein (Kuriyama *et al.*, 1971). Na$^+$-K$^+$ pump activity has also been shown to play important roles in maintaining the membrane potential of rabbit ear artery (Hendrickx and Casteels, 1974) and rabbit main pulmonary artery (Casteels *et al.*, 1977). In a resistant aretery, the second order branch of rat mesenteric artery, ouabain induced contraction, which was associated with membrane depolarization, an inhibition of ^{22}Na efflux, and an increase in [^{3}H]ouabain binding, indicating that ouabain contracted the artery through membrane depolarization by inhibition of Na$^+$-K$^+$ pump activity (Aalkjaer

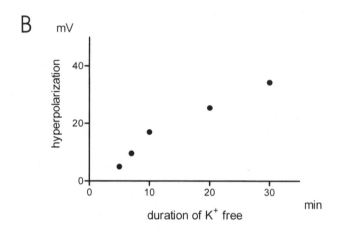

Figure 1. (A) A typical trace showing intracellularly recorded membrane potential (resting membrane potential -45 mV) and effect of reintroduction of K^+ in K^+-free medium. Membrane potential was recorded from a circular muscle cell of rat stomach fundus soaked in K^+-free medium for 10 min. Perfusion with medium containing 5.4 mM K^+ (arrow) which immediately caused transient membrane hyerpolarization with an amplitude of around 20 mV. (B) Summarized graph showing the relationship between time of incubation in K^+-free medium and the amplitude of maximal hyperpolarization by 5.4 mM K^+ at different incubation periods in K^+-free medium. Data were collected from 8 animals.

and Mulvany, 1985). Thus, the Na^+-K^+ pump plays a significant role in the regulation of vascular contraction (Mulvany, 1985). Na^+-K^+ pump currents were examined by patch clamp experiment in rabbit cerebral pial membrane precapillary arterioles (Quinn *et al.*, 2000). Inhibition of the Na^+-K^+ pump has been considered to elevate intracellular free Ca^{2+} through the Na^+/Ca^{2+} exchanger (Baker *et al.*, 1969). However, in arterial smooth muscle, ouabain increased intracellular Ca^{2+} concentration without intracellular Na^+ concentration increase (Arnon *et al.*, 2000). This mechanism may be related to a finding that cardiac glycoside increased inward current in canine coronary artery (Belardinelli *et al.*, 1979).

Involvement of neurotransmitters

When inhibition of the Na^+-K^+ pump induces vascular tissue contraction (Van-houtte and Lorenz, 1984), inhibition of the Na^+-K^+ pump excites non-smooth muscle cells as well as smooth muscle cells in the tissue. Experimental results of Na^+-K^+ pump inhibition need to be interpreted carefully because Na^+-K^+ pump inhibition has been reported to induce the release of neurotransmitters. For example, inhibition of the Na^+-K^+ pump by ouabain, digitoxigenin, or K^+-free medium causes endogenous catecholamine release from adrenergic nerve endings in the portal and saphenous veins of dogs (Tsuru and Shigei, 1976), and in the rat aorta (Karaki *et al.*, 1978) and rat tail artery (Bonaccorsi *et al.*, 1977b). In these studies, the contractile responses to ouabain were inhibited by α-adrenergic antagonists. Ouabain increased noradrenaline release by extracellular Ca^{2+}-dependent mechanism in non-vascular smooth muscle tissues such as guinea-pig vas deferens (Nakazato *et al.*, 1978). These studies indicate that in the tissue-level experiments, response mediated by Na$^+$-K$^+$ pump inhibition may involve the effects of neurotransmitters.

Intracellular mechanisms in regulation of Na$^+$-K$^+$ pump activity

The Na$^+$-K$^+$ pump is composed of α and β subunits. Different α subunits which are tissue- and cell type-specific show different affinities for ouabain and Na$^+$ (Clausen, 2003). For example, among four subunit isoforms, the α1-isoform is dominant in rat resistance artery, which is almost insensitive to cardiac glycosides (Blanco and Mercer, 1998).

The intracellular mechanisms involved in the regulation of Na$^+$-K$^+$ pump activity by humoral factors have been reviewed recently. Briefly, regulation of Na$^+$-K$^+$ pump activity is mediated by second messengers via the phosphorylation of protein kinase A by cAMP, diacylglycerol-induced activation of protein kinase C, and Ca^{2+}-induced calmodulin kinase (Ewart and Klip, 1995). For example, adrenaline induced hyperpolarization of the rat diaphragm possibly via cAMP mechanism, which was abolished by ouabain or low temperature (Kuba and Nohmi, 1987). Stimulation of protein kinase C by dioctanoyl glycerol reduced Na$^+$-K$^+$ pump currents, while injection of cAMP to Xenopus oocytes activated Na$^+$-K$^+$ pump currents (Vasilets and Schwarz, 1992).

In rat skeletal muscle (soleus muscle), salbutamol, a selective β$_2$-agonist, and adrenaline hyperpolarized plasma membranes and stimulated ^{22}Na-efflux and ^{42}K-uptake, indicating that Na$^+$-K$^+$ pump activation is involved in hyperpolarization (Clausen and Flatman, 1980). Similarly, isoproterenol, forskolin and 8Br-cAMP induced ouabain-sensitive hyperpolarization in rat soleus muscle, indicating the involvement of the Na$^+$-K$^+$ pump. Activation of protein kinase C by phorbol ester also hyperpolarized membrane and the hyperpolarization was inhibited by ouabain (Li and Sperelakis, 1993). Isoproterenol and 8Br-cAMP activated Na$^+$-K$^+$ pump current in patch clamp study of rat soleus muscle myoball (Li and Sperelakis, 1994). In non-smooth muscle tissues, for example, in rat renal cortex, phosphorylation of the Na$^+$-K$^+$ pump by cAMP-dependent protein kinase A or protein kinase C inhibited Na$^+$-K$^+$ pump activity (Bertorello *et al.*, 1991). Therefore, intracellular signal transduction mechanisms regulate Na$^+$-K$^+$ pump in different ways in different tissues and animals.

Furthermore, other transmitters and second messengers are involved in Na$^+$-K$^+$ pump regulation. In the rat renal tubules, α-adrenergic agonists stimulated Na$^+$-K$^+$ pump (Ibarra *et al.*, 1993), while dopamine inhibited Na$^+$-K$^+$ pump (Ibarra *et al.*, 1993; Takemoto *et al.*, 1992). In choroids plexus, carbachol-induced nitric oxide release inhibited the Na^+-K^+ pump through the cGMP pathway (Ellis et al., 2000). In rabbit aorta, nitric oxide increased ouabain-sensitive ^{86}Rb uptake, but the response did not depend on cGMP increase (Gupta et al., 1994). In mouse gastric fundus smooth muscle, relaxation by nitric oxide was inhibited by ouabain, indicating that activation of the Na^+-K^+ pump was involved in the relaxation (Yaktubay et al., 1999). Inhibition of cAMP-induced hyperpolarization by cGMP was reported in rat soleus muscle, indicating that cAMP and cGMP regulate Na^+-K^+ pump activity bidirectionally (Li and Sperelakis, 1993). Lastly, in human uterus, slow hyperpolarization induced by prostaglandins E_2 and $F_{2\alpha}$ were inhibited by ouabain or K^+-free medium, indicating that hyperpolarization was mediated by Na^+-K^+ pump activation (Parkington et al., 1999).

β-adrenoceptor and Na+-K+ pump in smooth muscles

It is widely accepted that β-adrenoceptor agonists dilate smooth muscle through an increase in cAMP production (Bülbring and Tomita, 1987). However, it has also been reported that isoproterenol relaxes pulmonary venous smooth muscle through cAMP-independent, as well as cAMP-dependent pathways (Gao and Raj, 2002). Regulation of Na^+-K^+ pump activity by β-agonists has been examined in different types of smooth muscle and animal species (Table 1).

Responses to β-agonists in gastrointestinal smooth muscles

In isolated smooth muscle cells from toad stomach, isoproterenol induced relaxation and increases in [42]K influx and [24]Na efflux (Scheid et al., 1979). Dibutyryl cAMP, a membrane permeable cAMP analogue, increased [42]K influx similarly (Scheid and Fay, 1984). There was a discrepancy between the results obtained from isolated cell-level experiments and tissue-level experiments. In isolated cells of toad stomach contracted by carbamylcholine, Na^+-K^+ pump inhibition was simulated by a loss of Na^+ gradient by low extracellular Na^+ (14 mM) or high intracellular Na^+ (46 mM). In this condition, β-agonist-induced relaxation was inhibited. But when tissue strips of toad stomach were contracted under this Na^+-K^+ pump inhibition, isoproterenol relaxed tissues to the same extent as in the control condition, suggesting that small numbers of isolated cells did not represent all cells in the tissue (Scheid, 1987). In isolated cells of toad stomach which had been loaded with SBFI, a Na^+-sensitive fluorescent dye, isoproterenol decreased intracellular Na^+ concentration and the decrease was inhibited by K^+-free medium. As forskolin and 8Br-cAMP mimicked the effects, it was suggested that isoproterenol activated adenylate cyclase to stimulate the Na^+-K^+ pump (Moore and Fay, 1993). Isoproterenol-induced hyperpolarization in isolated toad stomach cells was associated with a decrease in the input resistance (which means an increase in ion channel opening). As input resistance was decreased by isoproterenol in the presence of ouabain, ouabain might have inhibited hyperpolarization because of a decrease in the driving force for K^+ efflux, rather than because of an inhibition of Na^+-K^+ pump. Inhibition

Table 1.
Effects of Na^+-K^+ pump inhibitions on β-agonist/cAMP-induced responses

Materials	β-agonist/cAMP	Response to β-agonist/ cAMP	Inhibitory treatments	Changes by treatments	Reference
Canine trachea	Salbutamol ($>3\ \mu M$)	Relaxation	Ouabain 15 min	Inhibited	Tamaoki et al., 1994
Rabbit trachea	ISO	Relaxation	K-free	Inhibited	Schram and Gustein, 1995
Toad stomach cell	ISO	Relaxation	Ouabain 30 s	Inhibited	Scheid et al., 1979
Toad stomach cell	ISO	K^+ influx increase	Ouabain 45 min	Inhibited	Scheid and Fay, 1984
Toad stomach cell	ISO	Na^+ efflux increase	K-free 5 min	Inhibited	Moore and Fay, 1993
Toad stomach cell	ISO	Hyperpolarization	Ouabain 10 min	No*	Yamaguchi et al., 1988
Guinea-pig stomach fundus	Theophylline	Relaxation	Ouabain/K-free	No	Huizinga and den Hertog, 1979
Guinea-pig proximal colon	ISO	Hyperpolarization	Ouabain	Inhibited	Watson et al., 1996
Rabbit colon	Caged cAMP	Relaxation	Ouabain/K-free 30 min	No*	Willenbucher et al., 1992
Guinea-pig teania coli	ISO	Hyperpolarization	Ouabain 10 min	No	Bülbring and den Hertog, 1980
Guinea-pig teania coli	ISO	Relaxation	Ouabain 10 min	Inhibited	Watanabe, 1976
Guinea-pig bladder	ISO	Hyperpolarization	Ouabain/K 0.5 mM	Inhibited	Nakahira et al., 2001
Guinea-pig lymphatic vessel	ISO	Hyperpolarization	Ouabain 10 min	No	von der Weide and Van Helden, 1996
Cat carotid artery	ISO	Relaxation	K-free	No	Bose and Innes, 1972
Canine saphenous vein, E(−)	ISO	Hyperpolarization	Ouabain 60 min	No	Nakashima and Vanhoutte, 1995
Porcine coronary artery	ISO	Relaxation/Ca_i decrease	Ouabain	No	Yamanaka et al., 2003
Rat aorta cell	8Br-cAMP	Na^+ efflux decrease	Ouabain 30 min	Inhibited	Borin, 1995
Rat/pig tail artery	ISO	Relaxation	Low K (1 mM) 15 min	Inhibited	Webb and Bohr, 1981

E(−): endothelium-removed. ISO: isoproterenol.
No*: not changed by treatment of indicated period but inhibited by longer period treatment.

of the Na^+-K^+ pump by ouabain may decrease E_K (equilibrium potential for K^+) and also decrease the chemical gradient for outward K^+ current. This explanation indicates that in toad stomach cells, not the Na^+-K^+ pump, but outward K^+ current is important for isoproterenol-induced hyperpolarization (Yamaguchi et al., 1988). These reports suggest that isoproterenol induces relaxation of toad stomach cells by activation of the Na^+-K^+ pump; however, activation of the Na^+-K^+ pump is not responsible for membrane hyperpolarization by isoproterenol.

In guinea-pig taenia coli, isoproterenol-induced relaxation was partly inhibited by ouabain or K^+-free medium (Watanabe, 1976). In circular smooth muscle of guinea-pig proximal colon, isoproterenol-induced hyperpolarization in the presence of tetraethylammonium was inhibited by ouabain, indicating the contribution of the Na^+-K^+ pump in isoproterenol-induced hyperpolarization (Watson et al., 1996).

Responses to β-agonists in arterial, vesical and bronchial smooth muscles

Pig and rat tail arteries, contracted by noradrenaline in low (1 mM) K^+-medium, 6 mM K^+-induced relaxation was enhanced by isoproterenol; and it was shown that ouabain inhibited isoproterenol and dibutyryl cAMP-induced relaxation in these arteries (Webb and Bohr, 1981). Therefore, it appears that isoproterenol or dibutyryl-cAMP induced relaxation through Na^+-K^+ pump activation in pig and rat tail arteries.

In guinea-pig detrusor smooth muscle, hyperpolarization by isoproterenol was inhibited by H89 or Rp-cAMPS, inhibitors of protein kinase A. The hyperpolarization was not inhibited by K^+ channel blockers (charybdotoxin, apamin, glibenclamide, 4-aminopyridine, or barium) but was inhibited by low (0.5 mM) K^+-medium or ouabain, indicating that β-agonist hyperpolarized bladder smooth muscle membrane through activation of the Na^+-K^+ pump (Nakahira et al., 2001).

In rabbit tracheal muscle, isoproterenol-induced relaxation was inhibited by K^+-free medium or ouabain. Pretreatment of tracheal tissues with isoproterenol in K^+-free medium enhanced the relaxation induced by Na^+-K^+ pump activation with K^+ (Schramm and Grunstein, 1995). In canine tracheal muscle, salbutamol, a $β_2$-agonist, induced relaxation at concentrations lower than 1 μM and the relaxation was inhibited by charybdotoxin, a Ca^{2+}-activated K^+ channel inhibitor. However, salbutamol-induced relaxation at concentrations greater than 3 μM was inhibited by ouabain (Tamaoki et al., 1994). These reports indicate that β-agonists stimulate the Na^+-K^+ pump in tracheal smooth muscle. Since $β_2$-agonists were found to increase cAMP, these reports may indicate that phosphorylation of the Na^+-K^+ pump by protein kinase A may stimulate Na^+-K^+ pump activity in tracheal smooth muscle. This is in contrast with the kidney, where phosphorylation of the Na^+-K^+ pump by protein kinase A inhibited Na^+-K^+ pump activity (Ewart and Klip, 1995).

Evidence against the contribution of Na^+-K^+ pump activity in the response to β-agonists

In contrast, some reports have shown that β-agonist-induced relaxation or hyperpolarization does not involve Na^+-K^+ pump activation. In guinea-pig stomach fundus, theophylline increased cAMP content and induced relaxation which was not inhibited by Na^+-K^+ pump inhibition (Huizinga and den Hertog, 1979). In guinea-pig taenia coli, isoproterenol-induced hyperpolarization was not inhibited by exposure to K^+-free medium or ouabain, indicating that the Na^+-K^+ pump was not involved in the hyperpolarization, but 30 min exposure to ouabain abolished the isoproterenol-induced hyperpolarization. It was discussed that the longer exposure decreased the K^+ gradient to minimize outward K^+ current (Bülbring and den Hertog, 1980). Circular muscle strips of rabbit colon were shown to be relaxed by the rapid release of cAMP from caged cAMP by UV exposure. The relaxation was not inhibited by ouabain or K^+-free medium. K^+ channel inhibitors, such as glibenclamide, charybdotoxin, and tetraethylammonium, also failed to inhibit the relaxation; thus, indicating that the Na^+-K^+ pump and K^+ channels did not play an important role in the relaxation (Willenbucher *et al.*, 1992).

In guinea pig mesentery lymphatic vessel, isoproterenol, forskolin, and 8Br-cAMP caused hyperpolarization of membranes and an increase in membrane conductance (von der Weid and van Helden, 1996). The hyperpolarization was not inhibited by K^+-free medium or ouabain, but was inhibited by K^+ channel blockers (tetraethylammonium, 4-aminopyridine, and charybdotoxin). Furthermore, isoproterenol-induced hyperpolarization was inhibited by BAPTA/AM, a membrane-permeable calcium chelator, but forskolin-induced hyperpolarization was not. Thus, the K^+ channel was important in the hyperpolarization in guinea-pig mesenteric lymphatic vessels, and there was a difference in the involvement of Ca^{2+}-activated K^+ channels in membrane hyperpolarization induced by isoproterenol and forskolin. In forskolin-induced hyperpolarization, other types of K^+ channels may play more important roles.

In cat carotid artery, isoproterenol-induced relaxation was not inhibited by ouabain or K^+-free medium (Bose and Innes, 1972). In endothelium-removed canine saphenous vein, isoproterenol-induced hyperpolarization was not inhibited by ouabain or by K^+-free medium, but was attenuated by glibenclamide, an ATP-sensitive K^+ channel blocker. Charybdotoxin, a Ca^{2+}-activated K^+ channel blocker, did not inhibit the hyperpolarization. Forskolin caused membrane hyperpolarization which was inhibited by glibenclamide. Thus, isoproterenol activated ATP-sensitive K^+ channels in canine saphenous vein through activation of adenylate cyclase and the Na^+-K^+ pump did not appear to play an important role (Nakashima and Vanhoutte, 1995). In porcine coronary artery, isoproterenol relaxed the tonic contractions induced by U46619, a thromboxane A_2 analogue. The relaxation was inhibited by 2′,4′-dichlorobenzamil, a Na^+/Ca^{2+} exchanger inhibitor, but not by ouabain. These results indicated a contribution of the Na^+/Ca^{2+} exchanger, but not of the Na^+-K^+ pump, in isoproterenol-induced relaxation (Yamanaka *et al.*, 2003).

In isolated smooth muscle cells from rat aorta, the effects of intracellular cAMP increase by isoproterenol or forskolin on intracellular Na^+ concentration were examined by measurement with fluorescent dye SBFI. Elevation of intracellular cAMP or 8Br-cAMP increased intracellular Na^+ concentration. The increase in intracellular Na^+ concentration was associated with decreases in ^{86}Rb uptake and Na^+ efflux rate (Borin, 1995). Similar results were obtained in aortic ring tissue, indicating that increased cAMP relaxes rat aortic smooth muscle and inhibits Na^+-K^+ pump activity (Borin, 1995).

In the rat stomach, as observed in a previous report (Shimamura *et al.*, 2000), forskolin caused membrane hyperpolarization in the presence of tetraethylammonium, glibenclamide and verapamil (Fig. 2A), indicating that increased cAMP can hyperpolarize membranes under conditions in which K^+ channels are inhibited by tetraethylammonium or glibenclamide. Therefore, it was expected that mechanisms other than those involving K^+ channels, such as the Na^+-K^+ pump, are involved in hyperpolarization. However, isoproterenol caused membrane hyperpolarization when the Na^+-K^+ pump was inhibited by K^+-free medium (Fig. 2B). Thus, the role of the Na^+-K^+ pump in isoproterenol-induced membrane hyperpolarization remains to be clarified. Furthermore, little information is available on the role of the Na^+-K^+ pump in β-agonist-induced relaxation or membrane hyperpolarization in human tissues. The contribution of Na^+-K^+ pump activity in β-agonist-induced responses has been evaluated by observing the effect of Na^+-K^+ pump inhibition on Na^+ and K^+ transport, relaxation, and hyperpolarization. However, as indicated in some prior studies (Bülbring and den Hertog, 1980; Yamaguchi *et al.*, 1988), Na^+-K^+ pump inhibition over longer period may change the transmembrane gradients of Na^+ and K^+ ions and, therefore, the equilibrium potentials of Na^+ and K^+.

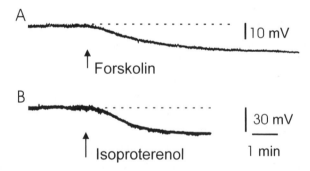

Figure 2. Effect of forskolin during K^+ channel inhibition (A) and effect of isoproterenol during Na^+-K^+ pump inhibition (B) on the membrane potential of smooth muscle cells of the circular muscle layer of rat stomach fundus. (A) Membrane potential recorded in the presence of 30 mM tetraethylammonium and 10 μM glibenclamide. The preparation showed increased electrical activity due to depolarization (resting membrane potential of -26 mV). To decrease electrical activity, voltage-dependent Ca^{2+} channels were inhibited by 10 μM verapamil. Forskolin (10 μM) was applied at the arrow. (B) Membrane potential recorded in K^+-free medium for 8 min. Isoproterenol (1 μM), applied at the arrow, caused membrane hyperpolarization with a magnitude of about 30 mV from the resting membrane potential of -47 mV.

Such changes can decrease K^+ currents and the experimental conditions inhibit hyperpolarization by outward K^+ currents as well as Na^+-K^+ pump. Furthermore, it has been shown that ouabain and K^+-free medium may affect Na^+-K^+ pump differently. The effects of ouabain and K^+-free medium differed in mouse pancreatic β cells (Meissner and Henquin, 1984). It is possible that in K^+-free medium, due to an increase in the transmembrane chemical gradient of K^+, increased K^+ currents can lead to underestimation of the contribution of the Na^+-K^+ pump.

Endogenous digitalis-like Na^+-K^+ pump inhibitors

Studies on endogenous digitalis-like factors have been reviewed elsewhere (Goto et al., 1992). Like plant cardiac glycosides, endogenous digitalis-like Na^+-K^+ pump inhibitors have been shown to regulate fluid and sodium balance in the animal body (Blaustein and Hamlyn, 1985; Haber and Haupert, 1987). Recently, an isomer of ouabain was isolated from bovine hypothalamus (Zhao et al., 1995). Injection of hypertonic saline to rats increased endogenous Na^+-K^+ pump inhibitor levels in plasma, pituitary gland and adrenal gland (Yamada et al., 1997). The amount of endogenous Na^+-K^+ pump inhibitors increased in the anteroventral third ventricle (AV3V) region of the hypothalamus when sympathoexcitatory and pressor responses to high sodium diet were observed in spontaneously hypertensive rats (SHR) and Dahl-salt sensitive rats (van Huysse and Leenen, 1998). Endogenous Na^+-K^+ pump inhibitors play roles in the development of hypertension. With regard to hypertension, the tissue and plasma levels of endogenous factors were reported to be increased in the Milan strain of hypertensive rats (Ferrandi et al., 1998). Endogenous Na^+-K^+ pump inhibitors have been shown to be released from adrenal cortex and hypothalamus and are responsible for blood pressure increases (Hamlyn et al., 1998). Hypoxia has been shown to trigger release of Na^+-K^+ pump inhibitors from midbrain and adrenal gland in rats (De Angelis and Haupert, 1998).

Na^+-K^+ pump and hypertension

Abnormal contractile force development has been reported in skeletal muscle of SHR, although its relation with Na^+-K^+ pump was not clear. Soleus muscle in SHR showed higher intracellular Na^+ and lower intracellular K^+ concentrations than those of normotensive control rats. $^{86}Rb^+$ uptake of the muscle, an indicator of K^+ uptake, was smaller in SHR than in normotensive rats. However, similar affinities to ouabain and a greater number of binding sites were observed in SHR (Carlsen et al., 1996). After incubation in K^+-free medium for 17 min, amplitude of K^+-induced relaxation was compared in rat tail artery. Tail artery from SHR showed less relaxation than that from normotensive Wistar Kyoto rats (WKY) (Webb and Bohr, 1979). The Na^+-K^+ pump number of aorta and caudal artery was examined by [^{3}H]-ouabain binding and it was found that Na^+-K^+ pump levels were not different between SHR and WKY (Wong et al., 1984). Pigs treated with deoxycorticosterone acetate (DOCA) for 55 days developed hypertension and had arterial pressures 37% higher than those of control pigs. Tail artery of the pig was contracted by noradrenaline in K^+-free medium, and it was found that the amplitude of K^+-induced relaxation of the artery from DOCA hypertensive pigs was greater than that from control pigs, indicating that Na^+-K^+ pump activity was increased in hypertensive pig (Webb, 1982). When the left femoral artery in DOCA hypertensive pigs was protected from high blood pressure by proximal ligature of the left iliac artery, the amplitude of relaxation was still increased when compared with that of control animals, indicating that the change in relaxation was not secondary change to an increase in intraluminal pressure. Hypothalamic Na^+-K^+ pump inhibitor has been shown to constrict the pulmonary arteries of rats. Arteries from SHR showed higher sensitivity to the hypothalamic Na^+-K^+ pump inhibitor than arteries from normotensive rats (Janssens *et al.*, 1993). The Milan strain of hypertensive rat is a genetic hypertensive model with primary renal dysfunction. An increase in Na^+-K^+ transport rate in the renal tubules and a 5- to 10-fold increase in the amount of hypothalamic Na^+-K^+-ATPase inhibitor have been reported in rats. Na^+-K^+ pump inhibitor seems to play a more important role in the development of hypertension in Milan rats (Ferrandi *et al.*, 1998); however, little is known about its role in smooth muscle.

CONCLUSION

The Na^+-K^+ pump contributes to β-adrenoceptor-mediated relaxation and hyperpolarization only in some smooth muscle tissues. Results in Na^+-K^+ pump inhibition study must be related to Na^+-K^+ pump with careful consideration of K^+ gradient. Influence of endogenous Na^+-K^+ pump inhibitors on β-adrenoceptor-mediated responses *in vivo* needs to be examined in future study. In hypertensive patients, Na^+-K^+ pump activity differs from that of normal subjects. The role of the Na^+-K^+ pump in β-agonist-induced relaxation in human smooth muscle tissues remains to be clarified in future studies.

Acknowledgement

This work was supported in part by the Academic Science Frontier Project of the Ministry of Education, Science, Sports and Culture of Japan.

REFERENCES

Aalkjaer, C. and Mulvany, M. J. (1985). Effect of ouabain on tone, membrane potential and sodium efflux compared with [^{3}H]ouabain binding in rat resistance vessels, *J. Physiol.* **362**, 215–231.

Arnon, A., Hamlyn, J. and Blaustein, M. P. (2000). Ouabain augments Ca^{2+} transients in arterial smooth muscle without raising cytosolic Na^+, *Am. J. Physiol.* **279**, H679–H691.

Baker, P. F., Blaustein, M. P., Hodgkin, A. L., *et al.* (1969). The influence of calcium ions on sodium efflux in squid axons, *J. Physiol.* **200**, 431–458.

Belardinelli, L., Harder, D., Sperelakis, N., et al. (1979). Cardiac glycoside stimulation of inward Ca^{++} current in vascular smooth muscle of canine coronary artery, J. Pharmacol. Exp. Ther. **209**, 62–66.

Bertorello, A. M., Aperia, A., Walaas, S. I., et al. (1991). Phosphorylation of the catalytic subunit of Na^+, K^+-ATPase inhibits the activity of the enzyme, Proc. Natl. Acad. Sci. USA **88**, 11359–11362.

Blanco, G. and Mercer, R. W. (1998). Isozymes of the Na-K-ATPase: heterogeneity in structure, diversity in function, Am. J. Physiol. **275**, F633–F650.

Blaustein, M. P. and Hamlyn, J. M. (1985). Sodium transport inhibition, cell calcium and hypertension. The natriuretic hormone/Na^+-Ca^{2+} exchange hypertension hypothesis, Am. J. Med. **77**, 45–59.

Bolton, T. B. (1973). Effects of electrogenic sodium pumping on the membrane potential of longitudinal smooth muscle from terminal ileum of guinea pig, J. Physiol. **228**, 693–712.

Bonaccorsi, A., Hermsmeyer, K., Aprigliano, O., et al. (1977a). Mechanism of potassium relaxation of arterial muscle, Blood Vessels **14**, 261–276.

Bonaccorsi, A., Hermsmeyer, K., Smith, C. B., et al. (1977b). Norepinephrine release in isolated arteries induced by K-free solution, Am. J. Physiol. **232**, H140–H145.

Borin, M. L. (1995). cAMP evokes a rise in intracellular Na^+ mediated by Na^+ pump inhibition in rat aortic smooth muscle cells, Am. J. Physiol. **269**, C884–C891.

Bose, D. and Innes, I. (1972). Isoprenaline-induced relaxation of smooth muscle not due to electrogenic sodium pumping, Can. J. Physiol. Pharmacol. **50**, 378–380.

Bülbring, E. and den Hertog, A. (1980). The action of isoprenaline on the smooth muscle of the guinea-pig taenia coli, J. Physiol. **304**, 277–296.

Bülbring, E. and Tomita, T. (1987). Catecholamine action on smooth muscle, Pharmacol. Rev. **39**, 49–96.

Carlsen, R. C., Gray, S. D. and Pickar, J. G. (1996). Na^+, K^+-pump activity and skeletal muscle contractile deficits in the spontaneously hypertensive rat, Acta Physiol. Scand. **156**, 237–245.

Casteels, R., Droogmans, G. and Hendrickx, H. (1971). Electrogenic sodium pump in smooth muscle cells of the guinea-pig's taenia coli, J. Physiol. **217**, 297–313.

Casteels, R., Kitamura, K., Kuriyama, H., et al. (1977). The membrane properties of the smooth muscle cells of the rabbit main pulmonary artery, J. Physiol. **271**, 41–61.

Clausen, T. (2003). Na^+-K^+ pump regulation and skeletal muscle contractility, Physiol. Rev. **83**, 1269–1324.

Clausen, T. and Flatman, J. A. (1980). β_2-adrenoceptors mediate the stimulating effect of adrenaline on active electrogenic Na-K-transport in rat soleus muscle, Br. J. Pharmacol. **68**, 749–755.

Connor, J. A. (1984). Physiological role of electrogenic pump in smooth muscle, in: Electrogenic Transport: Fundamental Principles and Physiological Implications, Blaustein, M. P. and Lieberman, M. (Eds), pp. 271–285. Raven Press, New York.

De Angelis, C. and Haupert, G. T., Jr. (1998). Hypoxia triggers release of an endogenous inhibitor of Na^+-K^+-ATPase from midbrain and adrenal, Am. J. Physiol. **274**, F182–F188.

De Weer, P., Gadsby, D. C. and Rakowski, R. F. (1988). Voltage dependence of the Na-K pump, Ann. Rev. Physiol. **50**, 225–241.

Ellis, D. Z., Nathanson, J. A. and Sweadner, K. J. (2000). Carbachol inhibits Na-K-ATPase activity in choroids plexus via stimulation of the NO cGMP pathway, Am. J. Physiol. **279**, C1685–C1693.

Ewart, H. S. and Klip, A. (1995). Hormonal regulation of the Na^+-K^+-ATPase: mechanisms underlying rapid and sustained changes in pump activity, Am. J. Physiol. **269**, C295–C311.

Ferrandi, M., Marunta, P., Rivera, R., et al. (1998). Role of the ouabain-like factor and Na-K pump in rat and human genetic hypertension, Clin. Exp. Hypertension **20**, 629–639.

Fleming, W. W. (1980). The electrogenic Na^+-K^+-pump in smooth muscle: physiologic and pharmacologic significance, Ann. Rev. Pharmacol. Toxicol. **20**, 129–149.

Gadsby, D. C. and Nakao, M. (1989). Steady-state current-voltage relationship of the Na/K pump in guinea pig ventricular myocytes, J. Gen. Physiol. **94**, 511–537.

Gao, Y. and Raj, J. U. (2002). Effects of SQ 22536, an adenylyl cyclase inhibitor, on isoproterenol-induced cyclic AMP elevation and relaxation in newborn ovine pulmonary veins, *Eur. J. Pharmacol.* **436**, 227–233.

Goto, A., Yamada, K., Yagi, N., *et al.* (1992). Physiology and pharmacology of endogenous digitalis-like factors, *Pharmacol. Rev.* **44**, 377–399.

Gupta, S., McArthur, C., Crady, C., *et al.* (1994). Stimulation of vascular Na$^+$-K$^+$-ATPase activity by nitric oxide: a cGMP-independent effect, *Am. J. Physiol.* **266**, H2146–H2151.

Haber, E. and Haupert, G. T., Jr. (1987). The search for a hypothalamic Na$^+$-K$^+$-ATPase inhibitor, *Hypertension* **9**, 315–324.

Hamlyn, J. M., Lu, Z.-R., Manunta, P., *et al.* (1998). Observations on the nature, biosynthesis, secretion and significance of endogenous ouabain, *Clin. Exp. Hypertens.* **20**, 523–533.

Hendrickx, H. and Casteels, R. (1974). Electrogenic sodium pump in arterial smooth muscle cells, *Pflügers Arch.* **346**, 299–306.

Huizinga, J. D. and den Hertog, A. (1979). Inhibition of fundic strips from guinea-pig stomach: the effect of theophylline on the membrane potential, muscle contraction and ion fluxes, *Eur. J. Pharmacol.* **57**, 1–11.

Ibarra, F., Aperia, A., Svensson, L.-B., *et al.* (1993). Bidirectional regulation of Na, K-ATPase activity by dopamine and an α-adrenergic agonist, *Proc. Natl. Acad. Sci.* **90**, 21–24.

Janssens, S. P., Kachoris, C., Parker, W. L., *et al.* (1993). Hypothalamic Na$^+$-K$^+$-ATPase inhibitor constricts pulmonary arteries of spontaneously hypertensive rats, *J. Cardiovasc. Pharmacol.* **22** (Suppl. 2), S42–S46.

Kao, C. Y. and Nishimura, A. (1969). Ion concentrations and membrane potentials of myometrium during recovery from cold, *Am. J. Physiol.* **217**, 525–531.

Karaki, H., Ozaki, H. and Urakawa, N. (1978). Effects of ouabain and potassium-free solution on the contraction of isolated blood vessels, *Eur. J. Pharmacol.* **48**, 439–443.

Kuba, K. and Nohmi, M. (1987). Role of ion conductance changes and of the sodium pump in adrenaline-induced hyperpolarization of rat diaphragm muscle fibres, *Br. J. Pharmacol.* **91**, 671–681.

Kuriyama, H., Ohshima, K. and Sakamoto, Y. (1971). The membrane properties of the smooth muscle of the guinea pig portal vein in isotonic and hypertonic solutions, *J. Physiol.* **217**, 179–199.

Li, K.-X. and Sperelakis, N. (1993). Isoproterenol- and insulin-induced hyperpolarization in rat skeletal muscle, *J. Cell Physiol.* **157**, 631–636.

Li, K.-X. and Sperelakis, N. (1994). Electrogenic Na-K pump current in rat skeletal myoballs, *J. Cell Physiol.* **159**, 181–186.

Matthews, E. K. and Sutter, M. C. (1966). Ouabain-induced changes in the contractile and electrical activity, potassium content, and response to drugs, of smooth muscle cells, *Can. J. Physiol. Pharmacol.* **45**, 509–520.

Meissner, H. P. and Henquin, J. C. (1984). The sodium pump of mouse pancreatic β-cells: electrogenic properties and activation by intracellular sodium, in: *Electrogenic Transport: Fundamental Principles and Physiological Implications*, Blaustein, M. P. and Lieberman, M. (Eds), pp. 271–285. Raven Press, New York, USA.

Miura, Y. and Kimura, J. (1989). Sodium-calcium exchange current, *J. Gen. Physiol.* **93**, 1129–1145.

Moore, E. D. W. and Fay, F. S. (1993). Isoproterenol stimulates rapid extrusion of sodium from isolated smooth muscle cells, *Proc. Natl. Acad. Sci. USA* **90**, 8058–8062.

Mulvany, M. J. (1985). Changes in sodium pump activity and vascular contraction, *J. Hypertension* **3**, 429–436.

Nakahira, Y., Hashitani, H., Fukuta, H., *et al.* (2001). Effects of isoproterenol on spontaneous excitations in detrusor smooth muscle cells of the guinea pig, *J. Urol.* **166**, 335–340.

Nakamura, Y., Ohya, Y., Abe, I., *et al.* (1999). Sodium-potassium pump current in smooth muscle cells from mesenteric resistance arteries of the guinea-pig, *J. Physiol.* **519**, 212–232.

Nakashima, M. and Vanhoutte, P. M. (1995). Isoproterenol causes hyperpolarization through opening of ATP-sensitive potassium channels in vascular smooth muscle of the canine saphenous vein, *J. Pharmacol. Exp. Ther.* **272**, 379–384.

Nakazato, Y., Ohga, A. and Onoda, Y. (1978). The effect of ouabain on noradrenaline output from peripheral adrenergic neurons of isolated guinea-pig vas deferens, *J. Physiol.* **278**, 45–54.

Oike, M., Droogmans, G. and Nilius, B. (1993). Electrogenic Na$^+$/K$^+$-transport in human endothelial cells, *Pflügers Arch.* **424**, 301–307.

Page, E., Goerke, R. J. and Storm, S. R. (1964). Cat heart muscle *in vitro*, IV. Inhibition of transport in quiescent muscles, *J. Gen. Physiol.* **47**, 531–543.

Parkington, H. C., Tonta, M. A., Davies, N. K., *et al.* (1999). Hyperpolarization and slowing of the rate of contraction in human uterus in pregnancy by prostaglandins E$_2$ and F$_{2\alpha}$: involvement of the Na$^+$ pump, *J. Physiol.* **514**, 229–243.

Quinn, K., Guibert, C. and Beech, D. J. (2000). Sodium-potassium-ATPase electrogenicity in cerebral precapillary arterioles, *Am. J. Physiol.* **279**, H351–H360.

Scheid, C. R. (1987). β-adrenergic relaxation of smooth muscle: differences between cells and tissues, *Am. J. Physiol.* **253**, C369–C374.

Scheid, C. R. and Fay, F. S. (1984). β-adrenergic stimulation of ^{42}K influx in isolated smooth mucle cells, *Am. J. Physiol.* **246**, C415–C421.

Scheid, C. R., Honeyman, T. W. and Fay, F. S. (1979). Mechanism of β-adrenergic relaxation of smooth muscle, *Nature* **277**, 32–36.

Schramm, C. M. and Grunstein, M. M. (1995). Mechanism of protein kinase C potentiation of airway β-adrenergic relaxation, *Life Sci.* **57**, 1163–1173.

Shimamura, K., Yamamoto, K., Sekiguchi, F., *et al.* (2000). Altered β-adrenoceptor-mediated responses in the gastric smooth muscle of hypertensive rats, *J. Smooth Muscle Res.* **36**, 1–12.

Takemoto, F., Cohen, H. T., Satoh, T., *et al.* (1992). Dopamine inhibits Na/K-ATPase in single tubules and cultured cells from distal nephron, *Pflügers Arch.* **421**, 302–306.

Tamaoki, J., Tagaya, E., Chiyotani, A., *et al.* (1994). Role of K$^+$ channel opening and Na$^+$-K$^+$ ATPase activity in airway relaxation induced by salbutamol, *Life Sci.* **55**, PL217–PL223.

Taylor, G. S., Paton, D. M. and Daniel, E. E. (1969). Evidence for an electrogenic sodium pump in smooth muscle, *Life Sci.* **8**, 769–773.

Tomita, T. and Yamamoto, T. (1971). Effects of removing external potassium on the smooth muscle of guinea-pig taenia coli, *J. Physiol.* **212**, 851–868.

Tsuru, H. and Shigei, T. (1976). Participation of catecholamine in the digitoxigenin-induced contraction of isolated dog veins, *Jpn. J. Pharmacol.* **26**, 120–123.

van Huysse, J. W. and Leenen, F. H. H. (1998). Role of endogenous brain 'ouabain' in the sympathoexcitatory and pressor effects of sodium, *Clin. Exp. Hypertens.* **20**, 657–667.

Vanhoutte, P. M. and Lorenz, R. R. (1984). Na$^+$, K$^+$-ATPase inhibitors and the adrenergic neuroeffector inteaction in the blood vessel wall, *J. Cardiovasc. Pharmacol.* **6**, S88–S94.

Vasilets, L. A. and Schwarz, W. (1992). Regulation of endogenous and expressed Na$^+$/K$^+$ pumps in Xenopus oocytes by membrane potential and stimulation of protein kinases, *J. Membrane Biol.* **125**, 119–132.

von der Weid, P.-Y. and van Helden, D. F. (1996). β-adrenoceptor-mediated hyperpolarization in lymphatic smooth muscle of guinea pig mesentery, *Am. J. Physiol.* **270**, H1687–H1695.

Watanabe, H. (1976). Inhibitory mechanisms of isoprenaline in the guinea-pig taenia coli, *Jpn. J. Pharmacol.* **26**, 217–226.

Watson, M. J., Bywater, R. A., Taylor, G. S., *et al.* (1996). Effects of nitric oxide (NO) and NO donors on the membrane conductance of circular smooth muscle cells of the guinea-pig proximal colon, *Br. J. Pharmacol.* **118**, 1605–1614.

Webb, R. C. (1982). Potassium relaxation of vascular smooth muscle from DOCA hypertensive pigs, *Hypertension* **4**, 609–619.

Webb, R. C. and Bohr, D. F. (1979). Potassium relaxation of vascular smooth muscle from spontaneously hypertensive rats, *Blood Vessels* **16**, 71–79.

Webb, R. C. and Bohr, D. F. (1981). Relaxation of vascular smooth muscle by isoproterenol, dibutyryl-cyclic AMP and theophylline, *J. Pharmacol. Exp. Ther.* **217**, 26–35.

Willenbucher, R. F., Xie, Y. N., Eysselein, V. E., et al. (1992). Mechanisms of cAMP-mediated relaxation of distal circular muscle in rabbit colon, *Am. J. Physiol.* **262**, G159–G164.

Wong, S. K., Westfall, D. P., Menear, D., et al. (1984). Sodium-potassium pump sites, as assessed by [^{3}H]-ouabain binding, in aorta and caudal artery of normotensive and spontaneously hypertensive rats, *Blood Vessels* **21**, 211–222.

Yaktubay, N., Ogulener, N., Onder, S., et al. (1999). Possible stimulation of Na^+-K^+-ATPase by NO produced from sodium nitrite by ultraviolet light in mouse gastric fundal strip, *Gen. Pharmacol.* **32**, 159–162.

Yamada, K., Goto, A., Nagoshi, H., et al. (1997). Elevation of ouabain like compound level with hypertonic sodium chloride load in rat plasma and tissues, *Hypertension* **30**, 94–98.

Yamaguchi, H., Honeyman, T. W. and Fay, F. S. (1988). β-adrenergic actions on membrane electrical properties of dissociated smooth muscle cells, *Am. J. Physiol.* **254**, C423–C431.

Yamanaka, J., Nishimura, J., Hirano, K., et al. (2003). An important role for the Na^+-Ca^{2+} exchanger in the decrease in cytosolic Ca^{2+} concentration induced by isoprenaline in the porcine coronary artery, *J. Physiol.* **549**, 553–562.

Zhao, N., Lo, L., Berova, C. N., et al. (1995). Na^+/K^+-ATPase inhibitors from bovine hypothalamus and human plasma are different from ouabain: nanogram scale CD structural analysis, *Biochem.* **34**, 9893–9896.

Advances in Neuroregulation and Neuroprotection (2005), pp. 491-502
Collin, C. *et al.* (Eds)
© VSP 2005

Cardiac tissue remodeling and renin-angiotensin system in hypertrophic heart

HIDEAKI KAWAGUCHI *

Department of Pathophysiological Science and Laboratory Medicine, Hokkaido University Graduate School of Medicine, Kita-5, Nishi-7, Kita-Ku, Sapporo 060-8638, Japan

Abstract—In the hearts of spontaneously hypertensive rats, mRNA of the renin-angiotensin system (renin, angiotensinogen, and angiotensin converting enzyme) was expressed. Level of angiotensinogen and renin mRNA expressed in ventricles, and angiotensinogen mRNA expressed in fibroblasts from SHR were higher than those from WKY. ACE mRNA was also more strongly expressed in the ventricles and fibroblasts from SHR compared with those of WKY.

We investigated the effects of angiotensin II on cardiac collagen synthesis in cardiac fibroblasts of 10-week-old spontaneously hypertensive rats and age-matched Wistar-Kyoto rats. Basal collagen synthesis in cardiac fibroblasts from spontaneously hypertensive rats was 1.6-fold greater than that in the cell of Wistar-Kyoto rats. Angiotensin II stimulated collagen synthesis in cardiac fibroblasts in a dose-dependent manner. The responsiveness of collagen production to angiotensin II was significantly enhanced in cardiac fibroblasts from spontaneously hypertensive rats (100 nM angiotensin II resulted in $185 \pm 18\%$ increase above basal levels, 185 ± 18 *versus* $128 \pm 19\%$ in Wistar-Kyoto rats, $p < 0.01$). This effect was receptor-specific, because the competitive inhibitor saralasin and MK 954 blocked it. These results indicate that collagen production was enhanced in cardiac fibroblasts from spontaneously hypertensive rats, that angiotensin II had a stimulatory effect on collagen synthesis in cardiac fibroblasts, and that cardiac fibroblasts from spontaneously hypertensive rats were hyper-responsive to stimulation by angiotensin II.

These findings suggest that the cardiac renin-angiotensin system may play an important role in collagen accumulation in hypertensive cardiac hypertrophy.

Keywords: Angiotensin II; collagen; cardiac fibroblasts; renin-angiotensin system; spontaneously hypertensive rats.

INTRODUCTION

The primary step in the function of many different hormones and neurotransmitters involves receptor-mediated stimulation of the breakdown of inositol phos-

---

*E-mail: hideaki@med.hokudai.ac.jp.

pholipids in plasma membrane (Baker and Singer, 1988; Brown et al., 1985). This phosphatidylinositol (PI)-turnover pathway generates two second messengers, inositol 1,4,5-trisphosphate (IP₃) and sn 1,2-diacylglycerol (DAG) (Berridge, 1984; Williamson et al., 1985). DAG stimulates membrane-bound phospholipid-dependent, Ca^{2+}-dependent protein kinase C (Nisizuka, 1983), whereas IP3 releases Ca^{2+} from endoplasmic reticulum stores (Ehrlich and Watras, 1988; Putney, 1987). This PI-turnover pathway may have an important role in inducing cardiac myocyte hypertrophy, even in adult rat myocytes. Recently, we showed that the PI turnover pathway in heart cells of cardiomyopathic hamsters was more active than in control cells (Kawaguchi et al., 1991a). We also reported that basal phosphatidylinositol and polyphosphoinositide metabolism (Shoki et al., 1992) and IP3 kinase activity (Kawaguchi et al., 1990) increased in spontaneously hypertensive rat heart without any stimulation. Recently, it has been reported (Sadoshima and Izumo, 1993a; 1993b), that angiotensin II (Ang II) stimulates PI-response in cardiac fibroblast and neonatal cardiac myocytes. Ang II induces the expression of immediate-early genes (c-fos, c-jun, jun B, Erg-1, and c-myc) in these cells.

Hypertensive cardiac hypertrophy is associated with the accumulation of collagen in the myocardial interstitium. Previous studies have demonstrated that this myocardial fibrosis accounts for impaired myocardial stiffness and ventricular dysfunction. Although cardiac fibroblasts are responsible for the synthesis of fibrillar collagen, factors that regulate collagen synthesis in cardiac fibroblasts are not fully understood. Several studies have shown the effects of antihypertensive drugs on cardiac collagen metabolism. In the spontaneously hypertensive rat (SHR), the concentration of cardiac collagen increases in the presence of α-methyldopa, remains unchanged with β-adrenoceptor blockers, and decreases with angiotensin-converting enzyme inhibitors during the regression of cardiac hypertrophy (Brilla et al., 1991; Motz et al., 1982; Sen and Bumpus, 1979). These data suggest that the renin-angiotensin system (RAS) may be involved in cardiac collagen remodeling in the regression of cardiac hypertrophy. We hypothesized that Ang II might regulate the proliferation and collagen production of cardiac fibroblasts. To investigate the effects of Ang II on the metabolism of cardiac collagen in the hypertrophic heart, we examined the collagen synthesis in cardiac fibroblasts from SHRs and age-matched Wistar-Kyoto (WKY) rats. We also attempted to determine the alteration of RAS in hypertrophic rat hearts.

MATERIALS AND METHODS

Materials

Ang II was donated by Ciba-Geigy Corporation, Edison, New Jersey. MK 954 (previous name, DuP 753) was supplied by Du Pont Merck Pharmaceutical, Wilmington, Delaware. [Sar1,Ala8]Angiotensin II (Saralasin) and collagenase type 1A (C-9891) for isolation of cardiac fibroblasts were purchased from Sigma

Chemical St. Louis, Missouri. Antibodies to smooth muscle a-actin, factor VIII and vimentin were from Boehringer Mannheim Corporation, Indianapolis, Indiana. Purified bacterial collagenase (collagenase form III) for collagen assay was from Advance Biofactures, Tokyo. L-[2,3-^{3}H] proline (1.11TBq/m mol) and [methyl-^{3}H] thymidine (925 GBq/m mol) were obtained from Amersham International, Buckinghamshire, UK Ready Value was obtained from Beckman, Tokyo.

Cell culture

The preparation of isolated cardiac fibroblasts was performed according to a modified procedure of Eghbali *et al.* (Eghbali *et al.*, 1988). Briefly, hearts were obtained from 10-week-old male SHRs and age- and sex-matched WKY rats. To prepare cardiac nonmyocytes, after removal of the atria and the right ventricles, the left ventricles were washed three times with ice-cold phosphate-buffered saline (PBS), minced with scissors and incubated with PBS containing 0.1% collagenase five times for 10 min at 37°C. The supernatant from each consecutive incubation was centrifuged at 37g for 5 min. The supernants were centrifuged at 200g for 5 min. Yields were 3 to 10 × 106 cells/heart. The cells were maintained in Dulbecco's modified Eagle's medium (DME) supplemented with 10% fetal bovine serum (FBS), penicillin (100 units/ml) and streptomycin (100 mg/ml) in 95% air and 5% CO_2 at 37°C (Kawaguchi and Yasuda, 1984). The purity of the cultured cell population was determined by immuno-histochemical staining of the cell layer with anti-factor VIII antibody for endothelial cells and smooth muscle anti-a-actin for vascular smooth muscle cells. On confluence, less than 1% of cultured cells stained positive with anti-factor VIII or anti-a-actin, but did stain positive with anti-vimentin antibody. The cultured cells appeared spindle-shaped with well-defined nuclei. For all studies, we used cells from passages 2–3 to avoid the possibility of differentiation of the cultured cells during long-term passages.

Collagen synthesis

In preparation for experiments, the cells were plated at a density of 2 × 10^5 in 35-mm Falcon plastic dishes. They were grown to confluence in DME containing 10% FBS and were made quiescent by washing twice with DME and by placing them in serum- free DME for 24 hours. The medium was aspirated and replaced with fresh DME containing 0.1 mM ascorbic acid and L-[2,3-3H] proline 74 kBq/ml plus Ang II for 24 h. The proteins in the medium and the cell layer were dialyzed against 0.05 M acetic acid, containing 25 mM EDTA, 10 mM N-ethylmaleimide, and 1 mM phenylmethylsulfonyl fluoride, and were then lyophilized. Collagen synthesis in the medium and the cell layer was determined by bacterial collagenase digestion (Advance Biofactures) (Peterkofsky and Diegelmann, 1971). Radioactivity in collagenase-sensitive protein was regarded as collagen synthesis, and radioactivity in collagenase-insensitive protein was regarded as noncollagen synthesis (Peterkofsky and Diegelmann, 1971). Relative rate of collagen synthesis compared with total protein synthesis (%) was calculated by the following formula: dpm in collagen $\times$ 100/[dpm in collagen + (dpm in noncollagen protein $\times$ 5.4)]. This formula is based on the fact that collagen has approximately 22.2% amino acids (proline and hydroxyproline) compared with an average of 4.1% in noncollagen proteins (Peterkofsky and Diegelmann, 1971).

DNA synthesis

The same procedure used to determine collagen synthesis was used to analyze DNA synthesis, except for the substitution of [methyl-^{3}H] thymidine (37 kBq/ml). The resulting medium was aspirated, and the cells were subsequently washed three times with ice-cold PBS. A 10% solution of trichloroacetic acid was added to dishes, and cells were kept on ice for 30 min. The trichloroacetic acid-insoluble material was collected on Whatman GF/C filters, and the radioactivity of the filters in Ready Value was determined by a liquid scintillation counter (Kawaguchi et al., 1991b).

Extraction of total RNA

Tissues from different hearts were homogenized individually, using a modified single-step method of RNA isolation by acid guanidinium thiocyanate-phenol-chloroform extraction with the RNA Zol (Cinna/Biotex Laboratories International, Inc., Friendswood, Texas) (Chomczynski and Sacchi, 1987). We added 0.2 ml chloroform per 2 ml homogenate and, after shaking, the specimen was stored on ice for 15 min. The suspension was centrifuged at 12 000g for 15 min, at which time an equal volume of isopropanol was added to the upper aqueous phase, and the specimen was stored for 45 min at $-20°C$. The sample was centrifuged for 15 min at 12 000g, and the precipitated RNA formed a white pellet. The pellet was washed twice with a solution of 75% ethanol/TEN (0.1 M NaCl, 10 mM Tris-HCl, 1 mM EDTA). The precipitated RNA was dissolved in TE (10 mM Tris-HCl, 1 mM EDTA, pH 8.0) and quantified using spectrophotometry at 260 nm.

Angiotensinogen, renin, and angiotensin converting enzyme DNA probes

Rat angiotensinogen cDNA cloned into pRag16 was excised with Bam HI (Kageyama et al., 1984). The 712-base-pair fragment was inserted into the paired-promoter pSPT19 (Amersham) by directional cloning. The DNA template was linearized with EcoRI restriction endonuclease, which cut at a single site, 712-base-pairs downstream from the SP6 promoter and 712-base-pairs upstream from the T7 promoter. Rat renin cDNA cloned into pUC19 was excised with Bam HI and Hin III (Imai et al., 1983). The 1433 base-pair fragment was inserted into the pSPT 19 promoter described above. The DNA template was linearized with BstEII restriction endonuclease, 880-base-pair downstream from the T7 promoter. Angiotensin converting enzyme (ACE) cDNA is a 191 base-pair fragment obtained by reverse transcription of 4 mg of total RNA from the ventricle of an adult rat, followed by polymerase chain reaction (PCR) amplification (Mullis *et al.*, 1986). The sense primer 1064-1083 (5'-CTGGAGAAGCCGGCCGACGG-3') from exon 7 and anti-sense primer 1236-1255 (5'-CCCCCGACGCAGGGAACGG-3') from exon 8 were used) (Hubert *et al.*, 1991). This PCR product was checked by southern blot analy-sis. The ACE cDNA fragment was inserted into pSPT19. The DNA template was linearized with EcoRI or Bam HI.

The cRNAs were labeled with a-[^{32}P] UTP. The antisense RNA probe was prepared using T7 RNA polymerase, and the sense RNA probe was prepared using SP6 RNA polymerase, *in vitro*. Probes were purified on a Quick Spin Column (Boehringer Mannheim Biochemicals). The specific activity of the probe used ranged from 0.8 to 1×109 cpm/mg.

Ribonuclease protection assay

The ^{32}P-labeled sense and antisense RNAs were mixed with total RNAs from human tissues and precipitated with ethanol. The pellets were resuspended in hybridization buffer (Ambion, Inc) and heated at 80°C for 3 min, then hybridized at 43°C overnight. All the RNAs except the control samples were treated with RNase A and T1 before treatment with SDS and protein kinase K (Sawa *et al.*, 1992). Samples were extracted with phenol/chloroform, precipitated with ethanol, resuspended in a loading buffer containing 80% formamide, heated 3 min at 80°C, and electrophoresed on 5% polyacrylamide gel containing 8 M urea. The gel was transferred to chromatography paper, covered with plastic wrap, and exposed to X-ray film (Kodak X-O mart) at −80° for 24 h. The relative amounts of a specific mRNA were quantified by laser densitometry of corresponding autoradiograms in the linear response range of the X-ray film. The hybridization signals of specific mRNA were normalized to those of glyceraldehyde-3-phosphate dehydrogenase mRNA to correct for differences in loading and/or transfer.

Statistical analysis

Six separate experiments in triplicate were analyzed in all the studies, and results were expressed as the mean ± S.E.M. Statistical significance was determined using the method of ANOVA with $p < 0.05$ as the limit of significance.

RESULTS

Blood pressure and heart weight

The ratio of left ventricular weight to body weight of SHR was larger as compared with WKY rats (Table 1). Systolic blood pressure also increased in SHR (Table 1).

mRNA expression of RAS

RNA was prepared from ventricles of both SHR and WKY heart. It was also extracted from cardiac fibroblasts from SHR and WKY. Level of angiotensinogen

Table 1.
Systolic blood pressure (SBP) and left ventricular weight/body weight (LV/BW)

	SHR ($n = 15$)	WKY ($n = 15$)
LV/BW (mg/100 g)	$300 \pm 12^*$	168 ± 5
SBP (mmHg)	$175 \pm 15^*$	120 ± 10

* $p < 0/001$, compared with age-matched WKY.

Table 2.
Distribution of collagen produced by cultured cardiac fibroblasts

	Medium	Cell
Control	$86.2 \pm 5.6\%$	$13.8 \pm 5.6\%$
Ang II (1 μM)	$84.5 \pm 4.7\%$	$15.5 \pm 4.7\%$

Ang II; stimulated by 1 μM of angiotensin II for 24 hour. Values are mean $\pm$ S.E.M.

and renin mRNA from SHR hearts strongly expressed in ventricles compared with WKY, and angiotensinogen mRNA expressed in fibroblasts from SHR were higher than those from WKY. ACE mRNA was also more strongly expressed in the ventricles and fibroblasts from SHR compared with those of WKY.

Effect of Ang II on collagen synthesis

The newly synthesized collagen was distributed mainly in the culture medium. The distribution remained unchanged with Ang II treatment (Table 2).

At 1 nM to 1 mM, Ang II induced an increase in collagen synthesis in a dose-dependent manner. The relative rate of collagen synthesis compared with total protein synthesis (calculated by the method described in the Materials and Methods section).

An elevation in collagen synthesis was noted 4 h after Ang II-treatment, and this effect persisted for 48 h.

To determine whether or not enhanced collagen synthesis was mediated by the Ang II receptor, we examined the effect of two specific competitive inhibitors on Ang II-induced collagen synthesis. The simultaneous addition of an equimolar concentration of saralasin or MK 954 and Ang II to the culture completely blocked the Ang II-induced increase in collagen synthesis.

Effect of Ang II on number of cells and synthesis of DNA

The number of cells remained unchanged for 64 h after Ang II. Significant effects on the incorporation of [^{3}H] thymidine by cardiac fibroblasts were observed in response to 100 pM–1 mM Ang II exposure under our experimental conditions.

Effect of Ang II on collagen synthesis in SHR

Basal (non-stimulated) collagen synthesis in cardiac fibroblasts from SHR was 1.6-fold greater than that in WKY rat cells. Collagen synthesis stimulated by Ang II increased in cardiac fibroblasts from SHR heart. Treatment with 100 nM Ang II resulted in an increase of $185 \pm 18\%$ in SHR cells and $128 \pm 19\%$ in WKY rat cells.

DISCUSSION

This study demonstrated the enhanced collagen production in SHR-derived cardiac fibroblasts, the stimulatory action of Ang II on collagen synthesis in cardiac fibroblasts, and the hyper-responsiveness of cardiac collagen synthesis to Ang II stimulation in SHR. The effect of Ang II on collagen synthesis was Ang II-receptor-mediated. We also reported that the expression of RAS mRNA was enhanced in SHR hearts.

Several studies have demonstrated increased collagen synthesis in the pressure-overloaded hypertrophied heart (Mukherjee and Sen, 1990; Weber, 1989; Weber and Brilla, 1991). Sen and Bumpus (1979) reported an augmented rate of collagen synthesis in the homogenates of ventricles from 4- and 8-week-old SHR compared with age-matched WKY controls. They also showed that differences between the two groups are less prominent at 10–15 weeks of age. In the present study, we measured enhanced collagen production in cultured cardiac fibroblasts from SHR. Our data are partly consistent with previous findings.

The reports on the ability of Ang II to promote collagen synthesis in cultured cardiac fibroblasts are rare. Recently, Brilla *et al.* (1994) reported that aldosterone is more potent than Ang II in the stimulation of collagen synthesis in cultured adult rat cardiac fibroblasts. Kato *et al.* (1991) showed that Ang II stimulated collagen synthesis in vascular smooth muscle cells. Studies have shown that Ang II stimulates cellular protein synthesis and induces hypertrophy in vascular smooth muscle cells but, has no mitogenic activity (Berk *et al.*, 1989; Geisterfer *et al.*, 1988). In embryonic chick myocytes, Ang II induced cellular hypertrophy that was associated with an increased rate of protein synthesis (Aceto and Naker, 1990). Ang II also stimulated protein synthesis in neonatal rat cardiomyocytes (Katoh *et al.*, 1989). As yet, no *in vivo* evidence exists to indicate that Ang II directly increases collagen synthesis. The relation appears to be indirect or by way of myocyte necrosis. Strong evidence indicates that elevated levels of plasma Ang II causes myocyte damage and necrosis and that the subsequent reparative process results in fibroblast proliferation and possibly increased collagen synthesis (Tan *et al.*, 1991). Our results conflict with the classic hypothesis that fibroblast proliferation and collagen deposition is followed by myocyte necrosis. In the present study, no substances were found to link myocyte necrosis and fibrosis; however, Ang II may independently influence cellular functions in myocytes and fibroblasts. This mechanism explains why volume-overloaded hypertrophy does not accompany the accumulation of collagen and heterogeneous regulations of growth of myocytes and fibroblasts.

The mechanisms by which Ang II stimulates collagen synthesis in cardiac fibroblasts are not clear. In vascular smooth muscle cells, earlier studies have illustrated that Ang II activates phosphoinositide turnover, augments intracellular calcium, and induces expression of platelet-derived growth factor (Naftilan *et al.*, 1989) and the proto-oncogenes c-fos (Kawahara *et al.*, 1988; Taubman *et al.*, 1989), c-myc (Naftilan *et al.*, 1989), and c-jun (Naftilan *et al.*, 1990). Indeed, it may be the case in cardiac fibroblasts that Ang II induces collagen synthesis by the same mechanisms (Sadoshima and Izumo, 1993b). Johnson and Aguilera showed that a progressive increase occurred in type 1-Ang II receptors in fetal skin fibroblasts of rats during culture, and that this responsible for stimulation of inositol phosphates and cAMP formation in these cells (Johnson and Aguilera, 1991). The inhibition of Ang II stimulated collagen synthesis by the type 1-Ang II receptor antagonist in this study, which suggests the exertion of the latter mechanism. Alternatively, it has been argued that other vasoactive substances and growth factors participate in this process. Transforming growth factor b1 is one candidate. This peptide stimulates collagen synthesis *in vivo* (Roberts *et al.*, 1986) and *in vitro* (Eghbali *et al.*, 1991). Ang II may induce gene expression of transforming growth factor b1 and stimulate collagen synthesis in cardiac fibroblasts.

Ang II slightly stimulated the incorporation of thymidine into cells but did not stimulate proliferation of cardiac fibroblasts in the present study. On the contrary, Ganten *et al.* (1975) reported that Ang II induced cell growth in 3T3 mouse fibroblasts in a culture containing 10% fetal calf serum. The discrepancy may be explained by either the presence of serum or cell-specific differences. In our experiment, we used confluent and serum-depleted cells, both of which were quiescent. However, we could not rule out the possibility that Ang II is mitogenic for cardiac fibroblasts during a logarithmic phase of growth. In vascular smooth muscle cells, several studies have shown that Ang II stimulates protein synthesis and induces hypertrophy but has no mitogenic activity (Berk *et al.*, 1989; Geisterfer *et al.*, 1988). Likewise, Ang II may cause cellular hypertrophy in cardiac fibroblasts, but its significance *in vivo* may be limited in cardiac myocytes or vascular smooth muscle cells. In our experiment, most of the newly synthesized collagen was distributed in the culture medium, which is in agreement with the fact that collagen is accumulated in the extracellular matrix in the heart.

Our experiment elicited the effects of Ang II on collagen synthesis in cardiac fibroblasts. However, additional experiments are required to determine whether Ang II-stimulated collagen synthesis is the results of an increase in the transcription of collagen genes or from a decrease in the degradation of newly synthesized collagen. Factors that stimulate collagenolytic activity may be inhibited in the development of cardiac hypertrophy.

This *in vitro* study permits us to conclude that Ang II can stimulate collagen synthesis in rat cardiac fibroblasts and thus may play an important role in the development and regression of myocardial fibrosis, which is associated with the development of cardiac hypertrophy.

Another important aspect of our findings is that cardiac fibroblasts from SHR exhibited enhanced collagen synthesis compared with fibroblasts from WKY controls. Two explanations can be offered: collagen may be produced through an intrinsic, genetic, or it may develop as a result of certain extracellular factors. In the latter case, growth factors may be produced by the fibroblast itself.

The present study also showed that the responsiveness of collagen synthesis to Ang II was enhanced in SHR cardiac fibroblasts from SHR. Several mechanisms may be involved, including an increased number of Ang II receptors or enhanced post-receptor signaling activated by Ang II, such as an increased activity of phospholipase C.

The concentrations of Ang II that affected the synthesis of collagen in our experiments were higher than the levels of plasma Ang II in physiological and pathologic states observed in man. Tissue concentrations of Ang II in the rhesus monkey heart ranged from 100 to 500 fmol per gram of wet tissue weight (Lindpaintner and Ganten, 1991). This discrepancy does not exclude the possibility that Ang II causes cardiac fibrosis *in vivo*. A long-term exposure to this vasoactive peptide may result in enhanced accumulation of collagen even at lower levels. Moreover, the levels of Ang II may be higher in the interstitium. Recently, the concept of a tissue renin-angiotesin system has been proposed (Lindpaintner and Ganten, 1991). Renin, angiotensinogen, and their mRNA expression have been identified in the heart (Kunapuli and Kumar, 1987; Paul *et al.*, 1988). This local system is supposed to exert its effects in an autocrine or paracrine manner. Implicit in this concept is the idea that local Ang II concentrations may exceed Ang II concentrations in the plasma. A local renin-angiotensin system may be present in cardiac fibroblasts. We attempted to prove the existence of cardiac RAS.

The hypertrophied heart caused by pressure overload showed an increased angiotensinogen gene expression even when plasma renin activity was not elevated (Baker *et al.*, 1990). An immunohistochemical study on the localization of angiotensinogen demonstrated that angiotensinogen was distributed mainly in the atrial muscles and the muscle fibers of the conduction system and was faint in those of the subendocardial layer of the ventricle in the normal human heart (Sawa *et al.*, 1992), whereas a wide-spread immunoreactivity was found in the failing hypertrophied ventricular muscles (Sawa *et al.*, 1994). Another study showed that either endogenous or exogenous Ang II was associated with altered sarcolemmal permeability, myocytolysis, fibroblast proliferation, and subsequent fibrosis (Aceto and Baker, 1990). These findings indicate that in the failing hypertrophied heart, the tissue renin-angiotensin system is activated. In our experiments, the expression of RAS mRNA was enhanced in SHR hearts. Although we could not specify the cellular distribution of RAS mRNA, our data support the existence of cardiac RAS. These findings suggest that cardiac RAS may play an important role in collagen accumulation in hypertensive cardiac hypertrophy.

REFERENCES

Aceto, J. F. and Baker, K. M. (1990). [Sar1] angiotensin II receptor-mediated stimulation of protein synthesis in chick heart cells, *Am. J. Physiol.* **258** (Heart Circ Physiol 27), H806–H813.

Baker, K. M. and Singer, H. A. (1988). Identification and characterization of guinea pig angiotensin II ventricular and atrial receptors: coupling to inositol phosphate production, *Circ. Res.* **62**, 896–904.

Baker, K. M., Chernin, M. I., Wixson, S. K., *et al.* (1990). Renin-angiotensin system involvement in pressure-overloaded cardiac hypertrophy in rats, *Am. J. Physiol.* **259** (Heart Circ Physiol 28), H324–H332.

Berk, B. C., Vekshtein, V., Gordon, H. M., *et al.* (1989). Angiotensin II-stimulated protein synthesis in cultured vascular smooth muscle cells, *Hypertension* **13**, 305–314.

Berridge, M. J. (1984). Inositol trisphosphate and diacylglycerol as second messengers, *Biochem. J.* **220**, 345–360.

Brilla, C. G., Janicki, J. S. and Weber, K. T. (1991). Cardioprotective effects of lisinopril in rats with genetic hypertension and left ventricular hypertrophy, *Circulation* **83**, 1771–1779.

Brilla, C., Zhou, G., Matubara, L., *et al.* (1994). Collagen metabolism in cultured adult rat cardiac fibroblasts: response to angiotensin II and aldosterone, *J. Mol. Cell. Cardiol.* **26**, 809–820.

Brown, J. H., Buxton, I. L. and Brunton, L. L. (1985). α1-adrenergic and muscarinic cholinergic stimulation of phosphoinositide hydrolysis in adult rat cardiomyocytes, *Circ. Res.* **57**, 532–537.

Chomczynski, P. and Sacchi, N. (1987). Single-step method of RNA isolation by acid guanidium thiocyanate-phenol-chloroform extraction, *Anal. Biochem.* **162**, 156–159.

Eghbali, M., Czaja, M. K., Zeydel, M., *et al.* (1988). Collagen chain mRNAs in isolated heart cells from young and adult rats, *J. Mol. Cell. Cardiol.* **20**, 267–276.

Eghbali, M., Tomek, R., Sukhatme, V. P., *et al.* (1991). Differential effects of transforming growth factor-b1 and phorbol myristate acetate on cardiac fibroblasts: Regulation of fibrillar collagen mRNA and expression of early transcription factors, *Circ. Res.* **69**, 483–490.

Ehrlich, B. E. and Watras, J. (1988). Inositol 1,4,5-trisphosphate activates a channel from smooth muscle sarcoplasmic reticulum, *Nature* **336**, 583–586.

Ganten, D., Schelling, P., Flugel, R. M., *et al.* (1975). Effect of angiotensin and the angiotensin antagonist P113 on iso-renin and cell growth in 3T3 mouse cells, *IRCS Med. Sci. Biochem.* **3**, 327.

Geisterfer, A. A. T., Peach, M. J. and Owens, G. K. (1988). Angiotensin II induces hypertrophy, not hyperplasia, of cultured rat aortic smooth muscle cells, *Circ. Res.* **62**, 749–756.

Hubert, C., Houot, A. N., Corvor, P., *et al.* (1991). Structure of the angiotensin I-converting enzyme gene: two alternate promotors correspond to evolutionary steps of duplicated gene, *J. Biol. Chem.* **266**, 15377–15383.

Imai, T., Miyazaki, H., Hirose, S., *et al.* (1983). Cloning and sequence analysis of cDNA for human renin precursor, *Proc. Natl. Acad. Sci. USA* **80**, 7405–7409.

Johnson, M. C. and Aguilera, G. (1991). Angiotensin-II receptor subtypes and coupling to signaling systems in cultured fetal fibroblasts, *Endocrinology* **129**, 1266–1274.

Kawaguchi, H. and Yasuda, H. (1984). Platelet-activating factor stimulates phospholipase in quiescent Swiss mouse 3T3 fibroblast, *FEBS Lett.* **176**, 93–96.

Kawaguchi. H., Iizuka, K., Takahashi, H., *et al.* (1990). Inositol trisphosphate kinase activity in hypertrophied rat heart, *Biochem. Med. Metab. Biol.* **44**, 42–50.

Kawaguchi, H., Shoki, M., Sano, H., *et al.* (1991a). Phospholipid metabolism in cardiomyopathic hamster heart cells, *Circ. Res.* **69**, 1015–1021.

Kawaguchi, H., Sano, H., Kudo, T., *et al.* (1991b). Effect of elastase on aortic smooth muscle cell proliferation, *Jpn. Heart J.* **32**, 131–138.

Kageyama, R., Ohkubo, H. and Nakanishi, S. (1984). Primary structure of human preangiotensinogen deduced from the cloned cDNA sequence, *Biochemistry* **23**, 3603–3609.

Kato, H., Suzuki, H., Tajima, S., *et al.* (1991). Angiotensin II stimulates collagen synthesis in cultured vascular smooth muscle cells, *J. Hypertens.* **9**, 17–22.

Katoh, Y., Komuro, I., Shibasaki, Y., *et al.* (1989). Angiotensin II induces hypertrophy and oncogene expression in cultured rat heart myocytes, *Circulation* **80** (Suppl. II), 450.

Kawahara, Y., Sunako, M., Tsuda, T., *et al.* (1988). Angiotensin II induces expressions of the c-fos gene through protein kinase C activation and calcium ion mobilization, *Biochem. Biophys. Res. Commun.* **150**, 52–59.

Kunapuli, S. P. and Kumar, A. (1987). Molecular cloning of human angiotensinogen cDNA and evidence for the presence of its mRNA in rat heart, *Circ. Res.* **60**, 786–790.

Lindpaintner, K. and Ganten, D. (1991). The cardiac renin-angiotensin system: An appraisal of present experimental and clinical evidence, *Circ. Res.* **68**, 905–921.

Motz, W., Ringswall, G., Goeldel, N., *et al.* (1982). Haemodynamics and connective tissue content in spontaneously hypertensive rat hearts under long-term antihypertensive therapy, in: *Hypertensive Mechanisms: The Spontaneously Hypertensive Rat as a Model to Study Human Hypertension*, Rascher, W., Glough, D. and Ganten, D. (Eds), pp. 715–718. Schatteur Verlag, Stuttgart.

Mukherjee, D. and Sen, S. (1990). Collagen phenotypes during development and regression of myocardial hypertrophy in spontaneously hypertensive rats, *Circ. Res.* **67**, 1474–1480.

Mullis, K. B., Faloona, F., Scharf, S. J., *et al.* (1986). Specific enzymatic amplification of DNA in vitro: the polymerase chain reaction, *Cold Spring Harbor Symp. Quant. Biol.* **51**, 335–350.

Naftilan, A. J., Pratt, R. E. and Dzau, V. J. (1989). Induction of platelet-derived growth factor A-chain and c-myc gene expressions by angiotensin II in cultured vascular smooth muscle cells, *J. Clin. Invest.* **83**, 1419–1424.

Naftilan, A. J., Gilliland, G. K., Eldridge, C. S., *et al.* (1990). Induction of the proto-oncogene c-jun by angiotensin II, *Mol. Cell. Biol.* **10**, 5536–5540.

Nisizuka, Y. (1983). Phospholipid degradation and signal translation for protein phosphorylation, *Trend. Biochem. Sci.* **8**, 13–16.

Paul, M., Wagner, D., Metzger, D., *et al.* (1988). Quantification of renin mRNA in various mouse tissues by a novel solution hybridization assay, *J. Hypertens.* **6**, 247–252.

Peterkofsky, B. and Diegelmann, R. (1971). Use of a mixture of proteinase-free collagenases for the specific assay of radioactive collagen in the presence of other proteins, *Biochemistry* **10**, 988–994.

Putney, J. W. (1987). Formation and actions of calcium-mobilizing messenger, inositol 1,4,5-trisphosphate, *Am. J. Physiol.* **252**, G149–G157.

Roberts, A. B., Sporn, M. B., Assoian, R. K., *et al.* (1986). Transforming growth factor type B: Rapid induction of fibrosis and angiogenesis in vivo and stimulation of collagen formation in vitro, *Proc. Natl. Acad. Sci. USA* **83**, 4167–4171.

Sadoshima, J. and Izumo, S. (1993a). Signal transduction pathways of angiotensin II-induced c-fos gene expression in cardiac myocytes in vitro. Role of phospholipid-derived second messengers, *Circ. Res.* **73**, 424–438.

Sadoshima, J. and Izumo, S. (1993b). Molecular characterization of angiotensin II-induced hypertrophy of cardiac myocytes and hyperplasia of cardiac fibroblasts. Critical role of the AT1 receptor subtype, *Circ. Res.* **73**, 413–423.

Sawa, H., Tokuchi, F., Mochizuki, N., *et al.* (1992). Expression of the angiotensinogen gene and localization of its protein in the human heart, *Circulation* **86**, 138–146.

Sawa, H., Kawaguchi, H., Mochizuki, N., *et al.* (1994). Distribution of angiotensinogen in diseased human hearts, *Mol. Cell. Biochem.* **132**, 15–23.

Sen, S. and Bumpus, F. M. (1979). Collagen synthesis in development and reversal of cardiac hypertrophy in spontaneously hypertensive rats, *Am. J. Cardiol.* **44**, 954–958.

Shoki, M., Kawaguchi, H., Okamoto, H., *et al.* (1992). Phosphatidylinositol and inositolphosphatide metabolism in hypertrophied rat heart, *Jpn. Circ. J.* **56**, 142–147.

Tan, L.-B., Jalil, J. E., Pick, R., *et al.* (1991). Cardiac myocyte necrosis induced by angiotensin II, *Circ. Res.* **69**, 1185–1195.

Taubman, M. B., Berk, B. C., Izumo, S., *et al.* (1989). Angiotensin II induces c-fos mRNA in aortic smooth muscle: Role of Ca^{2+} mobilization and protein kinase C activation, *J. Biol. Chem.* **264**, 526–530.

Weber, K. T. (1989). Cardiac interstitium in health and disease: the fibrillar collagen network, *J. Am. Coll. Cardiol.* **13**, 1637–1652.

Weber, K. T. and Brilla, C. G. (1991). Pathological hypertrophy and cardiac interstitium: Fibrosis and renin-angiotensin-aldosterone system, *Circulation* **83**, 1849–1865.

Williamson, J. R., Cooper, R. H., Joseph, N. E., *et al.* (1985). Inositol trisphosphate and diacylglycerol as intracellular second messengers in liver, *Am. J. Physiol.* **248**, C203–C216.

Advances in Neuroregulation and Neuroprotection (2005), pp. 503-516
Collin, C. *et al.* (Eds)
© VSP 2005

Biofeedback training in clinical settings

IWAO SAITO [1,*] and YASUKO SAITO [2]

[1] *Medical Service Center, Muroran Institute of Technology, 27-1 Mizumoto-cho, Muroran, Hokkaido 050-8585, Japan*
[2] *Sapporo Hukujuji Clinic, Sapporo 060-0808, Japan*

Abstract—Self-control of body functions by biofeedback training (BFT) has been developed and BFT is now employed in treatment over all medical sections as a complementary and alternative medicine (CAM). BFT is also useful in education and sports training. BFT uses electrical equipment to monitor very small signals from the body which are unrecognizable in usual life. Subjects are requested to learn physiology and the mechanism of BFT and then they make a voluntary effort to accomplish the stepwise task goal of BFT. This mini-review includes history and current application of BFT and also discusses future usefulness of BFT.

Keywords: Biofeedback training (BFT); psychophysiology; complementary and alternative medicine (CAM).

INTRODUCTION

Miller and DiCara (1967) found that heart rate (HR) can be changed voluntarily in rats that received operant conditioning. Soon after the first report, they succeeded in control of intestinal movement. Basmajian (1971) reported on voluntary control of single motor unit activity of skeletal muscle with a help of originally developed electromyography (EMG). They applied the EMG-biofeedback training (BFT) method to patients with flaccid and spastic paralysis by sequelae of cerebrovascular accidents, and succeeded in rehabilitation, which was the first successful clinical result by application of BFT. Since then, self-control of physical functions has been demonstrated and evaluated on a scientific and clinical basis.

The term 'biofeedback control' was coined about 40 years ago and its definition was proposed by several scientists in different ways. Olson summarized the accumulated definitions into three groups. Schwartz and Beatty, psychophysiologists, proposed that Biofeedback refers to a group of experimental procedures in which

---

*To whom correspondence should be addressed. E-mail: yaiku@u01.gate01.com

an external sensor is used to provide a subject with information about his body processes, to regulate body function.

According to Basmajian, biofeedback may be defined as a technique to tell subjects their internal physiological events to help them to manipulate involuntary and imperceptable events by using visual or auditory signal equipment (Basmajian, 1978). Green *et al.* (1980) defined the term from a psychosomatic point of view, that is, BFT is a tool for learning psychosomatic self-regulation. From the chronological point of view, Kamiya (1969) indicated three points that are required in the procedure for BFT: 'First, the physiological function to be controlled must be continuously monitored with sufficient sensitivity to detect moment-by-moment changes. Second, changes in the physiological measurement must be reflected immediately to the subject. Third, the subject must be motivated to learn the physiological changes under study'.

BFT is classified into two categories of self-control groups: one is BFT of prompt information index of the neuromuscular system, such as skin temperature, GSR, HR, BP, EEG, etc. The other is BFT of slow information index of the humoral system, such as the endocrine and immune system. Because of difficulty in making the measurements, the slow information system was hardly studied until recently.

BFT helps subjects to control physiological events within the body and it is useful for health and education. BFT is now regarded more favorably. In the United States and Europe, alternative medical treatment is considered as useful as a complement to orthodox medicine, and alternative medical treatments have been introduced so as to decrease the high cost of medical care.

The principles of BFT will be introduced in the early part of this review, while in the latter part, the current state and the image of the future of BFT in each clinical area will be introduced.

FEATURES OF BIOFEEDBACK TRAINING

BFT supports subjects' self control

BFT is convenient for the clients to understand the correlation of mind and body, and clients will usually accept the challenge of regulating their body function in a trial-and-error manner. With the improvement in instruments providing high accuracy and ease of operation, clients can understand their symptoms easily and can be encouraged in their efforts at self-help treatment.

Mechanism of BFT as a causal treatment

There was a criticism for a certain period that BFT was only a symptomatic therapy and not a causal therapy. However, it has been shown gradually that the mechanism of BFT shows excellent correlation with the pathophysiology of disease and its effect can be comparable to that of pharmacological treatment.

For example, in the treatment of oral dyskinesia, neither surgical operation nor pharmacological treatment work well and, in this case, EMG-BFT may provide one successful intervention. According to the facilitation theory, the mechanism of BFT is: (1) to substitute the function of degenerated CNS neuronal cells with remaining healthy neuronal cells; (2) to improve coordination between the CNS and peripheral nervous system. Therefore, BFT can be considered as a re-education with electronics.

Variations in BFT

Although pharmacological treatment has far more variation, BFT provides not a few indexes for self-control. Moreover, variations in BFT increase with progress in measurement devices and methodology. It is also expected that BFT could contribute to treatment and prevention of the life-style related diseases.

Treatment method of growth model (shaping)

In BFT, the therapist divides the goal into small steps of treatments and makes each patient's achievement easier. The therapist encourages patients to accumulate small successes one by one, so that the final goal will be achieved. The clients are led through the states of treatment, which might be called a 'success game'. The therapist uses praise and supporting talk and rewards to reinforce the patient's will.

Complementary and Alternative Medicine (CAM)

Because the costs involved in each nation's medical treatment are continually increasing, a cheap and effective treatment method is valued. Recently, alternative or complementary medicine has become a keyword to future wellness and treatment. Eisenberg *et al.* (1993) of Harvard University reported that, at that time, 33.8% of US citizens received 16 kinds of CAM for treatment, and in 1997, 42.1% of US citizens. Fifty percent of these patients graduated from College or University, and the number of high-income citizens is greater than those on low incomes.

According to MEDLINE, the number of references about CAM was about 2000 (0.4–0.5%) in 1963–1998. In the upper ranking, traditional alternative medicine (Homeopathy, Naturopathy and Herbalism) stands at 25%; electric stimulation treatment, 21%; acupuncture, 16%; biofeedback training, 9%; and relaxation, 9%. Both BFT and relaxation share the 4th place when documentation of CAM treatment is retrieved with MEDLINE. The most popular CAM treatment in Japan is Kampo (Chinese herbal) medicine.

Comparison between pharmacological treatment and BFT

Pharmacological treatment is an excellent choice with rapid effect and reliability, there are many different preparations available and it is popular with public and the medical profession. BFT is useful in the treatment of chronic symptoms and

Table 1.

Comparison of treatment modalities of essential hypertension between pharmacological treatment and biofeedback training (BFT)

	Pharmacological treatment	BFT
(A) Blood volume	diuretics	BFT of BP ?
(B) Cardiac rhythm	β-blockers for HR and stroke volume	HR-BFT for HR and BFT-Pulse volume pressure
(C) Blood vessel	α-blockers for control of blood vessel	BFT of skin temperature
	Ca-antagonists for vascular dilation	BFT of skin temperature
:	ACEIs	
(D) Psychological	Neuroleptics	EMG-BFT of the forehead and stress
:		BFT of GSR

ACEIs: Angiotensin converting enzyme inhibitors, GSR: Gulvano skin reflex.

diseases when other methods are less effective. BFT is also useful in the field of the education, welfare, and sports. As general medicine and surgery cooperate, so it may be possible to maintain cooperation between ordinary medical treatments and BFT.

We compared the pharmacological treatment of hypertension with BFT. In the physiology of blood pressure (BP), BP is shown as follows: 'BP = heart rate (HR) × total peripheral resistance (TPR)' or 'BP = ejection volume × HR × TPR'.

In pharmacological treatment of hypertension, loop diuretic hypotensives, α-blockers, β-blockers, Ca antagonists, angiotensin converting enzyme inhibitors and neuroleptic medication, etc. are used. Treatment modalities of essential hypertension in pharmacological treatment and BFT are summarized in Table 1.

Clinical BFT in the United States

The scientific basis of biofeedback control was fostered in the USA in 1960. Development of electronics and monitoring technology supported this establishment. The former Biofeedback Association of the United States and the present American Association for Applied Psychophysiology and Biofeedback (AAPB) has a 40-year history; it accumulated clinical experiences in medical fields, rehabilitation, pedagogy, sports, and other areas. In 1995, the US Government admitted insurance claims with BFT in chronic muscle contraction headache and insomnia (BF Research Letter from AAPB **23**: 1996) (A doctor's prescription or recommendation is necessary for an insurance claim; so, private practice of clinical psychologists and registered nurses increased). Later, BFT for urinary incontinence and neurofeedback training of ADHD and learning disorders was admitted. The AAPB is directing their next effort on insurance claims for essential hypertension. Recently, two new academic associations, the Neurofeedback Society and the Incontinence Society, were established in the USA.

BFT APPLICATIONS

BFT is applied mainly in the fields of medicine and rehabilitation, sports and education. We introduce minimum resources here. Original documents may be retrieved from the references given.

BFT in education

BFT is used in various ways in the field of education. In the education of young people, BFT is available as a tool for nonverbal training of physical function. BFT may be useful in the treatment of stuttering (Lanyon, 1976). Ray *et al.* (1979) introduced the following studies with BFT: voice quality and intensity, dysarthia (Netsell and Cleeland, 1973), teaching speech to the deaf (Nickerson *et al.*, 1976), sub-vocalization during reading (Hardyck and Petrovich, 1969). Pepper and Gibney (2000) applied EMG-BFT to the problem of VDT and lumbago for healthy computing.

Application of BFT to a patient with cerebral palsy was reported by Finley (1977). In early stage of BFT, the apparatus was elaborated for child's bed-wetting although it was not so effective as expected. Lubar (1979) applied EEG-BFT to children with ADHD. In alcoholism and drug intoxication, Crabtree and Lacey (1981) reported BF and relaxation training helped these patients control anxiety for more than 6 months, although it was not effective in reducing symptoms. Koyama *et al.* (1976) applied EMG-BFT for language training of students to pronounce English consonants unfamiliar to Japanese.

In emotional education, Webb (1977) developed the use of myoelectric feedback in teaching facial expression to the blind. Significant increase in successful expressions of emotion was observed in pre- and post-treatment assessments in clinical practice. In clinical situations, psychologists apply BFT of temperature and/or GSR to the college examinees for relaxation.

These reports indicate that education is an area where effectiveness of BFT will be expected more in the future.

BFT in sports

BFT used in sports is classified into three categories: (1) to enhance the performance ability (Cox and Matyas, 1983; Horino and Yamazaki, 1995; Yoshizawa *et al.*, 1983); (2) to regulate psychological problems such as stage fright; and (3) physical problems of chronic pain and other diseases. There are many reports that use EMG-BFT to treat chronic pain in sports players, and thermal BFT for relaxation. Furedy (1977) developed a BFT method so that biathlon participants significantly decrease their HR to practice exact shooting. It was requested for running-players to decrease their HR from 180 beats per minutes (bpm) to 120 bpm instantly.

Olympics and BFT

In the United States, BFT is one of the tools used for major sports events, such as the Olympic and other professional Games. It is known that, under mental tension, finger-tip temperature is 0.5–4°C lower than in the relaxed condition due to mental tension. To get better neuromuscular function and ability, players are asked to practice thermal BFT by their coach, 15–30 min before the game or ordinary training for relaxation and mental endurance. In the annual meeting of the Biofeedback Society of America just after the Los Angeles Olympics, a symposium 'BFT and Olympics' was held and it was shown how a US player had won a gold medal using BFT.

BFT IN MEDICINE

Although there are many kinds of BFT, several of them are most widely used in clinical situations. In this section, we discuss EMG-BFT, BFT of skin temperature, BFT of electroencephalogram (EEG), BFT of urinary and/or fecal function, and BFT of hypertension.

Electromyography (EMG)-BFT

It has been shown that EMG-BFT is beneficial in the treatment of sequelae of cerebrovascular accidents. Basmajian *et al.* (1975) applied EMG-BFT to patients with flaccid and spastic paralysis due to cerebrovascular accidents. Saito and Yasui (1992) applied EMG-BFT to patients with tongue spastic paralysis and achieved improvement with tongue relaxation and recovery of speech. Cerebral palsy was one area where BFT might be expected to be relevant. Finley *et al.* (1976) applied EMG-BFT in 6 children to reduce their spasticity.

Involuntary movement

In spasmodic torticollis, Cleeland (1973) applied EMG-BFT with joint-use of negative operant as electric stimulation for punishment. Brudny (1974) succeeded in application of BFT to a patient with spasmodic torticollis using the ordinary method. There is a BFT device for spasmodic torticollis for home practice using a paper speaker.

Application of BFT to a patient with writer's cramp was reported by Uchiyama *et al.* (1977). Application of BFT to a patient with blephalospasm was also reported.

Temporomandibular joint disease

Application of BFT to a patient with bruxism was reported by Berry and Wilmot (1977). Application of BFT to a patient with oral dyskinesia was reported by Saito *et al.* (1985).

Chronic pain syndromes, tension headache and lowback pain, etc.

Budzynski *et al.* (1973) was the first to report clinical effectiveness of EMG-BFT in treating tension headache. Kentsmith *et al.* (1977) discussed the effect of BF-induced plasma dopamine-β-hydroxylase activity upon suppression of migraine symptoms. EMG-BFT was most popular in treatment of chronic muscle contraction headache. Health insurance treatment has been authorized in the treatment of chronic muscle contraction headache in the USA (BF Research Letter from AAPB **23**: 1996).

Keef and Hoelscher (1987) evaluated 15 clinical studies of EMG-BFT of low back pain. Fourteen of 15 studies showed that the treatment resulted in increased mobility and exercise tolerance with decreased EMG level. EMG-BFT may also be effective to patients with trigeminal nerve pain.

Temporomandibular joint (TMJ) dysfunction and myalgia or dysfunction of the extremities

Use of EMG-BFT for treatment of TMJ (Berry and Wilmot, 1977) and for myalgia in the limbs are two of most popular treatment modalities in psychosomatic medicine.

Hauri (1981) applied EMG-BFT to subjects with insomnia. Hyperarousal due to anxiety is eased by relaxation. Health insurance treatment has been authorized in the treatment of chronic muscle contraction headache and insomnia in the USA. In parkinsonism, Shumaker (1973) tried frontal EMG-BFT for muscle relaxation on patients with tongue paralysis. Surwit *et al.* (1983) used frontal muscle EMG-BFT for chronic muscle contraction type headache, then he tried self-control of mental uneasiness, and succeeded in the dietary guidance of diabetic patients.

BFT of skin temperature

BFT of skin temperature is one of the most popular methods in education, medicine and sports. Application of BFT to a patient with Raynaud's disease was reported by Freedman and Ianni (1983). Application of BFT to a patient with essential hypertension was reported by Green *et al.* (1980). Application of BFT to a patient with migraine was reported by Sargent *et al.* (1975).

Electroencephalographic (EEG)-BFT

Kamiya (1969), a pioneer in EEG-BFT, proved that self-control of brain waves was possible. The United States National Space Development Agency (NASA) supported this research to prevent space intoxication of astronauts. By making good use of NASA technology, which is high in accuracy and reliability, Lubar (1979) succeeded in a clinical application of EEG-BFT to attention deficit hyperactive disorder (ADHD) and led to the proliferation of EEG-BFT today.

As for BFT in epilepsy, clinical application of BFT had been expected and BFT has succeeded experimentally in treatment of epilepsy (Finley, 1977). However, the necessity for use BFT in epilepsy is low, because excellent antiepileptic drugs have been developed. A neurofeedback academy has been established by the AAPB.

BFT of urinary and fecal incontinence

Urinary and fecal incontinence is an indispensable area in the clinical contribution of BFT. Elderly people tend to have trouble in controlling the excretion of urine and feces in the later stages of life due to central nervous degeneration. As we are facing a global expansion of aging society, economic and social demands are increasing. Incontinence remarkably limits social activity of aged people, hurts their pride and causes withdrawal from ordinary social activities. Up till now, there has been no good treatment for incontinence.

The self control method for treatment of urinary incontinence was practiced by Engel and group (Burgio *et al.*, 1986). When the physiological mechanism of urinary incontinence was elucidated, a concise treatment which did not use BFT was derived. Recently a new Academy of urinary incontinence was established by the AAPB. Incontinence treatment is an advanced area of BFT application in the USA.

BFT has become a powerful treatment for defecation problems in older people. BFT is not merely a symptomatic treatment but a causal treatment, and BFT belongs in a group of latest medical interventions in fecal and urinary incontinence. The principle is that patients learn muscle tension and function of the anus with EMG-monitoring, and it re-educates them to have a normal sense of the outer and inner constrictor muscles of the anus with the aid of visual and auditory signals in a trial-and-error manner (Engel *et al.*, 1974).

BFT in cardiovascular diseases: Essential hypertension

As psychophysiological approaches in dealing with essential hypertension, the following treatments have been applied: (1) Progressive relaxation, (2) Autogenic training, (3) Hypnotic relaxation, (4) Transcedental meditation, (5) Benson's relaxation response, (6) Yogic meditation, (7) Direct and indirect BP biofeedback.

As shown above, in the period 1970 to 1985, four major group controlled studies of hypertension treatment were conducted (Engel *et al.*, 1983; Goebel *et al.*, 1980; Green *et al.*, 1980; Patel, 1975). One interesting but shocking conclusion of the studies was that they pointed out the potential error in ordinary blood pressure measurement in a medical setting. This kind of measurement technique was introduced later to ordinary hypertensive treatment. These results are that a multicomponent non-drug approach works on regulation of high blood pressure when systematic component interventions are attempted but fail or succeed only to a limited extent.

Blanchard *et al.* (1993) indicated that this work has raised the possibility of psychological treatment of cardiovascular diseases as either an adjunct to standard pharmacological therapies, or as an alternative to them. At present, hypertension is the largest target AAPB aims at in insurance applications. Recently, Nakao *et al.* (2003) conducted a meta-analysis of non-drug treatment of essential hypertension and reported that, although the BFT group decrease BP more intentionally than non-treatment groups, BFT did not have a significant difference with non-specific behavioral medical intervention.

Other cardiovascular diseases

BFT has been studied in following disorders: premature ventricular contraction (PVC; Weiss and Engel, 1971), ventricular parasystole (Pickering and Gorham, 1975), atrial fibrillation (Blecker and Engel, 1973) and Raynaud's disease (Freedman and Ianni, 1983).

Brucker and Ince (1977) reported a postural hypotension in a patient with a spinal cord lesion at T3: he was trained to increase and decrease his BP, and after 11 sessions, the BP at standing was from 50 mmHg to 88 mmHg without falling. The results of this study and other studies suggested that BFT may be a cost-effective and contributory treatment for postural hypotension. Currently, BFT of cardiac failure is under research.

OTHER BF APPLICATIONS IN MEDICINE

Pediatrics

Application of BFT to a patient with bed-wetting was reported by Schwartz and Swash (1982) and Schwartz (1982). In congenital anal atresia and megacolon disease, BFT is used for control of the anal sphincter after operation. If the child succeeds in function, a toy frog jumps out for reward; if not, the frog does not appear. Application of BFT to a patient with stuttering was reported by Lanyon (1976). Application of BFT to a patient with learning disorders and ADHD was reported by Lubar (1979).

Ophthalmology

In blepharospasm, nystagmus, strabismus, myopia, chronic functional ophthalgia, BFT is applied when medication or surgery does not work well enough. BFT in blepharospasm is a popular application. We experienced a case of BFT used in a case of acquired strabism due to a traffic accident.

Otolaryngology

House (1978) tried EMG-BFT and thermal BFT of the skin to 41 tinnitus patients, and about 75% of patients showed some improvement.

Obstetrics, gynecology and urology

Applications are wide because the subjects are healthy in the case of delivery. Reported results are available on application of BFT to a patient with dysmenorrhea (Sedlacek and Perry, 1977) and sexual problems (Eversaul, 1974; Rosen, 1973; Rosen *et al.*, 1974).

Surgery, orthopedics, neurosurgery and rehabilitation

There are many reports of the application of BFT to hand and feet dysfunctions, and chronic pain problems due to surgical sequelae. For example, BFT was applied to a patient with chronic pain in postoperative rehabilitation. Application of BFT to a patient with fecal incontinence after surgery was reported by Engel *et al.* (1983).

Application of BFT to a patient with sequelae after surgery, cerebrovascular accident, hand and feet dysfunction has been reported. In posture training, to prevent toddling and lack of balance, there is a walk-training device that introduces part of the principles of BFT. There are some reports of BFT in cases of scoliosis. BFT also has possibilities as supplementary treatment to pharmacological intervention in Parkinson's disease (Shumaker, 1980; and other references).

Miscellaneous

One important point in the use of BFT on older patients is that such patients are normal subjects with integrated mind and body. They understand BFT unexpectedly well and quickly. Therapists and patients' families are often moved by their BFT performance.

Internal medicine, neurology, gerontology and psychosomatic medicine

Application of BFT to a patient with respiratory disorders has been reported. Kostes and Glaus (1981) summarized that BFT for respiratory resistance was effective in bronchial asthma; however, from the point of view of cost performance, ordinary treatments do better. Thus, generally BFT is not much used in clinical practice of bronchial asthma.

Application of BFT to a patient with emphysema and hyperventilation syndrome has been reported.

Digestive system

There is famous research by Miller (1969) that demonstrated voluntary control of a rat's intestinal peristalsis. Self-regulation of the intestinal rumbling in irritable colon syndrome has been discussed with feedback audio-signals using a special microphone, and using stethoscope (Saito, 1995, unpublished data). Shuster (1974) applied the BF method to patients with esophageal reflux who completed training within 6 sessions with good results.

Psychiatry and psychology

In BF approach for anxiolytic senses, BFT of HR, EMG-BFT of the forehead muscle (Fahrion, 1976) and BFT of skin temperature (Fahrion, 1976) have been widely studied. There are many reports that have proved the effectiveness of BFT in treating anxiety. However, as a variety of excellent psychotropic drugs were already developed, BFT of anxiety tends to be used only in refractory cases. Application of BFT to sleep problems has been reported (Hauri, 1981). Application of BFT to ADHD and learning disorders has been reported (Lubar, 1979).

Application of BFT to sexual dysfunction has been reported (Eversaul, 1974). In the treatment of male sexual problems in homosexual behavior, Rosen (1973) tried BFT to suppress tumescence in response to male nude pictures, then to female nude pictures by instrumental conditioning. Rosen *et al.* (1974) showed that voluntary control of penile tumescence was possible. However, Rosen cautioned that in most cases of erectile dysfunction, it seems necessary to consider crucial relationship factors.

In bereavement, the subjects often experience mind-body symptoms as hypertension, tachycardia, etc. Application of BFT to essential tremor has been reported. A BFT approach in depression and integrated malfunction syndrome (schizophrenia) is considered to be contraindicative; however, a definitive view should emerge in the future.

Odontology

BFT has been applied in appropriate cases here also: to bruxism (Casas *et al.*, 1982) and to temporomandibular pain dysfunction (Berry and Wilmot, 1977).

Expectations for future applications of BFT

In cancer interventions, although it is not possible to use BFT as yet, with progress in new technology, and new types of BFT, for instance, BFT involving immunological factors such as cancer antigens may become practicable. Norris's (1985) familial report was the first in natural healing with biofeedback and imagery training. Sibuya and Saito (2003) showed five instances of natural regression in cancer cases.

Direct BFT of glucose level is expected soon because new blood glucose measurement instruments are available that can report glucose level in couple of minutes.

REFERENCES

Basmajian, J. V. (1971). Neuromuscular facilitation techniques, *Arch. Phys. Med. Rehabil.* **52**, 40–42.

Basmajian, J. V. (1978). Biofeedback in medical practice, *Can. Med. Assoc. J.* **119**, 8–10.

Basmajian, J. V., Kukulka, C. G., Narayan, M. G., *et al.* (1975). BFT of foot drop after stroke compared with standard rehabilitation technique, *Arch. Phys. Med. Rehabil.* **56**, 231–236.

Berry, D. C. and Wilmot, G. (1977). The use of biofeedback techniques in the treatment of mandibular dysfunctional pain, *J. Oral Rehabilitation* **4**, 255–260.

Blanchard, E. B., Greene, B., Scharff, L., *et al.* (1993). Relaxation training as a treatment for irritable bowel syndrome. *Biofeedback and self-Regulation* **18**, 125–132.

Blecker, E. R. and Engel, B. E. (1973). Learned control of cardiac rate and cardiac conduction in the Wolf-Parkinson-White syndrome, *New Engl. J. Med.* **288**, 560–562.

Brudny, J., Grynbaum, B. B. and Korein, J. (1974). Spasmodic torticollis: treatment by feedback display of the EMG, *Arch. Phys. Med. Rehabil.* **55**, 403–408.

Brucker, B. C. and Ince, L. P. (1977). Biofeedback as an experimental treatment for postural hypotension in a patient a spinal-cord lesion, *Archives Physical Medicine Rehabilitation* **58**, 49–53.

Budzynski, T., Stoyva, J., Adler, C., *et al.* (1973). EMG biofeedback and tension headache: a controlled study, *Psychosom. Med.* **35**, 484–496.

Burgio, K. L., Robinson, J. C. and Engel, B. T. (1986). The role of biofeedback in Kegel exercise for stress urinary incontinence, *Am. J. Obstr. Gynecol.* **154**, 58–64.

Casas, J. M., Beemsterboer, P. and Clark, G. T. (1982). A comparison of stress reduction behavioral counseling and contingent nocturnal EMG feedback for treatment of bruxism, *Behav. Res. Ther.* **20**, 9–15.

Cleeland, C. S. (1973). Behavioral technics in the modification of spasmodic torticollis, *Neurology* **23**, 1241–1247.

Cox, J. R. and Matyas, T. A. (1983). Myoelectric and force feedback in the facilitation of isometric strength training: a controlled comparison, *Psychophysiology* **20**, 35–40.

Crabtree, M. and Lacey, R. (1981). Effectiveness of BF and relaxation training with drug addicts and alcoholics; a 6 month follow up, in: *Proceedings of 12th Annual Meeting of the BF Society of America*, pp. 32–34.

Eisenberg, D. M., Kessler, R. C., Foster, C., *et al.* (1993). Unconventional medicine in the United States. Prevalence, costs and patterns of use, *New Engl. J. Med.* **328**, 246–252.

Engel, B. T., Nikoomanesh, P. and Schuster, M. M. (1974). Operant conditioning of retro-sphincteric responses in the treatment of fecal incontinence, *New Engl. J. Med.* **290**, 646–649.

Engel, B. T., Glasgow, M. S. and Gaarder, K. R. (1983). Behavioral treatment of high blood pressure: III. Follow up results and treatment recommendation, *Psychosom. Med.* **45**, 23–29.

Eversaul, G. A. (1974). Psychophysiology and behavior training of premature ejaculation, in: *Annual Meetings of BF Research Society of America*.

Fahrion, S. (1976). Short term biofeedback program, in: *Handbook of Physiological Feedback, Vol. 1.* Autogenic Systems, Inc., Berkeley.

Finley, W. W. (1977). Operant conditioning of the EEG in two patients with epilepsy. *Pavlov J. Biol. Sci.* **12**, 93–111.

Finley, W. W., Niman, C., Stanley, J. and Ender, P. (1976). Frontal EMG biofeedback training of athetoid cerebral palsy, *Biofeedback Self-Regulation* **1**, 169–182.

Freedman, R. R. and Ianni, P. (1983). Self-control of digital temperature: Physiological factors and transfer effect, *Psychophysiology* **20**, 682–689.

Furedy, J. J. (1977). Biofeedback control of large magnitude of human heart rate decrease; its method and application, in: *Proceedings of the 4th Congress of the International College of Psychosomatic Medicine*, Kyoto, Japan.

Goebel, M., Viol, G. W., Lorenz, G. J., *et al.* (1980). Relaxation and biofeedback in essential hypertension: A preliminary report of a six year project, *Am. J. Clinical Biofeedback*, **3**, 20–29.

Green, E. E., Green, A. M. and Norris, P. (1980). Self-regulation training for control of hypertension, *Primary Cardiology* **6**, 126–137.

Hardyck, C. D. and Petrovich, L. F. (1969). Treatment of subvocal speech during reading, *J. Reading* **1**, 3–11.

Hauri, P. (1981). Treating psychosomatic insomnia with biofeedback, *Arch. Gen. Psychiatry* **38**, 752–758.

Horino, H. and Yamazaki, K. (1995). EMG-biofeedback and muscle control during exercise of the whole body, in: *Biobehavioral Self-Regulation*, Kikuchi, T., *et al.* (Eds), pp. 699–703. Springer-Verlag, Tokyo.

House, J. W. (1978). Treatment of severe tinnitus with biofeedback training, *The Laryngoscope* **88**, 406–412.

Kamiya, J. (1969). Operant control of the EEG alpha rhythm and some of its reported effects on consciousness, in: *Altered State of Consciousness*, Tart, C. J. (Ed.), pp. 507–517. John Wiley, New York.

Keef, F. J. and Hoelscher, T. J. (1987). Biofeedback in the management of chronic pain syndromes, in: *Biofeedback Studies in Clinical Efficacy*, Hatch, J. P., Fischer, J. G. and Rugh, J. D. (Eds), pp. 211–253. Biofeedback Society of America.

Kentsmith, D., Stridor, F., Copenhaver, J., *et al.* (1977). Effect of biofeedback upon suppression of migraine symptoms and plasma dopamine-beta-hydroxylase activity, *Headache* **16**, 173–177.

Kostes, H. and Glaus, K. (1981). Applications of biofeedback to the treatment of asthma: a critical review, *Biofeedback Self Regulation* **6**, 573–593.

Koyama, S., Okamoto, T., Yoshizawa, M., *et al.* (1976). An electromyographic study on training to pronounce English consonants unfamiliar to Japanese, *J. Human Ergol.* **5**, 51–60.

Lanyon, R. I. (1976). Modification of stuttering through biofeedback. *Behavior Therapy* **7**, 96–103.

Lubar, J. (1979). Neurological application of EEG biofeedback training, in: *Biofeedback, Principles and practice for clinicians*, 2nd edn, Basmajian (Ed.). Williams & Wilkins, Baltimore MD.

Miller, N. E. (1969). Learning of visceral and glandular responses. *Science*, **163**, 434–445.

Miller, N. E. and DiCara, B. S. (1967). Instrumental learning of HR changes in curarized rats: Shaping, and specificity indiscriminate stimulus, *J. Comp. Physiol. Psychol.* **63**, 12–19.

Nakao, M., Yano, E., Nomura, S., *et al.* (2003). Blood pressure-lowering effects of biofeedback treatment in hypertension: a meta-analysis of randomized controlled trials, *Hypertension Res.* **26**, 37–46.

Netsell, R. and Cleeland, C. S. (1973). Modification of lip hypertonia in dysarthria using EMG feedback, *J. Speech Hear. Disord.* **38**, 131–140.

Nickerson, R. S., Kalikow, D. N. and Stevens, K. N. (1976). Computer-aided speech training for the deaf, *J. Speech Hear Disord.* **41**, 120–132.

Patel, C. H. (1975). Yoga and biofeedback in the treatment of hypertension, *Lancet* **2**, 1152–1156.

Pepper, E. and Gibney, K. H. (2000). *Healthy Computing with Muscle Biofeedback*. Biofeedback Foundation of Europe, Woerden, The Netherlands.

Pickering, T. and Gorham, G. (1975). Learned heart-rate control by a patient with a ventricular parasystolic rhythm, *Lancet* **1**, 252–253.

Porter, G. and Norris, P. (1985). *Why me?: Learning to Harness the Healing Power of the Human Spirit*. Stillpoint Publishing, Walpole, NH, USA.

Ray, W., Raczynski, J. M., Rogers, T., *et al.* (1979). *Evaluation of Clinical Biofeedback*. Plenum Press, New York.

Rosen, R. C. (1973). Suppression of penile tumescence by instrumental conditioning, *Psychosom. Med.* **35**, 509–514.

Rosen, R. C., Shapiro, D. and Schwarz, G. E. (1974). Voluntary control of penile tumescence, *Psychophysiology* **11**, 230–231.

Saito, I. (1995). Relation of drug development and BF research, A friendly rivalry and concurrence, in: *Biobehavioral Self-Regulation*, Kikuchi, T., *et al.* (Eds), pp. 534–539. Springer-Verlag Tokyo.

Saito, I. and Yasui, K. (1992). EMG-BFT of spastic tongue paresis due to cerebrovascular sequelae, in: *Current BF Research in Japan 1992*, pp. 81–84. Shinko Igaku Shuppan, Tokyo.

Saito, I., Minami, M. and Saito, H. (1985). EMG-BFT of oral dyskinesia in geriatric patients. *Biofeedback Research* (Japanese) **12**, 28–32.

Sargent, J. D., Green, E. E. and Walters, E. D. (1975). Preliminary report on the use of autogenic biofeedback techniques in the treatment of migraine and tension headaches, *Psychosom. Med.* **35**, 129–137.

Schwartz, M. S. (1982). *Biofeedback A Practioner's Guide*. The Guilford Press, New York.

Schwartz, M. S. and Swash, M. (1982). Pattern of involvement in the cervical segments in the early stage of motor neurone disease: a single fiber EMG study, *Acta Neural. Scand.* **65**, 424–431.

Sedlacek, K. and Perry, C. (1977). Specific biofeedback treatment for dysmenorrhea, in: *Proceedings of the Biofeedback Research Society*, Orlando, FL.

Shumaker, R. G. (1980). The response of manual functioning in Parkinsonians to frontal EMG biofeedback and progressive relaxation, *Biofeedback Self Regulation* **5**, 229–234.

Shuster, M. M. (1974). Operant conditioning in gastrointestinal dysfunction, *Hospital Practice*, 135–143.

Sibuya, S. and Saito, S. (2003). Self-control interventions in cancer patients: 5 cases of natural regression, presented in the *44th Japanese Society of Psychosomatic Medicine, Proceedings*, pp. 111–112.

Surwit, R. S., Feinglos, M. and Scovern, A. (1983). Effect of relaxation training on non-insulin dependent diabetes mellitus, in: *Proceedings of the Biofeedback Society of America, 12th Annual Meeting*, pp. 82–83.

Uchiyama, I., Kudo, M. and Tada, R. (1977). Nursing of patients with a cervical spinal cord injury associated with persistent decubitus ulcer — a study of aged patients with total paralysis after an injury at the C5 level and sufficient recovery for home care, *Kango Gijutsu* **23**, 46–52.

Webb, N. C. (1977). The use of myoelectric feedback in teaching facial expression to the blind. *Biofeedback Self-Regulation* **2**, 147–160.

Weiss, T. and Engel, B. T. (1971). Operant conditioning of heart rate in patients with premature ventricular contraction, *Psychosom. Med.* **33**, 301–321.

Yoshizawa, M., Okamoto, T. and Kumamoto, M. (1983). Effects of EMG-BFT on swimming, in: *Biomechanics VIII*, Vol. 4B, Matsui, H. and Kobayashi, K. (Eds), pp. 828–832. Human Kinetics Publishers, Champaign, Illinois, USA.

Advances in Neuroregulation and Neuroprotection (2005), pp. 517-541
Collin, C. *et al.* (Eds)
© VSP 2005

Conformational diseases and the protein folding problem: Evidence for species-independent link between protein folding and genetic code degeneracy in ribosomal proteins

JEAN SOLOMOVICI [1], THIERRY LESNIK [1], RICARDO EHRLICH [2], HASAN PARVEZ [3], SIMONE PARVEZ [3] and CLAUDE REISS [1,*]

[1] *Alzheim'R&D, 2 rue de la Noue, F91190 Gif, France*
[2] *Seccion Bioquimica, Facultad de. Ciencias, Universidad de la Republica, 4225 Igua, U11400 Montevideo, Uruguay*
[3] *Laboratoire de Neuroendocrinologie, IAF-CNRS F91198 Gif, France*

Abstract—The family of conformational diseases is expanding rapidly, including not only neurodegenerative conditions like Alzheimer's and Parkinson's diseases, but also certain autoimmune diseases and cancers. The corresponding pathologies have in common their link to misfolded proteins, which accumulate in the cytosol, the rough ER or built up extracellularily, thereby inducing apoptosis, abolishing cell to cell communication, eliciting the immune response or cancelling proper enzymatic (i.e. tumour suppressor) activities. Understanding the mechanism leading to protein misconformation is essential for early diagnosis, efficient therapy and reliable prognosis of these devastating conditions. In previous articles, we showed that, in addition to the primary sequence, an important determinant of the protein structure is the local translation rate, controlled mainly by the choice of synonymous codons along the coding sequence. We found that in coding sequences of *E. coli*, amino acids having a propensity to appear in particular secondary structures are located in surroundings strongly constraint in fast, or slowly translated codons. To extend this observation to other species, we analyse here all ribosomal protein genes in four species (*H. sapiens, S. cerevisiae, B. subtilis and E. coli*), for constraints of particular amino acids within sequences of 5 to 21 codons, translated fastest or slowest on average. Although the species investigated make rather different usage of synonymous codons, each species utilizes its synonymous codon repertoire so as to achieve amino-acid constraints in these sequences similar to those observed in *E. coli*. This suggests that the amino-acid constraints within gene sequences translated at extreme rates are biological invariants to which the species have to conform. Although the constraints are more relaxed in sequences translated at less extreme rates, the general conclusion is that the local translation rate makes, at places at least, important contributions to the protein conformation. Conversely, modifications the local translation rate can at places interfere with native folding and lead to misfolding. Such modifications could be the consequence of spontaneous defects in the translation machinery of the cell, or of its unscheduled, excessive solicitation. This may be one of the routes to conformational diseases.

---

*To whom correspondence should be addressed. E-mail: cjreiss@yahoo.com

Keywords: Amino-acid constraints; ribosomal proteins; conformational diseases.

INTRODUCTION

Conformational diseases, which include neurodegenerative conditions like Alzheimer's, Parkinson's, prion diseases, and possibly also certain autoimmune diseases and cancer, are linked to protein misfolding (DeMager *et al.*, 2002; Johnson, 2000). It is known (Schubert *et al.*, 2000) that up to 70% of nascent proteins are detected as misfolded while still attached to the ribosome, meaning that misfolding can occur co-translationally and is a rather common event.

Several experimental results appear to link cotranslational protein folding, or misfolding, to the local translation rate (Cortazzo *et al.*, 2002). Based on appropriate biometric data obtained in species like *E. coli* (Ikemura, 1981a, b), the frequency with which a given codon is used in a species appears to be a measure of the average rate at which it is translated (Berg and Kurland, 1997; Dong *et al.*, 1996). The local translation rate could therefore be derived from the local average of codon frequency (Komar *et al.*, 1999). A protein coding sequence may then be considered a succession of codon clusters, each characterized, for a given cluster size, by the average frequency of the codons present in the cluster, inversely related to the cluster translation rate. Therefore, cluster analysis could help to solve the protein folding (and misfolding) problem.

For cluster sizes ranging from 5 to 21 codons, we observe in an accompanying paper that the clusters having the highest, or lowest, average codon frequency in each *E. coli* gene are characterized by biased use of synonymous codons. We further observe that the protein sequences corresponding to both clusters have characteristic amino-acid constraints. Surprisingly, amino acids favoured in clusters translated slowly have a propensity to enter beta strands, bulges or hinges, and are avoided in sequences translated fast, whereas amino acids favoured in the latter have a propensity to enter alpha helices and are avoided in clusters translated slowly.

From this, we infer that one mechanism leading to protein misfolding would depend on an inappropriate modification of the local translation rate. This could be due to a change in the population of the tRNA pool, synthase activity, yield of amino-acid synthesis, shortage in elongation factors, and impairment of other cellular activities involved in protein synthesis, such as foldases, etc.

If this finding in *E. coli* were of general validity, it would provide new approaches to the study of mechanisms leading to conformational diseases, hence to their early diagnosis, efficient therapy and meaningful prognosis. To assess the general validity of this conclusion, we reasoned that we should investigate the constraints in species for which the genome has been sequenced to a large part, so that the species-specific synonymous codon usage is established. Among such species, we should select those having synonymous codon usage as different as possible. Furthermore, rather than examining the coding sequences of the whole genomes of the selected species, we could restrict ourselves to a set of genes highly conserved in evolution, expressed most heavily in almost all cells in all species and sharing highly similar individual conformations and interactions.

These criteria are best met by the set of ribosomal proteins from *E. coli* (EC), *B. subtilis* (BS), *S. cerevisiae* (SC) and *H. sapiens* (HS). We report here that the constraints observed in the protein sequences translated fastest or slowest in the coding sequences of *E.coli* are not only conserved in the selected species and protein sets, but are even expressed more vigorously in the latter.

MATERIALS AND METHODS

The ribosomal protein gene sequences used in this work were taken from the EMBL DNA sequence base. The computer programmes used were written in Microsoft Quickbasic and can be provided on request to JS.

RESULTS

Codon translation rate

For a cell in a steady growth state, the global mass of protein synthesis varies linearly with time, at a constant average rate. Therefore, per unit time, the average number of codons translated in the cell is constant also and corresponds to the average codon demand. It follows that, at the level of an individual codons (i), the time t_i it takes to translate that codon is at average proportional to the inverse of its demand, the actual codon usage v_i (not the frequency with which it appears in the genome) of codon (i) in all the genes of the species under consideration). Hence $t_i \propto K/v_i$. Within a given cell in a given, steady state, to a good approximation the rate of translation of a given codon is proportional to the frequency with which that codon is used in the global protein synthesis.

Codon usage

We selected as codon usage (CU below) in each species the frequency (%) with which the codon is found in the ribosomal proteins of the species.

Searching genes for segments with highest, or lowest, codon usage

In order to see whether, within coding sequences, some amino acids tend to be associated with particular local translation rates, we set out looking for such correlations in sequence clusters of highest and lowest average CU values. To this end, a window of defined size z (ranging from 5 to 21 codons, as stated) is scanned through the coding sequence of the ribosomal genes by steps of one codon. The arithmetic average of the CU of the z codons in the window is computed for each window position, and the positions of the windows with the highest, zMAX, (and the lowest, zMIN) scores are memorized, together with the codons, and amino acids, present in the windows. Below, we refer to TMAX = 1/zMIN (*highest* average translation *time*, or lowest average CU) and TMIN = 1/zMAX (*lowest* average translation *time* or highest average CU).

Strong amino-acid constraints in both TMAX and TMIN windows, and effect of window size

For each species, we compute the ratio B of the average value of occurrence of the amino acid inside z corresponding to TMAX and TMIN, to its average occurrence in all analysed ribosomal protein genes of the species. B represents the constraints (if any) experienced by the amino acid (avoidance or preference) within the TMAX (or TMIN) window. We consider significant constraints departing from neutrality (B = 1) by 20% at least. Constraints below B = 0.8 are considered as 'discriminating' the codon, those above B = 1.2 are said to be 'favouring' the codon.

To study the effect of z, the window size, on amino-acid constraint, the latter were studied in TMIN and TMAX windows of 5 to 21 consecutive codons. The results are displayed in Fig. 1. The amino acids were classified according to their general characteristics, i.e. hydrophobic, polar uncharged, charged, special and aromatic.

(a) Special amino acids. In all species investigated, they are (with the exception of Gly for z < 13 SC) repressed in TMIN windows, especially Cys (not found for z < 13) and Pro. In contrast, in TMAX windows Cys is strongly favoured, even for z = 21, Pro is favoured in HS and BS, Gly in BS.

(b) Aromatic amino acids. In all species investigated, they are discriminated in TMIN windows, especially Trp, and strongly for z < 13. Conversely, in TMAX, Tyr and Trp are favoured in all four species, and Phe is almost neutral except for z = 5 in HS and BS, where it is avoided.

(c) Polar amino acids. All are repressed in TMIN except Gln in HS, where it is mostly neutral. In TMAX, Ser is strongly favoured in all species, even for z = 21. Gln tends to be repressed, in particular for large windows in EC, but is favoured in BS, Thr and Asn are neutral except Thr in HS where it is slightly favoured, and Asn, discriminated in small windows in all species except EC.

(d) Charged amino acids. Asp and Glu display a strikingly symmetric behaviour in both TMAX and TMIN. Glu is favoured in TMIN (except in BS for z > 9), whilst Asp is repressed (weakly in BS again). In TMAX, Asp is neutral, but Glu is repressed. Lys behaves like Glu in being strongly favoured in TMIN windows in the four species, as does, to a lesser extent, Arg (except in HS). In contrast, His and Arg are repressed in HS, SC, EC, especially in small windows. In TMAX windows, the five amino acids are either repressed or neutral, with a significant exception for His in SC and EC, where it is favoured.

(e) Hydrophobic amino acids. Ala and His, which have the smaller side chains, are generally avoided or neutral in TMAX windows, with the exception of Met in

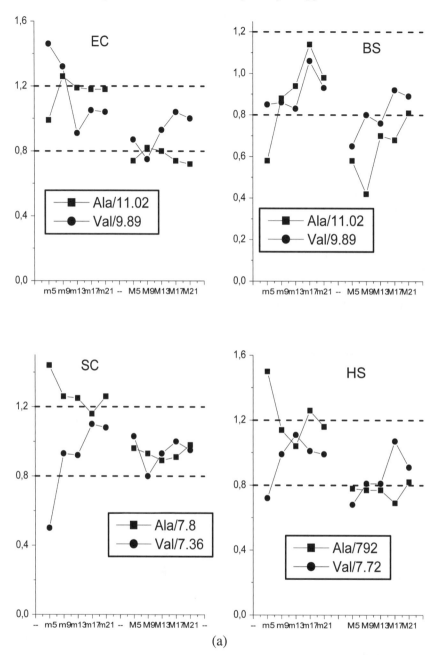

Figure 1. Dependence of constraint B on window size z for different amino acids in the species *E. coli* (EC); *B. subtilis* (BS); *S. cerevisiae* (SC) and *H. sapiens* (HS). Ordinate: constraint B; abscissa: window size z (5, 9, 13, 17 and 21 codons), for TMIN (mz, left half of each graph) and TMAX (Mz, right half of each graph) windows. Amino acids are denoted in the usual 3-letter code, together with their relative (%) abundance in the ribosomal proteins of the species. The dotted horizontal lines, at ordinates 0.8 and 1.2, delimit the range of significant constraint (B below 0.8, or above 1.2) — see text. (a) for the hydrophobic amino acids Ala and Val.

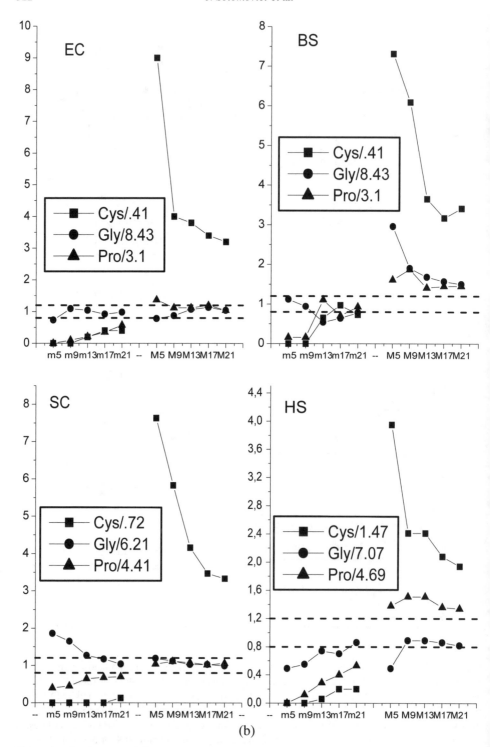

Figure 1. (Continued). (b) for the special amino acids Cys, Gly and Pro.

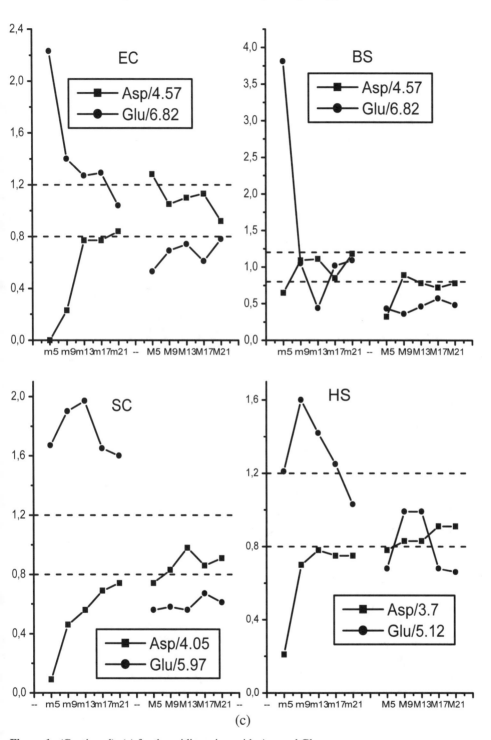

Figure 1. (Continued). (c) for the acidic amino acids Asp and Glu.

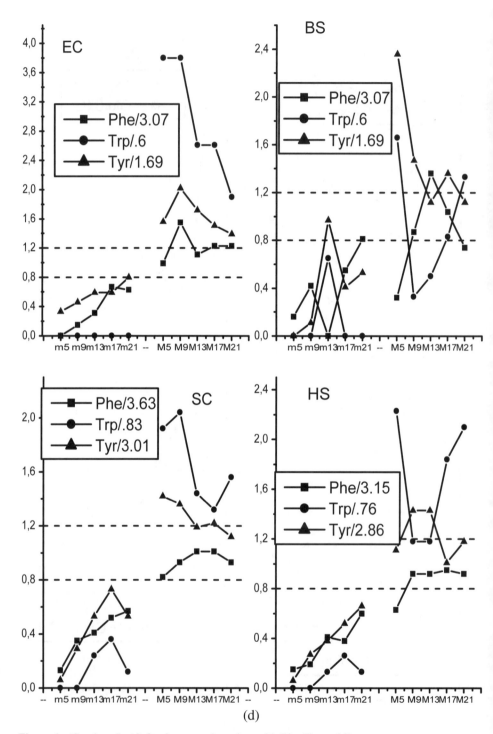

Figure 1. (Continued). (d) for the aromatic amino acids Phe, Trp and Tyr.

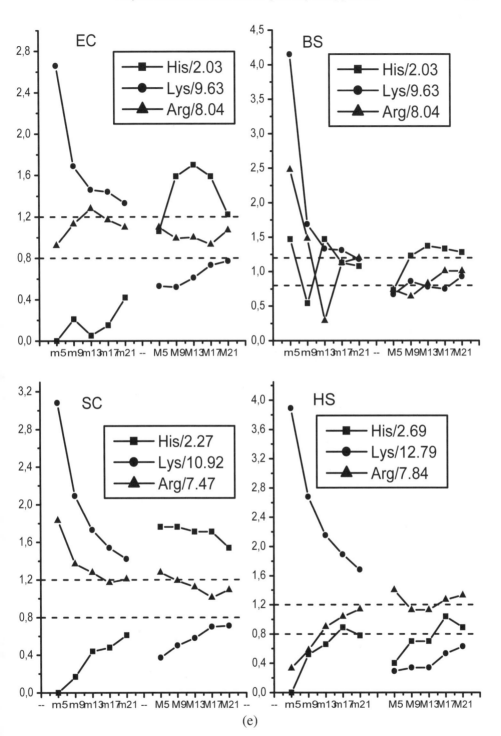

Figure 1. (Continued). (e) for the basic amino acids His, Lys and Arg.

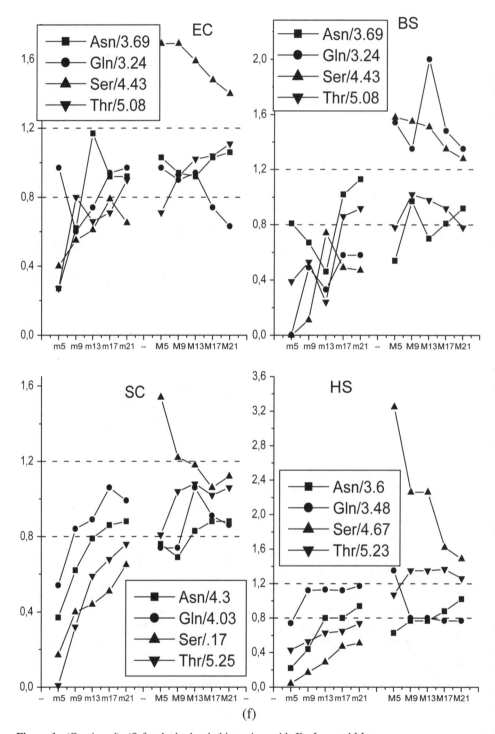

Figure 1. (Continued). (f) for the hydrophobic amino acids Ile, Leu and Met.

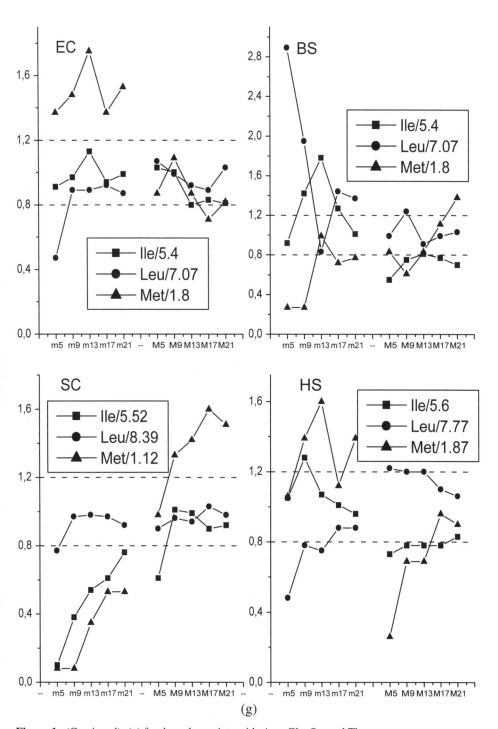

(g)

Figure 1. (Continued). (g) for the polar amino acids Asn, Gln, Ser and Thr.

SC, for z > 5. Their occurrence in TMIN windows is more disparate. Ala is favoured in TMIN in HS, EC, and SC, for z = 5, as is Met in HS and EC, and Leu in SB.

In most instances, despite the small size of the gene set used (50 to 70 sequences, average size about 200 codons), amino-acid constraints vary fairly monotonously with z (some hydrophobic amino acids in TMIN windows excepted). With few exceptions, the constraints (whether favouring or discriminating the amino acid) increase as z is reduced. For z = 5, the constraints become usually rather strong.

Taking as significant constraints those departing from neutrality (B = 1) by 20% at least, in summary, in the four species investigated, the TMAX window (translated slowest) favours most Cys, Trp, Ser, and discriminates most against Lys and Glu. The TMIN window (translated fastest) discriminates most Cys, Pro, Trp, Phe, Tyr, Ser, Thr, Asn, His and Asp.

Strategy of species-specific, synonymous codon usage in TMAX and TMIN windows

Each of the four species investigated makes a particular use of synonymous codon in its ribosomal protein genes. Since the TMAX and TMIN windows are the consequence of a local bias in synonymous, rare or frequent codons, respectively, these biases must be species-specific. We therefore investigated the codon usage in these windows, for each amino acid and window size ranging from 21 codons to 5 codons (Fig. 2).

The examination of the codon constraints in TMIN and TMAX in general reveals, as expected, that rare codons are favoured in TMAX windows, and conversely frequent codons are favoured in TMIN windows. These constraints are of course the stronger, the smaller the window. However, in many instances, strong constraints are already observed in the largest window (21 codons), in which on average not all codons are the most frequent (or rare). Hence these constraints result from a systematic strategy to localize the concerned codons in a sequence context globally translated 'fast' or 'slowly': the constraint is 'extended'. In cases where the constraint appears for smaller windows only, the corresponding modulation of the translation rate is restricted to areas close to the codon considered: the constraint is 'local'.

We chose to compare the synonymous codon selection in TMIN and TMAX, for ribosomal protein genes of the eukaryotic species, HS and SC. We will comment on the selection strategies for a representative set of amino acids only, and stress the comparison on mainly three features: the range covered by the constraint ('local', i.e. 5 or 9 codon windows, or 'extended', i.e. 5 to 21 codon windows), the intensity of the constraint, and the correlation of the codon selection with its usage.

For Leu, both species display similar behaviour in TMIN, the most used codon (CGT in HS, TTG in SC) experiences no significant constraint, while all others suffer extended and intense repression. In TMAX , the three less used codons (TTA, CTA and TTG) in HS are massively favoured in all windows, but this is shared in SC

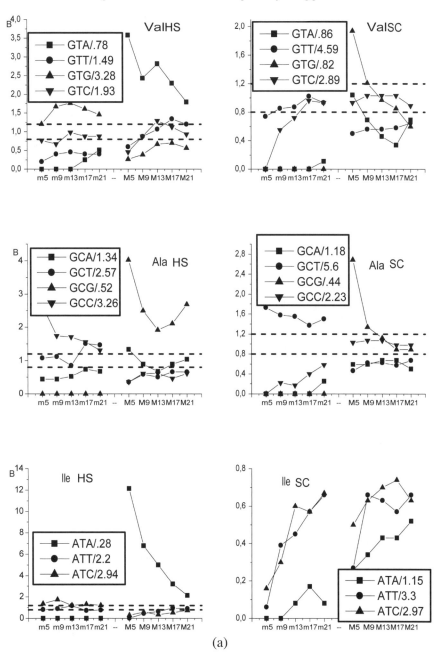

Figure 2. Dependence of constraint B on window size z for the codons of different amino acids in the species *H. sapiens* (HS) and *S. cerevisiae* (SC). Ordinate: constraint B; abscissa: window size z (5, 9, 13, 17 and 21 codons), for TMIN (mz, left half of each graph) and TMAX (Mz, right half of each graph) windows. Amino acids are denoted in the usual 3-letter code, together with the relative (%) abundance of the codons in the ribosomal proteins. The dotted horizontal lines, at ordinates 0.8 and 1.2, delimit the range of significant constraint (B below 0.8, or above 1.2) — see text. (a) for the hydrophobic amino acids Val, Ala, and Ile.

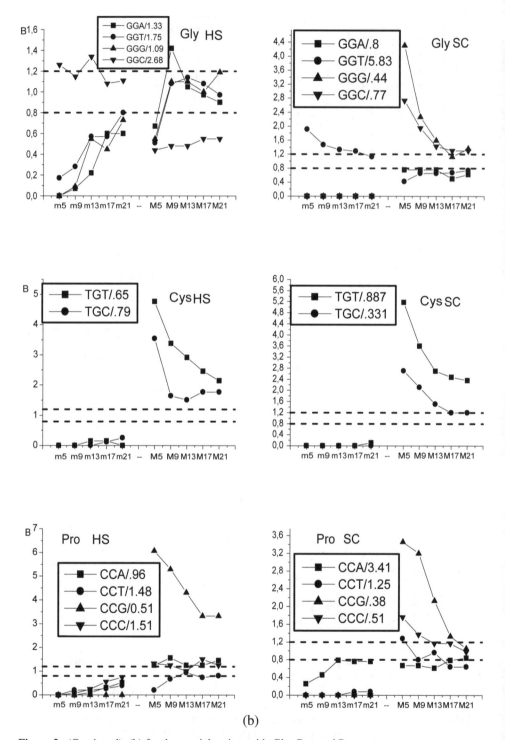

Figure 2. (Continued). (b) for the special amino acids Gly, Cys, and Pro.

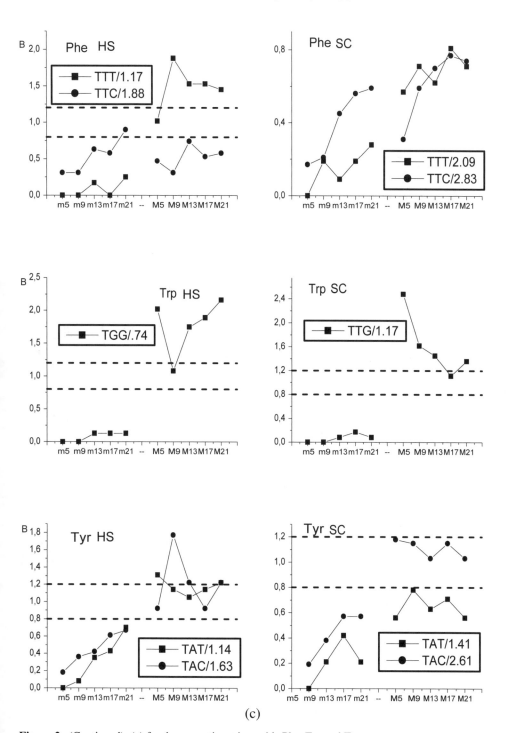

Figure 2. (Continued). (c) for the aromatic amino acids Phe, Trp and Trp.

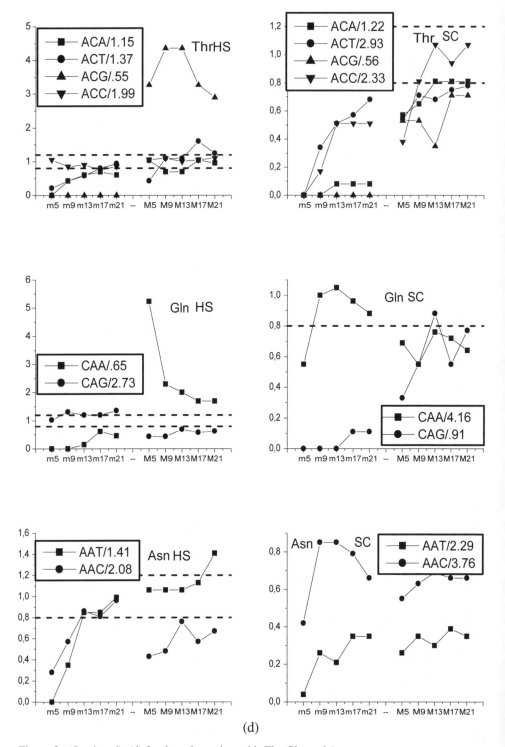

(d)

Figure 2. (Continued). (d) for the polar amino acids Thr, Gln, and Asn.

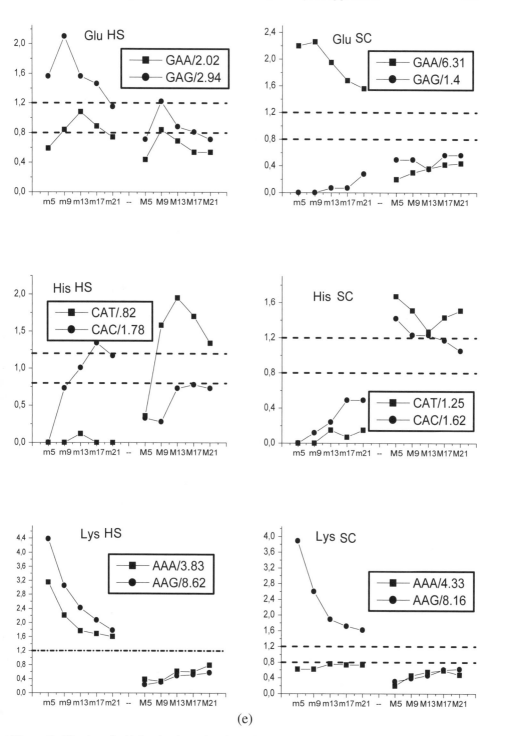

(e)

Figure 2. (Continued). (e) for the charged amino acids Glu, His and Lys.

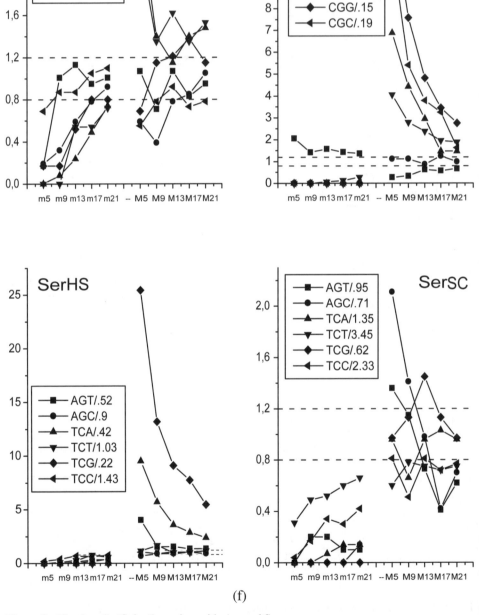

(f)

Figure 2. (Continued). (f) for the amino acids Arg and Ser.

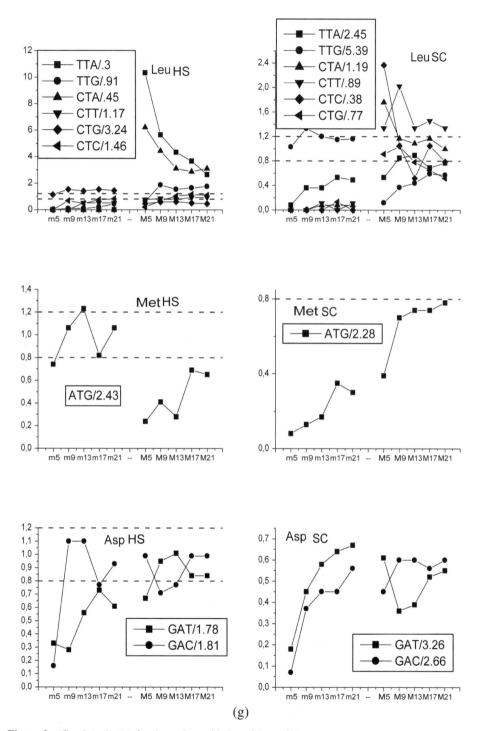

Figure 2. (Continued). (g) for the amino acids Leu, Met and Asp.

by one codon (CTT) only, which is not the less used, and its constraint is significant but less.

Glu is encoded by two codons with comparable usage in HS, four-fold different in SC. For TMIN, the most abundant codons in both species share strong and extended favourable selection, the less abundant are both discriminated (discrimination is strong and extended in SC). For TMAX, both codons share extended and strong discrimination in SC, and the same holds for the less abundant codon in HS, but for the most abundant, discrimination is barely significant.

Lys is encoded by two codons with comparable usage in both species, the usage of one codon being about half that of the other. For TMAX, in both species, both codons share strong and extended discrimination. For TMIN, in HS both codons experience strong and extended selection, but in SC this holds for the most abundant codon only, as the other suffers extended discrimination.

Pro is encoded by four codons. Their usage differs three-fold in HS, nine-fold in SC. In TMIN and in both species, all codons share strong and extended discrimination. In TMAX and in both species, only the less used codon benefits from strong and extended selection; the others are barely constrained.

These results show that each species, characterized by its distribution of synonymous codons, has its own strategy in selecting synonymous codons in the structural genes coding for its ribosomal proteins. The strategies used in individual species do however share a common feature: within codon sequences constrained on average in rare or frequent codons, codons for most amino acids are significantly favoured or discriminated.

DISCUSSION

In a previous report (Solomovici *et al.*, 2004), we have analysed in each coding sequence of *E. coli* the codon sequences having on average the highest contents in frequent or rare codons (averages were taken over sequences of $z = 5$ to 21 codons).

We first noticed that biometric studies in this species establish a strong correlation between the relative codon usage and the relative abundance of the cognate transfer RNA. Furthermore, the latter appears to limit the codon translation rate. Hence, codon usage correlates with codon translation rate. Within a given coding sequence, codon sequences having on average the highest contents in frequent or rare codons are those translated fastest or slowest; in other words their translation time is minimum (TMIN) or maximum (TMAX), respectively.

We also cited experimental evidence indicating that the local translation rate appears to take part in cotranslational protein folding. In addition, it is known that the amino acids favoured in TMAX are preferentially found in beta structures, while those favoured in TMIN enter preferentially alpha helices. This is consistent with the rationale stating that the latter involve near-neighbour amino acids and can therefore form fast. Beta sheets in contrast involve more or less remote amino-acid sequences, which must be maintained in the required folding capacity for extended periods of time. This would need the assistance of dedicated foldases, which would require time to 'react' with the strands involved and keep them in a competent structural state. Slow translation of these strands would allow for these reactions to occur.

In summary, in *E. coli*, the local selection of synonymous codons specifying TMIN and TMAX, favours, or discriminates against, most amino acids. Amino acids encoded in local sequences translated at extreme rates appear to be critical for the architecture of the protein.

Although this finding and the associated hypothesis are of interest, they are derived from a single species and for the whole set of coding sequences of this species, corresponding to proteins of rather diverse conformations. It remained to be seen whether these observations are of general value, holding in other species as well, and for coding sequences corresponding to homologous protein structures in these species. We further reasoned that the constraints should be exacerbated in genes expressed most heavily, for which the coding sequence would be expected to be optimised in terms of expression 'economy', i.e. usage of most abundant codons to enable maximal over-all rate of protein synthesis, and limited (or no) requirement for foldases, etc.

We therefore selected four species having rather different strategies in using synonymous codons, and in each of them the genes for ribosomal proteins. The number of the latter ranges from 50 to 70 in the four species, which provides reasonable statistical significance to our data. In all species, ribosomal protein genes are the most heavily expressed in almost all cells and they share gross structural homologies.

The results show that the amino-acid constraints in TMAX and TMIN windows observed in *E. coli* are strikingly similar to those in the ribosomal protein genes of EC, BS, SC and HS. These four species are rather diverse (Gram-negative and Gram-positive prokaryotes, fungal and mammalian eukaryotes). The usage of synonymous codons can be rather different (CGC for instance is 20 times less used in SC than in HS, CAA is over six times more frequent in SC compared to HS, but the synonymous CAG is 3 times more frequent in HS than in SC).

As already observed in *E. coli*, most amino acids display 'antisymmetric' behaviour in TMIN and TMAX, being discriminated in one and favoured in the other, sometimes with comparable strength. Also, we have observed that TMIN or TMAX do not necessarily involve the same sequence as the size z of the window is reduced from 21 to 5. For some amino acids, the variation of the constraint B with z is smooth, but for others it appears chaotic. For the few cases investigated in some detail, a smooth variation is associated with a fixed location of TMAX or TMIN sequences within the gene. The chaotic behaviour can be explained by the low number of ribosomal protein sequences in each species, and the paucity of certain amino-acids species in these genes.

We conclude that the amino-acid constraints observed in TMIN and TMAX are mainly species-independent, despite the species-specific usage of synonymous codons that encode TMIN and TMAX in the mRNA. Characteristic amino-acid constraints within TMIN and TMAX appear as biological invariants, which may be valid in general across species barriers (the results reported here, derived from sequences coding for ribosomal proteins, remains essentially valid in whole genome analysis of the four species (JS, unpublished results)).

The remarkable invariance of amino-acid constraints in TMAX and TMIN implies that within the latter, each species must have adapted its synonymous codon usage accordingly. A given species is free to set its own usage of synonymous codons, but it must provide adapted synonymous codon sequences at places where amino acids must be incorporated in the nascent protein at a specific rate. If indeed the local translation rate contributes to local protein folding, as observed in some experiments, the biological invariance is then a simple consequence of the physical and chemical constraints imposed by protein folding, which indeed must be species-independent.

The detailed analysis of individual codon constraints in HS and SC shows that minor codons tend to be preferred in TMAX, frequent codons in TMIN. The constraint on the corresponding amino acid is the average of individual codon constraints. For instance, in HS the six codons synonymous for Arg are of comparable frequency; in TMAX, two codons are preferred, two are neutral and two are discriminated, the amino acid is slightly preferred; in TMIN, two codons are neutral, four are discriminated, the amino acid is discriminated for $z < 13$, neutral above.

The local extent of the constraint can be derived from the B *vs.* z plot. Cys for instance is strongly discriminated in TMIN, and copiously favoured in TMAX, in all four species. The strength is already robust for $z = 21$, and rises steeply as z decreases. This means that the constraint on Cys results from long-range synonymous codon selections which creates an extended 'translation rate' domain. Fast-rated domains almost completely exclude Cys, but in contrast this amino acid is found up to 9-fold more frequently than average in slow-rated domains. We may conjecture that the purpose of this constraint is to allow the concerned Cys to enter disulfide bonds, which need enough time for the action of dedicated enzymes. Other amino acids exhibit strong constraints within small windows only. It could be that the latter require the extreme translation rate at a short-range only. Alternatively, the strong local constraint could be a simple consequence of the incidence of the amino acid's codon frequency on TMAX or TMIN, which increases as the window size reduces.

The findings reported here strongly suggest that at places of the mRNA, the local translation rate assists folding of secondary structures, and even the completion of tertiary and quaternary structures. Amino acids favoured in TMIN are barely exposed at the tip of the nascent protein to foldase 'reaction' and could therefore easily enter local secondary structure (mainly helices), whilst amino acids avoided in TMIN are prevented from entering such structures. In contrast, amino acids favoured in TMAX can react with diverse foldases, or several times with a given foldase (case of S—S bond rearrangement), and as a consequence either remain in a local conformation competent for long-range interactions (beta-sheet) instead of entering a short-range structure (like alpha helix), or remain available to enzymes like Pro *cis-trans* isomerase at places where a kink involving Pro is required (Pro is avoided in TMIN), or glutathione binding to those Cysteines which later enter S—S bond formation in the rough endoplasmic reticulum, etc. It is remarkable that amino acids having a general propensity to enter alpha helices are preferred in TMIN, those found in beta structures are favoured in TMAX, and those preferred in one are avoided in the other.

The findings could be summarized in form of a variant of the Anfinsen principle. The latter states that the conformation of a protein in defined by its amino-acid sequence. The principle holds for small proteins, which indeed usually can refold (as a rule rather slowly) to their native structure, but refolding can be problematic for larger proteins. This could be due to the fact that the refolding experiments lack the kinetic information provided cotranslationally by the synonymous codon usage. The same holds for the expression of exogenous genes, which results frequently in the accumulation of inclusion bodies made of misfolded gene products. This shows that even *in vivo*, the Anfinsen principle is violated because probably the synonymous codon selections in the donor species and the host are different. Based on our findings, we propose an amended version of the principle: the native conformation of a protein is defined by the *codon* sequence of its mRNA *and the synonymous codon usage in the species from which the protein is derived.*

As a corollary to this statement, modifications of the local translation rate determined by the local synonymous codon selection could in some places modify the local or distant cotranslationnal folding and result in a non-native conformation of the protein product. A coding sequence is associated with the map of the local translation rate, determined by the usage of the codons along the coding sequence. No misfolding would then be expected, provided:

(1) The codon usage correlates with the amount of charged, cognate tRNA, which set the codon translation rate.

(2) All other devices needed for protein synthesis (elongation factors, synthases, amino acids . . .) are available as required or in excess.

(3) All foldases (cytosolique or in the ER) needed to assist co- or post-translationnal folding are available as required or in excess.

(4) The folding devices born by the ribosome are operative.

Any defect in one of these conditions, or in the capability of the enzymes involved, due to mutation or otherwise defective biochemical activity, could produce misfolded proteins. For instance, unscheduled massive protein synthesis, due to untimely, extracellular or environmental signalling events (hormone analogs), could exceed the cellular resources needed for protein synthesis and folding, and result in misfolded proteins.

Declining production of some of the cellular resources needed for translation, due to senescence or some pathologic state, could also be responsible for protein misfolding. One may suspect the latter process to be responsible for conformational diseases occuring usually in the last quarter of human life, in particular, certain dementia, autoimmune diseases and types of cancers. These could be linked to progressive shortages in amino acids, synthases, tRNAs, elongation factors, foldases, etc, which could gradually increase the amount of misfolded protein species and favour their aggregation, by inhibiting their proteolysis.

This could in particular affect ribosomal proteins, which are the most abundant proteins in the cell. Misconformation of some ribosomal protein could affect the capacity of the ribosome to give access to, or release, tRNAs, or interfere with mRNA or nascent protein translocation rate, or modify the interaction with ribosomal components triggering local conformation events in, or at top of the protein exit tunnel. Hence minor defects in a component of translation machinery could promote a major cellular production of misfolded proteins, simply by misfolding some critical ribosomal protein. It would for instance be enough to reduce the production of an amino acid, or of a particular tRNA, or even of the modified bases at or close to its anticodon (which participates in the decoding rate), to perturb the local protein translation rate and give rise to massive production of various mifolded proteins.

Based on this mechanism leading to protein misfolding and eventually to conformational diseases, ways to early diagnostic, efficient therapy and reliable prognostic can be envisioned.

REFERENCES

Berg, O. G. and Kurland, C. G. (1997). Growth rate-optimised tRNA abundance and codon usage, *J. Mol. Biol.* **270**, 544–550.

Cortazzo, P., Cervenansky, C., Marin, M., Reiss, C., Ehrlich, R. and Deana, A. (2002). Silent mutations affect in vivo protein folding in *Escherichia coli*, *Biochem. Biophys. Res. Commun.* **293**, 537–541.

DeMager, P. P., Penke, B., Walter, R., Harkany, T. and Hartignny, W. (2002). Pathological peptide folding in Alzheimer's disease and other conformational disorders, *Curr. Med. Chem.* **9**, 1763–1780.

Dong, H., Nilsson, L. and Kurland, C. G. (1996). Co-variation of tRNA abundance and codon usage in *Escherichia coli* at different growth rates, *J. Mol. Biol.* **260**, 649–663.

Ikemura, T. (1981a). Correlation between the abundance of *Escherichia coli* transfer RNAs and the occurrence of the respective codons in its protein genes, *J. Mol. Biol.* **146**, 1–21.

Ikemura, T. (1981b). Correlation between the abundance of *Escherichia coli* transfer RNAs and the occurrence of the respective codons in its protein genes: a proposal for a synonymous codon choice that is optimal for the *E. coli* translational system, *J. Mol. Biol.* **151**, 389–409.

Johnson, W. G. (2000). Late-onset neurodegenerative diseases — the role of protein insolubility, *J. Anat.* **196**, 609–616.

Komar, A. A., Lesnik, T. and Reiss, C. (1999). Synonymous codon substitutions affect ribosome traffic and protein folding during in vitro translation, *FEBS Lett.* **462**, 387–391.

Schubert, U., Anton, L. C., Gibbs, J., Norbury, C. C., Yewdell, J. W. and Bennink, J. R. (2000). Rapid degradation of a large fraction of newly synthesized proteins by proteasomes, *Nature* **404**, 770–774.

Solomovici, J. (2004). *Biogenic Amines* **18** (3-6), 207.

Advances in Neuroregulation and Neuroprotection (2005), pp. 543-584
Collin, C. *et al.* (Eds)
© VSP 2005

Long-term stem cell gene therapy: zinc finger nuclease-boosted gene targeting and synergistic transient regenerative gene therapy

ROGER BERTOLOTTI*

*CNRS, Gene Therapy and Regulation Research, Faculty of Medicine,
University of Nice Sophia Antipolis, 06107 Nice, France*

Abstract—Stem cells have both self-renewing and homing/differentiative capabilities and thereby provide for life-long cell replacement in tissues and organs. They are therefore the ideal targets for most gene therapy protocols, i.e. both for long-term and transient gene expression protocols where they are either the long term carriers of the therapeutic gene (inherited/degenerative disease) or the mobilized/recruted targets of a transient regenerative process such as the formation of new blood vessels. Long-term gene therapy is thus amenable to synergistic combinations where *ex vivo/in vivo* genetic engineering of stem cells can be associated with *in vivo* transient topical expression of minigenes encoding various factors such as homing, regenerative and differentiative ones. Such a strategy is still hampered by many hurdles of which random integration of therapeutic DNA is a major safety concern. Emerging technologies are however aimed at efficient site-specific integration of therapeutic transgenes and at endogenous gene repair/modification. Promising site-specific integration vectors are at the pre-clinical stage and rely on an adeno-associated virus (AAV) *rep* platform or on phage phiC31 integrase. However, unlike these approaches, gene targeting is driven by homologous recombination and has thus target flexibility. It mediates DNA exchanges between chromosomal DNA and transfecting/transducing DNA, thereby providing the means to modify at will the sequence of target chromosomal DNA. Gene targeting stands thus as the ultimate process both for gene repair/alteration and targeted (i.e. site-specific) transgene integration. Such a process is however highly inefficient unless target chromosomal DNA is struck by a double-strand break (DSB). In addition, it is overwhelmed by random integration. In order to increase gene targeting frequency and eliminate random integration, we devised an approach that relies on the transfer into target cells of premade presynaptic filaments, i.e. the very complexes between recombinase protein and single-stranded DNA (ssDNA) that mediate the key reaction from homologous recombination (homologous DNA pairing-strand exchange with double-stranded [ds] DNA). Upon publication of the enzymatic properties of recombinase RAD51, we invented chimeric presynaptic filaments with a dsDNA core, and shifted from gene conversion to 'true' gene targeting. In these dsDNA-cored presynaptic filaments, therapeutic heterologous sequences and repetitive elements are comprised in the recombination-locked dsDNA core, and homologous recombination is driven by both recombination-activated ssDNA tails that are 100% identical to their respective target chromosomal sequences.

---

*E-mail: Roger.Bertolotti@unice.fr

Chimeric zinc finger nucleases are now available that create sequence-specific DSBs in target chromosomal DNA and stimulate gene targeting as expected. Such designed nucleases open exciting potentialities for standard gene targeting (non-viral) but also for emerging AAV gene targeting that has been shown to be boosted by DSB too. In these approaches, gene targeting frequency is raised to the random-integration level, i.e. ~1% of transfected/transduced cells. Our current approach is thus discussed in terms of synergistic combinations in which random-integration is blocked by the use of pre-made presynaptic complexes (stockiometric coating of transfecting DNA by RAD51 protein) and homologous recombination promoted both by transfecting pre-made presynaptic complexes and by sequence-specific DSB of target chromosomal DNA. Long-term gene therapy is thus amenable to sophisticated protocols in which stem-cell gene-targeting is combined with transient regenerative gene therapy, and could therefore in the future apply to the treatment of the neurologic symptoms of the Lesch–Nyhan disease in which the target is our paragon model, the hypoxanthine-guanine phosphoribosyl transferase (HPRT) gene.

Keywords: Stem cell gene therapy; transient regenerative gene therapy; zinc-finger nucleases; double strand break-boosted gene targeting; recombinase-DNA nucleoprotein vectors; AAV-mediated gene targeting.

INTRODUCTION

Stem cells as targets for both long-term and transient gene therapy protocols

Stem cells have both self-renewing and differentiative capabilities, thereby providing for life-long cell replacement in tissues and organs. They are thus the ideal targets for most gene therapy protocols, i.e. both for long-term and transient gene expression protocols (Bertolotti, 1998c, 2000b) where the therapeutic gene product is either a permanent need or an inducing factor, respectively (see Bertolotti, 2001, 2003a, c).

In long-term gene therapy protocols, targeted stem cells are expected to provide for life-long transgenic cell replacement as illustrated with hematopoietic stem cells (HSCs) for X-linked severe combined immunodeficiency (X-SCID) patients (Cavazzano-Calvo *et al.*, 2000). In this seminal gene therapy clinical trial, therapeutic stem cells have integrated in their genome a copy of a retroviral vector engineered to comprise a γC minigene encoding the common cytokine-receptor γ-chain that is defective in X-SCID patients.

In transient gene therapy protocols, stem/progenitor cells are mobilized and recruited in a regenerative process such as the formation of new blood vessels (see Bertolotti, 2001). By its transient nature relying on the safe use of non-integrating DNA, early therapeutic angio/vasculogenic gene therapy for critical limb ischemia (1994–1998) has been instrumental in the establishment of gene therapy as an effective medical practice (see Bertolotti, 2002a, b). The clinical outcomes of these pioneering Phase I/Phase II trials conducted by Isner and co-workers (Baumgartner *et al.*, 1998; Isner *et al.*, 1996, 1998) have been promising enough to prompt the recruitment of new patients (Baumgartner *et al.*, 2000; see Baumgartner, 2003) and to promote the extension of the approach to myocardial ischemic diseases both by

Isner and co-workers (Losordo *et al.*, 1998, 2002; Symes *et al.*, 1999, 2003; Vale *et al.*, 2000, 2001) and other groups (Rosengart *et al.*, 1999; Grines *et al.*, 2002; see Penny and Hammond, 2003).

Long-term gene therapy: stem cells and life-long therapeutic gene expression paradigm

The aforementioned seminal clinical gene therapy trial for an inherited disease conducted by Fischer and co-workers (Cavazzana-Calvo *et al.*, 2000) on X-SCID patients stands as the first unequivocal success for gene therapy and as the very paradigm of life-long therapeutic gene expression mediated by targeted autologous stem cells. Indeed, engraftment and selective repopulating ability of *ex vivo* γC-transduced HSCs have been instrumental in the success of this gene therapy trial where efficient retroviral transduction of bone marrow (BM) HSCs is synergized by the homing ability of these transduced stem cells and the selective growth advantage of their lymphoid progenitor derivatives over mutant X-SCID cognates. The self-renewing and differentiative capability of engrafted γC-transduced stem cells secures thus long-term transgenic lymphoid cell turn-over (Hacein-Bey-Abina *et al.*, 2002) and is thus expected to result in a life-long cure for this inherited disease. The same holds true for ADA-SCID in which the pathology results from an adenosine deaminase (ADA) deficiency (Aiuti *et al.*, 2002).

Interestingly, adult stem cells have been identified recently in tissues and organs such as brain (reviews: McKay, 1997; Gage, 2000; Momma *et al.*, 2000; Alvarez-Buylla and Garcia-Verdugo, 2002) and myocardium (Kajstura *et al.*, 1998; Beltrami *et al.*, 2003; review: Anversa and Nadal-Ginard, 2002) in which cell turn-over was reputedly absent, suggesting that stem cell targeting is most likely a quasi-universal requirement for life-long therapeutic gene expression (Bertolotti, 2003c). In this respect, the recent report of dopaminergic neurogenesis in the *substantia nigra* of adult rodent brain (Zhao *et al.*, 2003) opens exciting promises for stem cell gene therapy of Parkinson's disease.

Stem cell transgenesis and long-term safety

Stem cells are also the driving force of cancer (Reya *et al.*, 2001; Bonnet and Dick, 1997) where uncontrolled self-renewal culminates in tumorigenesis. Cancer stem cells have been identified in leukemia (Lapidot *et al.*, 1994; Bonnet and Dick, 1997) and in solid tumors (Al-Hajj *et al.*, 2003; Singh *et al.*, 2003), and shown to (1) both self-renew and differentiate giving rise to a heterogenous clonal population in which their unique proliferative ability drives both tumor cell replacement and unlimited growth, and (2) to engraft like normal stem cells (e.g. HSCs) into experimental hosts where they are the sole initiator of tumor propagation (Reya *et al.*, 2001; Dick, 2003).

Stem cells stand thus as the critical target for cancer transformation since (1) their conversion to cancer stem cells appears straightforward compared to putative reverse-differentiation of proliferation-restricted progenitor/mature cells, and (2) their long-lasting potential provides time flexibility for accumulation/inforcement of neoplastic mutations (Reya *et al.*, 2001; Dick, 2003). Consistent with these observations, efficient retroviral transduction and engraftment of therapeutic HSCs resulted both in (1) the very first unequivocal success for gene therapy (Cavazzana-Calvo *et al.*, 2000) and in the first cases (2 out of 9 successfully treated patients) of retroviral vector-mediated random insertional carcinogenesis (Hacein-Bey-Abina *et al.*, 2003a, b), and in (2) delayed pathologic symptoms that appeared almost three years after successful therapeutic gene transfer.

Stem cells are thus two-faceted targets with which the corollary of efficient retroviral transduction and selective engraftment is significant risk of random-insertional carcinogenesis. Like in the aforementioned two SCID-cases, random integration of transgenic DNA into host chromosomal DNA is an insertional mutagenesis event that can hit a cancer-prone gene (e.g. oncogene, tumor suppressor or DNA-repair gene), thereby promoting a true long-term safety hazard. Importantly, random integration is a common feature of current long-term gene therapy clinical vectors, i.e. replication-defective retroviruses and first generation Adeno-Associated Virus (AAV) vectors (see Bertolotti, 2003b). On the other hand, non-integrating vectors (e.g. naked DNA and replication-defective adenoviral vectors) are lost during stem/progenitor cell proliferation and are thus restricted to transient expression protocols (e.g. aforedescribed angio/vasculogenic gene therapy).

Site-specific integrating vectors and gene repair technologies

As shown above, although a theoretically low-probability risk, random insertional mutagenesis is a true potential drawback of retroviral vectors and other random-integrating vectors, in particular with repopulating stem cell targets. In addition, random integration into host chromosomal DNA does not provide for optimal transgene expression or regulation, a major concern when, unlike with SCID patients, we deal with a tightly regulated function (see Bertolotti, 1998a, 2000c). Site-specific integrating vectors are thus under intensive investigations and are now emerging both for viral and non-viral gene therapy protocols (see Bertolotti, 2003b). They are based on phage phiC31 integrase (see Olivares and Calos, 2003) and on a platform relying on AAV Rep78 or Rep68 protein (see below and Owens, 2002; Allen and Samulski, 2003; Fraefel *et al.*, 2003). On the other hand, the two cases of random-insertional pathology described above (Hacein-Bey-Abina *et al.*, 2003a, b) have rekindled interest in the genesis of efficient chimeric site-specific retroviral integrases based on the fusion of native enzyme to sequence-specific DNA-binding domains (review: Bushman, 2002).

Although promising, these site-specific integrating vectors lack chromosomal target flexibility and might thus not be optimal for transgene expression in various cell differentiation backgrounds. Unlike these approaches, integration of exogenous DNA mediated by gene targeting is driven by homologous recombination with chromosomal DNA (see Capecchi, 1989) and has thus no target restriction since the integration site is not specified by the DNA binding domain of a protein but by the very DNA sequence of the targeting vector that is homologous to target chromosomal DNA. Gene targeting mediates thus flexible DNA exchanges between chromosomal DNA and transfecting/transducing DNA, thereby providing the means to modify at will the sequence of target chromosomal DNA (see Capecchi, 1989). Gene targeting stands thus not only as the ultimate process for site-specific (i.e. targeted) transgene integration, but also for gene repair or alteration.

Repairing the dysfunctional gene in appropriate somatic cell targets is of course the most obvious approach to tackle an inherited disease: reestablishing the wild type sequence of the mutant gene and/or of its normal genomic background would restore genetic homeostasis and would prevent or reverse the genetic disorder (see Bertolotti, 2000a, c). Prevention is of course the best way since, after the onset of the disease, complete phenotypic reversion is not likely to be possible in many cases. Gene repair is thus the focus of intensive investigations that include oligonucleotide-based approaches (see Igoucheva *et al.*, 2001; Liu *et al.*, 2001a; Seidman and Glazer, 2003; Yoon, 2000) and the small fragment homologous replacement (SFHR) technique (see Gruenert *et al.*, 2003) in addition to emerging gene targeting technology (see below).

Emerging gene targeting vectors: toward efficient gene repair/alteration and non-random integration of therapeutic transgenes

Although ideal both for gene repair/alteration and for targeted integration of therapeutic transgenes, standard gene targeting is a highly inefficient process (Capecchi, 1989; Yanez and Porter, 1998) unless a double-strand break (DSB) hits target chromosomal DNA (Rouet *et al.*, 1994b; Smih *et al.*, 1995). In addition, it is overwhelmed by random integration (Capecchi, 1989; Merrihew *et al.*, 1996; Yanez and Porter, 1998). For these reasons, we devised in 1991 a new a new approach to gene targeting aimed at increasing its efficiency to a level compatible with gene therapy protocols. Our approach is based on the transfer of premade presynaptic filaments, i.e. the active recombinase-DNA nucleoprotein complexes that mediate the key reaction from homologous recombination (Bertolotti R., Grant application to French MRT, 1991; Bertolotti, 1996a, b, 1998a; review: Bertolotti, 2000c). Chimeric zinc finger nucleases are now available that create sequence-specific DSBs in target chromosomal DNA and stimulate gene targeting as expected (Bibikova *et al.*, 2003; Porteus and Baltimore, 2003). Our current approach is thus discussed in terms of synergistic combinations in which random-integration is blocked by the use of pre-made presynaptic complexes (stoichiometric coating of transfecting DNA by recombinase RAD51 protein) and homologous recombination promoted both by pre-made presynaptic complexes and by sequence-specific DSBs in target chromosomal DNA (Bertolotti, 2004). Additional features that might possibly be gained from AAV inverted terminal repeat (ITR) structure are also presented in view of the gene targeting ability that has been recently established for AAV vectors (Russell and Hirata, 1998).

HPRT GENE: TARGET OF LESCH-NYHAN DISEASE AND MODEL FOR GENE TARGETING

Lesch–Nyhan disease: from HPRT-deficiency metabolic impairment to brain dopaminergic deficit

Complete or virtually complete deficiency in hypoxanthine-guanine phosphoribosyl transferase (HPRT) enzyme activity results in Lesch–Nyhan syndrome (Seegmiller *et al.*, 1967) with hyperuricemia, choreoathetosis, compulsive self-mutilation and mental retardation (Lesch and Nyhan, 1964; Rossiter and Caskey, 1995).

The main metabolic feature of Lesch–Nyhan patients is hyperuricemia with uricosuria, tophaceous gouty arthritis, urinary tract calculi and urate nephropathy (Lesch and Nyhan, 1964; Rossiter and Caskey, 1995). It results from HPRT enzyme deficiency (Seegmiller *et al.*, 1967) and a subsequent increase in the rate of *de novo* purine synthesis (Lesch and Nyhan, 1964; Nyhan *et al.*, 1965) that leads to an overproduction of hypoxanthine upon HPRT recycling deficiency (Rossiter and Caskey, 1995). Hypoxanthine is subsequently converted to xanthine and uric acid mainly by liver xanthine oxidase (Rossiter and Caskey, 1995). Allopurinol, a xanthine oxidase inhibitor, is an effective medication for Lesch–Nyhan hyperuricemia and associated symptoms (Rossiter and Caskey, 1995). Unfortunatelly, allopurinol has no effect on the neurologic symptoms of the disease. This is consistent with the fact that these neurologic symptoms are not causally related to uric acid, since its concentration in the cerebrospinal fluid was shown to be the same in patients and control subjects (Rosenbloom *et al.*, 1967).

Unlike typical neurodegenerative diseases such as Parkinsonism and Alzheimer's disease that primarily affect the elderly, Lesch–Nyhan disease appears to be developmental in origin. Its neurological features, which include delayed motor development, severe spasticity, choreoathetosis, compulsive self-mutilation and mental retardation, appear to be a consequence of alterations in brain dopamine systems (Lloyd *et al.*, 1981; Wong *et al.*, 1996; Ernst *et al.*, 1996). Lesch–Nyhan patients have a reduced number of dopaminergic nerve terminal and cell bodies (Wong *et al.*, 1996; Ernst *et al.*, 1996). Such a dopaminergic deficit has been described in HPRT-deficient transgenic mice where it is mainly localized in basal ganglia and emerges during the first two months of postnatal development (Jinnah *et al.*, 1994). In Lesch–Nyhan patients, the dopaminergic deficit appears to involve all dopaminergic pathways (Ernst *et al.*, 1996) although it was first identified in basal ganglia (Lloyd *et al.*, 1981); it is pervasive and appears to be stable with age (patients were 10 to 20 years old; Ernst *et al.*, 1996). Such a stability implies that we are not dealing with a life-long degenerative process (Ernst *et al.*, 1996). It is consistent with the clinical course of the Lesch–Nyhan neurobehavioural syndrome that is fully expressed and constant by the age of 2 years (Mizuno, 1986; Anderson *et al.*, 1992).

There is not yet a cure to the neurologic symptoms of the Lesch–Nyhan syndrome (Nyhan and Wong, 1996; Rossiter and Caskey, 1995). Although the Lesch–

Nyhan disease has been proposed to be a target for pioneering gene therapy trials (see Anderson, 1984 and Friedmann, 1985) because the HPRT gene does not require tight physiological regulation and wild-type lymphocytes appear to have a selective survival advantage over HPRT-deficient ones in human heterozygotes (Dancis *et al.*, 1968; Nyhan *et al.*, 1970; McDonald and Kelley, 1972), the approach has been postponed because it became obvious that the neurologic symptoms of the disease will not be reversed by autologous HPRT-engineered lymphocytes. Consistent with this idea, normal HPRT activity in peripheral blood cells has been obtained upon BM transplantation in an adult Lesch–Nyhan patient (Nyhan *et al.*, 1986) and in a mouse model of the disease (Wojcik *et al.*, 1989) without a concomitant improvement in the neurological manifestations of the disease. The disease might however be amenable in the future to emerging sophisticated synergistic combinations associating long-term stem cell gene therapy and transient regenerative gene therapy protocols involving both HSCs and neural stem cells (see below and Bertolotti, 2001, 2003a, c).

The HPRT gene as a 'single-copy' X-linked selective marker

The HPRT gene (Fig. 1) is a classic marker of somatic cell genetics (Ephrussi, 1972; Szybalski, 1992). It is expressed in all cells and, to a notably higher level in the brain (Rossiter and Caskey, 1995; Jiralerspong and Patel, 1996). It provides both for positive and negative selections of cultured cells (Szybalska and Szybalski, 1962; Ephrussi, 1972; Szybalski, 1992). In addition, it is a 'single-copy' X-linked gene, i.e. it is hemizygous in males and it has only a single functional copy in adult females (X chromosome inactivation; Lyon, 1996). It is therefore an ideal target for gene repair/inactivation experiments where each targeting event can be readily scored upon positive or negative selection of transfectant cells in HPRT-specific growth media.

Selection of HPRT-deficient cells in vitro

HPRT (inosine monophosphate : pyrophosphatase phosphoribosyl transferase; EC 2.4.2.8) is a key purine salvage enzyme that catalyzes the transfer of the ribose phosphate moiety from phosphoribosyl pyrophosphate to hypoxanthine or guanine (Fig. 2). It thus recycles catabolic purine bases yielding inosine monophosphate (IMP) or guanosine monophosphate (GMP), respectively.

In addition, HPRT binds and phosphoribosylates toxic purine analogues such as azahypoxanthine, azaguanine or thioguanine (Szybalski, 1992; Szybalska and Szybalski, 1962). This ability has been used to isolate HPRT-deficient mutant cells upon selection in drug-containing media (Szybalska and Szybalski, 1962; Szybalski, 1992). Under these conditions, HPRT-positive cells die (counter-selection) while HPRT-negative mutants survive because they are unable to metabolize toxic purine analogs (negative selection).

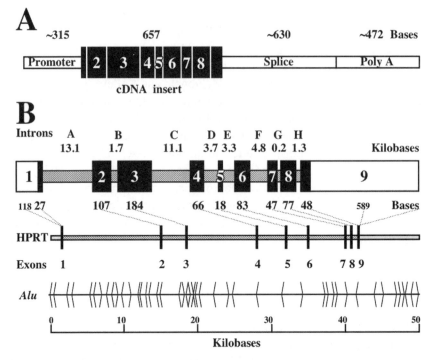

Figure 1. HPRT: structure of a minigene (A) and of the human gene (B; based on Edwards *et al.*, 1990, and EMBL/GeneBank M26434) (Bertolotti, 1998a, 1999). (A) The shortest full-length HPRT cDNA corresponding to complete coding sequence (657 base-long) is inserted in an SV40 expression cassette (Mulligan and Berg, 1981) comprising (1) the early promoter-enhancer sequence (Promoter) of the SV40 virus on the 5' end of the cDNA for transcription initiation, and (2) on the 3' end, the small t-antigen intervening sequence for mRNA splicing (Splice), and the early transcript termination/polyadenylation sequence (Poly A) for efficient termination of transcription and subsequent polyadenylation of the mRNA. Such a synthetic gene of about 2 kilobases (kb) is very small as compared to its normal human counterpart (~40 kb). (B) The HPRT gene comprises 9 exons (boxes) and 8 introns (hatched areas). Introns and flanking sequences (stippled areas) contain repetitive elements such as *Alu* (46 repeats, positioned with orientation) and LINE (not shown). Enlarged at the upper part of the figure, exons, not in scale with introns, span about 1364 bases out of a ~40 kb genomic sequence. The coding sequence is 657 base-long (filled boxes). Exons 1 and 9 comprise 5' and 3' non-coding regions (empty boxes), respectively.

Growth medium containing 15 μg/ml of thioguanine provided for efficient selection of HPRT-deficient variants from the human hepatic HepG2 line (Lutfalla *et al.*, 1989). Selection experiments were performed on a well-differentiated HepG2 subclone (Lutfalla *et al.*, 1989; Armbruster *et al.*, 1992), an excellent target for transactivating expression libraries (Bertolotti *et al.*, 1995) and for our recombinase-mediated gene therapy approach (Bertolotti, 1996b, 2000c). Under our experimental conditions, HPRT-deficient variants arose at a frequency of 2–5 × 10^{-6} (Lutfalla *et al.*, 1989).

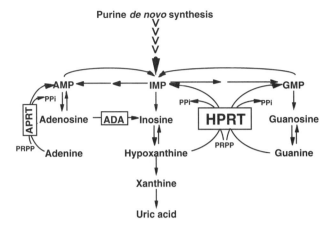

Figure 2. Simplified metabolic pathways for *de novo* synthesis and salvage of purines (based on Rossiter and Caskey, 1995; Nyhan and Wong, 1996; Simmonds *et al.*, 1995). Purine nucleotides (IMP, AMP and GMP) are produced upon either *de novo* purine synthesis or recycling of catabolic purines. The recycling of catabolic or exogenous purine bases is mediated by HPRT and adenine phosphoribosyl transferase (APRT) upon transfer of the phosphoribosyl moiety from 5-phosphoribosyl-1-pyrophosphate (PRPP) to hypoxanthine, guanine and adenine, respectively. PPi denotes inorganic pyrophosphate, IMP inosine monophosphate, AMP adenosine monophosphate and GMP guanosine monophosphate. Adenosine deaminase (ADA) deficiency results in severe combined immunodeficiency (SCID) syndrome (see above). Reprinted from Bertolotti, 1998b with permission.

Transfection of HPRT-deficient cells with XGPRT or HPRT minigenes

HPRT-deficient cells cannot grow in HAT medium where *de novo* synthesis of purine (and pyrimidine) nucleotides is inhibited by aminopterin (Szybalska and Szybalski, 1962; Littlefield, 1964). HAT stands for Hypoxanthine, Aminopterin and Thymidine, indicating that upon aminopterin inhibition of *de novo* synthesis pathways, growth in HAT medium relies on both purine and pyrimidine salvage enzymes: HPRT for phosphoribosylation of hypoxanthine and thymidine kinase for phosphorylation of thymidine (Szybalska and Szybalski, 1962; Littlefield, 1964; Szybalski, 1992).

Efficient survival of Lesch–Nyhan cells in HAT medium (positive selection) was first achieved upon transfection with XGPRT minigenes where the *Escherichia coli* DNA sequence encoding bacterial xanthine-guanine phosphoribosyltransferase (XGPRT) was placed under the control of an SV40 expression cassette (Mulligan and Berg, 1980; see Fig. 1). Unlike its mammalian HPRT counterpart, XGPRT phosphoribosylates both xanthine and hypoxanthine; it was therefore used as a universal bacterial selective marker for mammalian cells either with HAT medium for HPRT mutants or with xanthine selective medium for normal cells (Mulligan and Berg, 1980, 1981). Full-length cDNA encoding mammalian HPRT were subsequently isolated and introduced as selectable minigenes into HPRT-deficient cells (Brennand *et al.*, 1983; Jolly *et al.*, 1983).

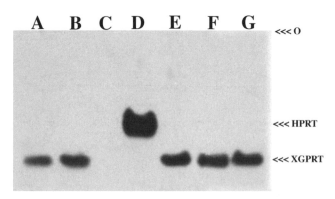

Figure 3. XGPRT and HPRT zymograms of transfected and control cells (Bertolotti, 1998b). Rat and mouse HPRT-deficient hepatoma cells were transfected with plasmids comprising a XGPRT minigene (Bertolotti and Lutfalla, 1983; Lutfalla *et al.*, 1985). Transfectants were selected in HAT medium and their XGPRT-HPRT content analyzed upon electrophoresis of cellular protein extracts on cellulose acetate gels (Shin *et al.*, 1971) and subsequent fluorography of purine phosphoribosyltransferase activity (Chasin and Urlaub, 1976; Lutfalla *et al.*, 1985). Unlike HPRT-deficient (lane C) and wild-type rat controls (lane D), rat (lanes A, B and F) and mouse (lanes E and G) transfectant cells expressed bacterial XGPRT enzyme as expected (Lutfalla *et al.*, 1985).

We have used XGPRT minigenes in our early transfection experiments with rat, mouse and human hepatoma cells (Bertolotti and Lutfalla, 1983; Lutfalla *et al.*, 1985). As illustrated in Fig. 3, growth of XGPRT transfectants in HAT medium was directly correlated with the production of active bacterial XGPRT enzyme in mammalian hepatoma cells (Lutfalla *et al.*, 1985). We then introduced an XGPRT minigene in a plasmid comprising the origin of replication of the Epstein-Barr virus (EBV) and showed that it was maintained as a self-replicating episome when HepG2 cells were engineered to produce trans-acting Epstein-Barr nuclear antigen 1 (Lutfalla *et al.*, 1989). A dramatic 200-fold increase in transfection efficiency was associated with the episomal maintenance of the XGPRT vector, suggesting that EBV shuttle vectors might be useful to the efficient transfection of proliferative tumor cells with a suicide gene.

HPRT as a paragon gene targeting model

With its selective or counter-selective properties together with its single-copy state, the HPRT gene is therefore perfectly fitted for gene targeting experiments where each targeting event is easily scored in the aforedescribed selective growth media.

Importantly, Lesch–Nyhan mutations are well-characterized at the molecular level (Gibbs *et al.*, 1989; Cariello and Skopek, 1993; Rossiter and Caskey, 1995). They were instrumental in the design of a series of DNA-backbones for nucleoprotein filaments aimed at repairing or inactivating the human HPRT gene (Bertolotti, 1996b; see below).

RECOMBINASE-DNA NUCLEOPROTEIN VECTORS

Transfecting DNA is subject to transient homologous recombination but integrates mainly upon illegitimate recombination

Transfected dividing cells give rise to stable transfectants when exogenous DNA is either integrated into host chromosomal DNA (Robins *et al.*, 1981; Folger *et al.*, 1982) or maintained as a self-replicating nuclear episome (see EBV-based shuttle plasmids described above and: DiMaio *et al.*, 1982; Yates *et al.*, 1985).

Under current transfection protocols, plasmid integration into host chromosomal DNA usually occurs as head-to-tail multimeric inserts. Such multimeric structures are generated by homologous recombination before integration into host chromosomal DNA (Folger *et al.*, 1982). Indeed, in most transfected cells, exogenous DNA molecules are subject to an intense homologous recombination activity (Capecchi, 1989; Folger *et al.*, 1982; Miller and Temin, 1983; Lutfalla *et al.*, 1985). Such a homologous recombination activity coincides with the early transfection phase when newly introduced DNA is not yet packaged into chromatin (Capecchi, 1989). It is transient and, in association with ligation and end-joining activities (Folger *et al.*, 1982; Miller and Temin, 1983; Lieber *et al.*, 2003), generates concatemers (pure plasmids) or concatenates (plasmids and carrier DNA) that give rise to stable transformants upon integration into host chromosomes (Perucho *et al.*, 1980; Robins *et al.*, 1981). Unlike concatemer genesis, concatemer integration into chromatinized chromosomal DNA is usually not the result of homologous recombination. Even with isogenic DNA (Te Riele *et al.*, 1992; Thomas *et al.*, 1992), gene targeting, i.e. integration of transfecting DNA upon homologous recombination with chromosomal DNA, is an infrequent event; its frequency is only a few percent of random integration upon illegitimate recombination (Capecchi, 1989; Thomas *et al.*, 1986; Merrihew *et al.*, 1996; Yanez and Porter, 1998). However, although infrequent, gene targeting in mammalian somatic cells is a highly significant process (Capecchi, 1989; Thomas *et al.*, 1992; Abuin and Bradley, 1996).

Somatic recombinases as RecA-like DNA repair enzymes

Such a transient homologous recombination activity of transfecting DNA molecules, either alone (Folger *et al.*, 1982; Lutfalla *et al.*, 1985; Capecchi, 1989) or with cognate chromosomal DNA (Thomas *et al.*, 1986; Capecchi, 1989), is in contrast with the absence of homologous chromosome recombination in most mammalian somatic cells and hybrids (Ephrussi, 1972; Panthier and Condamine, 1991; Johnson and Jasin, 2001), indicating that somatic recombinases do not usually mediate homologous genetic recombination (Bertolotti, 1996a). Like the bacterial RecA protein (Roca and Cox, 1990), they appear to be mainly involved in DNA repair (Bertolotti, 1996a; see RAD51 below). Therefore, in December 1991, our new approach to gene therapy (Bertolotti R., Grant application to French MRT, 1991) started with the search for such a mammalian homolog of RecA (Borchiellini *et al.*, 1997) that could be used to produce *in vitro* nucleoprotein complexes analogous to the bacterial RecA presynaptic filament. The premade recombinase-DNA complexes should mimic, upon transfection, recombinational DNA repair that occurs in most dividing mammalian cells and thereby promote homologous recombination between transfecting DNA and cognate chromosomal DNA (Bertolotti, 1996a, b).

Standard presynaptic filaments as gene conversion vectors

The eukaryotic RAD51 gene was cloned, sequenced and shown to encode a putative RecA-like protein in 1992 (Shinohara *et al.*, 1992; Aboussekhra *et al.*, 1992; Basile *et al.*, 1992). However, the enzymatic activity of the RAD51 protein was demonstrated in 1994 only (Sung, 1994). In fact, the search for an eukaryotic recombinase has been such an intensive disappointing task (see: Heyer, 1994; Borchiellini *et al.*, 1997) that RAD51 protein had to be the subject of an additional detailed enzymatic characterization (Sung and Robberson, 1995) before it could be established as a true recombinase. Indeed, like bacterial RecA (Kowalczykowski *et al.*, 1994; Roca and Cox, 1990), the eukaryotic RAD51 protein is able to mediate *in vitro* the key reaction from homologous recombination, i.e. ATP-dependent homologous DNA pairing-strand exchange with double-stranded DNA (Sung, 1994; Baumann *et al.*, 1996; Gutpa *et al.*, 1997). However, to be active under current experimental conditions, RAD51 requires the presence of replication protein A (RPA) (Sung, 1994) or eventually RAD52 (Benson *et al.*, 1998). Like in bacteria (Kowalczykowski *et al.*, 1994; Roca and Cox, 1990), the eukaryotic active molecule is a right-handed helical nucleoprotein (Fig. 4) formed, in the presence of ATP and RPA, upon recombinase protein polymerization onto single-stranded DNA (ssDNA) at a stochiometry of 3 nucleotides per monomer (Sung and Robberson, 1995). Adjunction of RAD52 protein may improve filament production (Sung, 1997; Benson *et al.*, 1998; Shinohara and Ogawa, 1998; New *et al.*, 1998). Other proteins, such as RAD54 that stimulates homologous pairing activity of RAD51 *in vitro* (yeast; Petukhova *et al.*, 1998), appear to be involved in other ancillary functions. They may be part of a putative recombinosome that includes a number of proteins such as RAD51, RPA and RAD52 (Firmenich *et al.*, 1995; Hays *et al.*, 1995).

As shown in Fig. 4, the final product of the homologous DNA pairing-strand exchange reaction is an heteroduplex, i.e. a duplex DNA in which the complementary strands originate from different parental molecules. Pairing of DNA strands that differ in some points of their sequences results in mismatched bases. *In vivo*, such mismatches are the targets of cellular DNA repair systems (Friedberg *et al.*, 1995; Modrich and Lahue, 1996) that include strand-specific mismatch repair (Modrich, 1997). Elimination of a mismatch involves long-patch or short-patch mismatch repair mechanisms; it is achieved upon replacement of one of the mismatched bases by the correct match of the other base. This process results in gene conversion when the incoming base is maintained, i.e. the sequence of target dsDNA is changed (gene repair or inactivation).

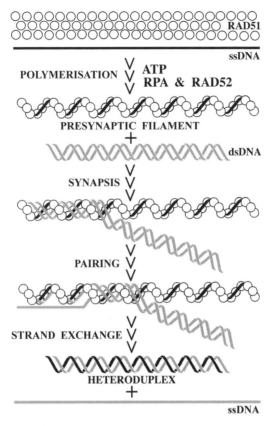

Figure 4. Homologous DNA pairing and strand exchange reaction mediated by RAD51 protein *in vitro* (Bertolotti, 1999). Under usual experimental conditions, production of active filaments requires the presence of replication protein A (RPA) (Sung, 1994; Sung and Robberson, 1995; Baumann and West, 1997) and is facilitated by RAD52 (Sung, 1997; Benson *et al.*, 1998; Shinohara and Ogawa, 1998; New *et al.*, 1998). RAD52 may obviate RPA requirement (Benson *et al.*, 1998). Like with RecA (Howard-Flanders *et al.*, 1984; Roca and Cox, 1990; Radding, 1991; Stasiak and Egelman, 1994), the reaction occurs within the filament where DNA is underwound and stretched by a factor 1.5 compared to B-form dsDNA.

Such a conversion of host genomic DNA may be achieved upon transfection of 'naked' ssDNA (Fujioka *et al.*, 1993; Yanez and Porter, 1998); its efficiency is however very poor with the potential exception of a new approach based on small PCR genomic fragments of an average length of about 500 nucleotides (see Gruenert *et al.*, 2003). On the other hand, efficient correction of mismatched bases in non-replicating heteroduplex injected into cultured mammalian cell nuclei has been described long ago (Folger *et al.*, 1985); in addition, efficient mismatch repair is essential to DNA replication fidelity and genomic stability of any replicating cell (Modrich and Lahue, 1996; Jiricny, 1998). Therefore, although we cannot formally exclude the possibility that gene conversion by transfecting ssDNA is impaired by a strand-specific mismatch repair mechanism (Modrich, 1997) that systematically removes mismatched bases from incoming transfecting DNA, the poor targeting efficiency of transfecting dsDNA into host chromosomes (see below) suggests that the limiting step of gene conversion by transfecting ssDNA is most likely the homologous DNA pairing-strand exchange reaction that is catalyzed by RAD51 *in vitro*.

Mammalian RAD51 is a ubiquitous protein that is expressed in most tissues (Shinohara *et al.*, 1993; Morita *et al.*, 1993), high level expression being restricted to tissues comprising actively dividing cells such as gonads, thymus, spleen, uterus, intestine and embryonic materials (Shinohara *et al.*, 1993; Morita *et al.*, 1993; Yamamoto *et al.*, 1996). In fact, RAD51 is essential for mammalian cell proliferation: RAD51 gene inactivation is lethal both to embryonic mice and to cultured cells (Tsuzuki *et al.*, 1996; Lim and Hasty, 1996). In addition, conditional transgene inactivation in proliferating chicken cells shows that RAD51 is indispensable to genome maintenance (Sonoda *et al.*, 1998). RAD51 stands thus as a basic recombinase that is essential to recombinational DNA repair. Consistent with this function, its expression is cell cycle-dependent and is associated to late G1-S-G2 phases (Yamamoto *et al.*, 1996; Flygare *et al.*, 1996). Interestingly enough, RAD51 expression is also well correlated to homologous recombination of transfecting DNA that peaks in early to mid-S phase (Wong and Capecchi, 1987).

Although present in dividing cells, RAD51 is active only as a stoichiometric complex with ssDNA and ATP (see above and Fig. 4; Sung and Robberson, 1995). Polymerization of RAD51 protein onto transfecting ssDNA *in vivo* is likely to be less efficient than under optimized conditions *in vitro*. In addition, transfecting ssDNA is subjected to nuclease degradation both in the cytoplasm and the nucleus of recipient cells. Therefore, the recombinase sheath from premade presynaptic filaments should protect transfecting ssDNA and prevent the binding of proteins that may interfere with homologous recombination (Bertolotti, 1996b, 1998a). In conclusion, premade presynaptic filaments should thus provide transfecting ssDNA with protection, with nuclear tropism (RAD51 is a nuclear protein; Haaf *et al.*, 1995) and with synapsing ability with chromosomal dsDNA (Bertolotti, 1996b, 1998a). Importantly enough, with or without the intermediate formation of recombinosomes (see above), RAD51 coating should provide for putative protein-protein interaction sites which are absent in bacterial RecA protein and which are supposed to be essential for an efficient spooling and processing of exogenous ssDNA into chromatinized mammalian chromosomal dsDNA (Bertolotti, 1996a, b, 1998a).

Upon transfection into human cells, human recombinase presynaptic filaments should therefore be able of carrying out genome-wide search on chromatinized human DNA (Bertolotti, 1996a, b, 1998a) in the same way *E. coli* RecA filaments do it *in vitro* on naked human DNA (Ferrin and Camerini-Otero, 1991). The ensuing directed homologous recombination should thus generate specific alterations to endogenous dsDNA sequences resulting in improved gene repair or inactivation (Bertolotti, 1996a, b, 1998a).

Chimeric dsDNA-cored filaments as gene targeting vectors

One of the obvious limitations of standard presynaptic filaments is the level of base pair heterology between transfecting ssDNA and host target DNA. RecA protein-mediated exchange of DNA strands can accomodate quite important DNA lesions and mismatched base-pair (Kowalczykowski *et al.*, 1994). Such information is missing for RAD51 *in vitro*. However, homologous recombination between transfecting dsDNA and cognate chromosomal DNA has been shown to be strongly inhibited by a few percents of base pair heterology (Deng and Capecchi, 1992; Te Riele *et al.*, 1992). In addition, many mutations result from deletions or insertions of multiple base sequences (see below and EMBL/GeneBank file No. M26434 for mutations of the human HPRT gene). This is why, upon publication of the enzymatic properties of RAD51 (Sung and Robberson, 1995), we devised a more sophisticated approach and invented dsDNA-cored filaments (Bertolotti, 1996b). As illustrated in Fig. 5, the idea is to eliminate mismatch problems by utilizing dsDNA as a therapeutic cassette and drive its targeted integration by using ssDNA tails that are 100% identical to host chromosomal dsDNA. We thus moved from gene conversion to 'true' gene targeting, i.e. integration of transfecting dsDNA into host chromosomal DNA upon homologous recombination (Capecchi, 1989).

The point is that, unlike bacterial RecA protein, RAD51 has rougthly the same affinity for dsDNA as for ssDNA and forms helical RecA-like filaments both with ssDNA and dsDNA (Ogawa *et al.*, 1993; Benson *et al.*, 1994; Sung and Robberson, 1995). However, dsDNA-nucleoprotein filaments are inactive (Sung and Robberson, 1995). Importantly enough, coating dsDNA with RAD51 prevents duplex DNA from participating in the pairing reaction but does not affect the ability of uncoated concurrent dsDNA to react with RAD51-ssDNA filaments (Sung and Robberson, 1995). The idea is thus to devise chimeric targeting filaments in which therapeutic heterologous bases are comprised in a dsDNA core (Fig. 5). Indeed, coating such DNA backbones with RAD51 should generate recombination-activated ssDNA tails and recombination-locked dsDNA cores. Having a ssDNA tail at each end, these filaments should initiate homologous pairing-strand exchange reactions on both flanking sides of the dsDNA core, resulting, like in a double crossing-over scheme, in the substitution of an endogenous dsDNA segment by an exogenous recombination-free dsDNA cassette (Fig. 5). Such a reaction will be all the more efficient as the sequences of the ssDNA tails are 100% identical to their respective cognate host targets. Perfect identity between these ssDNA tails and their respective targets eliminates mismatch problems (see Fig. 5) and optimizes homologous DNA pairing-strand exchange reactions on both sides of the therapeutic dsDNA cassette. The dsDNA cassette is hermetic to homologous recombination; therefore, heterologous therapeutic bases, comprised in such a dsDNA cassette, should not interfere with homologous recombination.

Another important feature of our dsDNA-cored presynaptic filaments is the opportunity to obviate unspecific targeting to repetitive chromosomal elements. Indeed, as illustrated with the human HPRT gene (Fig. 1), mammalian genomic DNA mainly comprises introns with repetitive elements such as *Alu* and LINE. Therefore, with our dsDNA-cored filaments, we should avoid unspecific targeting by excluding such ubiquitous intervening sequences from the ssDNA tails and inserting them in the dsDNA cassette when necessary (Fig. 6; Bertolotti, 1996b).

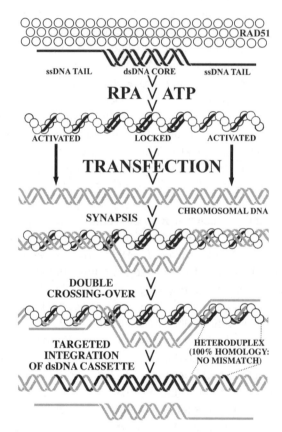

Figure 5. Targeting recombination-locked dsDNA cassettes to chromosomal DNA with dsDNA-cored RAD51 filaments: gene repair/inactivation or targeted integration of minigenes (chromatinian structures are not shown; Bertolotti, 1997, 1998a, 1999). The method consists in (1) producing *in vitro* dsDNA-cored presynatic filaments by polymerizing RAD51 protein onto appropriate DNA backbones (Fig. 6), and (2) transferring them (transfection) into recipient cells where recombination-activated ssDNA tails can drive homologous recombination with chromosomal cognate DNA. Substitution of an endogenous dsDNA segment by an exogenous recombination-free dsDNA cassette (targeted integration) should thus result from a double crossing-over mechanism involving both ssDNA tails from the targeting vector. The driving forces are therefore (1) the RAD51-coated ssDNA tails that are 100% identical to their respective targets and (2) the no-mismatch heteroduplexes resulting from homologous pairing of these activated ssDNA tails with cognate chromosomal strands. Importantly, adjunction of RAD52 protein to ATP and RPA may improve *in vitro* filament production (Sung, 1997; Benson *et al.*, 1998; Shinohara and Ogawa, 1998; New *et al.*, 1998).

DNA backbones for gene repair/inactivation

A series of DNA-backbones for nucleoprotein filaments aimed at inactivating then repairing the HPRT gene from human cells were designed (Bertolotti, 1996b) using EMBL/GenBank Data file No. M26434. This file comprises the nucleotide sequence from the human HPRT gene, its structure and its molecular alterations in a series of Lesch–Nyhan mutants (Edwards *et al.*, 1990). Some DNA backbones corresponding to two such mutants dealing with exon 3 are shown both for standard presynaptic filaments and dsDNA-cored ones (Fig. 6). Other selected mutations involve either introns or exons.

DNA backbones shown in Fig. 6 have been discussed previously (Bertolotti, 1996b, 1998a). Backbones for standard presynaptic filaments are illustrated for a mutation resulting from a deletion of two base-pairs (bp) in exon 3. Two sets of sens and antisens strands designed for the GT/CA deletion of mutant Lesch–Nyhan RJK 1332 are thus shown on the right part of Fig. 6. Short oligonucleotide-based filaments should be the best choice both for production and cell-internalization. However, although *in vitro* pairing can involve less than one helical repeat of DNA with RecA (Hsieh *et al.*, 1992), targeting efficiency upon transfection of RAD51-ssDNA filaments into mammalian cells may require oligonucleotide stretches longer than 50 nucleotides (nt). Aside from strand length, targeting efficiency

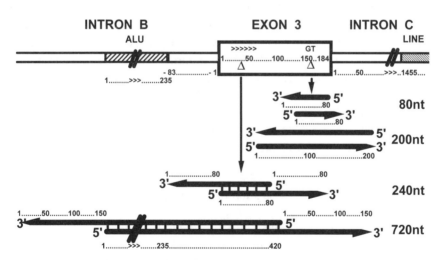

Figure 6. Designing HPRT DNA-backbones for standard and dsDNA-cored presynaptic filaments (Bertolotti, 1996b). Mutations Lesch–Nyhan RJK 2108 (>>>>>>) and 1332 (GT) are deletions (Δ) of either 40 base-pairs (bp) or 2 bp (GT/CA) from exon 3. Flanking exon 3 (184 bp), the 3′ end of intron B (1.7 kb) and the 5′ end of intron C (11.1 kb) are shown with their respective terminal *Alu* and proximal LINE repeat sequences. Single-stranded DNA stretches (ssDNA; thick horizontal arrows) are either derived from mutant (inactivating filaments) or wild-type DNA (reconstructing filaments). Backbones for chimeric presynaptic filaments comprise a double-stranded DNA (dsDNA) core for therapeutic heterologous bases and repetitive elements. Both ssDNA tails of dsDNA-cored backbones are 3′ overhanging ones. Small arabic numbers refer to nucleotide (nt) length.

may depend on strand type (sens or antisens) and concentration, on the position of heterologous bases on the strand and on properties inherent to recipient cell types. In addition, DNA methylation and discrete chromatinian structures may also interfere.

Both standard and dsDNA-cored presynaptic filaments may be used for point-mutations or very short lesions. However, illustrations for dsDNA-cored filaments have been restricted to backbones designed for a larger DNA lesion, i.e. a 40 bp deletion ($> > > > > >$; Fig. 6) in exon 3. The DNA backbones of two dsDNA-cored filaments corresponding to mutant Lesch–Nyhan RJK 2108 are thus presented in Fig. 6. The first one is 240 nt long. It comprises an 80 bp dsDNA core including a 40 bp cluster corresponding to deletion RJK 2108 and two 80 nt ssDNA tails. The second one is 720 nt long. It comprises a 420 bp dsDNA core including a 235 bp *Alu* repeat and the aforementioned 40 bp cluster; its tails are 150 nt long. In these cases, targeting efficiency may depend on the total length of the nucleoprotein filament as well as on the length of each component, i.e. tails and core. As above, potential parameters dealing with ssDNA tails include methylation of target host DNA, chromatinian structures and cell type. However, the new and main parameter is the single-stranded tail type, i.e. $3'$ or $5'$ overhang. Importantly enough, actual designs are based on RAD51 protein that has been shown to mediate *in vitro* the ATP-dependent homologous DNA pairing-strand exchange reaction with a polarity opposite to that of RecA (Sung and Robberson, 1995; Baumann and West, 1997). Therefore, $3'$ overhanging ssDNA tails are shown in Fig. 6 for dsDNA-cored filaments that will be used in conjunction with control filaments of the converse $5'$ overhang design (Bertolotti, 1996b). Such controls are all the more important as recent experiments have described a bidirectional strand exchange polarity both for yeast and human RAD51 (Namsaraev and Berg, 1998; McIlwraith *et al.*, 2000). In addition, bidirectionality implies that the properties of the two backones in which one ssDNA strand (the coding or non-coding one) provides both a $3'$ and $5'$ overhangs deserve to be evaluated too.

DNA backbones for targeted integration of minigenes

Chimeric dsDNA-cored filaments offer broad DNA exchange potentialities and should therefore accomodate minigenes (Bertolotti, 1999). In this case, the minigene of interest is inserted into the dsDNA core of appropriate recombinase-DNA nucleoprotein filaments. Such a new design should fit both sequence replacement vectors and sequence insertion vectors (Capecchi, 1989). Targeted integration of a minigene into a specific host chromosome is thus driven by genomic ssDNA tails. These tails are 100% identical to their respective chromosomal targets; they specify the integration site of the minigene. The HPRT gene is again a paragon target since each targeted insertional event disrupts the gene and can thereby be easily scored in HPRT counter-selective medium (see Capecchi, 1989, and above).

Gene targeting with conventional vectors or with dsDNA-cored RAD51 nucleoprotein filaments

Conventional gene targeting vectors are plain dsDNA (Capecchi, 1989; Thomas *et al.*, 1992; Te Riele *et al.*, 1992; Abuin and Bradley, 1996). By contrast to our dsDNA-cored presynaptic filaments, their design is not optimized for homologous recombination. First, they have to be processed in transfected cells to yield ssDNA-tailed molecules similar to our dsDNA-cored DNA-backbones (Fig. 6). Such a 5′ to 3′ exonuclease resectioning of dsDNA to produce 3′ overhangs has been shown to occur on chromosomal DNA ends when a double-strand break is repaired by the homologous recombination pathway (Sun *et al.*, 1991; see Pâques and Haber, 1999; West, 2003). In addition, this processing is in competition with unspecific nuclease degradation and with neutralizing effects of other DNA-binding proteins (Bertolotti, 1998a) such as the end-joining ones (Lieber *et al.*, 2003). Then, presynaptic filaments have to form. As discussed above, *in vivo* polymerization of RAD51 on ssDNA overhangs is not likely to be as efficient as under optimized *in vitro* conditions.

With our premade dsDNA-cored filaments where DNA is protected by a RAD51 protein sheath and ssDNA tails ready for synapsis, we thus escape the initial degradative step of dsDNA processing and ssDNA-recombinase nucleation, thereby promoting gene targeting while minimazing random-integration (Table 1). In addition, in our dsDNA-cored filaments, ssDNA tails are 100% identical to their host DNA targets which is not the case with current dsDNA vectors even isogenic DNA ones. Moreover, unlike most current dsDNA vectors, the ssDNA tails from our vectors do not comprise repetitive elements that may hamper targeting efficiency. We thus believe that our premade dsDNA-cored presynaptic filaments should be very effective at mediating the core homologous recombination reaction with target chromosomal DNA, thereby raising gene targeting efficiency while reducing the optimal

Table 1.
Putative effects of recombinase-coating on integration of transfecting DNA into mammalian host chromosomes (Bertolotti, 1997)

'Naked'* DNA	
– Gene targeting (homologous recombination):	rare
– Random integration ('illegitimate' recombination):	frequent
DNA-recombinase nucleoprotein filaments	
– Putative direct effect:	increase of gene targeting
– Putative indirect effects:	reduction/suppression of random integration
	reduction/suppression of concatemer formation

(recombinase coating should prevent homologous recombination between transfecting DNA molecules and should protect ssDNA and dsDNA-cores from a variety of DNA-binding proteins such as end-joining ones**)

*Transfection protocols may include formation of precipitates with calcium-phosphate (Graham and Van der Eb, 1973) or of various complexes (Bertolotti, 2000b).

**In addition, end-to-end ligation described for linear dsDNA (Baumann *et al.*, 1996) should be impossible with standard or dsDNA-cored ssDNA filaments.

homology length from 1–7 kb (conventional gene targeting: Thomas *et al.*, 1992; Deng and Capecchi, 1992) to handier ssDNA tails (see Bertolotti, 2000c).

CHIMERIC ZINC FINGER NUCLEASES AS BASIC GENE TARGETING BOOSTERS

Double-strand break (DSB) in target chromosomal DNA as a dramatic stimulator of gene targeting

The genesis of a double-strand break (DSB) in target chromosomal DNA is an efficient way to dramatically increase the frequency of gene targeting (Smih *et al.*, 1995; Choulika *et al.*, 1995; Donoho *et al.*, 1998). Such a stimulation was however initially achieved in artificial systems in which prior modification of the chromosomal target was required to introduce the 18-base pair recognition site of the rare-cutting endonuclease I-SceI. An 18 bp sequence provides a remarkable degree of specificity in a mammalian genome (3×10^9 bp) since its predicted occurrence is once in 6.9×10^{10} (i.e. 4^{18}) bp. Consistent with this calculation, expression of this yeast meganuclease is non-toxic to mammalian cells presumably because there are no endogenous sites in the genome of the mouse (Rouet *et al.*, 1994a, b) and of the other mammalian species that have been used in these experiments. Although essential to the proof of principle, I-SceI is of course not an option for gene targeting applications; the same holds true for the few other natural rare-cutting endonucleases that are currently available. Fortunately, chimeric zinc finger endonucleases are now emerging that provide the means to cleave DNA at virtually any site of interest.

Genesis of sequence-specific DSBs with designed zinc-finger nucleases

Chimeric zinc-finger nucleases are artificial proteins that comprise a designed zinc-finger domain for site-specific DNA binding linked to the non-specific cleavage domain from the restriction enzyme *Fok*I (see Fig. 7 and Kim *et al.*, 1996). Like for the original *Fok*I enzyme, the active form of the chimeric nuclease is a dimer (Smith *et al.*, 2000). Such a dimerization through the *Fok*I cleavage domain is promoted by the binding of the two zinc-finger domains to their respective DNA targets (see Fig. 7 and Smith *et al.*, 2000). There are therefore steric requirements both in the spacing of the two binding sites on target DNA and in the length of the linker peptide that joins the zinc-finger domain to the nucleasic one. Optimization of the design of zinc finger nuclease prototypes has been performed by Carroll, Chandrasegaran and co-workers (Smith *et al.*, 2000; Bibikova *et al.*, 2001) and is illustrated in Fig. 7 with the yA-yB set that has been designed for a target site in the drosophila *yellow* (*y*) gene (Bibikova *et al.*, 2002). The yA-yB set has been shown to mediate efficient targeted chromosomal clivage in drosophila (Bibikova *et al.*, 2002) and to enhance gene targeting as expected (Bibikova *et al.*, 2003), thereby validating the zinc-finger nuclease approach in a true biological context (i.e. the target is a real endogenous

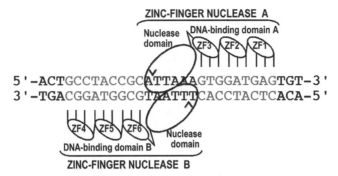

Figure 7. Genesis of a site-specific DSB by a set of designed zinc-finger nucleases. The prototypic pair of zinc-finger nucleases yA and yB designed to clive at a unique site the *yellow* (*y*) gene of Drosophila is schematized together with its target genomic sequence (modified from Fig. 1 of Bibikova *et al.*, 2002). The logic of the chimeric nuclease is to associate the DNA-cleavage domain of the *Fok*I endonuclease with a designed DNA binding domain. The active form of the enzyme being a dimer, the unique 18 bp recognition site is designed to be specified by a pair of DNA binding domains. A 9 bp specificity is achieved by the modular combination of three zinc fingers. The zinc-fingers recognize the target sequence in the major groove of dsDNA while making contacts primarily with bases in one strand, and are thus designed to provide close and anti-parallel binding to the two chimeric nucleases. Upon dimerization of the nuclease domain, the DNA is cleaved as indicated by carats.

gene in which the nuclease binding sites are not of transgenic origin). The same validation has been concurrently achieved in human cells in an artificial genomic context where gene targeting is directed at an engineered green fluorescent protein (GFP) transgene stably integrated in the genome of a human cell line (Porteus and Baltimore, 2003).

DSB-boosted gene targeting with designed zinc-finger nucleases in human cells

Consistent with the optimization of Bibikova *et al.* (2001; see Fig. 7), the sites for the two zinc-finger nucleases are spaced by a 6 bp insert in target GFP transgene (Porteus and Baltimore, 2003). Gene targeting is thus performed with a set of appropriate zinc-finger nucleases and transfecting dsDNA comprising a stretch of 2700 bp homologous to target GFP gene sequences. This transfecting dsDNA is the substrate of homologous recombination with the genomic target; it carries an heterologous core sequence for gene repair/modification. The reaction is initiated by the co-transfection of substrate DNA and minigenes encoding the two zinc-finger nucleases, and culminates in the genesis of easily scorable GFP gene targeting events. As anticipated, a dramatic increase in gene targeting frequencies is associated with the zinc-finger nucleases and fairly matches the effect of endonuclease I-*Sce*I in concurrent control experiments (Porteus and Baltimore, 2003). Optimization on control I-*Sce*I experiments indicates that gene repair/modification rates of 3–5% are achievable with this emerging DSB-dependent gene targeting technology (Porteus and Baltimore, Supporting Online Material, 2003).

Zinc-finger design as the critical point of chimeric nucleases

The DNA-binding domain of the aforedescribed zinc-finger nucleases is composed of three Cys_2His_2 zinc fingers, thereby recognizing a 9 bp sequence in the major groove of target DNA (Fig. 7). The effective recognition site of a pair of zinc finger nucleases is thus an 18 bp sequence, i.e. a statistically unique sequence in human/mammalian genomes (see above calculation for the binding site of I-*Sce*I).

Each finger module comprises about 30 amino acids that adopt a compact DNA-binding structure on chelating a zinc ion (Miller *et al.*, 1985) and is able to recognize 3–4 bp of DNA through the binding of its α helix into the major groove of dsDNA (Pavletich and Pabo, 1991). Individual zinc fingers with various DNA-binding specificities have been selected mainly with phage-display libraries and have been used to design multifinger proteins with pre-determined sequence-specificity (see: Choo and Isalan, 2000; Pabo *et al.*, 2001; Beerli and Barbas, 2002). A panel of individual zinc fingers specific to each of the 64 possible DNA base triplets should theoretically cover all tandem-array assembling needs. However, modular assembly of synthetic multifinger proteins is accompanied by cooperative and context-dependent contacts between neighboring fingers/neighboring DNA sites that interfere with the binding specificity of each finger module (Isalan *et al.*, 1997, 1998; Elrod-Erickson *et al.*, 1996; Wolfe *et al.*, 2001). Therefore, optimization strategies using random zinc-finger libraries have been devised to produce three-finger domains with the anticipated binding specificity that can be eventually combined into six-finger proteins in order to achieve an 18 bp DNA-binding specificity (Greisman and Pabo, 1997; Isalan *et al.*, 2001; Dreier *et al.*, 2001). Although effective (e.g. Beerli *et al.*, 2000; Liu *et al.*, 2001b; Rebar *et al.*, 2002; Reynolds *et al.*, 2003; Tan *et al.*, 2003), these approaches do not provide for a simultaneous randomized selection of the zinc finger modules comprising the three-finger domains and therefore request a careful characterization of the specificity and affinity of the final products (see: Pabo *et al.*, 2001; Segal *et al.*, 2003; Bae *et al.*, 2003). A new strategy is however emerging that might efficiently sustain such a direct simultaneous selection (Hurt *et al.*, 2003). Emerging zinc-finger technology should thus succeed in the production of the very DNA-binding domains that will allow efficient targeting of nearly any genomic sequence.

AAV GENE TARGETING VECTORS: SINGLE-STRANDED DNA AND SYNERGISTIC DNA DSB

From current episomal/random-integrating AAV vectors to site-specific integrating ones

Recombinant AAV vectors are currently used in pre-clinical and clinical gene therapy trials where they are either maintained as free monomeric/concatemeric episomes or random-integrated in host chromosomal DNA (see Kearns *et al.*, 1996; Schnepp *et al.*, 2003; Nakai *et al.*, 2003a, b). Importantly, to date, this human virus has not been associated with any disease (Review: Berns and Linden, 1995).

However, by contrast to current first generation AAV vectors, retroviruses and other known viruses, wild-type AAV has the unique ability to site-specifically integrate its DNA into the human genome: the target is the AAVS1 site on human chromosome 19 (Kotin *et al.*, 1990; review: Linden *et al.*, 1996). The site-specific integrating ability of the virus is still under investigations (see Allen and Samulski, 2003); it involves the two inverted terminal repeats (ITRs) and is dependent on AAV Rep78 or Rep68 protein (Balague *et al.*, 1997; Surosky *et al.*, 1997), i.e. the two large forms of the four overlapping Rep proteins that are encoded by the AAV *rep* gene. Due to the small packaging capability (~4.6 kb) of AAV virions, the viral backbone of current AAV vectors is restricted to the two flanking ITRs, thereby resulting in vectors that have lost the site-specific integration ability of wild type AAV (Kearns *et al.*, 1996). Rep78/Rep68 can however be provided in *trans* (Surosky *et al.*, 1997) and are now part of a co-transfer strategy that should confer site-specific integrating capability to first generation AAV vectors (see Allen and Samulski, 2003). In addition, these Rep proteins and AAV ITRs together with a newly identified p5 integration efficiency element (p5IEE) from the p5 promoter region of the *rep* gene (Philpott *et al.*, 2002) are currently used to confer site-specific integration to non-viral gene therapy vectors (Surosky *et al.*, 1997; Pieroni *et al.*, 1998) or to non-integrating viruses with large packaging capabilities such as herpes simplex virus type 1 (HSV1; Johnston *et al.*, 1997), baculovirus (Palombo *et al.*, 1998) or adenovirus (Recchia *et al.*, 1999).

Such a Rep-dependent platform of AAV-based site-specific integrating vectors is under intensive investigation and has exciting promises (see Owens, 2002; Bertolotti, 2003b; Allen and Samulski, 2003; Fraefel *et al.*, 2003). However, one of the most attractive features of recombinant AAV vectors is their gene targeting capability (see below).

Recombinant AAV as a gene targeting vector

First described in 1998 by Russell and Hirata with first generation AAV vectors on the very HPRT gene (Russell and Hirata, 1998) of our recombinase-mediated gene therapy model (Bertolotti, 1996b), AAV-mediated gene targeting was shown to be much more efficient than standard technology relying on dsDNA. With a frequency of up to 1% with normal human cells, such a viral gene targeting approach appears to be locus-independent and has been shown to drive 1- and 2-bp substitutions, small deletions, and insertions of up to 1.5 kb (Russell and Hirata, 1998; Inoue *et al.*, 1999, 2001; Hirata and Russell, 2000; Hirata *et al.*, 2002). Such a high gene targeting efficiency is however achieved at very high multiplicity of infection (MOI), from 10,000 to 400,000 viral genomes/cell, and is therefore not directly amenable to gene therapy protocols. In addition, it is overwhelmed by random integration (Hirata *et al.*, 2002) and is thereby associated with genomic rearrangements even at low MOI (Miller *et al.*, 2002).

Interestingly, the active substrate for the gene-targeting reaction appears to be the single-stranded genomic DNA from recombinant AAV (Hirata and Russell,

2000). The same holds true for the minute virus of mouse (MVM), the second parvovirus that has been used in gene targeting experiments (Hendrie *et al.*, 2003).
Such a situation is reminiscent of our pre-made presynaptic filaments in which
the form activated for homologous recombination is the RAD51-ssDNA complex
(Bertolotti, 1996a, b, 2000c). Moreover, like in our dsDNA-cored presynaptic filaments (Bertolotti, 1996b), the structure of the AAV ITRs and of the MVM palin-
drome sequences might confer to AAV/MVM genomic ssDNA a dsDNA-tailed
structure that has been shown to be the preferred substrate for RAD51 protein-
mediated homologous pairing (Mazin *et al.*, 2000). Such structural features could
thus provide to this viral approach a substantial advantage over conventional gene-
targeting technology where transfecting dsDNA has to be processed *in vivo* to be
converted into an active substrate for homologous recombination (see Bertolotti,
2000c).

AAV-mediated gene targeting and synergistic DNA DSBs: toward zinc finger
nuclease-boosted protocols

Like for conventional gene targeting, creating a DSB in target chromosomal DNA
produces a dramatic increase in viral gene targeting frequency (Miller *et al.*, 2003;
Porteus *et al.*, 2003). Two artificial genomic models have been used in these
pioneering experiments in which the target is an easily scorable transgene (either
the bacterial *lac Z* gene or a GFP minigene) that has been engineered to contain
the I-*Sce*I recognition site and then stably integrated in the genome of human
cells. Like in the AAV gene targeting experiments described above, standard first
generation AAV vectors (AAV sequences are restricted to the two viral ITRs) are
used to transduce the substrate for homologous recombination into target human
cells. On the other hand, expression of the I-*Sce*I endonuclease is either driven by an
integrated retroviral vector (Miller *et al.*, 2003) or by co-transduction/transfection of
a second AAV vector or of plasmid DNA (Porteus *et al.*, 2003). In both cases, AAV-
mediated gene targeting frequency is increased 60- to 100-fold and, under optimal
conditions, is raised to the random-integration level, i.e. ∼1–3% of transduced cells.
Importantly, the gene targeting frequency achieved with relatively low MOIs shows
that this approach is promising enough to be amenable to zinc-finger nucleases
and suggests that the resulting rAAV-targeted nuclease combination might be soon
compatible with gene therapy protocols (Porteus *et al.*, 2003; Miller *et al.*, 2003).
Toward this goal, new approaches might eventually involve a transient inhibition of
random-integration.

RECOMBINASE-DNA NUCLEOPROTEINS AS ZINC FINGER
NUCLEASE-BOOSTED GENE TARGETING VECTORS

Recombinase-DNA nucleoproteins as DSB-boosted gene targeting vectors and
random-integration inhibitors

Either with dsDNA transfection or AAV transduction, the genesis of a DSB on target chromosomal DNA has been shown to increase gene targeting frequency up

to the random-integration level, i.e. about 1–3% of transfected or transduced cells under current optimal conditions (Porteus *et al.*, 2003; Porteus and Baltimore, 2003; Miller *et al.*, 2003). Such approaches thus dramatically raise gene targeting efficiency but do not inhibit random-integration. In order to move toward clinical gene therapy, both a drastic reduction of random-integration and a further improvement in gene-targeting efficiency is therefore a highly desirable goal. To this end, our current approach is thus discussed in terms of synergistic combinations in which random-integration is blocked by the use of pre-made presynaptic complexes (stockiometric coating of transfecting DNA by RAD51 protein) and homologous recombination promoted both by transfecting pre-made presynaptic complexes and by sequence-specific double-strand breakage of target chromosomal DNA. Additional features might eventually be gained from AAV ITR structure.

DNA backbones for DSB-boosted gene targeting nucleoprotein vectors

Creating a DSB in the middle of chromosomal DNA target establishes a very different genomic context as compared to the previously described situation for homologous recombination mediated by recombinase-nucleoprotein vectors (see above), in particular for standard presynaptic filaments. Importantly, in this case, the result of the homologous DNA pairing-strand exchange reaction with chromosomal dsDNA mediated by both vector ssDNA tails is not a displacement loop but chromosomal ssDNA tails (see Fig. 8 for illustration with a standard presynaptic vector). On the other hand, the main product of the strand exchange reaction between the vector and target chromosomal DNA is an heteroduplex with a single-stranded gap of an unknown length located on the host strand. The size of the gap depends on the original DSB resection processing of the broken chromosomal dsDNA and on the length and nature of the heterologous core sequence of the recombinase-DNA nucleoprotein vector. Such a heteroduplex should be a perfect substrate for the DNA synthesis phase of standard homologous recombination repair (HRR) processing of DSBs (see Szostak *et al.*, 1983; Pâques and Haber, 1999; West, 2003), thereby boosting the very gene targeting reaction. The same appears to hold true for DSB-boosted AAV-mediated gene targeting (Miller *et al.*, 2003). However, providing an active RAD51-ssDNA complex should increase the efficiency of the gene targeting reaction by the swift initiation of the strand exchange reaction with chromosomal dsDNA, thereby shunting the initial steps of the DSB repair process of broken chromosomal dsDNA, i.e. either a 5′ to 3′ exonuclease resectioning of both ends to produce 3′ ssDNA overhangs (HRR and single-strand annealing [SSA] pathways; see Pâques and Haber, 1999; West, 2003; Van Dyck *et al.*, 2001) or Ku protein binding and subsequent non-homologous end-joining processing (NHEJ pathway; see Lieber *et al.*, 2003).

As detailed above, the active substrate for the AAV-mediated gene-targeting reaction appears to be the single-stranded genomic DNA from recombinant AAV (Hirata and Russell, 2000; Miller *et al.*, 2003). This observation implies that, if AAV gene targeting does not involve the SSA pathway (see below and Miller *et al.*,

2003), the homologous DNA pairing-strand exchange reaction with chromosomal dsDNA is initiated on both sides of chromosomal DSB by each end of AAV ssDNA. Such a pairing is consistent with the bidirectional strand exchange polarity that has been described for RAD51 (Namsaraev and Berg, 1998; McIlwraith *et al.*, 2000). Therefore, standard recombinase-ssDNA nucleoprotein complexes should work in the same way and thus behave as true gene targeting vector when a DSB is provided in chromosomal dsDNA target (Fig. 8). The AAV ITRs might confer some stimulating feature to viral gene targeting (see above) and should therefore be evaluated in the recombinase-DNA nucleoprotein vectors.

Importantly, additional benefits might be associated with our chimeric dsDNA-cored presynaptic vectors since (1) they are amenable to tail polarity optimization by comparing backbones with either both 3′ or 5′ overhangs or with a set of 3′-5′ overhangs (see discussion on dsDNA-cored backbones in the section on recombinase-DNA nucleoprotein vectors) and (2) their dsDNA-core might further improve gene targeting efficiency by eliminating potential mismatch problems and interference with repetitive elements (heterologous and repetitive sequences are in the recombination-locked dsDNA core) and by reducing the impact of DNA synthesis in the recombination procedure, in particular with a long stretch of heterologous core sequence such as a minigene.

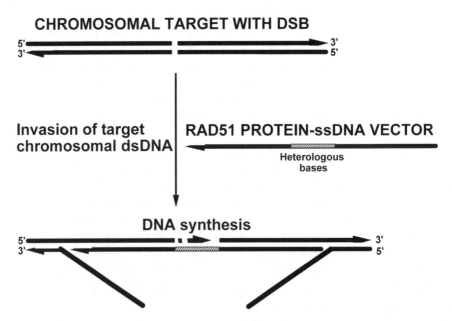

Figure 8. Schematic of the putative homologous DNA pairing and strand exchange reaction between a standard RAD51 protein-ssDNA nucleoprotein vector and unresected broken chromosomal dsDNA. As shown, the design is aimed at integrating a heterologous sequence at the site of the sequence-specific DSB.

DSB-boosted gene targeting vectors and DSB repair

During standard homologous recombination-mediated DSB repair (HHR pathway), the invading strands are the $3'$ overhangs from resected broken chromosomal dsDNA that have been converted into presynaptic filaments upon stoichiometric co-polymerisation with RAD51 protein (see West, 2003). The converse might hold true for DSB-boosted gene targeting since with AAV vectors the reaction appears to be mediated by genomic ssDNA (see above and Miller *et al.*, 2003). Therefore, even though RAD51 and ancillary proteins accumulate in foci at or near the site of DNA damage (Haaf *et al.*, 1995; Tashiro *et al.*, 2000), premade recombinase-DNA nucleoprotein vectors should improve the gene targeting efficiency, and prevent random integration as discussed previously (see above, and Bertolotti, 1996b, 2000c).

Regarding the homology length with target chromosomal DNA, there are currently no data available for the AAV-nuclease combination, i.e. a situation in which a sequence-specific DSB is provided and vector DNA does not need a $5'$ to $3'$ resection of its tails. Although both pioneering experiments rely on targeting vectors with a $\sim$3 kb homology length with target chromosomal DNA (based on data gained with standard AAV gene targeting; see Hirata and Russell, 2000), *in vitro* experiments have shown that the homologous pairing-strand exchange reaction mediated by human RAD51 protein is very efficient with $3'$ and $5'$ tails of approximately 600 and 500 nt in length, respectively, but is still detectable with 50 nt overhangs (McIlwraith *et al.*, 2000). Therefore, providing a vector with premade recombinase-ssDNA tails and creating the sequence-specific DSB in the very center of the chromosomal target sequence might significantly reduce optimal homology length. Such a length should however be adjusted to the average size of the resected portion of target chromosomal DNA.

Discussion: Rad52 and dsDNA-cored backbones as effectors of putative SSA-mediated gene targeting

Both yeast (Mortensen *et al.*, 1996) and human (Reddy *et al.*, 1997; Van Dyck *et al.*, 2001) Rad52 have been shown to mediate single-strand annealing *in vitro*. In addition to facilitating the efficient polymerisation of RAD51 on ssDNA, Rad52 stands therefore as a key enzyme of the SSA repair pathway (see: Van Dyck *et al.*, 2001; West, 2003) and as a potential SSA gene targeting protein. It might therefore be instrumental in DSB-boosted gene targeting mediated by ssDNA vectors including AAV and could thus be used to devise a premade nucleoprotein vector strategy similar to the presynaptic RAD51 filament one. Importantly, DSB-boosted SSA-mediated gene targeting stands as a symetric processing and should thus be directly boosted by a dsDNA-cored backbone similar to our chimeric presynaptic nucleoprotein vector one (see Fig. 9).

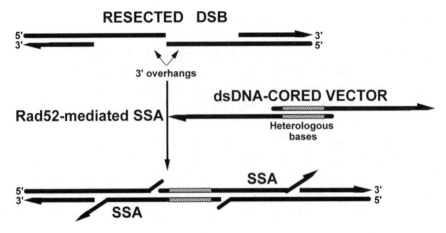

Figure 9. Schematic of the putative single-strand annealing (SSA) reaction mediated by Rad52 between the ssDNA tails of the dsDNA-cored vector and the 3′ overhangs of resected broken chromosomal dsDNA. Whether branch migration can follow SSA as shown *in vitro* with short oligonucleotides on a circular partial-duplex phage model (Reddy *et al.*, 1997) remains to be evaluated.

CONCLUSION: COMBINING STEM-CELL GENE-TARGETING WITH TRANSIENT REGENERATIVE GENE THERAPY

Long-term stem cell gene therapy is amenable to synergistic combinations with transient regenerative gene therapy

DSB-boosted gene targeting mediated by designed zinc-finger nucleases and either recombinase-DNA nucleoprotein vectors or AAV vectors associated to a transient inhibition of end-joining activity (see above) open exciting promises for long-term stem cell gene therapy. Such an approach is all the more promising as it is amenable to synergistic combinations with transient gene therapy protocols aimed at magnifying the homing/regenerative potential of transplanted engineered stem cells (see Fig. 10 and Bertolotti, 2001, 2003a, c).

Amplification of the proliferative and regenerative potential of stem cells by transient or regulated ex vivo *gene therapy*

Transient amplification of the regenerative or neovascularizative potential of *ex vivo* expanded endothelila progenitor cells (EPCs) has been pioneered by Isner and co-workers with a replication-defective adenoviral vector carrying a VEGF minigene (Iwaguro *et al.*, 2002; see Bertolotti, 2002b). A similar approach has been used for adult HSCs for which the rate of expansion and the *in vivo* repopulating ability have been shown to be dramatically increased by *ex vivo* HOXB4 gene therapy (Antonchuk *et al.*, 2001, 2002). However, instead of using non-integrating adenoviral vectors that are diluted and lost in proliferating cells, Antonchuk *et al.* focused on integrative retroviral vectors because transient expression is not required

STEM CELL GENE THERAPY BREAKTHROUGH

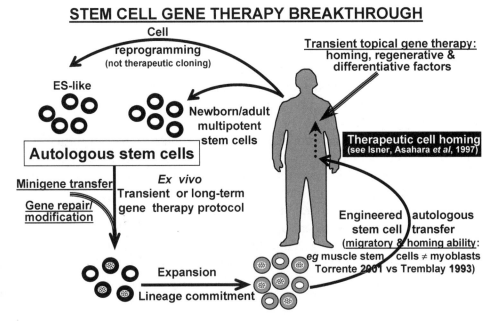

Figure 10. Stem cell gene therapy: combining *ex vivo* protocols with transient topical gene therapy (Poster Exhibition of June 6, 2002, *ASGT 5th Annual Meeting*, Boston: Bertolotti, 2002c). A schematic overview of the stem cell gene therapy breakthrough in which emerging autologous adult multipotent/pluripotent stem cells are displayed together with potential future autologous embryonic stem (ES)-like cells that might result from intensive cell reprogramming investigations (see Bertolotti, 2001). Adapted from Fig. 3 of Bertolotti, 2003a, and reprinted from Bertolotti, 2003c, with permission.

in this case (HOXB4 does not override extrinsic physiological mechanisms that control the population size of HSCs *in vivo*: Antonchuk *et al.*, 2001).

Transient topical gene therapy as a booster for the therapeutic homing/regenerative potential of target stem cells

Another way to increase the regenerative potential of transplanted stem cells is to couple *ex vivo* protocols to synergistic *in vivo* transient topical gene therapy (Bertolotti, 2001, 2002b).

Topical production of a growth factor may of course increase the proliferative/differentiative index of transplanted stem cells and differentiative progeny in the same way it acts on endogenous stem/progenitor cells (e.g. the angiogenic and vasculogenic gene therapy described above). Yet, transient local overproduction of a homing factor may have a more dramatic effect by chanelling curative cells to the very therapeutic site (see Bertolotti, 2001). Interestingly, topical injection of stroma cell-derived factor-1 (SDF-1) has been recently used to increase the recruitment of transplanted EPCs into ischemic muscle tissues and was shown to efficiently boost neovascularization in an experimental hind limb ischemia model (Yamaguchi *et al.*,

2003). Like for VEGF, FGF and other growth factors (e.g. Isner, 2002; Grines et al., 2002; Baumgartner, 2003), transient topical SDF-1 gene therapy should be more efficient than the mere topical infusion of the purified protein. Indeed, topical transplantation of cells engineered to constitutively express SDF-1 increases therapeutic homing of BM-derived c-kit$^+$ stem cells in the infarcted zone of a rat model (Askari et al., 2003), thereby confirming the broad chemoattractive ability of SDF-1 for BM stem cells (Mohle et al., 1998).

In addition to therapeutic homing, the curative efficiency of transplanted stem cells requires an optimal commitment to the target tissue or organ function. Such a functional integration is crucial for multipotent or pluripotent stem cells and may rely on transient topical expression of lineage-commitment and specific growth or differentiative factors (see Bertolotti, 2001). This process could also benefit from transient expression of a 'regenerative' inducer (Bertolotti, 2001) such as msx1 for skeletal muscle (Odelberg et al., 2000) or IGF-1 for many tissues including cardiac and skeletal muscles (e.g. Reiss et al., 1996; Barton-Davis et al., 1998; see Bertolotti, 2003a).

Combining stem-cell gene-targeting with transient regenerative gene therapy

As illustrated in Fig. 10, autologous stem cells are thus amenable to sophisticated cell gene therapy protocols. Upon potential mobilization by recombinant protein application, autologous stem cells are first subjected to *ex vivo* protocols where they can be purified, expanded, genetically engineered for transient amplification of their proliferative or regenerative potential and for long-term transgene expression or gene repair (DSB-boosted gene targeting involves transient expression of minigenes encoding the set of designed zinc-finger nucleases) and, eventually commited to a specific cell-lineage before being returned to the patient. Such *ex vivo* manipulations can then be combined to synergistic transient topical gene therapy culminating in a timely magnification of the regenerative, homing or differentiative capability of therapeutic stem cells and subsequent efficient targeted repopulation dynamics in relevant tissues and organs. Importantly, transient gene therapy may rely on exogenous drug control when fine tuning and transient gene expression have to be inforced in a very tight way (see Bertolotti, 2002b, 2003a).

Such a stem cell gene therapy in which stem cell therapy is combined with gene targeting and transient regenerative gene therapy protocols stands thus as a powerful means of combatting both inherited diseases and degenerative or acquired disorders (Bertolotti, 2001, 2002b, 2003a, c). It could therefore in the future be instrumental in the treatment of the Lesch–Nyhan disease where the metabolic disorder could be handled by HPRT-corrected circulating lymphoid-myeloid cells (HSC gene therapy) while the neurologic symptoms resulting from the dopaminergic deficit could be cured by stereotaxic regenerative brain stem cell gene therapy mediated by HPRT-corrected neural stem cells.

REFERENCES

Aboussekhra, A., Chanet, R., Adjiri, A. and Fabre, F. (1992). Semidominant suppressors of Srs2 helicase mutations of *S. cerevisiae* map in the *RAD51* gene, whose sequence predicts a protein with similarities to procaryotic RecA proteins, *Mol. Cell. Biol.* **12**, 3224–3234.

Abuin, A. and Bradley, A. (1996). Recycling selectable markers in mouse embryonic stem cells, *Mol. Cell. Biol.* **16**, 1851–1856.

Aiuti, A., Slavin, S., Aker, M., Ficara, F., Deola, S., Mortellaro, A., Morecki, S., Andolfi, G., Tabucchi, A., Carlucci, F., Marinello, E., Cattaneo, F., Vai, S., Servida, P., Miniero, R., Roncarolo, M. G. and Bordignon, C. (2002). Correction of ADA-SCID by stem cell gene therapy combined with nonmyeloablative conditioning, *Science* **296**, 2410–2413.

Al-Hajj, M., Wicha, M. S., Benito-Hernandez, A., Morrison, S. J. and Clarke, M. F. (2003). Prospective identification of tumorigenic breast cancer cells, *Proc. Natl. Acad. Sci. USA* **100**, 3983–3988.

Allen, N. A. and Samulski, R. J. (2003). The pros and cons of using the mechanism of AAV site-specific recombination in gene delivery, *Gene Therapy and Regulation* **2**, 121–138.

Alvarez-Buylla, A. and Garcia-Verdugo, J. M. (2002). Neurogenesis in adult subventricular zone, *J. Neurosci.* **22**, 629–634.

Anderson, L. T., Ernst, M. and Davis, S. V. (1992). Cognitive abilities of patients with Lesch–Nyhan disease, *J. Autism Dev. Disord.* **22**, 189–203.

Anderson, W. F. (1984). Prospects for human gene therapy, *Science* **226**, 401–419.

Antonchuk, J., Sauvageau, G. and Humphries, R. K. (2001). HOXB4 overexpression mediates very rapid stem cell regeneration and competitive hematopoietic repopulation, *Exp. Hematol.* **29**, 1125–1134.

Antonchuk, J., Sauvageau, G. and Humphries, R. K. (2002). HOXB4-induced expansion of adult hematopoietic stem cells *ex vivo*, *Cell* **109**, 39–45.

Anversa, P. and Nadal-Ginard, B. (2002). Myocyte renewal and ventricular remodelling, *Nature* **415**, 240–243.

Armbruster, L., Cavard, C., Briand, P. and Bertolotti, R. (1992). Selection of variant hepatoma cells in liver-specific growth media: regulation at the mRNA level, *Differentiation* **50**, 25–33.

Askari, A. T., Unzek, S., Popovic, Z. B., Goldman, C. K., Forudi, F., Kiedrowski, M., Rovner, A., Ellis, S. G., Thomas, J. D., DiCorleto, P. E., Topol, E. J. and Penn, M. S. (2003). Effect of stromal-cell-derived factor 1 on stem-cell homing and tissue regeneration in ischaemic cardiomyopathy, *Lancet* **362**, 697–703.

Bae, K. H., Kwon, Y. D., Shin, H. C., Hwang, M. S., Ryu, E. H., Park, K. S., Yang, H. Y., Lee, D. K., Lee, Y., Park, J., Kwon, H. S., Kim, H. W., Yeh, B. I., Lee, H. W., Sohn, S. H., Yoon, J., Seol, W. and Kim, J. S. (2003). Human zinc fingers as building blocks in the construction of artificial transcription factors, *Nature Biotechnol.* **21**, 275–280.

Balague, C., Kalla, M. and Zhang, W. W. (1997). Adeno-associated virus Rep78 protein and terminal repeats enhance integration of DNA sequences into the cellular genome, *J. Virol.* **71**, 3299–3306.

Barton-Davis, E. R., Shoturma, D. I., Musaro, A., Rosenthal, N. and Sweeney, H. L. (1998). Viral mediated expression of insulin-like growth factor I blocks the aging-related loss of skeletal muscle function, *Proc. Natl. Acad. Sci. USA* **95**, 15603–15607.

Basile, G., Aker, M. and Mortimer, R. K. (1992). Nucleotide sequence and transcriptional regulation of the yeast recombinational repair gene *RAD51*, *Mol. Cell. Biol.* **12**, 3235–3246.

Baumann, P. and West, S. C. (1997). The human Rad51 protein: polarity of strand transfer and stimulation by hRP-A, *EMBO J.* **17**, 5198–5206.

Baumann, P., Benson, F. and West, S. C. (1996). Human Rad51 protein promotes ATP-dependent homologous pairing and strand transfer reactions *in vitro*, *Cell* **87**, 757–766.

Baumgartner, I. (2003). Vascular endothelial growth factor (VEGF) gene therapy for critical limb ischemia, in: *Progress in Gene Therapy — 2: Pioneering Stem Cell/gene Therapy Trials*, Bertolotti, R., Ozawa, K. and Hammond, H. K. (Eds), pp. 65–106. VSP Publishers, Utrecht, The Netherlands.

Baumgartner, I., Pieczek, A., Manor, O., Blair, R., Kearney, M., Walsh, K. and Isner, J. M. (1998). Constitutive expression of phVEGF165 after intramuscular gene transfer promotes collateral vessel development in patients with critical limb ischemia, *Circulation* **97**, 1114–1123.

Baumgartner, I., Rauh, G., Pieczek, A., Wuensch, D., Magner, M., Kearney, M., Schainfeld, R. and Isner, J. M. (2000). Lower-extremity edema associated with gene transfer of naked DNA encoding vascular endothelial growth factor, *Ann. Intern. Med.* **132**, 880–884.

Beltrami, A. P., Barlucchi, L., Torella, D., Baker, M., Limana, F., Chimenti, S., Kasahara, H., Rota, M., Musso, E., Urbanek, K., Leri, A., Kajstura, J., Nadal-Ginard, B. and Anversa, P. (2003). Adult cardiac stem cells are multipotent and support myocardial regeneration, *Cell* **114**, 763–776.

Beerli, R. R. and Barbas, C. F. (2002). Engineering polydactyl zinc-finger transcription factors, *Nature Biotechnol.* **20**, 135–141.

Beerli, R. R., Dreier, B. and Barbas, C. F. (2000). Positive and negative regulation of endogenous genes by designed transcription factors, *Proc. Natl. Acad. Sci. USA* **97**, 1495–1500.

Benson, F. E., Stasiak, A. and West, S. C. (1994). Purification and characterization of the human Rad51 protein, an analogue of *E. coli* RecA, *EMBO J.* **13**, 5764–5771.

Benson, F. E., Baumann, P. and West, S. C. (1998). Synergistic actions of Rad51 and Rad52 in recombination and DNA repair, *Nature* **391**, 401–404.

Berns, K. I. and Linden, R. M. (1995). The cryptic life style of adeno-associated virus, *BioEssays* **17**, 237–245.

Bertolotti, R. (1996a). Recombinase-mediated gene therapy: targeting single stranded DNA to chromosomal DNA with human RAD51 presynaptic fibers, *Hepatology* **24**, 484A.

Bertolotti, R. (1996b). Recombinase-mediated gene therapy: strategies based on Lesch–Nyhan mutants for gene repair/inactivation using human RAD51 nucleoprotein filaments, *Biogenic Amines* **12**, 487–498.

Bertolotti, R. (1997). Recombinase-mediated gene therapy: strategies for gene repair/inactivation using human RAD51 nucleoprotein filaments, in: *Advances in Molecular Toxicology*, Reiss, C. and Parvez, H. (Eds), pp. 445–466. VSP Publishers, Utrecht, The Netherlands.

Bertolotti, R. (1998a). Recombinase-DNA nucleoprotein filaments as gene therapy vectors, *Biogenic Amines* **14**: 41–65.

Bertolotti, R. (1998b). Prospects of recombinase-mediated HPRT gene therapy for Lesch–Nyhan disease, in: *Neurochemical Markers of Degenerative Nervous Diseases and Drug Addiction*, Qureshi, G. A. Parvez, H., Caudy, P. and Parvez, S. (Eds), pp. 133–170. VSP Publishers, Utrecht, The Netherlands.

Bertolotti, R. (1998c). Gene therapy 1998: transient or stable minigene expression and gene repair/inactivation, *Biogenic Amines* **14**, 389–406.

Bertolotti, R. (1999). Recombinase-DNA nucleoprotein filaments as vectors for gene repair/inactivation and targeted integration of minigenes, *Biogenic Amines* **15**, 169–195.

Bertolotti, R. (2000a). RNA and gene repair/alteration: from inherited diseases to acquired disorders and tantalizing applications for non-disease conditions, *Gene Therapy and Regulation* **1**, 115–122.

Bertolotti, R. (2000b). Gene therapy: transient or long-term minigene expression and gene repair/inactivation, in: *Progress in Gene Therapy — Basic and Clinical Frontiers*, Bertolotti, R., Parvez, H. and Nagatsu, T. (Eds), pp. 1–34. VSP Publishers, Utrecht, The Netherlands.

Bertolotti, R. (2000c). Gene therapy: dsDNA-cored presynaptic filaments as vectors for gene repair and targeted integration of transgenes, in: *Progress in Gene Therapy — Basic and Clinical Frontiers*, Bertolotti, R., Parvez, H. and Nagatsu, T. (Eds), pp. 513–549. VSP Publishers, Utrecht, The Netherlands.

Bertolotti, R. (2001). Adult and embryonic-like stem cells: toward a major gene therapy breakthrough relying on autologous multipotent stem cells, *Gene Therapy and Regulation* **1**, 207–212.

Bertolotti, R. (2002a). In memoriam: Jeffrey M. Isner, stem cell/gene therapy pioneer, *Gene Therapy and Regulation* **1**, 285.

Bertolotti, R. (2002b). From therapeutic angiogenesis to myocardium regeneration, *Gene Therapy and Regulation* **1**, 287–295.

Bertolotti, R. (2002c). Gene repair/modification and targeted integration of minigenes: toward improved *ex vivo/in vivo* stem cell gene therapy, *Mol. Ther.* **5**, S133 (Abstract).

Bertolotti, R. (2003a). Stem cell gene therapy: a breakthrough combination magnified by therapeutic stem cell homing, in: *Progress in Gene Therapy — 2: Pioneering Stem Cell/gene Therapy Trials*, Bertolotti, R., Ozawa, K. and Hammond, H. K. (Eds), pp. 1–31. VSP Publishers, Utrecht, The Netherlands.

Bertolotti, R. (2003b). Site-specific integration, *Gene Therapy and Regulation* **2**, 1–5.

Bertolotti, R. (2003c). Stem cell gene therapy: breakthrough culminating in combination of *ex vivo* protocols with transient topical gene therapy, *Gene Therapy and Regulation* **2**, 91–102.

Bertolotti, R. (2004). Stem cell gene therapy: synergizing improved gene repair/modification & targeted integration of transgenes to transient in vivo topical minigene expression, *Mol. Ther.* (Abstract) (in press).

Bertolotti, R. and Lutfalla, G. (1983). Genes programming cell differentiation, *J. Cell. Biochem.* **7A**, 149.

Bertolotti, R., Armbruster-Hilbert, L. and Okayama, H. (1995). Liver fructose-1,6-bisphosphatase cDNA: trans-complementation of fission yeast and characterization of two human transcripts, *Differentiation* **59**, 51–60.

Bibikova, M., Carroll, D., Segal, D. J., Trautman, J. K., Smith, J., Kim, Y. G. and Chandrasegaran, S. (2001). Stimulation of homologous recombination through targeted cleavage by chimeric nucleases, *Mol. Cell. Biol.* **21**, 289–297.

Bibikova, M., Golic, M., Golic, K. G. and Carroll, D. (2002). Targeted chromosomal cleavage and mutagenesis in Drosophila using zinc-finger nucleases, *Genetics* **161**, 1169–1175.

Bibikova, M., Beumer, K., Trautman, J. K. and Carroll, D. (2003). Enhancing gene targeting with designed zinc finger nucleases, *Science* **300**, 764.

Bonnet, D. and Dick, J. E. (1997). Human acute myeloid leukemia is organized as a hierarchy that originates from a primitive hematopoietic cell, *Nature Med.* **3**, 730–737.

Borchiellini, P., Angulo, J. F. and Bertolotti, R. (1997). Genes encoding mammalian recombinases: cloning approach with anti-RecA antibodies, *Biogenic Amines* **13**, 195–215.

Brennand, J., Konecki, D. S. and Caskey, C. T. (1983). Expression of human and chinese hamster hypoxanthine-guanine phosphoribosyltransferase cDNA recombinants in cultured Lesch–Nyhan and chinese hamster fibroblasts, *J. Biol. Chem.* **258**, 9593–9596.

Bushman, F. (2002). Targeting retroviral integration? *Mol. Ther.* **6**, 570–571.

Capecchi, M. (1989). The new mouse genetics: altering the genome by gene targeting, *Trends Genetics* **5**, 70–76.

Cariello, N. and Skopek, T. (1993). Analysis of mutations occurring at the human *hprt* locus, *J. Mol. Biol.* **231**, 41–57.

Cavazzana-Calvo, M., Hacein-Bey, S., de Saint Basile, G., Gross, F., Yvon, E., Nusbaum, P., Selz, F., Hue, C., Certain, S., Casanova, J. L., Bousso, P., Deist, F. L. and Fischer, A. (2000). Gene therapy of human severe combined immunodeficiency (SCID)-X1 disease, *Science* **288**, 669–672.

Chasin, L. A. and Urlaub, G. (1976). Mutant alleles for hypoxanthine phosphoribosyltransferase: codominant expression, complementation, and segregation in hybrid Chinese hamster cells, *Somatic Cell Genet.* **2**, 453–467.

Choo, Y. and Isalan, M. (2000). Advances in zinc finger engineering, *Curr. Opin. Struct. Biol.* **10**, 411–416.

Choulika, A, Perrin, A., Dujon, B. and Nicolas, J. F. (1995). Induction of homologous recombination in mammalian chromosomes by using the I-SceI system of Saccharomyces cerevisiae, *Mol. Cell. Biol.* **15**, 1968–1973.

Dancis, J., Berman, P. H., Jansen, V. and Balis, M. E. (1968). Absence of mosaicism in the lymphocyte in X-linked congenital hyperuricosuria, *Life Sci.* **7**, 587–591.

Deng, C. and Capecchi, M. (1992). Reexamination of gene targeting frequency as a function of the extent of homology between the targeting vector and the target locus, *Mol. Cell. Biol.* **12**, 3365–3371.

Dick, J. E. (2003). Breast cancer stem cells revealed, *Proc. Natl. Acad. Sci. USA* **100**, 3547–3549.

DiMaio, D., Treisman, R. and Maniatis, T. (1982). Bovine papilloma-virus vector that propagates as a plasmid in both mouse and bacterial cells, *Proc. Natl. Acad. Sci. USA* **79**, 4030–4034.

Donoho, G., Jasin, M. and Berg, P. (1998). Analysis of gene targeting and intrachromosomal homologous recombination stimulated by genomic double-strand breaks in mouse embryonic stem cells, *Mol. Cell. Biol.* **18**, 4070–4078.

Dreier, B., Beerli, R. R., Segal, D. J., Flippin, J. D. and Barbas, C. F. (2001). Development of zinc finger domains for recognition of the 5′-ANN-3′ family of DNA sequences and their use in the construction of artificial transcription factors, *J. Biol. Chem.* **276**, 29466–29478.

Edwards, A., Voss, H., Rice, P., Civitello, A., Stegemann, J., Schwager, C., Zimmermann, J., Erfle, H., Caskey, C. and Ansorge, W. (1990). Automated DNA sequencing of the human *hprt* locus, *Genomics* **6**, 593–608.

Elrod-Erickson, M., Rould, M. A., Nekludova, L. and Pabo, C. O. (1996). Zif268 protein-DNA complex refined at 1.6 A: a model system for understanding zinc finger-DNA interactions, *Structure* **4**, 1171–1180.

Ephrussi, B. (1972), *Hybridization of Somatic Cells*. Princeton University Press, NJ.

Ernst, M., Zametkin, A. J., Matochik, J. A., Pascualvaca, D., Jons, P. H., Hardy, K., Hankerson, J. G., Doudet, D. J. and Cohen, R. M. (1996). Presynaptic dopaminergic deficits in Lesch–Nyhan disease, *New Engl. J. Med.* **334**, 1568–1572.

Ferrin, L. and Camerini-Otero, R. D. (1991). Selective cleavage of human DNA: RecA-assisted restriction endonuclease (RARE) cleavage, *Science* **254**, 1494–1497.

Firmenich, A. A., Elias-Arnanz, M. and Berg, P. (1995). A novel allele of Saccharomyces cerevisiae RFA1 that is deficient in recombination and repair and suppressible by RAD52, *Mol. Cell. Biol.* **15**, 1620–1631.

Flygare, J., Benson, F. and Hellgren, D. (1996). Expression of the human RAD51 gene during the cell cycle in primary human peripheral blood lymphocytes, *Biochim. Biophys. Acta* **1312**, 231–236.

Folger, K., Wong, E., Wahl, G. and Capecchi, M. (1982). Patterns of integration of DNA microinjected into cultured mammalian cells: evidence for homologous recombination between injected plasmid DNA molecules, *Mol. Cell. Biol.* **2**, 1372–1387.

Folger, K. R., Thomas, K. and Capecchi, M. R. (1985). Efficient correction of mismatched bases in plasmid heteroduplex injected into cultured mammalian cell nuclei, *Mol. Cell. Biol.* **5**, 70–74.

Fraefel, C., Heister, T. and Ackermann, M. (2003). Herpes simplex virus type 1/adeno-associated virus hybrid vectors, *Gene Therapy and Regulation* **2**, 7–28.

Friedberg, E. C., Walker, G. C. and Siede, W. (1995). *DNA Repair and Mutagenesis*. American Society for Microbiology, Washington, DC.

Friedmann, T. (1985). HPRT gene transfer as a model for gene therapy, in: *Genetic Engineering — Principles and Methods*, Setlow, J. and Hollaender, A. (Eds), Vol. 7, pp. 263–282. Plenum Press, New York.

Fujioka, K., Aratani, Y., Kusano, K. and Koyama, H. (1993). Targeted recombination with single-stranded DNA vectors in mammalian cells, *Nucleic Acids Res.* **21**, 407–412.

Gage, F. H. (2000). Mammalian neural stem cells, *Science* **287**, 1433–1438.

Gibbs, R. A., Nguyen, P. N., McBride, L. J., Koepf, S. M. and Caskey, C. T. (1989). Identification of mutations leading to the Lesch–Nyhan syndrome by automated direct sequencing of *in vitro* amplified cDNA, *Proc. Natl. Acad. Sci. USA* **86**, 1919–1923.

Graham, F. and van der Eb, J. (1973). A new technique for the assay of infectivity of human adenovirus 5 DNA, *Virology* **52**, 456–467.

Greisman, H. A. and Pabo, C. O. (1997). A general strategy for selecting high-affinity zinc finger proteins for diverse DNA target sites, *Science* **275**, 657–661.

Grines, C. L., Watkins, M. W., Helmer, G., Penny, W., Brinker, J., Marmur, J., West, A., Rade, J., Marrott, P., Hammond, H. K. and Engler, R. E. (2002). Angiogenic gene therapy trial (AGENT): intracoronary delivery of adenovirus encoding FGF4 for patients with stable angina pectoris, *Circulation* **105**, 1291–1297.

Gruenert, D. C., Bruscia, E., Novelli, G., Colosimo, A., Dallapiccola, B., Sangiuolo, F. and Goncz, K. K. (2003). Sequence-specific modification of genomic DNA by small DNA fragments, *J. Clin. Invest.* **112**, 637–641.

Gupta, R., Bazemore, L. R., Golub, E. and Radding, C. (1997). Activities of human recombination protein RAD51, *Proc. Natl. Acad. Sci. USA* **94**, 463–468.

Haaf, T., Golub, E., Reddy, G., Radding, C. and Ward, D. (1995). Nuclear foci of RAD51 recombination protein in somatic cells after DNA damage and its localization in synaptonemal complexes, *Proc. Natl. Acad. Sci. USA* **92**, 2298–2302.

Hacein-Bey-Abina, S., Le Deist, F., Carlier, F., Bouneaud, C., Hue, C., De Villartay, J. P., Thrasher, A. J., Wulffraat, N., Sorensen, R., Dupuis-Girod, S., Fischer, A., Davies, E. G., Kuis, W., Leiva, L. and Cavazzana-Calvo, M. (2002). Sustained correction of X-linked severe combined immunodeficiency by *ex vivo* gene therapy, *New Engl. J. Med.* **346**, 1185–1193.

Hacein-Bey-Abina, S., von Kalle, C., Schmidt, M., Le Deist, F., Wulffraat, N., McIntyre, E., Radford, I., Villeval, J. L., Fraser, C. C., Cavazzana-Calvo, M. and Fischer, A. (2003a). A serious adverse event after successful gene therapy for X-linked severe combined immunodeficiency, *New Engl. J. Med.* **348**, 255–256.

Hacein-Bey-Abina, S., Von Kalle, C., Schmidt, M., McCormack, M. P., Wulffraat, N., Leboulch, P., Lim, A., Osborne, C. S., Pawliuk, R., Morillon, E., Sorensen, R., Forster, A., Fraser, P., Cohen, J. I., de Saint Basile, G., Alexander, I., Wintergerst, U., Frebourg, T., Aurias, A., Stoppa-Lyonnet, D., Romana, S., Radford-Weiss, I., Gross, F., Valensi, F., Delabesse, E., Macintyre, E., Sigaux, F., Soulier, J., Leiva, L. E., Wissler, M., Prinz, C., Rabbitts, T. H., Le Deist, F., Fischer, A. and Cavazzana-Calvo, M. (2003b). LMO2-associated clonal T cell proliferation in two patients after gene therapy for SCID-X1, *Science* **302**, 415–419.

Hays, S., Firmenich, A. and Berg, P. (1995). Complex formation in yeast double-strand break repair: participation of Rad51, Rad52, Rad55 and Rad57 proteins, *Proc. Natl. Acad. Sci. USA* **92**, 6925–6929.

Hendrie, P. C., Hirata, R. K. and Russell, D. W. (2003). Chromosomal integration and homologous gene targeting by replication-incompetent vectors based on the autonomous parvovirus minute virus of mice, *J. Virol.* **77**, 13136–13145.

Heyer, W. D. (1994). The search for the right partner: homologous pairing and DNA strand exchange proteins in eukaryotes, *Experientia* **50**, 223–233.

Hirata, R. K. and Russell, D. W. (2000). Design and packaging of adeno-associated virus gene targeting vectors, *J. Virol.* **74**, 4612–4620.

Hirata, R., Chamberlain, J., Dong, R. and Russell, D. W. (2002). Targeted transgene insertion into human chromosomes by adeno-associated virus vectors, *Nature Biotechnol.* **20**, 735–738.

Howard-Flanders, P., West, S. C. and Stasiak, A. (1984). Role of RecA protein spiral filaments in genetic recombination, *Nature* **309**, 215–220.

Hsieh, P., Camerini-Otero, C. and Camerini-Otero, R. D. (1992). The synapsis event in the homologous pairing of DNAs: RecA recognizes and pairs less than one helical repeat of DNA, *Proc. Natl. Acad. Sci. USA* **89**, 6492–6496.

Hurt, J. A., Thibodeau, S. A., Hirsh, A. S., Pabo, C. O. and Joung, J. K. (2003). Highly specific zinc finger proteins obtained by directed domain shuffling and cell-based selection, *Proc. Natl. Acad. Sci. USA* **100**, 12271–12276.

Igoucheva, O., Alexeev, V. and Yoon, K. (2001). Targeted gene correction by small single-stranded oligonucleotides in mammalian cells, *Gene Ther.* **8**, 391–399.

Inoue, N., Hirata, R. K. and Russell, D. W. (1999). High-fidelity correction of mutations at multiple chromosomal positions by adeno-associated virus vectors, *J. Virol.* **73**, 7376–7380.

Inoue, N., Dong, R., Hirata, R. K. and Russell, D. W. (2001). Introduction of single base substitutions at homologous chromosomal sequences by adeno-associated virus vectors, *Mol. Ther.* **3**, 526–530.

Isalan, M., Choo, Y. and Klug, A. (1997). Synergy between adjacent zinc fingers in sequence-specific DNA recognition, *Proc. Natl. Acad. Sci. USA* **94**, 5617–5621.

Isalan, M., Klug, A. and Choo, Y. (1998). Comprehensive DNA recognition through concerted interactions from adjacent zinc fingers, *Biochemistry* **37**, 12026–12033.

Isalan, M., Klug, A. and Choo, Y. (2001). A rapid, generally applicable method to engineer zinc fingers illustrated by targeting the HIV-1 promoter, *Nature Biotechnol.* **19**, 656–660.

Isner, J. M. (2002). Myocardial gene therapy, *Nature* **415**, 234–239.

Isner, J. M., Pieczek, A., Schainfeld, R., Blair, R., Haley, L., Asahara, T., Rosenfield, K., Razvi, S., Walsh, K. and Symes, J. F. (1996). Clinical evidence of angiogenesis following arterial gene transfer of phVEGF165 in a patient with ischemic limb, *Lancet* **348**, 370–374.

Isner, J. M., Baumgartner, I., Rauh, G., Schainfeld, R., Blair, R., Manor, O., Razvi, S. and Symes, J. F. (1998). Treatment of thrombangiitis obliterans (Buerger's disease) by intramuscular gene transfer of vascular endothelial growth factor: preliminary clinical results, *J. Vasc. Surg.* **28**, 964–975.

Iwaguro, H., Yamaguchi, J., Kalka, C., Murasawa, S., Masuda, H., Hayashi, S., Silver, M., Li, T., Isner, J. M. and Asahara, T. (2002). Endothelial progenitor cell vascular endothelial growth factor gene transfer for vascular regeneration, *Circulation* **105**, 732–738.

Jinnah, H. A., Wojcik, B. E., Hunt, M., Narang, N., Lee, K. Y., Goldstein, M., Wamsley, J. K., Langlais, P. J. and Friedmann, T. (1994). Dopamine deficiency in a genetic mouse model of Lesch–Nyhan disease, *J. Neurosci.* **14**, 1164–1175.

Jiralerspong, S. and Patel, P. I. (1996). Regulation of the hypoxanthine phosphoribosyltransferase gene: *in vitro* and *in vivo* approaches, *Proc. Soc. Exp. Biol. Med.* **212**, 116–127.

Jiricny, J. (1998). Replication errors: challenging the genome, *EMBO J.* **17**, 6427–6436.

Johnson, R. D. and Jasin, M. (2001). Double-strand-break-induced homologous recombination in mammalian cells, *Biochem. Soc. Trans.* **29** (Pt. 2), 196–201.

Johnston, K. M., Jacoby, D., Pechan, P. A., Fraefel, C., Borghesani, P., Schuback, D., Dunn, R. J., Smith, F. I. and Breakefield, X. O. (1997). HSV/AAV hybrid amplicon vectors extend transgene expression in human glioma cells, *Human Gene Ther.* **8**, 359–370.

Jolly, D. J., Okayama, H., Berg, P., Esty, A., Filpula, D., Bohlen, P., Johnson, G. G., Shively, J. E., Hunkapiller, T. and Friedmann, T. (1983). Isolation and characterization of a full-length expressible cDNA for human hypoxanthine phosphoribosyltransferase, *Proc. Natl. Acad. Sci. USA* **80**, 477–481.

Kajstura, J., Leri, A., Finato, N., Di Loreto, C., Beltrami, C. A. and Anversa, P. (1998). Myocyte proliferation in end-stage cardiac failure in humans, *Proc. Natl. Acad. Sci. USA* **95**, 8801–8805.

Kearns, W. G., Afione, S. A., Fulmer, S. B., Pang, M. C., Erikson, D., Egan, M., Landrum, M. J., Flotte, T. R. and Cutting, G. R. (1996). Recombinant adeno-associated virus (AAV-CFTR) vectors do not integrate in a site-specific fashion in an immortalized epithelial cell line, *Gene Ther.* **3**, 748–755.

Kim, Y. G., Cha, J. and Chandrasegaran, S. (1996). Hybrid restriction enzymes: zinc finger fusions to Fok I cleavage domain, *Proc. Natl. Acad. Sci. USA* **93**, 1156–1160.

Kotin, R. M., Siniscalco, M., Samulski, R. J., Zhu, X. D., Hunter, L., Laughlin, C. A., McLaughlin, S., Muzyczka, N., Rocchi, M. and Berns, K. I. (1990). Site-specific integration by adeno-associated virus, *Proc. Natl. Acad. Sci. USA* **87**, 2211–2215.

Kowalczykowski, S., Dixon, D., Eggleston, A., Lauder, S. and Rehrauer, W. (1994). Biochemistry of homologous recombination in *E. coli*, *Microbiol. Rev.* **58**, 401–465.

Lapidot, T., Sirard, C., Vormoor, J., Murdoch, B., Hoang, T., Caceres-Cortes, J., Minden, M., Paterson, B., Caligiuri, M. A. and Dick, J. E. (1994). A cell initiating human acute myeloid leukaemia after transplantation into SCID mice, *Nature* **367**, 645–648.

Lesch, M. and Nyhan, W. L. (1964). A familiar disorder of uric acid metabolism and central nervous system function, *Am. J. Med.* **36**, 561–570.

Lieber, M. R., Ma, Y., Pannicke, U. and Schwarz, K. (2003). Mechanism and regulation of human non-homologous DNA end-joining, *Nature Rev. Mol. Cell. Biol.* **4**, 712–720.

Lim, D. S. and Hasty, P. (1996). A mutation in mouse RAD51 results in an early embryonic lethal that is suppressed by a mutation in p53, *Mol. Cell. Biol.* **16**, 7133–7143.

Linden, R. M., Ward, P., Giraud, C., Winocour, E. and Berns, K. I. (1996). Site-specific integration by adeno-associated virus, *Proc. Natl. Acad. Sci. USA* **93**, 11288–11294.

Littlefied, J. (1964). Selection of hybrids from mating of fibroblasts *in vitro* and their presumed recombinants, *Science* **145**, 709–710.

Liu, L., Rice, M. C. and Kmiec, E. B. (2001a). *In vivo* gene repair of point and frameshift mutations directed by chimeric RNA/DNA oligonucleotides and modified single-stranded oligonucleotides, *Nucleic Acids Res.* **29**, 4238–4250.

Liu, P. Q., Rebar, E. J., Zhang, L., Liu, Q., Jamieson, A. C., Liang, Y., Qi, H., Li, P. X., Chen, B., Mendel, M. C., Zhong, X., Lee, Y. L., Eisenberg, S. P., Spratt, S. K., Case, C. C. and Wolffe, A. P. (2001b). Regulation of an endogenous locus using a panel of designed zinc finger proteins targeted to accessible chromatin regions. Activation of vascular endothelial growth factor A, *J. Biol. Chem.* **276**, 11323–11334.

Lloyd, K. G., Hornykiewicz, O., Davidson, L., Shannak, K., Farley, I., Goldstein, M., Shibuya, M., Kelley, W. N. and Fox, I. H. (1981). Biochemical evidence of dysfunction of brain neurotransmitters in the Lesch–Nyhan syndrome, *New Engl. J. Med.* **305**, 1106–1111.

Losordo, D. W., Vale, P. R., Symes, J. F., Dunnington, C. H., Esakof, D. D., Maysky, M., Ashare, A. B., Lathi, K. and Isner, J. M. (1998). Gene therapy for myocardial angiogenesis: initial clinical results with direct myocardial injection of phVEGF165 as sole therapy for myocardial ischemia, *Circulation* **98**, 2800–2804.

Losordo, D. W., Vale, P. R., Hendel, R. C., Milliken, C. E., Fortuin, F. D., Cummings, N., Schatz, R. A., Asahara, T., Isner, J. M. and Kuntz, R. E. (2002). Phase 1/2 placebo-controlled, double-blind, dose-escalating trial of myocardial vascular endothelial growth factor 2 gene transfer by catheter delivery in patients with chronic myocardial ischemia, *Circulation* **105**, 2012–2018.

Lutfalla, G., Blanc, H. and Bertolotti, R. (1985). Shuttling of integrated vectors from mammalian cells to *E. coli* is mediated by head-to-tail multimeric inserts, *Somat. Cell. Mol. Genet.* **11**, 223–238.

Lutfalla, G., Armbruster, L., Dequin, S. and Bertolotti, R. (1989). Construction of an EBNA-producing line of well-differentiated human hepatoma cells and of appropriate Epstein-Barr virus-based vectors, *Gene* **76**, 27–39.

Lyon, M. F. (1996). X-chromosome inactivation: pinpointing the centre, *Nature* **379**, 116–117.

Mazin, A. V., Zaitseva, E., Sung, P. and Kowalczykowski, S. C. (2000). Tailed duplex DNA is the preferred substrate for Rad51 protein-mediated homologous pairing, *EMBO J.* **19**, 1148–1156.

McDonald, J. A. and Kelley, W. N. (1972). Lesch–Nyhan syndrome: absence of the mutant enzyme in erythrocytes of a heterozygote for both normal and mutant hypoxanthine-guanine phosphoribosyltransferase, *Biochem. Genet.* **6**, 21–26.

McIlwraith, M. J., Van Dyck, E., Masson, J. Y., Stasiak, A. Z., Stasiak, A. and West, S. C. (2000). Reconstitution of the strand invasion step of double-strand break repair using human Rad51, Rad52 and RPA proteins, *J. Mol. Biol.* **304**, 151–164.

McKay, R. (1997). Stem cells in the central nervous system, *Science* **276**, 66–71.

Merrihew, R., Marburger, K., Pennington, S., Roth, D. and Wilson, J. (1996). High-frequency illegitimate integration of transfected DNA at preintegrated target sites in a mammalian genome, *Mol. Cell. Biol.* **16**, 10–18.

Miller, C. and Temin, H. (1983). High-efficiency ligation and recombination of DNA fragments by vertebrate cells, *Science* **220**, 606–609.

Miller, D. G., Rutledge, E. A. and Russell, D. W. (2002). Chromosomal effects of adeno-associated virus vector integration, *Nat. Genet.* **30**, 147–148.

Miller, D. G., Petek, L. M. and Russell, D. W. (2003). Human gene targeting by adeno-associated virus vectors is enhanced by DNA double-strand breaks, *Mol. Cell. Biol.* **23**, 3550–3557.

Miller, J., McLachlan, A. D. and Klug, A. (1985). Repetitive zinc-binding domains in the protein transcription factor IIIA from Xenopus oocytes, *EMBO J.* **4**, 1609–1614.

Mizuno, T. (1986). Long-term follow-up of ten patients with Lesch–Nyhan syndrome, *Neuropediatrics* **17**, 158–161.

Modrich, P. (1997). Strand-specific mismatch repair in mammalian cells, *J. Biol. Chem.* **272**, 24727–24730.

Modrich, P. and Lahue, R. (1996). Mismatch repair in replication fidelity, genetic recombination, and cancer biology, *Annu. Rev. Biochem.* **65**, 101–133.

Mohle, R., Bautz, F., Rafii, S., Moore, M. A., Brugger, W. and Kanz, L. (1998). The chemokine receptor CXCR-4 is expressed on CD34+ hematopoietic progenitors and leukemic cells and mediates transendothelial migration induced by stromal cell-derived factor-1, *Blood* **91**, 4523–4530.

Momma, S., Johansson, C. B. and Frisen, J. (2000). Get to know your stem cells, *Curr. Opin. Neurobiol.* **10**, 45–49.

Morita, T., Yoshimura, Y., Yamamoto, A., Murata, K., Mori, M., Yamamoto, H. and Matsushiro, A. (1993). A mouse homolog of the *Escherichia coli* recA and Saccharomyces cerevisiae RAD51 genes, *Proc. Natl. Acad. Sci. USA* **90**, 6577–6580.

Mortensen, U. H., Bendixen, C., Sunjevaric, I. and Rothstein, R. (1996). DNA strand annealing is promoted by the yeast Rad52 protein, *Proc. Natl. Acad. Sci. USA* **93**, 10729–10734.

Mulligan, R. and Berg, P. (1980). Expression of a bacterial gene in mammalian cells, *Science* **209**, 1422–1427.

Mulligan, R. and Berg, P. (1981). Selection for animal cells that express the *Escherichia coli* gene coding for xanthine-guanine phosphoribosyltransferase, *Proc. Natl. Acad. Sci. USA* **78**, 2072–2076.

Nakai, H., Montini, E., Fuess, S., Storm, T. A., Grompe, M. and Kay, M. A. (2003a). AAV serotype 2 vectors preferentially integrate into active genes in mice, *Nat. Genet.* **34**, 297–302.

Nakai, H., Storm, T. A., Fuess, S. and Kay, M. A. (2003b). Pathways of removal of free DNA vector ends in normal and DNA-PKcs-deficient SCID mouse hepatocytes transduced with rAAV vectors, *Human Gene Ther.* **14**, 871–881.

Namsaraev, E. A. and Berg, P. (1998). Branch migration during Rad51-promoted strand exchange proceeds in either direction, *Proc. Natl. Acad. Sci. USA* **95**, 10477–10481.

New, J., Sugiyama, T., Zaitseva, E. and Kowalczykowski, S. (1998). Rad52 protein stimulates DNA strand exchange by Rad51 and replication protein A, *Nature* **391**, 407–410.

Nyhan, W. L. and Wong, D. F. (1996). New approaches to understanding Lesch–Nyhan disease, *New Engl. J. Med.* **334**, 1602–1604.

Nyhan, W. L., Olivier, W. J. and Lesch, M. (1965). A familial disorder of uric acid metabolism and central nervous system function, *J. Pediat.* **67**, 257–263.

Nyhan, W. L., Bakay, B., Connor, J. D., Marks, J. F. and Keele, D. (1970). Hemizygous expression of glucose-6-phosphate dehydrogenase in erythrocytes of heterozygotes for the Lesch–Nyhan syndrome, *Proc. Natl. Acad. Sci. USA* **65**, 214–218.

Nyhan, W. L., Parkman, R., Page, T., Gruber, H. E., Pyati, J., Jolly, D. and Friedmann, T. (1986). Bone marrow transplantation in Lesch–Nyhan disease, *Adv. Exp. Med. Biol.* **195A**, 167–170.

Odelberg, S. J., Kollhoff, A. and Keating, M. T. (2000). Dedifferentiation of mammalian myotubes induced by msx1, *Cell* **103**, 1099–1109.

Ogawa, T., Yu, X., Shinohara, A. and Egelman, E. H. (1993). Similarity of the yeast RAD51 filament to the bacterial RecA filament, *Science* **259**, 1896–1899.

Olivares, E. C. and Calos, M. P. (2003). Phage ϕC31 integrase-mediated site-specific integration for gene therapy, *Gene Therapy and Regulation* **2**, 103–120.

Owens, R. A. (2002). Second generation adeno-associated virus type 2-based gene therapy systems with the potential for preferential integration into AAVS1, *Curr. Gene Ther.* **2**, 145–159.

Pabo, C. O., Peisach, E. and Grant, R. A. (2001). Design and selection of novel Cys2His2 zinc finger proteins, *Annu. Rev. Biochem.* **70**, 313–340.

Palombo, F., Monciotti, A., Recchia, A., Cortese, R., Ciliberto, G. and La Monica, N. (1998). Site-specific integration in mammalian cells mediated by a new hybrid baculovirus-adeno-associated virus vector, *J. Virol.* **72**, 5025–5034.

Panthier, J. and Condamine, H. (1991). Mitotic recombination in mammals, *BioEssays* **13**, 351–356.

Pâques, F. and Haber, J. E. (1999). Multiple pathways of recombination induced by double-strand breaks in Saccharomyces cerevisiae, *Microbiol. Mol. Biol. Rev.* **63**, 349–404.

Pavletich, N. P. and Pabo, C. O. (1991). Zinc finger-DNA recognition: crystal structure of a Zif268-DNA complex at 2.1 A, *Science* **252**, 809–817.

Penny, W. F. and Hammond, H. K. (2003). Gene transfer via intracoronary delivery of fibroblast growth factors in experimental and clinical myocardial ischemia, in: *Progress in Gene Therapy — 2: Pioneering Stem Cell/gene Therapy Trials*, Bertolotti, R., Ozawa, K. and Hammond, H. K. (Eds), pp. 133–159. VSP Publishers, Utrecht, The Netherlands.

Perucho, M., Hanahan, D. and Wigler, M. (1980). Genetic and physical linkage of exogenous sequences in transformed cells, *Cell* **22**, 309–317.

Petukhova, G., Stratton, S. and Sung, P. (1998). Catalysis of homologous DNA pairing by yeast RAD51 and Rad54 proteins, *Nature* **393**, 91–94.

Philpott, N. J., Giraud-Wali, C., Dupuis, C., Gomos, J., Hamilton, H., Berns, K. I. and Falck-Pedersen, E. (2002). Efficient integration of recombinant adeno-associated virus DNA vectors requires a p5-rep sequence in cis, *J. Virol.* **76**, 5411–5421.

Pieroni, L., Fipaldini, C., Monciotti, A., Cimini, D., Sgura, A., Fattori, E., Epifano, O., Cortese, R., Palombo, F. and La Monica, N. (1998). Targeted integration of adeno-associated virus-derived plasmids in transfected human cells, *Virology* **249**, 249–259.

Porteus, M. H. and Baltimore, D. (2003). Chimeric nucleases stimulate gene targeting in human cells, *Science* **300**, 763.

Porteus, M. H., Cathomen, T., Weitzman, M. D. and Baltimore, D. (2003). Efficient gene targeting mediated by adeno-associated virus and DNA double-strand breaks, *Mol. Cell. Biol.* **23**, 3558–3565.

Radding, C. M. (1991). Helical interactions in homologous pairing and strand exchange driven by RecA protein, *J. Biol. Chem.* **266**, 5355–5358.

Rebar, E. J., Huang, Y., Hickey, R., Nath, A. K., Meoli, D., Nath, S., Chen, B., Xu, L., Liang, Y., Jamieson, A. C., Zhang, L., Spratt, S. K., Case, C. C., Wolffe, A. and Giordano, F. J. (2002). Induction of angiogenesis in a mouse model using engineered transcription factors, *Nat. Med.* **8**, 1427–1432.

Recchia, A., Parks, R. J., Lamartina, S., Toniatti, C., Pieroni, L., Palombo, F., Ciliberto, G., Graham, F. L., Cortese, R., La Monica, N. and Colloca, S. (1999). Site-specific integration mediated by a hybrid adenovirus/adeno-associated virus vector, *Proc. Natl. Acad. Sci. USA* **96**, 2615–2620.

Reddy, G., Golub, E. I. and Radding, C. M. (1997). Human Rad52 protein promotes single-strand DNA annealing followed by branch migration, *Mutation Res.* **377**, 53–59.

Reiss, K., Cheng, W., Ferber, A., Kajstura, J., Li, P., Li, B., Olivetti, G., Homcy, C. J., Baserga, R. and Anversa, P. (1996). Overexpression of insulin-like growth factor-1 in the heart is coupled with myocyte proliferation in transgenic mice, *Proc. Natl. Acad. Sci. USA* **93**, 8630–8635.

Reya, T., Morrison, S. J., Clarke, M. F. and Weissman, I. L. (2001). Stem cells, cancer, and cancer stem cells, *Nature* **414**, 105–111.

Reynolds, L., Ullman, C., Moore, M., Isalan, M., West, M. J., Clapham, P., Klug, A. and Choo, Y. (2003). Repression of the HIV-1 5′ LTR promoter and inhibition of HIV-1 replication by using engineered zinc-finger transcription factors, *Proc. Natl. Acad. Sci. USA* **100**, 1615–1620.

Robins, D., Ripley, S., Henderson, A. and Axel, R. (1981). Transforming DNA integrates into the host chromosome, *Cell* **23**, 29–39.

Roca, A. and Cox, M. (1990). The RecA protein: structure and function, *Crit. Rev. Biochem. Mol. Biol.* **25**, 415–455.

Rosenbloom, F. M., Kelley, W. N., Miller, J., Henderson, J. F. and Seegmiller, J. E. (1967). Inherited disorder of purine metabolism: correlation between central nervous system dysfunction and biochemical defects, *J. Amer. Med. Assoc.* **202**, 175–177.

Rosengart, T. K., Lee, L. Y., Patel, S. R., Sanborn, T. A., Parikh, M., Bergman, G. W., Hacha-movitch, R., Szulc, M., Kligfield, P. D., Okin, P. M., Hahn, R., Devereux, R., Post, M., Hackett, N., Foster, T., Grasso, T., Lesser, M., Isom, O. and Crystal, R. G. (1999). Angiogenesis gene therapy: phase I assessment of direct intramyocardial administration of an adenovirus vector expressing VEGF121 cDNA to individuals with clinically significant severe coronary artery disease, *Circulation* **100**, 468–474.

Rossiter, B. J. and Caskey, C. T. (1995). Hypoxanthine-guanine phosphoribosyl transferase deficiency: Lesch–Nyhan syndrome and gout, in: *The Metabolic and Molecular Bases of Inherited Disease,* Scriver, C., Beauded, A., Sly, W. and Valle, D. (Eds), pp. 1679–1706. McGraw-Hill, New York.

Rouet, P., Smih, F. and Jasin, M. (1994a). Expression of a site-specific endonuclease stimulates homologous recombination in mammalian cells, *Proc. Natl. Acad. Sci. USA* **91**, 6064–6068.

Rouet, P., Smih, F. and Jasin, M. (1994b). Introduction of double-strand breaks into the genome of mouse cells by expression of a rare-cutting endonuclease, *Mol. Cell. Biol.* **14**, 8096–8106.

Russell, D. W. and Hirata, R. K. (1998). Human gene targeting by viral vectors, *Nature Genetics* **18**, 325–330.

Schnepp, B. C., Clark, K. R., Klemanski, D. L., Pacak, C. A. and Johnson, P. R. (2003). Genetic fate of recombinant adeno-associated virus vector genomes in muscle, *J. Virol.* **77**, 3495–3504.

Seegmiller, J. E., Rosenbloom, F. M. and Kelley, W. N. (1967). Enzyme defect associated with a sex-linked human neurological disorder and excessive purine synthesis, *Science* **155**, 1682–1687.

Segal, D. J., Beerli, R. R., Blancafort, P., Dreier, B., Effertz, K., Huber, A., Koksch, B., Lund, C. V., Magnenat, L., Valente, D. and Barbas, C. F. (2003). Evaluation of a modular strategy for the construction of novel polydactyl zinc finger DNA-binding proteins, *Biochemistry* **42**, 2137–2148.

Seidman, M. M. and Glazer, P. M. (2003). The potential for gene repair via triple helix formation, *J. Clin. Invest.* **112**, 487–494.

Shin, S., Meera Khan, P. and Cook, P. R. (1971). Characterization of hypoxanthine-guanine phosphoribosyl transferase in man-mouse somatic cell hybrids by an improved electrophoretic method, *Biochem. Genet.* **5**, 91–99.

Shinohara, A. and Ogawa, T. (1998). Stimulation by Rad52 of yeast Rad51-mediated recombination, *Nature* **391**, 404–407.

Shinohara, A., Ogawa, H. and Ogawa, T. (1992). Rad51 protein involved in repair and recombination in *S. cerevisiae* is a RecA-like protein, *Cell* **69**, 457–470.

Shinohara, A., Ogawa, H., Matsuda, Y., Ushio, N., Ikeo, K. and Ogawa, T. (1993). Cloning of human, mouse and fission yeast recombination genes homologous to RAD51 and recA, *Nature Genetics* **4**, 239–243.

Simmonds, H. A., Sahota, A. S. and Van Acker, K. J. (1995). Adenine phosphoribosyltransferase deficiency and 2,8-dihydroxyadenine lithiasis, in: *The Metabolic and Molecular Bases of Inherited Disease*, Scriver, C., Beauded, A., Sly, W. and Valle, D. (Eds), pp. 1707–1724. McGraw-Hill, New York.

Singh, S. K., Clarke, I. D., Terasaki, M., Bonn, V. E., Hawkins, C., Squire, J. and Dirks, P. B. (2003). Identification of a cancer stem cell in human brain tumors, *Cancer Res.* **63**, 5821–5828.

Smih, F., Rouet, P., Romanienko, P. J. and Jasin, M. (1995). Double-strand breaks at the target locus stimulate gene targeting in embryonic stem cells, *Nucleic Acids Res.* **23**, 5012–5019.

Smith, J., Bibikova, M., Whitby, F. G., Reddy, A. R., Chandrasegaran, S. and Carroll, D. (2000). Requirements for double-strand cleavage by chimeric restriction enzymes with zinc finger DNA-recognition domains, *Nucleic Acids Res.* **28**, 3361–3369.

Sonada, E., Sasaki, M. S., Bluerstedde, J.-M., Bezzubova, O., Shinohara, A., Ogawa, H., Takata, M., Yamaguchi-Iwai, Y. and Takeda, S. (1998). Rad51-deficient vertebrate cells accumulate chromosomal breaks prior to cell death, *EMBO J.* **17**, 598–608.

Stasiak, A. and Egelman, E. H. (1994). Structure and function of RecA-DNA complexes, *Experientia* **50**, 192–203.

Sun, H., Treco, D. and Szostak, J. W. (1991). Extensive 3′-overhanging, single-stranded DNA associated with the meiosis-specific double-strand breaks at the ARG4 recombination initiation site, *Cell* **64**, 1155–1161.

Sung, P. (1994). Catalysis of ATP-dependent homologous DNA pairing and strand exchange by yeast RAD51 protein, *Science* **265**, 1241–1243.

Sung, P. (1997). Function of yeast Rad52 protein as a mediator between replication protein A and the Rad51 recombinase, *J. Biol. Chem.* **272**, 28194–28197.

Sung, P. and Robberson, D. (1995). DNA strand exchange mediated by a RAD51-ssDNA nucleoprotein filament with polarity opposite to that of RecA, *Cell* **82**, 453–461.

Surosky, R. T., Urabe, M., Godwin, S. G., McQuiston, S. A., Kurtzman, G. J., Ozawa, K. and Natsoulis, G. (1997). Adeno-associated virus Rep proteins target DNA sequences to a unique locus in the human genome, *J. Virol.* **71**, 7951–7959.

Symes, J. F., Losordo, D. W., Vale, P. R., Lathi, K. G., Esakof, D. D., Mayskiy, M. and Isner, J. M. (1999). Gene therapy with vascular endothelial growth factor for inoperable coronary artery disease, *Ann. Thorac. Surg.* **68**, 830–836.

Symes, J. F., Vale, P. R. and Losordo, D. W. (2003). Angiogenic VEGF gene therapy for refractory angina pectoris, in: *Progress in Gene Therapy — 2: Pioneering Stem Cell/gene Therapy Trials*, Bertolotti, R., Ozawa, K. and Hammond, H. K. (Eds), pp. 107–131. VSP Publishers, Utrecht, The Netherlands.

Szostak, J. W., Orr-Weaver, T. L., Rothstein, R. J. and Stahl, F. W. (1983). The double-strand-break repair model for recombination, *Cell* **33**, 25–35.

Szybalska, E. and Szybalski, W. (1962). Genetics of human cell lines, IV. DNA-mediated heritable transformation of a biochemical trait, *Proc. Natl. Acad. Sci. USA* **48**, 2026–2034.

Szybalski, W. (1992). Use of the HPRT gene and the HAT selection technique in DNA-mediated transformation of mammalian cells: first step toward developing hybridoma techniques and gene therapy, *BioEssays* **14**, 495–500.

Tan, S., Guschin, D., Davalos, A., Lee, Y. L., Snowden, A. W., Jouvenot, Y., Zhang, H. S., Howes, K., McNamara, A. R., Lai, A., Ullman, C., Reynolds, L., Moore, M., Isalan, M., Berg, L. P., Campos, B., Qi, H., Spratt, S. K., Case, C. C., Pabo, C. O., Campisi, J. and Gregory, P. D. (2003). Zinc-finger protein-targeted gene regulation: genomewide single-gene specificity, *Proc. Natl. Acad. Sci. USA* **100**, 11997–12002.

Tashiro, S., Walter, J., Shinohara, A., Kamada, N. and Cremer, T. (2000). Rad51 accumulation at sites of DNA damage and in postreplicative chromatin, *J. Cell. Biol.* **150**, 283–291.

Te Riele, H., Maandag, E. and Berns, A. (1992). Highly efficient gene targeting in embryonic stem cells through homologous recombination with isogenic DNA constructs, *Proc. Natl. Acad. Sci. USA* **89**, 5128–5132.

Thomas, K. and Capecchi, M. (1987). Site-directed mutagenesis by gene targeting in mouse embryo-derived stem cells, *Cell* **51**, 503–512.

Thomas, K., Folger, K. and Capecchi, M. (1986). High frequency targeting of genes to specific sites in the mammalian genome, *Cell* **44**, 419–428.

Thomas, K., Deng, C. and Capecchi, M. (1992). High-fidelity gene targeting in embryonic stem cells by using sequence replacement vectors, *Mol. Cell. Biol.* **12**, 2919–2923.

Torrente, Y., Tremblay, J. P., Pisati, F., Belicchi, M., Rossi, B., Sironi, M., Fortunato, F., El Fahime, M., D'Angelo, M. G., Caron, N. J., Constantin, G., Paulin, D., Scarlato, G. and Bresolin, N. (2001). Intraarterial injection of muscle-derived CD34(+) Sca-1(+) stem cells restores dystrophin in mdx mice, *J. Cell Biol.* **152**, 335–348.

Tremblay, J. P., Malouin, F., Roy, R., Huard, J., Bouchard, J. P., Satoh, A. and Richards, C. L. (1993). Results of a triple blind clinical study of myoblast transplantations without immunosuppressive treatment in young boys with Duchenne muscular dystrophy, *Cell Transplant.* **2**, 99–112.

Tsuzuki, T., Fujii, Y., Sakumi, K., Tominaga, Y., Nakao, K., Sekiguchi, M., Matsushiro, A., Yoshimura, Y. and Morita, T. (1996). Targeted disruption of the RAD51 gene leads to lethality in embryonic mice, *Proc. Natl. Acad. Sci. USA* **93**, 6236–6240.

Vale, P. R., Losordo, D. W., Milliken, C. E., Maysky, M., Esakof, D. D., Symes, J. F. and Isner, J. M. (2000). Left ventricular electromechanical mapping to assess efficacy of phVEGF$_{165}$ gene transfer for therapeutic angiogenesis in chronic myocardial ischemia, *Circulation* **102**, 965–974.

Vale, P. R., Losordo, D. W., Milliken, C. E., McDonald, M. C., Gravelin, L. M., Curry, C. M., Esakof, D. D., Maysky, M., Symes, J. F. and Isner, J. M. (2001). Randomized, single-blind, placebo-controlled pilot study of catheter-based myocardial gene transfer for therapeutic angiogenesis using left ventricular electromechanical mapping in patients with chronic myocardial ischemia, *Circulation* **103**, 2138–2143.

Van Dyck, E., Stasiak, A. Z., Stasiak, A. and West, S. C. (2001). Visualization of recombination intermediates produced by RAD52-mediated single-strand annealing, *EMBO Rep.* **2**, 905–909.

West, S. C. (2003). Molecular views of recombination proteins and their control, *Nat. Rev. Mol. Cell. Biol.* **4**, 435–445.

Wojcik, B. E., Jinnah, H. A., Muller-Sieburg, C. E. and Friedmann, T. (1989). Bone marrow transplantation does not ameliorate the neurologic symptoms in mice deficient in hypoxanthine guanine phosphoribosyl transferase (HPRT), *Metab. Brain Dis.* **14**, 57–65.

Wolfe, S. A., Grant, R. A., Elrod-Erickson, M. and Pabo, C. O. (2001). Beyond the "recognition code": structures of two Cys2His2 zinc finger/TATA box complexes, *Structure (Camb)* **9**, 717–723.

Wong, E. A. and Capecchi, M. R. (1987). Homologous recombination between coinjected DNA sequences peaks in early to mid-S phase, *Mol. Cell. Biol.* **7**, 2294–2295.

Wong, D. F., Harris, J. C., Naidu, S., Yokoi, F., Marenco, S., Dannals, R. F., Ravert, H. T., Yaster, M., Evans, A., Rousset, O., Bryan, R. N., Gjedde, A., Kuhar, M. J. and Breese, G. R. (1996). Dopamine transporters are markedly reduced in Lesch–Nyhan disease *in vivo*, *Proc. Natl. Acad. Sci. USA* **93**, 5539–5543.

Yamaguchi, J., Kusano, K. F., Masuo, O., Kawamoto, A., Silver, M., Murasawa, S., Bosch-Marce, M., Masuda, H., Losordo, D. W., Isner, J. M. and Asahara, T. (2003). Stromal cell-derived factor-1 effects on *ex vivo* expanded endothelial progenitor cell recruitment for ischemic neovascularization, *Circulation* **107**, 1322–1328.

Yamamoto, A., Taki, T., Yagi, H., Habu, T., Yoshida, K., Yoshimura, Y., Yamamoto, K., Matsushiro, A., Nishimune, Y. and Morita, T. (1996). Cell cycle-dependent expression of the mouse RAD51 gene in proliferating cells, *Mol. Gen. Genet.* **251**, 1–12.

Yanez, R. J. and Porter, A. C. (1998). Therapeutic gene targeting, *Gene Therapy* **5**, 149–159.

Yates, J., Warren, N. and Sugden, B. (1985). Stable replication of plasmids derived from Epstein-Barr virus in various mammalian cells, *Nature* **313**, 812–815.

Yoon, K. (2000). Single-base conversion of mammalian genes by an RNA-DNA oligonucleotide, in: *Progress in Gene Therapy — Basic and Clinical Frontiers*, Bertolotti, R., Parvez, H. and Nagatsu, T. (Eds), pp. 475–512. VSP Publishers, Utrecht, The Netherlands.

Zhao, M., Momma, S., Delfani, K., Carlen, M., Cassidy, R. M., Johansson, C. B., Brismar, H., Shupliakov, O., Frisen, J. and Janson, A. M. (2003). Evidence for neurogenesis in the adult mammalian substantia nigra, *Proc. Natl. Acad. Sci. USA* **100**, 7925–7930.

Advances in Neuroregulation and Neuroprotection (2005), pp. 585-594
Collin, C. *et al.* (Eds)
© VSP 2005

SHRSP: A novel field of study

H. SAITO *

Department of Basic Sciences, Japanese Red Cross Hokkaido College of Nursing, 664-1 Akebono chou, Kitami 090-0011, Japan

Keywords: Catecholamine; stroke-prone spontaneously hypertensive rats (SHRSP); stroke-related behaviour; vascular dementia.

INTRODUCTION

After graduation from Hokkaido University School of Medicine, in Northern Japan, I started out as a cardiovascular pharmacologist there at the Department of Pharmacology under Professor Tsuneyoshi Tanabe, MD, PhD, the head of the department. Dr. Tanabe had attended the Department of Pharmacology at the University of Michigan where his mentor Dr. Maurice H. Seevers had been noted for his studies on drug dependence. Professor Tanabe participated in the digitalis symposium at the First International Congress of Pharmacology.

I do not know why I decided to go into basic research, but my decision may have been influenced by my one-year internship at the Bibai City Hospital; I felt that I had more of an aptitude for basic science than clinical medicine.

After studying the effects of digitalis on active cation transport in the human red cell membrane for four years, I went abroad in 1964 to join Dr. Thomas E. Gaffney as a research fellow of the Division of Clinical Pharmacology, Department of Pharmacology, University of Cincinnati College of Medicine in Ohio. He asked me to measure human plasma catecholamine concentrations in patients with congestive heart failure. Although the plasma catecholamine assay is now a common laboratory test, in those days a sensitive method for the determination of epinephrine and norepinephrine in human plasma had not been established. After 8 months of energetic effort with our technician, Mr. Seraphim Woronkow using the Aminco-Bowman spectro-photofluorometer, we developed a sensitive method for determining human plasma catecholamines (Saito *et al.*, 1969) that finally made it possible to measure plasma catecholamine levels in patients with congestive heart

---

*E-mail: hideyas@rchokkaido-cn.ac.jp

failure and other diseases (Loggie *et al.*, 1967). Although I left the University of Cincinnati after one year due to family circumstances, the experience greatly influenced my research direction when I returned to Japan, and I continued in the clinical pharmacological research of biogenic amines. During the past 39 years, we have measured catecholamine concentrations in biological materials using five different methods: the THI method using a spectrophotofluorometer (Saito *et al.*, 1969), the radioenzymatic assay (Saito *et al.*, 1983), *in vivo* voltammetry (Minami *et al.*, 1985a), the HPLC plus THI method (Minami *et al.*, 1984) as well as the HPLC plus ECD method (Togashi *et al.*, 1990; Matsumoto *et al.*, 1995).

PERSONAL MOTIVES FOR PARTICIPATING IN SHR AND SHRSP STUDY

On returning to Hokkaido University, I was fortunate to be able to continue catecholamine research with an Aminco-Bowman spectrophotofluorometer that our laboratory had purchased two years earlier. At that time, Professor Tanabe and I were interested in the interaction between cardiac glycosides and endogenous catecholamines. Dr. Tanabe gave his last special lecture entitled 'Adrenergic Contribution to Digitalis Action' at the18th annual meeting of the Western Pharmacology Society at Hawaii in 1975 (Tanabe *et al.*, 1975).

After this meeting, I started my studies of SHR. Thus, using a colony of SHR which had been bred by the late Professor Hirofumi Sokabe, I began to study the mechanism of clonidine withdrawal hypertension. This mechanism had not been adequately confirmed or explained through laboratory experimentation (Yomaida *et al.*, 1979a, b). I presented our results at the Satellite Symposium of the 8th International Congress of Pharmacology at Sapporo in 1981 (Saito, 1981; Saito *et al.*, 1982).

For further clarification, we examined the effect of clonidine on efferent discharge rates of the adrenal sympathetic nerve and on catecholamine secretion from the adrenal medulla into the adrenal venous blood (Shimamura *et al.*, 1981). Our results revealed that clonidine (30 μg/kg i.v.) induced a significant decrease both in adrenal sympathetic nerve activity and in adrenal catecholamine secretion rate. These results suggest that clonidine depressed catecholamine secretion from the adrenal medulla by a centrally mediated mechanism in the rat. We also reported the central and peripheral effects of clonidine on the adrenal medullary function in SHR (Togashi, 1983; Togashi *et al.*, 1984).

Apart from these clonidine studies, we studied the effects of centrally acting drugs including a clonidine-like drug (Koike *et al.*, 1981), morphine (Togashi *et al.*, 1985), serotonin antagonist (Yoshioka *et al.*, 1987; Matsumoto *et al.*, 1988), α_1-adrenoceptor antagonists (Yoshioka *et al.*, 1990) and substance P antagonist (Togashi *et al.*, 1987) on the efferent discharges of adrenal sympathetic nerve activity in rats. I was invited to present this research work entitled 'A Comparative Study of the Effects of α_1-Adrenoceptor Antagonists on Sympathetic Function in

Rats' to the Satellite Symposium of the *10th Annual Scientific Meeting of the American Society* of Hypertension at New York in 1995 (Saito *et al.*, 1996).

In 1979, we started working with Stroke-Prone Spontaneously Hypertensive Rats (SHRSPs) that were kindly donated by the late Professor Kozo Okamoto. From this stock, we started breeding SHRSP at Hokkaido University. Thereafter, we studied the behavioral and neurochemical characteristics of SHRSP while evaluating drug reactions using SHRSP.

BEHAVIORAL AND NEUROCHEMICAL EVALUATION OF STROKE-PRONE SPONTANEOUSLY HYPERTENSIVE RATS FOR VASCULAR DEMENTIA ANIMAL MODEL

The SHRSP is known to be a unique model of stroke because the lethal time course of SHRSP coincides well with that of patients with cerebrovascular lesions. The aim of our study was to clarify the proposed possibility that SHRSP may serve as a vascular dementia animal model. First, a study was carried out to evaluate stroke-related behavioral changes and learning ability using the passive avoidance response (Minami *et al.*, 1985b). An attempt was also made to investigate the pathophysiology of SHRSP as an animal model for cerebrovascular lesions by determining neurotransmitter levels in the cerebrospinal fluid (CSF) (Togashi *et al.*, 1994).

In the stroke-related behavioral study, age-matched male SHRSP and Wistar Kyoto rats (WKY) were subjected to a 12-hour light and dark alternation cycle. Ambulation and drinking activity counts were determined simultaneously with an Ambulo-Drinkometer (O'Hara & Co., Ltd., Tokyo). Ambulatory and drinking activity data were obtained at hourly intervals and subjected to various statistical analyses: unpaired t-test, analysis of variance and two-tailed t-test. The data were also subjected to autocorrelation and power spectral analysis. The average rhythm period (tau) was determined by power spectral analysis.

In the learning ability study, 12-week-old SHRSP and Wistar normotensive rats (NWR) were used. Rats were bred under a 12-hour light and dark alternation cycle and subjected to a passive avoidance task. Each rat was allowed to habituate a light chamber for 60 s and to enter the dark chamber by removing a guillotine door. Three seconds after entering the dark chamber, the rat received a foot shock (75 V, 0.2 ms) for 3 s. Twenty four hours before and after the acquisition trial, latency times were recorded up to a maximum of 600 s. Retention trials were carried out for 5 days.

For the determination of neurotransmitters and their metabolite levels in the CSF, 15-week-old SHRSP and WKY rats were used. Rats were anesthetized with α-chloralose and urethane. CSF was collected via a polyethylene cannula which was inserted into the cisterna magna via the atlanto-occipital membrane under a microscope. CSF was allowed to flow from the cisterna magna into an iced microhematocrit tube. CSF catecholamines (CA), 5-HT and their metabolites (3,4-dihydroxyphenylacetic acid, DOPAC; 4-hydroxy-3-methoxy-phenylacetic acid, HVA; 5-hydroxyindole-3-acetic acid, 5-HIAA) levels were measured using high performance liquid chromatography with an electrochemical detector (HPLC-ECD). Total CSF volume was less than 150 μl for each rat. Ten μl of unprocessed CSF was used for simultaneous determination of monoamines and their metabolites. When necessary, alumina extraction was performed for catecholamine assay. For acetylcholine (ACh) and choline (Ch) determination, CSF was collected into polyethylene tubings containing an acetylcholine esterase inhibitor, eserine, and an internal standard, ethylhomocholine. CSF was injected directly into the HPLC-ECD with an attached immobilized enzyme column.

Stroke-related behavior in SHRSPs

From an early age, SHRSP systolic blood pressure is significantly higher than that of WKY rats. SHRSPs died from cerebral infarction, cerebral hemorrhage and their combination at an age of around 35 weeks. In the control WKY rats, whose average life span was 83 weeks old, on the other hand, senility was the most frequent cause of death. Before stroke (15 weeks), ambulation and drinking counts of SHRSP in the dark phase (82%) were higher than those in the light phase (18%). Both parameters were well synchronized with the light and dark alternation cycle. With aging, daily ambulation decreased while daily drinking activity increased in both SHRSP and WKY rats. Daily ambulation and drinking activity in 15- and 40-week-old SHRSPs were significantly greater than those of WKY.

In the SHRSPs that died of cerebral infarction, the ambulatory activity in the light period increased abruptly followed by a desynchronization with the light and dark alternation cycles. Moreover, the SHRSP ambulatory activity was desynchronized with their water drinking activity. With regard to the behavioral changes in the SHRSP whose deaths were caused by cerebral hemorrhage, desynchronization with light and dark alternation cycles and with water drinking activity was also observed.

The behavioral changes of the SHRSPs were analyzed by power spectral analysis. SHRSPs before stroke (15 weeks) had a 24-hour tau-value for both ambulation and drinking activity. However, at the onset of stroke, a much longer periodicity was observed in addition to the 24-hour periodicity in the SHRSPs that died from cerebral hemorrhage. On the other hand, a circadian rhythm persisted in the WKY that died from senility after reaching an age of 100 weeks.

These behavioral changes in ambulation and drinking activity, including the disturbance of circadian rhythms before death in SHRSPs, may correspond to behavioral changes such as the delirium-state observed in patients with dementia (Minami et al., 1985b). They point to the possibility that SHRSP may be a suitable vascular dementia animal model (Togashi et al., 1990; Saito et al., 1995).

Evaluation of learning ability using the passive avoidance test

In 12-week-old NWR, the response latency to entering the dark compartment was 545.9 ± 36.0 s (mean ± SE, $n = 7$) in the first retention test which was carried out

24 h after the acquisition trial. As compared with NWR, the response latency in the SHRSP was significantly reduced (355.3 ± 71.71 s, $n = 6$). A significant difference in latency time was noted until the second retention test, and the response latency declined to that before the acquisition trial (23.1 ± 7.1 s in NWR and 30.8 ± 0.6 s in SHRSPs) with successive retention tests. The significant impairment in passive avoidance response observed in SHRSPs may reflect an impairment of memory. However, further study is needed to evaluate the learning ability of SHRSPs, since there is a possibility that SHRSPs may have a different threshold for pain than WKY.

Determination of neurotransmitters and their metabolite levels in CSF

CSF monoamines, ACh and metabolites were determined with good reproducibility and high sensitivity. In 15-week-old SHRSPs, CSF norepinephrine (NE) concentration was significantly higher and CSF HVA and 5-HT levels were significantly lower than those in age-matched WKY rats. On the other hand, CSF ACh concentration was significantly lower in SHRSPs than that in WKY rats.

The changes in CSF NE and ACh may reflect central noradrenergic and cholinergic activity, although the pathophysiological significance of these findings was uncertain. It has been reported that an increase in CSF NE level was observed in patients with primary hypertension. On the other hand, CSF ACh levels were found to correlate negatively with the degree of dementia evaluated by the Memory and Information Test. The changes in CSF NE and ACh observed in SHRSPs might be involved in the pathogenesis of cerebrovascular lesions associated with hypertension. This study demonstrated that behavioral changes both in activities and rhythms and in memory impairment as evaluated by the passive avoidance response were observed in SHRSPs. Changes in central noradrenergic and cholinergic activity were also suggested. These changes might reflect a pathogenesis of cerebrovascular lesions caused by high blood pressure. The present behavioral and neurochemical study suggests that the SHRSP may be a suitable model for vascular dementia of the type caused by cerebrovascular lesions (Togashi *et al.*, 1990, 1994).

CHOLINERGIC CHANGES IN THE HIPPOCAMPUS OF SHRSP

Thereafter, our group investigated age-related changes in the central cholinergic systems of SHRSPs to examine whether regional and progressive cholinergic changes occurred and how these correlate with behavioral changes in the passive avoidance task (Togashi *et al.*, 1996).

We found that 15- to 20-week-old SHRSPs demonstrated a markedly lower level of hippocampal choline than age-matched WKY rats. A decrease in the choline level in 15- to 20-week-old SHRSPs was observed in all regions examined; however, a significant difference from WKY was subsequently observed in the hippocampus at age 30–40 weeks. The hippocampal ACh release was markedly decreased by repetitive stimulation with high concentrations of K^+ in 15- to 20-week-old SHRSPs. Behavioral impairment in the passive avoidance task was observed in both age groups of SHRSP, with significant and positive correlations between the hippocampal ACh levels and the response latency. In conclusion, a decrease in hippocampal choline level was observed in both 15- to 20-week-old and 30- to 40-week-old SHRSPs, accompanied by performance failure in the passive avoidance task. The abnormal release of hippocampal ACh in response to repetitive K^+ stimulation was also noted in 15- to 20-week-old SHRSPs. Thus, cholinergic dysfunction in the hippocampal system may be responsible for behavioral abnormalities in the passive avoidance task in SHRSPs. These findings support our hypothesis that the SHRSP may be a suitable animal model of vascular dementia.

PATHOLOGICAL CHANGES IN THE SHRSP BRAIN

The SHRSP brain was compared to a brain obtained from the autopsy of a patient with multiple cerebral infarcts to determine whether or not similar pathological changes occurred (Saito et al., 1995; Kimura et al., 2002). In the cerebral cortex of both the human case and the SHRSP, there was enlargement of the perivascular space and rarefaction of the parenchyma with scattered neuronal loss and astrocytic proliferation. In the SHRSP cerebral cortex, the population of neurons as well as the width of the cortical layer was smaller than that in the human brain. However, this feature of neuronal depopulation in the SHRSP cerebral cortex was very similar to that seen in the human case. In the white matter of the SHRSP, focal cystic changes with proliferation of macrophages and reactive astrocytes were observed. However, these changes were not common in SHRSP brain and were similar to changes observed in the human case. In the hippocampus or Ammon's horn of the SHRSP brain, the density of the neurons in the pyramidal cell layer was originally lower than that of human brain. In the pyramidal cell layer, nearly half of the nerve cells showed shrinkage, vacuolar formation, faintness of staining or other degenerative changes. However, so called 'senile changes' observed in the human case were not detected in the rat brain. This pathological study suggests that cerebrovascular disorder in the SHRSP is associated with lesions in the brain similar to those seen in the typical human case of multiple cerebral infarction. On the other hand, so called senile changes (Hirano bodies and senile plagues), as seen in the hippocampus of the human case, were not observed in SHRSP brains.

PERSPECTIVES OF SHRSP STUDY

The door to SHRSP studies has been opened to pathologists and pharmacologists by Dr. Okamoto, Dr. Yamori and Dr. Nagaoka for almost 30 years now, but SHRSP still continue to serve as an interesting and stimulating topic for research. In 1990, our group proposed the possibility of using the SHRSP as an animal model of vascular dementia.

In Japan, it is said that 2.5 million people are suffering from senile dementia. Dr. Yamori reported that nearly half of all the cases of senile dementia in Japan are due to cerebrovascular dementia. In contrast, Alzheimer-type dementia is more prevalent in western countries. This difference seems to be related to the greater prevalence of stroke in Japan than in occidental countries (Yamori *et al.*, 1989). Unfortunately, we do not have any rational therapy for the patients with vascular dementia. Although brain metabolic stimulants have recently been developed in Japan, in particular, four promising drugs — idebenone, bifemelane, indeloxazine and propentofylline — did not display evidence of ameliorating memory and cognition impairments in patients with vascular dementia. For this reason, The Japanese Government stopped the production of these four drugs.

A fundamental problem for basic research on vascular dementia, including the development of new therapeutic agents, is the lack of an appropriate animal model that mimics all aspects of this disease. The ideal model for vascular dementia would be one that exhibits histopathological, behavioral and biochemical abnormalities similar to those found in patients with vascular dementia.

The desired model should allow for a detailed evaluation of neuropathological, behavioral and neurochemical sequelae of the cerebrovascular lesions associated with this disease. Such a model also would be useful in testing new drugs to determine whether they lessen the extent of cerebrovascular lesions and restore cognitive functions. This animal model could permit one to conduct studies that cannot easily be performed on patients with vascular dementia. Thus, an appropriate animal model of vascular dementia that reproduces the specific cerebrovascular lesions and cognitive dysfunction associated with the human disorder might offer insight into the mechanisms that underlie the vascular dementia state. Then, we focused attention on behavioral, biochemical and pathological data obtained with SHRSPs. Our findings suggest that the SHRSP might serve as a suitable animal model for vascular dementia in humans caused by cerebrovascular lesions.

Thereafter, we evaluated drug reactions using SHRSPs. First, our group (Kimura *et al.*, 1998a) reported that methylcobalamin (vitamin-B_{12}) increases cerebral ACh levels and improves passive avoidance response in SHRSP.

In the late 1990s, we became interested in docosahexaenoic acid (DHA). Several investigators proposed the possibility that fish oil containing icosapentaenoic acid (EPA) and DHA produces an antithrombotic effect and prevents arteriosclerosis. Our group (Minami *et al.*, 1997a) reported that dietary DHA suppresses the development of hypertension and stroke-related behavioral changes, resulting in prolongation of the SHRSP's life span. We also reported that dietary DHA increases cerebral ACh levels and improves passive avoidance performance in SHRSP (Minami *et al.*, 1997b). Our group (Kimura *et al.*, 1998b) reported that DHA decreases blood viscosity in the SHRSP. We postulated that increased blood viscosity facilitates the formation of thrombosis, which is an important risk factor in the occurrence of cerebral infarctions. We undertook to elucidate whether DHA lowers blood viscosity, hematocrit and fibrinogen in the disease animal model

SHRSP. The blood viscosity, hematocrit and fibrinogen of non-treated SHRSP increased significantly when compared with levels in age-matched non-treated WKY rats. SHRSPs that received DHA for 5 weeks displayed significant decreases in blood viscosity, hematocrit and fibrinogen when compared with the values in non-treated SHRSPs. The blood pressure of DHA-treated SHRSPs was significantly lower than that of non-treated SHRSPs. There was a correlation between blood pressure and blood viscosity. These findings suggest that decreased blood viscosity induced by DHA appears to be associated with reduction of thrombosis formation and a hypotensive action in the SHRSP.

Our group (Kimura *et al.*, 1998c) also reported the ameliorative effects of DHA on serum lipid changes in the SHRSP. Our study was undertaken to elucidate the effects of long-term administration of DHA on the serum lipid concentrations in SHRSPs. SHRSPs were selected because serum lipid derangement is one of the primary risk factors in the development and maintenance of hypertension. DHA-treated SHRSP showed significantly lower blood pressure when compared with that of non-treated SHRSP; total cholesterol, triglyceride, low density lipoprotein and lipid peroxide levels were significantly decreased in DHA-treated SHRSP. On the other hand, high density lipoprotein concentrations tended to increase in DHA-treated SHRSP as compared with those in non-treated SHRSPs. These findings suggest that long-term administration of DHA has a protective effect against serum lipid derangement in SHRSPs. This DHA-induced amelioration of serum lipid changes in SHRSP might be associated with mechanisms involving the prevention of arteriosclerosis induced by DHA.

These results suggest that both the DHA-induced decrease in blood viscosity and the DHA- induced protective effect against serum lipid derangement may be related to the ameliorating effect on the symptoms of patients with vascular dementia.

The SHRSP is an animal model of vascular dementia that shows close correlation with the disease in humans. With basic studies of DHA in SHRSPs, one can make reasonable predictions with respect to the clinical significance of DHA for the treatment of vascular dementia in man. These findings suggest that the SHRSP would be useful in testing new drugs to determine whether they lessen the extent of the cerebrovascular lesions and restore cognitive functions.

REFERENCES

Kimura, S., Minami, M., Endo, T., *et al.* (1998a). Methylcobalamin (Vit-B$_{12}$) increases cerebral acethylcholine levels and improves passive avoidance response in stroke-prone spontaneously hypertensive rats, *Biog. Amines* **14**, 15–24.

Kimura, S., Tamayama, M., Minami, M., *et al.* (1998b). Docosahexaenoic acid inhibits blood viscosity in stroke-prone spontaneously hypertensive rats, *Res. Comm. Mol. Pathol. Pharmacol.* **100**, 351–361.

Kimura, S., Minami, M., Hata, N., *et al.* (1998c). Ameliorative effects of docosahexaenoic acid on serum lipid changes in stroke-prone spontaneously hypertensive rats, *Res. Comm. Mol. Pathol. Pharmacol.* **100**, 53–64.

Kimura, S., Saito, H., Minami, M., *et al.* (2002). Docosahexaenoic acid attenuated hypertension and vascular dementia in stroke-prone spontaneouslyhypertensive rats, *Neurotoxicol. Teratol.* **24**, 683–693.

Koike, Y., Togashi, H., Shimamura, K., *et al.* (1981). Effects of abrupt cessation of treatment with clonidine and guanfacine on blood pressure and heart rate in spontaneously hypertensive rats, *Clin. Exp. Hypert.* **3**, 103–120.

Loggie, J. M., Saito, H., Kahn, I., *et al.* (1967). Accidental reserpine poisoning: clinical and metabolic effects, *Clin. Pharmacol. Ther.* **8**, 692–695.

Matsumoto, M. (1988). Effect of ketanserin on adrenal sympathetic nerve activity in rats, *Jpn. J. Pharmacol.* **48**, 57–66.

Matsumoto, M., Yoshioka, M., Togashi, H., *et al.* (1995). Modulation of norepinephrine asmeasured by *in vivo* microdialysis, *J. Pharmacol. Exp. Ther.* **272**, 1044–1051.

Minami, M., Sano, M., Togashi, H., *et al.* (1984). The factors affecting plasma catecholamine concentration in rat and man, *Folia Pharmacol. Jpn.* **83**, 17–31.

Minami, M., Togashi, H., Morii, K., *et al.* (1985a). Changes in DOPAC and 5-HIAA after acute cerebral hemorrhage induced by hypertonic glucose solution (i.p.). *In vivo* voltammetry study, *Hokkaido Med. J.* **60**, 885–892.

Minami, M., Togashi, H., Koike, Y., *et al.* (1985b). Changes in ambulation and drinking behavior related to stroke in stroke-prone spontaneously hypertensive rats, *Stroke* **16**, 44–48.

Minami, M., Kimura, S., Endo, T., *et al.* (1997a). Effects of dietary docosahexaenoic acid on survival time and stroke-related behavior in stroke-prone spontaneously hypertensive rats, *Gen. Pharmacol.* **29**, 401–407.

Minami, M., Kimura, S., Endo, T., *et al.* (1997b). Dietary docosahexaenoic acid increases cerebral acethylcholine levels and improves passive avoidance performance in stroke-prone spontaneously hypertensive rats, *Pharmacol. Biochem. Behav.* **58**, 1123–1129.

Saito, H. (1981). Clonidine withdrawal hypertension in spontaneously hypertensive rats, *Trends Pharmacol. Sci.* **2**, 176–177.

Saito, H., Woronkow, S. and Myers, B. (1969). A rapid and sensitive method for simultaneous determination of epinephrine and norepinephrine in human plasma, *Jpn. Circ. J.* **33**, 677–683.

Saito, H., Yomaida, I., Togashi, H., *et al.* (1982). Clonidine withdrawal hypertension in spontaneously hypertensive rats, *Clin. Exp. Hypert.* **A4**, 316–317.

Saito, H., Minami, M., Togashi, H., *et al.* (1983). Radioenzymatic assays for catecholamine determination, in: *Methods in Biogenic Amine Research*, Parvez, S. and Nagatsu, T. (Eds), pp. 285–294. Elsevier, Amsterdam.

Saito, H., Togashi, H., Yoshioka, M., *et al.* (1995). Animal models of vascular dementia with emphasis on stroke-prone spontaneously hypertensive rats, *Clin. Exp. Pharmacol. Physiol.* **Suppl. I**, S257–S259.

Saito, H., Togashi, H. and Yoshioka, M. (1996). A comparative study of the effects of α_1-adrenoceptor antagonists on sympathetic function in rats, *Am. J. Hypertens.* **9**, 160S–169S.

Shimamura, K., Togashi, H. and Saito, H. (1981). Effect of clonidine on the function of the adrenal medulla in rats, *Res. Com. Chem. Pathol. Pharmacol.* **31**, 189–192.

Tanabe, T., Shudo, I. and Saito, H. (1975). Adrenergic contribution to digitalis action, *Proc. West. Pharmacol. Soc.* **18**, 20–22.

Togashi, H. (1983). Central and peripheral effects of clonidine on the adrenal medullary function in spontaneously hypertensive rats, *J. Pharmacol. Exp. Ther.* **225**, 191–197.

Togashi, H., Minami, M., Saito, I., *et al.* (1984). Guanfacine and clonidine: the effects on adrenal medullary function in spontaneously hypertensive rats, *Arch. Int. Pharmacodyn. Ther.* **272**, 79–87.

Togashi, H., Kurosawa, M., Minami, M., *et al.* (1985). Effect of morphine on the adrenal sympathetic reflex elucited by mechanical stimulation of the skin in rats, *Neurosci. Lett.* **62**, 81–87.

Togashi, H., Yoshioka, M., Minami, M., *et al.* (1987). Effect of the substance P antagonist spantaide in adrenal sympathetic nerve activity in rats, *Jpn. J. Pharmacol.* **43**, 253–261.

Togashi, H., Matsumoto, M., Yoshioka, M., *et al.* (1990). Behavioral and neurochemical evaluation of stroke-prone spontaneously hypertensive rats for vascular dementia-animal model, in: *Basic, Clinical, and Therapeutic Aspects of Alzheimer's and Parkinson's Diseases,* T. Nagatsu *et al.* (Eds), Vol. 1, pp. 473–476. Plenum Press, New York.

Togashi, H., Matsumoto, M., Yoshioka, M., *et al.* (1994). Neurochemical profiles in cerebrospinal fluid of stroke-prone spontaneously hypertensive rats, *Neurosci. Lett.* **166**, 117–120.

Togashi, H., Kimura, S., Matsumoto, M., *et al.* (1996). Cholinergic changes in the hippocampus of stroke-prone spontaneously hypertensive rats, *Stroke* **27**, 520–526.

Yamori, Y. (1989). Predictive and preventive pathology of cardiovascular diseases, *Acta Pathol. Jpn.* **39**, 683–705.

Yomaida, I., Murao, M., Togashi, H., *et al.* (1979a). Effects of long-term administration and withdrawal of clonidine on activity of sympathetic efferent nerve unit in spontaneously hypertensive rats, *Neurosci. Lett.* **15**, 249–251.

Yomaida, I., Koike, Y., Shimamura, K., *et al.* (1979b). Effect of long term oral administration of clonidine on blood pressure and heart rate in SHR, *Jpn. Heart J.* **20** (Suppl. 1), 204–206.

Yoshioka, M., Matsumoto, M., Togashi, H., *et al.* (1987). central sympathoinhibitory action of ketanserin in rats, *J. Pharmacol. Exp. Ther.* **243**, 1174–1178.

Yoshioka, M., Togashi, H., Abe, M., *et al.* (1990). Central sympathoinhibitory action of a new type of α-1 adrenoceptor antagonist, YM-617, in rats, *J. Pharmacol. Exp. Ther.* **253**, 427–431.

Advances in Neuroregulation and Neuroprotection (2005), pp. 595-614
Collin, C. *et al.* (Eds)
© VSP 2005

Serotonin-mediated neurotransmission in the pyramidal cells of the rat hippocampus

MACHKO MATSUMOTO [1], HIROKO TOGASHI [1,*], TAKU KOJIMA [2],
KAORI TACHIBANA [2], SATOSHI OHASHI [1] and MITSUHIRO YOSHIOKA [1]

[1] *Department of Pharmacology, Hokkaido University Graduate School of Medicine, Kita-15, Nishi-7,
Kita-ku, Sapporo 060-8638, Japan*
[2] *Department of Anesthesiology, Hokkaido University Graduate School of Medicine,
Sapporo 060-8638, Japan*

Abstract—Serotonin (5-HT)-mediated neurotransmission in the pyramidal cell layer of the rat hippocampus was reviewed, with special reference to our findings on a selective 5-HT reuptake inhibitor (SSRI) fluvoxamine. In the hippocampus, fluvoxamine increased the extracellular levels of 5-HT and increased the evoked population spike (PS) amplitudes in the CA1 and CA3 pyramidal cells. Fluvoxamine-induced synaptic facilitation was augmented by the 5-HT_{1A} receptor antagonist NAN 190, while it was prevented by the 5-HT_4 receptor antagonist GR 113808 and the 5-HT_7 receptor antagonist DR4004 in the CA1. They were abolished by either NAN 190 or DR 4004 in the CA3. A 5-HT_{1A} receptor agonist, tandospirone, mimicked the fluvoxamine-induced synaptic effects in the pyramidal cells in a NAN 190-sensitive manner. These results suggest that endogenous 5-HT-mediated neurotransmission is regulated by 5-HT_{1A} and $5\text{-HT}_4/5\text{-HT}_7$ receptors in an opposite manner in the CA1 field. On the contrary, both 5-HT_{1A} and 5-HT_7 receptors appear to be positively coupled with 5-HT-mediated synaptic responses in the CA3 field. These regional differences in 5-HT receptor-mediated neurotransmission in the pyramidal cells, may play a significant role in the hippocampal functions, such as fear/anxiety expression and emotional memory processing, and may be partly responsible for anxiolytic effects of SSRIs.

Keywords: 5-HT_{1A} receptors; 5-HT_4 receptors; 5-HT_7 receptors; hippocampal neurotransmission; CA1; CA3.

INTRODUCTION

The hippocampus is known to be an important structure for the expression of fear and anxiety, and emotional memory processing. The hippocampal formation consists of three principal fields that are connected as the trisynaptic input pathway,

---

*To whom correspondence should be addressed. E-mail: thiro@med.hokudai.ac.jp

namely, the entorhinal cortex to dentate gyrus granule cells, mossy fiber to CA3, and Schaffer collateral to CA1 pyramidal cells. Numerous *in vitro* studies have reported that the electrophysiological properties such as input resistance (Beck *et al.*, 1992), receptor-effector pathways (Okuhara and Beck, 1994) and the mechanism of long-term potentiation (LTP) (Weisskopf *et al.*, 1994; Villacres *et al.*, 1998) are different between the CA1 and CA3 pyramidal cells.

The hippocampal regions receive abundant 5-HTergic inputs from the raphe magnus, which have been implicated in some emotional disturbances such as depression and anxiety. Differences in the 5-HT receptor density and distribution and/or the 5-HT receptor-mediated responses have also been reported in these pyramidal regions. For instance, autoradiographic studies showed that the density of 5-HT$_{1A}$ receptors was higher in the CA1 than that in the CA3, and that the distribution of 5-HT$_{1A}$ receptors was found in all of the layers in the CA1 field, whereas in only parts of the CA3 field (Pazos and Palacios, 1985; Welner *et al.*, 1989). Beck *et al.* (1992) reported, using intracellular recordings of hippocampal slices, that the 5-HT-induced hyperpolarization was larger in the CA3 than in the CA1 field, which was blocked by the 5-HT$_{1A}$ receptor antagonist in the CA1 in a competitive manner, but not in the CA3. These findings possibly indicate that some differences in efficacy for effector-coupling, structure of the recognition site of 5-HT$_{1A}$ receptors and/or mediation by other 5-HT receptors exist in the pyramidal cell layers, CA1 and CA3 subfields. Recently, we reported that the 5-HT-induced accumulation of cAMP in the rat hippocampus was mediated via 5-HT$_{1A}$, 5-HT$_4$ and 5-HT$_7$ receptors in a cummulative manner (Markstein *et al.*, 1999). Thus, it is possible that these 5-HT receptors play a critical role in modulating hippocampal neurotransmission to incoming stimuli and contribute to the differential responses in each pyramidal layer. In turn, endogenous 5-HT might affect the synaptic transmission in the CA1 and CA3 fields by activation of these 5-HT receptors and consequently exert physiological functions.

Recently, selective 5-HT reuptake inhibitors (SSRIs) as well as 5-HT$_{1A}$ receptor agonists have been focused on their anxiolytic properties (Schreiber *et al.*, 1998; Figgitt and McClellan, 2000). The precise mechanisms remain to be clarified; however, SSRIs are considered to modulate 5-HT neurotransmission via a blockade of 5-HT transporters located in the cell bodies and terminal regions. It is assumed, therefore, that the synaptic 5-HT availability increased by SSRIs influences the pyramidal neurotransmission through activation of 5-HT receptors and consequently modulates the hippocampal functions.

We review here endogenous 5-HT-mediated neurotransmission in the pyramidal cell layer (CA1 and CA3 fields) of the rat hippocampus, with special reference to our findings on a SSRI, fluvoxamine (Matsumoto *et al.*, 2002; Ohashi *et al.*, 2002; Kojima *et al.*, 2003), and briefly discuss their implication with regard to anxiolytic effects. Thus, the present review was intended to summarize the following three issues: (1) To elucidate the effects of fluvoxamine on neurotransmission by monitoring the population spike (PS) amplitude; (2) To identify the 5-HT receptors involved in the fluvoxamine-induced changes in neurotransmission; (3) To assess whether 5-HT receptor agonists could mimic the fluvoxamine-induced synaptic responses, focusing on three 5-HT receptor candidates, 5-HT$_{1A}$, 5-HT$_4$ and 5-HT$_7$ receptors.

SEROTONERGIC PERTURBATION BY A SSRI, FLUVOXAMINE

Neuronal activity in the dorsal raphe nuclei

SSRIs act on the 5-HT transporters located not only at nerve terminals but also on cell bodies (Tao-Cheng and Zhou, 1999). To examine the effects of fluvoxamine on cell bodies, the neuronal activity was measured in the dorsal raphe nuclei, a major source of 5-HTergic inputs to the hippocampus. A stainless steel electrode (diameter: 0.1 mm) was inserted into the dorsal raphe with a posterior angle of 22° at the following coordinates: posterior, 10.3 mm, lateral, 0.1 mm, ventral, 6.4 mm from the bregma and dural surface, under ketamine (100 mg kg^{-1}, i.p.) anesthesia. Two days after surgery, neuronal activity was measured under non-anesthesia: the spontaneous firing was monitored using a telemetry system (Physiotel; TA11CTA-F40, Primetech, Co., Ltd., Tokyo, Japan) of which sensor was intraperitoneally implanted, and the neuronal signals obtained were amplified and displayed on an oscilloscope (AD-5141, A and D, Tokyo, Japan).

Systemic administration of fluvoxamine (30 mg kg^{-1}, i.p.) decreased the firing activity of the dorsal raphe nuclei of freely moving rats. A typical neurogram is shown in Fig. 1A. These findings indicate that fluvoxamine produced an inhibitory influence on 5-HTergic nerves via somatodendritic 5-HT$_{1A}$ autoreceptors, and consequently act to decrease the extracellular levels of 5-HT in the nerve terminal regions.

Endogenous 5-HT levels in the hippocampus

Synaptic 5-HT availability was examined by measuring extracellular levels of hippocampal 5-HT using *in vivo* microdialysis. Rats were anesthetized with ketamine (100 mg kg^{-1}, i.p.) and a 3-mm concentric guide cannula was stereotaxically implanted into the hippocampus (5.8 mm posterior, 4.8 mm lateral to the bregma, 4.0 mm ventral to the dura). Two days after surgery, a dialysis probe was inserted through the guide cannula and perfused at a flow rate of 1 μl min^{-1} with Ringer's solution (KCl 2.7, NaCl 147, CaCl$_2$ 2.3 mmol). Ringer's solution was used to obtain the stable baseline for 120–180 min and successive 20 μl samples were collected at 20 min intervals. Extracellular levels of 5-HT were measured using high-performance liquid chromatography with an electrochemical detector (HPLC-ECD), as described previously (Yoshioka *et al.*, 1995). Values obtained by use of *in vivo* microdialysis were expressed as a percentage of the baseline level before drug administration.

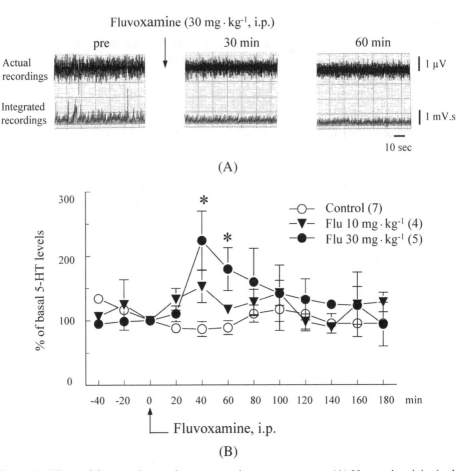

Figure 1. Effects of fluvoxamine on the serotonergic nervous system. (A) Neuronal activity in the dorsal raphe nuclei. Specimen recordings indicated spontaneous firing recorded in the dorsal raphe nuclei, before and 30 min and 60 min after fluvoxamine administration. (B) Extracellular levels of 5-HT in the hippocampus. 5-HT levels were measured by *in vivo* microdialysis. Values are expressed as a percentage of the basal levels obtained before fluvoxamine (Flu) administration. Each point represents the mean ± SEM. *$p < 0.05$ *vs.* control. The number of rats tested is given in parentheses.

The extracellular levels of hippocampal 5-HT were elevated by fluvoxamine (10 $mg\,kg^{-1}$ and 30 $mg\,kg^{-1}$, i.p.) in a dose-dependent manner. Significant increases in the 5-HT levels were observed after administration of a high dose (30 $mg\,kg^{-1}$, i.p.) to a maximum response of $237.4 \pm 42.1\%$. The peak response was obtained 40 min after drug administration, and then returned to the basal level (Fig. 1B). A low dose of fluvoxamine (10 $mg\,kg^{-1}$, i.p.) slightly, but not significantly, increased the hippocampal 5-HT levels.

The fact that systemically administered fluvoxamine (30 $mg\,kg^{-1}$, i.p.) decreased the neuronal firing in the dorsal raphe, while extracellular 5-HT levels in the hippocampus were increased, indicates that fluvoxamine-induced inhibition of 5-HT reuptake in terminal regions overcame the inhibitory influence of somatodendritic

5-HT$_{1A}$ autoreceptors on 5-HT release, and consequently increased the extracellular 5-HT levels of a nerve terminal region, hippocampus. It is conceivable, therefore, that synaptic 5-HT elevated by fluvoxamine modulates the neurotransmission in the hippocampus.

SYNAPTIC EFFECTS OF ENDOGENOUS 5-HT IN THE PYRAMIDAL CELLS

Synaptic efficacy

To evaluate the 5-HTergic influences on the synaptic efficacy in the pyramidal regions, fluvoxamine-induced changes in the population spike (PS) amplitude were evaluated in the hippocampal CA1 and CA3 fields of the anesthetized rat. The synaptic efficacy in the CA1 and CA3 fields was evaluated by monitoring the evoked PS amplitude under anesthesia with 1% halothane in a mixture of 20% O$_2$ and 80% N$_2$. A monopolar glass-coated recording electrode was placed in the ipsilateral pyramidal cell layer of the CA1 (5.0 mm posterior, 3.0 mm lateral to the bregma, approximately 2.3 mm ventral to the dura) and a bipolar stainless steel electrode was inserted to stimulate the Schaffer collateral (3.0 mm posterior, 1.5 mm lateral to the bregma, 2.8 mm ventral to the dura). The PS amplitude in the CA3 field (2.9 mm posterior, 2.4 mm lateral to the bregma, approximately 2.9 mm ventral to the dura) was evoked by stimulation of mossy fibers (5.7 mm posterior, 2.4 mm lateral to the bregma, a posterior angle of 15°, 2.8 mm ventral to the dura). The potentials evoked by test stimulation (frequency 0.1 Hz, pulse duration 250 μs, stimulus interval 30 s) was amplified and monitored with an oscilloscope (AD-5141, A and D, Tokyo, Japan). The latencies of evoked potentials in the CA1 and CA3 fields were around 10 ms and 5 ms, respectively. Integrated PS obtained from five successive stimuli was recorded every 5 min with a data-analysis system (ATAC-450, Nihon Kohden, Japan) and the amplitude of the PS height was measured. The intensity of the test stimulation was adjusted for each rat to elicit approximately 50% maximum amplitude. The PS amplitude was expressed as a percentage of the baseline level obtained before drug administration. The area under the curve (AUC) of the time course responses was determined to evaluate the ensemble effects on the PS amplitude.

Fluvoxamine (10 and 30 mg kg^{-1}, i.p.) produced increases in the PS amplitude in the CA1 and CA3 fields in a concentration-dependent manner (Fig. 2). Significant increment of the PS amplitude was found after a high dose of fluvoxamine (30 mg kg^{-1}, i.p.) in both fields. The fluvoxamine-induced excitatory effect was greater in the CA3 field than in the CA1 field. This difference may be due to the physiological properties of the pyramidal cells and/or associated 5-HT receptor subtypes. A number of *in vitro* studies have provided evidence for different characteristics of the pyramidal cells in the CA1 and CA3 fields. For instance, electrophysiological properties such as input resistance of the CA3 pyramidal cells are larger than those of the CA1, indicating the possibility that the synaptic input

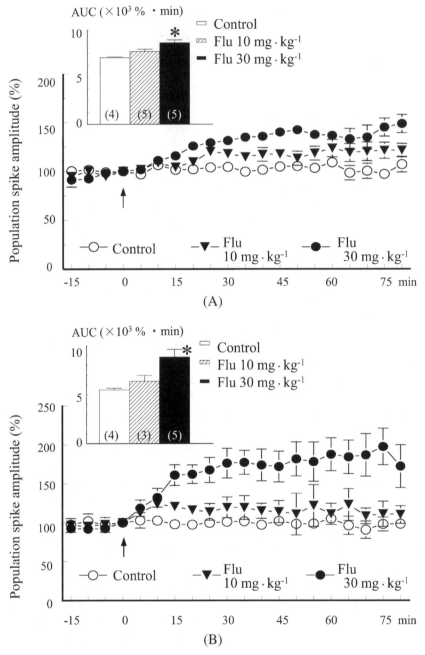

Figure 2. Time-course responses of the evoked population spikes (PS) amplitude after fluvoxamine (Flu; 10 and 30 mg kg^{-1}, i.p.) administration in the CA1 (A) and CA3 (B) fields. Values are expressed as a percentage of the basal PS amplitude obtained before fluvoxamine administration. Each point represents the mean $\pm$ SEM. Inset shows the area under the curve (AUC; $\times 10^3\%$ min) of the time-course responses in the PS amplitude during the 60-min period after fluvoxamine or saline (Control) administration. The number in each column indicates the number of rats tested. $^* p < 0.05$ *vs.* control.

to the CA3 region produces greater effects than that to the CA1 region (Beck *et al.*, 1992). The different sensitivity to fluvoxamine observed in the CA1 and CA3 fields, therefore, might be due to the synaptic properties of each pyramidal cell layer.

Synaptic plasticity

Based on the possible relationship between the anxiety and emotional memory processes, long-term potentiation (LTP), the electrophysiological basis of memory (Bliss and Collingridge, 1993), was analyzed in the CA1 and CA3 fields, focusing on the anxiolytic properties of fluvoxamine. To induce LTP, high-frequency, i.e. tetanic stimulation (Tetanus; 5 trains at 1 Hz each composed of 8 pulses at 400 Hz) was given at the same intensity as the test stimuli. Tetanic stimulation was given 20 min after the drug administration.

Tetanic stimulation of the Schaffer collaterals induced LTP formation in the CA1 field, enhancing the PS amplitude to a maximum of $210.3 \pm 24.1\%$. Pretreatment with fluvoxamine (10 and 30 mg kg^{-1}, i.p.) significantly suppressed the LTP formation in this field (Fig. 3). Areas under the curve (AUC) ($\times 10^3\%$ min) calculated during 60 min after the tetanic stimulation were 5.45 ± 0.15 and 6.93 ± 0.62 in the presence of 10 mg kg^{-1} and 30 mg kg^{-1} of fluvoxamine, respectively. The LTP formation was significantly reduced, when compared with controls (9.88 ± 0.72). In the CA3 field, the PS amplitude was elevated by tetanic stimulation of mossy fibers to a maximum of $221.4 \pm 23.2\%$ ($n = 7$). LTP formation was not affected by fluvoxamine (30 mg kg^{-1}, i.p.) in this field (Fig. 4).

PHARMACOLOGICAL CHARACTERIZATION OF THE ENDOGENOUS 5-HT-MEDIATED SYNAPTIC EFFECTS IN THE PYRAMIDAL CELLS

To determine 5-HT receptors responsible for the fluvoxamine-induced synaptic effects, 5-HT receptor agonists and/or antagonists, focusing on three 5-HT receptor candidates — 5-HT$_{1A}$, 5-HT$_4$ and 5-HT$_7$ receptors — were administered intraperitoneally (i.p.) or intracerebroventricularly (i.c.v.). A probe for i.c.v. injection was inserted at the following coordinates relative to the bregma and dural surface; caudal, 0.8 mm; lateral, 1.4 mm; ventral, 3.3 mm through a guide cannula and 10 μl of the drug was injected. As controls, saline (0.9% NaCl) or artificial cerebrospinal fluid (aCSF) (KCl 2.7, NaCl 140, CaCl$_2$ 1.2, MgCl$_2$ 1.0, NaH$_2$PO$_4$ 0.3, Na$_2$HPO$_4$ 1.7 mmol) were administered i.p. and i.c.v., respectively.

Synaptic efficacy

In the CA1 field, fluvoxamine (30 mg kg^{-1}, i.p.)-induced increase in the PS amplitude was further augmented by pretreatment with the 5-HT$_{1A}$ receptor antagonist NAN 190 (0.5 mg kg^{-1}, i.p.) (Fig. 5). The facilitatory effects of fluvoxamine (30 mg kg^{-1}, i.p.), on the other hand, were prevented by the 5-HT$_4$ receptor antagonist GR 113808 (20 μg rat^{-1}, i.c.v.) and the 5-HT$_7$ receptor antagonist

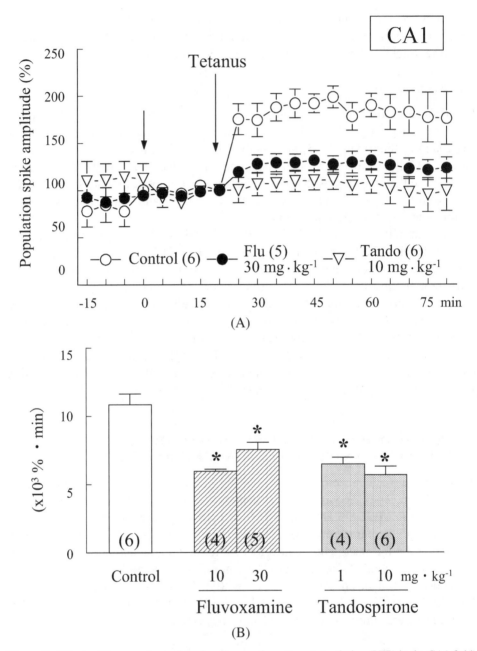

Figure 3. Effects of fluvoxamine and tandospilone on long-term potentiation (LTP) in the CA1 field. (A) LTP formation after high frequency stimulation (Tetanus) and time-course effects of fluvoxamine (Flu; 10 and 30 mg kg^{-1}, i.p.) and tandospilone (Tando; 1 and 10 mg kg^{-1}, i.p.). Fluvoxamine or tandospilone was applied 20 min before titanic stimulation as indicated by the short arrow. The number of rats tested is given in parentheses. (B) The area under the curves (AUC; $\times 10^3\%$ min) of the PS amplitude during the 60-min period after drug administration. The number in each column indicates the number of rats tested. *$p < 0.05$ *vs.* control.

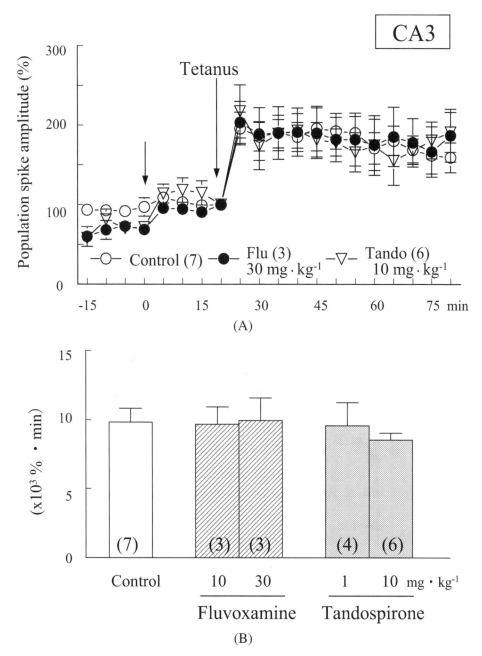

Figure 4. Effects of fluvoxamine and tandospilone on long-term potentiation (LTP) in the CA3 field. (A) LTP formation after high frequency stimulation (Tetanus) and time-course effects of fluvoxamine (Flu; 10 and 30 mg kg^{-1}, i.p.) and tandospilone (Tando; 1 and 10 mg kg^{-1}, i.p.). Fluvoxamine or tandospilone was applied 20 min before titanic stimulation as indicated by the short arrow. The number of rats tested is given in parentheses. (B) The area under the curves (AUC; $\times 10^3 \%$ min) of the PS amplitude during the 60-min period after drug administration. The number in each column indicates the number of rats tested.

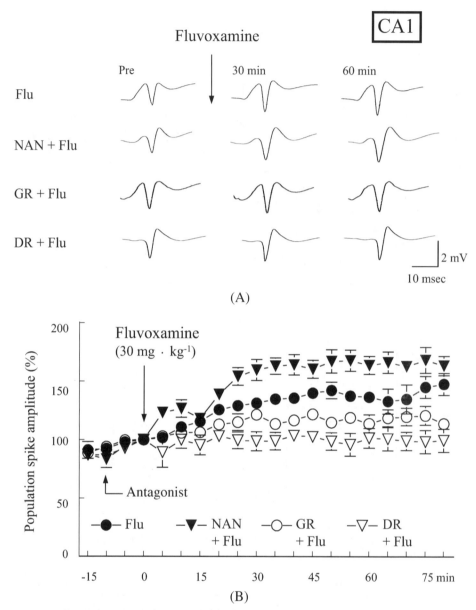

Figure 5. Effects of 5-HT receptor antagonists on the fluvoxamine (Flu; 30 mg kg^{-1}, i.p.)-induced changes in the evoked population spikes (PS) amplitude in the CA1 field. The 5-HT$_{1A}$ receptor antagonist NAN 190 (NAN; 0.5 mg kg^{-1}, i.p.), the 5-HT$_4$ receptor antagonist GR 113808 (GR; 20 μg rat^{-1}, i.c.v.) or the 5-HT$_7$ receptor antagonist DR 4004 (DR; 10 μg rat^{-1}, i.c.v.) was applied 10 min before fluvoxamine administration. (A) Specimen recordings show traces of the evoked PS before and after fluvoxamine administration in the presence or absence of 5-HT receptor antagonists. (B) Time-course responses of the PS amplitude induced by fluvoxamine in the presence or the absence of 5-HT receptor antagonists. Values are expressed as a percentage of the basal PS amplitude obtained before fluvoxamine administration. Each point represents the mean ± SEM. The number of rats tested is presented in Table 1.

Table 1.

5-HT receptor agonisms and antagonisms on fluvoxamine-induced neurotransmission in the pyramidal cells of the rat hippocampus

Treatment	CA1 field		CA3 field	
	n	AUC	n	AUC
Control (saline)	4	5.66 ± 0.08	4	5.51 ± 0.17
Control (aCSF)	3	5.54 ± 0.10	3	5.62 ± 0.24
Flu (30 mg kg^{-1}, i.p.)	5	6.94 ± 0.22^a	5	8.86 ± 0.80^a
NAN 190 (0.5 mg kg^{-1}, i.p.) alone	5	6.02 ± 0.12	4	5.61 ± 0.24
NAN 190 (0.5 mg kg^{-1}, i.p.) + Flu	5	8.06 ± 0.25^b	3	6.26 ± 0.20^b
GR 113808 (20 μg rat^{-1}, i.c.v.) alone	4	5.61 ± 0.11	3	5.86 ± 0.28
GR 113808 (20 μg rat^{-1}, i.c.v.) + Flu	6	6.21 ± 0.21^b	3	9.16 ± 0.16
DR 4004 (1 μg rat^{-1}, i.c.v.) alone	4	5.52 ± 0.03	3	5.59 ± 0.23
DR 4004 (1 μg rat^{-1}, i.c.v.) + Flu	5	7.36 ± 0.32	5	7.11 ± 0.40
DR 4004 (10 μg rat^{-1}, i.c.v.) alone	4	5.78 ± 0.26	3	5.66 ± 0.47
DR 4004 (10 μg rat^{-1}, i.c.v.) + Flu	4	5.42 ± 0.40^b	4	6.35 ± 0.27^b
SB 269970 (1 μg rat^{-1}, i.c.v.) + Flu			3	6.11 ± 0.33^b
SB 269970 (10 μg rat^{-1}, i.c.v.) + Flu	4	5.86 ± 0.29^b		
Tandospirone (10 mg kg^{-1}, i.p.)	5	4.35 ± 0.18^a	6	9.17 ± 0.54^a
NAN 190 (0.5 mg kg^{-1}, i.p.) + Tando	4	5.64 ± 0.07^c	4	6.88 ± 0.25^c
SC 53116 (10 μg rat^{-1}, i.c.v.)	6	7.37 ± 0.39^a	5	5.60 ± 0.17
GR 113808 (20 μg rat^{-1}, i.c.v.) + SC 53116 (10 μg rat^{-1}, i.c.v.)	5	4.83 ± 0.20^d		

Values are means $\pm$ SEM. Significant difference ($p < 0.05$) from saline control.
[a] Fluvoxamin (Flu; 30 mg kg^{-1}, i.p.).
[b] Tandospirone (Tando; 10 mg kg^{-1}, i.p.).
[c] SC 53116.
[d] Alone treated rats.
n: Number of rats tested. AUC: area under the curves ($\times 10^3\%$ min) during the 60-min period after administration of fluvoxamine or 5-HT receptor agonists.

DR 4004 (Kikuchi *et al.*, 1999) (10 μg rat^{-1}, i.c.v.) (Fig. 5). Low doses of GR 113808 (1 μg rat^{-1}, i.c.v.) and DR 4004 (1 μg rat^{-1}, i.c.v.) did not influence the fluvoxamine-induced facilitatory effects (Table 1). The possible involvement of the 5-HT$_7$ receptors in the fluvoxamine-induced synaptic responses was further confirmed by additional experiments using the selective 5-HT$_7$ receptor antagonist SB 269970 (Hagan *et al.*, 2000). As shown in Table 1, pretreatment with SB 269970 (10 μg rat^{-1}, i.c.v.), which did not alter the PS amplitude by itself, abolished the fluvoxamine-induced enhancement in the CA1 field.

In the CA3, fluvoxamine (30 mg kg^{-1}, i.p.)-induced increase in the PS amplitude was prevented by the 5-HT$_{1A}$ receptor antagonist NAN 190 (0.5 mg kg^{-1}, i.p.). Pretreatment with the 5-HT$_7$ receptor antagonists DR 4004 (10 μg rat^{-1}, i.c.v.) and SB 269970 (1 μg rat^{-1}, i.c.v.) also abolished this excitatory response. A low dose of DR 4004 (1 μg rat^{-1}, i.c.v.) exerted no significant effect by itself. In contrast, the 5-HT$_4$ receptor antagonist GR 113808 (20 μg rat^{-1}, i.c.v.) did not influence the fluvoxamine-induced synaptic facilitation (Fig. 6 and Table 1).

Involvement of 5-HT receptor subtypes was further confirmed by 5-HT agonists. Application of the 5-HT$_{1A}$ receptor agonist tandospirone (10 mg kg^{-1}, i.p.) decreased the PS amplitude in the CA1, but increased them in the CA3 (Fig. 7). The opposite effects of tandospirone, i.e. inhibition and facilitation, were abolished by pretreatment with the 5-HT$_{1A}$ receptor agonist NAN 190 (0.5 mg kg^{-1}, i.p.) in each field (Table 1). The 5-HT$_4$ receptor agonist SC 53116 (1 and 10 μg rat^{-1}, i.c.v.) produced increases in the PS amplitude in the CA1 field in a concentration-dependent manner. SC 53116 (10 μg rat^{-1}, i.c.v.)-induced facilitation was prevented by the 5-HT$_4$ receptor antagonist GR 113808 (20 μg rat^{-1}, i.c.v.) in this field. However, in the CA3 field, a high dose of SC 53116 (10 μg rat^{-1}, i.c.v.) did produce any changes in the PS amplitude (Fig. 7 and Table 1).

Synaptic plasticity

The LTP formation in the CA1 field was prevented by pretreatment with tandospirone (10 mg kg^{-1}, i.p.) as well as fluvoxamine (Fig. 3 and Fig. 8), whereas LTP was induced in the presence of SC 53116 (10 μg rat^{-1}, i.c.v.). In contrast, LTP formation in the CA3 field was observed after tandospirone (10 mg kg^{-1}, i.p.) as well as fluvoxamine administration (Fig. 4). We could not examine the possible involvement of the 5-HT$_7$ receptor mechanism in the LTP formation, because of the lack of selective 5-HT$_7$ receptor agonists. Therefore, effects of the 5-HT$_7$ receptor antagonist DR 4004 on the fluvoxamine-induced LTP formation were examined. As shown in Fig. 8, DR 4004 (10 μg rat^{-1}, i.c.v.) did not affect the fuluvoxamine-induced inhibition of LTP formation in the CA1 field (Fig. 8). Thus, it appears unlikely that the 5-HT$_7$ receptors are involved in the inhibitory effects of fluvoxamine on LTP formation.

ELECTROPHYSIOLOGICAL AND PHARMACOLOGICAL CONSIDERATIONS

Fluvoxamine produced an increase in the extracellular 5-HT and a concomitant 'LTP-like' enhancement of neurotransmission in the hippocampal pyramidal regions. In the CA1 field, the synaptic effect elicited by fluvoxamine was further augmented by the 5-HT$_{1A}$ receptor antagonist NAN 190. NAN 190 alone did not affect the PS amplitude. Thus, 5-HT$_{1A}$ receptor mechanisms are likely to contribute to the fluvoxamine-induced synaptic responses in an inhibitory manner. The finding that the 5-HT$_{1A}$ receptor agonist tandospirone produced a decrease in the PS amplitude in the CA1 region further supported this possibility. Interestingly, in the CA3 fields, the synaptic response mediated via 5-HT$_{1A}$ receptors showed opposite directions; fluvoxamine-induced facilitation was prevented by pretreatment with NAN 190. Thus, the activation of 5-HT$_{1A}$ receptors appears to be involved in the fluvoxamine-induced excitable changes in neurotransmission in a facilitatory manner. These observations were strengthened by the findings that tandospirone also

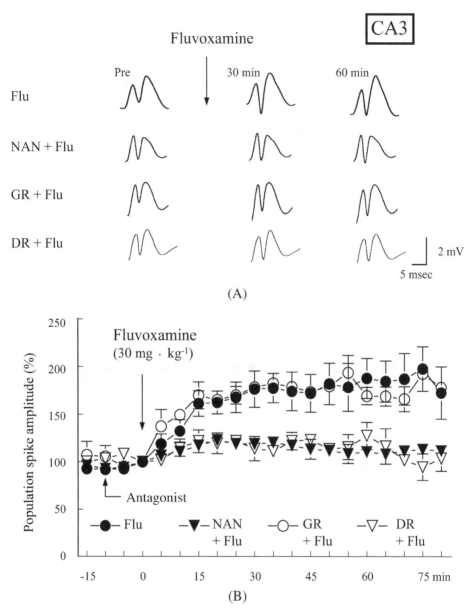

Figure 6. Effects of 5-HT receptor antagonists on the fluvoxamine (Flu; 30 mg kg^{-1}, i.p.)-induced changes in the evoked population spikes (PS) amplitude in the CA3 field. The 5-HT$_{1A}$ receptor antagonist NAN 190 (NAN; 0.5 mg kg^{-1}, i.p.), the 5-HT$_4$ receptor antagonist GR 113808 (GR; 20 μg rat^{-1}, i.c.v.) or the 5-HT$_7$ receptor antagonist DR 4004 (DR; 10 μg rat^{-1}, i.c.v.) was applied 10 min before fluvoxamine administration. (A) Specimen recordings show traces of the evoked PS before and after fluvoxamine administration in the presence or absence of 5-HT receptor antagonists. (B) Time-course responses of the PS amplitude induced by fluvoxamine in the presence or the absence of 5-HT receptor antagonists. Values are expressed as a percentage of the basal PS amplitude obtained before fluvoxamine administration. Each point represents the mean ± SEM. The number of rats tested is presented in Table 1.

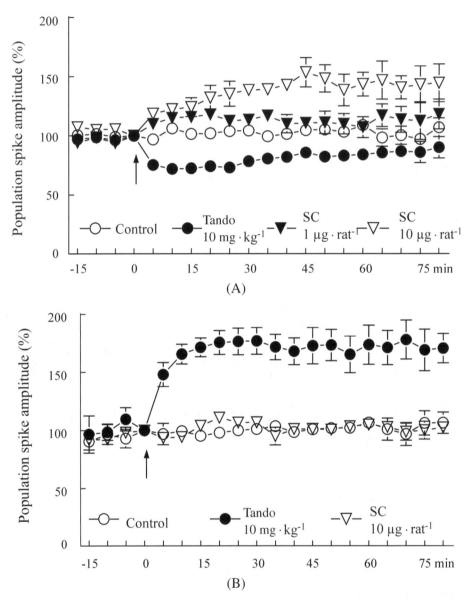

Figure 7. Effects of exogenously applied 5-HT receptor agonists on the evoked population spike (PS) amplitude in the pyramidal cells. Time-course changes in the PS amplitude were determined during the 60-min period after the 5-HT$_{1A}$ receptor agonist tandospirone (Tando; 10 mg kg^{-1}, i.p.) and the 5-HT$_4$ receptor agonist SC 53116 (SC; 1 or 10 μg rat^{-1}, i.c.v.) administration in the CA1 (A) and CA3 (B) fields. Values are expressed as a percentage of the basal PS amplitude obtained before agonist administration as indicated by the arrow. Each point represents the mean ± SEM. The number of rats tested is presented in Table 1.

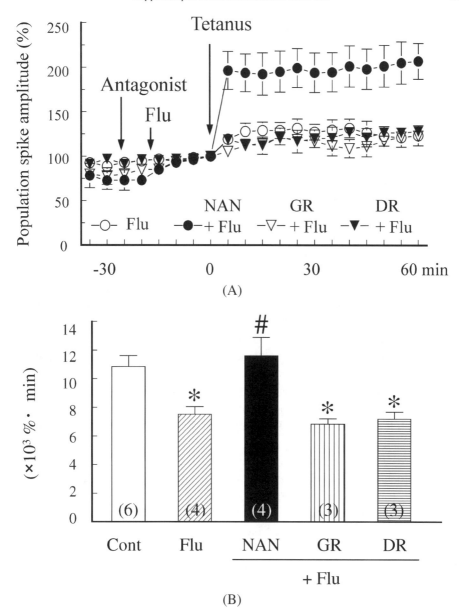

Figure 8. Effects of 5-HT receptor antagonists on the fluvoxamine (Flu; 30 mg kg^{-1}, i.p.)-induced changes in long-term potentiation (LTP) in the CA1 field. The 5-HT$_{1A}$ receptor antagonist NAN 190 (NAN; 0.5 mg kg^{-1}, i.p.), the 5-HT$_4$ receptor antagonist GR 113808 (GR; 20 μg rat^{-1}, i.c.v.) or the 5-HT$_7$ receptor antagonist DR 4004 (DR; 10 μg rat^{-1}, i.c.v.) was applied 10 min before fluvoxamine, which was applied 20 min before titanic stimulation (Tetanus). (A) Time course effects of 5-HT receptor antagonists on fluvoxamine-induced changes in LTP formation. Values are expressed as a percentage of the baseline level before titanic stimulation. The number of rats tested is presented in Table 1. (B) The area under the curves (AUC; $\times 10^3\%$ ·min) of the PS amplitude during the 60-min period after titanic stimulation. The number in each column indicates the number of rats tested. $^*p < 0.05$ *vs.* control. $^\#p < 0.05$ *vs.* fluvoxamine alone.

produced increases in the PS amplitude in the CA3 fields, which was completely blocked by NAN 190.

The different responses mediated via 5-HT$_{1A}$ receptors in the CA1 and the CA3 subfields of the hippocampus may be explained by possible involvement of neuronal circuits such as inhibitory interneurons: The CA3 field and mossy fibers are densely innervated by GABAergic neurons (Vida and Frotscher, 2000) and the synaptic transmission at the mossy fiber input to the CA3 was depressed by synaptically released GABA (Vogt and Nicoll, 1999). In the dentate gyrus (Matsuyama *et al.*, 1997) and amygdala (Kishimoto *et al.*, 2000), GABAergic neuronal activity is inhibited by activation of 5-HT$_{1A}$ receptors. It is, therefore, plausible that, in the CA1 region, the predominant effect mediated via 5-HT$_{1A}$ receptors is the inhibition of pyramidal neurons, whereas in the CA3 field it may be the inhibition of GABAergic interneurons. Another possibility may be due to the different mechanism of receptor-effector couplings including cAMP-dependent processes (Raymond *et al.*, 1999). Previous reports (Barbaccia *et al.*, 1983; Shenker *et al.*, 1985) and our recent study (Markstein *et al.*, 1999) showed that positive coupling of adenylyl cyclase (AC) to 5-HT$_{1A}$ receptors exist in the hippocampus. Blier *et al.* (1993) demonstrated that endogenous 5-HT regulates the pyramidal neuronal activity via cholera toxin sensitive 5-HT$_{1A}$ receptors in the CA3 region. These findings suggest that the cAMP-dependent processes may underlie the facilitatory neurotransmission via 5-HT$_{1A}$ receptors.

A previous study showed that 5-HT$_{1A}$ receptors are co-localized with 5-HT$_4$ receptors on the same pyramidal neurons in the CA1 field (Roychowdhury *et al.*, 1994). Colino and Halliwell (1987) reported that the 5-HT$_{1A}$ receptor-mediated hyperpolarization of the CA1 pyramidal cells masks the blockade of potassium conductance, which is probably mediated via 5-HT$_4$ receptors. The present finding that the blockade of 5-HT$_{1A}$ receptors by NAN 190 caused the augmentation of synaptic transmission is, therefore, likely to reflect the excitatory responses mediated via 5-HT$_4$ receptors. Indeed, the 5-HT$_4$ receptor antagonist GR 113808 prevented the fluvoxamine-induced facilitation and the 5-HT$_4$ receptor agonist SC 53116 (Flynn *et al.*, 1992) enhanced the PS amplitude. These findings using exogenously applied 5-HT receptor agonists were consistent with those of *in vitro* studies (Schmitz *et al.*, 1995; Bijak *et al.*, 1997; Pugliese *et al.*, 1998). Thus, in the CA1 field, 5-HT$_{1A}$ and 5-HT$_4$ receptors appear to contribute to endogenous 5-HT-mediated synaptic transmission in inhibitory and facilitatory manners, respectively. In the CA3 field, however, fluvoxamine-induced excitability was not changed by pretreatment with GR 113808. In addition, SC 53116 did not cause any changes in the PS amplitude in this field. Although the reason for the regional differences of the 5-HT$_4$ receptor-mediated responses is still unclear, it may be due to the 5-HT$_4$ receptor isoforms reported by Bender *et al.* (2000). They showed that the expression of these isoforms differ between tissues and at times even between samples from the hippocampus. Their findings lead us to speculate the existence of different isoforms in the CA1 and CA3; however, further studies will be required.

Table 2.
Involvement of 5-HT receptor subtypes in fluvoxamine-induced neurotransmission in the pyramidal cells of the rat hippocampus

	Fluvoxamine-induced neurotransmission	$5\text{-}HT_{1A}$	$5\text{-}HT_4$	$5\text{-}HT_7$
CA1	↑	↓	↑	↑
CA3	↑↑	↑	no effect	↑

↑: facilitation, ↓: inhibition.

Autoradiographic and *in situ* hybridization studies revealed that the CA3 field contains a high density of $5\text{-}HT_7$ receptors (Gustafson *et al.*, 1996; Heidmann *et al.*, 1998). A recent study (Bacon and Beck, 2000) reported that the reduction of the slow after hyperpolarization in the CA3 pyramidal cells was mediated via $5\text{-}HT_7$ receptors. In the present experiment, fluvoxamine-induced facilitation was prevented by two $5\text{-}HT_7$ receptor antagonists, DR 4004 (Kikuchi *et al.*, 1999) and SB 269970 (Hagan *et al.*, 2000), in the CA1 and the CA3 subfields. The precise role of $5\text{-}HT_7$ receptors in the hippocampus has not yet been clarified; however, the present finding that $5\text{-}HT_7$ receptors contribute to endogenous 5-HT-mediated synaptic transmission in both fields indicates the possibility that the $5\text{-}HT_7$ receptor subtype plays a functional role in the hippocampus. The different magnitude of synaptic efficacy produced by fluvoxamine in each field, therefore, may result from the differential mediation via these 5-HT receptor subtypes (Table 2).

CLINICAL IMPLICATIONS AND FUTURE DIRECTIONS

The present study revealed the different properties of endogenous 5-HT-mediated synaptic transmission in the CA1 and CA3 fields. In the CA1 field, $5\text{-}HT_{1A}$, $5\text{-}HT_4$ and $5\text{-}HT_7$ receptors are associated with the synaptic effects mediated by 5-HT in an opposite and independent manner. In the CA3 field, both $5\text{-}HT_{1A}$ and $5\text{-}HT_7$ receptors are attributable to the 5-HT mediated neurotransmission in a stimulatory manner. The regional difference in 5-HT receptor mechanisms observed in the pyramidal neurons may be significant to regulate the responses to incoming stimuli by a flexible machinery, which may play a functional role in the hippocampus.

The hippocampus is thought to be an essential region in associative fear memories evoked by contextual stimuli (Kim and Fanselow, 1992). LTP formation elicited by tetanic stimulation, which is proposed to be the electrophysiological basis of memory, was prevented by fluvoxamine pretreatment in the CA1 field. This suppression was also observed after administration of $5\text{-}HT_{1A}$ receptor agonist tandospirone. In the CA1 field, the NMDA component is thought to be essential for LTP formation (Bliss and Collingridge, 1993). Pugliese *et al.* (1998) reported that 5-HT decrease the NMDA components of the excitatory postsynaptic potentials (e.p.s.p.) through $5\text{-}HT_{1A}$ receptor-mediated action. Thus, increases in potassium conductance and hyperpolarization mediated by 5-HT$_{1A}$ receptors reduce the synaptic excitability and ultimately may participate in the blockade of LTP in the CA1 field. There has been increasing evidence that the cAMP-dependent mechanism plays a key role in LTP formation in the CA3 field (Weisscopf *et al.*, 1994; Villacres *et al.*, 1998). In this field, LTP was induced in the presence of either fluvoxamine or tandospirone. Interestingly, not only LTP but also 'LTP-like' response, i.e. an increased magnitude of the PS amplitude without tetanic stimulation was observed after fluvoxamine or tandospirone administration. This may indicate the possibility that the facilitation of neurotransmission mediated via 5-HT$_{1A}$ receptors is attributable to the cAMP-dependent processes in the CA3 field. Together, our findings that the effects of fluvoxamine on hippocampal LTP formation were mimicked by tandospirone suggest that the 5-HT$_{1A}$ receptors are predominantly responsible for the mechanism of the anxiolytic property of fluvoxamine.

We have recently reported that repetitive treatments with fluvoxamine enhanced the synaptic plasticity of the hippocampo-medial prefrontal pathway (Ohashi *et al.*, 2002). However, it is still uncertain whether the repetitive fluvoxamine would affect neurotransmission in the pyramidal cell regions of the hippocampus and, if so, what a 5-HT receptor subtype is responsible for. These are great concerns to be solved, from the viewpoint of the implication of SSRI-induced synaptic effects and their therapeutic effects on psychiatric disorders.

In conclusion, 5-HTergic neurotransmission in the pyramidal cells is regulated via 5-HT$_{1A}$, 5-HT$_4$ and 5-HT$_7$ receptors in a region-specific manner. The regional difference in 5-HT receptor-mediated neurotransmission may play a significant role in the hippocampal functions, including the expression of fear/anxiety and emotional memory processing, and may be partly responsible for anxiolytic effects of SSRIs.

Acknowledgements

The authors thank various drug companies for donations as follows: Fluvoxamine maleate and DR 4004 (2a-[4-(4-phenyl-1,2,3,6-tetrahydropyridyl)butyl]-2a,3,4,5-tetrahydro-benzo [*cd*]indol-2(1*H*)-one) by Meiji Seika Kaisha, Ltd. (Yokohama, Japan); Tandospirone citrate ((1R*, 2S*, 3R*, 4S*)-N-[4-[4-(2-pyrimidinyl)-1-piperazinyl] butyl]-2,3-bicyclo [2.2.1] heptanedicarboximide dihydrogen citrate) by Sumitomo, Co., Ltd. (Tokyo, Japan); SC 53116 ([1-(S)-1-exo-4-amino-5-chloro-N[(hexahydro-1H-pyrrolizin-1-yl)methyl]-2-methoxy-benzamide hydrochloride, monohydrate]) by G.D. Searle and Co., Ltd. (IL, USA); GR 113808 ([1-[2-(methyl sulphonylamino)ethyl]-4-piperidinyl]methyl-1-methyl-1H-indole-3-carboxylate maleate salt) by Glaxo Wellcome Co., Ltd. (Greenford, UK).

REFERENCES

Bacon, W. L. and Beck, S. G. (2000). 5-hydroxytryptamine$_7$ receptor activation decreases slow afterhyperpolarization amplitude in CA3 hippocampal pyramidal cells, *J. Pharmacol. Exp. Ther.* **294**, 672–679.

Barbaccia, M. L., Brunello, N., Chuang, D. M., *et al.* (1983). Serotonin-elicited amplification of adenylate cyclase activity in hippocampal membrans from adult rat, *J. Neurochem.* **40**, 1671–1679.

Beck, S. G., Choi, K. C. and List, T. J. (1992). Comparison of 5-hydroxytryptamine$_{1A}$-mediated hyperpolarization in CA1 and CA3 hippocampal pyramidal cells, *J. Pharmacol. Exp. Ther.* **263**, 350–359.

Bender, E., Pindon, A., Van Oers, I., *et al.* (2000). Structure of the human serotonin 5-HT$_4$ receptor gene and cloning of a novel 5-HT$_4$ splice variant, *J. Neurochem.* **74**, 478–489.

Bijak, M., Tokarski, K. and Maj, J. (1997). Repeated treatment with antidepressant drugs induces subsensitivity to the excitatory effects of 5-HT$_4$ receptor activation in the rat hippocampus, *Naunyn-Schmiedeberg's Arch Pharmacol.* **355**, 14–19.

Blier, P., Lista, A. and De Montigny, C. (1993). Differential properties of pre- and postsynaptic 5-hydroxytryptamine$_{1A}$ receptors in the dorsal raphe and hippocampus: II. Effect of pertussis and cholera toxins, *J. Pharmacol. Exp. Ther.* **265**, 16–23.

Bliss, T. V. P. and Collingridge, G. L. (1993). A synaptic model of memory: Long-term potentiation in the hippocampus, *Nature* **361**, 31–39.

Colino, A. and Halliwell, J. V. (1987). Differential modulation of three separate K-conductances in hippocampal CA1 neurons by serotonin, *Nature* **328**, 73–77.

Figgitt, D. P. and McClellan, K. J. (2000). Fluvoxamine. An updated review of its use in the management of adults with anxiety disorders, *Drugs* **60**, 925–954.

Flynn, D. L., Zabrawski, D. L., Becker, D. P., *et al.* (1992). SC-53116: The first selective agonist at the newly identified serotonin 5-HT$_4$ receptor subtype, *J. Med. Chem.* **35**, 1486–1489.

Gustafson, E. L., Durkin, M. M., Bard, J. A., *et al.* (1996). A receptor autoradiographic and in situ hybridization analysis of the distribution of the 5-HT$_7$ receptor in rat brain, *Br. J. Pharmacol.* **117**, 657–666.

Hagan, J. J., Price, G. W., Jeffrey, P., *et al.* (2000). Characterization of SB-269970-A, a selective 5-HT (7) receptor antagonist, *Br. J. Pharmacol.* **130**, 539–548.

Heidmann, D. E., Szot, P., Kohen, R., *et al.* (1998). Function and distribution of three rat 5-hydroxytryptamine 7 (5-HT7) receptor isoforms produced by alternative splicing, *Neuropharmacol.* **37**, 1621–1632.

Kikuchi, C., Nagaso, H., Hiranuma, T., *et al.* (1999). Tetrahydrobenzindoles: selective antagonists of the 5-HT$_7$ receptor, *J. Med. Chem.* **42**, 533–535.

Kim, J. J. and Fanselow, M. S. (1992). Modality-specific retrograde amnesia of fear, *Science* **256**, 675–677.

Kishimoto, K., Koyama, S. and Akaike, N. (2000). Presynaptic modulation of synaptic gamma-aminobutyric acid transmission by tandospirone in rat basolateral amygdala, *Eur. J. Pharmacol.* **407**, 257–265.

Kojima, T., Matsumoto, M., Tachibana, K., *et al.* (2003). Fluvoxamine suppresses the long-term potentiation in the hippocampal CA1 field of anesthetized rats: An effect mediated via 5HT$_{1A}$ receptors, *Brain Res.* **956**, 165–168.

Markstein, R., Matsumoto, M., Kohler, C., *et al.* (1999). Pharmacological characterisation of 5-HT receptors positively coupled to adenylyl cyclase in the rat hippocampus, *Naunyn-Schmiedeberg's Arch. Pharmacol.* **359**, 454–459.

Matsumoto, M., Kojima, T., Togashi, H., *et al.* (2002). Differential characteristics of endogenous serotonin-mediated synaptic transmission in the hippocampal CA1 and CA3 fields of anaesthetized rats, *Naunyn-Schmiedeberg's Arch. Pharmacol.* **366**, 570–577.

Matsuyama, S., Nei, K. and Tanaka, C. (1997). Regulation of GABA release via NMDA and 5-HT$_{1A}$ receptors in guinea pig dentate gyrus, *Brain Res.* **761**, 105–112.

Ohashi, S., Matsumoto, M., Otani, H., *et al.* (2002). Changes in synaptic plasticity in the rat hippocampo-medial prefrontal cortex pathway induced by repeated treatments with fluvoxamine, *Brain Res.* **949**, 131–138.

Okuhara, D. Y. and Beck, S. G. (1994). 5-HT$_{1A}$ receptor linked to inward-rectifying potassium current in hippocampal CA3 pyramidal cells, *J. Neurophysiol.* **71**, 2161–2164.

Pazos, A. and Palacios, J. M. (1985). Quantitative autoradiographic mapping of serotonin receptors in the rat brain. I. Serotonin-1 receptors, *Brain Res.* **346**, 205–230.

Pugliese, A. M., Passani, M. B. and Corradetti, R. (1998). Effect of the selective 5-HT$_{1A}$ receptor antagonist WAY 100635 on the inhibition of e.p.s.ps produced by 5-HT in the CA1 region of rat hippocampal slices, *Br. J. Pharmacol.* **124**, 93–100.

Raymond, J. R., Mukhin, Y. V., Gettys, T. W., *et al.* (1999). The recombinant 5-HT$_{1A}$ receptor: G protein coupling and signalling pathways, *Br. J. Pharmacol.* **127**, 1751–1764.

Roychowdhury, S., Haas, H. and Anderson, E. G. (1994). 5-HT$_{1A}$ and 5-HT$_4$ receptor colocalization on hippocampal pyramidal cells, *Neuropharmacology* **33**, 551–557.

Schmitz, D., Empsom, R. M. and Heinemann, U. (1995). Serotonin reduces inhibition via 5-HT$_{1A}$ receptors in area CA1 of rat hippocampal slices in vitro, *J. Neurosci.* **15**, 7217–7225.

Schreiber, R., Melon, C. and De Vrt, J. (1998). The role of 5-HT receptor subtypes in the anxiolytic effects of selective serotonin re-uptake inhibitors in the rat ultrasonic vocalization test, *Psychopharmacol.* **135**, 383–391.

Shenker, A., Maayani, S., Weinstein, H., *et al.* (1985). Two 5-HT receptors linked to adenylate cyclase in guinea pig hippocampus are discriminated by 5-carboxamidotryptamine and spiperone, *Eur. J. Pharmacol.* **109**, 427–429.

Tao-Cheng, J. H. and Zhou, F. C. (1999). Differential polarization of serotonin transporters in axons versus soma-dendrites: an immunogold electron microscopy study, *Neuroscience* **94**, 821–830.

Vida, I. and Frotscher, M. (2000). A hippocampal interneuron associated with the mossy fiber system, *Nerobiology* **97**, 1275–1280.

Villacres, E. C., Wong, S. T., Chavkin, C., *et al.* (1998). Type I adenylyl cyclase mutant mice have impaired mossy fiber long-term potentiation, *J. Neurosci.* **18**, 3186–3194.

Vogt, K. E. and Nicoll, R. A. (1999). Glutamate and gamma-aminobutyric acid mediate a heterosynaptic depression at mossy fiber synapses in the hippocampus, *Proc. Natl. Acad. Sci.* **96**, 1118–1122.

Weisskopf, M. G., Castillo, P. E., Zalutsky, R. A., *et al.* (1994). Mediation of hippocampal mossy fiber long-term potentiation by cyclic AMP, *Science* **265**, 1878–1882.

Welner, S. A., De Montigny, C., Desroches, J., *et al.* (1989). Autoradiographic quantification of serotonin$_{1A}$ receptors in rat brain following antidepressant drug treatment, *Synapse* **4**, 347–352.

Yoshioka, M., Matsumoto, M., Togashi, H., *et al.* (1995). Effects of conditioned fear stress on 5-HT release in the rat prefrontal cortex, *Pharmacol. Biochem. Behav.* **51**, 515–519.

Advances in Neuroregulation and Neuroprotection (2005), pp. 615-633
Collin, C. *et al.* (Eds)
© VSP 2005

The involvement of 5-HT$_3$ and 5-HT$_4$ receptors in anticancer drug-induced emesis

MASARU MINAMI [1,*], TORU ENDO [1], MASAHIKO HIRAFUJI [1], TSUTOMU HIROSHIGE [1], EMIKO FUKUSHI [1], YOUNGNAM KANG [1], HIDEYA SAITO [2] and HASAN S. PARVEZ [3]

[1] *Research Institute of Personalized Health Sciences, Health Sciences University of Hokkaido, Ishikari-Tobetsu, Hokkaido 061-0293, Japan*
[2] *Department of Basic Sciences, Japanese Red Cross Hokkaido College of Nursing, Kitami, Hokkaido 090-0011, Japan*
[3] *Neuroendocrinologie de Neuropharmacologie du Développement, Institut Alfred Fessard, Bât 5, Parc Chateau CNRS, 91190 Gif Sur Yvette, France*

Abstract—Emesis caused by cytotoxic drugs is associated with an increase in the concentration of serotonin (5-hydroxytryptamine: 5-HT) in the intestine and the brainstem. 5-HT receptors in the central nervous system (CNS) and the gut participate in the induction of emesis. 5-HT$_1$ receptors are negatively coupled to adenylate cyclase, but the 5-HT$_{2C}$ subtype is linked to phospholipase (PL) C activation. The 5-HT$_4$ receptor, by contrast, is positively coupled to adenylate cyclase. 5-HT$_{1A}$ and 5-HT$_{2A}$/5-HT$_{2C}$ agonists exhibit antiemetic properties. The 5-HT$_{1D}$ agonist reduces the emesis accompanying migraine headaches. Through use of 5-HT$_3$ receptor antagonists, which selectively antagonize 5-HT$_3$ receptors on the abdominal vagal afferent fibers, as much as 60% of the initial emetic response can be prevented. Inhibition of acute emesis also appears to be attained by blocking the initiation of the emetic reflex induced via 5-HT$_3$ receptors by 5-HT released from enterochromaffin (EC) cells in the small intestine. Although 5-HT$_4$ receptor antagonists have an antiemetic activity in some experimental models, the precise role of 5-HT$_4$ receptors has not been fully elucidated. The present review aims to compare the involvement of 5-HT$_4$ receptors with that of the 5-HT$_3$ receptors in anticancer drug-induced emesis.

Keywords: 5-HT$_3$ receptors; 5-HT$_4$ receptors; anticancer drug-induced emesis; 5-HT; vagal afferent nerve.

INTRODUCTION

The gut is estimated to contain more than 80% of the 5-HT in the body, and of this 90% is located within the EC cells (Erspamer, 1966). When the gastrointestinal

---

*To whom correspondence should be addressed. E-mail: minami@hoku-iryo-u.ac.jp

(GI) tract is totally resected, the urinary excretion of 5-hydroxyindole acetic acid (5-HIAA), a 5-HT metabolite, is abolished (Bertaccini, 1960). The time course of the increase in plasma chromogranin A, which is localized in the EC storage granules with 5-HT, matched that of emesis and of urinary 5-HIAA excretion (Cubeddu et al., 1995). These findings indicate that urinary 5-HIAA excretion provides an index of 5-HT release from the gastrointestinal tract (Cubeddu et al., 1992). With regard to the concentration of 5-HT and emesis, a cisplatin-induced increase in urinary 5-HIAA concentration paralleled the development of emesis in cancer patients (Cubeddu et al., 1992). Cisplatin-induced emesis in ferrets is also associated with an increase in the concentration of 5-HT in the intestine (Endo et al., 1990) and the brainstem (Endo et al., 1992). Cubeddu et al. (1992) reported that platelet and plasma 5-HT levels failed to increase in cyclophosphamide-treated patients, whereas chronic administration of cyclophosphamide significantly increased urinary 5-HIAA excretion in rats (Minami et al., 1997a).

Because of the importance of 5-HT derived from the EC cells, the effects of many substances on 5-HT release and their mechanisms have been investigated extensively. Numerous pharmacological approaches using receptor agonists or antagonists have revealed that 5-HT release from the EC cells is triggered or modulated via multiple autoreceptors or heteroreceptors present on the EC cells (Racké and Schwörer, 1991; Hirafuji et al., 2001). It is now widely accepted that the stimulation of 5-HT$_3$ receptors plays a pivotal role in the acute emesis induced by anticancer drugs (Andrews and Davis, 1995). Even so, the stimulus-secretion coupling and the cellular signaling pathways in the EC cells are still unclear. Using the vascularly perfused small intestine of guinea-pigs, Gebauer et al. (1993) reported that stimulation of 5-HT$_4$ receptors caused inhibition of 5-HT release. On the contrary, in the rat and ferret ileum dissected according to Milano's method (1991), 5-methoxytryptamine (5-MT), a 5-HT$_4$ receptor agonist, induced a concentration-dependent increase of 5-HT (Minami et al., 1995a). Thus, the mechanisms of receptor-mediated release of 5-HT from the EC cells are complicated. These discrepancies could be explained by a combination of the above mentioned factors including species difference.

Emesis is an instinctive defense mechanism caused by the somato-autonomic nerve reflex that is integrated in the medulla oblongata. The increased 5-HT induced by paraneuron (such as the EC cells) stimulation may release vasoactive intestinal polypeptides (VIP)s and diarrhea may occur (Andrews et al., 1990). Simultaneously, 5-HT may stimulate the 5-HT$_3$ receptors on the adjacent vagal afferent nerves. This vagal afferent nerve depolarization may evoke the vomiting reflex. The abdominal vagus nerve is the major afferent pathway involved in the detection of emetic stimuli (Andrews et al., 1990). We previously reported that cisplatin or copper sulfate increased abdominal afferent vagal nerve activity in a time course parallel with vomiting in ferrets (Endo et al., 1995). The administration of 5-HT$_3$ receptor antagonists suppresses both vagal nerve activity and vomiting. In the peripheral nervous system, the myenteric nerves are a therapeutic target for benzamide 5-HT$_4$ receptor agonists (Bockaert *et al.*, 1994). These 5-HT$_4$ receptor agonists act as gastroprokinetic drugs, increasing gastric peristalsis (Bockaert *et al.*, 1992). Both the mechanisms of action of 5-HT$_4$ receptor agonists and antagonists as well as the distribution of 5-HT$_4$ receptors may be useful therapeutic targets for the treatment of emesis. Recently, we demonstrated that N-3389, a new 5-HT$_3$ and 5-HT$_4$ receptor antagonist, significantly reduced cisplatin-induced emesis in parallel with the reduction of increased cisplatin-induced afferent vagus nerve activity in another group of ferrets (Minami *et al.*, 1997b).

This review compares the role of 5-HT$_4$ receptors with that of 5-HT$_3$ receptors on anticancer drug-induced emesis from the viewpoint of 5-HT release and vagus nerve activity.

5-HT RECEPTOR RELATED EMESIS

As a neurotransmitter and neuromodulator, 5-HT is involved in many physiological functions and pathological disorders including emesis. Additional 5-HT subtypes have been described using molecular techniques. Most 5-HT receptors belong to the G-protein-coupled receptor family (5-HT$_1$, 5-HT$_2$ and 5-HT$_4$ receptors), whereas 5-HT$_3$ receptors are a member of the ligand-gated ion-channel receptor family. 5-HT$_1$ receptors are negatively coupled to adenylate cyclase. The 5-HT$_4$ receptors, by contrast, are positively coupled to adenylate cyclase. Most 5-HT receptors have now been cloned, but their physiological roles are not completely understood (Raymond *et al.*, 2001).

5-HT receptors in the CNS and gut participate in the induction of emesis. 5-HT$_{1A}$ and 5-HT$_{2A}$/5-HT$_{2C}$ agonists exhibit antiemetic properties (Lucot and Crampton, 1989; Wolff and Leander, 1994). The 5-HT$_{1D}$ receptor agonist reduces the emesis accompanying migraine headaches (Ladabaum and Hasler, 1999; Cipolla *et al.*, 2001). Paradoxically, sumatriptan is reported both to relieve the nausea of a migraine attack and to have nausea as a side effect (Cipolla *et al.*, 2001). 5-HT$_{1D}$ receptors can inhibit 5-HT release in several brain regions. 5-HT efflux in the ventral lateral geniculate nucleus of the rat is modulated by presynaptic 5-HT$_{1D}$ receptors (Davidson and Stamford, 1996). Cloned 5-HT$_{1D}$ receptors transfected into C$_6$ glial cells promote cell growth, so there is little doubt that 5-HT$_{1D}$ receptor can be mitogenic (Pauwels *et al.*, 1996). Various selective 5-HT$_3$ receptor antagonists have been developed after the 5-HT receptors were classified by Bradley *et al.* (1986). There is now much evidence to confirm the pharmacological role of 5-HT$_3$ receptors in the control of anticancer drug-induced emesis (Costall *et al.*, 1987; Higgins *et al.*, 1989). Selective 5-HT$_3$ antagonists are a major advance in the treatment of chemotherapy- and radiotherapy-induced emesis in cancer patients. 5-HT$_4$ receptor antagonists have an antiemetic activity in some experimental models (Fukui *et al.*, 1994; Horikoshi *et al.*, 2001). Out of the seven recognized classes of 5-HT receptors (5-HT$_1$-5-HT$_7$), those involved in radiation- or chemotherapy-induced emesis belong principally to the 5-HT$_3$ receptors (Hasler, 1999), although it has been claimed that 5-HT$_4$ receptors participate in this process (Bhandari and Andrews, 1991).

5-HT$_4$ RECEPTOR INVOLVEMENT IN EMESIS

5-HT$_4$ receptors, which are positively coupled to adenylate cyclase (AC), were first analyzed in cell cultures of mouse embryo colliculi (Ford and Clark, 1993). Neural and non-neural 5-HT$_4$ receptors have been demonstrated in the GI tract. Non-neural receptors, which are located on excitatory pathways where they evoke transmitter release, are those involved in the GI prokinetic action of benzamide such as cisapride and zacopride. Zacopride, although it inhibits cisplatin-induced emesis, is emetic when given alone. This effect was originally described as 5-HT$_4$ receptor stimulation (Bhandari and Andrews, 1991).

The emesis induced by ipecac syrup

Ipecac syrup, prepared from galenical ipecac, contains the nauseate alkaloids cephaeline and emetine. Hasegawa *et al.* (2002) investigated the involvement of receptors and serotonin- and dopamine-metabolizing enzymes in the emesis induced by ipecac syrup (TJN-119). In the receptor binding assays of TJN-119, cephaeline and emetine had a distinct affinity to 5-HT$_4$ receptors, but displayed either weak or no affinity to 5-HT$_{1A}$, 5-HT$_3$, nicotine, M$_3$, beta$_1$, NK$_1$ and D$_2$ receptors. In ferrets, the selective 5-HT$_3$ receptor antagonist ondansetron prevented each emesis induced by TJN-119, cephaeline and emetine, but the intraperitoneal administration of the selective D$_2$-receptor antagonist sulpiride failed to significantly suppress the TJN-119-induced emesis. Cephaeline and emetine did not affect the metabolic enzyme activity of 5-HT and dopamine in the same way as monoamine oxidase (MAO)-A, MAO-B, tryptophan 5-hydroxylase (TPH) and tyrosine hydroxylase *in vitro*. Oral administration of TJN-119, cephaeline or emetine produced a significant increase in afferent abdominal vagus nerve activity which lasted approximately 1 h (Endo *et al.*, 2000). The maximum response induced by TJN-119 was estimated to be 219% of the pre-injection level. TJN-119 increased the 5-HT content in the ileum but not in the area postrema, suggesting that a possible mechanism for ipecac syrup may be in action at peripheral sites. These results obtained by Hasegawa *et al.* (2002) and Endo *et al.* (2000) suggest that 5-HT$_4$ receptors may be involved in the emetic action of TJN-119.

Emesis induced by a high dose of cisplatin

The emesis induced by a high dose of cisplatin (50 mg/kg) in the *suncus murinus* was reduced by oral pretreatment with tropisetron, which is known as a 5-HT$_3$- and 5-HT$_4$-receptor dual antagonist, with the ID$_{50}$ value of 0.52 mg/kg (Horikoshi *et al.*, 2001). Granisetron, a selective 5-HT$_3$ receptor antagonist, on the contrary, did not inhibit the emesis at up to 30 mg/kg. GR125487, a selective 5-HT₄-receptor antagonist, also did not inhibit the emesis. However, co-administration of GR125487 and granisetron significantly reduced the number of emetic episodes, suggesting that both the 5-HT₃ and 5-HT₄ receptors are involved in the emesis induced by a high dose of cisplatin.

Copper sulfate-induced vomiting

The vomiting induced by copper sulfate (100 mg/kg) was inhibited by blocking 5-HT₄ receptors with high doses (1 and 3 mg/kg, i.v.) of ICS205-930. On the other hand, blocking 5-HT₃ receptors with MDL72222 (0.5 and 5 mg/kg, i.v.) or low doses (0.01 mg/kg, i.v.) of ICS205-930 had no apparent effect on the vomiting induced by copper sulfate in beagle dogs (Fukui *et al.*, 1994). We previously reported that copper sulfate at a dose of 20 mg/kg (3/6) and 40 mg/kg (6/6) produced a dose-related increase in the number of emetic episodes in ferrets (Endo *et al.*, 1991). N-3389, a newly synthesized 5-HT₃ and 5-HT₄ receptor antagonist, halved the number of retching and vomiting episodes induced by copper sulfate in ferrets (Minami *et al.*, 1997b). Oral administration of the 5-HT₄ receptor agonist 5-MT caused vomiting, and the vomiting was inhibited by abdominal visceral nerve section or a high dose (1 mg/kg) of ICS205-930 but not by a low dose (0.1 mg/kg). These results suggest that peripheral 5-HT₄ receptors play an important role in the vomiting induced by oral administration of copper sulfate in dogs.

Zacopride-induced emesis

Bhandari and Andrews (1991) demonstrated that zacopride-induced emesis can be blocked by a high dose (1 mg/kg) but not by a low dose (0.1 mg/kg) of ICS205-930 or high doses (1 mg/kg) of another more selective 5-HT₃ receptor antagonist, granisetron. As high doses of ICS205-930 are reported to be a 5-HT₄ receptor antagonist, it appears likely that activation of the 5-HT₄ receptors contributes to the emesis induced by zacopride. High doses of ICS205-930, but not granisetron or ondansetron, can also block the vagally mediated emesis induced by oral copper sulfate, suggesting that the 5-HT₄ receptors involved in emesis are closely associated with abdominal vagal afferents.

ROLE OF 5-HT IN EMESIS

5-HT in the gastrointestinal tract

Over 80% of 5-HT in the body localizes in the gut and 90% of this is derived from the EC cells (Erspamer, 1966). 5-HT is synthesized by tryptophan hydroxylase (TPH), which is a rate-limiting enzyme for 5-HT biosynthesis at the endoplasmic reticulum, localized at the lumen site of the EC cells and the Golgi apparatus (Nilsson *et al.*, 1985). Many secretory granules, which are predominantly located in the supranuclear region (Golgi area), move towards the base of the EC cells and are released from the basal surface (Minami *et al.*, 1995b). Cisplatin (Gunning *et al.*, 1987; Endo *et al.*, 1990) or copper sulfate (Endo *et al.*, 1991) administration increased the level of 5-HT in the ileal tissue, but not in the gastric mucosa in ferrets. What is the functional role of the 5-HT in the EC cells? The presence of 5-HT might be an indication of defensive or protective reactions by the body to the outer environment (Andrews, 1991; Sanger, 1992). The vomiting or diarrhea caused by 5-HT seems to be a defense reaction, as both purge ingested toxic substances from the body. Afferent vagal nerve terminals are distributed in the areas adjacent to the EC cells. Therefore, the EC cell-vagal afferent nerve unit appears to be the anatomical mucosal chemoreceptor of the GI tract (Newson *et al.*, 1982). Cisplatin-administered ferrets showed increases in ileal 5-HT, 5-HIAA and norepinephrine (NE) concentrations, but no increases in ileal dopamine (DA) levels were observed (Gunning *et al.*, 1987; Endo *et al.*, 1990). Under the background of this 5-HT increase, cisplatin- and cyclophosphamide-administered ferrets demonstrate significant increases in TPH activity in the ileum (Endo *et al.*, 1993). Aromatic L-amino acid decarboxylase (AADC) activity also increased after the administration of these drugs. On the other hand, cytotoxic drugs induced a significant decrease in MAO activity in the ileum (Endo *et al.*, 1993). Thus, cytotoxic drugs may activate ileal 5-HT biosynthesis and further enhance the 5-HT concentration by reducing its degradation. Through this mechanism, the life span of 5-HT might be prolonged.

5-HT in the brainstem

Stimulation of afferent vagal nerve fibers appears to be related to increased levels of 5-HT in the brainstem. In fact, the electrical stimulation of abdominal vagal nerves produced an increase of 5-HT levels in the area postrema (AP), including the nucleus tractus solitarii (NTS), and evoked vomiting in anesthetized animals. We found that TPH activity in the AP also increased significantly in cisplatin-treated ferrets as compared with that in non-drug control ferrets (Minami *et al.*, 1995b). Pretreatment with ondansetron, a 5-HT$_3$ receptor antagonist, or abdominal vagotomy significantly inhibited the cytotoxic drug-induced increase in the 5-HT concentration of the AP area (Endo *et al.*, 1992). Although platelets could release 5-HT during their passage through the AP, vagotomy clearly inhibited the increase of 5-HT in the AP. These findings suggest that the AP-5-HT increase induced by cisplatin is due to emetic stimuli originating in the vagal afferent nerves.

5-HT$_3$ and 5-HT$_4$ receptors in the brainstem

The central terminal of the abdominal vagal afferent nerves projects into the brainstem. The vagal afferent nerve itself does not release 5-HT, but activates the 5-hydroxytryptaminergic neurons. 5-HT is distributed at the dorsal vagal complex, AP, NTS and fibers or varicosity of the dorsal motor vagal nucleus of the brainstem

(Leslie, 1985). With regard to the sources of 5-HT in the AP, numerous possibilities exist: the neuronal cell bodies of the AP itself, or the vagal afferent fibers which terminate at the AP. Another major source of neuronal groups containing 5-HT in the brainstem is the raphe nucleus. The NTS receives input from the raphe nucleus. Kilpatrick *et al.* (1987) first reported specific 5-HT_3 receptor binding sites in membranes derived from different regions of rat brain. Waeber *et al.* (1988) described dense binding of [^{3}H] tropisetron in the NTS, the dorsal motor nucleus of the vagus nerve, and in the spinal trigeminal nucleus in the mouse. The most dense region of 5-HT_3 receptor binding in the human brain was reported within the dorso-medial NTS, in the area that corresponds to the subnucleus gelatinosus and not the AP (Reynolds *et al.*, 1989). The observation that the highest density of 5-HT_3 receptor binding sites is in the subnucleus gelatinosus of the NTS may support the idea that the site of vomiting center which was put forward by Waeber *et al.* (1988). A vagal terminal distribution of 5-HT_3 receptors is also supported by the observation that 5-HT_3 receptor mRNA cannot be detected in the dorsal vagal complex using *in situ* hybridization (Tecott *et al.*, 1993). The neurons within the brainstem were also excited by histamine, neurotensin, substance P and enkephalin (Carpenter *et al.*, 1983). It has been reported that substance P may be associated with the depolarization of the 5-HT_3 receptors and that neurokinin (NK) B may excite 5-HT_4 receptors (Ramirez *et al.*, 1994). Substance P induces physiological activity via the tachykinin NK_1 receptors. The non-peptide NK_1 receptor antagonist, CP-99,994 has proved to have broad-spectrum antiemetic activity against cytotoxics, central and peripheral emetic agonists and motion sickness (Wason *et al.*, 1995). We reported that the 2-methyl-5-HT-induced 5-HT release was significantly inhibited by administration of sendide, a peptide NK_1 receptor antagonist, or granisetron (Minami *et al.*, 1998). Furthermore, the change from pre-injection level of the afferent nerve activity induced by substance P was significantly reduced by pretreatment with either sendide or granisetron. Together with the findings of afferent vagal nerve activity and 5-HT release from the isolated ileum, our experimental results indicated a close relation between tachykinin NK_1 and 5-HT_3 receptors (Minami *et al.*, 1998, 2001).

MODULATION OF 5-HT RELEASE

5-HT release from the intestinal mucosa

It has been reported that total body irradiation by X-ray produced a decrease in the 5-HT contents of the intestinal mucosa of mice, rats (Pentila and Kormano, 1971) and ferrets. This decreased 5-HT content appears to be associated with the increased release of 5-HT from the intestinal mucosa. Actually, using electron microscopy, examination of the ileum in cisplatin-administered ferrets demonstrated that the granules (5-HT store) of the EC cells were vacant of 5-HT for 6 hours after cisplatin administration (Minami *et al.*, 1995b). The most probable mode of 5-HT release may depend upon calcium influx (Racké and Schwörer, 1991), although the mechanisms and receptors involved have not been fully clarified.

5-HT$_3$ receptors

2-Methyl-5-HT, a selective 5-HT$_3$ receptor agonist, which acts as a ligand-gated cation channel causing action potential (Albuquerque *et al.*, 1995), facilitates 5-HT release. 2-Methyl-5-HT also increases 5-HT release from incubated strips of pig and human small intestine (Schwörer and Ramadori, 1998). Tetrodotoxin (TTX) blocks the cisplatin-activated neuronal input to the EC cells but does not prevent the EC autoreceptor-modulation of 5-HT release in guinea-pig tissue (Racké and Schwörer, 1991; Gebauer *et al.*, 1993). Therefore, the 5-HT release initiated via activation of the 5-HT$_3$ receptors in the gut would be mostly derived from the 5-HT$_3$ receptors located on the EC cells. All of the 5-HT$_3$ receptor antagonists marketed at the present time, such as granisetron, azasetron, ondansetron and ramosetron, were shown to inhibit the 2-methyl-5-HT-induced 5-HT release from the isolated ileum of ferrets, although the degree of inhibition varied among the antagonists (Endo *et al.*, 1999). These results agree with previous reports on the vascularly perfused guinea pig small intestine (Gebauer *et al.*, 1993) and the rat isolated ileum (Minami *et al.*, 1995a).

Butler *et al.* (1990) reported that granisetron, tropisetron and MDL72222 inhibited 5-HT release from the ileum, whereas ondansetron did not. Endo *et al.* (1999) demonstrated that ondansetron does inhibit 2-methyl-5-HT-induced 5-HT release from the ferret ileum. When the concentration of ondansetron was increased from 10^{-7} M to 10^{-6} M, inhibition of 5-HT was reduced, indicating that the affinity for the EC cells or neuronal 5-HT$_3$ receptors may be shifted according to the dose of ondansetron. Ramosetron, which has a greater binding affinity to the 5-HT$_3$ receptors of the rat brain than any other antagonists (Miyata *et al.*, 1991), inhibited the 5-HT release from the ferret ileum at a relatively high concentration of 10^{-6} M (Endo *et al.*, 1999). According to the results of Ito *et al.* (1995), ramosetron (pA$_2$: 10.27 and 10.48) showed more potent 5-HT$_3$ receptor antagonistic activity than granisetron (pA$_2$: 9.44 and 9.15) or ondansetron (pA$_2$: 8.63 and 8.70) in the isolated vagus nerve and cerebral cortex of rats, respectively. These studies showing that ondansetron or ramosetron elicit different responses to 5-HT release in different species suggest the possibility of 5-HT$_3$ receptor subtypes in various central and peripheral tissues.

In addition to the human 5-HT$_{3A}$ receptor subunit which shows 84% amino acid identity with the murine short isoform (Miyake *et al.*, 1995), the human 5-HT$_{3B}$ receptor subunit was identified (Davies *et al.*, 1999). Furthermore, two human 5-HT$_{3A}$ splice variants were found in the brain as well as in the intestine (Bruss *et al.*, 1998). Differences in these pharmacological studies could be due to differences in the amino acid sequence of the 5-HT$_3$ subunit and/or to differences in the subunit composition of the receptor complex among the species and/or the tissue.

5-HT₄ receptors

The 5-HT$_4$ receptors prototypically stimulate AC when expressed endogenously in tissues (Dumuis *et al.*, 1988; Bockaert *et al.*, 1992). As would be expected for 5-HT$_4$ receptors whose primary effect is to increase cAMP levels, most of the effects of 5-HT$_4$ receptors are due to the activation of protein kinase A (PKA) (Raymond *et al.*, 2001). Therefore, activation of L-type Ca^{2+} channels (Kaumann, 1991; Ouadid *et al.*, 1992) is probably responsible for the prokinetic actions of 5-HT$_4$ receptors in the gut. Bley *et al.* (1994) failed to find an increase in tissue cAMP generated by 5-HT$_4$ receptor activation in the guinea pig ileum or the rat vagus nerve. However, others have reported that 5-HT$_4$ receptors are linked to AC in the guinea pig ileum (Galzin and Delahaya, 1997). In the vascularly perfused isolated guinea pig small intestine, several 5-HT$_4$ receptor agonists concentration-dependently inhibited 5-HT release in the presence of TTX, suggesting that the EC cells of the guinea pig ileum are also endowed with inhibitory 5-HT$_4$ autoreceptors (Gebauer *et al.*, 1993).

The selective 5-HT$_4$ receptor agonist 5-MT also reduces 5-HT release from incubated strips of pig and human small intestine in the presence of TTX (Schwörer and Ramadori, 1998). In the absence of TTX, 5-MT produced a concentration-dependent increase in 5-HT release from the isolated rat (Minami *et al.*, 1995a) and ferret (Endo *et al.*, 1998a) ileum. SB204070, a selective 5-HT$_4$ receptor antagonist, significantly inhibited the increase in 5-HT release induced by 5-MT (Endo *et al.*, 1998b).

In the human small intestine, the electrogenic ion transport stimulated by 5-HT is mediated by 5-HT$_4$ receptors, probably located at the mucosal level (Burleigh and Borman, 1993). 5-HT$_3$ receptors act dominantly on the visceral neurons, which project from the gut to the central nervous system, and 5-HT$_4$ receptors act dominantly in the motility and secretory functions of the gut (Sanger, 1996). Since 5-HT$_4$ receptors are expressed in several parts of the alimentary tract including the enteric neurons (Bockaert *et al.*, 1994), the stimulation of 5-HT$_4$ receptors on those neurons may facilitate 5-HT release via stimulatory neurotransmitters which act on the EC cells in the small intestine of the rat and ferret. SB204070 alone at higher concentrations (10^{-6} to 10^{-5} M) increased the 5-HT release (Endo *et al.*, 1998b).

These results suggest that endogenous 5-HT could activate 5-HT$_3$ and 5-HT$_4$ receptors on the EC cells of the ferret or rat ileum. An experiment using isolated vascularly perfused guinea-pig small intestine showed that cisplatin produced a dose-dependent increase in both 5-HT and 5-HIAA (Schwörer *et al.*, 1991). The cisplatin-induced 5-HT release was blocked by TTX, hexamethonium and/or scopolamine. These findings suggest that the cisplatin-induced EC cell effects may not be due to direct action, but may be due to indirect action via a cholinergic mechanism, such as the stimulation of the enteric neurons by cisplatin to release acetylcholine (ACh). This may produce an increase in 5-HT release from the EC cells (Racké and Schwörer, 1993).

The increase in 5-HT release from the rat ileum induced by anticancer drug cyclophosphamide (10^{-6} M) was significantly suppressed by concomitant administration with TTX (10^{-6} M), granisetron (10^{-7} M) or GR113808 (10^{-7} M), a selective 5-HT$_4$ receptor antagonist (Ogawa *et al.*, unpublished data). Combined administration of granisetron with GR113808 showed an additive inhibitory effect on 5-HT release induced by cyclophosphamide. Cyclophosphamide stimulated 5-HT$_4$ receptors on the serotonergic interneurons and induced 5-HT release from the serotonergic interneuron. Since TTX significantly inhibited the 5-MT-induced increase in 5-HT release but not the 2-methyl-5-HT, 5-HT$_4$ receptors act dominantly on the interneurons and 5-HT$_3$ receptors act dominantly on the EC cells to release 5-HT. Recently, attention has been focused on the functional heterogeneity of 5-HT$_4$S and 5-HT$_4$L as the splice variants of 5-HT$_4$ receptors (Bonhaus *et al.*, 1997).

Intracellular Ca^{2+} dynamics in the EC cells

The release of 5-HT from the EC cells is largely dependent on extracellular Ca^{2+} (Satoh *et al.*, 1995; Racké *et al.*, 1996). A digital imaging analysis of intracellular Ca^{2+} dynamics in the ileal epithelial cells including the EC cells has been successfully reported in mouse isolated ileal crypts (Satoh *et al.*, 1995). When EC cells were identified by immunohistochemical staining with anti-5-HT antibody followed by confocal microscopy, 0-4 EC cells were found to be dispersed in a crypt (Hirafuji *et al.*, 2000). It has been demonstrated that ATP, but not adenosine, induces an increase in the [Ca^{2+}]$_i$ of the epithelial cells including the EC cells in the crypt isolated from the mouse ileum (Satoh *et al.*, 1995). However, ATP and adenosine are reported to inhibit 5-HT release from porcine EC cells (Racké *et al.*, 1996). This discrepancy may indicate important species differences or receptor subtype heterogeneity. Electrical depolarization with high K$^+$ solution increases [Ca^{2+}]$_i$ probably by causing Ca^{2+} influx through L-type voltage-dependent Ca^{2+} channels on the EC cells from guinea-pig and human duodenal crypts (Lomax *et al.*, 1999), although it has no effect on the mouse ileal crypts (Satoh *et al.*, 1995). High K$^+$ solution and FRL64176, a L-type Ca^{2+} channel agonist, also cause an increase in 5-HT release from guinea-pig crypts (Lomax *et al.*, 1999). In strips of porcine small intestine, 5-HT release induced by high K$^+$ solution was not influenced by TTX, but was reduced by the L-type and N-type Ca^{2+} channel blockers (Racké and Schwörer, 1993). These results suggest that the EC cells possess multiple voltage-dependent Ca^{2+} channels. It is well known that the G-protein-linked increase in [Ca^{2+}]$_i$ plays an important role in the stimulus-secretion coupling in a variety of cell types (Zucker, 1996). BON cells are a human carcinoid cell line derived from a metastasis of a pancreatic carcinoid tumor of enterochromaffin cell origin (Evers *et al.*, 1994). It has been reported that Ca^{2+} ionophore A23187 and forskolin stimulate adenylate cyclase to increase cAMP. To a lesser extent, isoproterenol and bethanechol also stimulate 5-HT release from BON cells (Kim *et al.*, 2001). NE-induced Ca^{2+} mobilization in mouse EC cells was not influenced by any antagonists of 5-HT$_1$, 5-HT$_2$, 5-HT$_3$ or 5-HT$_4$ receptors, suggesting that endogenous 5-HT has no role in the NE-induced increase in $[Ca^{2+}]_i$ (Hirafuji, unpublished data). Furthermore, cisplatin and cyclophosphamide had no direct effect on the $[Ca^{2+}]_i$ of mouse crypt cells (Hirafuji, unpublished data).

ABDOMINAL VAGAL NERVE ACTIVITY

Vagal fibers, which play a pivotal role in the transmission of emetogenic stimuli to the CNS, have been shown to be depolarized not only by 5-HT₃, but also by 5-HT₄ receptor stimulation (Rhodes *et al.*, 1992). Unlike 5-HT₃ receptor-induced depolarization, the depolarization evoked in the rat isolated vagus nerve by 5-HT₄ receptor stimulation was small and prolonged (Rhodes *et al.*, 1992; Nemoto *et al.*, 2001).

In vivo *afferent abdominal vagal nerve activity*

It is proposed that anticancer drugs cause 5-HT release from the EC cells and that the released 5-HT stimulates the 5-HT receptors on the afferent vagal nerve fibers. The vomiting center receives input from the afferent discharges of the vagal nerve fibers which evoke an emetic reflex (Hawthorn *et al.*, 1988). 5-HT has a potent depolarization action on the vagal afferent nerve (Round and Wallis, 1986; Ireland and Tyers, 1987). The central terminal of the abdominal vagal afferent nerves projects into the brainstem. Approximately 85% of the vomiting induced by cisplatin was inhibited by abdominal vagotomy (Endo *et al.*, 1992). The abdomen contains approximately 80% of the vagal afferent fibers (Andrews, 1992). Thus, the vagus nerve is the major nerve involved in the detection of emetic stimuli induced by anticancer drugs. In animals, electrical stimulation of the abdominal vagal afferents is capable of inducing emesis in ferrets (Andrews and Davidson, 1990) and cats (Milano *et al.*, 1990). The increase of abdominal afferent vagal nerve activity coincided well with the mode of emetic response induced by these emetics (Endo *et al.*, 1995). The time course of cisplatin-induced emesis in another group of ferrets paralleled the changes in the afferent vagal nerve activity (Endo *et al.*, 1995). Thus, 5-HT-mediated excitation of the afferent fibers might be responsible for vomiting.

5-HT₃ and 5-HT₄ receptors

Intravenous injection of 5-HT or 2-methyl-5-HT, a selective 5-HT₃ receptor agonist, produced a dose-dependent increase in abdominal afferent vagal nerve activity in ferrets (Endo *et al.*, 1995). Cisplatin also produced a significant increase in afferent abdominal vagal nerve activity and this increase in afferent vagal nerve activity was significantly inhibited by the intravenous injection of 5-HT₃ receptor antagonists (Endo *et al.*, 1995). The excitatory response of the vagal afferents might be mediated by 5-HT via the 5-HT₃ receptors of the afferent vagal fibers (Fukui *et al.*, 1992; Endo *et al.*, 1995). The issue of the involvement of 5-HT₄ receptors in emesis is controversial. 5-MT failed to increase the discharge of vagal afferent fibers in rats *in vitro* (Tonini, 1995). 5-MT (i.p.) produced a dose-dependent increase in kaolin intake (pica) in rats (Tamakai *et al.*, 1996). Furthermore, a 5-MT (i.v.)-induced increase in abdominal afferent vagal nerve activity in anesthetized rats was inhibited by GR113808, a selective 5-HT$_4$ receptor antagonist, but not by granisetron, a selective 5-HT$_3$ receptor antagonist (Tamakai *et al.*, 1996). These results suggest that abdominal afferent vagal nerve activity might be increased by the administration of 5-HT$_4$ receptor agonists and inhibited by antagonists of the 5-HT$_4$ receptors.

5-HT$_3$ and 5-HT$_4$ receptors in in vitro *isolated vagus nerve measurement*

It has been reported that 5-HT-induced vagal nerve depolarization is associated with both the 5-HT$_3$ and 5-HT$_4$ receptors (Bley *et al.*, 1994; Coleman and Rhodes, 1995). The 5-HT-induced depolarization reaction of isolated cervical vagus nerves in rats, rabbits and ferrets has been reported previously (Ireland, 1987; Elliott and Wallis, 1990; Newberry *et al.*, 1992; Bentley and Banner, 1998). We demonstrated that the 5-HT-induced depolarization response of the isolated rat abdominal vagal nerves was approximately 170% higher in value that that observed in the cervical vagus nerves (Nemoto *et al.*, 2001). These reports suggest that the 5-HT-induced depolarization occurs mainly due to a reaction mediated by 5-HT$_3$ receptors (Peters *et al.*, 1995), and, at least in part, a reaction mediated by 5-HT$_4$ receptors. The depolarization mediated by the 5-HT$_4$ receptors, however, was difficult to observe because there is no selective 5-HT$_4$ receptor agonist (Rhodes *et al.*, 1992; Bley *et al.*, 1994). 5-MT was originally used as a 5-HT$_4$ receptor agonist, but it has been reported that this compound acts as both 5-HT$_{1A}$ and 5-HT$_7$ receptors (Eison *et al.*, 1992; Tsou *et al.*, 1994; Martin, 1998). For these reasons, numerous investigators have used the residual response of 5-HT-induced depolarization in the presence of 5-HT$_3$ antagonist to evaluate depolarization reactions via the 5-HT$_4$ receptors (Rhodes *et al.*, 1992; Bley *et al.*, 1994). Depolarization of the vagus nerves by 5-HT may be mediated predominantly by 5-HT$_3$ receptors, since in the rat vagus nerve the depolarization affecting 5-HT$_4$ receptors was 13–19% of the maximum evoked by 5-HT (Rhodes *et al.*, 1992).

CONCLUSION

Cytotoxic drugs initially act within the gut to produce histological changes, activation of the biosynthesis of 5-HT and an increase in the concentration of 5-HT in the GI tract. 5-HT is released from the EC cells of the intestinal mucosa, which in turn stimulates 5-HT$_3$ receptors on the vagal afferent fibers. This stimulation of the vagal afferent fibers results in an increase in 5-HT in the brainstem and leads to emesis. Since the cisplatin-induced 5-HT release is significantly inhibited by TTX, the possible involvement of interneurons in the cisplatin-induced mechanism was suggested regarding the ileal tissue. The fact that abdominal vagotomy inhibited 85%

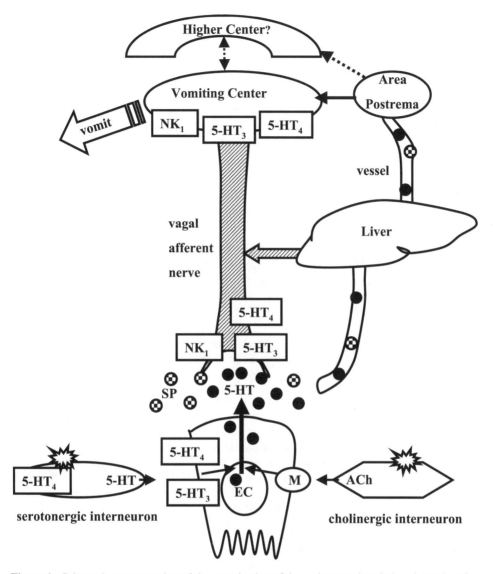

Figure 1. Schematic representation of the organization of the anticancer drug-induced emetic reflex involving 5-HT₃ and 5-HT₄ receptors. NK₁: tachykinin NK₁ receptors; M: muscarinic receptors; 5-HT 5-hydroxytryptamine (serotonin); ACh: acetylcholine; SP: substance P; E: enterochromaffin cells.

of the cisplatin-induced emesis indicates that abdominal vagal afferent nerves seem to be the major pathway for cisplatin-induced emesis. Pretreatment with 5-HT₃ receptor antagonists diminishes the brainstem 5-HT increase. The increased 5-HT level induced by anticancer drugs in the vomiting center might be inhibited by the chemical vagotomy caused by 5-HT₃ receptor antagonists. These studies confirmed that the cytotoxic drugs induced emesis mainly through action on the GI tract. The initial trigger of this pathway may be due to 5-HT release from the EC cells. The stimulation of 5-HT$_3$ and 5-HT$_4$ receptors on the EC cells and interneurons might be associated with 5-HT release from the EC cells.

5-HT$_4$ receptors exist in several parts of the alimentary tract. In the human small intestine, the electrogenic ion transport stimulated by 5-HT is mediated by 5-HT$_4$ receptors, probably located at the mucosal level. 5-HT$_3$ receptors act dominantly on the visceral neurons, which project from the gut to the CNS, and 5-HT$_4$ receptors act dominantly in the motility and secretory functions of the gut. The cisplatin-induced 5-HT release was blocked by TTX, hexamethonium and/or scopolamine. The cyclophosphamide-induced 5-HT release was blocked by TTX, granisetron and/or 5-HT$_4$ receptor antagonist. These findings suggest that the anticancer drug-induced EC cell effect may not be due to direct action, but may be due to indirect action via 5-HT$_4$ receptors in the cholinergic or serotonergic interneurons, such as the stimulation of the enteric neurons by anticancer drugs to release ACh or 5-HT. These studies suggest that ACh or 5-HT released from 5-HT$_4$ receptors in the enteric neurons stimulate 5-HT$_3$ receptors on the EC cells and on the vagal afferent nerves in anticancer drug-induced emesis (Fig. 1).

Acknowledgements

This study was partially supported the Academic Science Frontier Project of the Ministry of Education, Culture, Sports, Science and Technology, Japan.

REFERENCES

Albuquerque, E. C., Pereira, E. F., Castro, N. G., *et al.* (1995). Nicotinic receptor function in the mammalian central nervous system, *Ann. N.Y. Acad. Sci.* **757**, 48–72.

Andrews, P. L. R. (1991). Modulation of visceral afferent activity as therapeutic possibility for gastrointestinal disorders, in: *Irritable Bowel Syndrome*, Read, N. W. (Ed.), pp. 91–121. Blackwell Scientific, London.

Andrews, P. L. R. (1992). Physiology of nausea and vomiting, *Br. J. Anaesth.* **69** (Suppl.), 2–19.

Andrews, P. L. R. and Davidson, H. I. M. (1990). A method for the induction of emesis in the conscious ferret by abdominal vagal stimulation, *J. Physiol.* **422**, 5.

Andrews, P. L. R. and Davis, C. J. (1995). The physiology of emesis induced by anti-cancer therapy, in: *Serotonin and the Scientific Basis of Anti-emetic Therapy*, Reynolds, D. J. M., Andrews, P. L. R. and Davis, C. J. (Eds), pp. 25–49. Oxford Clinical Communications, Oxford.

Andrews, P. L. R., Davis, C. J., Bingham, S., *et al.* (1990). The abdominal visceral innervation and the emetic reflex: pathways, pharmacology and plasticity, *Can. J. Physiol. Pharmacol.* **68**, 325–345.

Bentley, K. R. and Barnes, N. M. (1998). 5-Hydroxytryptamine$_3$ (5-HT$_3$) receptor-mediated depolarization of the rat isolated vagus nerve: modulation by trichloroethanol and related alcohols, *Eur. J. Pharmacol.* **354**, 25–31.

Bertaccini, G. (1960). Tissue 5-hydroxytryptamine and urinary 5-hydroxyindole acetic acid after partial or total removal of the gastrointestinal tract in the rat, *J. Physiol.* **153**, 239–249.

Bhandari, P. and Andrews, P. L. (1991). Preliminary evidence for the involvement of the putative 5-HT$_4$ receptor in zacopride- and copper sulfate-induced vomiting in the ferret, *Eur. J. Pharmacol.* **204** (3), 273–280.

Bley, K. R., Eglen, R. M. and Wong, E. H. (1994). Characterization of 5-hydroxy-tryptamine-induced depolarizations in rat isolated vagus nerve, *Eur. J. Pharmacol.* **260**, 139–147.

Bockaert, J., Fozard, J. R., Dumuis, A., *et al.* (1992). The 5-HT$_4$ receptor — a place in the sun, *Trends Pharmacol. Sci.* **13**, 141–145.

Bockaert, J., Ansanay, H., Waeker, C., *et al.* (1994). 5-HT$_4$ receptors, *CNS Drugs* **1**, 6–15.

Bradley, P. B., Engel, G., Feniuk, W., *et al.* (1986). Proposal for the classification and nomenclature of functional receptors for 5-hydroxytryptamine, *Neuropharmacol.* **25**, 563–576.

Bruss, M., Göthert, M., Hayer, M., *et al.* (1998). Molecular cloning of alternatively spliced human 5-HT$_3$ receptor cDNAs, *Ann. N.Y. Acad. Sci.* **861**, 234–235.

Burleigh, D. E. and Borman, D. A. (1993). Short-circuit current responses to 5-hydroxytryptamine in human ileal mucosa are mediated by a 5-HT$_4$ receptor, *Eur. J. Pharmacol.* **241**, 125–128.

Butler, A., Elswood, C. J., Burridge, J., *et al.* (1990). The pharmacological characterization of 5-HT$_3$ receptors in these isolated preparations derived from guinea-pig tissues, *Br. J. Pharmacol.* **101**, 591–598.

Carpenter, D. O., Briggs, D. B. and Strominger, N. (1983). Responses of neurons of canine area postrema to neurotransmitters and peptides, *Cell. Mol. Neurobiol.* **3**, 113–126.

Cipolla, G., Sacco, S., Crema, F., *et al.* (2001). Gastric motor effects of triptans: open questions and further perspectives, *Pharmacol. Res.* **43** (3), 205–210.

Coleman, J. and Rhodes, K. F. (1995). Further characterization of the putative 5-HT$_4$ receptor mediating depolarization of the rat isolated vagus nerve, *Naunyn-Schmiedeberg's Arch. Pharmacol.* **352**, 74–78.

Costall, B., Domeney, A. M., Naylor, R. J., *et al.* (1987). Emesis induced by cisplatin in the ferret as a model for the detection of antiemetic drugs, *Neuropharmacology* **26**, 1321–1326.

Cubeddu, L. X., Hoffman, I. S., Fuenmayor, N. T., *et al.* (1992). Changes in serotonin metabolism in cancer patients: its relationship to nausea and vomiting induced by chemotherapeutic drugs, *Br. J. Cancer* **66** (1), 198–203.

Cubeddu, L. X., O'Connor, D. T., Hoffman, I., *et al.* (1995). Plasma chromogranin A marks emesis and serotonin release associated with dacarazine and nitorogen mustard but not with cyclophosphamide-based chemotherapies, *Br. J. Cancer* **72** (4), 1033–1038.

Davidson, C. and Stamford, J. A. (1996). Serotonin efflux in the rat ventral lateral geniculate nucleus assessed by fast cyclic voltammetry is modulated by 5-HT$_{1B}$ and 5-HT$_{1D}$ autoreceptors, *Neuropharmacology* **35**, 1627–1634.

Davies, P. A., Pistis, M., Hanna, M. C., *et al.* (1999). The 5-HT$_{3B}$ subunit is a major determinant of serotonin-receptor function, *Nature* **397**, 359–363.

Dumuis, A., Bouhelal, R., Sebben, M., *et al.* (1988). A 5-HT receptor in the central nervous system, positively coupled with adenylate cyclase, is antagonized by ICS 205-930, *Eur. J. Pharmacol.* **146**, 187–188.

Eison, A. S., Wright, R. N. and Freeman, R. (1992). Peripheral 5-carboxamidotryptamine induces hindlimb scratching by stimulating 5-HT$_{1A}$ receptors in rats, *Life Sci.* **51**, 95–99.

Elliott, P. and Wallis, D. I. (1990). Analysis of the actions of 5-hydroxytryptamine on the rabbit isolated vagus nerve, *Naunyn-Schmiedeberg's Arch. Pharmacol.* **341**, 494–502.

Endo, T., Minami, M., Monma, Y., *et al.* (1990). Effect of GR38032F on cisplatin- and cyclophosphamide-induced emesis in the ferret, *Biogenic Amines* **7**, 525–533.

Endo, T., Minami, M., Monma, Y., *et al.* (1991). Effect of ondansetron, a 5-HT$_3$ antagonist, on copper sulfate-induced emesis in the ferret, *Biogenic Amines* **8**, 79–86.

Endo, T., Minami, M., Monma, Y., *et al.* (1992). Vagotomy and ondansetron (5-HT$_3$ antagonist) inhibited the increase of serotonin concentration induced by cytotoxic drugs in the area postrema of ferrets, *Biogenic Amines* **9**, 163–175.

Endo, T., Takahashi, M., Minami, M., *et al.* (1993). Effects of anti-cancer drugs on enzyme activities and serotonin release from ileal tissue in ferrets, *Biogenic Amines* **9**, 479–489.

Endo, T., Nemoto, M., Minami, M., *et al.* (1995). Changes in the afferent abdominal vagal nerve activity induced by cisplatin and copper sulfate in the ferret, *Biogenic Amines* **11**, 399–407.

Endo, T., Ogawa, T., Hamaue, N., et al. (1998a). Granisetron, a 5-HT₃ receptor antagonist, inhibited cisplatin-induced 5-hydroxytryptamine release in the isolated ileum of ferrets, Res. Commun. Mol. Pathol. Pharmacol. **100**, 243–253.

Endo, T., Teramoto, Y., Hamaue, N., et al. (1998b). Effects of SB204070, a new 5-HT₄ receptor antagonist, on 5-hydroxytryptamine release from the isolated ileum of the ferret, Biogenic Amines **14**, 645–654.

Endo, T., Minami, M., Kitamura, N., et al. (1999). Effects of various 5-HT₃ receptor antagonists, granisetron, ondansetron, ramosetron and azasetron on serotonin (5-HT) release from the ferret isolated ileum, Res. Commun. Mol. Pathol. Pharmacol. **104**, 145–155.

Endo, T., Nemoto, M., Ogawa, T., et al. (2000). Pharmacological aspects of ipecac syrup (TJN-119)-induced emesis in ferrets, Res. Commun. Mol. Pathol. Pharmacol. **108** (3/4), 187–200.

Erspamer, V. (1966). Occurrence of indolalkyl amines in nature, in: Handbook of Experimental Pharmacology Vol. 19, Feichler, O. and Farah, A. (Eds), pp. 132–181. Springer, Berlin.

Evers, B. M., Ishizuka, J., Townsend, C. M., Jr., et al. (1994). The human carcinoid cell line, BON. A model system for the study of carcinoid tumors, Ann. N.Y. Acad. Sci. **733**, 393–406.

Ford, A. P. D. W. and Clark, D. E. (1993). The 5-HT₄ receptor, Med. Res. Rev. **13**, 633–662.

Fukui, H., Yamamoto, M., Sasaki, S., et al. (1994). Possible involvement of peripheral 5-HT₄ receptors in copper sulfate-induced vomiting in dogs, Eur. J. Pharmacol. **257** (1/2), 47–52.

Galzin, K. H. and Delahaya, M. (1992). Interaction of forskolin with the 5-HT₄ receptor mediated potentiation of twiches in guinea-pig ileum, in: Proc. 2nd Int. Symposium on Serotonin; From Cell Biology to Pharmacology and Therapeutics, p. 51.

Gebauer, A., Merger, M. and Kilbinger, H. (1993). Modulation by 5-HT₃ and 5-HT₄ receptors of the release of 5-hydroxytryptamine from the guinea-pig small intestine, Naunyn-Schmiedeberg's Arch. Pharmacol. **347**, 137–140.

Gunning, S. J., Hagan, R. M. and Tyers, M. B. (1987). Cisplatin induced biochemical and histological changes in the small intestine of the ferret, Br. J. Pharmacol. **90**, 135.

Hasegawa, M., Sasaki, T., Sadakane, K., et al. (2002). Studies for the emetic mechanisms of ipecac syrup (TJN-119) and its active components in ferrets: involvement of 5-hydroxytryptamine receptors, Jpn. J. Pharmacol. **89** (2), 113–119.

Hasler, W. L. (1999). Serotonin receptor physiology: relation to emesis, Dig. Dis. Sci. **44** (Suppl. 8), 108S–113S.

Hawthorn, J., Ostler, K. J. and Andrews, P. L. R. (1988). The role of the abdominal visceral innervation and 5-hydroxytryptamine M-receptors in vomiting induced by the cytotoxic drugs cyclophosphamide and cisplatin in the ferret, Quart. J. Exp. Physiol. **173**, 7–21.

Higgins, G. A., Kilpatrick, G. J., Bunce, K. T., et al. (1989). 5-HT₃ receptor antagonists injected into the area postrema inhibit cisplatin-induced emesis in the ferret, Br. J. Pharmacol. **97**, 247–255.

Hirafuji, M., Minami, M., Endo, T., et al. (2000). Intracellular regulatory mechanisms of 5-HT release from enterochromaffin cells in intestinal mucosa, Biogenic Amines **16**, 29–52.

Hirafuji, M., Ogawa, T., Kato, K., et al. (2001). Noradrenaline stimulates 5-hydroxytryptamine release from mouse ileal tissues via α₂-adrenoceptors, Eur. J. Pharmacol. **432**, 149–152.

Horikoshi, K., Yokoyama, T., Kishibayashi, N., et al. (2001). Possible involvement of 5-HT₄ receptors, in addition to 5-HT₃ receptors, in the emesis induced by high-dose cisplatin in Suncus murinus, Jpn. J. Pharmacol. **85** (1), 70–74.

Ireland, S. J. (1987). Origin of 5-hydroxytryptamine-induced hyperpolarization of the rat superior cervical ganglion and vagus nerve, Br. J. Pharmacol. **92**, 407–416.

Ireland, S. J. and Tyers, M. B. (1987). Pharmacological characterization of 5-hydroxytryptamine-induced depolarization of the rat isolated vagus nerve, Br. J. Pharmacol. **90**, 229–238.

Ito, H., Akuzawa, S., Tsutsumi, T., et al. (1995). Comparative study of affinities of the 5-HT₃ receptor antagonists, YM060, YM114 (KAE-393), granisetron and ondansetron in rat vagus nerve and cerebral cortex, Neuropharmacology **34**, 631–637.

Kaumann, A. J. (1991). 5-HT₄-like receptors in mammalian atria, *J. Neurol. Transm.* **34** (Suppl.), 195–201.

Kilpatrick, G. J., Jones, B. J. and Tyers, M. B. (1987). Identification and distribution of 5-HT₃ receptors in rat brain using radioligand binding, *Nature* **330**, 746–748.

Kim, M., Javad, N. H., Yu, J.-G., *et al.* (2001). Mechanical stimulation activates Gq signaling pathways and 5-hydroxytryptamine release from human carcinoid BON cells, *J. Clin. Invest.* **108**, 1051–1059.

Ladabaum, V. and Hasler, W. L. (1999). Novel approaches to the treatment of nausea and vomiting, *Dig. Dis.* **17** (3), 125–132.

Leslie, R. A. (1985). Neuroactive substances in the dorsal vagal complex of the medulla oblongata: Nucleus of the tractus solitarius, area postrema and dorsal motor nucleus of the vagus, *Neurochem. Int.* **7**, 191–211.

Lomax, R. B., Gallego, S., Novalbos, J., *et al.* (1999). L-Type calcium channels in enterochromaffin cells from guinea pig and human duodenal crypts: an in situ study, *Gastroenterology* **117**, 1363–1369.

Lucot, J. B. and Crampton, G. H. (1989). 8-OH-DPAT suppresses vomiting in the cat elicited by motion, cisplatin or xylazine, *Pharmacol. Biochem. Behav.* **33**, 627–631.

Martin, G. R. (1998). 5-Hydroxytryptamine receptors, in: *The IUPHAR Compendium of Receptor Characterization and Classification*, Girdestone, D. (Ed.), pp. 167–185. IUPHAR Media, UK.

Milano, S., Grelot, L., Chen, Z., *et al.* (1990). Vagal-induced vomiting in decerebrate cat is not suppressed by specific 5-HT₃ receptor antagonists, *J. Autonom. Nerv. Syst.* **31**, 109–118.

Milano, S., Simon, C. and Grelot, L. (1991). In vitro release and tissue levels of ileal serotonin after cisplatin-induced emesis in the cat, *Clin. Auton. Res.* **1**, 275–280.

Minami, M., Tamakai, H., Ogawa, T., *et al.* (1995a). Chemical modulation of 5-HT₃ and 5-HT₄ receptors affects the release of 5-hydroxytryptamine from the ferret and rat intestine, *Res. Commun. Mol. Pathol. Pharmacol.* **89**, 131–142.

Minami, M., Endo, T., Nemoto, M., *et al.* (1995b). How do toxic emetic stimuli cause 5-HT release in the gut and brain? in: *Serotonin and the Scientific Basis of Anti-emetic Therapy*, Reynolds, D. J. M., Andrews, P. L. R. and Davis, C. J. (Eds), pp. 68–76. Oxford Clinical Communications, Oxford.

Minami, M., Ogawa, T., Endo, T., *et al.* (1997a). Urinary 5-HIAA excretion in rats after chronic administration of cyclophosphamide, *Biogenic Amines* **13** (6), 579–590.

Minami, M., Endo, T., Tamakai, H., *et al.* (1997b). Antiemetic effects of N-3389, a newly synthesized 5-HT₃ and 5-HT₄ receptor antagonist, in ferrets, *Eur. J. Pharmacol.* **321**, 333–342.

Minami, M., Endo, T., Kikuchi, K., *et al.* (1998). Antiemetic effects of sendide, a peptide tachykinin NK₁ receptor antagonist, in the ferret, *Eur. J. Pharmacol.* **363**, 49–55.

Minami, M., Endo, T., Yokota, H., *et al.* (2001). Effects of CP-99994, a tachykinin NK₁ receptor antagonist, on abdominal afferent vagal activity in ferrets: evidence for involvement of NK₁ and 5-HT₃ receptors, *Eur. J. Pharmacol.* **428**, 215–220.

Miyake, A., Mochizuki, S., Takemoto, Y., *et al.* (1995). Molecular cloning of human 5-hydroxy-tryptamine₃ receptor: heterogeneity in distribution and function among species, *Mol. Pharmacol.* **48**, 407–416.

Miyata, K., Kamato, T., Yamano, M., *et al.* (1991). Serotonin (5-HT)₃ receptor blocking activities of YM060, a novel 4,5,6,7-tetrahydrobenzimidazole derivative, and its enantiomer in anesthetized rats, *J. Pharmacol. Exp. Ther.* **259**, 815–819.

Nemoto, M., Endo, T., Minami, M., *et al.* (2001). 5-Hydroxy-tryptamine (5-HT)-induced depolariza-tion in isolated abdominal vagus nerves in the rat: involvement of 5-HT₃ and 5-HT₄ receptors, *Res. Commun. Mol. Pathol. Pharmacol.* **109**, 217–230.

Newberry, N. R., Watkins, C. J., Reynolds, D. J., *et al.* (1992). Pharmacology of the 5-hydroxy-tryptamin-induced depolarization of the ferret vagus nerve in vitro, *Eur. J. Pharmacol.* **221**, 157–160.

Newson, B., Ahlman, H., Dahlström, A., et al. (1982). Ultrastructural observations in the rat ileal mucosa of possible epithelial 'taste cells' and submucosal sensory neurons, Acta Physiol. Scand. **114**, 161–164.

Nilsson, O., Ericson, L. E., Dahlström, A., et al. (1985). Subcellular localization of serotonin immunoreactivity in rat entero-chromaffin cells, Histochemistry **82**, 351–355.

Ouadid, H., Seguin, J., Dumuis, A., et al. (1992). Serotonin increases calcium current in human atrial myocytes via the newly described 5-hydroxytryptamine4 receptors, Mol. Pharmacol. **41**, 346–351.

Pauwels, P. J., Palmic, C., Wurch, T., et al. (1996). Pharmacology of cloned human 5-HT$_{1D}$ receptor-mediated functional responses in stably transfected rat C$_6$-glial cell lines: further evidence differentiating human 5-HT$_{1D}$ and 5-HT$_{1B}$ receptors, Naunyn-Schmiedeberg's Arch. Pharmacol. **335**, 144–156.

Pentila, A. and Kormano, M. (1971). Effect of X-irradiation of the exteriorized jejunal loop on the morphology and 5-hydroxytryptamine content of the enterochromaffin cells in the mouse, Strahlentherapie **142**, 238–247.

Peters, J. A., Lawbert, J. J., Hope, A. G., et al. (1995). The electrophysiology of vagal and recombinant 5-HT$_3$ receptors, in: Serotonin and the Scientific Basis of Anti-emetic Therapy, Reynolds, D. J. M., Andrews, P. L. R. and Davis, C. J. (Eds), pp. 95–105. Oxford Clinical Communications, Oxford.

Racké, K. and Schwörer, H. (1991). Regulation of serotonin release from the intestinal mucosa, Pharmacol. Res. **23**, 13–25.

Racké, K. and Schwörer, H. (1993). Characterization of the role of calcium and sodium channels in the stimulus secretion coupling of 5-hydroxytryptamine release from porcine enterochromaffin cells, Naunyn-Schmiedeberg's Arch. Pharmacol. **347**, 1–8.

Racké, K., Reimann, A., Schwörer, H., et al. (1996). Regulation of 5-HT release from enterochromaffin cells, Behav. Brain Res. **73**, 83–87.

Ramirez, M. J., Cenarruzabeitia, E., Ri, J. D., et al. (1994). Involvement of neurokinins in the non-cholinergic response to activation of 5-HT$_3$ and 5-HT$_4$ receptors in guinea pig ileum, Br. J. Pharmacol. **111**, 419–424.

Raymond, J. R., Mukhim, Y. V., Gelasco, A., et al. (2001). Multiplicity of mechanisms of serotonin receptor signal transduction, Pharmacol. Ther. **92**, 179–212.

Reynolds, D. J. M., Leslie, R. A., Graham-Smith, D. G., et al. (1989). Localization of 5-HT$_3$ receptor binding sites in human dorsal vagal complex, Eur. J. Pharmacol. **174**, 127–130.

Rhodes, K. F., Coleman, J. and Lattimer, N. (1992). A component of 5-HT-evoked depolarization of the rat isolated vagus nerve is mediated by a putative 5-HT$_4$ receptor, Naunyn-Schmiedeberg's Arch. Pharmacol. **346**, 496–503.

Round, A. and Wallis, D. I. (1986). The depolarization of 5-hydroxytryptamine on rabbit vagal afferent and sympathetic neurons in vitro and its selective blockade by ICS 205-930, Br. J. Pharmacol. **88**, 485–494.

Sanger, G. J. (1992). The involvement of 5-HT$_3$ receptors invisceral function, in: Central and Peripheral 5-HT$_3$ Receptors, Hamon, M. (Ed.), pp. 207–255. Academic Press, London.

Sanger, G. J. (1996). 5-Hydroxytryptamine and functional bowel disorders, Neuro-gastroenterol. Motil. **8**, 319–331.

Satoh, Y., Habara, Y., Ono, K., et al. (1995). Carbamylcholine- and catecholamine-induced intracellular calcium dynamics of epithelial cells in mouse ileal crypts, Gastroenterol. **108**, 1345–1356.

Schwörer, H. and Ramadori, G. (1998). Autoreceptors can modulate 5-hydroxytryptamine release from porcine and human small intestine in vitro, Naunyn-Schmiedeberg's Arch. Pharmacol. **357**, 548–552.

Tamakai, H., Sugahara, J., Ogawa, T., et al. (1996). Effects of 5-HT$_4$ receptor agonist on abdominal afferent nerve activity in rats (abstract), Jpn. J. Pharmacol. **71** (Suppl. 1), 245.

Tecott, L. H., Maricqi, A. V. and Julius, D. (1993). Nervous system diatribution of the serotonin 5-HT3 receptor mRNA, Proc. Natl. Acad. Sci. USA **90**, 1430–1434.

Tonini, M. (1995). 5-HT$_4$ receptor involvement in emesis, in: *Serotonin and the Scientific Basis of Anti-emetic Therapy*, Reynolds, D. J. M., Andrews, P. L. R. and Davis, C. J. (Eds), pp. 192–199. Oxford Clinical Communications, Oxford.

Tsou, A. P., Kosaka, A., Bach, C., *et al.* (1994). Cloning and expression of a 5-hydroxytryptamine 7 receptor positively coupled to adenylyl cyclase, *J. Neurochem.* **63**, 456–464.

Waeber, C., Dixon, K., Hoyer, D., *et al.* (1988). Localization by autoradiography of neuronal 5-HT3 receptors in the mouse CNS, *Eur. J. Pharmacol.* **151**, 351–352.

Wolff, M. C. and Leander, J. D. (1994). Antiemetic effects of 5-HT$_{1A}$ agonists in the pigeon, *Pharmacol. Biochem. Behav.* **49**, 385–391.

Zucker, R. S. (1996). Exocytosis: a molecular and physiological perspective, *Neuron* **17**, 1049–1055.

Advances in Neuroregulation and Neuroprotection (2005), pp. 635-644
Collin, C. *et al.* (Eds)
© VSP 2005

Urinary 5-hydroxyindoleacetic acid excretion and kaolin ingestion after a single administration of cisplatin in the delayed emesis rat model

YANXIA LIU, NAOYA HAMAUE, TORU ENDO, MASAHIKO HIRAFUJI
and MASARU MINAMI *

Department of Pharmacology, Faculty of Pharmaceutical Sciences, Health Sciences Universtiy of Hokkaido, Ishikari-Tobetsu, Hokkaido 061-0293, Japan

Abstract—In this study, rat kaolin ingestion (pica) and serial urinary 5-hydroxyindoleacetic acid (5-HIAA) excretions were determined for three days after a single administration of low-dose cisplatin (5 mg/kg, intraperitoneal, i.p.). We found that cisplatin induced acute (first 24 hours after chemotherapy) and delayed phases (24–72 hours after chemotherapy) of kaolin ingestion in rats. An increase in urinary 5-HIAA excretion was also observed in the cisplatin-administered group. These results suggest that 5-HT is involved in delayed emesis, and that urinary 5-HIAA excretion can be used as a marker in the study of delayed emesis.

Keywords: Delayed emesis; pica; cisplatin; 5-hydroxytryptamine (5-HT); 5-hydroxyindoleacetic acid (5-HIAA).

INTRODUCTION

Emesis induced by anticancer chemotherapeutic agents is classified into acute emesis during the first 24 hours and delayed emesis occurring 1 to 4 days after administration of cancer agents. The main mechanism of cancer chemotherapeutic agent-induced acute emesis is thought to be the stimulation of the vomiting center of the medulla oblongata by 5-hydroxytryptamine (5-HT), which is released from intestinal enterochromaffin (EC) cells following cisplatin administration, via the afferent vagal nerve fibers (Minami *et al.*, 1995). By the use of 5-HT$_3$ receptor antagonists, which selectively antagonize 5-HT$_3$ receptors on the abdominal vagal afferent fibers, as much as 60% of the initial emetic response can be prevented (Ruff

---

*To whom correspondence should be addressed. E-mail: minami@hoku-iryo-u.ac.jp

et al., 1994). Agents such as ondansetron and granisetron have been observed to be almost completely effective for 4 to 6 hours in inhibiting emesis in animal models (Bermudez *et al.*, 1988; Higgins *et al.*, 1989; Endo *et al.*, 1990). Delayed emesis, which occurs more than 24 hours after cancer chemotherapy, was first described in detail by Kris *et al.* (1985). The delayed emesis lasts for several days in humans, and there is a population of patients resistant to therapy with 5-HT$_3$ receptor antagonists (Kris *et al.*, 1992; Gandara *et al.*, 1993; Markman, 2002). The mechanism that causes delayed emesis remains uncertain (Andrews, 1992; Fukunaka, *et al.*, 1998). Delayed emesis was recently recognized to be a separate emetic entity with a pathological mechanism that differs from acute emesis (Kris *et al.*, 1992; Grunberg and Hesketh, 1993; Kris *et al.*, 1998).

The gut is estimated to contain more than 80% of the 5-HT in the body, of which 90% is located within EC cells (Erspamer, 1966). When the gastrointestinal tract is totally resected, the urinary excretion of 5-HIAA, a 5-HT metabolite, is abolished (Bertaccini, 1960). Chromogranin A (CgA) is co-localized with 5-HT in high concentration in EC cells. In patients who receive cisplatin treatment, the time course for the increases in plasma CgA parallels those of urinary 5-HIAA and the period of intense emesis (Cubeddu *et al.*, 1995). Taken together, these findings indicate that urinary excretion of 5-HIAA provides an index of 5-HT released from the gastrointestinal tract (Cubeddu *et al.*, 1992).

Rats do not vomit, but display a behavior known as 'pica' (Mitchell *et al.*, 1976; Takeda *et al.*, 1993). Mitchell *et al.* (1976, 1977) suggested that pica — eating of non-nutritive substances such as kaolin — is a naturally occurring response of the rat to illness that may be used profitably to investigate the etiology of motion sickness and to screen for the emetic properties of drugs. An increasing number of studies have shown that pica is analogous to emesis and can be used in the study of emesis (Takeda *et al.*, 1993; Rudd *et al.*, 2002).

We have found that the 5-HT level in the ileum and the 5-HIAA level in the area postrema are associated with the cisplatin-induced delayed emesis in the ferret (Fukunaka *et al.*, 1998). A continuous 72-hour observation revealed that acute and delayed emesis was induced by low-dose cisplatin administration (5 mg/kg, i.p.) in a ferret model (Rudd *et al.*, 1994). It is also known that cisplatin-induced acute and delayed-phase emesis in animals show different sensitivities to antiemetics. The 5-HT$_3$ receptor antagonists ondansetron and alosetron markedly antagonized or abolished the emesis during the acute phase and, whilst antagonizing the emesis during the delayed phase, also revealed a 5-HT$_3$ receptor antagonist resistant component (Rudd and Naylor, 1994). In the present study, in order to evaluate a suitable model for delayed emesis, we observed the kaolin ingestion in rats for 72 hours following a low dose and single injection of cisplatin (5 mg/kg, i.p.). We also examined whether 5-HT relating to the onset of emesis in the acute phase is also related to the mechanism of onset of delayed emesis.

MATERIALS AND METHODS

Animals

Male Wistar rats weighing 180–200 g were purchased from Sankyo Laboratory Service Co., Ltd. (Shizuoka, Japan). They were housed singly in metabolic cages (Natsume Co., Ltd., Japan) with a light-dark cycle (light 8:00–20:00) in an air conditioned room (22 ± 2°C, 50 ± 5% humidity). Food, water and kaolin were available *ad libitum*.

Kaolin pellet preparation

Kaolin pellets were prepared according to the method of Mitchell *et al.* (1976, 1977) with slight modifications. Kaolin (Wako Pure Chemical industries, Ltd., Japan) was mixed with 1% gum arabic (Suzu Pharmaceutical Co., Ltd., Japan) in distilled water to form a thick paste, which was extruded through a syringe made in the shape of a food column and completely dried at room temperature. Kaolin pellets and food were placed in individual containers. The kaolin ingestion was observed for 72 hours following a single administration of cisplatin.

Measurement of urinary 5-HIAA excretion

After 3 days for adaptation, the experimental group of rats were given a single i.p. injection of cisplatin (Nippon Kayaku, Japan) (5 mg in 10 ml/kg). The control group received a single injection of physiological saline (10 ml/kg). Kaolin, food intake and rat weight were measured at 15:00 each day. The urine was collected at 24 hours before cisplatin administration and at 24 h, 48 h, and 72 h after the injection. After being centrifuged for 10 min (3000 rpm, 2°C), the supernatant of the urine samples was stored at −80°C until assay.

Urinary 5-HIAA concentration was measured by high performance liquid chromatography (HPLC) (EP-10, Eicom, Japan) with electrochemical detection (ECD) (ECD-300, Eicom, Japan) (Minami *et al.*, 1988; Shintani *et al.*, 1993). HPLC was performed with a column (Eicompack, SC-5ODS: 3.0 mm × 150 mm, Eicom, Kyoto, Japan) containing EDTA-2Na (5 mg/l) (Sigma Chemical Co., St. Louis, MO, USA), sodium-1-octanesulfonate (190 mg/l) (Nacalai Tesque Co., Kyoto, Japan) and methanol (17%) as the mobile phase. The oxidation potential of the ECD was set at +750 mV. Isoproterenol · HCl (Sigma, USA) was used as the standard.

Statistical analysis

All values were expressed as mean ± SD. Data were subjected to one-way analysis of variance (ANOVA), followed by Dunnetts multicomparison test to compare more than two groups. The significance of the difference between treatments was assessed using Student's t-test (Wallenstein *et al.*, 1980).

RESULTS

Effect of cisplatin on kaolin ingestion

Figure 1 demonstrated the effect of cisplatin administration on kaolin ingestion in rats. Cisplatin (5 mg/kg i.p.) significantly increased kaolin ingestion in rats, during the 0–24 h ($p < 0.001$), and 24–48 h periods ($p < 0.05$), as compared with the control group. The peak kaolin ingestion was observed within the first 24 h. The kaolin ingestion remained elevated for 2 days following a single administration of cisplatin.

Effect of cisplatin on the changes of urinary 5-HIAA excretions

Cisplatin also induced increases in the urinary levels of 5-HIAA. As demonstrated in Fig. 2, urinary 5-HIAA excretion was significantly increased during the 0–72 h period after a single administration. The total urinary 5-HIAA excretion significantly correlated with the total kaolin ingestion during 72 h after cisplatin administration (Fig. 3).

As shown in Fig. 4, there were no changes in 5-HT turnover during the 0–24 h period, while there were evident changes during the 24–72 h period ($p < 0.01$) after cisplatin administration.

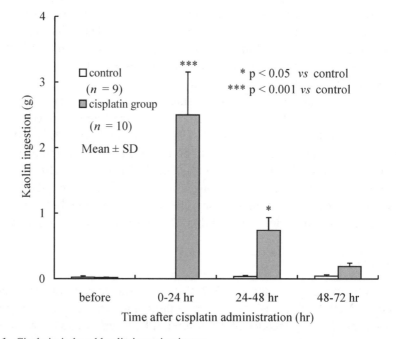

Figure 1. Cisplatin-induced kaolin ingestion in rats.

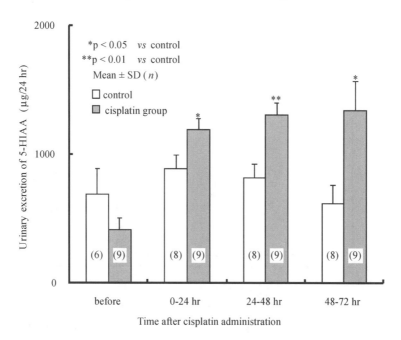

Figure 2. Effect of cisplatin on urinary 5-HIAA excretion in rats. 5-HIAA: 5-hydroxyindoleacetic acid.

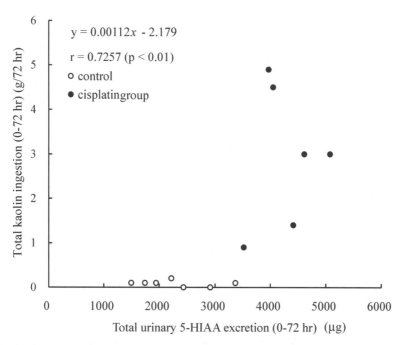

Figure 3. Correlation of kaolin ingestion with urinary 5-HIAA excretion in rats after cisplatin administration. 5-HIAA: 5-hydroxyindoleacetic acid.

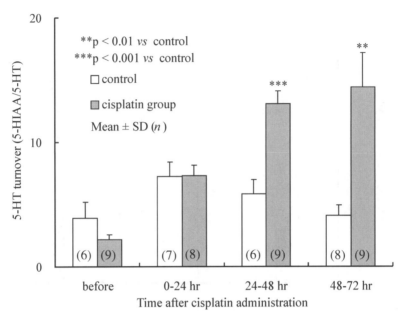

Figure 4.   Effects of cisplatin on the turnover of urinary 5-HT (5-HIAA/5-HT). 5-HIAA: 5-hydroxyindoleacetic acid; 5-HT: 5-hydroxytryptamine (serotonin).

DISCUSSION

We found that a single injection of low-dose cisplatin (5 mg/kg, i.p.) could induce pica (kaolin ingestion) during the acute and delayed emesis phase in rats. Simultaneously, an increase of urinary 5-HIAA excretion was observed in cisplatin-administered rats. Furthermore, a significant elevation of 5-HT turnover (5-HIAA/5-HT) was observed during the 24–72 h period after cisplatin administration. These results suggest that urinary 5-HIAA excretion can be used as a marker in the study of delayed emesis.

Cisplatin is a widely used antineoplastic agent with proven efficacy in the treatment of advanced ovarian cancer and testicular cancer. It is also useful in treating tumors of the head, neck, bladder and endometrium and small cell carcinoma of the lung (Bhat et al., 2002). Cisplatin-induced delayed emesis (60–70%) (Kris et al., 1992) has emerged as a vomiting problem for patients receiving this drug, while the pathogenesis of delayed emesis remains poorly understood in comparison to the recently improved understanding of acute emesis.

The acute phase (18–24 h) of nausea and vomiting induced by cisplatin has been demonstrated to correlate with the release of 5-HT, as measured by urinary 5-HIAA (Cubeddu et al., 1990) or by microdialysis in the ileal mucosa of the dog (Fukui et al., 1993). In man, the number of emetic episodes closely correlated with the peak urinary 5-HIAA levels (Cubeddu et al., 1990). Furthermore, in ferrets, depletion of tissue 5-HT by administration of para-chlorophenylalanine (PCPA), reduced cisplatin- and radiation-induced emesis (Barnes et al., 1988; Andrews,

1992). In cancer patients, PCPA reduced the spontaneous urinary excretion of 5-HIAA (Alfieri and Cubeddu, 1995). Furthermore, PCPA inhibited the increase in urinary 5-HIAA induced by cisplatin and markedly attenuated the acute period of nausea and vomiting associated with the cytotoxic drug.

In the present study, urinary 5-HIAA excretion levels remained elevated during the 3 days following cisplatin administration, while the kaolin ingestion did not increase during the 48–72 h period. It has been observed in the clinical study that, after a single administration of cisplatin, vomiting does not occur during the time when the urinary excretion of 5-HIAA is increasing in patients during the later days of delayed emesis (Suminaga *et al.*, 1995). Examination of the ileum of cisplatin-treated ferrets using election microscopy has demonstrated that 5-HT is stored in large electron-dense granules, which move towards the base of the EC cells and are released from the basal surface (Minami *et al.*, 1995). We examined the EC cells 6 hours after cisplatin administration and found them empty of 5-HT. These results suggest that the discrepancy between urinary 5-HIAA excretion and emetic episodes may be due to the timing of sampling. Nakajima *et al.* (1996) have reported no change in the plasma concentration of 5-HIAA after cisplatin administration, but a increase in the plasma concentration of 5-HT, although not significantly different from that of the control rats. Insertion of a probe into the blood stream to collect blood sample, however, may activate the coagulation system involving the platelets. The activated platelets may subsequently induce aggregation via the release of stored 5-HT, adenosine nucleotide and thromboxane A_2 (Verstraete, 1985). Under these circumstances, accurate determination of the circulating 5-HT content in the blood stream would be difficult. In fact, the concentration of 5-HT in the portal vein decreases gradually after insertion of a microdialysis probe but fluctuates markedly throughout the entire perfusion, while 5-HIAA values do not fluctuate in the same manner (Yoshioka *et al.*, 1993). An electron microscopic study of the post-dialysis probe membrane has shown platelet adhesion and aggregation on part of the membrane.

The present study examined the effect of a low dose of cisplatin (5 mg/kg) over a period of 72 h for the evaluation of the delayed emesis reported by Rudd and Naylor (1996). We observed that cisplatin 5 mg/kg (i.p.) induced acute and delayed emesis (pica) in rats compared with the vehicle-administered control group. These results coincided with the acute and delayed emesis observed in the ferret (Fukunaka *et al.*, 1998). The urinary 5-HIAA excretion also significantly increased correlaing with kaolin ingestion. 5-HIAA is the main metabolite of 5-HT, so the increase of urinary 5-HIAA concentration indicates an increase of 5-HT levels in delayed emesis. Thus, 5-HT is considered to be involved in delayed emesis as well as in acute emesis. This also could be supported by the increase in the turnover of 5-HT in this experiment. Taken together, these results suggest that rats are suitable and practical for use in the study of delayed emesis.

Contrary to acute emesis, delayed emesis is poorly controlled with antiemetic regimens, including 5-HT$_3$ receptor antagonists (Tavorath and Hesketh, 1996;

Gralla *et al.*, 1996; Gregory and Ettinger, 1998; Kris *et al.*, 1998). Perhaps, the mechanism in delayed emesis may differ from that of acute emesis and it may involve receptors other than 5-HT$_3$ receptors. Wilder-Smith *et al.* (1993) have shown that 5-HT levels do not correlate with delayed emesis. The basic mechanism in delayed emesis might be the toxic action of cancer chemotherapeutic agents on the rapidly dividing cells in the gastrointestinal tract (Esseboom *et al.*, 1995). Also, the sensitivity of the emotogenic center to gut-derived stimuli in the delayed phase is possible. Assuming the serum half-life of cisplatin in the human to be 100 hours, the stimulation of EC cells in the intestinal mucosa by a single administration of cisplatin may continue during the 72-h observation period and 5-HT may be continuously secreted. Fukunaka *et al.* (1998) reported that 5-HT levels in the ileum and 5-HIAA levels in the area postrema were significantly increased by cisplatin. In view of these results together with the present study, it is difficult to rule out the participation of 5-HT in delayed emesis.

In conclusion, even though the present dose of cisplatin was low, it could induce kaolin ingestion (pica), a significant increase in urinary 5-HIAA excretion and 5-HT turnover. These results suggest that 5-HT may be involved in delayed emesis, and urinary 5-HIAA can be used as a sensitive marker in the study of emesis.

Acknowledgement

This work was supported in part by High Technology Research Program (1998–2003) of the Ministry of Education, Culture, Sports, Science and Technology of Japan (to the project of Drs. Minami, Endo and Hirafuji).

REFERENCES

Alfieri, A. B. and Cubeddu, L. X. (1995). Treatment with para-chlorophenylalanine antagonises the emetic response and the serotonin-releasing actions of cisplatin in cancer patients, *Br. J. Cancer* **71**, 629–632.

Andrews, P. L. R. (1992). Physiology of nausea and vomiting, *Br. J. Anaesth.* **69**, 2S–19S.

Barnes, N. M., Costall, B., Naylor, R. J., *et al.* (1988). Identification of 5-HT$_3$ recognition sites in the ferret area postrema, *J. Pharm. Pharmacol.* **40**, 586–588.

Bermudez, J., Boyle, E. A., Miner, W. D., *et al.* (1988). The anti-emetic potential of the 5-hydroxytryptamine3 receptor antagonist BRL 43694, *Br. J. Cancer* **58**, 644–650.

Bertaccini, G. (1960). Tissue 5-hydroxytryptamine and urinary 5-hydroxyindoleacetic acid after partial or total removal of the gastro-intestinal tract in the rat, *J. Physiol.* **153**, 239–249.

Bhat, S. G., Mishra, S., Mei, Y., *et al.* (2002). Cisplatin up-regulates the adenosine A$_1$ receptor in the rat kidney, *Eur. J. Pharmacol.* **442**, 251–264.

Cubeddu, L. X., Hoffmann, I. S., Fuenmayor, N. T., *et al.* (1990). Efficacy of ondansetron (GR 38032F) and the role of serotonin in cisplatin-induced nausea and vomiting, *New Engl. J. Med.* **322**, 810–816.

Cubeddu, L. X., Hoffman, I. S., Fuenmayor, N. T., *et al.* (1992). Changes in serotonin metabolism in cancer patients: its relationship to nausea and vomiting induced by chemotherapeutic drugs, *Br. J. Cancer* **66**, 198–203.

Cubeddu, L. X., O'Connor, D. T. and Parmer, R. J. (1995). Plasma chromogranin A: a marker of serotonin release and of emesis associated with cisplatin chemotherapy, *J. Clin. Oncol.* **13**, 681–687.

Endo, T., Minami, M., Monma, Y., *et al.* (1990). Effect of GR38032F on cisplatin- and cyclophosphamide-induced emesis in the ferret, *Biogenic Amines* **7**, 525–533.

Erspamer, V. (1966). Occurrence of indolalkyl amines in nature, in: *Handbook of Experimental Pharmacology*, Eichler, O. and Farah, S. (Eds), Vol. 19, pp. 132–181. Springer, Berlin.

Esseboom. E. V., Rojer, R. A., Borm, J. J., *et al.* (1995). Prophylaxis of delayed nausea and vomiting after cancer chemotherapy, *Netherlands J. Med.* **47**, 12–17.

Fukui, H., Yamamoto, M., Sasaki, S., *et al.* (1993). Emetic effects of anticancer drugs and involvement of visceral afferent fibers and 5-HT$_3$ receptors in dogs, *Eur. J. Pharmacol.* **250**, 281–287.

Fukunaka, N., Sagae, S., Kudo, R., *et al.* (1998). Effects of granisetron and its combination with dexamethasone on cisplatin-induced delayed emesis in the ferret, *Gen. Pharmacol.* **31**, 775–781.

Gandara, D. R., Harvey, W. H., Monaghan, G. G., *et al.* (1993). Delayed emesis following high-dose cisplatin: a double-blind randomised comparative trial of ondansetron (GR 38032F) versus placebo, *Eur. J. Cancer* **29A** (Suppl. 1), S35–S38.

Gralla, R. J., Rittenberg, C., Peralta, M., *et al.* (1996). Cisplatin and emesis: aspects of treatment and a new trial for delayed emesis using oral dexamethasone plus ondansetron beginning at 16 hours after cisplatin, *Oncology* **53** (Suppl. 1), 86–91.

Gregory, R. E. and Ettinger, D. S. (1998). 5-HT$_3$ receptor antagonists for the prevention of chemotherapy-induced nausea and vomiting: a comparison of their pharmacology and clinical efficacy, *Drugs* **55**, 173–189.

Grunberg, S. M. and Hesketh, P. J. (1993). Control of chemotherapy-induced emesis, *New Engl. J. Med.* **329**, 1790–1796.

Higgins, G. A., Kilpatrick, G. J., Bunce, K. T., *et al.* (1989). 5-HT$_3$ receptor antagonists injected into the area postrema inhibit cisplatin-induced emesis in the ferret, *Br. J. Pharmacol.* **97**, 247–255.

Kris, M. G., Gralla, R. J., Clark, R. A., *et al.* (1985). Incidence, course, and severity of delayed nausea and vomiting following the administration of high-dose cisplatin, *J. Clin. Oncol.* **3**, 1379–1384.

Kris, M. G., Tyson, L. B., Clark, R. A., *et al.* (1992). Oral ondansetron for the control of delayed emesis after cisplatin: report of a phase-II study and a review of completed trials to manage delayed emesis, *Cancer* **70** (Suppl.), 1012–1016.

Kris, M. G., Roila, F., De Mulder, P. H., *et al.* (1998). Delayed emesis following anticancer chemotherapy, *Support. Care Cancer* **6**, 228–232.

Markman, M. (2002). Progress in preventing chemotherapy-induced nausea and vomiting, *Cleve. Clin. J. Med.* **69**, 609–617.

Minami, M., Sano, M., Togashi, H., *et al.* (1988). Stroke-related plasma noradrenaline, angiotensin II, arginin-vasopressin and serotonin concentrations in stroke-prone spontaneously hypertensive rats, in: *Progress in Hypertension*, Saito, H., Parvez, S. H. and Nagatsu, T. (Eds), Vol. 1, pp. 89–114. VSP, Utrecht, The Netherlands.

Minami, M., Endo, T., Nemoto, M., *et al.* (1995). How do toxic emetic stimuli cause 5-HT release in the gut and brain? in: *Serotonin and the Scientific Basis of Anti-emetic Therapy,* Reynolds, D. J. M., Andrews, P. L. R. and Davis, C. J. (Eds), pp. 68–76. Oxford Clinical Communications, Oxford, UK.

Mitchell, D., Wells, C., Hoch, N., *et al.* (1976). Poison induced pica in rats, *Physiol. Behav.* **17**, 691–697.

Mitchell, D., Krusemark, M. L. and Hafner, D. (1977). Pica: a species relevant behavioral assay of motion sickness in the rat, *Physiol. Behav.* **18**, 125–130.

Nakajima, Y., Yamamoto, K., Yamada, Y., *et al.* (1996). Effect of cisplatin on the disposition of endogenous serotonin and its main metabolite, 5-hydroxyindole-3-acetic acid, in rats and dogs, *Biol. Pharm. Bull.* **19**, 318–322.

Rudd, J. A. and Naylor, R. J. (1994). Effects of 5-HT$_3$ receptor antagonists on models of acute and delayed emesis induced by cisplatin in the ferret, *Neuropharmacol.* **33**, 1607–1608.

Rudd, J. A. and Naylor, R. J. (1996). An interaction of ondansetron and dexamethasone antagonizing cisplatin-induced acute and delayed emesis in the ferret, *Br. J. Pharmacol.* **118**, 209–214.

Rudd, J. A., Jordan, C. C. and Naylor, R. J. (1994). Profiles of emetic action of cisplatin in the ferret: a potential model of acute and delayed emesis, *Eur. J. Pharmacol.* **262**, R1–R2.

Rudd, J. A., Yamamoto, K., Yamatodani, A., *et al.* (2002). Differential action of ondansetron and dexamethasone to modify cisplatin-induced acute and delayed kaolin consumption ('pica') in rats, *Eur. J. Pharmacol.* **454**, 47–52.

Ruff, P., Paska, W., Goedhals, L., *et al.* (1994). Ondansetron compared with granisetron in the prophylaxis of cisplatin-induced acute emesis: a multicentre double-blind, randomised, parallel-group study, *Oncology* **51**, 113–118.

Shintani, F., Kanba, S., Nakaki, T., *et al.* (1993). Interleukin-1 beta augments release of norepinephrine, dopamine, and serotonin in the rat anterior hypothalamus, *J. Neurosci.* **13**, 3574–3581.

Suminaga, M., Akasaka, Y., Nukariya, N., *et al.* (1995). Pharmacokinetic investigation of 5-day multiple dose of tropisetron capsule in patients who had received cisplatin and the usefulness of tropisetron capsule in the treatment of nausea and vomiting, *Jpn. J. Cancer Chemother.* **22**, 1209–1221.

Takeda, N., Hasegawa, S., Morita, M., *et al.* (1993). Pica in rats is analogous to emesis: an animal model in emesis research, *Pharmacol. Biochem. Behav.* **45**, 817–821.

Tavorath, R. and Hesketh, P. J. (1996). Drug treatment of chemotherapy-induced delayed emesis, *Drugs* **52**, 639–648.

Verstraete, M. (1985). Platelet activation in patients with atherosclerosis of the arteries of the limbs, in: *Serotonin and the Cardiovasallar System*, P. M. Vanhoutte (Ed.), pp. 171–177. Raven Press, New York.

Wallenstein, S., Zucker, C. L. and Fleiss, J. L. (1980). Some statistical methods useful in circulation research, *Circ. Res.* **47**, 1–9.

Wilder-Smith, O. H., Borgeat, A., Chappuis, P., *et al.* (1993). Urinary serotonin metabolite excretion during cisplatin chemotherapy, *Cancer* **72**, 2239–2241.

Yoshioka, M., Kikuchi, A., Matsumoto, M., *et al.* (1993). Evaluation of 5-hydroxytryptamine concentration in portal vein measured by microdialysis, *Res. Commun. Chem. Pathol. Pharmacol.* **79**, 370–376.

Advances in Neuroregulation and Neuroprotection (2005), pp. 645-655
Collin, C. *et al.* (Eds)
© VSP 2005

Differences in rotational asymmetry in rats caused by single intranigral injections of 6-hydroxydopamine, 1-methyl-4-phenylpyridinium ion and rotenone

M. INDEN [1,*], J. KONDO [1,*], Y. KITAMURA [1,†], K. TAKATA [1], D. TSUCHIYA [1], K. NISHIMURA [1], T. TANIGUCHI [1], H. SAWADA [2] and S. SHIMOHAMA [2]

[1] *Department of Neurobiology, Kyoto Pharmaceutical University, Misasagi, Yamashina-ku, Kyoto 607-8412, Japan*
[2] *Department of Neurology, Graduate School of Medicine, Kyoto University, Kyoto 606-8507, Japan*

Abstract—Recently, 6-hydroxydopamine (6-OHDA), 1-methyl-4-phenylpyridinium ion (MPP$^+$) and rotenone have been shown to be dopaminergic neurotoxins. However, their neurotoxicities in rat brains *in vivo* are not fully understood. In the present study, we compared the *in vivo* neurotoxicities of 6-OHDA, MPP$^+$ and rotenone using a single intranigral injection. The injection of 6-OHDA caused the greatest loss of dopaminergic neurons in both the substantia nigra and striatum, and apomorphine induced contralateral rotation. In contrast, apomorphine-induced rotational behavior was in the ipsilateral direction with MPP$^+$, and was not observed with rotenone. Although MPP$^+$ and rotenone caused a loss of dopaminergic neurons in the substantia nigra, striatal neurodegeneration varied. These results suggest that intranigral injections of these neurotoxins produce different degrees of dopaminergic neurotoxicity in rats *in vivo*.

Keywords: 6-hydroxydopamine; MPP$^+$; rotenone; apomorphine; substantia nigra; striatum.

INTRODUCTION

The pathological hallmark of Parkinson's disease (PD), a major neurodegenerative disorder, is a loss of dopamine (DA) neurons in the substantia nigra pars compacta (SNpc). There is evidence that oxidative stress and free radical injury are pathogenic factors in PD (Dunnett and Björklund, 1999). A previous study suggested that the intracerebral injection of 6-hydroxydopamine (6-OHDA) induced unilateral degeneration of the DA system in rats (Ungerstedt, 1971). Recent studies of PD

---

*The first two authors (M. I. and J. K.) contributed equally to this work.

†To whom correspondence should be addressed. E-mail: yo-kita@mb.kyoto-phu.ac.jp

were prompted by the discovery of 1-methyl-4-phenyl-1,2,3,6-tetrahydropyridine (MPTP), which causes parkinsonism associated with the selective degeneration of nigrostriatal DA neurons in primates such as humans and monkeys (Langston and Irwin, 1986). Its metabolite, 1-methyl-4-phenylpyridinium ion (MPP$^+$), the formation of which is catalyzed by monoamine oxidase B, causes oxidative stress and DA neurodegeneration (Dunnett and Björklund, 1999; Kitamura et al., 2000). More recently, rotenone, a plant-derived pesticide that is a specific inhibitor of mitochondrial complex I, produced effects in rats which closely resemble PD (Betarbet et al., 2000). However, differences in the in vivo neurotoxicities of 6-OHDA, MPP$^+$ and rotenone are not fully understood. It is well known that axons of DA neurons in the SNpc project to the striatum (Schwarting and Huston, 1996; Dunnett and Björklund, 1999). In the present study, we performed a single intranigral injection of these neurotoxins and assessed the ensuing rotational behavior and immunoreactivities for tyrosine hydroxylase (TH) and α-synuclein.

MATERIALS AND METHODS

Materials

6-OHDA, MPP$^+$, rotenone and apomorphine were purchased from Sigma (St. Louis, USA). Primary antibodies included rabbit anti-TH antibody from Chemicon (Temecula, USA) and mouse monoclonal antibodies to TH from Sigma or α-synuclein from Transduction Lab. (Lexington, USA). The ABC Elite kit from Vector Laboratories (Burlingame, USA), enhanced chemiluminescent detection system kit (ECL kit) from Amersham Pharmacia Biotech (Buckinghamshire, UK), and Bradford protein assay from BioRad Laboratories (Hercules, USA) were used.

Rat models

Male Wistar rats weighing approximately 280 g were used. The rats were fasted overnight with free access to water. For stereotaxic microinjection, rats were anesthetized (sodium pentobarbital, 50 mg/kg, i.p.) and immobilized in a Kopf stereotaxic frame. The rats were then injected with 6-OHDA (8 μg), MPP$^+$ (8 μg), or rotenone (10 ng), in a final volume of 4 μl of sterilized physiological saline containing 1% dimethyl sulfoxide (DMSO), into the substantia nigra, where the bregma was −4.8 mm caudal, 1.8 mm left lateral, −7.8 mm ventral, via a motor-driven 10 μl-Hamilton syringe, using a 26-gauge needle. Coordinates were set according to a rat brain atlas (Paxinos and Watson, 1986). In cases of transplantation, approximately 4000 cells/2 μl of human neuroblastoma SH-SY5Y cells were grafted into the ipsilateral striatum at 2 weeks after the injection of 6-OHDA into the left substantia nigra.

Assay of rotational behavior

After the intranigral injection of a neurotoxin, drug-induced rotational asymmetry was assessed every week. Rotational behavior was tested in rotometer bowls (Ungerstedt, 1971). The total number of full 360° rotations in the ipsilateral and contralateral directions was counted for 60 min after the intraperitoneal administration of apomorphine (0.6 mg/kg, i.p.).

Immunohistochemistry

After 5 or 12 weeks, treated rats were perfused through the aorta with 150 ml of 10 mM phosphate-buffered saline (PBS), followed by 300 ml of a cold fixative consisting of 4% paraformaldehyde, 0.35% glutaraldehyde and 0.2% picric acid in 100 mM phosphate buffer (PB), under deep anesthesia with pentobarbital (100 mg/kg, i.p.). After perfusion, the brain was quickly removed and postfixed for 2 days with paraformaldehyde in 100 mM PB and then transferred to 15% sucrose solution in 100 mM PB containing 0.1% sodium azide at 4°C. Nigral sections were cut 20 μm-thick using a cryostat and collected in 100 mM PBS containing 0.3% Triton X-100 (PBS-T). After several washes, nigral slices were incubated with rabbit polyclonal TH antibody (diluted 1 : 10 000) for 3 days at 4°C. The antibody was detected by an ABC Elite kit using diaminobenzidine with nickel enhancement. The number of TH-immunopositive neurons in nigral sections was then counted.

Immunoblotting

After 1 or 5 weeks, treated rat striatum was rapidly removed and homogenized with 4 volumes of 50 mM Tris-HCl buffer (pH 7.4) containing 1 mM EDTA and 0.1 mM phenylmethylsulfonyl fluoride. After centrifugation at 50 000×g for 30 min, the supernatant and pellet were used as the cytosolic and membranous fractions, respectively. Aliquots of these fractions containing 10 μg of protein were subjected to sodium dodecyl sulfate–polyacrylamide gel electrophoresis (SDS-PAGE) and then immunoblotting using mouse monoclonal antibody to TH (diluted 1 : 5000) or α-synuclein (1 : 500). For semi-quantitative analysis, the bands of these proteins on radiographic films were scanned with a CCD color scanner (DuoScan, AGFA, Leverkusen, Germany) and then analyzed. Densitometric analysis was performed using the public-domain NIH Image 1.56 program (written by Wayne Rasband at the US National Institutes of Health and available from the Internet by anonymous ftp from zippy.nimh.nih.gov).

Statistical evaluation

Apomorphine-induced rotational asymmetry, the number of TH-immunopositive neurons in the substantia nigra, and the densitometric density of 60-kDa TH protein in the striatal fractions are given as mean ± standard error of the mean (S.E.M.). The significance of differences was determined by an analysis of variance (ANOVA).

Further statistical analysis for *post hoc* comparisons was performed using the Bonferroni/Dunn test.

RESULTS

Changes in rotational behavior produced by 6-OHDA, MPP+ and rotenone

Since rotenone is only slightly soluble in physiological saline, we preliminarily examined the influence of DMSO solution and rotenone when injected into the rat substantia nigra. The intranigral injection of sterilized physiological saline containing over 30% DMSO reduced the number of TH-immunopositive neurons in the SNpc. In addition, the injection of 4 μl of 100% DMSO containing 8 μg of rotenone caused death in rats within 1 week. Over 100 ng of rotenone did not dissolve in 4 μl of 1% DMSO saline. On the other hand, although 6-OHDA and MPP+ readily dissolved in the vehicle, injections of these reagents at less than 4 μg produced negligible behavioral changes. Based on these observations, we used sterilized physiological saline containing 1% DMSO as the vehicle, and injected 8 μg of 6-OHDA or MPP+, and 10 ng of rotenone per 4 μl of vehicle solution into the left substantia nigra.

After the injection of 6-OHDA, apomorphine induced marked contralateral rotation after 1 week, and the maximal number of rotations was reached after 2 weeks (Fig. 1). Thus, rotational asymmetry was maintained after 2 weeks. The injection

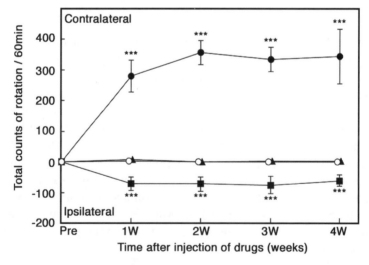

Figure 1. Apomorphine-induced rotational behavior after a unilateral injection of DA neurotoxin. Wistar rats received a single intranigral injection of vehicle (○), 6-OHDA (8 μg, ●), MPP+ (8 μg, ■), or rotenone (10 ng, ▲). Subsequently, rotational behavior induced by apomorphine (0.6 mg/kg, i.p.) was assessed every week. The number of rotations is given as the mean ± S.E.M. of 6 rats (in each group). After 1 week, the injection of 6-OHDA caused marked contralateral rotation, while MPP+ caused ipsilateral rotation. However, rotenone, like the vehicle control, did not induce either contralateral or ipsilateral rotation. *** $p < 0.001$ *vs.* the vehicle control.

of MPP$^+$ caused spontaneous and continuous rotation without apomorphine for several days. Briefly, the numbers of contralateral rotations per 60 min were 140 ± 55, 88 ± 38, 35 ± 18 and 14 ± 4 at 1, 2, 3 and 4 days after injection. Although spontaneous rotation gradually decreased and disappeared after 5 days, apomorphine induced ipsilateral rotation after 1 week (Fig. 1). In contrast, spontaneous rotation and apomorphine-induced rotational behavior were not induced by an injection of 10 ng of rotenone or by the vehicle control (Fig. 1).

Changes in TH immunoreactivity in the substantia nigra and striatum

After 5 weeks, we further assessed immunoreactivity for TH in the SNpc and striatum of rats injected with vehicle and neurotoxins (Figs 2 and 3). At 5 weeks after the injection of 6-OHDA into the left substantia nigra, TH immunoreactivity was markedly reduced in both the left substantia nigra and striatum (Fig. 2B and F). By a stoichiometric assessment, both the number of TH-immunopositive neurons in the SNpc and the level of TH protein in the striatum decreased by an average of over 95% (Fig. 3A and B). The injection of MPP$^+$ reduced the number of SNpc TH-immunopositive neurons by about 80% and the striatal TH protein level by about 30%. Although the injection of rotenone reduced the number of TH-immunopositive neurons in the SNpc by about 60%, it did not significantly reduce TH protein in the striatum.

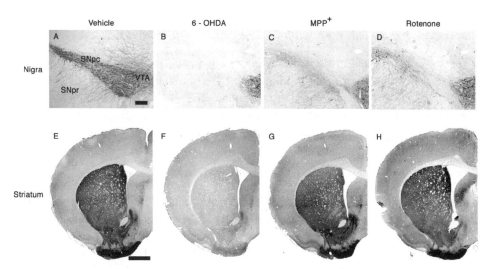

Figure 2. TH immunoreactivity in the substantia nigra and striatum. After 5 weeks, treated rats were fixed and brain slices were prepared. Sections of the substantia nigra (A–D) and striatum (E–H) were immunostained by anti-TH antibody. In upper panels: the pars compacta (SNpc), pars reticulata (SNpr) and ventral tegmental area (VTA) in the substantia nigra. Scale bars: 200 μm (A–D) and 1 mm (E–H).

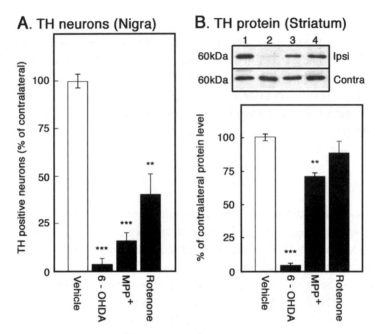

Figure 3. Changes in TH expression in the substantia nigra and striatum at 5 weeks after injection. (A) The number of TH-immunopositive neurons in the substantia nigra was counted. (B) The expression level of TH protein in the striatum was measured by immunoblot assay. Data are given as the mean ± S.E.M. (%) of three determinations, with the number of TH-immunopositive neurons in the SNpc and the density of 60-kDa TH in the ipsilateral striatum considered to be 100% in (A) and (B), respectively. $**p < 0.01$; $***p < 0.001$ *vs.* the vehicle control.

Changes in α-synuclein expression in the substantia nigra and striatum

After 1 week, the level of α-synuclein protein in the ipsilateral substantia nigra was markedly and significantly increased by a unilateral injection of 6-OHDA or MPP⁺, and slightly increased by rotenone (Fig. 4A). On the other hand, the injection of MPP⁺, but not 6-OHDA or rotenone, induced a significant increase in the α-synuclein level in the ipsilateral striatum (Fig. 4B).

Transplantation of human SH-SY5Y cells into 6-OHDA-lesioned rat

Human neuroblastoma SH-SY5Y cells can differentiate into DA neurons (Ross and Biedler, 1985; Kitamura *et al.*, 1998). Therefore, we further examined the transplantation of human SH-SY5Y cells (approximately 4000 cells) into the ipsilateral striatum of 6-OHDA-lesioned rats. Human SH-SY5Y cells survived even at 12 weeks after transplantation into rat striatum and had differentiated to TH-immunopositive neuron-like cells (Fig. 5C). In addition, apomorphine-induced contralateral rotation was markedly reduced after 6 weeks (Fig. 5A). In contrast, an injection of the vehicle (sterilized PBS) did not affect the apomorphine-induced contralateral rotation (Fig. 5A).

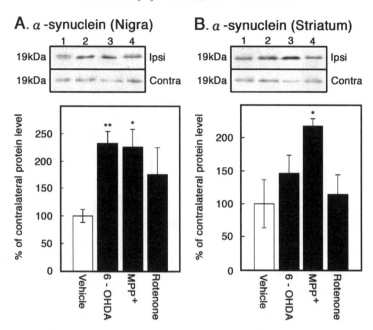

Figure 4. Changes in the α-synuclein level in the substantia nigra and striatum at 1 week after injection. The expression level of α-synuclein protein in the substantia nigra (A) and striatum (B) was measured by immunoblot assay. Data are given as the mean $\pm$ S.E.M. (%) of three determinations, with the density of 19-kDa α-synuclein in the ipsilateral tissues considered to be 100%. $^*p < 0.05$; $^{**}p < 0.01$ *vs.* the vehicle control.

DISCUSSION

Since PD is characterized by the loss of nigral DA neurons, it is thought that an animal with neurodegeneration induced by a DA neurotoxin may be a useful model of PD. In the present study, unilateral injections of 6-OHDA, MPP⁺ and rotenone caused different degrees of neurodegeneration in the substantia nigra and striatum. There were also differences in apomorphine-induced rotational asymmetry. Among these neurotoxins, 6-OHDA was more neurotoxic in the *in vivo* rat nigro-striatal DA pathway than MPP⁺ and rotenone. In brief, the injection of 6-OHDA reduced both the number of TH-immunopositive neurons in the SNpc and the level of TH protein in the striatum by more than 95%. Marked DA depletion in the striatum induces supersensitivity in DA receptors (Ungerstedt, 1971; Schwarting and Huston, 1996). Therefore, with the injection of 6-OHDA, apomorphine induced contralateral rotation.

Although unilateral rotation was not spontaneously observed after the injection of 6-OHDA, MPP⁺ caused spontaneous and continuous rotation in the contralateral direction for several days after injection. Subsequently, ipsilateral rotation was induced by the administration of apomorphine. In MPP⁺-treated rats, there was a marked decrease in DA neurons in the SNpc, but TH immunoreactivity was slightly reduced in the striatum. At 1 week after injection, the level of α-synuclein in the

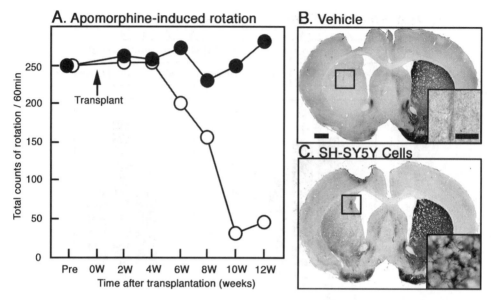

Figure 5. Transplantation of human SH-SY5Y cells into the striatum of 6-OHDA-lesioned rats. (A) Time-courses of changes in numbers of apomorphine-induced contralateral rotation in rats with vehicle injection (●) and transplanted SH-SY5Y cells (○). At 12 weeks after behavioral assay, the rat striatum with vehicle injection (PBS, B) or transplanted SH-SY5Y cells (about 4000 cells, C) was immunostained by anti-TH antibody. Although TH immunoreactivity was not detected around the site of injection with vehicle in the ipsilateral striatum (inset, B), marked TH immunoreactivity was observed in transplanted SH-SY5Y cells (inset, C). Scale bars: 1 mm (B, C) and 20 μm (inset).

striatum was significantly increased by MPP$^+$, but not by 6-OHDA. On the other hand, the α-synuclein level in the substantia nigra was increased by either MPP$^+$ or 6-OHDA. Since it is known that MPP$^+$, instead of DA, is taken up through both presynaptic DA transporter (DAT) and vesicular monoamine transporters (VMAT) and then accumulated in DA neurons (Przedborski and Jackson-Lewis, 1998), DA may be markedly released and then depleted in the striatum (Irwin et al., 1987; Kitamura et al., 1994). In addition, α-synuclein enhances the degeneration of DA neuron-like SH-SY5Y cells by oxidative stress (Kanda et al., 2000; Vila et al., 2000). Based on these observations, we consider that while the degeneration of DA neurons induced by a single intranigral injection of MPP$^+$ does not markedly denervate the rat striatum, postsynaptic DA receptors in the striatum may be downregulated by MPP$^+$-induced DA release and/or experience dysfunction due to increased α-synuclein. Therefore, an administration of apomorphine may induce the ipsilateral rotation after an injection of MPP$^+$.

A recent study has suggested that the continuous infusion of rotenone by an osmotic minipump produces effects in rats that closely resemble PD (Betarbet et al., 2000). However, a PD-like pathology was seen in only 12 (48%) of 25 Lewis rats, and these symptoms were not observed in other animals such as Sprague-Dawley rats or C57BL/6 mice (Ferrante et al., 1997; Betarbet et al., 2000; Thiffault et al.,

2000). In this study, although the injection of rotenone caused a degeneration of DA neurons in the SNpc, neither the level of striatal TH protein nor rotational behavior changed even after 4 weeks in Wistar rats. On the other hand, with regard to the effective concentration *in vitro*, rotenone was cytotoxic at the lowest concentration in mesencephalic DA neurons (Lotharius *et al.*, 1999; Lotharius and O'Malley, 2000) and human SH-SY5Y cells (Kitamura *et al.*, 1998, 2002). Furthermore, the *in vivo* neurotoxicity of MPTP is greatest in C57BL/6 among mouse strains but negligible in all strains of rats (Langston and Irwin, 1986; Kitamura *et al.*, 2000; Przedborski *et al.*, 2001). Thus, the neurotoxic potencies of MPP⁺ (or MPTP) and rotenone may be influenced by the experimental procedures such as *in vitro* and *in vivo* conditions, and also by the species and strains of the animals used. Considering these observations together, 6-OHDA may be the most efficient for generating an *in vivo* hemiparkinsonian model in rats.

Recent studies have suggested that the transplantation of DA-producing human cells improves the symptoms of PD (Lang and Lozano, 1998; Björklund and Lindvall, 2000). In addition, embryonic stem (ES) cells from mouse and human can differentiate into DA neurons *in vitro* (Lee *et al.*, 2000; Zhang *et al.*, 2001). In this study, the intrastriatal transplantation of human SH-SY5Y cells, DA neuron-like cells, led to the recovery of apomorphine-induced rotational asymmetry in 6-OHDA-lesioned rats. Thus, unilateral 6-OHDA-lesioned rats are suitable for evaluating *in vivo* behavioral recovery in studies of neuroprotective drugs, gene therapy and the regeneration of differentiated DA neurons by the transplantation of neuronal progenitor cells or embryonic stem cells.

In conclusion, we compared the *in vivo* neurotoxicities of 6-OHDA, MPP⁺ and rotenone caused by a single unilateral injection into the rat substantia nigra. 6-OHDA was the most efficient compound for generating the hemiparkinsonian rat model, which is suitable for evaluating behavioral recovery in studies of the transplantation of DA neuron-like cells.

Acknowledgements

This study was supported in part by the Frontier Research Program (T. T.), Grants-in-Aid from the Ministry of Education, Culture, Sports, Science and Technology of Japan (Y. K., T. T., H. S., S. S.), and a Grant-in-Aid for the promotion of the advancement of education and research in graduate schools (Special Research) from the Promotion and Mutual Aid Corporation for Private Schools of Japan (T. T.).

REFERENCES

Betarbet, R., Sherer, T. B., MacKenzie, G., *et al.* (2000). Chronic systemic pesticide exposure reproduces features of Parkinson's disease, *Nat. Neurosci.* **3**, 1301–1306.
Björklund, A. and Lindvall, O. (2000). Cell replacement therapies for central nervous system disorders, *Nat. Neurosci.* **3**, 537–544.

Björklund, L. M., Sánchez-Pernaute, R., Chung, S., *et al.* (2002). Embryonic stem cells develop into functional dopaminergic neurons after transplantation in a Parkinson rat model, *Proc. Natl. Acad. Sci. USA* **99**, 2344–2349.

Dunnett, S. B. and Björklund, A. (1999). Prospects for new restorative and neuroprotective treatments in Parkinson's disease, *Nature* **399** (Suppl.), A32–A39.

Ferrante, R. J., Schulz, J. B., Kowall, N. W., *et al.* (1997). Systemic administration of rotenone produces selective damage in the striatum and globus pallidus, but not in the substantia nigra, *Brain Res.* **753**, 157–162.

Irwin, I., Langston, J. W. and DeLanney, L. E. (1987). 4-Phenylpyridine (4PP) and MPTP: The relationship between striatal MPP$^+$ concentrations and neurotoxicity, *Life Sci.* **40**, 731–740.

Kanda, S., Bishop, J. F., Eglitis, M. A., *et al.* (2000). Enhanced vulnerability to oxidative stress by α-synuclein mutations and C-terminal truncation, *Neuroscience* **97**, 279–284.

Kitamura, Y., Itano, Y., Kubo, T., *et al.* (1994). Suppressive effects of FK-506, a novel immunosuppressant, against MPTP-induced dopamine depletion in the striatum of young C57BL/6N mice, *J. Neuroimmunol.* **50**, 221–224.

Kitamura, Y., Kosaka, T., Kakimura, J., *et al.* (1998). Protective effects of the antiparkinsonian drugs talipexole and pramipexole against 1-methyl-4-phenylpyridinium-induced apoptotic death in human neuroblastoma SH-SY5Y cells, *Mol. Pharmacol.* **54**, 1046–1054.

Kitamura, Y., Shimohama, S., Akaike, A., *et al.* (2000). The parkinsonian models: Invertebrates to mammals, *Jpn. J. Pharmacol.* **84**, 237–243.

Kitamura, Y., Inden, M., Miyamura, A., *et al.* (2002). Possible involvement of both mitochondria- and endoplasmic reticulum-dependent caspase pathways in rotenone-induced apoptosis in human neuroblastoma SH-SY5Y cells, *Neurosci. Lett.* **333**, 25–28.

Lang, A. E. and Lozano, A. M. (1998). Parkinson's disease: First of two parts, *New Engl. J. Med.* **339**, 1044–1053.

Langston, J. W. and Irwin, I. (1986). MPTP: current concepts and controversies, *Clin. Neuropharmacol.* **9**, 485–507.

Lee, S.-H., Lumelsky, N., Studer, L., *et al.* (2000). Efficient generation of midbrain and hindbrain neurons from mouse embryonic stem cells, *Nat. Biotechnol.* **18**, 675–679.

Lotharius, J. and O'Malley, K. L. (2000). The parkinsonism-inducing drug 1-methyl-4-phenylpyridinium triggers intracellular dopamine oxidation: A novel mechanism of toxicity, *J. Biol. Chem.* **275**, 38581–38588.

Lotharius, J., Dugan, L. L. and O'Malley, K. L. (1999). Distinct mechanisms underlie neurotoxin-mediated cell death in cultured dopaminergic neurons, *J. Neurosci.* **19**, 1284–1293.

Paxinos, G. and Watson, C. (1986). *The Rat Brain in Stereotaxic Coordinates*, 2nd edn. Academic Press, North Ryde, Australia.

Przedborski, S. and Jackson-Lewis, V. (1998). Mechanisms of MPTP toxicity, *Mov. Disord.* **13**, 35–38.

Przedborski, S., Jackson-Lewis, V., Naini, A. B., *et al.* (2001). The parkinsonian toxin 1-methyl-4-phenyl-1,2,3,6-tetrahydropyridine (MPTP): a technical review of its utility and safety, *J. Neurochem.* **76**, 1265–1274.

Ross, R. A. and Biedler, J. L. (1985). Presence and regulation of tyrosine activity in human neuroblastoma cell variants in vitro, *Cancer Res.* **45**, 1628–1632.

Schwarting, R. K. W. and Huston, J. P. (1996). The unilateral 6-hydroxydopamine lesion model in behavioral brain research: analysis of functional deficits, recovery and treatments, *Prog. Neurobiol.* **50**, 275–331.

Thiffault, C., Langston, J. W. and Di Monte, D. A. (2000). Increased striatal dopamine turnover following acute administration of rotenone to mice, *Brain Res.* **885**, 283–288.

Ungerstedt, U. (1971). Postsynaptic supersensitivity after 6-hydroxydopamine induced degeneration of the nigro-striatal dopamine system, *Acta Physiol. Scand.* **82** (Suppl. 367), 69–93.

Vila, M., Vukosavic, S., Jackson-Lewis, V., *et al.* (2000). α-Synuclein up-regulation in substantia nigra dopaminergic neurons following administration of the parkinsonian toxin MPTP, *J. Neurochem.* **74**, 721–729.

Zhang, S.-C., Wernig, M., Duncan, I. D., *et al.* (2001). *In vitro* differentiation of transplantable neural precursors from human embryonic stem cells, *Nat. Biotechnol.* **19**, 1129–1133.

Advances in Neuroregulation and Neuroprotection (2005), pp. 657-676
Collin, C. *et al.* (Eds)
© VSP 2005

Juvenile stroke-prone spontaneously hypertensive rats as an animal model of attention-deficit/hyperactivity disorder

KEN-ICHI UENO [1,*], HIROKO TOGASHI [1], MACHIKO MATSUMOTO [1], HIDEYA SAITO [2] and MITSUHIRO YOSHIOKA [1]

[1] *Department of Pharmacology, Hokkaido University School of Medicine, Kita-15, Nishi-7, Kita-ku, Sapporo 060-8638, Japan*
[2] *Department of Basic Sciences, Japanese Red Cross Hokkaido College of Nursing, Kitami 090-0011, Japan*

Abstract—Attention-deficit/hyperactivity disorder (AD/HD) is defined as a developmental disorder, manifested by inattention and/or hyperactivity-impulsivity. Neuropsychological studies suggest impaired cognitive function including working memory. 3–5% of school-aged children show signs of this disorder with male predominance. This article provides an overview of the symptomatic relevance of juvenile stroke-prone spontaneously hypertensive rats (SHRSP) as an animal model of AD/HD. To characterize behavioral alterations, i.e. hyperactivity-impulsivity and/or inattention, SHRSP were diagnosed according to motor activity, as well as emotional and cognitive behaviors with or without methylphenidate, a first choice drug for AD/HD therapy. Ambulatory and rearing activities in the open-field environment were significantly higher in SHRSP than in Wistar-Kyoto rats. In the elevated plus-maze task, anxiety-related behavior as an index of impulsivity was significantly increased in SHRSP. In the Y-maze task, spontaneous alternation behavior as an index of attention was significantly lowered in male, but not in female SHRSP, indicating gender specificity. Methylphenidate significantly attenuated locomotor hyperactivity at low doses, and dose-dependently improved the spontaneous alternation deficit in SHRSP. On the basis of these behavioral and pharmacological features, we have presented here that juvenile SHRSP are an appropriate animal model of AD/HD, for providing insights into the pathogenesis and developing therapeutic strategies.

Keywords: Attention-deficit/hyperactivity disorder (AD/HD); stroke-prone spontaneously hypertensive rats (SHRSP); methylphenidate; open-field; elevated plus-maze; Y-maze.

*To whom correspondence should be addressed. E-mail: ken-ueno@med.hokudai.ac.jp

INTRODUCTION

Attention-deficit/hyperactivity disorder (AD/HD) defined as a developmental disorder is manifested by inattention and/or hyperactivity-impulsivity (American Psychiatric Association, 1994). Among school-aged children, 3–5 % show signs of this disorder (Szatmari *et al.*, 1989; Barkley, 1990, 1998). Neuropsychological studies suggest impaired cognitive function, including working memory (Tannock *et al.*, 1995). The ratio of male to female children with AD/HD is currently 9:1 (Lahey *et al.*, 1994; Taylor, 1998). Although the pathogenesis of AD/HD, including the reason why it is predominant among males, remains unclear, inherent or genetic mechanisms are postulated (Barkley, 1998).

Neurochemical features of AD/HD are concerned on the overexpression of dopamine transporter (DAT); DAT density is increased in the striatum of AD/HD patients (Dougherty *et al.*, 1999; Krause *et al.*, 2000). Reduced regional cerebral blood flow (rCBF), which might result from cholinergic dysfunction (Honer *et al.*, 1988; Prohovnik *et al.*, 1997), is also implicated in AD/HD (Lou *et al.*, 1989; Powell *et al.*, 1997; Gustafsson *et al.*, 2000; Schweitzer *et al.*, 2000). In addition, omega-3 (n-3) and omega-6 (n-6) polyunsaturated fatty acid (PUFA), especially docosahexaenoic acid (DHA; C22: 6n-3) and arachidonic acid (AA; C20: 4n-6) deficiency, and a significant increase in the n-6/n-3 ratio have been observed in children with AD/HD (Colquhoun and Bunday, 1981; Mitchell *et al.*, 1987; Stevens *et al.*, 1995; Burgess *et al.*, 2000). A relationship between socially inappropriate behaviors, including aggression, impulsivity, learning disabilities and inattention, and serotonergic dysfunction indicated by lowered levels of serotonin (5-HT) and its metabolite 5-hydroxyindole-3-acetic acid (5-HIAA) in blood and cerebrospinal fluid (CSF), has also been noted in AD/HD patients (Saul and Ashby, 1986; Kruesi *et al.*, 1990; Comings, 1993; Hornig, 1998; Spivak *et al.*, 1999).

AD/HD is commonly treated with psychostimulants such as d-amphetamine, methylphenidate and pemoline. Among them, methylphenidate is currently the first choice psychostimulant for AD/HD medication. Such medication improves inattention and reduces hyperactivity-impulsivity (Bradley, 1937; Conners and Eisenberg, 1963; Sykes *et al.*, 1971). Methylphenidate also improves cognitive performance (Pliszka, 1989; Tannock *et al.*, 1995; Kempton *et al.*, 1999), and increases CBF to hypoperfused brain regions in patients with AD/HD (Lou *et al.*, 1989).

Several animal models of AD/HD, including those with chemical lesions (Luthman *et al.*, 1989; Kostrzewa *et al.*, 1994) and DAT gene knockout (KO) (Giros *et al.*, 1996; Gainetdinov *et al.*, 1999), have been proposed. Above all, spontaneously hypertensive rats (SHR) (Okamoto and Aoki, 1963) have been widely investigated because they exhibit behavioral similarities to AD/HD such as hyperactivity (Knardahl and Sagvolden, 1979; Myers *et al.*, 1982; Cierpial *et al.*, 1989; Wultz *et al.*, 1990), and impulsivity/inattention (Sagvolden *et al.*, 1992, 1993a, b; Sagvolden, 2000). In addition, central dopaminergic dysfunctions such as a decrease in dopamine (DA) release (Tsuda *et al.*, 1991; Russell *et al.*, 1995) and an increase in

DAT density (Watanabe *et al.*, 1997) have been found in SHR as in AD/HD. However, SHR do not fulfill the behavioral and pharmacological features as an AD/HD animal model. For instance, the behaviors are not male preponderance in SHR, as pointed out by Sagvolden and Berger (1996) and Berger and Sagvolden (1998). Female, not male, SHR exhibit impulsivity/inattention, although both are hyperactive. Furthermore, the effectiveness of psychostimulants in SHR is not consistent with the behavioral paradigms examined. Namely, the hyperactive behavior in SHR is ameliorated by high doses of d-amphetamine and methylphenidate (Myers *et al.*, 1982; Wultz *et al.*, 1990), whereas impulsivity/inattention are not (Sagvolden *et al.*, 1992; Evenden and Myerson, 1999). Recently, Bull *et al.* (2000) questioned the utility of SHR as an AD/HD animal model, since SHR do not exhibit impaired acquisition and performance in the differential low rate test. Furthermore, neurochemical or pathophysiological features do not always coincide with those in AD/HD, since rCBF in the frontal cortex is not reduced in SHR (Yamori and Horie, 1977).

Stroke-prone spontaneously hypertensive rats (SHRSP) are isolated from the SHR sub-strains (Okamoto *et al.*, 1974). SHRSP exert higher motor activity than SHR (Togashi *et al.*, 1982; Minami *et al.*, 1985). In addition, neurochemical evidence of similarities with AD/HD patients is accumulating. As in SHR, dopaminergic hypofunction such as decreased neural transmission was observed in the prefrontal cortex, nucleus accumbens shell, striatum, and basolateral amygdala of SHRSP (Nakamura *et al.*, 2001). The CSF 5-HT levels are significantly decreased in SHRSP, but not in SHR, as compared with Wistar-Kyoto rats (WKY) (Togashi *et al.*, 1994). In addition, cognitive impairment accompanied with central cholinergic dysfunction has been noted in SHRSP (Togashi *et al.*, 1996; Minami *et al.*, 1997). Moreover, rCBF in the frontal cortex is significantly reduced in SHRSP, but not in SHR or WKY (Yamori and Horie, 1977).

We review here juvenile SHRSP as an animal model of AD/HD with regard to the behavioral and pharmacological relevance. Behavioral features of juvenile SHRSP were evaluated using behavioral paradigms: locomotor activity in an open-field environment as an index of hyperactivity, anxiety-related behavior in the elevated plus-maze as an index of impulsivity, and spontaneous alternation behavior in the Y-maze as an index of inattention. Relevance as an AD/HD animal model was further evaluated on the pharmacological basis of the effect of a first choice psychostimulant methylphenidate upon the behavioral alteration of juvenile SHRSP.

BEHAVIORAL RELEVANCE OF JUVENILE SHRSP AS AN ANIMAL MODEL OF AD/HD

Hyperactivity in the open-field environment

The original SHRSP and normotensive control Wistar-Kyoto rats (WKY) were donated in 1979 by the late professor Kozo Okamoto, Department of Pathology, Kinki University School of Medicine, Osaka, Japan and are inbred in our laboratory (current generation F56). Juvenile (aged 6 weeks) male- and female-SHRSP and age- and sex-matched WKY were used. To examine the behavioral hyperactivity, horizontal (ambulatory) and vertical (rearing) activities of the rats were measured using the open-field apparatus (90 × 90 × 40 cm, 81 squares). Rats were placed at the central square and allowed to move freely in the field for a 60-min test session.

Horizontal ambulatory and vertical rearing activities of WKY and SHRSP are presented in Fig. 1. When exposed to an open-field environment, SHRSP and WKY immediately explored, then became habituated to the new environment. Although the number of square crossings (horizontal ambulatory activity) did not differ between male SHRSP and male WKY at the exploratory phase in the initial 0–15 min period, ambulatory activity significantly increased in male SHRSP at the habituated phases of 15–60 min (Fig. 1A, left). The numbers of rearings (vertical activity) of male SHRSP was also significantly greater than that of sex-matched WKY during the habituated phases (Fig. 1B, lower). Moreover, the numbers of square crossings and rearings of female SHRSP were also significantly higher than those of sex-matched WKY (Fig. 1A, right and B, right). Thus, SHRSP were hyperactive as compared with WKY. The hyperactivity of SHRSP was obvious at the habituated phases, but not at the exploratory phase. In addition, female SHRSP and WKY were more hyperactive than age-matched male SHRSP and WKY in open-field performance.

Thus, juvenile SHRSP showed locomotor hyperactivity when exposed to the open-field environment. Both horizontal (ambulation) and vertical rearing activities between male and female SHRSP were significantly more numerous than those of sex-matched WKY. Hyperactivity was noted from 15–60 min, but not during the initial 0–15 min period. In general, locomotor activity during the first few minutes of exposure to an open-field is thought to reflect the exploratory behavior or the ability to habituate to a novel environment. Therefore, the hyperactivity observed in juvenile SHRSP is probably not due to impaired habituation to new surroundings. This is supported by our previous finding that SHRSP exhibit a tonic increase in daily ambulation, as demonstrated by means of a consecutive recording apparatus (Minami et al., 1985).

Impulsivity in the elevated plus-maze

It has been reported that AD/HD children without anxiety exhibit more impulsive behavior than those with anxiety in the memory-scanning test (Pliszka et al., 1989). To examine the impulsivity, anxiety-relate behavior of the rats was evaluated in the elevated plus-maze paradigm (Pellow et al., 1985; Pellow and File, 1986). The plus-maze, consisting of two opposite open arms (50 × 10 cm) and two enclosed arms (50 × 10 × 40 cm), was elevated 50 cm above the floor. The arms extended from a central platform (10 × 10 cm). Rats were placed individually in the central platform of the maze and allowed to enter freely in the maze for a 10-min test session. The number of entries into open arms and the time spent in open arms were calculated as the percentage of total arm entries and of the time spent in all

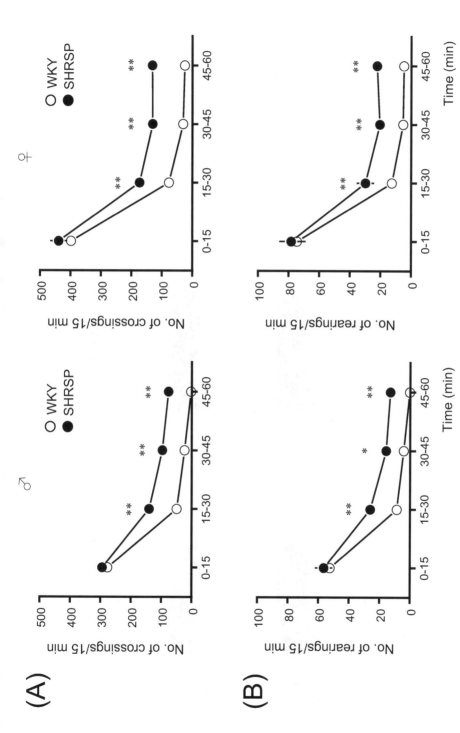

Figure 1. Time course changes in horizontal ambulatory activity (A) and vertical rearing activity (B) of WKY and SHRSP in the open-field. Data are expressed as the means ± SEM. * $p < 0.05$, ** $p < 0.01$ *versus* WKY.

four arms, respectively. These parameters were taken as measures of situational fear (Pellow and File, 1986).

Figure 2 shows anxiety-related behavior of SHRSP and WKY. The number of open arm entries was significantly increased in both male and female SHRSP as compared with sex-matched WKY. Also, SHRSP spent significantly more time in the open arms.

The elevated plus-maze is a task that evaluates the anxiolytic properties of drugs or animals exposed to unconditioned fear. In this task, juvenile SHRSP entered significantly more often into the open arms than WKY, implying that SHRSP exhibited less anxiety to unconditioned fear. The anxiety-related behavior that increases the number of entries into danger-zone open arms might reflect intense impulsivity in SHRSP. It has been reported that AD/HD children without anxiety

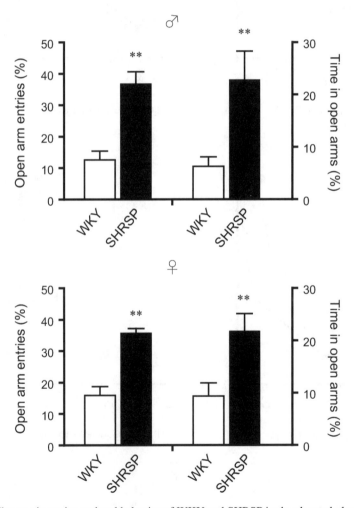

Figure 2. Changes in anxiety-related behavior of WKY and SHRSP in the elevated plus-maze. Data are expressed as the means ± SEM. ** $p < 0.01$ *versus* WKY.

show more impulsive behavior than those with anxiety in the Memory Scanning Test (Pliszka, 1989). Thus our findings that juvenile male and female SHRSP were insensitive to unconditioned fear, suggested that their impulsive-like behavior is based on their having less anxiety, as observed in children with AD/HD.

Inattention in the Y-maze

Cognitive function was assessed as a spontaneous alternation behavior in the Y-maze task, which is generally accepted as a paradigm evaluating working memory (or short-term memory) (Sarter *et al.*, 1988). Each arm of the black-painted Y-maze was 45 cm long, 10 cm wide, 35 cm high, and both arms were positioned at an equal angle. Each rat was placed at the end of an arm and allowed to freely enter the maze for an 8-min test session without reinforcers such as food, water or electric shock. Arm entry was defined as the entry of all four paws into one arm. The sequence of arm entries was recorded by video camera. The alternation behavior (actual alternations) was defined as the consecutive entry into three arms, i.e. the combination of three different arms, with stepwise combinations in the sequence. The number of maximum alternations was defined as the total number of arms entered minus 2, and the percent alternation behavior was calculated as (actual alternations/maximum alternations) × 100.

Figure 3A shows that spontaneous alternation of male SHRSP was significantly lower than that of sex-matched WKY. In contrast, female SHRSP did not show a significant decrease in alternation behavior. Thus, spontaneous alternation behavior was gender-specific in SHRSP. In contrast, the number of total arm entries of SHRSP significantly increased in both males and females, compared with that of sex-matched WKY.

Spontaneous alternation behavior in the Y-maze is implicated in working memory, which is classified as short-term memory (Sarter *et al.*, 1988; Parada-Turska and Turski, 1990). This is affected by several psychological factors such as attention (Katz and Schmaltz, 1980). In the Y-maze task, percent alternation behavior was significantly lowered in males, but not in female SHRSP or sex-matched WKY. The emphasis is mostly placed on the fact that spontaneous alternation deficit is gender-specific in juvenile SHRSP; i.e. males predominantly exhibit impaired alternation behavior. On the other hand, the total number of arm entries in the Y-maze performance was significantly increased in both male and female SHRSP as compared with sex-matched WKY, further supporting hyperactivity in SHRSP.

Based on the possible relationship between the attention and the memory processes, long-term potentiation (LTP), an electrophysiological basis of memory (Bliss and Lømo, 1973), was examined in the perforant path-dentate gyrus synapses. To induce LTP formation, high-frequency stimulation (i.e. tetanic stimulation, 10 trains at 1 Hz composed each of 8 pulses at 400 Hz) was given at the same intensity as the test stimuli.

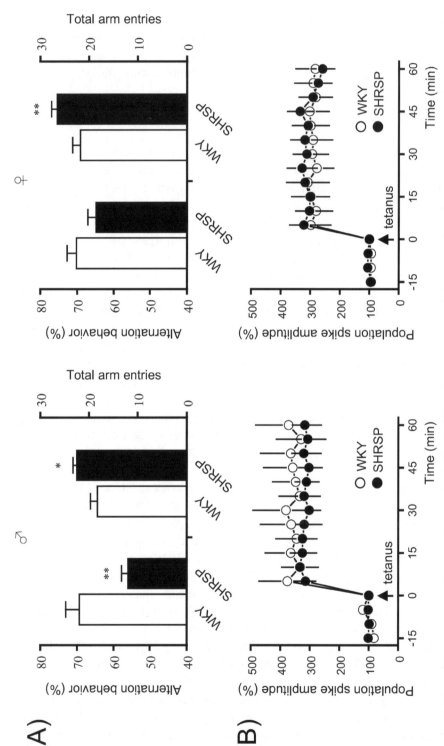

Figure 3. Changes in spontaneous alternation behavior in the Y-maze (A) and long-term potentiation in the dentate gyrus synapses (B) of WKY and SHRSP. Data are expressed as the means $\pm$ SEM. $*$ $p < 0.05$, $**$ $p < 0.01$ *versus* WKY.

In the hippocampal perforant path-dentate gyrus synapses, neither male nor female SHRSP exhibited impairment of hippocampal LTP formation, as compared with age- and sex-matched WKY (Fig. 3B).

It has been reported that spontaneous alternation behavior requires attention (Katz and Schmaltz, 1980) and working memory (Sarter *et al.*, 1988). Moreover, it is well known that the behavioral parameter of spontaneous alternation behavior in a Y-maze positively correlates with the hippocampal LTP in the perforant path-dentate gyrus synapses (Mori *et al.*, 2001; Nakao *et al.*, 2001).

To assess whether the spontaneous alternation deficit in juvenile male SHRSP was due to an attention-deficit (i.e. inability to sustain concentration) or learning and memory disabilities based on synaptic transmission impairment, we recorded LTP, a cellular mechanism of learning and memory (Bliss and Lømo, 1973), in juvenile SHRSP. Neither male nor female SHRSP exhibited impairment of hippocampal LTP formation in the perforant path-dentate gyrus synapses, which correlated well with a behavioral parameter of spatial working memory, spontaneous alternation behavior in the Y-maze (Nakao *et al.*, 2001). Our evidence that LTP formation in both male and female SHRSP was comparable to that in sex-matched WKY, whereas Y-maze performance was impaired in male SHRSP, suggests that the lowered spontaneous alternation behavior in juvenile male SHRSP is not due to a disability of hippocampal LTP formation. In other words, impairment of spontaneous alternation in juvenile male SHRSP might be due to an inability to sustain concentration or to attention-deficit.

PHARMACOLOGICAL RELEVANCE OF SHRSP AS AN ANIMAL MODEL OF AD/HD EVALUATED BY METHYLPHENIDATE

Effects on hyperactivity

Male SHRSP intraperitoneally administered with methylphenidate attenuated the hyperactivity observed in the open-field. Figure 4A shows that horizontal ambulatory activity in male SHRSP treated with methylphenidate at low and medium doses of 0.01 and 0.1 mg/kg was significantly decreased compared with that of vehicle-treated SHRSP during the habituated phases of 30–60 and 45–60 min. Furthermore, vertical rearing activity of male SHRSP was also significantly attenuated by methylphenidate, as compared with vehicle-treated SHRSP (Fig. 4B). Thus hyperactivity in SHRSP during the habituation phase in an open-field environment was paradoxically calmed by low (0.01 mg/kg) and medium (0.1 mg/kg) doses of a psychostimulant, methylphenidate. However, methylphenidate at high dose of 1 mg/kg did not produce a calming effect. In addition, an even higher dose of methylphenidate (3 and 30 mg/kg, i.p.) dose-dependently and significantly exacerbated both ambulatory and rearing activities in SHRSP (Fig. 5).

The present study found that intraperitoneal methylphenidate significantly attenuated hyperactivity in juvenile male SHRSP. Although the calming effect of

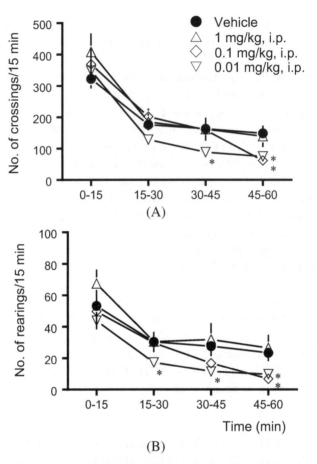

Figure 4. Effects of lower doses of methylphenidate on horizontal ambulatory hyperactity (A) and vertical rearing hyperactivity (A) of male SHRSP in the open-field. Data are expressed as the means ± SEM. *$p < 0.05$ *versus* vehicle-treated groups.

methylphenidate was not dose-dependent (0.01 – 1 mg/kg), locomotor hyperactivity was significantly improved at low (0.01 mg/kg) and medium (0.1 mg/kg) doses. Psychostimulant-induced calming effects in SHR have also been reported (Myers *et al.*, 1982; Wultz *et al.*, 1990), but the effects were inconsistent with those in SHRSP. Myers *et al.* (1982) reported that hyperactivity in SHR was attenuated by d-amphetamine only at the initial phase of the open-field performance. In contrast, methylphenidate induced calming effects in SHRSP during the habituation phases, but not in the initial exploratory phase. Our findings coincide with evidence that, in juvenile male and female SHRSP and WKY, methylphenidate (0.01 to 1 mg/kg, i.p.) did not affect total arm entries in the Y-maze (ambulatory activity) over an 8-min exploratory period, even when given at doses that significantly ameliorated the impaired spontaneous alternation behavior in juvenile male SHRSP.

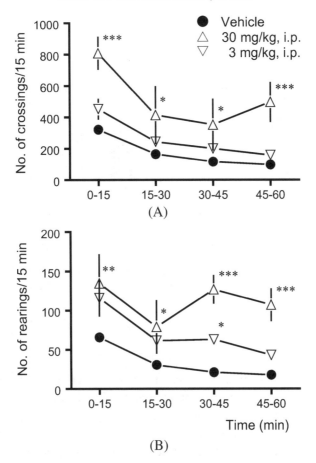

Figure 5. Effects of higher doses of methylphenidate on horizontal ambulatory hyperactity (A) and vertical rearing hyperactivity (B) of male SHRSP in the open-field. Data are expressed as the means ± SEM. * $p < 0.05$, ** $p < 0.01$, *** $p < 0.001$ *versus* vehicle-treated groups.

Effects on impulsivity

The effects of methylphenidate on the anxiety-related behavior of SHRSP in elevated plus-maze performance are presented in Fig. 6. Methylphenidate (0.01–1 mg/kg, i.p.) failed to ameliorate either the increased open arm entries or the increased time spent in open arms in both male and female SHRSP.

Methylphenidate (0.01–1 mg/kg, i.p.) was not effective in ameliorating impulsive-like behavior (i.e. the increased number of entries into open arms that are a danger zone) of juvenile SHRSP observed in the elevated plus-maze paradigm. The finding that methylphenidate has no effect against intense impulsivity of SHRSP coincides well with previous reports that methylphenidate is not effective in ameliorating premature responses (i.e. impulsivity) in SHR (Sagvolden *et al.*, 1992; Evenden and Meyerson, 1999) and with the clinical observations that it is difficult to inhibit intense impulsivity in AD/HD (Weiss, 1996).

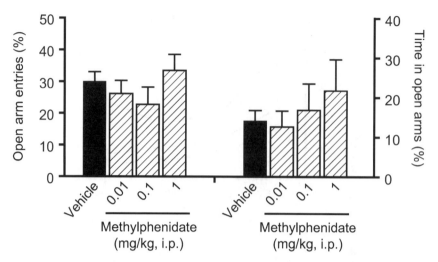

Figure 6. Effects of methylphenidate on anxiety-related behavior of male SHRSP in the elevated plus-maze. Data are expressed as the means ± SEM obtained from the SHRSP.

Effects on inattention

The effects of methylphenidate on the spontaneous alternation behavior of male SHRSP in Y-maze performance are summarized in Fig. 7. Methylphenidate (0.01–1 mg/kg, i.p.) significantly and dose-dependently improved the impaired alternation behavior that was specifically observed in male SHRSP (Fig. 7, left). At the high dose of 1 mg/kg, the methylphenidate-induced an increase in the accuracy of alternation behavior was accompanied by a concomitant and a significant increase in the total arm entries in male SHRSP (Fig. 7, right). On the contrary, methylphenidate had no effect upon either alternation behavior or total arm entries in female SHRSP (data not shown).

Methylphenidate dose-dependently (0.01–1 mg/kg, i.p.), improved spontaneous alternation behavior in male SHRSP. The drug enhances cognitive performance, including sustained attention, short-term memory, spatial memory and higher-order learning in children diagnosed with AD/HD (Aman *et al.*, 1991; Rapport *et al.*, 1993; Tannock *et al.*, 1995; Kempton *et al.*, 1999). Methylphenidate more significantly improves the working memory deficits and behavioral abnormalities displayed in AD/HD children without, rather than with anxiety (Tannock *et al.*, 1995). In our study, juvenile SHRSP exhibited distinctly less-anxious behavior than WKY, since SHRSP entered the open arms more often and stayed there longer in the elevated plus-maze task. Our present finding that methylphenidate significantly ameliorated the impaired spontaneous alternation behavior seen in juvenile male SHRSP, which exhibit less anxious behavior, coincides well with consistent clinical observations of non-anxious AD/HD children medicated with methylphenidate.

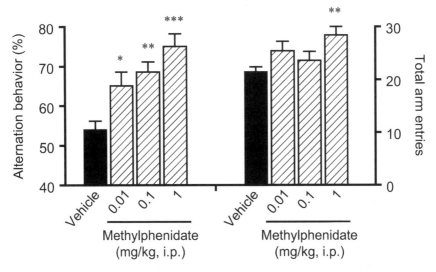

Figure 7. Effects of methylphenidate on spontaneous alternation behavior of male SHRSP in the Y-maze. Data are expressed as the means ± SEM. $*p < 0.05$, $**p < 0.01$, $***p < 0.001$ *versus* vehicle-treated groups.

Therapeutic dosage of methylphenidate

The clinically optimal dose range to obtain the therapeutic effect of methylphenidate is 0.1 to 1 mg/kg orally in children with developmental disabilities and AD/HD (Handen *et al.*, 1999; Ghuman *et al.*, 2001). However, the effective intraperitoneal doses of methylphenidate that elicited the paradoxical calming effect in SHR and DAT-KO mice were 24 and 30 mg/kg, respectively (Wultz *et al.*, 1990; Gainetdinov *et al.*, 1999). In contrast, we administered methylphenidate in the range of 0.01 to 1 mg/kg intraperitoneally to the juvenile SHRSP. This dosage was over two or three orders of magnitude lower than the dosage given to the SHR and DAT-KO mice, and was one order of magnitude lower than the oral dosage given to children with developmental disabilities and AD/HD. However, methylphenidate significantly ameliorated the impaired spontaneous alternation behavior in a dose-dependent manner, and significantly attenuated hyperactivity in juvenile male SHRSP at low (0.01 mg/kg) and medium (0.1 mg/kg) doses. Concerning the one-order difference of the methylphenidate dosage between juvenile male SHRSP (0.01–0.1 mg/kg, intraperitoneally) and children with developmental disabilities and AD/HD (0.1–1 mg/kg, orally), the administration routes should be considered since methylphenidate applied intraperitoneally is more available to the brain than when given orally (Gerasimov *et al.*, 2000). We therefore suggest that the dosage of methylphenidate given to the juvenile SHRSP in the present study is rational and in close agreement with the clinical dosage. In other words, the relevance of juvenile SHRSP as an animal model of the developmental disorder AD/HD is supported from the pharmacological viewpoint of the effectiveness of the psychostimulant methylphenidate.

CONSIDERATIONS FOR THE SHRSP AS AN ANIMAL MODEL OF AD/HD

Several animals have been proposed as animal models of AD/HD. SHR are popular models of AD/HD because of some behavioral similarities with those of AD/HD patients (Knardahl and Sagvolden, 1979; Myers *et al.*, 1982; Cierpial *et al.*, 1989; Wultz *et al.*, 1990; Sagvolden *et al.*, 1992, 1993a, b; Sagvolden, 2000). However, some disadvantages have also been pointed out. Indeed, male preponderance is not observed in SHR (Sagvolden and Berger, 1996; Berger and Sagvolden, 1998). Female, not male, SHR exhibit impulsivity/inattention, although both are hyperactive. Furthermore the effectiveness of psychostimulants in SHR is not consistent with behavioral paradigms used (Myers *et al.*, 1982; Wultz *et al.*, 1990; Sagvolden *et al.*, 1992; Evenden and Myerson, 1999). Thus, SHR is unlikely to fulfill the behavioral and pharmacological features as an AD/HD animal model.

A key concern about SHRSP as well as SHR as an AD/HD animal model is whether the behavioral alterations are associated with hypertension. However, the relationship between hyperactive behavior and hypertension in SHR has already been genetically separated by the establishment of the Wistar-Kyoto strain of hyperactive rats (WKHA) without hypertension and the Wistar-Kyoto strain of hypertensive rats (WKHT) without hyperactivity, derived from selective hybrid backcross breeding with WKY and SHR (Hendley *et al.*, 1986). The possibility that the behavioral alterations are implicated in hypertension is also denied by our observations that there is no correlation in juvenile SHRSP between systolic blood pressure and behavioral parameters such as ambulatory and rearing activity counts in the open-field task, percent alternation behavior in the Y-maze task or the number of open arm entries in the elevated plus-maze task (data not shown). In addition, SHRSP used are in the prehypertensive stage. Thus, we do not think that hypertension restrict the use of juvenile SHRSP as an AD/HD behavioral animal model.

How methylphenidate ameliorates the impaired spontaneous alternation and hyperactive behavior in juvenile SHRSP remains unknown. Methylphenidate has been reported to affect dopaminergic system via DAT blockade (Seeman and Madras, 1998). For the monoamine hypothesis, in particular, the DA system in the prefrontal cortex has been deeply involved in brain function in AD/HD (Barkley *et al.*, 1992; Pliszka *et al.*, 1996). Furthermore, Dougherty *et al.* (1999) reported that striatal DAT density was increased in patients with AD/HD and over expression of DAT may be the major cause of AD/HD. Enhancement of DA transmission by methylphenidate might have a key role, since central dopaminergic hypofunction such as lowered extracellular DA levels was found in the prefrontal cortex, nucleus accumbens shell, striatum, and basolateral amygdala of SHRSP (Nakamura *et al.*, 2001).

The n-3 PUFA (especially DHA and EPA) deficiency and an increase in the n-6/n-3 ratio have been reported not only in the plasma, but also in the cerebral cortex and the hippocampus of SHRSP (Minami *et al.*, 1997; Kimura *et al.*, 2000), as seen in children with AD/HD (Colquhoun and Bunday, 1981; Mitchell *et al.*, 1987; Stevens *et al.*, 1995; Burgess *et al.*, 2000). Therefore, it is likely that n-3 PUFA deficiency is, at least in part, explainable in the pathogenesis of AD/HD, which may in turn indicate an importance of the n-6/n-3 balance as the pathophysiological indication of AD/HD. The DHA deficiency in the central nervous system leads to cognitive dysfunction (Greiner *et al.*, 1999). It has been reported that the dietary n-3 fatty acid deficiency induces a decrease in hippocampal nerve growth factor (NGF) levels (Ikemoto *et al.*, 2000). It is also known that intracerebral infusion of NGF ameliorates cholinergic cell body atrophy and spatial memory impairment (Fischer *et al.*, 1987). Minami *et al.* (1997) have reported that dietary DHA supplementation increased the DHA and EPA levels and decreased AA levels in the plasma and brain, and consequently normalized the n-6/n-3 balance in SHRSP. The DHA supplementation certainly increases ACh levels in the cerebral cortex and hippocampus, and ameliorates cognitive dysfunction in SHRSP (Minami *et al.*, 1997). Tsukada *et al.* (2000) reported that DHA supplementation improved the decreased rCBF response in aged monkey. Hibbeln *et al.* (2000) have also reported that CSF 5-HIAA level is positively correlated with plasma PUFA levels in human. Thus, dietary DHA supplementation will become the beneficial strategy for ameliorating the behavioral abnormalities in AD/HD.

It is well known that serotonergic dysfunction such as 5-HT and 5-HIAA deficiency has been deeply implicated in aggressive and impulsive behaviors in patients with AD/HD (Saul and Ashby, 1986; Kruesi *et al.*, 1990; Comings, 1993; Hornig, 1998; Spivak *et al.*, 1999). Furthermore, AD/HD children with 5-HT deficiency have a good response to the psychostimulant pemoline, while AD/HD children without 5-HT deficiency have a poor response (Saul and Ashby, 1986). Thus, peripheral or central 5-HT levels may be a useful marker in deciding upon psychostimulant medication in treating hyperactivity/impulsivity and inattention. Togashi *et al.* (1994) have been reported that CSF 5-HT level was significantly lower in SHRSP, but not in SHR, than that of WKY. Therefore, it seems likely that serotonergic hypofunction due to lowered CSF 5-HT level may be deeply involved in hyperactivity/impulsivity and inattention in SHRSP, since serotonergic agents, such as a selective serotonin reuptake inhibitor, fluoxetine, a precursor of serotonin, tryptophan, and a mixed serotonin receptor agonist, quipazine, counteracted hyperlocomotion in hyperdopaminergic mutant mice inflicted with DAT-KO insult (Gainetdinov *et al.*, 1999). Therefore, it is noteworthy that SHRSP with CSF 5-HT level also have a good response to the psychostimulant methylphenidate, as seen in AD/HD children with 5-HT deficiency.

We cannot explain the mechanisms involving in the gender difference in spontaneous alternation behavior observed in juvenile SHRSP. Taken together previous reports (Fader *et al.*, 1999; Yamada *et al.*, 1999; King *et al.*, 2000), neurosteroidal mechanisms including the disruptive effects of androgens and the protective effects of estrogen on the working memory components or the ability to sustain attention might underlie the gender-specific behavioral features in juvenile SHRSP. It is of interest whether the mechanisms underlying the gender difference may be implicated in the neurochemical features in SHRSP as well as AD/HD patients mentioned above.

In conclusion, we have presented that juvenile SHRSP manifest behavioral features, hyperactivity/impulsivity and/or inattention, similar to the children with AD/HD. The attention-deficit was gender-specific; it was predominantly observed in male SHRSP. Clinical dosages of methylphenidate ameliorated these symptomatic behaviors. Thus, juvenile SHRSP are a behavioral animal model of AD/HD with some behavioral and pharmacological advantages. Further studies on juvenile SHRSP promises to provide insights into the pathogenesis and to develop new therapeutic strategies for AD/HD.

Acknowledgements

The authors would like to thank Novartis Pharma (Basel, Switzerland) for donation of the drug methylphenidate hydrochloride.

REFERENCES

Aman, M. G., Marks, R. E., Turbott, S. H., Wilsher, *et al.* (1991). Methylphenidate and thioridazine in the treatment of intellectually subaverage children: effects on cognitive-motor performance, *J. Am. Acad. Child Adolesc. Psychiatry* **30**, 816–824.

American Psychiatric Association (1994). Attention-deficit and disruptive behavior disorders, in: *Diagnostic and Statistical Manual of Mental Disorders*, pp. 78–85. DSM-IV, APA, Washington, DC.

Barkley, R. A. (1990). Attention deficit hyperactivity disorder, in: *A Handbook for Diagnosis and Treatment*, Barkley, R. A. (Ed.), pp. 135–137. Guilford Press, New York.

Barkley, R. A. (1998). Attention-deficit hyperactivity disorder, *Sci. Am.* **279**, 43–50.

Barkley, R. A., Grodzinsky, G. and DuPaul, G. J. (1992). Frontal lobe functions in attention deficit disorder with and without hyperactivity: a review and research report, *J. Abnorm. Child Psychol.* **20**, 163–188.

Berger, D. F. and Sagvolden, T. (1998). Sex differences in operant discrimination behaviour in an animal model of attention-deficit hyperactivity disorder, *Behav. Brain Res.* **94**, 73–82.

Bliss, T. V. and Lømo, T. (1973). Long-lasting potentiation of synaptic transmission in the dentate area of the anaesthetized rabbit following stimulation of the perforant path, *J. Physiol.* **232**, 331–356.

Bradley, C. (1937). The behavior of children receiving Benzedrine, *Am. J. Psychiatry* **94**, 577–585.

Bull, E., Reavill, C., Hagan, J. J., *et al.* (2000). Evaluation of the spontaneously hypertensive rat as a model of attention deficit hyperactivity disorder: acquisition and performance of the DRL-60s test, *Behav. Brain Res.* **109**, 27–35.

Burgess, J. R., Stevens, L., Zhang, W., *et al.* (2000). Long-chain polyunsaturated fatty acids in children with attention-deficit hyperactivity disorder, *Am. J. Clin. Nutr.* **71** (Suppl.), 327S–330S.

Cierpial, M. A., Shasby, D. E., Murphy, C. A., *et al.* (1989). Open-field behavior of spontaneously hypertensive and Wistar-Kyoto normotensive rats: effects of reciprocal cross-fostering, *Behav. Neural Biol.* **51**, 203–210.

Colquhoun, I. and Bunday, S. (1981). A lack of essential fatty acids as a possible cause of hyperactivity in children, *Med. Hypotheses* **7**, 673–679.

Comings, D. E. (1993). Serotonin and the biochemical genetics of alcoholism: lessons from studies of attention deficit hyperactivity disorder (ADHD) and Tourette syndrome, *Alcohol* **2** (Suppl.), 237–241.

Conners, C. K. and Eisenberg, L. (1963). The effects of methylphenidate on symptomatology and learning in disturbed children, *Am. J. Psychiatry* **120**, 458–464.

Dougherty, D. D., Bonab, A. A., Spencer, T. J., *et al.* (1999). Dopamine transporter density in patients with attention deficit hyperactivity disorder, *Lancet* **354**, 2132–2133.

Evenden, J. and Meyerson, B. (1999). The behavior of spontaneously hypertensive and Wistar-Kyoto rats under a paced fixed consecutive number schedule of reinforcement, *Pharmacol. Biochem. Behav.* **63**, 71–82.

Fader, A. J., Johnson, P. E. and Dohanich, G. P. (1999). Estrogen improves working but not reference memory and prevents amnestic effects of scopolamine of a radial-arm maze, *Pharmacol. Biochem. Behav.* **62**, 711–717.

Fischer, W., Wictorin, K., Bjorklund, A., *et al.* (1987). Amelioration of cholinergic neuron strophy and spatial memory impairment in aged rats by nerve growth factor, *Nature* **329**, 65–68.

Gainetdinov, R. R., Wetsel, W. C., Jones, S. R., *et al.* (1999). Role of serotonin in the paradoxical calming effect of psychomotor stimulants on hyperactivity, *Science* **283**, 397–401.

Gerasimov, M. R., Franceschi, M., Volkow, N. D., *et al.* (2000). Comparison between intraperitoneal and oral methylphenidate administration: a microdialysis and locomotor activity study, *J. Pharmacol. Exp. Ther.* **295**, 51–57.

Ghuman, J. K., Ginsburg, G. S., Subramaniam, G., *et al.* (2001). Psychostimulants in preschool children with attention-deficit/hyperactivity disorder: clinical evidence from a developmental disorder institution, *J. Am. Acad. Child Adolesc. Psychiatry* **40**, 516–524.

Giros, B., Jaber, M., Jones, S. R., *et al.* (1996). Hyperlocomotion and indifference to cocaine and amphetamine in mice lacking the dopamine transporter, *Nature* **379**, 606–612.

Greiner, R. S., Moriguchi, T., Hutton, A., *et al.* (1999). Rats with low levels of brain docosahexaenoic acid show impaired performance in olfactory-based and spatial learning tasks, *Lipids* **34** (Suppl.), S239–S243.

Gustafsson, P., Thernlund, G., Ryding, E., *et al.* (2000). Associations between cerebral blood-flow measured by single photon emission computed tomography (SPECT), electro-encephalogram (EEG), behaviour symptoms, cognition and neurological soft signs in children with attention-deficit hyperactivity disorder, *Acta Paediatr.* **89**, 830–835.

Handen, B. L., Feldman, H. M., Lurier, A., *et al.* (1999). Efficacy of methylphenidate among preschool children with developmental disabilities and ADHD, *J. Am. Acad. Child Adolesc. Psychiatry* **38**, 805–812.

Hendley, E. D., Wessel, D. J. and Van Houten, J. (1986). Inbreeding of Wistar-Kyoto rat strain with hyperactivity and without hypertension, *Behav. Neural Biol.* **45**, 1–16.

Hibbeln, J. R., Umhau, J. C., George, D. T., *et al.* (2000). Plasma total cholesterol concentrations do not predict cerebrospinal fluid neurotransmitter metabolites: implications for the biophysical role of highly unsaturated fatty acids, *Am. J. Clin. Nutr.* **71** (Suppl.), 331S–338S.

Honer, W. G., Prohovnik, I., Smith, G., *et al.* (1988). Scopolamine reduces frontal cortex perfusion, *J. Cereb. Blood Flow Metab.* **8**, 635–641.

Hornig, M. (1998). Addressing comorbidity in adults with attention-deficit/hyperactivity disorder, *J. Clin. Psychiatry* **59** (Suppl. 7), 69–75.

Ikemoto, A., Nitta, A., Furukawa, S., *et al.* (2000). Dietary n-3 fatty acid deficiency decreases nerve growth factor content in rat hippocampus, *Neurosci. Lett.* **285**, 99–102.

Katz, R. J. and Schmaltz, K. (1980). Dopaminergic involvement in attention: a novel animal model, *Prog. Neuropsychopharmacol.* **4**, 585–590.

Kempton, S., Vance, A., Maruff, P., *et al.* (1999). Executive function and attention deficit hyperactivity disorder: stimulant medication and better executive function performance in children, *Psychol. Med.* **29**, 527–538.

Kimura, S. (2000). Antihypertensive effect of docosahexaenoic acid in stroke-prone spontaneously hypertensive rats, *Yakugaku Zasshi* (in Japanese) **120**, 607–619.

King, J. A., Barkley, R. A., Delville, Y., *et al.* (2000). Early androgen treatment decreases cognitive function and catecholamine innervation in an animal model of ADHD, *Behav. Brain Res.* **107**, 35–43.

Knardahl, S. and Sagvolden, T. (1979). Open-field behavior of spontaneously hypertensive rats, *Behav. Neural Biol.* **27**, 187–200.

Kostrzewa, R. M., Brus, R., Kalbfleisch, J. H., *et al.* (1994). Proposed animal model of attention deficit hyperactivity disorder, *Brain Res. Bull.* **34**, 161–167.

Krause, K. H., Dresel, S. H., Krause, J., *et al.* (2000). Increased striatal dopamine transporter in adult patients with attention deficit hyperactivity disorder: effects of methylphenidate as measured by single photon emission computed tomography, *Neurosci. Lett.* **285**, 107–110.

Kruesi, M. J. P., Rapoport, J. L., Hamburger, S., *et al.* (1990). Cerebrospinal fluid monoamine metabolites, aggression, and impulsivity in disruptive behavior disorders of children and adolescents, *Arch. Gen. Psychiatry* **47**, 419–426.

Lahey, B. B., Applegate, B., McBurnett, K., *et al.* (1994). DSM-IV field trials for attention deficit hyperactivity disorder in children and Adolescents, *Am. J. Psychiatry* **151**, 1673–1685.

Lou, H. C., Henriksen, L., Bruhn, P., *et al.* (1989). Striatal dysfunction in attention deficit and hyperkinetic disorder, *Arch. Neurol.* **46**, 48–52.

Luthman, J., Fredriksson, A., Lewander, T., *et al.* (1989). Effects of d-amphetamine and methylphenidate on hyperactivity produced by neonatal 6-hydroxydopamine treatment, *Psychopharmacology* **99**, 550–557.

Minami, M., Togashi, H., Koike, Y., *et al.* (1985). Changes in ambulation and drinking behavior related to stroke in stroke-prone spontaneously hypertensive rats, *Stroke* **16**, 44–48.

Minami, M., Kimura, S., Endo, T., *et al.* (1997). Dietary docosahexaenoic acid increases cerebral acetylcholine levels and improves passive avoidance performance in stroke-prone spontaneously hypertensive rats, *Pharmacol. Biochem. Behav.* **58**, 1123–1129.

Mitchell, E. A., Aman, M. G., Turbott, S. H., *et al.* (1987). Clinical characteristics and serum essential fatty acid levels in hyperactive children, *Clin. Pediatr.* **26**, 406–411.

Mori, K., Togashi, H., Ueno, K.-i., *et al.* (2001). Aminoguanidine prevented the impairment of learning behavior and hippocampal long-term potentiation following transient cerebral ischemia, *Behav. Brain Res.* **120**, 159–168.

Myers, M. M., Musty, R. E. and Hendley, E. D. (1982). Attenuation of hyperactivity in the spontaneously hypertensive rat by amphetamine, *Behav. Neural Biol.* **34**, 42–54.

Nakamura, K., Shirane, M. and Koshikawa, N. (2001). Site-specific activation of dopamine and serotonine transmission by aniracetam in the mesocorticolimbic pathway of rats, *Brain Res.* **897**, 82–92.

Nakao, K., Ikegaya, Y., Yamada, M. K., *et al.* (2001). Spatial performance correlates with long-term potentiation of the dentate gyrus but not of the CA1 region in rats with fimbria-fornix lesions, *Neurosci. Lett.* **307**, 159–162.

Okamoto, K. and Aoki, K. (1963). Development of a strain of spontaneously hypertensive rat, *Jap. Circ. J.* **27**, 282–293.

Okamoto, K., Yamori, Y. and Nagaoka, A. (1974). Establishment of the stroke-prone spontaneously hypertensive rat (SHR), *Circ. Res.* **34/35** (Suppl. I), 143–153.

Parada-Turska, J. and Turski, W. A. (1990). Excitatory amino acid antagonists and memory: effect of drugs acting at N-methyl-d-aspartate receptors in learning and memory tasks, *Neuropharmacology* **29**, 1111–1116.

Pellow, S. and File, S. E. (1986). Anxiolytic and anxiogenic drug effects on exploratory activity in an elevated plus-maze : a novel test of anxiety in the rat, *Pharmacol. Biochem. Behav.* **24**, 525–529.

Pellow, S., Chopin, P., File, S. E., *et al.* (1985). Validation of open:closed arm entries in an elevated plus-maze as a measure of anxiety in the rat, *J. Neurosci. Methods* **14**, 149–167.

Pliszka, S. R. (1989). Effect of anxiety on cognition, behavior, and stimulant response in ADHD, *J. Am. Acad. Child Adolesc. Psychiatry* **28**, 882–887.

Pliszka, S. R., McCracken, J. T. and Maas, J. W. (1996). Catecholamines in attention-deficit hyperactivity disorder: current perspectives, *J. Am. Acad. Child Adolesc. Psychiatry* **35**, 264–272.

Powell, A. L., Yudd, A., Zee, P., *et al.* (1997). Attention deficit hyperactivity disorder associated with orbitofrontal epilepsy in a father and a son, *Neuropsychiatry Neuropsychol. Behav. Neurol.* **10**, 151–154.

Prohovnik, I., Arnold, S. E., Smith, G., *et al.* (1997). Physostigmine reversal of scopolamine-induced hypofrontality, *J. Cereb. Blood Flow Metab.* **17**, 220–228.

Rapport, M. D., Carlson, G. A., Kelly, K. L., *et al.* (1993). Methylphenidate and desipramine in hospitalized children: separate and combined effects on cognitive function, *J. Am. Acad. Child Adolesc. Psychiatry* **32**, 333–342.

Russell, V., de Villiers, A., Sagvolden, T., *et al.* (1995). Altered dopaminergic function in the prefrontal cortex, nucleus accumbens and caudate-putamen of an animal model of attention-deficit hyperactivity disorder — the spontaneously hypertensive rat, *Brain Res.* **676**, 343–351.

Sagvolden, T. (2000). Behavioral validation of the spontaneously hypertensive rat (SHR) as an animal model of attention-deficit/hyperactivity disorder (AD/HD), *Neurusci. Biobehav. Rev.* **24**, 31–39.

Sagvolden, T. and Berger, D. F. (1996). An animal model of attention deficit disorders: The female shows more behavioral problems and is more impulsive than the male, *Eur. Psychologist* **1**, 113–122.

Sagvolden, T., Metzger, M. A., Schiorbeck, H. K., *et al.* (1992). The spontaneously hypertensive rat (SHR) as an animal model of childhood hyperactivity (ADHD): changed reactivity to reinforcers and to psychomotor stimulants, *Behav. Neural Biol.* **58**, 103–112.

Sagvolden, T., Metzger, M. A. and Sagvolden, G. (1993a). Frequent reward eliminates differences in activity between hyperkinetic rats and controls, *Behav. Neural Biol.* **59**, 225–229.

Sagvolden, T., Pettersen, M. B. and Larsen, M. C. (1993b). Spontaneously hypertensive rats (SHR) as a putative animal model of childhood hyperkinesis: SHR behavior compared to four other rats strains, *Physiol. Behav.* **54**, 1047–1055.

Sarter, M., Bodewitz, G. and Stephens, D. N. (1988). Attenuation of scopolamine-induced impairment of spontaneous alternation behavior by antagonist but not inverse agonist and agonist β-carbolines, *Psychopharmacology* **94**, 491–495.

Saul, R. C. and Ashby, C. D. (1986). Measurement of whole blood serotonin as a guide in prescribing psychostimulant medication for children with attentional deficits, *Clin. Neuropharmacol.* **9**, 189–195.

Schweitzer, J. B., Faber, T. L., Grafton, S. T., *et al.* (2000). Alternations in the functional anatomy of working memory in adult attention deficit hyperactivity disorder, *Am. J. Psychiatry* **157**, 278–280.

Seeman, P. and Madras, B. K. (1998). Anti-hyperactivity medication: methylphenidate and amphetamine, *Mol. Psychiatry* **3**, 386–396.

Spivak, B., Vered, Y., Yoran-Hegesh, R., *et al.* (1999). Circulatory levels of catecholamines, serotonin and lipids in attention deficit hyperactivity disorder, *Acta Psychiat. Scand.* **99**, 300–304.

Stevens, L. J., Zentall, S. S., Deck, J. L., *et al.* (1995). Essential fatty acid metabolism in boys with attention-deficit hyperactivity disorder, *Am. J. Clin. Nutr.* **62**, 761–768.

Sykes, D. H., Douglas, V. I., Weiss, G., *et al.* (1971). Attention in hyperactive children and the effect of methylphenidate (ritalin), *J. Child Psychol. Psychiatry* **12**, 129–139.

Szatmari, P., Boyle, M. N. and Offord, D. R. (1989). ADHD and conduct disorder: degree of diagnostic overlap and differences among correlates, *J. Am. Acad. Child Adolesc. Psychiatry* **28**, 865–872.

Tannock, R., Ickowicz, A. and Schachar, R. (1995). Differential effects of methylphenidate on working memory in ADHD children with and without comorbid anxiety, *J. Am. Acad. Child Adolesc. Psychiatry* **34**, 886–896.

Taylor, E. (1998). Clinical foundations of hyperactivity research, *Behav. Brain Res.* **94**, 11–24.

Togashi, H., Minami, M., Bando, Y., *et al.* (1982). Effects of clonidine and guanfacine on drinking and ambulation in spontaneously hypertensive rats, *Pharmacol. Biochem. Behav.* **17**, 519–522.

Togashi, H., Matsumoto, M., Yoshioka, M., *et al.* (1994). Neurochemical profiles in cerebrospinal fluid of stroke-prone spontaneously hypertensive rats, *Neurosci. Lett.* **166**, 117–120.

Togashi, H., Kimura, S., Matsumoto, M., *et al.* (1996). Cholinergic changes in the hippocampus of stroke-prone spontaneously hypertensive rats, *Stroke* **27**, 520–526.

Tsuda, K., Tsuda, S., Masuyama, Y., *et al.* (1991). Alteration in catecholamine release in the central nervous system of spontaneously hypertensive rats, *Jpn. Heart J.* **32**, 701–709.

Tsukada, H., Kakiuchi, T., Fukumoto, D., *et al.* (2000). Docosahexaenoic acid (DHA) improves the age-related impairment of the coupling mechanism between neuronal activation and functional cerebral blood flow response: a PET study in conscious monkeys, *Brain Res.* **862**, 180–186.

Watanabe, Y., Fujita, M., Ito, Y., *et al.* (1997). Brain dopamine transporter in spontaneously hypertensive rats, *J. Nucl. Med.* **38**, 470–474.

Weiss, G. (1996). Attention deficit hyperactivity disorder, in: *Child and Adolescent Psychiatry: A Comprehensive Textbook*, 2nd edn, Lewis, M. (Ed.), pp. 544–563. Williams and Wilkins, Baltimore, MD, USA.

Wultz, B., Sagvolden, T., Moser, E. I., *et al.* (1990). The spontaneously hypertensive rat as an animal model of attention-deficit hyperactivity disorder: effects of methylphenidate on exploratory behavior, *Behav. Neural Biol.* **53**, 88–102.

Yamada, K., Tanaka, T., Zou, L. B., *et al.* (1999). Long-term deprivation of oestrogens by ovariectomy potentiates beta-amyloid-induced working memory deficits in rats, *Br. J. Pharmacol.* **128**, 419–427.

Yamori, Y. and Horie, R. (1977). Developmental course of hypertension and regional cerebral blood flow in stroke-prone spontaneously hypertensive rats, *Stroke* **8**, 456–461.

Advances in Neuroregulation and Neuroprotection (2005), pp. 677-684
Collin, C. *et al.* (Eds)
© VSP 2005

Intrathecal spermine and spermidine at high-doses induce antinociceptive effects in the mouse capsaicin test

K. TAN-NO [1,*], A. ESASHI [1], A. TAIRA [1], O. NAKAGAWASAI [1], F. NIIJIMA [1], C. SAKURADA [2], T. SAKURADA [2] and T. TADANO [1]

[1] *Department of Pharmacology, Tohoku Pharmaceutical University, 4-4-1 Komatsushima, Aoba-ku, Sendai 981-8558, Japan*
[2] *Department of Biochemistry, Daiichi College of Pharmaceutical Sciences, 22-1 Tamagawa-cho, Minami-ku, Fukuoka 815-8511, Japan*

Abstract—The antinociceptive effects of intrathecally (i.t.) administered high doses of spermine and spermidine, endogenous polyamines, were examined in the mouse capsaicin test. Spermine (2.5–10 nmol) and spermidine (10 and 15 nmol), when administered i.t. 5 min before the injection of capsaicin solution (1600 ng) into the plantar surface of a hindpaw, produced a significant reduction of the nociceptive behavioral response. Ifenprodil (2 and 4 nmol), an antagonist of the polyamine recognition site on N-methyl-D-aspartate (NMDA) receptor ion-channel complex, co-administered with spermine or spermidine significantly inhibited antinociception induced by both polyamines. These results indicate that spermine- and spermidine-induced antinociception may be mediated via the polyamine recognition site on NMDA receptor ion-channel complex in the mouse spinal cord.

Keywords: Spermine; spermidine; ifenprodil; capsaicin test; intrathecal administration; mouse.

INTRODUCTION

The polyamine recognition site has been found to exist as a part of the N-methyl-D-aspartate (NMDA) receptor ion-channel complex (for reviews, see Williams *et al.*, 1991; Rock and Macdonald, 1995). Spermine and spermidine, endogenous polyamines, modulate functions of the NMDA receptor ion-channel complex through the polyamine recognition site on the NR1 subunit of the receptor complex (for review, see Chizh *et al.*, 2001). In cultured cortical neurons, at low concentrations ($1-10\ \mu$M) spermine enhances NMDA-elicited currents by increasing channel opening frequency, and at higher concentrations ($>10\ \mu$M) it additionally produces a voltage-dependent decrease in NMDA receptor single-channel conductance and

---

*To whom correspondence should be addressed. E-mail: koichi@tohoku-pharm.ac.jp

average open time that limits its enhancing action (Rock and Macdonald, 1992). Spermidine also significantly increases hypertension induced by NMDA microinjected into the latero-caudal periaqueductal gray area at a low dose (0.01 μg/rat), but at higher doses (0.1 and 1 μg/rat) it significantly inhibits the NMDA-induced cardiovascular effect (Maione et al., 1994). These findings indicate that both spermine and spermidine have dual effects on the NMDA response; low concentrations (or doses) are stimulatory and high concentrations (or doses) are inhibitory. We have reported that intrathecally (i.t.) administered spermine at extremely low doses (0.1–10000 fmol) produces nociceptive behavior mainly consisting of biting and/or licking of the hindpaw along with a slight hindlimb scratching directed toward the flank in mice that is inhibited by several NMDA receptor antagonists including ifenprodil (Tan-No et al., 2000), which has an antagonistic activity at the polyamine recognition site (Carter et al., 1990). This observation suggests that spermine-induced nociceptive behavior is mediated through the activation of the NMDA receptor ion-channel complex by acting on the polyamine recognition site (Tan-No et al., 2000). These reports suggest that high dose polyamines can produce the antinociceptive effect through the polyamine recognition site in the spinal cord. However, this aspect remains to be elucidated. Therefore, we employed the mouse capsaicin test (Sakurada et al., 1992, 1998; Tan-No et al., 1998), which was developed in our laboratory as a model of chemogenic nociception to determine whether spermine and spermidine administered i.t. at high doses can produce the antinociceptive effects.

MATERIALS AND METHODS

Animals

Experiments were performed on male ddY-strain mice weighing 22–25 g (Japan SLC, Hamamatsu, Japan) and maintained on a 12 h light–dark cycle (light 8:00 a.m.–8:00 p.m.) with constant temperature (23 ± 1°C) and relative humidity (55 ± 5%). Groups of 10 mice were used only once for each experiment.

Drugs and chemicals

The following drugs and chemicals were purchased from commercial sources: spermine tetrahydrochloride, spermidine trihydrochloride (Nacalai Tesque, Kyoto, Japan), ifenprodil (Research Biochemical Inc., Natick, MA, USA) and capsaicin (Merck, West Point, PA, USA). All other chemicals were of the purest grade commercially available.

Administration procedure

Intrathecal injections were made in unanaesthetized mice at the L5, L6 intervertebral space as described by Hylden and Wilcox (1980). Briefly, a volume of 5 μl was administered i.t. with a 28 gauge needle connected to a 50 μl Hamilton microsyringe, with the animal lightly restrained to maintain the position of the needle. Puncture of the dura was indicated behaviorally by a slight flick of the tail. For i.t. administrations, both polyamines were dissolved in artificial cerebrospinal fluid (CSF), containing NaCl 7.4 g, KCl 0.19 g, $MgCl_2$ 0.19 g, $CaCl_2$ 0.14 g/1000 ml of distilled and sterilized water. Ifenprodil was dissolved in artificial CSF containing 20% dimethylsulfoxide. Capsaicin (1600 ng) was dissolved in physiological saline containing 7.5% dimethylsulfoxide and was injected under the skin of the plantar surface of the right hindpaw. Spermine and spermidine were administered i.t. 5 min before intraplantar injection of capsaicin. Ifenprodil was co-administered i.t. with each polyamine in a total volume of 5 μl.

Procedure of capsaicin test

Approximately 60 min before the i.t. administration, the mice were adapted to an individual cage (22.0 $\times$ 15.0 $\times$ 12.5 cm) which was also used as the observation chamber after injection of capsaicin solution. The capsaicin test was performed as previously described (Sakurada *et al.*, 1992, 1998; Tan-No *et al.*, 1998). Briefly, 20 μl of capsaicin solution (1600 ng) was injected under the skin of the plantar surface of the right hindpaw, using a microsyringe with a 26 gauge needle. Licking or biting of the injected hindpaw was defined as a nociceptive behavioral response and the total duration of the response was measured with a stopwatch. The animals were observed individually for 5 min, immediately after the intraplantar injection of capsaicin solution. The measurement of the behavioral response was performed blind, i.e. the observer had no information as to group designation.

Analyses of data

The results are presented as the means and S.E.M. The ID_{50} values with 95% confidence limits were calculated for reduction in capsaicin-induced nociceptive behavior by the method of Litchfield and Wilcoxon (1949). Significant differences between groups were determined by Tukey's test for multiple comparisons after analysis of variance (ANOVA). In all statistical comparisons, $p < 0.05$ was used as the criterion for statistical significance.

RESULTS

The injection of capsaicin solution (1600 ng) into the plantar surface of a hindpaw caused an acute nociceptive behavioral response that lasted about 5 min. When administered i.t. 5 min before the injection of capsaicin, spermine (2.5–10 nmol) and spermidine (10 and 15 nmol) produced a significant reduction of the nociceptive behavior with ID_{50} values of 4.2 (2.5–7.1) nmol and 10.5 (8.5–13.0) nmol, respectively (Fig. 1). Judging from the ID_{50} values, spermine was approximately

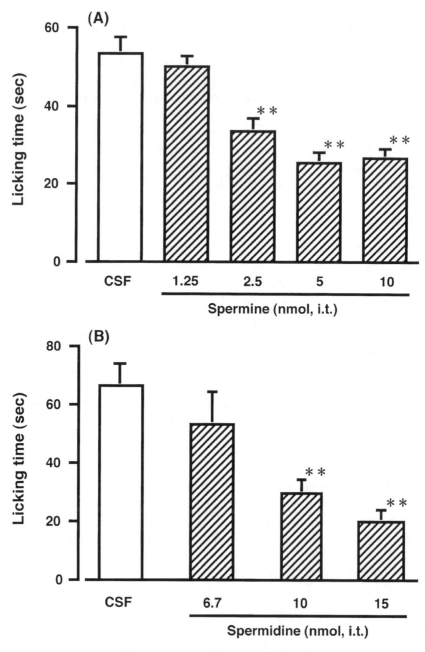

Figure 1. Antinociceptive effects of intrathecally administered spermine (A) and spermidine (B) in the mouse capsaicin test. The duration of licking and biting induced by capsaicin solution (1600 ng/20 μl) was determined over a 5 min period, starting immediately after intraplantar injection of capsaicin. Both polyamines were administered intrathecally 5 min before the injection of capsaicin. The data are given as means and S.E.M. for groups of 10 mice. ** $p < 0.01$ when compared to CSF-treated controls.

2.5 times more potent than spermidine in inducing the antinociceptive effect. Both polyamines applied caused no motor impairment such as convulsion and hindlimb paralysis even at the highest doses used.

As shown in Fig. 2, spermine (5 nmol)- and spermidine (15 nmol)-induced antinociception was significantly antagonized by co-administration of ifenprodil (2 and 4 nmol). A single administration of ifenprodil (4 nmol) did not induce an antinociceptive effect in the assay (data not shown).

DISCUSSION

Several lines of evidence indicate that NMDA receptors are involved in nociception at the spinal cord level. Behavioral experiments have shown that i.t. administration of NMDA at doses of 0.1–0.4 nmol in mice induces a behavioral response consisting of hindlimb scratching directed toward the flank, biting and/or licking of the hindpaw and the tail, which are thought to be nociceptive behavior (Sakurada *et al.*, 1990, 1991, 1996, 1999, 2000). We reported that i.t. administered spermine, at extremely low doses (0.1–10000 fmol) induces nociceptive behavior resembling that of NMDA (Tan-No *et al.*, 2000). The spermine-induced nociceptive behavior is dose-dependently inhibited by i.t. co-administration of ifenprodil, $(5R,10S)$-$(+)$-5-methyl-10,11-dihydro-5H-dibenzo[a,b]cycloheptene-5,10-imine hydrogen maleate (MK-801), an NMDA ion-channel blocker, and D$(-)$-2-amino-5-phosphonovaleric acid (D-APV) and 3-$((\pm)$-2-carboxypiperazin-4-yl)-propyl-1-phosphonic acid (CPP), the competitive NMDA receptor antagonists, but not by 6-cyano-7-nitroquinoxaline-2,3-dione (CNQX), a non-NMDA receptor antagonist, indicating that spermine-induced nociceptive behavior may be mediated through the activation of NMDA receptor ion-channel complex by acting on the polyamine recognition site (Tan-No *et al.*, 2000). We have recently found that big dynorphin (1–10 fmol), which is a prodynorphin-derived peptide and polycationic compound similar to spermine can also induce nociceptive behavior through the same mechanism as that of spermine (Tan-No *et al.*, 2002). These observations lead us to speculate that the polycationic compounds at extremely low doses can induce nociceptive behavior through the activation of the NMDA receptor ion-channel complex by acting on the polyamine recognition site. Moreover, i.t. administration of ifenprodil induces the antinociceptive effect with an ID$_{50}$ value of 13.8 (9.07–21.0) nmol in the mouse capsaicin test, though 7-chlorokynurenic acid, a competitive antagonist of the glycine recognition site on the NMDA receptor ion-channel complex, does not induce any antinociceptive effect even at the high dose of 20 nmol (Sakurada *et al.*, 1998). These reports suggest that the polyamine recognition site on the NMDA receptor ion-channel complex plays a significant role in the excitatory activities of NMDA receptors at the spinal cord level, which are involved in nociceptive processing.

On the other hand, as described in the Introduction, spermine and spermidine have dual effects on the NMDA response; facilitation at low concentrations and in-

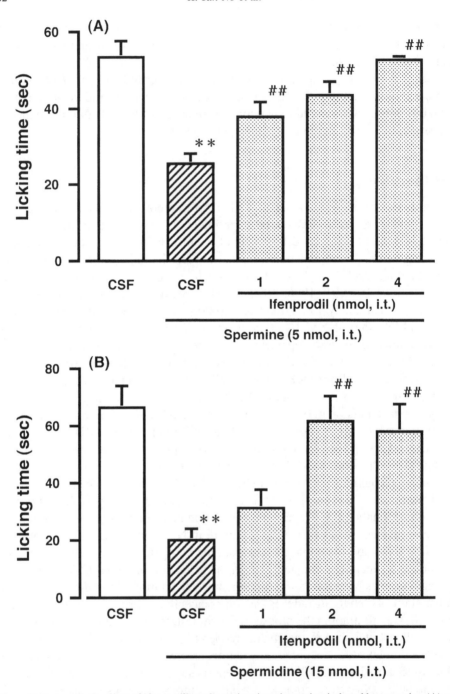

Figure 2. Antagonistic effect of ifenprodil on the antinociceptive action induced by spermine (A) and spermidine (B) in the mouse capsaicin test. Ifenprodil was co-administered intrathecally with each polyamine. Capsaicin was injected 5 min after intrathecal administration. The data are given as means and S.E.M. for groups of 10 mice. ** $p < 0.01$ when compared to CSF-treated controls. ## $p < 0.01$ when compared to each polyamine alone. For other details see Fig. 1.

hibition at high concentrations (Rock and Macdonald, 1992; Maione *et al.*, 1994). Based on those findings, it is suggested that high doses of polyamines can induce the antinociceptive effects through the polyamine recognition site in the spinal cord. Therefore, in the present study, we examined whether or not i.t. administration of spermine and spermidine at high doses can induce antinociceptive effects using the mouse capsaicin test; as a result, spermine (2.5–10 nmol) and spermidine (10 and 15 nmol) significantly induced antinociceptive effects. Spermine (5 nmol)- and spermidine (15 nmol)-induced antinociception was significantly antagonized by co-administration of ifenprodil at doses of 2 and 4 nmol which produces no antinociceptive effect in a single administration. Spermidine significantly decreases hypertension induced by NMDA microinjected into the latero-caudal periaqueductal gray area at doses of 0.1 and 1 μg/rat, but at 0.01 μg/rat, it significantly increases NMDA-induced hypertension (Maione *et al.*, 1994). The decrease of NMDA-induced hypertension produced by spermidine (0.1 and 1 μg/rat) is completely inhibited by pretreatment with arcain, another antagonist of the polyamine recognition site (Maione *et al.*, 1994). The present findings taken together with the report suggest that high doses of spermine and spermidine may negatively modulate the NMDA response through the polyamine recognition site. The mechanism by which high doses of polyamines exert negative modulation on the NMDA response remains to be elucidated. Polyamines dually modulate NMDA receptor functions through two different polyamine recognition sites. Rock and Macdonald (1992), and Maione *et al.* (1994) have speculated that on the outer side of the NMDA receptor ion-channel complex there is a high affinity and specific polyamine recognition site which may positively modulate the gating of the ion-channel associated with NMDA receptors. At high concentrations, however, polyamines have affinity with another recognition site near or in the ion-channel pore and exert negative modulatory effects by blocking the channel. Therefore, it seems likely that the antinociceptive effects produced by high dose polyamines may be mediated through the blockage of the ion-channel by acting on the polyamine recognition site near or in the ion-channel pore.

In conclusion, the present results indicate that i.t. administered spermine and spermidine at high doses induce antinociceptive effects, and that the antinociception induced by both polyamines may be mediated via the inhibition of the NMDA receptor ion-channel complex by acting on the polyamine recognition site.

Acknowledgements

This work was partly supported by a Grant-in-Aid for Scientific Research (14572062) from the Japanese Ministry of Education, Culture, Sports, Science and Technology to K. T.

REFERENCES

Carter, C. J., Lloyd, K. G., Zivkovic, B., *et al.* (1990). Ifenprodil and SL 82.0715 as cerebral antiischemic agents. III. Evidence for antagonistic effects at the polyamine modulatory site within the N-methyl-D-aspartate receptor complex, *J. Pharmacol. Exp. Ther.* **253**, 475–482.

Chizh, B. A., Headley, P. M. and Tzschentke, T. M. (2001). NMDA receptor antagonists as analgesics: focus on the NR2B subtype, *Trends Pharmacol. Sci.* **22**, 636–642.

Hylden, J. L. K. and Wilcox, G. L. (1980). Intrathecal morphine in mice, a new technique, *Eur. J. Pharmacol.* **67**, 313–316.

Litchfield, J. T. and Wilcoxon, F. (1949). A simplified method of evaluating dose-effect experiment, *J. Pharmacol. Exp. Ther.* **96**, 99–113.

Maione, S., Berrino, L., Pizzirusso, A., *et al.* (1994). Effects of the polyamine spermidine on NMDA-induced arterial hypertension in freely moving rats, *Neuropharmacology* **33**, 789–793.

Rock, D. M. and Macdonald, R. L. (1992). The polyamine spermine has multiple actions on N-methyl-D-aspartate receptor single-channel currents in cultured cortical neurons, *Mol. Pharmacol.* **41**, 83–88.

Rock, D. M. and Macdonald, R. L. (1995). Polyamine regulation of N-methyl-D-aspartate receptor channels, *Annu. Rev. Pharmacol. Toxicol.* **35**, 463–482.

Sakurada, T., Manome, Y., Tan-No, K., *et al.* (1990). The effects of substance P analogues on the scratching, biting and licking response induced by intrathecal injection of N-methyl-D-aspartate in mice, *Br. J. Pharmacol.* **101**, 307–310.

Sakurada, T., Tan-No, K., Manome, Y., *et al.* (1991). Aversive response produced by intrathecal injection of NMDA in mice: effects of substance P and 5-HT antagonists, in: *NMDA Receptor Related Agents: Biochemistry, Pharmacology and Behavior*, Kameyama, T., Nabeshima, T. and Domino, E. F. (Eds), pp. 219–225. NPP Books, Ann Arbor, Michigan.

Sakurada, T., Katsumata, K., Tan-No, K., *et al.* (1992). The capsaicin test in mice for evaluating tachykinin antagonists in the spinal cord, *Neuropharmacol.* **31**, 1279–1285.

Sakurada, T., Sugiyama, A., Sakurada, C., *et al.* (1996). Involvement of nitric oxide in spinally mediated capsaicin- and glutamate-induced behavioural responses in the mouse, *Neurochem. Int.* **29**, 271–278.

Sakurada, T., Wako, K., Sugiyama, A., *et al.* (1998). Involvement of spinal NMDA receptors in capsaicin-induced nociception, *Pharmacol. Biochem. Behav.* **59**, 339–345.

Sakurada, T., Yuhki, M., Inoue, M., *et al.* (1999). Opioid activity of sendide, a tachykinin NK1 receptor antagonist, *Eur. J. Pharmacol.* **369**, 261–266.

Sakurada, T., Sakurada, S., Katsuyama, S., *et al.* (2000). Evidence that N-terminal fragments of nociceptin modulate nociceptin-induced scratching, biting and licking in mice, *Neurosci. Lett.* **279**, 61–64.

Tan-No, K., Taira, A., Inoue, M., *et al.* (1998). Intrathecal administration of *p*-hydroxymercuribenzoate or phosphoramidon / bestatin-combined induces antinociceptive effects through different opioid mechanisms, *Neuropeptides* **32**, 411–415.

Tan-No, K., Taira, A., Wako, K., *et al.* (2000). Intrathecally administered spermine produces the scratching, biting and licking behaviour in mice, *Pain* **86**, 55–61.

Tan-No, K., Esashi, A., Nakagawasai, O., *et al.* (2002). Intrathecally administered big dynorphin, a prodynorphin-derived peptide, produces nociceptive behavior through an N-methyl-D-aspartate receptor mechanism, *Brain Res.* **952**, 7–14.

Williams, K., Romano, C., Dichter, M. A., *et al.* (1991). Modulation of the NMDA receptor by polyamines, *Life Sci.* **48**, 469–498.

Advances in Neuroregulation and Neuroprotection (2005), pp. 685-698
Collin, C. *et al.* (Eds)
© VSP 2005

An experimental rat model of Parkinson's disease induced by Japanese encephalitis virus

AKIHIKO OGATA *, SEIJI KIKUCHI and KUNIO TASHIRO

*Departments of Neurology, Hokkaido University Graduate School of Medicine, Kita-15, Nishi-7,
Kita-ku, Sapporo 060-8638, Japan*

Abstract—In adult Fischer rats sacrificed 12 weeks after infection with Japanese encephalitis virus
(JEV) at the age of 13 days, neuronal loss with gliosis was confined mainly to the zona compacta
of the bilateral substantia nigra, without lesions in the cerebral cortex and cerebellum. Furthermore,
the most severe lesions were in the central part of zona compacta; the lateral cell groups were less
affected. In addition, immunohistochemical study with anti-tyrosine hydroxylase (TH) showed that
the number of TH-positive neurons was decreased significantly in the substantia nigra compared with
that in controls, while comparable TH-positive neurons were found in the basal ganglia in the JEV-
treated rats and age-matched controls. Thus, the distribution of the pathologic lesions in infected
rats resembled those found in Parkinson's disease. The immunohistochemical studies failed to detect
JEV antigens in any region of the rat brain and the JEV genome was undetectable in the substantia
nigra and the cerebral cortex by reverse transcription-polymerase chain reaction. JEV-infected rats
showed marked bradykinesia. Significant behavioral improvement was observed upon administration
of L-DOPA. The findings suggest that JEV infection of rats under the conditions described may serve
as a model of virus induced Parkinson's disease. We could evaluate the possibilities of some new
drugs using this rat model.

Keywords: Parkinson's disease; dopamine; Japanese encephalitis virus; Parkinson model.

INTRODUCTION

The most widely used model of Parkinson's disease is that induced by the metabolites of 1-methyl-4-phenyl 1,2,3,6-tetrahydropyridine (MPTP) (Langston *et al.*,
1983; Burns *et al.*, 1983; Ballard *et al.*, 1985; Vingerhoets *et al.*, 1994). The
neuropathologic findings in MPTP-treated monkeys are similar to those of Parkinson's disease, in that neurons in the substantia nigra are primarily involved, while
other basal ganglia structures are spared. Despite these remarkable similarities,

---

*To whom correspondence should be addressed. E-mail: a-ogata@med.hokudai.ac.jp

MPTP-induced parkinsonism is different from Parkinson's disease in several respects (Langston et al., 1983; Ballard et al., 1985; Langston, 1989). Furthermore, MPTP acts only minimally or not at all in several animal species, including rats (Chiueh et al., 1983; Heikkila et al., 1984). Another model of parkinsonism, induced by stereotactic injection of 6-hydroxydopamine into the substantia nigra, is limited by technical difficulties (Kelly et al., 1975). We have reported the distribution of Japanese encephalitis virus (JEV) antigens in developing rat brain after intracerebral virus inoculation (Ogata et al., 1991). We concluded that the susceptibility to JEV infection in the rat brain was closely associated with neuronal immaturity; the neurons of basal ganglia and substantia nigra remained susceptible to infection longer than did those of the cerebral cortex. Developing immature neurons also showed the highest rate of infection by JEV in primary cultures of fetal rat brains (Kimura-Kuroda et al., 1993). In 10-day-old rats infected with JEV, JEV antigen was almost undetectable in the cerebral cortex, and in 14-day-old rats JEV antigen was confined to the basal ganglia and substantia nigra (Ogata et al., 1991). The analysis of the neuropathologic findings in rats infected with JEV on days 12, 13, and 14 after birth suggested that it would be possible to induce pathologic lesions similar to those of Parkinson's disease. These results strengthen the hypothesis that JEV might be a causal agent for some cases of Parkinson's disease.

MATERIALS AND METHODS

Animals and virus

Albino rats of the Fischer strain were obtained from SLC Japan. The virus strain used was the JaGAr-01 strain of JEV. The supernate from 10% homogenates of infected mouse brains (10^9 PFU/ml) were diluted with 20% Hemaccell (Hoechst) in Eagle's minimum essential medium and stored at $-70°C$ until use. Virus (0.03 ml containing 3×10^6 PFU) was inoculated intracerebrally with a specially designed two-step thin 27 gauge needle (Hoshimori Iryoki KK, Tokyo, Japan) with a stopper 3 mm from the tip. The site of inoculation was located at the midpoint of the line connecting the left eye to the midpoint between the right and left ears. Hemaccel (20%) in Eagle's minimum essential medium was injected into control rats.

Neuropathologic study

The topographical distribution of JEV antigen in the developing rat brain was determined 3 days after JEV inoculation. The neuropathologic changes in rats infected with JEV on days 12, 13, and 14 after birth were examined. Animals were sacrified 3 days, 10 days, 12 weeks, or one year after inoculation under ether anesthesia by perfusion fixation via the aorta with 4% freshly prepared paraformaldehyde in 0.1 M phosphate buffer. Coronal brain sections were taken from the frontal tip to the medulla, embedded in paraffin, and stained with hematoxylin-eosin, Luxol fast blue-cresyl violet (Klüver-Barrera method), anti-JEV antibody, and anti-tyrosine hydroxylase (TH) monoclonal antibody (Chemicon). The avidin-biotin-peroxidase comlex (ABC) method was used in this immunohistochemical study (Hsu *et al.*, 1981). After deparaffinization, the specimens were treated with 0.3% H_2O_2-methanol to suppress endogenous peroxidase activity, incubated with 10% normal goat serum, and allowed to react with anti-JEV rabbit serum or anti-TH monoclonal antibody, diluted in 1% bovine serum albumin (BSA) at 4°C overnight. Incubation with 1% BSA only was used as a negative control. The sections reacted with anti-JEV antibody and anti-TH monoclonal antibody were reacted with biotinylated goat anti-rabbit IgG and biotinylated goat anti-mouse IgG (Vecstain) respectively. ABC reaction products were visualized with 3,3'-diaminobenzidine tetrahydrochloride (Sigma), and counterstained with hematoxylin.

Reverse transcription-polymerase chain reaction (RT-PCR)

The primers used were 5'-5739 AGAGCGGGGAAAAAGGTCAT 5758-3' (sense) and 3'-5900 TTTCACGCTCTTTCTACAGT 5881-5' (antisense) for NS3 region amplification of the JEV genome (Sumiyoshi *et al.*, 1987). These were synthesized on a Cyclone Plus DNA/RNA synthesizer (Millipore Co., Japan). Prior neuropathological and immunohistochemical studies (Ogata *et al.*, 1991) demonstrated that rats infected with JEV on days 5 after birth had a severe panencephalitis with readily detectable JEV antigens. Therefore, brains taken from 3 rats inoculated with JEV on days 5 after birth and sacrificed 5 days later were used as positive controls. The cerebral cortex taken from an uninfected normal rat brain was used as a negative control. Rats infected with JEV on the day 13 of life were sacrificed 5 days (3 rats) or 12 weeks (3 rats) later. Pieces of the cerebral cortex and the ventral region of the midbrain were taken from each animal using a magnifying microscope. One hundred milligrams of brain tissue was added to 1 ml PBS and homogenized with a Polytron. Five microliters of the homogenate was incubated with an equal volume of detergent mix (2% NP-40, 10 U of RNase inhibitor (Takara Co., Kyoto, Japan) in PBS) in a 500-μl Eppendorf-type tube for 1 min at room temperature. This was added to 90 μl of RT-PCR mix (100 pmol of each primer, 0.2 mM deoxynucleoside triphosphate, 10 mM Tris (pH 8.9), 1.5 mM $MgCl_2$, 80 mM KCl, 0.5 mg of BSA per ml, 0.1% sodium cholate, 0.1% Triton X-100, 10 U of reverse transcriptase (Takara Co., Kyoto, Japan), and 2 U of Tth-DNA polymerase, a thermostable DNA polymerase (Toyobo Co., Osaka, Japan)). The reaction mixture was incubated for 10 min at 53°C for RT. PCR amplification (92°C for 60 s, 53°C for 90 s, and 72°C for 120 s by thermal cycler; PC-700, Astec Co., Fukuoka, Japan) was started immediately after the RT and performed for 35 cycles (Morita *et al.*, 1991). The PCR products were then separated on a 3% agarose gel and were visualized by ethidium bromide staining. Amplification yielded a 142-bp fragment.

Assessment of motor function

A pole test (Ogawa *et al.*, 1985) was performed as a way of evaluating bradykinesia in the rats. The time for the rats to descend from the top of a rough-surfaced pole (2.5 cm in diameter and 100 cm in height) to the floor was recorded in JEV-infected adult rats and control adult rats. To assess the efficiency of L-DOPA, the same procedure was repeated immediately after the successive intraperitoneal injection of L-DOPA (10 mg/kg/day) for 7 days. Comparisons among mean values for groups were done by Student's *t*-test.

RESULTS

The mortality of the rats infected with JEV

The mortality of rats infected with JEV is shown in Fig. 1. All rats infected when they were from 2 to 12 days old died. However, the mortality rate of animals infected when they were older than 12 days decreased with age, with 13-day-old rats having 50% mortality and 14-day-old rats having 8.3% mortality. Animals inoculated when they were older 17 days showed 0% mortality. This showed that the older the rat was at the time of infection, the longer it survived.

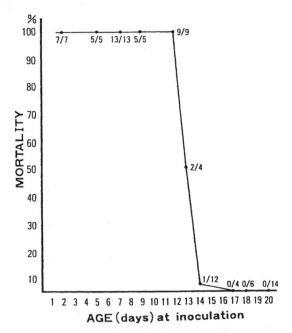

Figure 1. The mortality of the rats infected with Japanese encephalitis virus (JEV). All rats infected when they were from 2 to 12 days old died. However, the mortality rate of animals infected when they were older than 12 days decreased with age, with 13-day-old rats having 50% mortality and 14-day-old rats having 8.3% mortality. Animals inoculated when they were older 17 days showed 0% mortality. This showed that the older the rat was at the time of infection, the longer it survived.

Neuropathologic study

Pathological examination at higher magnifications clearly showed that in rats infected with JEV at the age of 7 days JEV antigen was present in the apical cytoplasmic portion of the neurons in the superficial areas of layer II (Fig. 2). Since the neuronal maturation of the cerebral cortex is known to proceed from the deeper layers (layers V and VI) to the upper layers (layers III and II), the results indicate that neurons in the cerebral cortex become resistant to JEV infection with maturation. In 15-day-old animals sacrificed 3 days after JEV infection, JEV antigen was found mainly in the substantia nigra and the caudate putamen in a distribution similar to that previously described (Table 1) (Ogata *et al.*, 1991). The expression of JEV antigen was strikingly localized to neurons in the substantia nigra. However, JEV antigen had disappeared from the brains of the rats sacrificed 10 days, 12 weeks, and one year after infection.

In the adult rats previously infected with JEV at the age of 13 days, neuronal loss was confined mainly to the zona compacta of the bilateral substantia nigra without lesions in the cerebral cortex or cerebellum (Figs 3A, 3B, arrows). Furthermore, the most severe lesions were found in the central part of the zona compacta, whereas the lateral cell groups were less affected (Fig. 4B). In the JEV-treated rats, marked neuronal loss with gliosis was found in the zona compacta of the substantia nigra

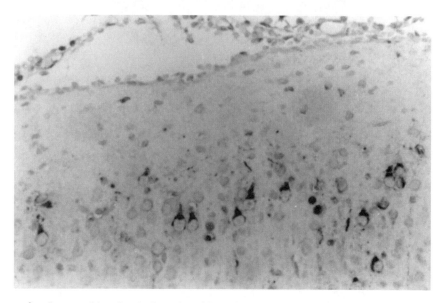

Figure 2. Immuno-histochemical study with anti-Japanese encephalitis virus (JEV) antibody. Pathological examination at higher magnifications clearly showed that in rats infected with JEV at the age of 7 days JEV antigen was present in the apical cytoplasmic portion of the neurons in the superficial areas of layer II (Fig. 2). Since the neuronal maturation of the cerebral cortex is known to proceed from the deeper layers (layers V and VI) to the upper layers (layers III and II), the results indicate that neurons in the cerebral cortex become resistant to JEV infection with maturation. The expression of JEV antigen was strikingly localized to neurons in the substantia nigra.

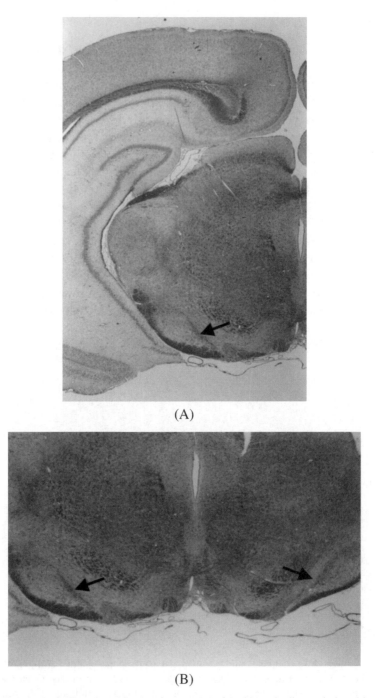

(A)

(B)

Figure 3. Pathologic findings in midbrain of rat model of Parkinson's disease induced by Japanese encephalitis virus (KB stain, ×40). In the adult rats previously infected with JEV at the age of 13 days, neuronal loss was confined mainly to the zona compacta of the bilateral substantia nigra without lesions in the cerebral cortex or cerebellum (Figs 3A, 3B, arrows).

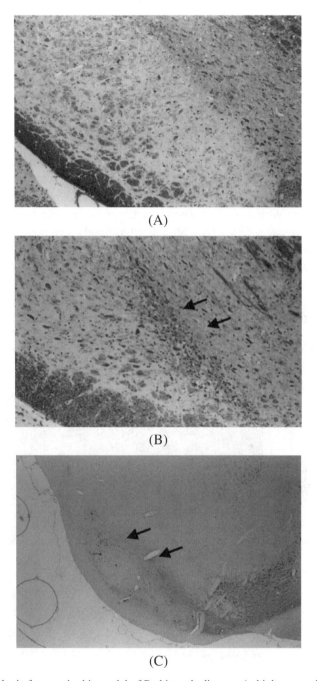

(A)

(B)

(C)

Figure 4. Pathologic features in this model of Parkinson's disease. At higher magnification (×400), the most severe lesions were found in the central part of the zona compacta, whereas the lateral cell groups were less affected (Fig. 4B). In the JEV-treated rats, marked neuronal loss with gliosis was found in the zona compacta of the substantia nigra (Fig. 4B), in contrast to the controls (Fig. 4A). Immunohistochemical analysis with anti-tyrosine hydroxylase (TH) antibody showed that TH-positive neurons were decreased in the substantia nigra (Fig. 4C, ×100).

(Fig. 4B), in contrast to the controls (Fig. 4A). In addition, immunohistochemical analysis with anti-tyrosine hydroxylase (TH) antibody showed that TH-positive neurons were decreased in the substantia nigra (Fig. 4C). However, many TH-positive neurons were found in the caudate putamen in JEV-infected rats as well as an age-matched control rat. These pathologic findings imply rather selective vulnerability of neurons of the substantia nigra and to a lesser extent caudate putamen of rats to the destructive effects of JEV following intracerebral inoculation.

In the JEV-infected rat brains, the same pathologic features were observed in more than 80% of the rats. In the rats sacrificed one year after JEV inoculation, almost no TH-positive neurons could be found in the substantia nigra (data not shown). This finding suggests that the function of the dopaminergic system may deteriorate further with age. Although no Lewy bodies could be observed and immunohistochemical staining with an ant-tau antibody was not seen (unpublished data), the pathologic distribution found in JEV-treated rats was similar to that found in Parkinson's disease.

Tissue distribution of the JEV genome by RT-PCR (Fig. 5)

The JEV genome was detected in the cerebral cortex (lane 1) from the rats sacrificed 5 days after JEV infection at the age of 5 days. In the rats infected with JEV at

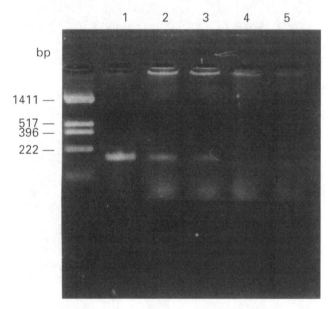

Figure 5. Tissue distribution of the JEV genome by RT-PCR. The JEV genome was detected in the cerebral cortex (lane 1) from the rats sacrificed 5 days after JEV infection at the age of 5 days. In the rats infected with JEV at the age of 13 days, the cerebral cortex (lane 2) and the substantia nigra (lane 3) taken from the rats sacrificed 5 days after infection contained the JEV genome, whereas the JEV genome was undetectable in both the cerebral cortex (lane 4) and the substantia nigra (lane 5) taken from rats sacrificed 12 weeks after infection.

the age of 13 days, the cerebral cortex (lane 2) and the substantia nigra (lane 3) taken from the rats sacrificed 5 days after infection contained the JEV genome, whereas the JEV genome was undetectable in both the cerebral cortex (lane 4) and the substantia nigra (lane 5) taken from rats sacrificed 12 weeks after infection. In both the negative control (not shown) and the cerebral cortex (lane 4) in which no abnormal neuropathologic change could be found, the JEV genome could not be detected. Moreover, in the affected substantia nigra (lane 5), the JEV genome had disappeared. These results indicated that there was no persistent infection.

Motor function

We measured the motor activity of the adult rats (12 weeks after infection) infected with JEV at the age of 13 days and age-matched control rats. The time required for the pole test was 18.3 ± 4.3 s ($n = 10$) for the JEV-infected rats before L-DOPA treatment, 6.5 ± 1.4 s ($n = 14$) for the control rats, and 8.3 ± 2.6 s ($n = 10$) for the JEV-infected rats after treatment with L-DOPA. The difference between the JEV-infected rats and the control rats was significant ($p < 0.001$, Fig. 6). The

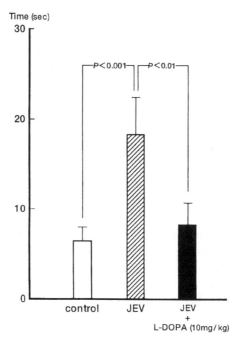

Figure 6. The evaluation of motor function by pole test. We measured the motor activity of the adult rats (12 weeks after infection) infected with JEV at the age of 13 days and age-matched control rats. The time required for the pole test was 18.3 ± 4.3 s ($n = 10$) for the JEV-infected rats before L-DOPA treatment, 6.5 ± 1.4 s ($n = 14$) for the control rats, and 8.3 ± 2.6 s ($n = 10$) for the JEV-infected rats after treatment with L-DOPA. The difference between the JEV-infected rats and the control rats was significant ($p < 0.001$, Fig. 6). The pole test results are consistent with marked bradykinesia of the JEV-infected rats. Significant behavioral improvement was observed after treatment with L-DOPA ($p < 0.01$, Fig. 6).

Table 1.
Topographical distribution of Japanese encephalitis virus (JEV) antigen in rat brains infected with (JEV) at each age

Sample location	Immunostaining* of samples from rats at the following age (days)											
	1	3	4	5	6	7	8	9	10	12	14	17
Caudate putamen	+	+	+	+	+	+	+	+	+	+	±	−
Substantia nigra	+	+	+	+	+	+	+	+	+	+	±	−
Thalamus	+	+	+	+	+	+	+	+	+	±	−	−
Amygdaloid nuclei	+	+	+	+	+	+	+	+	+	±	−	−
Mamillary body	+	+	+	+	+	+	+	+	−	−	−	−
Retrosplenial cortex	+	+	+	+	+	+	+	+	−	−	−	−
Dorsal cortex of inferior colliculus	+	+	+	+	+	+	+	+	−	−	−	−
Ammon's horn	+	+	+	+	+	+	+	±	−	−	−	−
Pontine nuclei	+	+	+	+	+	+	−	−	−	−	−	−
Nucleus solitary tract	+	+	+	+	+	+	−	−	−	−	−	−
Cerebellum	+	+	+	+	±	±	−	−	−	−	−	−
Olfactory bulb	+	+	+	−	−	−	−	−	−	−	−	−

*Intensity for immunostaining: +, definitely positive; ±, faintly positive; −, negative.

In 10-day-old rats infected with JEV, JEV antigen was almost undetectable in the cerebral cortex, and in 14-day-old rats JEV antigen was confined to the basal ganglia and substantia nigra. The analysis of the neuropathologic findings in rats infected with JEV on days 12, 13, and 14 after birth suggested that it would be possible to induce pathologic lesions similar to those of Parkinson's disease.

pole test results are consistent with marked bradykinesia of the JEV-infected rats. Masked faces or tremor could not be assessed in the rats. Significant behavioral improvement was observed after treatment with L-DOPA ($p < 0.01$, Fig. 6).

DISCUSSION

The pathogenesis of Parkinson's disease currently is thought to depend upon hereditary, aging, and environmental factors (Calne and Langston, 1983; Burns et al., 1983; Ballard et al., 1985; Cohen, 1986; Riederer et al., 1989; Dexter et al., 1989; Adams and Odunze, 1991; Nagatsu and Yoshida, 1988; Yoshida et al., 1990). Among the environmental factors, attention has largely focused on the role of toxic and infectious agents. Among the toxic factors, exogenous toxins such as MPTP and endogenous toxins such as free radicals or tetrahydroisoquinoline (Yoshida et al., 1990) have been implicated. Viruses can also selectively attack the substantia nigra and induce parkinsonism (Duvoisin and Yahr, 1965; Kristensson, 1992). Post-encephalitic parkinsonism was well documented (von Economo, 1917; Yahr, 1978). The world pandemic of encephalitis lethargica, von Economo encephalitis, from 1916 to 1927 resulted in the death or disability of approximately half a million people. Walters (1960) described a 54-year-old woman who manifested symptoms of parkinsonism while convalescing from meningoencephalitis due to Coxsackie B virus. Influenza A virus (Hudson and Rice, 1990), poliovirus (Bojinov, 1971), and measles virus (Alves *et al.*, 1992) have also been suspected from case reports as possible viral causes. Fishman *et al.* (1980) reported an experimental model with a selective attack on the substantia nigra and subthalamic nucleus by a strain of mouse hepatitis virus which is a corona virus that causes persistent CNS infection. Recently, similar features have been described in rats infected with influenza virus (Takahashi *et al.*, 1995).

JEV is a positive-stranded enveloped RNA virus that belongs to the family of the flavi viruses and is the most common cause of arthropod-borne human encephalitis worldwide. JEV is also occasionally the cause of post-encephalitis parkinsonism. Dickerson *et al.* (1952), in a study of 200 acute cases of Japanese encephalitis, noted tremors chiefly involving the fingers, tongue, and eyelids in 90%, muscular rigidity in more than 40%, and a mask-like face in 75% of the cases. Goto (1962) detected parkinsonian sequelae in 11.6% of 143 unselected patients five years after they had Japanese encephalitis.

The parkinsonian syndrome after Japanese encephalitis differs from that which follows encephalitis lethargica in several respects. In general, parkinsonism after Japanese encephalitis is mild, develops in the acute phase, and occasionally improves slightly over a long period. Recently, it was reported that MRI abnormalities were seen mainly in the substantia nigra and putamen in a case of typical parkinsonism after Japanese encephalitis (Shoji *et al.*, 1993). Patients without a clear history of encephalitis who follow the clinical course and have pathologic findings consistent with postencephalitic parkinsonism have been reported (Gibb and Lees, 1987; Geddes *et al.*, 1993). Although there were no Lewy bodies found in the substantia nigra in the JEV-treated rats, the pathologic findings otherwise resembled those of idiopathic Parkinson's disease. Furthermore, the immunohistochemical data using anti-TH antibody suggested that the function of the dopaminergic system might have deteriorated with age in the absence of ongoing or persistent JEV infection. McGeer *et al.* (1988) showed that the rate of neuronal cell degeneration was considerably higher in parkinsonian patients than could be accounted for on the basis of normal age-related neuronal degeneration alone. It seems likely that neuronal cell degeneration progressed more rapidly in the JEV-treated rats than in the controls. This observation raises the possibility that post-encephalitic parkinsonism as well as Parkinson's disease is a continuing degenerative process rather than an acute illness on which the effects of aging or decompensation are superimposed. Such late deterioration might be due to a resurgence of viral-mediated damage (Appel *et al.*, 1992), though our model does not support this view.

The complete nucleotide sequence of JEV genome RNA has been determined (Sumiyoshi *et al.*, 1987). Our RT-PCR study for NS3 region amplification (Morita *et al.*, 1991) of the JEV genome showed that JEV genome was undetectable in rats sacrificed 12 weeks after JEV infection at the age of 13 days. Moreover, JEV antigen as well as the JEV genome disappeared from the brain. These findings indicate that there is no persistent infection in the brain, and suggest that following the acute phase, JEV-infected rats are a safe model for researchers.

Thus far, no virus has been isolated from patients with Parkinson's disease, and there are no data that directly link known viruses to idiopathic Parkinson's disease. However, our findings support the possibility that as yet unidentified specific pathogens could cause similar pathologic lesions in man resulting in Parkinson's disease. Why neurons of the subtantia nigra remain susceptible to JEV infection longer than in other parts of the brain is unclear. One possibility is that virus receptors on the substantia nigra neurons persist longer. Certainly, the capacity of viruses to attack specific tissues selectively depends on an interaction between viral genes or proteins and host factors. Immune mechanism following an infection or other factors could be associated with the destruction of the substantia nigra. A more detailed understanding of JEV tropism for the substantia nigra in this experimental model might reveal mechanisms that aid in unraveling the degeneration of nigral dopaminergic neurons that is central to Parkinson's disease.

The JEV-induced parkinsonism in rats is characterized by selective destruction of neurons in the bilateral substantia nigra, especially in the zona compacta of the substantia nigra, similar to the lesions found in Parkinson's disease. Furthermore, significant behavioral improvement was observed upon administration of L-DOPA (Ogata *et al.*, 1997). Accordingly, this JEV-induced parkinsonian model is expected to be useful in the assessment of new anti-parkinsonian drugs (Ogata *et al.*, 1998, 2003; Hamaue *et al.*, 2001) the efficiency of neuronal transplantation therapy, and in the study of the pathophysiology of parkinsonism.

REFERENCES

Adams, J. D. and Odunze, I. N. (1991). Oxigen free radicals and Parkinson's disease, *Free Radical Biol. Med.* **10**, 161–169.

Alves, R. S. C., Barbosa, E. R. and Scaff, M. (1992). Postvaccinial parkinsonism, *Movement Disorders* **7**, 178–180.

Appel, S. H., Le, W.-D., Tajti, J., *et al.* (1992). Nigral damage and dopaminergic hypofunction in mesencephalon-immunized guinea pigs, *Ann. Neurol.* **32**, 494–501.

Ballard, P. A., Tetrud, J. W. and Langston, J. W. (1985). Permanent human parkinsonism due to 1-methyl-4-phenyl-1,2,3,6-tetrahydropyridine (MPTP): seven cases, *Neurology* **35**, 949–956.

Bojinov, S. (1971). Encephalitis with acute parkinsonian syndrome and bilateral inflammatory necrosis of the substantia nigra, *J. Neurol. Sci.* **12**, 383–415.

Burns, R. S., Chiueh, C. C., Markey, S. P., *et al.* (1983). A primate model of parkinsonism: Selective destruction of dopaminergic neurons in the pars compacta of the substantia nigra by N-methyl-4-phenyl-1,2,3,6-tetrahydropyridine, *Proc. Natl. Acad. Sci. USA* **80**, 4546–4550.

Calne, D. B. and Langston, J. W. (1983). Aetiology of Parkinson's disease, *Lancet* **ii**, 1457–1459.

Chiueh, C. C., Markey, S. P. and Burns, R. S. (1983). N-methyl-4-phenyl-1,2,3,6-tetrahydro-pyridine, a parkinsonian syndrome causing agent in man and monkey, produces different effects in guinea pig and rat, *Pharmacologist* **25**, 131.

Cohen, G. (1986). Monoamine oxidase, hydrogen peroxidase and Parkinson's disease, *Adv. Neurol.* **45**, 119–125.

Dexter, D. T., Wells, F. R., Lees, A. J., *et al.* (1989). Increased nigral iron content and alterations in other metal ions occurring in brain in Parkinson's disease, *J. Neurochem.* **52**, 1830–1836.

Dickerson, R. B., Newton, J. R. and Hansen, J. E. (1952). Diagnosis and immediate prognosis of Japanese B encephalitis, *Am. J. Med.* **12**, 277–288.

Duvoisin, R. C. and Yahr, M. D. (1965). Encephalitis and parkinsonism, *Arch. Neurol.* **12**, 227–239.

Fishman, P. S., Gass, J. S., Swoveland, P. T., *et al.* (1985). Infection of the basal ganglia by a murine coronavirus, *Science* **229**, 877–879.

Geddes, J. F., Hughes, A. J., Lees, A. J., *et al.* (1993). Pathological overlap in cases of parkinsonism associated with neurofibrillary tangles. A study of recent cases of postencephalitic parkinsonism and comparison with progressive supranuclear palsy and Guamanian parkinsonism-dementia complex, *Brain* **116**, 281–302.

Gibb, W. R. G. and Lees, A. J. (1987). The progression of idiopathic Parkinson's disease is not explained by age-related changes. Clinical and pathological comparisons with post-encephalitic parkinsonian syndrome, *Acta Neuropathol.* **73**, 195–201.

Goto, A. (1962). Follow-up study of Japanese B encephalitis, *Psychiat. Neurol. Jpn.* **64**, 236–266.

Hamaue, N., Ogata, A., Terado, M., *et al.* (2001). Selegiline effects on bradlykinesia and dopamine levels in a rat model of Parkinson's disease induced by the Japanese encephalitis virus, *Biogenic Amines* **16**, 523–530.

Heikkila, R. E., Hess, A. and Duvoisin, R. C. (1984). Dopaminergic neurotoxicity of 1-methyl-4-phenyl-1,2,5,6-tetrahydropyridine in mice, *Science* **224**, 1451–1453.

Hsu, S. M., Raine, L. and Fanger, H. (1981). Use of avidin-biotin-peroxidase complex (ABC) in immunoperoxidase techniques: a comparison between ABC and unlabeled antibody (PAP) procedures, *J. Histochem. Cytochem.* **29**, 577–580.

Hudson, A. J. and Rice, G. P. A. (1990). Similarities of Guamanian ALS/PD to post-encephalitic parkinsonism/ALS: possible viral cause, *Can. J. Neurol. Sci.* **17**, 427–433.

Kelly, P. H., Seviour, P. W. and Iversen, S. D. (1975). Amphetamine and apomorphine responses in the rat following 6-OHDA lesions of the nucleus accumbens septi and corpus striatum, *Brain Res.* **94**, 507–522.

Kimura-Kuroda, J., Ichikawa, M., Ogata, A., *et al.* (1993). Specific tropism of Japanese encephalitis virus for developing neurons in primary rat brain culture, *Arch. Virol.* **130**, 477–484.

Kristensson, K. (1992). Potential role of viruses in neuro-degeneration, *Mol. Chem. Neuropathol.* **16**, 45–58.

Langston, J. W. (1989). Current theories on the cause of Parkinson's disease, *J. Neurol. Neurosurg. Psychiat.* **52**, 13–17.

Langston, J. W., Ballard, P., Tetrud, J. W., *et al.* (1983). Chronic parkinsonism in humans due to a product of meperidine-analog synthesis, *Science* **219**, 979–980.

McGeer, P. L., Itagaki, S., Akiyama, H., *et al.* (1988). Rate of cell death in parkinsonism indicates active neuropathological process, *Ann. Neurol.* **24**, 574–576.

Morita, K., Tanaka, M. and Igarashi, A. (1991). Rapid identification of dengue virus serotypes by using polymerase chain reaction, *J. Clin. Microbiol.* **29**, 2107–2110.

Nagatsu, T. and Yoshida, M. (1988). An endogenous substance of the brain, tetra-hydroisoquinoline, produces parkinsonism in primate with decreased dopamine, tyrosine hydroxylase and biopterin in the nigrostriatal regions, *Neurosci. Lett.* **87**, 178–182.

Ogata, A., Nagashima, K., Hall, W. W., *et al.* (1991). Japanese encephalitis virus neurotropism is dependent on the degree of neuronal maturity, *J. Virol.* **65**, 880–886.

Ogata, A., Tashiro, K., Nukuzuma, S., *et al.* (1997). A rat model of Parkinson's disease induced by Japanese encephalitis virus, *J. Neuro. Virol.* **3**, 141–147.

Ogata, A., Nagashima, K., Yasui, K., *et al.* (1998). Sustained release dosage of thyrotropin-releasing hormone improves experimental Japanese encephalitis virus-induced parkinsonism in rats, *J. Neurol. Sci.* **159**, 135–139.

Ogata, A., Hamaue, N., Terado, M., *et al.* (2003). Isatin, an endogenous MAO inhibitor, improves bradykinesia and dopamine levels in a rat model of Parkinson's disease induced by Japanese encephalitis virus, *J. Neurol. Sci.* **206**, 79–83.

Ogawa, N., Hirose, Y., Ohara, T., *et al.* (1985). A simple quantitative bradykinesia test in MPTP-treated mice, *Res. Commun. Chem. Pathol. Pharmacol.* **50**, 435–441.

Richter, R. W. and Shimojyo, S. (1961). Neurologic sequelae of Japanese B encephalitis, *Neurology* **11**, 553–559.

Riederer, P., Sotic, E., Rausch, W. D., *et al.* (1989). Transition metals ferritin, glutathione and ascorbic acid in Parkinsonian brain, *J. Neurochem.* **52**, 515–520.

Shoji, H., Watanabe, M., Itoh, S., *et al.* (1993). Japanese encephalitis and parkinsonism, *J. Neurol.* **240**, 59–60.

Sumiyoshi, H., Mori, C., Fuke, I., *et al.* (1987). Complete nucleotide sequence of the Japanese encephalitis virus genome RNA, *Virology* **161**, 497–510.

Takahashi, M., Yamada, T., Nakajima, S., *et al.* (1995). The substania nigr is a major target for neurovirulent influenza A virus, *J. Exp. Med.* **181**, 2161–2169.

Vingerhoets, F. J. G., Snow, B. J., Tetrud, J. W., *et al.* (1994). Positron emission tomographic evidence for progression of human MPTP-induced dopaminergic lesions, *Ann. Neurol.* **36**, 765–770.

von Economo, C. (1917). Encephalitis lethargica, *Wien. Klin. Wschr.* **30**, 581–585.

Walters, J. H. (1960). Post-encephalitic parkinson syndrome after meningoencephalitis due to coxsackie virus group B, type 2, *New Eng. J. Med.* **263**, 744–747.

Yahr, M. D. (1978). Encephalitis lethargica (Von Economo's disease, epidemic encephalitis), in: *Handbook of Clinical Neurology,* Vinken, P. J. and Bruyn, G. W. (Eds), Vol. 34, pp. 451–457. Elsevier, Amsterdam.

Yoshida, M., Miwa, T. and Nagatsu, T. (1990). Parkinsonism in monkeys produced by chronic administration of an endogenous substance of the brain, tetrahydroisoquinoline: the behavioral and biochemical changes, *Neurosci. Lett.* **119**, 109–113.

Advances in Neuroregulation and Neuroprotection (2005), pp. 699-712
Collin, C. *et al.* (Eds)
© VSP 2005

Blood-brain barrier carrier-mediated transport and metabolism of L-histidine

E. SAKURAI [1,*], J. YAMAKAMI [2], T. SAKURADA [2], Y. OCHIAI [1]
and Y. TANAKA [1]

[1] *Department of Pharmaceutics I, Tohoku Pharmaceutical University, Aoba-ku, Sendai 981-8558, Japan*
[2] *Department of Pharmacy, Sapporo National Hospital, Sapporo 003-0804, Japan*

Abstract—To elucidate the functional importance of histamine at the blood-brain barrier (BBB), we discuss recent findings regarding the transport characteristics and metabolism of L-histidine, a precursor of histamine, in cultured rat brain microvascular endothelial cells (BMECs) which are major structural components of the BBB. L-Histidine was uptaken by rat BMECs via both Na^+-dependent system N and Na^+-independent system L transporters. Zinc ion had an enhancing effect on the BBB transport of L-histidine. L-Histidine is biotransformed to histamine by L-histidine decarboxylase (HDC). The presence of HDC protein and the expression of HDC mRNA were confirmed in rat BMECs, and the HDC activity of the BMECs was estimated to be 0.14 ± 0.05 pmol/mg protein/min. These findings indicated that L-histidine uptaken by rat BMECs was shown to be converted to histamine, suggesting that HDC may play an important role in the regulation of paracellular permeability through tight junctions in BBB.

Keywords: L-histidine; histamine; L-histidine decarboxylase; transport; metabolism; blood-brain barrier.

INTRODUCTION

The presence of a tight junction between the microvascular endothelial cells that make up the blood-brain barrier (BBB) limits the distribution of drug molecules from the blood into the brain intracellular spaces. In the discovery and development of new drugs that act on the central nervous system (CNS), it is essential to deliver the drugs into the CNS via the BBB. Thus, specific transporters are required for transcellular transport. However, acute potential alterations (within seconds to minutes) of tight junctions will be able to control the endothelial paracellular

---

*To whom correspondence should be addressed. E-mail: sakuraie@tohoku-pharm.ac.jp

permeability of drugs to the CNS through the tight junction. Histamine increases microvascular endothelial paracellular permeability (Niimi et al., 1992), and opens the inter-endothelial-cell tight junction (Majno and Palade, 1961). Histamine-induced endothelial cell retraction was also found both in vivo and in cultured endothelial cells (Majno et al., 1969; Meyrick and Brigham, 1984). This biogenic amine is known to be present in the walls of a variety of blood vessels of the rat (Howland and Spector, 1972; El-Ackad and Brody, 1975). Both H_1 and H_2 histamine-receptor stimulations increase microvascular permeability, depending on the circulatory bed. H_1 receptors increase permeability in cultured endothelial umbilical-vein cells, and H_2 receptors increase brain capillary permeability (Lo and Fan, 1987; Boertje et al., 1989). On the other hand, at least seven protein constituents of the tight junction have been identified (Anderson and Van Itallie, 1995). Of these, ZO-1, a 220 kDa phosphoprotein, is present in the BBB (Stevenson et al., 1986), and increased expression in parallel with trans-epithelial electrical resistance in cultured rat brain capillary endothelial cells (Krause et al., 1991). ZO-1 protein levels, therefore, respond to physiological influences and participate in the regulation of paracellular permeability through the tight junction. Histamine causes a reversible concentration-dependent reduction of this ZO-1 protein content, mediated by both H_1 and H_2 receptors, and also reduces ZO-1 expression within the time associated with increased paracellular permeability (Gardner et al., 1996). However, it was evident that histamine crossed the rat BBB with difficulty. Therefore, if the BBB has the ability to form histamine, histamine biosynthesized in brain microvascular endothelial cells may microregulate the paracellular permeability through the tight junction.

To elucidate this functional importance of histamine at the BBB, we first discussed the transport characteristics of L-histidine, a precursor of histamine, through the BBB using cultured rat brain microvascular endothelial cells (BMECs). Moreover, in this review, we discuss whether or not L-histidine decarboxylase (HDC: EC4.1.1.22), which catalyses the formation of histamine from L-histidine, is present in rat BMECs.

Isolation and culture of rat BMECs

Numerous works using cultured bovine or porcine brain microvascular endothelial cells have been described to investigate drug transport through the BBB (Pardridge et al., 1990; Terasaki et al., 1991; Walker et al., 1994). However, most in vivo studies on drug transport through the BBB have been done with small laboratory animals, such as rats. Thus, it is important to establish the cell culture system for laboratory animals which are commonly utilized in in vivo studies. We succeeded in the development of rat BMECs isolation with some modification of the technique described previously (Bowman et al., 1981; Gordon et al., 1991; Ichikawa et al., 1996; Yamakami et al., 1998). Isolated rat BMECs were cultured at 37°C with 95% air and 5% CO_2. Subculture was performed when the cells reached confluence, after approximately 10–12 days. Cells were trypsinized at a ratio of 1:3 after

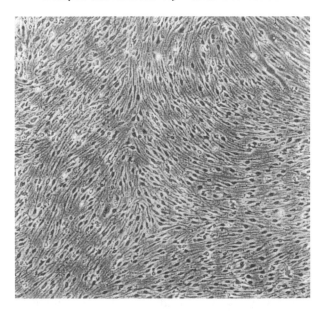

Figure 1. Phase contrast micrographs of cultured monolayers of rat BMECs at 12 days. Scale bar = 100 μm.

reaching confluence using 0.0025% trypsin in HBSS containing 0.02% EDTA. Secondary subcultured cells were grown on collagen-coated 25 cm² tissue culture flasks. Figure 1 shows a phase photomicrograph of the BMECs using rat collagen type 1 coated flask at 12 days. Identification of cells possessing Factor VIII antigen was done using immunofluorescence with goat anti-human von Willebrand factor antiserum as the primary antibody. Cellular uptake of Dil-Ac-LDLb was assessed by the method of Voyta *et al.* (1984). The culture also becomes progressively homogenous by the third passage, and all cells have this morphology (Magee *et al.*, 1994). Furthermore, it is expected that monolayers of BMECs remain polarized because several membrane functions have been reported to localize in the manner that is consistent with the brain microvascular endothelial cells *in vivo* (Audus *et al.*, 1990; Ichikawa *et al.*, 1996; Komura *et al.*, 1997). These features of the cultured cells enable us to study the polarized transport characteristics at luminal or antiluminal membrane.

Blood-brain barrier L-histidine transport

The plasma membrane of endothelial cells has been shown to be the site of several carrier-mediated transport systems (Pardridge, 1995; Tamai and Tsuji, 1996) including those for glucose, monocarboxylic acid and amino acids. System-N amino acid transport was first described by Kilberg *et al.* (1980) as mediating L-glutamine uptake in hepatocytes. This transport system has since been identified in skeletal muscle, placenta, lymphocytes, astrocytes and neurons. Three amino acids, L-glutamine, L-histidine and L-asparagine, all of which have nitrogen in their side group, are the natural substrates for this transporter. In particular, L-histidine, an essential amino acid, is a precursor of histamine. Histamine initiates transitory increases in endothelial permeability *in situ* and *in vitro* (Moy *et al.*, 1996). Under *in situ* conditions, increased permeability is associated with the development of small gaps between adjacent endothelial cells, and restored barrier function is associated with the reapposition of adjacent cells (Leach *et al.*, 1995; Wu and Baldwin, 1992). Therefore, it is of interest to investigate the transport mechanism of L-histidine from blood to brain and the metabolism of L-histidine in the microvascular endothelial cells. Xiang *et al.* (1998) reported that uptake of L-histidine into the rat choroids plexus is mediated not only via the System-L amino acid transporter described by Segal *et al.* (1990), but also via a System-N amino acid transporter similar to that described in liver and skeletal muscle.

The normal L-histidine concentration was 80 μM in rat plasma. When 100 μM of L-histidine was added in Krebs buffer, L-histidine uptake was linear over a 3 min incubation period. The uptake of L-histidine at 25°C was significantly lower than that measured at 37°C, indicating that the L-histidine uptake rate is temperature dependent. However, a remarkable decrease in the uptake of L-histidine was not observed during incubation at 0°C for 2 min, and there was no large difference in L-histidine uptake between 0 and 25°C.

An Eadie-Hofstee plot suggested that uptake of L-histidine is a saturable process. Nonlinear least-squares regression analysis of this data based on equation (1) yielded the following kinetic parameters: K_{m1} = 0.097 mM, V_{max1} = 960 pmol/mg protein/min for the high-affinity process and K_{m2} = 0.285 mM, V_{max2} = 1762 pmol/mg protein/min for the low-affinity process for L-histidine.

$$V = V_{max1}S/(K_{m1} + S) + V_{max2}/(K_{m2} + S), \tag{1}$$

where V_{max} is the maximum uptake rate for the carrier-mediated process, S is the concentration of substrate, and K_m is the half-saturation concentration (Michaelis constant); subscript integers indicate the saturable processes (1, high affinity; 2, low affinity).

Further, the addition of metabolic inhibitor (2,4-dinitrophenol (DNP) or rotenone) reduced the uptake rate of L-histidine, demonstrating that L-histidine uptake is metabolic energy-dependent. Ouabain, an inhibitor of (Na^+, K^+)-ATPase, which is localized in the antiluminal membrane, also inhibited uptake of L-histidine, possibly by reducing the sodium gradient and membrane potential. The effect of ouabain suggested that one of the driving forces for L-histidine transport is Na^+, as choline did not substitute for Na^+ (Table 1). These results suggested that L-histidine is actively taken up by a carrier-mediated mechanism with energy supplied by Na^+. However, L-histidine uptake was reduced in the presence of the Na^+-independent System-L substrate, 2-amino-2-norbornanecarboxylic acid (BCH), suggesting the coexistence of facilitated diffusion (Na^+-independent process) by

Table 1.
Effects of metabolic inhibitors and ion replacement in the Krebs buffer on the uptake of L-histidine

	Concentration (μM)	Uptake (n mol/mg protein/min)
Control		0.400 ± 0.034
DNP	250	$0.260 \pm 0.012^{**}$
Rotenone	25	$0.274 \pm 0.023^{*}$
Ouabain	100	$0.195 \pm 0.018^{**}$
Na$^+$ free		$0.252 \pm 0.015^{**}$

Rat BMECs were preincubated at 37°C for 15 min in 2 ml of Krebs buffer (pH 7.4) with 2,4-dinitrophenol (DNP), rotenone and ouabain. In sodium ion-replacement studies, incubation solution without Na$^+$ was prepared by substitution of choline chloride for NaCl and choline bicarbonate for NaHCO$_3$. The uptake of L-histidine (0.1 mM) was measured for 2 min at 37°C. Each value is the mean $\pm$ S.E. of 4 experiments. $^{*}p < 0.05$, $^{**}p < 0.01$, compared with control.

a carrier-mediated mechanism into the BMECs and an active transport system consuming Na$^+$.

The Na$^+$-dependent process for L-histidine uptake appears to involve System-N transporters as uptake was inhibited by glutamine, asparagine and L-glutamic acid γ-monohydroxamate, System-N substrates but not substrates for System-A and ransporters. In this experiment, in the presence of 143 mM Na$^+$ and 1 mM BCH, L-glutamic acid γ-monohydroxamate (1 mM) resulted in a progressive decrease in L-histidine uptake. This result suggested that System-N transport plays an important role in L-histidine uptake into rat BMECs (Fig. 2). Oldendorf *et al.* (1988) demonstrated pH dependence of histidine affinity for BBB carrier transport systems for neutral and cationic amino acids with the single-pass carotid injection technique. Reductions in pH markedly inhibited Na$^+$-dependent but not Na$^+$-independent transport, indicating that System-N mediated transport at the choroids plexus was also pH sensitive (Xiang *et al.*, 1998). Thus, such pH sensitivity may contribute to the derangement of brain amino acid composition during cerebral acidosis. However, the uptake of L-histidine into the rat BMECs was reported previously not to be affected by pH (5.0–7.4) in the absence or presence of BCH (Yamakami *et al.*, 1998). In this study, the System-N mediated transport system was also not pH sensitive in rat BMECs, suggesting that the transport mechanism of L-histidine is very similar in both the lung and brain microvascular endothelial cells (Sakurai *et al.*, 2002). Moreover, analysis by Eadie-Hofstee plots demonstrated at least two types of transport processes. The ($V_{\max}/K_m$) value of the high-affinity process was $\sim$1.6-fold greater than that of the low-affinity process. Further investigations would be necessary to confirm whether the high-affinity process dominates Na$^+$-dependent or Na$^+$-independent uptake of L-histidine in rat BMECs.

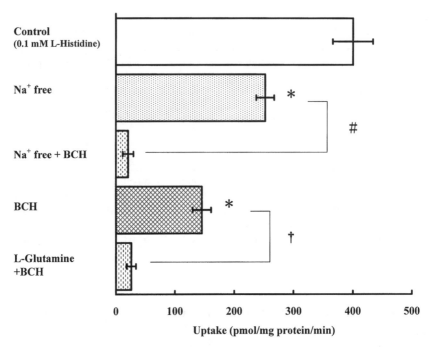

Figure 2. Rates of uptake of L-histidine into rat BMECs. Measurements were made in 143 mM Na$^+$ (control), Na$^+$ free, Na$^+$ free + 1 mM BCH, 143 mM Na$^+$ + 1 mM BCH, and 1 mM BCH + 1 mM L-glutamine to inhibit uptake via the Na$^+$-dependent System-N in the presence of 143 mM Na$^+$. Each point represents the mean ± S.E. of 4 determinations. *$p < 0.01$, significant difference from control. #$p < 0.01$, significant difference between Na$^+$ free and Na$^+$ free + 1 mM BCH; †$p < 0.01$, significant difference between 1 mM BCH and 1 mM L-glutamine + 1 mM BCH.

Enhancing effect of zinc on L-histidine uptake

Zinc has traditionally been identified in biology as a trace element, detectable in several biological systems, but the concentration of zinc in the brain is actually quite high. The typical concentration of zinc in the gray matter (about 0.15–0.20 mM) is 10–100 times higher than that of classical neurotransmitters such as acetylcholine and the monoamines. Huszti *et al.* (2001) clearly showed that zinc exposure enhances the astroglial and the cerebral endothelial uptake of histamine *in vitro* and it might be considered that zinc produces similar effects *in vivo*. However, although the mechanism by which the zinc ion affects histamine uptake is uncertain, they consider the possibility of a binding site for histamine on Zn^{2+} carrier, or the possibility of a Zn^{2+}-histamine complex transporter that modulates the operation of the carrier rather than the free histamine. On the other hand, the presence of zinc within the soluble matrix of secretory vesicles raised the possibility that zinc is released into the extracellular space during exocytosis (Assaf and Chung, 1984; Charton *et al.*, 1985). Glutamatergic nerve terminals of the hippocampal mossy fibers and the nerve endings of the neurons in the cortex have been demonstrated to contain zinc in high concentrations and it has been shown that zinc is co-released

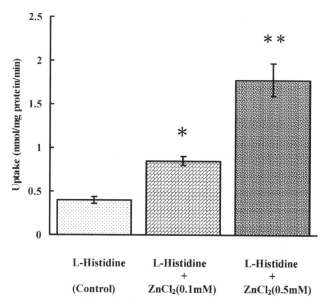

Figure 3. Effect of zinc ion on L-histidine uptake by cultured monolayers of rat BMECs. L-histidine uptake for 2 min at pH 7.4 was determined in the absence (control) and presence of 0.1 and 0.5 mM zinc chloride at 37°C. Each result represents the mean ± S.E. of 4 determinations. $^*p < 0.05$, $^{**}p < 0.01$, compared with control.

with glutamate from the presynaptic nerve terminal (Federickson, 1989). Moreover, the released metal ion may modulate both the uptake and the release of the amino acid by astroglial cells (Spiridon *et al.*, 1998) and the activity of the post-synaptic NMDA-receptor (Chadwick and Choi, 1990).

L-Histidine is an essential amino acid, and the imidazole rings of histamine and histidine are tautomeric. Therefore, zinc may have an influence on the BBB transport of L-histidine. Zinc displayed a dramatic dose-dependent enhancing effect in concentrations of 0.1 and 0.5 mM on L-histidine uptake by cultured BMECs of rat (Fig. 3). To characterize the effect of zinc ion, we followed the addition of metabolic inhibitors for the uptake of L-histidine. DNP, rotenone and ouabain did not affect the L-histidine uptake by zinc ion. The initial L-histidine uptake rate by addition of $ZnCl_2$ (0.5 mM) was also not reduced by the substitution of Na^+ with choline chloride and choline bicarbonate in the incubation buffer. Moreover, the system N substrate, L-glutamine, did not inhibit the enhancing effect of the zinc ion on uptake of L-histidine. However, the zinc-enhanced uptake of L-histidine was inhibited by the Na^+-independent system L substrate, BCH (Fig. 4). The Eadie-Hofstee analysis of the data revealed one component for L-histidine uptake by addition of $ZnCl_2$, suggesting that only system-L transport plays a role in zinc-enhanced uptake of L-histidine. We consider that free endogenous or free exogenous zinc might be favored to facilitate histamine synthesis from L-histidine at BBB, and delivery of the drugs into the CNS.

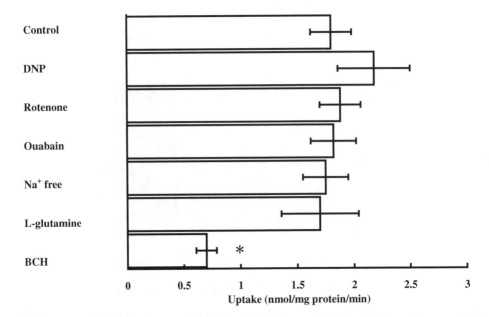

Figure 4. Effects of metabolic inhibitors, ion replacement and amino acids on the uptake of L-histidine in the presence of zinc chloride (0.5 mM) into cultured rat BMECs. The rat BMECs were preincubated at 37°C for 15 min in 2 ml of Krebs buffer (pH 7.4) with 2,4-dinitrophenol (DNP, 0.25 mM), rotenone (0.25 mM), ouabain (0.10 mM), L-glutamine (1.0 mM) and BCH (0.5 mM). The uptake of L-histidine (0.1 mM) was measured for 2 min at 37°C in the presence of zinc chloride (0.5 mM). In sodium ion-replacement studies, incubation solution without Na$^+$ was prepared by substitution of choline chloride for NaCl and choline bicarbonate for NaHCO$_3$. Each value is the mean ± S.E. of 4 experiments. *p < 0.01, compared with control. Control: rate of uptake of L-histidine increased by the presence of zinc chloride (0.5 mM).

Blood-brain barrier L-histidine metabolism

HDC catalyzes the formation of histamine from L-histidine. In rodents, this enzyme is primarily localized in the brain (hypothalamic neurons and projections to other brain regions) (Watanabe *et al.*, 1984), the glandular regions of the stomach (Savany and Cronenberger, 1982), and fetal liver (Taguchi *et al.*, 1984). Moreover, in addition to mast cells and basophils, several types of cells are reported to have the ability to form histamine (Aoi *et al.*, 1989). However, in monocytes histamine and HDC are colocalized in the cytoplasm, indicating a subcellular distribution different from mast cells and basophils. The endothelial cells of the brain microvasculature, generally considered to constitute the BBB, function as both a permeability barrier and a metabolic barrier to the passage of chemicals from the blood into the CNS (Pardridge, 1983). In this study, therefore, we examined whether rat BMECs have the ability to form histamine, and whether the protein and mRNA for HDC is present in rat BMECs.

The HDC activity in rat BMECs was estimated to be 0.14 ± 0.05 pmol/min/mg protein. Rat BMECs also contained histidine of 3.00 ± 0.14 nmol/mg protein

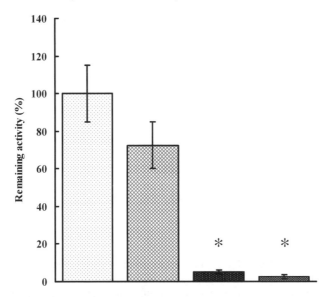

Figure 5. Inactivation of HDC activity by (S)-α-fluoromethylhistidine (FMH) in rat BMECs. Samples containing HDC activity were preincubated for 30 min with various concentrations of FMH, then the remaining HDC activity was assayed.

Figure 6. Indirect immunofluorescent histochemistry of rat BMECs using antibody against HDC purified from fetal rat liver. Immunofluorescence study was performed with primary antibody, anti-rat-HDC antibody and stained with secondary antibody, anti-rabbit IgG antibody conjugated with FITC (×200).

and histamine of 2.18 ± 0.10 pmol/mg protein. Since the BMECs constitute a non-neuronal and non-mast cellular histamine pool (Robinson-White and Beavean, 1982), the finding presented here provides evidence of histamine biosynthesis in rat BMECs. Moreover, the HDC activity was completely inhibited by FMH, a specific inhibitor of HDC which strongly inhibited histamine formation *in vitro* and *in vivo* (Fig. 5). When rat BMECs were also cultured with FMH (50 μM), the concentration of histamine decreased 97.8%. Figure 6 shows the result of indirect immunofluorescence analysis of HDC protein distribution in rat BMECs using the polyclonal anti HDC antibody. The presence of HDC protein was confirmed by fluorescence microscopy. Moreover, RT-PCR products of RNA amplified with specific primers of rat HDC cDNA in rat BMECs are shown in Fig. 7. HDC mRNA

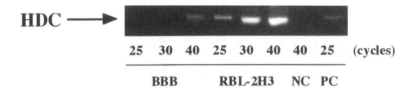

HDC ➔

25   30   40   25   30   40   40   25   (cycles)
_____   _____   _____
BBB   RBL-2H3   NC PC

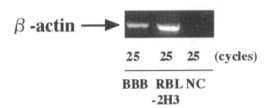

β-actin ➔

25   25   25   (cycles)
_____
BBB RBL NC
     -2H3

Figure 7. Expression of HDC mRNA in rat BMECs and rat basophilic leukemia cells (RBL-2H3). Total RNA isolated from rat BMECs and RBL-2H3 was analyzed by reverse transcriptase-PCR with specific primers for rat HDC cDNA. As positive control, rat HDC cDNA was used as template, and as negative control, water was used as template. The β-actin gene was used for evaluation of the amount of cDNA synthesized.

is expressed in rat BMECs, although the level of HDC mRNA was low compared with that in RBL-2H3.

These results suggested that HDC is present in the endothelial cells of rat microvessels. Thus, L-histidine uptaken by BMECs is biotransformed to histamine, demonstrating that HDC in BMECs plays an important role in BBB. Therefore, intracellular histamine might be involved in the regulation of brain microcirculation and/or in the regulation of the events occurring at the BBB level. However, it is not yet well understood if histamine synthesized in the endothelial cells affects the intracellular receptors, or if there is some releasing pathway effect on histamine H_1 and H_2 receptors on the surface of endothelial cells themselves. Further investigation is required.

HDC activity has also been reported to be increased by various stimuli. Inflammatory stimulation of Th2 cells induces HDC expression. The resulting histamine produced may act on histamine H_2-receptors in these cells to modulate IL-12 and IL-10 production (Elenkov et al., 1998). Maeyama et al. (1988) showed that HDC of rat basophilic leukemia cells (RBL-2H3) cells could be induced by phorbol myristate acetate (PMA). However, the activity of HDC in rat BMECs was not affected by PMA. Both the activity and mRNA for HDC in GMI 6-3 cells were induced by lipopolysaccharide (LPS) treatment, and the induction was sensitive to a calmodulin-dependent kinase II inhibitor, KN62 (Katoh et al., 2001). LPS also induced histamine production in the mouse macrophase-like cell line RAW 264.7, and the histamine production induced by LPS was inhibited by PD98059, a specific inhibitor of MEK-1 which phosphorylates p44/p42 MAP kinase (Shiraishi et al.,

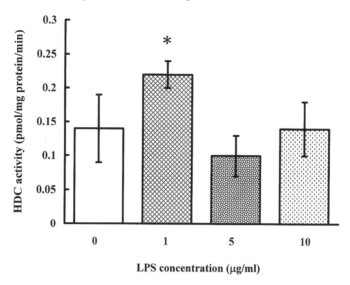

Figure 8. Effect of lipopolysaccharide (LPS) on HDC activity in rat BMECs. BMECs were treated with LPS (1, 5 and 10 μg/ml) for 12 h at 37°C, then the HDC activity was measured. Each value is the mean ± S.E. of 4 experiments. $^*p < 0.05$, compared with control.

2000). In our experiments, the HDC activity in cultured rat BMECs was induced by LPS treatment (Fig. 8), suggesting a cell type producing histamine in rat BMECs.

Blood-brain barrier histamine metabolism

Biosynthesized and released histamine is thought to be catabolized predominantly *in vivo*, mainly through two metabolic pathways. One is transmethylation to N-tele-methylhistamine, catalyzed by histamine N-tele-methyltransferase (HMT; EC 2.1.1.8), and the other is oxidative deamination to imidazole acetaldehyde, catalyzed by diamine oxidase (DAO; EC1.4.3.6). Hirata *et al.* (1999) demonstrated in the dissected inferior turbinate mucosa that expression of HMT mRNA was positively correlated to the expression of HDC mRNA in the normal subjects. Therefore, further investigation will be necessary on metabolism of histamine produced in the BBB, or uptaken from the blood.

CONCLUSION

Our findings suggest that the endothelial cells of the brain microvasculature which constitute the BBB have the ability to form histamine from L-histidine uptaken via both amino acid transporters System-N and -L (Fig. 9). Therefore, the increase in histamine production in the rat brain microvascular endothelial cells will be of advantage to the acute and reversible opening of the tight junction. We expect that the elucidation of this enhancing mechanism will assist the development of drug delivery system into the CNS via the BBB.

E. Sakurai et al.

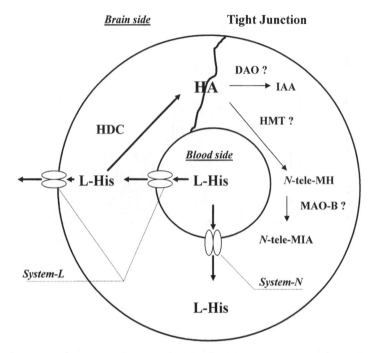

Figure 9. Blood-brain barrier carrier-mediated transport and metabolism of L-histidine in rats. L-His, L-histidine; HDC, L-histidine decarboxylase; HA, histamine; HMT, N-tele-methyltransferase; MAO, monoamine oxidase; DAO, diamine oxidase; N-tele-MH, N-tele-methylhistamine; N-tele-MIA, N-tele-methylimidazoleacetic acid; IAA, imidazoleacetic acid.

REFERENCES

Anderson, J. M. and Van Itallie, C. M. (1995). Tight junctions and the molecular basis for regulation of paracellular permeability, *Am. J. Physiol.* **269**, G467–475.

Aoi, R., Nakashima, I., Kitamura, Y., *et al.* (1989). Histamine synthesis by mouse T-lymphocytes through induced histidine decarboxylase, *Immunology* **66**, 219–223.

Assaf, S. Y. and Chung, S. H. (1984). Release of endogenous Zn^{++} from brain tissue during activity, *Nature* **308**, 734–738.

Audus, K. L., Bartel, R. L., Hidalgo, I. J., *et al.* (1990). The use of cultured epithelial and endothelial cells for drug transport and metabolism studies, *Pharm. Res.* **7**, 435–451.

Boertje, S. B., Le Beau, D. and Williams, C. (1989). Blockade of histamine-stimulated alterations in cerebrovascular permeability by the H$_2$-receptor antagonist cimetidine, *Neuropharmacol.* **28**, 749–752.

Bowman, P. D., Betz, A. L., Ar, D., *et al.* (1981). Primary culture of capillary endothelium from rat brain, *In Vitro* **17**, 353–362.

Chadwick, W. C. and Choi, D. W. (1990). Effect of zinc on NMDA receptor-mediated channel currents in cortical neurons, *J. Neurosci.* **10**, 108–161.

Charton, G., Rovira, C., Ben-Ari, Y., *et al.* (1985). Spontaneous and evoked release of endogenous Zn^{++} in the hippocampal mossy fiber zone of the rat in situ, *Exp. Brain Res.* **58**, 202–205.

El-Ackad, T. M. and Brody, M. J. (1975). Evidence for non-mast cell histamine in vascular wall, *Blood Vessels* **12**, 181–191.

Elenkov, I. J., Webster, E., Papanicolaou, D. A., *et al.* (1998). Histamine potently suppresses human IL-12 and stimulates IL-10 production via H2 receptors, *J. Immunol.* **161**, 2586–2593.

Federickson, C. J. (1989). Neurobiology of zinc and zinc-containing neurons, *Int. Rev. Neurobiol.* **31**, 145–238.

Gardner, T. W., Lesher, T., Khin, S., *et al.* (1996). Histamine reduces ZO-1 tight junction protein expression in cultured retinal microvascular endothelial cells, *Biochem. J.* **320**, 717–721.

Gordon, E. L., Danielsson, P. E., Nguyen, T.-S., *et al.* (1991). A comparison of primary cultures of rat cerebral microvascular endothelial cells to rat aortic endothelial cells, *In Vitro Cell Dev. Biol.* **27A**, 312–326.

Hirata, N., Takeuchi, K., Ukai, K., *et al.* (1999). Expression of histidine decarboxylase messenger RNA and histamine N-methyltransferase messenger RNA in nasal allergy, *Clin. Exp. Allergy* **29**, 76–83.

Howland, R. D. and Spector, S. (1972). Disposition of histamine in mammalian blood vessels, *J. Pharmacol. Exp. Ther.* **182**, 239–245.

Huszti, Z., Horvath-Sziklai, A., Noszal, B., *et al.* (2001). Enhancing effect of zinc on astroglial and cerebral endothelial histamine uptake, *Biochem. Pharmacol.* **62**, 1491–1500.

Ichikawa, N., Naora, K., Hirano, H., *et al.* (1996). Isolation and primary culture of rat cerebral microvascular endothelial cells for studing drug transport in vitro, *Pharmacol. Toxicol. Meth.* **36**, 45–52.

Katoh, Y., Niimi, M., Yamamoto, Y., *et al.* (2001). Histamine production by cultured microglial cells of the mouse, *Neurosci. Lett.* **305**, 181–184.

Kilberg, M. S., Handlogten, M. E. and Christensen, H. N. (1980). Characteristics of an amino acid transport system in rat liver for glutamine, asparagine, histidine and closely related analogs, *J. Biol. Chem.* **255**, 4011–4019.

Komura, J., Tamai, I., Senmaru, M., *et al.* (1997). Brain-to-blood active transport of β-alanine across the blood-brain barrier, *FEBS Lett.* **400**, 131–135.

Krause, D., Mischeck, U., Galla, H. J., *et al.* (1991). Correlation of zonula occludens ZO-1 antigen expression and transendothelial resistance in porcine and rat cultured cerebral endothelial cells, *Neurosci. Lett.* **128**, 301–304.

Leach, L., Eaton, B., Wescott, D., *et al.* (1995). Effect of histamine on endothelial permeability and structure and adhesion molecules of the paracellular junctions of perfused human placental microvessels, *Microvasc. Res.* **50**, 323–337.

Lo, W. W. and Fan, T. P. (1987). Histamine stimulates inositol phosphate accumulation via the H_1-receptor in cultured human endothelial cells, *Biochem. Biophys. Res. Commun.* **148**, 47–53.

Maeyama, K., Taguchi, Y., Sasaki, M., *et al.* (1988). Induction of histidine decarboxylase of rat basophilic leukemia (2H3) cells stimulated by higher oligomeric IgE or phorbol myristate, *Biochim. Biophys. Res. Commun.* **151**, 1402–1407.

Magee, J. C., Stone, A. E., Oldham, K. T., *et al.* (1994). Isolation, culture, and characterization of rat lung microvascular endothelial cells, *Am. J. Physiol. Lung Mol. Physiol.* **267**, L433–L441.

Majno, G. and Palade, G. E. (1961). The effect of histamine and serotonin on vascular permeability: an electron microscopy study, *J. Biophys. Biochem. Cytol.* **11**, 571–605.

Majno, G., Shea, S. M. and Leventhal, M. (1969). Endothelial contraction induced by histamine-type mediators: an electron microscopic study, *J. Cell Biol.* **42**, 647–672.

Meyrick, B. and Brigham, K. L. (1984). Increased permeability associated with dilation of endothelial cell junctions caused by histamine in intimal explants from bovine pulmonary artery, *Exp. Lung Res.* **6**, 11–25.

Moy, A. B., Van Engelenhoven, J., Bodmer, J., *et al.* (1996). Histamine and thrombin modulate endothelial focal adhesion through centripetal and centrifugal forces, *J. Clin. Invest.* **97**, 1020–1027.

Niimi, N., Noso, N. and Yamamoto, S. (1992). The effect of histamine on cultured endothelial cells. A study of the mechanism of increased vascular permeability, *Eur. J. Pharmacol.* **221**, 325–331.

Oldendorf, W. H., Crane, P. D., Braun, L. D., *et al.* (1988). pH dependence of histidine affinity for blood-brain barrier transport system for neutral and cationic amino acids, *J. Neurochem.* **50**, 857–861.

Pardridge, W. M. (1983). Brain metabolism: a perspective from the blood-brain barrier, *Physiol. Rev.* **63**, 1481–1535.

Pardridge, W. M. (1995). Transport of small molecules through the blood-brain barrier: biology and methodology, *Adv. Drug Delivery Res.* **15**, 5–36.

Pardridge, W. M., Triguero, D., Yang, J., *et al.* (1990). Comparison of in vitro and in vivo models of drug transcytosis through the blood-brain barrier, *J. Pharmacol. Exp. Ther.* **253**, 884–891.

Sakurai, E., Sakurada, T., Ochiai, Y., *et al.* (2002). Stereoselective transport of histidine in rat lung microvascular endothelial cells, *Am. J. Physiol. Lung Cell Mol. Physiol.* **282**, L1192–L1197.

Savany, A. and Cronenberger, L. (1982). Properties of histidine decarboxylase from rat gastric mucosa, *Eur. J. Biochem.* **123**, 593–599.

Segal, M. B., Preston, J. E., Collis, C. S., *et al.* (1990). Kinetics and Na independence of amino acid uptake by blood side of perfused sheep choroids plexus, *Am. J. Physiol. Renal Fluid Electrolyte Physiol.* **258**, F1288–F1294.

Shiraishi, M., Hirasawa, N., Kobayashi, Y., *et al.* (2000). Participation of mitogen-activated protein kinase in thapsigargin- and TPA-induced histamine production in murine macrophage RAW 264.7 cells, *Br. J. Pharmacol.* **129**, 515–524.

Spiridon, M., Kamm, D., Billups, B., *et al.* (1998). Modulation by zinc of the glutamate transporters in glial cells and cones isolated from the tiger salamander retina, *J. Physiol.* **506**, 363–376.

Stevenson, B. R., Siliciano, J. D., Mooseker, M. S., *et al.* (1986). Identification of ZO-1: a high molecular weight polypeptide associated with the tight junction (zonula occludens) in a variety of epithelia, *J. Cell Biol.* **103**, 755–766.

Taguchi, Y., Watanabe, T., Kubota, H., *et al.* (1984). Purification of histidine decarboxylase from the liver of fetal rats and its immunochemical and immunohistochemical characterization, *J. Biol. Chem.* **259**, 5214–5221.

Tamai, I. and Tsuji, A. (1996). Drug delivery through the blood-brain barrier, *Adv. Drug Delivery Res.* **19**, 410–424.

Terasaki, T., Takakuwa, S., Moritani, S., *et al.* (1991). Transport of monocarboxylic acids at the blood-brain barrier: studies with monolayers of primary cultured bovine brain capillary endothelial cells, *J. Pharmacol. Exp. Ther.* **258**, 932–937.

Voyta, J. C., Via, D. P., Butterfield, C. E., *et al.* (1984). Identification and isolation of endothelial cells based on their increased uptake of acetylated low density lipoprotein, *J. Cell Biol.* **99**, 2034–2040.

Walker, I., Nicholls, D., Irwin, W. J., *et al.* (1994). Drug delivery via active transport at the blood-brain barrier: II. Investigation of monocarboxylic acid transport in vitro, *Int. J. Pharm.* **108**, 225–232.

Watanabe, T., Taguchi, Y., Shiosaka, S., *et al.* (1984). Distribution of the histaminergic neuron system in the central nervous system of rats: A fluorescent immunohistochemical analysis with histidine decarboxylase as a marker, *Brain Res.* **295**, 13–25.

Wu, N. Z. and Baldwin, A. L. (1992). Transient venular permeability increase and endothelial gap formation induced by histamine, *Am. J. Physiol. Heart Circ. Physiol.* **262**, H1238–H1247.

Xiang, J., Fowkes, R. L. and Keep, R. F. (1998). Choroid plexus histidine transport, *Brain Res.* **783**, 37–43.

Yamakami, J., Sakurai, E., Sakurada, T., *et al.* (1998). Stereoselective blood-brain barrier transport of histidine in rats, *Brain Res.* **812**, 105–112.

Advances in Neuroregulation and Neuroprotection (2005), pp. 713-723
Collin, C. *et al.* (Eds)
© VSP 2005

Intracerebral serotonin and catecholamine metabolite concentrations in SART-stressed rats treated with diazepam and fluvoxamine

E. KIRIME [1,*], M. TABUCHI [2], H. HIGASHINO [2] and K. HITOMI [1]

[1] *Department of Neuropsychiatry, Kinki University School of Medicine, Osaka-Sayama 589-8511, Japan*
[2] *Department of Pharmacology, Kinki University School of Medicine, Osaka-Sayama 589-8511, Japan*

Abstract—It was investigated whether the anti-anxiety effects produced by benzodiazepine, diazepam, and a SSRI, fluvoxamine, treatments were related to the changes of intracerebral 5-HT and other catecholamine concentrations using SART-stressed rats. As a result, findings partially different from the previous reports were observed: (1) SART stress decreases the 5-HT concentration in the raphe nuclei region, diazepam partially induced recovery of the decrease, and fluvoxamine directly increases utilization of 5-HT without changing the 5-HT concentration; (2) diazepam and fluvoxamine caused recovery of the decreased DA concentration in the substantia nigra region and induce mental stability; and (3) SART stress increased the NE concentration in the region containing the ceruleus nucleus.

Keywords: SART-stress; serotonin; catecholamine; diazepam; fluvoxamine.

INTRODUCTION

In addition to benzodiazepines, the anti-anxiety efficacy of a selective serotonin reuptake inhibitor (SSRI) developed as an antidepressant has been confirmed in recent years, and it has been increasingly used for treatment of panic disorder, obsessive-compulsive disorder, post-traumatic stress disorder, and social phobia (Davidson *et al.*, 1998; Stein *et al.*, 1999; Ballenger, 2000; Figgitt and McClellan, 2000). Basic medical studies using anxiety animal models have suggested that many neurotransmitters regulate anxiety. In this study, we attempted to elucidate the relationship between anxiety and neurotransmitters based on neuropharmacology

---

*To whom correspondence should be addressed. E-mail: a-ji@med.kindai.ac.jp

using SART (Specific Alternation of Rhythm in Temperature)-stressed rats. SART-stressed rats are severely stressed by hourly changes in environmental temperature (Kita *et al.*, 1975). This model was originally designed as an autonomic imbalance model, but later studies have clarified that SART-stressed rats are in an anxiety state (Hata *et al.*, 2001). Thus, we administered a benzodiazepine, diazepam, and an SSRI, fluvoxamine, to SART-stressed rats, and measured the intracerebral concentrations of serotonin (5-HT) and catecholamine metabolites to investigate the actions of neurotransmitters under SART stress-induced anxiety.

MATERIALS AND METHODS

Animals

Male Wistar rats (160–185 g) aged six weeks were purchased from CLEA Japan, Osaka. The animals were housed at four animals per cage and acclimatized for four days in an animal room controlled at $23 \pm 0.5°C$ and $55 \pm 10\%$ humidity under a lighting cycle with lights on from 7:00 a.m. to 7:00 p.m. and lights off from 7:00 p.m. to 7:00 a.m. the next morning. Animals were given tap water and solid chow *ad libitum* (CLEA Japan, Inc., Osaka) throughout the acclimatization period and SART stress administration period. Animals were handled in accordance with the 'Guiding Principles for the Care and Use of Laboratory Animals approved by The Japanese Phamacological Society'.

SART stress administration

The temperature conditions of the method reported by Kita *et al.* (1975) was slightly modified and used for administration of SART stress to rats. Rats were exposed alternately to 24°C and 4°C hourly from 10:00 a.m. to 5:00 p.m., and then 4°C from 5:00 p.m. to 10:00 a.m. the following day. This SART stress was administered for five consecutive days, and the animals were individually kept in metal cages $(15(W) \times 21(D) \times 15(H)$ cm) under a lighting cycle with lights on from 10:00 a.m. to 5:00 p.m. and lights off from 5:00 p.m. to 10:00 a.m. during the stress period. In the original procedure, SART stress was administered to rats at 24°C and −3°C.

Drugs

The test drugs, diazepam (Cercine®, Takeda Chemical Industries, Osaka) and fluvoxamine maleate (Luvox®, Fujisawa Pharmaceutical Co., Osaka), were purchased from the market. The doses of diazepam and fluvoxamine maleate were 3 mg and 30 mg/kg body weight, respectively. These drugs were dissolved in 0.5 ml of 10% acacia (Wako Pure Chemical Industries, Osaka) solution on the administration day, and administered orally using a gastric tube at 4:00 p.m. daily during the SART stress period. To the control group, 0.5 ml of 10% Acacia solution was administered orally.

Experimental groups

The following experimental groups were setted based on the presence and absence of SART stress and drug administration.

(1) No-stress group ($n = 15$): a group without SART stress and medication.

(2) Control group ($n = 15$): a group with SART stress without medication.

(3) Diazepam group ($n = 15$): a group with SART stress treated with diazepam.

(4) Fluvoxamine group ($n = 15$): a group with SART stress treated with fluvoxamine.

Measurement schedule of intracerebral concentrations of various neurotransmitters

After rats were acclimatized for four days in an animal room, SART stress was loaded for six days on the rats. The rats were decapitated at 10:00 a.m. on the 6th day of SART stress. The brain tissue was immediately excised and frozen in liquid nitrogen. The frontal lobe, hypothalamus, brain stem including the raphe nuclei, substantia nigra, and ceruleus nucleus, and amygdala were excised promptly from the stock brain tissues and homogenized in 1.5 ml of 0.5 N perchloric acid (PCA) containing 75 ng of 3,4-dihydroxybenzylamine hydrobromide (DHBA) using a Polytron® homogenizer (Kinematica, AG, Switzerland). The supernatant after centrifugation was filtered through an ultrafiltration membrane tube (ULTRAFREE®-MC 0.1 μm Filter Unit, Millipore Corporation, USA), and the concentrations of serotonin (5-HT), 5-hydroxyindole acetic acid (5-HIAA), dopamine (DA), homovanillic acid (HVA), normetanephrine (NMN), 3,4-methoxy-hydroxyphenyl glycol (MHPG), metanephrine (MN), and vanillic acid (VA) were measured using high performance liquid chromatography (HPLC; Neurochem®, ESA Inc., MA, USA). The concentrations are presented as those per protein in the brain tissues (ng/mg protein).

Statistical analysis

The results are presented as the mean ± SE. Non-repeated measures ANOVA or Kruskal−Wallis H-test and Student−Newman−Keuls test were used for comparison of values among the groups. Differences between the results were considered as significant for $p < 0.05$.

RESULTS

5-HT concentrations in the regions containing the raphe nuclei, hypothalamus, and amygdala (Fig. 1)

As shown in Fig. 1A and B, the 5-HT concentrations in the regions containing the raphe nuclei and hypothalamus were significantly lower in the Control, Diazepam,

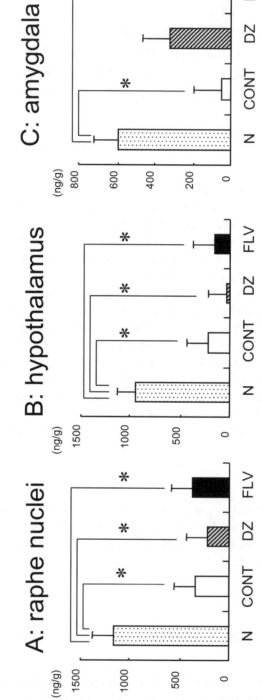

Figure 1. 5-HT concentrations in brain (A: raphe nuclei, B: hypothalamus, C: amygdala). Results are presented as mean $\pm$ SE ($n = 15$/group). N: No-stress group, CONT: Control group, DZ: Diazepam group, FLV: Fluvoxamine group. Significant differences: $*p < 0.05$ *vs.* Control group (Student–Newman–Keuls test).

and Fluvoxamine groups than in the No-stress group. As shown in Fig. 1C, the 5-HT concentration in the region containing the amygdala was significantly lower in the Control and Fluvoxamine groups than in the No-stress group. When diazepam was administered to rats that may have been anxious due to SART stress, 5-HT tended to increase, although the difference was not significant.

5-HIAA concentrations in the regions containing the raphe nuclei, hypothalamus, and amygdala (Fig. 2)

As shown in Fig. 2A, B, and C, the 5-HIAA concentrations in the regions containing the raphe nuclei, hypothalamus, and amygdala were significantly lower in the Control, Diazepam, and Fluvoxamine groups than in the No-stress group. Although it is not presented in the figure, SART stress tended to decrease 5-HT and 5-HIAA in the frontal lobe region in the final projection region of the 5-HT nervous system, and neither diazepam nor fluvoxamine counteracted the decreases. That is, 5-HT concentrations in the frontal lobe region were 117.2 ± 68.4 ng/g in the No-stress group, 45.2 ± 41.6 ng/g in the Control group, 2.69 ± 2.30 ng/g in the Diazepam group, and 9.60 ± 9.60 ng/g in the Fluvoxamine group, respectively, and 5-HIAA concentrations in the frontal lobe region were 231.1 ± 97.9 ng/g in the No-stress group, 84.2 ± 41.1 ng/g in the Control group, 26.7 ± 17.4 ng/g in the Diazepam group, and 63.4 ± 31.1 ng/g in the Fluvoxamine group.

Dopamine (DA) concentrations in the regions containing the substantia nigra and amygdala (Fig. 3A, B)

The DA concentration in the region containing the substantia nigra (Fig. 3A) was lowest in the stressed Control group and as high in the No-stress group as in the Fluvoxamine group. In the region containing the amygdala, the DA concentration in the Diazepam group was as high as that in the Fluvoximane group (Fig. 3B). A similar result was also obtained in the frontal lobe region, the final projection region of the DA nerve. That is, DA concentrations in the frontal lobe region were 13704 ± 4020 ng/g in the No-stress group, 8271 ± 2221 ng/g in the Control group, 4888 ± 1072 ng/g in the Diazepam group, and 9276 ± 1554 ng/g in the Fluvoxamine group.

Homovanillinic acid (HVA) concentrations in the regions containing the substantia nigra and amygdala (Fig. 4A, B)

The concentration of HVA (the final metabolite of DA, Fig. 4A) in the region containing the substantia nigra was similar to the change in the DA concentration. The concentration was lowest in the Control group loaded with SART stress, and as high in the No-stress group as in the Fluvoxamine group. In the region containing the amygdala, the concentration in the Diazepam group was equal to that in the Fluvoxamine group (Fig. 4B). In the frontal lobe region, the final projection region

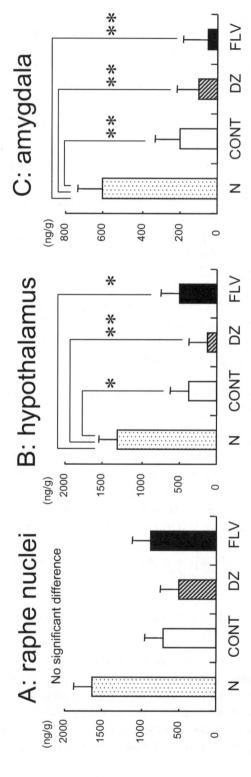

Figure 2. 5-HIAA concentrations in brain (A: raphe nuclei, B: hypothalamus, C: amygdala). Results are presented as mean ± SE ($n = 15$/group). N: No-stress group, CONT: Control group, DZ: Diazepam group, FLV: Fluvoxamine group. Significant differences: * $p < 0.05$, ** $p < 0.01$ *vs.* Control group (Student–Newman–Keuls test).

DOPAMINE

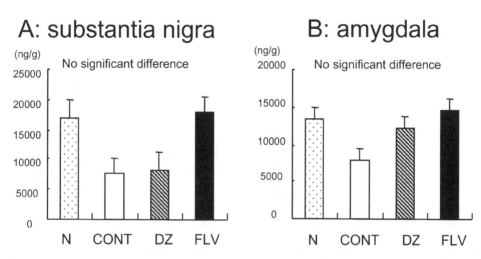

Figure 3. Dopamine concentrations in brain (A: substantia nigra, B: amygdala). Results are presented as mean ± SE (*n* = 15/group). N: No-stress group, CONT: Control group, DZ: Diazepam group, FLV: Fluvoxamine group. No significant difference (Student–Newman–Keuls test).

HVA

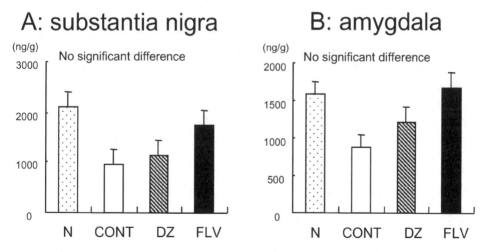

Figure 4. HVA concentrations in brain (A: substantia nigra, B: amygdala). Results are presented as mean ± SE (*n* = 15/group). N: No-stress group, CONT: Control group, DZ: Diazepam group, FLV: Fluvoxamine group. No significant difference (Student–Newman–Keuls test).

NMN

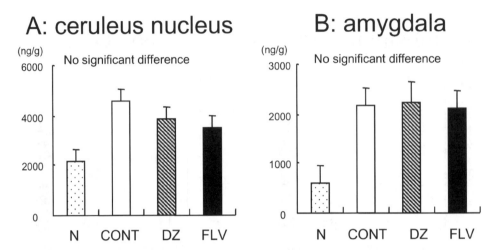

Figure 5. NMN concentrations in brain (A: ceruleus nucleus, B: amygdala). Results are presented as mean ± SE (*n* = 15/group). N: No-stress group, CONT: Control group, DZ: Diazepam group, FLV: Fluvoxamine group. No significant difference (Student–Newman–Keuls test).

of the DA nerve, the results shown below were obtained. HVA: No-stress group, 1595 ± 340 ng/g; Control group, 1176 ± 193 ng/g; Diazepam group, 648 ± 119 ng/g; Fluvoxamine group, 1127 ± 157 ng/g.

Normetanephrine (NMN) concentrations in the regions containing the ceruleus nucleus and amygdala (Fig. 5A, B)

The concentration of NMN (a norepinephrine metabolite, Fig. 5A) in the region containing the ceruleus nucleus was completely different from the changes in the 5-HT, 5-HIAA, and DA concentrations. The NMN concentration was higher in all three SART-stressed groups than in the No-stress group. Drug administrations did not affect these changes. Changes in the NMN concentration in the region containing amygdala were similar to those in the region containing the ceruleus nucleus (Fig. 5B).

3,4-Methoxy-hydroxyphenyl glycol (MHPG) concentrations in the regions containing the ceruleus nucleus and amygdala (figure not shown)

No differences in the MHPG concentrations (metabolite of norepinephrine and epinephrine) were observed in the presence or absence of SART stress among the groups in the region containing the ceruleus nucleus or amygdala. In the region containing the frontal area, the final projection region of the NE nervous system, there was no significant difference in the NMN concentration among the groups as follows. NMN: No-stress group, 17730 ± 4977 ng/g; Control group, 12809 ±

2849 ng/g; Diazepam group, 7515 ± 862 ng/g; Fluvoxamine group, 12622 ± 2498 ng/g. There were no significant differences in the concentration of the epinephrine metabolite, metanephrine, or the final metabolite, VA, among the groups. As for the concentrations of various neurotransmitters other than those described above, no significant differences were observed in any intracerebral region.

DISCUSSION

It has been reported that the anxiety state of SART-stressed animals was improved by benzodiazepines and SSRIs in a behavioral pharmacological study (Hata *et al.*, 2001; Kirime, 2002). However, we considered that it was necessary to investigate this improvement based on neuropharmacology. It has been reported that SART stress decreased 5-HT and 5-HIAA in various cerebral regions in laboratory animals (Hata *et al.*, 1991), and we obtained similar results. In this study, we administered a benzodiazepine, diazepam, and an SSRI, fluvoxamine, and investigated whether the exhibited anti-anxiety effects were related to intracerebral 5-HT and other catecholamines. SART stress decreased or tended to decrease the brain tissue concentrations of 5-HT and its metabolite, 5-HIAA, in the brain stem region containing the raphe nuclei. In the hypothalamus and amygdala, the projection regions of 5-HT nerve, the 5-HT and 5-HIAA concentrations were lower or tended to be lower in the stressed groups than in the No-stress group regardless of administration of the two drugs. However, in the amygdala, the 5-HIAA concentrations tended to be higher in the Diazepam group than in the other stress groups. In the frontal lobe, the final projection region of 5-HT, the 5-HT and 5-HIAA concentrations were lower in the stress groups than in the No-stress group. The 5-HT nervous system originates from the raphe nuclei in the lower brain stem and widely projects in cerebral regions such as the hypothalamus, amygdala, hippocampus, and cerebral cortex, and functions in mental stabilization (inhibition of fighting behavior), induction of sleep, inhibition of pain, and elevation of body temperature.

Based on the above findings, SART stress decreased the 5-HT concentration in the rat brain regions, suggesting that SART stress made rats anxious. In a behavioral pharmacological study using the elevated plus-maze test, fluvoxamine induced more active behavior than diazepam (Kirime, 2002). Although the measurements of 5-HT and 5-HIAA did not support this finding, diazepam administration increased the 5-HT concentration in the amygdala region, suggesting the possibility that diazepam, which exhibits an anti-anxiety effect via GABA receptors, acts in relation to the intracerebral 5-HT concentration. Since the marked efficacy of fluvoxamine on behavioral activity is an action of SSRI, it might be understood that the drug itself increased utilization of 5-HT in the 5-HT receptor region, although the drug did not directly affect the 5-HT concentration.

With regard to DA, it has been reported that increase in intracerebral DA reduces anxiety (Stein *et al.*, 2002), DA was increased in the hypothalamus and cerebral cortex in SART-stressed rats, and DA metabolites, 3,4-dihydroxy phenyl acetic acid (DOPAC) and HVA, were increased in a stress state. In contrast, in our study, DA and its metabolite, HVA, decreased in the substantia nigra and amygdala regions in SART-stressed rats. Diazepam and fluvoxamine tended to inhibit the decreases and the effect of an SSRI, fluvoxamine, tended to be higher than that of a tranquilizer, diazepam. The major intracerebral DA system consists of (1) action of the extrapyramidal tract system extending from the substantia nigra to the corpus striatum, and (2) pleasant action extending from the dorsal tegmental area (A10) to the amygdala and frontal area. Pharmacologically, greater SART stress-induced reduction of motility in the Diazepam group than in the Fluvoxamine group reported previously may have been related to the following: SART stress decreased the DA concentration in the regions containing the substantia nigra, amygdala, and frontal lobe. Diazepam and fluvoxamine caused recovery of the decrease, but the recovery by fluvoxamine was greater, suggesting that activation of the DA system via A10 fibers contributes to recovery from SART stress-induced anxiety.

The concentrations of NE metabolites, NMN and MHPG, were increased by SART stress in the cerebral regions containing the ceruleus nucleus and amygdala, and none of the drugs changed the increases. The NE nervous system projects from the ceruleus nucleus into the cerebral cortex and medulla oblongata/spinal cord, and is considered to be involved in anxiety, anger, and arousal. Accordingly, the NE nervous system may have been related to the result of SART stress-induced elevation of NE in the ceruleus nucleus region, that is, anxiety. Although stress is considered to increase the blood epinephrine (Ep) concentration, since Ep does not transfer to the brain tissue, no significant differences were observed in the intracerebral MN or VA concentration among the groups.

The above findings support the following possibility: (1) SART stress decreases the 5-HT concentration in the raphe nuclei region, and diazepam partially induces recovery of the decrease while fluvoxamine directly increases utilization of 5-HT without changing the 5-HT concentration. (2) Diazepam and fluvoxamine cause recovery of the decreased DA concentration in the substantia nigra region and induce mental stability. (3) SART stress increases the NE concentration in the region containing the ceruleus nucleus.

Clinical experience shows that SSRI exhibits an anti-anxiety effect, and its therapeutic usefulness for anxiety disorders has been reported as described above. The relation between anxiety and 5-HT has been discussed in many studies, but the pharmacological mechanism has not yet been elucidated. In this study, we investigated the relationship between various intracerebral neurotransmitters and drug administrations using SART-stressed animals with different properties from the previous models. As a result, findings partially different from the previous reports were observed. The pharmacological detailed mechanism on the relationship between anxiety and various neurotransmitters has not been elucidated. Studies of anxiety states in various anxiety animal models are awaited.

REFERENCES

Ballenger, J. C. (2000). Selective serotonin reuptake inhibitors (SSRIs) in panic disorder, *Jpn. J. Psychosom. Med.* **40**, 275–281.

Davidson, J. R. T., Weisler, R. H., Malik, M., *et al.* (1998). Fluvoxamine in civilians with post-traumatic stress disorder, *J. Clin. Psychopharmacol.* **18**, 93–95.

Figgitt, D. P. and McClellan, K. J. (2000). Fluvoxamine. An update review of its use in the management of adults with anxiety disorders, *Drugs* **60**, 925–954.

Hata, T., Itoh, E. and Kawabata, A. (1991). Changes in CNS levels of serotonin and its metabolite in SART-stressed (repeatedly cold-stressed) rats, *Jpn. J. Pharmacol.* **56**, 101–104.

Hata, T., Nishikawa, H., Itoh, E., *et al.* (2001). Anxiety-like behavior in elevated plus-maze tests in repeatedly cold-stressed mice, *Jpn. J. Pharmacol.* **85**, 189–196.

Kirime, E. (2002). Effects of diazepam and fluvoxamine on SART-stressed rats in the elevated plus-maze test: investigation of intracerebral serotonin action in anxiety, *Acta Med. Kinki Univ.* **27**, 57–65.

Kita, T., Hata, T., Yoneda, R., *et al.* (1975). Stress state caused by alteration of rhythm in environmental temperature, and the functional disorders in mice and rats, *Folia Phamacol. Jpn.* **71**, 195–210.

Stein, D. J., Westenberg, H. G. M. and Liebowitz, M. R. (2002). Social anxiety disorder and generalized anxiety disorder: serotonergic and dopaminergic neurocircuitry, *J. Clin. Psychiatry* **63** (Suppl. 6), 12–19.

Stein, M. B., Fyer, A. J., Davidson, J. R. T., *et al.* (1999). Fluvoxamine treatment of social phobia (social anxiety disorder): A double-blind, placebo-controlled study, *Am. J. Psychiatry* **156**, 756–760.

Advances in Neuroregulation and Neuroprotection (2005), pp. 725-736
Collin, C. *et al.* (Eds)
© VSP 2005

Protective effect of docosahexaenoic acid against development of hypertension in young stroke-prone spontaneously hypertensive rats. Focus on the sympathoadrenal system and lipid metabolism

H. SAITO [1,*], S. KIMURA [2], M. MINAMI [3], H. HAYASHI [2], K. SHIMAMURA [2], M. NEMOTO [1] and S. H. PARVEZ [4]

[1] *Department of Basic Sciences, Japanese Red Cross Hokkaido College of Nursing, Akebono chou 664-1, Kitami city, 090-0011 Japan*
[2] *Department of Clinical Pharmacology, Faculty of Pharmaceutical Sciences, Health Sciences University of Hokkaido, Ishikari-Tobetsu, Hokkaido 061-0212, Japan*
[3] *Department of Pharmacology, Faculty of Pharmaceutical Sciences, Health Sciences University of Hokkaido, Ishikari-Tobetsu, Hokkaido 061-0212, Japan*
[4] *Neuroendocrinologie et Neuropharmacologie du Development, CNRS-Institut Alfred Fessard de Neurosciences, IFR 2118, Bat 5, 91190 Gif Sur Yvette, France*

Abstract—The present study was undertaken to elucidate the effects of acute and subacute administration of docosahexaenoic acid (DHA) on blood pressure in SHRSP between the ages of 6 and 11 weeks. The continuous administration of DHA for 5 weeks in SHRSP significantly inhibited blood pressure rise as compared with levels in non-treated SHRSP. DHA-treated SHRSP showed significantly decreased urinary epinephrine contents as compared with those in non-treated SHRSP. DHA-treated SHRSP produced a significant decrease in adrenal epinephrine concentration as compared with those in non-treated SHRSP at the age of 11 weeks. DHA-treated SHRSP also demonstrated significant decreases in serum triglycerides, LDL and lipid peroxide levels as compared with levels in non-treated SHRSP. However, a single intraperitoneal administration of DHA did not exert any influence on blood pressure in SHRSP at the age of 11 weeks. These findings suggest that the anti-hypertensive effect induced by a 5-week, continuous administration of DHA appears to be associated with the amelioration of increased sympathoadrenal activity and serum lipid derangement during the developmental stage of hypertension in young SHRSP. The present study clarified that a 5-week, continuous administration of DHA had a protective effect against the development of hypertension in SHRSP.

Keywords: SHRSP; DHA; catecholamine; lipid metabolism.

---

*To whom correspondence should be addressed. E-mail: hideyas@rchokkaido-cn.ac.jp

INTRODUCTION

Stroke-prone spontaneously hypertensive rats (SHRSP) were developed by Oka-moto *et al.* in 1974. At present, SHRSP is considered the best animal model of essential hypertension and stroke (Okamoto *et al.*, 1974).

Similar to eicosapentaenoic acid (EPA; C20: 5n-3), docosahexaenoic acid (DHA; C22: 6n-3) is an unsaturated fatty acid derived from fish oil (Dyerberg *et al.*, 1975). DHA has effects on the cardiovascular system (Haifa *et al.*, 1992; Peter *et al.*, 1996) and central nervous system (Nishikawa *et al.*, 1994; Yoshida *et al.*, 1997). The precise mechanism of these biological actions induced by DHA, however, remains unknown.

Our previous studies revealed that chronic administration of DHA to SHRSP be-tween the ages of 6 and 20 weeks produced an inhibitory effect on the development and maintenance of high blood pressure (Kimura *et al.*, 1995). We also found that chronic DHA treatment inhibited the decline in the passive avoidance response in SHRSP and also produced an ameliorative effect on the impairment of spontaneous motor activity in SHRSP. The life span of chronic DHA-treated SHRSP was sig-nificanly prolonged as compared with that of non-treated SHRSP (Minami *et al.*, 1997a, b). We surmised that the ameliorative effect on renal function induced by chronic treatment with DHA might be related to the DHA-induced antihypertensive action in SHRSP between the ages of 6 and 20 weeks.

No report is available in the literature, however, concerning the effects of acute and subacute administration of DHA on the blood pressure of SHRSP. During the normotensive period, young SHRSP showed increased sympathetic nerve activity. Yamori *et al.* (1989) proposed the possibility that this increased sympathetic nerve activity during the developmental period of hypertension may stimulate protein synthesis in the vascular wall, leading to an increase in medial thickness of the blood vessels in SHR. During the development of hypertension, serum lipid derangement and abnormalities in membrane permeability were also reported in SHR (Yamori *et al.*, 1980; Okamoto *et al.*, 1989).

The present study was undertaken to elucidate whether or not acute and subacute administration of DHA induces protective effects against the development of hypertension. An attempt was also made to clarify whether DHA ameliorates the increased sympathoadrenal system and serum lipid derangement in SHRSP between the ages of 6 to 11 weeks.

MATERIALS AND METHODS

Animals and feeding

SHRSP were inbred at our own laboratory from the original stock donated by the late Professor Kozo Okamoto, Department of Pathology, Kinki University School of Medicine, Osaka, Japan. Rats were kept at a room temperature of $22 \pm 2°C$, and at a humidity of 50% throughout the experiment. After body weights and blood pressure measurements were taken, the SHRSP were divided into two groups with approximately equal mean body weights and systolic blood pressures. We used aged-matched Wistar Kyoto rats (WKY) as controls. The study was conducted in accordance with 'The Guideline for the Care and Use of Laboratory Animals Research Committee' at the Health Sciences University of Hokkaido.

Determination of blood pressure

In an acute DHA administration study, we used 11-week-old SHRSP. DHA ethyl ester (95% pure, Harima Chemicals, Inc., Ibaraki, Japan) at a dose of 300 mg/kg was injected intraperitoneally into the SHRSP. Blood pressures of DHA-treated SHRSP were measured directly from the femoral artery at 0, 30, 60, 180, 360 and 720 min after DHA administration.

SHRSP between the ages of 6 and 11 weeks old were used for the study of the continuous administration of DHA at a dose of 900 mg/kg/day, DHA ethyl ester (95% pure) was administrated orally for 5 continuous weeks. Both non-treated WKY and non-treated SHRSP received distilled water (1 ml/kg/day).

Before administration, the blood pressures of SHRSP were measured by the tail-cuff method (Natsume Co. Ltd., Tokyo, Japan: KN-0090). After five weeks of continuous DHA administration, the blood pressures of DHA-treated SHRSP were measured directly from the femoral artery, after which blood samples were withdrawn from the femoral artery for the determination of biochemical parameters.

Determination of biochemical parameters

Method for the determination of urinary catecholamine concentrations and adrenal catecholamine concentrations. Urinary samples for catecholamine assay were collected in metabolic cages and the urine volume of each sample was measured. Urinary catecholamine contents and adrenal catecholamine concentrations were determined by high performance liquid chromatography combined with electrochemical detector.

Determination of serum lipid concentrations. Serum total cholesterol concentration was determined by the enzymatic method developed by Allain *et al.* (1974). Serum triglycerides were determined by the method developed by Matsumiya (1983). Serum high density lipoprotein (HDL) concentration was determined by the method of Irie *et al.* (1992). Serum low density lipoprotein (LDL) was calculated using Friedwald's equation as follows: LDL cholesterol (mg/dl) = (total cholesterol) − (HDL cholesterol) − (triglycerides/5).

Determination of lipid peroxide concentration. Lipid peroxide was measured by the method of Yagi (1976).

Determination of renal function. Serum creatinine concentration was determined by the method of Cook (1971). Blood urea nitrogen concentration (BUN) was measured by the urease GLDH method (Kaltwasser *et al.*, 1966).

Statistical analysis

Values are expressed as the mean ± S.E. Comparison between two groups was performed using Student's *t*-test. Comparison of more than two groups was performed by analysis of Wallenstein *et al.* (1980).

RESULTS

Single intraperitoneal injection of DHA

No significant differences were observed in the response of blood pressure to a single intraperitoneal injection of 300 mg/kg DHA between DHA-treated SHRSP and non-treated SHRSP at 0, 30, 60, 180, 360 and 720 min after administration.

Five weeks continuous oral administration of DHA

No significant differences were observed in body weights 5 weeks after continuous oral administration of DHA at a dose of 900 mg/kg/day between non-treated WKY, non-treated SHRSP and DHA-treated SHRSP groups. After 5 weeks of continuous oral DHA administration, the blood pressure of non-treated WKY was 125.6 ± 3.7 mmHg, while the blood pressure of non-treated SHRSP was 226.1±2.0 mmHg. The blood pressure of DHA-treated SHRSP was 184.0±6.5 mmHg. Compared with that of non-treated SHRSP, the blood pressure of DHA-treated SHRSP decreased significantly (Table 1).

Table 1.
Effects of DHA administration on systolic blood pressures in SHRSP

	Systolic blood pressures (mmHg)		
	Non-treated WKY	Non-treated SHRSP	DHA-treated SHRSP
6-week-old	112.6 ± 3.7 (9)	126.1 ± 1.0 (12)	124.9 ± 3.0 (8)
11-week-old	125.6 ± 3.7 (9)	226.1 ± 2.0*** (12)	184.0 ± 6.5***, ### (8)

Mean ± S.E., (n), ***$p < 0.001$ *vs.* 11-week-old non treated WKY, ###$p < 0.001$ *vs.* 11-week-old non-treated SHRSP. DHA was administered orally at a dose of 900 mg/kg/day for 5 weeks from 6 to 11 weeks of age. Blood pressures were measured by the tail-cuff method at the age of 6 weeks. Blood pressures were measured directly from the femoral artery at the age of 11 weeks.

Effects of DHA on urinary catecholamine concentrations in SHRSP

At the ages of 6, 7 and 11 weeks, urinary epinephrine contents of non-treated SHRSP increased significantly as compared with those in non-treated WKY. At 10 and 11 weeks, DHA-treated SHRSP displayed significantly decreased urinary epinephrine contents as compared with those in non-treated SHRSP (Fig. 1). With regards to urinary norepinephrine and dopamine contents, DHA did not produce any significant effects in SHRSP at 7–11 weeks.

Effects of DHA on catecholamine concentrations in the adrenal glands of SHRSP

DHA-treated SHRSP demonstrated significantly decreased adrenal epinephrine concentrations as compared with those in non-treated SHRSP at the age of 11 weeks (Fig. 2). On the other hand, DHA did not influence the adrenal dopamine or the adrenal norepinephrine concentrations of SHRSP.

Effects of DHA on renal function and lipid metabolism in SHRSP

No significant differences were observed in serum creatinine and BUN levels five weeks after continuous oral administration of DHA between non-treated WKY, non-treated SHRSP and DHA-treated SHRSP groups (Table 2). Similarly, non-treated

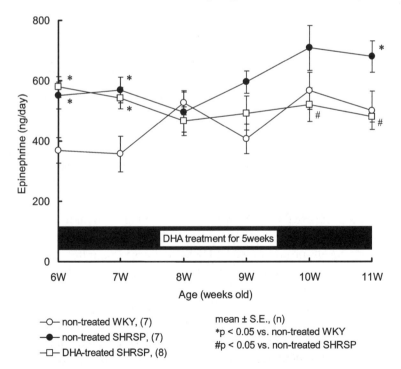

Figure 1. Effects of DHA on urinary excretion of epinephrine in WKY and SHRSP during the developmental stage of hypertension.

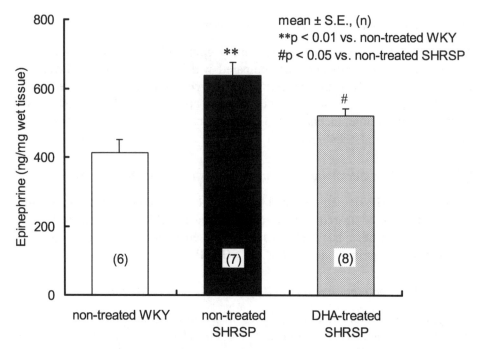

Figure 2. Effects of DHA on adrenal epinephrine levels in WKY and SHRSP at the age of 11 weeks.

Table 2.
Effects of DHA administration on serum creatinine and blood urea nitrogen levels in SHRSP

	Non-treated WKY	Non-treated SHRSP	DHA-treated SHRSP
S-creatinine	0.44 ± 0.02	0.48 ± 0.01	0.43 ± 0.01
(mg/dl)	(9)	(10)	(6)
BUN	20.3 ± 1.2	23.3 ± 1.4	23.8 ± 1.8
(mg/dl)	(7)	(11)	(6)

Mean ± S.E., (n), S-creatinine: serum creatinine, BUN: blood urea nitrogen. DHA was administered orally at a dose of 900 mg/kg/day for 5 weeks from 6 to 11 weeks of age.

WKY, non-treated SHRSP and DHA-treated SHRSP groups showed no significant differences in total cholesterol or HDL levels.

Triglycerides levels in non-treated SHRSP increased significantly as compared with those in non-treated WKY. A significant decrease in triglycerides was observed in DHA-treated SHRSP as compared with levels in non-treated SHRSP (Fig. 3). The LDL concentrations of non-treated SHRSP tended to be higher than those of non-treated WKY. LDL levels of DHA-treated SHRSP, however, decreased significantly as compared with those of non-treated SHRSP (Fig. 4). Similarly, the lipid peroxide levels in non-treated SHRSP increased significantly as compared with those in non-treated WKY. DHA-treated SHRSP showed a significant decrease in lipid peroxide levels as compared with levels in non-treated SHRSP (Fig. 5).

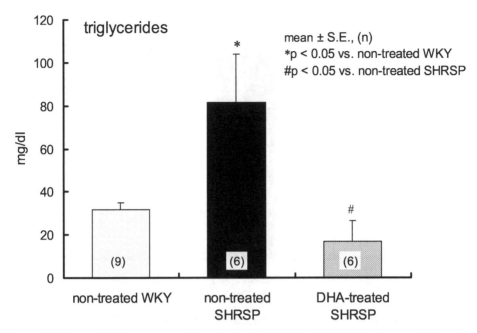

Figure 3. Effects of DHA on serum triglyceride levels in WKY and SHRSP at the age of 11 weeks.

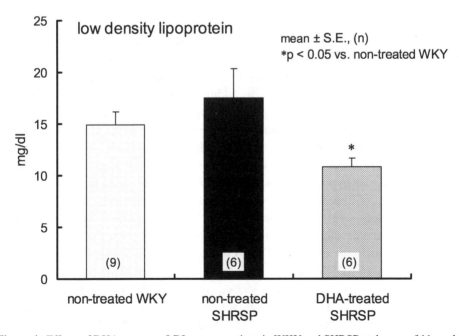

Figure 4. Effects of DHA on serum LDL concentrations in WKY and SHRSP at the age of 11 weeks.

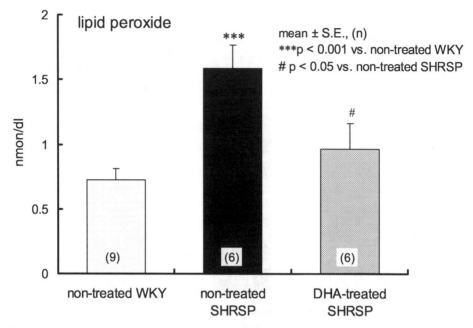

Figure 5. Effects of DHA on serum lipid peroxide levels in WKY and SHRSP at the age of 11 weeks.

DISCUSSION

Our previous studies revealed that SHRSP developed renal impairment. This renal impairment was not observed in SHRSP that received DHA for 14 weeks between the ages of 6 and 20 weeks (Kimura *et al.*, 1995). Thus, the ameliorative effect of DHA on renal impairment appears to be related to its antihypertensive action in SHRSP at the age of 20 weeks. In addition, one possible explanation of this DHA-induced ameliorative effect on renal functions would involve a secondary phenomenon induced by the antihypertensive action of DHA. Factors affecting the development of hypertension in both SHR and SHRSP include increased sympathetic nerve activity, increased protein synthesis such as collagen in vascular tissues and vascular hypertrophy. Serum lipid derangement may also participate in the development of hypertension (Ooshima *et al.*, 1979; Yamori *et al.*, 1979, 1981). In this experiment, we studied the effect of DHA on elevated sympathoadrenal activity during the development of hypertension in SHRSP.

In order to clarify the effects of DHA on blood pressure during the development of hypertension in SHRSP, DHA was administered to SHRSP between the ages of 6 and 11 weeks for 5 continuous weeks. Another purpose of this study was to elucidate the effects of DHA on serum lipid levels, which are a major risk factor for the development of hypertension. The blood pressure of DHA-treated SHRSP significantly decreased as compared with that in non-treated SHRSP at the ages of 10 and 11 weeks.

It has been realized that the increase in blood pressure in SHRSP may be due to the high catecholamine type of essential hypertension. In this study, we clarified that the urinary epinephrine levels and adrenal epinephrine concentrations of DHA-treated SHRSP decreased as compared to those of non-treated SHRSP.

One possible explanation would concern the decreased amount of epinephrine released from the adrenal glands. However, neither dopamine or norepinephrine concentrations in the urine and the adrenal glands changed in DHA-treated SHRSP. These findings suggest that DHA does not have an effect on the biosynthetic or the degradation enzymes involved in catecholamine metabolism, including TH, DBH, MAO and COMT. DHA might, instead, produce a decrease in PNMT activity in the adrenal glands. There is also a possibility that DHA might influence epinephrine release from the adrenal glands in SHRSP.

It has been reported that increased epinephrine release occurred in the adrenal glands of SHRSP. Moreover, the fact that DHA produces a decrease in Ca^{2+} influx into the cells suggests that DHA might inhibit epinephrine release due to the inhibition of excess Ca^{2+} influx into adrenal medullary cells. In line with the above, the present results suggest that decreased sympathoadrenal activity induced by DHA is, at least in part, associated with DHA-induced antihypertensive action.

Eleven-week-old SHRSP revealed an increase in serum LDL, triglycerides and lipid peroxide levels compared with those of normotensive non-treated WKY. Both LDL and triglycerides may diffuse from the tissues and invade the area beneath the vascular endothelium. Should this happen, both would accumulate in the vascular wall. These changes in the vascular wall might trigger vascular wall impairment (Ross and Glomset, 1976a, b). It may be such alterations in the vascular wall that produce hypertension and arteriosclerosis in SHRSP.

DHA-treated SHRSP showed significant decreases in serum LDL, triglycerides and lipid peroxide as compared with values of non-treated SHRSP. It has been reported that the n-3 unsaturated fatty acid, EPA, inhibited LDL absorption from the intestine (Mizuguchi *et al.*, 1992, 1993). In general, fatty acid produces an increase in LDL, but the administration of n-3 fatty acid produces a significant increase in the production of VLDL, which through rapid excretion has a high body clearance rate as compared with LDL (Huff and Telford, 1989). Furthermore, Livar *et al.* (1996) reported that EPA decreased cholesterol synthesis due to inhibition of HMG CoA reductase. For these reasons, we presupposed that DHA may produce the decreases in LDL and triglycerides levels mediated by decreased absorption, decreased biosynthesis and increased excretion of both LDL and triglycerides in SHRSP. Thus, DHA-induced decreases in LDL and triglycerides may produce the decreases in the occurrence of vascular structural impairment such as vascular hypertrophy, leading to inhibition of the development of hypertension.

Nishida and Kummerow (1960) reported that, similar to LDL and triglycerides, serum lipid peroxide accumulated in the vascular wall, causing vascular wall impairment. Cellular membrane derangement induced by lipid peroxide was also reported (Kikuchi and Ohishi, 1980). It has been realized that cell membranes of both SHRSP and patients with hypertension are markedly vulnerable as compared with those in both normotensive rats and normal subjects (Monique *et al.*, 1980; Yamori *et al.*, 1980; Kojima *et al.*, 1982).

The fact that a DHA-induced decrease in lipid peroxide was observed in this study suggests that DHA might contribute to membrane stability due to the inhibition of lipid peroxide biosynthesis.

With regard to renal function, non-treated 11-week-old SHRSP did not show any remarkable impairment of renal function as compared with levels in aged-matched normotensive WKY controls. However, in our previous study, the renal function impairment observed in SHRSP at the age of 20 weeks seems to be a secondary phenomenon that is due to the long-term continuation of hypertension. Our previous study, in which SHRSP received continuous administration of DHA for 14 weeks between the ages of 6 and 20 weeks, led us to conjecture that DHA-induced amelioration of renal function impairment might be related to DHA-induced antihypertensive action.

If this were so, functional impairment such as increased sympathoadrenal activity and serum lipid derangement may occur before and after the development of hypertension in SHRSP. This functional abnormality might relate to vascular structural changes, such as vascular endothelium impairment and vascular hypertrophy. Finally, these abnormalities appear to be related to blood pressure rise. Therefore, DHA might produce (1) an inhibition of functional impairment, (2) an ameliorative effect on vascular structural changes, and (3) an inhibition of blood pressure rise.

In this study, acute administration of DHA did not influence blood pressure in SHRSP during the development of hypertension. One explanation might pertain to the fact that a 5-week DHA administration may inhibit the risk factors for blood pressure rise in SHRSP between the ages of 6 and 11 weeks.

The present study demonstrated that a 5-week administration of DHA during pre-hypertensive and posthypertensive periods produced an antihypertensive action in SHRSP. Moreover, the DHA-induced amelioration of increased sympathoadrenal activity and serum lipid abnormalities before and after the development of hypertension appears to be related to its antihypertensive action in SHRSP.

Acknowledgements

This work was supported in part by grants for High-Tech Research Center from the Ministry of Education, Science, Sports and Culture of Japan.

REFERENCES

Allain, C. C., Poon, L. S., Chan, C. S. G., *et al.* (1974). Enzymatic determination of total serum cholesterol, *Clin. Chem.* **20**, 470–475.
Cook, J. G. (1971). Creatinine assay in the presence of protein, *Clin. Chim. Acta* **32**, 485–486.
Dyerberg, J., Bang, H. O. and Hjorne, N. (1975). Fatty acid composition of the plasma lipids in Greenland Eskimos, *Am. J. Nutr.* **28**, 958–966.

Haifa, H., Thomas, W. S. and Alexander, L. (1992). Modulation of dihydropyridine-sensitive calcium channels in heart cells by fish oil fatty acids, *Proc. Natl. Acad. Sci. USA* **89**, 1760–1764.

Huff, M. W. and Telford, D. E. (1989). Dietary fish oil increases conversion of very low density lipoprotein apoprotein B to low density lipoprotein, *Arteriosclerosis* **9**, 58–66.

Irie, T., Fukunaga, K. and Josef, P. (1992). Hydroxypropylcyclodextrins in parenteral use. I: Lipid dissolution and effects on lipid transfers in vitro, *J. Pharmacol.* **81**, 521–523.

Kaltwasser, H. and Schlegel, H. G. (1966). NADH-dependent coupled enzyme assay for urease and other ammonia-producing systems, *Anal. Biochem.* **16**, 132–138.

Kikuchi, S. and Ohishi, S. (1980). Lipoperoxide, *Rinsyou-kensa* **4**, 59–67.

Kimura, S., Minami, M., Togashi, H., *et al.* (1995). Antihypertensive effect of dietary docosahexaenoic acid (22: 6n-3) in stroke-prone spontaneously hypertensive rats, *Biogenic Amines* **11**, 195–203.

Kojima, S., Ito, K., Satani, M., *et al.* (1982). Abnormal electrolytes transport in red cells of essential hypertensive patients — Studies in Japanese, *Igaku-no-ayumi* **112**, 964–965.

Livar, F., Hege, V., Daniel, K. A., *et al.* (1996). Chronic administration of eicosapentaenoic acid and docosahexaenoic acid as ethyl esters reduced plasma cholesterol and changed the fatty acid composition in rat blood and organs, *Lipids* **31**, 169–178.

Matsumiya, K. (1983). Triglycerides (T-G), *Medical Technology* **11**, 1201–1206.

McLennan, P., Howe, P., Abeywardena, *et al.* (1996). The cardiovascular protective role of docosahexaenoic acid, *Eur. J. Pharmacol.* **300**, 83–89.

Minami, M., Kimura, S., Endo, T., *et al.* (1997a). Effetcs of dietary docosahexaenic acid on survival time and stroke-related behavior in stroke-prone spontaneously hypertensive rats, *Gen. Pharmacol.* **29**, 401–407.

Minami, M., Kimura, S., Endo, T., *et al.* (1997b). Dietary docosahexaenoic acid increases cerebral acetylcholine levels and improves passive avoidance performance in stroke-prone spontaneously hypertensive rats, *Pharmacol. Biochem. Behavior* **58**, 1123–1129.

Monique, D. M., Marie, L. G., Ricardo, P. G., *et al.* (1980). Abnormal net Na$^+$ and K$^+$ fluxes in erythrocytes of three varieties of genetically hypertensive rats, *Proc. Natl. Acad. Sci. USA* **77**, 4283–4286.

Mizuguchi, K., Yano, T. and Kojima, M. (1992). Hypolipidemic effect of ethyl all-cis-5,8,11,14,17-icosapentaenoate (EPA-E) in rats, *Jpn. J. Pharmacol.* **59**, 307–312.

Mizuguchi, K., Yano, T., Ishibashi, M., *et al.* (1993). Ethyl all-cis-5,8,11,14,17-icosapentaenoate modifies the biochemical properties of rat very low-density lipoprotein, *Eur. J. Pharmacol.* **235**, 221–227.

Nishida, T. and Kummerow, F. A. (1960). Interaction of serum lipoproteins with the hydroperoxide of methyl linoleate, *J. Lipid Res.* **1**, 450–458.

Nishikawa, M., Kimura, S. and Akaike, N. (1994). Facilitatory effect of docosahexaenoic acid on N-methyl-D-aspartate response in pyramidal neurons of rat cerebral cortex, *J. Physiol.* **475**, 83–93.

Okamoto, H., Kawaguchi, H., Minami, M., *et al.* (1989). Lipid alterations in renal membrane of stroke-prone spontaneously hypertensive rats, *Hypertension* **13**, 456–462.

Okamoto, K., Yamori, Y. and Nagaoka, A. (1974). Establishment of the stroke-prone spontaneously hypertensive rats (SHR), *Circ. Res.* **34/35** (Suppl. I), I-143–153.

Ooshima, A., Yamori, Y., Horie, R., *et al.* (1979). Vascular protein metabolism in hypertensive models in relation to the prevention of hypertensive diseases, in: *Prophylactic Approach to Hypertensive Diseases*, Yamori, Y. (Ed.), pp. 233–240. Raven Press, New York.

Ross, R. and Glomset, J. A. (1976a). The pathogenesis of atherosclerosis (First of two parts), *New Eng. J. Med.* **295**, 369–377.

Ross, R. and Glomset, J. A. (1976b). The pathogenesis of atherosclerosis (Second of two parts), *New Eng. J. Med.* **295**, 420–425.

Wallenstein, S., Zucker, C. L. and Fleiss, J. L. (1980). Some statistical methods useful in circulation research, *Circ. Res.* **47**, 1–9.

Yagi, K. (1976). A simple fluorometric assay for lipoperoxide in blood plasma, *Biochem. Med.* **15**, 212–216.

Yamori, Y. (1989). Predictive and preventive pathology of cardiovascular diseases, *Acta Pathologica Japonica* **39**, 683–705.

Yamori, Y., Horie, R., Akiguchi, I., *et al.* (1979). Pathophysiology and prevention of stroke-prone spontaneously hypertensive rats, in: *Prophylactic Approach to Hypertensive Diseases*, Yamori, Y. (Ed.), pp. 173–183. Raven Press, New York.

Yamori, Y., Nara, Y., Horie, R., *et al.* (1980). Abnormal membrane characteristics of erythrocytes in rat models and men with predisposition to stroke, *Clin. Exp. Hypertension* **2**, 1009–1021.

Yamori, Y., Kihara, M., Nara, Y., *et al.* (1981). New models for thrombogenesis and atherogenesis — spontaneous thrombogenic rats (STR)-myocardial lesioned rats (MLR), in: *Proc. of II German–Japanese Congress of Angiology*, pp. 154–158.

Yoshida, S., Miyazaki, M., Takeshita, M., *et al.* (1997). Functional changes of rat brain microsomal membrane surface after leaning task depending on dietary fatty acids, *J. Neurochem.* **68**, 1269–1277.

Advances in Neuroregulation and Neuroprotection (2005), pp. 737-750
Collin, C. *et al.* (Eds)
© VSP 2005

Expression of acetylcholine in lymphocytes and modulation of an independent lymphocytic cholinergic activity by immunological stimulation

TAKESHI FUJII, YOSHIHIRO WATANABE, KAZUKO FUJIMOTO
and KOICHIRO KAWASHIMA *

Department of Pharmacology, Kyoritsu College of Pharmacy, 1-5-30 Shibakoen, Minato-ku, Tokyo 105-8512, Japan

Abstract—Although acetylcholine (ACh) is classically thought of as a neurotransmitter in mammalian species, lymphocytes possess most of the components needed to constitute an independent non-neuronal cholinergic system. These include ACh itself, choline acetyltransferase (ChAT), high-affinity choline transporter, acetylcholinesterase, and both muscarinic and nicotinic ACh receptors (mAChRs and nAChRs, respectively). Activation of mAChRs and nAChRs on lymphocytes elicits increases in the intracellular Ca^{2+} concentration and stimulates c-fos gene expression and nitric oxide synthesis. Which subpopulations of T lymphocytes are the source of blood ACh is not yet known, though expression of ChAT mRNA was observed in human peripheral $CD4^+$ (helper) but not in $CD8^+$ (cytotoxic) T cells. Stimulation of lymphocytes with phytohemagglutinin, a T-cell activator, or with monoclonal antibodies that bind the CD7 and/or CD11a cell surface molecules activates lymphocytic cholinergic activity, as evidenced by increased synthesis and release of ACh and up-regulation of expression of ChAT and M_5 mAChR mRNA. Abnormalities in the lymphocytic cholinergic system have been detected in spontaneously hypertensive rats and MRL-*lpr* mice, two animal models of immune disorders. Taken together, these data present a compelling picture in which immune function is, at least in part, under the control of an independent non-neuronal lymphocytic cholinergic system.

Keywords: Acetylcholine; choline acetyltransferase; $CD4^+$ T cell; $CD8^+$ T cell; muscarinic receptor; nicotinic receptor.

INTRODUCTION

Acetylcholine (ACh) is most often thought of as a classical neurotransmitter synthesized by choline acetyltransferase (ChAT, E.C.2.3.1.6) from acetyl coenzyme A and choline (Tucek, 1988). However, we recently showed that ACh is ubiquitously

---

*To whom correspondence should be addressed. E-mail: kawashima-ki@kyoritsu-ph.ac.jp

expressed in life (Horiuchi *et al.*, 2003), and particular attention has been focused on the widespread expression of ACh in the non-neuronal tissues and organs of mammalian species — e.g. immune cells and epithelial, endothelial, mesothelial and parenchymal cells (Wessler *et al.*, 1998, 1999, 2001; Kawashima and Fujii, 2000; Fujii and Kawashima, 2001b; Sakuragawa *et al.*, 2001; Grando *et al.*, 2003) — where ACh acts in an autocrine and/or paracrine manner, serving as a local signaling molecule or cytotransmitter.

Using selective receptor binding assays and molecular biological techniques, both muscarinic and nicotinic ACh receptors (mAChRs and nAChRs, respectively) have been identified in various blood cell types, including lymphocytes; moreover, stimulation of lymphocytes with exogenous mAChR and nChR agonists elicits a variety of functional and biochemical effects (see reviews by Kawashima and Fujii, 2000; Fujii and Kawashima, 2001b; Okuma and Nomura, 2001). On the basis of these findings, it was initially postulated that the parasympathetic nervous system is directly involved in immune-neurohumoral crosstalk (Maslinski, 1989; Rinner and Schauenstein, 1991; Toyabe *et al.*, 1997; Abo and Kawamura, 2002). At odds with this interpretation, however, is the presence of ACh in the blood and plasma of a variety of animals, including humans (Kawashima *et al.*, 1987, 1989, 1993, 1997, 1998; Fujii *et al.*, 1995b, 1997; Yamada *et al.*, 1997). Indeed, subsequent investigation of its origin and function revealed that lymphocytes possess most of the essential components needed to constitute an independent, non-neuronal cholinergic system, and that ACh synthesized in and released from lymphocytes acts as an immunomodulator via both mAChRs and nAChRs (Kawashima and Fujii, 2000, 2003).

This review focuses primarily on (1) expression of the cholinergic components in lymphocytes; (2) modulation of lymphocytic cholinergic activity by immunological stimulation; (3) modulation of lymphocyte function by AChR agonists; and (4) lymphocytic cholinergic activity in animal models of immune disorders.

THE ORIGIN OF ACH IN LYMPHOCYTES

Detection of ACh and ChAT activity

Because ACh is rapidly hydrolyzed by acetylcholinesterase (AChE) and other cholinesterases (ChE), the presence of ACh in blood had long been believed to be unlikely, as blood contains high levels of ChE activity. Still, estimates made as early as 1930 (Kapfhammer and Bischoff, 1930) of ACh levels in the blood and plasma of animals have varied substantially (see a review by Kawashima and Fujii, 2000). Using a highly sensitive (3 pg/tube), specific radioimmunoassay (RIA) for ACh (Kawashima *et al.*, 1980; Oosawa *et al.*, 1999), we found ACh to be present in the blood and plasma of several mammals, including humans (Kawashima *et al.*, 1987, 1989, 1993, 1997, 1998; Fujii *et al.*, 1997; Yamada *et al.*, 1997). Among the species tested, the highest levels of blood ACh were found in cattle and horses

Table 1.

ACh concentrations in the blood and plasma of various mammalian species

Species	Number of samples	ACh (pmol/ml)		P/B ratio
Cattle (Holstain)	7	360.5 ± 59.7	7.14 ± 2.55	0.02
Chimpanzee	6	21.5 ± 2.60	1.26 ± 0.07	0.06
Dog (Beagle)	10	1.37 ± 0.23	0.13 ± 0.04	0.02
Goat	5	4.05 ± 0.93	0.22 ± 0.06	0.05
Horse	5	93.8 ± 16.3	0.66 ± 0.23	0.01
Human (Japanese)	30–32	8.66 ± 1.02	3.12 ± 0.36	0.38
Pig	5	11.7 ± 1.5	15.4 ± 1.58	1.32
Rat (Wistar)	10	1.43 ± 0.20	0.75 ± 0.11	0.57
Sheep	5	2.04 ± 0.20	0.27 ± 0.05	0.14

Values are means ± S.E.M. Strains are in parentheses. P/B ratio, ratio of the plasma and blood ACh concentrations. Adopted from Kawashima and Fujii (2000).

(Table 1). Furthermore, the ratios of the ACh contents of plasma and whole blood showed that, with the exception of pigs and rats, most of the ACh was present within blood cells. In humans, the ACh content of circulating mononuclear leukocytes (MNL), which consist mainly of lymphocytes, was 5.01 ± 0.59 pmol/ml blood ($n = 30$), corresponding to about 60% of that found in whole blood (Kawashima *et al.*, 1993). No ACh was detected in polymorphonuclear leukocytes. That ACh is predominantly localized within MNLs, as well as the significant correlation between the ACh content of MNLs and that of whole blood, suggests that most plasma ACh originates from MNLs, probably lymphocytes.

To verify the origin of blood ACh, we examined various human leukemic cell lines as models of lymphocytes. The MOLT-3, HSB-2 and CCRF-CEM (CEM) human leukemic T cell lines all contained significant amounts of ACh, along with corresponding ChAT activity. By contrast, the Daudi human leukemic B cell line and the U937 human monocytic cell line did not synthesize ACh (Fujii *et al.*, 1996b). These findings are consistent with those of Rinner and Schauenstein (1993), who found that HSB-2 cells contain the highest levels of ChAT activity among the various leukemic lymphoma cell lines they studied. Later, ChAT activity was detected in human MNLs, human leukemic T-cell lines and rat lymphocytes (Fujii *et al.*, 1998, 1999; Rinner *et al.*, 1998; Kawashima and Fujii, 2000).

Detection of ChAT mRNA and protein in T-lymphocytes

Using reverse transcription-polymerase chain reaction (RT-PCR) with specific sense and antisense primers (5′-AAGACGCCCATCCTGGAAAAG-3′ and 5′-TGAGACGGCGGAAATTAATGAC-3′, respectively) that corresponded to nucleotide positions 322-973 of human brain ChAT cDNA (Oda *et al.*, 1992), we were able to demonstrate expression of ChAT mRNA in MOLT-3 cells (Fig. 1A), which contain the highest levels of both ChAT activity and intracellular ACh among

(A) RT-PCR   (B) Western blot

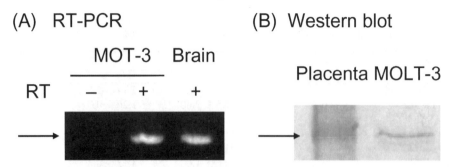

Figure 1. Expression of ChAT mRNA and protein in the MOLT-3 human leukemic T cells. (A) RT-PCR analysis of ChAT mRNA expression. (B) Detection of ChAT protein by Western blot analysis. Modified from Fujii *et al.* (1995a).

the cell types we tested (Fujii *et al.*, 1995a). Subsequently, expression of ChAT mRNA was similarly confirmed in human MNLs, rat lymphocytes and other human leukemic T-cell lines, as well as in lymphocytes from rat thymus, spleen and peripheral blood (Fujii *et al.*, 1998, 1999; Rinner *et al.*, 1998). B-cell lines, by contrast, did not express ChAT mRNA, indicating that only T-lymphocytes have the ability to synthesize ACh via ChAT.

That ChAT mRNA is translated to a corresponding protein was confirmed by Western blot analysis of MOLT-3 cells (Fig. 1B) (Fujii *et al.*, 1995a). This was the first demonstration of ChAT protein expression in a human T-cell line. Furthermore, the involvement of lymphoid ChAT in the synthesis of ACh was confirmed by the finding that bromoacetylcholine, a specific ChAT inhibitor, significantly diminished the ACh content of T-lymphocytes (Fujii *et al.*, 1996b).

Multiple ChAT mRNA species in T-lymphocytes

Multiple ChAT mRNA species (R-, N0-, N1-, N2-, and M-types), having identical coding regions but different 5′-noncoding regions, are expressed in the human brain and spinal cord (Misawa *et al.*, 1997). These mRNAs likely differ with respect to their translation efficiency and stability, reflecting differences in their promoter regions and alternative splicing. The types of ChAT mRNA expressed in T-lymphocytes remain unclear. However, we found that CEM and MOLT-3 human leukemic T-cells express the same ChAT mRNA as that expressed in the nervous system. CEM cells mainly express N2-, M-, and to a lesser extent, N1-type ChAT mRNA, whereas MOLT-3 cells expressed only N2-type mRNA (Ogawa *et al.*, 2003). Neither CEM nor MOLT-3 cells expressed R-type mRNA. Expression of the N1-, N2- and M-type mRNAs, but not the R-type, has also been detected in human MNLs (Fujii *et al.*, 1996a). How the inability to express R-type mRNA affects the capacity of lymphocytes to synthesize ACh, if at all, remains unknown.

OTHER CHOLINERGIC COMPONENTS IN LYMPHOCYTES

Expression of AChRs

Lymphoid expression of both mAChRs and nAChRs has been demonstrated using radioligand binding assays, RT-PCR and Western blot analyses (see reviews by Kawashima and Fujii, 2000; Fujii and Kawashima, 2001b). In addition, Fujii and Kawashima (1999) showed that some lymphocytes have the potential to simultaneously express both mAChRs and nAChRs.

mAChRs. Five mAChR subtypes (M_1-M_5) have been cloned (Bonner *et al.*, 1987, 1988; Alexander *et al.*, 2001), and mRNAs encoding several of these subtypes have been detected in rat and human lymphocytes, as well as in various human leukemic cell lines serving as models of lymphocytes. Sato *et al.* (1999) detected expression of mRNAs encoding the M_4 and M_5 receptor subtypes in all human MNLs; expression of mRNAs encoding the M_1, M_2 and M_3 subtypes varied, however. Thus, it may be that a wide array of combinations of mAChR subtypes is expressed among lymphocytes within individual subjects.

nAChRs. Radiolabeled ligand binding has been used to identify nAChRs in mouse, rat and human lymphocytes (Adem *et al.*, 1986; Gordon *et al.*, 1978). Five muscle-type ($\alpha 1, \beta 1, \gamma, \varepsilon$ and δ) and 12 neuronal-type ($\alpha 2$-$\alpha 10$ and $\beta 2$-$\beta 4$) subunits have been cloned (Alexander *et al.*, 2001). Sato *et al.* (1999) analyzed mRNAs encoding nAChR subunits in human MNLs and leukemic cell lines, and detected the expression of only neuronal-type subunits, though the specific expression pattern varied. Human MNLs expressed mRNAs encoding the $\alpha 2, \alpha 5$ and $\alpha 7$ subunits, but not the $\alpha 1, \beta 1$ or ε subunits.

Expression of high-affinity choline transporter (CHT)

In cholinergic neurons, choline taken up by the CHT is utilized exclusively for ACh synthesis; indeed choline uptake via the CHT is the rate-limiting step in ACh synthesis catalyzed by ChAT. Although cDNAs encoding rat and human CHT (CHT1) have been cloned recently (Okuda and Haga, 2000; Okuda *et al.*, 2000), the mechanism for choline uptake, and whether CHT1 is present in T-lymphocytes, remains unknown. Fujii *et al.* (2003a) used RT-PCR analysis to detect expression of hCHT1 mRNA in MOLT-3 human leukemic T cells. In addition, specific binding of [^{3}H]hemicholinium-3 (1.01 ± 0.55 fmol/mg protein, $n = 4$), an inhibitor of CHT1, and hemicholinium-3-sensitive [^{3}H]choline uptake (0.43 ± 0.01 pmol/10^6 cells per 1 h, $n = 3$) were both observed in MOLT-3 cells. Together, these data constitute the first evidence that non-neuronal cells express CHT1, and suggest that choline taken up via the hCHT1 is utilized, at least in part, for ACh synthesis in T-lymphocytes.

Vesicular acetylcholine transporter (VAChT)

The ACh synthesized in the cytosol of cholinergic neurons is then transported into synaptic vesicles by the VAChT. The VAChT gene is situated within the first intron of the ChAT gene, enabling coordinated regulation of VAChT and ChAT gene expression (Berrard *et al.*, 1995; Misawa *et al.*, 1995). Whether ACh in lymphocytes is stored in vesicles has not yet been confirmed, but expression of VAChT mRNA has not been detected in human MNLs or in leukemic T-cell lines, which argues against vesicular storage of ACh in lymphocytes (Fujii *et al.*, 1998; Ogawa *et al.*, 2003). Instead, these findings suggest that ACh is synthesized by T-lymphocytes when necessary and then directly released without storage.

Expression of acetylcholinesterase (AChE)

The available evidence clearly indicates that lymphocytes express AChE (Paldi-Haris *et al.*, 1990). RT-PCR analysis has confirmed the expression of mRNAs encoding all three AChE forms (hydrophilic, phosphoinositide-linked and readthrough) in human peripheral blood MNLs and in various leukemic T-cell (CEM and MOLT-3) and B-cell (Daudi and BALL-1) lines (Ando *et al.*, 1999), suggesting all have the capacity to express AChE. It is therefore reasonable to presume that ACh released from activated T-lymphocytes is hydrolyzed by AChE after interaction with AChRs and transmission of signals to target cells.

MODULATION OF AN INDEPENDENT LYMPHOCYTIC CHOLINERGIC ACTIVITY

In vivo stimulation of T- and B-lymphocytes leads to up-regulation of ACh synthesis and M_5 mAChR expression (Fujii *et al.*, 1998, 2003b). This suggests that immunological stimulation enhances cholinergic signal transmission between T-lympho cytes and target cells, including B-lymphocytes, by enhancing synthesis and release of ACh by T-lymphocytes and expression of M_5 mAChR by both T- and B-lymphocytes.

Phytohemagglutinin (PHA), a T-cell activator

Stimulation of T-lymphocytes with PHA induces activation of phospholipase C-mediated production of inositol-1,4,5-trisphosphate via the T-cell receptor (TCR)/CD3 molecule complex (Imboden *et al.*, 1986).

Enhancement of ACh synthesis and release and ChAT mRNA expression. Stimulation of MOLT-3 and HSB-2 cells with PHA increases both intracellular ACh content and its subsequent release (Fujii *et al.*, 1996b; Kawashima *et al.*, 1997), induces ChAT mRNA expression, and potentiates ChAT activity and ACh synthesis in human peripheral blood MNLs and MOLT-3 leukemic T cells (Fujii *et al.*, 1996b,

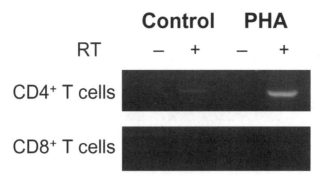

Figure 2. Expression of ChAT mRNA in human peripheral blood CD4$^+$ and CD8$^+$ T cells. The expected sizes of the ChAT mRNA RT-PCR product is 652 bp. The experiments were repeated at least three times, yielding essentially identical results.

1998; Ogawa *et al.*, 2003) and rat T cells (Rinner *et al.*, 1998). Since the increases in intracellular ACh content and release elicited by activation of T-lymphocytes are apparently mediated, at least in part, by increased ChAT activity, it would seem reasonable to suggest that immunologically activated T-lymphocytes synthesize and release ACh, which in turn acts on lymphocytes via mAChRs or nAChRs to modulate immune function.

Expression of ChAT mRNA in subpopulation of T-lymphocytes. We have shown that T-lymphocytes are the primary sources of blood ACh (Fujii *et al.*, 1999). However, which subpopulation of T-lymphocytes synthesizes ACh remains unclear. To address this question, we first separated human CD4$^+$ and CD8$^+$ T cells from MNLs using a magnetic cell sorting system. The resultant preparations usually consisted of >95% CD4$^+$ or CD8$^+$ T cells, as determined by flow cytometry using fluorescent-labeled antibodies.

Expression of ChAT mRNA was detected in unstimulated CD4$^+$ T cells (Fig. 2A, lane 2), but not in unstimulated CD8$^+$ T cells (Fig. 2B, lane 2). In addition, PHA up-regulated ChAT mRNA expression only in CD4$^+$ T cells (Fig. 2A and B, lane 4). On the other hand, both CD4$^+$ T cells and CD8$^+$ T cells expressed mRNA encoding the M$_3$, M$_4$ and M$_5$ mAChR subtypes (Fig. 3A and B). These data provide the first evidence that helper T cells are the major source of blood ACh, and that T cell activation modulates intracellular signaling processes via mAChRs expressed by themselves and other blood cells, with ACh serving as an immunomodulator.

Antithymocyte globulin-Fresenius (ATG-F)

ATG-F is a rabbit globulin against human thymocytes used clinically as an immunosuppressant following renal transplantation (Bock *et al.*, 1995; Passweg *et al.*, 1996). It binds to the CD2, CD7 and CD11a cell surface molecules, which are also respectively known as lymphocyte function-associated antigen 2, an Ig super-family member, and LFA-1 α-chain. Stimulation of CD7 and CD11a with monoclonal antibodies activated intracellular signaling pathways leading to increases in

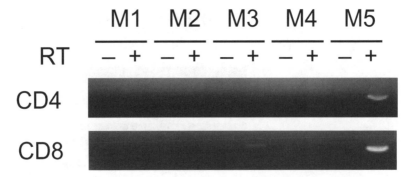

Figure 3. Expression of mAChR subtypes in human peripheral blood CD4$^+$ and CD8$^+$ T cells. The expected sizes of the mAChR mRNA RT-PCR products are 573, 469, 846, 648 and 800 bp. The experiments were repeated at least three times, yielding essentially identical results.

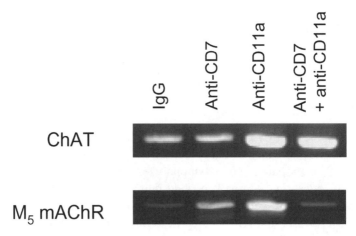

Figure 4. Up-regulation of ChAT and M$_5$ AChR in CEM cells after stimulation of the CD7 and CD11a surface molecules using monoclonal antibodies. The expected sizes of the ChAT and mAChR mRNA RT-PCR products are 652 and 800 bp.

the intracellular free Ca^{2+} concentration ([Ca^{2+}]$_i$), up-regulation of ChAT mRNA expression, and modulation of T-lymphocyte function (Fig. 4; Fujii *et al.*, 2002). In addition, CD11a stimulation also increased expression of M$_5$ mAChR mRNA (Fig. 4; Fujii and Kawashima, 2001c). ATG-F enhanced ACh release from CEM cells, most likely through transient increases in [Ca^{2+}]$_i$ mediated by CD7, which led to declines in intracellular ACh content. Within 48 h, however, the ACh content had increased as compared to control due to up-regulation of ChAT expression mediated by CD11a (Fig. 5).

INTRACELLULAR SIGNALING VIA MACHRS AND NACHRS

Consistent with the presence of both mAChRs and nAChRs on the surface of T- and B-lymphocytes, ACh and various mAChR and nAChR agonists were able

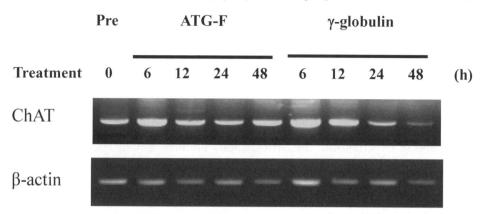

Figure 5. Effects of ATG-F on ChAT mRNA expression in CEM cells. Cells were exposed to ATG-F (100 μg/ml) or γ-globulin (control) for the indicated times. The expected sizes of the RT-PCR products are 652 (ChAT) and 1128 (β-actin) bp. Modified from Fujii *et al.* (2002).

to modulate lymphocyte function and metabolism (Maslinski, 1989), suggesting cholinergic mechanisms are involved in the regulation of immune function.

Effects of mAChR stimulation on lymphocytes

Fujii and Kawashima (2000a, b, 2001a) investigated the effects of ACh and mAChR agonists on $[Ca^{2+}]_i$ in human T-cell (CEM) and B-cell (Daudi) lines, and found that mAChR agonists induce atropine-sensitive increases in $[Ca^{2+}]_i$ followed by Ca^{2+} oscillations in both cell types, and that oxotremorine-M (Oxo-M, mAChR agonist) up-regulates c-fos mRNA expression. The effect of Oxo-M was blocked by 4-DAMP, but not by pirenzepine or AF-DX 116, providing the first evidence that ACh released from T-lymphocytes triggers nuclear signaling and up-regulation of gene expression in T- and B-lymphocytes via M_3 mAChRs (Fujii and Kawashima, 2000a). It is also notable that stimulation of mAChRs with Oxo-M induced an increase of nitric oxide synthesis in CEM cells (Kamimura *et al.*, 2003).

Effects of nAChRs stimulation on lymphocytes

Stimulation of nAChRs with nicotine or epibatidine, two nAChR agonists, elicited transient increases in $[Ca^{2+}]_i$ in CEM cells mediated by an influx of extracellular Ca^{2+} into the cytosol (Kimura *et al.*, 2003). These effects were abolished by α-bungarotoxin, an $\alpha7$ subtype nAChR antagonist, but not by pancuronium or d-tubocurarine, muscle-type nAChR antagonists. Stimulation of $\alpha7$ nAChR did not up-regulate c-fos mRNA expression (Ushiyama *et al.*, 2001), however, indicating that Ca^{2+} signaling via nAChRs differs from that via mAChRs. Prolonged exposure of CEM cells to nicotine down-regulated nAChR expression. Further study will be required to identify the precise role of nAChRs in the regulation of lymphocyte function.

LYMPHOCYTIC CHOLINERGIC ACTIVITY IN ANIMAL MODELS OF IMMUNE DISORDERS

The relationship between lymphocytic cholinergic activity and immune function was investigated in two animal models of immune disorders.

Spontaneously hypertensive rat (SHR)

In addition to hypertension, the SHR exhibits immune deficiencies related to the emergence of a natural thymocytotoxic autoantibody, to an age-related decline in T-cell function, and to morphological changes in immune organs (Takeichi *et al.*, 1988; Takeichi, 1995). Fujimoto *et al.* (2001) found significantly lower levels of ACh in the blood, thymus and spleen of SHRs and lower levels of ChAT mRNA expression in circulating MNLs, as compared with those of the immunologically normal Wistar Kyoto rat from which the SHR was derived. The diminished synthesis and release of ACh in MNLs and lymphoid organs could reflect an immune deficiency related to the T-cell dysfunction seen in SHRs.

MRL/lpr mouse

MRL/MpJ-*lpr*/*lpr* (MRL-*lpr*) mice develop autoimmune diseases such as lupus nephritis due to production of antinuclear autoantibodies (Watanabe *et al.*, 2000). Preliminary data show that the ACh content of the blood, thymus and spleen of MRL-*lpr* mice increases with age and is significantly greater than in age-matched control mice. Again, this is consistent with the idea that ACh derived from T-lymphocytes is involved in regulating immune activity (Fujimoto *et al.*, 2003; Fujimoto and Kawashima, in preparation).

CONCLUSION

The findings presented in this review provide clear evidence that lymphocyte function is regulated by an independent cholinergic system in the lymphocytes themselves. Stimulation of T-lymphocytes by PHA via TCR/CD3 or by monoclonal antibodies via CD7 and/or CD11a activates lymphocytic cholinergic activity, as evidenced by increased synthesis and release of ACh, increased AChE activity, and expression of mRNAs encoding ChAT and ACh receptors. Because lymphocytes can make direct contact with their target cells, we would anticipate that even small amounts of ACh released in the vicinity of their receptor sites would raise the ACh concentration enough to elicit a response before its hydrolysis by AChE. While the specific functions of the lymphoid cholinergic system are not yet precisely defined, the presence of both mAChRs and nAChRs on lymphocytes, together with the observed M_3 mAChR-mediated increases in $[Ca^{2+}]_i$ and c-fos gene expression, strongly suggests ACh synthesized and released by T-lymphocytes acts as an autocrine and/or paracrine factor regulating immune system function.

Acknowledgements

We are grateful to the present and past staff members and graduate students in the Department of Pharmacology, Kyoritsu College of Pharmacy for their encouragement and technical assistance. Human leukemic cell lines used in the studies cited were from the Fujisaki Cell Center, Hayashibara Biochemical Labs Inc. (Okayama, Japan). This work was supported by the Uehara Memorial Foundation (T. F.), the Naito Foundation (T. F.), by the Smoking Research Foundation (K. K.) and by a Grant-in Aid for Scientific Research from the Ministry of Education, Science, Sports, Culture and Technology (No. 14370037, K. K.).

REFERENCES

Abo, T. and Kawamura, T. (2002). Immunomodulation by the autonomic nervous system: Therapeutic approach for cancer, collagen diseases, and inflammatory bowel diseases, *Ther. Apher.* **6**, 348–357.

Adem, A., Nordoberg, A. and Slanina, P. (1986). A muscarinic receptor type in human lymphocytes: a comparison of ^{3}H-QNB binding to intact lymphocytes and lysed lymphocyte membranes, *Life Sci.* **38**, 1359–1368.

Alexander, S. P. H., Mathie, A. and Peters, J. A. (2001). in: *The 2001 Tips Nomenclature Supplement*, 12th edn, pp. 6–12. Current Trends, London.

Ando, T., Fujii, T. and Kawashima, K. (1999). Expression of three acetylcholinesterase mRNAs in human lymphocytes, *Jpn. J. Pharmacol.* **79** (Suppl. I), 289P.

Berrard, S., Varoqui, H., Cervini, R., *et al.* (1995). Coregulation of two embedded gene products, choline acetyltransferase and the vesicular acetylcholine transporter, *J. Neurochem.* **65**, 939–942.

Bock, H. A., Gallati, H., Zurcher, R. M., *et al.* (1995). A randomized prospective trial of prophylactic immunosuppression with ATG-fresenius versus OKT-3 after renal transplantation, *Transplantation* **59**, 830–840.

Bonner, T. I., Buckley, N. J., Young, A. C., *et al.* (1987). Identification of a family of muscarinic acetylcholine receptor genes, *Science* **237**, 527–532.

Bonner, T. I., Young, A. C., Brann, M. R., *et al.* (1988). Cloning and expression of the human and rat m5 muscarinic receptor genes, *Neuron* **1**, 403–410.

Fujii, T. and Kawashima, K. (1999). Presence of two distinct intracellular calcium signaling pathways via muscarinic and nicotinic receptors in human T- and B-cell lines, *Jpn. J. Pharmacol.* **79** (Suppl. I), 287P.

Fujii, T. and Kawashima, K. (2000a). Calcium signaling and c-fos gene expression via M_3 muscarinic acetylcholine receptors in human T- and B-cells, *Jpn. J. Pharmacol.* **84**, 124–132.

Fujii, T. and Kawashima, K. (2000b). Ca^{2+} oscillation and c-fos gene expression induced via muscarinic acetylcholine receptor in human leukemic T- and B-cell lines, *Naunyn-Schmiedeberg's Arch. Pharmacol.* **362**, 14–21.

Fujii, T. and Kawashima, K. (2001a). YM905, a novel M_3 antagonist, inhibits Ca^{2+} signaling and c-fos gene expression mediated via muscarinic receptors in human T-cells, *Gen. Pharmacol.* **35**, 71–75.

Fujii, T. and Kawashima, K. (2001b). The non-neuronal cholinergic system: an independent, non-neuronal cholinergic system in lymphocytes, *Jpn. J. Pharmacol.* **85**, 11–15.

Fujii, T. and Kawashima, K. (2001c). Activation of cholinergic system in T-lymphocytes by stimulation of CD11a molecule, *Neurosci. Res.* **25** (Suppl.), S147.

Fujii, T., Yamada, S., Misawa, H., *et al.* (1995a). Expression of choline acetyltransferase mRNA and protein in T-lymphocytes, *Proc. Jpn. Acad.* **71B**, 231–235.

Fujii, T., Yamada, S., Yamaguchi, N., *et al.* (1995b). Species differences in the concentration of acetylcholine, a neurotransmitter, in the whole blood and plasma, *Neurosci. Lett.* **201**, 207–210.

Fujii, T., Misawa, H. and Kawashma, K. (1996a). Induction of choline acetyltransferase mRNA in human T-lymphocytes during the immune responses, *Neurosci. Res.* **20** (Suppl.), S110.

Fujii, T., Tsuchiya, T., Yamada, S., *et al.* (1996b). Localization and synthesis of acetylcholine in human leukemic T cell lines, *J. Neurosci. Res.* **44**, 66–72.

Fujii, T., Mori, Y., Tominaga, H., *et al.* (1997). Maintenance of constant acetylcholine content before and after feeding in young chimpanzees, *Neurosci. Lett.* **227**, 21–24.

Fujii, T., Yamada, S., Watanabe, Y., *et al.* (1998). Induction of choline acetyltransferase mRNA in human mononuclear leukocytes stimulated by phytohemagglutinin, a T-cell activator, *J. Neuroimmunol.* **82**, 101–107.

Fujii, T., Tajima, S., Yamada, S., *et al.* (1999). Constitutive expression of mRNA for the same choline acetyltransferase as that in the nervous system, an acetylcholine-synthesizing enzyme, in human leukemic T-cell lines, *Neurosci. Lett.* **259**, 71–74.

Fujii, T., Ushiyama, N., Hosonuma, K., *et al.* (2002). Effects of human antithymocyte globulin on acetylcholine synthesis, its release and choline acetyltransferase transcription in a human leukemic T-cell line, *J. Neuroimmunol.* **128**, 1–8.

Fujii, T., Okuda, T., Haga, T., *et al.* (2003a). Detection of the high-affinity choline transporter in the MOLT-3 human leukemic T-cell line, *Life Sci.* **72**, 2131–2134.

Fujii, T., Watanabe, Y., Inoue, T., *et al.* (2003b). Up-regulation of mRNA encoding the M_5 muscarinic acetylcholine receptor in human T- and B-lymphocytes during immunological responses, *Neurochem. Res.* **28**, 423–429.

Fujimoto, K., Matsui, M., Fujii, T., *et al.* (2001). Decreased acetylcholine content and choline acetyltransferase mRNA expression in circulating mononuclear leukocytes and lymphoid organs of spontaneously hypertensive rat, *Life Sci.* **69**, 1629–1638.

Fujimoto, K., Fujii, T. and Kawashima, K. (2003). Increased acetylchoine contents in blood and lymphoid organs of MRL/MpJ-*lpr/lpr* (MRL-*lpr*) mice, an immune accelerated model, *J. Pharamacol. Sci.* **91** (Suppl. I), 286.

Gordon, M. A., Cohen, J. J. and Wilson, I. B. (1978). Muscarinic cholinergic receptors in murine lymphocytes: demonstration by direct binding, *Proc. Natl. Acad. Sci. USA* **75**, 2902–2904.

Grando, S., Kawashima, K. and Wessler, I. (2003). The non-neuronal cholinergic system in humans, *Life Sci.* **72**, 2009–2012.

Horiuchi, Y., Kimura, R., Kato, N., *et al.* (2003). Evolutional study on acetylcholine expression, *Life Sci.* **72**, 1745–1756.

Imboden, B. J., Shoback, M. D., Pattison, G., *et al.* (1986). Cholera toxin inhibits the T-cell antigen receptor-mediated increases in inositol trisphosphate and cytoplasmic free calcium, *Proc. Natl. Acad. Sci. USA* **83**, 5673–5677.

Kamimura, Y., Fujii, T., Kojima, H., *et al.* (2003). Induction of nitric oxide (NO) synthase expression and NO production via muscarinic acetylcholine receptor-mediated pathways in the CEM human leukemic T-cell line, *Life Sci.* **72**, 2151–2154.

Kapfhammer, J. and Bischoff, C. (1930). Acetylcholin und Cholin aus tierischen Organen, *Z. Physiol. Chem.* **191**, 179–182.

Kawashima, K. and Fujii, T. (2000). Extraneuronal cholinergic system in lymphocytes, *Pharmacol. Ther.* **86**, 29–48.

Kawashima, K. and Fujii, T. (2003). The lymphocytic cholinergic system in lymphocytes, *Life Sci.* **72**, 2101–2109.

Kawashima, K., Ishikawa, H. and Mochizuki, M. (1980). Radioimmunoassay for acetylcholine in the rat brain, *J. Pharmacol. Methods* **3**, 115–123.

Kawashima, K., Oohata, H., Fujimoto, K., *et al.* (1987). Plasma concentration of acetylcholine in young women, *Neurosci. Lett.* **80**, 209–212.

Kawashima, K., Oohata, H., Suzuki, T., *et al.* (1989). Extraneuronal localization of acetylcholine and its release upon nicotine stimulation, *Neurosci. Lett.* **104**, 336–339.

Kawashima, K., Kajiyama, K., Fujimoto, K., *et al.* (1993). Presence of acetylcholine in blood and its localization in circulating mononuclear leukocytes of humans, *Biogenic Amines* **9**, 251–258.

Kawashima, K., Fujii, T., Misawa, H., *et al.* (1997). Presence of acetylcholine in the blood and its production by choline acetyltransferase in lymphocytes, in: *Neurochemistry: Cellular, Molecular, and Clinical Aspects*, Teelken, A. and Korf, J. (Eds), pp. 813–819. Plenum, New York.

Kawashima, K., Fujii, T., Watanabe, Y., *et al.* (1998). Acetylcholine synthesis and muscarinic receptor subtype mRNA expression in T-lymphocytes, *Life Sci.* **62**, 1701–1705.

Kimura, R., Ushiyama, N., Fujii, T., *et al.* (2003). Nicotine-induced Ca^{2+} signaling and down-regulation of nicotinic acetylcholine receptor subunit gene expression in the CEM human leukemic T-cell line, *Life Sci.* **72**, 2155–2158.

Maslinski, W. (1989). Cholinergic receptors of lymphocytes, *Brain Behav. Immun.* **3**, 1–14.

Misawa, H., Takahashi, R. and Deguchi, T. (1995). Coordinate expression of vesicular acetylcholine transporter and choline acetyltransferase in sympathetic superior cervical neurons, *Neuroreport* **6**, 965–968.

Misawa, H., Matsuura, J., Oda, Y., *et al.* (1997). Human choline acetyltransferase mRNAs with different 5′-region produce a 69-kDa major translation product, *Mol. Brain Res.* **44**, 323–333.

Oda, Y., Nakanishi, I. and Deguchi, T. (1992). A complementary DNA for human choline acetyltransferase induces two forms of enzyme with different molecular weights in cultured cells, *Mol. Brain Res.* **16**, 287–294.

Ogawa, H., Fujii, T., Watanabe, Y., *et al.* (2003). Expression of multiple species for choline acetyltransferase mRNA in human T-lymphocytes, *Life Sci.* **72**, 2127–2130.

Okuda, T. and Haga, T. (2000). Functional characterization of the human high-affinity choline transporter, *FEBS Lett.* **484**, 92–97.

Okuda, T., Haga, T., Kanai, Y., *et al.* (2000). Identification and characterization of the high-affinity choline transporter, *Nat. Neurosci.* **3**, 120–125.

Okuma, Y. and Nomura, Y. (2001). Roles of muscarinic acetylcholine receptors in interleukin-2 synthesis in lymphocytes, *Jpn. J. Pharmacol.* **85**, 16–19.

Oosawa, H., Fujii, T. and Kawashima, K. (1999). Nerve growth factor increases the synthesis and release of acetylcholine and the expression of vesicular acetylcholine transporter in primary cultured rat embryonic septal cells, *J. Neurosci. Res.* **57**, 381–387.

Paldi-Haris, P., Szelenyi, J. G., Nguyen, T. H., *et al.* (1990). Changes in the expression of the cholinergic structures of human T lymphocytes due to maturation and stimulation, *Thymus* **16**, 119–122.

Passweg, J., Thiel, G. and Bock, H. A. (1996). Monoclonal gammopathy after intense induction immunosuppression in renal transplant patients, *Nephrol. Dial. Transplant.* **11**, 2461–2465.

Rinner, I. and Schauenstein, K. (1991). The parasympathetic nervous system takes part in the immuno-neuroendocrine dialogue, *J. Neuroimmunol.* **34**, 165–172.

Rinner, I. and Schauenstein, K. (1993). Detection of choline-acetyltransferase activity in lymphocytes, *J. Neurosci. Res.* **35**, 188–191.

Rinner, I., Kawashima, K. and Schauenstein, K. (1998). Rat lymphocytes produce and secrete acetylcholine in dependence of differentiation and activation, *J. Neuroimmunol.* **81**, 31–37.

Sakuragawa, N., Elwan, M. A., Uchida, S., *et al.* (2001). Non-neuronal neurotransmitters and neurotrophic factors in amniotic epithelial cells: expression and function in humans and monkey, *Jpn. J. Pharmacol.* **85**, 20–23.

Sato, K. Z., Fujii, T., Watanabe, Y., *et al.* (1999). Diversity of mRNA expression for muscarinic acetylcholine receptor subtypes and neuronal nicotinic acetylcholine receptor subunits in human mononuclear leukocytes and leukemic cell lines, *Neurosci. Lett.* **266**, 17–20.

Takeichi, N. (1995). Age-related immunological disorders in an animal model for hypertension, SHR rats, in: *Progress in Hypertension,* Vol. 3, *New Advances in SHR Research: Pathology and Harmacology*, Saito, H. and Yamori, Y., Minami, M. and Parvez, S. H. (Eds), pp. 33–43. VSP, Utrecht.

Takeichi, N., Hamada, J., Takimoto, M., *et al.* (1988). Depression of T cell-mediated immunity and enhancement of autoantibody production by natural infection with microorganisms in spontaneously hypertensive rats (SHR), *Microbiol. Immunol.* **32**, 1235–1244.

Toyabe, S., Iiai, T., Fukuda, M., *et al.* (1997). Identification of nicotinic acetylcholine receptors in lymphocytes in the periphery as well as thymus in mice, *Immunology* **92**, 201–205.

Tucek, S. (1988). Choline acetyltransferase and the synthesis of acetylcholine, in: *Handbook of Experimental Pharmacology, The Cholinergic Synapse,* Whittaker, V. P. (Ed.), Vol. 86, pp. 125–165. Springer Verlag, Berlin.

Ushiyama, N., Fujii, T. and Kawashima, K. (2001). A transient increase of intracellular free calcium ion concentration and up-regulation of alpha7 nicotinic acetylcholine receptor subunit gene expression by nicotine in CEM, a human leukemic T-cell line, *Jpn. J. Pharmacol.* **85** (Suppl.), 127P.

Watanabe, H., Garnier, G., Circolo, A., *et al.* (2000). Modulation of renal disease in MRL/*lpr* mice genetically deficient in the alternative complement pathway factor, *Br. J. Immunol.* **164**, 786–794.

Wessler, I., Kirkpatrick, C. J. and Racké, K. (1998). Non-neuronal acetylcholine, a locally acting molecule widely distributed in biological system: expression and function in humans, *Pharmacol. Ther.* **77**, 59–79.

Wessler, I., Kirkpatrick, C. J. and Racké, K. (1999). The cholinergic 'pitfall': acetylcholine, a universal cell molecule in biological systems including humans, *Clin. Exp. Pharmacol. Physiol.* **26**, 198–205.

Wessler, I., Kilbinger, H., Bittinger, F., *et al.* (2001). The biological role of non-neuronal acetylcholine in plants and humans, *Jpn. J. Pharmacol.* **85**, 2–10.

Yamada, S., Fujii, T. and Kawashima, K. (1997). Oral administration of KW-5092, a novel gastroprokinetic agent with acetylcholinesterase inhibitory and acetylcholine release enhancing activities, causes a dose-dependent increase in the blood acetylcholine content of beagle dogs, *Neurosci. Lett.* **225**, 25–28.

Advances in Neuroregulation and Neuroprotection (2005), pp. 751-763
Collin, C. *et al.* (Eds)
© VSP 2005

Central and peripheral sympathetic characteristics in hypertension considered from differences of catecholamine metabolite concentrations in plasma before and after adrenalectomy in male 2-month-old SHRSP and WKY

H. HIGASHINO* and K. OOSHIMA

Department of Pharmacology, Kinki University School of Medicine, Osaka-Sayama, 589-8511, Japan

Abstract—To clarify the detailed role of the adrenal medulla in 2-month-old SHRSP, the catecholamine concentrations in the plasma before and after bilateral adrenalectomy (Adrex) were measured with Neurochem®-HPLC system comparing with normotensive WKY. L-dopa, 3-OMDOPA (3-o-methyl-dopa), HVA (homovanillic acid), NE (norepinephrine) levels were not different between before and after Adrex in both of SHRSP and WKY. MHPG (3,4-methoxy-hydroxyphenyl glycol) and VA (vanillic acid) levels in WKY were not changed by Adrex except lower levels as compared with SHRSP, while those in SHRSP were remarkably decreased by Adrex. On the other hand, Ep (epinephrine) and MN (metanephrine) levels tended to decrease in WKY, and to increase in SHRSP after Adrex. Blood pressures were not changed by Adrex in either rat group. These findings indicate that (1) almost all of the MHPG and VA in blood are released from the adrenal glands in SHRSP; (2) the final catecholamine products of E and MN must be secreted by larger quantities of release from the accessory adrenal glands or extra-adrenal glands mainly in SHRSP than original release from the adrenal glands; and (3) as a result, the blood pressure might not be affected by Adrex in either SHRSP or WKY.

Keywords: Catecholamine metabolites; adrenal glands; SHRSP; adrenalectomy; sympathetic activity.

INTRODUCTION

Our previous studies (Maeda *et al.*, 1999) concerning catecholamine (CA) metabolites in plasma and urine between stroke-prone spontaneously hypertensive rats (SHRSP) (Okamoto *et al.*, 1974) and normotensive Wistar Kyoto rats (WKY) concluded as follows. Excessive doses of norepinephrine (NE) and epinephrine (Ep)

---

*To whom correspondence should be addressed. E-mail: pharm@med.kindai.ac.jp are rapidly released in the blood stream from the sympathetic nerve endings and adrenal medulla when SHRSP are active not only in the nighttime but also in the daytime, and under a cold stress. That is because our previous findings showed that three major enzymes related to the CA production, namely phenylalanine-4-monooxygenase (the enzyme that converts phenylalanine to tyrosine), tyrosine hydroxylase and phenylethanolamine-N-methyltransferase, may increase more in SHRSP than in WKY.

To clarify the detailed role of the adrenal medulla in these changes might be a key to solving the difference of CA metabolism between these two rat groups. Therefore, using a Neurochem®-HPLC system that could detect 21 CA metabolites at once, CA metabolite concentrations in the plasma and the urine before and after bilateral adrenalectomy (Adrex), and also CA metabolite contents in the isolated adrenal glands of male SHRSP and WKY at the age of 2-months, when hypertensive changes are at their most remarkable stage, were measured.

MATERIALS AND METHODS

Animals and blood pressure detection

Male 6-week-old SHRSP and WKY were taken from the common institute of the animal center in our school. These were raised in our laboratory with commercial chow (SP, Funahashi, Chiba, Japan) and tap water *ad libitum* at 22 ± 2°C, 60% humidity, and a light cycle consisting of 7:00–19:00 in the dark and 19:00–7:00 in the light for 2 weeks until use and a following experimental period of 2 days. These rats were at the 121st to 122nd generation in SHRSP and the 85–86th generation in WKY, respectively, from the establishment of the SHRSP strain by Okamoto *et al.* (1974). Rats in both strains were divided into two groups, each of 16–20 individuals. One was a bilateral adrenalectomy group (Adrex), and the other was a sham operated vehicle group (Non-Adrex). Systolic blood pressure (SBP) and pulse rate (HR) were measured with a volume-oscillometric manometer (TK-370A, UNICOM, Japan) as follows. The untreated and sham operated Non-Adrex or operated Adrex rats in conscious state were pre-warmed at 35°C for 5 min in a warming box, then put into the folder made of stainless steel wires, and the tail was inserted into a ring-shaped cuff with a laser blood pulse sensor. SBP and HR were measured in 1 min by an automatic up and down control of cuff pressure.

Procedure of adrenalectomy and a method of catecholamine determination in the samples

Bilateral adrenalectomies were performed through skin incision with scissors on both sides of the back in rats under sodium pentobarbital anesthesia (50 mg/kg i.p.) at 9:00 in the dark in one group and 21:00 in the light in the other group. The adrenal glands isolated from the rats were immediately homogenized for 20 s at a middle speed with a Polytron-homogenizer (Kinematica AG, Switzerland) in 1.5 ml of cold 0.5 N perchloric acid (PCA) solution with 75 ng 3,4-dihydroxybenzylamine hydrobromide (DHBA) for recovery calculation. After centrifugation with a conventional centrifuge at 3500 rpm for 10 min, the top layer of the solution in the tube was collected and ultrafiltered with Ultrafree-MC® (Millipore Co., US) by using a centrifuge. The filtrate was assayed as adrenal gland samples for 21 CA metabolites detection with a Neurochem®-HPLC Coullochem® electrode assay system (ESA, Inc., MA, USA). With regard to the blood samples, 200 μl of plasma in the blood taken from a tail vein at 9:00 in the dark in the conscious condition before and 24 h after Adrex was acidified by adding the same volume of 1.0 N PCA solution containing 10 ng of DHBA, and the top layer in the tube was filtered through Ultrafree-MC®. The filtrate was assayed in the same way as the adrenal gland samples. The urine samples were collected from a urine reservoir containing 0.1 N HCl solution in metabolic cages that each rat was put into for 24 hours. After diluting the filtrate samples with 0.1 N PCA solution by 400 times, CA metabolites were detected.

CA metabolite concentrations were expressed as ng/mg protein in adrenal glands, ng/ml in plasma, and μg/mg creatinine (Cre) in urine, respectively.

All the data were presented as mean $\pm$ S.E.M., and compared by the method of multiple statistical analysis of ANOVA-Scheffe between the groups. Statistical difference between the groups was set at $p < 0.05$.

RESULTS

Changes in blood pressures and pulse rates before and after Adrex

Before Adrex, SBP values were 137.5 ± 2.6 mmHg in WKY and 206.9 ± 5.4 mmHg in SHRSP, and HR were 448.1 ± 21.5 bpm in WKY and 414.8 ± 10.9 bpm in SHRSP, respectively. After Adrex, SBP changed only slightly in either SHRSP and WKY, but HR increased by 26% in SHRSP but not in WKY (Fig. 1).

Changes in catecholamine contents in plasma and urine before and after Adrex

The first metabolite, L-dopa, synthesized by tyrosine 3-monooxygenase (tyrosine hydroxylase) from tyrosine was higher in the plasma of WKY than in SHRSP at both situations before and after Adrex. In WKY, L-dopa showed an increasing tendency after Adrex, and no L-dopa was detected either before or after Adrex in SHRSP (Fig. 2A). L-dopa in urine increased both in SHRSP and WKY after Adrex compared with those before Adrex: from 5.25 ± 0.43 to 29.9 ± 4.2 μg/mg Cre in SHRSP, and from 10.9 ± 0.9 to 38.3 ± 5.7 μg/mg Cre in WKY. 3-OMDOPA (3-o-methyl-dopa), a direct metabolite converted by catechol-o-methyl transferase (COMT) from L-dopa, increased after Adrex compared with those before Adrex in WKY as shown from 21.1 ± 0.78 to 25.2 ± 0.79 ng/ml. In SHRSP, both values of 3-OMDOPA were higher by 24–55% than those of WKY, and no change was detected by Adrex treatment (Fig. 2B).

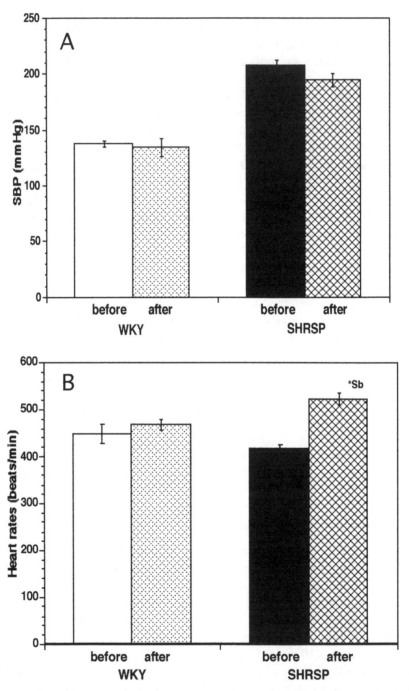

Figure 1. Systolic blood pressures (mmHg, A) and heart rates (bpm, B) in WKY and SHRSP before and after the bilateral adrenalectomy (Adrex). *Sb represents the significant difference between the values before and after Adrex in SHRSP.

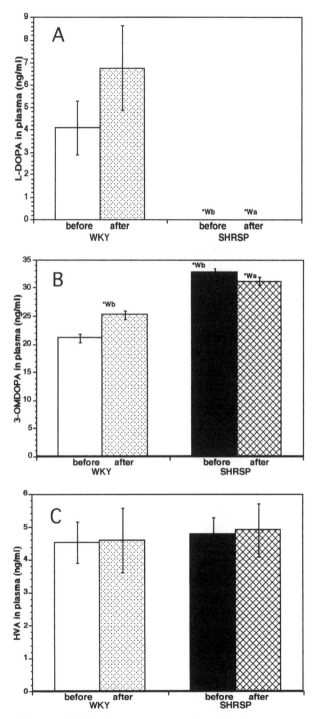

Figure 2. L-dopa (A), 3-OMDOPA (B), and HVA (C) levels in plasma before and after Adrex in WKY and SHRSP. *Wb and *Wa represent the significant differences *vs.* the values before Adrex and after Adrex in WKY, respectively.

The final excretion product in urine from L-dopa and dopamine, homovanillic acid (HVA), in plasma did not completely change between before and after Adrex in either WKY or SHRSP (from 4.52 ± 0.63 to 4.59 ± 0.98 ng/ml in WKY, and from 4.79 ± 0.51 to 4.90 ± 0.82 ng/ml in SHRSP, Fig. 2C). HVA in urine also did not significantly change between before and after Adrex in either WKY or SHRSP (5.67 ± 0.13 to 7.96 ± 1.20 μg/mg Cre in WKY, and 5.98 ± 0.48 to 6.51 ± 0.36 μg/mg Cre in SHRSP).

NE converted by dopamine β-hydroxylase (DBH) from dopamine showed an increasing tendency in plasma of SHRSP compared with WKY before Adrex, but no significant change was observed between before and after Adrex, and also between those for NE in WKY and SHRSP (from 4.06 ± 1.37 to 5.00 ± 1.05 ng/ml in WKY, and from 7.59 ± 2.77 to 4.91 ± 1.12 ng/ml in SHRSP, Fig. 3A). NE in urine changed non-significantly from 2.32 ± 0.25 to 8.08 ± 1.95 μg/mg Cre in WKY, and from 3.09 ± 0.22 to 2.17 ± 0.29 μg/mg Cre in SHRSP. MHPG (3,4-methoxy-hydroxyphenyl glycol), intermediate from NE and E, in plasma of SHRSP was remarkably higher compared with WKY before Adrex. In spite of a slight increase in WKY after Adrex, MHPG in plasma of SHRSP dramatically declined under detectable limit after Adrex. That is, the values changed from 0.33 ± 0.21 to 3.92 ± 1.07 ng/ml in WKY, and from 14.55 ± 2.52 ng/ml to values under the detectable limit in SHRSP (Fig. 3B). In the urine, however, the same change was observed as shown from 4.44 ± 0.58 to 1.08 ± 0.15 μg/mg Cre in SHRSP, and as shown from 2.20 ± 0.17 to 3.75 ± 0.39 μg/mg Cre in WKY. In WKY, vanillic acid (VA), the final excreting product from NE and E, did not change between before and after Adrex. On the other hand, higher VA level in plasma of SHRSP before Adrex decreased by 74% after Adrex (from 2.68 ± 0.46 to 3.05 ± 1.66 ng/ml in WKY, and from 8.10 ± 1.36 to 2.12 ± 0.35 ng/ml in SHRSP, Fig. 3C). Urine VA changed in a similar way to that in plasma: from 7.45 ± 0.39 to 7.10 ± 0.74 μg/mg Cre in WKY, and the data from 16.3 ± 0.7 to 5.82 ± 1.02 μg/mg Cre in SHRSP.

Ep and metanephrine (MN) are CA metabolites produced mainly in the adrenal glands, but not in the central and peripheral sympathetic nerves. Although Ep showed a decreasing tendency after Adrex in WKY, an increasing tendency was shown in SHRSP after Adrex (from 4.02 ± 2.00 to 1.30 ± 1.30 ng/ml in WKY, and from 8.40 ± 1.40 to 10.8 ± 2.4 ng/ml in SHRSP). MN converted by COMT from Ep also showed the same tendency in SHRSP as Ep did (from 0.93 ± 0.76 to 1.02 ± 0.27 ng/ml in WKY, and from 1.58 ± 0.24 to 2.06 ± 0.11 ng/ml in SHRSP, Fig. 4A and 4B).

Changes in catecholamine contents in adrenal glands

When the Adrex was performed in the dark, every catecholamine content in the adrenal glands was not different between WKY and SHRSP as shown at the part of 'in the dark' in Fig. 5A and 5B. When the Adrex was performed in

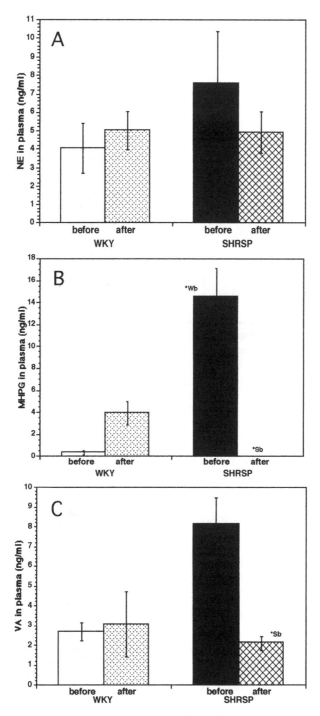

Figure 3. NE (A), MHPG (B), and VA (C) levels in plasma before and after Adrex in WKY and SHRSP. *Wb and *Sb represent the significant differences *vs.* the values before Adrex in WKY and SHRSP, respectively.

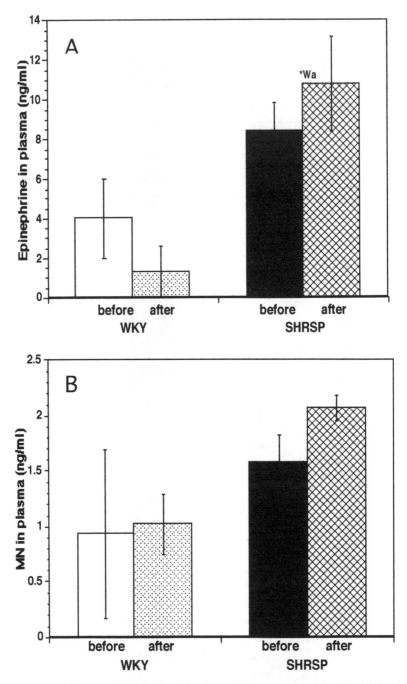

Figure 4. Epinephrine (A) and MN (B) levels in plasma before and after Adrex in WKY and SHRSP. *Wa represents the significant differences *vs.* the value after Adrex in WKY.

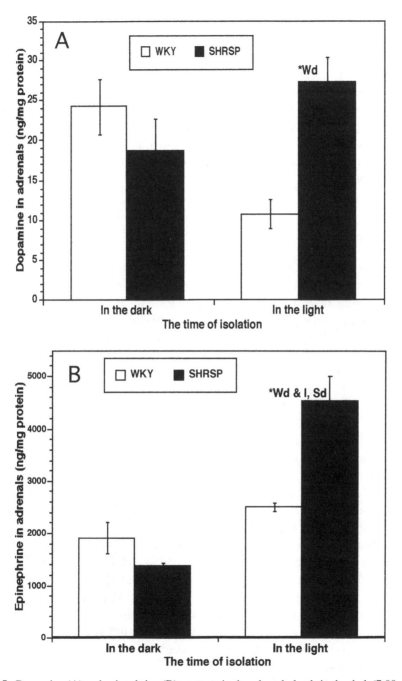

Figure 5. Dopamine (A) and epinephrine (B) contents in the adrenal glands in the dark (7:00–19:00) and in the light (19:00–7:00) in WKY and SHRSP. *Wd, *Wl and *Sd represent the significant differences *vs.* the values in the dark and in the light in WKY, and in the dark in SHRSP, respectively.

the light, however, every catecholamine content in the adrenal glands in SHRSP was significantly higher than in WKY. For example, 3-OMDOPA converted from L-dopa was 0.42 ± 0.12 ng/mg protein in WKY, and 1.66 ± 0.25 ng/mg protein in SHRSP. Dopamine in the adrenal glands was 10.8 ± 1.9 ng/mg protein in WKY, and 27.2 ± 3.2 ng/mg protein in SHRSP (Fig. 5A). NMN (normetanephrine) contents converted by COMT from NE were 1.02 ± 0.23 ng/mg protein in WKY, and 3.92 ± 0.57 ng/mg protein in SHRSP. Ep in the adrenal glands was 2492 ± 82 ng/mg protein in WKY, and 4538 ± 468 ng/mg protein in SHRSP (Fig. 5A and 5B). The final metabolite VA in the catecholamine pathway was 0.68 ± 0.07 ng/mg protein in WKY, and 1.35 ± 0.39 ng/mg protein in SHRSP adrenal glands.

DISCUSSION

From our previous studies (Maeda et al., 1999) concerning CA metabolites in plasma and urine between SHRSP and WKY, it was concluded that three major enzyme activities related to the catecholamine production such as phenylalanine-4-monooxygenase (the enzyme to produce tyrosine from phenylalanine), tyrosine hydroxylase, phenylethanolamine-N-methyltransferase might increase more in SHRSP than in WKY. So far, many reports (Pak, 1981; Togashi et al., 1984) showed that NE and Ep levels in plasma or urine went up more in young SHR than in normotensive WKY. So our interest was focused on the higher central sympathetic activity and the higher peripheral CA release mainly from the medulla of adrenal glands in SHR or SHRSP. The simplest way to investigate the role of adrenal glands was to perform demedullation or bilateral adrenalectomy, and to measure the CA concentrations in plasma or urine before and after the operation. While Borkowski (1991) showed that Ep levels in plasma of SHR decreased after adrenalectomy, Borovsky et al. (1998) showed that NE levels remained unchanged. Ep, NE and dopamine contents in the adrenal glands of SHR or SHRSP were higher than those of WKY (Shober et al., 1989; Kumai et al., 1994; Uchida et al., 1995). Transmural electrical stimulation to the adrenal glands of SHR increased more NE output in perfusate than those of WKY (Nagayama et al., 1999). NE contents in the kidney and adrenal glands were higher in SHR than in WKY, and bilateral renal denervation produced an almost complete reduction of NE content in the kidney, and then prevented the development of hypertension in SHR (Yoshida et al., 1995). Treatment with tyrosine hydroxylase antisense oligonucleotides in SHR decreased SBP with reductions of Ep and NE levels in plasma through inhibition of tyrosine hydroxylase in the adrenal glands (Nagayama et al., 1999).

On the other hand, bilateral adrenomedullectomy or chronic Ep infusion caused no significant blood pressure change in SHR (Jablonskis and Howe, 1994). However, although SHR with Adrex subjected to pulsed foot shocks elevated the blood pressure, SHR without Adrex did not, probably due to β-adrenoceptor-mediated dilatation of skeletal muscle vasculature (Knardahl *et al.*, 1988). Yamori *et al.* (1985) concluded that the adrenal glands of SHR work to develop and maintain the hypertension at young age, but not a direct pressor organ.

Consequently, it was considered to be a very important study to clarify the detailed role of the adrenal glands in the hypertension in SHR or SHRSP. In the study before and after Adrex in SHRSP, we found the following results. (1) L-dopa levels were higher in the plasma of WKY than SHRSP at both situations, before and after Adrex. L-dopa in urine increased both in SHRSP and WKY after Adrex compared with those before Adrex. (2) Values of 3-OMDOPA in SHRSP were higher by 24–55% than those of WKY, and no change was detected with Adrex. (3) HVA in plasma and urine did not change between before and after Adrex in either WKY or SHRSP. (4) NE levels in plasma and urine of SHRSP did not change before and after Adrex. Therefore, it will be concluded that NE synthesis from tyrosine is performed in the sympathetic nerve systems except the adrenal glands in the rats, since NE synthesis process via dopamine from L-dopa in plasma did not change after Adrex in either SHRSP or WKY. (5) MHPG in plasma of SHRSP was remarkably higher compared with WKY before Adrex, and it dramatically declined in plasma and urine after Adrex. (6) At the same time, higher VA level in plasma and urine of SHRSP before Adrex decreased after Adrex. (7) Ep and MN levels in plasma showed an increasing tendency in SHRSP, but not in WKY. (8) All catecholamine metabolite contents downstream of 3-OMDOPA in the adrenal glands were not different between WKY and SHRSP in the dark, at physiologically active stage, but were higher in SHRSP than in WKY in the light, at physiologically inactive sleeping stage.

Findings (5)–(7) explain that (a) every CA metabolite content in the adrenal glands of SHRSP was not different from that in WKY in the dark, since the adrenal glands of SHRSP released much more catecholamines in the fluid in the dark, (b) MHPG and VA, metabolites from Ep, in the urine decreased after Adrex, because the total amount of Ep secretion in the fluid was decreased by the Adrex, (c) Ep synthesis and release in SHRSP must occur in some of the accessory or extra-adrenal glands such as Zuckerkandle's body (Frezza *et al.*, 2002), in order to compensate for the lack of them. In the human, the same pathological situations as in SHRSP were often reported in diseases such as pheochromocytoma and myelolipoma (Katai *et al.*, 1998; Kageyama *et al.*, 1989). A summary is shown in Fig. 6 and in a published report (Maeda *et al.*, 2000).

In this experiment, SBP did not change after Adrex compared with before Adrex, and HR increased only in SHRSP after Adrex. As Jablonskis and Howe (1994) and Lee *et al.* (1991) reported, these results suggest that the adrenal glands in SHR or SHRSP are promoting factors for development and maintaining of hypertension, but not the causative ones for hypertension.

In essential hypertension in humans, the same situation with regard to the accessory adrenal glands as in SHRSP may be considered. More detailed investigation in these fields will be necessary to fully understand the nature of hypertension.

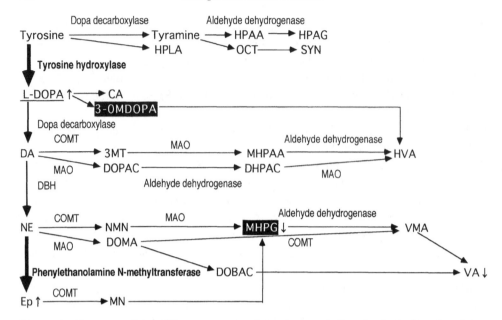

Figure 6. Summary of the differences on catecholamine metabolism in the bodies after Adrex between 2-month-old SHRSP and WKY. The thick arrows show the facilitated pathways in the catecholamine metabolism of SHRSP comparing WKY, and underlined or reversed words represent decreased or increased metabolites in plasma in SHRSP comparing WKY after Adrex, respectively. The arrows attached on the right of metabolite name show the increasing or decreasing metabolites excreted in urine after Adrex.

REFERENCES

Borkowski, K. R. (1991). Effect of adrenal demedullation and adrenaline on hypertension develop-ment and vascular reactivity in young spontaneously hypertensive rats, *J. Auton. Pharmacol.* **11**, 1–14.

Borovsky, V., Herman, M., Dunphy, G., *et al.* (1998). CO_2 asphyxia increases plasma norepinephrine in rats via sympathetic nerves, *Am. J. Physiol.* **274**, R19–R22.

Frezza, E. E., Ikramuddin, S., Gourash, W., *et al.* (2002). Laparoscopic resection of a large periadrenal nonmalignant pheochromocytoma, *Surg. Endosc.* **16**, 362–363.

Jablonskis, L. T. and Howe, P. R. (1994). Lack of influence of circulating adrenaline on blood pressure in normotensive and hypertensive rats, *Blood Press.* **3**, 112–119.

Kageyama, T., Doke, Y., Takahashi, M., *et al.* (1989). Computed tomography of myelolipoma in the accessory adrenal gland, *Urol. Radiol.* **11**, 153–155.

Katai, M., Sakurai, A., Ichikawa, K., *et al.* (1998). Pheochromocytoma arising from an accessory adrenal gland in a patient with multiple endocrine neoplasia type 2A: transient development of clinical manifestations after hemorrhagic necrosis, *Endocr. J.* **45**, 329–334.

Knardahl, S., Sanders, B. J. and Johnson, A. K. (1988). Effects of adrenal demedullation on stress-induced hypertension and cardiovascular responses to acute stress, *Acta Physiol. Scand.* **133**, 477–483.

Kumai, T., Tanaka, M., Watanabe, M., *et al.* (1994). Elevated tyrosine hydroxylase mRNA levels in the adrenal medulla of spontaneously hypertensive rats, *Jpn. J. Pharmacol.* **65**, 367–369.

Kumai, T., Tateishi, T., Tanaka, M., *et al.* (2001). Tyrosine hydroxylase therapy causes hypotensive effects in the spontaneously hypertensive rats, *J. Hypertens.* **19**, 1769–1773.

Lee, R. M., Borkowski, K. R., Leenen, F. H., *et al.* (1991). Combined effect of neonatal sympathetectomy and adrenal demedullation on blood pressure and vascular changes in spontaneously hypertensive rats, *Circ. Res.* **69**, 714–721.

Maeda, K., Azuma, M., Nishimura, Y., *et al.* (1999). Comparison of catecholamine metabolite levels in plasma drawn at daytime, nighttime, and acute cold stress between 2-month-old SHRSP and WKY, *Clin. Exp. Hypertens.* **21**, 464–465.

Maeda, K., Azuma, M., Nishimura, Y., *et al.* (2000). Catecholamine metabolites levels in plasma, and its synthetic and secretary activities in adrenal glands between 2-month-old SHRSP and WKY, *Clin. Exp. Hypertens.* **22**, 359–360.

Nagayama, T., Matsumoto, T., Yoshida, M., *et al.* (1999). Role of cholinergic receptors in adrenal catecholamine secretion in spontaneously hypertensive rats, *Am. J. Physiol.* **277**, R1057–R1062.

Okamoto, K., Yamori, Y. and Nagaoka, A. (1974). Establishment of the stroke-prone spontaneously hypertensive rats (SHR), *Circ. Res.* **34/35** (Suppl. 1), 143–153.

Pak, C. H. (1981). Plasma adrenaline and noradrenaline concentrations of the spontaneously hypertensive rat, *Jpn. Heart J.* **22**, 987–995.

Schober, M., Howe, P. R., Sperk, G., *et al.* (1989). An increased pool of secretary hormones and peptides in adrenal medulla of stroke-prone spontaneously hypertensive rats, *Hypertension* **13**, 469–474.

Togashi, H., Minami, M., Saito, I., *et al.* (1984). Guanfacine and clonidine: the effects on adrenal medullary function in spontaneously hypertensive rats, *Arch. Int. Pharmacodyn. Ther.* **272**, 79–87.

Uchida, T., Nishimura, Y. and Suzuki, A. (1995). Age-related changes in cerebral and peripheral monoamine contents in stroke-prone spontaneously hypertensive rats, *Clin. Exp. Pharmacol. Physiol.* **22** (Suppl.), S80–S82.

Yamori, Y., Ikeda, K., Kulakowski, E. C., *et al.* (1985). Enhanced sympathetic-adrenal medullary responses to cold exposure in spontaneously hypertensive rats, *J. Hypertens.* **3**, 63–66.

Yoshida, M., Yoshida, E. and Satoh, S. (1995). Effect of renal nerve denervation on tissue catecholamine content in spontaneously hypertensive rats, *Clin. Exp. Pharmacol. Physiol.* **22**, 512–517.

Advances in Neuroregulation and Neuroprotection (2005), pp. 765-771
Collin, C. *et al.* (Eds)
© VSP 2005

Effects of salicylate on the pharmacokinetics of valproic acid after oral administration of sodium valproate in rats

S. OHSHIRO [1], N. HOBARA [1], M. SAKAI [1], N. HOKAMA [1], H. KAMEYA [1] and M. SAKANASHI [2,*]

[1] *Department of Hospital Pharmacy, Faculty of Medicine, University of the Ryukyus, 207 Uehara, Nishihara-cho, Okinawa 903-0215, Japan*
[2] *Department of Pharmacology, School of Medicine, Faculty of Medicine, University of the Ryukyus, 207 Uehara, Nishihara-cho, Okinawa 903-0215, Japan*

Abstract—Effects of salicylate on the pharmacokinetics of valproic acid (VPA) after oral administration (p.o.) of sodium valproate were investigated in rats. When salicylic acid was administered orally after sodium valproate p.o., the plasma VPA concentrations including maximum plasma concentration (C_{max}), area under the plasma concentration–time curve up to 3 h (AUC_{0-3}) and the elimination half-life ($t_{1/2}$) of VPA were lower than control values. In addition, when salicylate was administered intraperitoneally or intravenously, the plasma VPA concentrations, AUC_{0-3} and $t_{1/2}$ of VPA were lower than those in the control group. These results suggest that decreases in the plasma VPA concentrations including C_{max} and AUC_{0-3} with simultaneous oral administration of sodium valproate and salicylic acid may be related to both reduction of absorption and increase in clearance of VPA.

Keywords: Valproic acid; salicylate; pharmacokinetics; rat.

INTRODUCTION

For the medical treatment of epilepsy, valproic acid (VPA) has been utilized in many patients. Some epileptic patients receiving long-term anticonvulsant therapy occasionally require co-administration with other medicines, e.g. aspirin. Feitman *et al.* (1980) observed that plasma protein binding sites of VPA are displaced by salicylate *in vitro*, and Orr *et al.* (1982), reported that, when sodium valproate was orally administered to epileptic children together with aspirin, total VPA concentrations in serum and free VPA concentrations increased as compared with the control group. On the other hand, Yu *et al.* (1990) reported a significant increase in total clearance and a significant decrease in area under the plasma

*To whom correspondence should be addressed. E-mail: sakanasi@med.u-ryukyu.ac.jp

concentration–time curve (AUC) of VPA when sodium valproate was injected intravenously to rats with a prior treatment of constant infusion of salicylate. In the present study, in order to clarify these discrepancies, effects of salicylate on the pharmacokinetics of VPA after oral administration of sodium valproate were investigated in rats.

MATERIALS AND METHODS

Chemicals

Depaken syrup of sodium valproate was purchased from Kyowa Hakko Kogyo (Tokyo, Japan). Salicylic acid and sodium salicylate were purchased from Nacalai Tesque (Kyoto, Japan). CMC-Na was purchased from Maruishi Pharmaceuticals (Osaka, Japan).

Animal experiments

This study was performed in accordance with the Guidelines for Animal Experimentation of University of the Ryukyus, and was approved by the Animal Care and Use Committee of this institution. Male Sprague-Dawley rats weighing 210 to 300 g were obtained from the Seac Yoshitomi, Ltd. (Fukuoka, Japan) and acclimatized for at least one week prior to the experiments. The rats were maintained in aluminum rat cages and housed in animal care facilities with a 12 h light/dark cycle, a temperature of 23–25°C, a humidity of 50 ± 15% and free access to food and water.

The rats were fasted from 19:00 on the previous day with free access to water, and were used for the experiment 12 h later. Under a light ether anesthesia, sodium valproate 100 mg/kg was administered orally using the polyethylene catheter to all rats. Salicylic acid 100 mg/kg suspended in 0.5% CMC-Na solution was administered orally or intraperitoneally immediately after oral administration of sodium valproate. In cases of the intravenous administration, sodium salicylate resolved in saline was administered directly into the jugular vein of rats. About 0.15 ml of blood samples were collected at 0.25, 0.5, 1, 2 and 3 h from the tail vein into a micro capillary and then centrifuged (model CT 12, Hitachi, Tokyo, Japan) at 12 000 rpm for 5 min to separate the plasma.

Determination of VPA

The plasma concentrations of VPA were determined by fluorescence polarization immunoassay (FPIA) using the TDXFLX analyzer (Dainabot, Tokyo, Japan).

Pharmacokinetic analysis

Maximum plasma concentration (C_{max}) of VPA and time to reach C_{max} (T_{max}) were estimated from the actual measurement. Elimination rate constant (K_{el})

was determined by a linear least square regression analysis using the plasma concentrations of VPA during the elimination phase. The $t_{1/2}$ of VPA was calculated from $0.693/K_{el}$. AUC_{0-3} values were calculated using the linear trapezoidal rule.

Statistical analysis

Data were expressed as the mean ± S.E. The statistical analysis was performed using unpaired Student's t-test. Significant difference was defined $p < 0.05$.

RESULTS

The plasma VPA concentration–time curve after oral administration of sodium valproate followed by oral or intraperitoneal administration of salicylic acid is shown in Fig. 1. Plasma VPA concentrations at 0.5 and 1 h after sodium valproate p.o. were significantly lower in the salicylic acid p.o. group than in the control group. In addition, plasma VPA concentrations at 0.25, 0.5, 1 and 3 h after sodium valproate p.o. were significantly lower in the salicylic acid i.p. group than in the control group.

Pharmacokinetic parameters of plasma VPA after sodium valproate p.o. are represented in Table 1. The oral administration of salicylic acid significantly

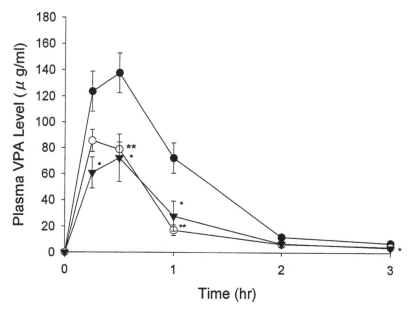

Figure 1. Time course of plasma concentration of VPA after oral administration (p.o.) of 100 mg/kg sodium valproate alone or with 100 mg/kg salicylic acid administered orally (p.o.) or intraperitoneally (i.p.) in rats. Filled circle: sodium valproate alone p.o. (control, $n = 9$), open circle: with salicylic acid p.o. ($n = 5$), filled triangle: with salicylic acid i.p. ($n = 5$). Each point represents the mean ± S.E. $^{*}p < 0.05$; $^{**}p < 0.01$, significantly different from the corresponding control value.

Table 1.

Pharmacokinetic parameters of plasma VPA after oral administration (p.o.) of 100 mg/kg sodium valproate alone (control) or with 100 mg/kg salicylic acid administered orally (p.o.) or intraperitoneally (i.p.) in rats

Treatment	n	T_{max} (h)	C_{max} (μg/ml)	$t_{1/2}$ (h)	AUC_{0-3} (μg·h/ml)
Sodium valproate alone p.o. (control)	9	0.44 ± 0.04	139.6 ± 15.9	0.53 ± 0.13	152.4 ± 14.3
Sodium valproate p.o. with salicylic acid p.o.	5	$0.30 \pm 0.05^*$	$86.1 \pm 8.3^*$	0.33 ± 0.04	$72.6 \pm 5.2^{***}$
Sodium valproate p.o. with salicylic acid i.p.	5	0.40 ± 0.06	$73.1 \pm 18.1^*$	0.40 ± 0.08	$72.0 \pm 19.2^{**}$

Data are mean values $\pm$ S.E. T_{max}: time to reach C_{max}. C_{max}: maximum plasma concentration. $t_{1/2}$: apparent elimination half-life. AUC: area under the plasma concentration–time curve. $^*p < 0.05$; $^{**}p < 0.01$; $^{***}p < 0.001$, significantly different from the corresponding control value.

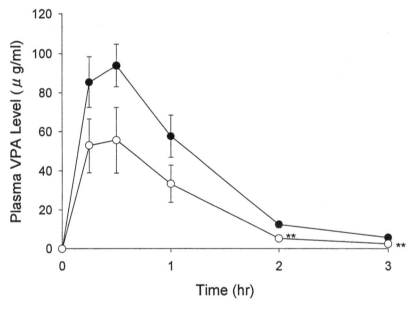

Figure 2. Time course of plasma concentration of VPA after oral administration (p.o.) of 100 mg/kg sodium valproate alone or with 100 mg/kg sodium salicylate administered intravenously (i.v.) in rats. Filled circle: sodium valproate alone p.o. (control, $n = 5$), open circle: with salicylic acid i.v. ($n = 5$). Each point represents the mean $\pm$ S.E. $^{**}p < 0.01$, significantly different from the corresponding control value.

decreased values of T_{max}, C_{max} and AUC_{0-3} of VPA by 32%, 38% and 52%, respectively. Salicylic acid p.o. also decreased the value of $t_{1/2}$, but not significantly compared with the control value. Additionally, the intraperitoneal administration of salicylic acid significantly decreased values of C_{max} and AUC_{0-3} of VPA by 48%

Table 2.

Pharmacokinetic parameters of plasma VPA after oral administration (p.o.) of 100 mg/kg sodium valproate alone (control) or with 100 mg/kg sodium salicylate administered intravenously (i.v.) in rats

Treatment	n	T_{max} (h)	C_{max} (μg/ml)	$t_{1/2}$ (h)	AUC$_{0-3}$ (μg·h/ml)
Sodium valproate alone p.o. (control)	5	0.45 ± 0.05	94.2 ± 11.1	0.49 ± 0.09	115.1 ± 10.0
Sodium valproate p.o. with sodium salicylate i.v.	5	0.35 ± 0.06	59.7 ± 15.7	0.36 ± 0.05	$65.6 \pm 15.9^*$

Data are mean values $\pm$ S.E. T_{max}: time to reach C_{max}. C_{max}: maximum plasma concentration. $t_{1/2}$: apparent elimination half-life. AUC: area under the plasma concentration–time curve. $^*p < 0.05$, significantly different from the corresponding control value.

and 53%, respectively. Salicylic acid i.p. also decreased values of T_{max} and $t_{1/2}$, but not significantly compared with control values.

The plasma VPA concentration–time curve after oral administration of sodium valproate followed by intravenous administration of sodium salicylate is shown in Fig. 2. Plasma VPA concentrations at 2 and 3 h after sodium valproate p.o. were significantly lower in the sodium salicylate i.v. group than in the control group.

Table 2 represents pharmacokinetic parameters of plasma VPA after sodium valproate p.o. The intravenous administration of sodium salicylate significantly decreased the value of AUC$_{0-3}$ of VPA by 43%, and induced decreases in C_{max}, T_{max} and $t_{1/2}$, which were not significant compared with control values.

DISCUSSION

Salicylic acid instead of acetylsalicylate was used in this study, because acetylsalicylic acid is rapidly hydrolyzed to salicylic acid in solution and *in vivo* (Iwamoto *et al.*, 1982).

It was reported that the therapeutic range of VPA in plasma was from 50 to 100 μg/ml (Gugler and Von Unruh, 1980) and that of salicylic acid was from 150 to 300 μg/ml (Flower *et al.*, 1985). In the present study, therefore, 100 mg/kg VPA and 100 mg/kg salicylic acid were used, respectively.

In this study, the simultaneous oral administration of VPA and salicylic acid was shown to decrease the plasma VPA concentrations including C_{max} and AUC$_{0-3}$ of VPA, suggesting a possibility that salicylic acid decreased the absorption of VPA. Intestinal absorption of VPA and salicylic acid has been proposed to occur not only through passive diffusion according to the pH-partition theory (Brodie and Hogben, 1957), but also through carrier-mediated mechanisms. In regard to the latter, it has been reported that VPA (Tsuji *et al.*, 1994) and salicylic acid (Takanaga *et al.*, 1994) are transported by monocarboxylic acid transporters (MCTs) in the Caco-2 cells. Moreover, the transport of salicylic acid across Caco-2 cells was significantly inhibited by VPA (Takanaga et al., 1994). Therefore, it might be possible that VPA transport by MCT was competitively inhibited by salicylic acid, and thus absorption of VPA was decreased.

In a clinical report (Orr et al., 1982), however, total serum concentration of VPA and the half-life for total serum VPA were increased after medication with aspirin. There was a discrepancy between this clinical report by Orr et al. (1982) and the present results. The former is supported by the observation that the gastrointestinal absorption of VPA was increased with salicylate-increased permeability in the small intestine (Kaiji et al., 1986). On the other hand, as mentioned above, VPA transport by MCT might be competitively inhibited by salicylic acid in the Caco-2 cells. Since Caco-2 cells, which originated from a human colon carcinoma (Fogh et al., 1977), have been shown to have same functions as observed in small intestinal cells (Higalgo et al., 1989), we proposed that salicylic acid might induce the reduction in absorption of VPA. Thus, increases in the serum concentration and the half-life of total VPA were considered to be due to changes in the metabolic pattern of VPA derived from salicylate (Abbott et al., 1986). However, in order to clarify these discrepancies of drug interactions, further detailed studies will be needed.

In this study, to remove the influence of salicylic acid on VPA absorption, salicylic acid was administered intraperitoneally. However, the plasma VPA concentration–time curve, C_{max} and AUC_{0-3} still showed the same tendency as oral administration of salicylic acid, suggesting the possibilities of decreased VPA absorption or changes in VPA metabolism and/or both of them.

On the other hand, in the present study, the intravenous administration of sodium salicylate significantly decreased AUC_{0-3} of VPA. Moreover, all three different administration methods of salicylate showed a tendency to decrease $t_{1/2}$ of VPA. These results suggested that the decrease in plasma VPA concentration including C_{max} would be due to the increase in free VPA followed by the increase in total clearance of VPA. This is supported by the proposal of Yu et al. (1990). They have reported that, when sodium valproate was injected intravenously to rats during constant infusion of sodium salicylate, a significant increase in total VPA clearance and a significant decrease in AUC and $t_{1/2}$ were observed. Thus, they surmised that salicylate would inhibit the plasma protein binding of VPA and, as a consequence, more free VPA might be available for distribution, metabolism and excretion, resulting in the decrease in AUC and the increase in total clearance of VPA.

In conclusion, this study demonstrated that simultaneous oral administration of sodium valproate and salicylic acid decreased the concentrations of VPA in plasma including C_{max} and AUC_{0-3} of VPA. Two possibilities can be considered: (1) competitive inhibition by salicylate on VPA absorption through MCT-mediated transport, (2) inhibition by salicylate on plasma protein binding of VPA followed by the decrease in AUC and increase in total clearance of VPA.

REFERENCES

Abbott, F. S., Kassan, J., Orr, J. M., *et al.* (1986). The effect of aspirin on valproic acid metabolism, *Clin. Pharmacol. Ther.* **40**, 94–100.

Brodie, B. B. and Hogben, C. A. M. (1957). Some physico-chemical factors in drug action, *J. Pharm. Pharmacol.* **9**, 345–380.

Feitman, J. S., Bruni, J., Perrin, J. H., *et al.* (1980). Albumin binding interactions of sodium valproate, *J. Clin. Pharmcol.* **20**, 514–517.

Flower, R. J., Moncada, S. and Vane, J. R. (1985). Analgesic-antipyretics and anti-inflammatory agents: drugs employed in the treatment of gout, in: *The Pharmacological Basis of Therapeutics,* Gilman, A. G., Goodman, L. S., Rall, T. W. and Murad, F. (Eds), 7th edn, pp. 674–715. Macmillan, New York.

Fogh, J., Fogh, J. M. and Orfeo, T. (1977). One hundred and twenty seven cultured human tumor cell lines producing tumors in nude mice, *J. Natl. Cancer Inst.* **59**, 221–226.

Gugler, R. and Von Unruh, G. E. (1980). Clinical pharmacokinetics of valproic acid, *Clin. Pharmacokinet.* **5**, 67–83.

Higalgo, I. J., Raub, T. J. and Borchardt, R. T. (1989). Characterization of the human colon carcinoma cell line (Caco-2) as a model system for intestinal epithelial permeability, *Gastroenterology* **96**, 736–749.

Iwamoto, K., Takei, M. and Watanabe, J. (1982). Gastrointestinal and hepatic first-pass metabolism of aspirin in rats, *J. Pharm. Pharmacol.* **34**, 176–180.

Kaiji, H., Horie, T., Hayasi, M., *et al.* (1986). Effect of salicylic acid on the permeability of the plasma membrane of the small intestine of the rat: a fluorescence spectroscopic approach to elucidate the mechanism of promoted drug absorption, *J. Pharm. Sci.* **75**, 475–478.

Orr, J. M., Abott, F. S., Farell, K., *et al.* (1982). Interaction between valproic acid and aspirin in epileptic children: Serum protein binding and metabolic effects, *Clin. Pharmacol. Ther.* **31**, 642–649.

Takanaga, H., Tamai, I. and Tsuji, A. (1994). pH-Dependent and carrier-mediated transport of salicylic acid across Caco-2 cells, *J. Pharm. Pharmacol.* **46**, 567–570.

Tsuji, A., Takanaga, H., Tamai, I., *et al.* (1994). Transcellular transport of benzoic acid across Caco-2 cells by a pH-dependent and carrier-mediated transport mechanism, *Pharm. Res.* **11**, 30–37.

Yu, H. Y., Shen, Y. Z., Sugiyama, Y., *et al.* (1990). Effects of salicylate on pharmacokinetics of valproic acid in rats, *Drug Metab. Dispos.* **18**, 121–126.

Advances in Neuroregulation and Neuroprotection (2005), pp. 773-784
Collin, C. *et al.* (Eds)
© VSP 2005

Possible mechanisms of low levels of plasma valproate concentration following simultaneous administration of sodium valproate and meropenem

NORIO HOBARA [1], NOBUO HOKAMA [1], SUSUMU OHSHIRO [1], HIROMASA KAMEYA [1] and MATAO SAKANASHI [2,*]

[1] *Department of Hospital Pharmacy, Faculty of Medicine, University of the Ryukyus, 207 Uehara, Nishihara-cho Okinawa 903-0215, Japan*
[2] *Department of Pharmacology, School of Medicine, Faculty of Medicine, University of the Ryukyus, 207 Uehara, Nishihara-cho Okinawa 903-0215, Japan*

Abstract—Drug interaction between meropenem (MEPM), a carbapenem of antibiotic agent, and sodium valproate (VPA), an anticonvulsant, was studied in rats and human volunteers. When sodium VPA and MEPM were administered simultaneously, the plasma VPA level was significantly lower and the blood cell/plasma VPA concentration ratio was significantly higher than after the administration of sodium VPA alone (control). Following the simultaneous administration of sodium VPA and MEPM, the pharmacokinetic parameters such as plasma VPA, volume of distribution (Vd) and clearance (CL) were significantly increased, while the maximum plasma concentration (Cmax) and area under the plasma concentration–time curve (AUC) till 3 h were significantly decreased. The ratio of free (unbound) VPA to protein-bound VPA was significantly increased after the administration of MEPM. The concentration of VPA in brain tissue was significantly decreased after the simultaneous administration of VPA and MEPM. Competition between VPA and MEPM for a protein binding site resulted in increased plasma levels of free (unbound) VPA, which then distributed to various tissues, in particular, the blood cells, liver and kidney. VPA which distributed to the liver was metabolized there and excreted. Subsequently, the levels of VPA in the plasma and brain would be significantly decreased.

Keywords: Sodium valproate; meropenem; protein binding; drug interaction.

INTRODUCTION

Sodium valproate (VPA) is a branched-chain fatty acid that is used in the treatment of absence seizures, myoclonus and, usually in combination with other traditional anticonvulsants, commonly generalized epilepsy (Meunier *et al.*, 1963).

---

*To whom correspondence should be addressed. E-mail: sakanasi@med.u-ryukyu.ac.jp

Meropenem (MEPM) is a carbapenem antibiotic, which is stable to human renal dehydropeptidase-I (Nakashima *et al.*, 1992) and also shows β-lactamase inhibitory activity. MEPM lacks a pro-convulsive activity and has no adverse actions in the central nervous system (Ohno *et al.*, 1992), but MEPM possesses broad antibacterial spectra similar to imipenem (IPM) (Fukasawa *et al.*, 1992). Clinically, there are some cases in which sodium VPA is administered together with MEPM. Therefore, we investigated a drug interaction between MEP and VPA in this study.

MATERIALS AND METHODS

Sodium VPA was prepared in the form of Depakene syrup (5% VPA) and Depakene tablets (200 mg VPA), which were donated from Kyowa Hakko Kogyo Co., Ltd., Tokyo, Japan. MEPM was prepared in the form of solution for intravenous drip infusion (Sumitomo Pharmaceutical Co., Ltd., Osaka, Japan). Twenty-five male Sprague-Dawley rats (230–300 g) were housed in groups of three or four to a plastic-walled cage (26 × 36 × 25 cm), and had free access to food and water expect for 12 h before the experiment. The animals were maintained on a 12-h light and 12-h dark cycle (light on from 8:00 to 20:00). The ambient temperature and humidity were kept at 22–24°C and *ca.* 60%, respectively.

A single dose of sodium VPA was given intragastrically to overnight-fasted rats ($n = 12$) at a dose of 100 mg/kg body weight (b.w.). Immediately after the administration of sodium VPA, MEPM was injected into the jugular vein of rats ($n = 6$) at a dose of 125 mg/kg b.w. As a control ($n = 6$), a physiological saline solution was similarly given to sodium VPA-treated rats at a volume of 2 ml/kg b.w. The blood samples were collected by tail-nick method. Small quantities of blood were drawn at 0.25, 0.5, 1, 2, and 3 h following treatment with sodium VPA. The quantitative analysis of VPA in blood, urine and tissue samples was conducted using an automated Abbott TDXFLX fluorescence polarization analyzer (Abbott Laboratories, Abbot Park, USA). Free serum VPA concentrations were determined after centrifuging the serum for 15 min at a speed of $1670 \times g$. The pharmaco-kinetic parameters were obtained from VPA concentrations using a personal computer program (PHAConet-one, System Wave Co., Ltd., Tokyo, Japan) for non-linear least-squares regression. Since VPA was reported to be absorbed approximately 100% from the digestive canal, VPA was assumed to have an absorption rate (F) of 1 (Klotz and Antonin, 1997). The maximum plasma concentration (Cmax) and its corresponding time (Tmax), the absorption rate constant (Ka), the elimination rate constant (Ke), clearance (CL), the volume of distribution (Vd), the elimination half-life ($T_{1/2}$), and the area under the plasma concentration-time curve (AUC) were estimated using the computer program.

In another 5 rats, blood samples were collected by hypodermic syringe from the jugular vein 39 min after the treatment with sodium VPA (100 mg/kg b.w.). After 1 min, MEPM was injected into the jugular vein at a dose of 125 mg/kg b.w. Blood samples were drawn again at 1 and 15 min following the injection of MEPM, and the free (unbound) and total VPA concentrations in the plasma were assayed.

In another series of rats ($n = 8$), sodium VPA (200 mg/kg b.w.) was administered intragastrically 1 min before the intravenous injection of MEPM at a dose of 125 mg/kg b.w. ($n = 4$) or 0.9% NaCl (control, $n = 4$). Rats were then anesthetized using ether and killed by exsanguination from the carotid artery 40 min after the sodium VPA administration. The liver, kidney and brain were removed and immediately frozen in liquid nitrogen. These frozen organs were pulverized in mortars cooled with liquid nitrogen. Tissue powders were suspended in 20% iced trichloric acid. VPA concentrations of the supernatant solution were assayed using the Abbott TDXFLX system described above. Average recovery using this method was found to be 100.1% for VPA.

The animals used in this study were handled in accordance with the Guidelines for Animal Experimentation of the University of the Ryukyus, and the experimental protocol was approved by the Animal Care and Use Committee of this institution.

Written consent for the study was obtained from all human volunteers before the experiments. Six healthy Japanese males (aged 37 to 54 years old and weighing 56 to 79 kg) were recruited. All of the subjects had abstained from alcohol for at least 24 hr before the experiment, and none was taking any drugs either regularly or occasionally. The subjects fasted after a last dinner and at 7:00 a.m. on the day of experiment took sodium VPA 3 tablets (600 mg) with 50 ml of water. Blood samples were obtained from antecubital vein 1 h after oral administration of sodium VPA. During this period, they did not eat or drink. After sera were separated from the blood samples, 100 μl of a 50 mg/ml solution of MEPM and 0.9 ml of fresh serum were mixed and kept for 30 min at room temperature. Sera were then centrifuged, and the concentration of free (unbound) VPA was measured. As a control, the same volume of physiological saline solution instead of MEPM solution was added to the sera of sodium VPA-treated subjects. The concentration of total protein, albumin, non-esterified fatty acid (NEFA), triglycerides, alanine aminotransferase (ALT, GPT), aspartate aminotransferase (AST, GOT), alkaline phosphatase, lactate dehydrogenase and total cholesterol were determined by routine laboratory methods.

One month later, blood was collected from the same normal healthy subjects after an overnight fast. VPA at 150 μg/ml and MEPM at 5 mg/ml were prepared for *in vitro* experiments. After that, VPA concentrations in total blood, plasma and blood cells were determined. As a control, we added physiological saline solution instead of MEPM to blood samples.

Results were expressed as mean $\pm$ SEM. The statistical significance ($p < 0.05$) was analyzed using Student's t-test.

RESULTS

The time courses of blood VPA concentrations following VPA ingestion in rats did not differ whether MEPM was administered or not (Fig. 1, left side). However,

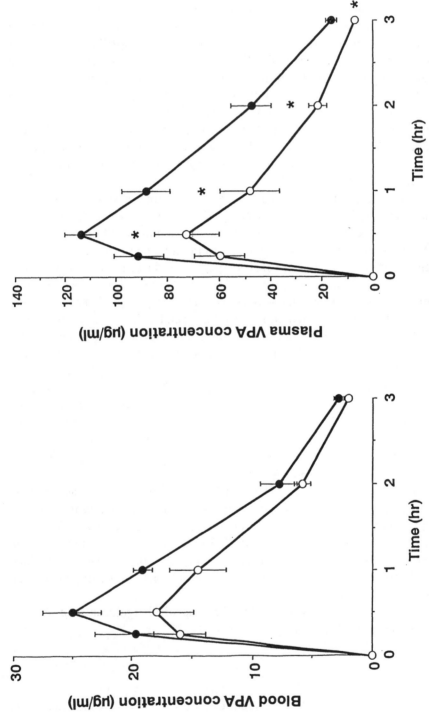

Figure 1. Time course of blood (left side) and plasma VPA concentrations (right side) following oral administration of sodium VPA 100 mg/kg plus intravenous administration of 0.9% NaCl or MEPM 125 mg/kg in rats. Result are expressed as mean ± SEM for 6 rats. An asterisk (*) denotes the significant difference ($p < 0.05$) between two groups. Vertical lines indicate SEM of the mean. (●) control (0.9% NaCl); (○) MEPM.

plasma VPA concentrations at 0.5, 1.0, 2.0 and 3.0 h after the simultaneous administration of sodium VPA and MEPM were significantly lower compared with those after the administration of VPA alone (Fig. 1, right side). On the other hand, the blood cell concentrations and blood cell/plasma VPA concentration ratios 1 h after the simultaneous administration of VPA and MEPM were markedly increased (Fig. 2). Figure 3 shows the available pharmaco-kinetic data for VPA after treatment of sodium VPA with and without MEPM. In MEPM-treated rats, the Cmax and AUC decreased significantly, while the Vd and CL increased significantly. In other rats, MEPM was injected into jugular vein 40 min after sodium VPA was administered. The ratio of free (unbound) VPA was significantly higher at 1 and 15 min after than before MEPM administration (Fig. 4).

Serum and brain VPA concentrations 40 min after sodium VPA (200 mg/kg p.o.) loading were significantly reduced in rats treated with MEPM (125 mg/kg i.v.), while the liver and kidney VPA/serum VPA concentration ratios were significantly higher in the MEPM-treated rats than in controls (Fig. 5).

When MEPM was added *in vitro* to the normal human sera following the oral administration of sodium VPA, the free (unbound) VPA concentration was significantly increased (Table 1). The clinical parameters of the healthy human subjects were within the normal ranges (Table 2). The experiment in which VPA and MEPM were added *in vitro* to the blood of normal human subjects showed a significant decrease in the plasma VPA concentration and a significant increase in the blood cell/plasma VPA concentration ratio (Table 3).

DISCUSSION

The time courses of plasma VPA concentration following the simultaneous administration of sodium VPA and MEPM showed lower levels than those following the administration of sodium VPA alone, and the Cmax and AUC were also decreased significantly. We have reported that plasma concentration levels of carvedilol, a β-blocker, following the simultaneous administration of carvedilol and MEPM or carvedilol and cefozopran were significantly lower than those following the administration of carvedilol alone in rats (Hobara *et al.*, 1998a, b).

Table 1.

Total and unbound (free) VPA concentration in normal human sera spiked with 5.0 mg/ml MEPM or 0.9% NaCl

	VPA concentration		
	Total	Unbound	Ratio of unbound
	(μg/ml)	(μg/ml)	(%)
0.9% NaCl	57.9 ± 3.3	4.7 ± 0.4	8.1 ± 0.2
MEPM	54.9 ± 2.8	6.2 ± 0.5*	11.3 ± 0.4*

Each value is mean ± SEM for 6 normal human subjects. An asterisk (*) denotes the significant difference ($p < 0.05$) between two groups.

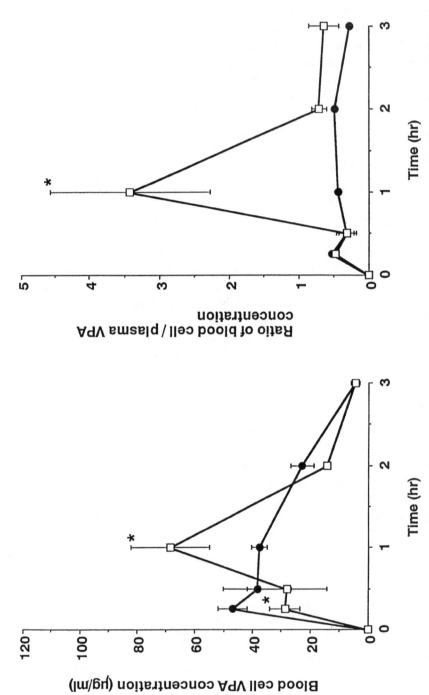

Figure 2. Blood cell VPA concentration (left side) and ratio of blood cell/plasma VPA concentration profile (right side) following oral administration of 100 mg/kg sodium VPA plus intravenous administration of 0.9% NaCl or 125 mg/kg MEPM in rats. Results are expressed as mean ± SEM for 6 rats. An asterisk (*) denotes the significant difference ($p < 0.05$) between two groups. Vertical lines indicate SEM of the mean. (●) control (0.9% NaCl); (□) MEPM.

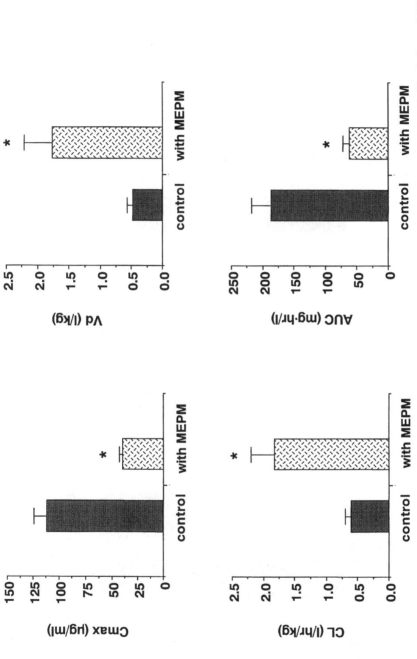

Figure 3. Pharmacokinetic parameters for VPA after simultaneous administration of sodium VPA 100 mg/kg p.o. plus 0.9% NaCl solution or MEPM 125 mg/kg i.v. in rats. Each column represents mean ± SEM for 6 rats. An asterisk (*) denotes the significant difference ($p < 0.05$) between two groups. Vertical lines indicate SEM of the mean. Vd: volume of distribution; CL: clearance; Cmax: maximum plasma concentration; AUC: area under the plasma concentration–time curve.

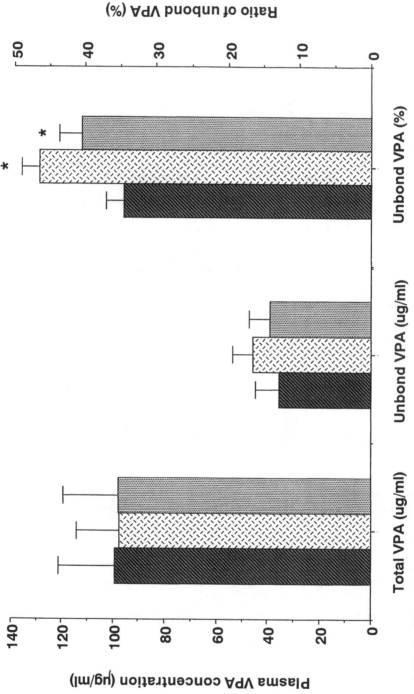

Figure 4. Total and free (unbound) plasma VPA concentration following oral administration of sodium VPA 100 mg/kg plus intravenous administration of MEPM 125 mg/kg in rats. Results are expressed as mean ± SEM for 5 to 4 rats. An asterisk (*) denotes the significant difference ($p < 0.05$) between 1 min before and 1 and 15 min after MEPM administration. Vertical lines indicate SEM of the mean. ▓ 1 min before MEPM; ▨ 1 min after MEPM; ▦ 15 min after MEPM.

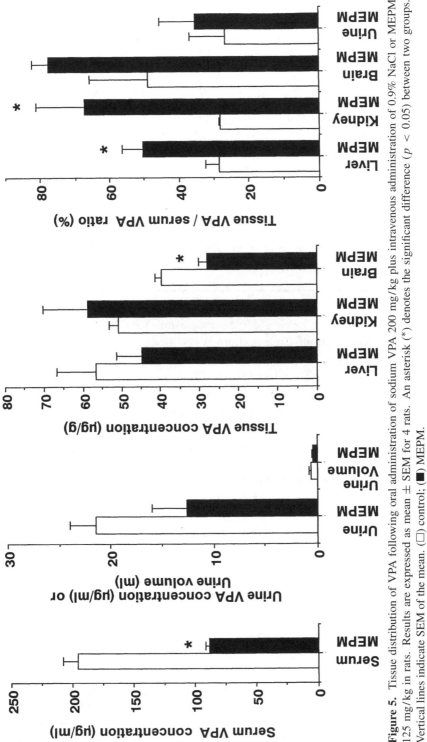

Figure 5. Tissue distribution of VPA following oral administration of sodium VPA 200 mg/kg plus intravenous administration of 0.9% NaCl or MEPM 125 mg/kg in rats. Results are expressed as mean ± SEM for 4 rats. An asterisk (*) denotes the significant difference ($p < 0.05$) between two groups. Vertical lines indicate SEM of the mean. (□) control; (■) MEPM.

Table 2.
Clinical parameters of 6 healthy human subjects

	Unit	Assay range	Mean ± SEM	Normal range
Age	year	37–54	44.5 ± 2.7	
Height	cm	162–175	167.5 ± 2.0	
Body weight	kg	56.8–80.8	66.9 ± 3.6	
Total protein	g/dl	7.21–7.91	7.55 ± 0.09	6.5–8.0
Albumin	g/dl	4.44–4.81	4.63 ± 0.05	3.7–5.2
Aspartate amino-transferase (AST, GOT)	IU	15–27	19.0 ± 1.8	0–38
Alanine amino-transferase (ALT, GPT)	IU	3–33	17.2 ± 4	0–44
Alkaline phospatase	U/l	121–206	157.8 ± 13.2	104–338
Lactate dehydrogenase	IU/l	177–355	271.2 ± 24.5	240–475
Total cholesterol	mg/dl	190–216	206.3 ± 4.3	130–230
Triglyceride	mg/dl	83–323	185.5 ± 39.8	30–160
Nonesterified fatty acid (NEFA)	μEq/l	0.31–0.66	0.43 ± 0.05	0.17–0.59

Araki *et al.* (1996) observed decreased plasma VPA concentration levels in patients who were administered sodium VPA along with injected MEPM and panipenem/betamipron, a carbapenem antibiotic drug derivative, and mentioned that convulsive fits observed in these patients suggest decreased plasma VPA levels. Since it was thought that MEPM might possibly delay the absorption of VPA, at 40 min after the administration of sodium VPA when the absorption of VPA had been made sufficient, MEPM was injected to rats. Then, the free (unbound) VPA concentration was measured 1 min before the injection of MEPM and after 1 and 15 min. The free VPA concentration ratio after MEPM injection significantly increased in comparison with that prior to MEPM injection. Thus, competition between VPA and MEPM for binding site on serum protein is considered to result in increased levels of free VPA.

In *in vitro* experiments using the serum of normal healthy-subjects, MEPM significantly increased the level of free (unbound) VPA concentration. On the other hand, the level of plasma VPA concentration decreased significantly, and the level of blood cell VPA concentration increased when MEPM was added to mixture of VPA and fresh blood obtained from normal healthy subjects. The possible reasons for these phenomena are discussed below.

The protein-binding rate of VPA at 20–60 μg/ml in human serum was 95.0% (Cramer and Mattson, 1979), while human and rat serum protein-binding rates of MEPM are reported to be 12.8% and 14.1%, respectively (Sumita *et al.*, 1992).

However, there are no reports of a protein-binding constant. When MEPM was added, the concentration of free (unbound) VPA increased in human sera, suggesting that the affinity of MEPM for protein binding in the plasma is stronger than that of VPA. The increase in free VPA after MEPM injection was considered to

Table 3.

Blood distribution of 150μg/ml VPA in human blood spiked with 5.0 mg/ml MEPM or 0.9% NaCl

	VPA concentration		
	Plasma (μg/ml)	Blood cell (μg/ml)	Ratio of blood cell/plasma (%)
0.9% NaCl	116.8 ± 2.7	30.9 ± 0.9	26.5 ± 1.3
MEPM	111.6 ± 1.4*	38.1 ± 2.5*	34.1 ± 1.9*

Each value is mean ± SEM for 4 normal human subjects. An asterisk (*) denotes the significant difference ($p < 0.05$) between two groups.

result in a redistribution to blood cells and tissues since there was an increase in Vd. However, when the VPA concentration in brain tissue was measured, the levels were significantly decreased in comparison with that in control rats. Free VPA distributes to the liver, too, and is metabolized there. Elevated levels of a metabolic substance of VPA and VPA itself were observed in the kidney and urine, possibly explaining the increased CL. Furthermore, large amounts of free (unbound) VPA that had been isolated by MEPM are considered to shift to the brain. There was a report that panipenem, which is a similar carbapenem-type antibiotic agent, prevented a shift of VPA to the transport process of brain (Murakami *et al.*, 1997). If MEPM has a similar action, a decrease in the concentration of brain VPA is also expected. When human plasma VPA concentration ranged from 26.8 to 150.4 μg/ml, the levels of brain VPA are reported to range from 5.7 to 26.7 μg/g brain tissue and the ratio of brain/plasma VPA concentration, from 6.8 to 27.9% (Vajda *et al.*, 1981). In the present experiment, the VPA concentration was 197.0 μg/ml in sera, and a brain VPA concentration was 39.7 μg/g brain tissue. Thus, the ratio of brain/plasma VPA concentration is calculated to be 20.2%. This experimental result agreed well with that mentioned above.

In humans, the majority of VPA exists as unchanged form in the blood. Though 2-propyl-E-2-pentenoic acid (E-2-en), 2-propyl-Z-2-pentenoic acid (Z-2-en), 2-propyl-3-pentenoic acid (3-en), 2-propyl-4-pentenoic acid (4-en), 3-hydroxy-2-propylpentanoic acid (3-OH), 4-hydroxy-2-propylpentanoic acid (4-OH), 4-hydro-xy-2-propyl-pentanoic acid (5-OH), 3-oxo-2-propylpentanoic acid (3-keto) and 4-oxo-2-propylpentanoic acid (4-keto) are detected as metabolic substances, their total concentration in sera amounts to only a little more than 10% of the total VPA (Muro and Tatsuhara, 1985; Tatsuhara and Muro, 1990).

The substance with the greatest excretion into urine is 3-keto, followed by VPA, E-2-en, 3-OH, 2-propylglutaric acid (PGA), 4-OH, 4-keto, 5-OH and Z-2-en substance. In these substances, VPA, E-2-en and Z-2-en substance were excreted mainly as glucuronides (Tatsuhara *et al.*, 1988). Thus, metabolites of VPA are excreted in large quantities into the urine. Though Muro and Tatsuhara (1985) reported that those do not reflect the true VPA concentration, we measured VPA in urine by TDXFLX, because metabolites of VPA can be measured with TDXFLX in the same way as VPA. In this study, the MEPM-treated group did not excrete metabolites of VPA or VPA into the urine more than the control group, and increases in the free VPA concentration and Vd were observed following the combined administration of sodium VPA and MEPM. Perhaps VPA would then redistribute in blood cells and body tissues, and VPA that shifted to the liver would be metabolized there. Metabolites would be excreted into the bile and urine, and CL would increase. These results could explain the fall in VPA concentration in the plasma and brain.

REFERENCES

Araki, A., Nakano, T., Yasuhara, T., *et al.* (1996). A report of decreased blood valpronic acid concentration while two patients admitted with panipenem/betamipron and meropenem. Extraction 4 in *The 128th Osaka Infant Academic Societies*, Osaka.

Cramer, J. A. and Mattson, R. H. (1979). Valpronic acid-in vitro plasma protein binding and interaction with phenytoin, *Therapeutic Drug Monitoring* **1**, 105–116.

Fukasawa, M., Sumita, Y., Tada, E., *et al.* (1992). In vitro antibacterial activity of meropenem, *Chemotherapy* **40** (S-1), 74–89.

Hobara, N., Hokama, N., Kameya, H., *et al.* (1998a). Influence of meropenem, a new carbapenem antibiotic, on plasma carvedilol concentration following simultaneous administration of carvedilol and meropenem in rats, *Kyusyhu Yakugaku Kaihou* **52**, 15–18 (in Japanese).

Hobara, N., Hokama, N., Kameya, H., *et al.* (1998b). Influence of cefozopran, a new carbapenem antibiotic, on plasma carvedilol concentration following simultaneous administration of carvedilol and cefozopran in rats, *Kyusyhu Yakugaku Kaihou* **52**, 9–13 (in Japanese).

Klotz, U. and Antonin, K. H. (1997). Pharmacokinetics and bioavailability of sodium valproate, *Clin. Pharmacol. Ther.* **21**, 736–743.

Meunier, H., Carraz, G., Neunier, Y., *et al.* (1963). Pharmacodynamic properties of N-dipropylacetic acid, *Therapie* **18**, 435–438.

Murakami, H., Fukayama, T., Matshuo, H., *et al.* (1997). Interaction of valproic acid and carbapenem antibiotic drug-effect of transport process on cultured cell, in: *The 117 Year Societies Speech 4*, 59. The Pharmaceutical Society of Japan.

Muro, H. and Tatsuhara, T. (1985). Effect of valproic acid metabolites on immunoassay, *J. Clin. Exp. Med.* **135**, 413–414.

Nakashima, M., Uematsu, T. and Kanamaru, M. (1992). Clinical phase I study of meropenem, *Chemotherapy* **40** (S-1), 258–275.

Ohno, Y., Hirose, A., Tsuji, R., *et al.* (1992). Behavioral and electroencephalographic studies on the central action of a novel carbapenem, meropenem, *Chemoterapy* **40** (S-1), 175–181.

Sumita, Y., Nouda, H., Tada, E., *et al.* (1992). Pharmacokinetics of meropenem, a new carbapenem antibiotic, parenterally administered to laboratory animals, *Chemotherapy* **40** (S-1), 123–131.

Tatsuhara, T. and Muro, H. (1990). Pharmacokinetics of valproic acid and its metabolites, *Jpn. J. Hosp. Pharm.* **16**, 329–342.

Tatsuhara, T., Muro, H., Matsuda, Y., *et al.* (1988). Pharmacokinetics of valproic acid and its metabolites after a single oral administration of sustained-release preparation of sodium valproate (KW-6066N), *Jpn. J. Clin. Pharmacol. Ther.* **19**, 749–757.

Vajda, F. J. E., Donnan, G. A., Phillips, J., *et al.* (1981). Human brain, plasma, and cerebrospinal fluid concentration of sodium valproate after 72 hours of therapy, *Neurology* **31**, 486–487.

Advances in Neuroregulation and Neuroprotection (2005), pp. 785-796
Collin, C. *et al.* (Eds)
© VSP 2005

Pressure stress stimulates aortic smooth muscle cell proliferation through angiotensin II receptor mediated signal transduction pathways

HIDEAKI KAWAGUCHI *, NORITERU MORITA, TAKESHI MURAKAMI
and KENJI IIZUKA

Laboratory Medicine, Hokkaido University School of Medicine, N-15 W-7, Kita-Ku, Sapporo 060-8638, Japan

Abstract—Mechanical forces related to pressure and flow are important for cell hypertrophy and proliferation. We hypothesized the presence of mechanosensors that were solely sensitive to pure atmospheric pressure in the absence of shear and tensile stresses. A pressure-loading apparatus was set up to examine the effects of atmospheric pressure on human aortic smooth muscle cells. Pressure application of 140 to 180 mmHg produced DNA synthesis in a pressure-dependent manner. In contrast, pressure of 120 mmHg or less produced no significant change. Pertussis toxin completely inhibited the pressure-induced increase of DNA synthesis, under high pressure of 200 mmHg. Both extracellular signal-related kinase and c-Jun N-terminal kinase activities, but not p38 activity, were stimulated by pressure of more than 160 mmHg. ACE inhibitor inhibited cell proliferation under the pressure. It also inhibited ERKs expression, but the addition of AII on ACE inhibitor under the pressure could not recover the cell proliferation. Furthermore, the inactive ACE inhibitor suppressed the cell proliferation. There is a possibility that these agents, not mediated by angiotensin II receptors, directly suppress cell proliferation system in our experiment. In summary, HASMC have a mechanosensing cellular switch for DNA synthesis which is sensitive to pure atmospheric pressure. The molecular switch is activated by pressure of more than 140 mmHg. The mechanism of the inhibitory effect of ACE inhibitor on cell proliferation stimulated by pure pressure is under way in our laboratory.

Keywords: Mechanical stress; JUN; ERK; p38; ACE inhibitor.

INTRODUCTION

Abnormal growth and proliferation of vascular smooth muscle cells have been implicated in the pathogenesis of atherosclerosis and hypertension (Safar *et al.*,

---

*To whom correspondence should be addressed. E-mail: hideaki@med.hokudai.ac.jp

1998). Mechanical stresses are likely to be involved in this process since it has been demonstrated that they regulate cell growth in many tissues (Vandenburgh, 1992). In arteries, for example, a mechanical stress related to pressure and flow is crucial in promoting blood vessel wall remodeling. Such vascular remodeling may have important clinical implications for the evolution of several vascular diseases, may alter vascular compliance in hypertension and atherosclerosis, and may cause vascular fragility and compensatory changes in atherosclerosis (Cowan and Langill, 1996).

Various stresses have been studied, including cytokines, mitogens, ultraviolet light, oxidant stress, hyperosmolarity, heat stress and mechanical stress (Kyriakis and Avruch, 1996). The vascular endothelial and vascular smooth muscle cells covering the inner surface of blood vessels are constantly exposed to such stress. The hemodynamic forces affecting endothelial cells include shear stress due to the frictional force of blood flow, and circumferential and pure pressure stress due to transmural pressure. Recently, a number of studies have shown that shear stress to the vessel wall, which is one of the mechanical stresses generated by blood flow, modulates endothelial morphology and function (Chien and Shyy, 1998). However, in an *in vitro* cell culture model, it has been shown that intracellular signalings produced by tensile and shear stresses may be influenced not only by pressure stress, but also by morphological and cytoskeletal changes of cells (Cucina *et al.*, 1995). Vascular smooth muscle cells make up the outer layer of the endothelium in blood vessels and are thus indirectly exposed to blood flow. Therefore, mechanosensing intracellular signal transductions in vascular smooth cells are defined as a cellular response to transmural pure pressure and tensile stress in the absence of shear stress. In previous studies, tensile stress was reported to accelerate DNA synthesis and activate stretch-activated cation channels in vascular smooth muscle cells (Weiser *et al.*, 1995; Setoguchi *et al.*, 1997). However, the molecular identities of these candidates for mechanosensitive receptors are currently unknown.

We and others reported that pressure stress without other mechanical stresses facilitated cell proliferation and cell migration of cultured human vascular smooth muscle cells and rat vascular smooth muscle cells and that pressure-induced pathological cellular responses resulted from activation of intracellular signal kinase C (PKC) and extracellular signal-regulated kinase (ERK). However, there are still very few studies that examined responses of human vascular smooth muscle cells to pure pressure stress, so it is unclear which mechanisms are involved in cell responses to pure pressure stress through activation of PKC and ERK. Previous reports have examined the involvement of angiotensin II via autocrine/paracrine secretion in pressure-induced responses, but the results were contradictory. It still remains to be determined whether renin angiotensin system in aortic smooth muscle cells is activated by pressure stress.

In the present study, we focused on the implication of the local renin–angiotensin system, especially angiotensin II that leads to activation of PKC and ERK, in pure pressure-induced proliferation of human aortic smooth muscle cells (HASMC).

The purpose of this study was to examine whether the local renin–angiotensin system is activated by pure pressure stress in HASMC, and we also determine the effects of an angiotensin converting enzyme (ACE) inhibitor on pressure-induced cell proliferation and the activation of MAP kinase.

METHODS

Cell culture

HASMC (Clonetics) were cultured in smooth muscle cell basal medium (SmBM; Clonetics) which was modified MCDB131 containing 5% fetal bovine serum, gentacine (50 μg/ml), amphotericine (50 μg/ml) and several growth factors: human epidermal growth factor (10 ng/ml), human fibroblast growth factor (2.0 ng/ml), and insulin (5.0 μg/ml). The cells were incubated at 37° in a humidified, 5% CO_2 atmosphere. The 4th through 8th passages of HASMC were plated on 6-well plates for further investigation.

Pure pressure-loading apparatus

An original pure pressure-loading apparatus was designed to expose HASMC to pure atmosphere pressure stress. The chamber allows for pumping air or nitrogen gas to raise the internal pressure (maximum at 300 mmHg), and can be sealed tightly by placing several clamps at the edge. Internal pressure levels were monitored with an aneroid barometer during the experiments. The compression chamber was established in the incubator and kept at 37°C, and a digital thermometer was mounted in the incubator to monitor the exact internal temperature. In the following series of experiments, the chamber was kept at 37°C, and the partial pressure of pO_2 and pCO_2 were theoretically preserved as constants according to the Boyle–Gay-Lussac law. The potential of hydrogen (pH) in the culture medium was supported at a constant level (7.41 ± 0.02) during the experiments. It was not possible to monitor the actual morphological changes of pressurized cells in our system set-up. However, the light microscopic investigations failed to find any changes in cell size or morphology during or after pressurization.

HASMC were cultured on 6-well plates for 48 hours (h). In 3 wells of each plate, the medium was changed to a starving medium (Dulbecco's modified Eagle's medium without serum). In the other 3 wells, the medium was changed to the same starving medium but containing PTx (0.1 μg/ml; Seikagaku, Japan). After 8 h of incubation, the plates were placed in the pure pressure-loading apparatus in the medium containing 10 mM HEPES (N-2-hydroxyethyl-piperazine-N-2-ethane-sulfonic acid, pH 7.4), exposed to various levels of atmospheric pressure (0 to 240 mmHg), and then incubated in an incubator for 1 or 3 h at 37°C.

DNA synthesis

[³H]-Thymidine (TdR) incorporation into DNA was studied as a marker for DNA synthesis acceleration by pure atmospheric pressure. [³H] TdR (2 μCi/ml, Amersham, UK) was added to the medium in all 6 wells of a 6-well plate. After pressure-loading procedures, the plate was incubated in a 37°C CO_2-incubator under normal pressure for 4 h. The cells were rinsed two times in ice-cold phosphate-buffered saline (PBS) followed by precipitation three times with ice-cold 10% trichloroacetic acid, and lysed in 0.5 N NaOH at 37°C by shaking for 30 min. The incorporation of [³H] TdR into DNA was quantified by pipetting the DNA hydrolysate into counting vials containing 4 ml of liquid scintillation cocktail (Ready Gel, Beckman). The protein concentration of HASMC was normalized by Lowry's method with bovine serum albumin as a standard (Lowry *et al.*, 1951). The counting results were expressed as disintegrations per protein and were expressed as a percent-increase compared to controls at the basal level (0 mmHg) for 1 and 3 h. All the experiments for each pressure were performed in triplicate, and were repeated 2 to 3 times.

Immunoblotting

HASMC were incubated on 100-mm dishes until 80% confluent. The medium was replaced with a starving medium. After 8 h of incubation, the dishes were placed in the pure pressure-loading apparatus in the medium containing 10 mM HEPES, and exposed to various levels of atmospheric pressure (0 mmHg, 120 through 180 mmHg, and 240 mmHg) for 3 h at 37°C. The cells were washed twice with ice-cold PBS, and harvested in a buffer containing 25 mM Tris-HCl (pH 6.8), 1% Triton X-100, 150 mM sodium chloride and protease inhibitors: benzamidine (100 μM), leupeptin (2 μM), aprotinin (0.15 μM), pepstatin A (1.5 μM), and phenylmethylsulfonyl fluoride (100 μM). The protein content of each sample was measured by Lowry's method. Samples (15 μg) were analyzed by SDS (sodium dodecyl sulfate)-polyacrylamide gel electrophoresis in a 12% gel using the Mini Gel Electrophoresis System (Marysol). Protein was transferred to a nitrocellulose membrane (Hybond ECL; Amersham, UK) by Western Blotting apparatus (Semi dry type; Marysol) with a buffer containing 20% method, 48 mM Tris-base, 78 mM glycine, and 0.375% SDS for 2 h. The membranes were submerged for 1 hour in 4% non-fat dry milk in TTBS (0.05% Tween-20, Tris-buffered saline, pH7.4), followed by an incubation for 2 h in TTBS containing the appropriate primary antibodies (anti-active extracellular signal-regulated kinase (ERK), -active c-Jun N-terminal kinase (JUN), and -active p38 anti-rabbit polyclonal antibodies, Promega) in concentrations recommended by the manufacturer. The membranes were incubated in TTBS containing horseradish peroxidase-labeled donkey anti-rabbit Ig's (1:2500; Amersham), followed by washing three times in TTBS. The detection of the protein was performed using a chemiluminescence method (ECL; Amersham, UK). Active phosphorylated forms of ERK, JNK and p38 were scanned with a digital image analyzing system and quantified NIH images.

Data analysis

All the results were expressed as means ± SEM, and statistical significance was assessed by Student's t-test. Values of $p < 0.005$ were considered statistically significant.

RESULTS

Pure pressure-dependent acceleration of DNA synthesis in HASMC

With the pure pressure-loading apparatus, an atmospheric pressure of 240 mmHg was applied on cultured HASMC. DNA synthesis was measured as [^{3}H]-TdR incorporation into DNA. After one hour of pressure application, the pure pressure of 240 mmHg induced an approximately 20% increase in [^{3}H]-TdR incorporation compared to the control HASMC. After more than 3 h, the degree of acceleration of DNA synthesis was similar.

Threshold of pure atmospheric pressure in acceleration of DNA synthesis

To determine the threshold of pure atmospheric pressure in acceleration of DNA synthesis, pure pressure was applied to HASMC for 1 or 3 h. Various levels of atmospheric pressure were applied to analyze the threshold of pure pressure in acceleration of DNA synthesis. The atmospheric pressure of less than 120 mmHg produced no significant change in [^{3}H]-TdR incorporation at 1 and 3 h. However, pressure of 140 mmHg produced an approximately 20% increase in [^{3}H]-TdR incorporation ($p < 0.005$ *vs.* control) at 1 and 3 h. Pressure of more than 140 mmHg also produced an increase (14 ± 6% to 27 ± 4%; $p < 0.005$ *vs.* control). At 3 h, the degree of increase of incorporation of [^{3}H]-TdR was dependent on pressure levels from 140 to 180 mmHg, with 180-mmHg pressure producing an increase of 49 ± 6% compared to control cells ($p < 0.01$ *vs.* control); 200 mmHg pressure significantly accelerated DNA synthesis at both 1 and 3 h. However, exposure to 240-mmHg pressure did not further increase the DNA synthesis.

PTx effect on pure pressure-dependent DNA synthesis

We examined whether Gi-proteins are involved in pure pressure-dependent acceleration of DNA synthesis in HASMC. DNA-synthesis in acceleration under 200 mmHg pressure was completely inhibited by PTx (0.1 μg/ml) at 3 h (increase of DNA synthesis *vs.* control: 18.8±2.7 at 1 h and 18.0±1.4% at 3 h in the absence of PTx; 4.4±2.7% at 1 h and 2.3±2.9% at 3 h in the presence of PTx). In contrast, an acceleration of DNA synthesis by 160 mmHg pressure was not significantly inhibited at this time point (increase of DNA synthesis *vs.* control: 21.6±2.5% at 1 h and 27.4±5.5% at 3 h in the absence of PTx; 16.5±6.9% at 1 h and 32.5±5.5% at 3 h in the presence of PTx); PTx also failed to significantly inhibit the acceleration produced by 180 mmHg pressure (increase of DNA synthesis *vs.* control:

24.6 ± 4.3%) at 1 h and 40.1 ± 11.4% at 3 h in the absence of PTx; 24.2 ± 3.9% at 1 h and 49.0 ± 6.0% at 3 h in the presence of PTx).

Signal transduction system under pure pressure

Pure atmospheric pressure stimulated ERK1/2 and JNK activities. Pressure of more than 160 mmHg induced an activation of ERK1 (p44), and ERK1 was activated to maximum level at pressure of 240 mmHg (increase in active form of ERK1 *vs.* control: 63.2 ± 12.1%). Pressures of more than 120 mmHg induced an activation of ERK2 (p42), and ERK2 was activated to maximum level at pressure of 180 mmHg (increase in active form of ERK2 *vs.* control: 49.5 ± 7.3%). JNK was activated at pressures of more than 160 mmHg in pressure-dependent manner and to a maximum level at pressure of 240 mmHg (increase in active form of JNK *vs.* control: 45.7 ± 6.9%). Application of pure atmospheric pressure induced no activation of p38.

Effect of ACE inhibitor on signal transduction system in HASMC stimulated by pressure

We studied the effect of ACE inhibitor on this system. ACE inhibitor perindoprilat was added to the culture medium, then DNA synthesis and ERK expression were determined. Perindoprilat suppressed DNA synthesis under 160 mmHg. Perindpril; inactive ACE inhibitor also suppressed DNA synthesis (Fig. 1). AII was added to this incubation system. ACE inhibitor + AII did not recover DNA synthesis, which was suppressed by ACE inhibitor (Fig. 2). This result shows that AII did not effect DNA synthesis under pressure, or ACE inhibitor directly inhibits the signal-pathway, which is activated by pure pressure. Figure 3 shows the effect of ACE inhibitor on MAP kinase. The ACE inhibitor, perindoprilat, suppressed

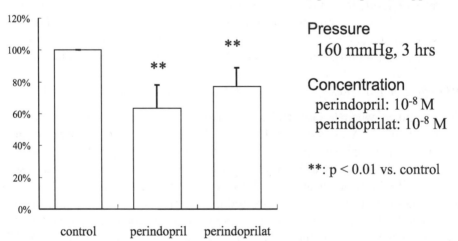

Figure 1. Effect of ACE inhibitor on pressure induced HASMC proliferation. Values are means ± SEM ($n = 6$–9). ** $p < 0.01$ *vs.* controls.

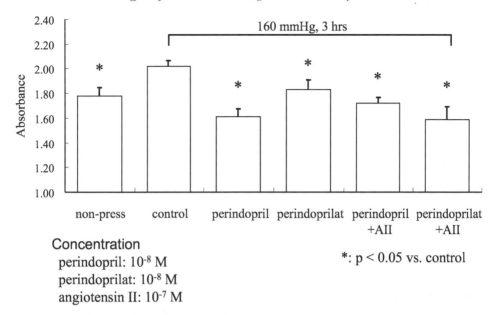

Figure 2. Effect of ACE inhibitor and angiotensin II on pressure-induced HASMC proliferation. Cells were incubated under an atmospheric pressure of 160 mmHg for 3 h. Values are means ± SEM ($n = 6–9$). $^*p < 0.05$ *vs.* control.

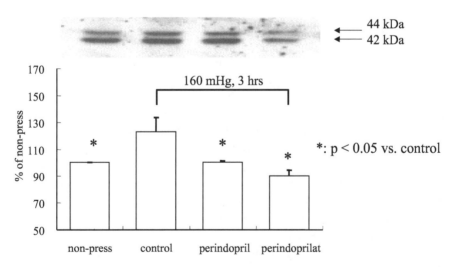

Figure 3. Effect of ACE inhibitor on active form of ERK. Cells were incubated under an atmospheric pressure of 160 mmHg for 3 h. Values are means ± SEM ($n = 3–5$). $^*p < 0.05$ *vs.* controls.

ERK expression, but perindoril, an inactive ACE inhibitor also suppressed ERK expression (Fig. 3). We used other ACE inhibitors in this system, for example captopril. Captopril also suppressed DNA synthesis (Fig. 4).

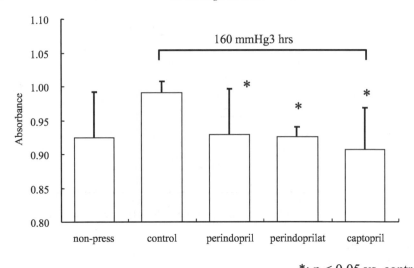

*: p < 0.05 vs. control

Figure 4. Effect of ACE inhibitors on pressure-induced HASMC proliferation. Cells were incubated under an atmospheric pressure of 160 mmHg for 3 h. Values are means ± SEM ($n = 3$–5). $^*p < 0.05$ vs. controls.

DISCUSSION

Mechanical forces are important modulators of cellular functions, and particularly in the cardiovascular system. Mechanical stress has various components, such as well shear stress, tensile stress and pure pressure stress. In the case of shear stress, a fluid shear stress of 1.3–4.1 dynes/cm^2 affected endothelial cell DNA synthesis during regeneration (Ando et al., 1990). It is also reported that shear stress stimulated ERK and JNK activity in a force-dependent manner in endothelial cells (Jo et al., 1997). Shear stress is also known to activate phospholipase C and to generate inositol triphosphate (IP$_3$) and diacylglycerol. IP$_3$ release Ca^{2+} from Ca^{2+} stores via IP$_3$ receptor, and diacylglycerol activates protein kinase C (PKC) (Helmlinger et al., 1995; Tseng et al., 1995). Several PKC isoenzymes have been suggested to be involved in several downstream signalings, such as ERK activation, NFκB-mediated gene transcription, erg-1 transcription and activation of c-Src families (Mohan et al., 1997; Jalali et al., 1998; Schwachtgen et al., 1998). Recently, it has been reported that shear stress induced changes in the morphology and cytoskeltal organization of endothelial cells, and that these changes may be correlated to the functional change after exposure to shear stress (Cucina et al., 1995). It is interesting to speculate about the role in shear stress-induced signalings, and that shear stress-induced signalings participate in intracellular cross-talk with integrine-coupled signal transductions (Wang et al., 1993; Miyamoto et al., 1995). In these studies, however, it is difficult to separate the direct effects of pure pressure from the indirect effects caused by the morphological change of cells. The pure stress is reported to promote DNA synthesis in rat cultured vascular smooth muscle cells in an original pressure-loading apparatus, as determined by immunocytochemical assay (Hishikawa *et al.*, 1994). This study suggests the possible presence of a mechanosensing cellular switch that is solely sensitive to pure pressure stress. However, in experiments with such apparatus, the threshold pressure for determining the on and off status of the mechanosensing cellular switch has not been clarified.

In this study, we demonstrated that pure pressure stress accelerated an increase of DNA synthesis in the absence of shear stress and tensile stress, and determined the threshold of pure pressure stress which presumably activates the mechanosensing cellular switch. Using an original pressure-loading apparatus, we investigated various pressure levels from 0 to 240 mmHg. Low atmospheric pressure of less than 120 mmHg had no significant effect on DNA synthesis, while pressures of more than 140 mmHg induced an acceleration of DNA synthesis. From these results, we conclude that a mechanosensing cellular switch for DNA synthesis that was solely sensitive to pure pressure was 'on' or 'off' at over or under 140 mmHg, respectively. Pressures of 140 to 180 mmHg promoted an increase of $[^3H]$-TdR incorporation in a pressure-dependent manner at 1 and 3 h. Pressure of more than 200 mmHg also induced approximately 20% increase of DNA synthesis. However, degree of acceleration of DNA synthesis at pressure of more than 200 mmHg was similar and pressure-independent.

We also demonstrated the differential pathway in HASMC in response to pure atmospheric pressures. ERK1 was activated at pressures of more than 160 mmHg, and to a maximum level at pressure 240 mmHg. ERK2 was activated at pressures of more than 120 mmHg, and to a maximum level at pressure of 180 mmHg. JNK was activated at pressures of more than 160 mmHg in a pressure-dependent manner. In contrast, p38 was not activated at pure atmospheric pressure. The mechanosensing mechanism stimulated by pure atmospheric pressure may consist of some differential pathways involving ERK and JNK signalings, and differential levels of pressure may activate each pathway. Moreover, differential activation of ERK and JNK in HASMC by pure atmospheric pressure may result in the selective phosphorylation and activation of transcription factors leading to selective gene regulatory events. It is not clear why the degrees of acceleration of DNA synthesis and activation of ERK2 were independent of pressure levels of more than 200 mmHg or why the pressure levels that accelerated DNA synthesis and activated ERK and JNK were different.

A role for heterotrimeric G-protein in mechanical stress-induced signal transductions in endothelial cells has recently emerged (Berthiaume and Frangos, 1992; Gudi *et al.*, 1996). It is known that both heterotrimeric G-proteins and small G-proteins are activated by shear stress. Shear stress-induced activation of G-proteins results in several flow-initiated endothelial responses that regulate vascular tone and that release such vasodilators as nitric oxide and prostaglandin I_2, and such vasoconstrictors as endothelin (Frangos *et al.*, 1985; Kuchan and Frangos, 1993; Ranjan *et al.*, 1995). Therefore, we also investigated whether mechanosensing signaling pathways that are sensitive to pure pressure in HASMC were related to Gi-dependent pathways or Gi-independent pathways. At atmospheric pressures of 160 mmHg, PTx demonstrated no significant effect on [^{3}H] TdR incorporation. However, at a high atmospheric pressure of 200 mmHg, PTx completely inhibited the increase of [^{3}H] TdR incorporation. Pure pressure-induced DNA synthesis was produced through intracellular signaling pathways, including both Gi-dependent and Gi-independent pathways. Although the molecular switch was 'on' at an atmospheric pressure of more than 140 mmHg, an atmospheric pressure of below 180 mmHg may activate intracellular signalings predominantly through Gi-independent pathways, and a pressure of more than 200 mmHg may activate DNA synthesis predominantly through Gi-dependent pathways.

In summary, we demonstrated that HASMC had a mechanosensing molecular switch for DNA synthesis which was solely sensitive to pure atmospheric pressure, and the switch was 'on' or 'off' at over or under 140 mmHg, respectively. Moreover, mechanosensing cellular mechanisms may consist of some mechanosensors or intracellular pathways activated by different levels of pure atmospheric pressure. This study showed a change in dominance of Gi-dependent or Gi-independent intracellular signaling pathways in a pressure range from 160 to 200 mmHg. Recently, it has been reported that minimum risk for cardiovascular mortality was reached at 138.8 mmHg for a mean systolic arterial pressure in randomized clinical study of patients with hypertension (Hansson *et al.*, 1998). Activation of

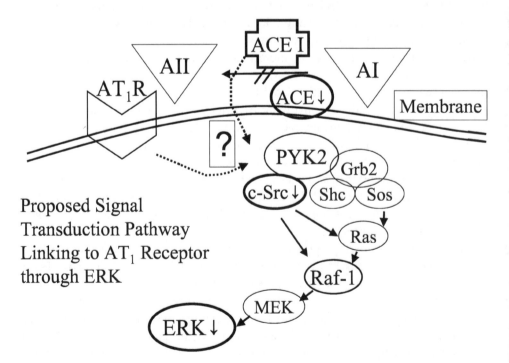

Figure 5. Proposed signal transduction pathway linking to AT$_1$ receptor through ERK.

mechanosensing switches in HASMC at 140 mmHg may be involved in clinical cardiovascular events.

ACE inhibitor suppressed cell proliferation under the pressure. But the addition of AII on ACE inhibitor under the pressure could not recover the cell proliferation. ERKs and JNK are increased under pressure, and inhibited by ACE inhibitors. In our recent studies, cSRC was activated by pressure and suppressed ACE inhibitors. But Raf1 did not changed by pressure. Usually AII stimulates cell proliferation but, in our experiments, the addition of AII to incubation system did not recover the DNA synthesis suppressed by ACE inhibitor. Furthermore, inactive ACE inhibitor also suppressed smooth muscle cells proliferation. These results suggest that ACE inhibitor and inactive ACE inhibitor directly inhibit the signal transduction system of cell proliferation. In our experiments, it was the MAP kinase pathway. At present time, we do not know how and at which portion ACE inhibitor and inactive ACE inhibitor suppressed DNA synthesis by pressure (Fig. 5). The mechanism of this inhibitory effect of ACE inhibitor on cell proliferation stimulated by pure pressure is under way in our laboratory.

REFERENCES

Ando, J., Komatsuda, T., Ishikawa, C., *et al.* (1990). Fluid shear stress enhanced DNA synthesis in cultured endothelial cells during repair of mechanical denudation, *Biorheology* **27**, 675–684.

Berthiaume, F. and Frangos, J. A. (1992). Flow-induced prostacyclin production is mediated by a pertussis toxin-sensitive G protein, *FEBS Lett.* **308**, 277–279.

Chien, S., Li, S. and Shyy, J. Y. J. (1998). Effects of mechanical forces on signal transduction and gene expression in endothelial cells, *Hypertension* **31**, 162–169.

Cowan, D. B. and Langill, B. L. (1996). Cellular and molecular biology of vascular remodeling, *Curr. Opi. Lipidol.* **7**, 94–100.

Cucina, A., Sterpetti, A. V., Pupelis, G., *et al.* (1995). Shear stress induces changes in the morphology and cytoskeleton organization of arterial endothelial cells, *Eur. J. Vas. Endovasc. Surg.* **9**, 86–92.

Frangos, J. A., Eskin, S. G., McIntire, L. V., *et al.* (1985). Flow effects prostacyclin production by cultured endothelial cells, *Science* **227**, 1477–1479.

Gudi, S. R. P., Clark, C. B. and Frangos, J. A. (1996). Fluid flow rapidly activates G proteins in human endothelial cells. Involvement of G proteins in mechanochemical signal transduction, *Circ. Res.* **79**, 834–839.

Hansson, L., Zanchetti, A., Carruthers, S. G., *et al.* (1998). Effect of intensive blood pressure lowering and low-dose aspirin in patients with hypertension: principal results of the hypertension optimal treatment (HOT) randomized trial, *Lancet* **351**, 1755–1762.

Helmlinger, G., Berk, B. C. and Nerem, R. M. (1995). Calcium responses of endothelial cell monolayers subjected to pulsatile and steady laminar flows differ, *Am. J. Physiol.* **269**, C367–375.

Hishikawa, K., Nakai, T., Marumo, T., *et al.* (1994). Pressure promotes DNA synthesis in rat cultured vascular smooth muscle cells, *J. Clin. Invest.* **93**, 1975–1980.

Jalali, S., Li, Y. S., Sotoudeh, M., *et al.* (1998). Shear stress activates p60src-Ras-MAPK signaling pathways in vascular endothelial cells, *Arterioscler. Thromb. Vas. Biol.* **2**, 227–234.

Jo, H., Sipos, K., Go, Y. M., *et al.* (1997). Differential effect of shear stress on extracellular signal-regulated kinase and N-terminal Jun kinase in endothelial cells. Gi2- and Gbeta/gamma-dependent signal pathways, *J. Biol. Chem.* **272**, 1395–1401.

Kuchan, M. J. and Frangos, J. A. (1993). Shear stress regulates endothelin-1 release via protein kinase C and cGMP in cultured endothelial calls, *Am. J. Physiol.* **264**, H150–H156.

Kyriakis, J. M. and Avruch, J. (1996). Sounding the alarm. Protein kinase cascades activated by stress and inflammation, *J. Biol. Chem.* **271**, 24313–24316.

Lowry, O. H., Rosenbrough, N. J., Farr, A. L., *et al.* (1951). Protein measurement with the folin phenol reagent, *J. Biol. Chem.* **193**, 265–275.

Miyamoto, S., Teramoto, H., Coso, A., *et al.* (1995). Integrin function: molecular hierarchies of cytoskeletal signaling molecules, *J. Cell Biol.* **131**, 791–805.

Mohan, S., Mohan, S. and Sprague, E. A. (1997). Differential activation of NF-kappa B in human aortic endothelial cells conditioned to specific flow environments, *Am. J. Physiol.* **273**, C572–578.

Ranjan, V., Xiao, Z. and Diamond, S. I. (1995). Constitutive NOS expression in cultured endothelial cells is elevated by fluid shear stress, *Am. J. Physiol.* **269**, H550–H555.

Safar, M. E., London. G., Asmar, R., *et al.* (1998). Recent advances on large arteries in hypertension, *Hypertension* **32**, 156–161.

Schwachtgen, J. L., Houston, P., Campbell, C., *et al.* (1998). Fluid shear stress activation of erg-1 transcription in cultured human endothelial and epithelial cells is mediated via the extracellular signal-related kinase 1/2 mitogen-activated protein kinase pathway, *J. Clin. Invest.* **101**, 2540–2549.

Setoguchi, M., Ohya, Y., Abe, I., *et al.* (1997). Stretch activated whole-cell currents in smooth muscle cells from mesenteric resistance artery of guinea pig, *J. Physiol.* **501**, 343–353.

Tseng, H., Peterson, T. E. and Berk, B. C. (1995). Fluid shear stress stimulates mitogen-activated protein kinase in endothelial cells, *Circ. Res.* **77**, 869–878.

Vandenburgh, H. H. (1992). Mechanical forces and their second messengers in stimulating cell growth in vitro, *Am. J. Physiol.* **262**, R350–R355.

Wang, N., Butler, J. P. and Ingber, D. E. (1993). Mechanotransduction across the cell surface and through the cytoskeleton, *Science* **260**, 1124–1127.

Weiser, M. C., Majack, R. A., Tucker, A., *et al.* (1995). Static tension is associated with increased smooth muscle cell DNA synthesis in rat pulmonary arteries, *Am. J. Physiol.* **268**, H1133–1138.

Advances in Neuroregulation and Neuroprotection (2005), pp. 797-809
Collin, C. *et al.* (Eds)
© VSP 2005

Pressure loading-induced mitogen-activated protein kinase activation is enhanced in prehypertensive spontaneously hypertensive rat vasculature

S. AIUCHI, H. HOSOKAWA, T. KAMBE, Y. HAGIWARA and T. KUBO*

Department of Pharmacology, Showa Pharmaceutical University, Machida, Tokyo 194-8543, Japan

Abstract—The vascular structural remodeling function may be altered in spontaneously hypertensive rats (SHR). To examine this possibility, we examined whether pressure loading-induced activation of mitogen-activated protein kinases (MAPKs) is enhanced in the vasculature of prehypertensive SHR and whether angiotensin and endothelin systems in the vasculature are involved in the enhanced MAPK activation in SHR vasculature. Male 4-week-old SHR and age-matched Wistar Kyoto rats (WKY) were used. Aortae were perfused with Tyrode solution. Increases in perfusion pressure caused an increase in p42/p44 MAPK activity in WKY and SHR aortae. MAPK activities in SHR aortae perfused at 100 and 200 mmHg were greater than those of WKY. The enhanced MAPK activation in SHR aortae was greatly inhibited by the angiotensin receptor antagonist losartan but minimally by the endothelin receptor antagonist BQ123. Cyclic stretching of aortic vascular smooth muscle cells from SHR and WKY aortae produced an increase in p42/p44 MAPK activity. The stretch-induced MAPK activation was almost the same in both cells. These results indicate that pressure loading-induced MAPK activation is enhanced in aortae from prehypertensive SHR. It appears that the enhancement of MAPK activation results partly from enhanced angiotensin system in SHR aortae.

Keywords: Angiotensin; mitogen-activated protein kinase; pressure loading; spontaneously hypertensive rats.

INTRODUCTION

Vascular hypertrophy occurs during hypertension and contributes to the elevation of peripheral vascular resistance in the established phase of hypertension (Folkow *et al.*, 1982; Mulvany *et al.*, 1985; Owens *et al.*, 1988). One of the possible biochemical events responsible for vascular hypertrophy is activation of mitogen-activated protein kinases (MAPKs), enzymes believed to be involved in the pathway for cell proliferation (Sturgill *et al.*, 1988; Pulverer *et al.*, 1991; Wilson *et al.*, 1993;

*To whom correspondence should be addressed. E-mail: t.kubo@cityfujisawa.ne.jp

Duff *et al.*, 1995; Li *et al.*, 1997; Matsusaki and Ichikawa, 1997; Kim and Iwao, 2000).

There are angiotensin systems and endothelin systems involved in MAPK activation in rat vasculature, and they are mediated via angiotensin AT_1 receptors and endothelin ETA receptors, respectively (Kubo *et al.*, 1998, 2001). We have previously demonstrated that mechanical strain of the vascular wall following pressure loading causes an increase of MAPK activity in isolated perfused rat aortae and that this MAPK activation is inhibited by the angiotensin AT_1 receptor antagonist losartan (Kubo *et al.*, 2000a). It has been reported that in aortae of hypertensive rats, MAPK activities are enhanced with the development of hypertension (Kim *et al.*, 1997) and losartan significantly decreases the enhanced MAPK activity. Xu *et al.* (1996) have demonstrated that an acute elevation of blood pressure induced by vasoconstrictors activates MAPKs in rat arteries. These findings clearly demonstrate that pressure loading increases MAPK activities in the vasculature and the mechanical strain-induced MAPK activation is mediated at least partly via the vascular angiotensin system.

If the pressure loading-induced MAPK activation is enhanced from prehypertensive stages in genetic hypertension, this may be one of the mechanisms of hypertension. We have previously demonstrated that MAPK activation reactivity to angiotensin II is enhanced at a prehypertensive stage in spontaneously hypertensive rats (SHR), a model of genetic hypertension (Kubo *et al.*, 1999b). In the present study, we examined whether pressure loading-induced MAPK activation is enhanced in the vasculature of prehypertensive SHR and whether angiotensin systems and/or endothelin systems in the vasculature are involved in the enhanced MAPK activation in prehypertensive SHR vasculature. In addition, we studied whether the enhanced pressure loading-induced MAPK activation in prehypertensive SHR vasculature results from genetic changes in vascular smooth muscle cells.

MATERIALS AND METHODS

Male 4-week-old SHR and age-matched WKY (Charles River, Japan) were used in this study. They were kept under alternating 12-h periods of dark and light, and given standard rat chow and tap water *ad libitum*. One day before the experiments, systolic blood pressure was measured indirectly by tail plethysmography. The procedure were in accordance with institutional on the National Research Council's guidelines.

Tissue preparation

Animals were killed with overdoses of ether. After sternotomy, 20 ml of saline containing 200 units of heparin was injected into the heart. A cannula (PE-60) was inserted into the thoracic aorta just next to the arcus aorta, and a second cannula (PE-60) was inserted into the aorta just next to the abdominal aorta as described previously (Kubo *et al.*, 2000a). The thoracic aorta was performed with Tyrode solution (mmol/l: NaCl 137, KCl 2.7, $CaCl_2$ 1.8, $MgCl_2$ 0.5, NaH_2PO_4 0.4, $NaHCO_3$ 11.9, glucose 5.5). All the branches of the thoracic aorta were ligated, and the aorta was isolated and placed in a dish with Tyrode solution maintained at 30°C.

The thoracic aorta was perfused with Tyrode solution maintained at 30°C and gassed with a mixture of 95% oxygen and 5% carbon dioxide. The aorta was perfused at a constant flow rate of 2 ml/min with a roller pump (Atto, Tokyo), and perfusion pressure was monitored by a pressure transducer connected to a polygraph. To elevate perfusion pressure, the end of the outlet cannula was connected to a silicon tube and the silicon tube was constricted by a clamp.

The thoracic aorta was first perfused without constriction of the silicon tube. After 60 min of the equilibration perfusion at basal pressure (10 mmHg), the perfusion pressure was elevated by constricting the silicon tube. Drugs were dissolved in the perfusion buffer solution and applied 5 min before and during elevation of perfusion pressure. We used losartan 10^{-6} mol/l and BQ123 10^{-6} mol/l, because these doses of the drugs produced a specific and maximal inhibition of MAPK activity in rat aortae (Kubo *et al.*, 1998, 1999a).

The tissues were homogenized and sonicated in 3 ml of an ice-cold buffer (10 mmol/l Tris, 150 mmol/l NaCl, 2 mmol/l ethylene glycol-O,O'-bis (2-aminoethyl)-N,N,N'N'-tetraacetic acid (EGTA), 2 mmol/l dithiothreitol, 1 mmol/l orthovanadate, 1 mmol/l (p-amidinophenyl) methansulphonyl fluoride, 10 μg/ml leupeptin and 10 μg/ml aprotinin) (pH 7.4). All further steps were performed at 4°C. Tissue homogenates were centrifuged at 15000 rpm for 30 min and the supernatant was retained to obtain cytoplasmic MAPKs.

Vascular smooth muscle cells

Rat aortic vascular smooth muscle cells were isolated from thoracic aortae of 4-week-old SHR and age-matched WKY using the method of Ross (1971) as described previously (Kubo *et al.*, 2001). In brief, striped aortae were placed on polyethylene dishes (diameter 60 mm). Dulbecco's modified Eagle Medium (DMEM) containing 20% fetal calf serum, 100 U/ml penicillin and 100 μg/ml streptomycin was added to the dishes to cover the aortae. The dishes were kept at 37°C in a humidified atmosphere of 95% air and 5% CO_2. The cells which grew from the explants reached confluence after about 14 days, were harvested by brief exposure to Hanks' medium supplemented with 0.25% trypsin and 0.02% ethylenediaminetetraacetic acid (EDTA)-sodium salt, and transferred to fresh dishes. The properties of the cells cultured showed a typical hills-and-valleys growth pattern and α-actin molecule upon immunohistochemical analysis using mouse anti-α-actin (Zymed Laboratories, Inc.), indicating that the cells were grown from vascular smooth muscle cells.

Primary cultures of vascular smooth muscle cells were plated in 60 mm culture dishes (1×10^5 cells per dish) containing DMEM with 10% fetal calf serum. The cells used in this study were passage 3.

Application of cyclic stretch

Cyclic stretching of vascular smooth muscle cells was performed according to the method of Kato *et al.* (1998). Vascular smooth muscle cells were removed from the dish with Hanks' medium supplemented with 0.25% trypsin and 0.02% EDTA-sodium salt, and were transferred onto a 10 cm^2 fibronectin-coated silicon chamber containing DMEM with 10% fetal calf serum, at a density of 2×10^5 cells/cm^2. The silicon chamber had a 200 μm-thick transparent bottom and the side wall was 400 μm thick to prevent narrowing at its bottom center. The silicon chamber was attached to a stretching apparatus that was driven by a computer controlled stepping motor (NS-100, Scalatec, Co., Osaka, Japan). Cells were allowed to attach to the chamber bottom for 48 h and rendered quiescent by a 48-hour serum deprivation period. The serum-free medium was replaced by 1 ml fresh serum free-DMEM, and after 3 h, uni-axial sinusoidal stretch was applied at 37°C, 5% CO$_2$. The relative elongation of the silicon membrane was uniform across the whole membrane area. Cells in some chambers were incubated under static conditions as control.

Cyclic stretching were performed at a condition of 20% elongation, 60 cycles/min for 20 min. Twenty min after cyclic stretching, the mechanical strain was stopped and the reaction was terminated by chilling the chamber on ice and washing twice with ice-cold phosphate-buffered saline. The cells were lysed and homogenized in 150 μl ice-cold buffer, consisting of 10 mM Tris, 150 mM NaCl, 2 mM EGTA, 2 mM dithiothreitol, 1 mM orthovanadate, 1 mM (p-amidinophenyl)methanesulphonyl fluoride, 10 μg/ml leupeptin and 10 μg/ml aprotinin (pH 7.4). Cellular debris was precipitated by centrifuging at 15 000 rpm for 30 min and the supernatant retained to obtain cytoplasmic MAP kinases.

Determination of MAPK activity

MAPK activity was assayed with the p42/p44 MAPK enzyme assay system (Amersham, UK) which is designed to detect MAPKs in lysed cells, as described elsewhere (Kubo *et al.*, 1998). In a previous study (Kubo *et al.*, 2002), we demonstrated that angiotensin II-induced MAPK activation measured by the p42/p44 MAPK enzyme assay system was similar to that measured by Western blot analysis with phospho-specific ERK antibodies, indicating that the enzyme assay system is a reasonable measure of MAPK activation. Protein was measured by the method of Lowry *et al.* (1951).

Reverse transcriptase/PCR for analysis of angiotensin AT$_1$ receptor mRNA Levels

Thoracic aortae of 4-week-old SHR and age-matched WKY were removed and immediately frozen in liquid nitrogen and stored at −80°C. Total RNA was extracted according to the guanidinium thiocyanate/cesium chloride centrifugation method (Chomczynski and Sacchi, 1987). The extracted RNA was suspended in RNase-free water and quantified by measuring the absorbance at 260 nm.

The relative levels of angiotensin AT_1 receptor mRNA were determined using reverse transcriptase/PCR (RT/PCR). Primers were synthesized by Japan Gene Research Laboratory (Tokyo, Japan). The sequences for the angiotensin AT_1 receptor primers were 5'-TACATATTTGTCATGATTCCT-3' and 5'-GTGAATATTTGGTG GGGAAC-3' (Cheng *et al.*, 1998). The sequences for the glyceraldehyde-3-phosphate-dehydrogenase (GAPDH) primers were 5'-GACCCCTTCATTGACC TCAAC-3' and 5'-CTCAGTGTAGCCCAGGAT-3' (Piechaczyk *et al.*, 1984). The sequences for the endothelin ETA receptor primers were 5'-GAAGTCGTCCGTGG GCATCA-3' and 5'-CTGTGCTGCTCGCCCTTGTA-3' (Shigematsu *et al.*, 1996).

For RT, 2 μg RNA was dissolved in 20 μl of a reaction mixture containing 2.5 mmol/l of dATP, dCTP, dTTP, and dGTP; 50 U of RNase inhibitor; 25 pmol of 3' primer; 75 mmol/l KCl; 50 mmol/l Tris-HCl (pH 8.3); 3 mmol/l $MgCl_2$; 50 U murine leukemia virus reverse transcriptase (M-MLVRT; Sawaday Tecnology, Tokyo, Japan). The samples were incubated 60 min at 42°C, boiled for 5 min at 95°C, and then quickly chilled on ice.

For amplification of the resulting cDNA, two μl of the RT-reaction mixture of the tissue total RNA were used. PCR amplification was done in 20 μl of 10 mmol/l Tris-HCl, 1.5 mmol/l $MgCl_2$, 50 mmol/l KCl, pH 8.3, 0.2 mmol/l dNTP, 10 pmoles of primers for angiotensin AT_1 receptor, 10 pmoles of primers for GAPDH, and 1 U of Taq DNA polymerase (Boeringer Mannheim). The reaction consisted of denaturation at 95°C for 60 s, annealing at 60°C for 60 s, and elongation at 72°C for 60 s. This cycle was repeated 25 times for GAPDH and 35 times for angiotensin AT_1 receptor and endothelin ETA receptor. Ten microliters of amplified cDNA were electrophoresed on 2% agarose gel and stained with ethidium bromide and then photographed. The intensity of each PCR band was analyzed using Image-Pro Plus TM (Medical Cybernetics, Tokyo).

Drugs were leupeptin hemisulfate, aprotinin (Sigma, St. Louis, MO), BQ123 (Research Biochemicals International, Natick, MA) and Dulbecco's modified Eagle's medium (DMEM) (Dainihon Pharmaceuticals, Osaka). Losartan was generously supplied by Dupon-Merck Pharmaceuticals (Wilmington, DE).

The results are expressed as means ± S.E.M. All results were analyzed by either Student's *t*-test or one-way analysis of variance combined with Dunnett's test for *post hoc* analysis for intergroup comparison. Differences were considered significant at $p < 0.05$.

RESULTS

Pressure loading-induced MAPK activation in aortae

The activity of p42/p44 MAPKs in 4-week-old SHR aortae placed in Tyrode solution without perfusion was almost the same with that of age-matched WKY (Fig. 1). In perfusion experiments, aortae were first perfused at basal pressure for equilibration for 60 min and then perfused further at the basal pressure (10 mmHg),

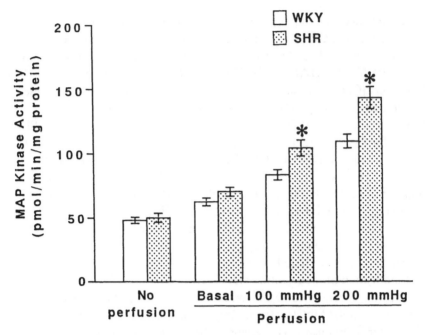

Figure 1. Effects of perfusion pressure on MAP kinase activity in isolated aortae from 4-week-old SHR and age-matched WKY. First, the aorta was perfused for 60 min without constriction of a silicon tube connected to the end of the outlet cannula. Then, the silicon tube was constricted for 20 min to elevate perfusion pressure to 100 mmHg or to 200 mmHg. In the basal perfusion pressure group (basal perfusion), the silicon tube was not constricted and the perfusion pressure was 10 mmHg. In the no perfusion group, aortae were isolated but not perfused. Values are means $\pm$ S.E.M. from six experiments using six animals. $*p < 0.05$, compared with WKY.

100 mmHg or 200 mmHg for 20 min. Pressure loading increased p42/p44 MAPK activities pressure-dependently in both WKY and SHR aortae (Fig. 1). MAPK activities were greater in SHR aortae perfused at 100 and 200 mmHg than those of WKY, while there was no significant difference in the enzyme activity between WKY and SHR aortae perfused at basal pressure.

The angiotensin AT_1 receptor antagonist losartan (10^{-6} mol/l) inhibited the pressure loading-induced increase of MAPK activity in WKY and SHR aortae perfused at 200 mmHg (Fig. 2). The losartan-induced inhibition of MAPK activation was greater in SHR aortae (-30.0 ± 2.8 pmol/min/mg protein) than in WKY aortae (-21.1 ± 2.1 pmol/min/mg protein) ($p < 0.05$). However, MAPK activities in the losartan-treated SHR aortae perfused at 200 mmHg were still greater than those of WKY. This antagonist minimally inhibited the MAPK activity in WKY and SHR aortae perfused at basal pressure.

The endothelin ETA receptor antagonist BQ123 (10^{-6} mol/l) inhibited the pressure loading-induced increase in MAPK activity in WKY and SHR aortae perfused at 200 mmHg, while it did not inhibit the MAPK activity in either of the aortae perfused at basal pressure (Fig. 3). The BQ123-induced inhibition of

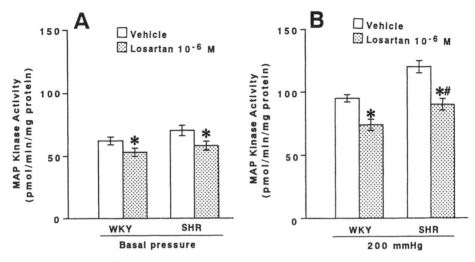

Figure 2. Effects of losartan on MAP kinase activity 20 min after perfusion at basal pressure (A) and at 200 mmHg (B) in isolated aortae from 4-week-old SHR and age-matched WKY. Saline (vehicle) and losartan were dissolved in the perfusion buffer and perfused 5 min before and during perfusion at basal pressure or at 200 mmHg. Values are means ± S.E.M. from five experiments. $^*p < 0.05$, compared with respective vehicle. $^{\#}p < 0.05$, compared with WKY-losartan.

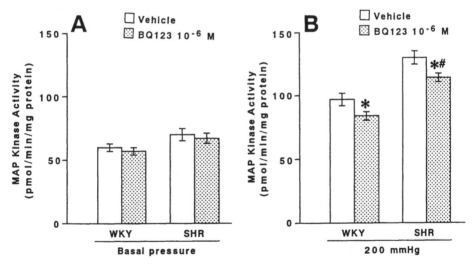

Figure 3. Effects of BQ123 on MAP kinase activity 20 min after perfusion at basal pressure (A) and at 200 mmHg (B) in isolated aortae from 4-week-old SHR and age-matched WKY. Saline (vehicle) and BQ123 were dissolved in the perfusion buffer and perfused 5 min before and during perfusion at basal pressure or at 200 mmHg. Values are means ± S.E.M. from five experiments. $^*p < 0.05$, compared with respective vehicle. $^{\#}p < 0.05$, compared with WKY-BQ123.

MAPK activation was almost the same in WKY and SHR aortae (-12.7 ± 1.5 and -15.8 ± 1.7 pmol/min/mg protein, respectively, $p > 0.05$).

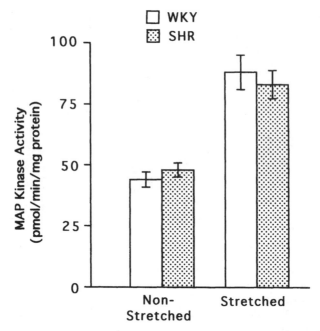

Figure 4. MAP kinase activity 20 min after cyclic stretching (60 cycles/min, 20% elongation) or non-stretching of aortic vascular smooth muscle cells from 4-week-old SHR and age-matched WKY. Values are means ± S.E.M. of five separate experiments (in triplicate) using five animals.

There was no significant difference in systolic blood pressure between 4-week-old SHR (113 ± 2 mmHg, $n = 10$) and age-matched WKY (109 ± 2 mmHg, $n = 10$).

Stretch-induced MAPK activation in vascular smooth muscle cells

To examine whether the enhanced pressure loading-induced MAPK activation observed in SHR aortae results from genetic changes in SHR vascular smooth muscle cells, we next measured MAPK activity in cultured aortic vascular smooth muscle cells from 4-week-old SHR and age-matched WKY. Cyclic stretching (20% elongation, 60 cycles/min for 20 min) of vascular smooth muscle cells from both kinds of rats produced an increase in p42/p44 MAPK activity (Fig. 4). The stretch-induced increase in MAPK activity was almost the same in both WKY and SHR vascular smooth muscle cells.

Angiotensin AT_1 Receptor mRNA Levels in aortae

Figure 5 shows RT-PCR analysis of angiotensin AT_1 receptor mRNA and endothelin ETA receptor mRNA in aortae from 4-week-old SHR and age-matched WKY. The AT_1 receptor mRNA level was increased in SHR aortae, whereas ETA receptor mRNA level was not different in aortae from both kinds of rats (Fig. 6).

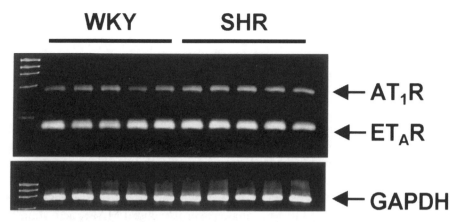

Figure 5. Angiotensin AT_1 receptor (AT_1R), endothelin ETA receptor (ETAR) and glyceraldehyde-3-phosphate-dehydrogenase(GAPDH) mRNA expression in aortae from 4-week-old SHR and age-matched WKY. Two micrograms of total RNA from aortae was subjected to 35 cycles (AT_1R and ETAR) or 25 cycles (GAPDH) of reverse transcriptase (RT)/polymerase chain reaction (PCR). Ten microliters of the PCR reaction mixture were electrophoresed through 2% agarose gel.

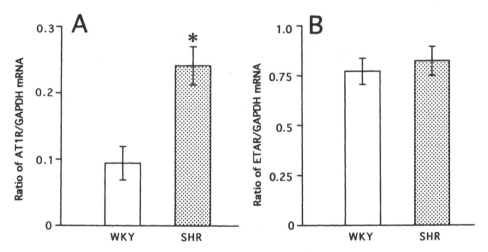

Figure 6. Relative amounts of angiotensin AT_1 receptor (AT_1R) (A) and endothelin ETA receptor (ETAR) (B) mRNA in aortae from 4-week-old SHR and age-matched WKY. Values are means $\pm$ S.E.M. from five animals. $^*p < 0.05$, compared with WKY.

DISCUSSION

In the present study, elevation of perfusion pressure in isolated aortae from 4-week-old SHR and age-matched WKY caused a pressure-dependent increase of MAPK activity. The pressure-induced increase of MAPK activity was greater in SHR aortae perfused at 100 or 200 mmHg than that of WKY, while the enzyme activity was almost the same in WKY and SHR aortae not perfused or perfused at basal pressure. There was no difference in systolic blood pressure between 4-week-old SHR and age-matched WKY. These findings suggest that pressure loading-induced

MAPK activation function is enhanced even at a prehypertensive stage in the SHR vasculature.

The angiotensin AT_1 rceptor antagonist losartan inhibited the pressure loading-induced MAPK activation in WKY and SHR aortae perfused at 200 mmHg. These findings are compatible with the idea that the pressure loading-induced MAPK activation is mediated via the production and release of angiotensin II in isolated aortae (Kubo *et al.*, 2000a). The losartan-induced inhibition of MAPK activation was enhanced in SHR aortae, suggesting that the vascular angiotensin system is involved in the enhanced pressure-induced MAPK activation in the SHR vasculature. Previously, we demonstrated that angiotensin II produced a concentration-dependent increase in MAPK activity in aortae from 4-week-old SHR and age-matched WKY, and that the angiotensin II-induced MAPK activation was greater in SHR aortae than in WKY aortae (Kubo *et al.*, 1999b). Furthermore, in the present study, we found that the expression of AT_1 receptor mRNA is increased in 4-week-old SHR aortae. Thus, it may be considered that the enhanced pressure loading-induced MAPK activation in SHR aortae is, at least in part, due to the enhanced reactivity to angiotensin II in SHR aortae.

In the present study, pressure loading-induced MAPK activation was also inhibited by the endothelin receptor antagonist BQ123 (10^{-6} mol/l) in WKY and SHR aortae. Previously, we found that BQ123 (10^{-6} mol/l) blocked the MAPK activation induced by endothelin-1 (10^{-8} mol/l) but not that induced by angiotensin II (10^{-8} mol/l), suggesting that BQ123 antagonizes endothelin receptors specifically. In the present study, BQ123 (10^{-6} mol/l) inhibited the pressure loading-induced MAPK activation to the same extent in both WKY and SHR aortae, suggesting that the vascular endothelin system is not mainly involved in the enhanced pressure-induced MAPK activation in the SHR vasculature. The results of the present study are compatible with our previous findings that the sensitivity to endothelin-1 for MAPK activation is almost the same in aortae (Kubo *et al.*, 1999a). Furthermore, in the present study, we found that ETA receptor mRNA levels were not different in aortae from 4-week-old SHR and age-matched WKY.

The vascular wall is composed of endothelial cells, smooth muscle cells and fibroblasts. We previously found that mechanical strain of vascular smooth muscle cells from rat aortae produced an increase in p42/p44 MAPK activity (Kubo *et al.*, 2002). In the present study, the stretch-induced MAPK activation was not different in vascular smooth muscle cells from 4-week-old SHR and age-matched WKY. These findings suggest that the enhanced pressure loading-induced MAPK activation observed in SHR aortae is not due to genetic changes in vascular smooth muscle cells from SHR.

In the present study, the pressure-induced MAPK activation in SHR aortae after losartan (10^{-6} mol/l) was still greater than that of WKY aortae after losartan (10^{-6} mol/l). The dose of losartan (10^{-6} mol/l) was enough to block angiotensin II-induced MAPK activation (Kubo *et al.*, 1998). Thus, other mechanisms, in addition to enhanced vascular angiotensin system, may be involved in the enhanced pressure-induced MAPK activation in SHR aortae. Stretching of vascular smooth muscle cells is reported to enhance Ca^{2+} influx through a voltage-dependent dihydropyridine-sensitive Ca^{2+} entry pathway (Ruiz-Velasco *et al.*, 1996) and thus the Ca^{2+} taken up into the cells activates MAPKs (Eguchi *et al.*, 1996). It is also reported that mechanical stresses can directly stretch the cell membrane and alter receptor or G protein conformation, thereby initiating signal pathways used by growth factors (Li and Xu, 2000). In addition, Iwasaki *et al.* (2000) have reported that Ca^{2+}-dependent activation of EGF receptor, mainly via SA ion channels, is required for the stretch-induced ERK1/2 activation. Whether some of these mechanisms are responsible for the enhanced MAPK activation in SHR aortae remains to be clarified.

Perfusion flow would cause both pressure stress and shear stress to the vascular wall. Since we always perfused at a constant flow rate of 2 ml/min in this study, shear stress would not be different in aortae perfused at basal pressure and at elevated pressures. Thus, the increase of aortic MAPK activity induced by elevation of perfusion pressure seems to be mainly due to pressure stress.

In the present study, pressure loading-induced MAPK activation was enhanced in prehypertensive SHR. Since MAPKs are believed to be involved in the pathway for cell proliferation and thus for vascular structural remodeling (Alvarez *et al.*, 1991; Tsuda *et al.*, 1992; Molloy *et al.*, 1993), the results of the present study are compatible with the idea that vascular structural remodeling function is enhanced in the genetically hypertensive rats, SHR, and that this may contribute to the development of hypertension in SHR.

In summary, pressure loading-induced MAPK activation is enhanced in perfused aortae from prehypertensive SHR, suggesting that vascular structural remodeling function may be enhanced in SHR. It appears that the enhancement of MAPK activation results partly from enhanced angiotensin system in SHR aortae.

Acknowledgements

This study was supported in part by a Grant-in-Aid for Scientific Research (No. 13672300) from Japan Society for the Promotion of Science.

REFERENCES

Alvarez, E., Northwood, I. C., Gonzalez, F. A., *et al.* (1991). Pro-Leu-Ser/Thr-Pro is a consensus primary sequence for substrate protein phosphorylation: Characterization of the phosphorylation of c-myc and c-jun proteins by an epidermal growth factor receptor threonine 669 protein kinase, *J. Biol. Chem.* **266**, 15277–15285.

Cheng, H. F., Wang, J. L., Vinson, G. P., *et al.* (1998). Young SHR express increased type 1 angiotensin II receptors in renal proximal tubule, *Am. J. Physiol.* **274**, F10–F17.

Chomczynski, P. and Sacchi, N. (1987). Single-step method of RNA isolation by acid guanidinium thiocyanate-phenol-chloroform extraction, *Anal. Biochem.* **162**, 156–159.

Duff, J. L., Monia, B. P. and Berk B. C. (1995). Mitogen-activated protein (MAP) kinase is regulated by the MAP kinase phosphatase (MKP-1) in vascular smooth muscle cells, *J. Biol. Chem.* **270**, 7161–7166.

Eguchi, S., Matsumoto, T., Motley, E. D., *et al.* (1996). Identification of an essential signaling cascade for mitogen-activated protein kinase activation by angiotensin II in cultured rat vascular smooth muscle cells, *J. Biol. Chem.* **271**, 14169–14175.

Folkow, B., Hallback, M., Lundgren, R., *et al.* (1982). Importance of adaptive changes in vascular design for establishment of primary hypertension studied in man and in spontaneously hypertensive rats, *Circ. Res.* **32** (Suppl. I), I-2–I-16.

Iwasaki, H., Eguchi, S., Ueno, H., *et al.* (2000). Mechanical stretch stimulates growth of vascular smooth muscle cells via epidermal growth factor receptor, *Am. J. Physiol.* **278**, H521–H529.

Kato, T., Ishiguro, N., Iwata, H., *et al.* (1998). Up-regulation of COX2 expression by uni-axial cyclic stretch in human lung fibroblast cells, *Biochem. Biophys. Res. Commun.* **244**, 615–619.

Kim, S. and Iwao, H. (2000). Molecular and cellular mechanisms of angiotensin II-mediated cardiovascular and renal diseases, *Pharmacol. Rev.* **52**, 11–34.

Kim, S., Murakami, T., Izumi, Y., *et al.* (1997). Extracellular signal-regulated kinase and c-jun NH2-terminal kinase activities are continuously and differentially increased in aorta of hypertensive rats, *Biochem. Biophys. Res. Commun.* **236**, 199–204.

Kubo, T., Saito, E., Hanada, M., *et al.* (1998). Evidence that angiotensin II, endothelins and nitric oxide regulate mitogen-activated protein kinase activity in rat aorta, *Eur. J. Pharmacol.* **347**, 337–346.

Kubo, T., Saito, E., Hosokawa, H., *et al.* (1999a). Local renin-angiotensin system and mitogen-activated protein kinase activation in rat aorta, *Eur. J. Pharmacol.* **365**, 103–110.

Kubo, T., Ibusuki, T., Saito, E., *et al.* (1999b). Vascular mitogen-activated protein kinase activity is enhanced via angiotensin system in spontaneously hypertensive rats, *Eur. J. Pharmacol.* **372**, 279–285.

Kubo, T., Hosokawa, H., Kambe, T., *et al.* (2000a). Angiotensin II mediates pressure loading-induced mitogen-activated protein kinase activation in isolated rat aorta, *Eur. J. Pharmacol.* **391**, 281–287.

Kubo, T., Ibusuki, T., Saito, E., *et al.* (2000b). Different activation of vascular mitogen-activated protein kinases in spontaneously and DOCA-salt hypertensive rats, *Eur. J. Pharmacol.* **400**, 231–237.

Kubo, T., Ibusuki, T., Chiba, S., *et al.* (2001). Mitogen-activated protein kinase activity regulation role of angiotensin and endothelin systems in vascular smooth muscle cells, *Eur. J. Pharmacol.* **411**, 27–34.

Kubo, T., Ibusuki, T., Chiba, S., *et al.* (2002). Altered mitogen-activated protein kinase activation in vascular smooth muscle cells from spontaneously hypertensive rats, *Clin. Exp. Pharmacol. Physiol.* **29**, 537–543.

Li, C. and Xu, Q. (2000). Mechanical stress-initiated signal transductions in vascular smoooth muscle cells, *Cellular Signal.* **12**, 435–445.

Li, Q., Muragaki, Y., Ueno, H., *et al.* (1997). Stretch-induced proliferation of cultured vascular smooth muscle cells and a possible involvement of local renin-angiotensin system and platelet-derived growth factor (PDGF), *Hypertension Res.* **20**, 217–223.

Lowry, O. H., Rosebrough, N. J., Farr, A. L., *et al.* (1951). Protein measurement with the Folin phenol reagent, *J. Biol. Chem.* **193**, 265–275.

Matsusaka, T. and Ichikawa, I. (1997). Biological functions of angiotensin and its receptors. *Ann. Rev. Physiol.* **59**, 395–412.

Molloy, C. J., Taylor, D. S. and Weber, H. (1993). Angiotensin II stimulation of rapid protein tyrosine phosphorylation and protein kinase activation in rat aortic smooth muscle cells, *J. Biol. Chem.* **268**, 7338–7345.

Mulvany, M. J., Baandrup, U. and Gundersen, H. J. G. (1985). Evidence for hyperplasia in mesenteric resistance vessels of spontaneously hypertensive rats using a three-dimensional disector, *Circ. Res.* **57**, 794–800.

Owens, G. K., Schwartz, S. M. and McCanna M. (1988). Evaluation of medial hypertrophy in resistance vessels of spontaneously hypertensive rats, *Hypertension* **11**, 198–207.

Piechaczyk, M., Bianchard, J. M., Marty, L., *et al.* (1984). Post-transcriptional regulation of glyceraldehyde-3-phosphate-dehydrogenase gene expression in rat tissues, *Nucleic Acids Res.* **12**, 6951–6963.

Pulverer, B. J., Kyriakis, J. M., Avruch, J., *et al.* (1991). Phosphorylation of c-jun mediated by MAP kinases, *Nature* **353**, 670–674.

Ross, R. (1971). The smooth muscle. II. Growth of smooth muscle in culture and formation of elastic fibers, *J. Cell. Biol.* **50**, 172–186.

Ruiz-Velasco, V., Mayer, M. B. and Hymel, L. J. (1996). Dihydropyridine-sensitive Ca^{2+} influx modulated by stretch in A7r5 vascular smooth muscle cells, *Eur. J. Pharmacol.* **296**, 327–334.

Shigematsu, K., Nakatani, A., Kawai, K., *et al.* (1996). Two subtypes of endothelin receptors and endothelin peptides are expressed in differential cell types of the rat placenta: in vitro receptor autoradiographic and in situ hybridization studies, *Endocrinology* **137**, 738–748.

Sturgill, T. W., Ray, L. B., Erikson, E., *et al.* (1988). Insulin-stimulated MAP-2 kinase phosphorylates and activates ribosomal protein S6 kinase II, *Nature* **334**, 715–718.

Tsuda, T., Kawahara, Y., Ishida, Y., *et al.* (1992). Angiotensin II stimulates two myelin basic protein/microtubule-associated protein 2 kinases in cultured vascular smooth muscle cells, *Circ. Res.* **71**, 620–630.

Wilson, E., Mai, Q., Sudhir, K., *et al.* (1993). Mechanical strain induces growth of vascular smooth muscle cells via autocrine action of PDGF, *J. Cell. Biol.* **123**, 741–747.

Xu, Q., Liu, Y., Gorospe, M., *et al.* (1996). Acute hypertension activates mitogen-activated protein kinases in arterial wall, *J. Clin. Invest.* **97**, 508–514.

Advances in Neuroregulation and Neuroprotection (2005), pp. 811-822
Collin, C. *et al.* (Eds)
© VSP 2005

Possible involvement of free radicals in hypertensive organ injury

H. ITO *

Department of Pathology, Kinki University School of Medicine, 377-2 Ohno-Higashi, Osaka-Sayama, 589-8511, Japan

Abstract—Hypertension is the major risk factor of various kinds of vascular changes and subsequent organ damage. In any kind of vascular change, endothelial injury, not only in large arteries but also in miscovessels, is the most important initial event; so the elucidation of the pathophysiological changes in endothelial cells in hypertension is undoubtedly important for the prevention and treatment the hypertensive patients. In hypertension, although physical stress such as high blood pressure and shear stress is important for endothelial changes, other factors must be involved besides these. Among several factors such as hypercholesterolemia, hyperglycemia, hyperinsulinemia, smoking, and others, free radicals may be a key factor, since a REDOX system could be affected by the above abnormal conditions. Thus, alteration of free radical generation or scavenging ability are essential events for development of hypertension as well as hypertension complications. In a series of experiments to elucidate the pathogenesis of hypertensive vascular injury and its sequelae, we revealed the important role of free radicals in endothelial injury, not only in the aorta but also in brain microvessels. In addition to genetic abnormalities such as mitochondrial DNA alteration, angiotensin II could be an important causative or, at least, stimulating factor of hypertensive endothelial injury. In other words, preexisting hypertension may act as a stimulating factor for free radical injury in the living body.

Keywords: Hypertension; microvascular injury; free radical; angiotensin II.

INTRODUCTION

As is well known, hypertension is the major risk factor for various kinds of arterial changes and subsequent organ damages such as stroke, myocardial infarction, nephrosclerosis and arteriosclerosis. Such diseases are still increasing as the population of aged people in Japan increases. To prevent or treat such patients is undoubtedly important in public health in modern society. Among the various types of hypertension, over 90% is essential hypertension which may be caused by genetic

---
*E-mail: hiroyuki@med.kindai.ac.jp

factors with association to life style. According to the recent advance in molecular technology, various kinds of genetic analysis are performed to clarify the pathogenesis of essential hypertension, but the gene responsible has not been determined. Of cause, although it is very important to elucidate the genetic abnormalities in human essential hypertension, there are limitations in performing human studies. In practical medicine, the treatment of hypertensive patients is focused on how to prevent cardiovascular events. Among the several animal models of genetic hypertension, spontaneously hypertensive rats (SHR) or stroke-prone SHR (SHRSP) is a suitable animal model for human essential hypertension and its consequences, since all of these rats suffer from severe hypertension (220–250 mmHg) and hypertensive organ damage (see the review by Ito and Suzuki, 1995) without any artificial treatments. Numerous studies have been conducted to elucidate the pathogenesis of hypertensive vascular changes using SHR and SHRSP, but precise mechanisms are still unknown. It has already been shown that membrane abnormality in hypertension is an important factor in the pathogenesis of hypertensive vascular changes (see the review by Ito, 1989). Although several factors could be involved in pathogenesis of hypertensive vascular changes, we are focusing our attention on the role of free radicals, since membrane abnormality may be caused by free radicals via peroxidation of membrane phospholipid, a major constituent of the membrane system of living cells. In this review, we summarize our previous experiments with regard to pathophysiological changes in cardiovascular and cerebrovascular system in SHR and SHRSP.

Cardiac and vascular changes in SHRSP

It is well known that hypertension is a major risk factor of cardiac hypertrophy and subsequent myocardial changes. To understand myocardial changes under such pathological conditions is particularly important in the treatment of hypertensive patients. Using SHR or SHRSP, it has been reported already that myocardial changes after administration of isoprotelenol or doxorubicin were much more intense in the hypertrophied heart than in normotensive Wistar-Kyoto rats (WKY) (Wexler, 1979; Tisne-Versailles, 1985). These results indicate that the SHR hypertrophied heart is myocardial cell-vulnerable when subjected to certain drugs. In order to clarify the pathogenesis of hypertrophied heart myocardial injuries, a number of enzymes were examined in both prehypertensive and hypertensive SHR, and compared with those in WKY (Torii and Ito, 1990). Results showed that at 6 weeks of age, isocitrate dehydrogenase (ICDH) activity in the mitochondrial fraction was higher in SHR than in WKY, but that at 16 weeks of age it was lower. Similar changes were observed in Na/K-ATPase activity in the microsomal fraction, one of the plasma membrane marker enzymes. These results indicate that mitochondria and plasma membrane alterations are important as causative factors in the initiation of SHR hypertrophied heart myocardial changes. To clarify the pathogenesis of such membrane alterations in SHR hypertrophied myocardium, we examined the myocardial characteristics of lipid peroxidation in SHR at the prehypertensive and hypertensive stages, in comparison with those in age-matched WKY (Ito *et al.*, 1992). The activity of glutathion peroxidase (GSH-Px), a well-known radical scavenger, was higher in SHR at 6 weeks of age, but lower at 16 weeks of age compared to that in WKY. The content of alpha-tocopherol, a potent radical scavenger, was lower in SHR at both 6 and 16 weeks of age than in WKY. On the other hand, *in vitro* formation of free malondialdehyde, a lipid peroxidation product, was more pronounced in SHR myocardium than in WKY. Coincidence of lower antiperoxidation ability and higher peroxidation of membrane phospholipid indicate myocardial cell vulnerability in SHR hypertrophied myocardium. To further investigate the membrane lipid peroxidation by free radicals, SHR hypertrophied hearts were examined histologically and biochemically on the first and fourth day after administration of doxorubicin (Torii *et al.*, 1992). Morphological examination of the SHR revealed focal myocytolysis on the first day and severe cardiomyopathy involving diffuse myocytolysis and vacuolar degeneration in the left ventricle on the fourth day. A thiobarbituric acid-reactant substance (TBARS), a lipid peroxidation marker, was significantly higher in treated SHR than in the treated WKY. Furthermore, in comparison with WKY, alpha-tocopherol in the left ventricle in SHR was significantly lower on the fourth day after administration. These results show that a proneness to lipid peroxidation in the membrane system is closely associated with severity of doxorubicin-induced cardiomyopathy in SHR and suggests that membrane lipid peroxidation may cause a higher degree of vulnerability in hypertrophied SHR myocardium. In order to investigate SHR hypertrophied heart myocardial vulnerability, superoxide dismutase (SOD), a potent scavenger of superoxide anion, was examined in myocardium of SHR and WKY in relation to aging, as well as under doxorubicin-induced cardiomyopathy (Ito *et al.*, 1995b). Superoxide anion production due to doxorubicin administration was also examined histochemically. SOD, either as Mn-SOD in mitochondria or as CuZn-SOD in cytoplasm, was found to be lower in aged SHR hypertrophied heart than in age-matched WKY heart. Under doxorubicin-induced cardiomyopathy, SOD activity was significantly lower in SHR hypertrophied heart than either control SHR or treated WKY myocardium. Superoxide anion generation, examined morphologically as formazan deposits, was much more intense in SHR myocardium and some degenerative changes were found in such formazan-containing cells. These results indicate that the myocardial vulnerability of SHR hypertrophied heart might be a result of a lowered free radical scavenging ability.

To further investigate the myocardial vulnerability in SHR hypertrophied heart, we examined the pathophysiological characteristics of mitochondria of myocardial cells using severely hypertensive SHR (SHRSP), because mitochondria is an important constituent of myocardial cells and free radicals are generated in mitochondria in the process of intracellular respiration (Tokoro *et al.*, 1996). The results showed that both ICDH and cytochrome C oxidase (COX), which are related to energy production or the respiratory chain in mitochondria, were significantly lower in SHRSP myocardium than in WKY. Furthermore, superoxide dismutase (SOD), a potent radical scavenger, was also lower in SHRSP myocardium. Restriction fragment length

polymorphism (RFLP) analysis of mitochondrial DNA by Rsa I revealed two deletions in the electrophoretic band in the SHRSP myocardium, but not in the liver.
These findings suggested that mitochondrial dysfunction, especially lower energy
production, could be an important factor for the pathogenesis of further myocardial
degeneration. The results also indicated that mitochondrial alterations, in the membrane system as well as mitochondrial DNA, could be caused by oxidative stress in
mitochondria because of decreased scavenging ability.

On the other hand, although severe vascular changes are major complications
of hypertension, the precise mechanisms of arterial changes are still obscure.
Endothelial changes such as cauliflower-like blebs, microvilli and crater-like holes,
which are frequently found in aged SHRSP aorta, are initial events in arterial
changes and could be related to the membrane changes in endothelial cells. To
investigate whether membrane changes could be induced by free radicals, the aorta
of 13-week-old male SHR and WKY were perfused by *tert*-butyl hydroperoxide
(t-BOOH) and the resulting hydrogen peroxide was morphologically examined
using cerium lanthanide (Torii *et al.*, 1994; Ito *et al.*, 1995a). The spin trap method
revealed a hydroxyl radical adduct at 5 min after t-BOOH perfusion. In t-BOOH-
perfused SHR, cauliflower-like blebs were frequently seen on the endothelial cell
surface, sometimes with prominent microvilli and marginal folds. Fine granular
materials were found clumped or scattered on the surface. Electron-microscopically,
in t-BOOH-perfused SHR, membrane blebs and intracytoplasmic edema were seen
in the endothelium. Electron-dense materials were abundant on the surface of the
intact and the degenerated endothelium. X-ray spectra showed the cerium peaks,
corresponding to the electron-dense material. These findings clearly indicate that
free radical injury, especially hydroxyl radical injury, is much more intense in the
endothelial cells of SHR than in those of WKY.

In conclusion, SHR and SHRSP have genetic defects for radical scavenging ability
and thus are prone to peroxidation of phospholipid and subsequent membrane
dysfunction. In addition to physical stress due to increasing of blood pressure, the
presence of free radicals must be an important causative factor in the pathogenesis
of hypertensive cardiovascular changes.

Microvascular injury in cerebral cortex of SHRSP brain

It is well known that hypertension is the major risk factor for various kinds of vascular injury, not only in large arteries but also in microvessels, and causes lethal
damage in several organs. In the brain, cerebral edema is a typical example of microvascular injury and is an essential event of stroke. Although cerebral edema is
caused with increased vascular permeability due to endothelial injury, the precise
mechanisms are still unknown. Ever since Demopoulos *et al.* (1977) reported the
important role of free radical formation and subsequent lipid peroxidation in ischemic cerebral injury, a number of experimental studies have been carried out on
oxygen free radicals and/or lipid peroxidation in ischemic brain injury (Flamm *et
al.*, 1978; Watson *et al.*, 1984; Abe *et al.*, 1988; Patt *et al.*, 1988). The results,

however, have not always matched. Also, because all of these studies were on acute artificial cerebral ischemia in experimental animals, the pathophysiological changes found in the brain may differ from those in chronic or repeated ischemia with hypertension, such as is seen in human cerebrovascular disorders. SHRSP suffer from stroke lesions, similar to those in humans, and are recognized as the best animal model available for studies related to human cerebrovascular disorders. Although many efforts have been made to elucidate the pathogenesis of stroke lesions in SHRSP, no studies on SHRSP brain free radical injury or cerebral lipid peroxidation have been performed to date. As a basic experiment designed to clarify the pathogenesis of cell death in cerebral ischemia, we examined free radical-induced lipid peroxidation in SHRSP cerebral cortex (Ito *et al.*, 1993). Results showed that *in vivo* formation of TBARS was higher in SHRSP at 20 weeks of age and *in vitro* generation of free malondialdehyde was greater in SHRSP brain, both at 5 and 20 weeks of age, as compared with those in WKY. Furthermore, membrane-associated enzymes such as Na/K-ATPase and 5′-nucleotidease activities were lower in 20-week-old SHRSP than in age-matched WKY. These results indicate how very prone the SHRSP brain is toward lipid peroxidation and subsequent membrane-related enzyme changes. It has already been reported that free radicals, mainly generated by the xanthine-xanthine oxidase system, may have an important role in endothelial damage in experimental studies using ischemia-reperfusion injury models (Patt *et al.*, 1988; Kinuta *et al.*, 1989). These models, however, may not be identical to human cerebral injury. However, the xanthine-xanthine oxidase system is not a major source of free radical generation in SHRSP cerebral cortex, because allopurinol, a well-known inhibitor of xanthine oxidase, did not have beneficial effects on cerebral injury (Maenishi *et al.*, 1997). Because our previous investigation revealed that the cerebral injury was more intensive in the SHRSP treated with allopurinol than in untreated rats. In such SHRSP, tissue uric acid content, a potent radical scavenger, was significantly lower than in non-treated rats as result of inhibition of xanthine degradation. These results indicated that tissue uric acid may act as an important radical scavenger and that the xanthine-xanthine oxidase system may not be a major source of free radical generation in the cortex of SHRSP.

With regard to free radical generation, it was previously reported that polymorphonuclear neutrophils (PMN) are the major source of free radicals via an activation of NADPH oxidase (Wexler *et al.*, 1997) and inducible nitric oxide synthase (iNOS) (Ideacola *et al.*, 1996), and that they play an important role in ischemia-reperfusion injuries in the brain. It is well known that adhesion of leukocytes to endothelial cells is essential to leukocyte migration in the inflammatory process (Seekamp *et al.*, 1998), and that Mac-1 (CD11b/CD18) in leukocytes and intercellular adhesion molecule-1 (ICAM-1) in endothelial cells are upregulated by the inflammatory cytokines, such as tumor necrosis factor-alpha (TNF-alpha) (Kupatt *et al.*, 1999) and interleukin-1 (Stanimirovic *et al.*, 1997). According to previous extensive studies by McCarron *et al.* (1994a, b), adhesion molecule expression and monocyte adhesion to endothelial cells were much more intense in SHR after administration of various

positive cells/HPF

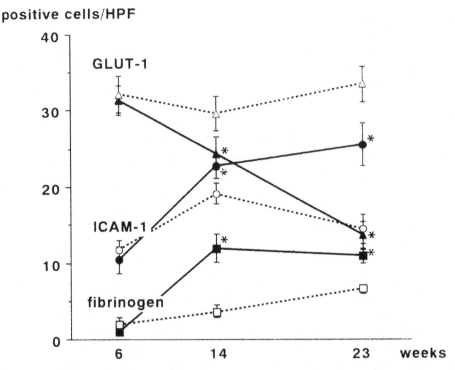

Figure 1. Immunohistochemical expression of ICAM-1, GLUT-1 and fibrinogen in cerebral cortex of SHRSP and WKY. Symbols connected by solid lines (—) and broken lines (- - - -) indicate SHRSP and WKY respectively.

cytokines. Based on these findings, it was speculated that PMN might even be involved in the pathogenesis of microvascular injuries of SHRSP. With regard to the role of leukocytes for vascular injury in SHRSP brain, Tagami and Yamori (1988) first reported that numerous monocytes adhered to the endothelium of the arteries having advanced medial damage, followed by cerebrovascular occlusions. These results clearly demonstrated the important role of leukocytes in the pathogenesis of cerebral injury in SHRSP.

To elucidate the role of PMN in microvascular injury in SHRSP cerebral cortex, we firstly investigated the expression of adhesion molecules in endothelial cells and also in blood-brain-barrier function. Furthermore, the effects of ACE inhibitor were also examined (unpublished data). Using male SHRSP and WKY, the expression of adhesion molecules such as ICAM-1 in endothelial cells was investigated at 6, 14, and 23 weeks of age by immunohistochemical examination and reverse transcription-polymerase chain reaction (RT-PCR) in relation to the expression of glucose transporter-1 (GLUT-1) in endothelial cells and fibrinogen around micro vessels. Angiotensin converting enzyme (ACE) inhibitor was administered to 19-week-old SHRSP for 4 weeks. As shown in Fig. 1, at 6 weeks of age, no significant differences between SHRSP and WKY were found in any of the parameters examined. At 23 weeks of age, brain weight was higher in SHRSP than in WKY, some of the former showing edematous changes. Furthermore, the expression of ICAM-1 in the cerebral cortex was much more intense in SHRSP than in WKY, both in immunohistochemical and RT-PCR examinations. In SHRSP, GLUT-1 expression was less intense, but fibrinogen expression was more intense in the microvessels of the cerebral cortex. Furthermore, in SHRSP, the altered expression of various molecules in endothelial cells were ameliorated by the administration of ACE inhibitor without lowering their blood pressure. These results indicate that the activation of endothelial cells in brain microvessels and possibly the subsequent enhanced adhesion of leukocytes are important initial events in hypertensive cerebral injury. Angiotensin II could be deeply involved in such processes.

With regard to the mechanisms of blood-brain barrier dysfunction by PMN, we hypothesized that free radicals derived from PMN could be an important causative factor. In order to clarify the causative role of cytotoxic NO in hypertensive cerebral injury, the effects of iNOS activation or inhibition on leukocyte and endothelial function were examined using SHRSP (Takemori *et al.*, 2000: unpublished data). For the iNOS activation or inhibition, lipopolysaccharide (LPS) or s-methyisothiourea (SMT) was administered to male SHRSP. Mac-1 expression in leukocytes was examined by adhesion on fibronectin and flow cytometric analysis. ICAM-1 (endothelial adhesion molecule), GLUT-1 (a marker of blood-brain-barrier) and fibrinogen (a marker of vascular permeability) were examined by immunohistochemical staining. The results showed that plasma NO metabolites level was extremely high in LPS group, but significantly lower in SMT group than in the respective control. Leukocyte adhesion to fibronectin as well as Mac-1 expression was greatly enhanced by LPS, but significantly inhibited by SMT. By LPS administration, ICAM-1 expression was enhanced, GLUT-1 was suppressed and fibrinogen was increased in cerebral cortex. In the SMT group, however, these changes were remarkably suppressed and brain weight was significantly lower than in control. These results indicated that iNOS-derived NO, mainly in activated PMN, could be an important causative factor for endothelial injury in hypertensive cerebral injury in SHRSP. Thus, we revealed the activation of leukocytes and enhanced expression of adhesion molecules both in leukocytes and endothelial cells in SHRSP after establishment of severe hypertension.

However, precise mechanisms as to how PMN and endothelial cells were activated in severe hypertension remain unknown. Among several humoral factors that are involved in blood pressure regulation, angiotensin II has multiple effects on vascular endothelial cells (Jeong *et al.*, 1996; Li *et al.*, 1999) and smooth muscle cells (Rajapopalan *et al.*, 1996; Kathy *et al.*, 1997). It has also been reported that plasma renin activity is elevated in SHRSP in accordance with the elevation of blood pressure (Kawashima *et al.*, 1980). Furthermore, since altered expression of several molecules related to endothelial function was ameliorated in SHRSP treated with ACE inhibitor, angiotensin II may be involved in activation of PMN and endothelial cells. To elucidate the possible involvement of angiotensin II in the pathogenesis of microvascular changes in severe hypertension, we investigated the effects of AT_1 receptor antagonist and ACE inhibitor on the expression of adhesion molecules of leukocytes and brain microvessels (Takemori et al., 2000a, b; Ito et al., 2000). Male SHRSP at 19 weeks of age were divided into three groups and age-matched WKY were used as the control group. AT_1 receptor antagonist (TCV-116, 0.5 mg/kg/day) and ACE inhibitor (captopril, 20 mg/kg/day) were administered to SHRSP for 4 weeks. Mac-1 expression in leukocytes was investigated by flow cytometric analysis. For endothelial cells, we examined the expression of ICAM-1, the AT_1 receptor and GLUT-1 using RT-PCR. Results showed that the blood pressure of AT_1 receptor antagonist and ACE inhibitor-treated groups was slightly lower than that of the control, but was still greater than 220 mmHg. Mac-1 expression, as well as ICAM-1 expression, was higher in control SHRSP than in WKY. Such enhanced expression of adhesion molecules in SHRSP was ameliorated by the administration of AT_1 receptor antagonist or ACE inhibitor, the former being more effective. AT_1 receptor expression was higher in control SHRSP than in WKY, and was lower in the AT_1 receptor antagonist group, whereas no difference was found in the ACE inhibitor group. No significant differences were found in GLUT-1 expression among all groups.

To confirm these findings, we further investigated the role of AT_1 receptor in microvascular injury in SHRSP cerebral cortex. To elucidate the mechanisms of activation of leukocytes and adhesion to endothelial cells in hypertensive cerebral injury, expression of AT_1 receptor in leukocytes was examined in relation to that in endothelial cells of brain microvessels using mature SHRSP (Ito et al., 2001). Effects of AT_1 receptor antagonist on leukocytes and endothelial cells were further investigated. To investigate the expression of AT_1 receptor, 23-week-old male SHRSP and age-matched WKY were used. For the effects of AT_1 receptor blockade, AT_1 receptor antagonist was orally administered at a dosage of 0.5 mg/kg/day for 4 weeks from 19 weeks of age. PMN-rich fraction was obtained by density gradient using Ficol-hyparc. AT_1 receptor expression in PMN was investigated by immunohistochemistry and RT-PCR. Mac-1 expression in PMN was examined by Flow cytometry. ICAM-1, GLUT-1 and fibrinogen expression in cerebral cortex (occipital region) were investigated by immunohistochemistry. We clearly demonstrated that AT_1 receptor was identified in PMN both in immunohistochemistry and RT-PCR. AT_1 receptor was also detected in the cerebral cortex. Such expression in both types of cells was much more intense in SHRSP than in WKY. Moreover, AT_1 receptor antagonist ameliorated the enhanced expression of Mac-1 in PMN. Again, it was confirmed that enhanced expression of adhesion molecules and increases of permeability in brain microvessels were decreased by AT_1 receptor antagonist.

As described above, we showed that (1) PMN as well as brain microvessel endothelial cells possess AT_1 receptor; (2) angiotensin II could activate both leukocytes and endothelial cells, followed by adhesion of these cells; (3) NO and related radicals cause endothelial dysfunction and increasing of vascular

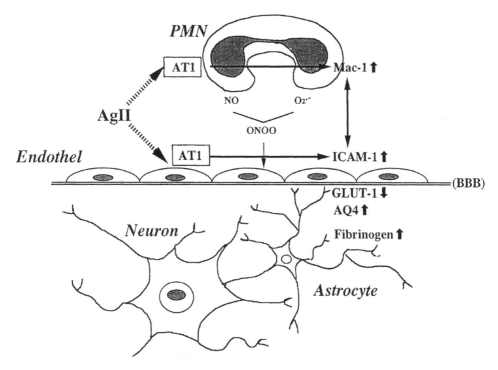

Figure 2. Schema of possible mechanisms of hypertensive microvascular injury in cerebral cortex.

permeability; (4) AT_1 receptor antagonist ameliorates the endothelial injury with pressure-independent manner. Our results suggest that angiotensin II must be a key factor for hypertensive endothelial injury. Figure 2 indicates the possible mechanisms of hypertensive microvascular injury in hypertension.

CONCLUSION

As is well known, hypertension is the major risk factor of various kinds of vascular changes and subsequent organ damage. In any kind of vascular changes, endothelial injury, not only in large arteries but also in microvessels, is the most important initial event and thus to elucidate the pathophysiological changes in endothelial cells in hypertension is undoubtedly important for the prevention of the condition and treatment of hypertensive patients. In hypertension, although physical stress such as high blood pressure and shear stress is important for endothelial changes, other factors must be involved. Among several factors such as hypercholesterolemia, hyperglycemia, hyperinsulinemia, smoking, and others, free radicals may be a key factor, since a REDOX system could be affected by the above abnormal conditions. Thus, alteration of free radical generation or scavenging ability are essential events for development of hypertension as well as hypertension complications. In a series of experiments to elucidate the pathogenesis of hypertensive vascular injury and its sequelae, we revealed the important role of free radicals in endothelial injury, not only in the aorta but also in brain microvessels. In addition to genetic abnormalities, such as mitochondrial DNA alteration, angiotensin II could be an important causative or, at least, stimulating factor in hypertensive endothelial injury. In other words, preexisting hypertension may act as a stimulating factor for free radical injury in the living body.

Acknowledgements

The author would like to express sincere thanks to Miss Kumiko Takemori for her great contribution during experimental work. The author also thanks many colleagues for their assistance in this research. This work was supported in part by a Grant of Aid from the Ministry of Education, Culture, Sports, Science and Technology of Japan.

REFERENCES

Abe, K., Yuki, S. and Kogure, K. (1988). Strong attenuation of ischemic and postischemic brain edema in rats by a novel free radical scavenger, *Stroke* **19**, 480–485.

Demopoulos, H., Flamm, E., Seligman, M., *et al.* (1977). Molecular pathology of lipid in CNS membrane, in: *Oxygen and Physiological Function*, F. F. Joebsis (Ed.), pp. 491–508. Professional Information Library, Dallas, TX, USA.

Flamm, E. S., Demopoulos, H. B., Seligman, H. L., *et al.* (1978). Free radicals in cerebral ischemia, *Stroke* **9**, 445–447.

Iadecola, C., Zhang, F., Casey, R., *et al.* (1996). Inducible nitric oxide synthase gene expression in vascular cells after transient focal cerebral ischemia, *Stroke* **27**, 1373–1380.

Ito, H. (1989). Pathophysiological changes of the membrane system of the arterial smooth muscle cells, in: *Membrane Abnormalities in Hypertension*, C. Y. Kwan (Ed.), Vol. II, pp. 1–21. CRC Press, Inc., Boca Ratón, Florida.

Ito, H. and Suzuki T. (1995). Pathophysiological overview of M-SHRSP, in: *Progress in Hypertension, New Advances in SHR Research. Pathophysiology & Pharmacology*, H. Saito, Y. Yamori, M. Minami and S. H. Parvez (Eds), Vol. 3, pp. 1–19. VSP, Utrecht.

Ito, H., Torii, M. and Suzuki, T. (1992). A comparative study on defense system for lipid peroxidation by free radicals in spontaneously hypertensive and normotensive rat myocardium, *Comp. Biochem. Physiol.* **103B**, 37–40.

Ito, H., Torii, M. and Suzuki, T. (1993). A comparative study on lipid peroxidation in cerebral cortex of stroke-prone spontaneously hypertensive and normotensive rats, *Int. J. Biochem.* **25**, 1801–1805.

Ito, H., Torii M. and Suzuki T. (1995a). Comparative study on free radical injury in the endothelium of SHR and WKY aorta, *Clin. Exp. Pharmacol. Physiol.* **1** (Suppl.), S157–S159.

Ito, H., Torii, M. and Suzuki, T. (1995b). Decreased superoxide dismutase activity and increased superoxide anion production in cardiac hypertrophy of spontaneously hypertensive rats, *Clin. Exp. Hypertens.* **17**, 803–816.

Ito, H., Takemori, K., Kawai, J., *et al.* (2000). AT_1 receptor antagonist prevents brain edema without lowering blood pressure, *Acta Neurochir.* **76** (Suppl.), 141–145.

Ito, H., Takemori, K. and Suzuki, T. (2001). Role of angiotensin II type 1 receptor in the leukocytes and endothelial cells of brain microvessels in the pathogenesis of hypertensive cerebral injury, *J. Hypertens.* **19**, 591–597.

Jeong, A. K., Judith, A. B. and Jerry, L. (1996). Angiotensin II increases monocyte binding to endothelial cells, *Biochem. Biophys. Res. Commun.* **226**, 862–868.

Kathy, K. G., Ushino-Fukai, M., Bernard, L. R., *et al.* (1997). Angiotensin II signaling in vascular smooth muscle, *Hypertension* **29**, 366–373.

Kawashima, K., Shiono, K. and Sokabe, H. (1980). Variation of plasma and kidney renin activities among substrains of spontaneously hypertensive rats, *Clin. Exp. Hypertens.* **2**, 229–245.

Kinuta, Y., Kikuchi, H., Ishikawa, M., *et al.* (1989). Lipid peroxidation in focal cerebral ischemia, *J. Neurosurg.* **71**, 421–429.

Komlos, M. and Seregi, A. (1975). Lipid peroxidation as the cause of the ascorbic acid induced decrease of adenosine triphosphate activities of rat brain microsomes and its inhibition by biogenic amines and psychotrophic drugs, *Biochem. Pharmacol.* **24**, 1781–1786.

Kupatt, C., Habazettl, H., Goedecke, A., *et al.* (1999). Tumor necrosis factor-alpha contributes to ischemia and reperfusion-induced endothelial activation in isolated hearts, *Circ. Res.* **84**, 392–400.

Li, D. Y., Zhang, Y. C., Philips, M. I., *et al.* (1999). Upregulation of endothelial receptor for oxidized low-density lipoprotein (LOX-1) in cultured human coronary artery endothelial cells by angiotensin II type 1 receptor activation, *Circ. Res.* **84**, 1043–1049.

McCarron, R. M., Wang, L., Siren, A. L., *et al.* (1994a). Monocyte adhesion to cerebromicrovascular endothelial cells derived from hypertensive and normotensive rats, *Am. J. Physiol.* **267**, H2491–H2497.

McCarron, R. M., Wang, L., Siren, A. L., *et al.* (1994b). Adhesion molecules on normotensive and hypertensive rat brain endothelial cells, *PSEBM* **205**, 257–262.

Maenishi, O., Ito, H. and Suzuki, T. (1997). Acceleration of hypertensive cerebral injury by the inhibition of xanthine-xanthine oxidase system in stroke-prone spontaneously hypertensive rats, *Clin. Exp. Hypertens.* **19**, 461–477.

Patt, A., Harken, A. H., Burton. L. K., *et al.* (1985). Cardiotoxicity of high dose of isoproterenol on cardiac hemodynamics and metabolism in SHR and WKY rats, *Arch. Int. Pharmacodyn.* **273**, 142–154.

Patt, A., Harken, A.H., Burton, L. K., *et al.* (1988). Xanthine oxidase-derived hydrogen peroxide contributes to ischemia reperfusion-induced edema in gerbil brains, *J. Clin. Invest.* **18**, 1556–1562.

Rajagopalan, S., Kurz, S., Münzel, T., *et al.* (1996). Angiotensin II-mediated hypertension in the rat increase vascular superoxide production via membrane NADH/NADPH oxidase activation, *J. Clin. Invest.* **97**, 1916–1923.

Seekamp, A., Jochum, M., Ziegler, M., *et al.* (1998). Cytokines and adhesion molecules on elective and accidental trauma-related ischemia/reperfusion, *J. Trauma* **44**, 874–882.

Stanimirovic, D., Shapiro, A., Wong, J., *et al.* (1997). The induction of ICAM-1 in human cerebromicrovascular endothelial cells (HCEC) by ischemia-like conditions promotes enhanced neutrophil/HCEC adhesion, *J. Neuroimmunol.* **76**, 193–205.

Tagami, M. and Yamori, Y. (1988). Morphological analysis of the pathogenesis of hypertensive cerebrovascular lesions: role of monocytes and platelets in intracerebral vessel occulusons, *Jpn. Circ. J.* **52**, 1351–1356.

Takemori, K., Ito, H. and Suzuki, T. (2000a). Effects of inducible nitric oxide synthase inhibition on cerebral edema in severe hypertension, *Acta Neurochir. (Suppl.)* **76**, 335–338.

Takemori, K., Ito, H. and Suzuki, T. (2000b). Effects of the AT_1 receptor antagonist on adhesion molecule expression in leukocytes and brain microvessels of stroke-prone spontaneously hypertensive rats, *Am. J. Hypertens.* **13**, 1233–1241.

Tokoro, T., Ito, H. and Suzuki, T. (1996). Alterations in mitochondrial DNA and enzyme activities in hypertrophied myocardium of stroke-prone SHRs, *Clin. Exp. Hypertens.* **18**, 595–606.

Torii, M. and Ito, H. (1990). Some enzyme characteristics of spontaneously hypertensive rats myocardium, *Jpn. Circ. J.* **54**, 688–694.

Torii, M., Ito, H. and Suzuki, T. (1992). Lipid peroxidation and myocardial vulnerability in hypertrophied SHR myocardium, *Exp. Mol. Pathol.* **57**, 29–38.

Torii, M., Inoue, S., Matsushima, S., *et al.* (1994). Free radical injury in endothelial cells of spontaneously hypertensive rat aorta, *J. Toxicol. Pathol.* **7**, 363–370.

Walder, C. E., Green, S. P., Darbonne, W. C., *et al.* (1997). Ischemic stroke injury is reduced in mice lacking a functional NADPH oxidase, *Stroke* **28**, 2252–2258.

Watson, B. D., Delivoria-Papadopoulos, M., Cahillane G., *et al.* (1990). Lipid peroxidation as the mechanism of modification of brain 5'-nucleotidase activity in vitro, *Neurochem. Res.* **15**, 237–242.

Wexler, B. C. (1979). Isoprenaline-induced myocardial infarction in spontaneously hypertensive rats, *Cardiovasc. Res.* **13**, 450–458.

Advances in Neuroregulation and Neuroprotection (2005), pp. 823-838
Collin, C. *et al.* (Eds)
© VSP 2005

Potential roles of interstitial fluid adenosine in the regulation of renal hemodynamics

AKIRA NISHIYAMA* and YOUICHI ABE

Department of Pharmacology, Kagawa Medical University, 1750-1 Ikenobe, Miki-cho, Kita-gun, Kagawa 761-0793, Japan

Abstract—Adenosine exerts membrane receptor-mediated effects on renal hemodynamics and function. Endogenously produced adenosine is proposed to be formed primarily by renal tubular epithelial cells and approaches the vascular smooth cells of the renal vasculature from the extracellular space, i.e. the interstitium. Hence, it is important to determine the dynamics of adenosine levels in renal interstitial fluid. Indeed, many studies have indicated that the effects of adenosine on renal hemodynamics depend on its concentration in the interstitium. Furthermore, *in vitro* studies reveal that adenosine in the renal interstitial fluid binds to membrane receptors from the adventitial side and initiates renal microvascular responses. Recently, we developed a renal microdialysis method to monitor the concentration of adenosine in renal interstitial fluid. This method is well suited for *in vivo* studies investigating alterations in renal interstitial adenosine levels. In this review, we summarize the literature pertaining to alterations in renal interstitial fluid concentrations of adenosine in some pathophysiological conditions. Potential roles of interstitial adenosine for regulating renal hemodynamics were also discussed.

Keywords: Adenosine; adenosine receptors; renal interstitial fluid; renal hemodynamics; renal microdialysis method.

INTRODUCTION

Adenosine is a byproduct of normal ATP hydrolysis or cellular energy-dependent processes. Adenosine secretion into the extracellular space involves facilitated diffusion via a nucleoside transporter that operates in a concentration dependent manner (Navar *et al.*, 1996; Jackson and Dubey, 2001). Adenosine can also be formed by conversion of extracellular ATP to adenosine via the ecto-nucleotidase cascade (Le Hir and Kaissling, 1993; Inscho *et al.*, 1994), an enzyme abundantly found in the brush border of proximal tubules or afferent and efferent arterioles (Dawson

---

*To whom correspondence should be addressed. E-mail: akira@kms.ac.jp

et al., 1989; Blanco *et al.*, 1993; Le Hir and Kaissling, 1993). Another biosynthetic route for adenosine involves the hydrolysis of S-adenosyl-L-homocysteine to L-homocysteine and adenosine (Lloyd *et al.*, 1988; Jackson and Dubey, 2001). These multiple pathways exist to produce appropriate extracellular levels of adenosine in the kidney (for reviews, see Refs Navar *et al.*, 1996; Jackson and Dubey, 2001).

Until recently, assessments of renal interstitial levels of adenosine have been limited to analyses of renal vein plasma, urine or renal lymph (Navar *et al.*, 1996). Data obtained from these indirect approaches are complex because of the ubiquitous uptake and metabolic pathways of adenosine. To more accurately determine the dynamics of adenosine levels in renal interstitial fluid, we developed an *in vivo* renal microdialysis method (Nishiyama *et al.*, 1999a–c, 2000, 2001b, c, 2002a–c). This novel technique allows us to collect samples directly from interstitial fluid and has many advantages over traditional measurements conducted in urine or lymph fluid. In particular, the molecular weight cut-off of the microdialysis membrane can function as a barrier separating small and large molecules, helping to exclude undesirable substances including degrading enzymes and carrier proteins (Nishiyama *et al.*, 1999c, 2002b, c). In this manuscript, we shall review some recent findings related to renal interstitial fluid adenosine with special emphasis on its hemodynamic roles and concentration changes under various pathophysiological conditions.

ADENOSINE RECEPTORS IN RENAL VASCULATURE

Measurements of total renal blood flow (RBF) in the whole kidney have demonstrated an effect of exogenously administered adenosine on RBF characterized by an initial transient renal vasoconstriction that wanes and becomes supplanted by a gradual vasodilation (Aki *et al.*, 1990, 1997, 2002). The mechanism of this biphasic response remains poorly understood, although it is presumed to involve the differential binding of adenosine to A_1 and A_2 receptors to evoke renal vasoconstriction and vasodilation, respectively (Spielman and Thompson, 1982; Nishiyama *et al.*, 2001a). Activation of adenosine A_1 receptors leads to inhibition of adenylate cyclase through a Gi protein mechanism and a concurrent elevation in cytosolic calcium (for reviews, see Refs Navar *et al.*, 1996; Jackson and Dubey, 2001). Adenosine A_2 receptors are subdivided into distinct A_{2a} and A_{2b} subtypes that are shown to be expressed in the kidney (Morton *et al.*, 1998; Zou *et al.*, 1999b). A_2 receptors stimulate adenylate cyclase activity through activation of a Gs protein, whereas adenosine A_3 receptors are not coupled to adenylate cyclase (for reviews, see Refs Navar *et al.*, 1996; Jackson and Dubey, 2001). Preliminary experiments by Jackson and Dubey (2001) showed that adenosine A_3 receptors are expressed in preglomerular microvessels; however, a role for adenosine A_3 receptors in the renal vascular function remains to be determined.

EFFECTS OF ALTERING INTESTINAL OR PERIVASCULAR ADENOSINE

Effects of superfusion with adenosine on renal microvascular function were examined by using an *in vitro* blood-perfused juxtamedullary nephron preparation (Carmines and Inscho, 1994; Nishiyama *et al.*, 2001a). Importantly, superfusing directly over the adventitial side of renal vasculature permits investigation of the actions of adenosine from the interstitial side as would occur if adenosine is serving as a paracrine agent. Superfusion with 10 μmol/l adenosine evokes afferent and efferent arteriolar vasoconstriction, whereas with higher concentrations of adenosine the vasoconstrictor effect is attenuated (Carmines and Inscho, 1994; Nishiyama *et al.*, 2001a). Similar results were observed in outer medullary descending vasa recta (Silldorff *et al.*, 1996). Further studies showed that adenosine elicits afferent and efferent arteriolar vasodilation during adenosine A_1 receptor blockade with 8-noradamantan-3-yl-1,3-dipropylxanthine (KW-3902) (Nishiyama *et al.*, 2001a). On the other hand, the adenosine-induced vasoconstrictor responses are markedly enhanced during adenosine A_{2a} receptor blockade with 1,3-dipropyl-7-methyl-8-(3,4-dimethoxystyryl) xanthine (KF17837). These data suggest that interstitially administered adenosine can influence renal microvascular tone via adenosine A_1 and A_{2a} receptors. Adenosine A_{2b} receptors are also suggested to be expressed in isolated preglomerular microvessels (Jackson and Dubey, 2001); however, their functional role in renal microvascular tone has not yet been investigated.

Early *in vivo* studies showed that RBF and glomerular filtration (GFR) rate were significantly decreased by the administration of dipyridamole, which inhibits cellular uptake of endogenous adenosine and thereby elevates interstitial adenosine levels (Arend *et al.*, 1985; Aki *et al.*, 2002). Studies using chronically implanted capsules to infuse drugs directly into the renal interstitium showed that interstitial administration of adenosine significantly decreased GFR, and that nonselective adenosine receptor antagonist theophylline completely blocked the effect of adenosine (Pawlowska *et al.*, 1987). Delivery of an adenosine A_1 receptor agonist, N^6-cyclopentyladenosine, into the renal interstitium at the corticomedullary junction significantly decreased cortical and medullary blood flow (Agmon *et al.*, 1993). It was also observed that interstitial administration of an adenosine A_2 receptor agonist 2-[p-(carboxyethyl) phenethylamino]-5'-N-ethylcarboxamido-adenosine significantly increased medullary blood flow (Agmon *et al.*, 1993). Further studies by Zou *et al.* (1999a) showed that renal medullary interstitial infusion of a selective adenosine A_2 receptor antagonist, 3,7-dimethyl-1-propagylxanthine, significantly decreased medullary blood flow. Collectively, these observations strongly support the concept that the effects of adenosine on renal hemodynamics depend on its concentration in the interstitium.

PHYSIOLOGICAL ROLES OF INTERSTITIAL ADENOSINE IN THE REGULATION OF RENAL HEMODYNAMICS

Possible physiological role of interstitial adenosine in renal microvascular tone

As mentioned above, both *in vitro* and *in vivo* studies reveal that exogenously administered adenosine causes an initial transient renal vasoconstriction that wanes and becomes supplanted by a gradual vasodilation (Aki *et al.*, 1990, 1997, 2002; Nishiyama *et al.*, 2001a). However, the contribution of endogenous interstitial adenosine to the basal renal vascular tone is unclear. Studies using juxtamedullary preparation showed that a significant vasoconstriction to interstitial administration of adenosine occurs at 10 μmol/l (Carmines and Inscho, 1994; Nishiyama *et al.*, 2001a). The data suggest the possibility that endogenous adenosine concentrations in renal interstitium (below a 1 μmol/l range) are lower than appears necessary for significant adenosine-mediated renal vasoconstriction under resting conditions. In support of this possibility, interstitial administration of adenosine A_1 or A_{2a} receptor antagonists did not alter basal renal microvascular diameters (Nishiyama *et al.*, 2001a). *In vivo* studies also showed that intrarenal arterial infusion of KW-3902, which prevents the vasoconstrictor influence of exogenous adenosine, did not alter basal RBF or GFR (Aki *et al.*, 1997, 2002). Thus, endogenous levels of renal interstitial fluid adenosine may not be sufficiently high to exert steady-state influences; however, they may be elevated to levels sufficient to elicit sustained effects by stimuli that increase the production and release of adenosine. In this regard, recent studies by Aki *et al.* (2002) have shown that during arterial infusion of dipyridamole, renal vasoconstriction is associated with increases in renal interstitial concentrations of adenosine (Fig. 1A).

Possible role of interstitial adenosine in renal vascular responsiveness of angiotensin II

Substantial experimental evidence supports the existence of a synergistic interaction between adenosine and the renin-angiotensin system for regulating renal hemo-dynamics (for review, see Navar *et al.*, 1996). In particular, several investigators have suggested that renal interstitial adenosine levels are an important determinant of vascular responsiveness of angiotensin II (Hall *et al.*, 1985; Carmines and Inscho, 1994; Weihprecht *et al.*, 1994; Navar *et al.*, 1996; Aki *et al.*, 2002). Aki *et al.* (2002) showed that the renal vasoconstrictor actions of both angiotensin II and norepinephrine were significantly augmented with increasing renal inter-stitial levels of adenosine induced by dipyridamole (Fig. 1B). Furthermore, the blockade of adenosine A_1 receptors by KW-3902 completely prevented the effects of dipyridamole on angiotensin II or norepinephrine-induced renal vasoconstric-tion (Fig. 1B). These data support the hypothesis that endogenous adenosine lev-els can influence angiotensin II and norepinephrine-mediated renal vasoconstric-tion via adenosine A_1 receptors. The authors also found that renal interstitial fluid concentrations of adenosine were not changed during infusion of angiotensin II

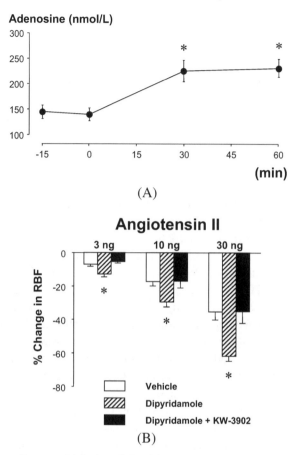

Figure 1. A: Effects of intrarenal infusion of dipyridamole (10 μg/kg/min) on renal interstitial concentrations of adenosine ($n = 7$). $^*p < 0.05$ vs. 0 min. Renal interstitial concentrations of adenosine were calculated by the equilibrium rate. B: Renal blood flow (RBF) responses to intrarenally administered angiotensin II during intrarenal infusion of vehicle, dipyridamole (10 μg/kg/min) or dipyridamole + 8-noradamantan-3-yl-1,3-dipropylxanthine (KW-3902: 10 μg/kg/min). $^*p < 0.05$ vs. vehicle, $n = 6$ [data from reference Aki *et al.*, 2002].

or norepinephrine. Thus, these data suggest that the responses of adenosine to angiotensin II and norepinephrine-induced renal vasoconstriction are not mediated through *de novo* intrarenal adenosine accumulation due to angiotensin II or norepinephrine-induced renal vasoconstriction. It has been suggested that synergistic interactions between adenosine and angiotensin II are mediated through crosstalk between adenosine A_1 receptors and receptors coupled to phospholipase C (Ardaillou *et al.*, 1992; Dickenson and Hill, 1994). Further studies have demonstrated that adenosine increases cytosolic free calcium concentrations along the entire length of the afferent arteriole from rabbit kidney (Gutierrez *et al.*, 1999). Clearly, further studies are needed to determine the intracellular interactions between adenosine and angiotensin II or norepinephrine.

Possible roles of adenosine in the tubuloglomerular feedback (TGF) mechanism and renal autoregulation

The ability of adenosine to inhibit renin release and the data indicating that adenosine might be derived from tubular cells including macula densa cells have led several investigators to suggest that adenosine may serve as the mediator of the TGF mechanism (for reviews, see Inscho *et al.*, 1994; Navar *et al.*, 1996; Schnermann, 2002). Indeed, several micropuncture studies (Franco *et al.*, 1989; Schnermann *et al.*, 1990) demonstrate that local administration of high doses of adenosine receptor antagonists reduces the magnitude of TGF-mediated reductions in stop-flow pressure and the single nephron filtration rate in response to increases in the distal nephron perfusion rate. Furthermore, these TGF responses were not observed in adenosine A_1 receptor deficient mice (Brown *et al.*, 2001; Sun *et al.*, 2001). However, this hypothesis has remained controversial because systemic administration of adenosine antagonists does not block RBF and GFR autoregulation (Premen *et al.*, 1985; Ibarrola *et al.*, 1991; Nishiyama and Navar, 2002). Because the TGF mechanism participates in the autoregulatory responses of the arteriolar vasculature to changes in perfusion pressure, it follows that the mediator of the TGF mechanism would also contribute to the changes in renal vascular resistance (RVR) associated with autoregulatory responses (Navar *et al.*, 1996; Nishiyama *et al.*, 2000). Furthermore, although the TGF mediator must exert selective actions on preglomerular arterioles (Inscho *et al.*, 1994; Navar *et al.*, 1996), several studies have demonstrated that adenosine or adenosine A_1 receptor agonists elicit significant constriction of efferent arterioles (Carmines and Inscho, 1994; Nishiyama *et al.*, 2001a) and vasa recta (Silldorff *et al.*, 1996). It is also recognized that TGF-mediated changes in afferent arteriolar resistance are sustained for long period (Navar *et al.*, 1996). In contrast, adenosine elicits transient vasoconstriction in the kidney that abates within a few minutes spontaneously (Aki *et al.*, 1990, 1997, 2002). Thus, adenosine as the predominant TGF mediator would be in a precarious role since its effects would wane as the TGF signal intensity increased.

The single most important criterion distinguishing between the mediator and modulators is that there should be a direct relationship between the change in the macula densa stimulus and changes in the release or concentration of the TGF mediator associated with changes in RVR (Navar *et al.*, 1996; Nishiyama *et al.*, 2000, 2001c; Nishiyama and Navar, 2002). In our recent studies evaluating renal interstitial concentrations of adenosine and ATP, we demonstrated that renal interstitial adenosine concentrations remained stable within the autoregulatory range, and did not show any significant relationship with either the autoregulatory or TGF related changes in RVR (Nishiyama *et al.*, 2000, 2001c). These data provide no support to the hypothesis that renal interstitial adenosine serves as a mediator of either the autoregulatory mechanism or the TGF response. In contrast, ATP clearly demonstrated such a relationship as discussed in other essays (Nishiyama and Navar, 2002). It should be noted that, although ATP can be metabolized to ADP,

AMP and adenosine (Le Hir and Kaissling, 1993; Inscho *et al.*, 1994), complete and immediate hydrolysis of all available ATP would still not yield sufficiently high levels of these substances to cause comparable vasoconstriction (Inscho *et al.*, 1994; Navar *et al.*, 1996). Thus, the collective data obtained using various approaches support the concept that adenosine serves as an important modulator rather than as the mediator of the TGF mechanism.

RENAL INTERSTITIAL CONCENTRATION OF ADENOSINE

Characteristics of a microdialysis probe

Microdialysis techniques have been applied for several tissues, but have not been widely used in the kidney because of the tissue injury induced by the insertion of the microdialysis probe. To minimize renal tissue injury, we developed a fiber type probe with a thinner diameter (Nishiyama *et al.*, 1999a–c, 2000, 2001b, c, 2002a–c). As shown in Fig. 2, the outer diameter of this probe is only 220–250 μm, which is one-third the diameter of commercially available probes. In addition, the length of the dialysis membrane is 1.0–1.5 cm, which is 2–4 times longer than that of a regular probe. Since the dialysis efficiency of this new probe was better than that of a regular probe, we can perfuse our probe at a high perfusion rate (3–10 μl/min) and shorten the sampling time (5–15 min). For determination of renal interstitial concentrations of adenosine, inosine and hypoxanthine, we used a microdialysis membrane made from cuprophan fiber with a 5500 Da transmembrane diffusion cut-off (Toyobo Co., Otsu, Japan). Thin stainless steel tubes (outer diameter: 190 μm, inner diameter: 100 μm) were inserted into both sides of the cuprophan fiber

Microdialysis Probe

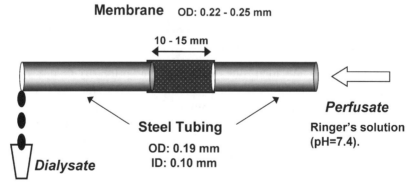

Figure 2. Schematic illustration of a microdialysis probe. The dialysis membrane is made from cuprophan fiber with a 5500 Da transmembrane diffusion cut-off (Toyobo Co., Ltd., Otsu, Japan). Thin stainless steel tubes (outer diameter: 190 μm, inner diameter: 100 μm) were inserted into both sides of the cuprophan fiber.

(Fig. 2). The inflow and outflow ends were inserted into the polyethylene tubes (PE-10) and sealed in place with glue. Negligible amounts of adenosine, inosine and hypoxanthine stick to the polyethylene tubes. The probes were connected to a CMA/100 microinfusion pump (Carnergie Medicine, Stockholm, Sweden) and perfused continuously at a rate of 3–10 μl/min. At a perfusion rate of 10 μl/min, the relative equilibrium rates of adenosine, inosine and hypoxanthine were $16 \pm 2\%$, $17 \pm 2\%$ and $25 \pm 3\%$, respectively (Nishiyama *et al.*, 1999a, 2001b). Renal interstitial concentrations of adenosine, inosine and hypoxanthine were estimated at each of the equilibrium rates.

Renal interstitial fluid concentrations of adenosine in physiological conditions

Several microdialysis experiments performed in rats indicate that resting renal interstitial adenosine, calculated from the relative equilibrium rates, normally remain in the 0.1 to 0.3 μmol/l range (Baranowski and Westenfelder, 1994; Siragy and Linden, 1996; Zou *et al.*, 1999a). Similar concentrations of adenosine in renal cortical interstitial fluid were observed in dog (Nishiyama *et al.*, 1999a, 2000, 2001b, c) and rabbit (Nishiyama *et al.*, 1999b). Basal adenosine effusion from the renal cortex exceeded that from the medulla in anesthetized rabbits (Nishiyama *et al.*, 1999b). However, the renal interstitial concentration of adenosine is higher in the renal medulla than that in the cortex in rats (Siragy and Linden, 1996; Zou *et al.*, 1999a). The reason for this discrepancy is not clear; however, it could be due to differences in species and experimental conditions. For example, it has been reported that during ischemia, the AMP concentration in the cortex is the highest in the rabbit kidney, whereas the AMP concentration in the outer medulla is the highest in the rat kidney (Zager *et al.*, 1990). Under resting conditions, renal cortical interstitial concentrations of other adenosine metabolites such as inosine and hypothanthine are in similar ranges to adenosine in both dogs (Nishiyama *et al.*, 2001b) and rats (Ekland *et al.*, 1991; Baranowski and Westenfelder, 1994).

Effects of inhibition of adenosine kinase and adenosine deaminase on renal interstitial fluid concentrations of adenosine

Both adenosine kinase and adenosine deaminase may primarily be intracellular enzymes (Spielman and Thompson, 1982), although recent studies have shown the activity of extracellular adenosine deaminase in renal proximal tubules (Blanco *et al.*, 1993). Renal cortical interstitial infusion of an adenosine kinase inhibitor, iodotubercidin, or an adenosine deaminase inhibitor, erythro-9-(2-hydroxy-3-nonyl)adenine (EHNA), by including these drugs in the perfusate, did not alter basal adenosine concentrations in renal interstitial fluid (Nishiyama *et al.*, 2001b). However, treatment of iodotubercidin plus EHNA significantly increased adenosine levels by approximately 2-fold (Nishiyama *et al.*, 2001b). These observations suggest that both adenosine kinase and adenosine deaminase have a sufficiently high enzyme activity to maintain the renal interstitial concentration of adenosine, despite the fact that one metabolic pathway was blocked. In support of this possibility, studies measuring adenosine kinase and adenosine deaminase activities have demonstrated the wide distribution of these enzymes in the renal cortex (Pawelczyk *et al.*, 1992). It was also observed that iodotubercidin did not alter renal interstitial concentrations of inosine and hypoxanthine, whereas EHNA significantly decreased these levels (Nishiyama *et al.*, 2001b). These data are consistent with the results from previous microdialysis studies performed in the heart (Manthei *et al.*, 1998), in which iodotubercidin present in perfusates was shown not to significantly alter pre-ischemic dialysate purine metabolite levels, whereas EHNA significantly decreases inosine and hypoxanthine levels. Nevertheless, it is still not clear why iodotubercidin had no effect on inosine and hypoxanthine levels under resting conditions. Other adenosine metabolic pathways cannot be ruled out and need further examination.

Renal interstitial fluid concentrations of adenosine in the ischemic kidneys

Increased extracellular adenosine, produced as a consequence of the energy-deficit state during ischemia and hypoxia, has been suggested as a regulator of renal hemodynamics in post-ischemia. Lin *et al.* (1986, 1988) reported that theophylline prevents ischemia-induced reductions in RBF and GFR during the initiation or maintenance phases of post-ischemic acute renal failure. It has also been shown that dipyridamole aggravated hemodynamic changes in post-ischemic acute renal failure, and that theophylline reverses the effects of dipyridamole (Lin *et al.*, 1987).

Early studies in dogs (Miller *et al.*, 1978) and rats (Osswald *et al.*, 1977) showed that tissue levels of adenosine increase only several-fold during global renal ischemia of 10–30 min duration. As shown in Fig. 3, our recent studies provide evidence that the elevation of renal interstitial adenosine levels induced by ischemia also remained only 3–4 fold (Nishiyama *et al.*, 2001b). During ischemia, the net increases of renal interstitial concentrations of adenosine, inosine and hypoxanthine were calculated to be 0.60, 1.82 and 21.5 μmol/l, respectively, which were not affected by treatment with iodotubercidin on its own. After treatment with iodotubercidin plus EHNA, net increases of adenosine, inosine and hypoxanthine during ischemia were 15.4, 0.70 and 9.7 μmol/l, respectively. These were similar to the increases with just EHNA. As shown in Fig. 3, the net increase of adenosine was significantly augmented by treatment with EHNA, while the net increases of inosine and hypoxanthine were inversely diminished (Nishiyama *et al.*, 2001b). These data suggest that the interstitial adenosine accumulated during ischemia was mainly degraded by adenosine deaminase, and that the rephosphorylation of adenosine via adenosine kinase is very small (Fig. 4). Pawelczyk *et al.* (1992) have demonstrated that under resting conditions, the activity of adenosine deaminase is much higher than that of adenosine kinase in glomeruli and cortical tubules. Thus, it is possible that in the renal cortex, the activity of adenosine deaminase is much more abundant basally than that of adenosine kinase. Another possibility is that ischemic conditions could change

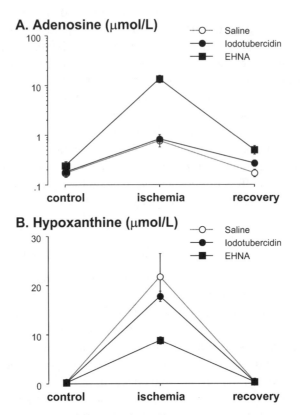

Figure 3. Changes in renal interstitial concentrations of adenosine (A) and hypoxanthine (B) during ischemia and recirculation in dogs. Microdialysis probes were perfused with the following solutions: (1) saline solution, (2) saline solution containing iodotubercidin (10 μmol/l), (3) saline solution containing erythro-9-(2-hydroxy-3-nonyl)adenine (EHNA: 100 μmol/l). Renal interstitial concentrations of adenosine and hypoxanthine were estimated at each equilibrium rate. *$p < 0.05$ *vs.* control, $n = 7$, respectively [data from reference Nishiyama *et al.*, 2001b].

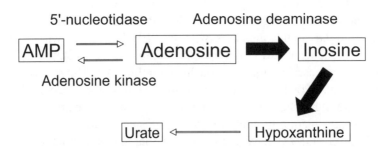

Figure 4. Schematic representation of adenosine metabolism. Renal interstitial adenosine accumulated during ischemia is mainly degraded by adenosine deaminase and the rephosphorylation of adenosine via adenosine kinase is very small.

the activities of adenosine kinase and/or adenosine deaminase. Studies performed in rats subjected to experimental myocardial infarction have demonstrated that adenosine deaminase activity is significantly increased in infarcted tissue (Saleem *et al.*, 1982). It has also been shown that global ischemia for 10 min significantly elevates adenosine deaminase activity in the rat heart (Choong *et al.*, 1987). In contrast, hypoxia decreases adenosine kinase activity in perfused guinea pig heart (Decking *et al.*, 1997). These observations suggest the possibility that under ischemic conditions, the activity of adenosine deaminase is elevated but adenosine kinase activity is reduced in the kidney.

Changes in renal interstitial fluid concentrations of adenosine in response to salt loading

It has been suggested that renal adenosine participates in the regulation of sodium excretion (Spielman and Thompson, 1982; Navar *et al.*, 1996). Therefore, the possibility exists that salt intake alters the adenosine metabolism for it to adapt to a higher salt loading. Early studies showed arterial, renal venous, and urinary adenosine levels were significantly higher in chronically sodium-loaded than in sodium-deprived dogs (Arend *et al.*, 1986). Similar results were observed for adenosine contents in renal rat tissues (Zou *et al.*, 1999b). Siragy *et al.* (1996) performed renal microdialysis studies in rats maintained on low-, normal, and high-salt diets for 5 days. The authors found that rats consuming a high salt diet had renal cortical and medullary dialysate adenosine concentrations that were both increased by 18-fold, compared with levels in rats on a low sodium diet. Based on these findings, they suggest that renal interstitial adenosine plays an important role in adaptation in the kidney to promote sodium excretion. To determine the mechanisms by which a high salt intake increases renal adenosine levels, Zou *et al.* (1999b) analyzed the enzyme activity of 5′-nucleotidase and adenosine deaminase. However, the authors found that neither activity was changed during chronic high salt-loading, indicating that increases in adenosine concentrations induced by salt loading are not associated with elevation of or inhibition of these enzyme activities. The authors also investigated the alterations in the expression of adenosine receptors and found that renal cortical and medullary adenosine A_1 receptors were downregulated in rats fed a high salt diet, compared with rats receiving a normal salt diet (Zou *et al.*, 1999b). However, adenosine A_{2a} and A_{2b} receptors were not changed, but adenosine A_3 receptors were substantially upregulated. Such observations are consistent with the concept that both increased adenosine production and altered adenosine receptors expression are important adaptive mechanisms against chronic salt loading; but the specific contributions of these alterations to salt-induced renal hemodynamic changes remain to be elucidated.

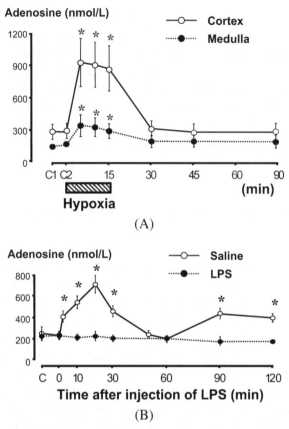

Figure 5. (A) Changes in renal interstitial concentrations of adenosine during hypoxemia in rabbits. Normocapnic systemic hypoxia resulted in severe renal vasoconstriction along with significant increases in renal interstitial concentrations of adenosine in both cortex and medulla. C; control. $^*p < 0.05$ vs. C, $n = 11$ [from Nishiyama et al., 1999b]. (B) Changes in renal interstitial concentrations of adenosine following the intravenous injection of either isotonic saline ($n = 5$) or lipopolysaccharide (LPS, $n = 7$) in dogs. Intravenous administration of LPS resulted in decreases in blood pressure and RBF, and increases in renal interstitial concentrations of adenosine. C; control, $^*p < 0.05$ vs. C [from Nishiyama et al., 1999a].

Renal interstitial fluid concentrations of adenosine during acute renal failure

Adenosine has been shown to play a critical role in the regulation of renal hemo-dynamic changes via adenosine A_1 receptors during various forms of acute renal failure (Churchill and Bidani, 1983). Administrations of adenosine A_1 receptor an-tagonists can attenuate renal injury induced by treatment with cisplatin (Knight et al., 1991) or glycerol (Bidani and Churchill, 1983) and during ischemia (Lin et al., 1986). However, there are few reports investigating how renal interstitial adenosine levels are altered during the progression of acute renal failure. We investigated the changes in renal interstitial adenosine levels during hypoxia in rabbits (Fig. 5A). We observed that normocapnic systemic hypoxia resulted in severe renal vasoconstriction along with significant increases in renal interstitial adenosine levels in both cortex and medulla (Nishiyama *et al.*, 1999b). During adenosine A_1 receptor blockade with KW-3902, hypoxemia caused similar increases in renal interstitial adenosine concentrations, but hypoxia-induced renal vasoconstriction was markedly attenuated by treatment with KW-3902 (Nishiyama *et al.*, 1999b). These data indicated that adenosine is involved in hypoxia-induced renal vasoconstriction via its effects on adenosine A_1 receptors. Renal interstitial concentrations were also measured in dogs subject to endotoxin shock (Fig. 5B). The results showed that the intravenous administration of lipopolysaccharide (LPS) resulted in decreases in blood pressure and RBF, and increases in renal interstitial concentrations of adenosine (Nishiyama *et al.*, 1999a). During treatment with an adenosine A_1 receptor antagonist, (E)-(R)-1-[3-(2-phenylpyrazolo [1,5-a] pyridin-3yl) acryloyl] pyperidin-2-ylacetic acid, LPS-induced reductions in blood pressure and RBF were significantly attenuated. Interestingly, the augmentation of renal interstitial adenosine production was also attenuated by treatment with FK352. Since adenosine concentrations were inversely proportional to RBF levels, it can be assumed that adenosine plays an important role as a mediator, but not as an initiator of renal hemodynamic changes during endotoxin shock.

CONCLUSIONS

We have briefly discussed the potential role of interstitial adenosine in the regulation of renal hemodynamics in pathophysiological conditions. Accumulating evidence is suggesting that renal interstitial adenosine levels are increased and contribute to the renal hemodynamic changes seen in some pathological conditions. Recent clinical studies have shown that treatment with an adenosine A_1 receptor antagonist protects against furosemide-induced decline in GFR in patients with congestive heart failure (Gottlieb *et al.*, 2000, 2002). Thus, it can be speculated that adenosine A_1 receptor antagonist is useful in therapies for renal failure.

Acknowledgements

This work was supported by grants-in-aid for scientific research from the Ministry of Education, Science and Culture of Japan (to Akira Nishiyama and Youichi Abe), The Salt Sciences Research Foundation (to Youichi Abe), and Mitsui Life Social Welfare Foundation and Uehara Memorial Foundation (to Akira Nishiyama).

REFERENCES

Agmon, Y., Dinour, D. and Brezis, M. (1993). Disparate effects of adenosine A_1- and A_2-receptor agonists on intrarenal blood flow, *Am. J. Physiol.* **265**, F802–F806.

Aki, Y., Shoji, T., Hasui, K., *et al.* (1990). Intrarenal vascular sites of action of adenosine and glucagon, *Jpn. J. Pharmacol.* **54**, 433–440.

Aki, Y., Tomohiro, A., Nishiyama, A., *et al.* (1997). Effects of KW-3902, a selective and potent adenosine A1 receptor antagonist, on renal hemodynamics and urine formation in anesthetized dogs, *Pharmacology* **55**, 193–201.

Aki, Y., Nishiyama, A., Miyatake, A., *et al.* (2002). Role of adenosine A_1 receptor in angiotensin II and norepinephrine-induced renal vasoconstriction, *J. Pharmacol. Exp. Ther.* **309**, 117–123.

Ardaillou, R., Chansel, D., Stefanovic, V., *et al.* (1992). Cell surface receptors and ectoenzymes in mesangial cells, *J. Am. Soc. Nephrol.* **2**, S107–S115.

Arend, L. J., Tompson, C. I. and Spielman, W. S. (1985). Dipyridamole decreases glomerular filtration in the sodium-depleted dog. Evidence for mediation by intrarenal adenosine, *Circ. Res.* **56**, 242–252.

Arend, L. J., Thompson, C. I., Brandt, M. A., *et al.* (1986). Elevation of intrarenal adenosine by maleic acid decreases GFR and renin release, *Kidney Int.* **30**, 656–661.

Bidani, A. K. and Churchill, P. C. (1983). Aminophylline ameliorates glycerol-induced acute renal failure in rats, *Can. J. Physiol. Pharmacol.* **61**, 567–571.

Blanco, J., Canela, E. I., Sayos, J., *et al.* (1993). Adenine nucleotides and adenosine metabolism in pig kidney proximal tubule membranes, *J. Cell. Physiol.* **157**, 77–83.

Brown, R., Ollerstam, A., Johansson, B., *et al.* (2001). Abolished tubuloglomerular feedback and increased plasma renin in adenosine A(1) receptor-deficient mice, *Am. J. Physiol. Regul. Integr. Comp. Physiol.* **281**, R1362–R1367.

Carmines, P. K. and Inscho, E. W. (1994). Renal arteriolar angiotensin responses during varied adenosine receptor activation, *Hypertension* **23**, I114–I119.

Choong, Y. S. and Humphrey, S. M. (1987). Differences in the regional distribution and response to ischemia of adenosine-regulating enzymes in the heart, *Basic. Res. Cardiol.* **82**, 576–584.

Churchill, P. C. and Bidani, A. K. (1983). Hypothesis: Adenosine mediates hemodynamic changes in renal failure, *Med. Hypothesis.* **8**, 275–285.

Dawson, T. P., Gandhi, R., Le Hir, M., *et al.* (1989). Ecto-5′-nucleotidase: localization in rat kidney by light microscopic histochemical and immunohistochemical methods, *J. Histochem. Cytochem.* **37**, 39–47.

Decking, U. K. M., Schlieper, G., Kroll, K., *et al.* (1997). Hypoxia-induced inhibition of adenosine kinase potentiates cardiac adenosine release, *Circ. Res.* **81**, 154–164.

Dickenson, J. M. and Hill, S. J. (1994). Interactions between adenosine A1- and histamine H1-receptors, *Int. J. Biochem.* **26**, 959–969.

Ekland, T., Wahlberg, J., Ungerstedt, U., *et al.* (1991). Interstitial lactate inosine and hypoxanthine in rat kidney during normothermic ischaemia and recirculation, *Acta Physiol. Scand.* **143**, 279–286.

Franco, M., Bell, P. D. and Navar, L. G. (1989). Effect of adenosine A1 analogue on tubuloglomerular feedback mechanism, *Am. J. Physiol.* **257**, F231–F236.

Gordon, E. L., Pearson, J. D., Dickinson, E. S., *et al.* (1982). The hydrolysis of extracellular adenine nucleotides by arterial smooth muscle cells. Regulation of adenosine production at the cell surface, *J. Biol. Chem.* **264**, 18986–18992.

Gottlieb, S. S., Skettino, S. L., Wolff, A., *et al.* (2000). Effects of BG9719 (CVT-124), an A1-adenosine receptor antagonist, and furosemide on glomerular filtration rate and natriuresis in patients with congestive heart failure, *J. Am. Coll. Cardiol.* **35**, 56–59.

Gottlieb, S. S., Brater, D. C., Thomas, I., *et al.* (2002). BG9719 (CVT-124), an A1 adenosine receptor antagonist, protects against the decline in renal function observed with diuretic therapy, *Circulation* **105**, 1348–1353.

Gutierrez, A. M., Kornfeld, M. and Persson, A. E. (1999). Calcium response to adenosine and ATP in rabbit afferent arterioles, *Acta. Physiol. Scand.* **166**, 175–181.

Hall, J. E., Granger, J. P. and Hester, R. L. (1985). Interactions between adenosine and angiotensin II in controlling glomerular filtration, *Am. J. Physiol.* **248**, F340–F346.

Ibarrola, A. M., Inscho, E. W., Vari, R. C., *et al.* (1991). Influence of adenosine receptor blockade on renal function and renal autoregulation, *J. Am. Soc. Nephrol.* **2**, 991–999.

Inscho, E. W., Mitchell, K. D. and Navar, L. G. (1994). Extracellular ATP in the regulation of renal microvascular function, *FASEB J.* **8**, 319–328.

Jackson, E. K. and Dubey, R. K. (2001). Role of the extracellular cAMP-adenosine pathway in renal physiology, *Am. J. Physiol. Renal Physiol.* **281**, F597–F612.

Knight, R. J., Collis, M. G., Yates, M. S., *et al.* (1991). Amelioration of cisplatin-induced acute renal failure with 8-cyclopentyl-1,3-dipropylxanthine, *Br. J. Pharmacol.* **104**, 1062–1068.

Le Hir, M. and Kaissling, B. (1993). Distribution and regulation of renal ecto-5′-nucleotidase: implications for physiological functions of adenosine, *Am. J. Physiol.* **264**, F377–F387.

Lin, J. J., Churchill, P. C. and Bidani, A. K. (1986). Effect of theophylline on the initiation phase of postischemic acute renal failure in rats, *J. Lab. Clin. Med.* **108**, 150–154.

Lin, J. J., Churchill, P. C. and Bidani, A. K. (1987). fect of dipyridamole on the initiation phase of postischemic acute renal failure in rats, *Can. J. Physiol. Pharmacol.* **65**, 1491–1495.

Lin, J. J., Churchill, P. C. and Bidani, A. K. (1988). Theophylline in rats during maintenace phase of post-ischemic acute renal failure, *Kidney Int.* **33**, 24–28.

Lloyd, H. G., Deussen, A., Wuppermann, H. and Schrader, J. (1988). The transmethylation pathway as a source for adenosine in the isolated guinea-pig heart, *Biochem. J.* **252**, 489–494.

Manthei, S. A., Reiling,C. M. and Van Wylen, D. G. L. (1998). Dual cardiac microdialysis to assess drug-induced changes in interstitial purine metabolites: adenosine deaminase inhibition versus adenosine kinase inhibition, *Cardiovas. Res.* **37**, 171–178.

Miller, W. L., Thames, D. A. and Berne, R. M. (1978). Adenosine production in the ischemic kidney, *Circ. Res.* **93**, 390–397.

Morton, M. J., Sivaprasadarao, A., Bowmer, C. J., *et al.* (1998). Adenosine receptor mRNA levels during post renal maturation in the rat, *J. Pharm. Pharmacol.* **50**, 649–654.

Navar, L. G., Inscho, E. W., Majid, D. S. A., *et al.* (1996). Paracrine regulation of the renal microcirculation, *Physiol. Rev.* **76**, 425–536.

Nishiyama, A. and Navar, L. G. (2002). ATP mediates tubuloglomerular feedback, *Am. J. Physiol. Regul. Integr. Comp. Physiol.* **283**, R273-R275; discussion, R278–R279.

Nishiyama, A., Miura, K., Miyatake, A., *et al.* (1999a). Renal interstitial concentration of adenosine during endotoxin shock, *Eur. J. Pharmacol.* **385**, 209–216.

Nishiyama, A., Miyatake, A., Aki, Y., *et al.* (1999b). Adenosine A_1 receptor antagonist KW-3902 prevents hypoxia-induced renal vasoconstriction, *J. Pharmacol. Exp. Ther.* **291**, 988–993.

Nishiyama, A., Miyatake, A., Kusudo, K., *et al.* (1999c). Effects of halothane on renal hemodynamics and interstitial nitric oxide in rabbits, *Eur. J. Pharmacol.* **367**, 299–306.

Nishiyama, A., Majid, D. S. A., Taher, K. A., *et al.* (2000). Relation between renal interstitial ATP concentrations and autoregulation-mediated changes in renal vascular resistance, *Circ. Res.* **86**, 656–662.

Nishiyama, A., Inscho, E. W. and Navar, L. G. (2001a). Interactions of adenosine A_1 and A_{2a} receptors on renal microvascular reactivity, *Am. J. Physiol. Renal Physiol.* **280**, F406–F414.

Nishiyama, A., Kimura, S., He, H., *et al.* (2001b). Renal interstitial adenosine metabolism during ischemia in dogs, *Am. J. Physiol. Renal Physiol.* **280**, F231–F238.

Nishiyama, A., Majid, D. S. A., Walker, M., III., *et al.* (2001c). Renal interstitial ATP responses to changes in arterial pressure during alterations in tubuloglomerular feedback activity, *Hypertension* **37**, 753–759.

Nishiyama, A., Kimura, S., Fukui, T., *et al.* (2002a). Blood flow-dependent changes in renal interstitial cyclic guanosine 3′,5′-monophosphate in rabbits, *Am. J. Physiol. Renal Physiol.* **282**, F238–F244.

Nishiyama, A., Seth, D. M. and Navar L. G. (2002b). Renal interstitial fluid concentrations Ang II concentrations during local ACE inhibition, *J. Am. Soc. Nephrol.* **13**, 2207–2212.

Nishiyama, A., Seth, D. M. and Navar L. G. (2002c). Renal interstitial fluid concentrations of angiotensins I and II in anesthetized rats, *Hypertension* **39**, 129–134.

Osswald, H., Schmitz, H. J. and Kemper, R. (1977). Tissue content of adenosine, inosine and hypoxanthine in the rat kidney after ischemia and postischemic recirculation. *Pflügers. Arch.* **371**, 45–49.

Pawelczyk, T., Bizon, D. and Angielski, S. (1992). The distribution of enzymes involved in purine metabolism in rat kidney, *Biochim. Biophys. Acta* **1116**, 309–314.

Pawlowska, D., Granger, J. P. and Knox, F. G. (1987). Effects of adenosine infusion into renal interstitium on renal hemodynamics, *Am. J. Physiol.* **252**, F678–F682.

Persson, A. E. and Wright, F. S. (1982). Evidence for feedback mediated reduction of glomerular filtration rate during infusion of acetazolamide, *Acta. Physiol. Scand.* **114**, 1–7.

Premen, A. J., Hall, J. E., Mizelle, H. L., *et al.* (1985). Maintenance of renal autoregulation during infusion of aminophylline or adenosine, *Am. J. Physiol.* **248**, F366–F373.

Saleem, Y., Niveditha, T. and Sadasivudu, B. (1982). AMP deaminase, 5'-nucleotidase and adenosine deaminase in rat myocardial tissue in myocardial infarction and hypothermia, *Experientia.* **38**, 776–777.

Schnermann, J. (2002). Adenosine mediates tubuloglomerular feedback, *Am. J. Physiol. Regul. Integr. Comp. Physiol.* **283**, R276-R277; discussion, R278–R279.

Schnermann, J., Weihprecht, H. and Briggs, J. P. (1990). Inhibition of tubuloglomerular feedback during adenosine1 receptor blockade, *Am. J. Physiol.* **258**, F553–F561.

Silldorff, E. P., Kreisberg, M. S. and Pallone, T. L. (1996). Adenosine modulates vasomotor tone in outer medullary descending vasa recta of the rat, *J. Clin. Invest.* **98**, 18–23.

Spielman, W. S. and Thompson, C. I. (1982). A proposed role for adenosine in the regulation of renal hemodynamics and renin release, *Am. J. Physiol.* **242**, F423–F435.

Sun, D., Samuelson, L. C., Yang, T., *et al.* (2001). Mediation of tubuloglomerular feedback by adenosine: evidence from mice lacking adenosine 1 receptors, *Proc. Natl. Acad. Sci. USA* **98**, 9983–9988.

Thomson, S., Bao, D., Deng, A. and Vallon, V. (2000). Adenosine formed by 5'-nucleotidase mediates tubuloglomerular feedback, *J. Clin. Invest.* **106**, 289–298.

Weihprecht, H., Lorenz, J. N., Briggs, J. P., *et al.* (1994). Synergistic effects of angiotensin and adenosine in the renal microvasculature, *Am. J. Physiol.* **266**, F227–F239.

Zager, R. A., Gmur, D. J., Bredl, C. R., *et al.* (1990). Regional responses within the kidney to ischemia: assessment of adenine nucleotide and catabolite profiles, *Biochim. Biophys. Acta* **1035**, 29–36.

Zou, A. P., Nithipatikom, K., Li, P. L., *et al.* (1999a). Role of renal medullary adenosine in the control of blood flow and sodium excretion, *Am. J. Physiol. Regul. Integr. Comp. Physiol.* **276**, R790–R798.

Zou, A. P., Wu, F., Li, P. L., *et al.* (1999b). Effect of chronic salt loading on adenosine metabolism and receptor expression in renal cortex and medulla in rats, *Hypertension* **33**, 511–516.

Advances in Neuroregulation and Neuroprotection (2005), pp. 839-847
Collin, C. *et al.* (Eds)
© VSP 2005

Renal pelvic pressure and the effect of hemorrhage in rats

K. SHIMAMURA [1,2,*], H. KUDO [1], H. SUZUKI [1], S. KIMURA [1], A. OHASHI [1] and H. SAITO [3]

[1] *Department of Clinical Pharmacology, Faculty of Pharmaceutical Sciences, Health Sciences University of Hokkaido, 1757 Kanazawa, Ishikari-Tobetsu, Hokkaido 061-0293, Japan*
[2] *The Research Institute of Personalized Health Sciences, Health Sciences University of Hokkaido, 1757 Kanazawa, Ishikari-Tobetsu, Hokkaido 061-0293, Japan*
[3] *Department of Basic Sciences, Japanese Red Cross Hokkaido College of Nursing, Kitami, Hokkaido 090-0011, Japan*

Abstract—Arterial pressure and renal pelvic pressure have been shown to modulate renal functions. We examined the effect of hemorrhage on renal pelvic motility in urethane-chloralose anesthetized rats. Renal pelvic pressure showed rhythmic oscillations and hemorrhage increased the frequency of pelvic pressure oscillations. In adrenomedullectomized rats, hemorrhage did not increase the frequency of oscillation. The frequency of spontaneous contraction in isolated ring preparations of renal pelvis was increased concentration-dependently by phenylephrine, clonidine, adrenaline and noradrenaline. Isoproterenol decreased the frequency at concentration below 10 μM. CGP-12177A, a β_3-agonist, was without effect. These results indicate that the increase of renal pelvic motility by hemorrhage in rats could be mediated by catecholamines released from the adrenal medulla.

Keywords: Renal pelvis; motility; hemorrhage; adrenal medulla; rat.

INTRODUCTION

The renal pelvis develops spontaneous motility, which propagates to the ureter and transports urine from the kidney to the bladder. Motility of the renal pelvis has been shown to be regulated by myogenic mechanisms (Lang *et al.*, 1998) and neurogenic factors (Santicioli and Maggi, 1998). The renal pelvis has been shown to be innervated by afferent rather than efferent nerve fibers. In the renal pelvis, adrenergic fibers innervate mainly blood vessels, rather than smooth muscle (Barajas and Liu, 1993). However, in anesthetized dogs and rabbits, the increased renal pelvic motility caused by electrical stimulation of the renal pedicle or pelvic wall was inhibited by α-adrenoceptor blockers, indicating that adrenergic receptors are involved in

---

*To whom correspondence should be addressed. E-mail: shimamu@hoku-iryo-u.ac.jp

the regulation of pyeloureteric motility (Boyarsky *et al.*, 1968; Gosling and Waas, 1971). Arterial pressure and renal pelvic pressure have been shown to modulate renal functions (Zanchetti and Stella, 1987; Santicioli and Maggi, 1998). Hypotensive hemorrhage is a typical stimulus for sympathoexcitation (Clement *et al.*, 1972). Hemorrhage decreases (Skoog *et al.*, 1985; Togashi *et al.*, 1990) or increases (Fujisawa *et al.*, 1999) efferent renal nerve activity; however, little information is available concerning the effect of hemorrhage on renal pelvic motility.

In the present study, we examined the effect of hemorrhage on pelvic motility in anesthetized rats and the effect of adrenoceptor agonists on spontaneous contractions of isolated pelvic preparations.

MATERIALS AND METHODS

Male Wistar rats weighing 200–300 g were used. They were treated according to the Guiding Principles for the Care and Use of Animals in the Field of Physiological Sciences approved by The Physiological Society of Japan (March 29, 2002). Rats were anesthetized with urethane and α-chloralose injected intraperitoneally at doses of 500 mg/kg and 50 mg/kg, respectively. Animals were tracheotomized and the femoral artery and vein were catheterized to record arterial pressure and administer drugs, respectively. Body temperature was maintained at 37°C by a regulated heater. Hemorrhage was induced by withdrawal of blood from a catheter inserted into the contralateral femoral artery. One week before the experiments, seven rats were adrenomedullectomized bilaterally while under anesthesia and were provided with physiological saline to drink postoperatively (Waynforth, 1980; Ariznavarreta *et al.*, 1989). Another seven rats were sham operated, which involved anesthesia and bilateral flank incision with manipulation of the adrenal glands (Cagampang *et al.*, 1999).

In experiments with isolated preparations, rats were anesthetized with carbon dioxide and exsanguinated. Kidneys were extracted and 1 mm wide ring preparations of pelvis were made. Preparations were mounted horizontally in organ baths filled with a modified Tyrode's solution at 37°C and isometric contractile force was measured by a force-displacement transducer (UL-10, Minebea, Karuizawa, Japan). After incubation for 1 h, the preparations were stretched to 1.3 times their original length and were allowed to develop stable spontaneous contraction. Modified Tyrode's solution consisted of (in mmol): NaCl 137, KCl 5.4, $CaCl_2$ 2.0, $MgCl_2$ 1.0, NaH_2PO_4 0.4, glucose 5.6; and was equilibrated with a gas mixture of 95% O_2 and 5% CO_2 (pH 7.3).

Drugs

Adrenaline, noradrenaline, isoproterenol, propranolol, phentolamine, clonidine, urethane, α-chloralose, forskolin and CGP-12177A were obtained from Sigma Chemical Co. (St. Louis MO, USA). All other drugs were from Wako Pure Chemical Industries, Ltd. (Osaka, Japan).

Statistics

Results are expressed as mean ± SEM with the number of animals indicated in parenthesis. Statistical analysis was performed by paired t-test or Student's t-test using StatView statistical software (Abacus Software, USA) and p-values less than 0.05 were considered significant.

RESULTS

In anesthetized rats, mean arterial pressure was 101 ± 5 mmHg ($n = 9$) and renal pelvic pressure showed rhythmic oscillatory changes from the basal level of 0 mmHg with an amplitude of 10 to 30 mmHg and a frequency of 20 ± 2 cycle per min ($n = 9$). When mean arterial pressure was decreased to 56 ± 4 mmHg ($n = 5$) by hemorrhage, the frequency of pelvic pressure oscillation increased gradually over a few minutes to reach a stable level of 28 ± 2 cpm ($n = 5$) (Fig. 1).

When mean arterial pressure was decreased from 93 ± 8 mmHg ($n = 7$) to 49 ± 3 mmHg ($n = 7$) in adrenomedullectomized rats, the frequency of pelvic pressure oscillation did not change remarkably (from 20 ± 1 cpm to 21 ± 1 cpm; $n = 7$ per group). When mean arterial pressure was decreased from 88 ± 5 mmHg ($n = 7$) to 57 ± 2 mmHg ($n = 7$) in sham operated rats, the frequency of pelvic pressure oscillation increased from 19 ± 2 cpm ($n = 7$) to 24 ± 2 cpm ($n = 7$). Thus, in adrenomedullectomized rats, the frequency of pelvic pressure oscillation was not increased by hemorrhage, while an increase was evident in sham-operated rats (Fig. 2).

To determine the effect of plasma catecholamines on pelvic pressure oscillation frequency, the effect of adrenergic agonists on contraction of isolated renal pelvis was examined. Ring preparations of isolated renal pelvis developed spontaneous rhythmic contractions at a frequency of 6.7 ± 0.2 cpm ($n = 40$) in the absence of drugs. Administration of adrenaline, noradrenaline, phenylephrine, clonidine or dopamine increased the contraction frequency concentration-dependently (Fig. 3).

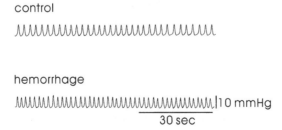

control

hemorrhage

10 mmHg

30 sec

Figure 1. Typical traces showing renal pelvic pressure recorded in an anesthetized Wistar rat. Upper trace: obtained from an anesthetized rat (mean blood pressure 117 mmHg) before hemorrhage. Bottom trace: obtained from the same rat 3 min after the start of hemorrhage induced by bleeding from the femoral artery (mean blood pressure 60 mmHg).

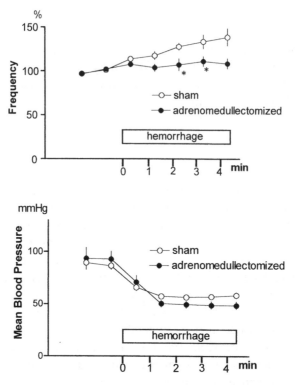

Figure 2. Contraction frequency of the renal pelvis and change in mean arterial blood pressure of anesthetized rats. Each symbol with bars indicates the mean ± SEM ($n = 7$ per group) of data obtained from sham operated rats (open circles) and bilaterally adrenomedullectomized rats (closed circles). Asterisks indicate a statistically significant difference ($p < 0.05$) in comparison with sham-operated rats.

Isoproterenol (0.1–10 μM) decreased the frequency of spontaneous contraction concentration-dependently. In the presence of propranolol (1 μM), the decrease in contraction frequency by isoproterenol was abolished. Isoproterenol (0.1 and 1 mM) increased the frequency of spontaneous contraction. In the presence of phentolamine (1 μM) or prazosin (1 μM), the increase in contraction frequency by isoproterenol was inhibited remarkably (data not shown). Forskolin did not increase contraction frequency and decreased contraction frequency concentration-dependently (0.1–10 μM).

DISCUSSION

Spontaneous contraction of the upper urinary tract has been shown to originate from atypical smooth muscle in the renal pelvis (Gosling and Dixon, 1974; Lang *et al.*, 2001). The excitation of renal pelvis is transferred to the ureter by cell-to-cell electrical coupling via gap junctions (Santicioli and Maggi, 2000). It has been reported that the frequency of ureteral peristalsis in anesthetized rats is 18–31 cpm

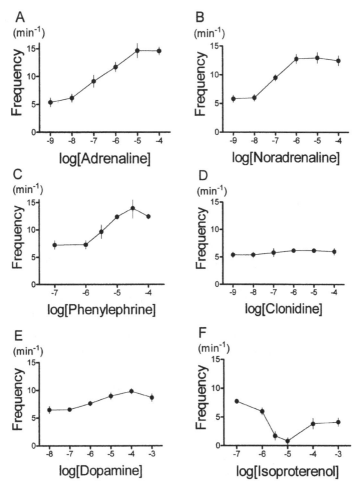

Figure 3. The effects of adrenaline, noradrenaline, phenylephrine, clonidine, dopamine, and isoproterenol on spontaneous contraction frequency of ring preparations of rat renal pelvis.

(Tindal, 1972). In isolated pyeloureter preparations, spontaneous contraction occurs at a frequency of 10–18 cpm (Finberg and Peart, 1970; Hannappel and Golenhofen, 1974a). Thus, the frequency of pelvic motility in the present study was in accordance with those values.

Adrenaline and noradrenaline increase the frequency and amplitude of contractions in the proximal ureter of rats (Hannappel and Golenhofen, 1974b; Qayum and Yusuf, 1983). Adrenaline increases the contraction frequency of isolated renal pelvis through stimulation of α-adrenoceptors in rabbits (Del Tacca *et al.*, 1981). In the present study, adrenaline and noradrenaline induced remarkable increases in the contraction frequency of rat renal pelvis. The effect of phenylephrine, an α_1-adrenoceptor agonist, was comparable in magnitude to that of adrenaline and noradrenaline.

There is little evidence available on the effect of α_2-adrenoceptor agonists on renal pelvic motility. In anesthetized rats, the contraction frequency of the upper urinary tract is reportedly decreased, following a transient increase, by intravenous administration of dexmedetomide, an α_2-adrenoceptor agonist (Harada *et al.*, 1992). However, in the present study, the α_2-adrenoceptor agonist clonidine caused only a slight increase in contraction frequency. This discrepancy may be due to differences between *in vivo* and *in vitro* experimental conditions.

Isoproterenol has been reported to decrease ureteral peristalsis in anesthetized dogs (MacLeod *et al.*, 1973; Rose and Gillenwater, 1974). In the renal pelvis of anesthetized pigs, intravenous or topical application of isoproterenol decreases contraction frequency (Danuser *et al.*, 2001). On the contrary, another report indicated that isoproterenol (1–100 μM) increases the contractile frequency of lower renal pelvic muscle strips from rabbits (Morita, 1986). The unexpected excitation may be explained by the facilitation of L-type voltage-dependent current by isoproterenol, as has been reported for portal vein (Xiong *et al.*, 1994). In the present study, low concentrations of isoproterenol decreased the contraction frequency of isolated rat renal pelvis. In the presence of propranolol, isoproterenol did not decrease contraction frequency. High concentrations of forskolin (10 μM), an activator of adenylate cyclase, decreased contraction frequency. The increase of contraction frequency by isoproterenol was inhibited by phentolamine (1 μM). These data may indicate that high concentrations of isoproterenol stimulate α-adrenoceptor to increase contraction frequency. Thus, stimulation of β-adrenoceptor decreases contraction frequency.

Spontaneous contraction induced by 40 mM potassium in guinea pig was inhibited concentration-dependently by CGP-12177A, a β_3 agonist (Yamamoto and Koike, 2000). However, CGP-12177A did not show any effect on tonic contraction of rat ureter induced by 80 mM potassium (Tomiyama *et al.*, 1998). In the present study, CGP-12177A at concentrations of up to 10 μM did not induce any change in spontaneous contraction of rat renal pelvis. This discrepancy may indicate species differences in β_3 adrenoceptors between guinea pigs and rats.

In the rat, sensory fibers containing substance P and calcitonin gene-related peptide are located in the renal pelvis (Ferguson and Bell, 1988). Application of guanethidine or phentolamine in combination with propranolol does not affect motility evoked by stimulation of the intramural nerves in the renal pelvis of rats or guinea pigs (Maggi and Giuliani, 1992). Thus, in renal pelvis, efferent neural control is less evident than that for afferent innervation (Santicioli and Maggi, 1998). Most reports indicate that adrenergic or cholinergic blockade does not affect ureteric peristalsis and it is believed that the efferent autonomic system does not have a tonic influence on ureteric motility (Amann, 1993). However, in anesthetized dogs and rabbits, increased motility evoked by electrical stimulation of the renal pedicle is inhibited by phenoxybenzamine or phentolamine (Boyarsky *et al.*, 1968; Gosling and Waas, 1971). Nociceptive cutanenous stimulation increases peristaltic movements of the ureter in anesthetized rats (Ohsawa *et al.*, 1988), and the increase is abolished by bilateral sectioning of the splanchnic nerves. These reports indicate the possible contribution of neural regulation in pyeloureteric motility.

In the present study, hemorrhage increased the frequency of pelvic pressure oscillations remarkably. However, an increase in efferent renal nerve activity may not be involved in the response, because it has been shown that hemorrhage decreases efferent renal nerve activity in anesthetized rats (Skoog *et al.*, 1985; Togashi *et al.*, 1990). Hemorrhage failed to increase the frequency of pelvic pressure oscillation in adrenomedullectomized rats. Therefore, the increase in frequency might be mediated by an increase in plasma adrenaline, since sympathetic adrenal nerve activity and plasma adrenaline are reportedly increased by hemorrhage in rats (Nishimura *et al.*, 1979; Ito *et al.*, 1984; Togashi *et al.*, 1990).

The present study indicated that the frequency of renal pelvic pressure oscillations can be increased by catecholamines from the adrenal medulla during hemorrhage in anesthetized rats.

Acknowledgement

This work was supported in part by High Technology Research Program and The Academic Science Frontier Project of the Ministry of Education, Culture, Sports, Science and Technology, Japan.

REFERENCES

Amann, R. (1993). Neural regulation of ureteric motility, in: *Nervous Control of the Urogenital System, The Autonomic Nervous System*, Vol. 3, Maggi, C. A. (Ed.), pp. 209–225. Harwood Academic Publishers, GmbH, Chur.

Ariznavarreta, C., Calderon, M. D., Tresguerres, J. A. F., *et al.* (1989). Effect of adrenomedullectomy and propranolol treatment on the response of gonadotrophins to chronic stress in male rats, *J. Endocrinol.* **120**, 275–279.

Barajas, L. and Liu, L. (1993). The renal nerve in the newborn rat, *Pediatr. Nephrol.* **7**, 657–666.

Boyarsky, S., Labay, P. and Pfautz, C. J. (1968). The effect of nicotine upon ureteral peristalsis, *Southern Medical J.* **61**, 573–579.

Cagampang, F. R. A., Strutton, P. H., Goubillon, M.-L., *et al.* (1999). Adrenomedullectomy prevents the suppression of pulsatile luteinizing hormone release during fasting in female rats, *J. Neuroendocrinol.* **11**, 429–433.

Clement, D. L., Pelletier, C. L. and Shepherd, J. T. (1972). Role of vagal afferents in the control of renal sympathetic nerve activity in the rabbit, *Circ. Res.* **31**, 824–830.

Danuser, H., Weiss, R., Abel, D., *et al.* (2001). Systemic and topical drug administration in the pig ureter: effect of phosphodiesterase inhibitors alpha1, beta and beta2-adrenergic receptor agonists and antagonists on the frequency and amplitude of ureteral contractions, *J. Urol.* **166**, 714–720.

Del Tacca, M., Constantinou, C. E. and Bernardini, C. (1981). The effects of drugs on pacemaker regions of isolated rabbit renal pelvis, *Eur. J. Pharmacol.* **71**, 43–51.

Ferguson, M. and Bell, C. (1988). Ultrastructural localization and characterization of sensory nerves in the rat kidney, *J. Comp. Neurol.* **247**, 9–16.

Finberg, J. P. M. and Peart, W. S. (1970). Function of smooth muscle of the rat renal pelvis — response of the isolated pelvis muscle to angiotensin and some other substances, *Br. J. Pharmacol.* **9**, 373–381.

Fujisawa, Y., Mori, N., Yube, K., *et al.* (1999). Role of nitric oxide in regulation of renal sympathetic nerve activity during hemorrhage in conscious rats, *Am. J. Physiol.* **277**, H8–H14.

Gosling, J. A. and Dixon, J. S. (1974). Species variation in the location of upper urinary tract pacemaker cells, *Inv. Urol.* **11**, 418–423.

Gosling, J. A. and Waas, A. N. C. (1971). The behaviour of the isolated rabbit renal calyx and pelvis compared with that of the ureter, *Eur. J. Pharmacol.* **16**, 100–104.

Hannappel, J. and Golenhofen, K. (1974a). Comparative studies on normal ureteral peristalsis in dogs, guinea-pigs and rats, *Pflugers Arch.* **348**, 65–76.

Hannappel, J. and Golenhofen, K. (1974b). The effect of catecholamines on ureteral peristalsis in different species (dogs, guinea-pig and rat), *Pflugers Arch.* **350**, 55–68.

Harada, T., Kigure, T., Yoshida, K., *et al.* (1992). The effects of alpha-2 agonists and antagonists on the upper urinary tract of the rat, *J. Smooth Muscle Res.* **28**, 139–151.

Ito, K., Sato, A., Shimamura, K., *et al.* (1984). Reflex changes in sympatho-adrenal medullary functions in response to baroreceptor stimulation in anesthetized rats, *J. Auton. Nerv. Syst.* **10**, 295–303.

Lang, R. J., Exintaris, B., Teele, M. E., *et al.* (1998). Electrical basis of peristalsis in the mammalian upper urinary tract, *Clin. Exp. Pharmacol. Physiol.* **25**, 310–321.

Lang, R. J., Takano, H., Davidson, M. E., *et al.* (2001). Characterization of the spontaneous electrical and contractile activity of smooth muscle cells in the rat upper urinary tract, *J. Urol.* **166**, 329–334.

MacLeod, D. G., Reynolds, D. G. and Swan, K. G. (1973). Adrenergic mechanisms in the canine ureter, *Am. J. Physiol.* **245**, 1054–1058.

Maggi, C. A. and Giuliani, S. (1992). Nonadrenergic noncholinergic excitatory innervation of the guinea-pig renal pelvis: Involvement of capsaicin-sensitive primary afferent neurons, *J. Urol.* **14**, 1394–1398.

Morita, T. (1986). Characteristics of spontaneous contraction and effects of isoproterenol on contractility in isolated rabbit renal pelvic smooth muscle strips, *J. Urol.* **135**, 604–607.

Nishimura, T., Nishio I., Ohtani H., *et al.* (1979). Plasma catecholamine determination using high pressure liquid chromatography and their roles in blood pressure regulation and experimental hypertension in rats, *Jpn. Circ. J.* **43**, 855–865.

Ohsawa, H., Nishijo, K. and Sato, Y. (1988). Reflex changes in ureter movements produced by noxious stimulation of the skin in anesthetized rats, *Zennippon Shinkyu Gakkai Zasshi* **38**, 271–280.

Qayum, A. and Yusuf, S. M. (1983). Effect of adrenaline and noradrenaline on isolated rat ureter preparations, *J. Pak. Med. Assoc.* **33**, 223–229.

Rose, J. G., and Gillenwater, J. Y. (1974). The effect of adrenergic and cholinergic agents and their blockers upon ureteral activity, *Invest. Urol.* **11**, 439–451.

Santicioli, P. and Maggi, C. A. (1998). Myogenic and neurogenic factors in the control of pyeloureteral motility and ureteral peristalsis, *Pharmacol. Rev.* **50**, 683–721.

Santicioli, P. and Maggi C. A. (2000). Effect of 18β-glycyrrhetinic acid on electromechanical coupling in the guinea-pig renal pelvis and ureter, *Br. J. Pharmacol.* **129**, 163–169.

Skoog, P., Mansson, J. and Thoren, P. (1985). Changes in renal sympathetic outflow during hypotensive haemorrhage in rats, *Acta Physiol. Scand.* **125**, 655–660.

Tindal, A. R. (1972). Preliminary observations of the mechanical and electrical activity of the rat ureter, *J. Physiol.* **223**, 633–647.

Togashi, H., Yoshioka, M., Tochihara, M., *et al.* (1990). Differential effects of hemorrhage on adrenal and renal nerve activity in anesthetized rats, *Am. J. Physiol.* **259**, H1134–H1141.

Tomiyama, Y., Hayakawa, K., Shinagawa, K., *et al.* (1998). α-Adrenoceptor subtypes in the ureteral smooth muscle of rats, rabbits and dogs, *Br. J. Pharmacol.* **352**, 369–278.

Waynforth, H. B. (1980). Specific surgical operation. in: *Experimental and Surgical Technique in the Rat.* Waynforth, H. B. (Ed.), pp. 124–208. Academic Press, London.

Xiong, Z., Sperelakis, N. and Fenoglio-Preser, C. (1994). Isoproterenol modulates the calcium channels through two different mechanisms in smooth-muscle cells from rabbit portal vein, *Pflugers Arch.* **428**, 105–113.

Yamamoto, Y. and Koike, K. (2000). The effects of α-adrenoceptor agonists on KCl-induced rhythmic contraction in the ureter of guinea-pig, *J. Smooth Muscle Res.* **36**, 13–19.

Zanchetti, A. and Stella, A. (1987). Sympatho-renal interactions, *Ital. J. Neural Sci.* **8**, 477–485.

Advances in Neuroregulation and Neuroprotection (2005), pp. 849-860
Collin, C. *et al.* (Eds)
© VSP 2005

Effects of dopamine, prostaglandins and digitalis on isolated abdominal vagus nerve in rats

MASAHIRO NEMOTO [1], TORU ENDO [2], MASARU MINAMI [2] and HIDEYA SAITO [1,*]

[1] *Department of Basic Sciences, Japanese Red Cross Hokkaido College of Nursing, 664-1 Akebono, Kitami, Hokkaido 090-0011, Japan*
[2] *Department of Pharmacology, Faculty of Pharmaceutical Sciences, Health Sciences University of Hokkaido, Ishikari-Tobetsu, Hokkaido 061-0293, Japan*

Abstract—The abdominal vagus is the major nerve involved in the detection of emetic stimuli. Although afferent vagus nerves are considered to have polymodal properties, depolarization of the vagus nerve is mainly mediated by 5-HT in emesis. We studied direct drug effects on rat abdominal vagus nerve using the grease-gap technique. In this study, we also investigated reactions to various emetic stimuli such as dopamine (DA), prostaglandins (PGs) and digitalis. The abdominal vagus nerves rapidly depolarize with potassium chloride application, rapidly declining towards basal levels by the end of the application. Concentration-related depolarization evoked by DA (1×10^{-8} M to 1×10^{-4} M) showed a pattern, in contrast to potassium, where depolarization occurred approximately 30 s later. PGE_1 and PGE_2 showed a maximum response to concentration of 1×10^{-5} M and 1×10^{-4} M, respectively. The concentration-related curves of PGE_1 are similar in PGE_2, in that 1 nM elicited depolarization. Ouabain (1×10^{-9} M to 1×10^{-4} M) induced concentration-dependent depolarization responses on the isolated abdominal vagus nerve. The application of 100 μM ouabain induced prolonged depolarization. We confirmed that the peripheral role of the vagus nerve may be involved in the appearance of emesis caused by DA, PGs and digitalis. The grease gap technique of the isolated abdominal vagus nerve is a useful technique for screening the emetic agents.

Keywords: Vagus nerve; depolarization; dopamine; prostaglandin; digitalis; emesis.

INTRODUCTION

The name 'vagus' nerve derives from the Latin word for 'wandering'. It is also well known as the Xth cranial nerve and it is widely distributed to peripheral organs, heart, lung and gut from the esophagus to the colon. The afferent vagus terminates at the chemoreceptor of the nucleus tranctus solitarius (van Giersbergen *et al.*, 1992).

---

*To whom correspondence should be addressed. E-mail: hideyas@rchokkaido-cn.ac.jp

The vagus is the major afferent pathway involved in the detection of emetic stimuli (Andrews, 1991). The vagus afferent nerve depolarization may evoke the vomiting reflex.

The grease-gap (or sucrose gap) technique is an *in vitro* technique for the estimation of the nerve depolarization (Neto, 1978). The nerve and its ganglion contain C-fibres and glia (e.g. Schwann cells) (Paintal, 1963). These express several receptors (serotonergic, nicotinic, purinergic, etc.) on the surface of the axons and ganglions (Ireland and Tyers, 1987). Therefore, application of these receptors activating agonist to the axon introduce an extracellular potential changes. This method has been widely used in the evaluation of 5-HT reactions on the cervical vagus nerve in several species (Ireland, 1987; Ireland and Tyers, 1987; Ito *et al.*, 1995). In order to evaluate drug-induced emesis, we previously developed a new technique for measurement of the abdominal vagus nerve depolarization (Nemoto *et al.*, 2001). We have reported that the reaction of the isolated abdominal vagus is significantly higher than that of the isolated cervical vagus nerve. It is well known that 5-HT$_3$ receptor antagonists block anticancer drug-induced emesis in several species (Costall *et al.*, 1986; Hawthorn *et al.*, 1988; Endo *et al.*, 1990). It has been reported that 5-HT-induced vagal nerve depolarization is associated with both 5-HT$_3$ and 5-HT$_4$ receptors (Bley *et al.*, 1994; Coleman and Rhodes, 1995; Nemoto *et al.*, 2001).

Despite many cervical vagus studies, no evidence has been found that excludes 5-HT from the reaction of the abdominal vagus nerve. In this study we investigated the effects of several emesis-related agents on the abdominal vagus nerves in rats. To determine whether the mechanism of drug-induced emesis involved the direct stimulation of the vagal nerve, we tested the membrane potential changes in the abdominal vagus nerve directly *in vitro*. First, we tested the reaction to dopamine (DA), as it is known have emetic actions via the 'central-mechanism' (Wang and Borison, 1952). We previously reported that dopamine D$_2$ receptor agonist induced depolarization responses on afferent vagus nerve activity *in vivo* (Minami *et al.*, 1995). The peripheral mechanism of DA is still unclear. The endogenous prostanoid participates in the appearance of pain or inflammation (Ferreira, 1979). Prostaglandin (PG) E$_2$ is especially potent as a nociceptive agent. The anti-inflammatory agent, dexamethasone, is effective against delayed emesis caused by anticancer chemotherapy. In order to clarify the peripheral action of PGs we studied the direct effect of PGs on the abdominal vagus nerve. We also tested the response to ouabain. Borison and Wang (1952) suggested that digitalis-induced emesis is associated with a central-mechanism. Interestingly, ouabain-induced emesis is inhibited in vagotomized animals (Kakimoto *et al.*, 1997). We previously reported that ouabain induced the increase in the abdominal vagus nerve activity *in vivo* (Endo *et al.*, 1998a). Ouabain-induced emesis may involve a peripheral mechanism via the abdominal vagus nerve activation as well as a central mechanism. To determine whether the mechanism of digitalis-induced emesis involves direct stimulation of the vagal nerves, we tested the depolarization response in the abdominal vagus nerve directly *in vitro*.

MATERIALS AND METHODS

In vitro *electrophysiological study*

Adult rats (Male, Wistar) weighing 190–250 g (purchased from Nippon SLC) anaesthetized with urethane (500 mg/kg, i.p.) and α-chloralose (50 mg/kg, i.p.) were killed by cardiac puncture. In order to isolate a bundle of abdominal vagus nerves, the esophagus was carefully isolated from the gastric cardia to the diaphragm. The isolated esophagus was placed in oxygenated ice-cold Krebs-bicarbonate buffer containing (in mmol): NaCl 118.2, KCl 4.6, $CaCl_2$ 2.5, KH_2PO_4 1.2, $NaHCO_3$ 24.8, $MgSO_4$ 1.2 and glucose 10.0. The vagus nerve was carefully removed under a binocular microscope. An *in vitro* depolarization study involving the abdominal vagal nerve was described previously (Nemoto *et al.*, 2001). De-sheathed vagus nerves were transferred into two-compartment Perspex baths to permit the extracellular recording of agonist-induced depolarization. Half of the vagus nerve was placed in the first compartment, while the remainder was projected through a vaseline (Dow Corning high vacuum grease) slot into the second chamber. The vagus nerves were superfused with gassed Krebs-bicarbonate solution at a rate of 1.5 ml/min and maintained at 23°C. The direct current potential between the two compartments was continuously recorded using Ag-AgCl electrodes (WPI, WK-1) mounted in 4% agar in saturated KCl. Potential changes induced by agonists were amplified (Nihon Kohden, MEZ7200), monitored on an oscilloscope (Nihon Kohden, VC11) and analyzed by a Power Lab system (ADI Instruments).

The isolated vagus nerves were then exposed to agonist for 1 min. The nerves were washed sufficiently for 25 min between every application. Nerves were also exposed to 3 mM KCl for 1.5 min before the initial application of agonist.

Drugs

DA, PGE_1 and PGE_2 were purchased from Sigma-Aldrich, USA. Since ouabain is water soluble and rapid-acting in isolated preparations, ouabain (Sigma-Aldrich Co., USA) was used as digitalis in this study. All the other reagents used were the highest quality or analytical grade. All chemicals were dissolved in physiological saline before each study.

Statistics

The data are expressed as the mean $\pm$ S.E. Statistical significance was assessed by the two-tailed, unpaired t-test using $p < 0.05$ as the level of significance.

RESULTS

Effects of potassium chloride

The Krebs solution application did not induce depolarization of any vagus nerves (data not shown). Figure 1 shows the typical pattern of depolarization with potassium chloride (KCl) (3 and 10 mM). The abdominal vagus nerves rapidly depolarized with KCl application (3 mM; 1024 ± 67 μV, 10 mM; 4758 ± 501 μV, $n = 5$ (Fig. 2)), it rapidly declined towards basal levels.

Effects of dopamine

The DA-induced depolarization of a typical experiment is shown in Fig. 3. Dop-amine evoked a concentration-related depolarization of the rat abdominal vagus

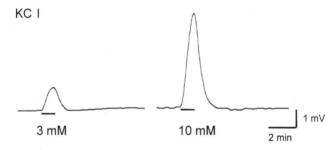

Figure 1. Depolarizing pattern of potassium chloride (KCl) on the isolated abdominal vagus nerve of the rat *in vitro*.

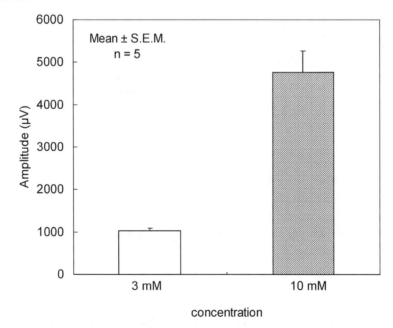

Figure 2. The effects of KCl on the isolated abdominal vagus nerves of the rat *in vitro*.

DA

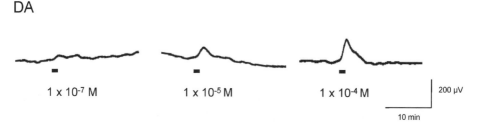

Figure 3. Depolarizing pattern of the dopamine (DA) on the isolated abdominal vagus nerve of rats *in vitro*.

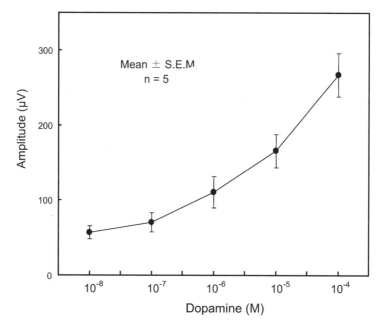

Figure 4. The effects of DA on the isolated abdominal vagus nerves of rats *in vitro*.

nerves ranging from 1×10^{-8} M to 1×10^{-4} M (Fig. 4). 1×10^{-4} M of DA elicited a maximum response (268 ± 29 μV, $n = 5$). The depolarization pattern of DA, in contrast to that of KCl, occurs approximately 30 s after DA application.

Effects of prostaglandins

The depolarization pattern of PGE_1 and PGE_2 is shown in Fig. 5. Application of PGs caused a large depolarization but not immediately, similar to that with KCl. PGE_1 and PGE_2 caused a maximum response at a concentration of 1×10^{-5} M and 1×10^{-4} M, respectively. The concentration-related curves in PGE_1 and PGE_2 are similar. Depolarization was elicited from a concentration of 1 nM (Fig. 6). A high concentration of PGs elicited a slow decline in each PG.

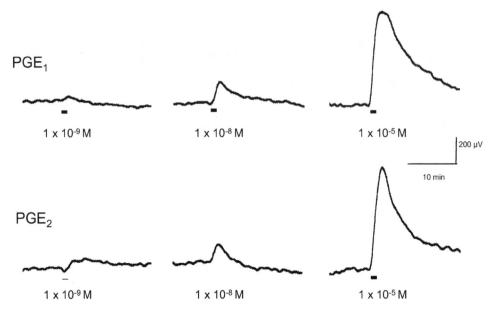

Figure 5. Depolarizing pattern of PGE$_1$ (upper) and PGE$_2$ (lower) on the isolated abdominal vagus nerves of rats *in vitro*.

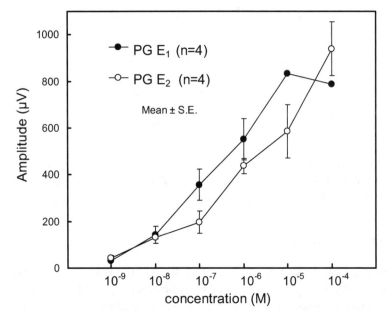

Figure 6. Concentration-response curves for depolarizing responses to PGE$_1$ and PGE$_2$ in the isolated abdominal vagus nerves of rats *in vitro*.

Effects of digitalis (study of ouabain)

As shown in Fig. 7, ouabain induced depolarization in a typical pattern. The depolarization developed slowly and in a sustained way as compared with that of KCl induced by ouabain.

Ouabain (1×10^{-9} M to 1×10^{-4} M) induced concentration-dependent depolarization responses in the isolated abdominal vagus nerve (Fig. 8). In this experiment, 10 μmol of ouabain evoked a maximum response (134 ± 16 μV). The application of 100 μM ouabain caused prolonged depolarization which required 30 min to return to the baseline.

ouabain

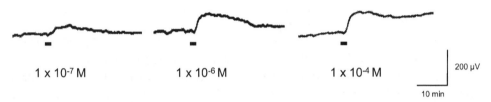

Figure 7. Depolarizing pattern of ouabain on the isolated abdominal vagus nerve of rats *in vitro*.

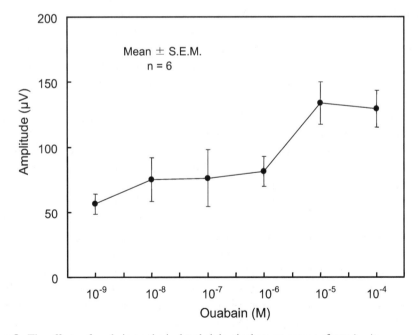

Figure 8. The effects of ouabain on the isolated abdominal vagus nerves of rats *in vitro*.

DISCUSSION

It has been demonstrated that exogenously administered DA, PGs and ouabain induced an increase in depolarization of the isolated abdominal vagus nerve in rats. In this study, we evaluated a new technique for the measurement of rat isolated abdominal vagus nerve depolarization *in vitro*.

The grease gap technique was used to investigate the reaction of the vagus nerves to various substances. Pharmacological evaluation of the cervical vagal nerves has been undertaken not only with 5-HT but also with PGs (Smith *et al.*, 1998), acetylcholine, GABA (Ito *et al.*, 1995), and cisplatin (Woods *et al.*, 1995). These studies have been undertaken mostly by using cervical vagal nerve preparations (Ireland 1987; Ireland *et al.*, 1987; Ito *et al.*, 1995). Nausea and vomiting are a defensive reflex providing the body with a means of ejecting toxicological substances (Parkes *et al.*, 1976; Blancquaert *et al.*, 1986; Sridhar and Donnelly, 1988; Andrews *et al.*, 1988; Nakanishi *et al.*, 1993). The electro-stimulation of afferents of the abdominal vagus nerve induced emetic reflex (Andrews *et al.*, 1995), vagotomy inhibited the emetic response induced by the cytotoxic drugs, digitalis and $CuSO_4$ (Carpenter *et al.*, 1988; Endo *et al.*, 1992; Kakimoto *et al.*, 1997). Thus, the abdominal vagal afferents play an important role in the peripheral mechanism of the emetic-episode. The abdominal vagus nerve contains about 90% afferent fibres and only about 10% efferent fibres. The 5-HT-induced depolarization response of the isolated abdominal vagus nerves using new technique was approximately 170% higher in value than that observed in the cervical vagus nerves (Nemoto *et al.*, 2001).

Not only anticancer drugs but also cardiac glycoside and anti-Parkinson's disease drugs induced nausea and emesis (Yanagisawa *et al.*, 1992; Nakanishi *et al.*, 1993; Minami *et al.*, 1997, 1999). The mechanism, however, remains elusive as to whether it involves central or peripheral action, or both (Andrews *et al.*, 1988; Kakimoto *et al.*, 1997). Thus, the present study was set up to produce reproducible and concentration related depolarization responses of the isolated abdominal vagus to KCl, DA, PGs and ouabain.

We tested the endogenous ligand DA by introducing concentration-related depolarization on the isolated abdominal vagus nerve. Our results suggested that DA acts both on the central medulla and the peripheral nerve endings. We previously reported that D_2 receptor agonist, bromocriptine, increased *in vivo* abdominal vagal afferent activity in a dose-dependent manner (Minami *et al.*, 1999). This reaction, however, was inhibited by $5-HT_3$ receptor antagonists in *in vivo* experiments which introduce the release of 5-HT from the enterochromaffin cells in the ileum. Thus, 5-HT seems to act on the abdominal vagus. In a nodose ganglia (cell body of vagus afferent) study, Lawrence *et al.* (1995) reported that DA elicited a concentration-dependent depolarization. However, in this preparation, at least 10 μmol of DA was required for minimum depolarization. In the present study, the abdominal vagus nerve demonstrated depolarization responses from a much smaller concentration (1 nM) than that of 5-HT (10 nM) (Nemoto *et al.*, 2001). In contrast, Birrell *et al.* (2002) reported that DA inhibited hypertonic-saline-induced or capsaicin-induced depolarization. D_2 receptors were visualized on the human inferior vagal ganglia (Lawrence, 1994). We also confirmed D_2 receptor mediated reactions using D_2 receptor agonist bromocriptine (unpublished data). Therefore, DA-induced depolarization is related to the D_2 receptors, but further information is necessary to clarify the mechanism of DA-related depolarization in the abdominal vagus nerves.

Anticancer agent-induced emesis is classified into acute emesis during the first 24 hours after anticancer chemotherapy, and delayed emesis, occurring 1 to 4 days after administration of cancer chemotherapeutive agents (Kris *et al.*, 1992; Rudd *et al.*, 1994). Although the acute phase of emesis is dramatically controlled by using 5-HT$_3$ receptor antagonists, there remain patients with delayed emesis resistant to therapy with 5-HT$_3$ receptor antagonists (Rudd *et al.*, 1994, 1996; Fukunaka *et al.*, 1998). Corticosteroids reduce the delayed emesis in chemotherapy when combined with 5-HT$_3$ receptor antagonists (Smith *et al.*, 1991, Roila, 1993). The mechanism of the antiemetic effect of corticosteroids is still unclear. Rich *et al.* (1980) suggested that corticosteroids initiate the inhibition of prostanoids synthesis in cancer chemotherapy. For this reason, we investigated the direct effect of the PGs on the abdominal vagus nerve. PGE$_1$ and PGE$_2$ exerted a potent agonistic activity on abdominal vagus nerve. No significant differences exist between PGE$_1$ and PGE$_2$. Recently, PG-derivative drugs have been used as oxytocic and anti gastro-ulcer agents (Lanza *et al.*, 1988; Creinin, 2000). Moreover, PG acts to one of the inflammatory mediators. Since, inflammation appeared not only during chemotherapy but also operatively, these substances affect the visceral organs and peripheral neurons and caused a depolarization response in the abdominal vagus nerves which might serve as a trigger to cause emetic episodes. PGE$_1$ and PGE$_2$ are well known to act on EP and IP receptors (Ushikubi *et al.*, 1995). Smith *et al.* (1998) suggested that a cervical vagus depolarizing response did not mediate the EP$_1$ receptor. Recently, Farrag *et al.* (2002) reported that PGE$_2$ did not show a measurable effect on action potential propagation in the cervical vagus nerve. This report also showed that PGE$_2$ induced very slight depolarization responses (74 μV at 20 μM in cervical vagus) compared to our present data (585 μV at 10 μM in abdominal vagus). Because EP receptors are characterized from EP$_1$ to EP$_4$ (Pierce *et al.*, 1995), we cannot use a specific agonist and antagonist of these receptors. Therefore, further investigations are needed to elucidate the PGs-induced depolarization of the isolated abdominal vagus nerves.

Borison and Wang (1953) suggested the existence of a chemoreceptor trigger zone (CTZ) in the medulla. They reported that digitalis-induced emesis is associated with a central-mechanism in this area. The dopaminergic agents (e.g. apomorphine) stimulate the D_2 receptor in the CTZ. Although digitalis is widely used as a cardiotonic drug in patients with heart failure, it has several side effects. In particular, the occurrence of digitalis-induced emesis was reported almost 50 years ago (Wang *et al.*, 1952). However, central administration of a D_2 receptor antagonist did not exert an antiemetic action in digitalis-induced emesis. Vagotomy abolished the vomiting reflex induced by ouabain (Uchiyama *et al.*, 1978; Kakimoto *et al.*,

1997). Moreover, 5-HT$_3$ receptor antagonist, granisetron, suppressed ouabain-induced emesis (Endo *et al.*, 1998). 5-HT$_3$ receptors are widely expressed in peripheral tissues, such as in enterochromaffin cells and the vagus nerve cell body and axons, as well as in the medulla of the central nervous system (Brown and Marsh, 1978). We previously reported that ouabain facilitated the release of 5-HT from the isolated ileum. Moreover, granisetron tended to suppress this release of 5-HT (Endo *et al.*, 1998). The released 5-HT may act to depolarize on the vagus nerves. In the present study, we investigated the direct action of ouabain on the abdominal vagus nerves. Ouabain induced a direct, concentration-related depolarization. Our results suggest that ouabain may act on the peripheral sensory nerve ends and participate through a peripheral mechanism in the vomiting reflex.

In conclusion, DA, PGs and ouabain induced depolarization concentration-dependently of the rat isolated abdominal vagus nerves. Their depolarization patterns and reactivity differed greatly. Among these substances, PGs induced the largest depolarization. These preliminary results suggest that the peripheral role of the vagus nerves may be involved in the appearance of emesis caused by dopaminergic agents, in the delayed phase of cancer chemotherapy and in digitalis therapeutics. The grease gap technique of the isolated abdominal vagus nerve is a useful technique for screening the various emetic substances.

REFERENCES

Andrews, P. L. R. (1991). Modulation of visceral afferent activity as therapeutic possibility for gastro intestinal disorders, in: *Irritable Bowel Syndrome*, Read N. W. (Ed.), pp. 91–121. Blackwell Scientific, London.

Andrews, P. L. R., Rapeport, W. G. and Sanger, G. J. (1988). Neuropharmacology of emesis induced by anti-cancer therapy, *Trends Pharmacol. Sci.* **9**, 334–341.

Andrews, P. L. R. and Davis, C. J. (1995). The physiology of emesis induced by anti-cancer therapy, in: *Serotonin and the scientific basis of anti-emetic therapy*, Reynolds, D. J. M., Andrews, P. L. R. and Davis, C. J. (Eds), pp. 25–49. Oxford Clinical Communications, Oxford.

Birrell, M. A., Crispino, N., Hele, D. J., *et al.* (2002). Effect of dopamine receptor agonists on sensory nerve activity: possible therapeutic targets for the treatment of asthma and COPD, *Br. J. Pharmacol.* **136**, 620–628.

Blancquaert, J. P., Lefebvre, R. A. and Willems, J. L. (1986). Emetic and antiemetic effects of opioids in the dog, *Eur. J. Pharmacol.* **128**, 143–150.

Bley, K. R., Eglen, R. M. and Wong, E. H. F. (1994). Characterization of 5-hydroxytryptamine-induced depolarizations in rat isolated vagus nerve, *Eur. J. Pharmacol.* **260**, 139–147.

Borison, H. L. and Wang, S. C. (1953). Physiology and pharmacology of vomiting, *Pharmacol. Rev.* **5**, 193–230.

Brown, D. A. and Marsh, S. J. (1978). Axonal GABA-receptors in mammalian peripheral nerve trunks, *Brain Res.* **156**, 187–191.

Carpenter, D. O., Briggs, D. B. and Strominger, N. L. (1988). Mechanisms of radiation-induced emesis in the dog, *Pharmacol. Ther.* **39**, 367–371.

Coleman, J. and Rhodes, K. F. (1995). Further characterization of the putative 5-HT$_4$ receptor mediating depolarization of the rat isolated vagus nerve, *Naunyn Schmiedebergs Arch Pharmacol.* **352**, 74–78.

Costall, B., Domeney, A. M., Naylor, R. J., *et al.* (1986). 5-Hydroxytryptamine M-receptor antagonism to prevent cisplatin-induced emesis, *Neuropharmacology* **25**, 959–961.

Creinin, M. D. (2000). Medical abortion regimens: historical context and overview, *Am. J. Obstet. Gynecol.* **183**, S3–9.

Endo, T., Minami, M., Monma, Y., *et al.* (1990). Emesis-related biochemical and histopathological changes induced by cisplatin in the ferret, *J. Toxicol. Sci.* **15**, 235–244.

Endo, T., Minami, M., Monma, Y., *et al.* (1992). Effect of G38032F on cisplatin- and cysclophosphamide-induced emesis in the ferret, *Biogenic Amines* **9**, 163–175.

Endo, T., Sugawara, J., Nemoto, M., *et al.* (1998). Effects of granisetron, a selective 5-HT$_3$ receptor antagonist, on ouabain-induced emesis in ferrets, *Res. Comm. Mol. Pathol. Pharmacol.* **102**, 227–239.

Farrag, K. J., Costa, S. K. P. and Docherty, R. J. (2002). Differential sensitivity to tetrodotoxin and lack of effect of prostaglandin E$_2$ on the pharmacology and physiology of propagated potentials, *Br. J. Pharmacol.* **135**, 1449–1456.

Ferreira, S. H. (1979). Site of action of aspirin-like drugs and opioids, in: *Mechanisms of Pain and Analgesic Compounds*, Beers, R. F., Jr. and Bassett, E. G. (Eds), pp. 309–321. Raven Press, New York.

Fukunaka, N., Sagae, S., Kudo, R., *et al.* (1998). Effects of granisetron and its combination with dexamethasone on cisplatin-induced delayed emesis in the ferret, *Gen. Pharmacol.* **31**, 775–781.

Hawthorn, J., Ostler, K. J. and Andrews, P. L. (1988). The role of the abdominal visceral innervation and 5-hydroxytryptamine M-receptors in vomiting induced by the cytotoxic drugs cyclophosphamide and cis-platin in the ferret, *Quart. J. Exper. Physiol.* **73**, 7–21.

Ireland, S. J. (1987). Origin of 5-hydroxytryptamine-induced hyperpolarization of the rat superior cervical ganglion and vagus nerve, *Br. J. Pharmacol.* **92**, 407–416.

Ireland, S. J. and Tyers, M. B. (1987). Pharmacological characterization of 5-hydroxytryptamine-induced depolarization of the rat isolated vagus nerve, *Br. J. Pharmacol.* **90**, 229–238.

Ito, H., Akuzawa, S., Tsutsumi, R., *et al.* (1995). Comparative study of the affinities of the 5-HT$_3$ receptor antagonists, YM060, YM114 (KAE-393), granisetron and ondansetron in rat vagus nerve and cerebral cortex, *Neuropharmacology* **34**, 631–637.

Kakimoto, S., Saito, H. and Matsuki, N. (1997). Antiemetic effects of morphine on motion- and drug-induced emesis in *Suncus murinus*, *Biol. Pharm. Bull.* **20**, 486–489.

Kamato, T., Ito, H., Nagakura, Y., *et al.* (1993). Mechanisms of cisplatin- and m-chlorophenylbiguanide-induced emesis in ferrets, *Eur. J. Pharmacol.* **20**, 369–376.

Kris, M. G., Tyson, L. B., Clark, R. A., *et al.* (1992). Oral ondansetron for the control of delayed emesis after cisplatin. Report of a phase II study and a review of completed trials to manage delayed emesis, *Cancer* **70**, 1012–1016.

Lanza, F., Peace, K., Gustitus, L., *et al.* (1988). A blinded endoscopic comparative study of misoprostol versus sucralfate and placebo in the prevention of aspirin-induced gastric and duodenal ulceration, *Am. J. Gastroenterol.* **83**, 143–146.

Lawrence, J. and Jarrott, B. (1995). Visualisation of dopamine D$_2$ binding sites on human inferior vagal ganglia, *Br. J. Pharmacol.* **114**, 1329–1334.

Lawrence, J., Krstew, E. and Jarrott, B. (1995). Functional dopamine D$_2$ receptors on rat vagal afferent neurons, *Br. J. Pharmacol.* **114**, 1329–1334.

Minami, M., Tamakai, H., Ogawa, T., *et al.* (1995). Chemical modulation of 5-HT$_3$ and 5-HT$_4$ receptors affects the release of 5-hydroxytryptamine from the ferret and rat intestine, *Res. Commun. Mol. Pathol. Pharmacol.* **89**, 131–142.

Minami, M., Nemoto, M., Endo, T., *et al.* (1997). Chemical modulation of 5-HT$_3$ and 5-HT$_4$ receptors affects the release of 5-hydroxytryptamine from the ferret and rat intestine, *Res. Comm. Mol. Pathol. Pharmacol.* **95**, 67–83.

Minami, M., Kohno, Y., Endo, T., *et al.* (1999). Differential effects of talipexole and bromocriptine on serotonin release from rat intestinal tissues — an in vitro study of the emetic response of antiparkinsonian dopamine agonists, *Res. Commun. Pathol. Pharmacol.* **104**, 3–12.

Nakanishi, T., Kowa, H., Mizuno, Y., *et al.* (1993). The clinical evaluation of B-HT 920 (talipexole hydrochloride) in Parkinson's disease, *Clin. Eval.* **21**, 59–110.

Nemoto, M., Endo, T., Minami, M., *et al.* (2001). 5-Hydroxytryptamine (5-HT)-induced depolarization in isolated abdominal vagus nerves in the rat: involvement of 5-HT$_3$ and 5-HT$_4$ receptors, *Res. Commun. Mol. Pathol. Pharmacol.* **109**, 217–230.

Neto, F. R., (1978). The depolarizing action of 5-HT on mammalian non-myelinated nerve fibers, *Eur. J. Pharmacol.* **49**, 351–356.

Paintal, A. S. (1963). Vagal afferent fibers, *Ergebn. Physiol.* **52**, 74–156.

Parkes, J. D., Debono, A. G. and Marsden, C. D. (1976). Bromocriptine in Parkinsonism: long-term treatment, dose response, and comparison with levodopa, *J. Neurol. Neurosurg. Psychiatry* **39**, 1101–1108.

Pierce, K. L., Gil, D. W., Woodward, D. F., *et al.* (1995). Cloning of human prostanoid receptors, *Trends Pharmacol. Sci.* **16**, 253–256.

Rich, N. M., Abdulhayoglu, G. and Disaia, P. J. (1980). Methylpredonisolone as an antiemetic during cancer chemotherapy; a pilot study, *Gynaecol. Oncol.* **9**, 193–198.

Roila, F. (1993). Ondansetron plus dexamethasone compared to the 'standard' metoclopramide combination, *Oncology* **50**, 163–167.

Rudd, J. A., Jordan, C. C. and Naylor, R. J. (1994). Profiles of emetic action of cisplatin in the ferret: a potential model of acute and delayed emesis, *Eur. J. Pharmacol.* **262**, R1–R2.

Rudd, J. A. and Naylor, R. J. (1996). An interaction of ondansetron and dexamethasone antagonizing cisplatin-induced acute and delayed emesis in the ferret, *Br. J. Pharmacol.* **118**, 209–214.

Smith, D. B., Newlands, E. S, Rustin, G. J., *et al.* (1991). Comparison of ondansetron and ondansetron plus dexamethasone as antiemetic prophylaxis during cisplatin-containing chemotherapy, *Lancet* **338**, 487–490.

Smith, J. A., Amagasu, S. M., Eglen, R. M., *et al.* (1998). Characterization of prostanoid receptor-evoked responses in rat sensory neurons, *Br. J. Pharmacol.* **124**, 513–523.

Sridhar, K. S. and Donnelly, E. (1988). Combination antiemetics for cisplatin chemotherapy, *Cancer* **61**, 1508–1517.

Uchiyama, T., Kaneko, A. and Ito, R. (1978). A simple method for the detection of emetic action using pigeons, *J. Med. Soc. Toho.* **25**, 912–914.

Ushikubi, F., Hirata, M. and Narumiya, S. (1995). Molecular biology of prostanoid receptors; an overview, *J. Lipid Mediators Cell Signal.* **12**, 343–359.

van Giersbergen, P. L., Palkovits, M. and De Jong, W. (1992). Involvement of neurotransmitters in the nucleus tractus solitarii in cardiovascular regulation, *Physiol. Rev.* **72**, 789–824.

Wang, S. C. and Borison, H. L. (1952). A new concept of organization of the central emetic mechanism: Recent studies on the site of action of apomorphine, copper sulfate and cardiac glycosides, *Gastroenterology* **22**, 1–12.

Woods, A. J. and Andrews, P. L. (1995). Cisplatin acutely reduces 5-hydroxytryptamine-induced vagal depolarization in the rat: protective action of dexamethasone, *Eur. J. Pharmacol.* **278**, 275–278.

Yanagisawa, N., Kowa, H., Mizuno, Y., *et al.* (1992). Clinical phase II study of talipexole dihydrochloride (B-HT 920) tablet in Parkinson's disease- results of multicenter open study (Step II), *Rinsho Iyaku* **8**, 2841–2867.

Advances in Neuroregulation and Neuroprotection (2005), pp. 861–864
Collin, C. *et al.* (Eds)
© VSP 2005

Index